RILEM Youth Symposium 2025 – RYS2025

RILEM Bookseries

Volume 67

RILEM, The International Union of Laboratories and Experts in Construction Materials, Systems and Structures, founded in 1947, is a non-governmental scientific association whose goal is to contribute to progress in the construction sciences, techniques and industries, essentially by means of the communication it fosters between research and practice. RILEM's focus is on construction materials and their use in building and civil engineering structures, covering all phases of the building process from manufacture to use and recycling of materials. More information on RILEM and its previous publications can be found on www.RILEM.net.

Indexed in SCOPUS, Google Scholar and SpringerLink.

Magda Posani · Surender Singh
Editors

RILEM Youth Symposium 2025 – RYS2025

1st RILEM Youth Symposium on Building Materials and Construction

Set 1

Springer

Editors
Magda Posani
Department of Civil Engineering
Aalto University
Espoo, Finland

Surender Singh
Department of Civil Engineering
IIT Madras
Chennai, Tamil Nadu, India

ISSN 2211-0844 ISSN 2211-0852 (electronic)
RILEM Bookseries
ISBN 978-3-032-14651-9 ISBN 978-3-032-14652-6 (eBook)
https://doi.org/10.1007/978-3-032-14652-6

This Springer imprint is published by the registered company Springer Nature Switzerland AG
The registered company address is: Gewerbestrasse 11, 6330 Cham, Switzerland

Contents

Innovative Low-Carbon Materials

Service Life and Durability

Hygrothermal Behaviour

RILEM Publications

The following list is presenting the global offer of RILEM Publications, sorted by series. Each publication is available in printed version and/or in online version.

RILEM Proceedings (PRO)

PRO 1: Durability of High Performance Concrete (ISBN: 2-912143-03-9; e-ISBN: 2-351580-12-5; e-ISBN: 2351580125); *Ed. H. Sommer*

PRO 2: Chloride Penetration into Concrete (ISBN: 2-912143-00-04; e-ISBN: 2912143454); *Eds. L.-O. Nilsson and J.-P. Ollivier*

PRO 3: Evaluation and Strengthening of Existing Masonry Structures (ISBN: 2-912143-02-0; e-ISBN: 2351580141); *Eds. L. Binda and C. Modena*

PRO 4: Concrete: From Material to Structure (ISBN: 2-912143-04-7; e-ISBN: 2351580206); *Eds. J.-P. Bournazel and Y. Malier*

PRO 5: The Role of Admixtures in High Performance Concrete (ISBN: 2-912143-05-5; e-ISBN: 2351580214); *Eds. J. G. Cabrera and R. Rivera-Villarreal*

PRO 6: High Performance Fiber Reinforced Cement Composites - HPFRCC 3 (ISBN: 2-912143-06-3; e-ISBN: 2351580222); *Eds. H. W. Reinhardt and A. E. Naaman*

PRO 7: 1st International RILEM Symposium on Self-Compacting Concrete (ISBN: 2-912143-09-8; e-ISBN: 2912143721); *Eds. Å. Skarendahl and Ö. Petersson*

PRO 8: International RILEM Symposium on Timber Engineering (ISBN: 2-912143-10-1; e-ISBN: 2351580230); *Ed. L. Boström*

PRO 9: 2nd International RILEM Symposium on Adhesion between Polymers and Concrete ISAP '99 (ISBN: 2-912143-11-X; e-ISBN: 2351580249); *Eds. Y. Ohama and M. Puterman*

PRO 10: 3rd International RILEM Symposium on Durability of Building and Construction Sealants (ISBN: 2-912143-13-6; e-ISBN: 2351580257); *Ed. A. T. Wolf*

PRO 11: 4th International RILEM Conference on Reflective Cracking in Pavements (ISBN: 2-912143-14-4; e-ISBN: 2351580265); *Eds. A. O. Abd El Halim, D. A. Taylor and El H. H. Mohamed*

PRO 12: International RILEM Workshop on Historic Mortars: Characteristics and Tests (ISBN: 2-912143-15-2; e-ISBN: 2351580273); *Eds. P. Bartos, C. Groot and J. J. Hughes*

PRO 13: 2nd International RILEM Symposium on Hydration and Setting (ISBN: 2-912143-16-0; e-ISBN: 2351580281); *Ed. A. Nonat*

PRO 14: Integrated Life-Cycle Design of Materials and Structures - ILCDES 2000 (ISBN: 951-758-408-3; e-ISBN: 235158029X); (ISSN: 0356-9403); *Ed. S. Sarja*

PRO 15: Fifth RILEM Symposium on Fibre-Reinforced Concretes (FRC) - BEFIB'2000 (ISBN: 2-912143-18-7; e-ISBN: 291214373X); *Eds. P. Rossi and G. Chanvillard*

PRO 16: Life Prediction and Management of Concrete Structures (ISBN: 2-912143-19-5; e-ISBN: 2351580303); *Ed. D. Naus*

PRO 17: Shrinkage of Concrete – Shrinkage 2000 (ISBN: 2-912143-20-9; e-ISBN: 2351580311); *Eds. V. Baroghel-Bouny and P.-C. Aïtcin*

PRO 18: Measurement and Interpretation of the On-Site Corrosion Rate (ISBN: 2-912143-21-7; e-ISBN: 235158032X); *Eds. C. Andrade, C. Alonso, J. Fullea, J. Polimon and J. Rodriguez*

PRO 19: Testing and Modelling the Chloride Ingress into Concrete (ISBN: 2-912143-22-5; e-ISBN: 2351580338); *Eds. C. Andrade and J. Kropp*

PRO 20: 1st International RILEM Workshop on Microbial Impacts on Building Materials (CD 02) (e-ISBN 978-2-35158-013-4); *Ed. M. Ribas Silva*

PRO 21: International RILEM Symposium on Connections between Steel and Concrete (ISBN: 2-912143-25-X; e-ISBN: 2351580346); *Ed. R. Eligehausen*

PRO 22: International RILEM Symposium on Joints in Timber Structures (ISBN: 2-912143-28-4; e-ISBN: 2351580354); *Eds. S. Aicher and H.-W. Reinhardt*

PRO 23: International RILEM Conference on Early Age Cracking in Cementitious Systems (ISBN: 2-912143-29-2; e-ISBN: 2351580362); *Eds. K. Kovler and A. Bentur*

PRO 24: 2nd International RILEM Workshop on Frost Resistance of Concrete (ISBN: 2-912143-30-6; e-ISBN: 2351580370); *Eds. M. J. Setzer, R. Auberg and H.-J. Keck*

PRO 25: International RILEM Workshop on Frost Damage in Concrete (ISBN: 2-912143-31-4; e-ISBN: 2351580389); *Eds. D. J. Janssen, M. J. Setzer and M. B. Snyder*

PRO 26: International RILEM Workshop on On-Site Control and Evaluation of Masonry Structures (ISBN: 2-912143-34-9; e-ISBN: 2351580141); *Eds. L. Binda and R. C. de Vekey*

PRO 27: International RILEM Symposium on Building Joint Sealants (CD03; e-ISBN: 235158015X); *Ed. A. T. Wolf*

PRO 28: 6th International RILEM Symposium on Performance Testing and Evaluation of Bituminous Materials - PTEBM'03 (ISBN: 2-912143-35-7; e-ISBN: 978-2-912143-77-8); *Ed. M. N. Partl*

PRO 29: 2nd International RILEM Workshop on Life Prediction and Ageing Management of Concrete Structures (ISBN: 2-912143-36-5; e-ISBN: 2912143780); *Ed. D. J. Naus*

PRO 30: 4th International RILEM Workshop on High Performance Fiber Reinforced Cement Composites - HPFRCC 4 (ISBN: 2-912143-37-3; e-ISBN: 2912143799); *Eds. A. E. Naaman and H. W. Reinhardt*

PRO 31: International RILEM Workshop on Test and Design Methods for Steel Fibre Reinforced Concrete: Background and Experiences (ISBN: 2-912143-38-1; e-ISBN: 2351580168); *Eds. B. Schnütgen and L. Vandewalle*

PRO 32: International Conference on Advances in Concrete and Structures 2 vol. (ISBN (set): 2-912143-41-1; e-ISBN: 2351580176); *Eds. Ying-shu Yuan, Surendra P. Shah and Heng-lin Lü*

PRO 33: 3rd International Symposium on Self-Compacting Concrete (ISBN: 2-912143-42-X; e-ISBN: 2912143713); *Eds. Ó. Wallevik and I. Níelsson*

PRO 34: International RILEM Conference on Microbial Impact on Building Materials (ISBN: 2-912143-43-8; e-ISBN: 2351580184); *Ed. M. Ribas Silva*

PRO 35: International RILEM TC 186-ISA on Internal Sulfate Attack and Delayed Ettringite Formation (ISBN: 2-912143-44-6; e-ISBN: 2912143802); *Eds. K. Scrivener and J. Skalny*

PRO 36: International RILEM Symposium on Concrete Science and Engineering – A Tribute to Arnon Bentur (ISBN: 2-912143-46-2; e-ISBN: 2912143586); *Eds. K. Kovler, J. Marchand, S. Mindess and J. Weiss*

PRO 37: 5th International RILEM Conference on Cracking in Pavements – Mitigation, Risk Assessment and Prevention (ISBN: 2-912143-47-0; e-ISBN: 2912143764); *Eds. C. Petit, I. Al-Qadi and A. Millien*

PRO 38: 3rd International RILEM Workshop on Testing and Modelling the Chloride Ingress into Concrete (ISBN: 2-912143-48-9; e-ISBN: 2912143578); *Eds. C. Andrade and J. Kropp*

PRO 39: 6th International RILEM Symposium on Fibre-Reinforced Concretes - BEFIB 2004 (ISBN: 2-912143-51-9; e-ISBN: 2912143748); *Eds. M. Di Prisco, R. Felicetti and G. A. Plizzari*

PRO 40: International RILEM Conference on the Use of Recycled Materials in Buildings and Structures (ISBN: 2-912143-52-7; e-ISBN: 2912143756); *Eds. E. Vázquez, Ch. F. Hendriks and G. M. T. Janssen*

PRO 41: RILEM International Symposium on Environment-Conscious Materials and Systems for Sustainable Development (ISBN: 2-912143-55-1; e-ISBN: 2912143640); *Eds. N. Kashino and Y. Ohama*

PRO 42: SCC'2005 - China: 1st International Symposium on Design, Performance and Use of Self-Consolidating Concrete (ISBN: 2-912143-61-6; e-ISBN: 2912143624); *Eds. Zhiwu Yu, Caijun Shi, Kamal Henri Khayat and Youjun Xie*

PRO 43: International RILEM Workshop on Bonded Concrete Overlays (e-ISBN: 2-912143-83-7); *Eds. J. L. Granju and J. Silfwerbrand*

PRO 44: 2nd International RILEM Workshop on Microbial Impacts on Building Materials (CD11) (e-ISBN: 2-912143-84-5); *Ed. M. Ribas Silva*

PRO 45: 2nd International Symposium on Nanotechnology in Construction, Bilbao (ISBN: 2-912143-87-X; e-ISBN: 2912143888); *Eds. Peter J. M. Bartos, Yolanda de Miguel and Antonio Porro*

PRO 46: ConcreteLife'06 - International RILEM-JCI Seminar on Concrete Durability and Service Life Planning: Curing, Crack Control, Performance in Harsh Environments (ISBN: 2-912143-89-6; e-ISBN: 291214390X); *Ed. K. Kovler*

PRO 47: International RILEM Workshop on Performance Based Evaluation and Indicators for Concrete Durability (ISBN: 978-2-912143-95-2; e-ISBN: 9782912143969); *Eds. V. Baroghel-Bouny, C. Andrade, R. Torrent and K. Scrivener*

PRO 48: 1st International RILEM Symposium on Advances in Concrete through Science and Engineering (e-ISBN: 2-912143-92-6); *Eds. J. Weiss, K. Kovler, J. Marchand and S. Mindess*

PRO 49: International RILEM Workshop on High Performance Fiber Reinforced Cementitious Composites in Structural Applications (ISBN: 2-912143-93-4; e-ISBN: 2912143942); *Eds. G. Fischer and V. C. Li*

PRO 50: 1st International RILEM Symposium on Textile Reinforced Concrete (ISBN: 2-912143-97-7; e-ISBN: 2351580087); *Eds. Josef Hegger, Wolfgang Brameshuber and Norbert Will*

PRO 51: 2nd International Symposium on Advances in Concrete through Science and Engineering (ISBN: 2-35158-003-6; e-ISBN: 2-35158-002-8); *Eds. J. Marchand, B. Bissonnette, R. Gagné, M. Jolin and F. Paradis*

PRO 52: Volume Changes of Hardening Concrete: Testing and Mitigation (ISBN: 2-35158-004-4; e-ISBN: 2-35158-005-2); *Eds. O. M. Jensen, P. Lura and K. Kovler*

PRO 53: High Performance Fiber Reinforced Cement Composites - HPFRCC5 (ISBN: 978-2-35158-046-2; e-ISBN: 978-2-35158-089-9); *Eds. H. W. Reinhardt and A. E. Naaman*

PRO 54: 5th International RILEM Symposium on Self-Compacting Concrete (ISBN: 978-2-35158-047-9; e-ISBN: 978-2-35158-088-2); *Eds. G. De Schutter and V. Boel*

PRO 55: International RILEM Symposium Photocatalysis, Environment and Construction Materials (ISBN: 978-2-35158-056-1; e-ISBN: 978-2-35158-057-8); *Eds. P. Baglioni and L. Cassar*

PRO 56: International RILEM Workshop on Integral Service Life Modelling of Concrete Structures (ISBN 978-2-35158-058-5; e-ISBN: 978-2-35158-090-5); *Eds. R. M. Ferreira, J. Gulikers and C. Andrade*

PRO 57: RILEM Workshop on Performance of cement-based materials in aggressive aqueous environments (e-ISBN: 978-2-35158-059-2); *Ed. N. De Belie*

PRO 58: International RILEM Symposium on Concrete Modelling - CONMOD'08 (ISBN: 978-2-35158-060-8; e-ISBN: 978-2-35158-076-9); *Eds. E. Schlangen and G. De Schutter*

PRO 59: International RILEM Conference on On Site Assessment of Concrete, Masonry and Timber Structures - SACoMaTiS 2008 (ISBN set: 978-2-35158-061-5; e-ISBN: 978-2-35158-075-2); *Eds. L. Binda, M. di Prisco and R. Felicetti*

PRO 60: Seventh RILEM International Symposium on Fibre Reinforced Concrete: Design and Applications - BEFIB 2008 (ISBN: 978-2-35158-064-6; e-ISBN: 978-2-35158-086-8); *Ed. R. Gettu*

PRO 61: 1st International Conference on Microstructure Related Durability of Cementitious Composites 2 vol., (ISBN: 978-2-35158-065-3; e-ISBN: 978-2-35158-084-4); *Eds. W. Sun, K. van Breugel, C. Miao, G. Ye and H. Chen*

PRO 62: NSF/ RILEM Workshop: In-situ Evaluation of Historic Wood and Masonry Structures (e-ISBN: 978-2-35158-068-4); *Eds. B. Kasal, R. Anthony and M. Drdácký*

PRO 63: Concrete in Aggressive Aqueous Environments: Performance, Testing and Modelling, 2 vol., (ISBN: 978-2-35158-071-4; e-ISBN: 978-2-35158-082-0); *Eds. M. G. Alexander and A. Bertron*

PRO 64: Long Term Performance of Cementitious Barriers and Reinforced Concrete in Nuclear Power Plants and Waste Management - NUCPERF 2009 (ISBN: 978-2-35158-072-1; e-ISBN: 978-2-35158-087-5); *Eds. V. L'Hostis, R. Gens and C. Gallé*

PRO 65: Design Performance and Use of Self-consolidating Concrete - SCC'2009 (ISBN: 978-2-35158-073-8; e-ISBN: 978-2-35158-093-6); *Eds. C. Shi, Z. Yu, K. H. Khayat and P. Yan*

PRO 66: 2nd International RILEM Workshop on Concrete Durability and Service Life Planning - ConcreteLife'09 (ISBN: 978-2-35158-074-5; ISBN: 978-2-35158-074-5); *Ed. K. Kovler*

PRO 67: Repairs Mortars for Historic Masonry (e-ISBN: 978-2-35158-083-7); *Ed. C. Groot*

PRO 68: Proceedings of the 3rd International RILEM Symposium on 'Rheology of Cement Suspensions such as Fresh Concrete (ISBN 978-2-35158-091-2; e-ISBN: 978-2-35158-092-9); *Eds. O. H. Wallevik, S. Kubens and S. Oesterheld*

PRO 69: 3rd International PhD Student Workshop on 'Modelling the Durability of Reinforced Concrete (ISBN: 978-2-35158-095-0); *Eds. R. M. Ferreira, J. Gulikers and C. Andrade*

PRO 70: 2nd International Conference on 'Service Life Design for Infrastructure' (ISBN set: 978-2-35158-096-7, e-ISBN: 978-2-35158-097-4); *Eds. K. van Breugel, G. Ye and Y. Yuan*

PRO 71: Advances in Civil Engineering Materials - The 50-year Teaching Anniversary of Prof. Sun Wei' (ISBN: 978-2-35158-098-1; e-ISBN: 978-2-35158-099-8); *Eds. C. Miao, G. Ye and H. Chen*

PRO 72: First International Conference on 'Advances in Chemically-Activated Materials – CAM'2010' (2010), 264 pp, ISBN: 978-2-35158-101-8; e-ISBN: 978-2-35158-115-5, *Eds. Caijun Shi and Xiaodong Shen*

PRO 73: 2nd International Conference on 'Waste Engineering and Management - ICWEM 2010' (2010), 894 pp, ISBN: 978-2-35158-102-5; e-ISBN: 978-2-35158-103-2, *Eds. J. Zh. Xiao, Y. Zhang, M. S. Cheung and R. Chu*

PRO 74: International RILEM Conference on 'Use of Superabsorbent Polymers and Other New Additives in Concrete' (2010) 374 pp., ISBN: 978-2-35158-104-9; e-ISBN: 978-2-35158-105-6; *Eds. O. M. Jensen, M. T. Hasholt and S. Laustsen*

PRO 75: International Conference on 'Material Science - 2nd ICTRC - Textile Reinforced Concrete - Theme 1' (2010) 436 pp., ISBN: 978-2-35158-106-3; e-ISBN: 978-2-35158-107-0; *Ed. W. Brameshuber*

PRO 76: International Conference on 'Material Science - HetMat - Modelling of Heterogeneous Materials - Theme 2' (2010) 255 pp., ISBN: 978-2-35158-108-7; e-ISBN: 978-2-35158-109-4; *Ed. W. Brameshuber*

PRO 77: International Conference on 'Material Science - AdIPoC - Additions Improving Properties of Concrete - Theme 3' (2010) 459 pp., ISBN: 978-2-35158-110-0; e-ISBN: 978-2-35158-111-7; *Ed. W. Brameshuber*

PRO 78: 2nd Historic Mortars Conference and RILEM TC 203-RHM Final Workshop – HMC2010 (2010) 1416 pp., e-ISBN: 978-2-35158-112-4; *Eds. J. Válek, C. Groot and J. J. Hughes*

PRO 79: International RILEM Conference on Advances in Construction Materials Through Science and Engineering (2011) 213 pp., ISBN: 978-2-35158-116-2, e-ISBN: 978-2-35158-117-9; *Eds. Christopher Leung and K. T. Wan*

PRO 80: 2nd International RILEM Conference on Concrete Spalling due to Fire Exposure (2011) 453 pp., ISBN: 978-2-35158-118-6, e-ISBN: 978-2-35158-119-3; *Eds. E. A. B. Koenders and F. Dehn*

PRO 81: 2nd International RILEM Conference on Strain Hardening Cementitious Composites (SHCC2-Rio) (2011) 451 pp., ISBN: 978-2-35158-120-9, e-ISBN: 978-2-35158-121-6; *Eds. R. D. Toledo Filho, F. A. Silva, E. A. B. Koenders and E. M. R. Fairbairn*

PRO 82: 2nd International RILEM Conference on Progress of Recycling in the Built Environment (2011) 507 pp., e-ISBN: 978-2-35158-122-3; *Eds. V. M. John, E. Vazquez, S. C. Angulo and C. Ulsen*

PRO 83: 2nd International Conference on Microstructural-related Durability of Cementitious Composites (2012) 250 pp., ISBN: 978-2-35158-129-2; e-ISBN: 978-2-35158-123-0; *Eds. G. Ye, K. van Breugel, W. Sun and C. Miao*

PRO 84: CONSEC13 - Seventh International Conference on Concrete under Severe Conditions – Environment and Loading (2013) 1930 pp., ISBN: 978-2-35158-124-7; e-ISBN: 978-2-35158-134-6; *Eds. Z. J. Li, W. Sun, C. W. Miao, K. Sakai, O. E. Gjorv and N. Banthia*

PRO 85: RILEM-JCI International Workshop on Crack Control of Mass Concrete and Related issues concerning Early-Age of Concrete Structures – ConCrack 3 – Control of Cracking in Concrete Structures 3 (2012) 237 pp., ISBN: 978-2-35158-125-4; e-ISBN: 978-2-35158-126-1; *Eds. F. Toutlemonde and J.-M. Torrenti*

PRO 86: International Symposium on Life Cycle Assessment and Construction (2012) 414 pp., ISBN: 978-2-35158-127-8, e-ISBN: 978-2-35158-128-5; *Eds. A. Ventura and C. de la Roche*

PRO 87: UHPFRC 2013 – RILEM-fib-AFGC International Symposium on Ultra-High Performance Fibre-Reinforced Concrete (2013), ISBN: 978-2-35158-130-8, e-ISBN: 978-2-35158-131-5; *Ed. F. Toutlemonde*

PRO 88: 8th RILEM International Symposium on Fibre Reinforced Concrete (2012) 344 pp., ISBN: 978-2-35158-132-2, e-ISBN: 978-2-35158-133-9; *Ed. Joaquim A. O. Barros*

PRO 89: RILEM International workshop on performance-based specification and control of concrete durability (2014) 678 pp, ISBN: 978-2-35158-135-3, e-ISBN: 978-2-35158-136-0; *Eds. D. Bjegović, H. Beushausen and M. Serdar*

PRO 90: 7th RILEM International Conference on Self-Compacting Concrete and of the 1st RILEM International Conference on Rheology and Processing of Construction Materials (2013) 396 pp, ISBN: 978-2-35158-137-7, e-ISBN: 978-2-35158-138-4; *Eds. Nicolas Roussel and Hela Bessaies-Bey*

PRO 91: CONMOD 2014 - RILEM International Symposium on Concrete Modelling (2014), ISBN: 978-2-35158-139-1; e-ISBN: 978-2-35158-140-7; *Eds. Kefei Li, Peiyu Yan and Rongwei Yang*

PRO 92: CAM 2014 - 2nd International Conference on advances in chemically-activated materials (2014) 392 pp., ISBN: 978-2-35158-141-4; e-ISBN: 978-2-35158-142-1; *Eds. Caijun Shi and Xiadong Shen*

PRO 93: SCC 2014 - 3rd International Symposium on Design, Performance and Use of Self-Consolidating Concrete (2014) 438 pp., ISBN: 978-2-35158-143-8; e-ISBN: 978-2-35158-144-5; *Eds. Caijun Shi, Zhihua Ou and Kamal H. Khayat*

PRO 94 (online version): HPFRCC-7 - 7th RILEM conference on High performance fiber reinforced cement composites (2015), e-ISBN: 978-2-35158-146-9; *Eds. H. W. Reinhardt, G. J. Parra-Montesinos and H. Garrecht*

PRO 95: International RILEM Conference on Application of superabsorbent polymers and other new admixtures in concrete construction (2014), ISBN: 978-2-35158-147-6; e-ISBN: 978-2-35158-148-3; *Eds. Viktor Mechtcherine and Christof Schroefl*

PRO 96 (online version): XIII DBMC: XIII International Conference on Durability of Building Materials and Components (2015), e-ISBN: 978-2-35158-149-0; *Eds. M. Quattrone and V. M. John*

PRO 97: SHCC3 – 3rd International RILEM Conference on Strain Hardening Cementitious Composites (2014), ISBN: 978-2-35158-150-6; e-ISBN: 978-2-35158-151-3; *Eds. E. Schlangen, M. G. Sierra Beltran, M. Lukovic and G. Ye*

PRO 98: FERRO-11 – 11th International Symposium on Ferrocement and 3rd ICTRC - International Conference on Textile Reinforced Concrete (2015), ISBN: 978-2-35158-152-0; e-ISBN: 978-2-35158-153-7; *Ed. W. Brameshuber*

PRO 99 (online version): ICBBM 2015 - 1st International Conference on Bio-Based Building Materials (2015), e-ISBN: 978-2-35158-154-4; *Eds. S. Amziane and M. Sonebi*

PRO 100: SCC16 - RILEM Self-Consolidating Concrete Conference (2016), ISBN: 978-2-35158-156-8; e-ISBN: 978-2-35158-157-5; *Ed. Kamal H. Kayat*

PRO 101 (online version): III Progress of Recycling in the Built Environment (2015), e-ISBN: 978-2-35158-158-2; *Eds. I. Martins, C. Ulsen and S. C. Angulo*

PRO 102 (online version): RILEM Conference on Microorganisms-Cementitious Materials Interactions (2016), e-ISBN: 978-2-35158-160-5; *Eds. Alexandra Bertron, Henk Jonkers and Virginie Wiktor*

PRO 103 (online version): ACESC'16 - Advances in Civil Engineering and Sustainable Construction (2016), e-ISBN: 978-2-35158-161-2; *Eds. T. Ch. Madhavi, G. Prabhakar, Santhosh Ram and P. M. Rameshwaran*

PRO 104 (online version): SSCS'2015 - Numerical Modeling - Strategies for Sustainable Concrete Structures (2015), e-ISBN: 978-2-35158-162-9

PRO 105: 1st International Conference on UHPC Materials and Structures (2016), ISBN: 978-2-35158-164-3, e-ISBN: 978-2-35158-165-0

PRO 106: AFGC-ACI-fib-RILEM International Conference on Ultra-High-Performance Fibre-Reinforced Concrete – UHPFRC 2017 (2017), ISBN: 978-2-35158-166-7, e-ISBN: 978-2-35158-167-4; *Eds. François Toutlemonde and Jacques Resplendino*

PRO 107 (online version): XIV DBMC – 14th International Conference on Durability of Building Materials and Components (2017), e-ISBN: 978-2-35158-159-9; *Eds. Geert De Schutter, Nele De Belie, Arnold Janssens and Nathan Van Den Bossche*

PRO 108: MSSCE 2016 - Innovation of Teaching in Materials and Structures (2016), ISBN: 978-2-35158-178-0, e-ISBN: 978-2-35158-179-7; *Ed. Per Goltermann*

PRO 109 (2 volumes): MSSCE 2016 - Service Life of Cement-Based Materials and Structures (2016), ISBN Vol. 1: 978-2-35158-170-4, Vol. 2: 978-2-35158-171-4, Set Vol. 1&2: 978-2-35158-172-8, e-ISBN: 978-2-35158-173-5; *Eds. Miguel Azenha, Ivan Gabrijel, Dirk Schlicke, Terje Kanstad and Ole Mejlhede Jensen*

PRO 110: MSSCE 2016 - Historical Masonry (2016), ISBN: 978-2-35158-178-0, e-ISBN: 978-2-35158-179-7; *Eds. Inge Rörig-Dalgaard and Ioannis Ioannou*

PRO 111: MSSCE 2016 - Electrochemistry in Civil Engineering (2016), ISBN: 978-2-35158-176-6, e-ISBN: 978-2-35158-177-3; *Ed. Lisbeth M. Ottosen*

PRO 112: MSSCE 2016 - Moisture in Materials and Structures (2016), ISBN: 978-2-35158-178-0, e-ISBN: 978-2-35158-179-7; *Eds. Kurt Kielsgaard Hansen, Carsten Rode and Lars-Olof Nilsson*

PRO 113: MSSCE 2016 - Concrete with Supplementary Cementitious Materials (2016), ISBN: 978-2-35158-178-0, e-ISBN: 978-2-35158-179-7; *Eds. Ole Mejlhede Jensen, Konstantin Kovler and Nele De Belie*

PRO 114: MSSCE 2016 - Frost Action in Concrete (2016), ISBN: 978-2-35158-182-7, e-ISBN: 978-2-35158-183-4; *Eds. Marianne Tange Hasholt, Katja Fridh and R. Doug Hooton*

PRO 115: MSSCE 2016 - Fresh Concrete (2016), ISBN: 978-2-35158-184-1, e-ISBN: 978-2-35158-185-8; *Eds. Lars N. Thrane, Claus Pade, Oldrich Svec and Nicolas Roussel*

PRO 116: BEFIB 2016 – 9th RILEM International Symposium on Fiber Reinforced Concrete (2016), ISBN: 978-2-35158-187-2, e-ISBN: 978-2-35158-186-5; *Eds. N. Banthia, M. di Prisco and S. Soleimani-Dashtaki*

PRO 117: 3rd International RILEM Conference on Microstructure Related Durability of Cementitious Composites (2016), ISBN: 978-2-35158-188-9, e-ISBN: 978-2-35158-189-6; *Eds. Changwen Miao, Wei Sun, Jiaping Liu, Huisu Chen, Guang Ye and Klaas van Breugel*

PRO 118 (4 volumes): International Conference on Advances in Construction Materials and Systems (2017), ISBN Set: 978-2-35158-190-2, Vol. 1: 978-2-35158-193-3, Vol. 2: 978-2-35158-194-0, Vol. 3: ISBN:978-2-35158-195-7, Vol. 4: ISBN:978-2-35158-196-4, e-ISBN: 978-2-35158-191-9; *Eds. Manu Santhanam, Ravindra Gettu, Radhakrishna G. Pillai and Sunitha K. Nayar*

PRO 119 (online version): ICBBM 2017 - Second International RILEM Conference on Bio-based Building Materials, (2017), e-ISBN: 978-2-35158-192-6; *Eds. Sofiane Amziane and Mohammed Sonebi*

PRO 120 (2 volumes): EAC-02 - 2nd International RILEM/COST Conference on Early Age Cracking and Serviceability in Cement-based Materials and Structures, (2017), Vol.

1: 978-2-35158-199-5, Vol. 2: 978-2-35158-200-8, Set: 978-2-35158-197-1, e-ISBN: 978-2-35158-198-8; *Eds. Stéphanie Staquet and Dimitrios Aggelis*

PRO 121 (2 volumes): SynerCrete18: Interdisciplinary Approaches for Cement-based Materials and Structural Concrete: Synergizing Expertise and Bridging Scales of Space and Time, (2018), Set: 978-2-35158-202-2, Vol.1: 978-2-35158-211-4, Vol. 2: 978-2-35158-212-1, e-ISBN: 978-2-35158-203-9; *Eds. Miguel Azenha, Dirk Schlicke, Farid Benboudjema and Agnieszka Knoppik*

PRO 122: SCC'2018 China - Fourth International Symposium on Design, Performance and Use of Self-Consolidating Concrete, (2018), ISBN: 978-2-35158-204-6, e-ISBN: 978-2-35158-205-3; *Eds. C. Shi, Z. Zhang and K. H. Khayat*

PRO 123: Final Conference of RILEM TC 253-MCI: Microorganisms-Cementitious Materials Interactions (2018), Set: 978-2-35158-207-7, Vol.1: 978-2-35158-209-1, Vol.2: 978-2-35158-210-7, e-ISBN: 978-2-35158-206-0; *Ed. Alexandra Bertron*

PRO 124 (online version): Fourth International Conference Progress of Recycling in the Built Environment (2018), e-ISBN: 978-2-35158-208-4; *Eds. Isabel M. Martins, Carina Ulsen and Yury Villagran*

PRO 125 (online version): SLD4 - 4th International Conference on Service Life Design for Infrastructures (2018), e-ISBN: 978-2-35158-213-8; *Eds. Guang Ye, Yong Yuan, Claudia Romero Rodriguez, Hongzhi Zhang and Branko Savija*

PRO 126: Workshop on Concrete Modelling and Material Behaviour in honor of Professor Klaas van Breugel (2018), ISBN: 978-2-35158-214-5, e-ISBN: 978-2-35158-215-2; *Ed. Guang Ye*

PRO 127 (online version): CONMOD2018 - Symposium on Concrete Modelling (2018), e-ISBN: 978-2-35158-216-9; *Eds. Erik Schlangen, Geert de Schutter, Branko Savija, Hongzhi Zhang and Claudia Romero Rodriguez*

PRO 128: SMSS2019 - International Conference on Sustainable Materials, Systems and Structures (2019), ISBN: 978-2-35158-217-6, e-ISBN: 978-2-35158-218-3

PRO 129: 2nd International Conference on UHPC Materials and Structures (UHPC2018-China), ISBN: 978-2-35158-219-0, e-ISBN: 978-2-35158-220-6;

PRO 130: 5th Historic Mortars Conference (2019), ISBN: 978-2-35158-221-3, e-ISBN: 978-2-35158-222-0; *Eds. José Ignacio Álvarez, José María Fernández, Íñigo Navarro, Adrián Durán and Rafael Sirera*

PRO 131 (online version): 3rd International Conference on Bio-Based Building Materials (ICBBM2019), e-ISBN: 978-2-35158-229-9; *Eds. Mohammed Sonebi, Sofiane Amziane and Jonathan Page*

PRO 132: IRWRMC'18 - International RILEM Workshop on Rheological Measurements of Cement-based Materials (2018), ISBN: 978-2-35158-230-5, e-ISBN: 978-2-35158-231-2; *Eds. Chafika Djelal and Yannick Vanhove*

PRO 133 (online version): CO2STO2019 - International Workshop CO2 Storage in Concrete (2019), e-ISBN: 978-2-35158-232-9; *Eds. Assia Djerbi, Othman Omikrine-Metalssi and Teddy Fen-Chong*

PRO 134: 3rd ACF/HNU International Conference on UHPC Materials and Structures - UHPC'2020, ISBN: 978-2-35158-233-6, e-ISBN: 978-2-35158-234-3; *Eds. Caijun Shi and Jiaping Liu*

PRO 135: Fourth International Conference on Chemically Activated Materials (CAM2021), ISBN: 978-2-35158-235-0, e-ISBN: 978-2-35158-236-7; *Eds. Caijun Shi and Xiang Hu*

PRO 136: Lime Based Materials for Repairing Historic Structures (2022), ISBN: 978-2-35158-240-4, e-ISBN: 978-2-35158-237-4; *Eds. Ioanna Papayianni and Jan Válek*

PRO 137 (online version): V International Conference Progress of Recycling in the Built Environment (2023), e-ISBN: 978-2-35158-238-1; *Eds. Ines Döring, Yury Villagrán Zaccardi and Anette Müller*

RILEM Reports (REP)

Report 19: Considerations for Use in Managing the Aging of Nuclear Power Plant Concrete Structures (ISBN: 2-912143-07-1); *Ed. D. J. Naus*

Report 20: Engineering and Transport Properties of the Interfacial Transition Zone in Cementitious Composites (ISBN: 2-912143-08-X); *Eds. M. G. Alexander, G. Arliguie, G. Ballivy, A. Bentur and J. Marchand*

Report 21: Durability of Building Sealants (ISBN: 2-912143-12-8); *Ed. A. T. Wolf*

Report 22: Sustainable Raw Materials - Construction and Demolition Waste (ISBN: 2-912143-17-9); *Eds. C. F. Hendriks and H. S. Pietersen*

Report 23: Self-Compacting Concrete state-of-the-art report (ISBN: 2-912143-23-3); *Eds. Å. Skarendahl and Ö. Petersson*

Report 24: Workability and Rheology of Fresh Concrete: Compendium of Tests (ISBN: 2-912143-32-2); *Eds. P. J. M. Bartos, M. Sonebi and A. K. Tamimi*

Report 25: Early Age Cracking in Cementitious Systems (ISBN: 2-912143-33-0); *Ed. A. Bentur*

Report 26: Towards Sustainable Roofing (Joint Committee CIB/RILEM) (CD 07) (e-ISBN 978-2-912143-65-5); *Eds. Thomas W. Hutchinson and Keith Roberts*

Report 27: Condition Assessment of Roofs (Joint Committee CIB/RILEM) (CD 08) (e-ISBN 978-2-912143-66-2); *Ed. CIB W 83/RILEM TC166-RMS*

Report 28: Final report of RILEM TC 167-COM 'Characterisation of Old Mortars with Respect to Their Repair (ISBN: 978-2-912143-56-3); *Eds. C. Groot, G. Ashall and J. Hughes*

Report 29: Pavement Performance Prediction and Evaluation (PPPE): Interlaboratory Tests (e-ISBN: 2-912143-68-3); *Eds. M. Partl and H. Piber*

Report 30: Final Report of RILEM TC 198-URM 'Use of Recycled Materials' (ISBN: 2-912143-82-9; e-ISBN: 2-912143-69-1); *Eds. Ch. F. Hendriks, G. M. T. Janssen and E. Vázquez*

Report 31: Final Report of RILEM TC 185-ATC 'Advanced testing of cement-based materials during setting and hardening' (ISBN: 2-912143-81-0; e-ISBN: 2-912143-70-5); *Eds. H. W. Reinhardt and C. U. Grosse*

Report 32: Probabilistic Assessment of Existing Structures. A JCSS publication (ISBN 2-912143-24-1); *Ed. D. Diamantidis*

Report 33: State-of-the-Art Report of RILEM Technical Committee TC 184-IFE 'Industrial Floors' (ISBN 2-35158-006-0); *Ed. P. Seidler*

Report 34: Report of RILEM Technical Committee TC 147-FMB 'Fracture mechanics applications to anchorage and bond' Tension of Reinforced Concrete Prisms – Round Robin Analysis and Tests on Bond (e-ISBN 2-912143-91-8); *Eds. L. Elfgren and K. Noghabai*

Report 35: Final Report of RILEM Technical Committee TC 188-CSC 'Casting of Self Compacting Concrete' (ISBN 2-35158-001-X; e-ISBN: 2-912143-98-5); *Eds. Å. Skarendahl and P. Billberg*

Report 36: State-of-the-Art Report of RILEM Technical Committee TC 201-TRC 'Textile Reinforced Concrete' (ISBN 2-912143-99-3); *Ed. W. Brameshuber*

Report 37: State-of-the-Art Report of RILEM Technical Committee TC 192-ECM 'Environment-conscious construction materials and systems' (ISBN: 978-2-35158-053-0); *Eds. N. Kashino, D. Van Gemert and K. Imamoto*

Report 38: State-of-the-Art Report of RILEM Technical Committee TC 205-DSC 'Durability of Self-Compacting Concrete' (ISBN: 978-2-35158-048-6); *Eds. G. De Schutter and K. Audenaert*

Report 39: Final Report of RILEM Technical Committee TC 187-SOC 'Experimental determination of the stress-crack opening curve for concrete in tension' (ISBN 978-2-35158-049-3); *Ed. J. Planas*

Report 40: State-of-the-Art Report of RILEM Technical Committee TC 189-NEC 'Non-Destructive Evaluation of the Penetrability and Thickness of the Concrete Cover' (ISBN 978-2-35158-054-7); *Eds. R. Torrent and L. Fernández Luco*

Report 41: State-of-the-Art Report of RILEM Technical Committee TC 196-ICC 'Internal Curing of Concrete' (ISBN 978-2-35158-009-7); *Eds. K. Kovler and O. M. Jensen*

Report 42: 'Acoustic Emission and Related Non-destructive Evaluation Techniques for Crack Detection and Damage Evaluation in Concrete' - Final Report of RILEM Technical Committee 212-ACD (e-ISBN: 978-2-35158-100-1); *Ed. M. Ohtsu*

Report 45: Repair Mortars for Historic Masonry - State-of-the-Art Report of RILEM Technical Committee TC 203-RHM (e-ISBN: 978-2-35158-163-6); *Eds. Paul Maurenbrecher and Caspar Groot*

Report 46: *S*urface delamination of concrete industrial floors and other durability related aspects guide - Report of RILEM Technical Committee TC 268-SIF (e-ISBN: 978-2-35158-201-5); *Ed. Valerie Pollet*

Cementitious Materials

Active Rheology Control of Cementitious Materials: Comparing Residual Materials with Potentially Magnetic Particles

P. Heik(✉), F. Ehle, T. Echt, and L. Göbel

Chair of Mechanics of Engineering Materials, F. A. Finger-Institute for Building Material Science, Bauhaus-Universität Weimar, Weimar, Germany

paula.heik@uni-weimar.de

Abstract. Active Rheology Control (ARC) represents an innovative concept to control the rheological behavior of cementitious materials. By incorporating magneto-responsive particles, rheological properties can be modified in a targeted manner through the application of an external magnetic field. However, many of the particles studied present challenges in terms of cost and availability, limiting their suitability for large-scale applications. Therefore, there is an increasing need to identify more economical and environmentally friendly alternatives. In this study, two different residual materials in the form of slags, representing promising candidates to exhibit magneto-responsive behavior, were investigated. Their magnetic properties were analyzed using vibrating sample magnetometry, while particle size distributions were determined by means of laser diffraction. Additionally, oscillatory rheometer tests with a magnetorheological device were used to evaluate the magnetorheological behavior of slag-including cement pastes. The results reveal that both slag materials exhibit a pronounced magnetorheological response, that is furthermore strongly dependent on the particle dosage. With these findings, the range of suitable materials for ARC expands, contributing to the development of more sustainable, eco-efficient cementitious systems with enhanced functional properties.

Keywords: Active Rheology Control · Cementitious Materials · Residual Materials · Magneto-Responsive · Oscillatory Rheometry

1 Introduction

The rheological behavior of cementitious materials is essential for their workability, buildability and mechanical performance [1]. Yet, the rheological properties must satisfy contradictory requirements in different manufacturing steps, for instance in 3D concrete printing, where the material should first exhibit a lower viscosity to be easily pumpable [2]. Afterwards, however, it should have an appropriate green strength to ensure the stability of the printed layers. Therefore, mechanisms which allow for on-demand influencing of the rheological properties of the cementitious materials are of great interest. One approach is Active Rheology Control (ARC), according to which the

M. Posani and S. Singh (Eds.): RYS 2025, RILEM Bookseries 67, pp. 3–11, 2026.
https://doi.org/10.1007/978-3-032-14652-6_1

rheological properties are altered by targeted external stimuli [3]. This can be achieved by vibration [4], microwaves [5], heat [6] or magnetic fields [7]. The present paper focuses on the magnetorheological activation in cement pastes, as this concept has proven to be promising. For this concept, magneto-responsive particles are added to the cement paste and excited by an external magnetic field [8]. The initially uniformly distributed magneto-responsive particles align in the cement paste matrix to form columnar structures upon magnetic stimulation [9] leading to an increase of the storage modulus [10]. This approach is derived from magnetorheological fluids and can even be performed reversibly [11].

To date, mostly pure materials such as carbonyl iron powder [12–15] or Fe_3O_4 nanoparticles [7–9, 11, 12, 16, 17] have been used as magneto-responsive particles for ARC. Nevertheless, they exhibit certain limitations in terms of cost and availability, which constrain their suitability for large-scale use. Therefore, it is of particular importance for the further enhancement of ARC techniques, to find economically and ecologically appropriate alternatives. This is where residual materials, some of which have already been used as cement fillers, come into play. First attempts have been made with fly ash [18–20] and fayalite slag [21], revealing that the materials are sufficiently magneto-responsive for ARC. The aim of the present study is to expand the selection of suitable materials, focusing particularly on influences of their dosage on the magnetorheological response in the cement paste.

To address this, two different types of slags were selected as residual materials, and their magnetic properties were investigated by vibrating sample magnetometry. Particle size distributions were measured by laser diffraction. Oscillatory rheometer tests in the form of Small Amplitude Oscillatory Shear tests (SAOS, time-sweep tests) were performed in the absence and presence of an external magnetic field.

2 Methodology

2.1 Materials

An ordinary Portland cement CEM I 42.5 R (Heidelberg Materials, Germany) with a density of 3.14 g/cm^3 was used for the investigations. As residual materials, two different types of slags were used: electric arc furnace (EAFS, steel plant Eisenhüttenstadt, Germany) and modified steel slag (MSS), referred to as LiDonite (Thyssenkrupp Steel Europe AG, Germany). EAFS exhibits a density of 4.13 g/cm^3 and MSS a considerably smaller density of 3.75 g/cm^3.

The two slags differ in their manufacturing process. While EAFS is formed during the production of crude steel from recycled steel, MSS originates when SiO_2 is blown into liquid basic oxygen furnace (BOF) slag [22]. The particle size distributions of the materials were determined by laser diffraction particle size analyser (Beckmann Coulter, LS 13320XR).

2.2 Methods

Magnetic Characterization The magnetic characterization of the residual materials in carefully compacted powder form was carried out at 25 °C using vibrating sample magnetometry (VSM, Mini Cryogen Free Magnet System equipped with Vibrating Sample Magnetometer Measurement Option, Cryogenic Ltd., United Kingdom) within a magnetic field range of ±2 T. Following the usual procedure, the samples were placed in a uniform magnetic field and vibrated sinusoidally, inducing a voltage in surrounding pickup coils that is proportional to its magnetic moment. By sweeping the applied magnetic field while recording the induced signal, the magnetization as a function of field (σ–H-curve) was obtained [23].

Preparation Procedures The cement pastes were prepared with a water-to-binder *(w/b)* ratio of 0.4 using deionized water. Two different dosages were studied: 70% by volume cement and 30% by volume residual material (EAFS or MSS, respectively) as well as 85% by volume cement and 15% by volume residual material. Before mixing with water, the dry components were manually mixed to ensure uniformity. The water was then added to the mixing tube of an Ultra-Turrax Tube Drive system (IKA), followed by the gradual addition of the cement–residual material blend over the course of 60 s while continuously stirring at approximately 600 rpm. Manual mixing was performed briefly in between to loosen any material adhering to the walls of the tube. The paste was then subjected to high-speed mixing at 6000 rpm for 180 s to ensure thorough dispersion.

Small Amplitude Oscillatory Shear Tests Rheological characterization was conducted using a rotational rheometer (Anton Paar MCR 102e) equipped with a 20 mm parallel plate geometry and a gap of 1 mm. A corresponding magneto-rheological device (MRD 70/1 T) was used to apply a magnetic field perpendicular to the face of the plates. For each sample type, the relationship between applied current and resulting magnetic flux density in the gap was calibrated to ensure consistent field strengths across materials. The structural evolution and viscoelastic behavior of the pastes were examined via small amplitude oscillatory shear (SAOS) testing within the linear viscoelastic region (LVER). The latter was identified through strain sweep tests ranging from 0.0001% to 10% strain amplitude at a frequency of 2 Hz. SAOS measurements were conducted to probe the linear viscoelastic behavior of the samples over time. Prior to oscillatory testing, the samples were subjected to a pre-shearing step at a shear rate of 100 s^{-1} for 100 s in order to erase previous shear history and ensure a reproducible microstructural state. Subsequently, the material was excited with a strictly sinusoidal shear strain at a constant frequency of 2 Hz and an amplitude of 0.004%, within the linear viscoelastic regime. The resulting shear stress response was recorded, and the storage modulus (G'), representing the material's elastic response to oscillatory strain within the LVER, was calculated from the phase shift between strain and stress. Temperature was controlled at 20 ± 0.5 °C throughout the experiments, and data were logged at 5-s intervals for 900 s. The magnetic flux density (0 or 0.5 T) was held constant throughout the entire SAOS measurement. Each formulation was tested twice, with a freshly prepared sample for every run.

3 Results and Discussion

3.1 Material Properties

The particle size distributions of the materials are given in Fig. 1. The MSS has the smallest particles with a median, d_{50}, of 5.51 µm, while the cement has the largest particles with a d_{50}-value of 13.06 µm. EAFS has a d_{50} of 6.36 µm. While EAFS exhibits two pronounced peaks in its particle size distribution, MSS displays a broader distribution. Notably, MSS contains a higher proportion of particles smaller than 2 µm, while EAFS contains more particles around 60 µm.

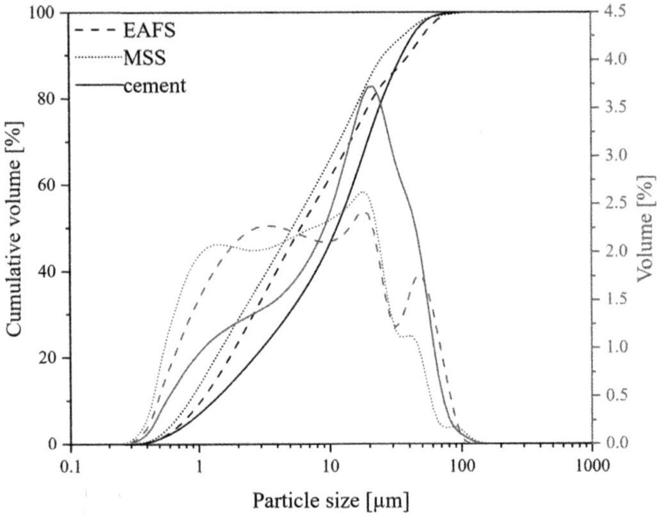

Fig. 1. Particle size distribution of the used materials.

The magnetic hysteresis loops, measured via VSM, of the materials used in this study are illustrated in Fig. 2. The cement exhibits negligible magnetic behavior with a saturation magnetization of 0.09 Am^2/kg and a remanence magnetization of 0.003 Am^2/kg. EAFS and MSS exhibit saturation magnetizations of 5.5 Am^2/kg and 3.9 Am^2/kg and remanence magnetizations of 1.0 Am^2/kg and 0.9 Am^2/kg, respectively. Thus, both slags have the same degree of magnetic orientation once the field is gone, EAFS shows a higher degree of magnetic orientation when a magnetic field is applied.

The coercive fields of EAFS, MSS and cement are 18.9 kA/m, 30.08 kA/m and 19.74 kA/m, respectively. This implies that EAFS is more easily demagnetized than MSS.

3.2 Small Amplitude Oscillatory Shear Tests

The results of the SAOS tests for the cement pastes ($w/b = 0.4$) with residual materials in different dosages at magnetic flux densities of 0 T and 0.5 T are provided in Fig. 3. Comparatively, a plain reference (Ref) is also shown, measured for zero flux density.

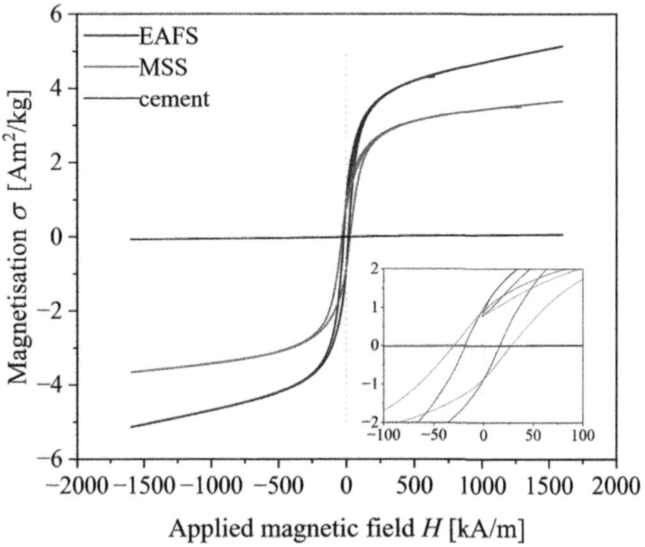

Fig. 2. Measured magnetization loops of the materials used in this study.

Due to the weak magnetic properties of cement, the respective curve at 0.5 T is not presented. All cement pastes that include residual materials exhibit a larger storage modulus at 0.5 T when comparing the storage modulus evolutions measured at 0.5 T and 0 T. Therefore, the magneto-responsive properties of both slags suggest their potential as candidates for ARC. For both materials, the magnetorheological response is expectedly more pronounced at 30 vol.% addition than at 15 vol.%.

In the absence of a magnetic field, the storage modulus evolutions differ for the three mixtures. In case of MSS, values of *G'* are below those of the reference irrespective of the added slag volume, indicating a reduced hydration reaction as a consequence of the partial cement substitution by the residual material. As a result, the development of more a solid-like behavior, reflected in an increase in storage modulus, is slower than in pure cement paste. For EAFS, in contrast, this is only observed at a slag content of 15 vol.%. Still, this explanation does not fully account for the behavior of the 30 vol.% EAFS sample, where the 0 T curve partially exceeds the reference line. This may result from altered hydration dynamics due to slag addition or from particle packing effects, as discussed in [24]. The complexity of rheological behavior is further highlighted in [7], where different magneto-responsive particles exhibited varying magnetorheological responses depending on their dosage. Notably, one of the materials studied showed a strong response at a low dosage but a diminished response at a higher dosage compared to the other materials. These observations highlight the complex interaction of material properties and dosage effects on rheology, emphasizing the importance of continued research to better understand these mechanisms.

When assessing the strength of the magnetorheological response, it is insufficient to consider the absolute value of the storage modulus with applied magnetic field due to the variations in storage modulus between the samples in the absence of a magnetic

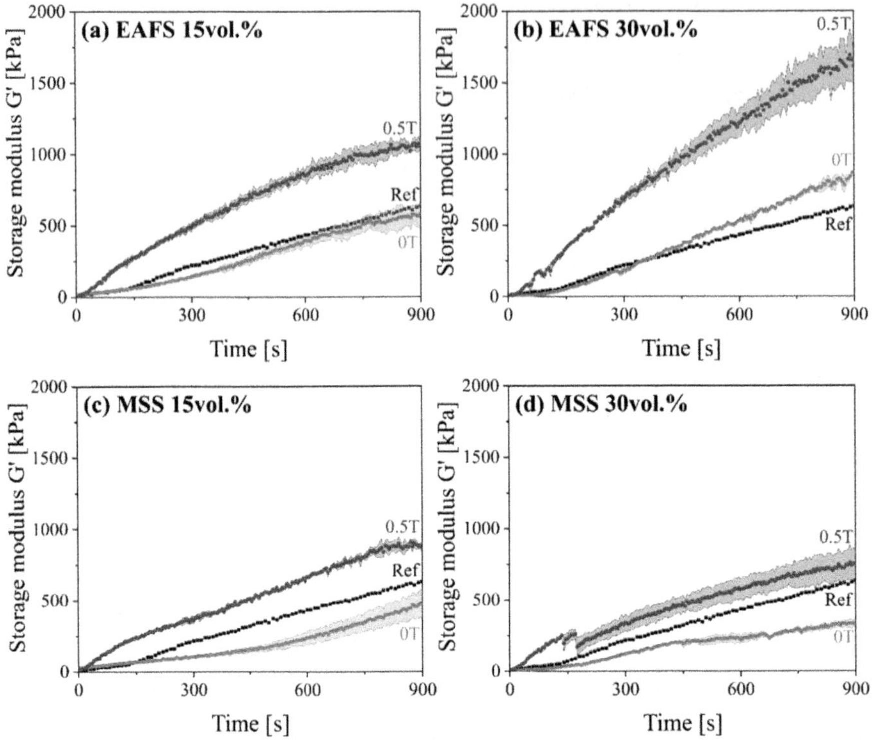

Fig. 3. Storage modulus evolutions measured during SAOS tests for the two different slag types ((a) EAFS 15 vol.%, (b) EAFS 30 vol.%, (c) MSS 15 vol.% and (d) MSS 30 vol.%) tested at flux densities of 0 T and 0.5 T. The reference (Ref) refers to the plain cement paste with no magneto-responsive particles, measured at 0 T. The error band presents the scatter between the two measurements.

field. As a consequence, the differences between the storage modulus values obtained from the measurement curves at 0.5 T and 0 T at the 300-s time instant were calculated and referred to as D_{300} [7], see Fig. 4. For both volume additions, EAFS shows the stronger magnetorheological response with 347 kPa for 15 vol.% and 486 kPa for 30 vol.%, respectively, presenting an increase of 40%. For MSS, the increase is only 23%, referring to 255 kPa at 15 vol.% and 313 kPa at 30 vol.%. Therefore, the magnetorheological response depends on the dosage of magneto-responsive particles, though not in a proportional manner and differs depending on the residual material.

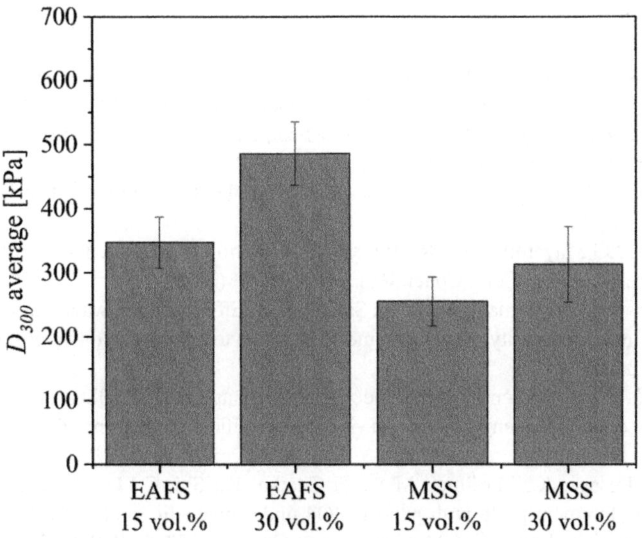

Fig. 4. Average difference between the storage modulus at 0 T and 0.5 T at 300 s (D_{300}) of the two slags with different dosages according to the SAOS measurements. The error bar refers to the sum of the mean deviations of the 0 T and 0.5 T measurements.

4 Conclusions

The present study examined the suitability of the residual materials electric arc furnace slag and modified steel slag (LiDonit) as magneto-responsive particles for ARC in cement pastes. SAOS measurements were performed to gain information on the structural build-up of the cement pastes with EAFS and MSS, comparing the storage modulus evolutions with and without an applied magnetic field. Both slags showed a distinct magnetorheological response, whereby the strength of the response depended on their added dosage in the cement paste. With increasing dosage, a more pronounced magnetorheological response was observed. The obtained results highlight the potential of EAFS and MSS as more sustainable and cost-effective alternatives to conventional magneto-responsive particles while further research is needed to clarify the influences of dosage and particle size on the magnetorheological behavior.

Acknowledgments. The authors gratefully acknowledge the financial support from the Bundesministerium für Forschung, Technologie und Raumfahrt (BMFTR) as part of the NanoMatFutur grant (grant no. 13XP5193). They also thank Dr.-Ing. Manuel Heidenreich and Prof. Dr. rer. nat. Jörg Töpfer (Department of SciTec, Ernst-Abbe-Hochschule Jena) for the performance of the VSM measurements.

References

1. Khayat, K.H., Meng, W., Vallurupalli, K., Teng, L.: Rheological properties of ultra-high-performance concrete—an overview. Cem. Concr. Res. **124**, 105828 (2019)
2. Jakathilakage, R., Rajeev, P., Sanjayan, J.: Rheometry for concrete 3D printing: a review and an experimental comparison. Buildings. **12**(8), 1190 (2022)
3. De Schutter, G., Lesage, K.: Active Rheology Control of Cementitious Materials. CRC Press, Boca Raton (2023)
4. Sanjayan, J.G., Jayathilakage, R., Rajeev, P.: Vibration induced active rheology control for 3D concrete printing. Cem. Concr. Res. **140**, 106293 (2021)
5. Muthukrishnan, S., Ramakrishnan, S., Sanjayan, J.: Effect of microwave heating on interlayer bonding and buildability of geopolymer 3D concrete printing. Constr. Build. Mater. **265**, 120786 (2020)
6. Zhang, Y., et al.: A potential active rheology control approach for 3D printable cementbased materials: coupling of temperature and viscosity modifiers. Cem. Concr. Compos. **149**, 105496 (2024)
7. Jiao, D., Lesage, K., Yardimci, M.Y., El Cheikh, K., Shi, C., De Schutter, G.: Structural evolution of cement paste with nano-Fe3O4 under magnetic field - effect of concentration and particle size of nano-Fe3O4. Cem. Concr. Compos. **120**, 104036 (2021)
8. Jiao, D., Lesage, K., Yardimci, M.Y., El Cheikh, K., Shi, C., de Schutter, G.: Rheological properties of cement paste with Nano-Fe3O4 under magnetic field: flow curve and nanoparticle agglomeration. Materials. **13**, 5164 (2020)
9. Jiao, D., Lesage, K., Yardimci, M.Y., El Cheikh, K., Shi, C., de Schutter, G.: Quantitative assessment of the influence of external magnetic field on clustering of nano-Fe3O4 particles in cementitious paste. Cem. Concr. Res. **142**, 106345 (2021)
10. Jiao, D., Lesage, K., Yardimci, M.Y., El Cheikh, K., Shi, C., de Schutter, G.: Rheological behavior of cement paste with nano-Fe3O4 under magnetic field: magneto-rheological responses and conceptual calculations. Cem. Concr. Compos. **120**, 104035 (2021)
11. Jiao, D., El Cheikh, K., Shi, C., Lesage, K., de Schutter, G.: Structural build-up of cementitious paste with nano-Fe3O4 under time-varying magnetic fields. Cem. Concr. Res. **124**, 105857 (2019)
12. Zhang, Y., Lesge, K., Zhang, Y., Tao, Y., van Tittelboom, K., de Schutter, G.: A comparison of magneto-responsive particles and testing protocols for active rheology control of cementitious materials. Cem. Concr. Compos. **146**, 105390 (2024)
13. Nair, F., Ferron, R.: Set-on-demand concrete. Cem. Concr. Res. **57**, 13–27 (2014)
14. Nair, S., Ferron, R.: Real time control of fresh cement paste stiffening: smart cement-based materials via a magnetorheological approach. Rheol. Acta. **55**, 571–579 (2016)
15. Kanagasuntharam, S., Ramakrishnan, S., Muthukrishnan, S., Sanjayan, J.: Effect of magnetorheological additives on the buildability of 3D concrete printing. J. Building Eng. **74**, 106814 (2023)
16. Chibulu, C., Jiao, D., Yardimci, M., de Schutter, G.: Magneto-rheology control of cement paste containing Fe3O4 nanoparticles in view of reducing or preventing formwork leakage. Cem. Concr. Compos. **142**, 105176 (2023)
17. Chibulu, C., Yardimci, M., de Schutter, G.: Effect of volume fraction of fine sand on magnetorheological response and blocking mechanisms of cementitious mixtures containing Fe3O4 nanoparticles. Appl. Sci. **12**(19), 10104 (2022)
18. Chibulu, C., Yradimci, M., Jiao, D., de Schryver, R., Lesage, K., de Schutter, G.: Active stiffening control by magnetically induced blocking in confined flow of fly ash pastes. Constr. Build. Mater. **313**, 125485 (2021)

19. Jiao, D., Lesage, K., Yardimci, M.Y., Shi, C., de Schutter, G.: Possibilities of fly ash as responsive additive in magneto-rheology control of cementitious materials. Constr. Build. Mater. **296**, 123656 (2021)
20. Jiao, D., Shi, C., de Schutter, G.: Magneto-responsive structural build-up of highly flowable cementitious paste in the presence of PCE superplasticizer. Constr. Build. Mater. **327**, 126925 (2022)
21. Yardimci, M., Jiao, D., de Schutter, G.: Magnetorheological properties of fayalite slag incorporated cement mixtures. In: RILEM Spring Convention and Conference, pp. 238–246. Springer Nature Switzerland (2024)
22. Pöllmann, H.: Industrial Waste - Characterization, Modification and Application. deGruyter, Berlin/Boston (2021)
23. Wecker, J., Bayreuther, G., Ross, G., Grössinger, R.: Magnetic Properties. In: Czichos, H., Saito, T., Smith, L.E. (eds.) Springer Handbook of Metrology and Testing, p. 552ff (2011)
24. Ahmadah, O., Bessaies-Bey, H., Yahia, A., Roussel, N.: A new mix design method for low-environmental-impact blended cementitious materials: optimization of the physical characteristics of powders for better rheological and mechanical properties. Cem. Concr. Compos. **128**, 104437 (2022)

Comparison between Different Methods of Determining Apparent Elasticity Modulus of Cementitious Materials

Mauro José da Silva Filho[1(✉)], Ana Karoliny Lemos Bezerra[2], Thamires Ximenes Cavalcante[3], Lucas Feitosa de Albuquerque Lima Babadopulos[1], Jean-Claude Carret[3], and Jorge Barbosa Soares[1]

[1] Federal University of Ceará, Fortaleza, Brazil
mauro.silva@det.ufc.br
[2] Ecole Centrale de Lyon, Écully, France
[3] École de Technologie Supérieure, Montréal, Canada

Abstract. This study aims to compare four different techniques for determining the apparent elasticity modulus of cementitious materials: Impact Resonance Testing (IRT), Ultrasonic Testing (UT), Static Modulus Testing (SM), and Dynamic Modulus Testing (DM). Tests were conducted on twelve cylindrical specimens comprising three conventional concretes (with and without superplasticizer), a high-performance concrete (HPC), and a cement mortar. The experimental campaign followed a sequential protocol: (i) initial characterization using UT; (ii) characterization using RM; (iii) after, evaluating the CM testing under sinusoidal loading; (iv) evaluating RM again, to check for microdamage associated with the tests and (v) final evaluating SM. All specimens were tested using the four techniques under the same conditions, allowing for direct comparison of the results. The IRT showed excellent agreement with the DM method, with differences of less than 1 GPa across all materials. The UT also presented good consistency, particularly for stiffer concretes, while the SM yielded lower average values and higher variability. Strong linear correlations were found (near-1 R^2 values) between IRT and the other testing methods analyzed. The findings confirm the potential of IRT as a reliable, fast, and non-destructive technique for evaluating the stiffness of cement-based materials. The methodology also proved effective in detecting modulus reductions caused by repeated loading cycles, especially in mortar specimens.

Keywords: Apparent Modulus · Impact Resonance Test · Ultrasonic Test · Cement Concrete · Non-destructive testing (NDT)

1 Introduction

The elasticity modulus of cementitious materials is fundamental for structural engineering, as it provides essential input for design, quality control and predicted performance in service [1]. To represent the mechanical behavior of these materials, especially in

M. Posani and S. Singh (Eds.): RYS 2025, RILEM Bookseries 67, pp. 12–25, 2026.
https://doi.org/10.1007/978-3-032-14652-6_2

analyses involving stresses and strains, the selection of suitable rheological models - such as the isotropic linear elastic model - requires the precise determination of model parameters such as the modulus of elasticity [2–4]. Various experimental methods can be used to obtain such material property, and its value may depend of the testing method, which would suggest it to be seen as a parameter. Tests may be conducted in moderate strains, in which case they are probably destructive (some of them requiring the rupture of the specimens to check for strength), while other apply lower strains and are considered non-destructive. Each method has its own characteristics in terms of application, sensitivity and repeatability.

Among the non-destructive techniques, Impact Resonance Testing (IRT) stands out for its simplicity, speed, and low cost [5]. By inducing vibrations in the specimen through a controlled impact and recording the dynamic response using accelerometers, it is possible to identify resonant frequencies associated with the vibrational modes of the material. From these frequencies, elastic moduli can be estimated and correlated with values obtained through conventional methods [6–8]. Ultrasonic Testing (UT), in turn, is based on the measurement of longitudinal wave propagation velocity through the material. This velocity, combined with the specimen's density, allows for estimating the dynamic modulus of elasticity. Recent studies have demonstrated the effectiveness of ultrasonic pulse velocity techniques for evaluating the stiffness and integrity of both asphalt mixtures and fiber-reinforced concretes [9, 10].

Between the destructive methods, ASTM C469 [11] provides a standard procedure for determining the modulus of elasticity of concrete through uniaxial compression tests, with loading limited to 40% of the characteristic strength (fc) and measurements of axial and transverse strains. Another relevant method is defined in D3497 [12], originally for asphalt mixtures, describes a method for determining the dynamic modulus of elasticity by applying a haversine compressive load to the specimen. This test yields the ratio between stress and strain amplitudes ($|E|$), typically treated as a scalar value, and is commonly applied to asphalt mixtures under conditions where viscoelastic effects are moderate. However, when the phase angle (φ) between stress and strain is considered, the material response is better described by the complex modulus (E^*), which incorporates both the elastic (storage) and viscous (loss) components. The absolute value of this quantity ($|E^*|$), represents the dynamic stiffness of the material, while the phase angle quantifies its viscoelastic nature [13, 14]. Other fields also use such kind of measurement of elasticity, where damped systems present deformation. Although initially developed for characterizing asphalt mixtures, this principle was adapted in the present study to assess stiffness degradation in cementitious materials subjected to quasi-static and cyclic loading.

Several studies have previously explored the characterization of cementitious materials using both destructive and non-destructive testing techniques. Mounier et al. [9] investigated the linear elastic properties of bituminous mixtures using ultrasonic wave propagation, highlighting its potential for dynamic modulus estimation [10]. applied ultrasonic pulse velocity methods to determine the modulus of elasticity and Poisson's ratio in fiber-reinforced concretes. Gudmarsson et al. [6], Carret et al. [7, 15] focused on IRT to evaluate stiffness degradation and monitor dynamic behavior in asphalt concrete. Bezerra et al. [16] proposed a comparative approach using S-wave measurements to align results from static, ultrasonic, and resonant tests. In addition, Carrasco et al. [17] evaluated mortars with iron ore tailings using both destructive and NDT methods, contributing to material sustainability discussions. Holanda [18] compared several methods and loading conditions to assess the modulus of elasticity in concrete, while Melo [19] examined moisture effects and self-heating in fatigue behavior. Norberto [20] also investigated the elastic modulus of mortars using multiple test methods with a focus on fatigue performance. These works underline the relevance and growing interest in comparing different techniques to evaluate stiffness properties in cementitious composites.

In this context, this study proposes a comparative analysis of four widely used techniques for obtaining the apparent modulus of cementitious materials: UT, IRT, Complex Modulus (CM) determined using a servo-hydraulic press and Static Testing (ST). The objective is to assess the consistency of the results, their limitations, and possible correlations among the methods used, considering their different natures (destructive and non-destructive). This study is aligned with the ongoing efforts of the RILEM Technical Committee MWP – *Mechanical Wave Propagation to Characterize Bituminous Mixtures*, officially launched in 2024. The committee aims to standardize the use of ultrasonic and impact resonance techniques to characterize stiffness and viscoelastic properties of asphalt and cementitious materials. As highlighted in its opening letter [21], one of the main challenges in this field is the lack of harmonized testing procedures and interpretation frameworks for non-destructive methods such as IRT and UT. By comparing multiple techniques (including IRT, UT, static and dynamic modulus) applied to cement-based materials, the present study contributes directly to the objectives of the committee, providing data and analysis that can support the development of robust protocols for laboratory and field applications.

2 Materials and Methods

The experimental campaign followed a sequential protocol: (i) initial characterization using UT; (ii) characterization using IRT; (iii) after, evaluating the CM testing under sinusoidal loading; (iv) evaluating the modulus again with IRT, to check for microdamage associated with the tests and (v) final evaluating SM (ASTM C469), which ends with the rupture of the specimen.

2.1 Materials

Four different types of Portland cement-based mixtures were used: Conventional Concrete (CC), Conventional Concrete with Superplasticizer Additive (CCS), High-Performance Concrete (HPC) with high-strength, and Mortars (M). For each mixture type, three different specimens were tested at ages ranging from 60 to 100 days. A total of 45 elastic modulus evaluations were performed for each test method considered (IRT, dynamic modulus, and static loading). Table 1 provides details regarding the specimen dimensions and Table 2 provides the composition of Portland cement-based mixtures considered in this study.

Table 1. Weights and geometries of Portland cement concretes and mortars.

Specimen	Mass (g)	Height (mm)	Diameter (mm)	Density (kg/m^3)
CC1	3591.3	195.34	99.42	2368.23
CC2	3555.7	193.78	99.55	2357.46
CC3	3545.1	195.35	98.95	2359.90
CCS1	3471.8	192.01	99.87	2308.19
CCS2	3526.0	192.98	99.57	2346.51
CCS3	3562.6	194.63	99.44	2356.92
HPC1	3701.3	193.25	100.36	2421.16
HPC2	3725.9	194.80	100.01	2434.81
HPC3	3745.7	197.97	098.99	2458.45
M1	2907.0	205.63	102.26	1721.30
M2	2851.5	201.04	101.32	1772.51
M3	2873.1	202.30	100.83	1765.26

Table 2. Composition of Portland cement-based mixtures

Component	CC (1,2,3)	CCS (1,2,3)	HPC (1,2,3)	M (1,2,3)
Cement PC V (kg/m^3)	368,02	368,02	573,89	–
Cement CP II-40 RS (kg/m^3)	–	–	–	223,00
Sand (kg/m3)	736,04	736.044	654,23	1331
Coarse aggregate (25 mm) (kg/m3)	–	–	860,83	–
Fine aggregate (12,5 mm) (kg/m3)	1104,07	1104,07	212,34	–
Water (kg/m3)	184,01	184,01	172,17	337
Silica fume	–	–	63,13	–
Superplasticizer (%)	2,21	2,58	6,89	–

2.2 Impact Resonance Test (IRT)

Figure 1a illustrates the schematic of the equipment used for conducting the IRT, as discussed in studies by [22–25]. The figure represents the prototype developed at the Federal University of Ceará (UFC) by [8]. In the method, the specimens are subjected to five controlled excitations using an instrumented impact hammer, while the response is captured as vibrational acceleration by a piezoelectric accelerometer. Both sensors (hammer and accelerometer) are integrated with load cells and connected to a signal conditioner, which filters and amplifies the analog signals prior to digital conversion. A Data Acquisition System (DAQ), operating at a 50 kHz sampling rate, converts the conditioned analog inputs into digital data. Test execution and data processing are managed via a custom Python-based interface. All measurements were conducted at room temperature (~25 °C). Figure 1b shows the average of the time-domain signals corresponding to the five impacts on specimen HPC2: input force in newtons (N) and output acceleration in m/s^2. These signals were transformed into the frequency domain using Fast Fourier Transform (FFT), as illustrated in Fig. 3c.

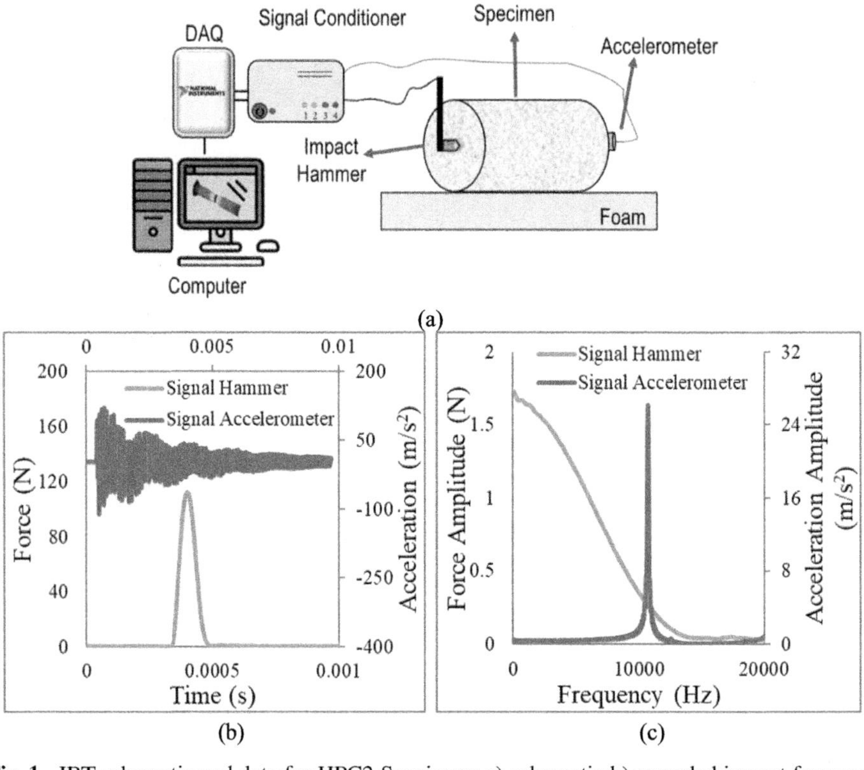

Fig. 1. IRT schematic and data for HPC2 Specimen: a) schematic b) recorded impact force and acceleration in the time domain; c)) force and frequency signals converted to the frequency domain.

Using the fundamental resonance frequency, which corresponds to the first peak of the FRF at maximum vibration, and the specimen's geometry and mass, it is possible to calculate the Resonant Modulus (RM) according to the guidelines of ASTM C215 [26], traditionally used for concrete specimens:

$$RM = D\left(L/d^2\right)M\left(n'\right)^2 \tag{1}$$

In such equation, D is equal to 5.093 for cylinders, L is the length (m), d the diameter (m), M the mass of the specimen (kg) and n' the fundamental resonance frequency (Hz).

2.3 Ultrasonic Test (UT)

Ultrasonic testing was conducted using a Pundit Lab device equipped with two types of transducers, as illustrated in Fig. 2. For the emission of P-waves, flat-surface transducers with a diameter of 50 mm and a nominal frequency of 54 kHz were used. In the case of S-waves, 84-mm diameter transducers operating at a nominal frequency of 40 kHz were employed. The measurements followed the direct transmission method, in which the emitter and receiver were positioned at opposite ends of the cylindrical specimens to ensure optimal signal transmission. Prior to each test, the equipment was calibrated by setting the delay to 0 μs when the emitter and receiver were placed in direct contact. Additionally, device settings such as gain and pulse voltage were adjusted to maximize the amplitude of the received wave signals [27].

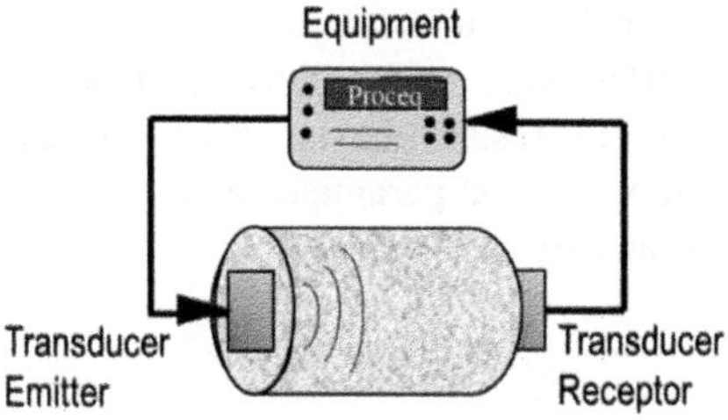

Fig. 2. UT schematic

For S waves, the intersection between the x-axis and the signal just before the waves with the lowest amplitude was adopted for ToF calculations. This is referred to as the intersection time before the arrival of the lower amplitude waves, as recommended by Rabelo et al. [29].

Following the completion of the test, the time-of-flight (ToF) for both P-waves and S-waves can be accurately determined. By applying Eqs. 2 and 3, which relate the specimen length (L) to the respective ToF values, the propagation velocities of the longitudinal and shear waves are calculated. These velocity values for P-waves and S-waves obtained using Eqs. 4 and 5, respectively, serve as the basis for computing the Ultrasonic Modulus (UM) and Poisson's ratio (v) through Eqs. 6 and 7 [28]. The use of these equations is justified by the assumption that the tested materials exhibit isotropic, homogeneous, and linear elastic behavior.

$$Cp = L/Tp \tag{2}$$

$$Cs = L/Ts \tag{3}$$

$$Cp = \left\{[(1-\nu)\ UM]/\left[(1+\nu)\ 1-2\nu\right)\rho\right]\right\}^{1/2} \tag{4}$$

$$Cs = \{[(UM/[2(1+\nu)\rho]\}^{1/2} \tag{5}$$

$$UM = Cp^2 \rho\,[(2\nu - 1)(\nu + 1)/(\nu - 1)] \quad (6)$$

$$\nu = \left[0,5(Cp/Cs)^2 - 1\right]/\left[(Cp/Cs)^2 - 1\right] \quad (7)$$

2.4 Complex Modulus (CM)

The complex elastic modulus (|E*|) quantifies the stiffness of materials subjected to cyclic loading and has been widely adopted for engineering applications - initially in France since the 1960s and later in the United States from the 1990s onward [30]. This modulus is a complex number composed of two components: the storage modulus (real part), which reflects the recoverable elastic energy stored during loading, and the loss modulus (imaginary part), which accounts for the energy dissipated as a result of viscous effects [23, 30]. As illustrated in Fig. 3, under sinusoidal loading at a given frequency (f), the applied stress (green) and the resulting axial strain (red) exhibit a phase difference. This phase angle (φ) is a fundamental parameter in viscoelastic analysis, representing the time lag between stress and strain responses.

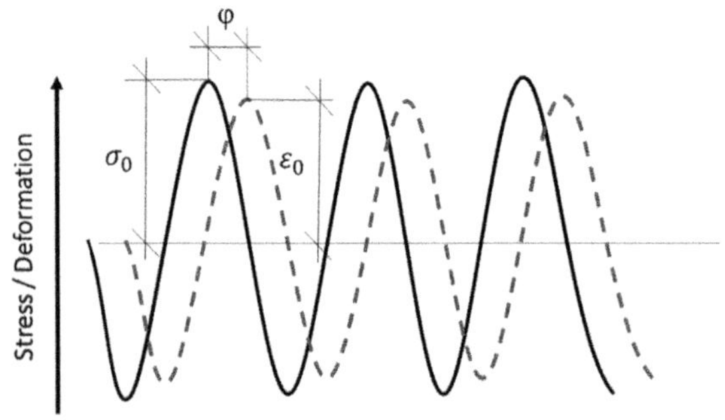

Fig. 3. Representation of stress–strain signals from the complex modulus test.

The complex modulus of elasticity of the cementitious mixtures was determined following the procedures outlined in [12], using a UTM-25 servo-hydraulic press (Fig. 4). Such tool is equipped with a high-speed data acquisition system, capable of accurately measuring mechanical responses under cyclic loading, with a maximum load capacity of 25kN (Table 3). For test execution, small metal inserts called “targets” are initially glued to the specimens - three pairs evenly spaced around the cylinder, in the middle third of the specimen. Each target pair is used to position objects used for fitting in the LVDTs. The actuator piston is responsible for applying the loads.

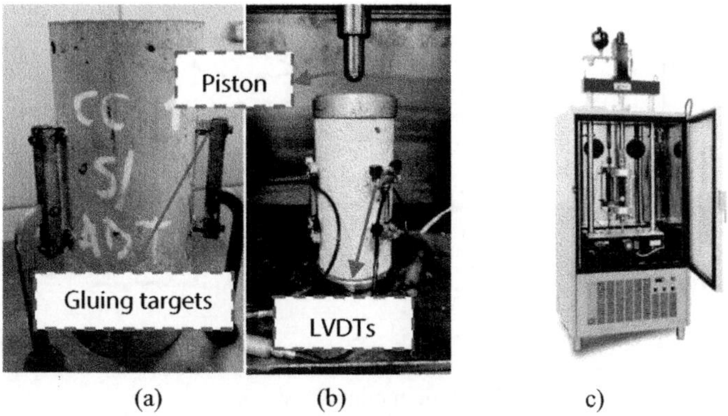

Fig. 4. Complex modulus test (a) gluing the targets, (b) test procedure and (c) UTM.

Table 3. Parameters for sinusoidal loading

Frequency	25	10	5	1	0.5	0.1
Number of cycles	20	20	15	10	5	5
Maximum Applied Stress (kPa)	2000	2000	2000	2000	2000	2000
Minimum Applied Stress (kPa)	200	200	200	200	200	200

2.5 Static Modulus (SM)

The static elastic modulus is determined by applying controlled monotonic loading while simultaneously measuring the deformation response. In experimental practice, two main approaches are distinguished [4]: (i) The initial tangent modulus (*Eci*), obtained from the slope of the tangent line at the origin of the stress-strain curve, representing stiffness in the linear elastic regime. (ii) The secant modulus, calculated as the slope of the line connecting the origin to a point corresponding to 30% of the compressive strength (*fe)*, a conventional value that precedes the significant development of microcracks in concrete.

The static modulus test (Fig. 5) was performed according to the specifications of the standard ASTM C469/C469M [11]. The procedure consists of applying a maximum compressive load equivalent to 30% of the specimen's compressive strength (*fc*), following an initial preload of 0.5 MPa, maintained for 60 s before the main loading stage, which is held for 90 s. Among the three specimens tested, one was used to determine fc. Strain measurements were recorded using two extensometers fixed to the specimen surface during the test.

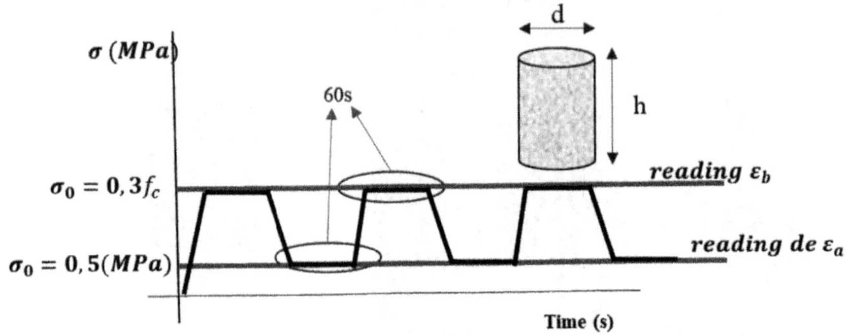

Fig. 5. Static modulus test.

3 Results and Discussion

3.1 Apparent Modulus

The average values obtained from all tests showed good agreement. The values from the Impact Resonance Test (IRT), when compared to those from the Complex Modulus (CM), differed by less than 1 GPa for all materials, reinforcing the similarity between the two methods under vibratory conditions. Ultrasonic Testing (UT), in turn, consistently produced slightly lower values than IRT, though still close, especially for materials with higher stiffness. In contrast, the Static Modulus (SM) yielded the lowest average values and the highest variability, reflecting its greater sensitivity to initial irreversible deformations.

As shown in Fig. 6, the materials exhibited the following behavior:

- CC: The average RM obtained by IRT was 40.10 $\pm$ 0.60 GPa, while UM indicated 37.42 $\pm$ 0.65 GPa. The CM reached 40.68 $\pm$ 0.29 GPa, nearly identical to the RM. The SM, however, recorded 31.99 $\pm$ 0.75 GPa, confirming the method's sensitivity to damage effects.
- CCS: Results were similar to those of CC, with RM indicating 39.00 $\pm$ 0.87 GPa and UM 35.96 $\pm$ 0.70 GPa. The CM had an average of 39.98 $\pm$ 1.20 GPa, while the SM showed greater dispersion at 33.15 $\pm$ 2.50 GPa, possibly due to increased microcracking associated with the testing procedure.
- HPC: This material presented the highest apparent modulus values. RM recorded 43.49 $\pm$ 1.18 GPa, UM 40.16 $\pm$ 0.67 GPa, and the CM 43.60 $\pm$ 1.77 GPa, with excellent consistency across all three methods. The SM was 38.44 $\pm$ 1.41 GPa – lower, but still compatible with the expected performance for this type of concrete.
- M: As expected, this group showed the lowest average values. RM resulted in 6.40 $\pm$ 0.19 GPa, UM 7.15 $\pm$ 0.72 GPa, DM 4.80 $\pm$ 0.78 GPa, and SM 5.83 $\pm$ 0.27 GPa. The greater discrepancy between methods may be attributed to the mortar's higher heterogeneity and porosity, which affect wave propagation and stress distribution.

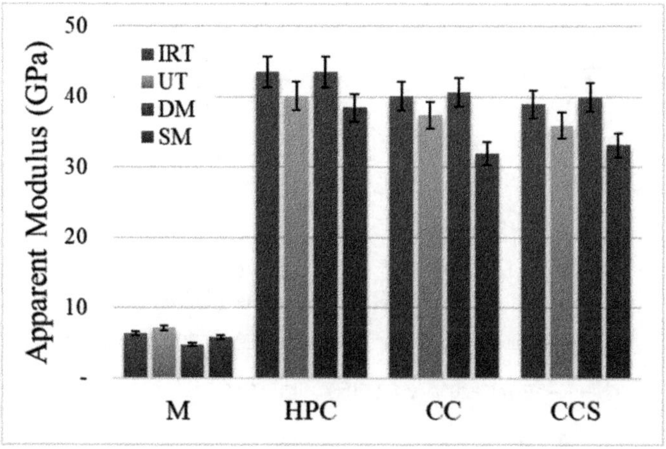

Fig. 6. Apparent modulus on different tests.

The lowest standard deviations were observed in the dynamic modulus methods, indicating greater repeatability of results, particularly for concrete (CC, CCS, and HPC). In contrast, the highest deviations were found in the SM method, especially for CCS (±2.50 GPa), suggesting that this procedure is more sensitive to testing conditions, such as extensometer placement and small variations in the surface characteristics of the specimens. Similar results were reported by [16, 17, and 18] when comparing the impact resonance test to other testing techniques.

As shown in Fig. 7a for mortar and Fig. 7b for concrete, the results revealed a strong linear correlation between the apparent modulus values obtained from IRT and those from the other methods analyzed. It is possible to see the results around values that varied up to 20% for the concretes and in the case of the mortars, the CM values were lower than this percentage. UT exhibited the highest coefficient of determination in relation to IRT, with an R^2 value of 0.9991, indicating excellent agreement between the estimated values. The CM also showed a high correlation ($R^2 = 0.9981$), confirming that both methods similarly capture the dynamic response of cementitious materials. Although the ST yielded lower average values compared to the other methods, its correlation with IRT remained high ($R^2 = 0.9867$), suggesting that despite different sensitivities, there is a consistent relationship among the results. These high coefficients of determination reinforce the reliability of IRT as a characterization technique, demonstrating its potential to replace or complement more traditional methods, especially in applications requiring speed, repeatability, and minimal destructive interference.

Moreover, it can be observed that the results obtained from the Static Modulus (SM) test provide lower values of elastic modulus, possibly due to small nonlinearities in the material behavior. Since dynamic tests (IRT, UT, and CM) are performed under significantly lower strain levels compared to traditional static tests, these results were expected and are consistent with previous investigations regarding the strain-level dependency of cementitious materials, as reported in studies by Bezerra et al. [16] and Holanda [18].

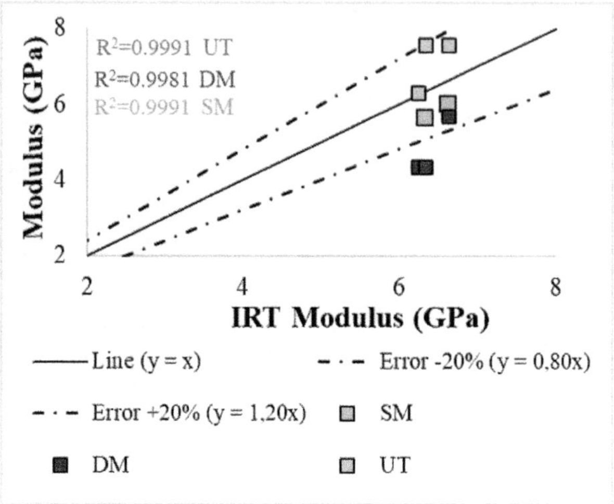

(a)

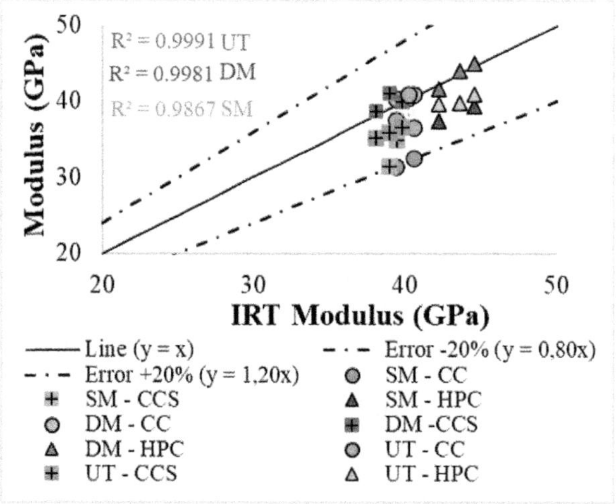

(b)

Fig. 7. Comparison between apparent modulus and impact resonant modulus (a) for Mortar and (b) for All Concretes.

After performing the tests to determine the complex modulus under cyclic loading (adapted from ASTM D3497–79), the IRT was reapplied to the same specimens (Fig. 8). The objective was to investigate possible changes in the RM, potentially associated with microdamage induced by the repeated loading cycles during the dynamic modulus test. For this purpose, the RM values obtained before and after the CM test were compared. This approach allowed the evaluation of IRT's sensitivity to detecting accumulated damage and provided insight into the structural stability of the materials under low-intensity cyclic loading. For the concrete specimens, the average variation in RM was 0.35%, indicating negligible stiffness loss. In contrast, the mortar specimens showed a significantly higher loss, with a 24.77% reduction in stiffness.

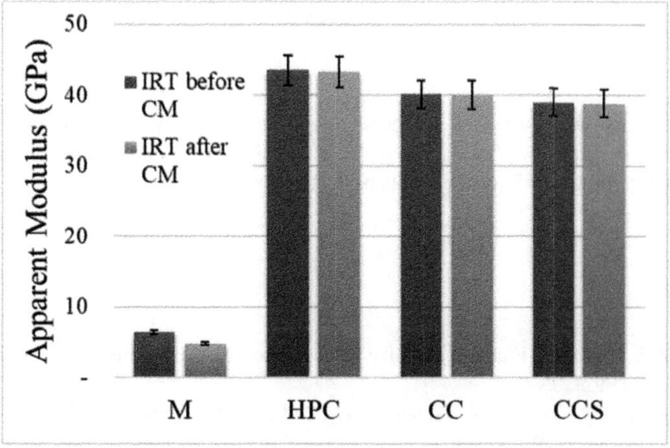

Fig. 8. Apparent modulus on IRT.

4 Conclusions

This study compared four different methods for determining the apparent modulus of cementitious materials: RM, UM, DM, SM. The experimental campaign applied all methods to the same specimens, three concretes (CC, CCS, HPC) and one mortar (M).

The results demonstrated good consistency between RM and DM, with differences below 1 GPa across all material types, reinforcing their compatibility under dynamic loading conditions. UM, while slightly lower than RM, still showed close agreement, especially for higher-stiffness concretes. In contrast, SM produced lower mean values and greater variability, indicating higher sensitivity to early microcracking and test configuration. Statistical analysis revealed strong linear correlations between RM and other methods. The highest correlation was observed with UM ($R^2 = 0.9991$), followed by DM ($R^2 = 0.9981$) and SM ($R^2 = 0.9867$). These high coefficients of determination confirm the reliability of IRT and suggest its potential as a practical and non-destructive alternative to traditional methods. Moreover, it is also evident that the SM, the most traditional testing method, yielded lower modulus values compared to the other techniques. This outcome is likely due to its inability to account for the small nonlinearities present in the specimens.

Additionally, the decrease in apparent modulus after repeated loading, particularly in mortar specimens, highlights that even quasi-static or "non-destructive" tests can induce measurable changes in material stiffness. The use of IRT before and after these loading protocols proved to be effective in identifying such variations.

Overall, the methodology developed in this work shows strong potential for future applications in:

- Non-destructive evaluation of structural integrity;
- Monitoring damage accumulation under service loads;

- Supporting the development of standards for dynamic testing of cement-based materials.

Acknowledgments. This work was financially supported by the Foundation for Scientific and Technological Development of the State of Ceará (FUNCAP). Authors also acknowledge the support of CNPq and CAPES for the grants of the authors.

References

1. Bazant, Z.P., Planas, J.: Fracture and Size Effect in Concrete and Other Quasibrittle Materials. CRC Press (1997)
2. Christensen, R.M.: Theory of Viscoelasticity, 2nd edn. Academic, New York (1982)
3. Darwin, D., Dolan, C.W., Nilson, A.H.: Design of Concrete Structures, 14th edn. McGraw-Hill Education, New York (2010)
4. Mehta, P.K., Monteiro, P.J.M.: Concrete: Microstructure, Properties, and Materials. McGraw-Hill (2014)
5. Boz I (2016) Characterization of asphalt mixtures and RAP binder properties through impact resonance test. PhD thesis, Pennsylvania State University.
6. Gudmarsson, A., Ryden, N., Birgisson, B.: Characterizing the low strain complex modulus of asphalt concrete specimens through optimization of frequency response functions. J. Acoust. Soc. Am. **132**(4), 2304–2312 (2012)
7. Carret, J.C., Pedraza, A., Di Benedetto, H., Sauzeat, C.: Comparison of the 3D linear viscoelastic behavior of asphalt mixes determined with tension-compression and dynamic tests. Constr. Build. Mater. **174**, 529–536 (2018)
8. Bezerra AKL (2023) Development of a Non-destructive Test for Viscoelastic Characterization of Asphalt Mixtures. MSc Dissertation, Federal University of Ceará.
9. Mounier, D., Di Benedetto, H., Sauzeat, C.: Determination of bituminous mixtures linear properties using ultrasonic wave propagation. Constr. Build. Mater. **36**, 638–647 (2012). https://doi.org/10.1016/j.conbuildmat.2012.04.136
10. Carrillo, J., Ramirez, J., Lizarazo-Marriaga, J.: Modulus of elasticity and Poisson's ratio of fiber-reinforced concrete in Colombia from ultrasonic pulse velocities. J Build Eng. **23**, 18–26 (2019). https://doi.org/10.1016/j.jobe.2019.01.016
11. ASTM C469/C469M-22: Standard Test Method for Static Modulus of Elasticity and Poisson's Ratio of Concrete in Compression. ASTM International (2022)
12. AASHTO D3497-79 (1979) Standard test method for dynamic modulus of asphalt mixtures. American Association of State Highway and Transportation Officials.
13. Ferry, J.D.: Viscoelastic Properties of Polymers, 3rd edn. Wiley, New York (1980)
14. Pritz, T.: Frequency dependences of complex moduli and complex Poisson's ratio of real solid materials. J. Sound Vib. **214**, 83–104 (1998)
15. Carret, J.C., Di Benedetto, H., Sauzeat, C.: Linear viscoelastic behavior of asphalt mixes from dynamic frequency response functions. Int. J. Numer. Anal. Methods Geomech. **44**(7), 1–13 (2020)
16. Bezerra, A.K.L., Melo, A.R.S., Freitas, I.L.B., Babadopulos, L.F.A.L., Carret, J.C., Soares, J.B.: Determination of modulus of elasticity and Poisson's ratio of cementitious materials using S-wave measurements to get consistent results between static, ultrasonic and resonant testing. Constr. Build. Mater. **398**, 132456 (2023). https://doi.org/10.1016/j.conbuildmat.2023.132456

17. Carrasco, E.V.M., Magalhães, M.D.C., Santos, W.J.D., Alves, R.C., Mantilla, J.N.R.: Characterization of mortars with iron ore tailings using destructive and nondestructive tests. Constr. Build. Mater. **131**, 31–38 (2017). https://doi.org/10.1016/j.conbuildmat.2016.11.065
18. Holanda ADC (2023) Caracterização do módulo de elasticidade de concretos por meio de diferentes métodos e condições de carregamento. BSc Thesis, Federal University of Ceará.
19. Melo ARS (2022) Avaliação da influência da umidade e investigação do autoaquecimento na fadiga de concreto. BSc Thesis, Federal University of Ceará.
20. Noberto CC (2022) Avaliação do módulo de elasticidade de argamassas utilizando diferentes métodos de ensaio com vistas a uma caracterização quanto à fadiga. MSc Thesis, Federal University of Ceará.
21. Carret, J.C., Babadopulos, L.F.A.L.: Opening letter of RILEM TC MWP: mechanical wave propagation to characterize bituminous mixtures. RILEM Tech Lett. **10**, 15–21 (2025)
22. Gudmarsson, A., Ryden, N., Birgisson, B., Di Benedetto, H., Sauzeat, C.: Characterizing the complex modulus and Poisson's ratio of asphalt concrete specimens through modal testing. In: Proceeding International Symposium Non-Destructive Testing in Civil Engineering (2015)
23. Carret JC (2018) Linear viscoelastic characterization of bituminous mixtures from dynamic tests back analysis. PhD Thesis, Université of Lyon (ENTPE).
24. Bezerra, A.K.L., et al.: Desenvolvimento de ensaio de ressonância por impacto para determinação de rigidez de diferentes materiais. Transport. **30**(3), 2757 (2022). https://doi.org/10.14295/transportes.v30i3.2757
25. Filho, M.J.S., Babadopulos, L.F.A.L., Carret, J.C., Soares, J.B.: Influence of test configuration parameters on the results of impact resonance tests for stiffness determination. In: Proceedings 38th Congresso de Pesquisa e Ensino Em Transportes, Florianópolis (2024)
26. ASTM C215-19 (2019) Standard test method for fundamental transverse, longitudinal, and torsional resonant frequencies of concrete specimens. ASTM International.
27. Proceq (2014) PUNDIT PL-200 Operating Instructions.
28. Mandel J (1966) Cours de mécanique des milieux continus.
29. Rabelo KR, Babadopulos LFAL, Soares JB (2020) Caracterização de rigidez de misturas asfálticas por meio de ensaios não destrutivos. In: Proc 34° Congresso de Pesquisa e Ensino em Transporte da ANPET, Online.
30. Lucas Júnior JLO (2018) Influência da adesividade agregado-ligante no trincamento por fadiga de misturas asfálticas. MSc Dissertation, Federal University of Ceará.

Correlating In-line Sensor Data to Hardened State Material Properties in Digital Fabrication with Concrete

Derk Bos(✉), Arjen Deetman, and Rob Wolfs

Eindhoven University of Technology, Eindhoven 5612 AZ, The Netherlands
d.h.bos@tue.nl

Abstract. High resolution quality control is essential for many digital concrete fabrication processes, since the systems used are often sensitive to disturbances and the objects that are produced more sensitive to defects when optimized towards material minimization. Furthermore, high-resolution quality control could decrease the uncertainty of the object's material properties and thereby reduce the cement consumption. In-line measurements are preferred over off-line or on-line measurements to obtain this high-resolution control, since it would require less labor, could be conducted at higher frequency, and would measure the same material that will later be placed in the product. However, the properties of the hardened state govern the quality of the finished concrete objects to a large extent. Since these properties are typically determined after 28 days or later, and tested using destructive, off-line methods, high-resolution quality control would be laborious and material intensive. Furthermore, making informed decisions based upon the results would be significantly delayed.

To close this gap, this contribution aims to investigate whether there exists a correlation between in-line measured sensor data and off-line measured hardened state properties obtained at a later age. For the current study, a 3D concrete printing system is used that is equipped with various sensors. The material's strength is determined at the deposition age by the slugs test, and at an age of 15 min, 1.5 h, 3 h, 1 day, 2 days, 7 days, and 28 days by compression tests.

For variations in the water-to-solids ratio, the results show that with increasing age the correlation between in-line quality indicators and the material's strength is decreasing. At deposition and ages of 15 min, 1.5 h, and 3 h, the R-squared adjusted is above 0.9. For 1 and 2 days, it is around 0.8, and for 7 and 28 days, a negligible correlation with and R-squared of around 0.2 is found, which can be attributed to a low sensitivity of the strength at these ages to variations in the water-to-solids ratio.

Keywords: Quality control · hardened state properties · 3d concrete printing · digital fabrication of concrete · in-line sensors

1 Introduction

In recent decades, digital fabrication with concrete (DFC) methods have been developed to address the need for more sustainable and productive concrete production [1]. These techniques eliminate the need for formwork partially or completely, and, thereby, allow

M. Posani and S. Singh (Eds.): RYS 2025, RILEM Bookseries 67, pp. 26–34, 2026.
https://doi.org/10.1007/978-3-032-14652-6_3

for an object shape that follows the function rather than the formwork. This, in turn, allows for optimized concrete products that reduce the material need and, thereby, could reduce the emissions that are associated with the construction industry. In addition, these production techniques are designed to be automated and could therefore improve the productivity of the production process.

With the optimization of the product towards material minimization comes also a cost in the form of reduced robustness. A small defect in the quality could, therefore, lead to a major defect in the product's resistance. To fully enable the potential that DFC has to offer in terms of material minimization, it is therefore essential to control the quality of the product at a high resolution [2]. Controlling this quality at a high resolution using traditional off-line experiments would be labor-intensive and cost-ineffective. Therefore, new quality control experiments have been proposed, of which the slugs test appears to be the most relevant for industrial applications [3]. Here, the droplet size is measured after separation from the nozzle, either by measuring its mass with a load cell [4] or its volume with a camera [5]. This droplet size is correlated to the tensile strength of the material, which is an important indicator of the quality in the fresh state. Droplets typically separate at a frequency of 1 Hz, and minimal labor is required to conduct the test; therefore, it complies with the needs of an industry-ready high-frequency measurement. However, it does not allow for measuring the quality of the product itself since it can only be performed before or after the production of the actual product. To close this gap and complete a quality control procedure for the fresh state of the material, it has been proposed to draw correlations between the tensile strength determined from the slugs test, and in-line readings of sensor data [2, 6]. Such a correlation could be robust, meaning it is governing for multiple system and material configurations, as long as the material is in a similar state during pumping and mixing as it is at deposition.

However, the quality of the product during the use phase is to a large extent governed by the hardened state properties. Therefore, this paper aims to investigate whether in-line sensor reading could also be used to obtain a high-resolution reading of these hardened state properties. When chemical reactions start dominating the strength of the material, a robust correlation between in-line sensors and the hardened-state properties is not likely to exist. However, for a given system and material, a correlation could be found and used to control the properties of the hardened state. Since this correlation is not necessarily robust, it should be checked at a frequency that is proportional to the handling volume of the production process, for example, the size of an on-site production plant, mixing truck, big bag, or mixing chamber. Then, it can be used to guarantee that the variations in hardened-state properties are within a given bound, or it can be used to detect defects when variations exceed the acceptable limits.

As a first investigation of the applicability of such a concept, the sensitivity of fresh and hardened state bulk compressive strength to variations in water-to-solids ratio is investigated, since it is expected that both material states are affected by these changes.

2 Methodology

A 3D concrete printing setup at the Structures Laboratory of Eindhoven University of Technology was used, as shown in Fig. 1. A commercially available premix (Weber 160–2) was stored in a small silo, from which it was fed to a MAI MULTIMIX through a rotary feeder. The material was then mixed with water in a semibatch operation and stored in the pumping chamber from which it was extracted continuously. The material was then pumped through a series of hoses, sensors, and tubes before it was extruded through the nozzle. Upon extrusion, either the slugs test was performed or the material was directly extruded in a mold, for the determination of the compressive strength.

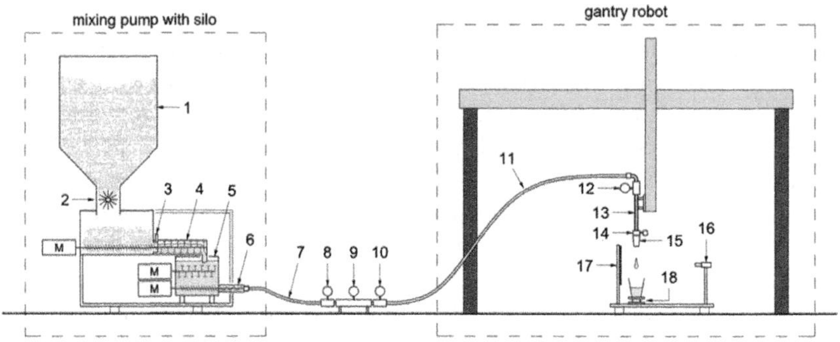

Fig. 1. Schematic overview of the 3D concrete printing system with the mixing pump, sensing station, and gantry robot with printhead. 1) Silo with dry premix. 2) Rotary feeder 3) Water inlet. 4) Horizontal mixing chamber. 5) Pumping chamber with agitating screw and temperature sensor. 6) Progressive cavity pump. 7) 1 m long 35 mm diameter hose. 8) Mortar pressure sensor P_1. 9) Coriolis-type sensor for temperature, density, and viscosity measurement. 10) Mortar pressure sensor P_2. 11) 13.3 m long 35 mm diameter hose. 12) Mortar pressure sensor P_3. 13) 1-m-long steel tube. 14) Temperature sensor. 15) Round nozzle reducing from 35 to 25 mm diameter. 16) Industrial camera. 17) Backlight. 18) Load cell.

Subsequently, the water content in the MAI MULTIMIX was set to 180, 170, and 190 L/h, resulting in water-to-solids ratios of 11.4, 10.7, and 12.1%, respectively. The tensile strength was tested at the deposition age by the slugs test [5] in 30-min intervals per water-to-solids ratio, during which a total of 1743 measurements were conducted. For the other ages, the material was cast into molds with internal dimensions of approximately 70 x 140 mm (diameter x height). The compressive strength of the samples was defined by unconfined uniaxial compression tests at ages of 15 min, 1.5 h, and 3 h, [7] and by standard compression tests at 1 day, 2 days, 7 days, and 28 days, according to EN 12390–3. For each water-to-solids ratio, 5 repetitions were sampled, including samples that were omitted due to imperfections. For the 15-min, 1.5-h, 3-h, and 1-day samples, the tests were conducted directly on the demoulded samples, while for the other ages, approximately 10 mm of the top side of the sample was sawn off to reduce the surface roughness.

3 Results

3.1 Results as a Function of Water Content

The strength as a function of water-to-solids ratio and material age is presented in Fig. 2. For all ratios, the strength increases approximately linearly up to a material age of 1.5 h, quadratically up to 3 h, and even faster up to the 1-day age. Thereafter, the strength development slows down significantly. The highest strength is obtained from the lowest water-to-solids ratio in the fresh state, while the highest strength is obtained from the middle ratio in the hardened state. At all ages, the lowest strength is obtained from the water-to-solids ratio. To move towards the aim of the current paper, we investigate whether this variation in strength can be explained or predicted by some of the in-line measurements.

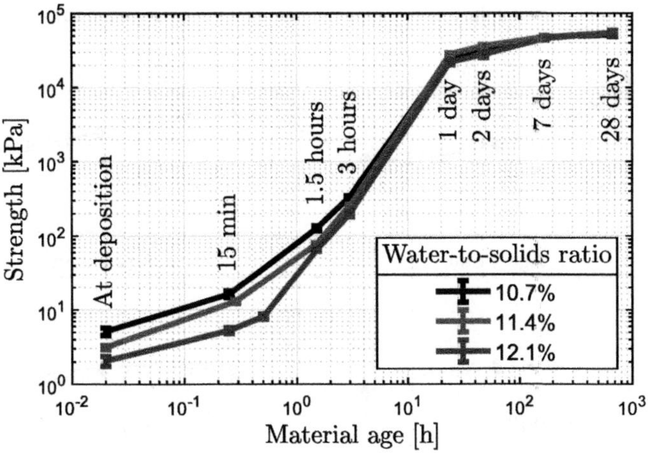

Fig. 2. Strength as a function of material age for water-to-solids ratios of 10.7, 11.4, and 12.1%.

Next, the normalized strengths per age are plotted in Fig. 3 to show how the strength varies as a function of the water-to-solids ratio. Furthermore, the trend is captured by a quadratic fit, and the adjusted R-squared is calculated to determine how much of the variation in strength could be explained by the variation in water-to-solids ratio. With increasing age, the adjusted R-squared is relatively constant above 0.9 up to an age of 3 h. At 1 and 2 days, the adjusted R-squared is around 0.8, and at 7 and 28 days it is around 0.2. This indicates that the percentage of the total variation that can be explained by the variation in water-to-solids ratio is decreasing with increasing age.

3.2 Results as a Function of In-line Indicators

In addition to the controlled variation that is induced on the material by changing the water-to-solids ratio, uncontrolled variations could have occurred as well. Therefore, it is of interest to consider the in-line indicator as predictors of the strength. This concept has been researched in depth [8] from which it has been concluded that from the four indicators *motor power of motor M3, pressure difference over the pipe* (P_2-P_1, *detail 9 and 7), torsional exciter current of the Coriolis sensor (detail 8),* and *pressure difference over the nozzle* (P_3, *detail 12)*, see Fig. 1, the nozzle pressure had the highest correlation with the bulk rheological properties. For conciseness, only the pressure difference over the nozzle is taken as a predictor for the current research.

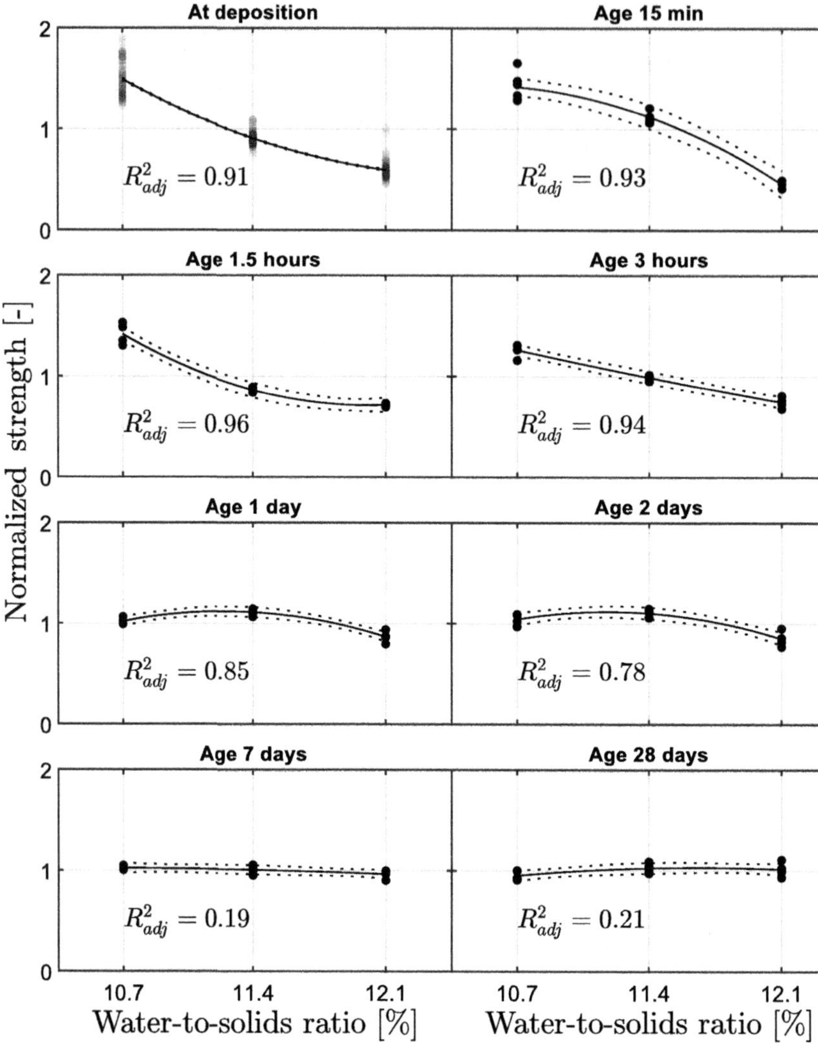

Fig. 3. Normalized strength as a function of water-to-solids ratio of the material at deposition (slugs test), at an age of 15 min, 1.5 h, 3 h (unconfined uniaxial compression test), and at an age of 1 day, 2 days, 7 days, and 28 days (compression tests).

The R-squared adjusted values are calculated as is done for the water-to-solids ratio in Fig. 3. The results are presented in Fig. 4, where it is observed that when taking the nozzle pressure as a predictor, similar R-squared adjusted values are obtained.

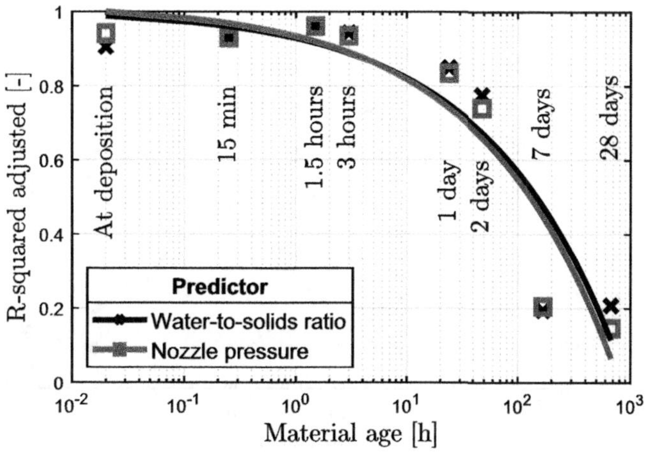

Fig. 4. R-squared adjusted as a function of material age with the water-to-solids ratio (black) and the nozzle pressure (red) as predictors of variations in strength.

4 Discussion

In principle, the decreasing R-squared values with increasing age, as shown in Fig. 4 could be obtained in two ways. First, the total variation could increase with increasing age, which would hide any effect of a variation in the water-to-solids ratio behind other sources of variation. Second, the hardened-state properties could be insensitive to variations in the water-to-solids ratio. To some extent, these two explanations coexist for any correlation, and the question reduces to: what is the magnitude of the random (unexplained) variation with respect to the effect size? This is exactly what is measured by the R-squared values that were calculated. Nevertheless, to gain more insight into the potential causes of the reduced R-squared values, it is useful to calculate the relative standard deviation of strength as a function of material age, which is shown in Fig. 5. It is shown to decrease with increasing material age, making the most likely explanation that the hardened-state material is relatively insensitive to the applied variations in water-to-solids ratio.

This is further supported by the fact that the fresh-state yield stress diverges at the volume fraction that marks the transition from a plastic to a frictional material. For a monodisperse suspension of perfect spheres, this volume fraction lies at 0.61, while it is typically slightly lower for monodisperse suspensions with irregularly shaped particles at around 0.59 [9, 10]. The hardened state strength diverges at the maximal packing density [11] which lies at a volume fraction of 0.64 for monodisperse spheres. For polydisperse suspensions, these diverging volume fractions lie higher, but the diverging volume fractions for the fresh-state yield stress will always lie lower than the diverging volume fraction for hardened-state strength. Since the sensitivity to changes in the water-to-solids ratio is the slope of the strength-volume fraction curve, as schematically shown in Fig. 6, a decreasing variation of properties with increasing age is generally to be expected.

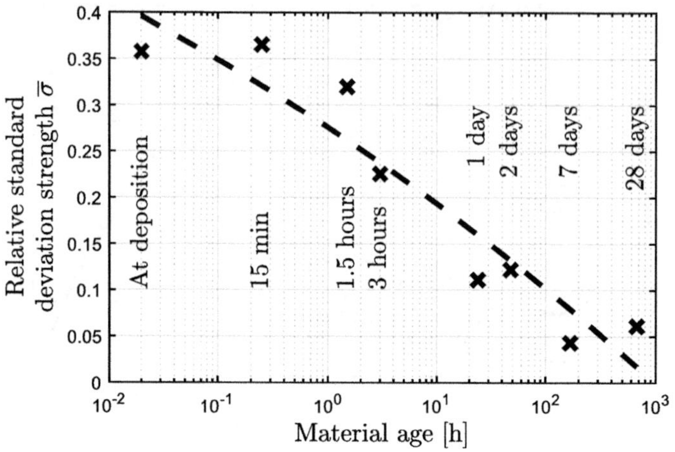

Fig. 5. Relative standard deviation of all measured strength values per material age as a function of material age.

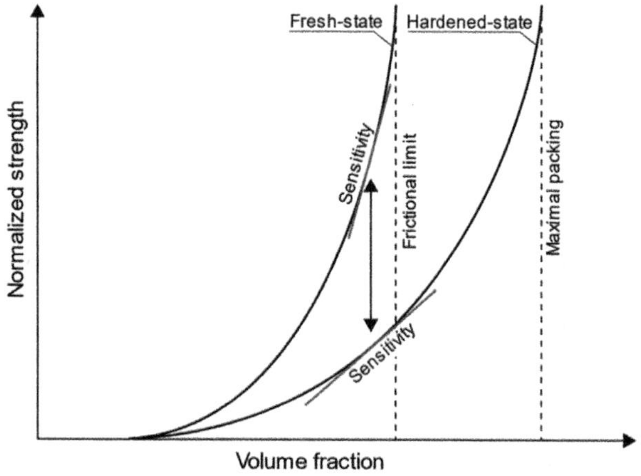

Fig. 6. Schematic representation of the diverging strength with increasing volume fraction in the fresh and hardened state, showing a decrease in sensitivity with increasing age.

5 Conclusions

In-line measurements of the system correlate well with the strength of the researched material up to an age of 3 h, indicated by an adjusted R-squared well above 0.9. For material ages of 1 and 2 days, the correlation was slightly weaker with adjusted R-squared values of around 0.8, while it was only around 0.2 for 7 and 28 days. This decreasing correlation can be attributed to the fact that in the fresh state, the material's strength diverges at the friction limit, while in the hardened state, it diverges at the maximum packing limit. Therefore, the hardened state strength is much less sensitive to the variation in water content that was applied in the current study.

Thus, the research suggests that there is a very high potential for using in-line measurement for quality control purposes in the production phase. For the use phase, a direct correlation between in-line measurements and bulk hardened state properties was relatively weak for the applied range variations in the water-to-solids ratio. Potentially, such a correlation could be established for other variations, such as changes in environmental conditions, other material properties, such as interface properties or durability aspects, or other in-line measurements. A similar methodology as presented in the current contribution could be adopted for determining the degree of correlation of such variations, conditions, and measurements.

6 Data Availability

A complete data set containing all measured values as well as the setup and configuration of the system is available in [12] following the structure described in [13].

References

1. Buswell, R.A., et al.: A process classification framework for defining and describing digital fabrication with concrete. Cem. Concr. Res. **134**, 106068 (2020). https://doi.org/10.1016/j.cemconres.2020.106068
2. Bos, D., Wolfs, R.: A quality control framework for digital fabrication with concrete. RILEM Tech. Lett. **8**, 106–112 (2023). https://doi.org/10.21809/rilemtechlett.2023.181
3. Roussel, N., et al.: Assessing the fresh properties of printable cement-based materials: high potential tests for quality control. Cem. Concr. Res. **158**, 106836 (2022). https://doi.org/10.1016/j.cemconres.2022.106836
4. Ducoulombier, N., et al.: The slug test: inline assessment of yield stress for extrusion-based additive manufacturing. In: Bos, F., Lucas, S., Wolfs, R., Salet, T. (eds.) Second RILEM International Conference on Concrete and Digital Fabrication, vol. 28, pp. 216–224. Springer, Cham (2020). https://doi.org/10.1007/978-3-030-49916-7_22
5. Bos, D., Wolfs, R.: Automated visual inspection of near nozzle droplet formation for quality control of additive manufacturing. In: RILEM Bookseries, vol. 37, pp. 453–459. Springer Science and Business Media B.V (2022). https://doi.org/10.1007/978-3-031-06116-5_67
6. Bos, D.H.: Quality Engineering and Control for Digital Fabrication with Concrete, p. 159. Eindhoven University of Technology, Eindhoven (2024) (Bouwstenen)
7. Wolfs, R.J.M.: Experimental Characterization and Numerical Modelling of 3D Printed Concrete: Controlling Structural Behaviour in the Fresh and Hardened State. Eindhoven University of Technology, Eindhoven (2019)
8. Bos, D., Deetman, A. & Wolfs, R. From on-line to in-line quality control of digitally fabricated concrete. Manuscript in preparation.
9. Flatt, R.J., Bowen, P.: Yodel: a yield stress model for suspensions. J. Am. Ceram. Soc. **89**, 1244–1256 (2006). https://doi.org/10.1111/j.1551-2916.2005.00888.x
10. Jacquet, Y., Perrot, A., Picandet, V.: Assessment of asymmetrical rheological behavior of cementitious material for 3D printing application. Cem. Concr. Res. **140**, 106305 (2021). https://doi.org/10.1016/j.cemconres.2020.106305
11. Yin, T., Liu, K., Fan, D., Yu, R.: Derivation and verification of multilevel particle packing model for ultra-high performance concrete (UHPC): modelling and experiments. Cem. Concr. Compos. **136**, 104889 (2023). https://doi.org/10.1016/j.cemconcomp.2022.104889

12. Bos D, Deetman A, Wolfs R. Data for correlating in-line sensor data to hardened state material properties in digital fabrication with concrete. Zenodo; 2025. doi:https://doi.org/10.5281/zenodo.16837346
13. Deetman,A., Bos,D., Lucas,S., Salet,T., Wolfs,R.: A database framework for digital fabrication with concrete. Results Eng. **29**, 108669 (2026). https://doi.org/10.1016/j.rineng.2025.108669

Development of an Ettringite-Rich Cementitious Material for Thermal Energy Storage

S. Bonnin[1,2](✉)

[1] Holcim Innovation Center, 95 rue du Montmurier, 38291 Saint Quentin Fallavier, France
salome.bonnin@holcim.com

[2] INSA Lyon, CNRS, CETHIL, UMR5008, 69621 Villeurbanne, France

Abstract. Ettringite is a mineral compound and one of the products of cement hydration. The Ettringite/Meta-ettringite couple is also an interesting candidate for thermochemical heat storage. A thermochemical storage solution has an interest in a heating system for network application if its storage capacity it is higher than that of sensitive water, within the same temperature range. Given that water has a storage capacity of 4.18 kJ/kg.K, it can be deduced that the thermochemical material must be capable of storing more than 50 kWh/m^3 for a ΔT of 40 °C.

These storage materials link the chemistry structure to a thermal storage performance. Several studies have found cementitious materials with a maximum ettringite content of 72 wt%. In this study, new cement mixes formulations were tested that could theoretically contain between 80 wt% and 100 wt% ettringite. The objective of this study is to quantify the ettringite content in cement pastes through thermogravimetric analysis (TGA) and to evaluate its correlation with the theoretically calculated value. Additionally, other thermochemical parameters were conducted to assess the potential of cement pastes as a thermochemical storage material.

The newly formulated solutions have been demonstrated to yield a matrix with 93 wt% of ettringite. However, thermochemical cycling analyses, which are employed to study the heat storage and release capacity, do not show a significantly higher result than the reference material "C80P20". The correlation between ettringite chemistry and its capacity as a heat storage agent is not contingent upon the quantity of this agent within the cement matrix.

Keywords: Long term energy storage · Thermochemical heat storage · Ettringite · Cement

Nomenclature:

OPC - Ordinary Portland cement
CEM I - Cement type I
CSA - Calcium sulphate aluminium
TGA - ThermoGravimetric Analysis
XRD - X-Ray Diffraction
XRF - X-Ray Fluorescence
MIP - Mercure intrusion porosity
TCHS - Thermochemical heat storage

M. Posani and S. Singh (Eds.): RYS 2025, RILEM Bookseries 67, pp. 35–52, 2026.
https://doi.org/10.1007/978-3-032-14652-6_4

LOI - Loss on ignition
IPCC - Intergovernmental panel on climate change
Symbols
ΔT - Temperature difference, °C
RH - Relative humidity, %
R - Rate, wt%
C - Thermal energy storage capacity, J/g
E - Energy density, kWh/m3
subscript
Ett - Ettringite
Met Meta-ettringite
CemCementitious material
Cement phase - Chemistry - Acronym
Gypse - $CaSO_4.2H_2O$ -CH_2$
Anhydrite -$CaSO_4$ -C$
Lime- CaO / $Ca(OH)_2$- C / CH
Mayenite-$12CaO.7Al_2O_3$-C12A7
Ettringite- $3CaO.Al_2O_3.3CaSO_4.32H_2O$-$C_6A\$_3H_{32}$ or AFt
Monosulfate- $3CaO.Al_2O_3.CaSO_4.12H_2O$ - $C_6A\$H_{12}$ or AFm
Aluminium hydroxideAl$(OH)_3$- AH3

1 Introduction

Ettringite is a mineral compound produced during the hydration of Portland cement (OPC) with the following chemical formula: $Ca_6Al_2(SO_4)_3(OH)_{12}\cdot 26H_2O$. It is also known in cement chemistry notation as $3CaO.Al_2O_3.3CaSO_4.32H_2O$, $C_6A\$_3H_{32}$, calcium trisulfoaluminate hydrate or AFt. While the reversible and exothermic reaction of ettringite was first highlighted by Struble and Brown in 1986 [1], it was not until the 2000s that ettringite truly emerged as an interesting material for thermochemical heat storage (TCHS) applications [2].

The crystal structure of ettringite is characterised by columns of $[Ca_6[Al(OH)_6)]_2.24H_2O]^{+2}$ and sulphate tetrahedra (SO_4^{2-}) linked by water molecules and a space group P31c (Fig. 1). Its structure allows a reversible dehydration and hydration reactions that can be exploited for heat storage and release by using the binding energy of water molecules in the molecule as represented in Eqs. (1), (2).

$$3CaO.Al_2O_3.3CaSO_4.32H_2O + Heat \rightarrow 3CaO.Al_2O_3.3CaSO_4.nH_2O + (32 - n)H_2O \quad (1)$$

$$3CaO.Al_2O_3.3CaSO_4.nH_2O + xH_2O \rightarrow 3CaO.Al_2O_3.3CaSO_4.32H_2O \quad (2)$$

During dehydration, ettringite undergoes a transformation into an amorphous phase, referred as meta-ettringite, as a result of the loss of hydrogen and oxygen molecules in the form of water (H_2O). Dent Glasser et al., (1962) described this process as a topotactic transformation, where the three-dimensional crystal structure is preserved despite the loss of essential structural constituents [3]. It is an endothermic reaction that occurs at low temperatures of under 100 °C and low humidity [4].

Conversely, the hydration reaction is exothermic, allowing the release of stored heat. This reversible process, combined with the low cost of the material and its low carbon footprint, makes ettringite an attractive alternative to existing TCHS materials [5–7].

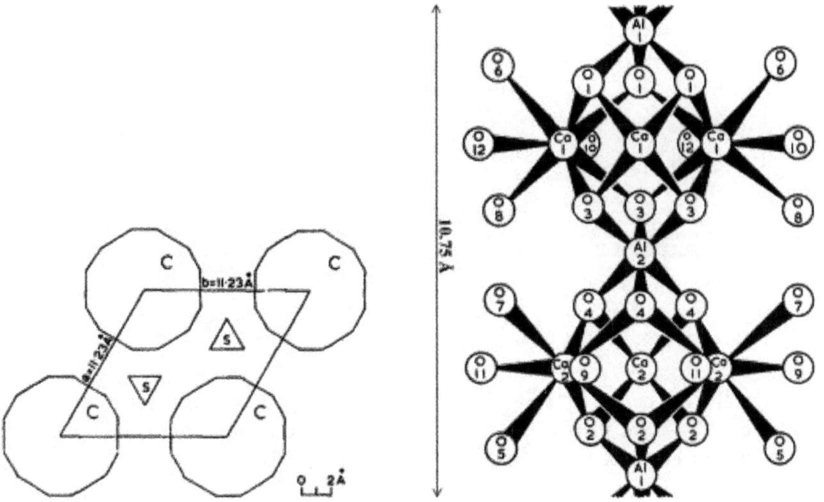

Fig. 1. Section of ettringite unit structure (left): the polygons and the triangles represent respectively the main column's structure ($[Ca_6[Al(OH)_6]_2{\bullet}24H_2O]^{6+}$) and $SO_4{}^{2-}$ or H_2O groups in channels. Basic unit crystal structure of ettringite in c-axis direction (right): the O represents OH or H_2O group, while Al and Ca atoms are showed by chemical symbols [2], adapted from [11].

Thermochemical heat storage represents a promising avenue within the wider landscape of thermal energy storage technologies (Fig. 3). These technologies, encompassing sensible heat storage, latent heat storage, and thermochemical heat storage, are useful tools for bridging the gap between energy supply and demand over time (Fig. 2).

Energy consumption in buildings is a major issue in the energy transition due to the reduction in fossil fuel reserves, the growth in energy demand and the increase in greenhouse gas emissions. As indicated in the Intergovernmental Panel on Climate Change (IPCC) report of 2023, the advancement of alternative renewable energy systems is imperative for the fulfilment of daily energy requirements while concurrently reducing CO_2 emissions [8].

Efficient storage of thermal energy represents a significant challenge in the use of intermittent, clean and renewable energy sources. While sensible and latent heat storage methods are well-established, thermochemical heat storage offers unique advantages, particularly for long-duration storage and high energy densities, positioning materials like ettringite at the forefront of advanced thermal storage research [9].

Figure 3 categorised storage mechanism: sensible heat, latent heat (using Phase Change Materials or PCMs), sorption, and chemical reactions. The left side of the figure visually represents the approximate volume required to store 6.7 GJ of energy

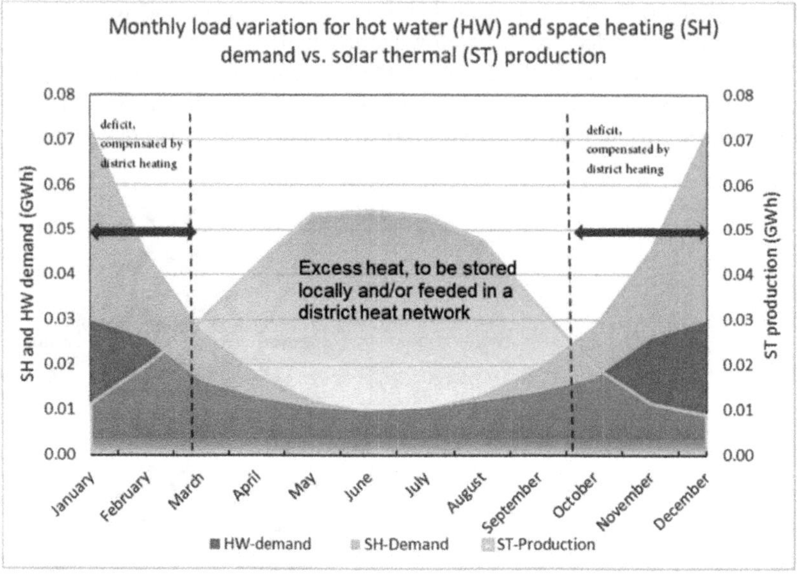

Fig. 2. Monthly load curves for HW—hot water and SH—space heating demand and the supply curve of local ST—solar thermal: demonstrating the period of heat excess and deficit of the conceived PED in a detached housing district in Vienna, Austria [10]

for four hypothetical scenarios, each corresponding to a different material category. This highlights the challenges in sizing heat storage solutions and the high potential of Thermochemical Heat Storage (TCHS) materials. The graph helps in identifying suitable materials for specific applications based on key parameters such as working temperature, energy density, size, and volume.

Ettringite occupies an unusual position on the graph because it is a thermochemical storage material that functions under lower operating conditions than materials in its category. This material opens a novel domain of application for thermal storage, offering advantages for low-temperature storage (< 100 °C) with higher energy density than PCM with stability over time due to the chemical storage process.

Clark et al., 2022 analysed several salt hydrates based on their stability, cost, and energy density [12]. The study highlights Strontium chloride ($SrCl_2$) and Strontium bromide ($SrBr_2$) as the most promising salts, but their cost remains a limiting factor. The Table 1 unequivocally demonstrates the disparity in cost and, consequently, the greater affordability of ettringite as a heat storage medium in comparison to other TCHS materials. Furthermore, a preliminary exploratory study [13] has demonstrated that ettringite, in a cementitious material for heat storage, exhibits a reduced carbon footprint when compared to its primary competitor: zeolite 13x.[1]

[1] Zeolite is a physical sorption storage material and is well known in the literature, different types of zeolites exist and depend on the chemical formula of the molecule and the pore size of the molecular sieve. Zeolite 13X have the following chemical structure: $Na_2O.Al_2O_3.(2.8 \pm 0.2)SiO_2.(6 \sim 7)H_2O$

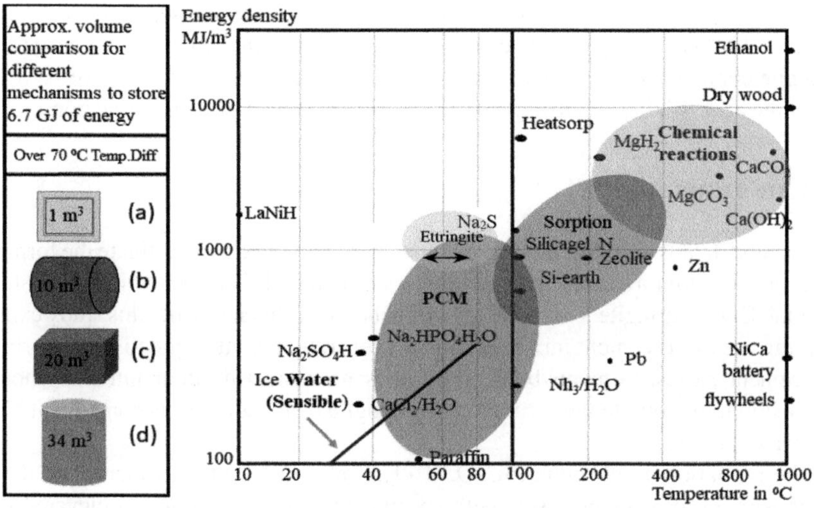

Fig. 3. Energy density of material for heat storage. The blue part represents latent storage materials (PCM), in gray, physical sorption storage and in green, chemical sorption storage (TCHS) [Adapted from 11].

The approach of generating ettringite within a cement paste offers the potential for a dual-functional material, where the cementitious matrix provides structural integrity and a low-cost process production while the embedded ettringite contributes to thermal energy storage. This in-situ formation could lead to more material-efficient and cost-effective TCHS solutions for large-scale applications.

Ndiaye et al., (2020) [18] developed an ettringite-rich foamed material from calcium sulfoaluminate (CSA) cement containing 68 wt % ettringite with a heat storage capacity of 117 kWh/m^3 [18]. Chen et al., (2020) also use a CSA cement to produce and compare three ettringite-based cementitious material. The study highlights one candidate: "C80P20" with 71.8 wt% ettringite and an energy storage capacity of 282 kWh/m^3 of original hydrated materials [19].

Table 1. Promising TCHS material [14–17]

Material	Theoretic energy density (kWh/m^3)	Material cost (€/m^3)	Dehydration temperature
MgSO4·7H2O/MgSO4·H2O	639	4870	150
SrBr2·6H2O/SrBr2·H2O	628	42000	80
LaCl3·7H2O/LaCl3·H2O	591	38400	100
$Ca_6Al_2(SO_4)_3(OH)_{12}\cdot 26H_2O$/ $Ca_6Al_2(SO_4)_3(OH)_{12}\cdot 3H_2O$	547	<1000	60–80

This study aims to evaluate the potential of new cement mixtures to produce ettringitic materials for heat storage. The selection of an optimal candidate will be based on its ettringite content and other thermochemical parameters. These criteria will be used to assess the accessibility of ettringite for the reversible and exothermic reaction.

2 Experimental Part

This section details the experimental methodology employed to investigate the formation of ettringite within a cement paste matrix as a potential thermochemical heat storage material. Given ettringite's promising thermochemical characteristic, this study explores the influence of the cement mix on the in-situ generation of ettringite to make ettringite-rich cement pastes. Cements/binders are chosen considering their intrinsic chemical composition, to promote the formation of ettringite by reaction of its constituent phases with water [20, 21].

The formation of ettringite ($3CaO.Al_2O_3.3CaSO_4.32H_2O$) was facilitated through the preparation of cement pastes with specific cements that exhibited elevated levels of aluminate (Al_2O_3) and calcium sulphate ($CaSO_4$). This controlled addition of sulphate ions is crucial as it provides the necessary sulphate for the formation of ettringite and instead of other hydrates such as monosulphate (AFm) through reactions with the aluminate phases present in the cement.

This section details the cement mixes used in the present study and the chemical hydration process leading to the formation of ettringite. In the extant literature, the C80P20 cement mix proposed by Chen et al. exhibited the best ettringite content applied to thermochemical storage (72–75 wt%) [19]. Furthermore, chemical and thermal data of the "C80P20" material are available in the literature; this cement mixture is then selected as the reference material.

2.1 Material

High calcium aluminate cement containing additional calcium sulphate (CSA) was used to obtain ettringite-rich cement pastes. To produce pastes with different ettringite contents, some CSA was mixed with OPC (CEM I 52.5 N), improving mechanical properties and decreasing the risk of self-breakage due to significant shrinkage. The addition of gypsum ($CaSO_4.2H_2O$), anhydrite ($CaSO_4$), and lime (CaO or $Ca(OH)_2$) was employed to regulate and optimise the ettringite formation within the cement paste.

The reactions involved in the formation of ettringite are listed with Eqs.(3, 4 and 5).

$$\text{C12A7} + 12\ \text{C\$} + 137\ H_2O \rightarrow 4\ \text{AFt} + 3\ \text{AH3} \quad (3)$$

$$\text{C12A7} + 12\ \text{C\$} + 9\ \text{CH} + 128\ H_2O \rightarrow 4\ \text{AFt} + 3\ \text{AH3} + 9\ \text{CaO} \quad (4)$$

$$\text{AFt} + \text{C12A7} \rightarrow 3\text{AFm} + 2.5\ \text{AH3} \quad (5)$$

Chemical compositions and mineralogical phases of CSA and OPC cements as well as mineral additions were analysed by X-ray fluorescence (XRF) and XRD and are

Table 2. FX analysis giving the composition of the cements and binders

	CEM I (OPC)	CSA	Gypse ($CaSO_4.2H_2O$)	Anhydrite ($CaSO_4$)	Lime (CaO)	Lime ($Ca(OH)_2$)
	P	**C**	**G**	**A**	**L**	**Lh**
LOI 950 °C (%)	1.6	–	22.5	3.1	2	25.7
SiO_2 (%)	20.2	1.8	–	–	–	–
Al_2O_3 (%)	4.8	23	–	–	–	–
Fe_2O_3 (%)	3	–	–	–	–	–
CaO (%)	64.5	43	32	40	97	73.4
MgO (%)	1.1	–	–	–	–	–
SO_3 (%)	3.2	29.5	44.3	55	–	–
Other (%)	1.6	1.6	1.2	1.9	1	0.9

shown in Table 2. Minor oxides and mineralogical phases under 1% are traces and are in the "other" category.

The Bogue calculation is a method of calculating the approximate proportions of the four main minerals in Portland cement clinker. The OPC cement utilised in this study consists of 72.5 wt% C3S, 3.3 wt% C2S, 7.6 wt% C3A and 9.1 wt% C4AF. CSA cement is highly pure, consisting of only two phases: C12A7 and C$. Their proportion can be calculated from the data of the FX pearls data. The composition of the substance is 52.9 wt% C$ and 47.1 wt% C12A7.

2.2 Preparation of Samples

The cement mix was manually mixed to prepare samples with different levels of adding phases. Tap water was used with water to solid ratio (w/s) between 0.7 and 1.1 to hydrate binders by a VMI/Rayneri mixer at 700 r/min for 1 min, then at 1000 r/min for 2 min. Before the addition of the blend into tap water, a small quantity of a superplasticizer (0.11 wt% of blend) and tartaric acid (0.06 wt% of bend) was pre-introduced to the water to assure good workability retention and setting time. After the addition of cement mixtures, a thickener (0.06 wt% of water) was added slowly to prevent agglomeration. Details of the different mixtures are summarised in Table 3.

One column of the table is dedicated to calculating the theoretical rate of ettringite formation from each proposed cement mix in this study. This calculation is based on a Holcim proprietary calculation tool. The workbook is based on the thermodynamic data of the involved species, as well as the application of the Gibbs phase rule.

The cementitious pastes were then poured into plastic bags and sealed for 24 h. Then, the samples were demoulded and stored in aluminum cups in an open air and controlled laboratory at 20 °C as shown in the photo in Fig. 4 for 7 days of hydration (1 day in sealed plastic bags and 6 days in a cup) at ~20 °C. After 7 days of curing, cement pastes were ground to a size less than 100 μm for analysis. The hydration of the samples was not

Table 3. Description of the cement mixes of the samples

		Cements (wt%)		Adding phases (wt%)					Theorical ettringite rate (wt%)
Sample designation	W/C	P	C	G	A	L	Lh	Zeolite 13x	
C80P20a	1.1	20	80	0	0	0	0	0	91
C80P20b	0.8	20	80	0	0	0	0	0	91
C80P20c	0.44	20	80	0	0	0	0	0	–
C80P20d	0.8	20	80	0	0	0	0	0.4	91
C100P0	0.8	0	100	0	0	0	0	0	87
C57L12G31	0.75	0	57	31	0	12	0	0	100
C44P28G27	0.7	28.5	44.2	27.3	0	0	0	0	87.5
C47P30A23	0.8	30	47	0	23	0	0	0	87.5
C63L10A27	0.9	0	63	0	27	0	10	0	100
C59L10G31	0.8	0	59	0	31.4	0	10	0	100

stopped by organic solutions, in order to reproduce a real-life energy storage scenario and also because the reaction kinetics of ettringite is very fast (a few hours) [20]. In a semi-humid environment (50% relative humidity) at 20 °C, ettringite is found in a stable form with between 30 and 32 water molecules in its chemical structure [22, 23].

Fig. 4. Picture of cement paste samples at day 1 (ref) C80P20a (a-b) C57L12G31 (c) C44P28G27 (d) C47P30A23 (e) C63L10A27 (f) C59L10G31

2.3 Analysis

X-ray diffraction analysis (XRD)

The X-ray diffraction qualitative analysis (XRD) method used in this contribution was to follow the phase assemblage. It was equipped with a DX Bruker D8 Advance with a with copper tube (CuKα radiation beam, 25 kV and 40 mA). The scanned range is between $2\theta = 3–65°$ by a step of 0,135 s for 10 min (4336 measurements) in the ambient environment (Appendix 3).

Thermogravimetry analysis (TGA)

TGA (Mettler TGA-DSC 3+) was principally used to quantify ettringite. An approximately 40 mg sample was added to the device in a half-closed capsule and heated from 30 to 600 °C at 2.5 K/min under a nitrogen flow of 20 mL/min. The error in the TGA measurement was approximately 1%.

A preliminary study was carried out on synthetic ettringite ($R_{ett} > 99wt$) to determine the decomposition temperature ranges of the 32 water molecules in ettringite. Indeed, according to the Fig. 1, the water molecules do not have the same location in the ettringite structure, the associated bond energies are therefore also different leading to decompositions spread between 30 °C and 600 °C [24]. In a cement paste, ettringite is not the only hydration product, which makes it challenging to directly use thermogravimetric analysis (TGA) data from 30 °C to 600 °C for quantification [25]. However, a preliminary study allowed us to hypothesize a predominant decomposition zone for ettringite. A loss of 24 water molecules per ettringite molecule was recorded between 40 °C and 130 °C under the previously described analysis conditions (Appendix 1 and 2). This finding makes it possible to determine the ettringite content in a cement paste using the Eq. (6):

$$R_{ett}\,(wt\%) = 100 \times wt\%loss\big[40°C - 130°C\big] \times \frac{1}{\left(\frac{30\times M_{H2O}}{M_{AFt30}} - \frac{6\times M_{H2O}}{M_{AFt6}}\right)} = 100 \times \frac{wt\%loss[40°C - 130°C]}{30.6} \quad (6)$$

Mercury Intrusion Porosimetry (MIP)

MIP tests were carried out on small pieces from the hardened pastes of 7 days (1–3 mm) using an Autopore IV (Micromeritics, USA). Before testing, samples were dried in an oven at 45 °C overnight. The suggested contact angle for ordinary cement pastes was presumed to be 130 °. On this basis, the pressures used enabled pore diameter coverage of approximately 3 nm–360 µm.

Thermogravimetric Analysis (TGA) and Differential Scanning Calorimetry (DSC)

To determine thermal energy storage capacity (C), powdered samples were characterised by TGA-DSC (Sensys Evo TG-DSC), which provided thermal analysis simultaneous with the loss of weight. Approximately 10 mg of specimen follows the cycle described in Fig. 5 comprising an 8-h loading (dehydration) phase, a 12-h storage phase and a 12-h unloading (hydration) phase under a nitrogen flow of 10 mL / min. Thermal analysis data were acquired by Calisto Acquisition and processed by Calisto Processing (both software) to give loss of weight and enthalpies. The systematic uncertainty in enthalpy measurement was approximately 1%.

The dehydration/hydration cycle program allows to ensure the reversible reaction between ettringite and meta-ettringite according to the study of Chen et al., [19] and Baquerizo et al., [22]. The analysis is carried out under nitrogen to avoid the phenomenon of carbonation during cycling (Eq. 7).

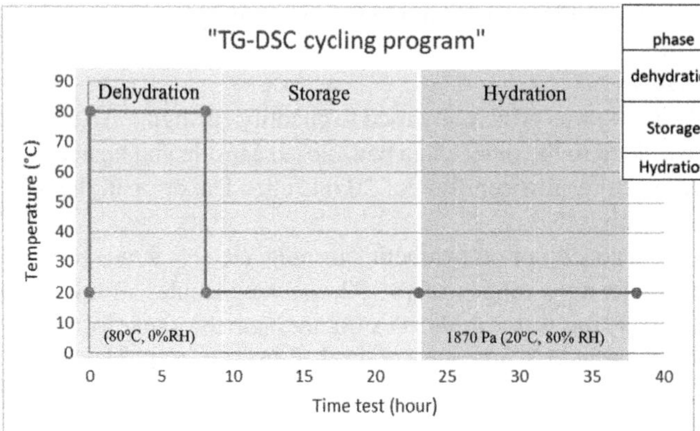

phase	step (K/min)	Time (h)	Ti (°C)	Tf (°C)
dehydration	10	0,1	20	80
	-	8	80	80
Storage	99	0	80	20
	-	15	20	20
Hydration	-	15	20	20

Fig. 5. TG-DSC cycling program in 3 steps: Dehydration of 8 h at 80 °C, storage of 12 h at 20 °C and low humidity, hydration of 12 h at 20 °C and 80% RH

3 Results and Discussion

Results of the detection and quantification of ettringite in the samples produced are summarised in the Table 4 using methods like XRD and TGA analysis. The detection of the crystalline structure ettringite is based on XRD analysis. The quantification of ettringite was carried out using TGA analysis and a calculation method detailed in the previous section. The XRD diffractogram and the TGA thermogram of the C80P20a W/C = 1.1 material are available in Appendix 2 to show an example of the raw results of the analyses before treatment on the "C80P20a" reference material.

Table 4 shows a difference between the theoretical and experimental measurements of the ettringite content. For instance, cement mix C59L10G31 was designed because it was predicted to have a 100 wt% theoretical ettringite content. However, the experimental results show a final 55% wt% of ettringite, with significant gypsum formation in the diffractogram at the expense of ettringite.

These first chemical analyses allow us to make an initial selection of interesting materials to be tested in thermochemical cycling with ATG-DSC according to the protocol described in the previous section. The interesting materials are the cement matrices presenting more ettringite than the reference material "C80P20a".

Table 5 presents the heat storage and release results called storage capacity and energy density respectively measured in J/g and extrapolated in kWh/m^3 by considering the bulk density of each material. These results are from the TGA-DSC cycling of cementitious ettringite powders based on different cement mix. Synthetic ettringite was

Table 4. The results of exploitation of XRD and TGA data for the detection of ettringite in samples

Sample designation	W/L	Ettringite detection?	Ettringite rate (wt%)	Theorical ettringite rate (%)
C80P20a	1.1	Yes	78.6	91
C80P20b	0.8	Yes	74.6	91
C80P20c	0.44	Yes	60	–
C80P20d	0.8	Yes	81	91
C100P0	0.8	Yes	77.6	87
C57L12G31	0.75	Yes – with gypsum	77.4	100
C44P28G27	0.7	Yes	83.8	87.5
C47P30A23	0.8	Yes	76.7	87.5
C63L10A27	0.9	Yes	92.8	100
C59L10G31	0.8	Yes – with gypsum	55.3	100

Table 5. Exploited results of TG-DSC cycling data for thermochemical heat storage

				Storage		Release	
Sample designation	W/C	Ettringite rate (wt%)	Bulk density (kg/m^3)	C (J/g)	E (kWh/m^3)	C (J/g)	E (kWh/m^3)
Synthetic ettringite *	/	> 95	/	1029	/	1092	/
C80P20 *	1.1	72	/	811	/	636	/
Synthetic ettringite	/	> 95	590	928	152	1189	195
C80P20b	0.8	74.6	555	838	130	502	77
C44P28G27	0.7	83.8	590	864	133	413	64
C63L10A27	0.9	92.8	520	719	111	458	71

*: Data from Chen et al., 2021 [17, 19]

also produced in the laboratory, the manufacturing process will not be detailed in this study.

The C80P20b material, exhibits an energy density of 502 J/g, which is the best result of the three selected matrices, although the results are similar. It is important to note that these new energy density results are similar to the data from Chen et al., 2021 [17, 19] using the same analytical method. Despite the development of a cement matrix with elevated ettringite rate, the storage capacities in J/g and kWh/m3 do not demonstrate superior performance in comparison to the C80P20a reference mix. News formulations do not appear to be better candidates for TCES in comparison to C80P20a.

According to the Fig. 7, increasing the W/C ratio increases the overall porosity but also impacts the pore distribution. Figure 6 illustrates the pore size distribution of cement pastes with various W/C ratios, based on the "C80P20" cement mix. The cement paste with an optimal W/C ratio of 0.8 for ettringite formation has a majority pore distribution around 0.25 µm, whereas the cement paste with a W/C ratio of 1.1 has a majority pore distribution around 0.7 µm. Therefore, excess water in the manufacturing process increases the overall porosity and opens the pore size of the cement matrix.

The pore structure of a hardened cement paste is largely governed by the water/cement (W/C) ratio used during mixing. This ratio determines the initial volume of water available for hydration and, consequently, the space occupied by pores within the hydrated cement matrix.

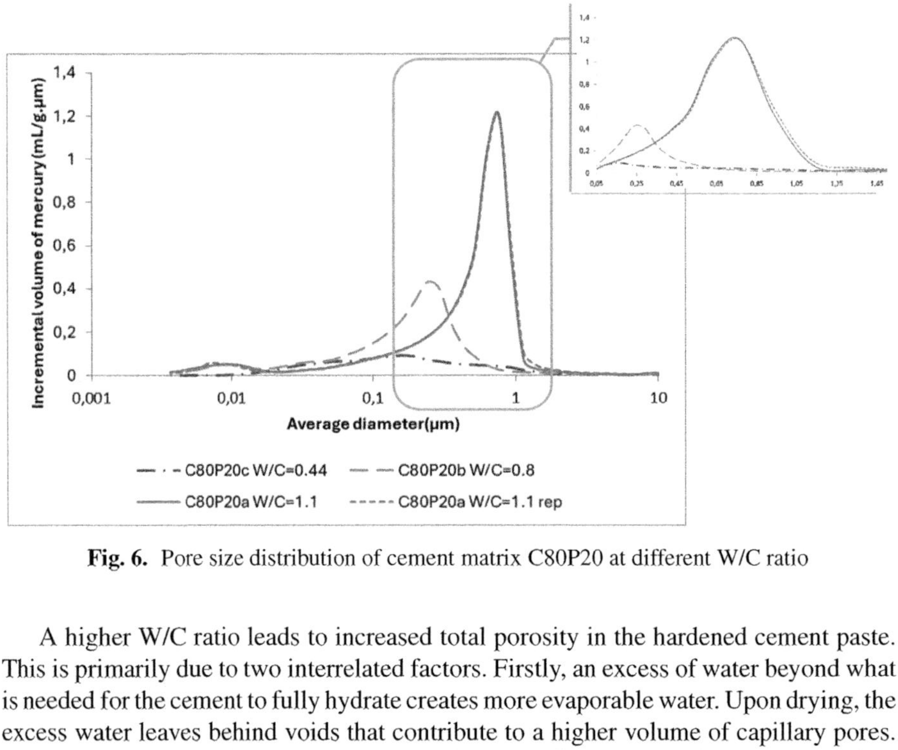

Fig. 6. Pore size distribution of cement matrix C80P20 at different W/C ratio

A higher W/C ratio leads to increased total porosity in the hardened cement paste. This is primarily due to two interrelated factors. Firstly, an excess of water beyond what is needed for the cement to fully hydrate creates more evaporable water. Upon drying, the excess water leaves behind voids that contribute to a higher volume of capillary pores. These larger pores typically have a more detrimental effect on strength and permeability. Secondly, a higher W/C ratio can result in a less dense packing of cement particles during mixing, leading to a more open structure even before significant hydration occurs. As hydration proceeds, the formation of hydration products may not fully infill these larger initial voids, thus contributing to higher porosity.

Conversely, a lower W/C ratio tends to produce a denser, less porous hardened cement paste. With less free water, the capillary pore network is finer and less interconnected. This reduces the total pore volume and refines the pore size distribution. The result is a higher proportion of smaller gel pores, which are less detrimental to strength, and a lower proportion of larger capillary pores. Consequently, cement pastes with a lower

W/C ratio generally exhibit increased compressive strength and reduced permeability to aggressive agents, such as CO_2.

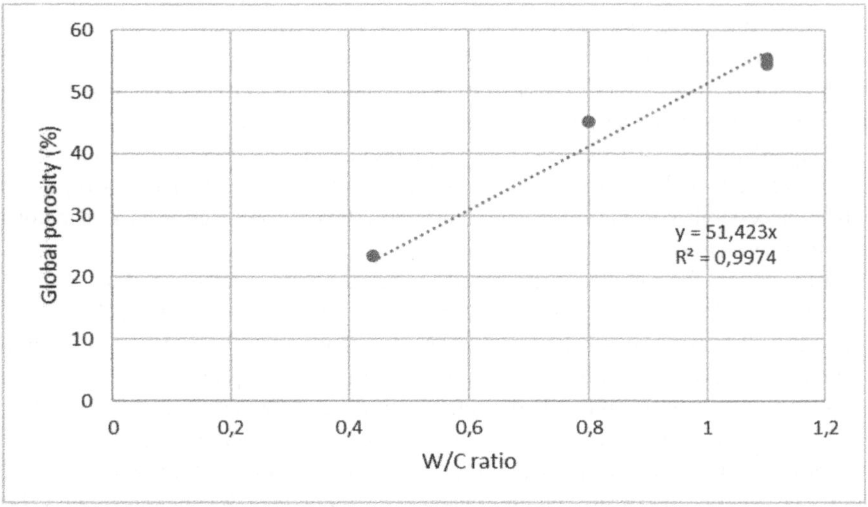

Fig. 7. Graph representing the overall porosity (%) from MIP analyses as a function of the W/C ratio of C80P20 cement pastes

It is important to note, however, that very low W/C ratios can lead to practical difficulties, such as reduced workability of the fresh paste, potentially compromising proper mixing and compaction. Furthermore, since ettringite is a molecule requiring a large amount of water in its structure compared to other cement hydrates (32 water molecules), high W/C ratios (minimum 0.7) are calculated to meet the water demand for ettringite formation.

Figure 7 suggests the presence of a linear relationship between the W/C ratio and the overall porosity of a cement paste cured using the "C80P20" cement mix.

Overall porosity (after 7 days curing) = 51.4 * W/C

In this study, high porosity is sought, including a network of well-interconnected capillary pores, which directly translates into increased permeability within the cement paste matrix. Indeed, this connection of the network within the matrix will influence the accessibility as well as the kinetics of the dehydration and hydration reaction described by Eq. (1) and Eq. (2) allowing thermochemical storage.

Conversely, the same pore network that allows efficient water penetration also provides pathways for atmospheric carbon dioxide (CO_2) to diffuse into the cement paste. This diffusion initiates the carbonation process, a chemical reaction between CO_2 and cement hydration products, particularly ettringite, through the following reaction:

$$3CaO.Al_2O_3.3CaSO_4.32H_2O + 3CO_2 \rightarrow 3CaCO_3 + CaSO_4.2H_2O + Al_2O_3.xH_2O + (26 - x)H_2O \quad (7)$$

While carbonation leads to the formation of calcium carbonate ($CaCO3$), which can densify the pore structure and reduce the porosity of the carbonate zone, its overall impact on durability and a decrease in the pH of the concrete (pH < 9). In the case of developing an ettringite-rich cementitious solution for heat storage, carbonation is a phenomenon that deteriorates the structure of the ettringite and, more specifically, the water molecules attached to it (Eq. 1). Since thermochemical storage relies on the chemical bonds of water molecules to the ettringite structure, carbonation can potentially impair thermal storage properties through various processes [19]:

- Degradation of ettringite (the active ingredient in thermochemical storage)
- Reduction in the porosity and permeability of the cement matrix, resulting in reduced accessibility to the reversible dehydration/hydration reaction (a key factor in heat storage/release)

In this study, the samples were stored in sealed bottles after seven days of curing and analysed using nitrogen to overcome problems caused by carbonation. Carbonation is not studied here; however, it is a phenomenon that should be considered in future research on the potential industrialisation of ettringite for heat storage.

While ettringite is theoretically the active ingredient of the material for thermochemical storage, no correlation of experimental results was identified between the ettringite content in the new cement mixes and the energy density results. Exploration of additional parameters is necessary to comprehensively understand the phenomenon of thermochemical storage of ettringite in a cement matrix. A comprehensive understanding of the predominant phenomena in the ettringite formation reaction during cement matrix production is also imperative, as theoretical ettringite content calculations often diverge from experimental values.

The results of this study underscore the necessity of considering a multitude of parameters in the development of an ettringite-rich cement matrix for heat storage. Indeed, the technical duality of the subject requires a cement structure that is protected from carbonation and mechanically stable over time, while offering a high and accessible ettringite content for the kinetics of the thermochemical storage reaction.

4 Conclusion

New tests on cement samples showed promising theoretical results, but the C80P20 cement mix remains more energy-efficient. While a 93% ettringite matrix was created, its storage capacity (111 kWh/m3) and energy density (71 kWh/m3) were surprisingly lower than the reference C80P20 matrix with 75% ettringite.

Furthermore, a non-correlation between theoretically calculated and experimentally observed ettringite levels is highlighted, underline the existence of unexplored variables in the reaction mechanisms of ettringite formation.

The inability to establish a correlation between the storage capacity of a cement paste powder and its ettringite content underscores the complexity of the system. The energy densities of synthetic ettringite (100 wt%) and cementitious ettringite (93 wt%) are 195 kWh/m^3 and 71 kWh/m^3, respectively, indicating that even with similar internal chemistry, the energy release from the meta-ettringite/ettringite transformation varies significantly. These findings reveal that the structure of formed ettringite and the reactivity of the ettringite/meta-ettringite pair for heat storage are not yet fully understood.

Few cement mixes for thermochemical storage have been reported in the literature. This study aimed to improve the composition of the cement mix for thermochemical heat storage. The tests carried out at ATG/DSC led to select the C80P20 cement mix as a high-performance cement mix. This observation can be explained by the purity of the CSA cement used, which alone allows the formation of high-quality ettringite.

Ultimately, C80P20b material is expected to be the most energy storage-efficient. In fact, a reduction in the water-to-cement ratio (W/C) from 1.1 to 0.8 results in an increase in the density of the cement matrix. This allows to maximise the energy density (kWh/m^3), which is an important factor in developing an industrial-scale storage solution (Fig. 8).

This study reveals the mismatch between ettringitic cement's inherent chemistry and its potential for heat storage. Future research needs to consider factors like BET specific surface area, thermal conductivity, and the importance of using thermally aged materials (an ettringitic cement matrix which has undergone several charge/discharge cycles) to accurately assess the ettringite/meta-ettringite couple's reactivity, rather than post-fabrication ettringite alone (Figs. 9 and 10).

Annexe

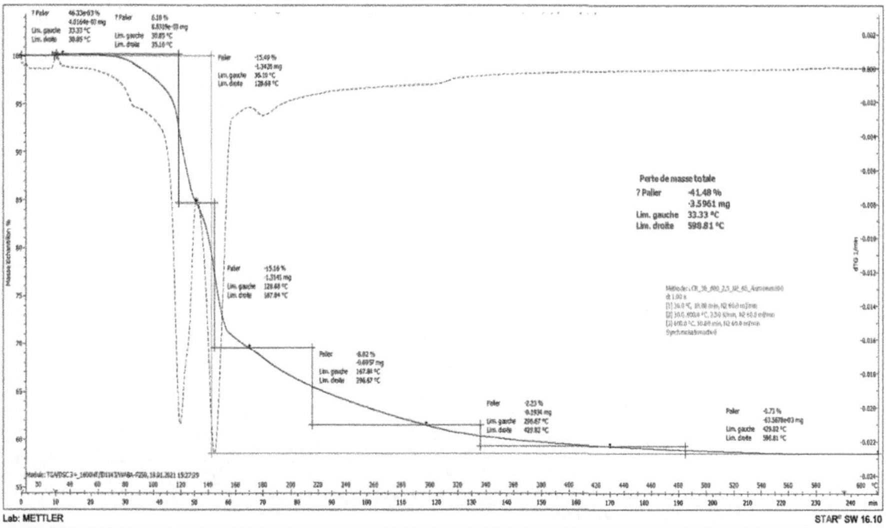

Fig. 8. Thermogram TGA of synthetic ettringite

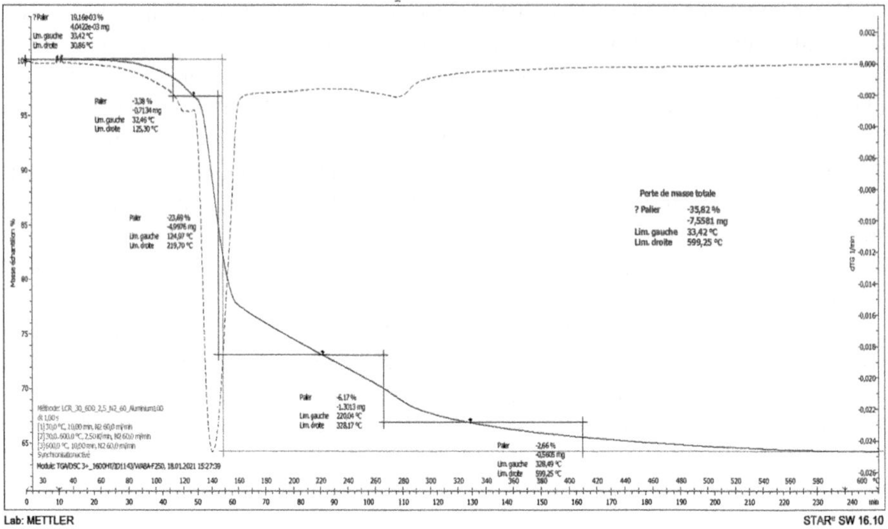

Fig. 9. Thermogram TGA of C80P20 W/C = 1.1

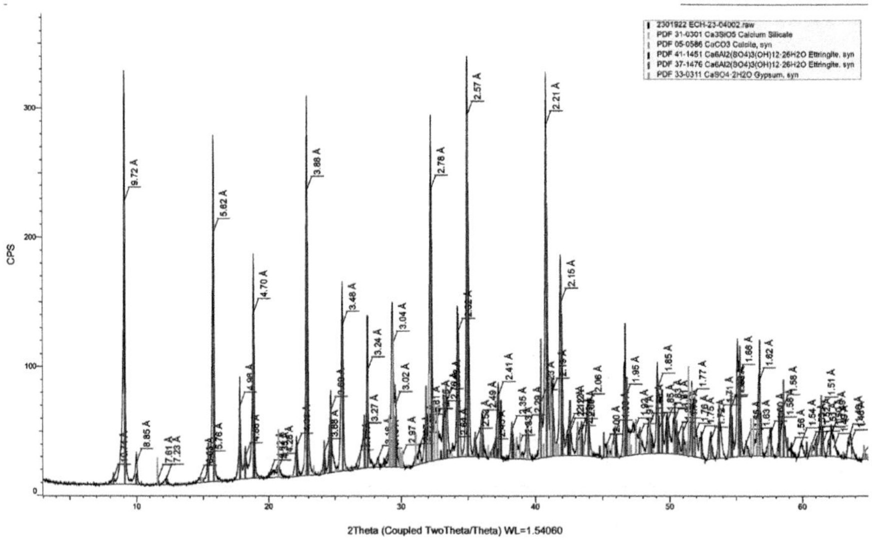

Fig. 10. Diffractogram XRD of C80P20 W/C = 1.1

References

1. Struble, L.J., Brown, P.W.: Heats of dehydration and specific heats of compounds found in concrete and their potential for thermal energy storage. Solar Energy Mater. **14**(1), 1–12 (1986)

2. Chen, B., Kuznik, F., Horgnies, M., Johannes, K., Morin, V., Gengembre, E.: Physicochemical properties of Ettringite/meta-Ettringite for thermal energy storage: review. Sol. Energy Mater. Sol. Cells. **193**, 320–334 (2019)
3. Dent Glasser, L., Glasser, F., Taylor, H.F.: Topotactic reactions in inorganic oxycompounds. Q. Rev. Chem. Soc. **16**, 343360 (1962)
4. Ndiaye, Khadim, Stéphane Ginestet, Martin Cyr: LMDC, Université de Toulouse, INSAT, UPS, France. Thermal energy storage based on cementitious materials: a review. AIMS Energy 6, 1 (2018): 97–120.
5. Skoblinskaya, N.N., Krasilnikov, K.G.: Changes in crystal structure of Ettringite on dehydration. Cem. Concr. Res. **5**(4), 381–393 (1975)
6. Winnefeld, F., Kaufmann, J.P.: Concrete produced with calcium Sulfoaluminate cement – a potential system for energy and heat storage. In: Proceedings of the First Middle East Conference on Smart Monitoring, Assessment and Rehabilitation of Civil Structures (SMAR 2011), pp. 8–10 (2011)
7. Van Berkel J (2000) Solar thermal storage techniques. Research commissioned by The Netherlands Agency for Energy and the Environment NOVEM, project # 143.620–935.8.
8. IPCC report 2023 - https://www.ecologie.gouv.fr/actualites/publication-du-6e-rapport-synthese-du-giec
9. Kuznik, F., Johannes, K.: A review on chemisorption heat storage in low-energy buildings. Energy Procedia. **57**, 2333–2341 (2014)
10. Neumann, H.-M., Hainoun, A., Stollnberger, R., Etminan, G., Schaffler, V.: Analysis and evaluation of the feasibility of positive energy districts in selected urban typologies in Vienna using a bottom-up district energy modelling approach. Energies. **14**(15), 4449 (2021)
11. Ayaz, H., Chinnasamy, V., Yong, J., Cho, H.: Review of technologies and recent advances in low-temperature sorption thermal storage systems. Energies. **14**(19), 6052 (2021)
12. Clark, R.-J., et al.: Experimental screening of salt hydrates for thermochemical energy storage for building heating application. J. Energy Storage. **51**, 104415 (2022)
13. Bonnin, S., Gengembre, E., Johannes, K., Kuznik, F., Malley-Ernewein, A.: Evaluation of the life cycle energy and greenhouse gas emissions of space heating systems using ettringite as an adsorbent material. J. Energy Storage. (2025)
14. N'Tsoukpoe, K.E., et al.: A systematic multi-step screening of numerous salt hydrates for low temperature thermochemical energy storage. Appl. Energy. **124**, 1–16 (2014)
15. Cabeza, L.F., et al.: Review on sorption materials and technologies for heat pumps and thermal energy storage. Renew. Energy. **110**, 3–39 (2017)
16. N'Tsoukpoe, K.E., et al.: The concept of cascade thermochemical storage based on multimaterial system for household applications. Energ. Buildings. **129**, 138–149 (2016)
17. Chen, B., Johannes, K., Ratel, L., Horgnies, M., Morin, V., Kuznik, F.: Investigation on Ettringite as a low-cost high-density thermochemical heat storage material: thermodynamics and kinetics. Sol. Energy Mater. Sol. Cells. **221**, 110877 (2021)
18. Ndiaye, K., Cyr, M., Ginestet, S.: Development of a cementitious material for thermal energy storage at low temperature. Constr. Build. Mater. **242**, 118130 (2020)
19. Chen, B., Horgnies, M., Huet, B., Morin, V., Johannes, K., Kuznik, F.: Comparative kinetics study on carbonation of Ettringite and meta-Ettringite based materials. Cem. Concr. Res. **137**, 106209 (2020)
20. Noor, L., Rost, P., Kirchberger, I., Goetz-Neunhoeffer, F., Ideker, J.H.: Hygrothermal stability of Ettringite in blended systems with CAC-OPC-CŠ. Cem. Concr. Res. **184**, 107590 (2024)
21. Poupelloz, Mme Estelle, & Mme Sandrine Gauffinet. Etude des processus de germination-croissance de l'ettringite, seule ou dans un système aluminate tricalcique/sulfate de calcium, 2019.
22. Baquerizo, L.G., Matschei, T., Scrivener, K.L.: Impact of water activity on the stability of ettringite. Cem. Concr. Res. **79**, 31–44 (2016)

23. Ogorodova, L.P., Gritsenko, Y.D., Kosova, D.A., et al.: Physicochemical and thermochemical study of Ettringite. Geochem. Int. **59**, 1188–1197 (2021)
24. Perkins, R.B., Palmer, C.D.: Solubility of Ettringite (Ca6[Al(OH)6]2(SO4)3 · 26H2O) at 5–75°C. Geochim. Cosmochim. Acta. **63**(13–14), 1969–1980 (1999)
25. Noor, Lamiya, Pauline Rost, Irina Kirchberger, Friedlinde Goetz-Neunhoeffer, & Jason H. Ideker. Hygrothermal stability of Ettringite in blended systems with CAC-OPC-CŠ (2024)

Effective Utilization of End-of-Life Concrete Via an Innovative Thermo-Mechanical Treatment for Sustainable Cement Mortar with Improved Mechanical Properties

Wisal Ahmed(✉), Muhammad Basit Khan, Guoyang Lu, and S. Thomas Ng

Department of Architecture and Civil Engineering, City University of Hong Kong, Tat Chee Avenue, Kowloon Tong, Hong Kong SAR, P.R. China
wahmed7-c@my.cityu.edu.hk

Abstract. Amidst the challenges of increasing urbanization and the pressing need for sustainable green building and circular economy practices, the construction industry is actively pursuing innovative and eco-friendly solutions to reduce environmental impact, enhance resource efficiency, and ensure long-term sustainability. Recycling End-of-Life (EOL) concrete to replace cement and natural sand in cement-based products offers a promising approach to reducing carbon emissions and valorizing construction waste. However, the low reactivity of recycled concrete fines often compromises the mechanical and durability performance of concrete. This study investigates a novel thermo-mechanical treatment to convert EOL concrete into ultrafine recycled concrete powder (RCP) and recycled fine aggregate (RFA) suitable for use in sustainable cement mortar. The treatment involves a three-stage process, which comprises initial ball milling (for 30 min), thermal activation at 800 °C, and secondary ball milling (for 30 min). Experimental results show that mortar mixes with 5–15% treated RCP and 20% treated RFA exhibit up to 12% higher compressive strength and 5% improved flexural strength. Microstructural analysis revealed enhanced pozzolanic activity, improved interfacial transition zones, and better overall mortar performance due to the treatment of RCP. However, the combined addition of treated RCP and RFA reduced mortar flowability by 4–18%, primarily due to the porous and angular surface texture of RCP. Overall, the findings demonstrate that thermo-mechanically treated RCP and RFA can produce a sustainable mortar, with optimal performance at 10% RCP and 20% RFA as cement and sand replacements, offering a viable path for eco-friendly construction practices.

Keywords: Recycled Concrete Fines · Thermo-Mechanical Treatment · Recycled Concrete Powder · Mechanical Properties · Microstructure Analysis

1 Introduction

The construction industry plays a critical role in making new buildings and other infrastructures. However, it faces significant challenges due to fast-growing urbanization, resource depletion, and environmental degradation. Concrete, the most essential and

M. Posani and S. Singh (Eds.): RYS 2025, RILEM Bookseries 67, pp. 53–66, 2026.
https://doi.org/10.1007/978-3-032-14652-6_5

widely employed construction material, relies heavily on Ordinary Portland Cement (OPC), which results in about 8% of global carbon emissions and 3% of global energy consumption due to its energy-intensive production process [1, 2]. With annual concrete production exceeding 30 billion tonnes, the high environmental footprint of cement manufacturing poses a significant barrier to achieving sustainability goals [1]. Concurrently, the industry generates substantial construction and demolition waste, estimated at 3.57 billion tonnes annually, much of which consists of End-of-Life (EOL) concrete [3, 4]. This dual challenge of high carbon emissions and waste accumulation demands urgent attention from both academia and industry to emphasize the utilization of sustainable and waste-derived materials in construction practices.

Over the last decades, supplementary cementitious materials (SCMs) have emerged as one of the most widely employed solutions to mitigate the environmental impact of OPC-based concrete [5–7]. By partially replacing OPC, SCMs reduce cement consumption, lower CO_2 emissions, and enhance concrete durability through pozzolanic reactions that form additional cementitious compounds. However, traditional SCMs like blast-furnace slag and fly ash are declining at a fast rate due to the phasing out of coal-based power plants and the adoption of greener technologies and recycling by the steel industry due to environmental and economic factors [8]. Consequently, researchers are investigating alternative materials, particularly recycled concrete fines (RCFs) derived from EOL concrete, which can serve as both cement and sand replacements in new concrete production [9–11]. Recycling EOL concrete not only conserves natural aggregates and reduces landfill waste but also aligns with global sustainability objectives [5, 12, 13].

RCFs typically exhibit high porosity and low pozzolanic reactivity due to their heterogeneous composition and residual hydration products, resulting in reduced compressive strength, increased water absorption, and compromised durability [14–18]. Numerous approaches including thermal, physical, and chemical activation methods have been investigated to enhance the reactivity of RCFs [19–21]. In thermal activation, RCFs are subjected to elevated temperatures to dehydrate hydration products, thereby generating more reactive constituents [22]. In physical activation, the RCFs are converted into finer particles using the ball milling technique, which disrupts the tetrahedral structure of α-SiO_2, converting it into amorphous SiO_2 [23]. This process also facilitates the release of additional calcium hydroxide and unhydrated cement particles, which further enhance its reactivity. In a recent study, various activation methods were explored to enhance concrete recycled fine's low activity, with thermal activation at 700 °C proving most effective, achieving a higher activity index (77.4%) and promoting hydration and structural compactness of cementitious materials [24]. Another study investigated the performance of recycled cement comprising thermally treated cement and concrete wastes [25]. Their finding demonstrates that mixes with recycled cement paste, particularly when heated at 650 °C, achieved comparative mechanical properties at up to 20% replacement ratios. Similarly, available literature also shows that thermal activation of RCFs at 800 °C is the most effective method which resulted in a high strength activity index of 80.47% [23].

This study introduces a novel three-stage thermo-mechanical activation process to enhance the material properties and pozzolanic reactivity of recycled concrete powder

(RCP) and recycled fine aggregate (RFA). It comprises an initial ball milling for 30 mins, thermal treatment at 800 °C, and secondary ball milling for 30 mins. This synergistic combination of mechanical and thermal treatments effectively overcomes the limitations of individual activation techniques, significantly enhancing the RCF's potential as SCM. For instance, thermal activation alone may cause agglomeration, and mechanical activation alone may not sufficiently alter the chemical composition of RCFs. To investigate the efficacy of this new technique, various mortar mixes were formulated by using 0–30% RCP and 20% RFA, and their performance was assessed in terms of flowability, compressive strength, flexural strength, and microstructure characteristics.

By developing a scalable, eco-friendly recycling method for EOL concrete, this research aims to improve the material properties of RCFs for use in sustainable mortar production, thus reducing the environmental impact of cement-based materials and supporting the global transition to green building and circular economy principles.

2 Materials and Methods

2.1 Materials

For the preparation of cement mortar, ordinary Portland cement (OPC) conforming to ASTM C150 [26] was utilized as the binder material. The OPC was obtained from Green Island Cement Ltd., Hong Kong, and its material properties are presented in Table 1. Natural sand with a fineness modulus of 2.6 and a particle size distribution adhering to ASTM C136 [27] standards was utilized as fine aggregate. A third-generation polycarboxylate-ether (PCE)-based superplasticizer (BASF HK), compliant with ASTM C494 [28], was used to ensure adequate flow of the mortar mixes. Treated RFA, processed via ball milling, and treated RCP, prepared through a combination of ball milling and thermal treatment, were used as partial replacements for natural sand and OPC in the modified mortar mixes, respectively. The EOL concrete was sourced from demolished residential buildings in Hong Kong, provided by Tiostone Holdings Ltd., and consisted primarily of OPC with limestone-based aggregates. The parent concrete compressive strength was 30–40 MPa.

The treatment approach is comprised of a three-stage process. The first step involves initial ball milling for 30 min, where RCF is mechanically processed to reduce particle size, improve specific surface area, and introduce micro-defects in the crystalline structure. This step is critical because finer particles with higher surface areas expose more reactive sites and thus enhance the RCF's susceptibility to subsequent thermal treatment. After this stage, treated RFA and RCP were separated through sieving. In the next step, the RCP was subjected to a thermal treatment at 800 °C in a controlled furnace environment. This temperature is selected based on prior research [23], which demonstrates that 800 °C effectively decomposes low-reactivity phases like calcite and forms highly reactive cementitious compounds such as calcium silicate, and calcium oxide [29]. In the final step secondary ball milling was performed for 30 min which further refined the thermally activated RCP by breaking down any agglomerates or partially sintered particles formed during high-temperature treatment. The treated RCP was sieved through a 75 μm sieve, and only the passed material was used as cement replacement. Figure 1(a) illustrates a sample of the treated RCP used in this study, while the SEM micrograph in Fig. 1(b) reveals the porous and rough surface morphology of the RCP. The energy-dispersive X-ray spectroscopy (EDX) results in Fig. 1(c) show the elemental distribution of the treated RCP, primarily consisting of Ca (46.6 wt%), O (35.3 wt%), Si (7.1 wt%), C (7.1 wt%), Al (3.1 wt%), and Mg (0.6 wt%).

Table 1. Material properties of OPC.

CaO (%)	SiO_2 (%)	Fe_2O_3 (%)	Al_2O_3 (%)	SO_3 (%)	MgO (%)	LOI	Specific Area (m^2/g)
65.5	21.1	3.4	5.8	2.42	0.8	2.85	0.359

Fig. 1. a) Sample of treated RCP, b) SEM image of treated RCP, c) EDX spectra and elemental composition of treated RCP.

2.2 Mix Proportions

This study considered a total of eight mortar mixes having their compositions as detailed in Table 2. The reference mortar denoted as RM, was prepared using OPC and natural sand. The mix designated MM-20RFA-0RCP incorporated 20% treated RFA as a natural sand replacement. The remaining six mixes were designed with a constant 20% RFA content and varying RCP levels, ranging from 5% to 30%, as a partial replacement for OPC. In all mixes a constant water-to-binder (w/b) ratio of 0.38 and a superplasticizer dosage of 0.38% by weight of the binder was used. The production process of treated RFA and treated RCP-based mortar mixes is illustrated in Fig. 2.

Table 2. Mix proportions of reference and RCP-based modified mortars

Mixture ID	Cement (kg/m^3)	Sand (kg/m^3)	RCP (%)	RFA (%)	Superplasticizer (%)
RM	576	864	0	0	0.38
MM-20RFA-0RCP	576	691	0	20	0.38
MM-20RFA-5RCP	547	691	5	20	0.38
MM-20RFA-10RCP	518	691	10	20	0.38
MM-20RFA-15RCP	490	691	15	20	0.38
MM-20RFA-20RCP	461	691	20	20	0.38
MM-20RFA-25RCP	432	691	25	20	0.38
MM-20RFA-30RCP	403	691	30	20	0.38

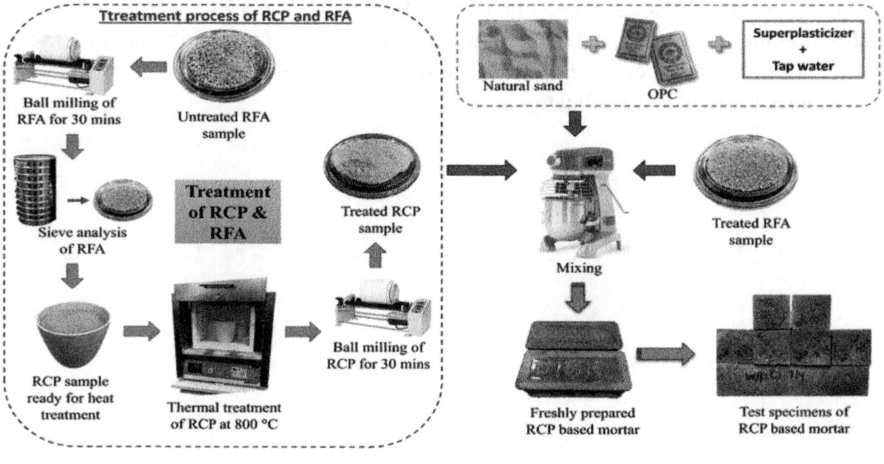

Fig. 2. Schematic diagram illustrating the treatment process of RFA and RCP, and production of modified mortar mixes.

2.3 Methods

The flowability of reference and modified mortar mixes was measured using a standard flow table as per ASTM C1437 [30]. For the compressive strength, tests were conducted on the cubical mortar samples cured for 3, 7, and 28 days, following the procedures outlined in ASTM C109 [31]. Additionally, the flexural strength of prismatic mortar specimens was evaluated through a three-point bending test, performed in compliance with ASTM C348 [32]. The microstructural characteristics of reference and modified mortar samples were studied with the help of a scanning electron microscope (SEM).

3 Results and Discussion

3.1 Flowability

The flowability of reference and modified mortar mixes is presented in Fig. 3. The RM exhibited a baseline fluidity of 216 mm. The MM-20RFA-0RCP mix, with 20% treated RFA replacing natural sand and no RCP, showed a reduced fluidity of 208 mm, reflecting a 3.7% decrease. This reduction is attributed to the angular particle shape and slightly higher surface area of treated RFA compared to natural sand. However, considering the fluidity of untreated RFA-based mortar mixes reported in previous studies, this loss is moderate [33, 34]. The treatment of RFA with ball milling improves particle packing and reduces water absorption relative to untreated RFA.

For mixes combining 20% treated RFA with varying RCP replacements (0–30%), flowability progressively decreased as RCP content increased. Specifically, the flowability values were 204 mm, 201 mm, 196 mm, 190 mm, 185 mm, and 177 mm for 5%, 10%, 15%, 20%, 25%, and 30% RCP replacements, respectively. These values correspond to a decrease of 5–18% in reference to RM. This reduction in fluidity is predominantly due to the increased fineness, angular shape, and porous surface texture of RCP as shown in Fig. 1(b). These results suggest that while treated RFA and RCP can be viable partial replacements for sand and cement, respectively, their incorporation at higher levels (beyond 20%) may necessitate increased superplasticizer dosage to achieve acceptable workability for practical applications.

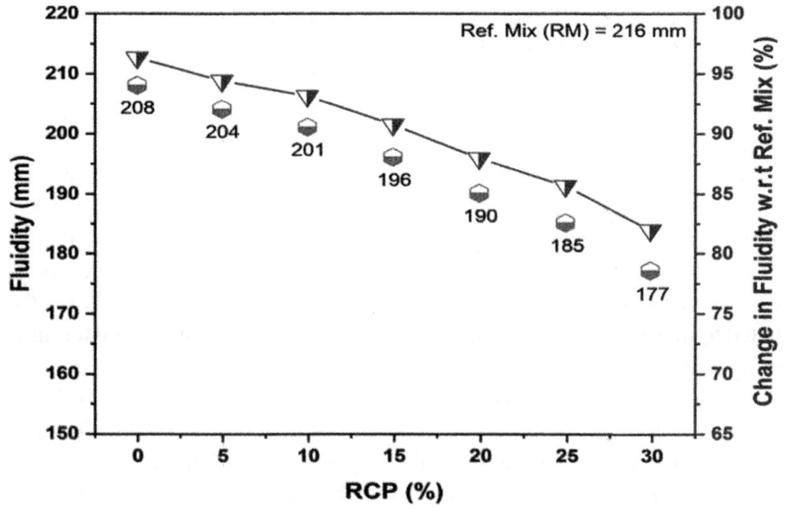

Fig. 3. Combined effect of treated RCP and RFA on the flowability of mortar mixes.

3.2 Compressive Strength

The compressive strength results of mortar mixes incorporating treated RCP and RFA evaluated at various curing periods are presented in Fig. 4. RM achieved compressive strengths of 33.7 MPa, 48.5 MPa, and 62.2 MPa at 3, 7, and 28 days, respectively. The strength development follows a typical OPC hydration curve, with approximately 54% and 78% of the 28-day strength achieved at 3 and 7 days, respectively, indicating robust early and later-age performance. The MM-20RFA-0RCP exhibited slightly lower strengths of 32.2 MPa, 45.9 MPa, and 58.4 MPa, corresponding to about 4–6% reduction compared to the RM. These results are in agreement with those reported in previous studies [35, 36]. This reduction is primarily due to the weak material properties of RFA, adherent cement mortar, and poor interfacial transition zone (ITZ) of RFA-cement paste.

For mixes combining 20% treated RFA with varying RCP replacements (0–30%), the compressive strength reveals first increasing and then decreasing trend possibly due to the RCP's filler effect, pozzolanic reactivity, and OPC dilution. The modified mix with 5% RCP showed enhanced strengths of 33.8 MPa, 49.8 MPa, and 65.5 MPa at 3, 7, and 28 days, respectively, surpassing the RM by about 5.3% at 28 days. This improvement is attributed to the fine particle size of treated RCP acting as a micro-filler, enhancing particle packing and densifying the cement matrix. Additionally, the thermal treatment at 800 °C likely forms reactive phases (e.g., calcium silicates), contributing minor pozzolanic activity that augments early and later-age strengths. With a further increase in the RCP content, the MM-20RFA-10RCP mix attained strengths of 34.9 MPa at 3 days, 52.4 MPa at 7 days, and 69.6 MPa at 28 days. This mix achieved the highest 11.9% improvement in 28-day strength. This reflects an optimal balance where RCP's filler effect and limited pozzolanic reactivity compensate for the OPC dilution and RFA's ITZ weakening. The fine RCP particles likely fill voids in the matrix, while the low w/b ratio ensures a dense microstructure.

The MM-20RFA-15RCP did not reveal any further increase and attained strength of 34.7 MPa at 3 days, 51.2 MPa at 7 days, and 68.3 MPa at 28 days. At 20% RCP (MM-20RFA-20RCP), strength decreased to 33.1 MPa, 48.2 MPa, and 63.4 MPa, remaining comparable to RM. This suggests that the dilution of OPC begins to offset the filler and pozzolanic benefits, reducing the formation of C-S-H and slightly compromising early-age hydration due to RCP's slower reactivity. This trend intensifies at 25% and 30% RCP, with strength dropping to 31.9 MPa, 45.2 MPa, and 58.5 MPa for 25% RCP, and 29.7 MPa, 43.6 MPa, and 55.2 MPa for 30% RCP. These strength values are 6% and 11% below RM at 28 days, respectively. The continued OPC replacement reduces the primary cementitious material, limiting the overall C-S-H formation. The filler effect of RCP still contributed to matrix densification, but its impact was insufficient to counterbalance the reduced hydration products.

The strength variations across the modified mixes highlight the dual role of utilizing treated RCP as a filler and partial cementitious material. The low w/b ratio (0.38) resulted in a dense matrix but amplified the impact of OPC dilution at higher RCP levels, as less water is available for hydration. Notably, the 5–15% RCP mixes outperform the RM, suggesting that low-level RCP replacement is a viable strategy for sustainable mortars with enhanced strength. However, higher RCP contents require careful consideration of workability and long-term performance, potentially necessitating adjustments in superplasticizer dosage or curing conditions to optimize strength development.

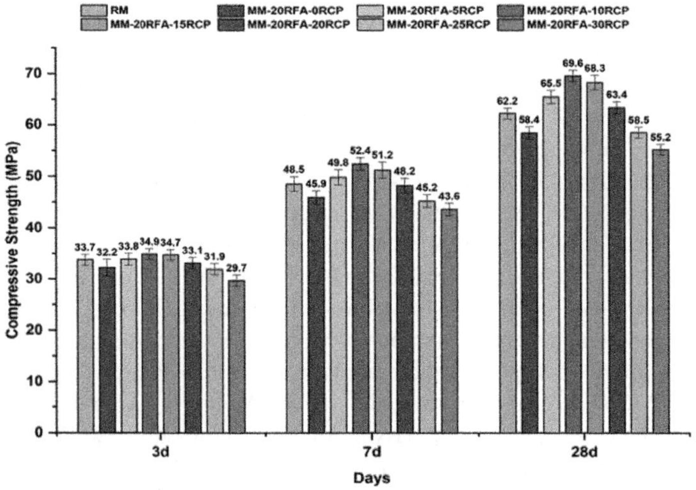

Fig. 4. Compressive strength of reference and modified mortar mixes at 3, 7, and 28 days.

3.3 Flexural Strength

The 28-day flexural strength results of reference and modified mortar mixes incorporating treated RCP and treated RFA are presented in Fig. 5. The RM achieved a flexural strength of 5.59 MPa, consistent with its high 28-day compressive strength of 62.2 MPa.

MM-20RFA-0RCP exhibited a slightly reduced flexural strength of 5.45 MPa, a 2.5% decrease compared to the RM. This correlates with the compressive strength results and is possibly due to the weak material properties of RFA.

For modified mixes incorporating 5–30% RCP, the flexural strength trends resemble closely that of the compressive strength patterns. The MM-20RFA-5RCP, MM-20RFA-10RCP, and MM-20RFA-15RCP mix showed an improved flexural strength of 5.72 MPa, 5.86 MPa, and 5.82 MPa, respectively, demonstrating an increase of 2.3%, 4.8%, and 4.1% over RM. This increase could be attributed to the RCP's filler effect and pozzolanic reactivity which enhanced the matrix density and crack resistance, offsetting the weak performance of 20% RFA. However, at 20% RCP, the flexural strength decreased to 5.58 MPa, comparable to that of RM. This decline indicates that OPC dilution begins to limit C-S-H formation, reducing the matrix's flexural strength capacity, though the filler effect still provides some benefit. The decreasing trend continues with 25% and 30% RCP, where flexural strength falls to 5.21 MPa and 4.93 MPa, respectively, decreased by 6.8% and 11.8% in comparison to the RM. This trend correlates well with the compressive strengths and could be the result of a relatively weak matrix quality of the mortar matrix possibly due to reduced C-S-H and other hydration products caused by the OPC dilution.

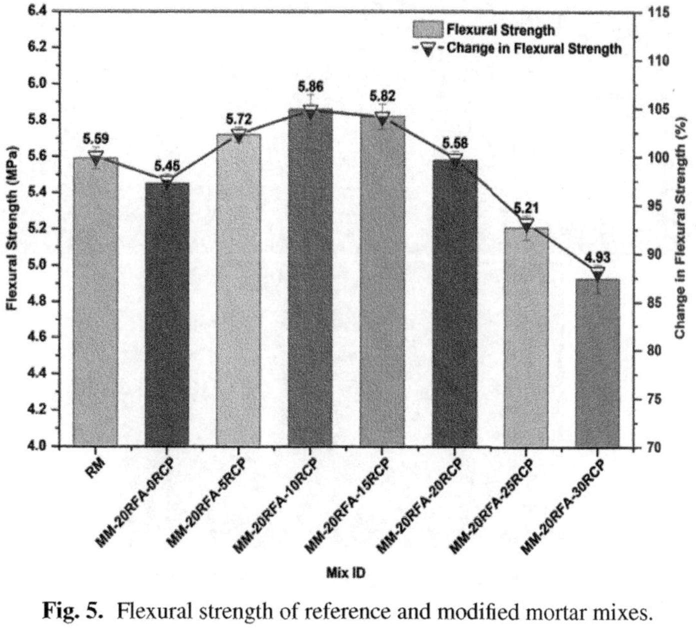

Fig. 5. Flexural strength of reference and modified mortar mixes.

3.4 Microstructural Analysis

The SEM image of the RM mix in Fig. 6(a) reveals a dense and well-compacted cementitious matrix. The microstructure exhibits a homogeneous distribution of C-S-H with minimal porosity and few microcracks. Similarly, the SEM micrograph of the MM-20RFA-10RCP mix in Fig. 6(b) displays a relatively denser matrix with enhanced microstructural characteristics. The C-S-H gel appears more compact, with finer RCP particles acting as micro-fillers that fill voids and improve particle packing. The ITZ between RFA and cement paste is notably improved, showing stronger bonding compared to RM, likely due to the pozzolanic reactivity of thermo-mechanically treated RCP. This enhanced ITZ and dense matrix correlate with the mix's peak improvements in mechanical properties. However, the SEM image of the MM-20RFA-30RCP mix in Fig. 6(c) reveals a compromised microstructure. The matrix exhibits voids, microcracks, and a weakened ITZ, with poor bonding between the cement paste and RFA. Loose adhered mortar is also observed, indicating incomplete integration of RFA. These features reflect the excessive OPC dilution, reducing C-S-H formation and leading to lower strength performance in mortar samples with higher replacement levels of RCP.

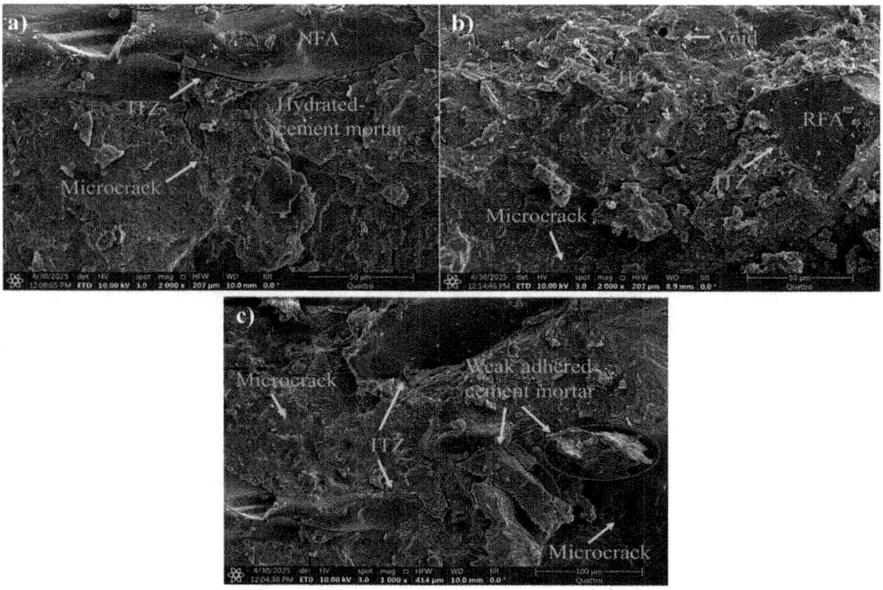

Fig. 6. a) SEM image of RM; b) SEM image of MM-20RFA-10RCP; c) SEM image of MM-20RFA-30RCP.

4 Conclusions

This study experimentally assessed the combined effect of treated RCP and treated RFA as partial alternatives for cement and natural sand, respectively in mortar mixes. The experimental findings reveal that a combination of 20% treated RFA with 5–15% treated

RCP enhances strength performance compared to the reference mortar, with the MM-20RFA-10RCP mix representing the highest 28-day compressive strength and flexural strength improvements of 11.9% and 4.8%, respectively. These improvements are primarily attributed to the filler effect and the pozzolanic reactivity of thermo-mechanically treated RCP particles, which densified the cement matrix and produced additional C-S-H as evidenced by the microstructural analysis of MM-20RFA-10RCP mix.

The higher RCP replacements (beyond 20%) lead to a decline in strength, with the MM-20RFA-30RCP mix exhibiting 11–12% reductions in 28-day compressive and flexural strengths, compared to RM. This decline is majorly due to excessive OPC dilution, which reduced C-S-H formation and compromised the matrix quality, as evidenced by increased voids, microcracks, and weakened ITZ in the SEM analysis. Additionally, the flowability of modified mixes decreased with increasing RCP content, with the MM-20RFA-30RCP mix showing about an 18% reduction in comparison to RM.

These findings suggest that treated RFA and RCP can be effectively utilized in sustainable mortar production, with optimal performance achieved at 20% RFA and 10% RCP replacements. This approach not only enhances the mechanical properties of cement-based products but also promotes the employment of recycled concrete fines as viable construction material for sustainable buildings and infrastructures. Future research should focus on optimizing superplasticizer dosages and evaluating long-term durability to further enhance the practical applicability of these RCFs-based sustainable mortar formulations.

5 Recommendations for Further Research

To further enhance the practical applicability of the proposed thermo-mechanical treatment for RCP and RFA, future research should focus on the following areas. Chemical analyses such as thermogravimetric analysis (TGA) and X-ray diffraction (XRD) should be performed to quantify phase transformations (e.g., dehydration of hydration products and formation of amorphous silica) induced by the thermo-mechanical treatment of RCP. In order, to quantify the CO_2 reduction potential of using treated RCP and RFA as partial replacements for cement and natural sand in mortar, a comprehensive life-cycle assessment (LCA) should be conducted. Furthermore, the variability of RCP derived from diverse end-of-life (EOL) concrete sources (e.g., different aggregate types such as limestone or siliceous, cement compositions, and exposure conditions) should be investigated to support the broader applicability of the proposed technique. Lastly, future works should also explore optimizing superplasticizer dosages and evaluating the long-term durability of RCP-based mortars to ensure their viability in sustainable construction practices.

Acknowledgements. The work described in this paper was supported by the Environment and Ecology Bureau of the Government of the Hong Kong Administrative Region through the Green Tech Fund (Grant No.: GTF202110158).

References

1. Monteiro, P.J.M., Miller, S.A., Horvath, A.: Towards sustainable concrete. Nat. Mater. **16**, 698–699 (2017). https://doi.org/10.1038/nmat4930
2. Miller, S.A., Horvath, A., Monteiro, P.J.M.: Readily implementable techniques can cut annual CO 2 emissions from the production of concrete by readily implementable techniques can cut annual CO 2 emissions from the production of concrete by over 20%. Environ. Res. Lett. **11**, 074029 (2016)
3. Patil, Y.R., Dakwale, V.A., Ralegaonkar, R.V.: Recycling construction and demolition waste in the sector of construction. Adv. Civ. Eng. **2024**, 6234010 (2024). https://doi.org/10.1155/2024/6234010
4. Ahmed, W., Lu, G., Ng, S.T., Liu, G.: Innovative valorization of solid waste materials for production of sustainable low-carbon pavement: a systematic review and scientometric analysis. Case Stud. Constr. Mater. **22**, e04541 (2025). https://doi.org/10.1016/j.cscm.2025.e04541
5. Snellings, R., Suraneni, P., Skibsted, J.: Future and emerging supplementary cementitious materials. Cem. Concr. Res. **171**, 107199 (2023). https://doi.org/10.1016/j.cemconres.2023.107199
6. Ahmed, W., Ye, C., Lu, G., Ng, S.T., Liu, G., Wang, Y.: Low-carbon concrete comprising high-volume pozzolan and recycled aggregate: evaluating mechanical performance, microstructure, environmental impact, and cost efficiency. J. Clean. Prod. **518**, 145796 (2025). https://doi.org/10.1016/j.jclepro.2025.145796
7. Hu, J., Ahmed, W., Jiao, D.: A critical review of the technical characteristics of recycled brick powder and its influence on concrete properties. Buildings. **14**, 3691 (2024)
8. E.E.S. Association, Low Carbon Roadmap: Pathways to a CO2-Neutral European Steel Industry, Final Rep. (2019) 1–18. https://www.eurofer.eu/assets/Uploads/EUROFER-Low-Carbon-Roadmap-Pathways-to-a-CO2-neutral-European-Steel-Industry.pdf.
9. Evangelista, L., de Brito, J.: Mechanical behaviour of concrete made with fine recycled concrete aggregates. Cem. Concr. Compos. **29**, 397–401 (2007). https://doi.org/10.1016/j.cemconcomp.2006.12.004
10. Oksri-Nelfia, L., Mahieux, P.Y., Amiri, O., Turcry, P., Lux, J.: Reuse of recycled crushed concrete fines as mineral addition in cementitious materials. Mater. Struct. Constr. **49**, 3239–3251 (2016). https://doi.org/10.1617/s11527-015-0716-1
11. Tang, Q., Ma, Z., Wu, H., Wang, W.: The utilization of eco-friendly recycled powder from concrete and brick waste in new concrete: a critical review. Cem. Concr. Compos. **114**, 103807 (2020). https://doi.org/10.1016/j.cemconcomp.2020.103807
12. Ahmed, W., Lim, C.W.: Coupling effect assessment of vacuum based pozzolana slurry encrusted recycled aggregate and basalt fiber on mechanical performance of fiber reinforced concrete. Constr. Build. Mater. **300**, 124032 (2021). https://doi.org/10.1016/j.conbuildmat.2021.124032
13. Jamil, S., Idrees, M., Akbar, A., Ahmed, W.: Investigating the mechanical and durability properties of carbonated recycled aggregate concrete and its performance with SCMs. Buildings. **15**, 1–23 (2025). https://doi.org/10.3390/buildings15020201
14. Nedeljković, M., Visser, J., Šavija, B., Valcke, S., Schlangen, E.: Use of fine recycled concrete aggregates in concrete: a critical review. J. Build. Eng. **38**, 102196 (2021). https://doi.org/10.1016/j.jobe.2021.102196
15. Evangelista, L., de Brito, J.: Durability performance of concrete made with fine recycled concrete aggregates. Cem. Concr. Compos. **32**, 9–14 (2010). https://doi.org/10.1016/j.cemconcomp.2009.09.005
16. Wu, H., Yang, D., Ma, Z.: Micro-structure, mechanical and transport properties of cementitious materials with high-volume waste concrete powder and thermal modification. Constr. Build. Mater. **313**, 125477 (2021). https://doi.org/10.1016/j.conbuildmat.2021.125477

17. Ahmed, W., Lim, C.W.: Production of sustainable and structural fiber reinforced recycled aggregate concrete with improved fracture properties: a review. J. Clean. Prod. **279**, 123832 (2021)
18. Ahmed, W., Lim, C.W.: Effective recycling of disposable medical face masks for sustainable green concrete via a new fiber hybridization technique. Constr. Build. Mater. **344**, 128245 (2022). https://doi.org/10.1016/J.CONBUILDMAT.2022.128245
19. Sousa, L.N., Zepper, J.C.O., Schollbach, K., Brouwers, H.J.H.: Improving the reactivity of industrial recycled concrete fines: exploring mechanical and hydrothermal activation. Constr. Build. Mater. **442**, 137594 (2024). https://doi.org/10.1016/j.conbuildmat.2024.137594
20. Wang, L., et al.: Eco-friendly treatment of recycled concrete fines as supplementary cementitious materials. Constr. Build. Mater. **322**, 126491 (2022). https://doi.org/10.1016/j.conbuildmat.2022.126491
21. Shen, P., et al.: Synthesis of amorphous nano-silica from recycled concrete fines by two-step wet carbonation. Cem. Concr. Res. **147**, 106526 (2021). https://doi.org/10.1016/j.cemconres.2021.106526
22. Sui, Y., Ou, C., Liu, S., Zhang, J., Tian, Q.: Study on properties of waste concrete powder by thermal treatment and application in mortar. Appl. Sci. **10**, 998 (2020). https://doi.org/10.3390/app10030998
23. Zhang, D., Zhang, S., Huang, B., Yang, Q., Li, J.: Comparison of mechanical, chemical, and thermal activation methods on the utilisation of recycled concrete powder from construction and demolition waste. J. Build. Eng. **61**, 105295 (2022). https://doi.org/10.1016/j.jobe.2022.105295
24. Chen, L., Wei, M., Lei, N., Li, H.: Effect of chemical–thermal activation on the properties of recycled fine powder cementitious materials. Case Stud. Constr. Mater. **20**, e02956 (2024). https://doi.org/10.1016/j.cscm.2024.e02956
25. Bogas, J.A., Carriço, A., Pereira, M.F.C.: Mechanical characterization of thermal activated low-carbon recycled cement mortars. J. Clean. Prod. **218**, 377–389 (2019). https://doi.org/10.1016/j.jclepro.2019.01.325
26. C. ASTM, 150, Standard specification for Portland cement, Annu. B. ASTM Stand. (2002).
27. ASTM, ASTM C136 / C136M - 19 Standard Test Method for Sieve Analysis of Fine and Coarse Aggregates, (2019). https://www.astm.org/Standards/C136 (Accessed 23 Mar 2021).
28. ASTM C494/494M, Standard Specification for Chemical Admixtures for Concrete, (2019). https://www.astm.org/Standards/C494.htm (Accessed 8 Dec 2020).
29. Kim, J., Nciri, N., Sicakova, A., Kim, N.: Characteristics of waste concrete powders from multi-recycled coarse aggregate concrete and their effects as cement replacements. Constr. Build. Mater. **398**, 132525 (2023). https://doi.org/10.1016/j.conbuildmat.2023.132525
30. ASTM C1437, Test Method for Flow of Hydraulic Cement Mortar, (2020). doi:https://doi.org/10.1520/C1437-20.
31. ASTM C109/C109M-20, Test Method for Compressive Strength of Hydraulic Cement Mortars (Using 2-in. or [50-mm] Cube Specimens), (2020). doi:https://doi.org/10.1520/C0109_C0109M-20.
32. ASTM C348–21, Test Method for Flexural Strength of Hydraulic-Cement Mortars, (2021). doi:https://doi.org/10.1520/C0348-21.
33. Jiang, Y., Li, B., Liu, S., He, J., Hernandez, A.G.: Role of recycled concrete powder as sand replacement in the properties of cement mortar. J. Clean. Prod. **371**, 133424 (2022). https://doi.org/10.1016/j.jclepro.2022.133424
34. Deng, X., et al.: Rheological and early hydration of cementitious material containing recycled concrete powders collected from recycled aggregates. Constr. Build. Mater. **393**, 132108 (2023). https://doi.org/10.1016/j.conbuildmat.2023.132108

35. Zhao, Z., Remond, S., Damidot, D., Xu, W.: Influence of fine recycled concrete aggregates on the properties of mortars. Constr. Build. Mater. **81**, 179–186 (2015). https://doi.org/10.1016/j.conbuildmat.2015.02.037
36. Fan, C.C., Huang, R., Hwang, H., Chao, S.J.: The effects of different fine recycled concrete aggregates on the properties of mortar. Materials (Basel). **8**, 2658–2672 (2015). https://doi.org/10.3390/ma8052658

Evaluating the Performance of Limestone Calcined Clay Cement (LC3) Incorporating Low-Kaolinite Clay from the Ceramic Industry

Nesil Ozbakan Orhan(✉) and Derya Over

Eskisehir Technical University Civil Engineering Department, 26555 Eskisehir, Turkey
nesil_ozbakan@eskisehir.edu.tr

Abstract. Clay, a globally abundant material, has the potential to serve as a cement substitute following calcination. Although the reactivity of kaolinitic clays in this process is well-established, further research is needed to fully understand the potential of low-grade clays with lower kaolinite content in cement production. This study investigated the potential of using clays with low kaolinite content (~22%) sourced from the ceramic industry in the production of limestone calcined clay cement (LC3) by calcining at 750 °C. The chemical composition of the clay was determined by X-ray fluorescence (XRF), and its mineralogical analysis was performed using X-ray diffraction (XRD). The mechanical properties of LC3 mortars were assessed at various water-to-binder ratios and different curing ages. Furthermore, the mineralogical composition of LC3 pastes was characterized using XRD analysis, while the microstructural development of LC3 mortars was investigated through scanning electron microscopy.

Keywords: LC3 · Calcined clay · Kaolinite · Compressive Strength

1 Introduction

The growing global demand for cement, driven by rapid urbanization and infrastructure development, poses a major environmental challenge due to carbon emissions. Cement production accounts for approximately 7–8% of global CO_2 emissions [1], largely due to the energy- and carbon-intensive process of clinker production. In this context, LC3 cement, which allows for up to a 50% reduction in clinker content, is a promising alternative to reduce CO_2 emissions and energy demand. The kaolinite content of the clay, along with the calcination process and temperature, significantly influences both the pozzolanic activity of the clay and the performance of the blended cement [2, 3]. The global availability of clays meeting the recommended kaolinite content for LC3 is limited [4]. While research on utilizing low-grade clays in concrete remains scarce, existing studies suggest their potential as a cement replacement material.

Alujas et al. [5] investigated the pozzolanic reactivity of low-grade kaolinitic clay as a supplementary cementitious material. The raw clay contained approximately 40% kaolinite and 40% 2:1 type clay minerals (illite and montmorillonite), and was calcined

M. Posani and S. Singh (Eds.): RYS 2025, RILEM Bookseries 67, pp. 67–75, 2026.
https://doi.org/10.1007/978-3-032-14652-6_6

at temperatures ranging from 500 °C to 1000 °C. The highest reactivity was observed at 800 °C, which provided an optimal balance between structural disorder and specific surface area. At this temperature, the calcined clay enhanced compressive strength and influenced ordinary Portland cement (OPC) hydration through both filler effects and pozzolanic reactions. The findings demonstrated that properly calcined low-grade kaolinitic clays offered high reactivity and potential for sustainable cement substitution.

Du and Pang [6] evaluated the pozzolanic reactivity of marine clay, containing approximately 20% kaolinite, after thermal activation at various calcination temperatures. The calcined marine clay was used to replace 30% of ordinary Portland cement in mortar. Despite the low kaolinite content, the thermally treated clay exhibited clear pozzolanic behavior, confirmed by calcium hydroxide consumption and increased heat release. The findings demonstrated the feasibility of converting marine clay excavation waste into a sustainable supplementary cementitious material.

Hu et al. [7] investigated the performance of LC3 cement incorporating low-grade marine clay sourced from Singapore. Despite the relatively low kaolinite content (16–20%) and a reduction of approximately 25% in 28-day compressive strength compared to OPC, the LC3 systems demonstrated remarkable resistance to chloride ion penetration. Notably, the LC3–50 formulation exhibited significantly lower chloride diffusion coefficients, underscoring its potential for durable and sustainable construction applications.

Furthermore, long-term investigations into the phase composition of LC3 systems have revealed substantial similarities despite variations in initial kaolinite content, although the reaction kinetics governing the formation of hydration products differ [8]. These observations suggest that clays with lower kaolinite content may still be viable for LC3 production. Given the limited availability and uneven distribution of high-quality clays, exploring the use of locally sourced, low-grade clays holds both practical and environmental significance. In this context, this study investigated the feasibility of utilizing low-kaolinite clays for the production of limestone calcined clay cement through calcination. The chemical composition of the clays was analyzed using XRF, while their mineralogical characteristics were examined via XRD. The mechanical performance of LC3 mortars was evaluated across various water-to-binder ratios and curing periods. In addition, the mineralogical composition of the LC3 pastes was further characterized using XRD, and the microstructural evolution of the mortars was investigated through scanning electron microscopy (SEM).

2 Materials and Test Procedures

Limestone calcined clay cement (LC3) was developed using 55% Portland cement, 30% calcined clay and 15% limestone powder. In this study, ordinary Portland cement (OPC), limestone, and low-grade clay were utilized in the production of limestone calcined clay cement (LC3). The chemical compositions of the materials used in the study were determined by X-ray fluorescence spectroscopy (XRF) analysis and are presented in Table 1. The clay was oven-dried at 105 $\pm$ 5 °C for 24 h and subsequently ground using a ring mill. Its particle size distribution was determined by a laser diffraction particle size analyzer (Mastersizer 2000, Malvern Instruments) and is illustrated in Fig. 1 (Table 1).

Table 1. Chemical compositions of the materials

Chemical Composition (%)	OPC	Clay	Limestone
SiO_2	18.36	56.50	0.42
Al_2O_3	4.88	27.70	-
Fe_2O_3	3.10	2.05	-
TiO_2	0.25	0.94	-
CaO	61.36	0.62	56.41
MgO	1.94	0.53	0.56
Na_2O	0.80	-	-
K_2O	0.87	1.74	-
P_2O_5	0.23	0.07	-
SO_3	2.89	0.06	-
BaO	-	-	-
LOI	3.21	-	42.61

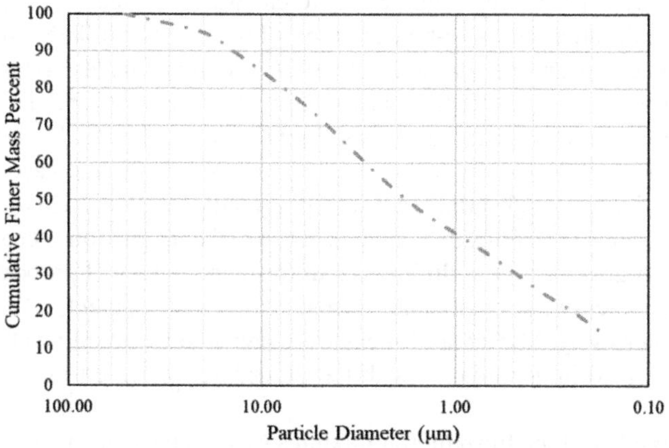

Fig. 1. Particle size distribution of clay

The mineralogical composition of the clay was analyzed by X-ray diffraction (XRD) technique using a Rigaku Rint 2000 diffractometer with Rietveld refinement, scanning a 2θ range of 5–70° at 0.5°/min. As a result of XRD analysis of the clay, quartz (SiO_2), kaolinite ($Al_2Si_2O_5(OH)_4$), muscovite ($KAl_2(Al_2\ Si_3\ O_{10})(OH)_2$) and microcline ($KAlSi_3O_8$) phases were detected (Fig. 2.). As a result of Rietveld analysis, the kaolinite content of the clay was determined to be approximately 22%.

The thermal activation was conducted at 750 °C for 1 h in a laboratory muffle furnace. Before calcination, the clay sample was dried in an oven at a controlled temperature of 105 ± 5 °C for 24 h to remove moisture. Subsequently, the dried clay was uniformly distributed as a thin layer on refractory plates and subjected to calcination within a muffle

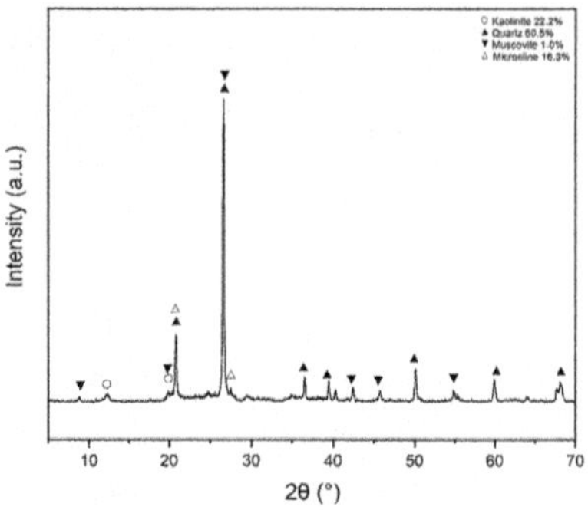

Fig. 2. XRD pattern of clay

furnace at a target temperature of 750 °C. The furnace temperature increased to the target temperature at a rate of 10 °C/min and maintained at that temperature for one hour. After the calcination period, the clay samples were allowed to cool inside the furnace.

LC3 cement was produced by mixing 55% ordinary Portland cement, 30% calcined clay, and 15% limestone. For comparative evaluation of the fresh and hardened properties of the LC3 mixtures, reference mortars (REF) containing only OPC were also prepared. All mixtures were produced using two different water-to-binder (w/b) ratios: 0.485 and 0.55. Two different water-to-binder (w/b) ratios (0.485 and 0.55) were adopted to assess the robustness and performance of the LC3 mixtures under varying workability conditions. The lower ratio (0.485) was selected to represent a relatively dense mix design commonly used in performance-based applications, while the higher w/b ratio was specifically chosen considering the potential of calcined clay to increase water demand of the mixture due to its high surface area and pozzolanic reactivity. This approach aimed to simulate practical variations that could occur in real construction scenarios. By using a lower and a higher w/b ratio, differences in fresh and hardened properties, including flowability, compressive strength, and microstructural development, could be evaluated more comprehensively. Mortar specimens were prepared by mixing water, binders and CEN standard sand in accordance with the procedures specified in the ASTM C305 standard [9].

The workability of the fresh mortar mixtures was evaluated using the flow table test, following the ASTM C1437 standard [10]. Before testing, both the surface of the flow table and the conical mold were cleaned and dried. The mold was then positioned in the center of the table and filled with mortar in two layers. Each layer was compacted with 20 uniform strokes using a tamping rod. After compaction, the mold was carefully removed, and the table was dropped 25 times. Subsequently, the spread diameter of the mortar was measured. This procedure was consistently applied to all mortar mixtures that were evaluated.

In order to assess the compressive strength of mortars, three specimens with dimensions of 50 mm × 50 mm × 50 mm were cured in water at room temperature until the day of testing, and then tested after 3, 7, and 28 days in accordance with ASTM C109 standard [11].

For the XRD analysis, paste specimens were prepared using LC3 cement and OPC according to ASTM C305. The samples were cured in water for a duration of 28 days. XRD measurements were performed in 2θ range of 5–70° with a scan rate of 2°/min.

The microstructural evolution of hydrated OPC and LC3 mortars at different water-to-binder ratios was examined using scanning electron microscopy (SEM). The morphology and microstructure of LC3 mortars were analyzed and compared with those of the reference mortar.

3 Results and Discussion

The changes in flow diameter of the mortars are illustrated in Fig. 3. In both reference and LC3 mixes, increasing the w/b ratio from 0.485 to 0.55 results in higher slump flow. Reference mixtures exhibited significantly higher slump flow compared to the LC3 mixes at both w/b ratios. This indicates that LC3 mixes are less workable than the reference mixes. The reduced workability of LC3 mixes can be attributed to the high surface area and water demand of calcined clay, as well as its tendency to increase the yield stress of the mix due to flocculation [12, 13]. These findings suggest that although LC3 mixes may be viable alternatives from a sustainability perspective, their lower workability requires the use of chemical admixtures or adjustments in mix design to achieve performance levels comparable to conventional systems.

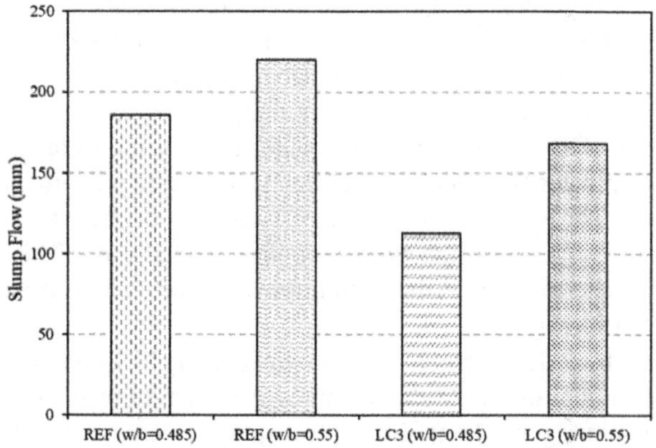

Fig. 3. Slump flow diameter values of the mixtures

Fig. 4. presents the compressive strength development of the reference and LC3 mortars, each tested at two w/b ratios (0.485 and 0.55) and measured at 3, 7, and 28 days. For all mixes, compressive strength demonstrates a positive correlation with curing duration (Fig. 4a). The most pronounced gains were recorded between 3 and 7 days, followed by continued but more moderate increases up to 28 days.

Among the mixtures, the reference mortar with a w/b ratio of 0.485 exhibited the highest overall strength, reaching approximately 44 MPa at 28 days. The corresponding reference mix with a higher w/b ratio of 0.55 attained a slightly lower strength of about

38 MPa, consistent with expectations regarding the influence of increased water content on matrix density and hydration efficiency.

LC3 mixes with water-to-binder ratios of 0.485 and 0.55 exhibit lower compressive strength compared to their corresponding reference mixes. LC3 with a w/b ratio of 0.485 reaches approximately 29 MPa at 28 days, while the mix with a w/b ratio of 0.55 achieves about 32 MPa. The LC3 mixture with a w/b ratio of 0.55 achieved approximately 85% of the compressive strength of the reference mixture after 28 days of curing. Considering that similar mixtures prepared with low-quality clays in the literature typically reach only 70–75% of the strength of the reference mixtures [4, 7], these results indicate that the LC3 (w/b = 0,55) mixture demonstrates a satisfactory performance in terms of compressive strength.

Fig. 4b further contextualizes these findings by illustrating the development of strength over time across all mixtures. The reference mortars exhibit a consistently strong relationship between curing time and compressive strength. Similarly, the LC3 mix with a w/b ratio of 0.55 displays a comparable strength evolution pattern, while the LC3 mix with a lower w/b ratio shows a more modest and less consistent rate of increase. This weaker response may result from decreased workability, incomplete hydration, or limitations in pozzolanic reaction kinetics at lower water content.

In summary, the results in Fig. 4 clearly demonstrate the influence of curing duration, binder type, and water-to-binder ratio on compressive strength. While reference mortars provide superior mechanical performance at all ages, the LC3 mixture with a w/b ratio of 0.55 offers a competitive alternative in terms of strength development, particularly considering its sustainability advantages and material efficiency.

The XRD patterns of reference and LC3 cement pastes after 28 days are illustrated in 5. The main phases observed in the reference paste are portlandite (CH), ettringite (AFt) and alite (C3S) and belite (C2S). The primary phases identified in the LC3 paste include portlandite (CH), ettringite (AFt), quartz (SiO_2), monocarboaluminate (Mc), and hemicarboaluminates (Hc), and calcite ($CaCO_3$). Calcite predominantly originating from raw materials. The formation of Mc and Hc phases results from the reaction between calcium carbonate in limestone and alumina in the calcined clays. The combined effect of limestone and metakaolin contributes to improved mechanical strength and reduced permeability [14, 15] (Fig. 5).

The microstructural development was examined using SEM on the surfaces of mortar specimens fractured during 28-day compressive strength tests. SEM images of the samples are shown in Fig. 6. The REF sample displayed a compact surface morphology with low porosity and minor indications of microcracking. The LC3 specimen with a water-to-binder ratio of 0.485 exhibited more elongated cracks than the LC3 specimen with a water-to-binder ratio of 0.55. A dense structure with microcracks was observed in the LC3 (w/b = 0.55) specimen.

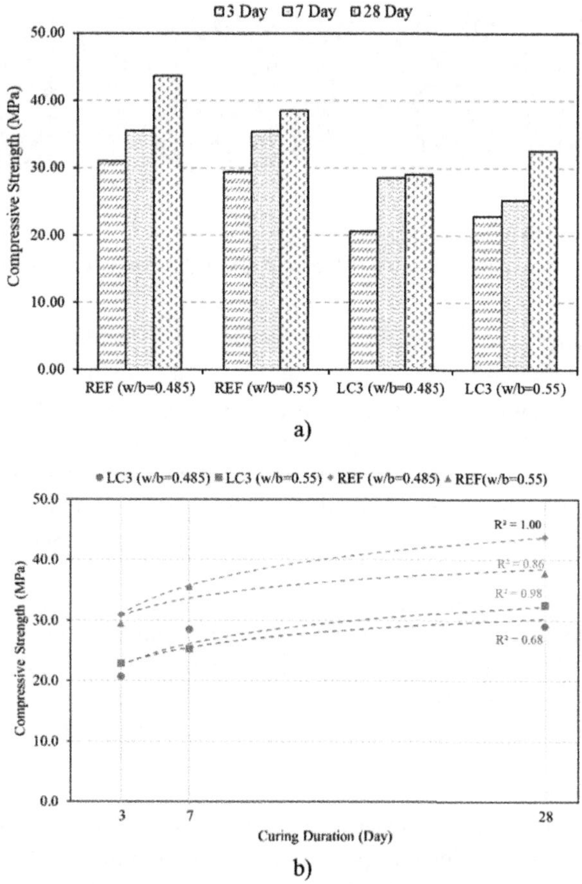

Fig. 4. a) Compressive strength results of mixtures, b) Variation of compressive strength with curing time

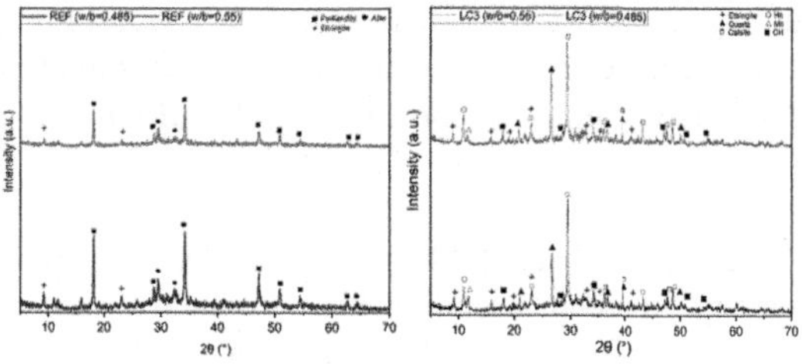

Fig. 5. XRD patterns of a) REF paste, b) LC3 paste

Fig. 6. SEM images of mortar samples: (a) REF mixture with w/b = 0.485, (b) REF mixture with w/b = 0.55, (c) LC3 mixture with w/b = 0.485, and (d) LC3 mixture with w/b = 0.55.

4 Conclusion

In this study, the influence of low-grade calcined clay on the flow characteristics, compressive strength development, phase assemblage, and microstructural evolution of limestone calcined clay cement was examined. The main findings are as follows:

It was observed that reference mixtures exhibited superior flowability compared to LC3 mixes, primarily due to the high surface area and increased water demand of calcined clay.

Despite the limitation in fresh properties, LC3 mixtures showed promising strength development trends over time. Notably, the LC3 mix with a w/b ratio of 0.55 achieved approximately 85% of the 28-day compressive strength of its corresponding reference mix. This result exceeds typical strength retention values reported in the literature (70–75%) for systems incorporating low-reactivity clays and suggests that adequate mechanical performance can be achieved even with low-grade clay resources when mixture design is optimized.

Microstructural analysis revealed that higher water-to-binder (w/b) ratios in LC3 systems promoted denser internal packing due to improved compaction. While it is acknowledged that LC3 mortars may require extended curing periods to fully match the strength levels of traditional cementitious systems, their significantly reduced clinker content and related carbon emissions position them as viable options for sustainable construction.

Overall, the findings demonstrate that with proper proportioning and curing, even low-grade calcined clays can be effectively utilized in LC3 formulations without compromising mechanical integrity, thereby supporting broader adoption of low-carbon cement technologies.

Acknowledgments. The authors would like to thank SAM (Ceramic Research Center-Eskisehir/Türkiye) for their support in laser diffraction particle size analysis, X-ray fluorescence (XRF) analysis, X-ray diffraction (XRD) analysis and Scanning Electron Microscopy Analysis (SEM).

References

1. Coffetti, D., Crotti, E., Gazzaniga, G., et al.: Pathways towards sustainable concrete. Cem. Concr. Res. **154**, 106718 (2022). https://doi.org/10.1016/j.cemconres.2022.106718

2. Avet, F., Scrivener, K.: Investigation of the calcined kaolinite content on the hydration of limestone calcined clay cement (LC3). Cem. Concr. Res. **107**, 124–135 (2018). https://doi.org/10.1016/j.cemconres.2018.02.016
3. Dixit, A., Du, H., Pang, S.D.: Performance of mortar incorporating calcined marine clays with varying kaolinite content. J. Clean. Prod. **282**, 124513 (2021). https://doi.org/10.1016/j.jclepro.2020.124513
4. Ram, K., Flegar, M., Serdar, M., Scrivener, K.: Influence of low- to medium-kaolinite clay on the durability of limestone calcined clay cement (LC3) concrete. Mater. Basel. **16**, 374 (2023). https://doi.org/10.3390/ma16010374
5. Alujas, A., Fernández, R., Quintana, R., et al.: Pozzolanic reactivity of low grade kaolinitic clays: influence of calcination temperature and impact of calcination products on OPC hydration. Appl. Clay Sci. **108**, 94–101 (2015). https://doi.org/10.1016/j.clay.2015.01.028
6. Du, H., Pang, S.D.: Value-added utilization of marine clay as cement replacement for sustainable concrete production. J. Clean. Prod. **198**, 867–873 (2018). https://doi.org/10.1016/j.jclepro.2018.07.068
7. Hu, Y., Xiong, L., Yan, Y., Geng, G.: Performance of limestone calcined clay cement (LC3) incorporating low-grade marine clay. Case Stud. Constr. Mater. **20** (2024). https://doi.org/10.1016/j.cscm.2024.e03283
8. Zunino, F., Scrivener, K.: Cement and concrete research factors influencing the sulfate balance in pure phase C3S/C3A systems. Cem. Concr. Res. **133**, 106085 (2020). https://doi.org/10.1016/j.cemconres.2020.106085
9. ASTM C305: Standard practice for mechanical mixing of hydraulic cement pastes and mortars of plastic consistency. In: ASTM Int West Conshohocken, PA, USA (2020). https://doi.org/10.1177/089033449801400218
10. ASTM C1437: Standard test method for flow of hydraulic cement mortar. In: ASTM Int West Conshohocken, PA, USA (2020). https://doi.org/10.1520/C1437-20.2
11. ASTM C109: Standard test method for compressive strength of hydraulic cement mortars. In: ASTM Int West Conshohocken, PA, USA (2021). https://doi.org/10.1520/C0109
12. Ez-zaki, H., Marangu, J.M., Bellotto, M., et al.: A fresh view on limestone calcined clay cement (LC3) pastes. Mater. Basel. **14**, 3037 (2021). https://doi.org/10.3390/ma14113037
13. Muzenda, T.R., Hou, P., Kawashima, S., et al.: The role of limestone and calcined clay on the rheological properties of LC3. Cem. Concr. Compos. **107**, 103516 (2020). https://doi.org/10.1016/j.cemconcomp.2020.103516
14. Blouch, N., Rashid, K., Ju, M.: Exploring low-grade clay minerals diving into limestone calcined clay cement (LC3): characterization – hydration – performance. J. Clean. Prod. **426**, 139065 (2023). https://doi.org/10.1016/j.jclepro.2023.139065
15. Zunino, F., Martirena, F., Scrivener, K.: Limestone calcined clay cements (LC3). ACI Mater. J. **118**, 49–60 (2021). https://doi.org/10.14359/51730422

Extrusion-Based 3D Printing of Fiber-Reinforced Concrete

Haodao Li(✉) and Kamal H. Khayat

Missouri University of Science and Technology, Rolla, MO 65401, USA
hlbb4@mst.edu

Abstract. Embedding conventional reinforcement in 3D concrete printing (3DCP) remains a critical challenge due to the constraints imposed by the layer-wise deposition process. A promising solution is the incorporation of discontinuous short fibers into printable mixtures, as they can enhance tensile strength and toughness, mitigate the risk of shrinkage-induced cracking, and reduce crack width. In this study, three types of short fibers, namely 6-mm steel fiber, 8-mm polyvinyl alcohol (PVA) fiber, and 6-mm polypropylene (PP) fiber were incorporated into a high-performance printable mixture. The effects of individual and hybrid fibers on fresh properties, extrudability, mechanical performance, and shrinkage resistance were systematically evaluated. Results indicate that increasing steel fiber volume or partially replacing it with synthetic fibers significantly increased air content and accelerated early slump flow loss, despite improved water retention. Mixtures with up to 2% steel fibers maintained excellent printability, whereas hybrid fiber combinations introduced surface defects while remaining acceptable extrudability. The incorporation of hybrid fibers also increased mechanical anisotropy due to compromised filament quality. Drying shrinkage was most effectively mitigated by increasing the steel fiber volume, while partial reductions were observed with the incorporation of PP or PVA fibers. Overall, steel fibers provided optimal performance in balancing flowability, mechanical strength, and filament stability, while hybrid fiber strategies requires careful optimization to avoid print quality degradation.

Keywords: 3D Concrete Printing (3DCP) · Anisotropic Mechanical Properties · Extrudability · Fiber Reinforcement · Shrinkage Resistance

1 Introduction

Traditional non-reinforced 3D-printed concrete elements often exhibit low tensile-to-compressive strength ratios due to the inherent difficulty of embedding steel rebars during printing. This limitation poses a significant challenge to the scalability and structural performance of large-scale 3D concrete printing (3DCP). To overcome this, fiber reinforcement has gained attention for its ability to enhance tensile strength, ductility, and crack resistance of printed elements [1–3]. Short fibers, typically less than 13 mm in length, are particularly suitable for 3DCP due to their compatibility with extrusion systems [2]. A wide variety of fiber types, including metal, synthetic, non-metallic, and

M. Posani and S. Singh (Eds.): RYS 2025, RILEM Bookseries 67, pp. 76–88, 2026.
https://doi.org/10.1007/978-3-032-14652-6_7

plant-derived fibers, have been successfully incorporated into printable mixtures [4–8]. Each type offers distinct benefits, allowing for customized reinforcement strategies tailored to specific application needs. However, the addition of fibers also introduces challenges to the extrusion process. On the one hand, high fiber volumes and longer fibers can substantially increase extrusion pressures and hinder material flow through the narrow barrel, leading to nozzle clogging, filament cracking, and compromised print quality [9]. Additionally, phase separation may occur if the mixture is unable to exhibit adequate water retention ability, further increasing the risk of blockage [10]. On the other hand, the convergent flow within the extrusion nozzle tends to induce preferential fiber alignment in the printing direction. While it can enhance strength along the extrusion path, it also amplifies mechanical anisotropy and weakens the performance in directions perpendicular to the layers due to reduced fiber bridging and poor interlayer bonding. These extrusion-related issues are highly influenced by fiber volume and length, mixture design, and printing parameters [6, 11–12].

This study aims to investigate the effect of fiber characteristics, including type, volume, and combination, on the key fresh and hardened properties of a high-performance printable mixture. The fresh properties evaluated include slump flow retention, air content, and water retention ability, while extrudability was assessed through the printing of a winding-line geometry using an extrusion-based 3D printer. In terms of hardened properties, the anisotropic compressive strength in three loading directions and drying shrinkage of 3D-printed elements were mainly examined.

2 Experimental Program

2.1 Materials and Mixture Design

The high-performance printable mixture adopted in this study is based on an optimized formulation developed by the authors in previous work [4, 13]. The detailed mixture proportion is shown in Table 1. The water-to-binder (w/b) ratio and sand-to-binder (s/b) ratio were fixed at 0.3 and 1.5, respectively. The binder materials include 55% Type III cement, 30% calcined clay (MK), and 15% limestone powder (LF), by volume. A high-fineness Type III cement (Blaine surface area of 625 m^2/kg) was selected to enhance early-age strength development and structural build-up during 3DCP. The MK contains 65% metakaolin, calcined at a maximum temperature of 650 °C. Fine LF, with a median particle size of 3.5 μm, was used as a filler to improve particle packing and adjust rheology. To optimize granular packing, a blend fine aggregate composed of 70% conventional concrete sand (CS) and 30% masonry sand (MS) was used. The maximum particle sizes of CS and MS are 4.75 and 2 mm, respectively, and their water absorptions are 0.14% and 0.06%.

Two chemical admixtures were incorporated to enhance printability: a polycarboxylate-based superplasticizer (SP) and a cellulose-type viscosity-modifying agent (VMA), with solid contents of 30% and 25%, respectively. The SP dosage was adjusted to achieve a target slump flow of 170 ± 10 mm after 25 jolts on a flow table, ensuring the mixture fell within the printable range characterized by adequate extrudability and buildability [14].

Table 1. Proportioning of high-performance printable mixtures with fibers (kg/m^3).

Mixture	Binder materials			Aggregates		Chemicals		Fiber combination		
	Cement	MK	LF	CS	MS	VMA	SP	Steel	PVA	PP
SL1.5	499	204	100	843	361	2	5.6	117	-	-
SL2							6.4	156	-	-
SL1.25PVA0.25							7.0	97.5	3.25	-
SL1.25PP0.25							6.7	97.5	-	3.25
SL1PVA0.25PP0.25							7.1	78	3.25	3.25

Three types of short fiber were selected, namely 6-mm straight steel fiber, 8-mm monofilament polyvinyl alcohol (PVA) fiber, and 6-mm monofilament polypropylene (PP) fiber. The physical characteristics of these fibers are listed in Table 2. Various fiber combinations were evaluated, with a constant total fiber volume fraction of 1.5% for all hybrid systems, as shown in Table 1. For the steel fiber-only mixture, an additional dosage level of 2% was also investigated.

Table 2. Physical characteristics of selected fibers.

Fiber type	Length (mm)	Diameter (mm)	Aspect ratio	Tensile strength (MPa)	Elastic modulus (GPa)
Steel fiber	6	0.2	30	2100	154
PVA fiber	8	0.038	210	1600	25–40
PP fiber	6	0.038	158	600–650	9.5

All mixtures were prepared at ambient temperature using a 19-L Hobart mixer, following a standardized batching and mixing protocol. The mixing procedure was as follows: (1) dry mixing. Cementitious materials and sand were mixed at a low speed of 1 rps for 2 min; (2) initial wet mixing. A large portion of the mixing water, premixed with half of the SP (the other half was premix with the VMA), was added and mixed at 2 rps until the mixture reached a flowable state; (3) final wet mixing. The remaining water containing the dispersed suspension of SP and VMA was introduced and mixed at the same speed; (4) fiber addition. Fibers were added sequentially at the end to ensure uniform dispersion throughout the matrix. The total mixing time for each batch was approximately 12 min.

2.2 Test Methods

Fresh Properties. The slump flow retention test, adapted from ASTM C230 (flow table test), was conducted to evaluate the flowability of fiber-reinforced mixtures and estimate their open time for printability. The initial slump flow after 25 drops were measured, followed by subsequent measurements at 5, 15, and 30 min after mixing to track the loss in fluidity over time. To prevent moisture loss during this period, the remaining mixture was stored in the mixing bowl and sealed with an impermeable film. Additionally, the air content of the fresh mixture was measured in accordance with ASTM C231.

To assess water retention ability, the fluid filtration test was performed under a constant pressure gradient, following the methodology described in [10, 15]. This test characterizes the mixture's resistance to forced bleeding during extrusion, which is closely associated with extrudability and interlayer bonding. An overhead pressure of 120 kPa was applied using inert gas, and the test was limited to 5 min to minimize the effect of structural build-up. The forced bleeding coefficient (D_e), defined as the slope of the filtrated water volume versus the square root of time, was used as an indicator of water retention ability. This relationship follows a linear relationship that can be expressed as Eq. (1).

$$V_c/V_w = D_e{}^* \, t^{1/2} \tag{1}$$

where V_c is the volume of collected bleed water, V_w is the total water volume in the sample; t is the elapsed time. A lower D_e value indicates better water retention and, therefore, a lower risk of phase separation during extrusion.

Printing Evaluation. Printability was evaluated directly using a gantry-based extrusion-type 3D printer, as illustrated in Fig. 1. The printer had a build volume of $1.5 \times 1.5 \times 1.5$ m^3. The extrusion system consisted of a motorized hopper, funnel, auger, and nozzle mounted on a three-axis gantry frame controlled by a programmable controller. The printhead travel speed was set at 50 mm/s, and the auger rotation speed was 1.0 rad/s, resulting in an average filament width of 45 ± 5 mm. To accommodate higher fiber volumes and lengths, a 40-mm nozzle was used.

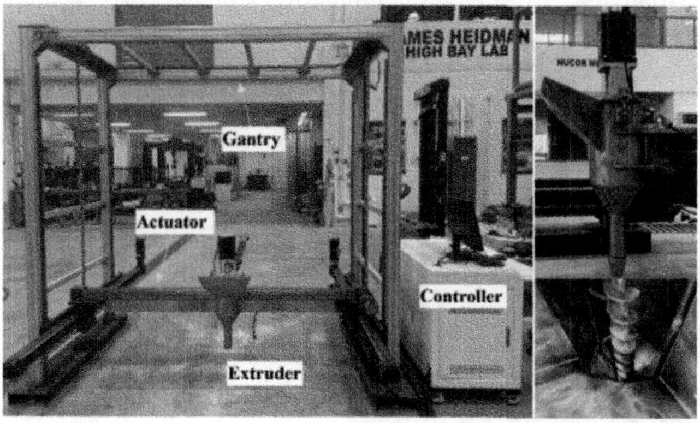

Fig. 1. Extrusion-based 3D printer equipped with a screw-type extruder.

Each printed element was built layer-by-layer with a layer height of 15 mm and a time interval of 10 s between layers. Extrudability was assessed through the successful deposition of a winding-line object composed of at least six straight and curved segments (each 300 mm in length). Given the increased risk of nozzle blockage and filament discontinuity due to fiber incorporation, three key criteria were considered: (1) smooth surface formation with no visible nozzle clogging; (2) continuous, break-free filament extrusion; and (3) uniform filament width across all segments.

In addition, quantitative evaluation was conducted using digital image analysis to assess voids and tears on the external filament surfaces, as exemplified in Fig. 2. A MATLAB code incorporating the Sauvola binarization algorithm was employed [16]. High-resolution images were first converted into binarized form, in which black pixels represented the filament surface, whereas white pixels denoted defects (e.g., voids, tears, and interfaces). The defect density was then defined as the ratio of white pixels (defects) to the total pixel count. A higher defect density corresponded to reduced extrudability, indicating greater occurrence of voids, tears, or weakened interfaces.

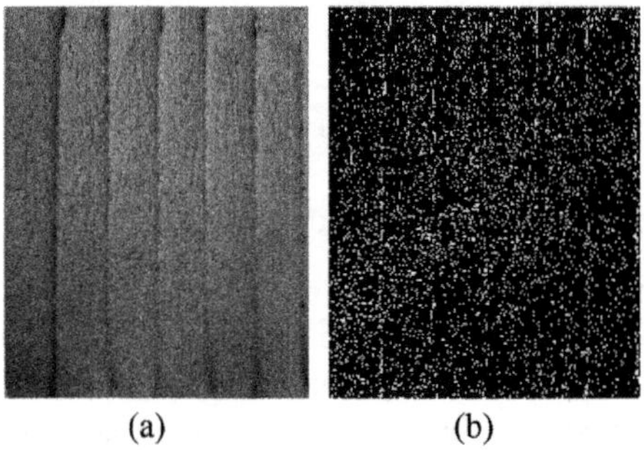

Fig. 2. Digital image processing of the printed element made with SL2 mixture. (a) original high-quality image and (b) binarized image.

Hardened Properties. To evaluate compressive strength, two winding-line objects were printed for each mixture to ensure repeatability. After printing, the elements were covered with wet burlap for 24 h and then cured in a moist environment for an additional 27 days. At 28 days, at least eight 50 × 50 × 50 mm^3 cubes were saw-cut from the central portion of each winding object (which consisted of five printed layers, 300 mm long), as shown in Fig. 3.

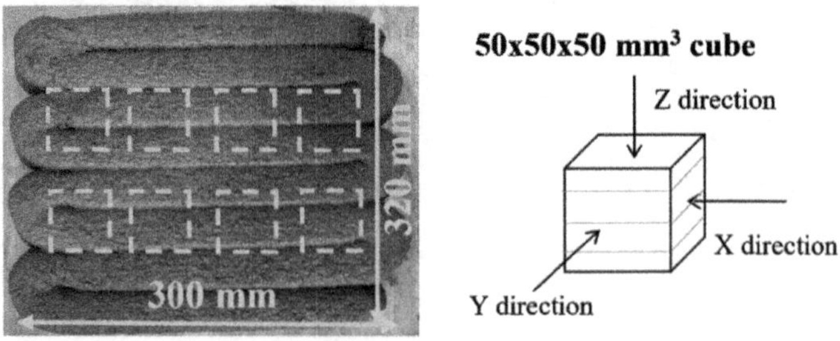

Fig. 3. Illustration of printed winding-lines and loading directions for compression test.

Compressive tests were conducted in three principal directions: X (longitudinal), Y (lateral), and Z (perpendicular to print layers), at a constant loading rate of 0.4 MPa/s to assess mechanical anisotropy. To ensure consistency, all specimens contained three uniform interlayer interfaces. Prior to testing, the loading faces of the cubes were sequentially ground using three sandpaper grades (each applied for 10 min) to ensure flat, parallel surfaces.

The drying shrinkage of the printed elements was evaluated using a dial gauge comparator setup. Two 50 × 50 × 300 mm^3 prisms were saw-cut from the printed winding-line after 7 days of moist curing, as illustrated in Fig. 4. These prisms were stored in a controlled environment at 50 ± 4% relative humidity and 23 °C. Each prism contained one interface on the top side. Two steel discs were glued along one longitudinal side using heavy-duty AB adhesive for length change measurement. Shrinkage readings were recorded immediately after saw-cutting, on a daily basis for the first week, and a weekly basis until the age of 63 days.

Fig. 4. Drying shrinkage measurement of printed object. (a) a 300 × 360 mm^2 winding-line object in five layers; (b) hand-held apparatus; and (c) extracted specimens.

3 Results and Discussion

3.1 Effect of Fiber Characteristics on Fresh Properties

Table 3 summarizes the air content and slump flow retention of mixtures containing individual or hybrid fibers. An increase in steel fiber volume and partial replacement with PVA or PP fibers led to higher air contents. For instance, increasing the steel fiber volume from 1.5% to 2% resulted in a 59% increase in entrapped air, while replacing 0.25% of steel fibers with PVA or PP fibers led to air content increases of 76% and 32%, respectively. This increase can be attributed to fiber agglomeration and flexible movement of synthetic fibers, both of which promote air entrapment. Excess air adversely affects filament surface quality and interlayer bonding.

Regarding slump flow loss over time, all mixtures were adjusted using SP to an initial target value of 170 ± 10 mm after 25 jolts. Two distinct phases of slump flow loss were observed: a rapid drop in flow within the first 5 min due to particle flocculation, followed by a linear decrease up to 30 min. Mixtures containing higher fiber volumes and hybrid fibers exhibited sharper initial reductions. For instance, incorporating 2% steel fiber resulted in a 15% slump flow loss at 5 min, while 1.5% steel fiber led to only a 5% reduction. Similarly, replacing 0.25% fiber volume with PVA or PP fibers resulted in 11% and 14% reductions, respectively. Furthermore, replacing 0.5% volume with PVA and PP fibers led to a 10% reduction. These reductions occurred despite the increased SP dosage, highlighting the flow resistance introduced by rigid steel fibers and flexible synthetic fibers. Notably, mixtures with higher fiber volume or hybrid combinations exhibited slower flow loss rates from 5 to 30 min, likely due to delayed hydration and reduced structural buildup induced by SP, which extended the mixture's extrudability window.

Table 3. Slump flow after 25 jolts, air content, and D_e of fiber-reinforced mixtures.

Mixture	Slump flow after 25 jolts (mm)				Air content (%)	D_e (10^{-3} $s^{-0.5}$)
	Initial	5 min	15 min	30 min		
SL1.5 (Ref.)	173	164	154	144	3.4	1.9
SL2	175	149	140	131	5.4	1.6
SL1.25PVA0.25	179	160	152	143	6	1.7
SL1.25PP0.25	174	150	142	132	4.5	1.8
SL1PVA0.25PP0.25	177	160	150	144	6	2.2

The water retention ability of fiber-reinforced mixtures, as evaluated via forced bleeding coefficient (D_e), is also listed in Table 3. Increasing the steel fiber volume of 0.5% significantly improved water retention, reducing D_e by 16% compared to the reference with 1.5% steel fiber. The effect was less pronounced when replacing steel fibers with 0.25% PVA or PP fibers. This can be attributed to the stronger steel fiber-matrix interaction and the flexible movement of synthetic fibers, which effectively reduce bleeding and segregation by improving matrix cohesion. However, replacing 0.5% steel fibers with a combination of PVA and PP fibers increased D_e by 16%, likely caused by reduced matrix cohesion and excessive SP dosage that promoted bleeding and segregation.

3.2 Effect of Fiber Characteristics on Extrudability

Figure 5 represents the printed winding-line elements made with various fiber-reinforced mixtures. As reported in [4], high-performance mixtures with 1.5% 6-mm steel fibers showed excellent extrudability and filament integrity. In this study, when the steel fiber volume increased to 2%, the printability remained superior without noticeable deficiencies. However, partial replacement with 0.25% PVA or PP fibers slightly compromised filament surface quality, resulting in voids and surface undulations, as also reflected in the quantitative results of Table 4. In particular, hybrid mixtures in which 0.5% of steel fibers were substituted with 0.25% PVA and 0.25% PP fibers retained adequate printability, with only a 23.4% increase in defect density compared with the mixture made with 2% steel fibers. It should be noted that the present work does not account for pumping systems. The incorporation of fibers, especially at relatively high volumes or when using flexible fibers, can significantly increase pumping pressure and elevate the risk of pipe clogging. To ensure smooth pumping, higher flowability and sufficient workability retention are required. Accordingly, adjusting the dosage of SP and VMA is necessary to balance pumpability, extrudability, and buildability.

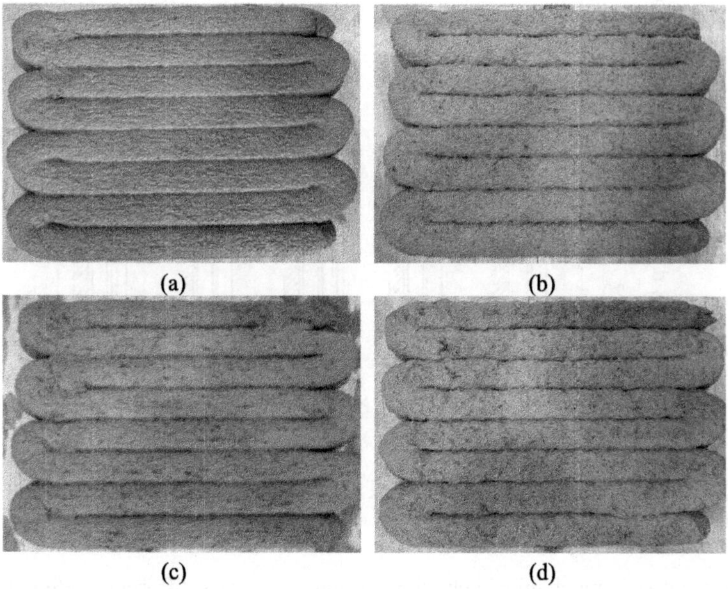

Fig. 5. Printing of winding-line objects. (a) SL2; (b) SL1.25PVA0.25; (c) SL1.25PP0.25; and (d) SL1PVA0.25PP0.25.

Table 4 Quantitative evaluation of printed winding-line objects made with various fibers.

Mixture	SL2	SL1.25PVA0.25	SL1.25PP0.25	SL1PVA0.25PP0.25
Defect density	10.37%	10.92%	11.89%	12.8%

3.3 Effect of Fiber Characteristics on Anisotropic Compressive Strength

Figure 6 illustrates the 28-d compressive strength of both cast and printed specimens tested in three loading directions (X, Y, and Z). The printed elements with 2% steel fibers exhibited an 8% increase in compressive strength over the 1.5% steel fiber mixture, though the enhancement was more pronounced in the cast counterparts. The incorporation of PVA and PP fibers led to greater strength reductions in printed elements, primarily due to print quality deficiencies such as surface voids, filament tears, and weakened interlayer interfaces, as observed in Fig. 5. Furthermore, the mixing of flexible fibers can entrap air that is detrimental to strength, and higher air content can be found in Table 3. Notably, replacing 0.5% steel fiber with equal parts PVA and PP fibers led to a 16% reduction in compressive strength in cast samples and the lowest strength values in printed specimens across all directions.

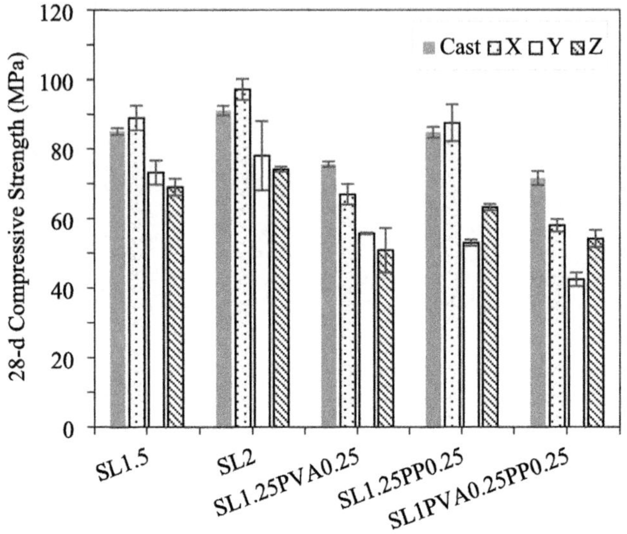

Fig. 6. The 28-d compressive strength of cast samples and printed elements in three loading directions.

The anisotropic mechanical behavior was evident across different loading directions in the printed element. The X-direction (parallel to layer deposition) exhibited the highest strength, occasionally surpassing cast samples, due to enhanced confinement and interlayer compaction. In contrast, the Y- and Z-directions (lateral and perpendicular) were more influenced by filament quality and bonding [17]. Poor dispersion of flexible fibers led to greater strength reductions in these directions. The Z-direction, in particular, was susceptible to shear failure, although improved fiber alignment could mitigate this effect [18].

To quantify this anisotropic compressive strength, the anisotropic index (I_a) adapted from [19] was employed. The I_a is calculated as follows:

$$I_a = \sqrt{\left((f_c - f_x)^2 + \left(f_c - f_y\right)^2 + (f_c - f_z)^2\right)}/f_c \tag{2}$$

where f_c is the compressive strength of the cast sample, and f_x, f_y, and f_z are the strength of the printed element measured in the X, Y, and Z directions, respectively. A higher I_a value indicates more significant anisotropic behavior. Fiber type and its dispersion, as well as overall print quality, strongly affect the anisotropic mechanical behavior of printed elements [20].

Table 5 shows the reduction rates of printed elements in three loading directions compared to their counterparts and I_a. The strength reduction rates across the three directions ranged from 11% to 41% for mixtures made with various fiber combinations.

Moreover, increasing steel fiber volume had a negligible effect on I_a, suggesting consistently good print quality. The improved interfacial bonding resulting from the bridging effect of steel fibers mitigated strength reductions in the Y direction. Notably, the adequate printability ensured an increased confinement when measured in the X direction that led to an increase in strength than the cast counterpart. In contrast, the incorporation of PVA and PP fibers significantly increased I_a, primarily due to compromised filament quality and weakened interfacial properties. The SL1PVA0.25PP0.25 mixture exhibited the highest I_a of 0.51, which is 113% higher than the reference mixture with pure 1.5% steel fibers.

Table 5. Compressive strength reduction rates and I_a of printed elements with various fiber combinations compared to cast counterparts (negative values denote strength increase).

Mixture	Strength reduction rate (%)			I_a
	Δf_x	Δf_y	Δf_z	
SL1.5 (Ref.)	-5%	14%	19%	0.24
SL2	-7%	14%	19%	0.24
SL1.25PVA0.25	11%	26%	33%	0.43
SL1.25PP0.25	-3%	37%	25%	0.45
SL1PVA0.25PP0.25	19%	41%	24%	0.51

3.4 Effect of Fiber Characteristics on Drying Shrinkage

Figure 7 shows the drying shrinkage development from 7 to 63 days of printed elements incorporating individual and hybrid fibers. In 3DCP, the layer-by-layer deposition provides interlayer confinement that can enhance shrinkage resistance compared to cast samples, particularly in fiber-reinforced mixtures [2]. However, inadequate print quality introduces defects at interlayer interfaces, creating stress concentration points that promote shrinkage cracking. This effect is more pronounced in drying shrinkage due to the high surface-to-volume ratio of printed elements, making them more susceptible to moisture loss.

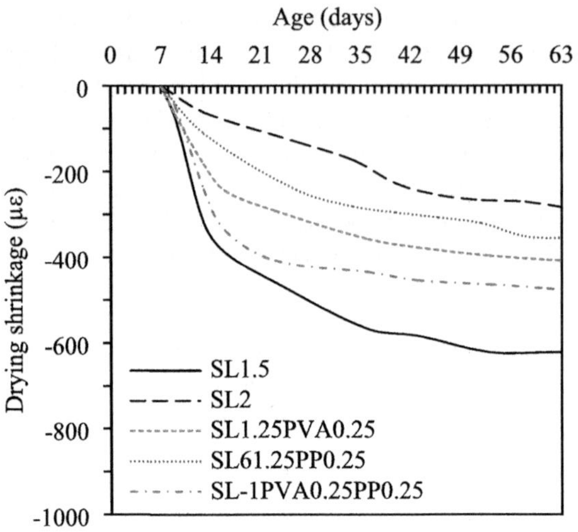

Fig. 7. The drying shrinkage of printed elements up to 63 days.

As shown in Fig. 7, increasing steel fiber volume consistently improved shrinkage resistance. Notably, the SL2 mixture exhibited the lowest drying shrinkage of −290 με at 63 days, representing a 53% reduction compared to the 1.5% steel fiber reference. Among flexible fibers, PP fiber demonstrated superior shrinkage control relative to PVA fiber. For example, the SL1.25PP0.25 mixture showed a 43% reduction, while the SL1.25PVA0.25 mixture exhibited 34%. However, combining 0.25% PVA and 0.25% PP fibers led to a limited modification in shrinkage resistance. This is likely due to the negative impact of higher PVA/PP fiber volume on printability, leading to inferior interlayer bonding and higher susceptibility to drying shrinkage.

4 Conclusions

This study systematically investigated the influence of fiber characteristics, including type, volume, and hybrid combinations, on the fresh properties, extrudability, mechanical anisotropy, and drying shrinkage of a high-performance 3D printable mixture. The key findings were summarized based on experimental results:

- Increasing fiber volume or introducing hybrid fibers increased air content and accelerated early-age slump flow loss, while improving water retention ability. Specifically, increasing steel fiber volume from 1.5% to 2% increased air content by 59%, while replacing 0.25% of steel with PVA or PP fibers elevated it by 76% and 32%, respectively.
- Mixtures containing up to 2% steel fibers maintained excellent printability with continuous and uniform filament extrusion. In contrast, partial replacement with PVA or PP fibers introduced surface defects and reduced filament integrity, though acceptable printability was retained at moderate replacement ratios.

- Fiber volume and type strongly affected the compressive strength and anisotropy. Steel fiber-reinforced elements achieved higher strength in the longitudinal (X-axis) direction due to improved interlayer confinement. The anisotropic index (I_a) increased from 0.24 (pure 1.5% steel fiber) to 0.51 with hybrid fibers, reflecting a 113% increase in anisotropic behavior linked to poor interlayer bonding and air entrapment.
- Enhanced shrinkage resistance was observed with increased steel fiber volume. The 2% steel fiber mixture exhibited the lowest 63-day drying shrinkage of −290 $\mu\varepsilon$, a 53% reduction compared to the 1.5% mixture. Replacing 0.25% with PP or PVA reduced shrinkage by 43% and 34%, respectively, whereas dual replacement showed diminished effectiveness, likely due to compromised print quality.

References

1. Mechtcherine, V., et al.: Integrating reinforcement in digital fabrication with concrete: a review and classification framework. Cem. Concr. Compos. **119**, 103964 (2021)
2. Zhou, Y.Y., Jiang, D., Sharma, R., Xie, Y.M., Singh, A.: Enhancement of 3D printed cementitious composite by short fibers: a review. Constr. Build. Mater. **362**, 129763 (2023)
3. Le, T.T., et al.: Hardened properties of high-performance printing concrete. Cem. Concr. Res. **42**, 558–566 (2012)
4. Li, H.D., Addai-Nimoh, A., Kreiger, E., Khayat, K.H.: Methodology to design eco-friendly fiber-reinforced concrete for 3D printing. Cem. Concr. Compos. **147**, 105415 (2024)
5. Chu, S.H., Li, L.G., Kwan, A.K.H.H.: Development of extrudable high strength fiber reinforced concrete incorporating nano calcium carbonate. Addit. Manuf. **37**, 101617 (2021)
6. Arunothayan, A.R., Nematollahi, B., Ranade, R., Bong, S.H., Sanjayan, J.G., Khayat, K.H.: Fiber orientation effects on ultra-high performance concrete formed by 3D printing. Cem. Concr. Res. **143**, 106384 (2021)
7. Pham, L., Tran, P., Sanjayan, J.: Steel fibres reinforced 3D printed concrete: influence of fibre sizes on mechanical performance. Constr. Build. Mater. **250**, 118785 (2020)
8. Luhar, S., et al: Sustainable and renewable bio-based natural fibres and its application for 3D printed concrete: a review. Sustainability. **12**, 10485 (2020)
9. Kudera, K.G., Shah, S.P.: Processing of high-performance fiber-reinforced cement-based composites. Constr. Build. Mater. **24**, 181–186 (2010)
10. Vallurupalli, K., Libre, N.A., Khayat, K.H.: Characterization of extrudability using rheology and desorptivity. ACI Mater. J. **121** (2024)
11. Abrishambaf, A., Pimentel, M., Nunes, S.: Influence of fibre orientation on the tensile behaviour of ultra-high performance fibre reinforced cementitious composites. Cem. Concr. Res. **97**, 28–40 (2017)
12. Carvalho, M.R., Barros, J.A.O., Zhang, Y., Dias-da-Costa, D.: A computational model for simulation of steel fibre reinforced concrete with explicit fibres and cracks. Comput. Methods Appl. Mech. Eng. **363**, 112879 (2020)
13. Li, H.D., Wei, J.J., Khyat, K.H.: 3D printing of fiber-reinforced calcined clay-limestone-based cementitious materials: from mixture design to printability evaluation. Buildings. **14**(6), 1666 (2024)
14. Tay, Y.W.D., Qian, Y., Tan, M.J.: Printability region for 3D concrete printing using slump and slump flow test. Compos. Part B Eng. **174**, 106968 (2019)
15. Rahul, A.V., Sharma, A., Santhanam, M.: A desorptivity-based approach for the assessment of phase separation during extrusion of cementitious materials. Cem. Concr. Compos. **108**, 103546 (2020)

16. Ahn, E., Han, S., Shin, M., Khayat, K.H., Popovics, J.S.: Novel method to evaluate 3-D printed concrete quality using ultrasonic scatter energy techniques. Nondestruct. Test. Eval., 1–17 (2025)
17. van den, H.M., Bester, F., Kruger, J., van, Z.G.: Mechanical characterisation for numerical simulation of extrusion-based 3D concrete printing. J. Build. Eng. **44**, 102944 (2021)
18. Panda, B., Chandra Paul, S., Jen Tan, M., Paul, S.C., Tan, M.J.: Anisotropic mechanical performance of 3D printed fiber reinforced sustainable construction material. Mater. Lett. **209**, 146–149 (2017)
19. Ma, G.W., Li, Z.J., Wang, L., Wang, F., Sanjayan, J.: Mechanical anisotropy of aligned fiber reinforced composite for extrusion-based 3D printing. Constr. Build. Mater. **202**, 770–783 (2019)
20. Yang, Y.K., Wu, C.Q., Liu, Z.X., Wang, H.L., Ren, Q.C.: Mechanical anisotropy of ultra-high performance fibre-reinforced concrete for 3D printing. Cem. Concr. Compos. **125**, 104310 (2022)

Fiber Reinforced Limestone Calcined Clay Cementitious (LC^3) Composites Produced with Construction and Demolition Waste (CDW) Clays

İmirza Okay Altuntaş(✉) and Oğuzhan Öztürk

Civil Engineering Department, Konya Technical University, Konya, Türkiye
e248122001003@ktun.edu.tr

Abstract. The high energy demand and CO_2 emissions associated with Portland cement (PC) production have intensified the need for more sustainable cementitious materials. In this context, low-clinker and blended cements have gained increasing attention. Limestone Calcined Clay Cement (LC^3) offers significant potential to mitigate environmental impacts, particularly using low-grade and waste-derived clays. Although numerous studies have explored the enhancement of engineering properties in cementitious systems using synthetic fibers, the fiber–matrix interactions in LC^3 systems—especially those incorporating clays derived from construction and demolition waste (CDW)—remain underexplored. This study investigates the processing and mechanical behavior of fiber-reinforced LC^3 composites incorporating hollow brick (HB) and ceramic waste tiles (CWT) clays. By incorporating polyvinyl alcohol (PVA) fibers, the cracking resistance of fiber-reinforced LC^3 composites (FRLC^3)—formulated with only 22.5% Portland cement by binder mass—was enhanced, leading to improved tensile properties, despite slight reductions in compressive strength and flowability. The findings demonstrate that CDW-based low-grade clays are viable alternatives for producing low-clinker containing FRLC^3 composites with superior tensile performance, offering a sustainable substitute for fly ash, which is increasingly scarce in many regions.

Keywords: Limestone Calcined Clay Cement (LC^3) · Construction and Demolition Waste (CDW) · Clays · Fiber-reinforced · Tensile Properties

1 Introduction

Concrete is the most widely used material in construction and its primary binder, Portland cement (PC), is produced by clinker after sintering of limestone and clay at high temperatures (~1450 °C). However, the processing of PC leads to significant environmental challenges due to high energy consumption and CO_2 emissions [1]. It is anticipated that PC accounts for approximately 8% of CO_2 emissions worldwide [2]. Consequently, lowering the clinker content and increasing the use of supplementary cementitious materials

M. Posani and S. Singh (Eds.): RYS 2025, RILEM Bookseries 67, pp. 89–103, 2026.
https://doi.org/10.1007/978-3-032-14652-6_8

(SCMs) have become increasingly important to reduce carbon footprints. The limited availability and quality control challenges of conventional SCMs such as fly ash and ground granulated blast furnace slag [3] have necessitated the investigation of other type of SCMs for blended binders. In this context, Limestone Calcined Clay Cement (LC^3), composed of limestone and calcined clay, has emerged as a promising blended cement system for reducing environmental impacts offering 50% of less clinker content [4]. While LC^3 mixtures produced with high-quality and kaolinite-rich clays exhibit good performance in terms of early and long-term strength, the limited availability and high cost of such clays have led to growing interest in utilizing low-grade and regionally available waste-derived clays. Notably, it has been demonstrated that clay-based components derived from construction and demolition waste (CDW) can develop binding properties through appropriate pretreatment and be effectively used in LC^3 systems [5]. Although there are numerous studies in the literature on LC^3 systems [6–8], their use in fiber-reinforced cementitious composites remains limited [9]. Fiber reinforcement is generally used to tailor crack resistance and overall engineering properties in cement-based systems, with various fibers such as polypropylene (PP), polyethylene (PE), and polyvinyl alcohol (PVA) offering different engineering performance [10, 11]. Study [12] demonstrated that replacing 75% of Portland cement with LC^3, along with a specially engineered cementitious matrix, can enable strain-hardening behavior in composites incorporating polyethylene (PE) fibers. Compressive strength were obtained within acceptable range after 28 days of curing. The specimens exhibited reasonable tensile strength exceeding 4% strain showing notable strain capacity at all tested ages, including 7, 14, 28 and 60 days. For LC^3-based composites, a 28-day tensile strength of 0.6 MPa and an ultimate strain of 4.8% were reported. In another study [13], LC^3-based Engineered Cementitious Composites (ECC) with PVA fibers required a 20% higher water-to-binder ratio to ensure workability and fiber dispersion, which accelerated early strength development. Although 28-day compressive strength was limited, the conventional ECC reached 55.3 MPa. Nevertheless, the LC^3-based ECC exhibited a tensile strain capacity exceeding 6%, with an initial cracking strength of ~2.5 MPa and a final tensile strength of ~4.2 MPa. The correlation between tensile and compressive strength confirmed that the LC^3-based ECC had lower matrix integrity than its conventional counterpart. As evidenced by the limited number of studies, the effect of LC^3 matrices on fiber–matrix behavior has not been sufficiently understood particularly for the fiber-reinforced Limestone Calcined Clay cementitious composites ($FRLC^3$). This underscores the importance of conducting comprehensive investigations into the engineering properties of $FRLC^3$ composites. Improving these low-clinker binders with fiber reinforcement can present significant potential for both sustainable binder utilization and enhanced tensile properties.

In this study, the engineering performance of $FRLC^3$ composites were evaluated. A reference mixture containing PC and Class F fly ash was produced. Subsequently, $FRLC^3$ composites were produced by partially replacing PC with an equivalent amount of limestone, calcined clay and gypsum. The developed LC^3 system with clays from CDW stream employed two different waste clays, hollow brick (HB) powder and waste ceramic tile (WCT). The raw materials were characterized using XRD and XRF analyses. The fresh-state properties of the produced $FRLC^3$ composites were assessed using the flow table test, while compressive and direct tensile strength tests were conducted on 28-days specimens to analyze the influence of LC^3 matrices on the performance of fiber-reinforced systems.

2 Materials and Method

2.1 Materials

The composition of LC^3 consisted of PC (CEM I 42.5R), two different CDW sources, hollow brick (HB) and ceramic waste tile (CWT), limestone (LS), and gypsum (GP). The HB and CWT, processed as clay sources, were fired during their original production processes at approximately 750 °C and 1000 °C, respectively, and exhibit partial pozzolanic potential due to their low crystallinity. In addition, limestone ($CaCO_3$) and calcium sulphate dihydrate ($CaSO_4 \cdot 2H_2O$) were used by 15% and 5% of total binder fraction. For the production of $FRLC^3$, the binder was combined with PVA fibers and fine silica sand. To compensate for the workability loss induced by the presence of calcined clay in LC^3 mixtures, two types of superplasticizers (SPs) were utilized: PC61, specifically formulated for clay-rich systems, and SF18, commonly applied in conventional PC-based mixtures to improve flowability. The use of these SPs aimed to mitigate the adverse effects of calcined clay on the fresh properties of LC^3 composites. In the reference mixtures, Class F fly ash was employed as a partial cement replacement.

2.2 Method

Hybrid Grinding of Raw Materials

In this study, a hybrid grinding method was employed in the production of the LC^3 dry mix to enhance material homogeneity. Based on the varying hardness of the constituent materials, an individual optimum grinding duration was determined for each raw material, aiming to improve energy efficiency [14]. The HB and CWT, waste clay sources from CDW, were first crushed using a mechanical crusher to a particle size of 40 mm, followed by oven drying at 104 °C for 24 h to eliminate moisture. The dried materials were then ground in a 22-l ball mill at 40 rpm until reaching a particle size of 73 μm. Fineness control was performed according to ASTM E-11 [15] using a No. 200 sieve at one-minute intervals, and the grinding duration was recorded once the target fineness was achieved. Fig. 1 presents the materials in both their bulk and ground forms. The same hybrid grinding process was applied to limestone (LS) and gypsum (GP) to determine the optimum grinding duration for each component. The blend was completed using a sequential feeding approach according to the hardness of the materials. Grinding performance was verified using laser diffraction analysis conducted in an aqueous medium based on the Mie scattering principle.

Characterization of Raw Materials

The mineralogical and chemical compositions of the materials in the LC^3 binder phase were determined by X-ray diffraction (XRD) and X-ray fluorescence (XRF) analyses, respectively. XRD analyses were conducted on <63 μm powder samples using

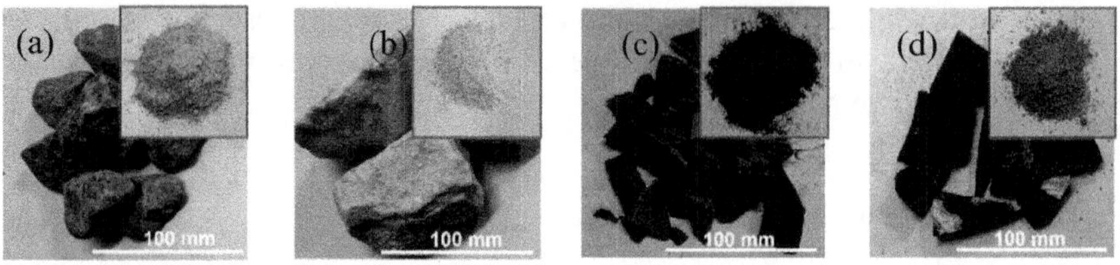

Fig. 1. Raw material composition of LC^3 blends (a) Limestone (LS), (b) Gypsum (GP), (c) Hollow Bricks (HB), (d) Ceramic Waste Tile (CWT).

a Rigaku Miniflex 600 diffractometer within the 5°–70° 2θ range at a scanning rate of 2°/min, and quantitative phase analysis was performed using the MAUD-Rietveld refinement method. The chemical compositions were analyzed using a Bruker Tiger S8 XRF spectrometer on samples ground to <63 μm and oven-dried at 105 °C for 4 h.

Production of $FRLC^3$ Composites

In the production of fiber-reinforced mixtures, LC^3 was used as the primary binder, replacing Portland cement at equivalent proportions. In all mixtures, the water-to-binder (w/b) ratio was fixed at 0.27, and the sand-to-binder (s/b) ratio was maintained at 0.33. Consequently, PC content-constituting 40% of the binder in Ref. mixtures-was reduced to half in LC^3 systems (Table 1). For fiber reinforcement, 6 mm long polyvinyl alcohol (PVA) fibers were incorporated at 2% by volume of the binder. Due to the high specific surface area resulting from the elevated fly ash content and fine LC^3 particles, the low w/b ratio limited workability. To compensate, a hybrid superplasticizer system was used at a total dosage of 2.5% (SP1: 1.75%, SP2: 0.75%). In the $FRLC^3$ mixtures, the gypsum used as a set regulator was pre-treated at 130 °C for 30 min to convert it into its hemihydrate/anhydrite form before use. During mixture preparation, the binder components, fly ash, and sand were first dry-mixed at low speed for 4 min. Subsequently, water and superplasticizer were added and mixed for another 6 min. Fibers were introduced at low speed over 1 min, followed by 2 min of homogenization and a final 2 min of high-speed mixing to achieve a fresh, uniform mix. The reference mixture included only Portland cement and fly ash as binders, with all other parameters kept constant. The LC^3 mixture incorporating hollow brick waste (HB) was designated as $FRLC^3$/HB, while the one using ceramic wall tile waste (CWT) was labeled as $FRLC^3$/CWT.

Table 1. Mixture proportions of $FRLC^3$ produced with clays from CDW

Mix-ID	PC	F-FA	LS	GP	HB	CWT	w/b	s/b	SP*
Ref.	0.95	1.2	-	0.05	-	-	0.27	0.33	2.5
$FRLC^3$/HB	0.5		0.15		0.3	-			
$FRLC^3$/CWT	0.5		0.15		-	0.3			

aBy weight of total binder (%)

Flowability, Compressive Strength and Tensile Properties

To evaluate the fresh properties of the mixtures, the flow table test was conducted in accordance with EN 1015-3 [16]. LC^3 composites were placed into a lubricated conical mold (bottom diameter: 100 mm, top diameter: 70 mm, height: 60 mm), and the mold was carefully lifted to allow the mix to spread freely. The initial flow diameter was measured after 15 s, followed by 15 flow table drops, after which the final diameter was recorded. For compressive strength, the LC^3 composites was placed into greased 50 mm edge steel cube molds in two layers, with each layer compacted by 25 rod strokes and 25 flow table drops. The molds were then covered with plastic sheets and kept undisturbed for 48 h. After demolding, the specimens were cured under controlled conditions at 24 °C and 95% relative humidity for 28 days. The compressive strength tests were performed using a compression test machine at a loading rate of 0.9 MPa/s. To assess the stress–strain behavior, in this study, dog bone-shaped specimens were prepared, and uniaxial tension tests were conducted in accordance with the specifications of the Japan Society of Civil Engineers [17]. The dog bone-shaped specimens produced for the tensile test are shown in Fig. 2.a. The tensile zone has a width of 10 mm, while the head section measures 20 mm by 15 mm. The thickness of the specimen is 6 mm. Three specimens were prepared for each mix, covered with plastic in the molds for 48 h, and subsequently cured under the same conditions for 28 days. Tensile tests were conducted at a displacement rate of 0.1 mm/min using a Shimadzu testing machine.

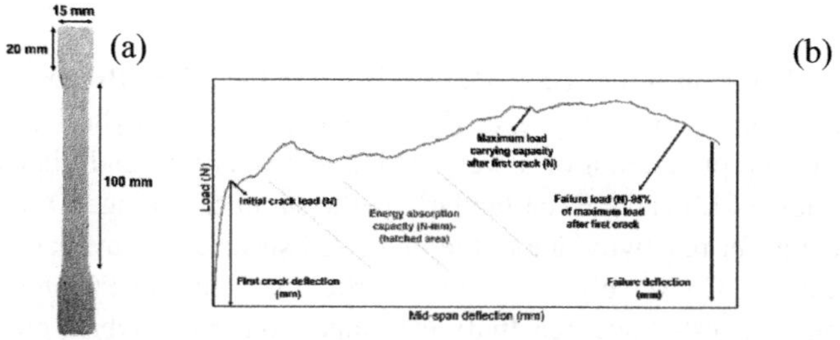

Fig. 2. Dog bone-shaped specimen (a) and typical load (N) – mid-span deflection (mm) curve and calculation parameters of $FRLC^3$ (b).

The engineering properties of the produced specimens were evaluated based on their yield and fracture points, ductility ratio, initial stiffness, and strain-hardening capacity ratio [17]. The direct tensile strain capacity of the specimens (at yield and ultimate strength) was recorded under the direct tensile test configuration. The ductility ratios were calculated by dividing the displacement value recorded at the fracture point (corresponding to 85% of the maximum load-bearing capacity) by the displacement value recorded at the yield point, where the linear elastic region ends (Fig. 2.b). The initial stiffness values of the specimens (N/mm) were assessed based on the slope of the initial linear elastic region in the load/mid-span displacement curves, which corresponds to the range where the load-displacement relationship remains linear and no visible cracking is observed [17, 18]. The strain-hardening capacity ratio was determined by the ratio of the displacement at maximum load to the displacement at the yield load. The energy absorption capacity was determined by calculating the total area under load–mid-span deflection curve, from the onset of loading up to the failure point.

3 Results and Discussion

3.1 Grindability of Raw Materials

In this study, a hybrid grinding method based on sequential feeding was applied to reduce energy consumption, ensure a homogeneous dry blend, and improve binder reactivity [19]. The grinding durations of each raw material were individually optimized. Construction and demolition waste (CDW) materials were first crushed to ~40 mm, then ground to below 73 μm. Among the materials, limestone (LS) required the longest grinding time (600 min), followed by gypsum (GP, 480 min), due to their mineralogical hardness and crystallinity [20, 21]. Based on the Mohs scale, LS (hardness 3) required more time than GP (hardness 2) [22]. CDW-based clays required longer grinding than natural clays due to their non-plastic nature and high quartz content [19]. HB reached target fineness in 120 min, while CWT required 150 min. The hybrid grinding process began with LS (120 min), followed by co-grinding with GP (360 min for HB mixtures, 330 min for CWT mixtures), and finally HB or CWT was added and co-ground for 120 or 150 min, respectively. This staged grinding ensured optimal fineness and blend homogeneity. The final hybrid blends were sieved through a No. 200 sieve (73 μm) in accordance with ASTM E-11.

Particle Size Distribution of Raw Materials After Grinding Process
Particle size distribution of raw materials suggests that the hybrid grinding method contributes to the production of a well-graded and finely ground binder system. The notable fineness of LS and HB, particularly in the submicron range (0–1 μm), implies a higher potential for reactivity due to the increased surface area available for hydration reactions (Fig. 3). Conversely, the coarser particle structure of PC and GP in the 10–100 μm range suggests lower reactivity in comparison to the hybrid-ground materials. The similar particle size characteristics observed between HB and CWT also indicate consistent grinding behavior among CDW-based clay materials, despite their differing origin. Additionally, the specific surface areas determined from the PSD analysis were 798 cm^2/g for PC, 1345 cm^2/g for HB, and 1041 cm^2/g for CWT. Although PSD alone cannot be directly associated with reactivity, an increase in fineness is generally expected to improve reactivity when similar raw materials are used [23]. Overall, the particle size distribution results confirm that the hybrid grinding process not only achieved the target fineness but also potentially improved the reactivity and performance of the alternative binder system.

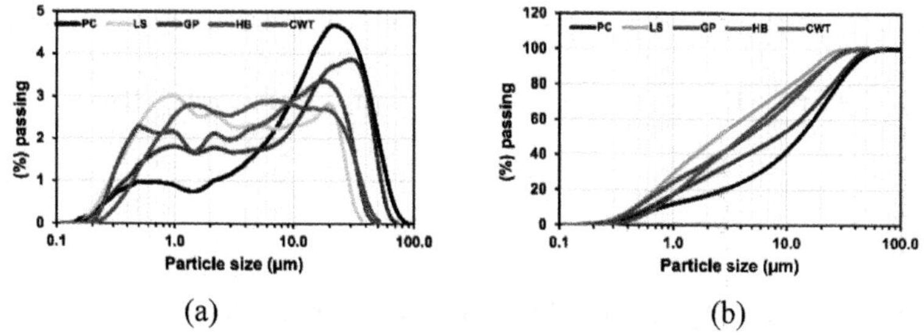

Fig. 3. PSD results (a) % volume; (b) % total passing.

3.2 Mineralogical and Chemical Compositions of Raw Materials

Similarly, the mineralogical phase compositions of Portland cement (PC), limestone (LS), and gypsum (GP) samples as determined by XRD-Rietveld analysis. The PC sample is notably rich in β-form larnite (Ca_2SiO_4, 51%), a critical phase contributing to its binding capacity. Additionally, brownmillerite ($Ca_2Al_4O_7$, 20.6%) was identified as the ferrite phase, along with calcite ($CaCO_3$, 14.6%) and basanite (13.2%). The LS sample consists of 97.3% calcite ($CaCO_3$) and 1.7% dolomite ($CaMg(CO_3)_2$), indicating that LS is a high-purity calcium carbonate source and can act effectively as a filler component in the LC^3 system. The GP sample predominantly contains gypsum (78.7%), supporting the necessity of a preheating process at 130 °C for 30 min to promote the conversion of gypsum into its anhydrite and/or hemihydrate. This thermal treatment is essential for enhancing the sulfate and reactivity of the material within the binder system.

Similarly, Fig. 4 presents the XRD patterns of HB and CWT samples. Both exhibit prominent quartz peaks, while HB shows a greater presence of crystalline mineral oxides compared to CWT. Notably, CWT displays more intense broad humps associated with the amorphous phase. Table 2 summarizes the mineralogical phase compositions determined by XRD-Rietveld analysis. The HB sample contains muscovite (15.6%) and kaolinite (16.3%) in a complex structure, along with 29.6% quartz. Its relatively low amorphous phase content (30.9%) suggests that HB may serve more as a filler than a reactive binder in the LC^3 system. In contrast, the CWT sample contains minerals such as mullite and albite and exhibits a high amorphous content (64.5%), which may be attributed to its origin from ceramic tile production and/or exposure to high-temperature processing.

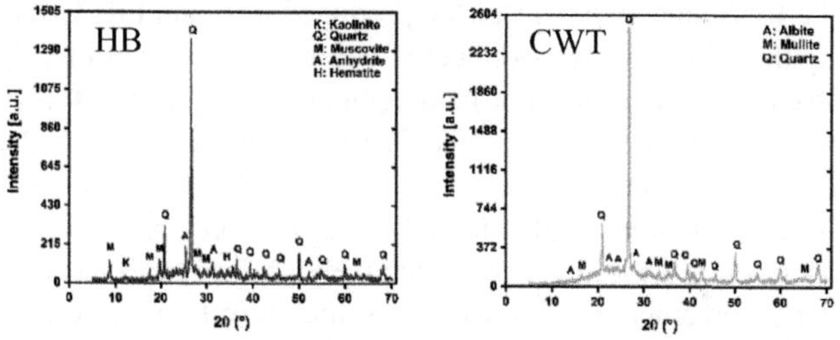

Fig. 4. XRD plots of the CDW clays without calcination.

Table 2. Mineralogical compositions (%) of HB and CWT obtained from XRD analysis.

Mineral	Quartz	Kao.	Mullite	Albite	Anhy.	Muscovit	Hematite	A.F.
HB	29.6	16.3	-	-	3.6	15.6	3.9	31.0
CWT	26.0	-	4.9	4.9	-	-	-	64.5

Kao.: Kaolinite, Anhy.: Anhydratie, A.F.: Amourphous phase

Similarly, Table 3 presents the chemical compositions of the raw material samples obtained through XRF analysis. The construction and demolition waste (CDW) exhibits varying chemical contents, likely due to the use of low-quality mineral sources during their initial production. The HB sample is notable for its high Fe_2O_3 content, which contributes to its orange-reddish coloration after calcination [24]. It also contains elevated levels of CaO, MgO, and SO_3. The high LOI (loss on ignition) value (>1%) of HB suggests incomplete amorphization and confirms the presence of kaolinite (16.3%). In contrast, the CWT sample, with its higher amorphous phase content and LOI value (<1%), indicates a more advanced level of amorphization.

Table 3. Chemical compositions (wt.%) of the raw materials determined by XRF.

Raw materials	Composition									
	SiO_2	**Al_2O_3**	**Fe_2O_3**	**CaO**	**MgO**	**SO_3**	**K_2O**	**Na_2O**	**TiO_2**	**LOI**
PC	16.4	4.7	3.4	58.0	0.9	2.5	0.7	0.5	0.2	4.4
LS	0.9	0.4	0.1	54.1	0.8	0.1	0.1	-	-	43.2
GP	1.7	-	0.2	38.9	0.7	43.7	-	-	-	14.8
HB	55.3	15.1	6.63	8.6	4.2	2.3	2.6	1.8	0.8	2.67
WCT	68.1	19.0	2.5	1.1	1.6	0.1	3.0	3.0	0.7	0.2

LOI: Loss on ignition

3.3 Flowability of Mixtures

In the study, the fresh-state behavior of the produced Ref. and $FRLC^3$ specimens were evaluated using the flow table test conducted before and after stroke (Fig. 5). The reference mix, produced solely with Portland cement (PC) and fly ash (FA), exhibited a spread diameter of 17 cm before shaking and 21 cm afterward. Among the $FRLC^3$ mixtures, the $FRLC^3$/HB specimen-containing HB-type CDW-achieved a spread of 10 cm before stroke and 15 cm after, while the $FRLC^3$/CWT specimen-containing CWT-type CDW-recorded 14.5 cm and 19.5 cm, respectively. Although the low water-to-binder ratio and fiber reinforcement are known in the literature to adversely affect flowability, this study also observed limited flowability despite the use of a high-range water reducer (superplasticizer, SP). The $FRLC^3$ mixtures exhibited lower flowability compared to the reference mix, which can be attributed to the higher specific surface area and lower maximum particle size (<73 μm) of the binder components used in the LC^3 systems, namely limestone (LS), gypsum (GP), and the clay sources (HB and CWT). When evaluating the LC^3 mixtures individually, the $FRLC^3$/HB specimen showed a lower spread diameter than $FRLC^3$/CWT. This difference is linked to the higher specific surface area of HB (1345 m^2/kg). As a result, the flowability performance was ranked as Ref > $FRLC^3$/CWT > $FRLC^3$/HB, consistent with the specific surface area values of the binder components.

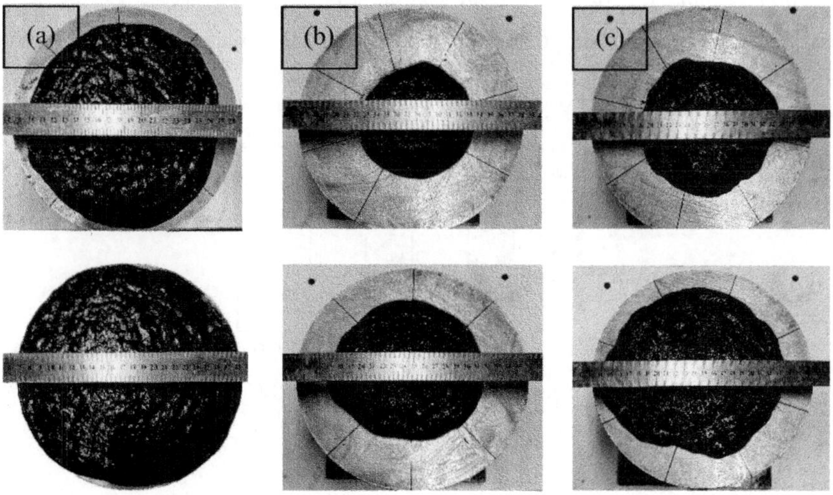

Fig. 5. The spread flow images of the LC^3 composites before (top row) and after (bottom row) shaking. (a) Ref., (b) $FRLC^3$/HB, (c) $FRLC^3$/CWT.

4 Mechanical Properties of LC^3 Mixtures

Compressive strength

According to the compressive strength results of the 28-days Ref. and $FRLC^3$ specimen presented in Fig. 6, the reference (Ref.) mixture achieved an average compressive strength of 29.8 MPa, the $FRLC^3$/HB mixture containing brick waste reached 26 MPa, and the $FRLC^3$/CWT mixture with ceramic tile waste demonstrated an average compressive strength of 22 MPa. Specifically, the increased use of FA, which was 20% higher than that of cement, is believed to have contributed to the low strength. This highlights quality issues in FA use and limitations when substituting it for cement [25]. Despite the binder volume of LC^3 mixtures being only 22.5% Portland cement (PC), particularly the $FRLC^3$/HB mixture exhibited strength close to that of the reference. This can be linked to the 31% amorphous phase found in HB, where the synergistic interaction of the clay with limestone and gypsum leads to pozzolanic reactions, filler effects, and the formation of carboaluminate, enhancing the strength.

The FRLC^3/CWT mixture showed the lowest average compressive strength. Despite the high amorphous content (61.4%) of CWT, the high quartz content and lack of calcination could have contributed to this result. Additionally, the reduced clinker content in LC^3 mixtures leads to less $Ca(OH)_2$ production, which, combined with limited CH (calcium hydroxide) due to the clays originating from FA and CDW, induces chemical competition and negatively affects strength [26]. The superior performance of the FRLC^3/HB mixture can be explained by the filler effect associated with its high specific surface area (1345 m^2/kg), as observed in the particle size distribution analyses [27]. Overall, despite the low PC content (22.5%), the LC^3 mixtures achieving a compressive strength range of 25–35 MPa demonstrate the potential of these systems for structural applications [12].

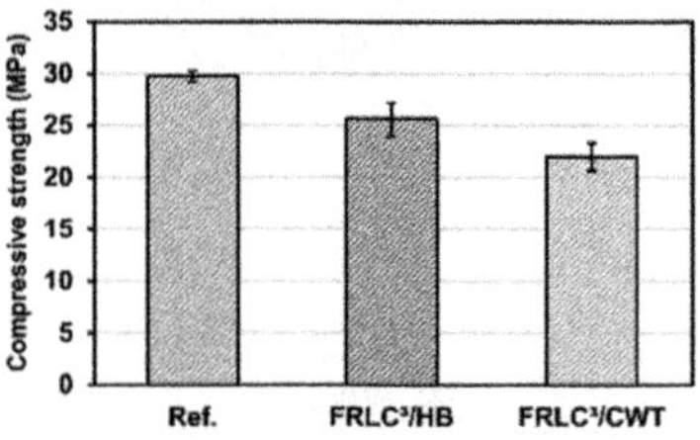

Fig. 6. Average compressive strength of FRLC^3 at 28 days.

Tensile properties

Fig. 7 presents the stress–strain curves obtained from the direct tensile tests conducted on dog-bone specimens. The reference (Ref.) mixture did not exhibit robust strain-hardening behavior, reaching a maximum tensile stress of only 0.5 MPa and a deformation capacity of approximately 1.4%. The FRLC^3/HB mixture exhibited higher engineering performance under tensile loadings and achieved a tensile strength of approximately 1.4 MPa. With a deformation capacity exceeding 1.4%, it exhibited a significant improvement in ductility compared to the Ref. This enhancement can be attributed to the strengthened fiber–matrix interfacial bond in the LC^3 system [28].

The FRLC^3/CWT mixture reached a maximum tensile stress of 0.7 MPa. Among all specimens, the minimum deformation was 1.5%, while the maximum reached up to 5.5%. Overall, the FRLC^3/HB mixture exhibited the best performance in terms of tensile strength, whereas FRLC^3/CWT showed superior ductility. These findings suggest that in LC^3 systems with low clinker and high SCM content, the deformation-hardening phase is extended, resulting in enhanced ductility [29].

The maximum tensile stress (MPa), stress corresponding to the maximum displacement (MPa), and maximum strain (%) values obtained from the direct tensile test conducted on specimens at 28 days of curing has been obtained from stress-strain graphs. According to the results, Ref. mixture, the maximum stress, stress at maximum displacement, and maximum strain was determined as 0.44 MPa, 0.36 MPa, and 1.32%, respectively. In contrast, for the FRLC^3/HB mixture incorporating hollow brick (HB) waste, the values were 1.42 MPa, 1.13 MPa, and 1.47%, respectively. Similarly, the

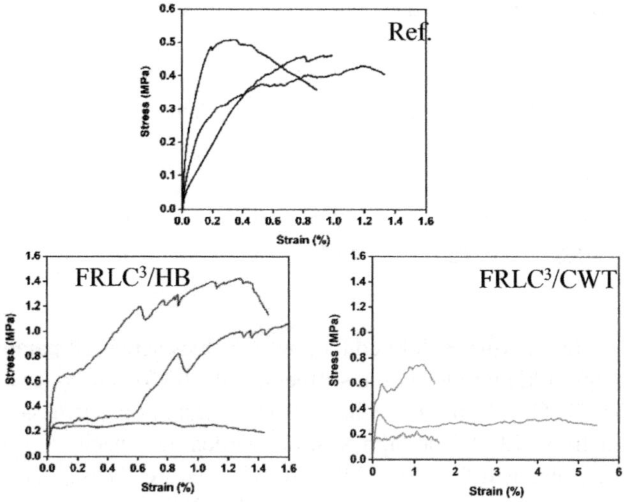

Fig. 7. Stress (MPa)–strain (%) graphs obtained from the direct tensile test of the $FRLC^3$.

$FRLC^3$/CWT mixture, produced with ceramic wall tile (CWT) waste, exhibited values of 0.80 MPa, 0.60 MPa, and 1.50%, respectively. As shown in Fig. 7, the stiffness values of the reference specimens exhibited considerable variability. The observed variability is likely associated with the small and delicate nature of the specimens prepared for the direct tensile test, which can lead to increased susceptibility to measurement inconsistencies and, consequently, higher standard deviations.

Upon evaluation, although the Ref. mixture demonstrated the highest 28-days compressive strength, it exhibited the weakest performance in terms of engineering performance under tensile loadings. On the other hand, the $FRLC^3$/HB mixture, which achieved a compressive strength close to that of the Ref. mixture after 28 days of curing, showed significantly superior performance in terms of both maximum tensile stress and stress at maximum displacement compared to the Ref. and other LC^3 mixtures. From a ductility standpoint, all LC^3 mixtures demonstrated greater ductility than the Ref. mixture. $FRLC^3$/HB and $FRLC^3$/CWT exhibited comparable displacement capacities, both surpassing the Ref. mixture. These findings suggest that the enhanced pozzolanic reaction potential and fine particle structure of LC^3 systems contribute to a denser matrix through a filler effect, improving mechanical performance. Moreover, the values of maximum stress, stress at maximum displacement, and maximum displacement (mm) observed in LC^3 mixtures produced using construction and demolition waste (CDW) indicate their high energy absorption capacity, which highlights their potential applicability in seismic design. Additionally, the use of recycled CDW enhances environmental sustainability, making these systems more advantageous from an ecological perspective. Also, Table 4 presents the results of dog-bone shape specimens under direct tensile loadings based on their load-displacement raw data.

To evaluate the engineering performance, the ductility ratio, initial stiffness, strain-hardening capacity and energy absorption capacity of the specimens were calculated. Fig. 8 represents the engineering performance of the $FRLC^3$ based on the data given

Table 4. Results of dog-bone specimens under direct tensile test.

Specimens	Yield		Failure		Max. load (N)	Disp. under max load (mm)
	Load (N)	Disp.(mm)	Load(N)	Disp.(mm)		
Ref.	31	0.35	36.6	0.36	43	0.88
FRLC^3/HB	64	0.097	125.8	0.62	148	1.28
FRLC^3/CWT	35.6	0.025	49.7	0.77	58.5	1.19

in Table 4. The Ref. mixture exhibited the lowest values across all parameters-2.51 for ductility ratio, 86.11 kN/mm for initial stiffness, and 1.03 for strain-hardening capacity. In contrast, the FRLC^3/HB mixture achieved the highest initial stiffness (103.2 kN/mm) and notable ductility (13.2). The highest strain-hardening capacity (47.6) and ductility ratio (30.8) were obtained from the FRLC^3/CWT mixture. These results indicate that FRLC^3 mixtures, particularly the one incorporating CWT, provide superior performance in terms of ductility and deformation capacity. Similarly, when evaluating the energy absorption capacities of the FRLC^3 mixtures, the Ref. mixture exhibited a value of 8.3 N.mm, whereas the FRLC^3/HB and FRLC^3/CWT mixtures showed significantly higher values of 136.4 N.mm and 39 N.mm, respectively. These results indicate that the FRLC^3 mixtures produced with clays from CDW exhibit markedly superior engineering performance compared to the reference mixture.

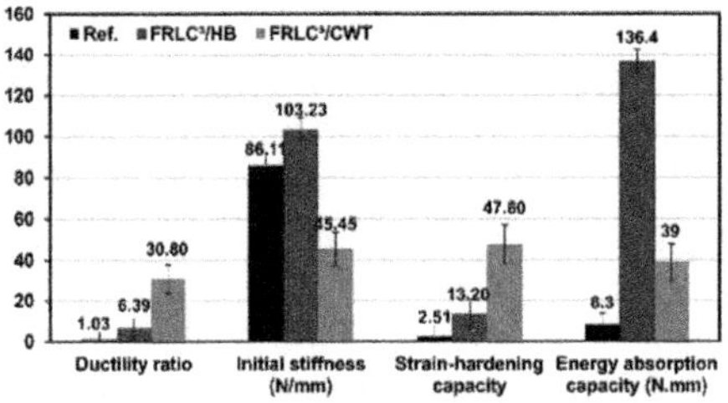

Fig. 8. Strain hardening capacity, initial stiffness, ductility ratios and energy absorption capacity of the FRLC^3.

5 Conclusions

LC^3 binders derived from two different construction and demolition waste (CDW) sources were utilized as replacements for Portland cement to produce fiber-reinforced composites at low water-to-binder ratios. The following can be inferred from the study findings:

- Through the hybrid grinding method, binder systems that are reactive, energy-efficient, and ground to a fineness appropriate to the physical properties of the components were successfully produced.

- Grinding and particle size distribution analyses confirmed that all components achieved the target fineness, with HB and CWT in particular exhibiting high specific surface areas.
- According to XRD analyses, HB exhibited limited binding capability due to its high quartz content and low amorphous phase proportion.
- CWT on the other hand, demonstrated higher pozzolanic potential owing to its high amorphous phase content.
- In terms of fresh-state properties, a decrease in flowability was observed in the LC^3 mixtures, which is attributed to the high specific surface area of the binder components.
- In terms of compressive strength, the $FRLC^3$/HB mixture exhibited strength comparable to the reference, whereas the $FRLC^3$/CWT mixture showed lower compressive strength.
- Both LC^3 mixtures exhibited engineering performance in tensile tests, with $FRLC^3$/CWT achieving the highest ductility and strain-hardening capacity. Despite their low clinker content, all LC^3 mixtures demonstrated superior ductility performance compared to the reference mixture.

Low-quality CDW-derived clays, with appropriate grinding and design strategies, can be used in fiber-reinforced composites, offering significant potential for the environmentally and economically sustainable production of fiber reinforced cementitious composites.

Concrete Committee, Japan Society of Civil Engineers, 2008, Recommendations for Design and Construction of High Performance Fiber Reinforced Cement Composites with Multiple Fine Cracks, Sub-Committee, Recommendations for Design and Construction of HPFRCC.

References

1. Farzadnia, N., Ali, A., Demirboga, R.: Development of nanotechnology in high performance concrete. Adv. Mater. Res. **364**, 115–118 (2011). https://doi.org/10.4028/www.scientific.net/AMR.364.115
2. Oss, H., Padovani, A.: Cement manufacture and the environment part II: environmental challenges and opportunities. J. Ind. Ecol. **7**, 93–126 (2003). https://doi.org/10.1162/108819803766729212
3. Worrell, E., Price, L., Martin, N., Hendriks, C., Ozawa-Meida, L.: Carbon dioxide emission from the global cement industry. Annu. Rev. Energy Environ. **26**, 303–329 (2001). https://doi.org/10.1146/annurev.energy.26.1.303
4. Kline, J., Kline, C.: Cement and CO2: what's happening? IEEE Trans. Ind. Appl. **67**, 1–1 (2014). https://doi.org/10.1109/TIA.2014.2339396
5. GCCA: Concrete future the GCCA 2050 cement and concrete industry roadmap for net zero concrete roadmap. In: The Global Cement and Concrete Association. Central (Ed.), p. 48 (2021)
6. Scrivener, K.: Options for the future of cement. Indian Concr. J. **88**(7), 11–21 (2014)
7. Snellings, R.: Assessing, Underst. Unlocking Supplementary Cementitious Materials. RILEM Tech. Lett. **1**(0), 50–55 (2016). https://doi.org/10.21809/rilemtechlett.2016.12
8. Snellings, R., Suraneni, P., Skibsted, J.R.: Future and emerging supplementary cementitious materials. Cem. Concr. Res. (2023). https://doi.org/10.1016/j.cemconres.2023.107199

9. Chen, W., Dang, J., Du, H.: Using low-grade calcined clay to develop low-carbon and lightweight strain-hardening cement composites. J. Build. Eng. **58**, 105023 (2022). https://doi.org/10.1016/j.jobe.2022.105023
10. Gamage, N, Liyanage, K, Fragomeni, S: Overview of Different Type of Fly Ash and their Use as a Building and Construction Material. (2011)
11. Dushimimana, A., Niyonsenga, A.A., Nzamurambaho, F.: A review on strength development of high performance concrete. Constr. Build. Mater. **307**, 124865 (2021). https://doi.org/10.1016/j.conbuildmat.2021.124865
12. Ahmed, A.H., Signorini, C., Chikhradze, M., Liebscher, M., Butler, M., Mechtcherine, V.: Employing limestone and calcined clay for preserving the strain-hardening response of PET fiber-reinforced cementitious composites. Constr. Build. Mater. **438**, 137166 (2024). https://doi.org/10.1016/j.conbuildmat.2024.137166
13. Zhang, D., Jaworska, B., Zhu, H., Dahlquist, K., Li, V.C.: Engineered cementitious composites (ECC) with limestone calcined clay cement (LC3). Cem. Concr. Compos. **114**, 103766 (2020). https://doi.org/10.1016/j.cemconcomp.2020.103766
14. Abdelaziz, G., Shoukry, H., Selim, A., Saif, M.: Recent Progress in limestone-calcined clay cement (LC3): a review. Mater. Sci. Forum. **1089**, 165–174 (2023). https://doi.org/10.4028/p-74p7so
15. ASTM: ASTM E11–24. Specification for Woven Wire Test Sieve Cloth and Test Sieves, Standard (2024)
16. EN, 1999, EN 1015–3: Methods of test for mortar for masonry - Part 3: Determination of consistence of fresh mortar (by flow table) (1999)
17. Engineers J S o C: Recommendations for design and construction of high performance fiber reinforced cement composites with multiple fine cracks (2008)
18. Yıldırım, G., Sarwary, M.H., Al-Dahawi, A., Öztürk, O., Anıl, Ö., Şahmaran, M.: Piezoresistive behavior of CF- and CNT-based reinforced concrete beams subjected to static flexural loading: shear failure investigation. Constr. Build. Mater. **168**, 266–279 (2018). https://doi.org/10.1016/j.conbuildmat.2018.02.124
19. Altuntas, I.O., Ozturk, O., Yildirim, G.: Exploring the viability of low-grade Turkish clays in the development of limestone calcined clay cement (LC3) and mortars. Constr. Build. Mater. **463**, 140077 (2025). https://doi.org/10.1016/j.conbuildmat.2025.140077
20. Graf, D.L., Lamar, J.E.: Properties of calcium and magnesium carbonates and their bearing on some uses of carbonate rocks. Econ. Geol. **50**, 639–713 (1955). https://doi.org/10.5382/AV50.17
21. Zhou, J., et al: Preparation of specific gypsum with advanced hardness and bending strength by a novel in-situ loading-hydration process. Cem. Concr. Res. **67**, 179–183 (2015). https://doi.org/10.1016/j.cemconres.2014.09.004
22. Hanein, T., et al.: Clay calcination technology: state-of-the-art review by the RILEM TC 282-CCL. Mater. Struct. **55**(1), 3 (2021). https://doi.org/10.1617/s11527-021-01807-6
23. Zaldívar-Cadena, A.A., et al.: Effect of milling time on mechanical properties of Fly ash incorporated cement mortars. Adv. Mater. Res. **787**, 286–290 (2013). https://doi.org/10.4028/www.scientific.net/AMR.787.286
24. Martirena Hernandez, J.F., Reyes, R.S., Zunino, F., Alujas, A., Scrivener, K.: Color control in industrial clay calcination. RILEM Tech. Lett. **5**, 1–7 (2020). https://doi.org/10.21809/rilemtechlett.2020.107
25. Little D N, Males E H, Prusinski J R, Stewart B: Cementitious Stabilization, Engineering, Environmental Science, Materials Science, A2J01: Committee on Cementitious Stabilization. Accession Number: 00784589 (2000)
26. Zhou, Y., Gong, G., Xi, B., Guo, M., Xing, F., Chen, C.: Sustainable lightweight engineered cementitious composites using limestone calcined clay cement (LC3). Compos. Part B. **243**, 110183 (2022). https://doi.org/10.1016/j.compositesb.2022.110183

27. Ahmed, A.H., et al.: Exploring the role of dilutive effects on microstructural development and hydration kinetics of limestone calcined clay cement (LC3) made of low-grade raw materials. J. Clean. Prod. **428**, 139438 (2023). https://doi.org/10.1016/j.jclepro.2023.139438
28. Wang, L., et al: On the use of limestone calcined clay cement (LC3) in high-strength strain-hardening cement-based composites (HS-SHCC). Cem. Concr. Res. **144**, 106421 (2021). https://doi.org/10.1016/j.cemconres.2021.106421
29. Qi, X., Liu, Z., Gao, X., Shui, Z.: Engineered cementitious composites with limestone calcined clay cement (LC3-ECC): effects of clinker content and Fly ash, journal of building. Engineering, 112828 (2025). https://doi.org/10.1016/j.jobe.2025.112828

From Waste to Value: Stubble Biochar for Sustainable and Low Carbon Construction

Sarmad Rashid(✉), Arpit Goyal, A. B. Danie Roy, and Manpreet Singh

Civil Engineering Department, Thapar Institute of Engineering and Technology, Patiala, Punjab 147005, India
srashid_phd22@thapar.edu

Abstract. This study presents an innovative approach to sustainable construction by utilizing stubble waste biochar as a cementitious composite material. Crop stubble burning is a critical environmental issue, contributing significantly to air pollution and carbon emissions. This research addresses this challenge by converting stubble waste into biochar, a carbon-rich material with substantial potential for carbon sequestration and construction applications. Comprehensive characterization of biochar revealed key properties, including a fine particle size (1.23 μm), complex pore structure for improved water absorption and retention, and the presence of silica and calcite phases, which contribute to enhanced cementitious behavior. Biochar, produced through the pyrolysis of rice stubble, was incorporated into mortar mixes at replacement levels of 0%, 5%, and 10% by weight of cement. The mixes underwent carbonation curing to enhance carbon capture and were evaluated for workability, compressive strength, flexural strength, and water absorption. While biochar incorporation initially reduced workability due to high water absorption, this was mitigated by adjusting superplasticizer dosages. Post-carbonation curing, the compressive strength of biochar mixes increased by 20.5% and 35.3% for 5% and 10% biochar, respectively. Flexural strength improvements were even more pronounced, with increases of 68.7% and 102.5% for the same biochar levels. Additionally, water absorption decreased by 9.9% and 5.2% for 5% and 10% biochar mixes, respectively. These results highlight the dual benefits of biochar in enhancing mechanical properties and contributing to carbon sequestration, showcasing its promise as a sustainable and carbon-negative material for construction applications.

Keywords: Bio-composites · Carbonation Curing · Sustainable Construction · Biomass Valorization · Carbon Sequestration · Stubble Waste Biochar

1 Introduction

Stubble burning in Northern India has emerged as a significant environmental issue, especially during the post-harvest season. Farmers often resort to burning leftover crop residue owing to a lack of cost-effective disposal alternatives. This practice emits considerable amounts of particulate matter, carbon dioxide (CO_2), and other toxic gasses, considerably escalating the region's severe air pollution levels [1]. The resulting health risks,

M. Posani and S. Singh (Eds.): RYS 2025, RILEM Bookseries 67, pp. 104–115, 2026.
https://doi.org/10.1007/978-3-032-14652-6_9

particularly respiratory and cardiovascular diseases, pose serious public health concerns [2]. At the same time, the construction sector, another major contributor to emissions, mainly depends on cement, the production of which alone accounts for around 8% of global CO_2 emissions [3, 4].

Amid these dual environmental challenges, there is an urgent need for synergistic solutions that address both agricultural residue management and the carbon footprint of the construction sector. Biochar, a carbon-rich porous material produced through the pyrolysis of organic biomass in an oxygen-limited environment, emerges as a promising solution [5]. Converting stubble waste into biochar provides a sustainable solution for mitigating the environmental impacts of stubble burning, while simultaneously producing a viable supplementary cementitious material [6, 7]. Biochar can partially replace cement in cementitious composites, thereby reducing CO_2 emissions associated with its production [8, 9].

Beyond sustainability, biochar exhibits beneficial impacts on the characteristics of cementitious systems [10, 11]. Its porous structure and surface functional groups facilitate cement hydration, resulting in denser microstructures and enhanced mechanical strength [12]. Furthermore, biochar's inherent ability to absorb and store CO_2 further contributes to carbon sequestration during the carbonation curing [13]. Numerous studies have demonstrated that incorporating biochar into cement mixes could reduce the workability of fresh cementitious composites owing to biochar's water-absorbing characteristics. Gupta et al. [14] observed a 9–13% decrease in workability with 0–5 by percentage weight (wt%) of biochar, whereas Yang and Wang [15] found an 8% decrease with 5 wt% rice husk biochar. To mitigate these effects, Sirico et al. [16] proposed the utilization of superplasticizers. Biochar can be produced from a range of feedstocks, rice stubble-derived biochar (RS biochar) is particularly notable owing to its silica-rich composition and affordability. Rashid et al. [17] indicated that RS biochar decreased workability due to increased water retention, while enhancing 28-day compressive strength by up to as 23.07%, while affecting flexural strength and water absorption trends. Subsequent study by Rashid et al. [18] revealed that the physical modification of RS biochar improved microstructural densification, mechanical properties, and durability in cementitious composites.

The mechanical characteristics of biochar-cement composites have been extensively studied. Mo et al. [19] observed a 3.4% increase in compressive strength at 28 days by incorporating 2 wt% of weed tree biochar. Similarly, Tan et al. [20] witnessed an 8.4% improvement in strength by adding 2–3 wt% of biochar. Liu et al. [21] discovered that adding up to 4 wt% bamboo chips biochar resulted in a 20% improvement in 28-day compressive strength. The effect of biochar on the durability and permeability of cementitious materials has also been thoroughly investigated. Aneja et al. [22] found that adding 5% rice husk biochar resulted in a 17.3% decrease in permeability. However, higher levels of replacement increased permeability due to increased porosity. Studies have shown that fine biochar particles effectively fill voids, decreasing permeability and improving water resistance [23, 24].

These findings indicate that biochar produced from stubble waste offers a sustainable solution for agricultural and construction pollution, while simultaneously improving the performance of cementitious composite. The aim of this study is to explore stubble waste biochar as a cementitious composite, assess its carbon sequestration potential through carbonation curing, and investigate the mechanical characteristics and durability of biochar-incorporated mortar. The significance of this study lies in its interdisciplinary approach in addressing real-world environmental issues. By transforming agricultural stubble waste into a viable construction material, this study paves the way for scalable, low-carbon solutions which promote global sustainability goals, especially in climate change mitigation and sustainable infrastructure.

2 Experimental Program

2.1 Cement, Sand, and Superplasticizer

Ordinary Portland Cement (OPC) of 43 grade, adhering to IS:269–2020 [25] standards, was utilized in all mortar compositions. The cement exhibited a specific gravity of 3.16. Locally sourced natural sand classified as zone II according to IS: 383–2016 [26] guidelines was utilized as the fine aggregate, exhibiting a specific gravity of 2.70 and a fineness modulus of 2.67. Portable water was utilized during the experimental work. A polycarboxylic acid-based superplasticizer was incorporated to improve the workability of the mortar mixes.

2.2 Biochar

Biochar derived from rice stubble was produced under controlled laboratory conditions through pyrolysis at 550 °C for 120 min. Subsequent to pyrolysis, the biochar was ground using a mechanical grinder with 1000watt motor for 90 s to reduce particle size, resulting in an average particle size of 1.23 μm as determined by Dynamic Light Scattering (DLS). The resultant ground biochar exhibited a fibrous surface texture and a complex porous framework, improving its water absorption and retention capabilities. Fig. 1 illustrates the external and internal morphological characteristics of the ground biochar, revealing a honeycomb-like pore structure formed by the volatilization of organic matter during pyrolysis.

X-ray Diffraction (XRD) analysis revealed a broad hump in the 2θ range of 15–30°, signifying the presence of amorphous phases. Moreover, distinct crystalline peaks were observed at 2θ values of 26.49°, 20.73°, and 68.24°, indicative of quartz (crystalline silica). Minor peaks at 29.26° and 39.31° corresponded to crystalline calcite and potassium chloride, respectively. The findings indicate a composite structure consisting of both amorphous and crystalline phases, primarily including silica, calcite, and KCl.

The Fourier Transform Infrared Spectroscopy (FTIR) further elucidated the chemical composition of the biochar. Key functional groups identified included C-H bending vibrations between 750–900 cm^{-1} (indicating the presence of hydrocarbons), C-O-C stretching in the 1030–1110 cm^{-1} range (suggesting ether or ester linkages), O-H bending at 1375 cm^{-1} (indicating hydroxyl groups), and C=C stretching between 1600–1630 cm^{-1} (characteristic of aromatic structures). The results offer significant insight on the chemical composition of biochar, confirming its suitability for use in cement-based applications.

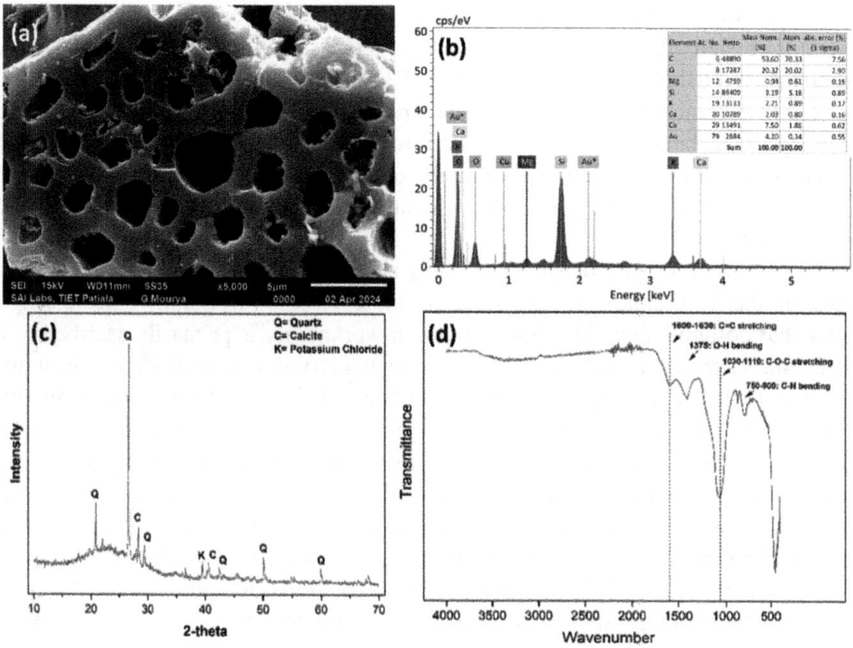

Fig. 1. Characteristics of Ground Biochar: (a-b) SEM and EDS micrographs, (c) XRD pattern, and (d) FTIR spectra

The pH of the biochar was assessed by mixing 2 g of ground biochar with 20 g of distilled water (1,10 w/w), equilibrating for 24 h before filtering. The filtrate exhibited an alkaline pH of 9.58, owing to the presence of basic oxides formed during pyrolysis, indicating no adverse impact on the alkalinity of the pore solution in cementitious systems. Additionally, the ground biochar exhibited a high-water retention capacity of 5.80 g/g, assessed by oven drying 2 g biochar for 24 h, soaking in 400 g of distilled water for 24 h with intermittent stirring, and filtering through pre-wetted filter paper. The retained mass was calculated from the difference between the wet and dry weights. This notable high water retention capacity can significantly impact the workability, internal curing, and hydration dynamics of mortar mixes.

2.3 Mix Design and Sample Preparation

Three mortar mixes were formulated for this study: a control mix and two experimental mixes which incorporated 5% and 10% rice stubble biochar as partial cement replacements by weight. The water-to-binder ratio was kept constant at 0.48, and the binder-to-sand ratio was maintained at 1:2.75. The superplasticizer dosage was altered to attain a consistent flow value of 115 ± 10 mm, as depicted in Table 1.

The mixing process encompassed dry mixing of cement, sand, and biochar in a Digimixer at a low speed of 145 ± 5 rpm for 60 s to ensure a uniform distribution of fine particles. Subsequently, water and superplasticizer were added, followed by high-speed mixing at 290 ± 10 rpm for 180 s to ensure uniform dispersion of the biochar and proper hydration of the binder. Fresh mortar was assessed for workability using flow table test and thereafter poured into molds as per testing specifications detailed in Table 2 and compacted using a vibration table for 45 s to remove trapped air. The specimens were left at the ambient temperature for 24 h prior to demolding. Following demolding, specimens were labelled and categorized into two groups: one group underwent 8 h of accelerated carbonation curing, while the other was directly immersed in water for curing. Carbonated samples were subsequently transferred to water curing until the specified testing age.

The incorporation of biochar significantly impacted the mortar's workability, determined by the flow table test (as shown in Table 1), with workability decreasing as the biochar dosage increased. This decrease in workability is primarily ascribed to the porous and water-absorbing character of biochar, which reduces the fluidity of the mortar. The superplasticizer dosage was altered to obtain a desired workability of the mixture. The control mix required a minimal quantity of superplasticizer, while the 5% and 10% biochar mixtures required 1.25 g and 2.25 g of superplasticizer, respectively. These modifications underscore the significance of maximizing the utilization of biochar in mortar, as shown by similar studies conducted by Gupta et al., 2018 [14, 27]. The consistent findings of reduced workability with biochar addition align with previous studies by Ofori-Boadu et al., 2017 [28] and Zhang and Islam, 2012 [29], emphasizing the crucial role of biochar characteristics in affecting mortar workability.

Table 1. Mix proportions and Flow table results.

Mix	Cement (g)	Biochar (g)	Sand (g)	Water (g)	Superplasticizer (g)	Flow (mm)
Control	250	0	687	120	0.75	112
B5	237.5	12.5	687	120	1.25	109
B10	225	25	687	120	2.25	120

Table 2. Test specifications and codes followed.

Test	Code	Samples (per mix)	Specifications
Flow	ASTM C1437 2007 [30]	-	Fresh mortar
Compressive Strength	ASTM C109 2020 [31]	6	Three cubes (50 x 50 x 50 mm) at 7 and 28 days
Flexural Strength	ASTM C348–02 2002 [32]	6	Three prisms (40 x 40 x 160 mm) at 7 and 28 days
Water Absorption	ASTM C642 [33]	3	Three cubes (50 x 50 x 50 mm) at 28 days

3 Results and Discussion

3.1 Compressive Strength

The compressive strength of the mortar mixes was evaluated at 7 and 28 days for both non-carbonated and carbonated specimens (as illustrated in Fig. 2). The results indicate significant variations in strength based on biochar dosage and curing conditions, highlighting the synergistic effects of biochar incorporation and carbonation curing.

At 7 days, the control mix had a compressive strength of 17.2 MPa for non-carbonated samples, increasing by 21.5% to 20.9 MPa following carbonation. The 5% biochar mix increased strength from 18.4 MPa to 22.1 MPa (19.5% gain), while the 10% biochar mix reduced strength to 16.1 MPa for non-carbonated samples but increased by 36.9% to 22 MPa after carbonation. At 28 days, the non-carbonated control mix achieved a strength of 24.2 MPa and improved by 16.1% to 28.2 MPa after undergoing carbonation. The 5% biochar mix exhibited the highest compressive strength, with a 20.5% increase from 31.5 MPa for non-carbonated samples to 37.9 MPa after carbonation. Furthermore, the 10% biochar mix demonstrated a non-carbonated strength of 27.5 MPa, which improved by 35.3% to reach 37.2 MPa following carbonation.

The retention and improvement of compressive strength, despite the partial substitution of cement with biochar, can be ascribed to several synergistic mechanisms. The finely ground biochar (average particle size 1.23 μm) serves as a micro-filler, enhancing packing density and optimizing pore structure. Its high specific surface area offers abundant nucleation sites for hydration and carbonation products, leading to matrix densification. Moreover, water retained within the biochar pores facilitates internal curing, maintaining hydration beyond the initial phases. These effects counterbalance the dilution of cement and, under carbonation curing, substantially improve strength.

The significant enhancement in compressive strength through carbonation curing primarily is due to the formation of calcium carbonate in the capillary pores and interfacial transition zones, therefore densifying the microstructure. The 5% biochar mix appears optimal as it maximizes these benefits without excessively increasing water demands or reducing early-age matrix cohesiveness. In contrast, the 10% biochar mix exhibited reduced early-age strength, most likely due to its higher water absorption, although this drawback was mitigated by densification after carbonation curing.

It is important to acknowledge that not all studies report strength enhancements with biochar incorporation; coarser, more porous particles have often been linked with increased matrix porosity, weakened interfacial zones, and inadequate hydration caused by excessive water absorption [34, 35]. This study demonstrated that the fine particle size and higher water retention lowered the free water-to-cement ratio and improved packing density, hence contributing to strength retention and enhancement. When compared with prior studies, the results from this study demonstrate a more significant strength improvement. Gupta et al. [27] reported a 12–15% increase in 28-day strength for rice husk biochar under conventional curing, whereas Tan et al. [36] observed an 18–21% improvement for high-temperature biochar. Conversely, the 5% biochar–carbonated mix in this study yielded a 20.5% increase, while the 10% mix achieved a 35.3% increase, demonstrating that the combination of fine particle size, micro-filler effect, and carbonation curing offers a substantial synergistic benefit over conventional curing methods.

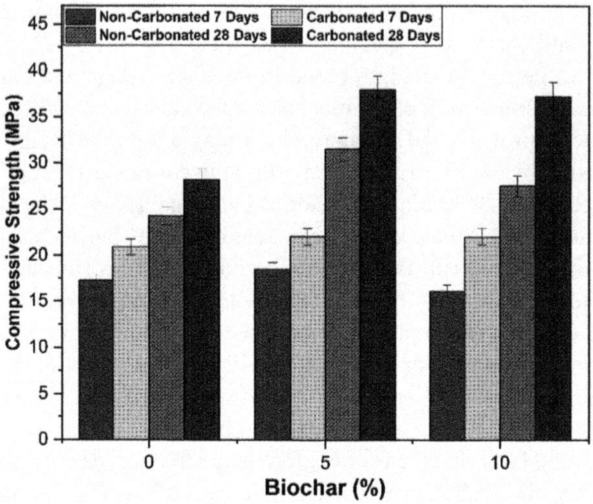

Fig. 2. Compressive strength test results of the mixes at 7 and 28 days.

3.2 Flexural Strength

The flexural strength of the mortar mixes was assessed at 7 and 28 days for both non-carbonated and carbonated samples (as shown in Fig. 3). The results demonstrate significant variations due to the incorporation of biochar and carbonation curing.

At 7 days, the control mix achieved a flexural strength of 7.1 MPa for non-carbonated samples, which increased by 15.8% to 8.2 MPa after carbonation. The 5% biochar mix showed no change in non-carbonated strength (7.1Mpa) but increased by 16.6% to 8.3 MPa following carbonation. The 10% biochar mix resulted in a reduction in non-carbonated strength to 5.5 MPa, which improved by 39.7% to 7.7 MPa after carbonation.

At 28 days, the non-carbonated control mix reached a strength of 8.7 MPa, which increased by 17.6% to 10.2 MPa following carbonation. The 5% biochar mix exhibited a reduction in non-carbonated strength to 7.6 MPa, but increased significantly by 68.7% to 12.9 MPa after carbonation. Similarly, the 10% biochar mix improved from 6 MPa to 12.2 MPa, a 102.5% increase after carbonation.

The improvement in flexural performance after carbonation curing can be attributed to enhanced microstructural integrity and stronger interfacial bonding. The fine particle size of biochar acts as a micro-filler, reducing flaw sizes while promoting stress distribution. During carbonation curing, calcium carbonate precipitates within capillary pores and at the biochar-matrix interface, further reinforcing these effects. While higher biochar dosage initially compromises flexural strength due to increased water absorption and reduced early hydration, carbonation curing mitigates these drawbacks by enhancing microstructural density.

Nevertheless, literature reveals inconsistent results. Numerous investigations have indicated a decline in flexural strength with the incorporation of biochar, particularly at higher dosages or with coarser biochar, where increased porosity and reduced matrix cohesiveness could outweigh the filler benefits [17, 37]. Comparatively, the findings of this study demonstrate superior flexural improvement compared to the 15–25% increase reported by Liu et al. [21] and Mo et al. [19] for biochar–cement composites under conventional curing conditions. This indicates that the combination of fine biochar particles with carbonation curing yields a synergistic effect, significantly improving tensile load resistance in mortar specimens.

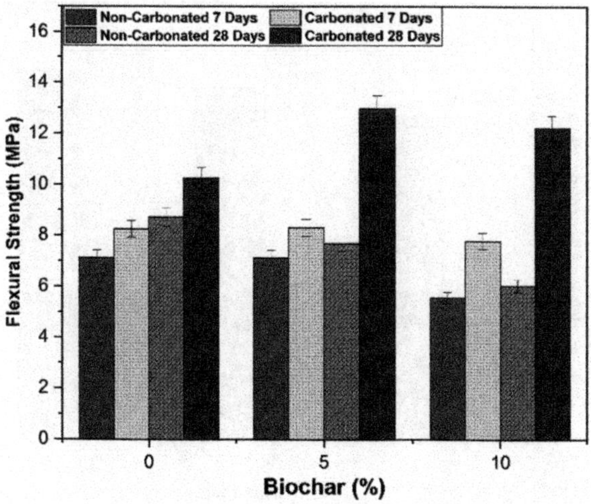

Fig. 3. Flexural strength test results of the mixes at 7 and 28 days.

3.3 Water Absorption

The water absorption of the mortar mixes was evaluated for both non-carbonated and carbonated samples (as depicted in Fig. 4). In the non-carbonated state, the control mix exhibited 8.2% absorption. The 5% and 10% biochar mixes recorded 9.10% and 9.15%, respectively, indicating that additional biochar beyond 5% led to only a negligible change in total water-accessible porosity. This minimal difference suggests a possible saturation threshold for capillary and interconnected pores introduced by biochar addition.

In the carbonated state, water absorption values were consistently lower. The control mix decreased to 7.5% (7.7% reduction), the 5% biochar mix to 8.2% (9.9% reduction), and the 10% biochar mix to 8.6% (5.2% reduction). This reduction is ascribed to the densification of the matrix by calcium carbonate precipitation, which enhances pore structure and restricts water infiltration.

Although increased water absorption in non-carbonated specimens is often related with decreased strength, the results this study indicate that carbonation curing mitigates this impact by reducing effective porosity. The substantial water retention capacity of biochar (5.80 g/g) also plays a role in a lowering the effective water-to-cement ratio, which may contribute to both the retention and the enhancement of strength in carbonated specimens.

When compared to prior research, Aneja et al. [22] found a 17.3% reduction in permeability with 5% rice husk biochar, whereas Chen et al. [23] observed reduced capillary water absorption following carbonation curing. This study corroborates these findings and also demonstrates that the reduction in water absorption could be attained at higher biochar dosages when combined with carbonation curing, underscoring the significance of the curing regime in enhancing durability.

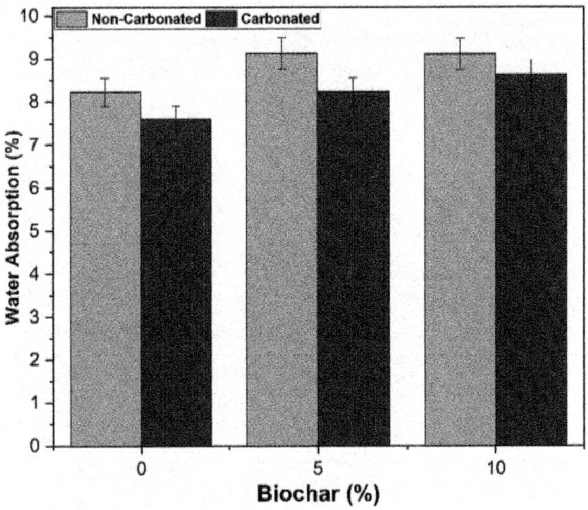

Fig. 4. Water absorption test results for the mixes.

4 Conclusion

This study explored the utilization of rice stubble biochar as a sustainable cementitious composite to address the environmental challenges caused by stubble burning and the construction sector. The study attempts to reduce the carbon footprint while also enhancing the overall performance of cementitious materials by incorporating biochar, known for its carbon sequestration capabilities. The main findings of the study are:

- The addition of biochar significantly reduced the workability of the mortar due to its high water absorption capacity, necessitating the addition of superplasticizers to attain a workable mix.
- The compressive strength improved with the biochar integration, especially after carbonation curing. The 5% carbonated biochar mix exhibited a 20.5% increase at 28 days, whereas the 10% carbonated biochar mix demonstrated a 35.3% increase.

- The inclusion of higher biochar dosages initially decreased the flexural strength but increased considerably after carbonation curing. The 5% carbonated biochar mix exhibited a 68.7% increase at 28 days, whereas the 10% biochar mix demonstrated a 102.5% increase.
- Higher biochar dosages increased water absorption in non-carbonated samples due to its porous nature. However, carbonation curing significantly decreased the water absorption. The carbonated control mix experienced a decrease of 7.7%, with 5% and 10% carbonated biochar mix exhibited a decrease of 9.9% and 5.2% respectively.

The study highlights the synergistic benefits of biochar and carbonation curing in improving the mechanical characteristics and durability of cementitious composites. Optimizing biochar content and curing methods can significantly enhance both sustainability and performance, showcasing the potential of agricultural waste valorization and advanced curing techniques in climate change mitigation.

Future research should focus on detailed biochar characterization, encompassing yield, elemental composition (CHNS/XRF), ash content, pore size distribution and total porosity (MIP), and specific surface area (BET). Standardized water uptake testing with repeatability checks is essential to more accurately correlate absorption behavior to effective w/c ratio and mechanical performance. Integrating such advanced characterization with extended mechanical and durability testing will strengthen correlations between biochar properties, microstructure, and composite performance, advancing both scientific understanding and practical application.

Acknowledgments. The authors gratefully acknowledge the financial support provided by the Center of Excellence for Emerging Materials under TIET-VT (Thapar Institute of Engineering and Technology, Patiala – Virginia Tech, USA, Sanction No. TIET/CEEMS/Regular/2022/027).

References

1. Abdurrahman, M.I., Chaki, S., Saini, G.: Stubble burning: effects on health & environment, regulations and management practices. Environ. Adv. **2**, 100011 (2020). https://doi.org/10.1016/j.envadv.2020.100011
2. Sharma, A.R., Kharol, S.K., Badarinath, K.V.S., Singh, D.: Impact of agriculture crop residue burning on atmospheric aerosol loading; a study over Punjab state, India. Ann. Geophys. **28**, 367–379 (2010). https://doi.org/10.5194/angeo-28-367-2010
3. Rashid, S., Singh, M.: An investigation on carbon dioxide incorporated sustainable ready-mix concrete using OPC and PPC. Arab. J. Sci. Eng. (2023). https://doi.org/10.1007/s13369-023-08106-y
4. Chadha, V., Rashid, S., Singh, M., Roy, A.B.D.: Enhancing sustainable concrete: a study on mechanical properties and durability of concrete with treated recycled aggregates and marble dust. Iran. J. Sci. Technol. Trans. Civ. Eng. (2024). https://doi.org/10.1007/s40996-024-01368-y
5. Yaashikaa, P.R., Kumar, P.S., Varjani, S., Saravanan, A.: A critical review on the biochar production techniques, characterization, stability and applications for circular bioeconomy. Biotech. Rep. **28**, e00570 (2020)
6. Ghani, W.A.W.A.K., Mohd, A., da Silva, G., et al.: Biochar production from waste rubber-wood-sawdust and its potential use in C sequestration: chemical and physical characterization. Ind. Crop. Prod. **44**, 18–24 (2013)

7. Sirico, A., Bernardi, P., Sciancalepore, C., et al.: Biochar from wood waste as additive for structural concrete. Constr. Build. Mater. **303**, 124500 (2021). https://doi.org/10.1016/j.conbuildmat.2021.124500
8. Dixit, A., Verma, A., Dai Pang, S.: Dual waste utilization in ultra-high performance concrete using biochar and marine clay. Cem. Concr. Compos. **120**, 104049 (2021)
9. Tan, K.-H., Wang, T.-Y., Zhou, Z.-H., Qin, Y.-H.: Biochar as a partial cement replacement material for developing sustainable concrete: an overview. J. Mater. Civ. Eng. **33**, 3121001 (2021)
10. Zaid, O., Alsharari, F., Ahmed, M.: Utilization of engineered biochar as a binder in carbon negative cement-based composites: a review. Constr. Build. Mater. **417**, 135246 (2024). https://doi.org/10.1016/j.conbuildmat.2024.135246
11. Mensah, R.A., Shanmugam, V., Narayanan, S., et al.: Biochar-added cementitious materials—a review on mechanical, thermal, and environmental properties. Sustainability. **13** (2021). https://doi.org/10.3390/su13169336
12. Haris Javed, M., Ali Sikandar, M., Ahmad, W., et al.: Effect of various biochars on physical, mechanical, and microstructural characteristics of cement pastes and mortars. J. Build. Eng. **57**, 104850 (2022). https://doi.org/10.1016/j.jobe.2022.104850
13. Murali, G., Wong, L.S.: A comprehensive review of biochar-modified concrete: mechanical performance and microstructural insights. Constr. Build. Mater. **425**, 135986 (2024). https://doi.org/10.1016/j.conbuildmat.2024.135986
14. Gupta, S., Kua, H.W., Dai Pang, S.: Biochar-mortar composite: manufacturing, evaluation of physical properties and economic viability. Constr. Build. Mater. **167**, 874–889 (2018)
15. Yang, X., Wang, X.Y.: Hydration-strength-durability-workability of biochar-cement binary blends. J. Build. Eng. **42**, 103064 (2021). https://doi.org/10.1016/j.jobe.2021.103064
16. Sirico, A., Belletti, B., Bernardi, P., et al.: Effects of biochar addition on long-term behavior of concrete. Theor. Appl. Fract. Mech. **122**, 103626 (2022). https://doi.org/10.1016/j.tafmec.2022.103626
17. Rashid, S., Raghav, A., Goyal, A., et al.: Biochar as a sustainable additive in cementitious composites: a comprehensive analysis of properties and environmental impact. Ind. Crop. Prod. **209**, 118044 (2024). https://doi.org/10.1016/j.indcrop.2024.118044
18. Rashid, S., Goyal, A., Danie Roy, A.B., Singh, M.: Effect of physical modification of biochar on its characteristics and cementitious properties. J. Mater. Civ. Eng. (2025). https://doi.org/10.1061/JMCEE7.MTENG-20957
19. Mo, L., Fang, J., Huang, B., et al.: Combined effects of biochar and MgO expansive additive on the autogenous shrinkage, internal relative humidity and compressive strength of cement pastes. Constr. Build. Mater. **229**, 116877 (2019). https://doi.org/10.1016/j.conbuildmat.2019.116877
20. Tan, K., Pang, X., Qin, Y., Wang, J.: Properties of cement mortar containing pulverized biochar pyrolyzed at different temperatures. Constr. Build. Mater. **263**, 120616 (2020). https://doi.org/10.1016/j.conbuildmat.2020.120616
21. Liu, W., Li, K., Xu, S.: Utilizing bamboo biochar in cement mortar as a bio-modifier to improve the compressive strength and crack-resistance fracture ability. Constr. Build. Mater. **327**, 126917 (2022). https://doi.org/10.1016/j.conbuildmat.2022.126917
22. Aneja, A., Sharma, R.L., Singh, H.: Mechanical and durability properties of biochar concrete. Mater. Today Proc. **65**, 3724–3730 (2022). https://doi.org/10.1016/j.matpr.2022.06.371
23. Chen, T., Zhao, L., Gao, X., et al.: Modification of carbonation-cured cement mortar using biochar and its environmental evaluation. Cem. Concr. Compos. **134**, 104764 (2022). https://doi.org/10.1016/j.cemconcomp.2022.104764
24. Garg, A., Huang, H., Cai, W., et al.: Influence of soil density on gas permeability and water retention in soils amended with in-house produced biochar. J. Rock Mech. Geotech. Eng. **13**, 593–602 (2021). https://doi.org/10.1016/j.jrmge.2020.10.007

25. IS: 269-2015: IS: 269-2015, Specification for Ordinary Portland Cement. Bur Indian Stand Delhi 2015 (2020)
26. BIS 383: Specification for coarse and fine aggregates from natural sources for concrete. Bur. Indian Stand. Delhi, 1–24 (2016)
27. Gupta, S., Kua, H.W., Koh, H.J.: Application of biochar from food and wood waste as green admixture for cement mortar. Sci. Total Environ. **619–620**, 419–435 (2018). https://doi.org/10.1016/j.scitotenv.2017.11.044
28. Ofori-Boadu, A.N., Kelley, R., Aryeetey, F., et al.: The influence of swine-waste biochar on the early-age characteristics of cement paste. Int. J. Eng. Res. Appl. **07**, 01–07 (2017). https://doi.org/10.9790/9622-0706010107
29. Zhang, M.-H., Islam, J.: Use of nano-silica to reduce setting time and increase early strength of concretes with high volumes of fly ash or slag. Constr. Build. Mater. **29**, 573–580 (2012). https://doi.org/10.1016/j.conbuildmat.2011.11.013
30. ASTM C1437: Standard test method for flow of hydraulic cement mortar. C1437 (2007)
31. ASTM C109: ASTM C 109-Standard Test Method for Compressive Strength of Hydraulic Cement Mortars. United States ASTM Int (2020)
32. ASTM C348-02: ASTM C348–02 Standard Test Method for Flexural Strength of Hydraulic-Cement Mortars. Annu. B, ASTM Stand (2002)
33. ASTM C642: Standard Test Method for Density, Absorption, and Voids in Hardened Concrete. ASTM, ASTM Int (2013)
34. Qin, Y., Pang, X., Tan, K., Bao, T.: Evaluation of pervious concrete performance with pulverized biochar as cement replacement. Cem. Concr. Compos. **119**, 104022 (2021). https://doi.org/10.1016/j.cemconcomp.2021.104022
35. Lyu, H., Gao, B., He, F., et al.: Effects of ball milling on the physicochemical and sorptive properties of biochar: experimental observations and governing mechanisms. Environ. Pollut. **233**, 54–63 (2018). https://doi.org/10.1016/j.envpol.2017.10.037
36. Tan, K., Qin, Y., Wang, J.: Evaluation of the properties and carbon sequestration potential of biochar-modified pervious concrete. Constr. Build. Mater. **314**, 125648 (2022). https://doi.org/10.1016/j.conbuildmat.2021.125648
37. Restuccia, L., Ferro, G.A.: Influence of filler size on the mechanical properties of cement-based composites. Fatigue Fract. Eng. Mater. Struct. **41**, 797–805 (2018). https://doi.org/10.1111/ffe.12694

Industrial-Scale Production of Low-Clinker LC3 Cements Using Low-Grade Clays

J. I. Galindo-Barajas[1,2(✉)], M. A. Perez[1], R. Gonzalez-Olmos[1], M. Guillem[3], and P. Perez-Cortes[3]

[1] IQS School of Engineering, Universitat Ramon Llull, Via Augusta, 390, 08017 Barcelona, Spain
jorgegalindob@iqs.url.edu, jorgeisaac.galindo@molins.es
[2] Cementos Molins Industrial, S.A., CN-340 nº2 Al 38 Km 1242.3, Sant Vicenç Dels Horts, 08620 Barcelona, Spain
[3] Cementos Molins, S.A., R&D Group, CN-340 nº2 Al 38 Km 1242.3, Sant Vicenç Dels Horts, 08620 Barcelona, Spain

Abstract. Clays and limestone are widely available materials used as supplementary cementitious material (SCM's) and filler, respectively, to reduce the clinker content in cement. The pozzolanic reactivity of clays is known to correlate directly with their kaolinite content; however, sourcing kaolinite-rich clays can be challenging in certain regions, despite the general abundance of clays deposits. This study investigates the feasibility of utilising low-kaolinite clays, in combination with limestone, for the production of low-clinker cements. The clays were calcined at 850 °C and 950 °C and subsequent characterisation was carried out via X-ray diffraction (XRD), thermogravimetric analysis (TGA), and X-ray fluorescence (XRF) to determine their mineralogical and chemical composition. Clay reactivity was assessed through the R^3 test using isothermal calorimetry; in addition, the reactive silica content of selected clays was quantified. Finally, low-clinker cements incorporating calcined clay and limestone were produced at industrial scale. Their mechanical strength, setting times, and workability were evaluated. The results indicate that clays containing only 45% kaolinite can significantly contribute to the development of low-clinker cements, providing a viable alternative in regions where high-kaolinite clays are not readily available. Moreover, this study presents a practical approach to clay characterisation for industrial cement applications, which may serve as a useful reference for cement manufacturers interested in incorporating calcined clays into new products.

Keywords: Clays characterisation · LC3 · low-grade clays · low-carbon cements · differential scanning calorimetry

1 Introduction

Cement production accounts for approximately 8% of global anthropogenic CO_2 emissions, primarily from clinker manufacturing. Reducing clinker content by incorporating supplementary cementitious materials (SCMs) is a key strategy for decarbonisation of

M. Posani and S. Singh (Eds.): RYS 2025, RILEM Bookseries 67, pp. 116–129, 2026.
https://doi.org/10.1007/978-3-032-14652-6_10

the sector. Although fly ash and slag have traditionally been used, their availability is decreasing [1]. Calcined clays offer a promising alternative due to their pozzolanic activity and wide availability; however, their reactivity depends largely on kaolinite content [2] which is not consistently accessible across regions. Most studies on calcined clays are limited to laboratory-scale formulations. In contrast, this work presents results from industrial-scale cement production using clays with 45% kaolinite, outlining the clay selection process and calcination parameters. These findings provide practical insight into the large-scale implementation of low-kaolinite clays in commercial cement manufacturing.

2 Materials and Methods

2.1 Clays Characterisation

Five clays from different quarries were characterised using thermogravimetric and differential scanning calorimetry (TGA-DSC), X-ray diffraction (XRD), R^3 test via isothermal calorimetry to assess the effect of the calcination temperature on the microstructure and reactivity of clays. The clays were calcined at the laboratory scale at 850 °C and 950 °C. Additionally, the reactive silica content was determined for selected clays. The clays used for cement production were industrially calcined in a rotary kiln at approximately 900 °C. The chemical composition by XRF of the clays and the industrially produced cements is shown in Table 1. The XRF analysis was carried out in an automated equipment (S8 Tiger).

XRD analysis was performed to evaluate the evolution of the crystalline structure of the clays at different calcination temperatures. The test was carried out using an automated diffractometer (D8 Endeavor) from 8° to 65° 2θ, with a step size of 0.019° 2θ every 0.1 s, using a voltage of 45 kV and a current of 30 mA, with a Cu tube ($\lambda = 1.5418$ Å).

TGA-DSC analysis was used to quantify the kaolinite content in raw clays and to observe phase changes in both raw and calcined clays. The equipment used was a Discovery SDT 650, with a heating rate of 10 °C/min under a nitrogen atmosphere with a constant flow rate of 100 mL/min. To better compare the endothermic and exothermic peaks between the clays, the baseline generated by the reference crucibles was subtracted. Additionally, the heat released below 200 °C was excluded due to overlapping signals from crucible heat flow and moisture evaporation.

For the R^3 test the clays were grounded to achieve a particle size distribution with 8% retained on a 45 μm sieve (EN 196-12). The reactive silica content of the clays was determined in accordance with the EN 196-2 standard.

Table 1. Chemical composition of clays (wt.%)

Sample	CaO	SiO_2	Al_2O_3	Fe_2O_3	MgO	K_2O	SO_3	Na_2O	Other	LOI*
Clay - H	0,30	55,40	23,50	5,09	1,93	1,89	0,31	0,74	1,76	9,08
Clay - R	0,11	51,60	26,70	7,53	1,36	2,66	0,26	0,67	1,2	7,91
Clay - F	0,38	52,90	26,50	4,70	1,72	3,18	0,22	0,78	1,18	8,44
Clay - L	0,12	50,00	29,80	5,07	1,04	0,97	0,44	0,63	1,53	10,4
Clay - C	0,36	45,30	31,90	5,90	1,25	0,68	0,55	0,7	1,86	11,5

*Loss on ignition

2.2 Industrial Cements Characterisation

Four different CEM II B-M (Q-L) type cements were produced on an industrial scale using selected clays and their mixtures, along with a reference cement containing natural pozzolan and limestone. The chemical composition by XRF of cements are shown in Table 2. The reference cement was selected because it corresponded to the CEM II B-M type routinely produced at the industrial plant where this study was conducted. Additionally, Table 3 presents the estimated dosage of the materials used in the cement formulation. Cement strength was assessed in mortar specimens following the EN 196-1 standard, using a mass sand-to-cement ratio of 3:1 and a water-to-cement ratio of 0.5. Compressive strength was measured at 1, 2, 7, and 28 days. Flowability, setting time and water demand were determined according to EN 196-3 and UNE-EN 1015-3 standards, respectively.

Table 2. Chemical composition of cements produced industrially (%wt).

Sample	CaO	SiO_2	Al_2O_3	Fe_2O_3	MgO	K_2O	SO_3	Na_2O	Other	LOI
CEM 68-12-13-60K	54,88	21,13	8,34	4,01	1,17	0,59	3,03	0,14	0,02	6,69
CEM 62-12-20-60K	54,23	19,58	7,99	3,87	1,11	0,58	2,93	0,12	0,03	9,56

(continued)

Table 2. (*continued*)

Sample	CaO	SiO_2	Al_2O_3	Fe_2O_3	MgO	K_2O	SO_3	Na_2O	Other	LOI
CEM 65-07-22-45K	58,43	17,54	5,08	3,34	1,35	0,68	2,84	0,12	0,02	10,6
CEM 57-15-22-45K	52,91	20,46	7,02	3,88	1,29	0,72	2,91	0,12	0,02	10,67
CEM REF B-M (P-L)	57,05	20,34	5,06	4,12	2,06	0,83	3,16	0,41	0,05	6,93

Table 3. Estimated dosage of cements produced at industrial scale (%wt).

Cement	Clinker	Clay	Limestone	Gypsum
CEM 68-12-13-60K	68	12	13	6
CEM 62-12-20-60K	62	12	20	6
CEM 65-7-22-45K	65	7	22	6
CEM 57-15-22-45K	57	15	22	6
CEM REF B-M (P-L)	70	10	14	6

3 Results

3.1 XRD Characterisation

Figure 1 shows XRD patterns of the five natural clays and those calcined at 850 °C, 950 °C, and 1100 °C. The natural clays (Fig. 1a) clearly exhibit the most intense kaolinite peak around 12,4° 2θ, which follows this order: Clay C > L > F > R > H, correlating well with the alumina content, as detailed in Table 1.

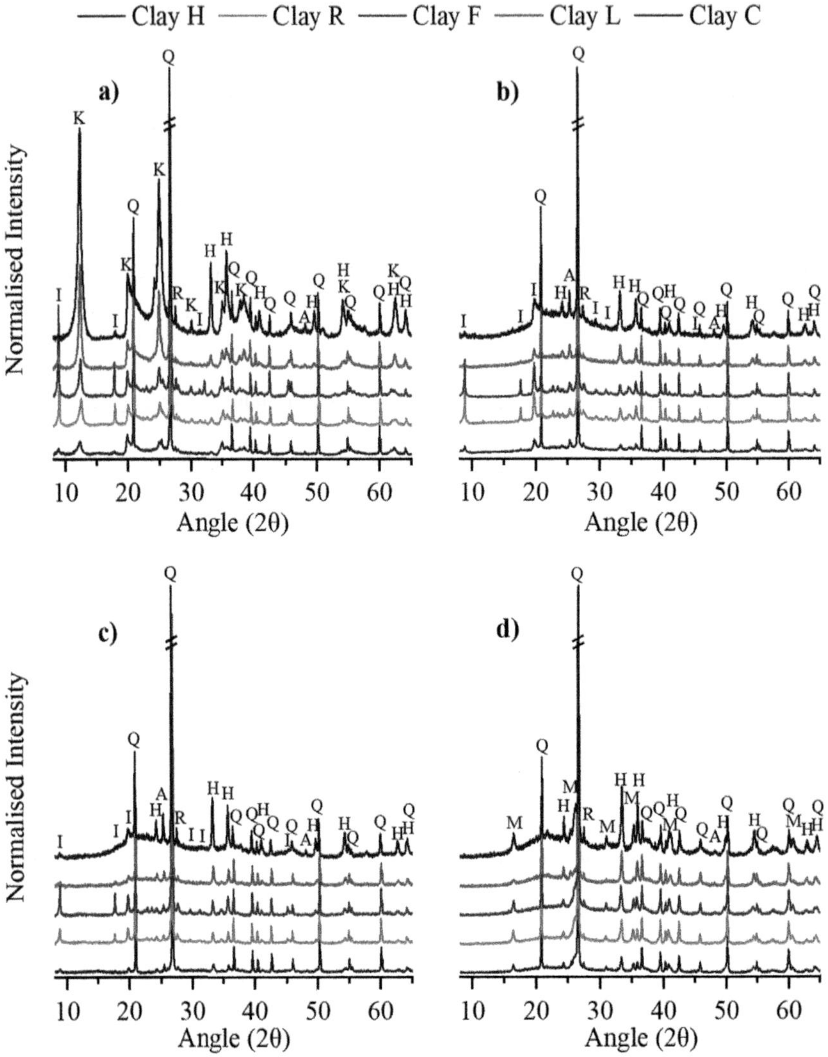

Fig. 1. XRD of clays calcined at different temperature: a) Raw, b) 850 °C, c) 950 °C and d) 1100 °C. K: Kaolinite, Q: Quartz, I: I llite/muscovite, H: Hematite, R: Rutile, A: Anatase, M: Mullite

After calcination at 850 °C (Fig. 1b), the main change observed is the disappearance of the kaolinite peaks and a reduction in the illite/muscovite peaks, mainly seen at around 8,7° and 19,7° 2θ. At 950 °C, no significant changes are observed in the diffractograms, except for a slight decrease in the intensity of the illite/muscovite peaks, compared to those at 850 °C. At first glance, this suggests that recrystallisation may not occur at this temperature, as no new crystalline phases are detected (Fig. 1c). In contrast, when the clays are calcined above 1100 °C, the appearance of a mullite phase is evident at around

16,5° and 26,4° 2θ, indicating recrystallisation of the amorphous phase. Simultaneously, illite/muscovite-related reflections disappear completely (Fig. 1d).

The formation of Al–Si spinel phase is typically expected within the temperature range of 925–1050 °C, accompanied by the presence of amorphous SiO_2 [3], while mullite generally forms above 1050 °C. However, in the current diffractograms, only the mullite phase was clearly identified. Dhar and Bishnoi [4], did not detect characteristic peaks of Al–Si spinel at 19.15°, 31.45°, or 36.95° 2θ in clays calcined up to 1000 °C. Zunino and Scrivener [5] observed that in clays calcined at 950 °C, no clear evidence or distinct peaks of the spinel phase were present. However, at 1050 °C, the spinel phase was clearly identifiable, with higher intensity in the kaolinitic clay compared to the natural clay, despite both having similar alumina contents. Since the Al–Si spinel phase results from the thermal transformation of aluminium-rich clays—particularly those with high kaolinite content—its formation is expected to be even more difficult to detect in low-grade clays. This is mainly the case for clays H, R, and F, as indicated by the XRF results in Table 1. Additionally, Al – Si spinel crystals typically form at the nanometric scale [6] which broadens the diffraction peaks and reduces their intensity, making detection by XRD more challenging.

According to the diffractograms, no evidence of recrystallisation is observed at 950 °C. A noticeable reduction in illite/muscovite peaks is present, which, from a preliminary standpoint, could be favourable for the reactivity of low-grade clays. Specifically, low-grade kaolinitic clays might become more reactive if the illitic phases are further amorphised at this temperature. To support the hypothesis of limited or absent recrystallisation at 950 °C, TGA-DSC analysis was carried out.

3.2 TGA/DSC Characterisation

Figure 2 presents the thermogravimetric analysis (TGA) curves for the five clays. All samples exhibit two distinct mass loss regions: the first below 200 °C, attributed to the evaporation of physically adsorbed water, and the second between 400 °C and 600 °C, corresponding to the dehydroxylation of kaolinite. Based on this, the kaolinite content in each clay was estimated using Eq. 1:

$$\text{Kaolinite content}(\%) = W_{600-400}/(100 - W_{200-RT}) \cdot 7.17 \cdot 100 \quad (1)$$

$W_{600–400}$ = Weight loss between 400 °C and 600 °C.

$W_{200\text{-}RT}$ = Weight loss between room temperature and 200 °C.

7.17 = This factor refers to the ratio of the molecular weight of kaolinite ($M_{kaolinite}$ = 258.16 g/mol) and two water molecules ($2xM_{water}$ = 36.03 g/mol).

The calculated kaolinite contents (Table 4) align with the relative intensity of the kaolinite peaks observed in the XRD patterns of raw clays (Fig. 1a), confirming consistency between both characterisation techniques.

Table 4. Kaolinite content of raw clays

Sample	Kaolinite content (%)
Clay H	31,3
Clay R	33,5
Clay F	35,9
Clay L	51,8
Clay C	64,6

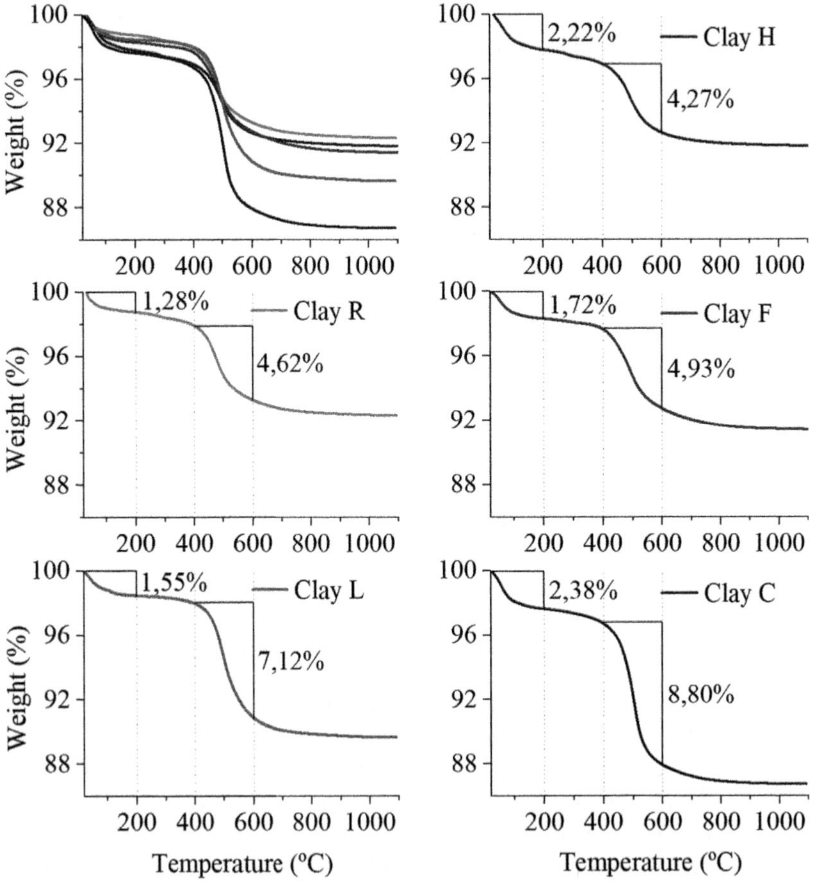

Fig. 2. TGA of the 5 raw clays from room temperature to 1100 °C.

Figure 3 shows the DSC results of the clays, both in their raw and calcined states. In the natural state, all clays exhibit a pronounced endothermic peak near 500 °C, attributed to the dehydroxylation of kaolinite. The peak intensity correlates with the kaolinite content calculated from TGA and XRD, reflecting the thermal breakdown of the kaolinite crystal structure into an amorphous phase. Between 900 °C and 1030 °C an exothermic peak emerges attributed to the recrystallisation of the amorphous phase into Al-Si spinel [4].

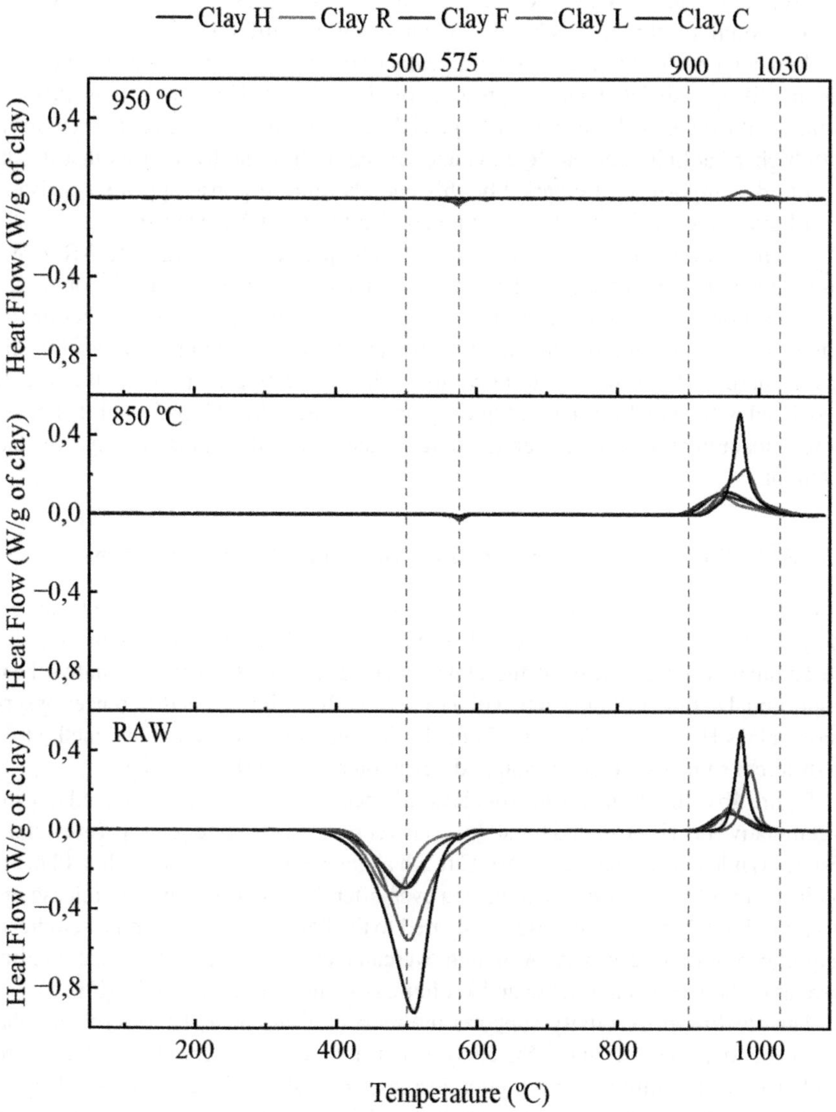

Fig. 3. DSC of clays at different temperatures: Raw (bottom), 850 °C (middle) and 950 °C (top).

For clays calcined at 850 °C, the dehydroxylation peak is not present, confirming the complete transformation of kaolinite. Interestingly, even in samples previously calcined at 850 °C, the recrystallisation peak remains visible in the DSC traces. This suggests that while kaolinite was fully dehydroxylated, the resulting amorphous phase had not yet been crystallised and remained thermally reactive. By contrast, DSC of clays calcined above 900 °C show no exothermic signals, indicating that recrystallisation has already occurred. The Clay - L shows a minor residual recrystallisation peak, which may be explained by its slightly higher crystallisation temperature compared to the other clays, as observed in the raw DSC profile. This subtle shift suggests a slower transformation of the amorphous phase, possibly due to compositional differences.

Another noteworthy feature is the appearance of an endothermic peak around 575 °C, which is likely associated with the presence of illitic phases. This peak is more prominent in clays with lower kaolinite content (e.g., H, R, and F), while it is weak or absent in clays with higher kaolinite content (e.g., C and L), supporting the hypothesis that it relates to illite decomposition. Remarkably, this endothermic response appears consistently, regardless of the calcination temperature applied prior to DSC analysis.

In summary, the DSC technique provides complementary insights to XRD. While Al-Si spinel phase was not detectable by XRD at 950 °C, the DSC analysis confirms that crystallisation processes did occur upon heating, highlighting the value of thermal techniques in the evaluating pozzolanic potential and transformation pathways of calcined clays. These findings align with previous studies indicating that the exothermic peak associated with spinel formation typically appears after 900 °C [3–5] and this phase is more evident in clays with higher kaolinite content due to their greater dehydroxylation potential.

3.3 R^3 Isothermal Conductivity and Reactive Silica Characterisation

To evaluate the reactivity of the clays and compare the findings from XRD and DSC analyses, the R^3 test was conducted using isothermal calorimetry. Figure 4a shows the cumulative heat release of the clays calcined at 850 °C. The cumulative heat is proportional to the kaolinite content of the clays, with no significant differences observed among clays H, R, and F. Tests performed using the same equipment and methodology on different samples yielded a standard deviation around 3.0 J/g of clay.

Figure 4b displays the results for clays calcined at 950 °C, where a marked reduction in cumulative heat is observed for all five clays compared to their counterparts calcined at 850 °C. While clays calcined at 850 °C reach a plateau in heat release within 48 h, those calcined at 950 °C show no such plateau even after 7 days. The total heat release drops by approximately 50% in all cases compared to the 850 °C calcined samples, indicating a significant loss in reactivity. Although overcalcined clays release less heat, the linear correlation between heat release and kaolinite content appears to persist (Fig. 5).

This decline in reactivity supports the recrystallisation of the amorphous phase, which was detected through DSC analysis but not clearly observed via XRD. Zunino and Scrivener [5] similarly reported a decrease in total heat release when a kaolinitic clay was calcined at 950 °C. In their study, the DSC recrystallisation peak also appeared around this temperature. However, for a clay in which recrystallisation did not occur until close to 1000 °C, the drop in reactivity measured by the R^3 test was less pronounced, suggesting that some amorphous content remained at 950 °C. This could explain why Clay – L, showed a similar total heat release to Clay – C at 950 °C, despite the lower kaolinite content, as a fraction of its amorphous phase may have been persevered (Fig. 3 top).

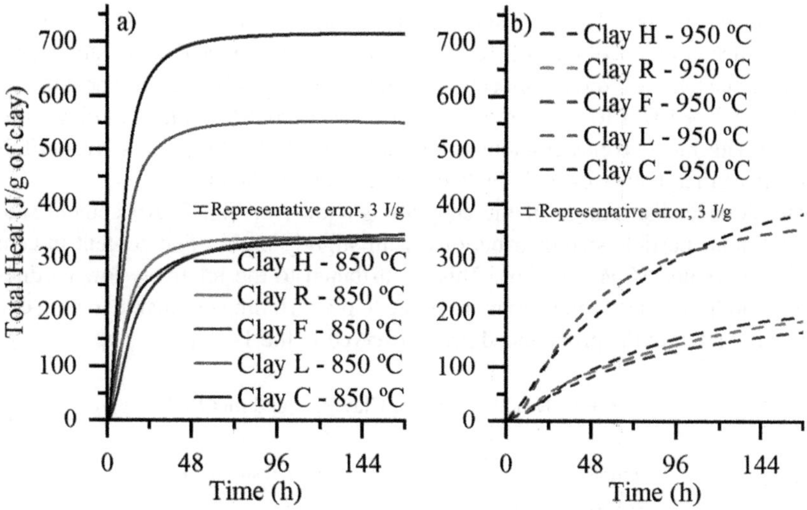

Fig. 4. Total heat during 7 days for the 5 clays calcined at: a) 850 °C and b) 950 °C.

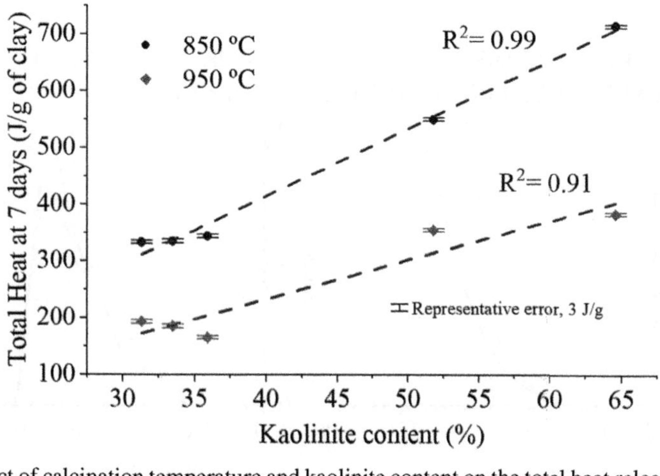

Fig. 5. Effect of calcination temperature and kaolinite content on the total heat released at 7 days.

These results highlight the capability of DSC to more accurately identify optimal calcination conditions by tracking the intensity of the recrystallisation peak and comparing it with the recrystallisation peak in the raw clay. This approach is particularly effective for assessing the preservation or loss of the amorphous phase, which is directly linked to the reactivity of the calcined material.

3.4 Cements Produced at Industrial Scale

Based on the characterisation of the clays, two were selected for industrial-scale testing: clays C and R. The selection of clay R was driven primarily by its greater availability, and the fact that clays H and F showed only marginal differences in reactivity. The reactive silica contents of clays R, C, and a blend of clays C and R are summarised in Table 5. Since Clay – R did not meet the minimum reactive silica requirement of 25 wt% specified by EN 197-1 for the use of calcined clays as SCMs, a blend of clays R and C was used. This blended clay contains approximately 45% kaolinite. As shown in Table 5, the reactive silica increases with the kaolinite content of the clay.

Particle size distribution of the cements is shown in Fig. 5. All cement samples exhibit similar particle size distributions, with slightly higher fine content in cements containing greater SCM fractions. This is attributed to the relatively low hardness of the clay and limestone, which are more prone to overgrinding compared to clinker. This effect is reflected in the pronounced peak observed in the 1–10 μm range.

Table 5. Reactive silica of Clay R, C and mix R/C.

Sample	Average reactive silica (%)
Clay R – 900 °C	22.1
Clay C – 900 °C	39.1
Mix 45K – Clay R/C	28.4

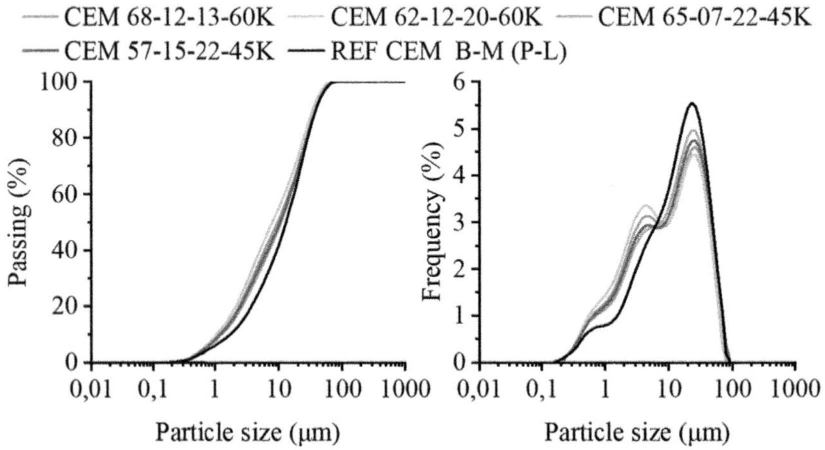

Fig. 6. Particle size distribution of industrial cements.

Compressive strength results for the four cements with calcined clay and limestone are presented in Fig. 6, alongside the reference B-M (P-L) cement. During the first two days, strength development was primarily influenced by clinker content and possibly by physical effects from SCMs, such as the nucleation provided by limestone and/or calcined clay fines. At this early stage, the specific kaolinite content of the clay appears to have minimal impact on strength development. By day 7, pozzolanic activity becomes more apparent. For a similar clinker content, cements incorporating higher-kaolinite clays show greater strength development. Notably, a formulation with 5% less clinker (65% clinker) and a 45% kaolinite clay surpassed the reference cement in strength at this age. At 28 days, all four cements formulated with calcined clays outperformed the reference, including the one with 13% less clinker (57% clinker), but with a higher content of calcined clay. The CEM 57-15-22-45K cement, in particular, achieved comparable performance to the CEM 65-07-22-45K, indicating that 8% of clinker could be effectively replaced by the clay blend without compromising long-term strength. However, lower clinker contents generally reduced early-age strength, which is often undesirable in commercial applications. Setting time, water demand, and mortar workability are shown in Fig. 7. Cements with higher clay content (12–15%) exhibited an approximate 10% increase in water demand compared to the reference. This increase is attributed to the high specific surface area and water absorption capacity of the calcined clays [7].

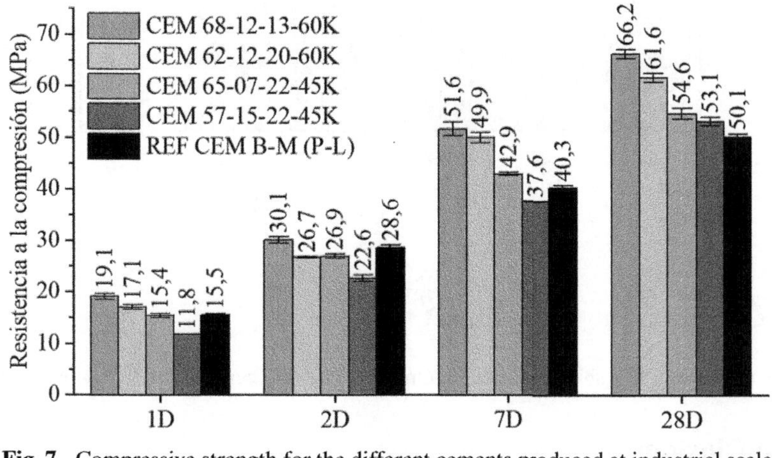

Fig. 7. Compressive strength for the different cements produced at industrial scale.

Correspondingly, these cements showed reduced flow, with the CEM 62-12-20-60K cement presenting the most significant reduction (19%) relative to the reference. Although this cement contains more limestone and the same clay content as CEM 68-12-13-60K, its slightly finer particle size distribution (Fig. 5) contributes to the increased water demand. It should also be noted that an amine-based grinding aid was used during cement preparation. In addition to improving grinding efficiency, this additive influences cement properties by producing finer particle sizes, enhancing compressive strength, and, in this case, improving rheology and flowability. While its dosage was intended to be consistent across samples, slight variations may have occurred, potentially affecting flow behaviour. Furthermore, no chemical admixtures were used to improve rheology during mortar preparation. However, admixtures based on polycarboxylates have been shown to perform adequately in cements containing calcined clays when optimized with different absorbing groups or combined with other admixtures to retain flow [8]. Nevertheless, the study of admixture effects on rheology improvement falls outside the scope of this work. Regarding setting behaviour, cements containing the 60% kaolinite clay exhibited slightly reduced initial and final setting times. Conversely, cements incorporating the 45% kaolinite blend tended to increase setting times, particularly CEM 65-07-22-45K, likely due to the higher limestone content and lower clay dosage. Similarly, cements incorporating the clay with 45% kaolinite exhibited a coarser particle size distribution, which could also help explain the delayed setting time (Fig. 8).

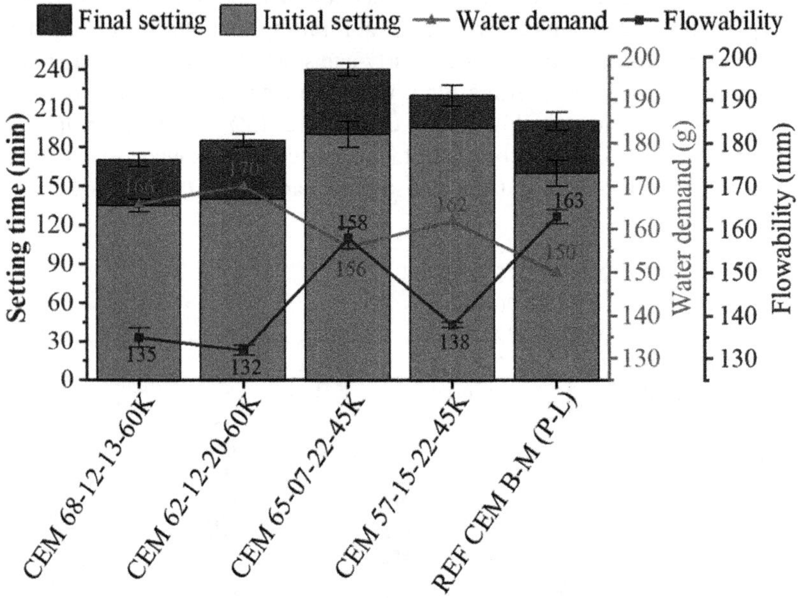

Fig. 8. Setting, flowability and water demand for the five cements produced industrially.

4 Conclusions

This study investigated the thermal behaviour and reactivity of low- to medium-grade clays subjected to different calcination temperatures, with a focus on their potential use as supplementary cementitious materials (SCMs) and their application in industrial cement production. The results show that calcination above 1050 °C leads to mullite formation, while Al–Si spinel, expected around 950 °C, was not detected by XRD—highlighting the limitations of this technique for complex mineralogies. In contrast, DSC effectively identified recrystallisation through exothermic signals and confirmed that clays calcined at 850 °C retained a reactive amorphous phase, which diminished at 950 °C due to crystallisation. This trend was corroborated by R^3 isothermal calorimetry, which revealed a significant reduction in heat release—and thus reactivity—at higher temperatures. Industrial-scale tests using a blend of clays with 45% kaolinite demonstrated promising performance. Despite increased water demand and lower workability, the resulting cements achieved higher 28-day strength than the reference, supporting the practical viability of calcined clays in low-clinker formulations. Overall, the combined use of XRD, DSC, and calorimetry provides a solid basis for assessing clay reactivity, optimising calcination parameters, and anticipating performance in cement formulations. These findings offer relevant insights for the development and implementation of low-clinker cements incorporating calcined clays at industrial scale.

References

1. Giergiczny, Z.: Fly ash and slag. Cem. Concr. Res. **124** (2019). https://doi.org/10.1016/j.cemconres.2019.105826
2. Avet, F., Snellings, R., Alujas Diaz, A., Ben Haha, M., Scrivener, K.: Development of a new rapid, relevant and reliable (R3) test method to evaluate the pozzolanic reactivity of calcined kaolinitic clays. Cem. Concr. Res. **85**, 1–11 (2016). https://doi.org/10.1016/j.cemconres.2016.02.015
3. Ptáček, P., Šoukal, F., Opravil, T., Nosková, M., Havlica, J., Brandštetr, J.: The kinetics of Al–Si spinel phase crystallization from calcined kaolin. J. Solid State Chem. **183**(11), 2565–2569 (2010). https://doi.org/10.1016/j.jssc.2010.08.030
4. Dhar, M., Bishnoi, S.: Influence of calcination temperature on the physical and chemical characteristics of kaolinitic clays for use as supplementary cementitious materials. Cem. Concr. Res. **178** (2024). https://doi.org/10.1016/j.cemconres.2024.107464
5. Zunino, F., Scrivener, K.: Reactivity of kaolinitic clays calcined in the 650 °C–1050 °C temperature range: towards a robust assessment of overcalcination. Cem. Concr. Compos. **146** (2023). https://doi.org/10.1016/j.cemconcomp.2023.105380
6. Chauffeton, C., Wallez, G.: Crystal-chemical investigation of the "cubic Al–Si spinel" issued from the thermal transformation of kaolinite and halloysite. J. Am. Ceram. Soc. **105**(7), 4986–4993 (2022). https://doi.org/10.1111/jace.18439
7. Nehdi, M.L.: Clay in cement-based materials: critical overview of state-of-the-art. Constr. Build. Mater. **51**, 372–382 (2013). https://doi.org/10.1016/j.conbuildmat.2013.10.059
8. Moghul, S., Zunino, F., Flatt, R.J.: Flow loss in superplasticized limestone calcined clay cement. J. Am. Ceram. Soc. **108**(5) (2025). https://doi.org/10.1111/jace.20344

Investigating Traditional Rainwater Harvesting Systems: A Study of the Tanka at Jaigarh Fort

Nigar Shaikh(✉) and Swathy Manohar

Department of Civil Engineering, Indian Institute of Technology Bombay, Mumbai, India
23D0314@iitb.ac.in

Abstract. This study explores the "Tanka", a traditional underground rainwater harvesting system situated within the historic Jaigarh Fort, on the outskirts of Jaipur, Rajasthan, India. The research focuses on understanding the functional role and engineering benefits of organic and mineral admixtures historically used in its construction, alongside insights into production techniques and layering mechanisms. Detailed analysis revealed the significant contribution of pozzolanic admixtures in enhancing hydraulic reactions, improving water resistance, and increasing durability within lime-based plaster systems. Additionally, the incorporation of organic fibers, such as straw, was identified in the plaster, demonstrating improved crack resistance. The results highlight the synergy between these traditional admixtures and local materials, refining porosity while contributing to long-term structural stability and performance. Furthermore, the study examines the vernacular production technology, including the sequential layering of plaster and the influence of aggregate size. The layering method facilitated optimized adhesion and resilience, while aggregate size impacted the permeability of the structure.

Keywords: Pozzolanic lime mortars · Cultural heritage · Microstructure · Hygric properties

1 Introduction

In construction, binding materials play a pivotal role, acting as the adhesive that unifies structural elements, ensuring strength and stability, and offering protection against water infiltration and environmental degradation [1]. Lime, as a binding material, has been historically celebrated for its versatility and durability, as evidenced by its extensive use in monuments and heritage structures worldwide [2]. However, the advent of ordinary Portland cement (OPC) in the 20th century led to a decline in the use of lime, often resulting in incompatible repairs of historic structures and further deterioration [3]. The emphasis on material compatibility in conservation, advocated by guidelines like ISCARSAH ICOMOS, highlights the need for minimal intervention and reversible methodologies.

M. Posani and S. Singh (Eds.): RYS 2025, RILEM Bookseries 67, pp. 130–140, 2026.
https://doi.org/10.1007/978-3-032-14652-6_11

Among the various architectural typologies constructed using lime-based materials, traditional rainwater harvesting systems such as tanks and stepwells remain relatively underexplored, especially when compared to more prominently studied structures like forts, palaces, and residential buildings. These systems, common in the arid and semi-arid regions of northwestern India, have played a vital role in water conservation, sustaining communities through prolonged dry spells [4].

Their design is marked by simplicity, cost-effectiveness, and adaptability, providing efficient and sustainable solutions to water scarcity [5]. Despite their functional and cultural significance, the materiality of these systems particularly the use of lime-based mortars and plasters has received limited scholarly attention, highlighting the need for detailed studies to inform compatible and sustainable conservation strategies. Traditional rainwater harvesting structures such as tanks and stepwells were built with lime-based plasters and locally sourced materials, chosen specifically for their durability and water-resistant properties [4]. However, many modern repair practices introduce materials incompatible with historic lime-based masonry, resulting in microstructural damage and accelerated decay. Understanding the original material including the type of lime used, its additives, and method of application is essential to formulate compatible repair strategies that preserve the structural and functional integrity of these heritage systems.

This study focuses on the material characterization of historic lime-based plaster from a large water tank located within the Jaigarh Palace complex in Jaipur, Rajasthan.

1.1 Traditional Rainwater Harvesting System at Jaigarh Fort

The Jaigarh Fort complex was built in 18th century by Maharaja Sawai Jai Singh II. The water management system of the fort comprises a series of masonry-lined channels constructed along the surrounding hill slopes to collect and convey rainwater into dedicated storage tanks within the fort precincts [6]. The largest tank, referred to locally as "Tanka", measures approximately 48 m in length, 42 m in width, and 12 m in depth, with an estimated storage capacity of about 22,700 cubic meters. It was strategically constructed beneath the central courtyard to minimize water loss through evaporation and to protect against external contamination (Fig. 1) [6]. Its roof system supported by a network of stone masonry columns and arches. Approximately eighty-one stone columns are arranged in a grid to support the load-bearing arches, which in turn carry the masonry roof.

(a.)

(b.)

Fig. 1. (a.) Location of large water tank in precinct of Jaigarh fort (Image source: Google earth, accessed on 1-05-2025); (b.) Tank from inside shows large stone masonry column and arched roof.

2 Materials and Methods

The selected large water tank "Tanka" is functional and used to store rainwater for use throughout the year. However, it could not retain water above a 10 ft height. The walls and columns of the Tanka showed signs of deterioration and cracks. Samples of plaster and bedding mortar were collected from accessible locations, as shown in Fig. 2.

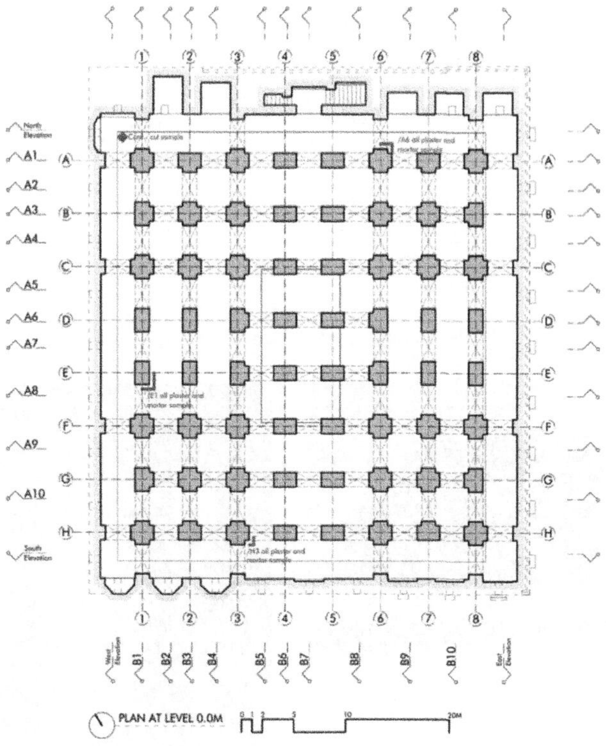

Fig. 2. Location of samples collected from the Tanka. (Base drawing: CHC team)

2.1 Sample Collection

The locations for sample collection were selected based on their accessibility, material used and visual observations. The samples were carefully collected using manual tools such as a scalpel, a small chisel and a little hammer and making sure contamination was avoided. Table 1 shows the list of samples which are lime plaster, along with their corresponding locations, and quantities.

Following a detailed sample documentation which included unique sample code for identification, visual observations of samples and other details related to sample collection. The plaster and mortar samples from location A6 were further selected for physical, chemical and microstructure analysis (Fig. 3).

Table 1. Sample Collection locations

Sr. no	Type	Column number	Location	Level from bottom to collect the sample (m)	Span of sample collection space (m)	Qty.		
						Nos. Surface plaster	Nos. Lime plaster	Nos. Bedding lime mortar
1	Lime plaster & mortar	A6	NE corner	0.53	1.6	2	3	2
2		E1	SE corner	0.69	1.07	-	2	1
3		H3	SE corner	0.53	0.91	-	1	1
4		S1	West surface	4.57	0.23	4		
5	Core cut	A1	SE Junction	-	-		1	

(a.) (b.)

Fig. 3. (a.) Documentation of collected samples; (b.) Location A6 plaster samples

2.2 Testing Methodology

Minimal intervention was ensured during sampling, with only the required quantity of material extracted for testing. For the plaster samples, each layer was tested individually to preserve the integrity of the stratigraphy. Selected samples underwent the following tests:

Visual Observations

Samples were carefully examined for visible features such as additives, contamination, thickness, color, and texture. These physical observations supported inferences drawn from subsequent chemical and microstructural analyses.

Chemical Tests

Chemical analysis, including an acid dissolution test based on Rilem Code 167 Part 2. This test determined the binder-to-aggregate ratio of mortars and provided insights into grain size distribution through sieve analysis. Understanding this ratio is crucial for estimating mix design and particle size distribution.

Mineralogical and Microstructural Assessment
For the microstructural analysis such as X-ray Diffraction (XRD), Fourier Transform Infrared (FTIR), and Thermogravimetric Analysis (TGA), a uniform sample preparation method was adopted. The selected samples were uniformly crushed in mortar and pestle to a size finer than 75 μm. This ensured uniform sample distribution, then freeze-dried to remove the water present in the sample. The freeze-dried samples were then taken for microstructural analyses.

3 Results and Discussion

Visual observation of the plaster samples revealed notable characteristics regarding material composition and layering. Plaster sample A603 exhibited three clearly distinguishable layers, each varying in color, texture, and inclusions (Fig. 4). Layer 2 showed evident use of fibrous material, possibly straw, suggesting intentional reinforcement for improved tensile strength (Fig. 5). The presence of unburnt lime lumps across the layers indicated incomplete calcination or inadequate mixing during preparation. A distinct pinkish hue in certain sections of the plaster pointed to the addition of surkhi (brick dust), a traditional pozzolanic additive known for enhancing strength and water resistance in lime-based mortars.

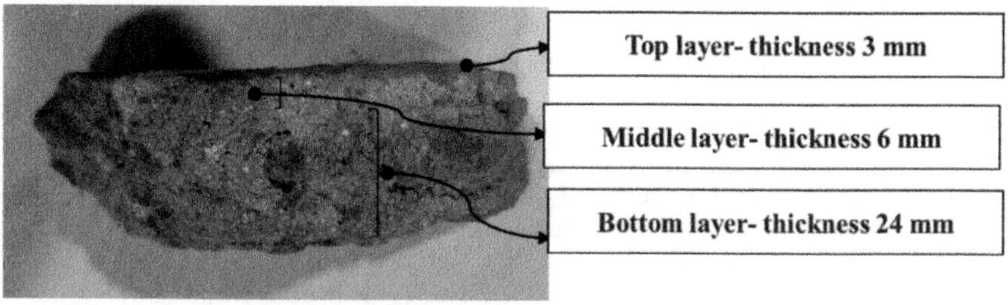

Fig. 4. Observed three layers of Plaster/A603.

Fig. 5. Observations from Plaster/A603 sample. (a) Unburnt lime; (b) straw and (c) brick bats.

Figure 6 (a.) shows infrared spectrum for bottom and middle layer of the plaster PA6031/2 as well as the bedding mortar BA603. Characteristic peaks around, 870 cm¯1, 1400–1450 cm¯1 corresponding to calcite, and 996 cm¯1, attributed to the O-Si-O bond. For sample the top layer(A6033), the infrared spectrum showed characteristic peaks at 824 cm¯1, corresponding to calcite out-of-plane bending, and 986 cm¯1, attributed to the O-Si-O bond (Fig. 6 (b.)). Additionally, a peak at 1638 cm¯1 indicates N–H bending vibrations associated with secondary amines, while a broad band at 3387 cm¯1 further confirms the presence of secondary amines.

The thermograms of bottom plaster layer (A6031) exhibited a prominent peak indicating carbonate decomposition at approximately 750 °C. A peak around 100 °C shows higher free water content. Wide endothermic shoulder around 500 °C highlights presence of hydraulic components that corresponding to the dehydration of the calcium silicate hydrates formed (Fig. 6).

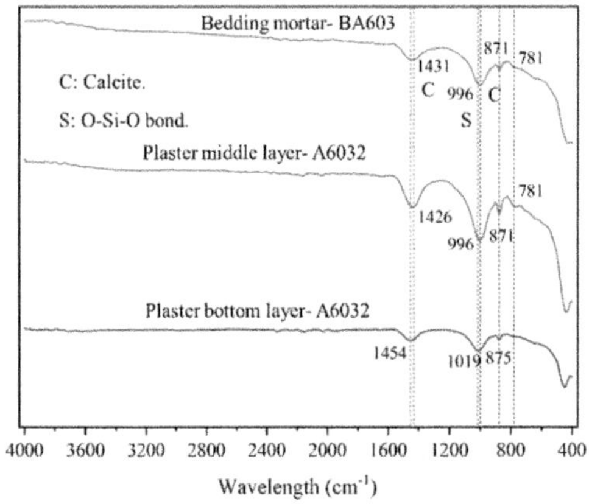

(a.)

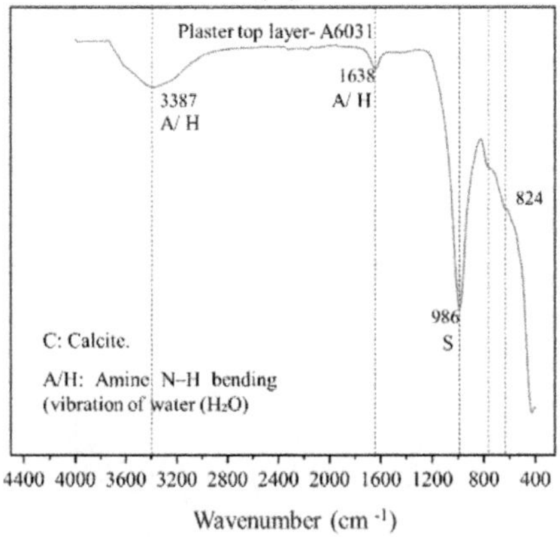

(b.)

Fig. 6. FTIR spectra (a.) for the bedding mortar sample (BA603) and plaster middle (A6032) and bottom layer (A6031); (b.) for top layer (A6033)

Thermograph for middle layer A6032, showed a prominent peak indicating carbonate decomposition at approximately 700 °C (Fig. 7). A peak around 100 °C shows higher free water content. Endothermic shoulder at 740 °C can also be suggestive of decomposition of magnesium carbonate. For top layer the prominent peak around 100 °C corresponds to high free water content whereas the broad peak between 300 °C and 600 °C can be caused due to presence of organic additives or $Mg(OH)_2$. This needs to be further confirmed with tests results of XRD and SEM analysis.

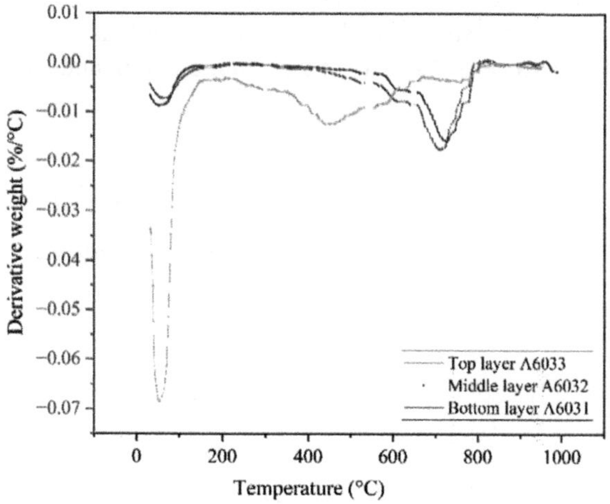

Fig. 7. TGA spectra for the three layers of plaster sample PA603.

For the top and middle layer of the plaster -A6033 and A6032 respectively, the TGA results indicated the presence of hydraulic components in air lime-pozzolan mortars. This was evidenced by a shoulder observed on the DTG (derivative thermogravimetry) curve, which corresponds to the dehydration of calcium silicate hydrates formed during pozzolanic reactions. This weak shoulder appeared in the temperature range of 600 °C, suggesting that reactions were occurring at this temperature. The CO_2/H_2O ratio was determined by comparing the weight loss of carbonates above 600 °C to the weight loss of hydraulic water between 200–600 °C, providing an assessment of the binder's hydraulic indices [7]. All samples indicate high hydraulicity and presence of pozzolanic material as shown in Table 2.

Table 2. Weight loss calculated at established temperature range with CO_2: bound H_2O

Sample ID	<120 °C	120–200 °C	200–600 °C	>600 °C	CO_2: bound H_2O
PA6031	2.75	0.51	4.06	9.99	2.46
PA6032	2.97	0.44	3.18	10.21	3.21
PA6033	4.55	0.49	3.02	1.12	0.37
BA603	2.34	0.66	3.83	8	2.09

The XRD spectrum results for the top layer (A6031) aligned with the observations from the TGA thermogram and FTIR spectrum, confirming the absence of calcite in its composition. The analysis revealed the presence of sepiolite, C–S–H phases, and a predominant quartz phase. Additionally, traces of M–S–H were identified, suggesting the possible use of dolomitic lime as a binder. It is also likely that both C–S–H and M–S–H phases exist in an amorphous form, not fully detectable by XRD. In contrast, the middle and bottom layers exhibited dominant peaks of quartz and calcite, with biotite and anorthite commonly present across all samples, confirming FTIR and TGA results. These layers also displayed C–S–H phases confirming the TGA peak. The bedding mortar samples produced a spectrum closely resembling that of the bottom layer but also indicated the presence of M–S–H and ettringite compounds not observed in the earlier layers (Fig. 8).

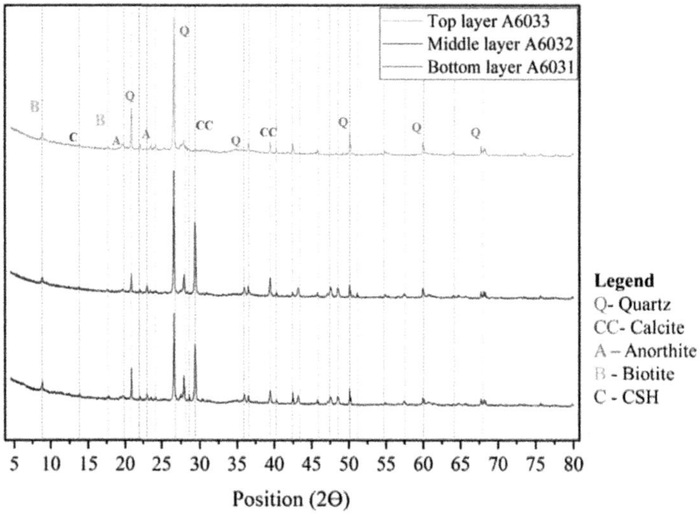

Fig. 8. XRD pattern o for the three layers of plaster sample PA603

The binder aggregate ratio derived from the sample is nearly 1:1.20 which means the sample is having almost 50% binder in the mixture (Table 3).

Table 3. Binder to Aggregate ratio for tested samples

Sr. No.	Sample code (Place name/Site name/ Sample no)	Ratio (Binder: Aggregate)	Approx ratio
1	JAI/Tanka/Mortar/A6-03	1:1.03	1:1
2	JAI/Tanka/Plaster/A6-03-01 (Bottom)	1:1.37	1:1.5
3	JAI/Tanka/Plaster/A6-03-02 (Middle)	1:1.44	1:1.5
4	JAI/Tanka/Plaster/A6-03-03 (Top)	1:0.45	1:0.5

Particle size distribution through sieve analysis reveals notable differences in gradation among the plaster layers and the bedding mortar sample. Top layer (A6033) exhibits the highest fineness, with over 80% of the particles finer than 300 μm, indicating a well-graded and fine aggregate mix, possibly contributing to a denser microstructure. In contrast, bottom layer (A6031) displays a coarser profile, with a more gradual increase in fineness across the sieve range, especially between 150 μm and 1.18 mm, suggesting a higher proportion of coarser particles. When compared to the middle layer (A6032), the bottom layer consistently shows lower % passing at equivalent sieve sizes, confirming that it is relatively coarser (Fig. 9). Bedding mortar A603 showed very similar particle distribution as the bottom layer. These variations in gradation suggest differences in the original mortar mix design, application layer, or function (e.g., finish vs. core plaster), potentially influencing the strength, porosity, and water resistance of each layer.

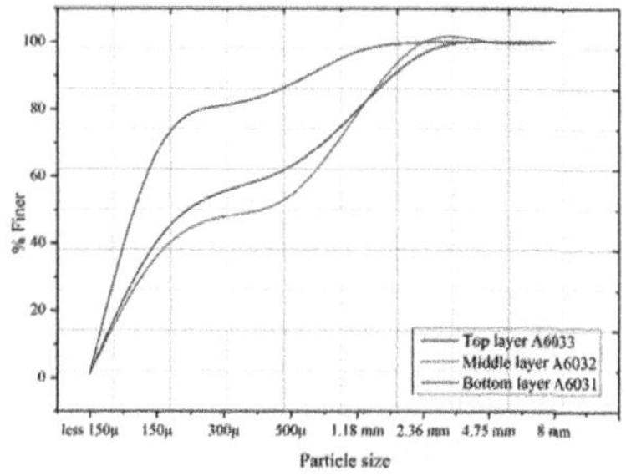

Fig. 9. Particle size distribution graph for the three layers of the plaster sample PA603

4 Conclusion

This study demonstrates that the historic Tanka plaster system at Jaigarh Fort was a product of informed and intentional material engineering. Through visual, mineralogical, and physical analyses, it was established that traditional masons employed a multi-layered lime plaster strategy in order to obtain the functional demands of a water-retaining structure. The use of pozzolanic additives, dolomitic lime, and organic fibers such as straw contributed to improved hydraulicity, crack resistance, and long-term performance. The results also strengthen the hypothesis that informed decisions were employed for grading, additives and application technique. Further SEM, EDS analysis can help strengthen the findings and chemical analysis can be carried out to identify the amine based organic additive detected through FTIR.

These findings highlight a nuanced understanding of lime technology in historic construction, where breathability and water resistance were selectively achieved through material composition and layering. For conservation practice, this underscores the need to base repair strategies on a deep understanding of original materials, ensuring compatibility in both physical behavior and long-term performance especially in structures exposed to water.

Acknowledgment. This research was supported by the Center for Heritage Conservation (CHC), CEPT Research and Development Foundation (CRDF), and the Jaigarh Public Charitable Trust (JPCT), under their MoU for the Sustainable Conservation of the Jaigarh Fort Precinct. We gratefully acknowledge both organisations for facilitating access to the site and samples for analysis. Their collaboration and commitment in enabling site-based research for heritage conservation have been valuable to this study.

References

1. Artioli, G., Secco, M., Addis, A.: The vitruvian legacy: mortars and binders before and after the Roman world. In: European Mineralogical Union Notes in Mineralogy. Mineralogical Society, pp. 151–202 (2019)
2. Sarda, K., et al.: Hunnarshala Foundation for Building Technology and Innovations Gaurav Dinodia | Content Structure (2020)
3. Lanas, J., et al.: Mechanical properties of natural hydraulic lime-based mortars. Cem. Concr. Res. **34**(12), 2191–2201 (2004)
4. Bhattacharya, S.: Traditional water harvesting structures and sustainable water management in India: a socio-hydrological review. Int. Lett. Nat. Sci. **37**, 30–38 (2015). https://doi.org/10.56431/p-a84p4z
5. More, H., More, H.B.: A review of Indian traditional method of rain water harvesting. Int. J. Innov. Res. Sci. Eng. Technol. [Preprint] (2021)
6. CyArk: Water Systems of Jaigarh. Tapestry (2024). https://tapestry.cyark.org/content/jaigarh-fort
7. Medeghini, L., et al.: The secret of ancient Roman hydraulic mortar: the lesson learnt from the past for future cements. Cement Concr. Compos. **148**, 105484 (2024). https://doi.org/10.1016/j.cemconcomp.2024.105484

Ionic Environment's Change of Cement Paste Blended with Slag and Their Effects on Dispersing Property of PCE

Bin Li[1,2], Ling Wang[1,2], Zhendi Wang[1,2,2](✉), Mingwei Wang[1,2], and Fei Wang[1,2]

[1] State Key Laboratory of Green Building Materials, China Building Materials Academy, Beijing 100024, China
wzhendi@163.com

[2] Tianjin Chengjian University, Tianjin 300384, China

Abstract. Ground granulated blast furnace slag (GGBS) is the main by-product of steel industry and has been widely utilized as mineral addition in concrete. Due to the difference of chemical composition and hydration reactivity of slag and cement particles, the ionic environment of cement paste changed a lot and influenced the dispersion performance of PCEs. For clarifying the ionic environment change and their effects on the dispersing ability of PCEs. The ICP-OES was used for clarifying ionic environment of cement paste, and their effects on dispersing performance of PCEs were determined by flowability and rheological experiments. Test results showed that ionic environment influenced the surface charge of slag, affected adsorption behavior and conformation of PCEs inducing the different dispersing properties. PCE with longer side chain length or smaller acid-ether ratio exhibited better dispersion performance.

Keywords: Ionic environment · PCE · Surface charge · Adsorption behavior · Conformation · Dispersing properties

1 Introduction

Slag as a mineral addition has been widely used during the concrete preparation. Through substituting cement to lower the carbon emission and the amount of clinker promotes the final products' eco-efficiency [1, 2]. The influence of slag on the hydration process and structural features of hydrates have been extensively investigated [3–6]. Pavía S. et al. [7] have pointed out that slag provides greater mechanical and durability properties as it significantly decreases both the content and the size of $Ca(OH)_2$ crystals in the aggregate-paste interface, which makes the microstructure of the transition zone aggregate/binder dense and strong. H.N. Atahan [8] illustrated the introduced slag effectively improved the sulfate resistance of mortars due to the pozzolanic reaction between calcium hydroxide and the slag, which reduces calcium hydroxide and improves the sulfate resistance. Lin Li et al. [9] found that the slag significantly improved the impermeability and air voids distribution of cement mortars. The reduction of $Ca(OH)_2$, gentler hydration rate,

M. Posani and S. Singh (Eds.): RYS 2025, RILEM Bookseries 67, pp. 141–156, 2026.
https://doi.org/10.1007/978-3-032-14652-6_12

and filling effect of fine slag particles induced the improvement of impermeability of cement mortars. Previous studies have demonstrated that slag has a positive impact on the hydration, mechanical properties and durability of cement-based materials.

However, as a kind of mineral addition, slag's chemical composition and hydration activity are different from cement. Therefore, the effect of slag on the ionic environment of cement pastes and their effects on the compatibility of PCE should be significantly considered.

The effect of mineral additions on the workability of cement paste are determined by its physical and chemical properties. When considering chemical properties, the ionic environment of cement paste blended with slag could not be ignored. The dispersing performance and mechanism of PCEs was sensitive to the ionic environment of cement paste solution [10–14]. Yamada et al. [15] have proved that the sulfate ions in pore solution would enhance the competition adsorption between PCE and sulfate ions. Moreover, while ions concentration was high in the liquid phase, the spatial size of PCE would be curled and the dispersion performance of PCE would be seriously affected. Hanehara S. [16] also proposed that the compatibility between PCEs and cement particles was affected by the adsorption behavior of PCE on the cement surface and the amount of alkaline sulfates in cement paste pore solution. Hongwei Tian et al. [14] found that Ca^{2+} ions in the interstitial solution induced the deleterious bridging effect among different PCE polymers and impacted PCEs' dispersing performance. These results indicated that the ionic type and concentration of the cement paste solution directly influenced the PCE's dispersing performance.

Aiming to understand the role of slag on the ionic environment of cement pastes and the impacts of ionic environment change on the compatibility between slag/cement and PCEs. The ionic environment of cement pastes blended with slag was investigated, and dispersion performance of PCEs with different molecular parameters were also studied. Furthermore, adsorption behaviors and conformation of PCE with various structural parameters in different ionic environments were quantified to illustrate the influence of the ionic environmental change on the dispersing mechanism of PCEs. The results from this study can serve as guidance for the practical application of slag in cementitious materials.

2 Materials and Test Method

2.1 Materials for Synthesizing PCEs

2–methyl allyl polyoxyenthylene (HPEG) with molecular weight 1000 g/mol, 3000 g/mol, and 5000 g/mol (Industrial-grade, Liaoning Oxiranchem Co., Ltd. of China); hydrogen peroxide (H_2O_2) (Beijing Chemical Works); acrylic acid (AA) (analytical grade, Aladdin); thioglycolic acid (TA) (analytical grade, Aladdin); sodium hydroxide (NaOH) (analytical grade, Aladdin); ascorbic acid (Vc) (analytical grade, Aladdin).

2.2 Cement and Slag

The chemical compositions of cement/slag are shown in Table 1. The cement used has a specific surface area of 341 m^2kg^{-1} and a strength grade of 42.5.

The slag in this work was commercially available, and the Nitrogen BET surface area is 3955 m^2kg^{-1}. And its activity index is S95 according to GB/T 18046-2017.

Table 1. Chemical composition of cement/slag

Sample	SiO_2	Al_2O_3	Fe_2O_3	CaO	MgO	SO_3	Na_2O_{eq}	Loss
Cement	25.10	6.38	4.19	54.87	2.61	2.66	0.56	2.18
slag	33.04	18.31	0.49	32.21	11.21	2.93	1.09	1.00

2.3 The Preparation of PCEs

The preparation of PCEs was carried out through free radical polymerization as shown in Fig. 1.

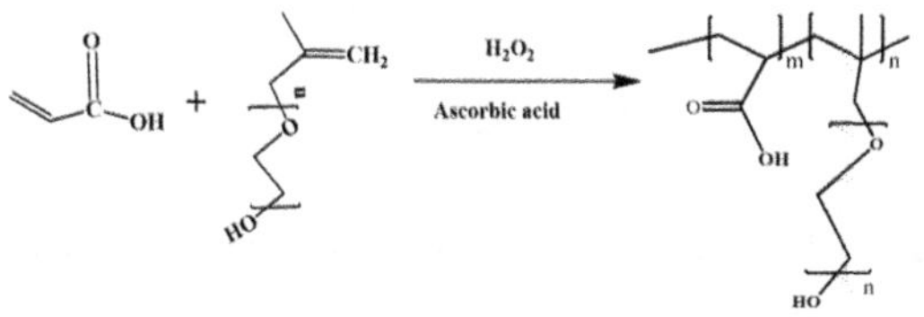

Fig. 1. The schema of the synthesis of PCEs

The chemical composition of PCEs was shown in Table 2. A certain amount of macromonomer HPEG was placed in a three-necked flask, and a certain amount of DI water was poured into the flask to form an HPEG solution. When the system was heated to 60 °C, the initiator H_2O_2 was added. Next, the acrylic acid solution AA and ascorbic acid (Vc)/thioglycolic acid (TA) solution were added dropwise into the HPEG solution for 3 h and 3.5 h, respectively. The system was reserved for a heat preservation reaction that lasted for 6.5 h. When the reaction was completed, the polymer solution's pH was adjusted to 7.0 by 30 wt.% NaOH.

Table 2. Chemical compositions of PCEs

Sample	$[AA]_0$:[HPEG)$]_0$	HPEG's molecular weight (g/mol)	H_2O_2 (wight of AA + HPEG, %)	$[Vc]_0:[H_2O_2]_0$	$[TA]_0:[AA]_0$
PCE-1	2:1	3000	0.5	1:5.8	0.015:1
PCE-2	4:1	3000	0.5	1:5.8	0.015:1
PCE-3	6:1	3000	0.5	1:5.8	0.015:1
PCE-4	4:1	1000	0.5	1:5.8	0.015:1
PCE-5	4:1	5000	0.5	1:5.8	0.015:1

2.4 Characterization of PCEs

For confirming the structure of synthesized PCEs, FT-IR, and GPC were used.

Figure 2 shows the FT-IR curves of PCEs, vibration absorption at 1646.20 cm^{-1} (corresponding to the absorption peaks of -C = C- group) almost disappeared, indicating that the macromonomer HPEG has copolymerized with AA. Meanwhile, the absorption peaks observed at 1109.87 cm^{-1} and 1725.04 cm^{-1} correspond to -C-O-C- and -C = O, respectively.

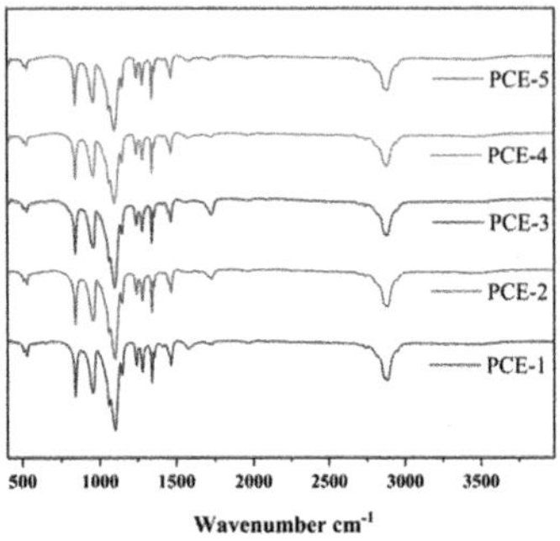

Fig. 2. The FT IR spectrum of PCEs with various parameters

The number-average molecular weight (Mn), weight-average molecular weight (Mw) and dispersity of PCEs were tested by GPC. And the test results were shown in Table 3, the number-average molecular weights (Mn) of the synthesized PCEs are almost the same, around 40000 g/mol. The PDIs of the PCEs are in the range of 1.47~1.57.

Table 3. Molecular weights information of PCEs

Sample	Mn (g/mol)	Mw (g/mol)	Dispersity (Mw/Mn)
PCE-1	18000	34750	1.92
PCE-2	25500	40200	1.57
PCE-3	32700	65500	2.00
PCE-4	34950	74200	2.13
PCE-5	36000	61000	1.69

The above findings demonstrated that the PCEs with different structural parameters were successfully synthesized.

2.5 Test Method of the Flowability and Rheology

Flowability Measurement: A certain amount of slag was used to replace cement particles at a replacement ratio of 20%-80%, to prepare cement blended slag. Then, the water containing 0.2% PCE (by weight of binder, calculated based on the solid content) was poured into the binder (the water/binder ratio was 0.29) and the mixture was stirred at a low speed of (140 ± 5) rpm for 2 min, then rested for 15 s and stirred at a high speed of (285 ± 10) rpm for 2 min. Next, the paste was poured into a cone (60 mm height, 36 mm top diameter, and 60 mm bottom diameter) on a glass plate, followed by vertically upwards removing the cone. The spread paste diameter was measured after 30 s, and then the second measurement was vertical to the first. The final paste flowability was the average of the two measurements, as depicted in the figure below:

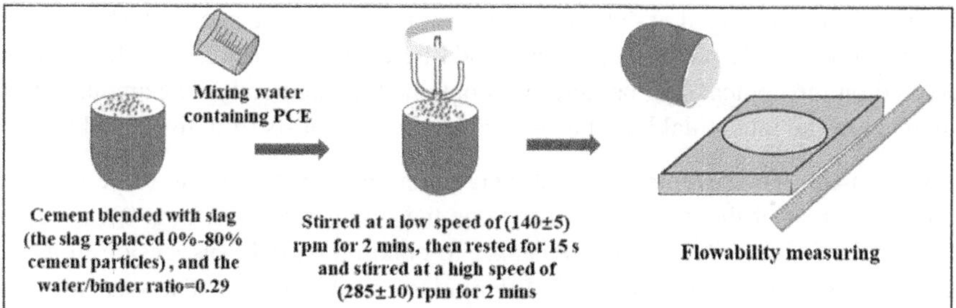

Schematic diagram of the measurement of flowability of paste

Rheology: Rheological properties of pastes were tested by Viskom at XL rheometer (Schleibinger Co., Germany). And the Bingham model (Eq. 1) was used to calculate the rheological parameters of pastes [17–19]. The preparation of cement paste was the same as the flowability measurement.

$$\tau = \tau_0 + \mu_p \dot{\gamma} \tag{1}$$

where τ, τ_0, η, and $\dot{\gamma}$ represent the shear stress (Pa), yield stress (Pa), plastic viscosity (Pa·s), and shear rate (s^{-1}), respectively.

ICP Measurement: The ionic environment of cement past solution was tested via the inductively coupled plasma (ICP-OES) method. A certain amount of cement and slag was stirred with water in a beaker for 4 min after the initial contact between cement and water. Then, the suspension was filled into a centrifuge tube and centrifuged in the centrifuge with a speed of 4000 rpm for 3 min. The obtained supernatant was filtered by 0.22 μm film and diluted with deionized water, and a certain amount of HNO_3 was added to adjust the pH value of the solution to 1 before ICP-OES measurement, as depicted in the figure below:

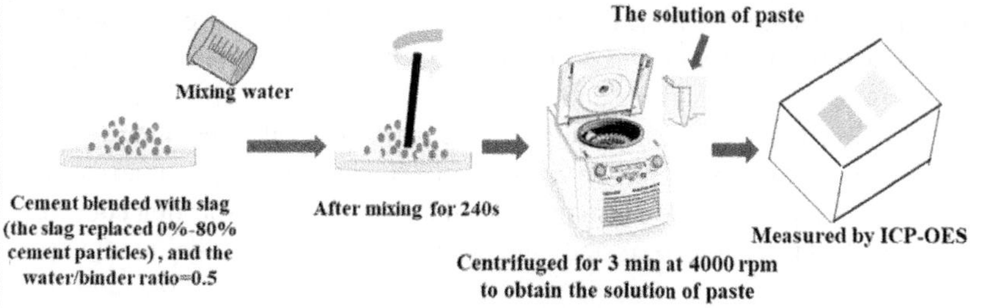

Schematic diagram of the measurement of ionic environment of paste

DLS Measurement: The Zetasizer Nano ZS90 (Malvern PANalytical, China) was used to measure the hydrodynamic radius of PCEs in different solutions. The concentration of PCEs was diluted to 1000 mg/L and dispersed for 10 min by ultrasound before measurements.

TOC Measurement: Adsorption measurements were performed to describe the adsorption behavior of PCEs on cement /slag particles via the TOC analyzer (Analytik Jena AG, German). During the experiment, the w/b was fixed as 4, immediately after mixing, the paste was centrifuged at 4000 rpm for 2 min to obtain the supernatant liquid. The supernatant was collected using a syringe and filtered through a 0.22 μm filter. The PCE adsorbed on the cement/slag particles was obtained by subtracting the amount of PCE remaining in the interstitial liquid measured by the TOC tests from the added PCE.

Zeta Potential Measurement: Batch zeta potential experiments were performed to clarify the effect of the type of electrolyte on the zeta potential of slag. 0.5 g slag was mixed with 50 mL of DI water in a beaker to obtain a particle suspension. Ions were then added to form suspensions with different ionic environment. The suspension was then agitated for 2 min to achieve complete mixing. Finally, the bottles suspension was placed in a zeta meter for testing.

3 Results

3.1 The Effect of Slag on the Ionic Environment of Cement Paste

Based on previous studies, it can be found that compared with cement, the hydration reactivity of slag is lower [20, 21]. The introduced slag would change the ionic concentration in the liquid of cement paste. ICP test results show a similar phenomenon. As the w/b ratio was fixed at 0.50, in the liquid phase, the ionic concentrations greatly varied, as shown in Fig. 3. When the mass ratio of slag increases from 0% to 80%, the concentration of Na^+, K^+, and SO_4^{2-} in the solution of blended pastes all decrease accordingly, and the ionic strength also decreased. However, the concentration of Ca^{2+} did not decrease as the mass ratio of slag increased.

For further clarifying this phenomenon, the ionic environment of cement paste solutions without slag in different water/cement ratio was measured. As shown in Table 4, it can be seen that with the increase of water/cement ratio, the dissolving Al^{3+}, K^+, Na^+ decreased accordingly. But the change of the dissolving concentration of Ca^{2+} was not obviously, indicating that the dissolving degree of calcium phase improved with the water/cement ratio increasing. It can be deduced that the difference in water-holding capacity between cement particle and slag powder affected the dissolution behavior of cement particle and the concentration of leached ions. Recent research has highlighted that the dynamic evolution of ionic strength and pH in the liquid phase not only affects early-age reaction pathways but also modifies the adsorption–desorption equilibrium of polycarboxylate superplasticizers and other polymers at the solid–liquid interface.

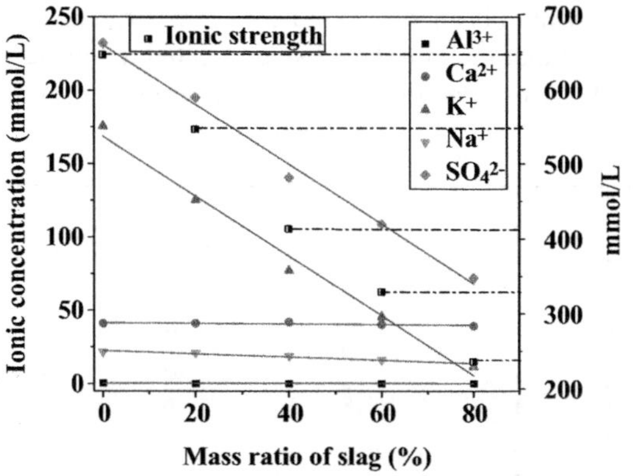

Fig. 3. The effect of mass ratio of slag on the concentration of Al^{3+}, Ca^{2+}, Na^{+}, K^{+}, SO_4^{2-} and the ionic strength of dissolving solution of blended pastes

Table 4. Ionic concentrations of the solution of cement paste without slag in the different water-cement ratios (mmol/L)

Sample	W/C ratio	Al^{3+}	Ca^{2+}	K^{+}	Na^{+}	SO_4^{2-}
1	0.5	0.39	40.89	175.30	21.48	231.88
2	1.0	0.38	77.20	76.23	12.06	216.32
3	1.5	0.28	72.14	48.18	6.74	181.60
4	2.0	0.21	68.88	33.34	5.93	153.01
5	2.5	0.08	65.67	27.04	4.48	137.05

3.2 The Effect of Introduced Slag on the Dispersion Performance of PCEs

For clarifying the effect of slag on the dispersion performance of PCE, the flowability of cement pastes in the presence of PCEs with different mass ratios of slag was investigated. During the experiment, the water/binder ratio was 0.29, and the dosage of PCE was 0.20% by the weight of binder.

Figure 4 shows flowability of cement pastes and the corresponding flowability increasing rate. Comparing with cement pastes without slag, cement pastes blended with slag exhibited larger flowability, with mass ratio of slag increasing, the flowability of cement pastes plasticized by PCE-1, PCE-2, PCE-4 and PCE-5 all showed an increasing trend. However, their flowability increasing rates were different, the order of the flowability increasing rates of cement pastes containing PCEs with different parameters was following: PCE-4>PCE-1>PCE-2>PCE-5>PCE-3. For instance, when the mass ratio of slag was 20%, 40%, 60% and 80%, respectively. Compared with cement paste without slag, flowability of cement pastes containing PCE-4 increased 5.26%, 23.68%, 31.58%, and 35.53%, respectively. However, under the same experimental conditions, the flowability of cement paste containing PCE-3 (with higher acid-ether ratio) showed a contrary tendency. As the mass ratio of slag was 20%, 40%, 60%, and 80%, respectively.

The corresponding flowability increased −1.79%, 0%, 0%, and 0.714%, respectively. These results indicated that the introduced slag affected the dispersing performance of PCEs and PCEs with various parameters exhibit different impact degrees.

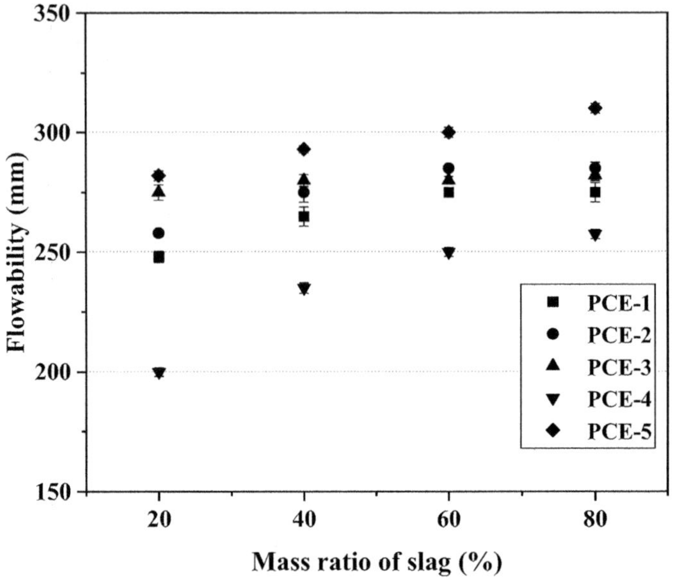

Fig. 4. The flowability of cement pastes with different mass ratio of slag in the presence of PCEs with different parameters

3.3 The Effect of Slag on the Rheological Property of Cement Pastes Containing PCEs

To further study the effect of slag on the workability of cement pastes, the rheological behavior was discussed and the results are shown in Fig. 5 and Fig. 6.

Figure 5 displayed the rheological curves of cement pastes with slag in the presence of PCEs, which were fitted by the Bingham model. The correlation coefficients (R^2) were larger than 0.97, indicating that the Bingham model fits their rheological curves well. According to the Bingham model, the yield stress and plastic viscosity can be calculated and the results were shown in Fig. 6. It can be seen that with the mass ratio of slag increasing, the yield stress and plastic viscosity of cement paste containing PCE-1, PCE-2, PCE-4 and PCE-5, decreased significantly. For example, the cement pastes without slag containing PCE-4 has the largest yield stress 16.57 Pa among all samples, corresponding plastic viscosity reached 0.68 Pa·s. With the increase of mass ratio of slag from 0 to 80 wt%, the cement paste's yield stress and plastic viscosity gradually decreased to 7.73 Pa and 0.19 Pa·s, which decreased by 53.3% and 72.0%, respectively.

However, for PCE-3, the yield stress and plastic viscosity of cement paste increased as the mass ratio of slag was below 60%, and when the mass of slag was at 40%, its yield stress reached 3.92 Pa, which increased by 5.7% compared with reference group without slag.

In the presence of PCEs, the increase of slag mass ratio decreased the yield stress and plastic viscosity of the cement pastes containing PCEs, besides PCE-3. As an important effect, the change in the adsorption behavior of PCE may explain this phenomenon [22]. With the increasing slag mass ratio, the dissolution of slag affects the adsorption behavior of PCEs on the surface of cement particles, resulting in a great difference in the rheological behavior of the system. The adsorption behavior and microstructure of PCE would be discussed in the following sections.

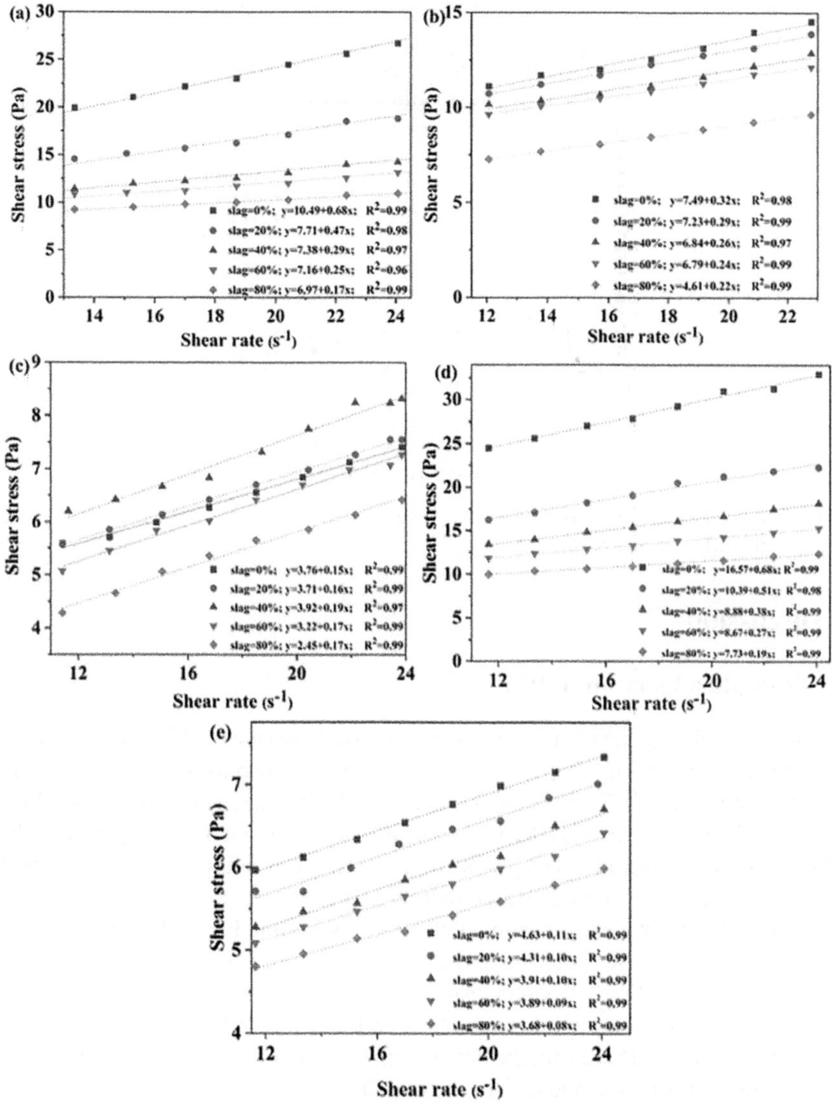

Fig. 5. The rheological curves of cement pastes with different slag dosages: (a) PCE-1; (b) PCE-2; (c) PCE-3; (d) PCE-4 and (e) PCE-5

(a) PCE-1; (b) PCE-2; (c) PCE-3; (d) PCE-4 and (e) PCE-5

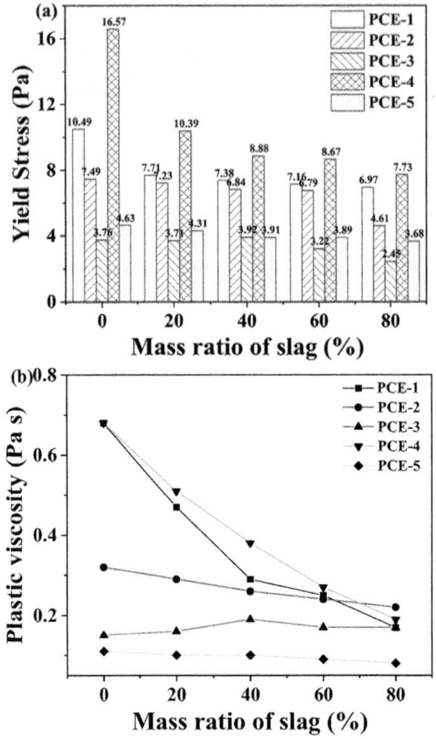

Fig. 6. The yield stress and plastic viscosity of cement pastes blended with slag in the presence of PCEs with different parameters

4 Discussion

4.1 Adsorption Behavior of PCE

The influence of slag mass ratio on the adsorption behavior of PCEs was tested, and results were shown in Fig. 7. With the mass ratio of slag increasing, compared with the adsorption amount of PCEs in cement paste, the adsorption amounts of PCE-1, PCE-2, PCE-4 and PCE-5 in cement pastes blended with slag increased obviously, especially for PCE-4. When the mass ratio of slag was 20%, its adsorption amount increased by 56.6%, compared with that of cement paste without slag. According to the ionic concentration of the solution of cement pastes blended with slag, as shown in Fig. 3. It can be found that with the mass ratio of slag increasing the concentration of Ca^{2+} was unchanged, while the concentration of $SO_4{}^{2-}$ was decreasing accordingly. Li et al. and Yamad have proved that the competitive adsorption between PCE and $SO_4{}^{2-}$ would induce poor dispersing performance of PCE. Therefore, the decline of $SO_4{}^{2-}$ concentration probably hindered the competitive adsorption and caused a larger amount of adsorption.

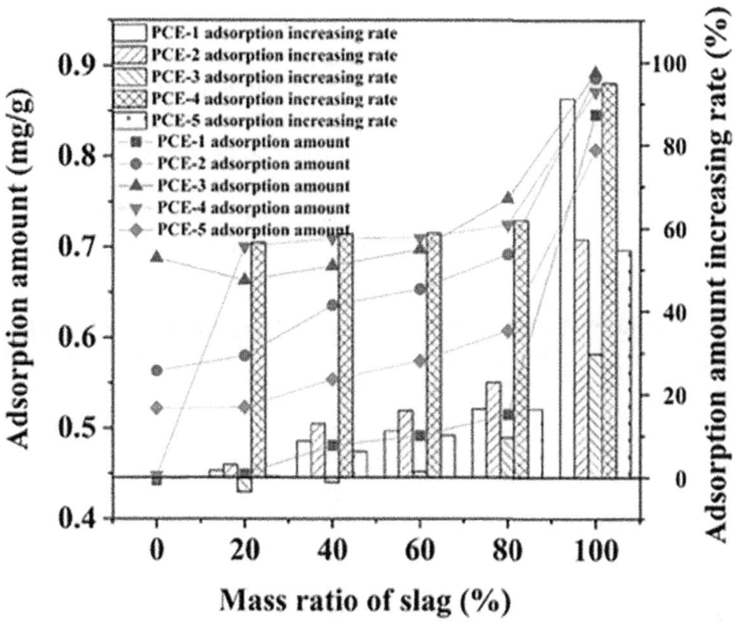

Fig. 7. The adsorption amount and adsorption amount increasing rate of PCEs with various parameters in cement pastes blended with slag

However, the PCE-3 exhibited contract results, when the mass ratio of slag was below 60%, its adsorption amount was less than that of cement paste without slag, despite a smaller concentration of SO_4^{2-} in cement paste solution. Thus, for further clarifying the adsorption behaviors of PCEs in cement paste blended with slag, the zeta potential of slag in different electrolyte solutions according to the ionic environment of cement paste solution was investigated.

As shown in Fig. 8, the zeta potential of slag in the DI water was −1.198 mv. However, when the slag was added in various solutions containing different electrolytes, its zeta potentials exhibited various values.

According to previous studies [23, 24], the zeta potential of suspensions depends on the kind of contact solution due to the adsorption of ions on the surface of solid. Figure 8a and Fig. 8b show the zeta potentials of slag as a function of $Ca(OH)_2$ concentration and Na_2SO_4 concentration. With the concentration of $Ca(OH)_2$ increasing, the zeta potential value of slag increased obviously (Fig. 8a), indicating that the surface charge of slag turned from negative to positive. And the zeta potential value of slag reached 12.88 mv when the calcium hydroxide was at 20 mmol/L. In contrast, the slag exhibits negative zeta potential values and the zeta potential value decreased significantly as the concentration of Na_2SO_4 increased (Fig. 8b).

Moreover, based on the ionic environment of the solution of cement pastes blended with slag (as shown in Fig. 3), the effect of Na^+ and K^+ on the zeta potential of slag could not be ignored. It can be found that the Na^+ concentration range was from 4.48 mmol/L to 21.48 mmol/L, and with the concentration of NaCl increasing, the zeta potential of slag changed accordingly. The maximum zeta potential of slag reached 7.85 mv, as the concentration NaCl was 15 mmol/L. As for K^+, with the concentration of KCl increasing, the zeta potential exhibited a similar trend as NaCl, the maximum zeta potential value reached 49.50 mv, when the KCl concentration was at 100 mmol/L.

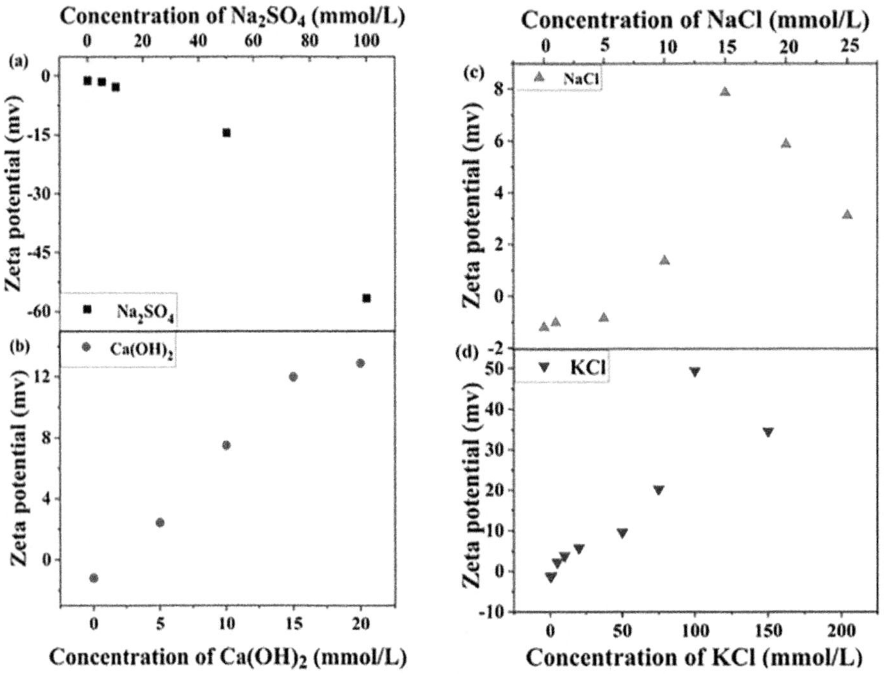

Fig. 8. Zeta potential values of slag in different electrolytes

Figure 8 shows, as noted above, that the ionic concentration of cement paste solution considerably influenced the zeta potential of slag. On the one hand, compared with Na^+, the concentration of K^+ was much larger (concentration reached 175.30 mmol/L) in the solution of cement paste with slag, and Ca^{2+} showed the strongest impact on the zeta potential of slag among cations. On the other hand, the effect of SO_4^{2-} could not be ignored, which significantly decreased the zeta potential of slag.

PCE is a kind of anionic surfactant, which is easier to adsorb on the particle surface with positive charges than negative charges. Therefore, the surface charge of slag would play an important role in the adsorption behavior of PCE. According to the zeta potential results, the zeta potential of slag depended on the type of electrolytes and their concentrations. And the slag surface charge would change with the ionic environment of the solution of cement paste with slag, which would induce different adsorption behavior of PCEs and lead to various adsorption amounts. For instance, in pure cement paste solution, it had the highest concentration of cations (Na^{+}, K^{+} and Ca^{2+}), the $SO_4{}^{2-}$ concentration was also the highest. In the solution of cement paste blended with slag, with the content of slag increasing, both $SO_4{}^{2-}$ and cations concentrations decreased significantly, but Ca^{2+} (which has the most significant effect on the surface charge of slag) did not change a lot. This result indicates that the surface charge of slag increased as the content of slag increases and induce to a higher adsorption amount.

4.2 Conformation of PCE

The dispersing mechanism of PCE has been extensively studied. When PCE is adsorbed by particles, the long side chains of PCE can provide a steric hindrance effect and disperse the particles [25], releasing the free water wrapped by cement particles and bringing the lubrication effect [26]. Therefore, PCE can effectively improve the workability of cement pastes. It should be mentioned that the dispersing performance of PCE cannot simply be attributed to the adsorption behavior. It is essential to clarify the conformation of PCE in the solution of cement paste.

The DLS was conducted to evaluate the hydrodynamic radius of PCEs in various paste solutions. Previous researchers have illustrated that anionic surfactant is more sensitive to counterions [27, 28], especially with higher charge numbers. According to the ionic environment of cement paste blended with slag, as shown in Fig. 3, with the mass ratio of slag increasing, the concentration of Ca^{2+} almost did not change. Thus, the radius diameters of PCEs in aqueous with different Ca^{2+} concentrations were studied. Figure 9 shows hydrodynamic radius of PCEs with various parameters in different ionic environments. It can be found that the hydrodynamic radius of PCEs exhibited various trends as the concentration of Ca^{2+} increasing. As for PCE-1 (Fig. 9a), the hydrodynamic radius values of the single PCE-1 molecule remained almost unchanged, but with Ca^{2+} further increasing, small peaks at 50.74 nm, 164.18 nm, 190.14 nm and 295.31 nm were observed. For PCE-2 (Fig. 9b), a series of small peaks at 91.28 nm, 141.77 nm, 220.19 nm and 295.31 nm could also be found. Moreover, more small peaks (at 50.75 nm, 91.28 nm, 105.71 nm, 255.10 nm, 458.67 nm and 511.67 nm) were found for PCE-3, as shown in Fig. 9c. And this phenomenon could be attributed to the bridge effect of Ca^{2+}, which crosslinked different PCE molecules and induced higher Rh values.

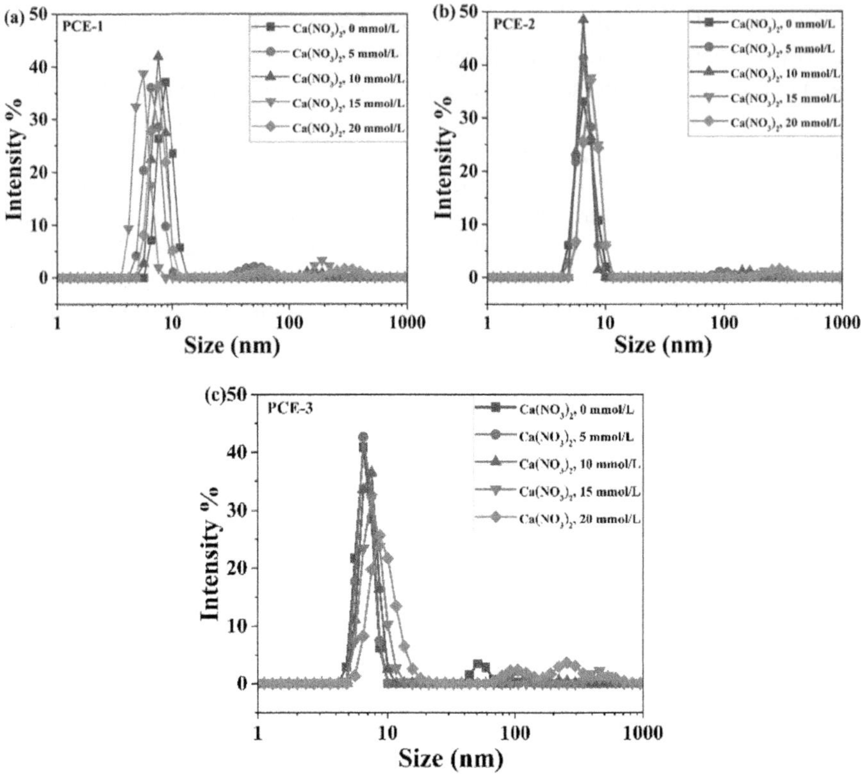

Fig. 9. The hydrodynamic diameter of PCEs in the different ionic environment

The PCE-1, PCE-2 and PCE-3 have different acid-ether ratios (Table 2). Their Rh values were compared to clarify the influence of PCE's charge density on the crosslinking phenomenon. It can be found that PCEs with higher acid-ether ratios generated smaller peaks and corresponded to stronger intensity despite their single molecular diameters almost unchanged. This result indicated that the PCE with a higher acid-ether ratio (PCE-3) was easier to crosslink in the presence of Ca^{2+}. And it would influence the adsorption behavior of PCEs, which explains why the adsorption amount decreases for PCE-3. Due to the crosslinked by Ca^{2+}, fewer PCE-3 molecules were able to adsorb on the cement/slag surface and this induced a lower adsorption amount for PCE-3.

5 Conclusions

The ionic concentration's change had an effect on the adsorption behavior, and conformation of PCEs and changed the surface charge of slag, which induced PCEs with various dispersing performances.

(1) The introduced slag changed the ionic concentration in the solution of cement paste. As the mass ratio of slag increases, the concentration of Na^+, K^+, and $SO_4{}^{2-}$ in the blended pastes all decrease accordingly. However, the concentration of Ca^{2+} almost not changed.
(2) The introduced slag influenced the adsorption behavior of PCEs. With the mass ratio of slag increasing, the adsorption amount of PCE on cement/slag particles increased for PCE-1, PCE-2, PCE-4 and PCE-5. However, the PCE-3 exhibited contrast results.
(3) The slag surface charge would change with the ionic environment of the solution of cement paste blended with slag, which would induce different adsorption behavior of PCEs and lead to various adsorption amounts.
(4) PCE with a higher acid-ether ratio was easier to be crosslinked in the presence of Ca^{2+}, which might influence the adsorption behavior of PCEs.

References

1. Lothenbach, B., Scrivener, K., Hooton, R.D.: Supplementary cementitious materials. Cem. Concr. Res. **41**(12), 1244–1256 (2011)
2. Chopperla, K.S.T., Ideker, J.H.: Using electrical resistivity to determine the efficiency of supplementary cementitious materials to prevent alkali-silica reaction in concrete. Cem. Concr. Compos. **125** (2022)
3. Miao, Z., Wu, J., Zhang, Y., Zhao, X., et al.: Chemical characterizations of different sized mineral-rich particles in fine slag from entrained-flow gasification. Adv. Powder Technol. **31**(9), 3715–3723 (2020)
4. Fayaz, M., Krishnaiah, R.V., Raju, K.V.B., et al.: Experimental study on mechanical properties of concrete using mineral admixtures. Mater. Today: Proc. (2023)
5. Liang, W., Lin, M., Yin, W., et al.: Experimental study and simulation of workability and compressive performance of SCC with high-volume mineral admixtures. Constr. Build. Mater. **409** (2023)
6. Zhao, G., Yan, D., Ren, G., et al.: Influence of ground granulated blast furnace slag on the dispersivity and mechanical property of dispersive soil. Constr. Build. Mater. **409** (2023)
7. Pavía, S., Condren, E.: Study of the durability of OPC versus GGBS concrete on exposure to silage effluent. J. Mater. Civ. Eng. **20**, 313–320 (2008)
8. Atahan, H.N., Dikme, D.: Use of mineral admixtures for enhanced resistance against sulfate attack. Constr. Build. Mater. **25**(8), 3450–3457 (2011)
9. Li, L., Ye, T., Li, Y., et al.: Improvement of microporous structure and impermeability of cement mortars using fly ash and blast furnace slag under low curing pressures. Constr. Build. Mater. **401** (2023)
10. Han, S., Plank, J.: Mechanistic study on the effect of sulfate ions on polycarboxylate superplasticisers in cement. Adv. Cem. Res. **25**(4), 200–207 (2013)
11. Song, I.-C., Lee, H.J., Kim, H.-J., et al.: A multicenter retrospective analysis of the clinical features of pernicious Anemia in a Korean population. J. Korean Med. Sci. **28**(2) (2013)
12. Alonso, M.M., Puertas, F.: Adsorption of PCE and PNS superplasticisers on cubic and orthorhombic C3A. Effect of sulfate. Constr. Build. Mater. **78**, 324–332 (2015)
13. Guan, J., Liu, X., Lai, G., et al.: Effect of sulfonation modification of polycarboxylate superplasticizer on tolerance enhancement in sulfate. Constr. Build. Mater. **273** (2021)
14. Tian, H., Kong, X., Miao, X., et al.: A new insight into the working mechanism of PCE emphasizing the interaction between PCE and Ca2+ in fresh cement paste. Constr. Build. Mater. **275** (2021)

15. Yamada, K., Ogawa, S., Hanehara, S.: Controlling of the adsorption and dispersing force of polycarboxylate-type superplasticizer by sulfate ion concentration in aqueous phase. Cem. Concr. Res. **31**, 375–383 (2001)
16. Yamada, K., Ogawa, S., Hanehara, S.: Controlling of the adsorption and dispersing force of polycarboxylate-type superplasticizer by sulfate ion concentration in aqueous phase. Cem. Concr. Res. **31**, 375–383 (2001)
17. Sposito, R., Maier, M., Beuntner, N., et al.: Physical and mineralogical properties of calcined common clays as SCM and their impact on flow resistance and demand for superplasticizer. Cem. Concr. Res. **154**, 106743 (2022)
18. Yao, H., Fan, M., Huang, T., et al.: Retardation and bridging effect of anionic polyacrylamide in cement paste and its relationship with early properties. Constr. Build. Mater. **306**, 124822 (2021)
19. Li, B., Gao, R., Wang, L.: Synthesis and performance of a novel polycarboxylate superplasticizer through grafting the β-cyclodextrin on the side chain by click reaction. J. Appl. Polym. Sci. **139**(3) (2021)
20. Zhang, T., Ma, B., Jiang, D., et al.: Comparative research on the effect of various mineral admixtures on the early hydration process of cement. Const. Build. Mater. **301** (2021)
21. Luo, F., Jiang, Y., Wei, C.: Potential of decarbonized coal gasification residues as the mineral admixture of cement-based material. Constr. Build. Mater. **269**, 121259 (2021)
22. Bessaies-Bey, H., Massoussi, N., Mulik, S., et al.: Polycarboxylate ester adsorption on cement grains: Influence of polydispersity. Cem. Concr. Res. **143**, 106383 (2021)
23. Ersoy, B., Dikmen, S., Uygunoğlu, T., et al.: Effect of mixing water types on the time-dependent zeta potential of Portland cement paste. Sci. Eng. Compos. Mater. **20**(3), 285–292 (2013)
24. Guo, Y., Zhao, Y., Wang, S., et al.: Relationship between the zeta potential and the chemical agglomeration efficiency of fine particles in flue gas during coal combustion. Fuel **215**, 756–765 (2018)
25. Sha, S., Wang, M., Shi, C., et al.: Influence of the structures of polycarboxylate superplasticizer on its performance in cement-based materials - a review. Constr. Build. Mater. **233**, 117257 (2020)
26. Lei, L., Hirata, T., Plank, J.: 40 years of PCE superplasticizers - history, current state-of-the-art and an outlook. Cem. Concr. Res. **157**, 106826 (2022)
27. Cui, X., Mao, S., Liu, M., et al.: Mechanism of surfactant micelle formation. Langmuir **19**(24), 10771–10775 (2008)
28. Feneuil, B., Pitois, O., Roussel, N.: Effect of surfactants on the yield stress of cement paste. Cem. Concr. Res. **100**, 32–39 (2017)

Performance Evaluation of Carbon Fibre-Reinforced Pervious Concrete Subjected to Drop Weight Impact Loading

Bright Singh Seeni(✉)

L&T Heavy Civil Infrastructure, L&T Construction, Chennai, Tamil Nadu 600089, India
bright.singh@lntecc.com

Abstract. Pervious concrete faces challenges in its practical applications owing to its lesser strength characteristics compared to conventional concrete. This study focuses on the evaluation of the performance enhancement of pervious concrete subjected to impact loading with the incorporation of carbon fibre reinforcement. Pervious concrete mix was designed with two distinct aggregate sizes, namely 12.5 mm and 20 mm, with a constant binder to aggregate ratio of 0.28 and water to binder ratio of 0.35. Cement and metakaolin constitute to 85% and 15% of the total binder, respectively. Discrete carbon fibre, accounting for 0.2% of the volume of the mix, was used to prepare the carbon fibre-reinforced pervious concrete. The crack resistance and energy absorption at service and ultimate conditions were determined. Also, the post-crack behaviour of the mix was determined using impact residual strength ratio. The outcome of the study indicated that the post-crack behaviour of the pervious mix was improved considerably with the incorporation of carbon fibre in the 12.5 mm aggregate-sized mix.

Keywords: Pervious concrete · metakaolin · carbon fibre · impact loading · post-crack behaviour

1 Introduction

Low-velocity impacts are usually characterized by a medium level of energy and relatively slow reaction rates and are commonly occurring phenomena in different construction sites and pavement surfaces [1]. These structures are exposed to impact loads under different circumstances such as water runoff, suspended particles and impact due to falling objects, etc. which affects their durability over prolonged time [2]. The repeated impacts generally reduce the stiffness of the slab. Experimental studies have revealed the improvement in the performance of the slabs under impact loading with the use of fibre reinforcements. Fibre reinforcements when used in concrete impacts better toughness properties and energy absorption capacities. Carbon fibre, when used in strips, was found to improve the ultimate bearing capacity better than the conventional steel reinforcements under static and impact loading [3]. Several studies exists with the evaluation of the low-velocity impact loading for ordinary concrete, reinforced concrete or composite specimens [4]. However, studies on the influence of fibres and supplementary

M. Posani and S. Singh (Eds.): RYS 2025, RILEM Bookseries 67, pp. 157–166, 2026.
https://doi.org/10.1007/978-3-032-14652-6_13

cementitious materials on the impact load resisting behaviour of pervious concrete are not well researched. This study evaluates the performance of pervious concrete mixes with two aggregate sizes and with metakaolin and carbon fibre in its composition in individual and combined forms.

2 Materials and Methodology

Ordinary Portland Cement (53 grade) and two single-size coarse aggregate namely 12.5 mm and 20 mm were used commonly for the preparation of the pervious concrete mixes.

2.1 Carbon Fibre

Discrete carbon fibre with a tensile strength of 3450 MPa and modulus of elasticity of 228 GPa was used for development of carbon fibre reinforced pervious concrete mixes. The dimensions of the carbon fibre include length of 12 mm and a diameter of 0.007.

2.2 Metakaolin

Metakaolin obtained through the calcination of kaolin material with a mean particle size of 17 μm, possessing a specific gravity and fineness value of 2.51 and 720 m^2/kg respectively, was used as a supplementary cementitious material.

2.3 Mix Preparation

The pervious mix was developed by maintaining a constant binder to aggregate ratio and water to binder ratio of 0.28 and 0.35, respectively. Table 1 presents the proportion of ingredients used for the mixes. In the mix ID mentioned in Table 1, 12.5 or 20 represents aggregate size and CPC, PCMK15 and CFRPC0.2 indicate Conventional pervious concrete, pervious concrete with 15% metakaolin and Carbon fibre reinforced pervious concrete with fibre content of 0.2% respectively. The proportion of the mix mainly the metakaolin and carbon fibre content were fixed based on our previous efforts [5, 6]. Slab elements with length, width and depth, namely 600 mm, 450 mm and 100 mm respectively, were cast with reinforcement provided in the form of wire mesh at 25 mm from the bottom. The elements were subjected to an impact load test after water curing for 28 days (Fig. 1).

Fig. 1. Pervious concrete slabs

Table 1. Mix proportions

Mix*	Cement (kg/m^3)	Metakaolin (kg/m^3)	Carbon fibre (%)	12.5 mm aggregate (kg/m^3)	20 mm aggregate (kg/m^3)	Water (kg/m^3)	Superplasticizers (kg/m^3)
20CPC	430	-	-	-	1521	150	4.3
12.5CPC	430	-	-	1521	-		
20PCMK15	365.5	64.5	-	-	1521		
12.5PCMK15	365.5	64.5	-	1521			
20CFRPC0.2	430	-	0.2	-	1521		
12.5CFRPC0.2	430	-	0.2	1521	-		
20CFRPC0.2 + MK15	430	64.5	0.2	-	1521		
12.5CFRPC0.2 + MK15	430	64.5	0.2	1521	-		

2.4 Test Method

Repeated drop-weight impact test was conducted on the pervious concrete slabs to investigate the impact resistance. To improve the visibility of the crack pattern, a coat of whitewash was applied all around the surface of the slabs. A dropping load of 6.5 kg was made to free fall from a height of 1.5 m over a steel ball of 64 mm diameter using a tripod pulley arrangement. To determine the impact load effect on road pavements, the slabs were placed in a compacted earthen base and confined on sides with soil. The blows required to induce the service and ultimate cracks on the slabs were recorded, and the dimensions of the cracks were also noted in both conditions. The mentioned test setup provides an easy approach to perform impact load test on field conditions without the requirement of a standard tripod and it has flexibility to vary the height of drop and mass of the drop which can be ultimately be converted to energy absorption. The service and ultimate crack resistance, and post-crack behavior were calculated based on the previous studies [7]. For each mix, three numbers of samples were tested and the average value is reported.

3 Results and Discussion

Carbon fibre was found to improve the mechanical strength properties, especially the tensile and flexural behaviour of pervious concrete to optimum values for a fibre content of 0.2% [5]. 0.2% of carbon fibre was found to improve the compressive strength by 11.59% and 13.21% in 12.5 mm and 20 mm aggregate-sized mixes, respectively. Similarly, in the case of cement replacement by metakaolin, the compressive strength increased to 39.25% and 43.41%, respectively [6].

The optimum mix proportion of pervious concrete was thus designed by the combination of the optimum content of fibres and mineral admixtures. Table 1 provides the mix proportion of the optimized pervious mix (for both 12.5 mm and 20 mm aggregate sized) obtained by 15% replacement of cement by metakaolin with an addition of 0.2% of carbon fibre. The other properties such as permeability, flexural strength, tensile strength and abrasion resistance for the mixes can be referred from our previous research [5, 6].

Table 2. Properties of mix and cracking-related parameters

Mix	ρ (kg/m^3)	n (%)	f_c (MPa)	N_s	N_u	lc (mm)	d_c (mm)	w_c (mm)
20CPC	1798	22.30%	27.02	78	89	1723	10	4.5
12.5CPC	1924	17.21%	29.76	87	102	1677	10	4.0
20PCMK15	1833	21.15%	38.75	122	159	1775	10	5
12.5PCMK15	1949	16.05%	41.44	133	182	1628	10	4
20CFRPC0.2	1764	21.87%	30.59	107	161	1585	10	6
12.5CFRPC0.2	1920	16.79%	33.21	120	193	1359	10	5
20CFRPC0.2 + MK15	1839	20.98%	39.66	120	181	1612	10	6
12.5CFRPC0.2 + MK15	1951	15.99%	43.86	149	226	1428	10	5

Note: ρ, n & f_c indicate density, porosity and compressive strength, respectively. N_s and N_u refer to the number of blows required for service and ultimate cracks under impact loading. l_c,d_c and w_c refers to the total length, maximum depth and maximum width of the cracks, respectively

3.1 Physical Properties and Compressive Strength

Table 2 provides the properties of the pervious concrete mixes with metakaolin and carbon fibre. The control pervious concrete achieved an average compressive strength of

27.02 MPa and 29.76 MPa for mixes with 12.5 mm and 20 mm aggregates, respectively. The porosity was observed to drop from 22.30% to 17.21% with the change in aggregate size from 20 mm to 12.5 mm for the same control mix. The drop in porosity also contributes to the improvement in the density of the concrete. The inclusion of metakaolin and carbon fibre contributed to the increase in compressive strength. Replacement of cement with 15% metakaolin improved compressive strength by 43% and 39% for 20 mm and 12.5 mm aggregate-sized mixes, respectively. However, this improvement is also associated with a decrease in the porosity values by 5.16% and 6.74%, respectively. Compared to the influence of carbon fibre, the influence of metakaolin was observed to be higher with respect to the compressive strength. The combined incorporation of metakaolin and carbon fibre imparted the maximum improvement in the compressive strength for both aggregate sizes (optimized mix).

3.2 Performance Under Impact Load

The performance of the pervious slabs subjected to impact load is determined mainly by the number of drops of the impact load and the cracking behaviour of the slab. Tables 2 and 3 list the details pertaining to the repetitions of impact load sustained by the pervious slabs of different composition under service and ultimate failure conditions.

The number of blows resisted by the control pervious slabs is higher in the case of 12.5 mm aggregates compared to 20 mm aggregates, which is influenced by the variation in the porosity values. This is similar for both service and ultimate failure conditions. In case of individual inclusion of carbon fibre and metakaolin, the number of blows to service and ultimate failure was observed to be higher in comparison with the control mixes for both the aggregate size. The combined use of metakaolin and carbon fibre improved the impact resistance to the highest among the investigated mix aided with the improvement in mechanical strengths.

Table 3. Impact resistance properties

	Energy absorption (kN.m)		Crack resistance (MPa)		Impact crack resistance ratio	Impact residual strength ratio
	Service	Ultimate	Service	Ultimate		
Notation	EA_S	EA_u	R_S	R_u	C_r	I_{rs}
12.5CPC	8.32	9.76	124.05	145.44	4.89	1.17
20CPC	7.46	8.51	96.22	109.79	4.06	1.14
12.5CFRPC0.2	11.48	18.46	168.91	271.67	8.18	1.61
20CFRPC0.2	10.23	15.40	107.62	161.93	5.29	1.50
12.5PCMK15	12.72	17.41	195.35	267.32	6.45	1.37
20PCMK15	11.67	15.21	131.48	171.36	4.42	1.30
12.5CFRPC0.2 + MK15	14.25	21.62	199.60	302.75	6.90	1.52
20CFRPC0.2 + MK15	11.48	17.31	118.67	178.99	4.58	1.51

3.3 Energy Absorption

Energy absorption at service and ultimate conditions was calculated by the product of the energy per blow to the number of blows. Energy per blow is calculated by the product of the mass of the dropping load, acceleration due to gravity and the height of the drop. Energy absorption provides a clear picture of the ability of the mixes to resist impact loads in comparison to the number of blows. The impact energy absorption at service and ultimate conditions for 12.5 is 8.32 and 9.76 kN.m respectively (Fig. 2). The impact energy at service and ultimate condition for 12.5 mm sized mixes is 7.46 and 8.51 kN.m respectively. The impact energy absorption of the 12.5CFRPC0.2 mix was 37.93% and 89.22% higher than the service and ultimate energy absorption of the 12.5CPC mix, respectively. The impact energy absorption of the 20CFRPC0.2 mix was 37.18% and 80.90% higher than the service and ultimate energy absorption of the 20CPC mix, respectively. The impact energy absorption of the 12.5PCMK15 mix was 52.87% and 78.43% higher than the service and ultimate energy absorption of the 12.5CPC mix, respectively. The impact energy absorption of the 20PCMK15 mix was 56.41% and 78.65% higher than the service and ultimate energy absorption of the 20CPC mix, respectively. 12.5CFRPC0.2 + MK15 mix has 71.26% and 121.57% higher service and ultimate energy absorption, respectively, in comparison to 12.5CPC mix. 20CFRPC0.2 + MK15 mix has 53.85% and 103.37% higher service and ultimate energy absorption, respectively, in comparison to 20CPC mix.

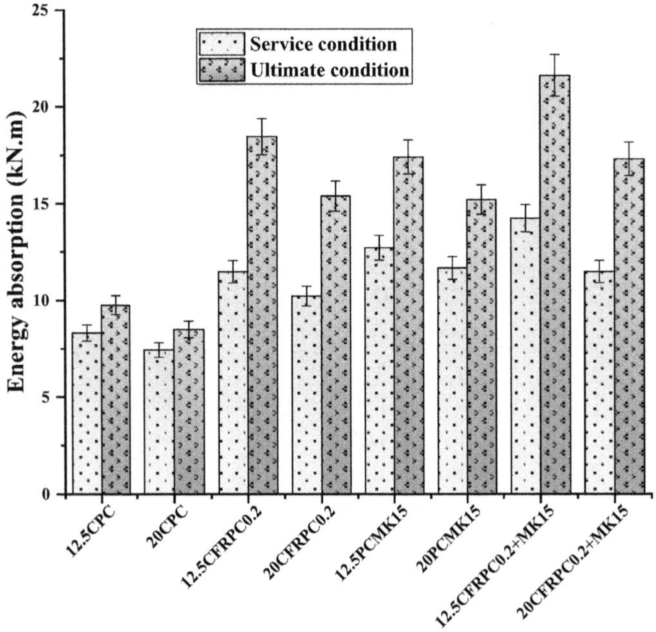

Fig. 2. Impact energy absorption

3.4 Crack Resistance

The crack resistance is estimated by dividing the energy absorption by the dimension of the cracks, namely length, width and depth in service or ultimate failure conditions. The crack resistance in ultimate condition is higher than the service crack resistance for all the investigated mixes. Moreover, the influence of aggregate size on the crack resistance is higher, i.e. smaller-sized aggregate tends to decrease the porosity, which improves the crack resistance of the mixes. The inclusion of metakaolin and carbon fibre made a significant contribution to the crack resistance values. However, the mix with carbon fibre possessed higher improvement in the ultimate crack resistance, indicating the possibility of improved post-crack behaviour. The optimum crack resistance was observed for 12.5CFRPC0.2 + MK15 mix at service and ultimate conditions, namely 199.60 and 302.75 MPa, respectively. This pertains to 60.90% and 108.16%, respectively, higher than the 12.5CPC mix (Fig. 3).

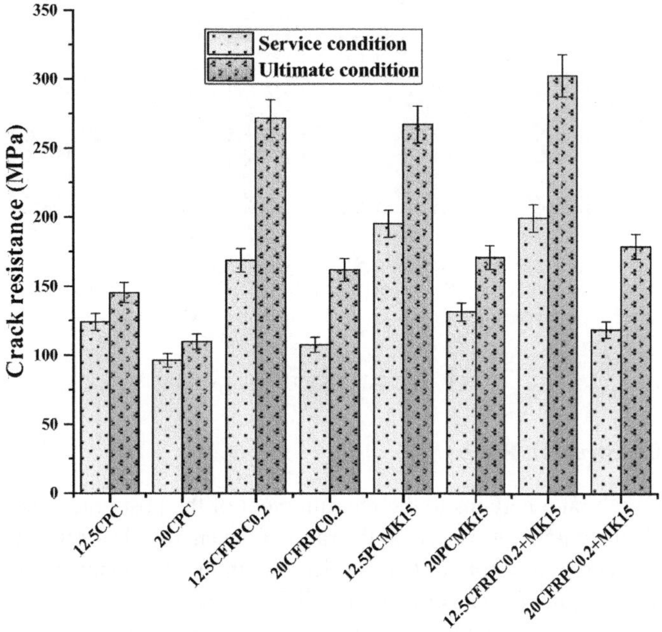

Fig. 3. Impact crack resistance

3.5 Crack Resistance Ratio

Crack resistance ratio is determined by the ratio of the crack resistance at the ultimate condition to the compressive strength. This ratio establishes a relationship between the crack resistance behavior with the compressive strength of the concrete. The higher range of crack resistance ratio indicates a better ability of the mixes to withstand the impact loads for the given compressive strength. The mixes with 20 mm aggregates were found to be superior in par with the 12.5 mm aggregate-sized mixes (Fig. 4). The reason behind this pattern is due to the increased compressive strength and impact resistance of the latter in comparison to the former. Moreover, the inclusion of metakaolin and carbon fibre improved the compressive strength and impact resistance considerably resulting in superior performance than the control mix. Also, the combined use of the metakaolin and carbon fibre resulted in the maximum crack resistance ratio.

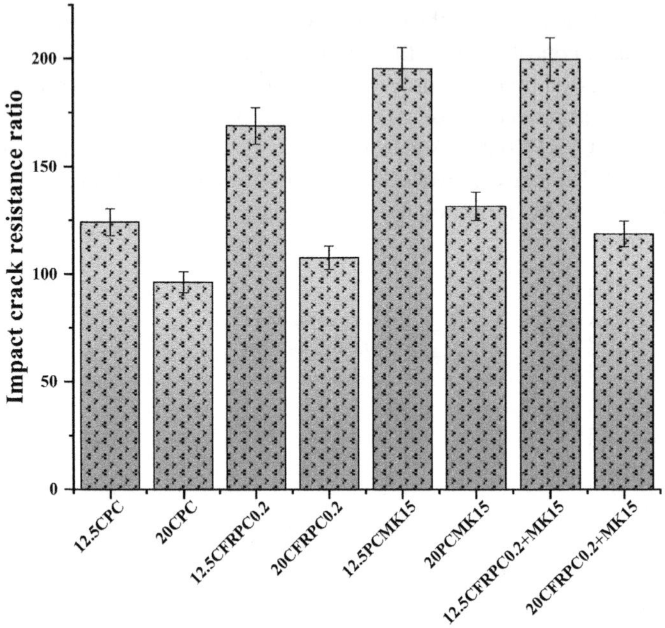

Fig. 4. Impact crack resistance ratio

3.6 Residual Strength Ratio

Residual strength ratio pertains to the measurement of the post-crack behaviour of the mixes. Residual strength ratio refers to the ratio of the energy absorption at the ultimate condition to the energy absorption at the service condition. The higher the ratio, the higher the post-crack strength of the mixes. Among all the investigated mixes, the control mixes were observed to have the lowest residual strength ratio owing to their brittle nature. The inclusion of metakaolin improved the post-crack behaviour slightly higher than the control mixes, but less than the carbon fibre included mixes (Fig. 5). Carbon fibre was observed to impart the maximum post-crack energy absorption owing to its superior flexural and tensile properties and the ability to withstand the widening of the cracks developed at the service stage. Further, carbon fibre acts as a bridge in connecting the cracks and also improves the residual ductility properties of the mix.

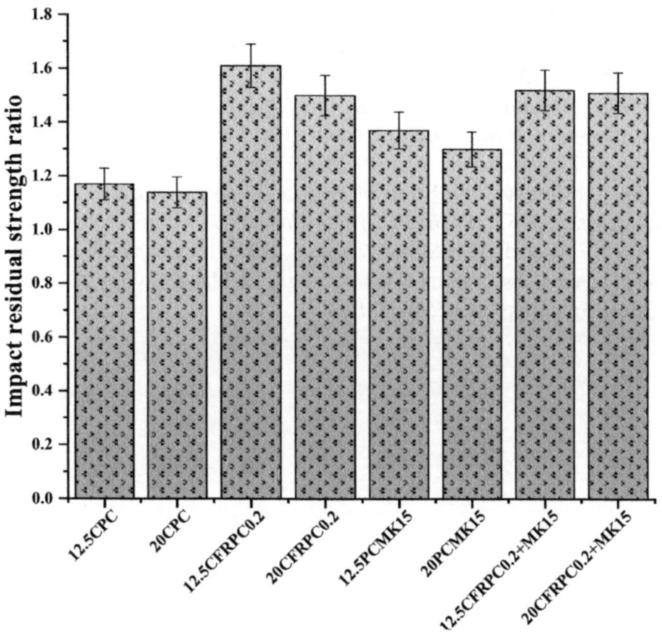

Fig. 5. Impact residual strength ratio

4 Conclusion

The study evaluated the performance of pervious mixes with the individual and combined use of metakaolin and carbon fibre under impact loads. The results indicated that the use of 12.5 mm aggregates instead of 20 mm aggregates resulted in better impact performance. Further, the use of metakaolin and carbon fibre tends to improve the impact performance considerably in individual and combination. The crack resistance at the service and ultimate condition of 12.5CFRPC0.2 + MK15 pertains to 60.90% and 108.16%, respectively, higher than the 12.5CPC mix indicating the optimal performance. The use of carbon fibre improved the ductility of the mixes and the post-crack behaviour of the mixes considerably, owing to the superior tensile and flexural properties.

References

1. Nassar, A.K., Kathirvel, P., Murali, G., Krishna, A.: Development and performance evaluation of novel sustainable one-part alkali-activated fibrous concrete subjected to drop weight impact loading: an experimental study. Constr. Build. Mater. **459**, 139754 (2025). https://doi.org/10.1016/j.conbuildmat.2024.139754
2. Li, Y., Huang, Y., Wang, R., et al.: Experimental investigation on the impact resistance of the concrete-polyurethane mortar composite structures under different influencing factors. J. Build. Eng. **100**, 111735 (2025). https://doi.org/10.1016/j.jobe.2024.111735
3. Radnić, J., Matešan, D., Grgić, N., Baloević, G.: Impact testing of RC slabs strengthened with CFRP strips. Compos. Struct. **121**, 90–103 (2015). https://doi.org/10.1016/j.compstruct.2014.10.033

4. Chakradhara Rao, M., Bhattacharyya, S.K., Barai, S.V.: Behaviour of recycled aggregate concrete under drop weight impact load. Constr. Build. Mater. **25**, 69–80 (2011). https://doi.org/10.1016/j.conbuildmat.2010.06.055
5. Bright Singh, S., Madasamy, M.: Investigation of aggregate size effects on properties of basalt and carbon fibre-reinforced pervious concrete. Road Mater. Pavement Des. **23**, 1305–1328 (2022). https://doi.org/10.1080/14680629.2021.1886158
6. Bright Singh, S., Murugan, M.: Effect of metakaolin on the properties of pervious concrete. Constr. Build. Mater. **346**, 128476 (2022). https://doi.org/10.1016/j.conbuildmat.2022.128476
7. Al Qadi, A.N.S., Al-Kadi, Q.N.S., Al-Zaidyeen, S.M.: Impact strength of oil-palm shell on lightweight concrete slabs reinforced with a geo-grid. J. Mater. Civ. Eng. **27**, 1–7 (2015). https://doi.org/10.1061/(asce)mt.1943-5533.0001142

Silicate Solidified Lunar Regolith: A Novel Scheme from Material Properties to Processing Methods

Zifan Geng(✉) and Guoqing Geng

Department of Civil and Environmental Engineering, National University of Singapore, Singapore 119077, Singapore
gengzifan_cee@nus.edu.sg

Abstract. Extraterrestrial exploration, represented by Lunar base construction, has been promoted worldwide as a multi-disciplinary, cutting-edge and strategic issue. Originated from lunar geopolymers, we herein proposed a novel extraterrestrial construction scheme, silicate solidified lunar regolith (SSLR). It contains only 4 wt% of silicates, which no longer undergo geopolymerization but bind the particles through vacuum dehydration. Under this mechanism, various silicate modulus and modifications (inorganic, organic and physical modifications) have been detected to optimize the silicate adhesive. Through a simulated lunar environment combining high vacuum and large temperature variation, SSLR can be produced in both thermal and cryogenic vacuum, and it presents comprehensive durability with stable microstructures and over 70% of residual strength. Based on these material properties, this study further designed two processing methods for SSLR, mold-pressing and roll-pressing. In the former, a press mold along with water condensation recycle system was verified, and the influences of lunar regolith type, molding pressure and mortise-tenon fit relations are systematically revealed to study the mechanical properties of SSLR bricks and their assembly. In the latter, a specialized 3D printhead was designed for powder extrusion and passive roll-pressing. Therein, the roll-press effects, as well as the flexural and interfacial strengths of print layers are comprehensively investigated considering various roller-spring parameters, lunar regolith types and printing modes. This novel scheme will bring more prospects in both material and processing aspects for future lunar constructions.

Keywords: Lunar construction · Silicate solidified lunar regolith · Solidification mechanism · Mechanical properties · Processing method

1 Introduction

A famous saying goes by Tsiolkovsky: "Earth is the cradle of humanity, but one cannot remain in the cradle forever". Previous lunar exploration projects have succeeded in automatic rover inspection, lunar regolith retrieval and manned landing [1, 2]. Based on these achievements, more ambitious plans on lunar base are proposed, such as the Artemis

M. Posani and S. Singh (Eds.): RYS 2025, RILEM Bookseries 67, pp. 167–182, 2026.
https://doi.org/10.1007/978-3-032-14652-6_14

Program from NASA [3] and the program of International Lunar Research Station from CNSA [4]. It can be seen lunar construction has become a multi-disciplinary, cutting-edge and strategic topic worldwide. However, lunar construction will face much more challenges than terrestrial ones, including rigorous space environments and extremely high cost to transport any material or astronaut onto the Moon [5, 6]. This means lunar regolith must be utilized as the major building material, and automatic processing methods are urgently demanded to in-situ build up required structures.

Several lunar construction schemes have been proposed recently, involving lunar regolith sintering, lunar soil welding, sulfur concrete and lunar geopolymer, etc. [7] Therein, lunar regolith sintering utilizes high temperature over 1000 °C to directly sinter fine particles as structural materials without any terrestrial material [8]. Through laser beam or focused sunlight, the exquisite components can be flexibly 3D printed but there are still some concerns on its weak strength, low processing efficiency for bulk production and complex energy supply [9, 10]. Another series of lunar construction schemes depend on more or less carried binders to acquire higher strength under relatively moderate processing conditions, such as the alkali to produce lunar geopolymers [11]. This material could reach 50 MPa of compressive strength based on appropriate composition and curing temperature [12, 13]. Nevertheless, its durability under lunar environments is still doubtable in view of the significant deterioration reported after only one cycle between −180 °C and 100 °C [14].

In a word currently, an inevitable contradiction emerges among regolith utilization (i.e., fewer carried materials), processing conditions and mechanical performances. Terrestrial binders tend to perform better on the second and third points compared with lunar regolith sintering but consequently introduce much higher transport and storage costs. In addition, the applicability of most adhesives proposed is still worrisome regarding their processing conditions or durability in space. Therefore, a novel extraterrestrial construction scheme, silicate solidified lunar regolith (SSLR), is herein developed to comprehensively handle these existing contradictions. Its binder and water consumption are expected to be strictly limited so carrying water from the earth or in-situ utilizing tiny amount of water in lunar regolith [15] can be acceptable.

2 Materials and Methods

2.1 Raw Materials

Lunar Regolith Simulant (LRS). The lunar soil samples collected from the moon are too scarce to support macro-scale research on construction materials. Therefore, the utilization of LRS is a common approach in this field. Two types of LRS were prepared herein as shown in Fig. 1a. The first LRS1 was made up of fine brown volcanic powders from Huinan, Jilin province, and the second LRS2C was composed of coarse grey basalt particles from Luhe, Jiangsu province. Both of them exhibit irregular and rough particles in micro scale. Their particle sizes are 44.6 μm and 122.0 μm in average respectively, almost within the particle size distribution of actual lunar soils (seeing Fig. 1c) [16]. Figure 1b shows the element composition of these LRS, mainly including oxygen, silicon, aluminum, iron, calcium, magnesium, sodium and potassium, etc. Therein, the composition of LRS1 is similar to the average of the lunar samples collected by Apollo-17 [17]. X-ray diffraction (XRD) phase analysis in Fig. 1d indicates various feldspars, olivine, amorphous glass and a small amount of quartz as its main minerals. As for LRS2C, its composition is close to the average of Apollo-14 samples [17]. Compared with LRS1, LRS2C has slightly higher silicon and aluminum contents and lower iron content. Its amorphous glass content is also relatively lower, but a new mineral diopside occurs. Overall, these two types of LRS ensure both diversity and representativeness. Before use, all the LRS mentioned above were dried at 105 °C until constant weight and stored in a vacuum desiccator.

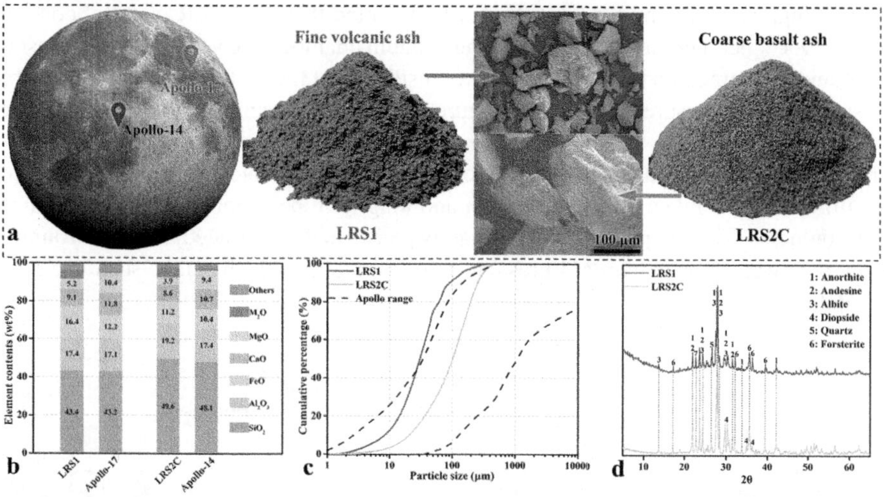

Fig. 1. Raw materials information of lunar regolith simulants LRS1 and LRS2C. (a) Visual and microscopic particle appearances. (b) Element composition. (c) Particle size distribution. (d) Mineral phases composition.

Silicate Binder. This paper used a commercial sodium silicate (SS) sol (34 wt%, modulus 3.3) as the precursor. Other chemicals included analytically pure modifiers and deionized water. The former was used to adjust the modulus (i.e., the molar ratio of SiO_2 to M_2O), cation/anion types, and the structure of silica tetrahedra, which will be discussed in Sect. 3.2. The latter was used to dissolve the chemicals and obtain the required concentration. Herein, silicate modulus was set among 0~3, and sodium hydroxide (SH) solution was seen as a special silicate binder with 0 modulus.

2.2 Simulated Lunar Environments

The simulated lunar environments platform was established based on a programmable oven/refrigerator, a freeze dryer, vacuum chambers, high-performance vacuum pumps and a Pirani vacuum gauge, which were used to generate a thermal and cryogenic vacuum environment as shown in Fig. 2a. The dotted lines in Fig. 2b show the typical diurnal temperature on the lunar surface at different latitudes derived from remote sensing [18]. Since the binder content in the SSLR studied herein is quite low, it can be assumed that the heating and cooling behaviors of this material are similar to those of surficial lunar soil. Therefore, the remote sensing temperature can be approximately used for the simulated lunar environments. In addition, the actual heating or cooling rate of lunar surface is relatively slow since 1 lunar day is around 27.3 Earth days. A stepped heating/cooling protocol is thus adopted to ensure the temperature of each step is consistent with the average temperature of lunar surface in the corresponding time period (seeing solid lines in Fig. 2b).

Apart from the large temperature variation, this platform could simultaneously simulate a high vacuum environment. It should be noted that the SSLR material contains water. Once the material was placed in the vacuum chamber, the vacuum pump must be continuously turned on to remove the water inside out of the system as much as possible to simulate the exposure on the moon. Figure 2c shows the vacuum degree in the chamber containing fresh SSLR samples. It can be seen that the chamber pressure drops sharply from 100 kPa and stabilizes to 10 Pa level after 1 h approximately. This vacuum degree is sufficient to simulate the solidification and long-term behavior of SSLR in the lunar environments. The samples were immediately placed in the simulated lunar environment once they were shaped by mold-pressing or roll-pressing, which can be seen in Sect. 4 for details.

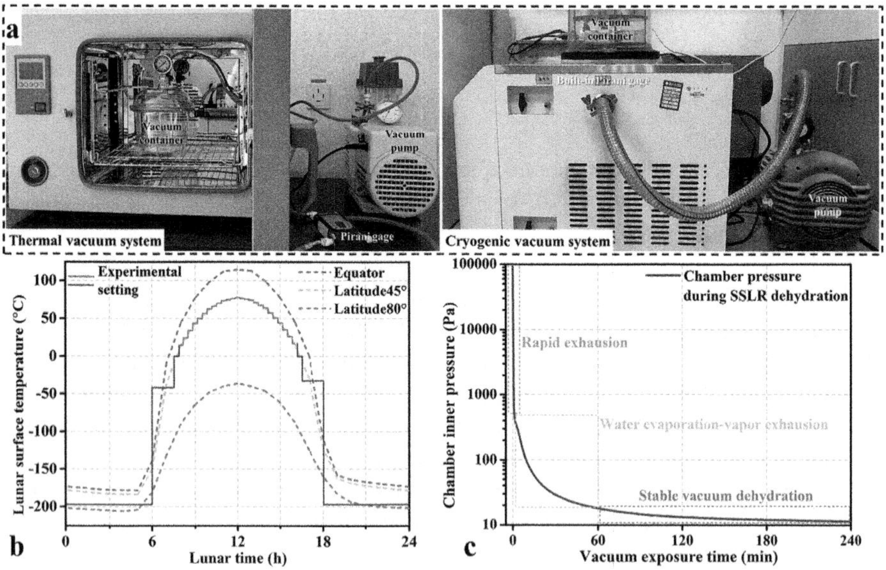

Fig. 2. Simulated lunar environments. (a) Thermal and cryogenic vacuum systems. (b) Lunar diel temperature cycle and experimental temperature setting. (c) Vacuum chamber pressure.

2.3 Test Methods

Mechanical Test. Mechanical tests mainly involve soil mechanics tests of fresh SSLR powders and three-point bending, uniaxial compression and 30° slant shear of solidified SSLR blocks. The soil mechanics test referred to the unconsolidated-undrained triaxial compression test in GB/T 50123. The mechanical tests of solid specimens were conducted using a high-precision universal testing machine. For the three-point bending, the loading rate is 1.5 N/s, the span is set as 25 mm, and the sample size is 40 × 10 × 10 mm^3. Regarding the uniaxial compression, the broken column was used according to ISO 679. The compressed area is around 10 × 10 mm^2, and the loading rate is 30 N/s. The 30° slant shear was specifically used to test the interfacial bonding strength of 3D printed specimens in Sect. 4.2 with a loading rate of 0.1 mm/min and 15 × 15 mm^2 of shear area approximately. Each group was repeated for multiple times (3 times for flexure and 6 times for compression and shear) to calculate the average and standard deviation.

Micro Characterization. X-ray fluorescence (XRF) spectrometry and X-ray diffraction (XRD) analysis were respectively used to detect the element and phase compositions of the LRS raw materials. The X-ray source applied CuKα radiation with 40 kV and 40 mA. The XRD scanning range was set as 5° to 65° with 0.02° and 0.45 s for each step. Scanning electron microscopy-energy dispersive spectroscopy (SEM-EDS) was used to observe SSLR's microstructure after solidification. The products were observed through secondary electron imaging mode with an accelerating voltage of 5~10 kV. For the particle-binder interface specifically, the internal microstructure and element distribution were detected using 2~5 kV of backscattered electron imaging mode and 15kV of EDS mapping. X-ray computed tomography (XCT) was employed for in-situ observing the mortise-tenon joints and 3D-printed interlayer of SSLR. The X-ray source is 130 kV and 60 μA, and the scanning step is 0.4° for 1 s. The voxel resolution is 15~16 μm. The data packet was reconstructed to obtain the 2D cross-section image of required internal structures.

3 Material Properties of Silicate Solidified Lunar Regolith

3.1 From Geopolymerization to Vacuum Hardening

Based on previous studies of lunar geopolymers [19], the LRS1-based geopolymer slurry was prepared with a water-to-regolith ratio of 0.3 and an alkali-to-regolith ratio of 0.075. As shown in Fig. 3a, the green body was sealed by aluminum foil bags after casting and placed in a vacuum chamber to evaluate their 24-h early strength. At 34 °C of average curing temperature, all the groups exhibit minimal solidification with compressive strength only ranging from 0.3~0.4 MPa. Only when further elevating the curing temperature to 82 °C, a significant early strength can be observed. Therein, the SH-solidified group achieves the highest flexural and compressive strengths of 3.46 ± 0.21 MPa and 12.8 ± 1.4 MPa, respectively.

It should be noted that the products through sealed thermal curing are insoluble. At 82 °C of curing temperature, the alkali-activation of SSLR proceeds sufficiently, generating abundant insoluble granules that fill the gaps among LRS particles (seeing Fig. 3b). Specifically, the SH-solidified group predominantly forms spherical products, whereas the products of SS-solidified groups display irregular clusters. These observations demonstrate that the primary products of SSLR through sealed thermal curing are sodium aluminosilicate hydrates (N-A-S-H), a characteristic geopolymer.

Fig. 3. (a) Mechanical properties of SSLR under seal curing. (b) Microstructures of SSLR under seal curing with insoluble products.

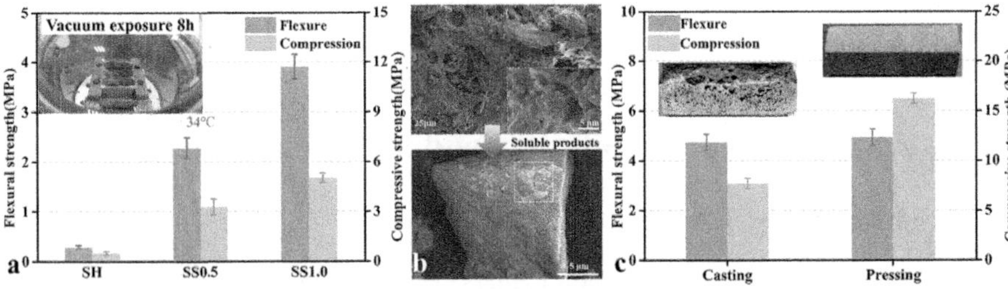

Fig. 4. (a) Mechanical properties of SSLR under vacuum exposure. (b) Microstructures of SSLR under vacuum exposure with soluble products. (c) Mechanical properties of SSLR through different forming methods.

However, it can be seen the geopolymerization of SSLR demands high temperature and suitable silicate-regolith combination. When it comes to vacuum exposure rather than sealed curing, the solidification behavior of SSLR will entirely change. As illustrated in Fig. 4a, the SSLR slurry underwent rapid water loss when exposed in thermal vacuum, resulting in 28~39% of volumetric expansion in the green body. After 8 h approximately, the specimen's mass stabilized and a porous foam-like structure was formed. Therein, the SH-solidified group exhibits negligible strength, while the SS1.0-solidified group achieves 3.90 ± 0.25 MPa of flexural and 5.03 ± 0.29 MPa of compressive strengths, respectively. Even though the SS-solidified groups generate amorphous membranes with notable strength in vacuum, these products exhibit significant water solubility, where they almost completely redissolve into powders as shown in Fig. 4b. This suggests that under vacuum conditions, silicates tend to undergo physical deposition on lunar regolith particles through dehydration instead of initiating geopolymerization.

In addition, cold-pressing was introduced to handle the expansion in vacuum. As shown in Fig. 4c, the green body was pressed under 9.5 MPa with reduced water-to-regolith ratio of 0.098 and silicate-to-regolith ratio of 0.042. This pressed SSLR enables the same flexural strength, over double compressive strength and much denser microstructure than casted one. As a result, the vacuum dehydration along with press method is more applicable for SSLR than geopolymerization in terms of processing conditions and material properties.

3.2 In-situ Hardening Properties

In this section, SSLR samples were mold-pressed according to the above paragraph, and the effects of multiple silicates and environmental temperature on their hardening properties were comprehensively investigated. As listed in the labels of Fig. 5a, silicate modulus ranges from 2~3. Cation/anion modifications refer to 50 mol% of potassium ions, 50 mol% of lithium ions, 10 wt% of sodium aluminate, 2.5 wt% of sodium tetraborate, and 2.5 wt% of sodium pyrophosphate accordingly. Organic modifiers involve 0.4 wt% of sodium dodecyl sulfate, 1 wt% of glycerol and 1 wt% of sorbitol, respectively. Physical modifications are correspondingly 110 °C, 0.7 h of hydrothermal treatment and 40 kHz, 2.5 h of ultrasonic treatment.

Figure 5a systematically illustrates the effects of various silicates on the strength of vacuum-hardened SSLR. Regarding silicate modulus, SS2.5 demonstrates optimal performance with flexural and compressive strengths of 5.48 ± 0.05 MPa and 18.2 ± 0.3 MPa. For cation modification, the introduction of potassium or lithium ions significantly reduces the strength. In contrast, anion modification shows divergent trends: aluminate severely weakens strength, tetraborate slightly degrades strength, while pyrophosphate enhances strength, yielding 6.23 ± 0.39 MPa of flexural strength and 20.3 ± 0.6 MPa of compressive strength. Organic modifiers, especially glycerol and sorbitol, improve the compressive strength by approximately 12%. Physical modifications also notably enhance strength. For example, hydrothermal treatment enables flexural and compressive strengths of 6.54 ± 0.39 MPa and 20.3 ± 0.7 MPa, respectively. As a result, a hybrid-modified silicate binder (prepared with 1 wt% of glycerol and 120 °C, 1 h of hydrothermal treatment) is selected for subsequent studies.

As shown in Figs. 5b–5c, the dehydration process and strength evolution of SSLR are investigated in thermal and cryogenic vacuum environments. The silicate binder exhibits an initial water content of 70%. When exposed to a 44 °C thermal vacuum, rapid dehydration occurs within 0.5 h, generating sufficient strength for subsequent construction. Over the following 4 h, the silicate binder continues to dehydrate slowly and further increases the flexural and compressive strengths to 7.13 ± 0.29 MPa and

23.0 ± 0.5 MPa, respectively. Remarkably, frozen SSLR remains formable even under −40 °C. The frozen mixture was pressed into a green body and then exposed to a cryogenic vacuum. The dehydration rate under freeze-drying is notably decelerated, leading to delayed solidification. After 4.5 h, the silicate binder retains 39% of water content. Over the subsequent 10 h, slow dehydration persists in the cryogenic vacuum, which gradually elevates the flexural and compressive strengths to 6.21 ± 0.24 MPa and 19.5 ± 1.0 MPa, respectively. These findings demonstrate the feasibility for in-situ forming SSLR even under extreme cryogenic conditions.

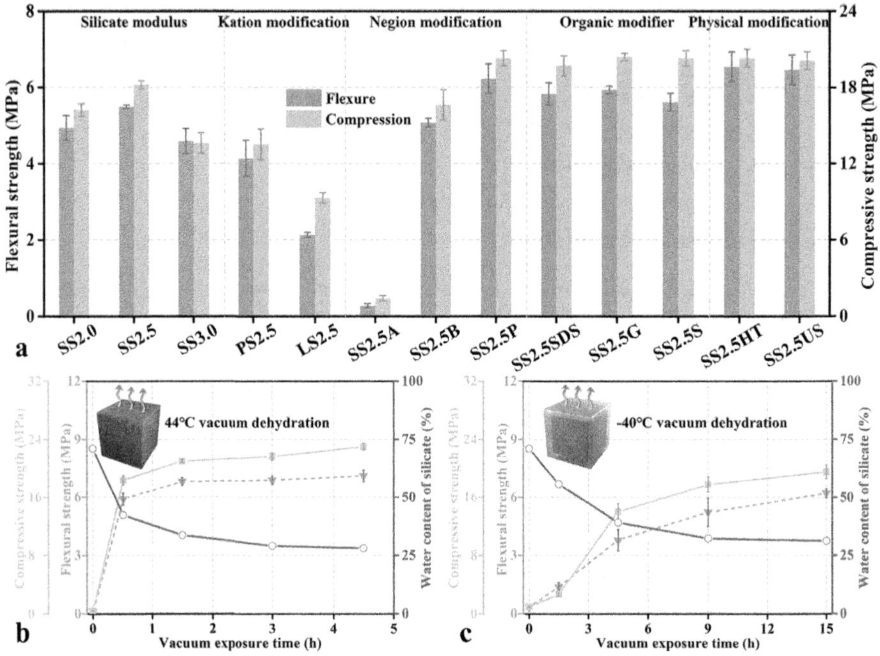

Fig. 5. (a) Mechanical properties of SSLR with different silicates. (b–c) Strength and dehydration evolution of SSLR during the vacuum hardening at 44 °C and −40 °C.

3.3 Long-Term Durability

This section investigates the long-term durability of SSLR in simulated lunar environments at four critical stages. (1) Initial hardening stage: materials are exposed to a thermal vacuum at 44 °C for 4.5 h. (2) Midday high-temperature stage: materials undergo 81 h of thermal vacuum exposure with rising temperature from 44 °C to 78 °C. (3) Midnight low-temperature stage: materials experience 540 h of vacuum exposure with decreasing temperature from 78 °C to −196 °C. (4) Long term stage: the vacuum exposure is accumulated to 108 d (4 lunar days).

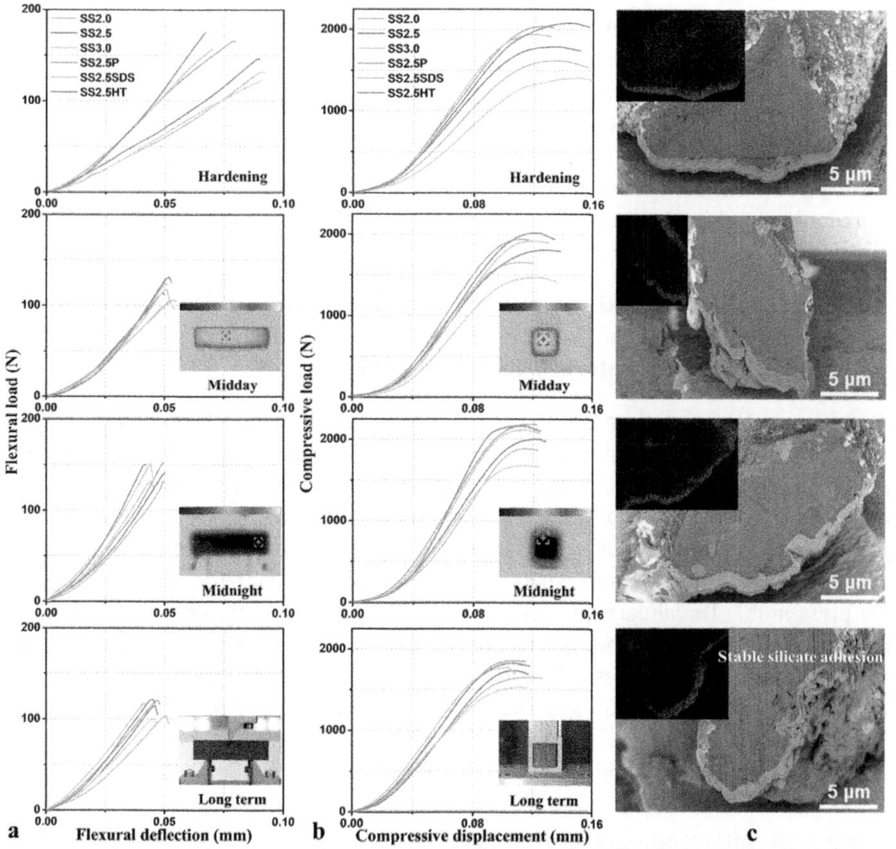

Fig. 6. (a–b) Flexural and compressive curves and (c) microstructural evolution of SSLR at different service stages.

Figures 6a–6b systematically present the load-displacement curves of representative SSLR across these stages. During the hardening stage, specimens exhibit quasi-brittle failure with ultimate deflection ranging 0.067~0.091 mm for flexure and ultimate deformation ranging 0.123~0.143 mm for compression. At the first midday, continued bound water evaporation induces embrittlement in all groups. The high-temperature flexural strength of 6 representative SSLR groups decreases by 18~26%, accompanied by a significant reduction in ultimate deflection (0.049~0.054 mm). However, the high-temperature compressive strength shows only a negligible decline. During the midnight stage, the dehydration further decelerates. The combined effects of dehydration and low temperature simultaneously embrittle and strengthen the material. After this stage, nearly all bound water in the silicate binder has been lost. Compared with the first midday, the long-term flexural strength of six SSLR groups declines by less than 4% with ultimate deflection stabilizing at 0.041~0.051 mm, while the long-term compressive strength decreases by only 2~8% with ultimate deformation maintaining at 0.107~0.116 mm.

These results indicate minimal mechanical degradation in SSLR after the first lunar temperature cycle, demonstrating its exceptional durability under lunar environments.

Additionally, Fig. 6c reveals the stable adhesion between silicate binder and regolith particles. The binder consistently forms membrane products tightly coating LRS particles with no visible defects. High grayscale regions with red EDS mapping (Na) around particles confirm the robust interfacial adhesion of the silicate binder during vacuum dehydration, and no detectable delamination or spalling are observed.

4 Processing Methods of Silicate Solidified Lunar Regolith

4.1 Mold-Pressing and Mortise-Tenon Assembling

According to Sect. 3, cold-pressing can be seen as an essential method to produce SSLR on the moon. In this section, a mold-pressed SSLR brick with mortise-tenon joints was firstly conceptualized as illustrated in Fig. 7a. It should be noted the water condensation recycle accessory is compatible with press mold. Since most water in SSLR does not participate in geopolymerization but serves as a dispersant, they can be naturally removed during the vacuum hardening. Therefore, a condenser can be integrated at the base of the press mold. To validate this concept, a condensation recycle system was designed as shown in Fig. 7b. The commercial condenser is set up between the vacuum chamber and pump. The temperatures at hot and cold ends are monitored using infrared thermal imaging (44 °C and −40 °C respectively). Vapors are condensed and recycled during the vacuum hardening to enable the fabrication of mortise-tenon SSLR bricks with minimal terrestrial binders consumption.

Fig. 7. (a) Conceptualization and (b) experimental settings of SSLR brick mold-pressing, water condensation recycling and mortise-tenon preparation.

The performance of this mold system is systematically validated. Figure 8a indicates the condensation rate during SSLR dehydration. At 10 min, the condenser captures nearly all the vapors. By 20 min, the condensation rate reaches a maximum of 52%. As vacuum dehydration continues however, ice sublimation occurs in the condenser, leading to a slightly reduced efficiency. Figures 8b–8c demonstrate that LRS2C bricks are generally superior to those made from LRS1. Additionally, the mechanical properties of both show nonlinear variations with the pressing stress. At low pressures, the performances of SSLR bricks seem more sensitive to molding pressure, where LRS1 displays a greater sensitivity than LRS2C. This implies LRS2C is more suitable for low-pressure molding.

These trends can be explained through lunar soil mechanics. Figures 8d–8e reveal that LRS2C (both in dry regolith and mixtures) exhibits lower cohesion but higher friction angles than LRS1. Consequently, LRS2C achieves higher shear strength during mold-pressing and forms a "stiffer" granular skeleton. This results in higher strength for LRS2C bricks compared to LRS1 ones.

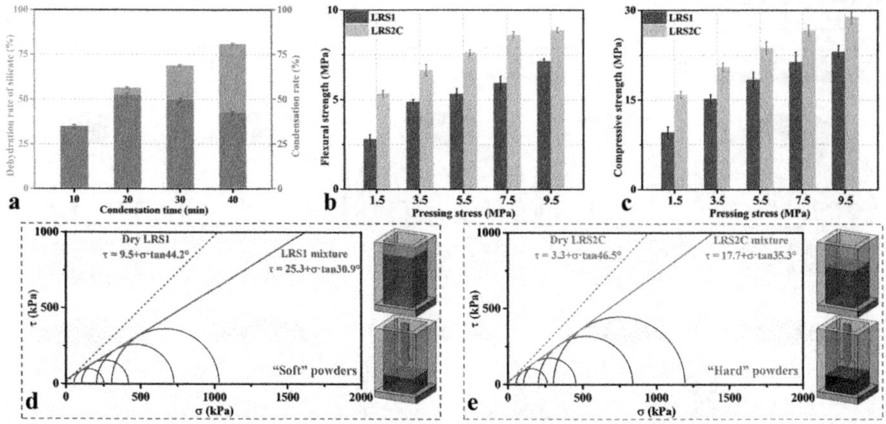

Fig. 8. (a) Water condensation rate during the vacuum hardening of SSLR brick. (b–c) Flexural and compressive strength of SSLR brick with different regolith under different pressing stresses. (d–e) Soil mechanical properties of LRS1 and LRS2C.

Finally, the mortise-tenon joints were assembled using LRS2C bricks (seeing Fig. 9). The mechanical properties of mortise-tenon joints under varying fit relations are first examined in this section. As shown in Figs. 9a–9b, tight joints (interference fit) adversely reduce the strength. The peak loads of two clearance-fit joints increase to 83.4 ± 8.9 N and 89.0 ± 6.4 N, respectively. On the other hand, joints gradient reinforcement leads to a more substantial enhancement. Figure 9b shows the peak load of gradient-reinforced joints reaches as high as 173.2 ± 6.2 N. Understandably, stress concentrations in mortise-tenon SSLR bricks primarily occur at the tenon and mortise corners, while the stress level is low within the matrix. This suggests a simple and effective optimization strategy to weaken the matrix and enhance the joints through a gradient binder distribution during molding.

Figures 9c–9d systematically present the XCT internal structures and failure modes of different joints to further confirm these findings. For interference-fit and transition-fit joints, fractures initiate at the tenon root, accompanied by pronounced cracking at the mortise periphery. These cracks are generated during assembly and compromise mechanical properties. In contrast, clearance-fit joints exhibit upward fracture locations in the tenon and reduced crack dimensions in the mortise periphery. Moreover, gradient-reinforced joints demonstrate a synergistic failure mode between the weakened matrix and reinforced interfaces, significantly improving the strength without additional silicate binder consumption.

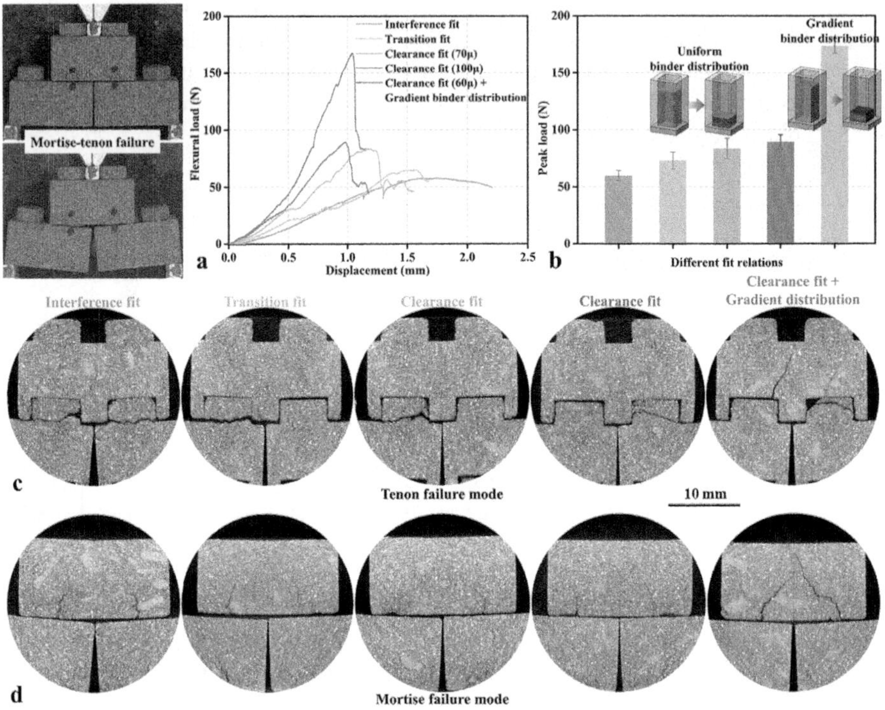

Fig. 9. (a–b) Mechanical properties of assembled SSLR bricks with different mortise-tenon fits. (c–d) Tenon and mortise failure modes under different fitting relations.

4.2 Roll-Pressing and Additive Manufacturing

Apart from mold-pressing, another process type, roll-pressing, is also verified for in-situ SSLR production in this section. Considering the difference between mold-press and roll-press, additive manufacturing (3D printing) is introduced herein. This section presents a novel 3D printing system specifically designed for SSLR powders, different from extrudable concrete 3D printing. Figures 10a–10c display the CAD model and physical prototype of the 3D printing system based on roll-pressing [20]. The specialized printhead consists of: (1) support module, (2) drive module, (3) screw-blade extrusion module, and (4) roller-spring pressing module. The screw-blade module in the printhead frame can extrude SSLR powders downward controllably. The roller-spring module is designed as the terminal nozzle. As the printhead moves, the roller will passively compact the loose SSLR powders into dense print layers.

This system further enables flexible selection of two printing modes as shown in Fig. 10d. (1) Premix printing: Premixed LRS and silicate solution form a powdered mixture to be extruded and compacted into layers. (2) Dry printing: Dry LRS powders are loaded, extruded and compacted into dense layers directly, while the silicate solution is subsequently dropped onto the powder layer through a rear-mounted dropper. The solution will spontaneously infiltrate and solidify the entire print layer.

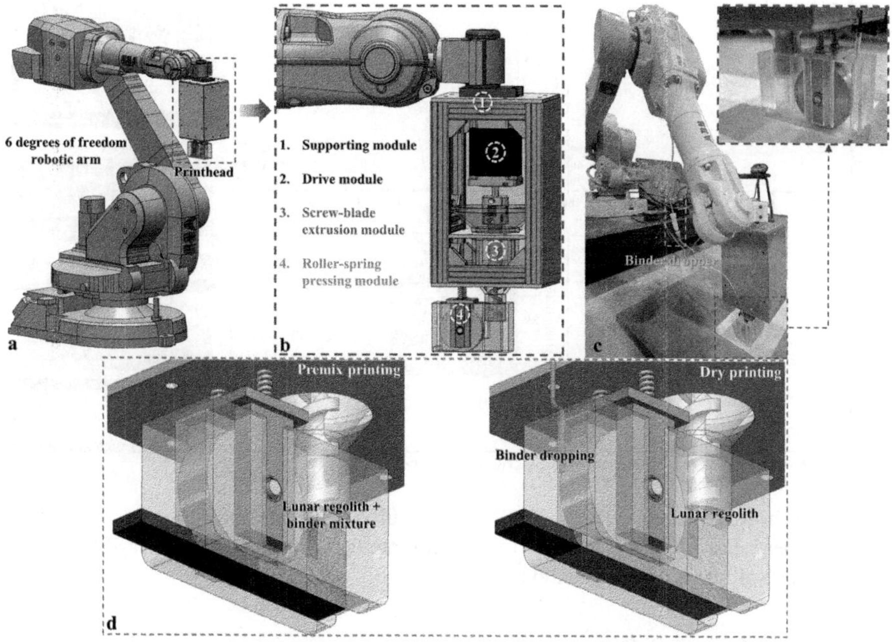

Fig. 10. (a-c) CAD models and photograph of SSLR 3D printing system. (d) Conceptualizations of premix printing and dry printing of SSLR.

The influences of roller-spring module parameters on printing outcomes are investigated firstly. As shown in Fig. 11a, a larger roller significantly increases layer thickness, where a 50 mm roller achieves a thickness of 2.07 ± 0.05 mm. In contrast, stiff springs show limited improvement in layer thickness. Layer density correlates with both spring stiffness and roller diameter positively (seeing Fig. 11b). This densification substantially enhances the layer strength: stiffer springs and larger rollers elevate the flexural strength of print layers from 1.30 ± 0.18 MPa to 1.99 ± 0.16 MPa as shown in Fig. 11c. Both stiff springs and large rollers expand the critical compaction angle (the capacity of printhead to compact powders into layers). Notably, LRS1-based SSLR demonstrates significantly lower layer thickness, density and strength compared to LRS2C, which is consistent with the results in Sect. 4.1.

Additionally, the influences of two printing modes are examined. Dry-printed layers exhibit superior mechanical properties to premix-printed ones, achieving a flexural strength as high as 5.15 ± 0.37 MPa (seeing Fig. 11c). Figures 11d–11e further demonstrate dry printing generates a denser interlayer with much higher interfacial strength, whereas the interfaces via premix printing contain visible pores. This difference arises from the spontaneous infiltration of the silicate solution during dry printing. However, this mechanism also increases the binder consumption in dry printing (around 0.16 of solution-to-regolith ratio) compared to premix methods (0.13~0.14).

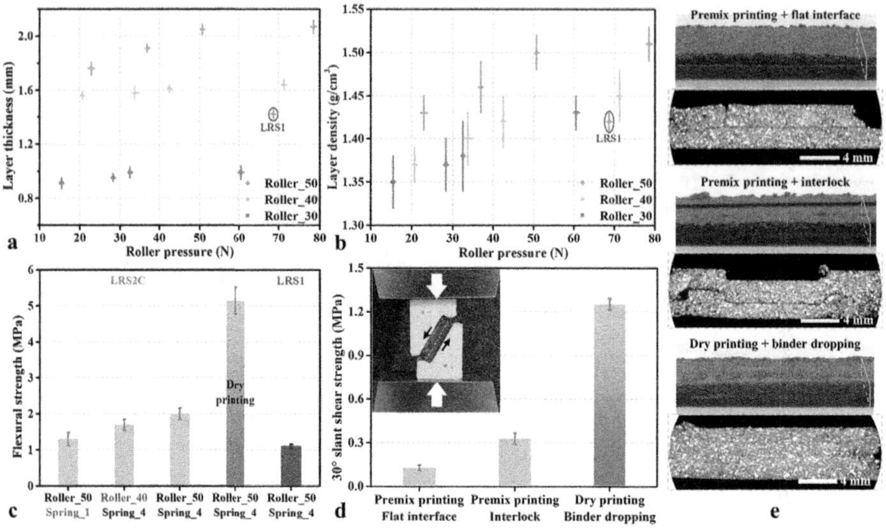

Fig. 11. (a–b) Roll-pressing effects of SSLR under different rollers and springs. (c–d) Flexural and slant-shear strength of SSLR printed layers. (e) Interfacial meso-structures of SSLR printed layers through different printing methods.

5 Conclusion and Prospect

Originated from lunar geopolymers, this paper proposed a novel extraterrestrial construction scheme, silicate solidified lunar regolith (SSLR). Using two types of lunar regolith simulants and mimicked lunar environments with high vacuum and large temperature variation, this scheme has been comprehensively investigated in both material and processing aspects. Specific findings are listed below.

In term of material properties, SSLR is composed of regolith particles and 4 wt% of silicate binders, and can be in-situ hardened through vacuum dehydration rather than geopolymerization. The former is fundamentally different with the latter since its products are soluble membranes. Various silicate modulus and modifications (inorganic, organic and physical modifications) are detected, which indicate the enhancements of 2.5 silicate modulus, pyrophosphate, polyhydric alcohols and hydrothermal reaction. The in-situ hardening of SSLR under both thermal and cryogenic vacuum are also verified. Furthermore, SSLR presents long-term durability after 4 lunar days with stable microstructures and over 70% of residual strength. A slight brittleness can be detected considering the long-term dehydration.

In term of processing methods, this study further design mold-pressing and roll-pressing, which are separately used for brick assembling and 3D printing. In the former, a press mold along with water condensation recycle system is verified, and the influences of lunar regolith type, molding pressure and mortise-tenon fit relations are systematically revealed. Therein, LRS2C with higher friction angle is more suitable for low-pressure molding, and clearance fit along with gradient-reinforced mortise-tenon joints achieves better performances. In the latter, a specialized 3D printhead is designed for powder extrusion and passive roll-pressing. Therein, the roll-press effects (layer thickness and density, etc.), and the flexural and interfacial strengths of print layers are comprehensively investigated considering various roller-spring parameters, lunar regolith types and printing modes.

This novel scheme is hopeful to bring more prospects in both material and processing aspects for lunar base constructions. In the future, more silicate modifications can be investigated to enhance the mechanical properties and further save the terrestrial materials. On the other hand, the automatic pressing equipment can be further designed for actual lunar environments, where some validations on space station may be required.

References

1. Lunar Sample Laboratory Facility, NASA Astromaterials Acquisition & Curation Office. https://curator.jsc.nasa.gov/lunar/lun-fac.cfm. Accessed May 2022
2. China's Chang'e-5 retrieves 1,731 grams of moon samples, China National Space Administration. http://www.cnsa.gov.cn/english/n6465652/n6465653/c6810963/content.html. Accessed July 2025
3. Patrinos, T.: Artemis Moon Program Advances – the Story So Far, National Aeronautics and Space Administration. https://www.nasa.gov/artemis-moon-program-advances. Accessed July 2025
4. International Lunar Research Station (ILRS) Guide for Partnership, China National Space Administration. http://www.cnsa.gov.cn/english/n6465645/n6465648/c6812150/content.html. Accessed July 2025
5. Ding, L., Xu, J., Luo, H., Cai, L.: Lunar construction: challenges and research progresses. Manned Spacefl. **25**, 277–285 (2019)
6. Advanced Space Transportation Program: Paving the Highway to Space, Marshall Space Flight Center. https://www.nasa.gov/centers/marshall/news/background/facts/astp.html. Accessed July 2025
7. Zhou, C., et al.: Properties and characteristics of regolith-based materials for extraterrestrial construction. Engineering **37**, 159–181 (2024)
8. Han, W., Ding, L., Cai, L., Zhu, J., Luo, H., Tang, T.: Sintering of HUST-1 lunar regolith simulant. Constr. Build. Mater. **324**, 126655 (2022)
9. Meurisse, A., Makaya, A., Willsch, C., Sperl, M.: Solar 3D printing of lunar regolith. Acta Astronaut. **152**, 800–810 (2018)
10. Fateri, M., et al.: Solar sintering for lunar additive manufacturing. J. Aerosp. Eng. **32**, 04019101 (2019)
11. Yao, Y., et al.: Influence of vacuum and high-temperature on the evolution of mechanical strength and microstructure of alkali-activated lunar regolith simulant. J. Build. Eng. **97**, 110709 (2024)
12. Zhou, S., Lu, C., Zhu, X., Li, F.: Preparation and characterization of high-strength geopolymer based on BH-1 lunar soil simulant with low alkali content. Engineering **7**, 1631–1645 (2021)
13. Wu, Z., Geng, Z., Pan, H., Zhang, L., She, W.: Determinants of lunar geopolymer synthesis based on experimental and statistical analysis: towards lunar regolith's diversity. Constr. Build. Mater. **450**, 138557 (2024)

14. Zhang, R., Zhou, S., Li, F.: Preparation of geopolymer based on lunar regolith simulant at in-situ lunar temperature and its durability under lunar high and cryogenic temperature. Constr. Build. Mater. **318**, 126033 (2022)
15. Lin, H., et al.: In situ detection of water on the Moon by the Chang'E-5 lander. Sci. Adv. **8**, eabl9174 (2022)
16. Carrier, W.D.: Particle size distribution of lunar soil. J. Geotech. Geoenviron. Eng. **129**, 956–959 (2003)
17. Heiken, G.H., Vaniman, D.T., French, B.M.: Lunar Sourcebook. Cambridge University Press (1991). https://www.lpi.usra.edu/publications/books/lunar_sourcebook/
18. Geng, Z., et al.: Silicate-lunar regolith composite: a vacuum self-hardened and comprehensively durable extraterrestrial construction material. Constr. Build. Mater. **449**, 138517 (2024)
19. Geng, Z., et al.: In-situ solidification of alkali-activated lunar regolith: Insights into the chemical and physical origins. J. Cleaner Prod. **391**, 136147 (2023)
20. Geng, Z., Wu, Z., Wang, X., Zhang, L., She, W., Tan, M.J.: A novel 3D printing scheme for lunar construction with extremely low binder utilization. Addit. Manuf. **99**, 104657 (2025)

Spalling Resistance of Engineered Cementitious Composites Under Fire: A Review of Mechanisms and Influencing Parameters

Sanket Rawat(✉)

School of Civil and Environmental Engineering, University of Technology Sydney (UTS), Sydney, NSW 2007, Australia
sanket.rawat@uts.edu.au

Abstract. Fiber-reinforced cementitious composites (FRCC), particularly engineered cementitious composites (ECC), are widely recognized for their exceptional tensile performance and durability. However, their dense microstructure, due to the absence of coarse aggregates, presents challenges under fire exposure. Among the various factors influencing fire performance, fiber type plays a critical role in determining spalling resistance. It is commonly believed that low-melting-point fibers improve spalling resistance by creating channels upon melting, which helps mitigate internal pressure build-up. Hybrid fiber combinations, which incorporate both low-melting-point and high-melting-point fibers, have also shown potential in maintaining strength after fiber melting and enhancing spalling resistance. However, a key debate in the literature revolves around which parameters most significantly contribute to spalling resistance. While lower melting point fibers (e.g. polyethylene (PE) fiber) are expected to improve performance, research does not consistently support this hypothesis. In fact, fibers like polyvinyl alcohol (PVA), which have much higher melting points (~250 °C for PVA vs ~140 °C for PE), are often found to provide better spalling resistance. Moreover, some studies have provided evidence of better spalling resistance in steel-FRCC (steel fiber melting point > 1000 °C). However, a clear analysis of the current findings and prevailing mechanism remains unclear. This paper critically examines the current body of research on cementitious composites reinforced with various fiber types to identify key factors influencing spalling resistance. By exploring the governing parameters and mechanisms, the study aims to provide guidance for future material selection to improve the fire performance of FRCCs.

Keywords: Elevated temperature performance · Engineered cementitious composites (ECC) · Fiber-reinforced cementitious composites (FRCC) · Spalling Resistance · Ultra-high-performance concrete (UHPC)

1 Introduction

Engineered Cementitious Composites (ECC) have attracted significant attention in recent years due to their exceptional mechanical properties and durability, leading to their application in various structural projects, including bridge link slabs and tunnel linings [1]. These high-performance fiber-reinforced cementitious composites (FRCCs)

M. Posani and S. Singh (Eds.): RYS 2025, RILEM Bookseries 67, pp. 183–197, 2026.
https://doi.org/10.1007/978-3-032-14652-6_15

are often referred to interchangeably as ultra-high-performance concrete (UHPC) or strain-hardening cementitious composites (SHCC), depending on their specific characteristics. Typically composed of binder materials, fine aggregates and fibers (typically $\leq$ 2% by volume), ECC are designed using micromechanics principle to achieve superior strain hardening behavior and high ultimate tensile strain. Due to the absence of coarse aggregates, the packing density in ECC is generally higher than in conventional concrete, resulting in enhanced mechanical properties along with reduced porosity. While this dense microstructure and low permeability contribute to its superior mechanical performance and durability, they also increase susceptibility to spalling under fire exposure [2].

Fire resistance is a critical durability parameter, particularly for buildings, underground structures, and tunnels. Current international standards typically adopt a prescriptive approach to account for fire resistance in structural design, specifying minimum dimensions for various structural members or cover thicknesses to achieve a certain fire resistance rating [3]. However, in the case of UHPC and ECC, such prescriptive measures may not be adequate. Their dense, low-permeability microstructure restricts moisture escape during fire exposure, causing rapid vapor pressure build-up that can lead to explosive spalling, despite cover thickness and dimensions meeting standard requirements. Even minor spalling can significantly compromise structural integrity by further reducing mechanical properties and exposing reinforcement to excessive heat. To mitigate this issue, use of polypropylene (PP) fibers is often recommended. For instance, Eurocode 2 [4] suggests incorporating more than 2 kg/m^3 of PP fibers to address the spalling risk in high-strength concrete.

However, spalling resistance and mechanical performance in FRCCs are influenced not only by fiber type, but also by factors such as fiber geometry and the cementitious matrix. While PP fibers (melting point ~ 175 °C) are effective in preventing spalling, they are not ideal when high tensile strain capacity is desired. Polyethylene (PE) fibers, known for their superior strain capacity, have been found to enhance the tensile strain of ECC to as much as 12% [5], far exceeding that of normal or conventional fiber-reinforced concrete. However, despite this advantage and despite PE fiber having lower melting point than PP fiber (~160 °C for PE vs ~175 °C for PP), ECCs reinforced solely with PE fibers have been criticized for their poor spalling resistance. On the other hand, better spalling resistance has been observed for polyvinyl alcohol (PVA) fiber (melting point ~ 250 °C) or steel fiber (melting point > 1000 °C) based FRCCs [6]. These findings contradict the commonly held theory that low-melting-point fibers should improve spalling resistance by forming interconnected channels during melting and highlight the need for a more detailed understanding of the factors that contribute to spalling resistance, especially in PE fiber-based ECC.

This paper seeks to address this gap by providing a succinct review of the spalling resistance of PE fiber-based ECC, alongside broader trends in FRCC spalling behavior. It critically examines existing theories on spalling resistance and correlates these theories with the performance of various FRCCs. The study integrates findings from the author's own extensive experimental investigations and existing literature. The aim is to help both new and established researchers gain a deeper understanding of the mechanisms behind spalling resistance, ultimately enabling the design and development of better fire-resistant FRCCs for future applications.

2 Types of Spalling in FRCCs

Explosive spalling remains a significant failure mode in cementitious composites exposed to fire. Traditionally, it was believed that spalling risk is dependent on the moisture content and could be minimal if the moisture content of the material is below 3% [7]. However, moisture content can vary throughout the cross-section of the concrete structure, making it an unreliable and inconsistent criterion for evaluating spalling risk. Spalling has also been observed in concrete with low moisture content [8], suggesting that moisture content alone cannot account for spalling. Therefore, a more comprehensive understanding of the different types of spalling is required to better identify the contributing factors.

Liu et al. [9] provided a detailed classification of fire-induced spalling, categorizing it into three types based on the contributing mechanisms:

- Thermo-hygral spalling: Caused by pore pressure and moisture clogging,
- Thermo-mechanical spalling: Driven by mechanical stresses such as compressive loads, thermal gradient, or restrained thermal expansion, and
- Thermo-chemical spalling: Triggered by the decomposition of hydration products and the resulting loss of bonding between hydration phases and aggregates.

Among these, thermo-hygral spalling is of primary interest in this study, as fiber reinforcement significantly mitigates this form of spalling by enhancing permeability and controlling internal stresses. Consequently, this paper primarily focuses on the mechanisms and mitigation strategies for thermo-hygral spalling in ECCs. For a broader context, relevant studies on UHPC, SHCC, and other FRCCs without coarse aggregates are also discussed.

2.1 Thermo-Hygral Spalling

Thermo-hygral spalling is primarily attributed to the interaction between heat, moisture, and pore pressure within a dense cementitious matrix under fire exposure. Several mechanisms have been proposed in the literature to explain this phenomenon.

One of the most widely accepted explanations involves the build-up of pore pressure due to moisture clogging [10]. As the internal temperature of the material rises, free water within the pores evaporates and attempts to migrate toward the cooler exterior surfaces. In low-permeability systems such as UHPC, this vapor flow often encounters resistance, leading to the formation of a saturated zone (commonly referred to as a *moisture clog*). This zone restricts further vapor movement, causing a rapid increase in pore pressure. When the internal vapor pressure surpasses the reduced tensile strength of the heated matrix, the material can fracture and spall explosively (Fig. 1).

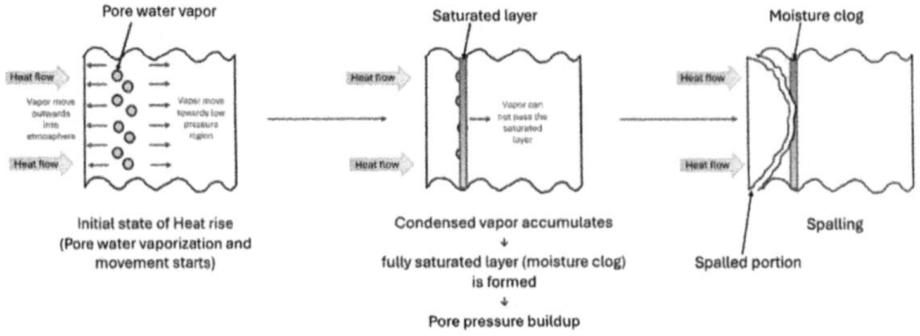

Fig. 1. Spalling due to moisture clog formation (adapted and redrawn from [11])

Another mechanism is based on a BLEVE (Boiling Liquid Expanding Vapor Explosion)-like phenomenon [12]. In this scenario, water trapped within the dense matrix becomes superheated, remaining in liquid form even above 100 °C due to the elevated internal pressure. If microcracking or sudden stress release exposes this superheated water to lower ambient pressures, it undergoes instantaneous phase change into steam, resulting in rapid volume expansion and energy release. This abrupt process can simulate the explosive characteristics observed in BLEVE events, offering a thermodynamic explanation for the violent nature of spalling (Fig. 2).

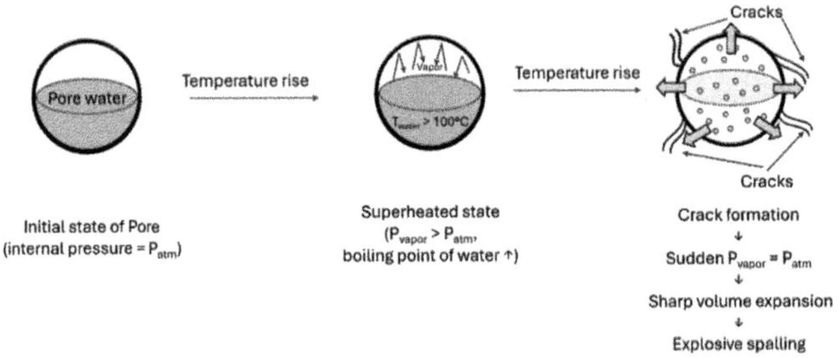

Fig. 2. Spalling due to phase change of water (adapted and redrawn from [12])

Both mechanisms can occur independently or simultaneously, and their relative contribution depends on factors such as heating rate, permeability, initial moisture content, and the presence of SCMs. In either case, this form of spalling is particularly significant for dense composites, where water evaporation and vapor buildup can happen rapidly. The role of fibers, especially low melting point polymer fibers, in mitigating thermo-hygral spalling is frequently highlighted in literature. However, the specific mechanisms explaining why some fibers are more effective than others in preventing spalling remain unclear. The following sections explore these knowledge gaps and examine key findings from recent studies to clarify the role of fiber type, test parameters, and matrix in mitigating spalling.

3 Main Research Questions Related to Spalling Resistance

3.1 Do Test Parameters Play a Role?

In fire performance studies of cementitious composites, experimental parameters such as curing regime, heating rate, and specimen size significantly influence the observed spalling behavior. These factors can affect not only the occurrence of spalling but also the underlying mechanisms responsible for it.

There is no clear consensus on how specimen size affects spalling sensitivity [6]. To address this, we recently investigated the spalling resistance of specimens of varying sizes under both low and high heating rates [13, 14]. Overall, smaller specimens, such as 50 mm cubes commonly used in FRCC testing, may not reliably indicate spalling resistance under high-temperature exposure. This is primarily due to their limited internal volume, which restricts the development of large thermal gradients and reduces the length of moisture migration paths. In contrast, larger specimens can be more vulnerable due to more pronounced thermal gradients and greater internal pore pressure under rapid heating. This size sensitivity was confirmed in the author's recent investigations. At low heating rates, even large specimens (e.g., 150 mm diameter × 300 mm height cylinders) showed no signs of spalling [13]. However, in a separate study conducted under ISO heating rate [14], 300 × 300 × 50 mm panel specimens exhibited severe spalling, whereas the smaller 100 × 100 × 50 mm panel specimens remained unaffected as shown in Fig. 3. It is important to note that material composition also plays a critical role. In some cases, even low heating rates (e.g., 1–10 °C/min) have been shown to trigger spalling, depending on the fiber content and mix design [15, 16]. Additionally, the curing regime has been found to significantly influence spalling severity. For instance, Lu et al. [17] reported improved spalling resistance in specimens that were steam-cured at 90 °C for 48 h and subsequently dry-heat cured at 250 °C for 8–24 h, compared to those that were only water- or steam-cured. Some studies have attempted to use oven drying to suppress spalling; however, for realistic spalling assessments, it is recommended that the curing regime reflects the intended application.

Fig. 3. Different sized ECC specimens (1.25% PE + 0.75% basalt fiber) under ISO heating rate (adapted from [14])

Therefore, to ensure a reliable assessment of spalling resistance, it is advisable to use large-scale specimens such as 300 × 300 panels, apply curing regimes consistent with the intended end-use, and adopt heating protocols that replicate real fire conditions, such as the ISO 834 standard or a project-specific fire curve. Where full-scale furnace setups are unavailable, comparable rapid heating profiles can be simulated using a blowtorch test [14, 18], which provides a practical and reasonably accurate representation of spalling behavior in structural applications.

3.2 Is Cementitious Matrix Important?

In ECC, composition of the cementitious matrix, particularly the type of binder used, plays a crucial role in affecting its spalling resistance under high-temperature conditions. This is especially significant as ECCs do not contain coarse aggregates.

Among commonly used SCMs, silica fume (SF) is often associated with poor performance under high-temperature exposure [6]. While SF enhances mechanical strength at room temperature by refining the pore structure, this densification can hinder moisture migration and may result in increased pore pressure and a higher likelihood of spalling. In our recent study, significant spalling was observed in SF-containing mixes, even at a relatively low heating rate of 2 °C/min [16]. These findings align with Eurocode 2 [4], which provides special recommendations when SF is included in concrete designed for fire resistance.

On the other hand, SCMs such as fly ash and ground granulated blast furnace slag (GGBFS) are generally found to offer better performance at high temperatures. These materials promote the formation of thermally stable hydration products and contribute to a more porous matrix, facilitating moisture migration during heating [19, 20]. Contradictory findings do exist, but such discrepancies are often due to additional influencing factors. For instance, Cao et al. [21] reported increased spalling when the fly ash content was raised from 30% to 70%. However, their mixes also contained hybrid steel-basalt fibers, which may have contributed to the observed behavior. Ternary or quaternary blends incorporating GGBFS, fly ash, dolomite powder, or magnesium oxide have also shown promise in enhancing thermal performance and mitigating spalling [15, 16, 19, 20]. Similarly, combined use of high-volume fly ash mixtures and the use of steel slag aggregates has also been associated with improved thermal performance [22–24].

Another promising approach involves the complete replacement of ordinary Portland cement with calcium aluminate cement (CAC) in UHPC. It was reported in a recent study that CAC-based UHPCs exhibit enhanced spalling resistance due to reduced vapor pressure buildup, less degradation of hydration products, and the formation of coarser pore structures [25]. However, these results were obtained under relatively slow heating rates (~3 °C/min), and further research is needed to confirm their effectiveness under more severe thermal exposures. Other alternative binders, such as calcined clay (e.g., in limestone calcined clay cement) [26], nano-silica [27], carbon nanotubes [28], and hollow glass or ceramic microspheres [2], have also been explored. However, the current body of evidence does not yet strongly support their effectiveness in reducing spalling. More research is required, particularly in the context of ECCs.

Overall, the particle size distribution and reactivity of SCMs are critical microstructural factors affecting spalling resistance. Fine SCMs, such as silica fume and metakaolin (calcined clay), typically refine the pore structure and densify the matrix, which can restrict vapor movement and increase spalling risk. In contrast, coarser SCMs like fly ash and GGBFS tend to offer better matrix permeability and thermal resilience, thereby reducing the likelihood of spalling under fire exposure. Therefore, if the particle size of reactive SCMs, such as calcined clay, can be optimized without compromising mechanical performance, improved spalling resistance may be achievable.

3.3 Role of Fibers

Fibers are integral to FRCCs for enhancing mechanical performance and mitigating spalling under elevated temperatures. However, the influence of fiber type on spalling resistance is complex and several questions need to be clarified as explained below.

Are PP Fibers Suitable?: PP fibers are commonly considered effective for applications where spalling resistance is a priority [29]. However, their suitability depends on specific dosage, fiber length, and aspect ratio. For instance, at fiber lengths of 9, 12, and 18 mm, even a fiber dosage of 0.33% (~3 kg/m^3 > Eurocode recommended dosage) has been shown to cause spalling when the fiber diameter is increased from 28 μm to 100 μm. Therefore, PP fibers can provide adequate spalling resistance only when the appropriate combination of length and diameter is used [30]. The equation (Eq. 1) proposed by Zhang et al. [31] can offer a general guide for determining the appropriate dosage of PP fibers for effective spalling resistance. It is important to note that PP fibers should be used in conjunction with other fiber types, as they alone cannot provide strain-hardening behaviour if that is a desired characteristic.

$$\%\ \text{PP fiber volume (for non} - \text{spalling)} = 55.12\left(\frac{\text{Length of fiber}}{\text{Diameter of fiber}}\right)^{-0.87} \quad (1)$$

Are PE Fibers Suitable?: The consensus is that PE fibers generally do not offer any improvement in spalling resistance. Studies by Liu and Tan [32], as well as Zhang et al. [33, 34], have reported spalling in ECC specimens containing PE fibers at volumes ranging from 1.3% to 1.8%. However, our recent study [14] demonstrated that 2% dosage of 12 mm PE fibers successfully prevented spalling in ECC specimens exposed to both one-directional (1D) and three-directional (3D) fire tests under ISO-equivalent heating conditions (Fig. 4). Although this dosage exceeds those reported in earlier studies, it consistently performed well across all tested specimen sizes.

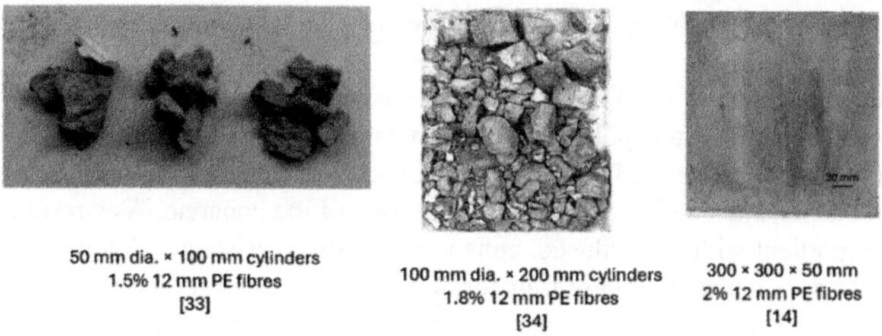

Fig. 4. PE fibres reinforced ECC after fire exposure (adapted from [14, 33, 34])

It should be noted that while the differences in outcomes may have stemmed from other variables, such as the cementitious matrix composition and specific test conditions, our investigation [14] indicated that PE fibers can enhance spalling resistance when used at a sufficiently high dosage and combined with a compatible matrix. Notably, the effective dosage of 2% aligns with typical ECC design, suggesting that PE fibers remain a practical and viable option. Therefore, it would be unfair to dismiss PE fibers altogether in fire-resistant applications considering the additional benefits they offer (e.g. better tensile performance).

What Other Types of Fibers are Suitable for Providing Spalling Resistance?: Several alternative fibers have been explored for improving spalling resistance in ECC and UHPC. Among them, PVA fibers are widely used in ECC design and have demonstrated effective spalling resistance at dosages of 1.5–2% by volume with 12 mm fiber length [35], despite their relatively high melting point compared to PE fibers. Liu and Tan [36] attributed this performance to changes in the interfacial transition zone (ITZ) between 105 °C and 200 °C, where a slight increase in ITZ improved matrix permeability. Moreover, at around 200 °C, shrinkage of the PVA fibers created disc-shaped voids, which served as vapor reservoirs and helped alleviate internal pressure buildup. Polyamide fibers are another promising alternative due to their very high coefficient of thermal expansion [37]. Their expansion under heat generates radial microcracks around the fibers, which enhances permeability and can suppress spalling.

Natural fibers, such as jute and flax, have also demonstrated effective spalling resistance [38–40]. Jute fibers have a straw-like structure that facilitates the release of vapor up to 390 °C. During heating, this structure may carbonize, creating gaps at the fiber-concrete interface to release vapor and reduce spalling. Flax fibers follow similar mechanism, shrinking at elevated temperatures and creating interfacial gaps with the surrounding cement matrix.

Though steel fibers are commonly used in FRCCs but are typically not effective on their own in preventing spalling. The theory behind their use is that their expansion under heat may generate microcracks, allowing for the release of vapor pressure. However, this mechanism is more effective when combined with other fibers, as high dosages of steel fibers alone (e.g., 2% by volume) may still result in spalling [41, 42]. Basalt fibers, which also have a high melting point, have also been explored for spalling resistance. However, these fibers suffer agglomeration issues [14] and therefore, may not be the best choice if added alone at high dosage. Spalling has been found when basalt fibers are present in the matrix [43].

Recently, Zhang et al. [44] developed flame-retardant polymer fibers incorporating additives like ammonium polyphosphate, melamine, magnesium hydroxide and aluminum hydroxide. These fibers expand when exposed to heat, filling thermal stress-induced cracks and slowing down the heating rate of the concrete. As a result, the temperature gradient within is reduced, enhancing spalling resistance. Additionally, chemical reactions, such as the phase transformation of PP resin and the foaming reactions of metallic organic flame retardants, can improve pore connectivity and help prevent explosive spalling.

Are Hybrid Fibers (Combined Especially with PE) Suitable?: Hybrid fiber composites, particularly when PE fibers are used in combination with high-melting-point fibers (e.g., steel or basalt), have been proposed to enhance spalling resistance. However, their effectiveness has been inconsistent. For instance, Li et al. [45] observed spalling at 600 °C in ultra-high-performance fiber-reinforced concrete containing 1% PE and 2% steel fibers, whereas Liu et al. [32] found that a combination of 1.25% PE and 0.75% steel fibers offered adequate spalling resistance. In our recent investigation, ECC reinforced with 12 mm 1.25% PE and 0.75% basalt fibers experienced significant spalling under both 1-D and 3-D exposure tests [14]. However, spalling was mitigated when the length of the PE fibers was increased to 18 mm or when 0.3% PP fibers were added. These findings suggest that hybrid fiber composites may be effective in preventing spalling, but the specific type and dosage of fibers must be carefully optimized. Notably, the inclusion of even small amounts of PP fiber generally improves performance [14, 33, 34].

Carbon fibers, when used alongside polymer fibers, have shown promise in improving mechanical performance and providing sufficient spalling resistance. Wang et al. [46] demonstrated that 0.5–1% carbon fiber, replacing PVA fiber, resulted in minimal spalling, while residual strength was significantly enhanced. Polyoxymethylene (POM) fibers are another alternative, often used in conjunction with PE fibers. Despite their high melting point (~280 °C) and vaporization temperature (~420 °C), they can contribute to pore coarsening and shrinkage, which can help with vapor dissipation [47]. Polybenzimidazole (PBI) fibers, which are stable up to 1050 °C, could also be potential alternatives to high-melting-point fibers [48]. However, research on these new types of fibers (except PP, PE and PVA) is still very limited, and their spalling performance requires further investigation.

3.4 Possible Mechanism for Spalling Mechanism

Based on the above review, it is evident that fiber dosage, type, and geometry significantly impact spalling resistance. However, several unconventional factors also play a crucial role as discussed below.

Fiber Melting, Vaporization and Melt Flow Index: Fiber melting is essential for forming interconnected porous networks in ECC, but the vaporization point of fibers is equally important. Only after vaporization is the fiber completely removed from the matrix, making the channels fully effective in aiding vapor release. Additionally, melting point of fibers should be considered in conjunction with their melt flow index (MFI), as the melting point alone does not provide a complete picture.

MFI is a key parameter for evaluating the processability of thermoplastics and indicates the melt viscosity of a polymer. PE fibers, despite having a lower melting point than PP fibers, have a lower MFI and higher vaporization point, which reduces their spalling resistance effectiveness. High-density PE fibers (HDPE) have an MFI of 0.1–3 g/10 min, significantly lower than that of low-density PE fibers (0.1–30 g/10 min) or PP fibers (1–50 g/10 min) [49]. Therefore, the melting of HDPE fibers is usually not as effective as that of PP fibers. The commonly used PE fiber type in ECC, ultra-high molecular weight PE fibers (UHMWPE), have even higher molecular weight than HDPE, and therefore they exhibit even higher melt viscosities and lower MFI. Knack [50] compared PP and PE fibers with different MFIs and found that spalling resistance improved as MFI increased, as this allows the polymer to flow more freely and diffuse more effectively. POM fibers may also offer a good alternative when used with PE fibers. However, POM fibers are available in different grades based on their MFIs (e.g., K30, K90, K130, K270, K520) [51], and therefore, should be carefully selected for optimal performance.

Fiber-Matrix Bonding and Thermal Expansion: Fiber-matrix bonding and thermal expansion significantly impact spalling resistance, particularly under 1-D exposure. Our recent study [14] demonstrated that, after 30 min of exposure reaching 800 °C under 1-D exposure, the temperature at the center and back of the specimen only reached about <120 °C and <50 °C, respectively (Fig. 5) indicating that melting alone is insufficient to explain spalling. Typically, PP and PE fibers bond weakly with the cement matrix, whereas basalt fibers exhibit stronger bonding. Weak fiber-matrix bonding creates spaces (known as *pressure-induced tangential space*) that facilitate water and vapor transport and help in pore pressure dissipation. In our study, ECC specimens incorporating 1.25% PE and 0.75% basalt (12 mm) fibers exhibited spalling. This was attributed to the PE fibers failing to form a continuous porous network for vapor transport, and the basalt fibers interrupting this network due to their stronger adhesion. However, increasing the PE fiber length to 18 mm improved the performance, likely because the longer fibers helped form a more effective network for pore pressure dissipation.

Thermal expansion coefficients (CTEs) of fibers also play a crucial role in spalling. When temperature mismatch between fibers and the matrix occurs, cracks can form, helping polymer fibers assist with pore pressure dissipation even without melting. PP fibers, for example, have a thermal expansion coefficient that is ten times higher than that of concrete [52]. As the temperature rises, the differing expansion rates between PP fibers and the matrix can cause microcracks, increasing the matrix permeability before fiber melting occurs. Additionally, the CTE of fibers increases with temperature which further aids in the increase in the permeability. For instance, the CTE of low-density PE fibers ranges from 0.2 to 2.1 $\times$ 10^{-3}/°C between 25 °C–105 °C, while for PP fibers, the CTE ranges from 0.1 to 3.6 $\times$ 10^{-3}/°C between 25 °C–150 °C [37]. This may also have been the reason why PP fibres show better spalling resistance compared to PE fibre. These properties, however, depend on the exact polymer composition and should be evaluated before use.

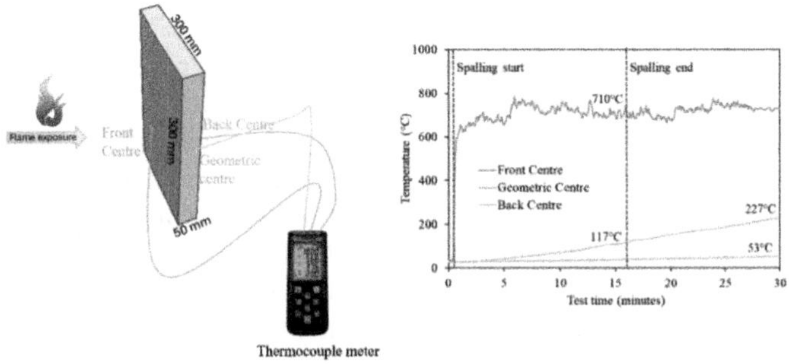

Fig. 5. Temperature at different points in ECC panel specimen under 1-D fire exposure (adapted from [14])

Fiber Distribution: Fiber distribution within the mix is another factor that is often overlooked [14]. Different fibers distribute differently in the cementitious matrix. Basalt fibers tend to clump together, while PP and PVA fibers generally distribute more evenly. PE and steel fibers can also be uniformly distributed with proper adjustments in the mix. A uniform distribution of polymer fibers helps form an interconnected network, which is crucial for effective spalling resistance.

4 Recommendations for Spalling Resistance Design of FRCC

Designing FRCCs for fire-resistant applications requires careful consideration of several interrelated factors. Thermo-hygral spalling, which is driven by the buildup of pressure during heating, can only be mitigated if a network forms within the matrix to facilitate vapor transfer. This depends on the combined influence of fiber properties, matrix characteristics, and test conditions.

- Test Parameters: Low heating rates and small specimen sizes may underestimate spalling risk. Therefore, to ensure a reliable assessment of spalling resistance, large-scale specimens, such as 300 × 300 × 20 mm panels, should be used. Curing conditions that replicate the intended end-use of the material should be applied, and heating protocols should closely mirror real fire conditions. Where full-scale furnace testing is not feasible, blowtorch test can offer a practical alternative for replicating rapid heating profiles.
- Cementitious Matrix: Particle size distribution and reactivity of SCMs are key microstructural factors influencing spalling. Finer SCM, such as silica fume, tend to densify the matrix, hindering vapor migration. In contrast, coarser SCMs like fly ash and GGBFS promote a more permeable and thermally resilient pore network, reducing spalling risk during fire exposure.
- Fibers: The ability of fibers to form an interconnected percolation network is essential for spalling resistance. Key factors such as fiber length, aspect ratio, and dosage are critical when selecting the appropriate fiber type for FRCC design. However, the selected fiber type should be thoroughly characterized before use for properties such as melting point, MFI, vaporization point, bonding with the matrix, thermal expansion, and ease of uniform distribution. Only when these parameters are carefully accounted for, spalling resistance can be significantly enhanced. A summary of these recommendations is further presented in Fig. 6.

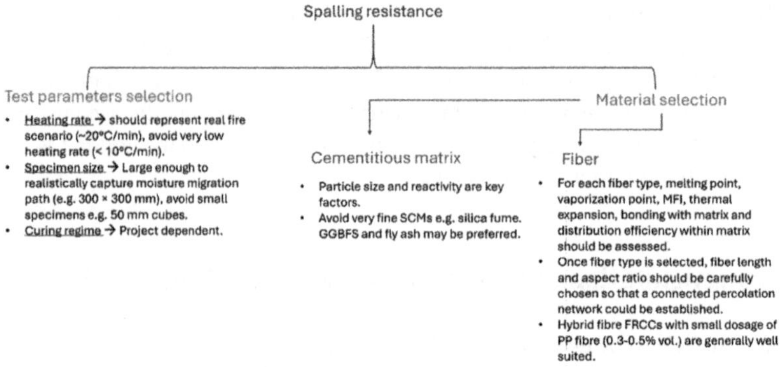

Fig. 6. Recommendations for material and test parameters selection to achieve better spalling resistance

5 Conclusions

This review critically evaluated current research on the spalling resistance of FRCCs and ECCs. While there is broad agreement that thermo-hygral spalling can be mitigated through the formation of vapor-transfer channels within the matrix, the precise mechanisms and design strategies to reliably achieve this remain unclear. By analyzing existing findings, this work identified the key matrix, fiber, and testing parameters that govern spalling performance. The recommendations outlined in this review further offer a practical framework to guide future research and the design of FRCCs with enhanced spalling resistance.

References

1. Li, V.C.: Engineered Cementitious Composites (ECC): Bendable Concrete for Sustainable and Resilient Infrastructure. Springer, Cham (2019)
2. Yang, J., Kang, H., Yang, J., Pae, J., Shi, C., Moon, J.: Enhancing fire resistance of lightweight high-performance cementitious composites using hollow microspheres. Cement Concr. Compos. **155**, 105845 (2025)
3. Upasani, N., Bansal, M., Satapathy, A., Rawat, S., Muthukumar, G.: Design and performance criteria for fire-resistant design of structures—an overview. Adv. Struct. Technol. Sel. Proc. CoAST **2019**, 277–294 (2020)
4. EN 1992-1-2: Eurocode 2: Design of concrete structures - Part 1–2: General rules - Structural fire design (2004)
5. Yu, K., Wang, Y., Yu, J., Xu, S.: A strain-hardening cementitious composites with the tensile capacity up to 8%. Constr. Build. Mater. **137**, 410–419 (2017)
6. Rawat, S., Lee, C.K., Zhang, Y.X.: Performance of fibre-reinforced cementitious composites at elevated temperatures: a review. Constr. Build. Mater. **292**, 123382 (2021)
7. Hertz, K.D.: Limits of spalling of fire-exposed concrete. Fire Saf. J. **38**(2), 103–116 (2003)
8. Boström, L., Jansson, R.: The age effect on fire spalling of concrete. In: 2nd International RILEM Workshop on Concrete Spalling due to Fire Exposure, vol. 1, pp. 33–41 (2011)
9. Liu, J.C., Tan, K.H., Yao, Y.: A new perspective on nature of fire-induced spalling in concrete. Constr. Build. Mater. **184**, 581–590 (2018)

10. Fu, Y., Li, L.: Study on mechanism of thermal spalling in concrete exposed to elevated temperatures. Mater. Struct. **44**, 361–376 (2011)
11. Zeiml, M., Lackner, R., Mang, H.A.: Experimental insight into spalling behavior of concrete tunnel linings under fire loading. Acta Geotech. **3**, 295–308 (2008)
12. Choe, G., Kim, G., Yoon, M., Hwang, E., Nam, J., Guncunski, N.: Effect of moisture migration and water vapor pressure build-up with the heating rate on concrete spalling type. Cem. Concr. Res. **116**, 1–10 (2019)
13. Rawat, S., Zhang, Y.X., Lee, C.K.: Effect of specimen size and shape on the compressive performance of high strength engineered cementitious composites at elevated temperatures. Innov. Infrastruct. Sol. **9**(8), 317 (2024)
14. Rawat, S., Zhang, L., Zhang, Y.X.: Fire-induced spalling in hybrid polyethylene fiber-reinforced engineered cementitious composite panels. Eng. Struct. **338**, 120589 (2025)
15. Rawat, S., Zhang, Y.X., Lee, C.K.: Spalling resistance of hybrid polyethylene and steel fiber-reinforced high-strength engineered cementitious composite. In: Nanotechnology in Construction for Circular Economy: Proceedings of NICOM7, 31 October–02 November 2022, Melbourne, Australia, vol. 356, p. 321 (2023)
16. Rawat, S., Vongsvivut, J., Zhang, L., Zhang, Y.X.: Mechanical performance and microstructure evolution of MgO-doped high volume GGBFS-based engineered cementitious composites at room and elevated temperatures. J. Build. Eng. **98**, 111437 (2024)
17. Lu, J.X., Shen, P., Sun, Y., Poon, C.S.: Strategy for preventing explosive spalling and enhancing material efficiency of lightweight ultra high-performance concrete. Cem. Concr. Res. **158**, 106842 (2022)
18. Rawat, S., Zhang, L., Zhang, Y.X.: A Cost-effective Testing Method for Spalling Resistance of Cementitious Composites Using Propane Torch, ICES2024, 8–10 November 2024, Guangzhou, China (2024)
19. Rawat, S., Lee, C.K., Zhang, Y.X.: Green engineered cementitious composites with enhanced tensile and flexural properties at elevated temperatures. Clean. Mater. **12**, 100240 (2024)
20. Rawat, S., Zhang, Y.X., Fanna, D.J., Lee, C.K.: Development of sustainable engineered cementitious composite with enhanced compressive performance at elevated temperatures using high volume GGBFS. J. Clean. Prod. **451**, 142011 (2024)
21. Cao, K., Liu, G., Li, H., Huang, Z., Ji, X.: Effects of high temperature and substitution rates of fly ash on the mechanical properties and microstructures of steel-basalt hybrid fibers reinforced cementitious composites. Constr. Build. Mater. **352**, 128895 (2022)
22. Tangirala, A., Rawat, S., Lahoti, M.: High volume fly ash and basalt-polypropylene fibres as performance enhancers of novel fire-resistant fibre reinforced cementitious composites. J. Build. Eng. **78**, 107586 (2023)
23. Tangirala, A., Rawat, S., Tan, K.H., Lahoti, M.: Enhanced thermal performance of fiber-reinforced cementitious composite with high-volume fly ash and steel slag aggregates. J. Mater. Civ. Eng. **37**(7), 04025211 (2025)
24. Liang, X., Wu, C., Su, Y., Chen, Z., Li, Z.: Development of ultra-high performance concrete with high fire resistance. Constr. Build. Mater. **179**, 400–412 (2018)
25. Khan, M., Lao, J., Ahmad, M.R., Kai, M.F., Dai, J.G.: The role of calcium aluminate cement in developing an efficient ultra-high performance concrete resistant to explosive spalling under high temperatures. Constr. Build. Mater. **384**, 131469 (2023)
26. Hassan, A.M.S., Shoukry, H., Perumal, P., Abd El-razik, M.M., Aly, R.M., Alzahrani, A.M.Y.: Evaluation of the thermo-physical, mechanical, and fire resistance performances of limestone calcined clay cement (LC3)-based lightweight rendering mortars. J. Build. Eng. **71**, 106495 (2023)

27. Mohammed, B.S., Achara, B.E., Liew, M.S., Alaloul, W.S., Khed, V.C.: Effects of elevated temperature on the tensile properties of NS-modified self-consolidating engineered cementitious composites and property optimization using response surface methodology (RSM). Constr. Build. Mater. **206**, 449–469 (2019)
28. Lu, H., Yao, Y.: Spalling mechanism of carbon nanotube concrete at elevated temperature. Constr. Build. Mater. **314**, 125594 (2022)
29. Hager, I., Mróz, K.: Role of polypropylene fibres in concrete spalling risk mitigation in fire and test methods of fibres effectiveness evaluation. Materials **12**(23), 3869 (2019)
30. Zhang, D., Tan, K.H.: Critical fibre dimensions for preventing spalling of ultra-high performance concrete at high temperature. In: The 11th International Conference on Structures in Fire, The University of Queensland, Australia, 30 November–2 December (2020)
31. Zhang, D., Zhang, Y., Dasari, A., Tan, K.H., Weng, Y.: Effect of spatial distribution of polymer fibers on preventing spalling of UHPC at high temperatures. Cem. Concr. Res. **140**, 106281 (2021)
32. Liu, J.C., Tan, K.H.: Fire resistance of ultra-high performance strain hardening cementitious composite: residual mechanical properties and spalling resistance. Cem. Concr. Compos. **89**, 62–75 (2018)
33. Zhang, D., Liu, Y., Tan, K.H.: Spalling resistance and mechanical properties of strain-hardening ultra-high performance concrete at elevated temperature. Constr. Build. Mater. **266**, 120961 (2021)
34. Zhang, Z., Li, Z., He, J., Qian, S., Shi, X.: Recycled mask polypropylene microfibers benefit tensile properties and prevent thermally induced spalling of high-strength engineered cementitious composite (HS-ECC). J. Clean. Prod. **457**, 142476 (2024)
35. He, J., Wang, Q., Yao, B., Ho, J.: Impact of elevated temperatures on the performance of high-strength engineered cementitious composite. J. Mater. Civ. Eng. **33**(9), 04021222 (2021)
36. Liu, J.C., Tan, K.H.: Mechanism of PVA fibers in mitigating explosive spalling of engineered cementitious composite at elevated temperature. Cem. Concr. Compos. **93**, 235–245 (2018)
37. Zhang, D., Tan, K.H.: Effect of various polymer fibers on spalling mitigation of ultra-high performance concrete at high temperature. Cem. Concr. Compos. **114**, 103815 (2020)
38. Zhang, D., Tan, G.Y., Tan, K.H.: Combined effect of flax fibers and steel fibers on spalling resistance of ultra-high performance concrete at high temperature. Cem. Concr. Compos. **121**, 104067 (2021)
39. Ozawa, M., Parajuli, S.S., Uchida, Y., Zhou, B.: Preventive effects of polypropylene and jute fibers on spalling of UHPC at high temperatures in combination with waste porous ceramic fine aggregate as an internal curing material. Constr. Build. Mater. **206**, 219–225 (2019)
40. Ridha, M.M.: Combined effect of natural fibre and steel fibre on the thermal-mechanical properties of UHPC subjected to high temperature. Cem. Concr. Res. **180**, 107510 (2024)
41. Li, Y., Tan, K.H., Yang, E.H.: Synergistic effects of hybrid polypropylene and steel fibers on explosive spalling prevention of ultra-high performance concrete at elevated temperature. Cem. Concr. Compos. **96**, 174–181 (2019)
42. Park, G.K., Park, G.J., Park, J.J., Lee, N., Kim, S.W.: Residual tensile properties and explosive spalling of high-performance fiber-reinforced cementitious composites exposed to thermal damage. Materials **14**(7), 1608 (2021)
43. Xie, D., Hu, Z., Wang, B., Zhao, K., Zhou, J.: Effect of PVA fiber and basalt fiber on residual mechanical and permeability properties of ECC after elevated temperatures. Constr. Build. Mater. **433**, 136610 (2024)
44. Zhang, T., Zhang, M., Shen, Y., Zhu, H., Yan, Z.: Mitigating the damage of ultra-high performance concrete at elevated temperatures using synergistic flame-retardant polymer fibres. Cem. Concr. Res. **158**, 106835 (2022)

45. Li, Y., Yang, E.H., Tan, K.H.: Flexural behavior of ultra-high performance hybrid fiber reinforced concrete at the ambient and elevated temperature. Constr. Build. Mater. **250**, 118487 (2020)
46. Wang, Q., Zhou, Y., Lai, M., Gu, M., Ho, J.C.M.: Carbon fiber to improve the resistance of high strength PVA-ECC to elevated temperatures. J. Build. Eng. **71**, 106475 (2023)
47. Yu, P., Yu, C., Chen, Z., Ren, Z.: Mechanical behavior of ultra-high toughness engineered cementitious composite using seawater coral sand exposed to elevated temperature. Constr. Build. Mater. **473**, 141016 (2025)
48. Fares, G., Khan, M.I., Mourad, S., Abbass, W.: Evaluation of PVA and PBI-based engineered-cementitious composites under different environments. Constr. Build. Mater. **85**, 109–118 (2015)
49. Cathetermelt.com. https://cathetermelt.com/melt-flow-index-of-various-polymer-materials/. Accessed 11 May 2025
50. Knack, I.: The use of PP fibers in tunnel construction to avoid explosive concrete spalling in case of fire. New test results for the clarification of the mode of action. In: Proceedings of the 2nd International RILEM Workshop on Concrete Spalling due to Fire Exposure, Delft, The Netherlands, pp. 5–7 (2011)
51. Add-tek.com. https://www.add-tek.com/product/advance-chemicals/polyoxymethylene.html. Accessed 11 May 2025
52. Khoury, G.A.: Polypropylene fibres in heated concrete. Part 2: pressure relief mechanisms and modelling criteria. Mag. Concr. Res. **60**(3), 189–204 (2008)

Synthesis and Processing of Dense Tricalcium and Dicalcium Silicates for Surface-Sensitive Characterization

Aidyn Tugelbayev[1(✉)], Carlos A. Fortulan[2], David Jarret Wright[3], Carolina Tallon[4], and Alexander S. Brand[1,4]

[1] The Charles E. Via, Jr. Department of Civil and Environmental Engineering, Virginia Polytechnic Institute and State University, Blacksburg, VA, USA
aidyn@vt.edu

[2] Department of Mechanical Engineering, University of São Paulo, São Carlos, SP, Brazil

[3] Institute for Critical Technology and Applied Science, Nanoscale Characterization and Fabrication Laboratory, Virginia Polytechnic Institute and State University, Blacksburg, VA, USA

[4] Department of Materials Science and Engineering, Virginia Polytechnic Institute and State University, Blacksburg, VA, USA

Abstract. Tricalcium silicate (Ca_3SiO_5, C_3S) and dicalcium silicate (Ca_2SiO_4, C_2S) are the most critical mineral components in portland cement, governing early-age and long-term hydration and strength development. Understanding their structural and surface behavior is critical for advancing cement chemistry and improving concrete mechanical performance and durability. Currently, most studies rely on powder forms of calcium silicates that limit advanced surface analyses of these materials. This study addresses the need for bulk, low-porosity calcium silicate samples suitable for high-resolution surface characterization by presenting a robust, reproducible method for synthesizing dense polycrystalline C_3S and β-C_2S bulk specimens with 5% to 8% porosity. Mixtures of calcium carbonate and amorphous silica were sintered at 1450 °C and 1600 °C, with and without stabilizing ions, to produce dense polycrystalline β-C_2S and monoclinic and triclinic C_3S bodies, respectively. The powders were granulated with a polyvinyl butyral binder in isopropanol and compacted via cold isostatic pressing, followed by a final sintering step to ensure homogeneity. The study presents detailed optimization procedures for sample preparation, including binder selection, pressing conditions, and sintering protocols, while offering practical recommendations and troubleshooting guidance to address potential challenges, such as phase uniformity and polishing requirements. The polishing of the product yields samples with smooth surfaces suitable for surface-sensitive analyses and characterization of crystallographic orientation, including scanning electron microscopy (SEM) and electron backscatter diffraction (EBSD). By enabling consistent synthesis of dense calcium silicates, this work supports advanced experimental investigations into the microstructural evolution of cement phases, and contributes to the design of more sustainable and durable construction materials.

Keywords: Solid-State Synthesis · Sintering · Cement · Tricalcium Silicate · Dicalcium Silicate · Alite · Belite · Cold Isostatic Pressing · Powder Processing

M. Posani and S. Singh (Eds.): RYS 2025, RILEM Bookseries 67, pp. 198–211, 2026.
https://doi.org/10.1007/978-3-032-14652-6_16

1 Introduction

Tricalcium silicate (C_3S) and dicalcium silicate (C_2S) are the fundamental silicate constituents of portland cement, comprising up to 85% of the mineralogical phases responsible for the cements' mechanical properties, hydration behavior, and long-term durability. The synthesis of the pure calcium silicates is crucial for cement chemistry studies, as their hydration reactions are pivotal in the development of concrete and other cementitious materials [1–3]. Consequently, there is growing interest in developing scalable synthesis routes that enable both fundamental research and practical application in cement formulation and sustainability. Although most studies have reported calcium silicate synthesis in powder form [1, 3–5], efficient reproducible methods for the synthesis of dense, low-porosity polycrystalline bodies, critical for surface analysis, remain underexplored. So far, Naber et al. [6] reported on the synthesis of monocrystalline alite using optical floating zone method, which produced a large crystal of C_3S after heat treating the sample for 16 days at 1600 °C. While being able to produce a monocrystalline sample, this method requires both advanced setup and significant amount of time to successfully yield the product. However, there is little literature on time- and cost-efficient synthesis of dense bodies of calcium silicates. The existing gap in literature hinders further insight into the crystallographic and surface characteristics of cement phases.

This study presents a novel method of producing dense polycrystalline solid bodies through the use of cold isostatic pressing of the C_3S and C_2S powders mixed with polyvinyl butyral (PVB) as a binder. Sintering protocols adapted from prior ceramic processing studies [7] were optimized to reduce porosity and increase grain size, enabling high-quality surface preparation. The study also offers guidelines for polishing procedures to prepare surfaces suitable for surface-sensitive analyses such as electron backscatter diffraction (EBSD).

The main objective of this study is to establish a reproducible route for preparing dense calcium silicate specimens tailored for advanced surface and crystallographic studies. This contributes to broader efforts in cement phase synthesis for experimental analyses to better understand the role of microstructure in hydration mechanisms, ultimately guiding the development of more durable and sustainable cementitious materials.

2 Materials and Methods

2.1 Materials

Calcium silicates were synthesized using calcium carbonate ($CaCO_3$, Aldon Corporation, USA) and amorphous fumed silica (SiO_2, BeanTown Chemical Corporation, USA) as raw base materials. The compositions of the C_3S and β-C_2S mixes are shown in Table 1.

For triclinic C_3S polymorph (C_3S-T), $CaCO_3$ and SiO_2 powders were wet-homogenized in a stoichiometric Ca/Si ratio of 3.15:1 in a ceramic jar with ZrO_2 balls for 24 h following the procedure by Li *et al.* [1]. Isopropyl alcohol was used as wet media to improve homogenization and avoid contamination that could result in the formation of monoclinic polymorph of C_3S. The homogenized slurry was placed into plastic cylindrical molds (Ø50 mm by 100 mm) and dried for 48 h in an oven at 105 °C. The resultant

Table 1. Mixture composition for C_3S and C_2S slurries

	Composition (wt. %)						C/S ratio
	CaO	SiO_2	Mn_2O_3	P_2O_5	B_2O_3	K_2SO_4	
C_3S-T	74.62	25.38					3.15
C_3S-M	74.32	25.28	0.20	0.20			
β-C_2SB	64.87	34.77			0.35		2.00
β-C_2SK	64.60	34.72				0.60	

dry shrunk mix cylinder was placed into a platinum crucible and sintered as discussed in Sect. 2.2. Since some portion of C_3S transforms into β-C_2S during cooling [1], the ratio of Ca/Si needed to be adjusted to increase the C_3S yield. From trial experiments, the amount of β-C_2S reduced from ~15% to 7% and 2% by increasing the Ca/Si ratio to 3.07 and 3.15, respectively. Thus, Ca/Si ratio of 3.15:1 was chosen due to lower yield of β-C_2S compared to the Ca/Si ratio of 3:1.

For monoclinic C_3S polymorph (C_3S-M), the procedure was the same as for the C_3S-T synthesis except for the addition of Mn_2O_3 (AK Scientific Inc., USA) and P_2O_5 (Thermo Fisher Scientific, USA) powders as stabilizing ions, where Mn^{2+} and PO_4^{3-} partially substitute Ca^{2+} and SiO_4^{4-}, respectively [4]. The mixture was wet-homogenized in deionized water and dried in the oven. The stabilizing manganese and phosphorus ions were chosen to avoid the formation of calcium aluminate, which could be formed when Al_2O_3 are added during C_3S-M synthesis [3].

For β-C_2S synthesis, the ratio of $CaCO_3$ to SiO_2 was 2:1, with the use of two different types of stabilizing ions, namely B_2O_3 (β-C_2S^B) (Thermo Fisher Scientific, USA) [3] and K_2SO_4 (β-C_2S^K) (BeanTown Chemical Corporation, USA) [8] to compare the β-C_2S polymorph formation. Similar to C_3S, the wet slurry was mixed for 24 h and dried for 48 h at 105 °C (Fig. 1).

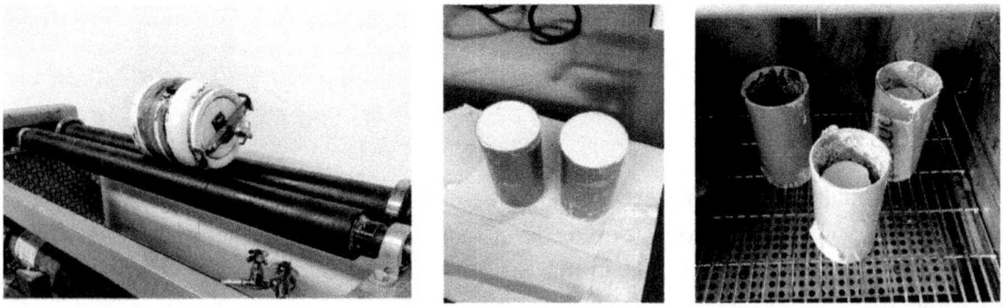

(a) (b) (c)

Fig. 1. Mixture preparation: (a) ball-milling jar with wet mixture, (b) cylinder molds filled with calcium silicate slurry, and (c) dried slurry ready for sintering.

2.2 Solid-State Synthesis *via* Sintering

The dried mixtures were thermally treated in a high temperature elevator furnace (Deltech, CO, USA). For C_3S, the temperature was ramped to 1600 °C at 5 °C/min, held for 3 h at 1600 °C (dwell time), followed by air-quenching with compressed air [1, 9]. For β-C_2S, the temperature was ramped to 1450 °C at 5 °C/min, held for 2 h, with the subsequent quenching the product with compressed air [3]. The images of cylinders before and after sintering are shown in Fig. 2.

2.3 Sample Preparation Using PVB Binder and Cold-Isostatic Pressing

The sample preparation using cold-isostatic pressing was adapted from the work reported by Fiocchi *et al.* [7]. The sintered calcium silicate was finely ground using ball milling in isopropyl alcohol as wet media, where the powder and IPA was mixed in 30 vol. % and 70 vol. %, respectively. To improve green body integrity and prevent cracking during pressing, polyvinyl butyral (PVB) was incorporated in two stages. First, 0.1 wt. % of PVB was added to the solution as a dispersant and mixed for 24 h. Next, 1.9 wt. % PVB was added to the solution as a binder, and the mixture was mixed for at least 2 h. The isopropanol was dried from the slurry using hot air from a hair dryer until the powder could be granulated and sequentially sieved to sizes passing #40 (425 μm), #50 (300 μm), and #80 (180 μm) mesh, achieving a free-flowing powder behavior. The progression of the procedure is shown in Fig. 3.

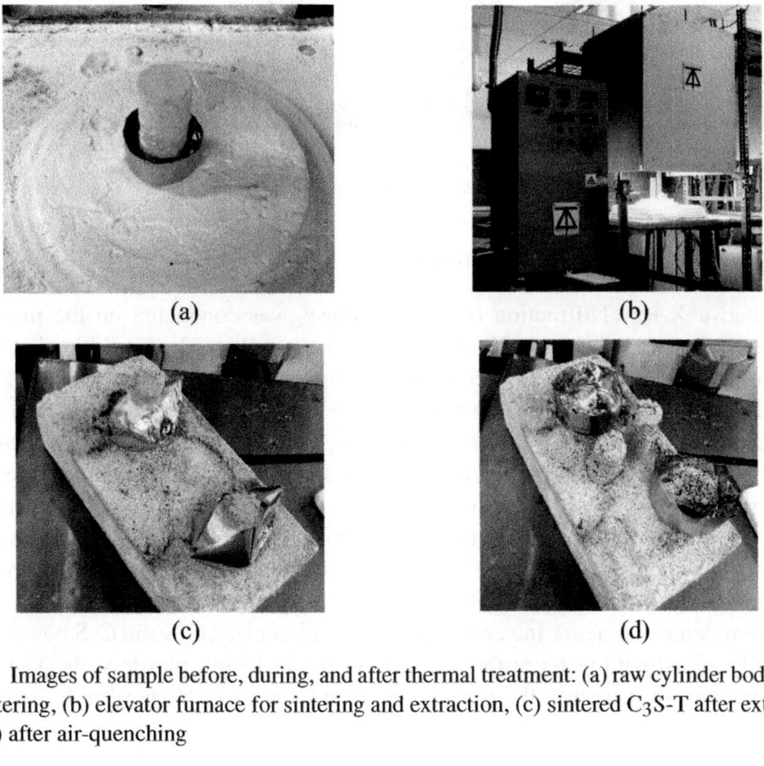

Fig. 2. Images of sample before, during, and after thermal treatment: (a) raw cylinder body ready for sintering, (b) elevator furnace for sintering and extraction, (c) sintered C_3S-T after extraction and (d) after air-quenching

The sample preparation through the use of cold isostatic pressing (CIP) is shown in Fig. 4. First, the powder with PVB was shape-formed into pellets using a uniaxial hand press (Ø9 mm by 8 mm and Ø25 mm by 6 mm) or a ceramic rod dye (Ø5 mm by 50 mm). The pellets were vacuum sealed into a glove and finally compacted in the cold isostatic press at 30 ksi (207 MPa) for 2 min. The pressed green bodies were sintered following the same protocol as in Sect. 2.1 and recommendations described in Sect. 4. The produced pellets exhibited density of 92% to 95% (*i.e.*, 8% to 5% porosity, respectively). The measurement of the skeletal (apparent) density was conducted using helium pycnometry, which was compared to the bulk density of the material measured with calipers and scales.

Fig. 3. Powder preparation procedure

2.4 Sample Characterization Methods

Quantitative X-Ray Diffraction (QXRD) analysis was conducted on the micronized powder of C_3S and C_2S with 50 wt. % rutile as internal standard, backloaded into the sample holder. The data were collected using Empyrean X-Ray Diffractometer (Malvern PANalytical Ltd., the Netherlands) working at 45 kV and 40 mA using copper Kα X-rays. Incident beam Soller slits of 0.04 rad were used, and the antiscatter and divergence slits were both fixed at 0.5°. The receiving Soller slits in the diffracted beam path were set at 0.04 rad. The measurement was conducted on a spinning sample stage between 8° and 65° with 0.013° 2θ step size for a total measurement time of about 120 min per sample. The data were analyzed through Rietveld refinement using Profex [10] software.

The heat of hydration of the calcium silicates was measured using I-Cal Flex isothermal calorimeter (Calmetrix Inc., MA, USA). The micronized C_3S and C_2S powders were manually mixed with water at the w/c ratio of 0.5 in 15 mL plastic vials. The heat of hydration was measured in the isothermal chamber at to 23 °C for 168 h. Since there were no changes observed in the heat of hydration after 40 h of the measurement, the analysis was conducted on the data starting from 20 min of the reaction and up until 40 h.

For electron backscatter diffraction (EBSD) analysis, the sintered pellet samples were epoxy impregnated and polished on a DualPrep 3 grinding/polishing system (Allied High Tech Products Inc., USA) using 400, 600, 800 and 1200 SiC grits and 3 μm, 1 μm and 0.25 μm diamond paste on an Imperial polishing microcloth pad (Allied High Tech Products Inc., USA). Propylene glycol was used as a lubricant during polishing. To ensure better quality of the surface flatness, the polishing time was 5 min on each SiC grit and 20 min on each microcloth pad with diamond paste. The sample rotation was kept at 50 RPM, whereas the platen was rotating at 100 RPM in the opposite direction.

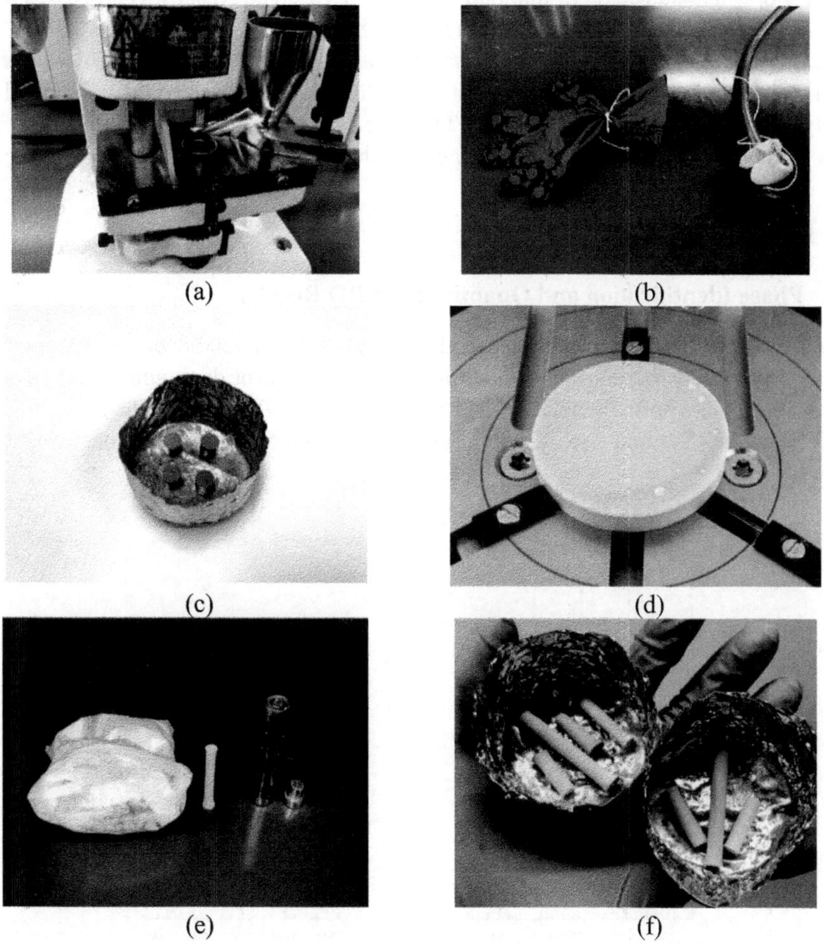

Fig. 4. Sample preparation for CIP. (a) Uniaxial hand press, (b) vacuum-sealed pellets for CIP submersion, (c) sintered C_3S-M pellets, (d) C_3S-T disc, (e) ceramic rod dye, and (f) C_3S-T and C_3S-M rod green bodies.

The polished samples were carbon coated using a Cressington 208 (Cressington Scientific Instruments Ltd., UK). The coating was applied in 3 bursts of 7 s (21 s total) for a total thickness of approximately 5 nm to 8 nm. Carbon was applied to samples in coatings from 2 nm to 20 nm thick as part of this study. Coatings below approximately 5 nm were generally not conductive enough to sufficiently reduce charging on the polished cross-sectional surfaces, particularly in areas that were relatively more porous. Coatings greater than 12 nm were generally observed to reduce the quality of the EBSD patterns that could be obtained on the samples.

The EBSD analysis was performed on a Helios 5 UC Scanning Electron Microscope and Focused Ion Beam (SEM-FIB) dual beam system (Thermo Fisher Scientific, USA) with EDAX Velocity Super EBSD camera (Gatan Inc., USA). The images were taken at accelerating voltage of 20 kV and probe current of 13 nA. The sample was analyzed at the working distance of 13 mm and stage tilt of 70°. The signal was processed using enhanced noise reduction setting through APEX software (AMETEK Inc., USA). The indexing of the crystallographic orientation was performed at a 0.5 μm step size per approximately 300 μm by 300 μm area for each sample. The scan data was plotted and analyzed using OIM9 software (AMETEK Inc., USA). Cleanup of the indexed areas was made by correlating the neighboring confidence indices (CI).

3 Results and Discussion

3.1 Phase Identification and Quantitative XRD Results

The qualitative XRD of the C_3S was conducted to confirm the stabilization of monoclinic polymorph, where triclinic and monoclinic polymorphs can be distinguished based on the peaks located at approximately 29° and 52° 2θ angle [1, 9]. Based on the diffractogram, the obtained C_3S-T and C_3S-M polymorphs are T1 and M3 (Fig. 5), respectively.

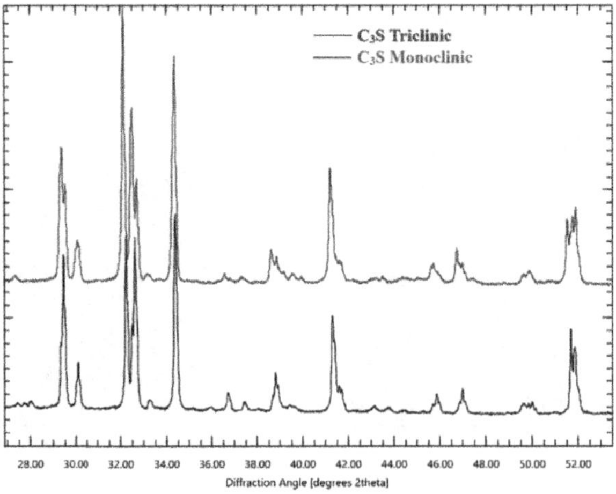

Fig. 5. XRD of C_3S polymorphs with characteristic peaks 29° and 52° 2θ angle.

The quantitative XRD with 50% rutile as internal standard was conducted to estimate the phase purity of calcium silicates. The results are summarized in Table 2. According to the results, the use of Mn^{3+} and P^{5+} stabilizing ions reduces the amount of β-C_2S formation in C_3S, whereas some of the C_3S-T product undergoes the conversion to β-C_2S [1, 3]. As for the C_2S stabilized by B_2O_3 and K_2SO_4, the use of K_2SO_4 seemed to be less stable and dusting of the product to form γ-C_2S was more pronounced compared to the use of B_2O_3. Also, while different stabilizers were used, and other proportions of the material have been used in the mixture design, the XRD pattern of both β-C_2S products revealed an unresolved peak at around 33.5° 2θ, which could affect the correct phase identification with currently available crystallographic information on C_2S (Fig. 6). The peak at ~33.5° was also present in other studies [11, 12], yet its presence was not as significant and could be omitted during Rietveld refinement. Cuesta *et al.* [13] attributed a similar peak to belong to α'$_H$-C_2S polymorph, while the diffractograms by Ghosh *et al.* [14] indicate that this peak is present for β-C_2S. Thus, this peak could indicate the resultant crystal distortion from the transformation from α-C_2S to β-C_2S phase [15].

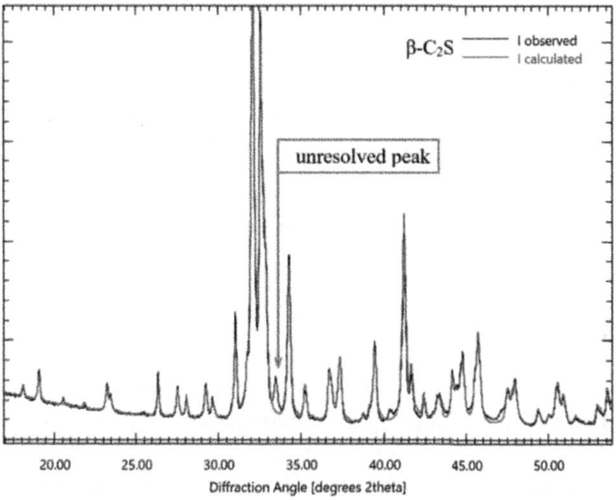

Fig. 6. XRD of β-C_2S with an unresolved peak at 33.4 degrees 2θ.

Table 2. Quantitative XRD Results using 50% rutile as internal standard. The uncertainty represents Rietveld refinement regression error.

Sample	Phase composition (%)					
	C_3S	β-C_2S	γ-C_2S	α-C_2S	CaO	Amorphous
C_3S-T	87.0 ± 0.6	7.6 ± 0.4	-	-	-	5.4 ± 0.7
C_3S-M	92.7 ± 0.4	0.2 ± 0.1	-	-	~0.1	7.0 ± 0.4
β-C_2S^B	-	77.1 ± 1.2	8.6 ± 0.5	4.9 ± 0.2	-	9.4 ± 0.9
β-C_2S^K	-	69.3 ± 0.7	11.1 ± 0.4	14.4 ± 0.4	-	5.3 ± 0.7

3.2 Isothermal Calorimetry

The isothermal calorimetry results are shown in Fig. 7. As expected, C_3S powders released higher heat in the early periods compared to C_2S, where C_3S-M exhibited slightly retarded rates probably due to Mn^{3+} and P^{5+} ion doping. Similar retardation behavior was observed by Neto *et al.* [16], where the slower hydration of monoclinic C_3S was due to the doping by Al^{3+} ions.

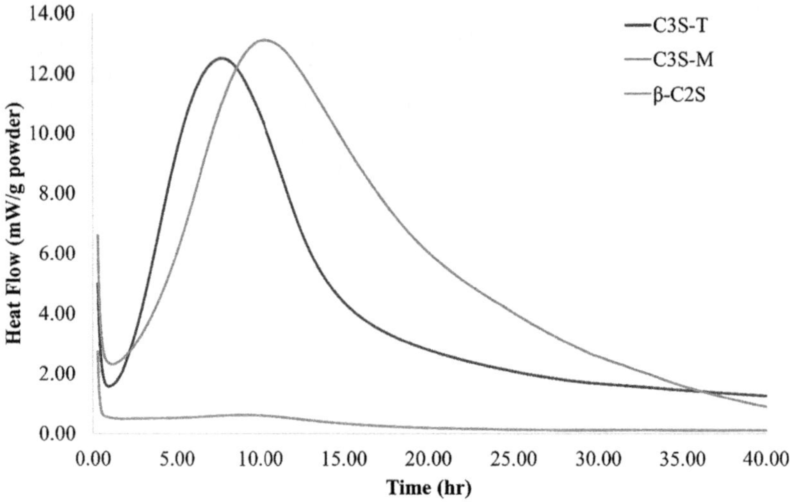

Fig. 7. Heat of hydration of calcium silicates.

3.3 EBSD Indexing

Figure 8 shows the obtained EBSD-indexed inverse pole figure (IPF) maps on the surfaces of triclinic and monoclinic C_3S using diffraction patterns reported by De la Torre *et al.* [5, 9]. The grain orientations are labelled based on the most repeated plane index with confidence index (CI) larger than 0.1 within the respective grain area [6, 17]. The samples shown in Fig. 8 were sintered at 8 h to 10 h of dwell time, which resulted in larger grain size, and the porosity measured for these samples ranged from 5% to 8%. Occasional isolated porosity within the larger grains could result from trapped air or insufficient particle packing during isostatic pressing. At shorter dwell time (3 h), the produced C_3S showed higher porosity (up to 10% to 12%) and many smaller grains.

Based on the EBSD IPF maps, both triclinic and monoclinic C_3S samples have grains with the size up to and larger than 100 μm by 100 μm, although the triclinic polymorph was more prone to smaller grain size overall. Thus, it is possible to produce large grains of the material using 8-h to 10-h sintering dwell time and obtain good EBSD patterns on the samples prepared by mechanical polishing using 0.25 μm diamond paste.

As for the β-C_2S samples, the obtained Kikuchi patterns appeared noisy even after thorough camera optimization. Weak EBSD patterns on these C_2S areas possibly appears due to small crystallite size, poor crystallinity, and lamellar twinning nature of this mineral [6, 17, 18]. All of these factors complicate the pattern indexing, and the Kikuchi bands were faint on β-C_2S surfaces. Another possible reason is the transformation from the α-phase to β-C_2S upon quenching, which may introduce crystal distortion, preventing the generation of clear EBSD patterns [19]. Such signals could not be resolved by EBSD pattern matching, where the software erroneously assigned α-C_2S, α'-C_2S, and β-C_2S patterns within a single grain of the sample. For this reason, EBSD analysis of C_2S sample surface was conducted on an ion cross-sectioning polished sample.

The sample with dimensions of 8 mm by 8 mm by 3 mm was polished using Argon Ion Cooling Cross Section Polisher™ (IB-19520, JEOL, Japan). The procedure was done in two steps with initial ion milling at 8 kV for 10 h and finishing with fine milling at 4 kV for 1 h. The polished β-C_2S sample, SEM image of the cross-sectioned surface, and an example of misindexed EBSD IPF map of C_2S are shown in Fig. 9. Even after ion-polishing and despite obtaining a good Kikuchi pattern signal (Fig. 10), the crystallographic orientation indexing of β-C_2S was not successful and a single grain could be assigned an α, β, and γ C_2S phase within a small area. It is possible that the crystallographic patterns of C_2S available today do not fully represent the structure of actual β-C_2S (*e.g.*, due to unresolved peak at 33.4° 2θ shown in Fig. 6) and require refinement. More importantly, solid-state synthesis of β-C_2S *via* sintering introduces distortion to the crystal, which complicates indexing crystallographic orientation through EBSD. Thus, alternative synthesis techniques, such as the Pechini process [19], could resolve this issue by producing higher purity β-C_2S phase.

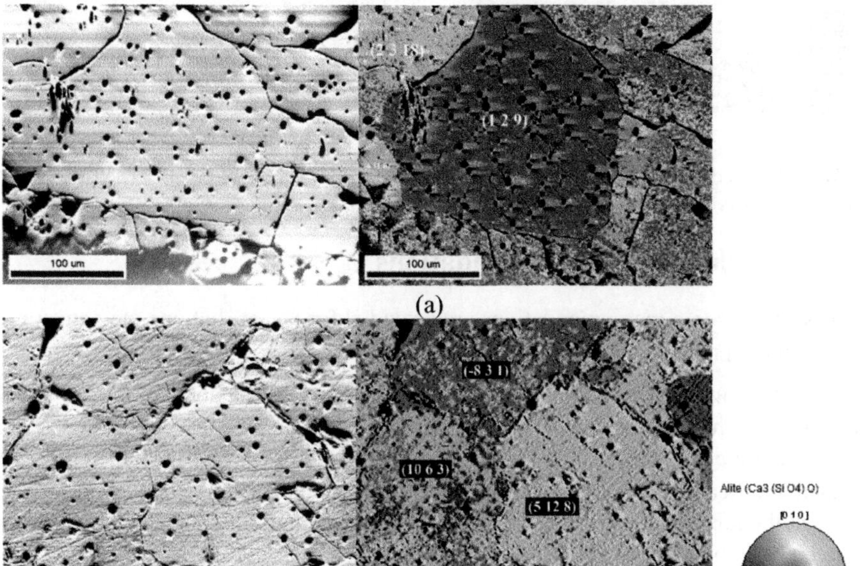

Fig. 8. SEM images (left) and inverse pole figure maps (right) overlaid on SEM image of the (a) C_3S-T and (b) C_3S-M area near epoxy (at bottom of images). Grains are labeled based on the most repeated index with a confidence index > 0.1.

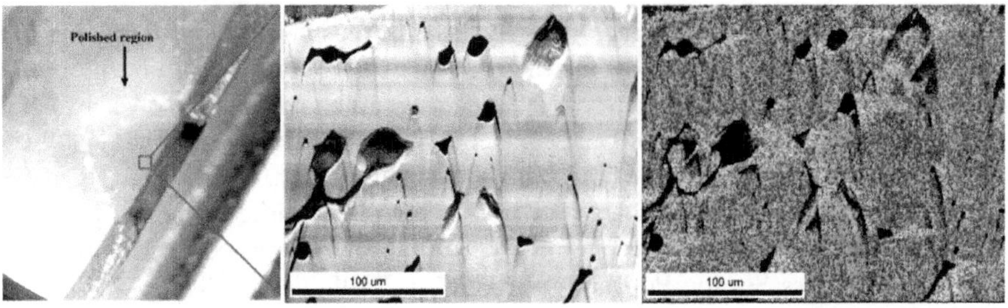

(a) (b) (c)

Fig. 9. Ion polished surface of β-C_2S *via* (a) optical and (b) SEM microscope, and (c) an example of a mis-indexed EBSD IPF map of β-C_2S.

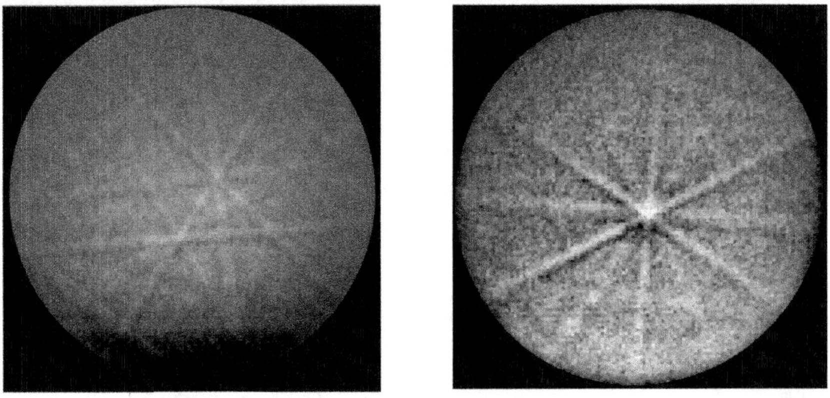

(a) (b)

Fig. 10. Observed examples of Kikuchi patterns from (a) C_3S and (b) β-C_2S. The C_3S pattern appears sharper compared to the slightly noisier C_2S signal.

4 Troubleshooting Recommendations

In this section, some recommendations are provided for issues that one may encounter during sample preparation described in the paper.

Green Body Powder Preparation. After mixing PVB in isopropanol solution with calcium silicates, the mixture should be dried on a flat surface (*e.g.*, on a sieving pan) with a hair dryer on a hot setting. The material drying should be controlled until it becomes granulated, otherwise if the material hardens, it is difficult to sieve. In such cases, one can grind the hardened powder with mortar and pestle, and the powder is still suitable for green body preparation.

Porosity Due to Trapped Air. During the preparation of the green body, it is important to make sure minimal air remains in the sample, as it is not fully compressible in the isostatic press. The residual air from dry pressing may lead to poor compression of the sample, which further will result in higher porosity after sintering. To minimize the residual air, one should fully fill the dye with the powder and tap or vibrate it, then fill the dye completely again, and vacuum seal the sample before submerging into CIP.

Sintering and Porosity. The porosity in the sample is also dependent on the sintering time, where longer sintering will reduce the porosity and facilitate grain growth. While regular sintering of C_3S and C_2S takes 2 h to 3 h of dwell time (*i.e.*, time after reaching 1600 °C and 1450 °C, respectively), this will result in a porous sample. Prolonged sintering (8 h to 10 h of dwell time) will reduce the porosity and form unform large grains. However, it was found that if the sintering is too long, this could lead to the formation of glass after 24 h of dwell time. For C_3S samples, especially monoclinic C_3S, slow cooling is permissible since stabilized C_3S did not convert into C_2S, while there was some conversion of triclinic C_3S. However, C_2S should always be quenched in order to avoid the large dusting of the sample into γ-C_2S.

Polishing and Coating. To obtain a proper surface polish just on SiC grits and microcloth with diamond paste, it is better to increase the polishing time on an autopolisher. For the grits up to 1200, five minutes of polishing for each of the grit is enough to remove scratches. However, for diamond paste polishing, even 10 min of polishing on each of the size (3 μm, 1 μm, 0.25 μm) leaves scratches that reduces the Kikuchi band clarity. By polishing the sample for 20 min, especially on the 1 μm and 0.25 μm, the desired level of polish can be achieved. Since no water can be used during polishing, abundant lubrication with propylene glycol is recommended. Starting after 1200 SiC grit polishing, the samples should be sonicated in alcohol between each polishing step to remove the residual particles that may cause scratching. To avoid polishing relief, the use of rod samples and longer epoxy hardening times are also recommended. Fast polishing on the autopolisher (platen and sample in the same rotation direction set to 200 and 100 RPM, respectively), could also be utilized with shorter polishing times. However, the surface should be monitored carefully for the scratches.

5 Conclusion

This work reports a method of synthesizing tricalcium and dicalcium silicate bulk bodies with density of 92% to 95% using cold isostatic pressing and PVB binder. The recommendations on the potential issues related to powder preparation and processing were discussed. After mechanical polishing, the produced C_3S phases exhibit good EBSD signal suitable for grain indexing, which greatly reduces the need for advanced polishing of C_3S due to its high crystallinity. As for β-C_2S, the mechanical polishing of the surface is not enough, and ion polishing is required to obtain a good signal. Still, even with a surface polished through ion milling, the indexing of β-C_2S grains is challenging due to the nature of this material.

Besides EBSD, the mechanically polished surfaces of the tricalcium and dicalcium silicates can be used in the studies related to surface-sensitive nanoscale characterization techniques, such as X-ray photoelectron spectroscopy, X-ray reflectivity, and microscopy. These findings will aid in surface analyses and deeper understanding of silicate phases in cement, potentially informing chemical and microstructural optimization strategies in cement-based construction materials.

Acknowledgement. This material is based upon work supported by the National Science Foundation under Grant No. 2237137. Any opinions, findings, and conclusions or recommendations expressed in this material are those of the author(s) and do not necessarily reflect the views of the National Science Foundation. This work used shared facilities at the Nanoscale Characterization and Fabrication Laboratory, which is funded and managed by Virginia Tech's Institute for Critical Technology and Applied Science. Additional support is provided by the Virginia Tech National Center for Earth and Environmental Nanotechnology Infrastructure (NanoEarth), a member of the National Nanotechnology Coordinated Infrastructure (NNCI), supported by NSF (ECCS 1542100 and ECCS 2025151).

References

1. Li, X., Ouzia, A., Scrivener, K.: Laboratory synthesis of C3S on the kilogram scale. Cem. Concr. Res. **108**, 201–207 (2018). https://doi.org/10.1016/j.cemconres.2018.03.019
2. Brand, A.S., Gorham, J.M., Bullard, J.W.: Dissolution rate spectra of β-dicalcium silicate in water of varying activity. Cem. Concr. Res. **118**, 69–83 (2019). https://doi.org/10.1016/j.cemconres.2019.02.014
3. Wesselsky, A., Jensen, O.M.: Synthesis of pure Portland cement phases. Cem. Concr. Res. **39**, 973–980 (2009). https://doi.org/10.1016/j.cemconres.2009.07.013
4. Diouri, A., Boukhari, A., Aride, J., Puertas, F., Vazquez, T.: Stable Ca3SiO5 solid solution containing manganese and phosphorus. Cem. Concr. Res. **27**, 1203–1212 (1997). https://doi.org/10.1016/S0008-8846(97)00110-5
5. De La Torre, Á.G., Bruque, S., Campo, J., Aranda, M.A.: The superstructure of C3S from synchrotron and neutron powder diffraction and its role in quantitative phase analyses. Cem. Concr. Res. **32**, 1347–1356 (2002). https://doi.org/10.1016/S0008-8846(02)00796-2
6. Naber, C., Goetz-Neunhoeffer, F., Göbbels, M., Rößler, C., Neubauer, J.: Synthesis of monocrystalline Ca3SiO5 using the optical floating zone method. Cem. Concr. Res. **85**, 156–162 (2016). https://doi.org/10.1016/j.cemconres.2016.04.008
7. Fiocchi, A.A., de Angelo Sanchez, L.E., Lisboa-Filho, P.N., Fortulan, C.A.: The ultra-precision Ud-lap grinding of flat advanced ceramics. J. Mater. Process. Technol. **231**, 336–356 (2016). https://doi.org/10.1016/j.jmatprotec.2015.10.003
8. Krishnan, S., Zunino, F., Bishnoi, S., Scrivener, K.: Characterisation and hydration kinetics of β-C2S synthesised with K2SO4 as dopant. Cem. Concr. Res. **167**, 107119 (2023). https://doi.org/10.1016/j.cemconres.2023.107119
9. De la Torre, Á.G., De Vera, R.N., Cuberos, A.J.M., Aranda, M.A.G.: Crystal structure of low magnesium-content alite: application to Rietveld quantitative phase analysis. Cem. Concr. Res. **38**, 1261–1269 (2008). https://doi.org/10.1016/j.cemconres.2008.06.005
10. Doebelin, N., Kleeberg, R.: Profex : a graphical user interface for the Rietveld refinement program BGMN. J. Appl. Crystallogr. **48**, 1573–1580 (2015). https://doi.org/10.1107/S1600576715014685
11. Rheinheimer, V., Casanova, I.: An X-ray photoelectron spectroscopy study of the hydration of C2S thin films. Cem. Concr. Res. **60**, 83–90 (2014). https://doi.org/10.1016/j.cemconres.2014.03.005
12. Sun, F., Pang, X., Wei, J., Zeng, J., Niu, J.: Synthesis of alite, belite and ferrite in both monophase and polyphase states and their hydration behavior. J. Mater. Res. Technol. **25**, 3901–3916 (2023). https://doi.org/10.1016/j.jmrt.2023.06.151
13. Cuesta, A., Losilla, E.R., Aranda, M.A.G., Sanz, J., De la Torre, Á.G.: Reactive belite stabilization mechanisms by boron-bearing dopants. Cem. Concr. Res. **42**, 598–606 (2012). https://doi.org/10.1016/j.cemconres.2012.01.006

14. Ghosh, S.N., Rao, P.B., Paul, A.K., Raina, K.: The chemistry of dicalcium silicate mineral. J. Mater. Sci. **14**, 1554–1566 (1979). https://doi.org/10.1007/BF00569274
15. Kocis, J., Rößler, C., Elhoweris, A., Abdalla, O., Ludwig, H.M.: The performance of α'-belite calcium sulfoaluminate (C$A) cement preserved under SO2 atmosphere compared to commercial products. Cem. Concr. Compos. **148** (2024). https://doi.org/10.1016/j.cemconcomp.2024.105468
16. Andrade Neto, J.S., et al.: Hydration and interactions between pure and doped C3S and C3A in the presence of different calcium sulfates. Cem. Concr. Res. **159**, 106893 (2022). https://doi.org/10.1016/j.cemconres.2022.106893
17. Rößler, C., Zimmer, D., Trimby, P., Ludwig, H.-M.: Chemical – crystallographic characterisation of cement clinkers by EBSD-EDS analysis in the SEM. Cem. Concr. Res. **154**, 106721 (2022). https://doi.org/10.1016/j.cemconres.2022.106721
18. Ghose, A., Chopra, S., Young, J.F.: Microstructural characterization of doped dicalcium silicate polymorphs. J. Mater. Sci. **18**, 2905–2914 (1983). https://doi.org/10.1007/BF00700771
19. Jawad Ahmed, M., Schollbach, K., van der Laan, S., Florea, M., Brouwers, H.J.: A quantitative analysis of dicalcium silicate synthesized via different sol-gel methods. Mater. Des. **213**, 110329 (2022). https://doi.org/10.1016/j.matdes.2021.110329

Thermodynamic Simulation and Challenges in the Deferrization of EAF Black Slags

Filippo Disconzi[1](✉) and Maurizio Bellotto[2]

[1] Geosciences Department, University of Padua, Padova, Italy
Filippo.disconzi@phd.unipd.it
[2] Opigeo S.R.L., Grisignano di Zocco, Italy

Abstract. Electric Arc Furnace (EAF) black slags are a major by-product of steelmaking, rich in iron oxides (FeO, Fe_2O_3) and containing silicates and other oxides.

Although EAF black slags are already widely recycled as aggregates in construction applications, as reported by Pellegrino et al. [1], this represents a form of downcycling. Despite their favourable properties, such as abrasion resistance, their high density limits broader application. Therefore, deferrization aimed at selectively removing iron from the slag remains crucial for maximizing resource recovery, reducing waste, and enhancing the potential for higher-value reuse of the residual material.

This study investigates deferrization using thermodynamic modeling with FactSage software, focusing on a post-slagging process applied directly to molten slag.

The approach evaluates the reduction of iron oxides using metallic silicon (Si) or aluminum (Al) as reducing agents, which also act as slag modifiers. Simulations were carried out to predict phase equilibria under varying temperature and slag composition conditions, highlighting the dual role of Si and Al in both reducing iron oxides and modifying the slag's properties.

The results show that, under optimized conditions, iron oxides can be effectively reduced to metallic iron or magnetite, enabling separation through magnetic or gravitational methods.

The use of Si and Al not only improves the reduction process but also modifies the slag's melting behavior and phase stability, allowing a better division between ferrous and non-ferrous fractions. The residual non-ferrous slag, after processing, has significant reuse potential, particularly as a clinker substitute in the cement industry. This reuse aligns with circular economy principles and can greatly reduce the carbon footprint of cement production.

This work emphasizes the need to integrate thermodynamic modeling with experimental validation to optimize deferrization strategies. By leveraging the properties of Si and Al and ensuring consistent slag quality, the findings contribute to resource-efficient and sustainable solutions for managing steelmaking by-products, supporting a more circular approach to slag utilization and valorization.

Keywords: Circular Economy · Deferrization · Electric Arc Furnace Slag · Resource Recovery · Thermodynamic Modeling

M. Posani and S. Singh (Eds.): RYS 2025, RILEM Bookseries 67, pp. 212–223, 2026.
https://doi.org/10.1007/978-3-032-14652-6_17

1 Introduction

Steelmaking is mainly based on two main production processes, the iron ore based, or the scrap based one.

In the scrap process, steel is produced directly by melting steel scrap in an electric arc furnace (EAF).

In 2021, 152.6 million tons of steel were produced in Europe, from which 67 million tons using the EAF furnace. About 32.9% of the worldwide steel production output is constituted by side streams, mainly consisting of slags that can be recycled in several processes inside and outside the steel production cycle [2].

EAF slag, coming from the de-carburization and oxidation step of iron refining, is a by-product rich in iron oxides (FeO, Fe_2O_3, Fe_3O_4) alongside silicates and other oxides. Although EAF slag contains a significant amount of iron oxides, these are not typically recovered as metallic iron. Instead, the slag is often recycled as an aggregate in construction. The presence of Fe oxides enhances abrasion resistance but also contributes to the high density of the material, which can limit its applications. Consequently, a significant fraction of iron remains trapped within the slag matrix. While, EAF black slag is rarely landfilled and is typically recycled as aggregate, environmental concerns remain due to the potential leaching of toxic elements, especially under prolonged exposure to weathering or acidic conditions [3].

Finding effective ways to recover iron from EAF slag and valorize the residual material is therefore crucial for resource efficiency and sustainable waste management.

Deferrization, or the selective removal of iron from EAF slag, can be achieved through thermodynamic control of phase equilibria, facilitating the formation of iron-rich phases that are more easily separable.

This study investigates an alternative approach to EAF slag deferrization using FactSage thermodynamic modeling to predict phase equilibria and process conditions for optimized Fe-recovery [4]. Specifically, the reduction of iron oxides using metallic silicon (Si) or aluminum (Al) as reducing agents is evaluated.

These elements not only promote the selective reduction of Fe_2O_3 but also act as slag modifiers, altering the phase stability of the residual non-ferrous slag.

More in depth, the study utilizes the FactSage 8.3 thermodynamic database, which includes data for metallurgical systems, slags, and oxides. Specifically, the FToxid and FSstel databases are employed, containing thermodynamic parameters for oxides, silicates, and metallic compounds relevant to EAF slags. These databases have been developed and validated based on experimental data published [4] and are regularly updated to ensure maximum accuracy in simulations of high-temperature metallurgical processes.

Through computational modeling, the study aims to define optimal temperature ranges, reducing agent dosages, and slag compositions that maximize iron recovery while ensuring that the remaining slag meets quality standards for industrial reuse. Thermodynamic simulations are conducted using the Equilib module of FactSage, which calculates thermodynamic equilibrium by minimizing the Gibbs free energy of the system. Calculations are performed at temperatures ranging from 1400 °C to 1700 °C, considering atmospheric pressure.

The valorization of deferrized EAF slag is particularly relevant, as the residual CaO-SiO_2-Al_2O_3-rich material shows promising potential as a supplementary cementitious material (SCM). In contrast to its common downcycling as aggregate, this application can be considered an upcycling route, since the production of cementitious binders adds higher value and functionality to the material. Substituting clinker with deferrized slag contributes to reducing the carbon footprint of cement manufacturing and supports circular economy goals [2]. Furthermore, under reducing conditions, not only are Fe oxides converted to metallic iron, but Cr is also partially reduced. This promotes the stabilization of Cr in spinel or silicate phases, lowering the risk of leaching and enhancing the environmental compatibility of the final material [5].

By integrating thermodynamic modeling with process optimization using, this work provides a pathway for improving iron recovery from EAF slag, while simultaneously enhancing the sustainability and economic viability of steelmaking by-products. Future work will focus on experimental validation of the predicted phase transformations and process efficiencies, ensuring that the proposed methodology is applicable at the industrial scale.

2 Materials and Method

The starting material used in this study is a black slag (EAF slag) named BL_SLAG from an Italian Electric Arc Furnace (EAF) steelmaking process.

Unlike white slag, which is produced during secondary refining and is mainly composed of CaO and Al_2O_3, black slag contains a significant amount of iron oxides making it a valuable candidate for deferrization and subsequent valorization.

The chemical composition of the black slag analyzed in this work is as follows on Table 1:

Table 1. BL_SLAG chemical composition (XRF analysis)

Oxide	CaO	SiO_2	Fe_2O_3	Al_2O_3	MgO
%	42.2	25.6	20.2	11.4	2.8

The high content of iron oxides indicates the potential for metal recovery, while the remaining matrix, rich in CaO-SiO_2-Al_2O_3, suggests possible applications in cementitious materials. The presence of MgO contributes to the stability and viscosity of the slag melt, affecting its behavior during processing.

EAF black slags vary in composition depending on raw materials, process conditions, and fluxing agents used during steel production. The slag considered in this study was collected immediately after tapping, ensuring that its composition reflects the high-temperature equilibrium state before cooling and solidification.

Aluminothermic-Silicothermic reaction to promote deferrization.

The deferrization of Electric Arc Furnace (EAF) black slag through aluminothermic or silicothermic reduction offers an efficient approach for recovering iron and modifying the slag composition for further applications. This process involves the selective reduction of iron oxides (Fe_2O_3, Fe_3O_4) using aluminum (Al) or silicon (Si) as reducing agents, facilitating the separation of metallic iron from the residual slag.

In the aluminothermic reaction, aluminum reacts with iron oxides to form metallic iron and alumina (Al_2O_3), as described by the reaction (Eq. 1):

$$Fe_2O_3 + 2Al \rightarrow 2Fe + Al_2O_3 + \Delta H \tag{1}$$

This process is highly exothermic, generating sufficient heat to sustain the reaction without external energy input. The metallic iron produced can be easily separated due to density differences, while the resulting Al_2O_3-rich slag may have potential applications in cementitious materials or refractory products. Aluminothermic reactions have been extensively studied for various metallurgical applications, including the reduction of silicate slags as demonstrated by Philipson et al. [6].

Similarly, the silicothermic reduction employs silicon as the reducing agent, yielding metallic iron and silica (SiO_2) as by-products (Eq. 2):

$$Fe_2O_3 + 3Si \rightarrow 2Fe + 3SiO_2 + \Delta H \tag{2}$$

This reaction also proceeds exothermically, ensuring a self-sustaining process, and results in a slag rich in SiO_2, which can be utilized as a supplementary cementitious material in blended cement formulations. The silicothermic reduction process has been applied effectively in various metallurgical contexts, as demonstrated by Chen et al. [7], who for example investigated the direct alloying of steel through silicothermic self-reduction of chromite ore using Si-containing solid waste.

The choice between Al and Si as reducing agents influences both the efficiency of iron recovery and the final slag composition, which determines its potential industrial applications.

Each method presents specific advantages and limitations. Aluminothermic reduction offers a highly exothermic reaction that ensures rapid iron recovery and self-sustaining process conditions. The resulting Al_2O_3-rich slag can enhance the mechanical properties of high-alumina cements or be considered for refractory applications. In this study, aluminum is assumed to be introduced in the form of aluminum scrap, which is more suitable for this process compared to conventional metallurgical recycling. Scrap Al, typically alloyed, oxidized, or polymer-coated, is difficult to remelt into pure metal, making its use in high-temperature reduction processes more feasible and less energy intensive. However, for the purpose of thermodynamic modeling, pure aluminum is used as an input to simplify the simulation and ensure reproducibility. Despite its advantages, the use of aluminum (scrap or pure) remains economically less favorable, and excessive alumina in the slag may reduce compatibility with standard cement formulations.

Silicothermic reduction represents a more cost-effective option, particularly when considering silicon scrap. Although less available than aluminum scrap, secondary silicon sources, such as those recovered from photovoltaic panel recycling, offer promising alternatives. As with aluminum, pure silicon is used as an input in simulations to standardize the thermodynamic analysis. The resulting SiO_2-rich slag displays high pozzolanic potential, making it attractive as a supplementary cementitious material and reinforcing the sustainability of this valorization pathway.

However, the reaction produces less heat than the aluminothermic process, which may require additional energy input to maintain optimal processing temperatures. Additionally, controlling the final slag viscosity is critical, as excessive SiO_2 content may lead to undesirable rheological properties.

3 Results

3.1 Heat Recovery Analysis

Aluminothermic reduction

In the aluminothermic deferrization of 100 g BL_SLAG black slag, approximately 7 g of aluminum is introduced as a reducing agent to selectively reduce Fe_2O_3 to metallic iron. A key aspect of process optimization is ensuring energy efficiency by minimizing external heat input.

Since EAF slag is already in a molten state at the end of steelmaking operations, it is logical to exploit this condition rather than applying unnecessary additional heat.

Therefore, in the simulation setup, aimed at minimizing resource consumption and energy waste, the BL_SLAG is assumed to be fully molten at 1430.00 °C, and aluminum is introduced at ambient temperature (25 °C) to initiate the deferrization reaction.

Introducing aluminum at a lower temperature into the molten slag has direct thermal implications. First, the aluminum must be heated from 25 °C to 1541.93 °C (new T liquidus of BL_SLAG + Al addition), requiring an energy input that must be accounted for in the overall energy balance. At the same time, the aluminothermic reaction is highly exothermic, with an enthalpy release of $\Delta H = -5.81 \times 10^4$ J, which contributes to maintaining the system temperature. The heat released by the reaction offsets the energy required to heat and react the aluminum, ensuring that, under optimal conditions, the process remains thermally self-sufficient. However, if aluminum addition significantly affects the slag temperature, it could influence phase separation efficiency and reaction kinetics.

The composition used in the simulation corresponds to the actual EAF black slag chemistry, without any additional fluxing agents aimed at lowering the melting point. However, the resulting residual slag after aluminothermic deferrization remains fully molten at 1541.93 °C, which is consistent with the temperature range of steelmaking operations and does not require further adjustment. Therefore, no external additives are introduced to modify the slag melting behavior.

To assess the energy efficiency of the aluminothermic deferrization process, a detailed thermal balance evaluation is conducted. The calculations consider the energy required to heat aluminum from ambient temperature, radiative heat losses, and the heat released by the aluminothermic reaction.

The energy required to heat the aluminum from 25 °C to the final molten slag temperature of 1541.93 °C is determined using the specific heat capacity (Eq. 3) obtaining a value of 9.54 kJ.

Radiative heat losses are estimated using Stefan-Boltzmann's law (Eq. 4), obtaining as results: 196.80 W.

In a minute, the total radiative heat losses are: 11.81 kJ.

The total energy balance equation integrates all contributions (Eq. 5) is −36.75 kJ. Since it is negative, the reaction produces excess thermal energy, confirming that the aluminothermic deferrization process is thermally self-sustaining. The heat generated by the reaction is sufficient to maintain the system in a fully molten state, ensuring efficient phase separation and slag processing without requiring additional external heating.

This analysis demonstrates that the aluminothermic approach is both energy-efficient and industrially viable, maximizing process efficiency while minimizing unnecessary energy consumption.

Heat Required to Raise Aluminum T **(Equation 3)**	$Q_{Al}=m_{Al}\cdot C_p\cdot \Delta T$
Stefan-Boltzmann's law **(Equation 4)**	$Q_{rad}=\epsilon\cdot\sigma\cdot A\cdot(T^4_{final,K}-T^4_{amb,K})$
Total energy balance **(Equation 5)**	$Q_{ext}=Q_{Al}+Q_{rad,total}-\lvert\Delta H_{reaction}\rvert$

Silicothermic reduction

In the silicothermic deferrization of 100 g of BL_SLAG, 10.5 g of silicon is introduced as a reducing agent to selectively reduce Fe_2O_3 to metallic iron.

BL_SLAG is assumed to be fully molten at 1430.00 °C, and silicon is introduced at room temperature (25 °C) to initiate the deferrization reaction.

Introducing silicon at a lower temperature into the molten slag has direct thermal implications. The silicon must be heated from 25 °C to 1430.00 °C (liquidus T of BL_SLAG + Si addiction), requiring an energy input that must be accounted for in the overall energy balance. At the same time, the silicothermic reaction is highly exothermic, with an enthalpy release of $\Delta H = -131.09$ kJ, which contributes to maintaining the system temperature. The heat released by the reaction offsets the energy required to heat and react the silicon, ensuring that, under optimal conditions, the process remains thermally self-sufficient. However, if silicon addition significantly affects the slag temperature, it may influence phase separation efficiency and reaction kinetics.

The simulation is based on the native composition of the EAF black slag, without introducing additional fluxing components. The final SiO_2-rich slag obtained after silicothermic deferrization remains fully molten at 1430.00 °C, which aligns with typical operating temperatures in industrial processes. Therefore, no modifications were made to tailor the melting behavior of the system. Silicon is introduced at 25 °C, and the reaction proceeds without external heat, relying entirely on the energy released by the silicothermic process. A thermal balance evaluation is performed to confirm whether the reaction-generated heat is sufficient to maintain the system in a fully molten state throughout the process.

The energy requires to heat the 10.5 g of silicon from ambient temperature (25 °C) to the operating temperature of 1430.00 °C is 10.34 kJ. In addition, the system experiences radiative heat losses to the surroundings, estimated at 11.81 kJ over a 60 s reaction interval.

Despite these energy demands, the highly exothermic nature of the reaction, which releases 131.09 kJ, ensures that the process remains thermally stable. The total energy balance, calculated as Q_ext = −109.95 kJ, indicates a significant excess of energy generated by the reaction, more than compensating for the heating requirements of the silicon and radiative losses.

In the energy calculations related to heating the reducing agents (Al, Si) from ambient temperature (25 °C) to process temperature (>1500 °C), both sensible heat and latent heat of fusion are considered.

As a result, no additional thermal input is necessary, as the reaction itself provides sufficient energy to maintain the molten state of the system. This finding supports the practical feasibility of silicothermic deferrization as an energy-efficient process, reinforcing its potential for industrial applications in slag valorization and resource recovery (Fig. 1).

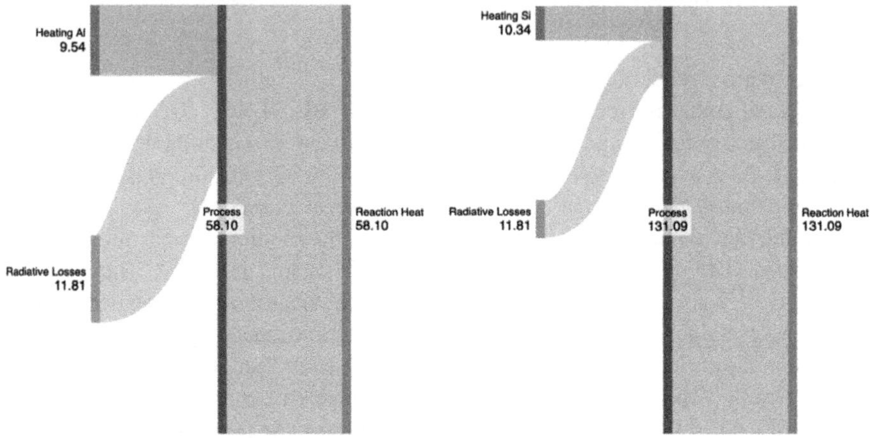

Fig. 1. Sankey diagrams for the aluminothermic and silicothermic deferrization processes. Each node represents a process stage ("Heating", "Radiative Losses", "Reaction Heat") and each link's width is proportional to the energy flow (in kJ). Positive values (inflows) correspond to energy inputs or losses, while the outgoing link indicates energy released by the exothermic reduction reaction.

3.2 Chemical/phases Analysis of Products

Aluminothermic reduction

The aluminothermic deferrization process applied to 100 g of BL_SLAG EAF black slag produces two main output phases: a metallic iron phase and a residual slag.

The metallic iron phase, representing 14.249 g of the total reaction products, consists primarily of 99.01 wt.% Fe, confirming that iron extraction is highly efficient. The remaining fraction includes 0.97 wt.% Si and 0.016 wt.% Al, suggesting minor interactions with the slag. The high iron purity indicates that the aluminothermic reduction successfully converts Fe_2O_3 into metallic iron, making this material potentially suitable for direct reuse in steel production or as an alloying component.

The residual slag, with a total mass of about 100 g, is reported on Table 2. The presence of high calcium and aluminum oxide concentrations makes this slag particularly interesting for potential applications in the cement industry, where such oxides are valuable in blended cements and supplementary cementitious materials (SCMs). The very low FeO and Fe_2O_3 content in the slag (0.036 wt.% and 0.00007 wt.%, respectively,

Table 3) confirms that nearly all iron has been successfully reduced, further supporting the efficiency of the process.

From a valorization perspective, the composition of the residual slag suggests that it could be stabilized and incorporated into cement formulations, reducing the need for clinker and thereby lowering the carbon footprint of cement production. However, before considering large-scale applications, further mineralogical characterization is necessary to assess the phase stability, hydraulic properties, and leaching behavior of the material.

Overall, the aluminothermic deferrization of EAF black slag demonstrates high iron recovery efficiency and produces a calcium-alumino-silicate-rich slag with promising potential for industrial reuse.

Table 2. Metallic Iron Phase Composition

Component	wt. %
Fe	99.0
Si	0.9
Al	0.02

Table 3. Residual Slag Composition

Component	wt. %
Al_2O_3	25.9
SiO_2	26.6
CaO	44.5
FeO	0.04
Fe_2O_3	<0.01

Silicothermic reduction

The silicothermic deferrization process applied to 100 g of BL_SLAG EAF black slag results in the formation of two main phases: a metallic silicon-iron alloy and a residual slag. The effectiveness of the reaction and the composition of these products are crucial in assessing their potential industrial applications and overall process feasibility.

At 1430 °C, the metallic phase consists of about 19 g of liquid alloy, with a composition as reported on Table 4.

The high silicon content confirms the successful reduction of Fe_2O_3 by Si, leading to the formation of Fe-Si intermetallic compounds, which could be useful as ferrosilicon alloys in steel production. These alloys serve as deoxidizers in steelmaking and are often employed in the manufacture of high-strength steels and cast irons. The negligible aluminum and calcium content indicates that these elements remain in the slag phase rather than dissolving into the metal, ensuring a high-purity Fe-Si alloy.

The residual slag phase, weighing 93 g, is primarily composed of SiO_2, and Al_2O_3, with minimal iron oxide content FeO and Fe_2O_3 (Table 5). The very low iron content confirms the high efficiency of the deferrization process, demonstrating that nearly all iron oxides have been successfully reduced and transferred to the metallic phase.

The high silica and calcium oxide content suggests that the slag could be valorized in the cement industry, particularly as a supplementary cementitious material (SCM) or clinker substitute. Phases such as gehlenite ($Ca_2Al_2SiO_7$) and wollastonite ($CaSiO_3$), which are commonly observed in high-temperature slags, contribute to pozzolanic activity and hydraulic reactivity, making the slag suitable for cement formulations (Fig. 2).

Table 4. Metallic Iron Phase Composition

Component	wt. %
Fe	73.2
Si	26.8
Al	0.01

Table 5. Residual Slag Composition

Component	wt. %
Al_2O_3	12.2
SiO_2	39.6
CaO	45.2
FeO	<0.01
$Fe2O_3$	<0.01
MgO	3.0

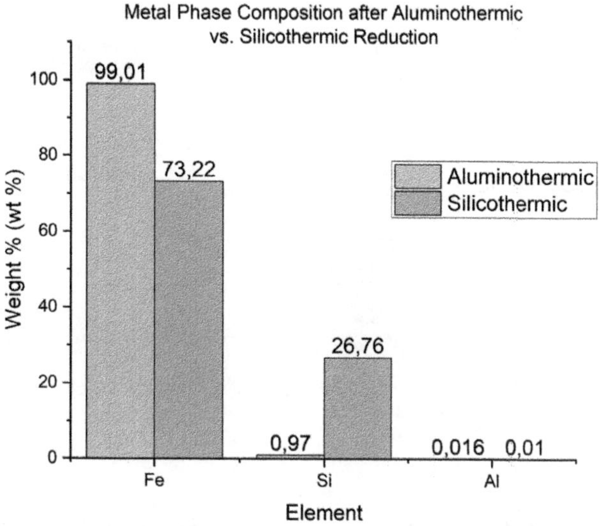

Fig. 2. Metal Phase Composition after Aluminothermic vs. Silicothermic Reduction

A critical issue of the silicothermic pathway is the inevitable enrichment of silicon in the obtained metallic product. Simulations show that reduced iron can contain up to 27% Si, limiting its direct use as a steelmaking raw material. This problem could be addressed through: (i) a secondary refining process to remove excess silicon; (ii) direct integration into Fe-Si alloy production processes, valorizing the silicon content as a useful component rather than a contaminant; (iii) optimization of process parameters to minimize Si transfer to the metallic phase, for example by operating at lower temperatures or with controlled stoichiometric ratios.

Figure 3 compares five critical performance indicators for the aluminothermic (red) and silicothermic (blue) deferrization routes in a radar chart format.

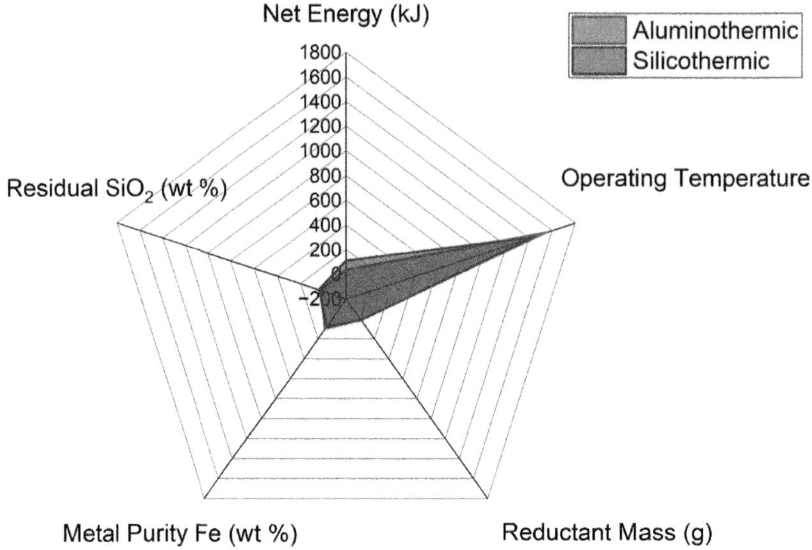

Fig. 3. Radar Chart of Key Process Metrics for Aluminothermic and Silicothermic reduction

The axes represent: net energy balance (kJ surplus), operating temperature (°C), reductant mass (g), iron purity in the metallic phase (wt % Fe), and silica content in the residual slag (wt % SiO_2). The colored areas illustrate key trade-offs:

- Net Energy: Silicothermic delivers a larger thermal surplus (~110 kJ vs. 37 kJ), indicating greater self-sufficiency.
- Operating Temperature: Aluminothermic requires higher temperatures (~1542 °C vs. 1430 °C), potentially affecting refractory life.
- Reductant Mass: Aluminothermic uses less reagent (7 g Al vs. 10.5 g Si), reducing material costs.
- Metal Purity: Aluminothermic yields purer iron (99 wt % Fe vs. 73 wt % Fe).
- Residual SiO_2: Silicothermic slag contains more silica (~40 wt % vs. 26 wt %), suggesting different valorization routes.

This visualization succinctly highlights the relative advantages of each method in terms of energy efficiency, product quality, and downstream slag utilization.

It is important to clarify that, despite their thermodynamic effectiveness, the use of metallic aluminum and silicon as reducing agents presents significant environmental challenges. Both primary aluminum and silicon production are energy-intensive processes with substantial carbon footprints. These high energy requirements and associated greenhouse gas emissions limit the industrial sustainability of these processes. However, the environmental impact assessment changes significantly when considering: (i) the use of aluminum and silicon scrap from industrial processes, which substantially reduces the embedded energy and carbon footprint; (ii) the resource recovery value of the extracted iron; (iii) the qualitative improvement of the residual slag, which can be valorized as a raw material for the cement industry, replacing clinker and generating CO_2 credits.

4 Conclusions

A comparative assessment between aluminothermic and silicothermic deferrization reveals notable differences in process efficiency and outcomes. The aluminothermic process requires less reductant (7.0 g of Al compared to 10.5 g of Si for 100 g of starting material considered) and operates at higher temperatures (1542 °C), potentially enhancing phase separation between molten iron and slag. However, its lower reaction enthalpy results in a more limited thermal surplus, making it more susceptible to heat losses in industrial settings.

The silicothermic process generates greater thermal surplus, improving self-sufficiency while operating at lower temperatures (1430 °C), which benefits process control and reduces refractory wear. Both processes are thermodynamically self-sustaining under the assumed conditions, with the choice between them depending on specific constraints and intended applications.

From an industrial viewpoint, the aluminothermic route appears more suitable for recovering high-purity metallic iron, whereas the silicothermic pathway may be preferred for Fe-Si alloy production. Both approaches yield CaO-SiO_2-Al_2O_3-rich residues that are chemically compatible with cement formulations, supporting circular economy principles by maximizing resource recovery.

The industrial adoption of either route will depend on the balance between product specifications, reductant availability, and the feasibility of integrating the resulting slag into sustainable construction materials. Future work should focus on experimental validation of these thermodynamic predictions, particularly regarding the behavior of deferrized slag during hydration and its performance in cementitious applications.

References

1. Pellegrino, C., Cavagnis, P., Faleschini, F., Brunelli, K.: Properties of concretes with Black/Oxidizing Electric Arc Furnace slag aggregate. Cem. Concr. Compos. **37**, 232–240 (2013). https://doi.org/10.1016/j.cemconcomp.2012.11.005
2. Colla, V., et al.: Future research and developments on reuse and recycling of steelmaking by-products. Metals **13**(4), 676 (2023). https://doi.org/10.3390/met13040676
3. Lim, J.W., et al.: Heavy metal recovery from electric arc furnace steel slag by using hydrochloric acid leaching. In: E3S Web of Conferences, vol. 34, p. 02007 (2018). https://doi.org/10.1051/e3sconf/20183402007
4. Bale, C.W., et al.: FactSage Thermochemical Software and Databases, 2010–2016, Calphad, vol. 54, pp. 35–53, set. 2016. https://doi.org/10.1016/j.calphad.2016.05.002
5. Andersson, A., Isaksson, J., Lennartsson, A., Engström, F.: Insights into the valorization of electric arc furnace slags as supplementary cementitious materials. J. Sustain. Metall. **9**(1), 96–109 (2023). https://doi.org/10.1007/s40831-023-00778-y
6. Philipson, H.G.R., Wallin, M., Einarsrud, K.E.: Aluminothermic reduction kinetics of calcium silicate slag for silicon alloy production. Metals **14**, 604 (2024). https://doi.org/10.3390/met14060604
7. Chen, Y., Xue, Z., Song, S.: The direct alloying of steel through silicothermic self-reduction of chromite ore utilizing Si-containing solid waste. Metals **14**, 138 (2024). https://doi.org/10.3390/met14020138

Transforming Crustacean Waste into an "Accelerating" Mineral Admixture for Cementitious Systems

Parisa Heidari(✉), Patrice Rivard, and William Wilson

Université de Sherbrooke, Sherbrooke, QC, Canada
Parisa.Heidari@USherbrooke.ca

Abstract. In the context of climate change and waste management, transforming locally available waste into valuable resources for construction is essential. The main objective of this study is to transform low-value crustacean waste (CW) into an accelerating admixture for cementitious systems. This study investigates the impact of ultrasonication on heat-treated, ground CW particles. The impact of incorporating 10% CW, ground using roller-mill and high-shear grinders, without and with ultrasonication, was evaluated in terms of slump flow, compressive strength, and heat of hydration. The results showed that undispersed CW reduced compressive strength by about 10%, whereas ultrasonicated CW led to a 30% increase in both early and later-age strengths. The system with 10% ultrasonicated CW exhibited accelerated hydration kinetics, including a 4% increase in the C_3S hydration and a 12% reduction in the time to reach maximum heat release. This was attributed to filler and nucleation effects. These findings highlight the importance of dispersing alternative fine materials within cementitious systems to improve the compatibility of these materials in developing sustainable cement-based systems.

Keywords: Calcium carbonate · Cementitious systems · Compressive strength · Crustacean waste · Hydration kinetics · Ultrasonication

1 Introduction

The cement industry is responsible for about 8% of global anthropogenic carbon emissions [1]. This significant carbon footprint has drawn increasing concern and calls for more sustainable practices in the cement and concrete sector.

One of the most extensively studied approaches to reducing the carbon footprint is incorporating limestone powder as a partial cement replacement [2]. In Europe, the EN 197-1 standard [3] allows up to 35% limestone replacement in cement, while standards in the United States and other countries, such as ASTM C595 [4], AASHTO M 240 [5], and CSA-A3000 [6], limit the limestone content in blended cements to a maximum of 15%.

The need to find alternative sources of construction materials is particularly critical in remote or island regions, such as the Magdalen Islands (Îles de la Madeleine) in the Gulf of St. Lawrence, Quebec. In these areas, most construction materials must

M. Posani and S. Singh (Eds.): RYS 2025, RILEM Bookseries 67, pp. 224–238, 2026.
https://doi.org/10.1007/978-3-032-14652-6_18

be transported from the mainland, significantly increasing both costs and the associated carbon footprint [7]. These islands also face environmental issues associated with marine waste generated by local fisheries. A large volume of untreated crustacean waste (CW), rich in $CaCO_3$, accumulates annually. This waste contains residual organic matter that decomposes, producing foul odors and releasing toxic gases like ammonia (NH_3) and hydrogen sulfide (H_2S) [8].

These intertwined challenges can be addressed by transforming locally sourced CW into a calcium carbonate-rich powder suitable for use in concrete. This strategy simultaneously addresses two issues critical for islands: marine waste management and material sustainability in construction. Processing CW into a $CaCO_3$-based mineral filler can reduce reliance on imported limestone, reduce transportation emissions, and offer an environmentally friendly solution to island waste management.

CW, composed of $CaCO_3$ (~80%) and chitin (~15–20%), has been widely studied for applications in pharmaceuticals, cosmetics, food processing, fertilizers, and biopolymer production [9]. In the construction sector, recent research has primarily focused on extracting chitin derivatives, such as chitosan, chitin nanocrystals (ChNC), and chitin nanofibers (ChNF), for use as high-performance admixtures in cementitious systems [10, 11]. These materials have shown beneficial effects on mechanical strength, crack resistance, and durability even at low dosages. For instance, Baykara et al. (2024) [10] utilized chitosan derived from shrimp waste as a biodegradable alternative to synthetic polymers for concrete reinforcement, aiming to reduce their environmental impact. Similarly, Haidar et al. (2022) [11] addressed the high disposal costs of crustacean waste by incorporating chitin as an admixture in mortar, demonstrating that chitin-based nanomaterials can significantly enhance the flexural strength and elastic modulus of mortar systems. However, the major component of the CW, $CaCO_3$, has been overlooked in the context of construction materials. The potential of using $CaCO_3$ derived from CW as a sustainable mineral filler has received limited attention in the literature.

In this study, we aim to bridge this knowledge gap by evaluating the use of $CaCO_3$ derived from CW as an alternative mineral additive in cementitious systems. Unlike the complex and chemical intensive processes required to extract chitin-based nanomaterials [12], $CaCO_3$ can be recovered through a simple thermal treatment to remove organic matter, without producing chemical waste. With its moderate power requirements and inherent scalability, this method is particularly suitable for regions with limited infrastructure. Moreover, unlike chitin derivatives, typically used in small dosages as admixtures, calcium carbonate powder can be applied as a bulk cement replacement or filler, contributing more substantially to clinker reduction and the overall mitigation of CO_2 emissions.

Uniform dispersion of alternative cement-based materials is essential for its effective use in engineering and construction applications [13, 14]. Traditional mixing methods [15, 16] often fail to effectively disperse fine or agglomerated particles, limiting the performance of alternative cementitious materials. Although techniques like double-blade, vibration, and secondary mixing have been proposed [17], they mainly act on macroscopic dispersion and are limited in their ability to act directly on individual particles [18]. This limitation is a key reason why current technologies fail to uniformly disperse materials in cementitious systems. However, ultrasound has been shown to effectively disperse solid particles at the microscopic level [19].

Therefore, this study proposes the use of ultrasonication (US) on heat-treated and ground CW particles to convert the waste into a high-value admixture for cementitious systems by dispersing the particles uniformly. This paper examines the physicochemical properties, particle characteristics, and the effects of $CaCO_3$ derived from CW on cement hydration, fresh-state properties, and mechanical behavior of cementitious mixtures, both without and with ultrasonication of the powder. This approach aims to demonstrate the environmental and technical viability of transforming a low-value waste into a functional component for low-carbon concrete production, with particular emphasis on remote island regions where access to locally sourced limestone can be limited.

2 Materials and Methods

2.1 Materials

A general use (GU) cement conforming to the CAN/CSA A3000 specification [6], was used to produce cement-based composites. Crustacean waste (CW), sourced from the Magdalen Islands in Québec, Canada, was used as a partial cement replacement. The physicochemical properties of GU cement and CW are presented in Table 1. Quantitative X-ray diffraction (QXRD) analysis indicated that the processed crustacean waste (CW) was composed of approximately 80% calcite ($CaCO_3$). The densities of the materials, measured using helium pycnometry, were found to be 3100 kg/m^3 for general use (GU) cement and 2800 kg/m^3 for CW.

Table 1. Chemical properties of raw materials used (wt. %)

Oxides	SiO_2	Al_2O_3	Fe_2O_3	CaO	MgO	Na_2O	K_2O	LOI
GU	19.3	4.8	2.9	62.1	2.7	0.3	0.7	2.2
CW	0.1	0.1	0.0	49.8	3.6	0.7	0.0	36.7

2.2 Experimental Program

CW Treatment and Characterization

A series of procedures were carried out to convert CW into a suitable mineral admixture in cementitious systems, followed by material characterization. The treatment and characterization procedures for CW are summarized in Table 2. CW samples were cleaned and dried in a laboratory oven at 105 °C ± 5 °C for 24 h. Thermogravimetric analysis (TGA) accompanied by mass spectrometry was performed to determine the optimum heat treatment of CW for the complete removal of organic matter. CW was ground using two different grinding methods, roller milling (RM) and high shear grinding (HS), with different grinding times. Roller milling was conducted in a stainless steel jar using steel balls of 20 mm, 10 mm, and 5 mm diameters, both without grinding agents (GAs) and with three different organic GAs: TIPA, V3, and V21. The compositions of V3 and V21 are confidential. For clarity, TIPA, V3, and V21 are referred to as GA1, GA2, and GA3, respectively, throughout this study. For further characterization, X-ray fluorescence (XRF), quantitative X-ray powder diffraction (QXRD), scanning electron microscopy (SEM), laser granulometry, and reactivity assessment by the R^3 method (ASTM 1897) [20] were performed.

Table 2. Summary of CW treatment and characterization with objective and description

Process	Objective	Description
Stage: As received		
TGA	Determine the optimum heat treatment temperature	Nitrogen gas flow of 75 mL/min, Temperature rise rate of 5 °C/min
XRF analysis	Determine oxide compositions	Anode power of 50 mA and 60 kV
Stage: After drying		
Heat treatment	Remove organic matter	Performed at 500 °C ± 10 for 3 h
Loss on Ignition measurement	Determine optimum heat treatment period	Heated at 500 °C for 30 min to 3 h in 30-min increments
Stage: After heat treatment		
Grinding	Prepare CW powder	RM and HS grinders
QXRD analysis	Quantify calcite content	Cu-Kα (λ = 1.5406 Å) radiation at 40 kV and 50 mA current
Stage: After grinding		
Dispersion	De-aggregation of adherent particles	Ultrasonic processor at a power of 130 watts and an operating frequency of 20 kHz for 5 min
Physical properties	Measure density, particle size distribution	Helium Pycnometer, Laser Granulometry
R^3 test	Assess degree of reactivity of CW	ASTM C1702 [21] and ASTM C1897 [20]
Stage: After heat treatment and mechanical dispersion of ground particles		
Imaging	Examine CW morphology	SEM at acceleration voltage of 10 kV and vacuum of 10 Pa

Mixture Proportions and Testing Procedures

Table 3 summarizes the formulation of the cement paste systems, all prepared with a constant water-to-binder (w/b) ratio of 0.4. Five mixtures were tested: one reference (OPC) and four with 10% CW replacing cement. The nomenclature indicates the CW content, the grinding method (RM or HS), and the dispersion condition (see Table 3). Superplasticizer dosage was adjusted to achieve a consistent slump flow of 60 ± 10 mm in the OPC reference and CW blended systems without ultrasonication.

Mixing was performed using a laboratory mixer operating at 1600 rpm for 2 min according to the mixing procedure used by Georget et al. [22]. For the undispersed CW, cement and CW were dry mixed before being added to the water. For the ultrasonicated CW, a portion of the mixing water was used for ultrasonication, while the remaining water was mixed with the SP, then cement was added to the water-SP mixture, followed by the addition of the dispersed CW.

Table 3. Formulation of cement paste systems in mass percentages

Labels	Description	Cement	CW	SP*
OPC	Reference	100	0	0.22
$10CW_{HS}$	$10\%CW_{\text{High shear (HS) without ultrasonication}}$	90	10	0.35
$10CW_{RM}$	$10\%CW_{\text{Roller mill (RM) without ultrasonication}}$	90	10	0.30
$10CW_{HS\text{-}US}$	$10\%CW_{\text{High shear (HS) with ultrasonication}}$	90	10	0.35
$10CW_{RM\text{-}US}$	$10\%CW_{\text{Roller mill (RM) with ultrasonication}}$	90	10	0.30

*SP is a polycarboxylate-based superplasticizer and it is calculated based on the % of binder.

Slump flow was measured using a mini-slump cone (40 mm base, 20 mm top). The compressive strength was tested on 30 × 60 mm cylindrical specimens of cement pastes at 1, 3, 7, 28, and 91 days, using a loading rate of 0.171 kN/s [23]. For isothermal calorimetry, fresh paste samples were sealed in plastic vials and placed in an 8-channel TAM Air calorimeter at 23 °C.

3 Results and Discussion

3.1 Optimal Heat Treatment of CW

Figure 1 shows thermogravimetric analysis (TGA) results coupled with mass spectrometry, which highlight the phase transformations of as-received CW over different temperatures. An initial mass loss of ~6% was observed between 25 °C and 230 °C, accompanied by water release, corresponding to the evaporation of physically adsorbed water. A subsequent mass loss of ~19% occurred between 230 °C and 410 °C, which is associated with the thermal degradation and combustion of organic matter, releasing water and CO_2. A mass loss of ~24% was recorded between 600 °C and 810 °C, indicating $CaCO_3$ decomposition into CaO and CO_2. Based on these findings, 500 °C was selected as the optimal heat treatment temperature to fully remove organic matter without decomposing $CaCO_3$.

To determine the optimum heat treatment duration at 500 °C, the loss on ignition (LOI) was measured using Eq. (1) at 30-min intervals, as specified in CSA A3003 [24].

Heating was continued until the LOI difference between two consecutive measurements was less than 0.1%. This condition was met after 3 h, confirming the complete removal of organic matter and establishing 500 °C for 3 h as the optimal heat treatment condition.

$$\text{Loss on ignition} = \left(\frac{\text{Mass of sample} - \text{Mass of ignited residue}}{\text{Mass of sample}}\right) \times 100 \quad (1)$$

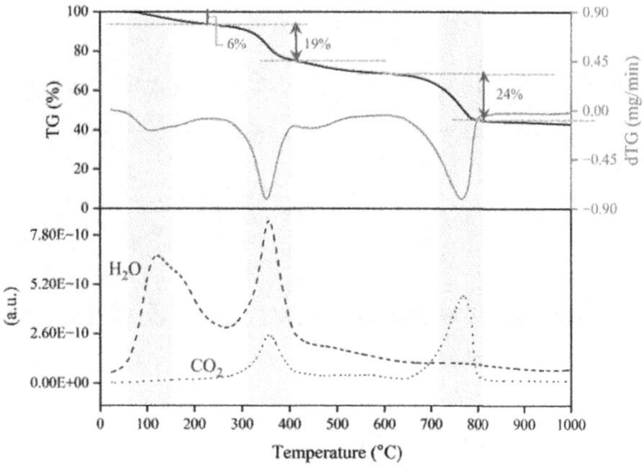

Fig. 1. Thermogravimetric analysis (TGA) coupled with mass spectrometry of raw CW

XRD patterns of CW before and after heat treatment are shown in Fig. 2. In the raw CW, crystal-like peaks at $2\theta \approx 10°$ and $20°$ indicated the presence of chitin [25]. These peaks disappeared after heat treatment at 500 °C, confirming the effective removal of organic matter. In addition, hydroxyapatite crystallization was observed in the heat-treated sample, attributed to the thermal transformation of amorphous calcium phosphate at elevated temperatures [26]. Overall, the XRD analysis confirmed that the chosen thermal condition (500 °C for 3 h) successfully removed the organic matter while preserving the mineral phases.

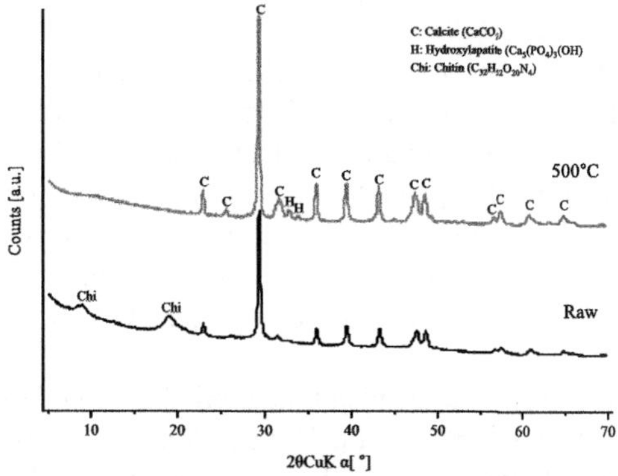

Fig. 2. XRD profiles of CW before and after heat treatment at 500 °C for 3 h

Figure 3 shows SEM images of CW after heat treatment at different magnifications. As is shown in Figs. 3(a), 3(b), and 3(c), the heat-treated CW exhibits distinct morphological features, including a loose and fractured surface structure, a layered arrangement, and a sponge-like porous network, respectively. The observed sponge-like morphology, particularly evident at higher magnification, suggests the development of an interconnected porous structure. This morphology likely increase the porosity of the particles, thereby increasing the overall specific surface area and water absorption of CW [27].

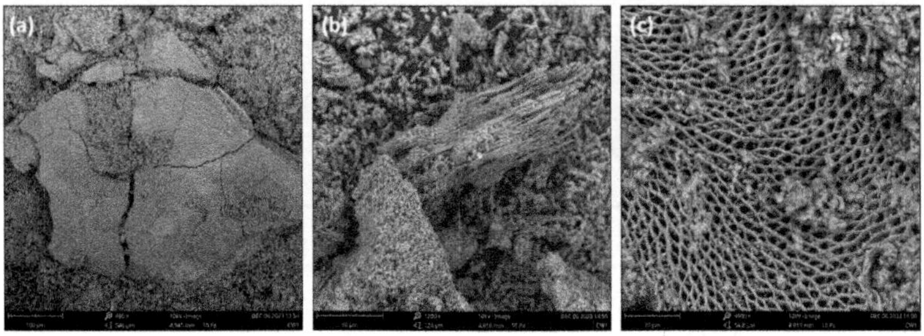

Fig. 3. SEM images of CW after heat treatment (a) × 490, (b) × 1200, (c) × 4900 magnifications

3.2 CW Grinding and Ultrasonication

Following heat treatment, two grinding approaches, roller milling (RM) and high shear (HS) grinding, were used to reduce the particle size of CW. To improve grinding efficiency and reduce particle adhesion to each other, the mill lining, and the grinding media during roller milling, three grinding agents (GAs), namely GA1, GA2, and GA3, were employed to enhance particle dispersion. These GAs function primarily via steric stabilization by forming a protective layer around the particles, which serves as a steric barrier to prevent close contact and aggregation of fine particles [28]. After grinding, a mechanical dispersion technique was used to break down particle aggregates. This was achieved through ultrasonic treatment in an aqueous suspension. The ultrasonic method operates by generating microbubbles in the liquid, which then undergo rapid growth and collapse, a process known as cavitation. This phenomenon produces intense shear forces and shock waves capable of disaggregating particles and transforming them into uniformly dispersed particles [19].

Figure 4. Shows the median particle size (D_{50}) of CW subjected to different grinding and dispersing techniques. Under RM and HS grinding without GAs, the D_{50} progressively decreased with longer grinding times. However, the HS grinder achieved a finer D_{50} of ~8 μm within 4 min, compared to the roller milled CW, which reached a D_{50} of ~34 μm after 90 min of grinding. Across all grinding durations, the presence of GAs (GA1, GA2, GA3) consistently outperformed dry RM grinding. Specifically, using GA2 for 90 min reduced D_{50} by ~40%, from 34 μm to 21 μm. GA1 and GA3 also contributed to finer particles at 60 min.

Mechanical dispersion via ultrasonication, applied after both RM and HS grinding, significantly reduced D_{50} by disrupting the Van der Waals interparticle forces. For example, Fig. 4a shows that US after roller milling reduced D_{50} by about 85%, from 34 μm to 5.3 μm, demonstrating that particle aggregation during grinding can be effectively reversed through ultrasonic treatment [28].

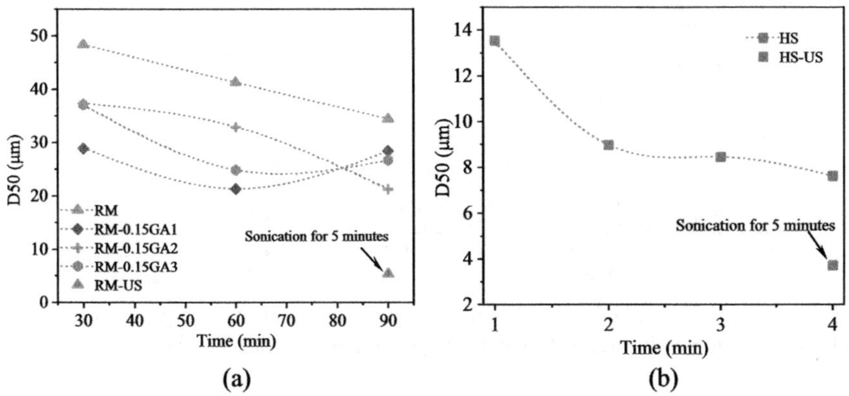

Fig. 4. Median particle size distributions of CW powder (a) roller-milled RM at different grinding durations with and without grinding agents and with post-grinding ultrasonication (b) ground with high-shear HS grinder for different grinding durations, with post-grinding ultrasonication

Figure 5 shows the volumetric particle size distribution of CW processed by different grinding and dispersion techniques, compared to GU cement. HS grinding produced finer particles in less time than RM. Samples ground with GAs had more fine particles (<5 μm) and fewer coarse ones (>100 μm) than dry-ground CW, confirming improved dispersion, consistent with lower D_{50} values in Fig. 4a.

As shown in Fig. 5, mechanical dispersion through sonication further reduced particle aggregation and resulted in a finer particle size distribution compared to dispersion achieved by using GAs during milling. CW particles showed two distinct size distributions, regardless of the grinding and dispersion techniques employed. These variations can be attributed to the differing grindability of calcite (Mohs of 3) and hydroxyapatite (Mohs of 5), with calcite forming fine particles and hydroxyapatite contributing to larger ones (>100 μm) [26]. This coexistence of very fine and relatively coarse particles promotes the formation of aggregates due to Van der Waals forces during grinding [28]. Ultrasonication has shown an effective post-grinding treatment to improve particle dispersion.

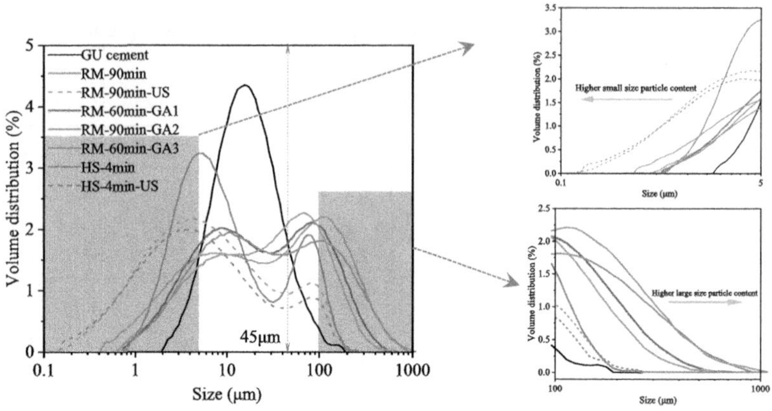

Fig. 5. Particle size distributions of CW ground with roller-mill RM and high-shear HS grinders, with and without grinding agents GA and with post-grinding ultrasonication US

3.3 Reactivity of CW

Table 4 shows the cumulative heat release of CW compared to quartz, based on the R^3 test (ASTM C1897) [20], used to evaluate material reactivity. Over a 7-day period, CW exhibited a cumulative heat release of 49 J/g, which is very similar to that of quartz. According to RILEM TC 267 [29], materials with a 7-day cumulative heat release below 160 J/g are considered inert, and CW falls well within this threshold. The R^3 method analysis thus confirms that CW has negligible reactivity and functions primarily as an inert filler in cementitious systems.

Table 4. Cumulative heat of hydration of quartz and $CW_{RM\text{-}US}$ based on R^3 test according to ASTM C1897 [20]

Samples	Cumulative heat release (J/g_{SCM})		
	1 day	3 days	7 days
CW	29.4	37.4	49
Quartz	15.7	27.8	43.1

3.4 Performance of CW in Cementitious Systems

Slump Flow

Figure 6 shows the slump flow of cement pastes with 10% ground CW using RM-90min and HS-4min grinders, with and without post-grinding ultrasonication, compared to a reference mixture. CW-blended pastes required more SP than the reference due to the finer particle size and higher porosity [30, 31] of CW particles (Fig. 3). SP demand increased by about 60% for HS-ground CW and about 30% for RM-ground CW, likely due to the finer and more angular particles produced in HS systems [32].

US significantly improved slump flow by about 15% for HS-ground CW and about 20% for RM-ground CW without additional SP, confirming its effectiveness in enhancing workability.

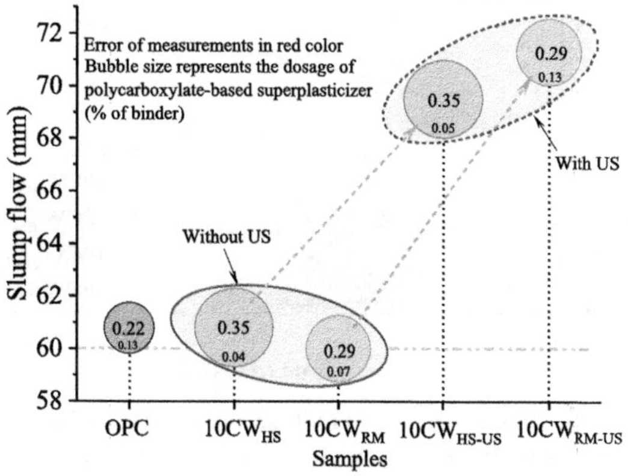

Fig. 6. Slump flow of OPC and systems containing 10% CW ground with roller mill RM and high shear HS grinders, without and with ultrasonication US

Compressive Strength

Figure 7 illustrates the compressive strength development of cement pastes containing 10% CW, ground by RM or HS methods, with and without post-grinding ultrasonication, up to 91 days. Strength increased over time for all systems, but CW pastes without dispersion showed about 10% lower strength than the reference at 28 days. Despite HS-ground CW having a much finer D_{50} (~8 μm) compared to RM-ground CW (~34 μm), as discussed in Sect. 3.2, both showed similar strength, suggesting re-aggregation of particles in paste systems. Also, the low solubility and reactivity of coarse hydroxyapatite particles was assumed not to have a significant influence on strength development [33].

Ultrasonic dispersion significantly enhanced the compressive strength of CW-containing cement pastes, with 30% improvement at both 7 and 28 days. This increase was mainly due to the filler and nucleation effects of finer fraction of CW particles (<5 μm) which is mainly composed of $CaCO_3$ promoting the precipitation of hydration products, resulting in a denser microstructure with reduced porosity [34–36]. Meanwhile, the coarser fraction (>100 μm) composed of hydroxyapatite remains largely inert. Sonication improves the distribution of $CaCO_3$ particles, maximizing their effectiveness and minimizing the dilution effect of CW. Overall, sonication is a practical method to enhance the strength and performance of CW-blended cementitious systems.

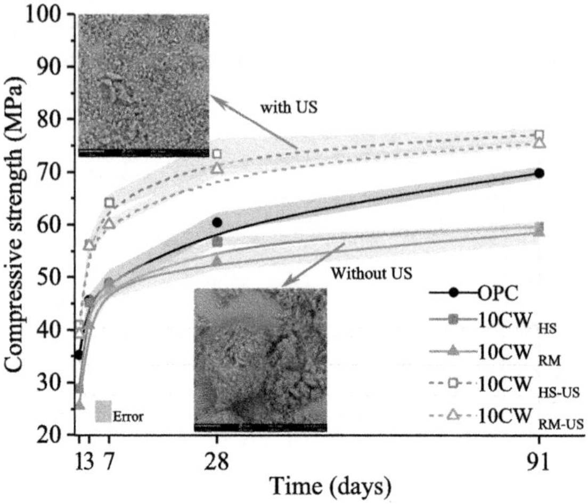

Fig. 7. Compressive strength development of OPC and systems containing 10% CW ground with roller mill RM and high shear HS grinders, without and with ultrasonication US up to 91 days

Heat of Hydration

Figure 8 presents isothermal calorimetry results for OPC and cement pastes with 10% CW, ground by RM or HS methods, without and with ultrasonication. As shown in Fig. 8a, systems containing CW exhibited comparable heat flow profiles regardless of the grinding technique, with both $10CW_{HS}$ and $10CW_{RM}$ exhibiting slightly lower early hydration peaks than OPC, due to CW's dilution effect and inert nature. With ultrasonication (dashed lines), both systems showed an increase in peak intensity, indicating improved particle dispersion and enhanced nucleation site formation, which slightly accelerated early hydration (and significantly impacted strength as shown in Fig. 7).

Figure 8b shows that sonication reduced the time to peak heat flow by about 12% and increased the peak intensity by about 4% for both CW systems. These improvements highlight the role of ultrasonication in promoting early hydration by better dispersion of fine particles. Similar findings have been reported in the literature, where ultrasonication was shown to increase the rate of cement hydration by facilitating the dispersion of fine particles and providing a greater surface area for hydration [37]. These findings show that ultrasonication can be an effective method to partially offset the dilution effect associated with partial cement replacement with CW through enhancing the hydration kinetics.

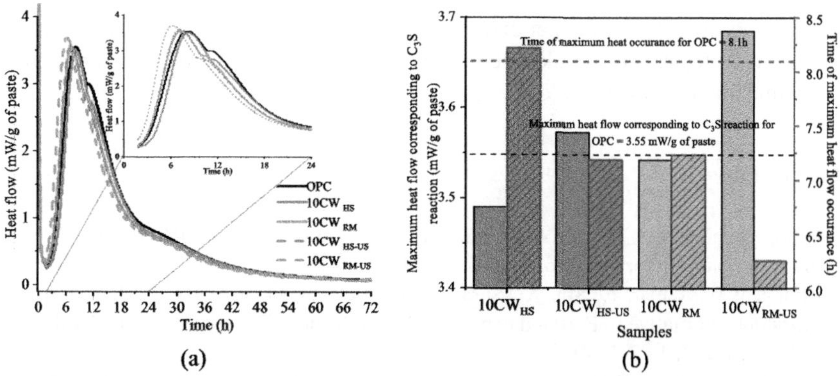

Fig. 8. Calorimetry results of OPC and systems containing10% CW ground with roller-mill RM and high-shear HS grinders, without and with ultrasonication (a) heat flow up to 3 days (b) maximum heat flow and its time of occurrence

4 Conclusion

This study introduces calcium carbonate derived from crustacean wastes (CW) as a mineral admixture for eco-friendly cementitious systems. The optimized processing of CW, combined with ultrasonic (US) dispersion, offers an attractive opportunity to divert waste from landfills for cement-based applications. The following conclusions can be drawn:

- Heat-treated CW mainly consists of calcite with trace amounts of hydroxyapatite.
- The difference in hardness between calcite and hydroxyapatite resulted in a bimodal particle size distribution in the ground CW, with finer calcite and coarser hydroxyapatite particles. This variation promoted particle aggregation and led to a broader overall particle size distribution.
- Mechanical dispersion of CW particles resulted in slump flow enhancement by about 14% and 19% for high-shear and roller-mill ground CW, respectively, compared to undispersed CW.
- While undispersed CW reduced the compressive strength by about 10%, the use of ultrasonicated CW as a partial cement replacement led to a 30% increase in both early and later-age strength.
- Isothermal calorimetry further confirmed that mechanical dispersion improved hydration kinetics, increasing heat flow associated with the C_3S reaction by about 4% and reducing the time to peak heat release by 12%.

Overall, the findings demonstrate that calcium carbonate derived from crustacean waste (CW) can function as an accelerating admixture in cement-based systems, particularly when processed through ultrasonication. While a 10% cement replacement with undispersed CW led to a reduction in strength, proper dispersion significantly improved both early and later-age strengths. These results highlight the critical importance of dispersing alternative fine materials within cementitious systems. Future research should optimize CW dosage and further explore the chemical effects of CW-derived $CaCO_3$ to clarify its role in strength development. This would help better understand both physical and chemical contributions in cementitious systems.

References

1. Global Cement and Concrete Association: Concrete future: GCCA 2050 cement and concrete industry roadmap for net zero concrete (2021)
2. Chong, B.W., Gujar, P., Shi, X., Suraneni, P.: Assessment of waste eggshell powder as a limestone alternative in Portland cement. J. Mater. Struct. **57**, 219 (2024). https://doi.org/10.1617/s11527-024-02478-9
3. EN 197-1, Cement - Part 1: Composition, specifications and conformity criteria for common cements (2011)
4. ASTM C595, Standard Specification for Blended Hydraulic Cements. ASTM International, West Conshohocken, PA. p. 9 (2024)
5. AASHTO 240, Standard specification for blended cement. American Association of State Highway and Transportation officials, AASHTO: Washington D.C, United States (2023)
6. CSA A3000, Cementitious materials compendium (2018)
7. Heidari, P., Wilson, W., Rivard, P.: Marine Dredged Sediments as a Partial Alternative to Sand in Cementitious Systems Using Particle Packing Optimization. ACI Special Publication, vol. 362, pp. 639–650 (2024). https://doi.org/10.14359/51741018
8. Belmokhtar, Z., Cousin, P., Elkoun, S., Robert, M.: Eco-friendly wall cladding panels from recycled fishing gear and clamshell waste. Compos. Sci. **8**, 484 (2024). https://doi.org/10.3390/jcs8110484
9. Yusoff, M.A., et al.: Valorization of seafood waste: a review of life cycle assessment studies in biorefinery applications. Sci. Total Environ. **952**, 175810 (2024). https://doi.org/10.1016/j.scitotenv.2024.175810
10. Baykara, H., et al.: Chitosan-cement composite mortars: exploring interactions, structural evolution, environmental footprint and mechanical performance. ACS Omega **9**, 24978–24986 (2024). https://doi.org/10.1021/acsomega.4c02040
11. Haider, M.M., et al.: Impact of chitin nanofibers and nanocrystals from waste shrimp shells on mechanical properties, setting time, and late-age hydration of mortar. Sci. Rep. **12**, 20539 (2022). https://doi.org/10.1038/s41598-022-24366-4
12. Mathew, G.M., et al.: Sustainable and eco-friendly strategies for shrimp shell valorization. Environ. Pollut. **267**, 115656 (2020). https://doi.org/10.1016/j.envpol.2020.115656
13. Dequiedt, A.-S., Coster, M., Chermant, L., Chermant, J.-L.: Study of phase dispersion in concrete by image analysis. Cement Concr. Compos. **23**, 215–226 (2001). https://doi.org/10.1016/S0958-9465(00)00060-3
14. Zhang, Y., Kong, X., Gao, L., Bai, Y.: Characterization of the mesostructural organization of cement particles in fresh cement paste. Constr. Build. Mater. **124**, 1038–1050 (2016). https://doi.org/10.1016/j.conbuildmat.2016.08.143
15. Abd Elhakam, A., Mohamed, A.E., Awad, E.: Influence of self-healing, mixing method and adding silica fume on mechanical properties of recycled aggregates concrete. Constr. Build. Mater. **35**, 421–427 (2012). https://doi.org/10.1016/j.conbuildmat.2012.04.013
16. Li, N., Ma, B., Wang, Y., Si, W., Qin, F.: Influence analyses of mixing approaches on properties of conventional and interlocking-dense concrete. Constr. Build. Mater. **122**, 465–472 (2016). https://doi.org/10.1016/j.conbuildmat.2016.06.093
17. Ferraris, C.: Concrete Mixing Methods and Concrete Mixers: State of the Art. Research of the National Institute of Standards and Technology. https://doi.org/10.6028/jres.106.016

18. Jézéquel, P.-H., Collin, V.: Mixing of concrete or mortars: dispersive aspects. Cem. Concr. Res. **37**, 1321–1333 (2007). https://doi.org/10.1016/j.cemconres.2007.05.007
19. Xu, Z., et al.: Strengthening mechanism of ultrasonic action on mechanical properties of cement-based materials. Constr. Build. Mater. **362**, 129788 (2023). https://doi.org/10.1016/j.conbuildmat.2022.129788
20. ASTM C1897, A. Standard Test Methods for Measuring the Reactivity of Supplementary Cementitious Materials by Isothermal Calorimetry and Bound Water Measurements (2020)
21. ASTM C1702, A. Standard Test Method for Measurement of Heat of Hydration of Hydraulic Cementitious Materials Using Isothermal Conduction Calorimetry (2017)
22. Georget, F., Sui, S., Wilson, W., Scrivener, K.L.: Reconciliation of pore structure characterization methods: the simple case of PC-limestone cement pastes. Cem. Concr. Res. **184**, 107624 (2024). https://doi.org/10.1016/j.cemconres.2024.107624
23. Fischer, I., Pichler, B., Lach, E., Terner, C., Barraud, E., Britz, F.: Compressive strength of cement paste as a function of loading rate: experiments and engineering mechanics analysis. Cem. Concr. Res. **58**, 186–200 (2014). https://doi.org/10.1016/j.cemconres.2014.01.013
24. CSA A3003, Chemical test methods for cementitious materials for use in concrete and masonry (2023)
25. Mohan, K., Muralisankar, T., Jayakumar, R., Rajeevgandhi, C.: A study on structural comparisons of α-chitin extracted from marine crustacean shell waste. Carbohyd. Polym. Technol. Appl. **2**, 100037 (2021). https://doi.org/10.1016/j.carpta.2021.100037
26. Basri, H.: Crystallographic and morphological studies of nanocrystalline hydroxyapatite synthesized from bovine bone at different calcination temperatures. Mechanics of Continua and Mathematical Sciences (2019). https://doi.org/10.26782/jmcms.spl.4/2019.11.00002
27. Roux, V., Wilson, W., Deneele, D., Paris, M., Tagnit-Hamou, A.: Linking Water Demand and Surface Area of Calcined Palygorskite. ACI Special Publication, p. 362 (2024). https://doi.org/10.14359/51740891
28. Balaz, P.: Mechanochemistry in Nanoscience and Minerals Engineering. Springer, Heidelberg (2008)
29. Li, X., et al.: Reactivity tests for supplementary cementitious materials: RILEM TC 267-TRM phase 1. J. Mater. Struct. **51**, 1–14 (2018). https://doi.org/10.1617/s11527-018-1269-x
30. Hunger, M., Brouwers, H.J.H.: Flow analysis of water–powder mixtures: application to specific surface area and shape factor. Cement Concr. Compos. **31**, 39–59 (2009). https://doi.org/10.1016/j.cemconcomp.2008.09.010
31. Sua-Iam, G., Makul, N.: Utilization of limestone powder to improve the properties of self-compacting concrete incorporating high volumes of untreated rice husk ash as fine aggregate. Constr. Build. Mater. **38**, 455–464 (2013). https://doi.org/10.1016/j.conbuildmat.2012.08.016
32. Sposito, R., Maier, M., Beuntner, N., Thienel, K.: Physical and mineralogical properties of calcined common clays as SCM and their impact on flow resistance and demand for superplasticizer. Cem. Concr. Res. **154**, 106743 (2022). https://doi.org/10.1016/j.cemconres.2022.106743
33. Barralet, J., Lilley, K., Grover, L., Farrar, D., Ansell, C., Gbureck, U.: Cements from nanocrystalline hydroxyapatite: selected papers from the 18th European Conference on Biomaterials (ESB2003), Stuttgart, Germany (2003). (Guest Editors: Michael Doser and Heinrich Planck). J. Mater. Sci. Mater. Med. **15** (2004). https://doi.org/10.1023/B:JMSM.0000021111.48592.ab
34. Berodier, E., Scrivener, K.: Understanding the filler effect on the nucleation and growth of C-S-H. Am. Ceram. Soc. **97**, 3764–3773 (2014). https://doi.org/10.1111/jace.13177
35. De Weerdt, K., Kjellsen, K.O., Sellevold, E., Justnes, H.: Synergy between fly ash and limestone powder in ternary cements. Cem. Concr. Compos. **33**, 30–38 (2011). https://doi.org/10.1016/j.cemconcomp.2010.09.006

36. Matschei, T., Lothenbach, B., Glasser, F.P.: The role of calcium carbonate in cement hydration. Cem. Concr. Res. **37**, 551–558 (2007). https://doi.org/10.1016/j.cemconres.2006.10.013
37. Vallurupalli, K., Meng, W., Liu, J., Khayat, K.H.: Effect of graphene oxide on rheology, hydration and strength development of cement paste. Constr. Build. Mater. **265**, 120311 (2020). https://doi.org/10.1016/j.conbuildmat.2020.120311

Innovative Low-Carbon Materials

Adaptation of Low Environmental Impact Mortar Formulations for 3D Printing

Sourour Elleuch(✉), Hélène Carré, and Christian La Borderie

Laboratoire des Sciences de l'Ingénieur Appliquées à la Mécanique et au Génie Electrique [SIAME], Université de Pau et des Pays de l'Adour [UPPA], Pau, France
Sourour.elleuch@univ-pau.fr

Abstract. The construction industry is experiencing a transformative shift with the integration of 3D printing technologies, revolutionizing traditional construction methods. The growing adoption of 3D printing offers a promising approach to minimizing material waste, reducing manufacturing costs, and improving overall efficiency. However, 3D printable mortars often require a very high cement content and high dosage of admixtures compared to conventional mortars or concretes, which increases their carbon footprint.

While 3D printing presents an opportunity to reduce waste and optimize material use, this challenge highlights the need for developing low environmental impact materials that maintain the required properties for successful additive construction.

In this context, our research explores the advantages of combining low environmental impact materials with 3D printing to maximize the benefits of this innovative approach while addressing the environmental concerns associated with high cement content. Specifically, we focus on designing mortars that meet the rheological, mechanical, and environmental requirements for additive construction. Key parameters such as pumpability, buildability, and extrudability are targeted by selecting eco-friendly cement, fine calcareous filler, silica fume and sand.

To ensure compatibility with 3D printing processes, we employ experimental methods such as rheological characterization, extrusion trials, and buildability tests. We adapt the formulation to ensure it possesses the necessary flow, workability, and setting properties required for extrusion and layer-by-layer construction. This adaptability to the 3D printing process is crucial, as it ensures the material can be successfully printed without compromising its performance or structural integrity.

This research proposes a low-impact mortar that meet the technical demands of additive manufacturing while helping to reduce the environmental footprint of the construction industry.

Keywords: 3D printing · Additive construction · Low environmental impact materials · Mortar formulation · Fresh properties · Mechanical performances

M. Posani and S. Singh (Eds.): RYS 2025, RILEM Bookseries 67, pp. 241–254, 2026.
https://doi.org/10.1007/978-3-032-14652-6_19

1 Introduction

Additive manufacturing in the construction sector offers significant advantages, including enhanced design freedom, reduced labour demands, and decreased material waste.

Nevertheless, the cement-based mortars commonly employed in this context pose substantial environmental challenges, particularly due to their contribution to CO_2 emissions. With cement production accounting for approximately 6% of global CO_2 emissions [1], and the formulation of 3D mortars typically requires a high cement content can reach 683 kg/m^3 [2] or even 863 kg/m^3 [3] making these materials particularly impactful from an environmental perspective. Consequently, there is a critical need to develop low-impact mortars specifically tailored for 3D printing applications.

Currently, printable concrete lacks standardized regulations, and 3D mortar formulations are not well established. Some researchers combine self-compacting concrete and shotcrete properties for extrudability [4], while others use experimental methods to optimize mix design by adjusting water content and admixtures based on extrusion, pumpability, and buildability tests [5]. Our research adopts the experimental approach to develop a low-impact 3D printable mortar tailored for additive construction.

In this study, we first define the experimental framework by specifying the key requirements "environmental, printability, and mechanical performances" and the corresponding testing protocols designed to validate these criteria.

Following this, we describe the materials and methods employed, including the equipment used, the selection of materials, and the detailed mix design methodology implemented to develop a 3D printable mortar with low environmental impact.

Finally, the results and discussion present the outcomes of the mix design formulations and the mechanical performance evaluations and the study concludes with an assessment of how the final mortar formulation meets the established requirements.

2 Experimental Framework: Requirements and Testing Protocols

The essential criteria required for developing a mortar that is both environmentally sustainable and compatible with 3D printing processes are organized into three main categories: environmental impact, printability, and mechanical performance. Building upon a comprehensive literature review and preliminary experimental investigations, a set of performance requirements was summarized along with the final results for our mortar in Table 3.

2.1 Environmental Requirements

Environmental performance is one of the key criteria guiding our formulation methodology. Although the term low-carbon concrete is widely used, it currently lacks a standardized definition grounded in regulatory or normative frameworks. To address this gap, the French national ready-mix concrete association (SNBPE) [6] proposes benchmark thresholds to classify concretes according to their carbon footprint.

In this context, the present study adopts these thresholds to distinguish between standard and low-carbon concretes. As a reference point, we compare our mortar formulation to a typical C35/45 concrete, which is widely used in structural application concrete designed for XC-XF1-XD1 exposure classes [7] (Fig. 1), the carbon footprint of a conventional mix ranges between 260 and 290 kg CO_2-eq/m^3, whereas a formulation may be classified as low-carbon if it exhibits emissions below 230 kg CO_2-eq/m^3.

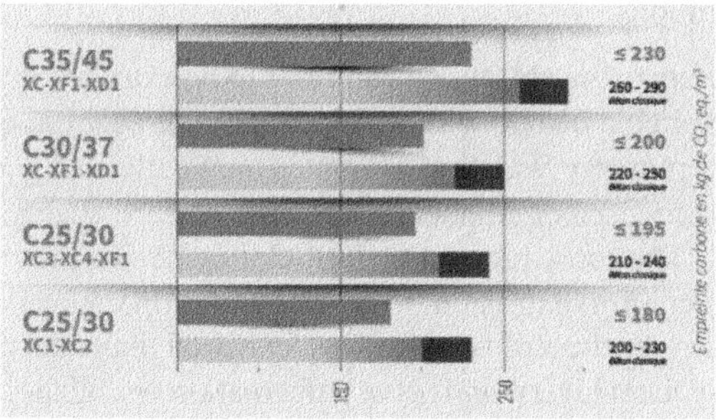

Fig. 1. Comparison of the carbon footprint of standard (grey) and low-carbon (mustard) concrete formulations [6]

To assess the environmental impact of the proposed mortar formulation, a carbon footprint calculation was conducted in accordance with [8], focusing on life cycle stages A1–A3, which include raw material extraction, transport, and production. All materials were assumed to be locally sourced and transported within a 100 km radius of the laboratory using a EURO 5 semi-trailer truck, in accordance with emission factors provided by the Ecoinvent database. This assumption reflects realistic sourcing conditions, as the selected constituents are widely available at most cement production facilities. The environmental impact of each formulation was calculated using Eq. (1) and Eq. (2).

- **Material emissions ME** (kg CO_2-eq/m^3):

$$ME = \sum \left[\text{Mass} \left(\text{kg/m}^3 \right) \times (A1 + A2 + A3) \right] (\text{kg } CO_2 \text{ - eq/kg}) \quad (1)$$

- **Transport emissions TE** (kg CO_2-eq/m^3):

$$TE = \text{Mass}\left(\text{kg/m}^3 \right) \times \text{Distance(km)} \times \text{Transport factor}\left(\text{kg } CO_2 \text{ - /kg\textbackslash km} \right) \quad (2)$$

2.2 Printability Requirements

The printability a multifaceted concept defined in [9] that must be optimized to ensure successful 3D printing. This optimisation must consider contradictory expectations, as the mixture must be fluid enough to be pumped and extruded, but also stiff enough to retain its shape after deposition and support subsequent layers [10].

Due to the lack of standardized testing protocols, these parameters were evaluated through a combination of visual observations and empirical tests.

- Pumpability, defined as the capacity of the mix to be conveyed through a pumping system over a distance, without inducing phase separation, loss of workability, or mechanical failure within the system [9].

Pumpability was initially assessed via the flow table test [11] which simulates the fluidization induced by pumping. A spread diameter in the range of 135–220 mm was considered acceptable, as recommended by several studies [3, 12–16]. The formulations were subsequently tested using an actual pumping system to confirm their suitability under realistic conditions.

- Extrudability, refers to the ability of the mortar to be extruded in a continuous, homogeneous flow through a nozzle without blockage, segregation, or excessive pressure system [9]. This property was first evaluated manually using a handheld extruder. Visual inspection focused on filament continuity, surface quality, and absence of segregation [4, 17]. The most promising formulations were then tested using the 3D printer nozzle to confirm extrusion behaviour under operational constraints.
- Buildability, describes the ability of freshly deposited filaments to retain their geometry, support subsequent layers, and resist deformation or collapse during the layer-by-layer deposition process [9]. Buildability was quantified by printing vertical elements and recording the maximum number of layers that could be deposited without significant deformation or collapse. The slump ratio, defined as the percentage reduction in height due to layer settlement, was also measured to characterize the material's capacity to sustain self-weight [4, 5]. Printing straight wall elements with minimum 10 layers is a practical benchmark for evaluating this property [17].
- Rheological retention (Open Time), refers to the material's ability to maintain its flowability and workability over time. It is essential to ensure printability throughout the printing process. The formulation must maintain its flow and structural characteristics for a period of 30 to 60 min [15, 18] to accommodate batching, transfer, and printing operations. To assess this stability, a flow test is performed every 15 min for a total duration of 90 min to evaluate the fluidity over time.

2.3 Mechanical Performance Requirements

Mechanical strength was evaluated using standardized tests conducted on [40 × 40 × 160 mm] prismatic specimens, in accordance with [19]. The specimens were sealed in plastic bags to minimize moisture loss. The results were compared to reference values typical of conventional concrete of class C35/45, to benchmark the performance of the developed formulations.

- Flexural strength (ft) was measured using the three-point bending test on the entire specimens prior to compression.
- Compressive strength (fc) was determined via uniaxial compression on the specimen halves.

3 Materials and Methods

3.1 Equipment

The 3D mortar formulation was validated through a series of targeted experimental set-ups, each designed to assess specific performance aspects.

A manual extrusion device (Fig. 2a) with two nozzle types (Ø5 cm and 1 × 4 cm rectangular) was used for the first extrusion trials to visually inspect flow, shape retention, and workability under low pressure.This setup allowed a preliminary assessment of flow behavior, continuity of extruded filament and shape retention.

To assess pumpability, a progressive cavity pump (Fig. 2b) with a maximum flow rate of 10 l/min was used, simulating realistic on-site material delivery.

For the printing trials, a StoneFlower 3D concrete printer (Fig. 2c) of gantry type with a maximum printing speed = 100 mm/s was employed. It can print object with maximum dimensions 50 × 120 × 80 cm^3.

(a) (b) (c)

Fig. 2. Equipment used for extrusion and 3D printing trials (a) manual extruder, (b) concrete pump, (c) 3D printer

3.2 Materials Selection

The selection of raw materials is a critical phase in the development of 3D printable mortars. These mixtures typically rely on a high cement content to ensure sufficient extrudability and optimal interlayer bonding [5]. However, cement-rich formulations present several drawbacks, including increased environmental impact, elevated cost, and a heightened risk of shrinkage-induced cracking. These issues stem largely from the high cement consumption and its substantial carbon footprint.

Cement is widely recognized as the most environmentally burdensome component in cementitious systems. Consequently, reducing the cement content is a key strategy for improving the environmental profile of 3D printable mortars.

To this end, many researchers have explored the partial replacement of cement with supplementary cementitious materials (SCM) that exhibit lower environmental impacts without compromising mechanical properties. Binders containing up to 70% cement combined with SCMs such as fly ash, silica fume, and limestone fillers have been shown to enhance mechanical strength and extrudability [14, 17, 20]. More ambitious formulations have reduced cement content to 50%, incorporating calcined clays and limestone fillers [21, 22] in an effort to optimize the balance between technical performance and environmental sustainability, using minimal chemical admixtures.

All raw materials used are listed in Table 1 along with their estimated CO_2 emissions.

A low-carbon cement "CEM III/C 32.5 N-LH/SR CE PM NF" supplied by EQIOM was selected as the primary binder. This cement consists of approximately 15% clinker and 82% ground granulated blast furnace slag (GGBFS), offering a significantly reduced CO_2 footprint compared to conventional Portland cement.

A natural fine sand (0–1 mm) was used as the granular skeleton, ensuring smooth extrusion, minimizing segregation, and preserving the geometry of printed layers.

Mineral additions were used to improve performance and reduce CO_2 emissions:

- A limestone filler (particle size < 63 µm) supplied by Omya, was used to improve particle packing, porosity and flowability.
- Silica fume, due to its high pozzolanic reactivity and fine particle size, was added to improve matrix compactness and long-term strength.

Two Chryso admixtures were added to control the mortar's fresh properties:

- EnviroMix 2091, a high-performance plasticizer designed for low-carbon concrete, was used to extend workability and improve rheology.
- Tard MTD, a set retarder, was included to regulate the setting time and ensure continuous and consistent layer deposition.

The synergistic use of these materials enabled the formulation of a 3D-printable mortar with a well-graded particle size distribution, ensuring dense packing, enhanced compactness, and desirable printability characteristics.

Table 1. Raw material characteristics and associated carbon footprint

Material	Description	Density (g/cm^3)	Granulometry	CO_2 Emission Footprint (A1-A3) (kg CO_2 eq/kg)
CEM III/C 32.5 N-LH/SR	Low heat of hydration, sulfate-resistant blast furnace cement	2.92	D50 ≈ 10 µm	0.228
Sand (0/1 mm)	Natural fine aggregate	2.64	0 – 1 mm	0.005
Limestone filler	Ultra-fine calcium carbonate,	2.70	D50 ≈ 1-3 µm, 100% < 63 µm	0.0542
Silica Fume	Reactive pozzolanic additive	2.20	D50 ≈ 0.15 µm	0.052
EnviroMix 2091	rheology-extending plasticizer	1.064	-	3.067
Tard MTD	set-retarder	1.090	-	2.5

3.3 Mix Design Methodology

The formulation strategy was established based on a literature review and the technical constraints of the laboratory equipment. The approach was structured into a sequence of experimental stages, each validated by practical tests.

Mixing Procedure

For the mixing process, we employed two types of mixers. For the initial small-scale trials, a 5 L laboratory mixer was used. The binder and sand were first mixed at low speed (62 rpm) for 60 s. Subsequently, two-thirds of the total water was added and mixed for another 60 s at the same low speed. After this, the sides of the bowl were scraped for 30 s to ensure homogeneity. Finally, the remaining one-third of the water, along with the admixtures, was added, and the mixture was mixed at high speed (125 rpm) for 60 s.

For the printing trials, a larger 50 L mixer was used following the same mixing procedure. All mixing was conducted at room temperature (approximately 20 °C).

Step 1 – Determination of Optimal Mass Ratios with Cement-Only Binder

Initial manual extrusion trials were conducted using a 5 cm diameter nozzle to identify suitable water-to-binder (W/B) and sand-to-binder (S/B) ratios that enable continuous and stable extrusion.

A wide range of S/B ratios has been reported in the literature, including S/B = 0.4 [15], S/B = 1.0 [23] up to S/B = 1.7 [17, 18] and even S/B = 2.3 [24]. In this study, S/B ratios ranging from 1.0 to 2.0 were evaluated. Similarly, literature reports W/B ratios from 0.26 [14] to 0.51 [17]. In our trials, W/B ratios between 0.25 and 0.40 were tested. We avoided using higher water content to prevent a decrease in mechanical performance.

Step 2 – Binder Substitution and Water Adjustment

In the second phase, partial substitution of CEM III/C cement with varying percentages of limestone filler and silica fume was carried out. Multiple combinations and replacement levels were tested to assess their impact on extrudability and rheological behavior. These substitutions aimed to ensure successful manual extrusion using a 1 × 4 cm rectangular nozzle. Water content was adjusted accordingly for each formulation to maintain suitable workability and extrusion performance.

Step 3 – Adjusting the Admixtures

The selection of a suitable plasticizer was based on preliminary flow-table tests conducted with various types and dosages of plasticizers. The most promising superplasticizer was then evaluated for its ability to retain rheological properties over time, both with and without the use of a set retarder.

In parallel, the environmental impact of the formulation was assessed through emissions calculations. The final formulation was optimized to ensure rheological retention between 30 and 60 min.

Step 4 – Validation of Final Mix

The final step consists of validating the optimized mortar formulation by assessing its compliance with all performance requirements. This includes verifying its environmental impact, fresh-state properties (extrudability, pumpability, buildability, and open time), as well as its early mechanical strength. To complete this validation, a full-scale 3D printing test is carried out using the actual 3D printer. The printed structure must demonstrate continuous and homogeneous extrusion, stable layer stacking without collapse, and sufficient build-up of layers ($\geq$ 10 layers) while maintaining workability for at least one hour. The formulation must also meet the compressive and flexural strength targets and fall below the predefined CO_2 emission threshold.

4 Results and Discussion

4.1 Formulation Mix Design Results

Step 1 – Determination of Optimal Mass Ratios with Cement-Only Binder
The optimal performance was achieved with W/B = 0.32 and S/B = 1.5, resulting in a continuous, void-free filament over a 20 cm length (maximum capacity of the manual extruder) as showed in Fig. 6a.

This confirmed the good extrudability of the cement-only binder and established a reference for incorporating mineral additions in the next steps.

Step 2 – Binder Substitution and Water Adjustment
After numerous formulation trials with varying addition-to-binder ratios, the optimal mix that achieved both low environmental impact and continuous extrusion without voids was obtained by replacing 50% of the cement in the binder. The final binder composition is as follows:

- 50% CEM III/C
- 30% limestone filler
- 20% silica fume

Initial tests with a W/B ratio of 0.4 yielded a flow-table spread of 145 mm. Although this formulation demonstrated some potential buildability, it was difficult to extrude manually and resulted in an inconsistent, discontinuous filament.

Step 3 – Adjusting the Admixtures
A W/B of 0.35 was selected to minimize the need for high dosages of plasticizers, thereby reducing the environmental impact of the formulation. While also maintaining a low enough water content to ensure good mechanical performance.

Figure 3 compares the performance of Enviromix 2091 with Enviromix P7055, highlighting that the Enviromix 2091 proved to be the most effective at 1% of the cement mass, demonstrating superior ability to maintain rheological properties over time.

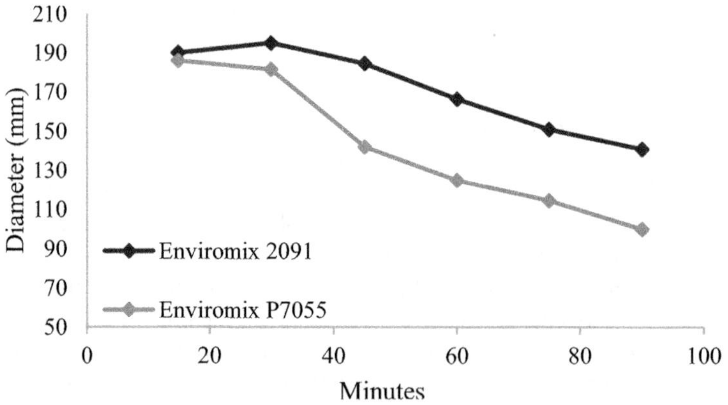

Fig. 3. Comparison of plasticizers based on flow table spread over time

However, despite the promising initial performance with the plasticizer, pumpability began to decline after approximately 45 min due to the onset of cement setting eventually leading to a complete interruption of flow, as shown in Fig. 5a.

To improve workability retention, a set retarder was incorporated into the mix. The impact of adding 0.5% retarder (by cement mass) with W/B ratio = 0.35 is illustrated in Fig. 4.

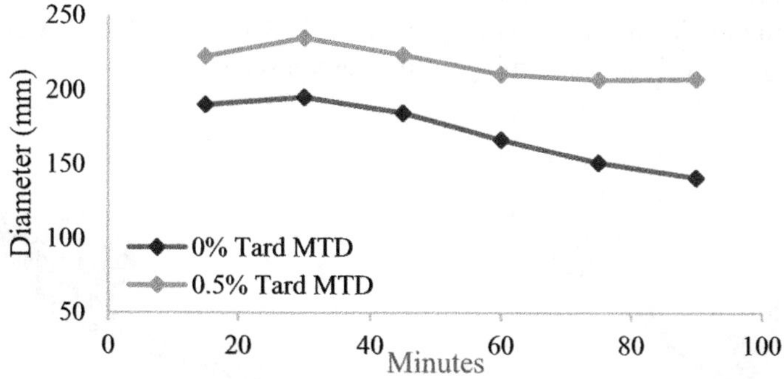

Fig. 4. Effect of 0.5% retarder on flow retention over time

The resulting fluidity exceeded 220 mm in the flowtable test, confirming excellent flowability.

Nevertheless, while the mortar remained fluid and pumpable, it exhibited poor buildability. Significant collapse occurred during the printing process, as shown in Fig. 5b. This indicates that while workability was improved, further optimization is required to enhance the structural stability during deposition. Notably, the addition of the retarder slightly increased the fluidity of the mix.

(a) (b)

Fig. 5. Printability issues observed with 3D printing: (a) discontinuous pumpability; (b) poor buildability with collapse.

Step 4 – Validation of Final Mix

After testing various water contents with the addition of the superplasticizer and retarder, we adjusted the water-to-binder ratio (W/B) to 0.3. This formulation proved to be manually extrudable with 1 × 4 cm nozzle as shown in Fig. 6b and to be sufficiently fluid for pumping and continuous extrusion with the 3D printer as shown in a Fig. 7a.

Buildability was confirmed by the successful deposition of 15 consecutive layers as shown in Fig. 7b without collapse, with a settlement of 27% of the initial height (which should be 22). The flow-table spread reached 196 mm, indicating appropriate flow behaviour while maintaining sufficient cohesion. The open time exceeded 60 min, satisfying the rheological retention requirements.

Furthermore, the mix's carbon footprint (156.14 kg CO_2 eq./m^3) is well below the 230 kg CO_2 eq./m^3 limit, confirming its low-carbon status.

(a) (b)

Fig. 6. Validation of printability with manual extruder: (a) valid extrusion with 5 cm diameter nozzle; (b) valid extrusion with 1 × 4 cm nozzle

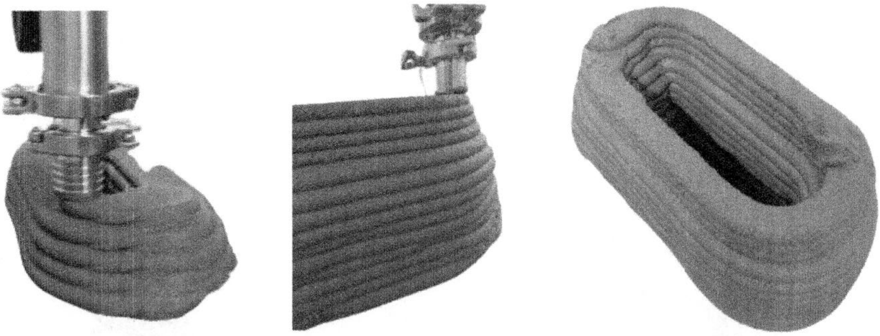

(a) (b) (c)

Fig. 7. Successful 3D printing results: (a) continuous Extrusion; (b) good buildability; (c) final printed object

The final mix, detailed in Table 2, shows all component proportions as mass ratios relative to the binder.

Table 2. 3D final mortar composition and carbon footprint

Materiel		Mass ratio (relative to binder)	Quantity (kg/m³)	Footprint (kg CO_2-eq/ m³)
Binder	CEM III/C	0.5	403,3	98,13
	limestone filler	0.3	242;0	16.82
	Silica Fume	0.2	161,3	10,86
Sand		1.5	1209,9	24,56
EnviroMix 2091		0.01	4,03	3,76
Tard MTD		0.005	2.02	1,34
Water		0.3	241,2	0,67
Total				156,14

4.2 Mechanical Performance Results

Figure 8 shows the results of the compressive and flexural strength tests for the final mortar at 1, 2, 7, and 14 days [19], where points represent individual results and the line represents the average. The flexural strength (ft) was measured on (40 × 40 × 160 mm) prisms, while the compressive strength (fc) was determined on (40 × 40 × 40 mm) cubes.

Our low-carbon mortar surpassed the expected performance at 28 days by achieving remarkable results at only 14 days. The compressive strength (Fc) reached 50 MPa, while the flexural strength (Ft) was 5.5 MPa at 14 days.

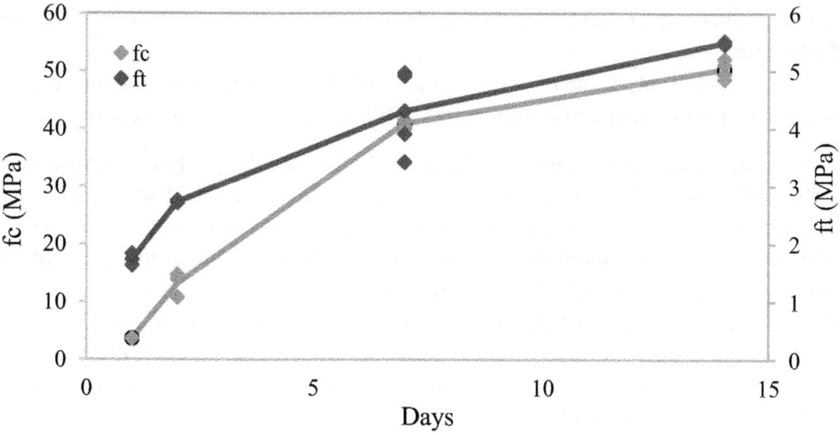

Fig. 8. Mechanical performance (days)

Table 3 summarizes all the criteria we set as targets "environmental impact, printability, and mechanical performance" along with our experimental results.

Table 3. Performance requirements and experimental validation

Category	Criterion	Target requirement	Experimental result	Validation
Environmental Impact	Carbon footprint	≤ 230 kg CO_2 eq./ m³	156 kg CO_2 eq./ m³	✓
Printability	Extrudability	Continuous, homogeneous filament (manual + 3D printer)	Achieved	✓
	Rheological retention	≥ 30–60 minutes of open time	60 min	✓
	Pumpability	Flow table spread [135-220] + continuous pumping	Flow table spread = 196 mm + continuous pumping	✓
	Buildability	≥ 10 layers without collapse	15 layers, stable geometry	✓
Mechanical Performance	Compressive strength (fc)	≥ 35 MPa at 7 days	41 MPa at 7 days	✓
		≥ 42 MPa at 14 days ≥ 45 MPa at 28 days	50 MPa at 14 days	✓
	Flexural strength (ft)	≥ 4 MPa at 7 days	4.5 MPa at 7 days	✓
		≥ 5 MPa at 14 days ≥ 5.5 MPa at 28 days	5.5 MPa at 14 days	✓

5 Conclusion

This study successfully developed and validated a low-carbon mortar formulation tailored for 3D printing applications. By adopting a one-component (1 K) approach based on a low-clinker binder combining low-carbon cement, calcium carbonate, silica fume and admixtures.

As detailed in Table 3, the final mix achieved full compliance with the multi-criteria requirements for environmental impact, printability and mechanical strength:

- Our mortar reached a carbon footprint of 156.14 kg CO_2 eq/m^3, representing a 32% reduction in CO_2 emissions compared to the green C35/45 concrete, and approximately a 40–46% reduction compared to the conventional C35/45 concrete.
- The optimized mix demonstrated excellent fresh-state behaviour, including pumpability, continuous extrudability, and buildability.
- It achieved early-age mechanical strengths of 41 MPa in compression after 7 days, representing a 17% increase over the reference concrete class C35/45. At 14 days, the compressive strength reached 50 MPa, marking a 19% improvement compared to the 42 MPa typically achieved by C35/45.

These results underline the potential of this low-impact, high-performance mortar for sustainable additive manufacturing in the construction industry.

6 Perspectives

This study highlights key aspects of 3D printable low-carbon mortars, yet further work is needed to fully quantify their mechanical and environmental performance.

- Environmental impact: Comprehensive life cycle assessment (LCA) incorporating structural performance metrics, energy consumption, material waste during printing, and overall carbon footprint, to provide a more accurate evaluation of the environmental benefits of low-carbon printable concrete.
- Mechanical performance: Strength assessment that accounts for irregular cross-sections, with systematic comparison to monolithically cast specimens of the same mix design, to distinguish intrinsic material properties from effects induced by the 3D printing process, including interlayer bonding and anisotropy.

Overall, combining refined mechanical assessment with comprehensive environmental analysis will provide a clear basis for optimizing printable concrete materials and promoting their sustainable adoption in construction.

References

1. Tiseo, I.: Cement industry emissions worldwide - Statistics & Facts. https://www.statista.com/topics/11056/cement-industry-emissions-worldwide/#topicOverview
2. Khalil, N.: Formulation et caractérisation chimique et rhéologique des mortiers imprimables en 3D à base de mélanges de ciments Portland et sulfoalumineux. Ecole nationale supérieure Mines-Télécom, Lille Douai (2018)
3. Hynek, E., Bulteel, D., Antoine, U., Rémond, S., David, B., Sébastien, R.: Formulation et caractérisation d'une encre cimentaire à faible impact carbone pour l'impression 3D (2022). https://hal.science/hal-03879722
4. Li, Z., et al.: Fresh and hardened properties of extrusion-based 3D-printed cementitious materials: a review. MDPI (2020). https://doi.org/10.3390/su12145628
5. Malaeb, Z., et al.: 3D Concrete Printing: Machine and Mix Design (2015)
6. SNBPE: BÉTON BAS CARBONE-Définition (2021). www.snbpe.org
7. AFNOR: NF EN 206+A2/CN- Concrete – Specification, performance, production and conformity (2022)
8. AFNOR: NF EN 15804+A2-CN - Sustainability of construction works – Environmental product declarations (2022)
9. AFNOR: NF EN ISO/ASTM 52939 - Additive manufacturing for construction—Qualification principles—Structural and infrastructure elements (ISO/ASTM 52939:2023) (2023)
10. Zhang, C., et al.: Mix Design Concepts for 3D Printable Concrete: A Review. Elsevier Ltd. (2021). https://doi.org/10.1016/j.cemconcomp.2021.104155
11. AFNOR: NF EN 1015-3 Methods of test for mortar for masonry - Part 3 : determination of consistence of fresh mortar (by flow table) (1999)
12. Paritala, S., Singaram, K.K., Bathina, I., Khan, M.A., Jyosyula, S.K.R.: Rheology and Pumpability of Mix Suitable for Extrusion-Based Concrete 3D Printing – A Review. Elsevier Ltd. (2023). https://doi.org/10.1016/j.conbuildmat.2023.132962
13. Baz, B., Aouad, G., Leblond, P., Al-Mansouri, O., D'hondt, M., Remond, S.: Mechanical assessment of concrete – steel bonding in 3D printed elements. Constr. Build. Mater. **256** (2020). https://doi.org/10.1016/j.conbuildmat.2020.119457

14. Rehman, A.U., Perrot, A., Birru, B.M., Kim, J.H.: Recommendations for quality control in industrial 3D concrete printing construction with mono-component concrete: a critical evaluation of ten test methods and the introduction of the performance index. Dev. Built Environ. **16** (2023). https://doi.org/10.1016/j.dibe.2023.100232
15. Ye, J., Cui, C., Yu, J., Yu, K., Xiao, J.: Fresh and anisotropic-mechanical properties of 3D printable ultra-high ductile concrete with crumb rubber. Compos. B Eng. **211** (2021). https://doi.org/10.1016/j.compositesb.2021.108639
16. Jo, J.H., Jo, B.W., Cho, W., Kim, J.H.: Development of a 3D printer for concrete structures: laboratory testing of cementitious materials. Int. J. Concr. Struct. Mater. **14**(1) (2020). https://doi.org/10.1186/s40069-019-0388-2
17. Baz, B., Aouad, G., Remond, S.: Effect of the printing method and mortar's workability on pull-out strength of 3D printed elements. Constr. Build. Mater. **230** (2020). https://doi.org/10.1016/j.conbuildmat.2019.117002
18. Paul, S.C., Tay, Y.W.D., Panda, B., Tan, M.J.: Fresh and hardened properties of 3D printable cementitious materials for building and construction. Arch. Civil Mech. Eng. **18**(1), 311–319 (2018). https://doi.org/10.1016/j.acme.2017.02.008
19. AFNOR: NF EN 196-1 - Methods of testing cement – Part 1: Determination of mechanical strengths (2016)
20. Le, T.T., Austin, S.A., Lim, S., Buswell, R.A., Gibb, A.G.F., Thorpe, T.: Mix design and fresh properties for high-performance printing concrete. Mater. Struct./Materiaux et Construct. **45**(8), 1221–1232 (2012). https://doi.org/10.1617/s11527-012-9828-z
21. Perrot, A., Rangeard, D., Pierre, A.: Structural built-up of cement-based materials used for 3D-printing extrusion techniques. Mater. Struct./Materiaux et Construct. **49**(4), 1213–1220 (2016). https://doi.org/10.1617/s11527-015-0571-0
22. Markandeya, A., Galvez Moreno, D., Ley Hernandez, A.M., Kupwade-Patil, K., Le Roux, A.: 3D printable Portland limestone clay-based mortar utilizing locally available materials. WO 2022/071992 A1 (2022)
23. Harbouz, I.: Comportement rhéo-mécanique des matériaux cimentaires adaptés pour une mise en place par impression 3D (2022)
24. Kazemian, A., Yuan, X., Cochran, E., Khoshnevis, B.: Cementitious materials for construction-scale 3D printing: laboratory testing of fresh printing mixture. Constr. Build. Mater. **145**, 639–647 (2017). https://doi.org/10.1016/j.conbuildmat.2017.04.015

Concrete Slurry Waste as Supplementary Cementitious Material—A Consolidated Review Report

Manigandan Nagarajan(✉) and Indu Siva Ranjani Gandhi

Department of Civil Engineering, Indian Institute of Technology Guwahati, North Guwahati , Assam 39, India

manigandan@rnd.iitg.ac.in, gindu@iitg.ac.in

Abstract. Concrete is the most used man-made material globally and its consumption is expected to be 25 giga tonnes every year. The fresh concrete waste is estimated to be 330 million metric tonnes. Allowing the slurry from the RMC to the settling tank and landfill often has a negative impact on society. Recent scientific reports on concrete slurry waste (CSW) suggest that CSW has the potential to be used as a building material and is capable of sequestering CO_2. Therefore, a lack of comprehensiveness in CSW has been observed regarding its material behaviour and fresh and hardened behaviour in paste/mortar. The current study aims to conduct a state-of-the-art review on the physico-chemical properties of both carbonated and non-carbonated CSW. This report assesses the morphology, mineralogy phases, and thermogravimetric analysis (TGA) on CSW powder. Along with this, the fresh properties, hydration mechanics, hardening behaviour, and porosity behaviour of CSW in paste and mortar are intellectually summarised as supplementary cementitious material (SCM). Consequently, the CSW fines possess the ability to uptake 17–24% of CO_2 and have a carbonation degree of 48.3%. Substituting 10% to 20% of CSW as SCM primarily results in a filler effect, dilution effect, and nucleus effect. Moreover, this innocuous and eco-friendly pathway suggests the CSW should be recommended as a performance-improving SCM.

Keywords: Concrete Slurry Waste · Physico-chemical Properties · Morphology · Fresh · Strength · Porosity

1 Background

The demand for concrete increases globally and more drastically in developing countries like India and China. A global quantity of concrete production is nearly 25 gigatonnes. Concrete is the most-used man-made product on Earth. Concrete is impossible without clinker cement [1, 2]. Figure 1 replicates the data from the US Geological Survey report of 2024 on clinker cement production, which is 4000+ million tonnes [3]. India alone contributes 10% of the entire global cement production, i.e., 400+ million tonnes. Clinker combustion and production processes alone contribute 7% of global CO_2 emissions [4, 5]. The alternative of clinker will reduce the emission and conserve the calcareous

M. Posani and S. Singh (Eds.): RYS 2025, RILEM Bookseries 67, pp. 255–266, 2026.
https://doi.org/10.1007/978-3-032-14652-6_20

and argillaceous materials. Generally, GGBS, fly ash (C and F), calcined clay, micro- and nano-silica, industrial reactive slags, volcanic ashes, and agro-ashes (RHA, SCBA, etc.) are well known as supplementary cementitious materials (SCM) [6–8]. The most used SCM like GBBS and fly ash, are expected to rigorously decrease in the upcoming days, so there is a need for finding possible pathways for novel and sustainable SCM. Adhered mortar from the chunks of recycled concrete is also a booming research area for alternative SCM. These adhered mortars are carbonated and utilised in place of cement [9, 10]. Similarly, research on fresh waste concrete (also known as returned/rejected concrete) and backwashed concrete slurry from ready-mix is a recently focused area of study. Noteworthy, 330 million metric tonnes of fresh concrete are wasted every year due to various reasons like unsatisfactory concrete as designed criteria, high transit time of mixers, weather conditions and miscellaneous problems. This wasted concrete is dumped in the yards of ready-mix concrete plants (RMC), and some RMC trucks unload the waste concrete in the unauthorised areas like roadsides, riverbed areas, etc. To avoid this, the proper reclaiming system is essential for all the RMC plants. The reclaiming system segregates and separates the coarser particles (coarse and fine aggregate) and finer particles (slurry) as schematised in Fig. 2. The reclaimed coarse aggregate is extensively reused in the next set of batching materials. Moreover, the slurries are allowed to settle in a sedimentation or settling tank. Once the tank is full, the slurries are simply dumped in the yard and play a negative role in the ecosystem [11, 12].

In 2006, the dewatered liquid from the settling tank, which was free of dissolved solids, was used for concrete production. This limited study indicates that the reused water, which has a specific gravity of 1.02, meets the strength and shrinkage performance standards of tap water concrete [13]. Countable articles are only published by exploring the RMC slurry for reuse. Table 1 presents a summary of CSW processing techniques, properties, CO_2 uptake and analysis. Kaliyavaradhan et al. (2020) collected this slurry cake by removing water with a filter press mechanism and used it as a binder by oven drying and sieving. The systematic processing sketch needs to be emphasised in the CSW processing methods [14]. Researchers and industrialists face a significant gap in understanding how to reuse CSW, whether through carbonation or non-carbonation. Haripan et al. (2024) explored the properties of CSW particles smaller than 75 microns, which have a carbonation degree of 48.3% and a CO_2 uptake of 176 kg per tonne. This degree of carbonation is superior to that of other waste concrete fines, like construction and demolition waste. The details on CO_2 sequestration need to be structured, specifying the precise humidity range, carbonation period, and percentage [15]. Therefore, this state-of-the-art technology highlights the nature of CSW and its role as SCM in paste and mortar.

2 Nature of Concrete Slurry Waste (CSW)

2.1 Physicochemical Properties and Leaching Characteristics on CWS Powder

The physical properties of CSW, like specific gravity and density, are comparable with fly ash and other natural SCM. The specific surface area (SSA) of CSW ranges from 200 to 349 m^2/kg, depending on the processing technique and activation. In the carbonated CSW, SSA will be reduced due to the formation of calcite crystals. From the particle size

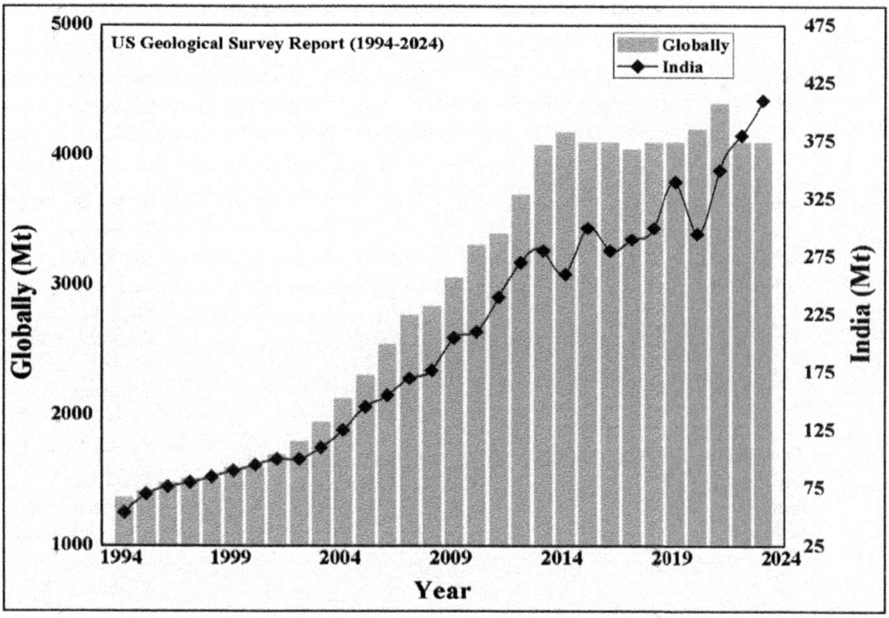

Fig. 1. Cement Production in Last Three Decades

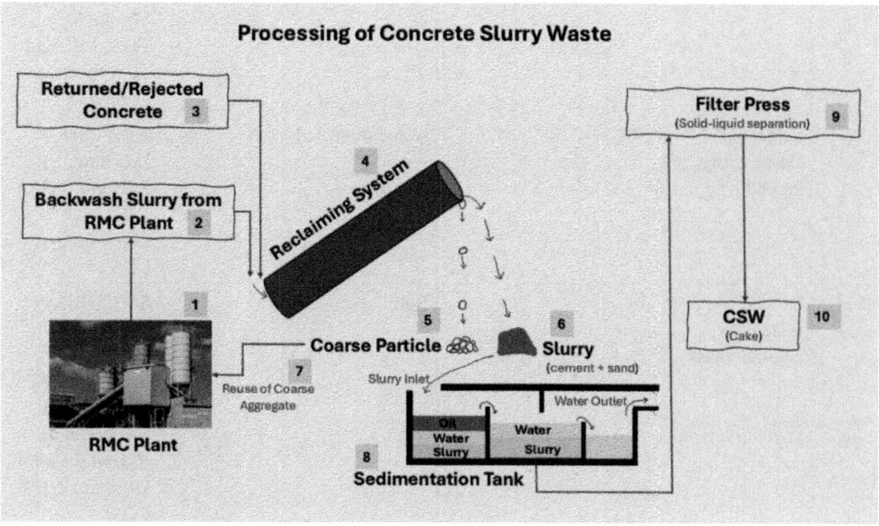

Fig. 2. Schematic Representation of CSW Processing

Table 1. Summary of the literature on CSW

Article	Processing of CSW	Predominant composition in CSW (%)			Cabonated or Non-carbonated	CO_2 Uptake	Area of Inspections
		CaO*	SiO_2	Al_2O_3			
[14]	Settlement, Filter press, oven dried and ground (<300 μm)	53	27	09	Both	20.4%	XRF, PSD, SSA, Morphology, XRD, TGA, Leaching, Strength, and Carbonated for 1–168 h. *(20% CO_2 with humidity of 65%)*
[15]	Atmospheric dried, and ground manually (<75 μm)	45	30	10	Both	17.6%	XRF, PSD, SSA, TGA, and Carbonated for 0,1,7, and 28 days *(3% CO_2 with humidity of 65%)*
[17]	Sieving, shearing and ball milling	03	82	05	Non-carbonated	-	XRF, SSA, PSD, XRD, Calorimetry, Strength, Morphology, TGA, MIP and LCA by three techniques
[18]	Settlement, De-watering, and ground	44	18	5	Non-carbonated	-	XRF, PSD, Leaching, Strength, TGA, XRD, and Morphology with EDS
[19]	Ground and oven dried until constant mass	-	-	-	Both	-	SSA, XRD, Calorimetry, Strength, Shrinkage, TGA, Pore solution concentration *(Carbonated for 90 min at 100% CO_2)*

(*continued*)

Table 1. (*continued*)

Article	Processing of CSW	Predominant composition in CSW (%)			Cabonated or Non-carbonated	CO_2 Uptake	Area of Inspections
		CaO*	SiO_2	Al_2O_3			
[20]	Settlement, and Filter press	35	29	9	Non-carbonated	-	XRF, PSD, Strength, Morphology, XRD, Calorimetry, and Embodied energy (*CSW blended with bamboo biochar*)
[21]	Settlement, and wet-grinding	44	27	14	Both	24.8%	XRF, PSD, XRD, TGA, FT-IR, Calorimetry, Strength, Morphology, MIP. (*Carbonated for 5 and 30 min at 99.9% CO_2*)
[23]	Sedimentation, drying and ball milling (30 min)	37	37	05	Non-carbonated	-	XRF, PSD, SSA, Strength, Morphology, XRD, and MIP (*Analysed in AAC block*)

Notes: XRF – X-ray fluorescence; PSD – Particle size distribution; SSA – Specific surface area; XRD – X-ray diffraction; FT-IR – Fourier-transformed infrared spectroscopy; MIP – Mercury intrusion porosimetry; AAC – Autoclaved aerated cellular concrete; TGA – Thermogravimetric analysis; LCA – Life cycle assessment and Chemical composition is expressed only for maximum CaO* content CSW

distribution (PSD), the D90 of non-carbonated CSW is 39.7 μm, whereas the carbonated has 48 μm. The pore volume of CSW in carbonated slurry will be less compared to non-carbonated slurry. The loss of ignition (LOI) of CSW in non-carbonated slurry is lesser, and carbonated CSW is evidently noted at up to 17%. More particles of carbon due to carbonation were observed by He et al. (2020) [16]. The chemical analysis of raw CSW indicates that it has CaO, SiO_2, and Al_2O_3, which are better for hydration. A considerable quantity of Fe_2O_3 (<10%) is also found in the CSW produced from fly ash concrete. The types and quality of fly ash determine the iron oxide content in CSW. According to the X-ray fluorescence (XRF) analysis, the CSW demonstrates reliability as a supplementary cementitious material (SCM). But the mineralogy phases of these oxide compounds will play a crucial role. Kaliyavaradhan et al. (2020) studied heavy metals in the CSW using a method called inductively coupled plasma optical emission spectrometry (ICP-OES). Metals like chromium (Cr), barium (Ba), lead (Pb), copper (Cu), and zinc (Zn) are at the allowable leaching limit for hazardous landfills. Therefore, the findings recommend that the CSW can be substituted for SCM in concrete. Ren et al., 2022 explored the particle size distribution of CSW and reported that the D50 values for sieving, ball milling, and shearing processes are 37.2 μm, 19.45 μm, and 20.88 μm, respectively [17]. Keppert et al. (2021) inspected the leaching behaviour of metals from raw CSW powder as per EN 12457–2. The total chromium (Cr) and arsenic (As) are classified in the other waste category, while the remaining metals are classified in the inert waste category. The CSW and modern cement have D50s of 7 μm and 18 μm, respectively. This study confirms the CSW can be recycled and used in the construction industry [18].

2.2 Microstructural Properties on CWS Powder

XRD Analysis on CSW Powder. According to the X-ray powder diffraction (XRD) of raw CSW, the mineral and glassy phases will influence the reactivity. In the non-carbonated CSW, the major phases are like portlandite, quartz calcite, dolomite, muscovite, ettringite, feldspar, gypsum, monocarbonates (Mc), and AFm. While carbonating the CSW with CO_2 under controlled humidity and temperature, the peaks for hydroxides, AFt, and AFm diminish, while the calcite peak increases tremendously, as noted by Winnefeld et al. (2024) [19]. According to He et al. (2020), the raw CSW contains SiO_2, calcite, and dolomite. Liu & Li (2025) observed portlandite, quartz, ettringite, and anhydrous C_2S and C_3S in CSW. Based on the raw materials used in making the concrete and processing techniques, the mineral phases in the CSW are influenced [20]. Yang et al., 2024 inspected the raw CSW and CSW with 5 min and 30 min of carbonation. The non-carbonated CSW contains portlandite, C_2S, C_3S, Mc, gypsum, AFm, AFt, and $CaCO_3$. Based on the QXRD, the anhydrous and portlandite phases are converted to carbonate phases. In the raw CSW, the $CaCO_3$ and Mc contents are 6.41 and 4.11%, respectively. Moreover, the carbonated slurry for 30 min has 47.01% and 7.74% of $CaCO_3$ and Mc, respectively [21].

Strnadel et al. (2024) examined the carbonated slurry (CSW) at various wet milling durations, including 0 min (dry CSW), 10 min, and 20 min. The quartz, calcite and AFt phases are noted in all the CSW. Specifically in wet milling CSW, the AFt peaks appear stronger and represent the hydration (i.e., CSW with water) [22]. Ren et al. (2022) observed the intense peaks of SiO_2 and calcite for sieved and sheared processed CSW. Whereas the ball milling process not only makes the particle size finer, it also exposes some anhydrite in cement like C_2S and C_3S. Feng et al. (2024) found that the non-carbonated CSW has anhydrites, quartz, gypsum, and calcite from the powder diffraction analysis using x-rays [23]. Keppert et al. (2021) observed that the raw CSW powder contains mineral phases such as quartz, portlandite, calcite, hydrotalcite, AFm, AFt, and clinker particles of C_2S and C_3S. Herein the hydrotalcite represents that there is a considerable amount of slag used in the concrete-making.

Morphology Analysis on CSW Powder. Feng et al. (2024) examined the morphology of CSW and confirmed that the flocculent CSH, which contains irregular crystals, represents quartz and calcite. Kaliyavaradhan et al. (2020) investigated and differentiated the morphology of carbonated and non-carbonated CSW. The non-carbonated

CSW contains unreacted fly ash particles, which are present as spherical-shaped crystals, well-defined hexagonal-corner-shaped crystals identified as portlandite, flocculated and acicular CSH, and needle-shaped microcrystal AFt. Once the CSW is carbonated by 20% CO_2 with a humidity of 65% the macro-calcite, micro-calcite and nano-calcite are witnessed from the morphology. The size and role of macro-calcite are greater than 1mm, and it is an inert filler. This macro-calcite has excellent thermostability in elevated exposure. Micro-calcite, sized between 1 mm and 1 μm, exhibits synergistic resistance to acid attacks while also assisting in accelerating yield stress and hydration. It primarily assists in extending the capillary pores within the matrix. The nano-calcite <1 μm contributes to the early hydration, thermal resistance, filler effect, and better long-term durability by densifying the microstructure. Cao et al. (2024) observed that micro-calcite and nano-calcite can act as SCM [24].

He et al. (2020) portrayed the morphology of CSW, which contains fly ash (spherical) and sand particles (SiO_2). Yang et al. (2024) compared the morphology of CSW with carbonated and non-carbonated samples. The non-carbonated CSW has a smooth texture of spherical fly ash particles. The 5-min carbonated CSW morphology reveals porous, layered calcite on the fly ash's periphery. The 30 min of carbonated CSW noted a new layer of flocculent $CaCO_3$ precipitated on the surface of the fly ash. Thus, the CO_2 sequestration of CSW leads to the formation of calcite. The morphology of dry CSW reveals the presence of aluminosilicates from the unreacted slag and fly ash, as reported by Strnadel et al. in 2024. The irregular particle shape becomes round after wet milling. The wet milling tends to break the fly ash and stimulate the kinetic reaction of CSW, as noted in the morphology. Ren et al. (2022) observed the hydrated products in all three processed CSWs. Feng et al. (2024) confirm that the morphology of CSW contains AFt and flocculent CSH.

TGA and FT-IT Analysis on CSW Powder. Thermogravimetric analysis (TGA) of the processed CSW helps predict the nature of the materials. Kaliyavaradhan et al. (2020) explored the TGA in both carbonated and non-carbonated CSW. The non-carbonated CSW contains dehydration, dehydroxylation and decarbonation peaks. The partial dehydration peak and complete dehydroxylation peaks are invisible in the carbonated CSW. The decarbonation peak in the non-carbonated sample shows a mass loss of 14%, while the same peak in the carbonated CSW exhibits a mass loss of 27%. The CO_2 with CSH and portlandite are converted into the $CaCO_3$ under the respective temperature and humidity as per Eq. 1.

$$[\![Ca(OH)]\!]_2 \quad (aq)+[\![CO]\!]\_2 \quad (g)\rightarrow [\![CaCO]\!]\_3 \quad (s)+H\_2\ O \qquad (1)$$

Haripan et al. (2024) observed that the similarity in calcium carbonate precipitation is significant in the continuous 7-day carbonated sample. Therefore, the CSW at 3% CO_2 with a humidity of 65% on 7 days has 48.3 degrees of carbonation and 176 kg/tonne of CO_2 uptake. Among the other recycled waste concrete fines, the slurry waste from RMC (i.e., <75 μm of CSW) has a higher degree of carbonation. Yang et al. (2024) published the results of CSW's TGA. The 30-min carbonated CSW has 19.79% mass loss in the decarbonisation region, whereas the 5-min carbonated CSW and non-carbonated CSW have 13.63% and 2.63% mass loss, respectively. The 30-min carbonated CSW has 24.76% CO_2 uptake and a 62.49% degree of carbonation. Fourier-transformed infrared spectroscopy (FT-IR) confirms that the peak at 1140 cm^{-1}, associated with the stretching of Si–O, indicates silica gel formation in the 30-min carbonated CSW, while an increase in CO_3^{2-} vibrational bending is observed at 875 cm^{-1}. A similarly intense peak on the C-O vibration is witnessed at 1470 cm^{-1} and 1417 cm^{-1}. Strnadel et al. (2024) observed that during wet milling, the decarbonation peak shows a mass loss of 5.49%, while the mass loss from dry CSW is 4.77%. Without the external CO_2, by adding the water and milling alone, the carbonisation can be increased by the decline of CSH. The portlandite peak (dehydroxylate) in all mixes of TGA is very minimal (<1% mass loss) in CSW. The long exposure of CSW with moist air (atmospheric CO_2) will convert the calcium hydroxide into calcite. Ren et al. (2022) observed only two ranges of exothermic peaks (dehydration and decarbonation). Even though the CSW is non-carbonated, the portlandite region is not visible from the TGA. The carbonated peak of the sieved CSW exhibits a higher mass loss, reaching up to 30.06%. Winnefeld et al. (2024) reveals from the TGA report of the dehydration peak that there is a partial conversion occurring during the carbonation of CSW. The Si and Al from the hydrated region will be transferred to amorphous silica alumina gel. Keppert et al., 2021 confirms that the calcite content, as determined from TGA mass loss, is 35.4%; they note that there is only a minimal amount of hydrotalcite, and it consists entirely of $CaCO_3$. The dehydroxylate peak (portlandite) of CSW has 13.6% mass loss.

3 CWS as SCM in Paste and Mortar

3.1 Fresh State Performance of CSW Matrix

He et al. (2020) investigated the setting nature of CSW. The plain blend has 180 min and 230 min of initial and final setting time. The 15% CSW blend exhibits a setting time of 110 min and 130 min, respectively. The micro calcite in CSW affects lubrication, while the pre-hydrated C_3S minimises the dormant period and accelerates further hydration, which causes the decrease in setting time. Strnadel et al. (2024) have discussed how wet milling CSW reduces the dormant period (initial setting time) by increasing its content. The main peak appears earlier (final setting time) after 20 min of wet milling with 10% CSW compared to other blends. Yang et al. (2024) investigated 30 min of wet grinding with carbonated CSW at 25%, which has decreased the initial setting time to 4.2 h (h) and the final hardening time to 6.8 h. The formation of micro-calcite crystals reduces the system's alkalinity, allows for earlier hydration, and contributes to the reduction in setting time. Ren et al. (2022) investigated the mini slump of 15%, 30% and 45% substitution of CSW. Increasing the content of CSW increases the water demand and restrains the flow. The CSW blend from ball milling has very little initial and final setting time due to the enhanced hydration from C_3S.

3.2 Strength, Hydration Mechanism and Porosity Analysis of CSW Matrix

Liu & Li (2025) observed a 3 h sedimentation time of CSW with 30% bio-bamboo char, showing the 45.9 MPa compressive strength. On the other hand, a similar blend, when sedimented for 5 h, achieves a compressive strength of 43 MPa within 28 days.

Therefore, the findings confirm that the settlement time and processing technique will also influence the CSW reaction process. The cumulative heat of 30% bio-bamboo char with 3 h sedimented CSW is 26.2% higher than plain CSW. Thus, the calorimetry result confirms the bamboo biochar with CSW accelerates the early hydration process and enhances the dense microstructure. He et al. (2020) noted that the strength activity index (SAI) of 15% non-carbonated CSW is 121% on day one when compared to plain cement blends. The filler effect and nucleus effect accelerate the initial day's strength in CSW. The 28-day strength of 15% CSW has 90% SAI. At later ages, plain CSW does not exhibit any adverse effects on strength. The isothermal calorimetry heat rate of the plain cement blend is 2.2 mW/g, and 15% CSW has 3 mW/g, where the main peak moves left upward from 11.3 h to 6.7 h. This observation confirms that the nucleus effect reduces the final setting time. The pre-hydrated C3S in the CSW will be transformed into CSH, which acts as a nucleus crystal, reduces the induction and dormant periods, and sparks the hydration known as the nucleus effect. The MIP investigation by He et al. (2020) shows that the critical pore diameter of the 1-day specimen provides a superior filler effect compared to that of the plain blend. In 28 days, the same 15% CSW has a similar critical pore diameter to a plain cement matrix. Strnadel et al., 2024 reported that CSW with wet milling accelerates the filler effect and nucleus effect. The other two beneficial effects offset the adverse dilution effect of replacing CSW. The mixture containing 10% CSW shows synergistic behaviour in compressive strength, pore refinement from MIP, and heat rate by calorimetry after 20 min compared to all other blends. The optimum blend has a pore volume of 0.097 cc/g, down from 0.134 cc/g for the referral blend. Similarly, the chloride ion diffusion is controlled by 65.4% more than the referral blend.

Kaliyavaradhan et al. (2020) reported that the SAI of non-carbonated CSW is 79%, while the carbonated CSW exhibits 85% activity after 28 days of curing. The calcite reacts with tricalcium aluminate to form microcrystalline phases, such as calcium carboaluminate hydrate, which contribute to the strength gain process of carbonated CSW. Yang et al. (2024) discussed the strength of 1, 3, 7 and 28 days of 25% wet ground carbonated CSW. The carbonated CSW systems exhibit a strength gain of 23.6% when compared to the non-carbonated CSW systems. The dilution effect, the nucleus effect of $CaCO_3$, and the pozzolanic effect from reactivating the silica gel enhance the performance of a 30-min carbonated 25% CSW matrix. Hydration kinetics by isothermal calorimetry study confirm there is a shortened induction period and the main peak occurs at 10.4 h. For the control blend, the maximum heat of the main peak is at 13.88 h. This value clearly states that the calcite favours growth on the surface and nucleates hydration kinetics. The shoulder peak at 20 h to 24 h in all the carbonated CSW confirms there is a reaction between calcite and other cement phases to form carboaluminates (Mc). The secondary shoulder peak is absent in plain Portland cement blends. The volume proportion of porosity in carbonated CSW decreases for pores larger than 100 nm. The gel pores of <10 nm of carbonated CSW are similar to the referral blend.

Ren et al. (2022) reported that the strength of 15% CSW produced by ball milling exhibits superior behaviour at the early and later stages, while higher replacement levels do not match the control group's performance during the initial curing days. The exothermic main peak of 15% CSW with ball milling has 1.62 mW/g of heat flow at 6.5 h. The dilution effect and pre-hydrated C_3S affect the heat flow and enhance the nucleation site earlier. The similar blend has only 18.12% pore volume when compared to the referral, which has 35.87% pore volume. Therefore, the filler effect and dilution effect of 15% substituted CSW from ball milling are recommended and suggested for concrete making. Feng et al. (2024) studied the porosity of autoclaved cellular aerated concrete (AAC) using 60% CSW along with nano silica and micro silica. The plain CSW blend (60%) in AAC has more harmful macropores from 0.1 μm to 0.8 μm. The combination of nano silica and micro silica with 60% CSW enhances the cementation and filler effects, shifting the critical pore size to less than 0.1 μm.

Winnefeld et al. (2024) conducted isothermal calorimetry on the CSW paste as scripted in ASTM C1897-20. The heat obtained from non-carbonated and carbonated CSW is 42 J/g and 79 J/g. The carbonated CSW shows little reactivity but not up to the limit of other natural minerals and fly ash. The compressive strength with 30% of carbonated CSW has a bit of a pozzolanic effect and a better filler effect than the non-carbonated CSW (30%) and inter-quartz powder (30%) blends. The micro calcite in the CSW accelerates the kinetic reaction in the early hours by supporting nucleation for CSH. Keppert et al. (2021) suggest the 10% substitution ratio of CSW has a comparable strength (compressive and flexural) and cost-effectiveness instead of landfill. Therefore, this consolidated review report concludes that the material nature of CSW is acceptable for SCM. The quantity of silica gel, alumina gel, and micro-calcite significantly influences the reactivity of CSW. By processing and carbonating, CSW can be utilized as a performance improver, and SCM with 10–20% clinker can enhance circularity. Furthermore, large-scale studies on processing, carbonating, and reactivity assessment should be conducted to evaluate their effectiveness in the built environment.

4 Conclusion

The rundown of the current work is succinctly presented below regarding concrete slurry waste (CSW) as a supplementary cementitious material (SCM).

- The processing and activating are playing a crucial role for matching the suitability of CSW as supplementary cementitious material. Generally, the dried CSW and wet milling CSW can be used as inter fillers by 10–20%, and the carbonated CSW by accelerated carbonation or wet grinding can accelerate the early hydration effect by forming a hydrate of calcium carbon aluminate. Heavy metal leaching from the raw CSW is also within acceptable limits. However, further research is required to finalise the clear processing and activation technique for CSW.
- The CaO, SiO_2 and Al_2O_3 are major predominant chemical compositions of CSW. The mineralogy phases, like portlandite, calcite, quartz, CSH, AFt, AFm, Mc, gypsum, dolomite, muscovite, and feldspar, are present in CSW, and the quantity depends on the concrete-making materials. The morphology of CSW confirms the presence of calcite, anhydrous clinker particles, and unreacted or partially reacted fly ash with fine aggregate particles. The TGA with carbonated CSW has a maximum mass loss of nearly 30%. The portlandite and partial amount of CSH are converted into calcite crystals.

- The dilution effect of both carbonated and non-carbonated CSW drastically reduces the setting time and flow behaviour. Isothermal calorimetry observes the CSW nucleus effect, revealing an earlier main peak. The 15% CSW releases a cumulative heat that is comparable to regular blends. The strength performance of both 10%–20% carbonated and non-carbonated CSW is comparable to that of a plain cement blend. The calcite particles of different sizes serve distinct functions by reducing nanosized pores and enhancing pozzolanic activity.
- The use of CSW as a supplementary cementitious material necessitates the assessments and regulations for processing and activation strategies. Since CSW has only been examined on paste, mortar, and AAC blocks, there is ample room to trial it on concrete, special concrete, structural members, and other lightweight concrete. There should be a focus on implementing CSW from lab scale to large scale in the built environment.

References

1. Lothenbach, B., Scrivener, K., Hooton, R.D.: Supplementary cementitious materials. Cem. Concr. Res. **41**(12), 1244–1256 (2011). https://doi.org/10.1016/j.cemconres.2010.12.001
2. Manigandan, N., Ponmalar, V.: M100 concrete probing on quinary cementitious matrix incorporated with industrial ferrochrome slag sand. Constr. Build. Mater. **420**, 135478 (2024). https://doi.org/10.1016/j.conbuildmat.2024.135478
3. US Geological survey report on cement (2024). https://pubs.usgs.gov/periodicals/mcs2024/mcs2024-cement.pdf
4. Nagarajan, M., Vijayan, P.: Some investigation on ternary powder (binder) technology incorporated with ferrochrome slag as fine aggregate in concrete. J. Mater. Cycles Waste Manage. **25**(5), 2822–2834 (2023). https://doi.org/10.1007/s10163-023-01710-y
5. Manigandan, N., Ponmalar, V.: M5 investigation on ternary binder incorporated with ferrochrome slag aggregate in concrete. Appl. Nanosci. **12**(12), 3925–3944 (2022). https://doi.org/10.1007/s13204-022-02615-2
6. Nagarajan, M., Vijayan, P., Ranganathan, S.: Unlocking the insightful performances of metakaolin and ferrochrome slag sand mortar at elevated core temperatures. Innov. Infrastruct. Solutions **9**(12) (2024). https://doi.org/10.1007/s41062-024-01771-4
7. Nagarajan, M., Vijayan, P.: Affordable Solid Block (SB) using industrial waste fly ash binder and ferrochrome slag fine aggregate. In: Jayalekshmi, B.R., Rao, K.S.N., Pavan, G.S. (eds.) LNCV, pp. 695–707 (2024). https://doi.org/10.1007/978-981-97-4844-0_55
8. Scrivener, K., Snellings, R., Lothenbach, B. (eds.). A Practical Guide to Microstructural Analysis of Cementitious Materials. CRC Press (2018). https://doi.org/10.1201/b19074
9. Duan, Z.H., Poon, C.S.: Properties of recycled aggregate concrete made with recycled aggregates with different amounts of old adhered mortars. Mater. Des. **58**, 19–29 (2014). https://doi.org/10.1016/j.matdes.2014.01.044
10. Prajapati, R., Stephen, S.J., Gettu, R., Singh, S.: Effect of thermomechanically beneficiated recycled concrete aggregates on the mechanical and durability characteristics of concrete. Ind. Concr. J. **97**(10), 9–19 (2023)
11. Xuan, D., Poon, C.S., Zheng, W.: Management and sustainable utilization of processing wastes from ready-mixed concrete plants in construction: a review. Resour. Conserv. Recycl. **136**, 238–247 (2018). https://doi.org/10.1016/j.resconrec.2018.04.007

12. Audo, M., Mahieux, P.-Y., Turcry, P.: Utilization of sludge from ready-mixed concrete plants as a substitute for limestone fillers. Constr. Build. Mater. **112**, 790–799 (2016). https://doi.org/10.1016/j.conbuildmat.2016.02.044
13. Low G.L., Ng K.Y., Ng W.L., Tam C.T., Heng R.B.W.: Use of recycled cement-based slurry water for making concrete. Inst. Eng. Malaysia **68**, 47–55 (2007)
14. Kaliyavaradhan, S.K., Ling, T.-C., Mo, K.H.: CO2 sequestration of fresh concrete slurry waste: optimization of CO2 uptake and feasible use as a potential cement binder. J. CO2 Util. **42**, 101330 (2020). https://doi.org/10.1016/j.jcou.2020.101330
15. Haripan, V., Gettu, R., Santhanam, M.: Assessment of the CO2 sequestration potential of waste concrete fines. Mater. Struct. **57**(10) (2024). https://doi.org/10.1617/s11527-024-02531-7
16. He, X., et al.: New treatment technology: the use of wet-milling concrete slurry waste to substitute cement. J. Cleaner Prod. **242**, 118347 (2020). https://doi.org/10.1016/j.jclepro.2019.118347
17. Ren, J., et al.: Pre-treatment of reclaimed concrete slurry waste for substituting cementitious materials: effect of treatment approach and substitution content. J. Cleaner Prod. **380**, 134987–134987 (2022). https://doi.org/10.1016/j.jclepro.2022.134987
18. Keppert, M., Davidová, V., Doušová, B., Scheinherrová, L., Reiterman, P.: Recycling of fresh concrete slurry waste as supplementary cementing material: characterization, application and leaching of selected elements. Construct. Build. Mater. **300**, 124061–124061 (2021). https://doi.org/10.1016/j.conbuildmat.2021.124061
19. Winnefeld, F., Tiefenthaler, J., Leemann, A.: Carbonated concrete slurry waste as supplementary cementitious material. J. Adv. Concr. Technol. **22**(8), 485–498 (2024). https://doi.org/10.3151/jact.22.485
20. Liu, H., Li, Q.: Preparation of green concrete from bamboo biochar (BB) and concrete slurry waste (CSW): preparation method and performance evaluation. Construct. Build. Mater. **462**, 139964–139964 (2025). https://doi.org/10.1016/j.conbuildmat.2025.139964
21. Yang, J., et al.: Achieving carbon utilization and storage (CUS) in cement-based materials with wet-grinding carbonated concrete slurry waste. Cement and Concrete Composites **152**, 105642 (2024). https://doi.org/10.1016/j.cemconcomp.2024.105642
22. Strnadel, B., et a.: A comparative study on concrete slurry waste: performance optimization from the wet-milling process. Mater. Struct. **54**(5) (2021). https://doi.org/10.1617/s11527-021-01771-1
23. Feng, W., Li, Z., Long, Q., Tang, S., Zhao, Y.: Study on the properties of autoclaved aerated concrete with high content concrete slurry waste. Dev. Built Environ. **17**, 100338–100338 (2024). https://doi.org/10.1016/j.dibe.2024.100338
24. Cao, M., Ming, X., He, K., Li, L., Shen, S.: Effect of macro-, micro- and nano-calcium carbonate on properties of cementitious composites—a review. Materials **12**(5), 781–781 (2019). https://doi.org/10.3390/ma12050781

Development of Alkali-Activated Lunar Regolith for Sustainable Space Construction

Anas Driouich[1,2](✉), Maria Chiara Dalconi[1], Claudia Esposito[3], and Luca Valentini[1,2]

[1] CISAS Centre for the Study of Space Activities, University of Padua, Padua, Italy
anas.driouich@unipd.it
[2] Department of Geosciences, University of Padua, Padua, Italy
[3] ASI Italian Space Agency, Rome, Italy

Abstract. With the increasing emphasis on space exploration, global space agencies are advancing toward the establishment of human settlements on the Moon, which is considered as a pivotal step for future extraterrestrial missions. In this context, the In Situ Resource Utilization (ISRU) strategy promotes the use of local lunar resources, such as regolith, to produce durable building materials. Alkali activation of lunar regolith emerges as a promising approach, minimizing dependence on terrestrial supply chains by generating cementitious binders capable of withstanding lunar environmental conditions.

This study explores the potential of using lunar regolith simulants to synthesize alkali-activated materials suitable for lunar construction, with a particular focus on optimizing their fresh and hardened properties. It was observed that the high content of reactive silica and alumina provided by metakaolin significantly enhances the formation of N-A-S-H gel phases, thereby increasing both yield stress and compressive strength. Response surface methodology revealed that under a defined optimal condition the material reached a yield stress of 19.77 Pa and a compressive strength of 27.6 MPa. Additionally, incorporating 3% urea as an admixture markedly improved workability and reactivity, enabling a reduced liquid-to-solid ratio and enhanced compressive strength. Preliminary vacuum-curing tests simulating lunar environments further emphasized the importance of optimizing water content and additive selection to minimize void content and maintain dimensional stability under low-pressure conditions. This work represents a critical advancement toward sustainable lunar infrastructure, contributing both to scientific knowledge and practical applications in extraterrestrial engineering.

Keywords: Regolith · alkali-activated materials · rheology · urea · design of experiments

1 Introduction

Space agencies around the world are more oriented toward investments in planetary research, enabling life support technologies, as well as the development of spacecraft and launch systems capable of carrying humans to Mars. For this matter, living and working on extraterrestrial lands is considered a solvable challenge, and the ability to construct human-habitable structures using planetary resources is a significant aspect of

M. Posani and S. Singh (Eds.): RYS 2025, RILEM Bookseries 67, pp. 267–278, 2026.
https://doi.org/10.1007/978-3-032-14652-6_21

the problem (Alexiadis et al., 2017). More than that, it is considered a radical solution, considering that the lunar regolith is a source of aggregate and binder, which allows it to be a useful solution for the ISRU building material by geopolymerization (Momi et al., 2021). Lunar soil is a term often used interchangeably with regolith; however, the soil is defined as the subcentimeter fraction of the regolith. Lunar dust has been defined in many ways by different researchers but generally refers to only the very finest fractions of the soil, less than ~10 or 20 microns (Noble, 2009).

Establishing long-term human settlements on the lunar surface requires a sufficient stock of lunar regolith to test the technologies for lunar buildings. However, due to limited availability of returned lunar regolith samples different types of lunar regolith simulants were developed for manufacturing processes and scientific studies (Long-Fox et al., 2023). Several studies have explored different lunar regolith simulants as precursors for geopolymer synthesis. For example, geopolymers based on JSC simulants have shown compressive strengths of about 8 MPa under vacuum and lunar daytime conditions (Davis et al., 2017). Neves et al. reported that JSC-1A could act as a low-reactivity SCM in blended systems, reaching strengths up to 30 MPa (Neves et al., 2020). Beyond the JSC series, KLS-1 has been shown to perform well under thermal cycling when microwave-sintered (Kim et al., 2021), BH-1 geopolymers have been tested under lunar high and cryogenic temperatures (Zhang et al., 2022), and comparative studies across multiple simulants further demonstrated differences in behavior under vacuum and extreme temperatures (Mills et al., 2022). Furthermore, as part of the NASA Solar System Research Virtual Institute (SSERVI) Center for Lunar and Asteroid Surface Science (CLASS), the LMS-1 lunar mare simulant was developed by Cannon and Britt (2019) to achieve mineralogical fidelity. In the same context, several works studied the LMS simulant to produce materials for lunar construction, focussing e.g. on the investigation of the shear characteristics, and the geochemistry and geomechanics properties (Javed et al., 2024; Long-Fox et al., 2023; Yin et al., 2023).

To achieve environmental and engineering objectives, it is necessary to use an efficient and environmentally friendly method of synthesis. Considered as an eco-friendly material, alkali-activated binders (AABs) consist of alkali-metal solutions, known as activators, and aluminosilicate-rich materials like metakaolin, fly ash, and slag. When synthesized correctly, AABs demonstrate superior fresh, mechanical, durability, and microstructural properties, comparable to those of ordinary Portland cement (OPC), while also reducing harmful component leaching (Provis et al., 2015). AABs are gaining interest due to their potential for a 20–80% reduction in carbon emissions, depending on the materials and strength grades used (Provis, 2014).

The importance of improving the yield of reactivity and achieving a high rate in terms of fresh and hardened properties is very important to optimize the process of alkali activation. Traditional methods of studying a process by holding other factors constant at unspecified levels do not accurately capture the combined effects of all variables involved. Additionally, this approach is time-consuming and requires numerous experiments to identify optimal conditions, often leading to unreliable results (Montgomery, 2017).To achieve the desired goals, a powerful tool has been widely used recently to optimize and understand the performance of complex systems. The Design of Experiments is a combination of mathematical and statistical techniques used to study the significance of affecting factors even the possible interactions between them (Montgomery, 2017; Weiß, 2007).

In the present work, we focus on assessing the performance of alkali-activated materials synthesized using LMS as a lunar regolith simulant with the lowest water content in an alkaline activator. By applying the In-Situ Resource Utilization (ISRU) approach, this research aims to optimize the formulation of LMS-based alkali-activated material using a design of experiments methodology. To quantify the optimization, it is crucial to evaluate both the fresh and hardened properties of the synthesized material. The strength of the alkali-activated materials is measured by compressive strength, while yield stress is assessed to understand the workability of the slurry. The findings aim to support the creation of 3D-printable lunar alkali-activated materials capable of withstanding the harsh lunar environment.

2 Materials and Methods

2.1 Lunar Regolith Simulants Characterization

The proper formulation of geopolymer binders from powdered raw materials and reactive agents requires a comprehensive understanding of the mineralogical and physicochemical characteristics of the source materials. The lunar regolith simulant used in the present study is LMS-1D (Lunar Mare Simulant 1 Dust), which is developed by Space Resource Technology (Oviedo, Florida, USA), and designed to simulate the mineralogical and chemical properties of lunar mare basalt regolith (Long-Fox et al., 2023). The mineralogical composition of the LMS-1D lunar regolith simulant was determined using X-ray powder diffraction (XRD) with a PANalytical X'Pert PRO diffractometer (Malvern Panalytical, Almelo, Netherlands), employing cobalt radiation ($K\alpha1 = 1.7890$ Å, $K\alpha2 = 1.7929$ Å). To quantify the amorphous content, 20 wt.% of zincite (ZnO) was used as an internal standard. The results revealed that LMS-1D contains typical crystalline phases found in lunar basalts such as plagioclase, pyroxene, olivine, and ilmenite alongside an amorphous content of approximately 10 wt.%.

Scanning Electron Microscopy (SEM) was performed using a TESCAN SOLARIS instrument (TESCAN, Brno, Czech Republic), and Energy-Dispersive X-ray Spectroscopy (EDS) was employed for the chemical microanalysis of the glassy phases. The SEM was operated at an accelerating voltage of 15 keV, a beam current of 3 nA, a spot size of 12 nm, and a working distance of 5 mm. The chemical composition of the amorphous phase, as determined by EDS, was compared to the theoretical composition calculated via mass balance. This calculation was based on the results of quantitative XRD phase analysis, the known stoichiometry of the crystalline phases, and the bulk chemical composition of the simulant. In addition to the phase composition identified by XRD, the particle size distribution of the LMS-1D simulant was also characterized by laser scattering (Malvern Mastersizer, Malvern Panalytical, Malvern, UK) and ranging from 0.04 to 32 μm with a median of 4 μm and a mean of 7 μm.

The oxide composition shows strong similarities between LMS-1D and lunar regolith samples returned by the Apollo missions (Levinson & Taylor, 2013), particularly in terms of alumina (Al_2O_3) and silica (SiO_2) content. As for the amorphous (glassy) fraction, which is considered the most chemically reactive component of the simulant, a similar composition was observed, except for minor deviations in MgO and CaO contents (Fig. 1).

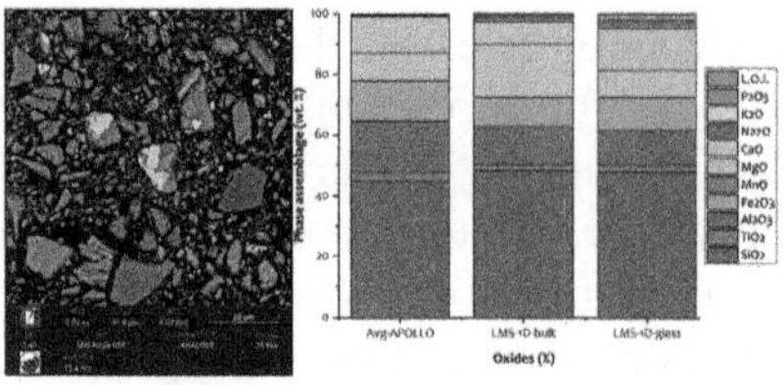

Fig. 1. Microscopy image of LMS-1D simulant powder, and the main oxides present in the bulk and glass part, comparing to composition of Apollo mission lunar regolith.

2.2 Experimental Procedures Design

The alkali-activated binder formulations investigated in this study are based on LMS-1D lunar regolith simulant, activated using a sodium silicate solution. The activator was prepared from a commercial sodium silicate with a SiO_2/Na_2O molar ratio of 2.6, which was adjusted by adding sodium hydroxide pellets to achieve the desired alkalinity. Prior to mixing with the alkaline solution, the lunar simulant was blended with varying proportions of commercial metakaolin powder. This addition was necessary to enhance the reactivity of the binder system by compensating for the low amorphous content of LMS-1D, which is significantly lower than that of actual lunar samples collected during the Apollo and Chang'E-5 missions, reported to contain approximately 25–40 wt.% and 12–20 wt.% of glassy phases, respectively (Li et al., 2022; Taylor et al., 2019).

The paste formulations were mixed using an IKA ULTRA-TURRAX Tube Drive system, operating at 1700 rpm for 3 min. Rheological measurements were performed using a stress-controlled Anton Paar MCR 92 rheometer, fitted with a serrated parallel plate geometry (plate gap: 1.5 mm), at a controlled temperature of 20 ± 1 °C. To minimize water evaporation during testing, a solvent trap was used. Rheological behavior was assessed using a rotational protocol to determine the yield stress. For compressive strength evaluation, the fresh pastes were cast into Teflon molds ($6 \times 1.5 \times 1.5$ cm^3) and cured for 14 days. Compressive strength was measured in triplicate, and the average value was used for statistical analysis.

Traditional experimental approaches typically involve varying one factor at a time while keeping all other parameters constant. However, this method is often inefficient, as it requires numerous experimental runs, leading to excessive consumption of time, energy, and raw materials, with limited insight into factor interactions. As a more effective alternative, the design of experiments (DoE) offers a structured and strategic methodology for planning, executing, and analyzing experiments. By systematically selecting and varying the input variables, DoE facilitates the identification of optimal process conditions, minimizes variability, and enhances overall product quality (Rigdon et al., 2022).

In this study, a central composite design (CCD) was employed to structure the experimental plan. The key independent variables investigated were: the silica modulus (SiO_2/Na_2O) ranging from a minimum value of 0.8 to a maximum value of 1.6; the liquid-to-solid ratio (L/S) varying within the range 0.4 to 0.44, which describes the ratio between the liquid (consisting of sodium silicate and sodium hydroxide solution) and the solid (consisting of the simulant with metakaolin as additive); the metakaolin content (wt%), ranging from 0 to 15%; the curing temperature, varying from 20 to 60 °C.

The selected response variables reflect both fresh and hardened properties of the alkali-activated materials, specifically yield stress (Pa) as an indicator of rheological behavior, and compressive strength (MPa) as a measure of mechanical performance. Optimization of the alkali-activation process was performed using response surface methodology (RSM), a statistical tool designed to identify the best combination of input factors that satisfies specific performance criteria. The primary objective was to simultaneously achieve a suitable yield stress for 3D printing applications and a high compressive strength in the final product. Statistical analysis was carried out using analysis of variance (ANOVA) to assess the significance of the model terms and their interactions. JMP and Origin software were utilized for regression modeling, data visualization, and parameter estimation, with a confidence level set at 95% (Santana-Quintero et al., 2010).

3 Results and Discussion

3.1 Design of Experiments for the LMS Alkali-Activation Optimization

The first step in DoE analysis consists of evaluating the analysis of variance and assessing the goodness of fit of the proposed models. For the compressive strength model, the fit is validated by a high coefficient of determination ($R^2 = 0.971$) and an adjusted R^2 of 0.935, indicating that the model explains a substantial portion of the variability in the response. This is further supported by a highly significant p-value (<0.0001) and an F-statistic value (26.646), which exceeds the critical threshold, confirming the model's overall statistical significance. In comparison, the rheology model exhibits a slightly lower, yet still acceptable, goodness of fit with an R^2 of 0.897. The model remains statistically significant, as demonstrated by a p-value of 0.0014 and a significant F-statistic at the 95% confidence level, although its predictive performance is somewhat less robust than the compressive strength model.

The optimization of alkali-activated lunar regolith simulant (LMS) was carried out using response surface methodology (RSM), aiming to identify the optimal combination of factors that would balance appropriate yield stress and high compressive strength. Firstly, compressive strength rose from 5 to 32 MPa as metakaolin content increased, due to its high amorphous material and Al_2O_3 content, which enhances the alkali activation. On the other hand, yield stress also increased with metakaolin content and curing temperature, driven by accelerated chemical reactions and microstructural changes at higher temperatures, which resulted in a denser, more cohesive matrix. In contrast, no significant impact on yield stress was observed from the SiO_2/Na_2O and liquid-to-solid ratios. According to the optimization results, a silica modulus of 1.5 proved to be an effective activator. A liquid-to-solid ratio of 0.4 delivered good results in terms of both fresh and hardened properties. By reducing the metakaolin content to 8.5 wt.%, the proportion of lunar regolith simulant in the precursor mix increased to approximately 91.5 wt.%, thereby maximizing its utilization, and a curing temperature of 20 °C was found to be most suitable. Under these optimal conditions, the compressive strength measured was 19.8 MPa, and the yield stress was 27.6 Pa. Verification experiments conducted under these predicted conditions, derived from ridge analysis of RSM, showed that the experimental values were reasonably close to the predicted ones, confirming the validity and reliability of the models (Fig. 2).

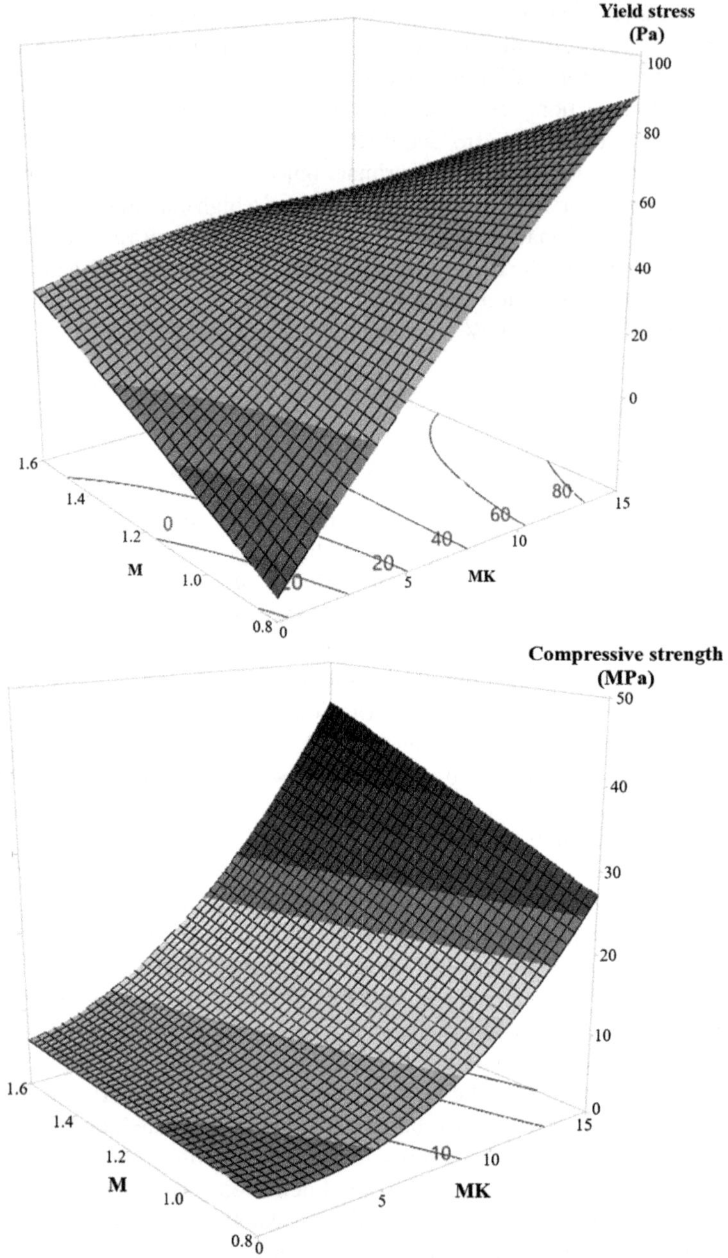

Fig. 2. Contour plots and response surface plots showing the effects of the Silica modulus and the metakaolin on the compressive strength (a) and the yield stress (b) of the LMS alkali-activated materials.

3.2 Rheology and Reduction of the Amount of Water

Based on the optimization results described in the previous section, the goal of this stage is to reduce water usage for several key reasons: first, the limited availability of water on the Moon, and second, the lunar vacuum environment, which poses additional challenges compared to Earth (Clinton Jr et al., 2021). Therefore, further experiments were conducted to optimize the process by reducing the liquid-to-solid (L/S) ratio, aiming to lower water demand while enhancing the extrusion process.

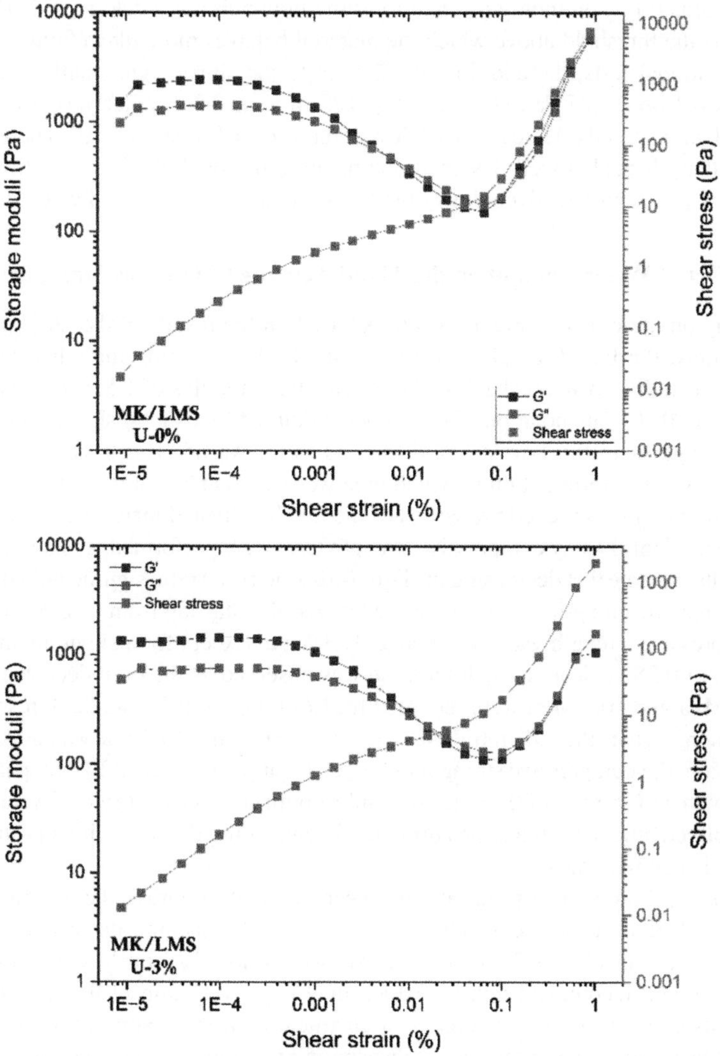

Fig. 3. Variation of the viscoelastic moduli G' and G" for sample "OPT2" before and after adding 3% wt. of urea.

The experiments were performed under previously established optimal conditions: a silica modulus of 1.5, 8.5% metakaolin substitution, and a curing temperature of 20 °C. The L/S ratio was initially set at approximately 0.4 (23.63 Pa), as further reduction in liquid content hindered homogeneous mixing between the solid precursor and the alkaline solution.

Preliminary tests focused on reducing water content by modifying only the L/S ratio while keeping all other parameters constant. Rheological properties were assessed using both oscillatory and rotational protocols. After characterizing the rheological behavior of the fresh pastes, the shear stress value at which the storage modulus (G') intersects the loss modulus (G") commonly referred to as the flow point was determined. This value represents the threshold above which the material behaves more like a fluid.

In rotational tests, the addition of 3% urea to the slurry significantly reduced the yield stress from 19.27 Pa to 0.32 Pa at an L/S ratio of 0.3. Similarly, oscillatory tests (Fig. 3) demonstrated a decrease in the flow point when 3% urea was used. These findings confirm that, through a well-designed optimization protocol, the incorporation of urea as an admixture can effectively lower the L/S ratio and reduce overall water demand.

3.3 Effect of Water Amount on the Alkali Activated Materials Strength

The study on the effect of water content is based on the results of the DOE part (3.1). In this study, the liquid-to-solid ratio was varied while keeping the other parameters constant, namely, 8.5% metakaolin addition, a silica modulus of 1.5, and a curing temperature of 20 °C. The notations F1 to F4 were defined to distinguish the four samples, from the highest to the lowest liquid-to-solid ratio. An additional sample, F5-U3%, corresponds to the formulation in which urea was added at 3% of the solid weight.

To evaluate the hardened properties of the alkali activated materials based on lunar regolith simulant, 14-day compressive strength tests were performed, as shown in Fig. 4. The results indicate that decreasing the liquid-to-solid (L/S) ratio significantly influences the compressive strength of the synthesized materials. Starting from an L/S ratio of 0.4, the compressive strength was measured at 31.1 MPa. Except for a slight anomaly at an L/S ratio of 0.38, a clear strength increase was observed as the ratio decreased: values of 35.5 MPa and 40.9 MPa were recorded for L/S ratios of 0.35 and 0.33, respectively. Furthermore, when the L/S ratio was reduced to 0.3 and 3% urea was added to the formulation, the compressive strength increased by approximately 22% compared to the sample with an L/S ratio of 0.33. These results confirm the critical role of water content and the effectiveness of urea as an admixture in enhancing the reactivity and strength of alkali-activated materials.

Reducing the liquid-to-solid ratio has been shown to improve the mechanical performance of alkali-activated systems by increasing the packing density and limiting porosity (Wang et al., 2023). However, excessively low L/S ratios can lead to poor workability and incomplete mixing, which may negatively impact the reaction kinetics. In this context, the use of urea as a chemical admixture appears to enhance the paste's flowability, allowing for better dispersion and more efficient geopolymerization at lower water contents. Urea likely acts as a plasticizing agent, disrupting inter-particle hydrogen bonding and promoting better particle suspension, which facilitates improved strength development (Pilehvar et al., 2020). The observed 22% increase in compressive strength at 0.3 L/S with urea supports the conclusion that an optimized water-to-binder balance, aided by additives, can significantly enhance the microstructural integrity and mechanical performance of alkali-activated binders.

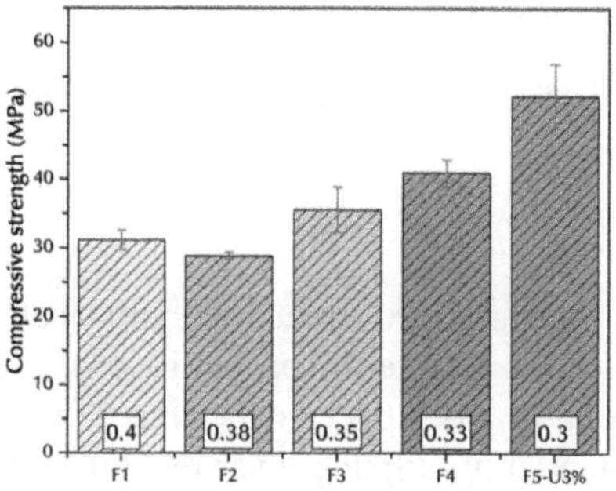

Fig. 4. Variation of compressive strength (MPa) as a function of liquid/Solid ratio

3.4 Vacuum Curing Tests for Simulating Lunar Conditions

To simulate the behavior of alkali-activated materials (AAMs) synthesized from lunar regolith simulants, a couple of tests were conducted in a vacuum environment. Unlike conventional curing conditions, vacuum curing replicates the low-pressure environment of the Moon, where water retention and evaporation control are crucial for the success of alkali-activation reactions.

The vacuum curing tests were applied to three samples. The first sample labelled "F1," was prepared under optimal synthesis conditions. The second sample 'F4' was synthesized under the same conditions as "F1" but with decreasing liquid-to-solid (L/S) ratio to 0.33. The third sample 'F4' was prepared with the same conditions as "F1" but with an L/S ratio of 0.3 and the addition of 3% urea, used as a superplasticizer.

Upon placing the samples in the vacuum environment, notable differences in the density of the hardened material were observed among the three samples as observed in Fig. 5. The most significant void content was seen in sample 'F1', highlighting the critical role of water content. The sample 'F4' with a lower L/S ratio (0.3) exhibited reduced void content, which was further minimized by the addition of urea. These observations underscore the importance of optimizing the water content and additive use in maintaining dimensional stability under vacuum conditions.

Compressive strength testing further confirms these findings. Sample "F5" with a L/S ratio equal to 0.3 and 3% of urea used demonstrated the highest compressive strength, reaching 3 MPa, while the sample 'F4' with a L/S ratio of 0.33 showed a compressive strength of 0.6 MPa. These results emphasize the interplay between water content, additives, and curing environment on the structural and mechanical properties of alkali-activated materials synthesized for extraterrestrial applications.

Fig. 5. Vacuum curing for one-week tests

4 Conclusion

In summary, the ability to manufacture structural components and spare parts directly on-site is a critical enabler for long-term space missions, offering significant advantages in terms of cost efficiency and reduced dependency on earth-based resupply. This study investigated the potential of utilizing lunar regolith simulants to produce alkali-activated materials (AAMs) suitable for construction applications, with a specific focus on optimizing their fresh and hardened properties. The key outcomes of this optimization process can be outlined as follows:

- The positive effect of metakaolin on both yield stress and compressive strength can be attributed to its high content of reactive silica and alumina, which contributes to a denser and more cohesive microstructure.
- The response surface curves indicate that, under the optimal conditions of 20 °C curing temperature, a liquid-to-solid ratio of approximately 0.4, 8.5% metakaolin content, and a silica modulus of 1.5, the yield stress reaches an optimal value of 19.77 Pa, while the compressive strength is 27.6 MPa.
- The addition of urea as an admixture significantly improved workability, reduced the liquid-to-solid ratio, and enhanced the reactivity of the amorphous phases, leading to notable increases in compressive strength up to 52.2 MPa for a L/S ratio of 0.3.
- The preliminary tests under vacuum curing conditions, simulating lunar environments, demonstrated the importance of optimizing water content and additives to minimize void content and maintain dimensional stability.

The approach adopted in this study holds significant promise for future lunar construction, as it utilizes in-situ resources to develop sustainable and environmentally responsible building materials. The next phase will focus on the optimization of urea using as a superplasticizer and simulating lunar conditions through vacuum curing. Durability will also be assessed under freeze-thaw cycle les to ensure the materials' suitability for lunar construction.

Acknowledgments. This work was funded by the Italian Space Agency (ASI) grant Research Days 2023-6-U.0 F93C23000200005.

References

Alexiadis, A., Alberini, F., Meyer, M.E.: Geopolymers from lunar and Martian soil simulants. Adv. Space Res. **59**(1), 490–495 (2017). https://doi.org/10.1016/j.asr.2016.10.003

Clinton Jr, R., et al.: Moon to Mars Planetary Autonomous Construction Technologies (MMPACT). ASCEND, Planetary Construction technologies (2021)

Davis, G., Montes, C., Eklund, S.: Preparation of lunar regolith based geopolymer cement under heat and vacuum. Adv. Space Res. **59**(7), 1872–1885 (2017). https://doi.org/10.1016/j.asr.2017.01.024

Javed, U., Shaikh, F.U.A., Rahman, A.S.: Synthesizing lunar regolith-geopolymer emulating lunar positive temperature regime. Planet. Space Sci. **244**, 105890 (2024)

Kim, Y.-J., Ryu, B.H., Jin, H., Lee, J., Shin, H.-S.: Microstructural, mechanical, and thermal properties of microwave-sintered KLS-1 lunar regolith simulant. Ceram. Int. **47**(19), 26891–26897 (2021). https://doi.org/10.1016/j.ceramint.2021.06.098

Levinson, A.A., Taylor, S.R.: Moon Rocks and Minerals: Scientific Results of the Study of the Apollo 11 Lunar Samples with Preliminary Data on Apollo 12 Samples. Elsevier (2013)

Li, C., et al.: Characteristics of the lunar samples returned by the Chang'E-5 mission. Nat. Sci. Rev. **9**(2), nwab188 (2022)

Long-Fox, J.M., Landsman, Z.A., Easter, P.B., Millwater, C.A., Britt, D.T.: Geomechanical properties of lunar regolith simulants LHS-1 and LMS-1. Adv. Space Res. **71**(12), 5400–5412 (2023). https://doi.org/10.1016/j.asr.2023.02.034

Mills, J.N., Katzarova, M., Wagner, N.J.: Comparison of lunar and Martian regolith simulant-based geopolymer cements formed by alkali-activation for in-situ resource utilization. Adv. Space Res. **69**(1), 761–777 (2022). https://doi.org/10.1016/j.asr.2021.10.045

Momi, J., Lewis, T., Alberini, F., Meyer, M.E., Alexiadis, A.: Study of the rheology of lunar regolith simulant and water slurries for geopolymer applications on the Moon. Adv. Space Res. **68**(11), 4496–4504 (2021). https://doi.org/10.1016/j.asr.2021.08.037

Montgomery, D.C.: Design and Analysis of Experiments. Wiley (2017)

Neves, J.M., Ramanathan, S., Suraneni, P., Grugel, R., Radlińska, A.: Characterization, mechanical properties, and microstructural development of lunar regolith simulant-portland cement blended mixtures. Construct. Build. Mater. **258** (2020). https://doi.org/10.1016/j.conbuildmat.2020.120315

Noble, S.: The lunar regolith. In: Lunor Regolith Simulant Workshop (2009)

Pilehvar, S., Arnhof, M., Pamies, R., Valentini, L., Kjøniksen, A.-L.: Utilization of urea as an accessible superplasticizer on the moon for lunar geopolymer mixtures. J. Clean. Prod. **247**, 119177 (2020)

Provis, J.L.: Geopolymers and other alkali activated materials: why, how, and what? Mater. Struct. **47**, 11–25 (2014)

Provis, J.L., Palomo, A., Shi, C.: Advances in understanding alkali-activated materials. Cem. Concr. Res. **78**, 110–125 (2015)

Rigdon, S.E., Pan, R., Montgomery, D.C., Freeman, L.: Design of Experiments for Reliability Achievement, vol. 1. Wiley (2022)

Santana-Quintero, L.V., Montano, A.A., Coello, C.A.C.: A review of techniques for handling expensive functions in evolutionary multi-objective optimization. Computational intelligence in expensive optimization problems, pp. 29–59 (2010)

Taylor, G.J., Martel, L.M., Lucey, P.G., Gillis-Davis, J.J., Blake, D.F., Sarrazin, P.: Modal analyses of lunar soils by quantitative X-ray diffraction analysis. Geochim. Cosmochim. Acta **266**, 17–28 (2019)

Wang, W., Fan, C., Wang, B., Zhang, X., Liu, Z.: Workability, rheology, and geopolymerization of fly ash geopolymer: role of alkali content, modulus, and water–binder ratio. Construct. Build. Mater. **367** (2023). https://doi.org/10.1016/j.conbuildmat.2023.130357

Weiß, C.: Statistica, version 8. AStA Adv. Stat. Anal. **91**, 339–341 (2007)
Yin, K., Cheng, Z., Liu, J., Chen, A.: Shear properties of LHS-1 and LMS-1 Lunar regolith simulants. Planetary Space Sci. **226** (2023). https://doi.org/10.1016/j.pss.2022.105630
Preparation of geopolymer based on lunar regolith simulant at in-situ lunar temperature and its durability under lunar high and cryogenic temperature. Construct. Build. Mater. **318** (2022). https://doi.org/10.1016/j.conbuildmat.2021.126033

Exploring Sustainable Alternatives in Concrete: A Study on Recycled Aggregates, Palm Oil Fibers, and Coconut Shell Waste

Mohammed Abdulkareem Adisa[1(✉)], Noah Ifeoluwapo Badmus[1], Kolawole Favour Wuraola[2], Olushola Emmanuel Olasunkanmi[1], and Teslim Opeyemi Onaolapo[3]

[1] Kwara State University, Malete, Nigeria
mickeya344@gmail.com
[2] Ladoke Akintola University of Technology, Ogbomosho, Nigeria
[3] University of Ilorin, Ilorin, Nigeria

Abstract. Concrete is the most used construction material, driven by industrialization and urbanization, with worldwide demand surpassing 20 billion tons annually. Increasing demand for sustainable construction materials has led to research into the use of recycled concrete aggregates (RCA), palm oil fiber (POF), and coconut shell (CS) waste for sustainability advancement and improvement of mechanical performance. RCA was incorporated as a partial replacement of natural coarse aggregate at 2.5–10% by weight of the coarse aggregate fraction. The mixes demonstrated optimal workability and strength at the 2.5% RCA level. The addition of fibers improved mechanical properties at 2.5%; however, it decreased workability as the fiber content increased. CS waste presented a sustainable option but reduced workability and compressive strength in the high proportions due to higher porosity and a weak interfacial transition zone. These findings indicate the potential of these materials for sustainable concrete applications with proper optimization.

Keywords: Concrete · Recycled concrete aggregates · Palm oil fiber · Coconut shell waste · Sustainable construction materials

1 Introduction

Concrete is the most popular construction material, with the increasing pace of industrialization and urbanization driving a global demand surge [1]. Approximately 70% of a concrete mix volume is comprised of coarse and fine natural aggregates. Current estimates indicate a worldwide demand for concrete aggregates exceeding 20 billion tons annually [2]. This heightened reliance on concrete leads to the depletion of natural aggregates and substantial energy consumption in the production, transportation, and use of raw materials. Simultaneously, the escalating demolition of old structures, due to limited new construction sites, generates substantial amounts of construction and demolition (C&D) wastes [3]. Traditional construction materials contribute to resource

M. Posani and S. Singh (Eds.): RYS 2025, RILEM Bookseries 67, pp. 279–294, 2026.
https://doi.org/10.1007/978-3-032-14652-6_22

depletion and waste generation. This project promotes sustainable alternatives by leveraging emerging technologies to minimize environmental impact and enhance economic viability. With the urgent need to reduce CO_2 emissions, the development of alternative materials aligns with global policies striving for net-zero carbon emissions [4]. Alternative materials may include pozzolans, waste byproducts, recycled aggregates, and agricultural residues. This study specifically examines the application of recycled aggregates, palm oil fibers, and coconut shell waste.

In the last twenty years, aggregates sourced from demolished concrete structures have become a sustainable substitute for traditional materials. Their use helps reduce reliance on natural resources while mitigating the environmental impact associated with C&D waste. The EPA [5] estimated that in 2018, The U.S. generated 600 million tons of C&D debris, over twice the municipal solid waste produced that year. Despite growing recycling efforts, much still ends up in landfills. Lei et al. [6] emphasized the environmental and economic benefits of reusing C&D waste in concrete and noted that curing methods enhance RAC performance. However, considering factors like cost-effectiveness and carbon emissions, the primary barriers to the widespread use of RAC are the replacement ratio of recycled aggregates and transportation distance. Also studies by Berredjem et al. [7], Skochek et al. [8] and Delsaute [9] revealed that natural aggregate concrete outperformed recycled aggregate in strength and durability. Recycled aggregates increased porosity and vulnerability.

Coconut shell aggregate, derived from agricultural waste, has recently been utilized in concrete production. Coconut shells are an abundant and renewable resource. According to UNESCAP [10], coconut-producing countries generated over 65.7 billion coconuts, equivalent to 11.7 million metric tons (MT) of copra in 2022. In 2021, Nigeria's coconut production was estimated at 229,710 metric tons. Since 1966, the country's coconut supply has grown steadily at an average annual rate of 0.2%, driven by increasing demand and gradual expansion in cultivation [11]. Hediye et al.[12] and Tomar et al. [13] stated that coconut shell is suitable for lightweight concrete, making it ideal for non-load-bearing structures, strip footings, and other non-structural elements. And as such, incorporating them into concrete as aggregates can help minimize the construction industry's environmental impact [14]. Coconut shell concrete can cut construction costs by 30%, benefiting low-income families [15]. Although studies has shown that the shells reduces workability [16] and increase the porosity of concrete [17].

Bio-fibers are increasingly used to reinforce concrete composites due to their flexibility, strength, and cost-effectiveness [18]. Their performance depends on factors such as origin, climate, and extraction methods. Natural fibers like hemp, flax, jute, sisal, and coir are particularly suitable for bituminous mixtures and cementitious composites [17]. The oil palm tree is a vital natural source of vegetable fibers. In 2023, Nigeria's palm oil production reached 1.4 million metric tons, with steady growth from 2009 to 2023, including a peak increase of 14% in 2010 [19]. However, in developing countries, the disposal of byproducts from palm oil extraction, cultivation, and residual tree components presents challenges. These waste materials are often discarded improperly or used as cooking fuel, raising significant environmental concerns [20].

Researchers have used Fibers like Kaliwon et al. [21] reinforced concrete roof slates with Empty Fruit Bunch Fibers (EFBF), observing improved flexural strength with 1%

EFBF, exceeding ASTM requirements. Odeyemi et al. [22] showed that adding 0.2% coconut fiber to self-compacting concrete enhanced its strength, while Odeyemi et al. [23] found that 3% coconut fiber inclusion significantly improved impact resistance. Adamu et al. [24] recommended less than 1% fiber for shrinkage control, and 2% for improving energy absorption, with a 3 cm fiber length as optimal.

The growing use of alternative materials in concrete production, such as recycled aggregates, palm oil fibers, and coconut shells, reflects a shift toward more sustainable construction practices. While the potential of these materials to reduce environmental impact is evident, challenges remain, particularly concerning workability, strength, and cost-effectiveness. Although recycled aggregates help conserve natural resources and reduce C&D waste, their use is limited by issues related to porosity and the interfacial transition zone, which can impact concrete performance. Similarly, while coconut shells and palm oil fibers offer benefits in terms of sustainability and cost savings, their use may reduce workability and increase porosity in concrete mixtures. These gaps in material performance and the need for optimization present a significant opportunity for future research. The effectiveness of these alternative materials, especially in enhancing both mechanical and durability properties, requires further exploration to better understand their long-term impacts and feasibility in large-scale construction applications. This study evaluated the fresh properties and compressive strength of concrete reinforced with these materials.

2 Materials and Methods

2.1 Materials

This study utilizes cement, fine and coarse aggregates, waste materials, and admixtures. CEM I 42.5R (BUA Cement Plc, Sokoto, Nigeria), conforming to NIS 444-1 [25] and BS EN 197-1 [26], was used. River sand sourced from Asa River, Ilorin, met BS 812 [27] standards, and its particle size distribution was tested. Potable water complying with BS EN 12620 [28] was utilized in all mixes. Properties of PLC, Fine Aggregates, and Coarse Aggregates are shown in Table 1. Palm oil fiber (POF) was obtained from a local mill in Ilorin, derived from crushed fruit bunches, shells, and husks. The length of fiber was trimmed to 30mm before use. Water-reducing admixture (Sika, Lagos State, Nigeria) was used to enhance workability, adhesion, and durability. Coconut shells (CS) were collected in sacks in Ilorin, Kwara State, they were sieved using a 12mm sieve to ensure they were prepared as coarse aggregates. The recycled concrete aggregates (RCA), sourced from demolished or renovated concrete blocks in Ilorin, Kwara State, were sieved through a 12 mm sieve. The properties of waste aggregates are shown in Table 2 following BS 812 standards. These materials were shown in Fig. 1.

(a) (b) (c)

Fig. 1. Materials - (a) Recycled Concrete Aggregates, (b) Palm Oil Fiber, (c) Coconut Shells

Table 1. Properties of Cement, Fine Aggregates (FA), and Coarse Aggregates (CA)

Property	Cement	FA	CA	RCA
Specific Gravity	2.93	2.65	2.7	2.6
Initial Setting Time	42 min	-	-	
Final Setting Time	580 min	-	-	
Fineness (%)	6.50%	-	-	
Standard Consistency (%)	30	-	-	
Fineness Modulus	-	2.67	7.1	6.7
Water Absorption (%)	-	1.1	0.75	3.6
Bulk Density (kg/m3)	-	1461	1423	1265

Table 2. Properties of Waste aggregates

Aggregate Type	Water Absorption (%)	Aggregate Crushing Value (%)
CS	18	26
RCA	7	21

2.2 Methodology

The concrete mixes were produced with Portland Limestone Cement (PLC), river sand, and a constant water-to-binder ratio of 0.35 (binder = PLC). Coconut shells (CS) and recycled concrete aggregate (RCA) replaced the natural coarse aggregate at 25%, 50%, 75%, 100% CS and 2.5%, 5.0%, 7.5%, 10% RCA, respectively, all by weight of the coarse-aggregate fraction. Palm oil fibre (POF) was incorporated at 2.5–10% by weight of cement, and a superplasticizer was added at 1% of the cement weight to enhance workability. Replacement levels were selected in small low-range increments (2.5%, 5%,

7.5%, 10%) to capture early performance inflection points while maintaining baseline mechanical properties. Literature justifies this for RCA (≤10% limits strength penalties, and palm-based fibers (benefits at 0.5–2.5%) [29–32]. Overall, these closely spaced sub-10% levels balance sustainability with mechanical performance.

As detailed in Table 3, different mixtures were prepared, including a control mix and variations with CS, POF, and RCA at selected replacement levels. The mixing process, illustrated in Fig. 2, was carried out using a mechanical mixer to ensure uniformity, and the hardened mixtures were cured for 7, 14, and 28 days (Fig. 3) before being evaluated for strength to assess the impact of CS, POF, and RCA on mechanical performance.

Table 3. Mix proportion (kg/m^3)

Mix ID	Cement	FA	CA	CS	POF	RCA
Control	400	700	1100	0	0	0
CS-25	400	700	1000	100	0	0
CS-50	400	700	550	550	0	0
CS-75	400	700	275	825	0	0
CS-100	400	700	0	1100	0	0
POF-2.5	400	700	1100	0	10	0
POF-5.0	400	700	1100	0	20	0
POF-7.5	400	700	1100	0	30	0
POF-10	400	700	1100	0	40	0
RCA-2.5	400	700	1072.5	0	0	27.5
RCA-5.0	400	700	1045	0	0	55
RCA-7.5	400	700	1017.5	0	0	82.5
RCA-10	400	700	990	0	0	110

Fig. 2. Concrete production using a mechanical drum mixer during the mixing process. (Civil Engineering Laboratory, Kwara state University).

Fig. 3. Concrete cube specimens submerged in a curing tank with water maintained at room temperature for specified curing ages (7, 14, 28 days). (Civil Engineering Laboratory, Kwara state University).

2.2.1 Workability

The concrete mixes were prepared and thoroughly mixed using a concrete mixer. The slump cone was placed on a clean base, filled in three layers, each tamped 25 times, and leveled. The cone was then lifted vertically, allowing the concrete to settle freely. The slump was measured as the height difference and recorded to assess workability per BS EN 12350-1 [33] at Kwara State University Civil Engineering Lab. It is illustrated in Fig. 4.

Fig. 4. Slump test of Concrete

2.2.2 Compressive Strength

Concrete specimens were cast in cube moulds of 150 × 150 × 150 mm, compacted, and cured for 7, 14, and 28 days. Compressive strength testing was performed using a 1560 kN capacity machine, applying a gradual load of 140 kg/cm^2 per minute until failure. The compressive strength was calculated using Eq. (1), in accordance with BS EN 12390-3 [34]. Kwara State University Civil Engineering Lab. The testing equipment is shown in Fig. 5.

$$\text{Compressive strength} = \frac{\text{Maximum load (kN)} \times 1000}{\text{Cross} - \text{ sectional Area (mm}^2\text{)}} \quad (1)$$

Compressive strength results are presented as mean values with error bars representing the standard deviation of replicate tests. For each curing age (7, 14, and 28 days), three cube specimens were tested per mixture, and the variability across these replicates is reflected in the plotted error bars.

Fig. 5. Compressive strength machine

2.2.3 Scanning Electron Microscopy (SEM).

SEM analysis examined the morphology, distribution, and interaction of coconut shells (CS) within the cement matrix. Concrete samples (control, 50% CS, and 100% CS) were cured, sectioned, and analyzed using SE imaging for surface morphology and BSE imaging for CS distribution. This test conformed to ASTM E1508 [35]. It was conducted in the material science laboratory in Kaduna State University.

3 Results and Discussion

3.1 Slump Results

The trend in Fig. 6 shows a consistent decrease in slump height as the dosage of RCA in the concrete mix increases, with the control sample having the highest slump height at 65 mm, followed by 2.5% RCA at 49 mm, 50% RCA at 37 mm, 7.5% RCA at 30 mm, and 10% RCA at 28 mm. This progressive decrease in slump height can be attributed to several factors: RCA typically has higher porosity and water absorption compared to natural aggregates, leading to reduced free water in the mix and thus a lower slump; the rougher surface texture of RCA increases internal friction within the mix, reducing workability. The findings is consistent with research by Rostami et al. [36].

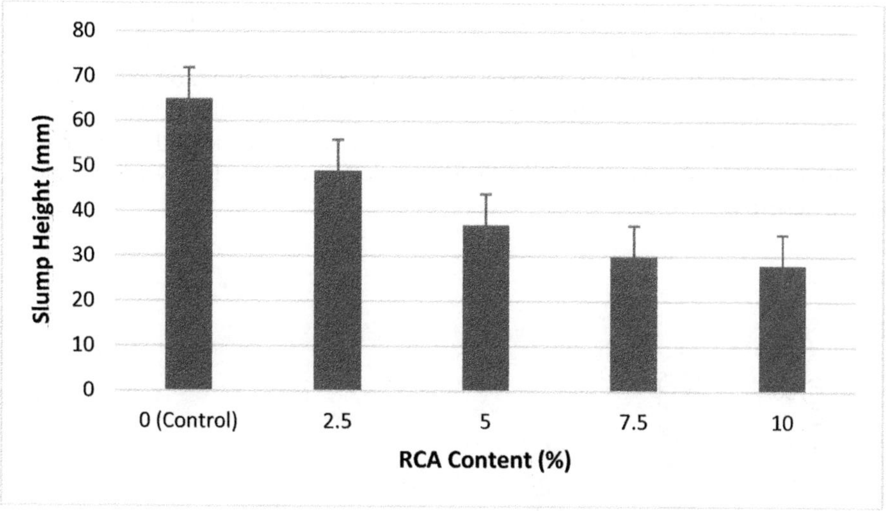

Fig. 6. Slump results for RCA samples

Figure 7 shows workability result of POF samples. The slump test results show a clear trend: as the percentage of Palm Oil Fiber (POF) increases, the workability decreases. The control had the Top-most slump height at 75 mm, but 25% POF reduced it to 67 mm, 50% POF to 50 mm, 75% POF to 48 mm, and 100% POF to 30 mm. This indicates that higher POF content leads to a significant decrease in workability due to increased internal friction and stiffness in the concrete mix. These findings align with previous research by Awal & Mohammadhosseini [37].

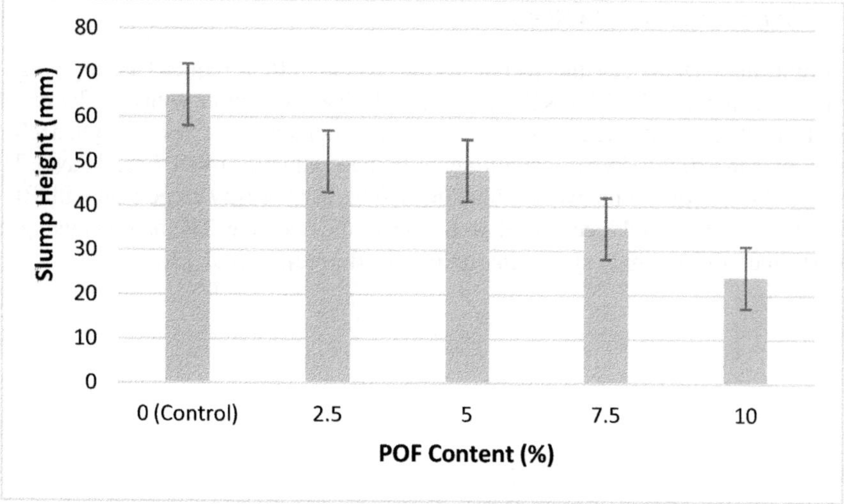

Fig. 7. Slump result of POF samples

The slump result of coconut shells (CS) samples in Fig. 8 shows a progressive decrease in workability as the percentage of CS increases in the mix. The control mix (0% CS) exhibits the highest slump at 65 mm, signifying good workability. As CS content increases to 25%, 50%, 75%, and 100%, the slump values drop to 51 mm, 44 mm, 39 mm, and 30 mm, respectively. This reduction was due to the higher water absorption and rough texture of CS, which reduces the available free water in the mix, making it stiffer.

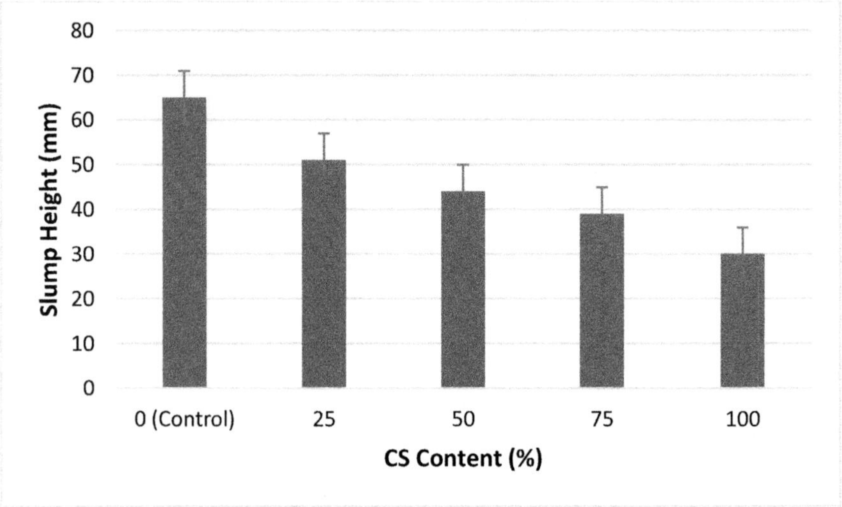

Fig. 8. Slump result of CS samples

3.2 Compressive Strength Results

The results in Fig. 9 indicate that increasing the content of RCA reduced the compressive strength of concrete. The 2.5% RCA mix shows the least reduction, with a 23.2% drop in strength at 28 days compared to the control. As RCA content increases, strength declines further, with the 10% RCA mix exhibiting the highest reduction (46.4% at 28 days). This decrease is likely due to the weaker bonding and higher porosity associated with RCA, which affects the overall mechanical properties of the concrete. The natural aggregates outperformed RCA, which align with findings of Berredjem et al. [7].

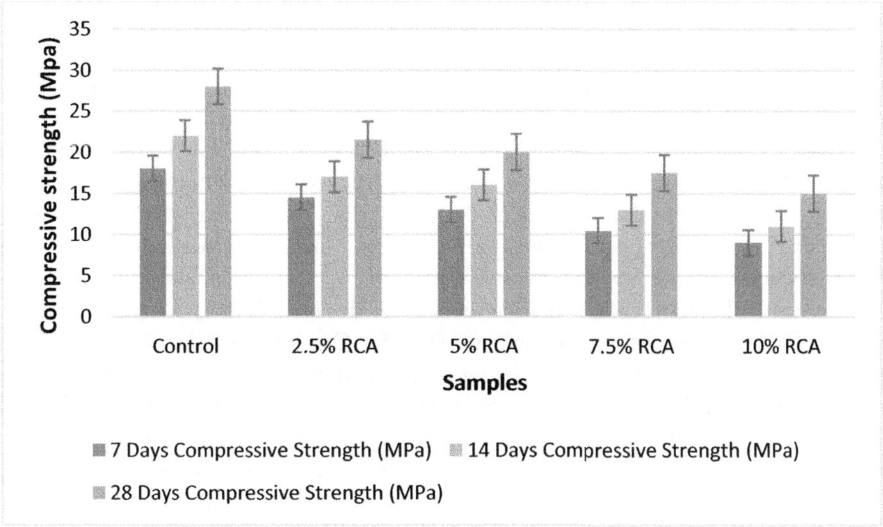

Fig. 9. Compressive strength of RCA samples

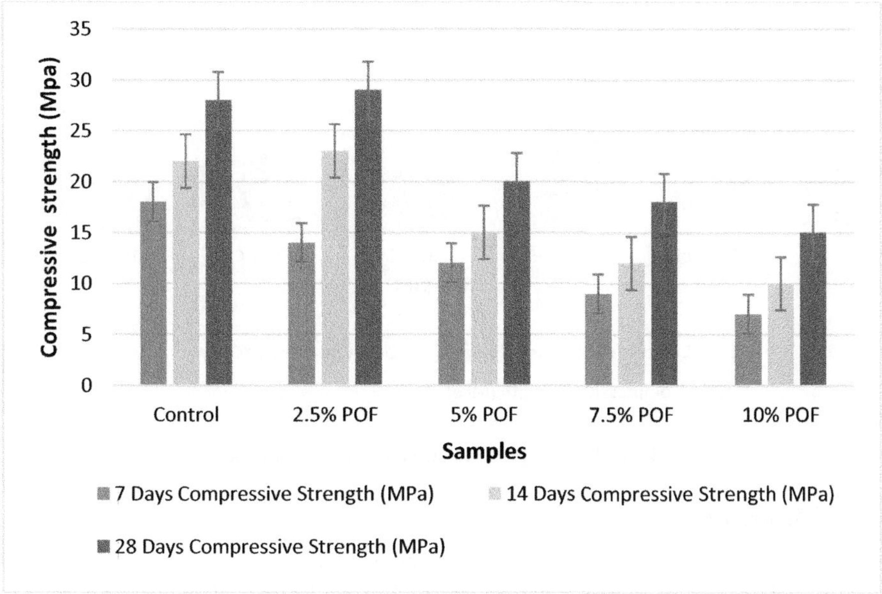

Fig. 10. Compressive strength of POF samples

Figure 10 show compressive strength of POF samples. It showed that the control sample achieved the greatest strength at 28 days, reaching 28 MPa. The addition of 2.5% POF slightly reduced the early strength at 7 days but led to a marginal increase in 28-day strength (29 MPa), suggesting an improvement in long-term performance. However, as the fiber content increased to 5%, 7.5%, and 10%, the compressive strength gradually declined at all curing ages. At 5% POF, the 28-day strength dropped to 20 MPa, indicating that excessive fibers may have disrupted the concrete matrix. The lowest strengths were observed at 7.5% and 10% POF, with 18 MPa and 15 MPa at 28 days, respectively. This suggests that excessive fiber content negatively affects the bonding and compaction of concrete, likely introducing voids and reducing the overall strength. While a small dosage of POF (2.5%) appears beneficial in the long run, higher fiber content weakens the matrix, leading to lower compressive strength.

The results in Fig. 11 show that increasing the dosage of CS as a replacement significantly reduces the strength of concrete. The 25% CS mix showed the least reduction among the modified samples, with a 26.8% drop in compressive strength at 28 days compared to the control. As the CS content increases, strength decreases further, with the 100% CS mix showing the highest reduction (63.2% at 28 days). This decline is likely due to the lower density and weaker interfacial bonding of CS, leading to increased porosity and reduced load-bearing capacity. This results align with the findings of kanojia et al. [38].

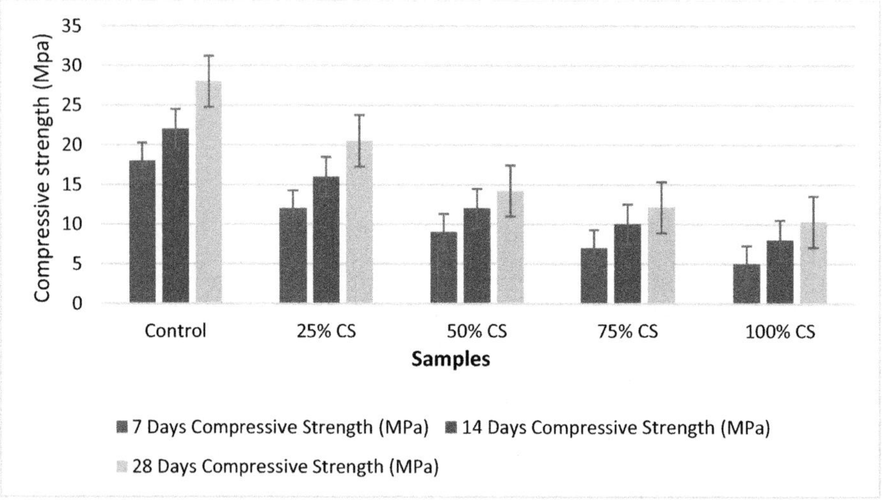

Fig. 11. Compressive strength of CS samples

3.3 SEM Results

SEM in Fig. 12 depicts the concrete with coconut shells after 28 days curing. The SEM images show that the control concrete has a dense and compact microstructure with fewer voids, while the 50% CS sample appears more porous with increased voids and weaker bonding. The 100% CS sample exhibits a highly porous and irregular structure with significant voids, indicating a more disconnected matrix that compromises overall integrity and strength. Increasing CS content results in a progressively more porous and less compact microstructure, aligning with the observed reduction in compressive strength (Table 4).

(a) SEM images of concrete with 50% CS

(b) SEM image of control concrete

(c) SEM image of concrete with 100% CS

Fig. 12. SEM images

Table 4. Comparison of pore structure, crack density, density

Parameter	Control (0% CS)	50% CS	100% CS
Pore Structure	Dense, compact ITZ; minimal pores; air 3.5%	Porous ITZ, 0.5–5 μm peaks; larger 50–100 μm pores; air 4.5%	Very high porosity, pores up to 100–120 μm; weak matrix
Crack Density	No cracks (7–90 days)	Microcracks after 5–10 cycles; fissures ~80–110 μm; mass loss 70–80 g	Severe cracking at all interfaces; 2–3 × control
Density	2227 kg/m3	2095 kg/m^3 (−12.7%)	1800 kg/m^3 (23%)
SEM Observations	Uniform C-S-H, no voids	Flakes of cement-sand, weak bonding, reduced ettringite	Poor adhesion, abundant voids, minimal hydration products

4 Conclusions

The conclusions drawn at end of study:

I. Incorporation of RCA, POF, and CS generally reduced compressive strength compared to the control mix. However, an optimal balance between strength, workability, and sustainability was achieved with 2.5% RCA, 2.5% POF and 25% CS.
II. The inclusion of fibers enhanced mechanical properties at 2.5% dosage, although increasing the fiber content progressively reduced workability.
III. The use of coconut shell (CS) waste indicates potential as a partial replacement material; however, higher CS content led to reduced workability and compressive strength, likely due to increased porosity and a weaker interfacial transition zone (ITZ).

4.1 Recommendations

I. Future studies should focus on optimizing the use of RCA, CS, and POF in combination with supplementary cementitious materials (SCMs) to mitigate strength loss while enhancing sustainability.
II. Surface treatment of CS and RCA should be explored to reduce porosity, improve the ITZ, and enhance mechanical properties.
III. Blending these waste-based additives with hybrid reinforcement strategies or modified mix designs could improve their performance, enabling broader application in structural and non-structural concrete.

References

1. Adisa, M.A., Odeyemi, S.O., Ogaji, C.O.: Effect of Various Cement Grades on the Properties of Self- Compacting Concrete. In: Raheem, A., Ikotun, B., Oluremi, J. (eds.) International Conference on Advances in Cement and Concrete Research. Ladoke Akintola University, Ogbomoso pp. 113–120 (2023)
2. Makul, N., et al.: Design strategy for recycled aggregate concrete: a review of status and future perspectives (2021). https://doi.org/10.3390/cryst11060695
3. Wang, B., Yan, L., Fu, Q., Kasal, B.: A comprehensive review on recycled aggregate and recycled aggregate concrete (2021). https://doi.org/10.1016/j.resconrec.2021.105565
4. Rissman, J., et al.: Technologies and policies to decarbonize global industry: Review and assessment of mitigation drivers through 2070 (2020). https://doi.org/10.1016/j.apenergy.2020.114848
5. EPA: Construction and demolition debris: material-specific data. https://www.epa.gov/facts-and-figures-about-materials-waste-and-recycling/construction-and-demolition-debris-material. Accessed 18 Feb 2025
6. Wu, L., Sun, Z., Cao, Y.: Modification of recycled aggregate and conservation and application of recycled aggregate concrete: a review. Constr. Build. Mater. **431**, 136567 (2024). https://doi.org/10.1016/J.CONBUILDMAT.2024.136567
7. Berredjem, L., Arabi, N., Molez, L.: Mechanical and durability properties of concrete based on recycled coarse and fine aggregates produced from demolished concrete. Constr. Build. Mater. **246** (2020). https://doi.org/10.1016/j.conbuildmat.2020.118421

8. Skocek, J., Ouzia, A., Vargas Serrano, E., Pato, N.: Recycled sand and aggregates for structural concrete: toward the industrial production of high-quality recycled materials with low water absorption. Sustainability **16**(2) (2024). https://doi.org/10.3390/su16020814
9. Delsaute, B., Staquet, S.: Impact of recycled sand and gravels in concrete on volume change. Constr. Build. Mater. **232** (2020). https://doi.org/10.1016/j.conbuildmat.2019.117279
10. UNESCAP: International coconut community. https://coconutcommunity.org/executive-director/optimizing-the-coconut-husk-potential. Accessed 18 Feb 2025
11. Report Linker: Nigeria coconut industry outlook 2022–2026. https://www.reportlinker.com/clp/country/3054/726344. Accessed 18 Feb 2025
12. Kumbasaroglu, H., Kumbasaroglu, A.: Applicability of agro-waste materials in structural systems for building construction: a scoping review. Appl. Sci. **15**(1), 71 (2025). https://doi.org/10.3390/APP15010071
13. Tomar, R., Kishore, K., Singh Parihar, H., Gupta, N.: A comprehensive study of waste coconut shell aggregate as raw material in concrete. Mater. Today: Proc. (2021). https://doi.org/10.1016/j.matpr.2020.09.754
14. Bhoj, S., Manoj, A., Bhaskar, S.: Usage potential and benefits of processed coconut shells in concrete as coarse aggregates. Mater. Today Proc. (2023). https://doi.org/10.1016/j.matpr.2023.03.529
15. Oparinde, A.O., et al.: Properties of coconut shells as coarse aggregate in concrete production: a productive waste usage. Sci. **10**(8) (2024).•ceeol.com, https://doi.org/10.22178/pos.107-9
16. Hasan, N.M.S., et al.: Investigation of lightweight and green concrete characteristics using coconut shell aggregate as a replacement for conventional aggregates. Int. J. Civ. Eng. (2023). https://doi.org/10.1007/s40999-023-00881-x
17. Nandiyanto, A.B.D., et al.: The effect of the addition of coconut fibers and coconut shells on the mechanical characteristics of porous concrete. J. Eng. Sci. Technol. **17**(3) (2022)
18. Raja, P., et al.: A review of sustainable bio-based insulation materials for energy-efficient buildings (2023). https://doi.org/10.1002/mame.202300086
19. Dokua, D.: Nigeria: production of palm oil 2023. Statista. https://www.statista.com/statistics/1134494/production-of-palm-oil-in-nigeria/. Accessed 18 Feb 2025
20. Singh, A., Yadav, B.P.: Sustainable innovations and future prospects in construction material: a review on natural fiber-reinforced cement composites. Environ. Sci. Pollut. Res. **31**(54), 62549–62587 (2024). https://doi.org/10.1007/S11356-024-35236-Z
21. Kaliwon, J., Ahmad, S.S., Aziz, A.A.: Performance of oil palm EFB fibre reinforced concrete roof slates. In: CSSR 2010 - 2010 International Conference on Science and Social Research (2010). https://doi.org/10.1109/CSSR.2010.5773834
22. Odeyemi, S.O., Adisa, M.A., Atoyebi, O.D., Adesina, A., Lukman, A., Olakiitan, A.: Variability analysis of compressive and flexural performance of coconut fibre reinforced self-compacting concrete. Int. J. Eng. Res. Africa (2023). https://www.scientific.net/JERA.66.19. Accessed 01 Oct 2023
23. Odeyemi, S.O., Shitu, A., Adisa, M.O.: Impact resistance of coconut fibre-reinforced concrete. J. Eng. Res. **29** (2024)
24. Adamu, M., Ibrahim, Y.E., Ahmed, O.S., Drmosh, Q.A.: Mechanical performance of date palm fiber-reinforced concrete modified with nano-activated carbon. Nanotechnol. Rev. **12**(1) (2023). https://doi.org/10.1515/ntrev-2022-0564
25. NIS 444-1: Composition, specification and conformity criteria for common cement Standard Organisation of Nigeria, Niger. Ind. Stand. Cent. (2003)
26. British Standard Institution BSI: BS EN 197-1:2011 Cement Part 1: Composition, specifications and conformity criteria for common cements. BSI Stand. Publ., no. November, p. 50, (2011)
27. British Standards Institution, BS 812-2:1995 Testing aggregates (1995)

28. British Standards Institution (BSI): BS EN 12620:2013 Aggregates for concrete Br. Stand. (2013)
29. Mahmood, A., Nanos, N., Begg, D.: An evaluation of the strength for recycled fine aggregate replacement in cementitious mortars. Buildings **14**(2) (2024). https://doi.org/10.3390/buildings14020470
30. Ramli: Effects of palm fiber on the mechanical properties of lightweight concrete crushed brick. Am. J. Eng. Appl. Sci. **3**(2) (2010). https://doi.org/10.3844/ajeassp.2010.489.493
31. Chinnu, S.N., Minnu, S.N., Bahurudeen, A., Senthilkumar, R.: Recycling of industrial and agricultural wastes as alternative coarse aggregates: a step towards cleaner production of concrete (2021). https://doi.org/10.1016/j.conbuildmat.2021.123056
32. Dabiri, H., Kioumarsi, M., Kheyroddin, A., Kandiri, A., Sartipi, F.: Compressive strength of concrete with recycled aggregate; a machine learning-based evaluation. Clean. Mater. **3**, 100044 (2022). https://doi.org/10.1016/J.CLEMA.2022.100044
33. BS-EN-12350-2:2019, Testing fresh concrete - Part 2: Slump test (2019)
34. BS EN 12390-3: Testing hardened concrete. Compressive strength of test specimens (2019)
35. ASTM: ASTM E1508: Standard guide for quantitative analysis by energy-dispersive Spectroscopy. ASTM Stand. (2012)
36. Rostami, J., Khandel, O., Sedighardekani, R., Sahneh, A.R., Ghahari, S.A.: Enhanced workability, durability, and thermal properties of cement-based composites with aerogel and paraffin coated recycled aggregates. J. Clean. Prod. **297** (2021). https://doi.org/10.1016/j.jclepro.2021.126518
37. Awal, A.S.M.A., Mohammadhosseini, H.: Green concrete production incorporating waste carpet fiber and palm oil fuel ash. J. Clean. Prod. **137** (2016). https://doi.org/10.1016/j.jclepro.2016.06.162
38. Kanojia, A., Jain, S.K.: Performance of coconut shell as coarse aggregate in concrete. Constr. Build. Mater. **140** (2017). https://doi.org/10.1016/j.conbuildmat.2017.02.066

Nanotechnology as a Feasible Strategy for Sustainable Precast Concrete Production

Simone Rapelli(✉), Denny Coffetti, and Luigi Coppola

Department of Engineering and Applied Sciences, University of Bergamo, Via Marconi 5, 24044 Dalmine, BG, Italy
simone.rapelli@unibg.it

Abstract. Nanotechnology, originally developed for high-tech applications, is increasingly being applied in the concrete sector. Among its various applications, it provides a cost-effective solution to accelerate early-age compressive strength development, a critical requirement in precast concrete production, without increasing environmental impact or compromising the long-term durability.

This study investigates the use of a C-S-H-based admixture to improve the economic and environmental sustainability, as well as manufacturing efficiency, of precast concrete. A conventional precast mixture was optimized by gradually reducing the cement content by up to 60 kg/m^3, while incorporating the seeding admixture. The experimental investigation included adiabatic calorimetry, compressive strength tests from 16 h to 28 days, and simulations of environmental impact and production costs.

The results demonstrate that the optimized mixtures achieved both early-age and long-term compressive strength targets, even with significantly lower cement content. Moreover, the optimized concretes exhibited a lower carbon footprint and production costs comparable to the reference mixture. These findings support the adoption of nanotechnology as a scalable and sustainable strategy to reduce the carbon footprint of precast concrete production.

Keywords: Precast concrete · Nanotechnology · C-S-H seeding · Sustainability

1 Introduction

Decarbonisation goals across all industrial sectors require the implementation of effective strategies to reduce the environmental impact. The concrete industry, responsible for approximately 7–8% of global man-made CO_2 emissions, is no exception and is called to adopt more sustainable practices [1, 2].

As highlighted in a recent study [3], about 85% of concrete's CO_2 emissions are related to Portland cement production; consequently, one of the most effective strategies to lower concrete's carbon footprint is the development of mixtures with a reduced cement factor.

In addition to improving sustainability, reducing the cement content can enhance durability by mitigating issues related to alkali-silica reaction (ASR), acid attack, and

M. Posani and S. Singh (Eds.): RYS 2025, RILEM Bookseries 67, pp. 295–304, 2026.
https://doi.org/10.1007/978-3-032-14652-6_23

shrinkage-induced cracking. However, it often results in limited early-age compressive strength development [4–6]. This limitation is especially critical in precast concrete production, where slower strength gain delays demolding and decreases overall manufacturing efficiency and economic viability [7].

Therefore, optimizing the production rate of low-carbon precast elements requires the adoption of strategies, such as steam curing or chemical admixtures, that accelerate early-age compressive strength development and shorten the curing time needed for safe demolding.

Steam curing promotes hydration by exposing fresh concrete to water vapor at 40 °C–60 °C for 8–16 h. While effective, this method requires expensive equipment and accounts for up to about 90% of the total energy consumption [8]. Moreover, high-temperature curing may negatively affect the microstructural development, resulting in a coarser and more porous matrix that compromises long-term compressive strength and durability [9].

As an alternative, accelerating admixtures enhance early-age strength by promoting tricalcium silicate (C_3S) dissolution. However, at high dosages, they may increase total porosity and reduce both long-term strength and durability. For instance, chloride-based accelerators, while being cheap and effective, are banned in several countries due to the increased risk of reinforcing steel corrosion induced by free chloride ions in the pore solution. Moreover, they can negatively affect long term durability by increasing drying shrinkage, reducing sulphate resistance, and increasing the risk of alkali-silica reaction (ASR). Chloride-free alternatives (calcium carbonate or lithium carbonate-based ones) are available but more expensive and their effectiveness is highly dependent on cement composition [6, 7, 10–13].

In this context, nanotechnology, once limited to medicine and electronics, has emerged as a promising solution for improving the economic and environmental sustainability of precast concrete production, without compromising long term compressive strength or durability. Notably, C-S-H seeding, through the addition of preformed nanoparticles with high specific surface area, can serve as a viable alternative to thermal curing in precast concrete. When used in place of steam curing, it has been shown to reduce CO_2 emissions by more than 30 kg per cubic meter, corresponding to about 10% in the total carbon footprint of precast elements [5, 8]. Beyond accelerating hydration kinetics, C-S-H nanoparticles promote a more uniform spatial distribution of hydration products, leading, especially at early ages, to a denser and more refined microstructure that enhances durability.

Experimental studies support this mechanism. For instance, Feng et al. [8] reported a reduction in threshold pore size from 823 nm to 76 nm after 6 h of curing when C-S-H nanoparticles were added. Similarly, another study [6] observed a 50% reduction in water penetration depth using a 3% dosage of C-S-H-based admixture [14]. These microstructural improvements are consistent with mechanical performance: compressive strength gains of 330%, 550% and 640% were achieved after 6 h of curing for mixtures containing 0.50%, 1.00% and 1.50% C-S-H nucleator, respectively [8].

Considering the viability of using C-S-H-based admixtures to improve early-age compressive strength development, this study assesses the feasibility of designing optimized low-carbon precast concrete mixtures by reducing cement content. The reference mixture, specifically tailored for precast applications, was proportioned to meet compressive strength classes of 20 MPa at 16 h and 55 MPa at 28 days, without the use of seeding admixture. To improve environmental sustainability the cement content was gradually reduced, while workability was kept constant by adjusting the dosage of superplasticizer at a fixed water/cement ratio.

The experimental program was divided into two parts: the first assessed the effectiveness of C-S-H seeding in enhancing hydration kinetics and compressive strength development across mixtures with different cement contents, while the second focused on assessing the environmental impact and cost effectiveness of the optimized mixtures, with the aim of validating this approach as a practical and scalable solution for more sustainable precast concrete production.

2 Materials and Methods

2.1 Materials

To better replicate industrial-scale precast concrete production and to account for cement variability, two commercially available cements were used: CEM I 52.5 R and CEM II/A-LL 42.5 R, sourced from different suppliers. The main physico-chemical properties are summarized in Table 1.

Table 1. Main physical properties of the cements.

	CEM I 52.5R	CEM II/A-LL 45.2R
Specific surface area [m^2/kg]	609.5	584.1
Particle density [kg/m^3]	2900	2900
Dv (50) [μm]	10.1	11.8
CO_2 content [%]	0.32	8.44
Sulphate content [%]	3.46	3.14

Concrete mixtures were formulated using a blend of natural (N) and industrial (I) aggregates (maximum size: 22 mm), classified into 5 grain size classes. The main properties of the aggregates are summarized in Table 2, while the granulometric curves are showed in Fig. 1.

To ensure consistent workability across all mixtures, a commercial polycarboxylate ether-based superplasticizer, specifically formulated for precast applications, was added. Additionally, a commercially available C-S-H seeding admixture was used to enhance hydration kinetics and promote early-age strength development. The main properties of the chemical admixtures are showed in Table 3.

Table 2. Main physical properties of the aggregates.

	N 0-2	N 2-8	N 8-22	I 8-16
Particle density [kg/m^3]	2620	2630	2630	3635
Water absorption [%]	1.2	1.5	1.4	1.4
Min-max size [mm]	0–2	2–8	8–22	8–16

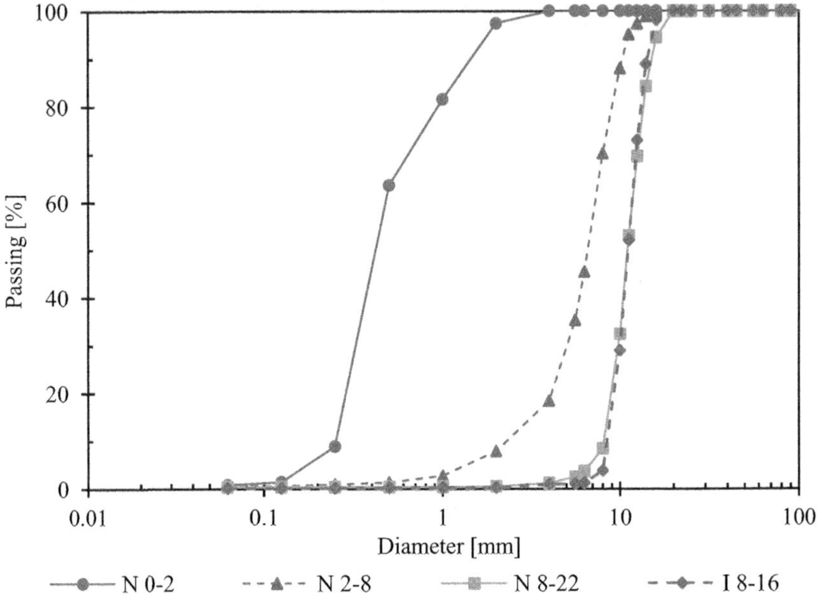

Fig. 1. Granulometric curves of the aggregates

Table 3. Main physico-chemical properties of the admixtures

	Super-plasticizer	C-S-H nucleator
Physic form	Brown liquid	Brown liquid
Dry residue [%]	34%	30%
Specific mass [kg/m^3]	1070	1100
pH	4.5	7.5

2.2 Mix Proportions

Four concrete mixtures were designed. The reference mix (C400), targeting a slump class S5 and compressive strength class of 20 MPa at 16 h and 55 MPa and 28 days, reflects typical precast production requirements. Cement was gradually reduced by up to 60 kg/m^3 and replaced with an equivalent volume of aggregates. In all mixtures, the water/binder ratio was kept constant at 0.45 and the superplasticizer dosage was adjusted to maintain consistent workability. The mix proportions are shown in Table 4.

Table 4. Mix design of the concretes produced

Component [kg/m^3]	C400 (Ref)	C370 + C-S-H	C350 + C-S-H	C340 + C-S-H
N 0 – 2	650	690	715	730
N 2 – 8	550	560	565	565
N 8 – 22	230	235	235	235
I 8 – 16	440	445	450	455
CEM I 52.5R	200	185	175	170
CEM II/A-LL 42.5R	200	185	175	170
Filler	30	30	30	30
Water	179	166	157	152
Super-plasticizer	3.0	2.8	3.0	3.0
C-S-H nucleator	--	3.7	3.5	3.4
water/cement ratio	0.45	0.45	0.45	0.45

2.3 Methods

All concrete mixtures were prepared in the laboratory under controlled environmental conditions and using a horizontal-axis mixer with a maximum capacity of 300 L. The mixing procedure was carried out according to the following steps:

- Step 1: Mixing of aggregates with part of the mixing water (equivalent to the aggregates' absorption capacity) for 2 min;
- Step 2: Resting phase with the mixer sealed for 2 min;
- Step 3: Addition of cement and water and mixing for 2 min;
- Step 4: Addition of the admixtures (superplasticizer and C-S-H nucleator);
- Step 5: Final mixing for 2 min.

Workability of the fresh concrete was assessed through slump tests (EN 12350-2). After fresh-state testing, 100 mm cubic specimens were cast in triplicate for compressive strength testing (EN 12390-3) at 16 h, 1, 3, 7 and 28 days. In addition, a 150 mm cubic specimen per mix was produced to perform adiabatic calorimetry testing (EN 12390-15).

In accordance with EN 12390-2, for the first 24 h all specimens were initially cured into their moulds, covered with a plastic sheet, and stored in a climatic chamber at 20 ± 2 °C. After demoulding, the specimens were cured in plain water at 20 ± 2 °C until testing.

The environmental impact of each concrete mixture was assessed in terms of Global Warming Potential (GWP), based on the concrete mix designs and raw data obtained from manufacturer-supplied Environmental Product Declarations (EPDs).

Production costs were estimated based on average unit prices of raw materials used in precast concrete plants located in Northern Italy. The cost per m^3 was obtained by multiplying the unit price of each constituent by its corresponding quantity in the mixture.

3 Experimental Results and Discussion

3.1 Adiabatic Calorimetry Test

Figure 2 shows the results of the adiabatic calorimetry tests, expressed as temperature evolution over a 48-h period.

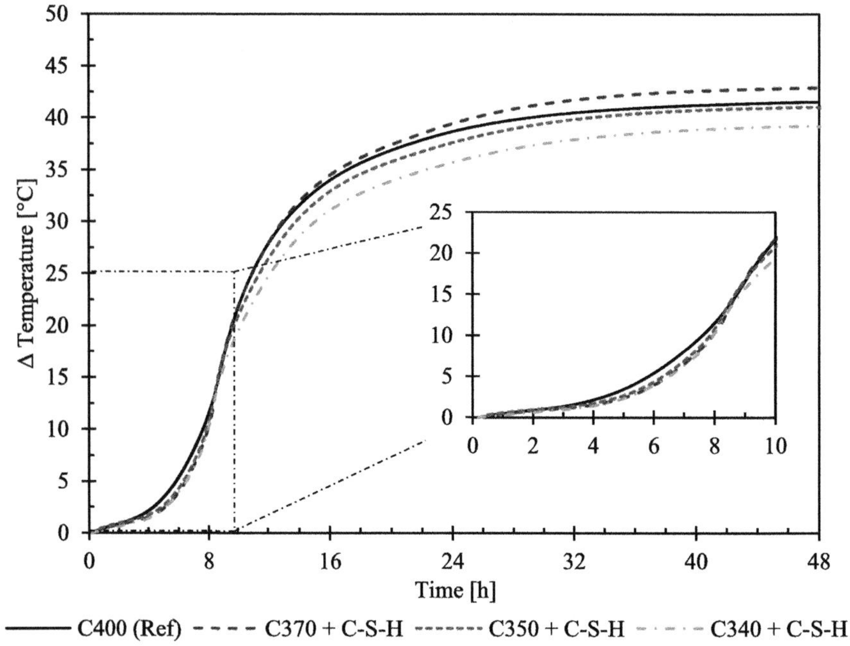

Fig. 2. Adiabatic calorimetry results

As expected, the reference mixture (C400) exhibited the fastest initial heat release, indicating a faster hydration rate. However, when cement content was reduced to 370 kg/m^3 and 350 kg/m^3, the addition of the C-S-H seeding admixture enabled temperature profiles that closely followed the reference mix, with only a slight delay observed within the first 8 h. Conversely, for the mixture containing 340 kg/m^3 of cement, a more pronounced delay in heat release was observed. These results indicate that, while the use of nucleation seeding-based admixtures effectively compensates for cement content reductions by up to 50 kg/m^3, its beneficial effect is no longer sufficient to maintain comparable hydration kinetics beyond this threshold.

3.2 Compressive Strength Development

The compressive strength development of the mixtures is shown in Fig. 3. All tested concretes successfully reached the target compressive strength classes of 20 MPa at 16 h and 55 MPa at 28 days, confirming their suitability for precast production. Specifically, the early-age strengths after 16 h ranged between 28 MPa and 32 MPa and continued to increase to values between 64 MPa to 75 MPa at 28 days.

Even at very-early ages, all optimized mixtures outperformed the reference, demonstrating the ability of C-S-H seeding to enhance early-age strength development in mixtures with a reduced cement content. This improvement ranged from 7% to 16% at 16 h and reached up to 17% at 28 days, highlighting the admixture's contribution not only to early hydration but also to later-age strength gain.

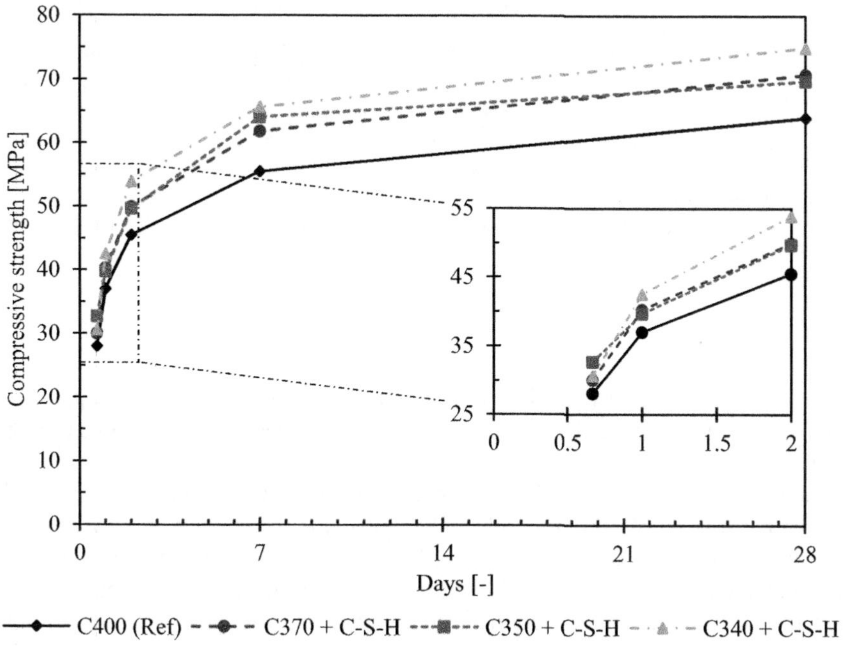

Fig. 3. Compressive strength development

3.3 Carbon Footprint Estimation and Cost Analysis

Figure 4 presents a comparative analysis of the global warming potential (GWP) of each mixture, divided into the contributions from cement, aggregates and admixtures.

As expected, regardless of composition, cement production is the main contributor to total greenhouse gases emission. The gradual reduction in cement content across the C370 + C-S-H, C350 + C-S-H and C340 + C-S-H mixtures led to a corresponding decrease in overall GWP of approximately 6%, 9% and 11%, respectively, when compared to the reference mix. Although the absolute environmental impact of the chemical admixtures remains limited (ranging from 3 to 7 kg_{CO2eq}/m^3), their relative contribution to the overall footprint increases in mixtures with lower cement content due to the use of the C-S-H nucleator. Nevertheless, this increase did not offset the substantial reduction in emissions achieved through cement reduction.

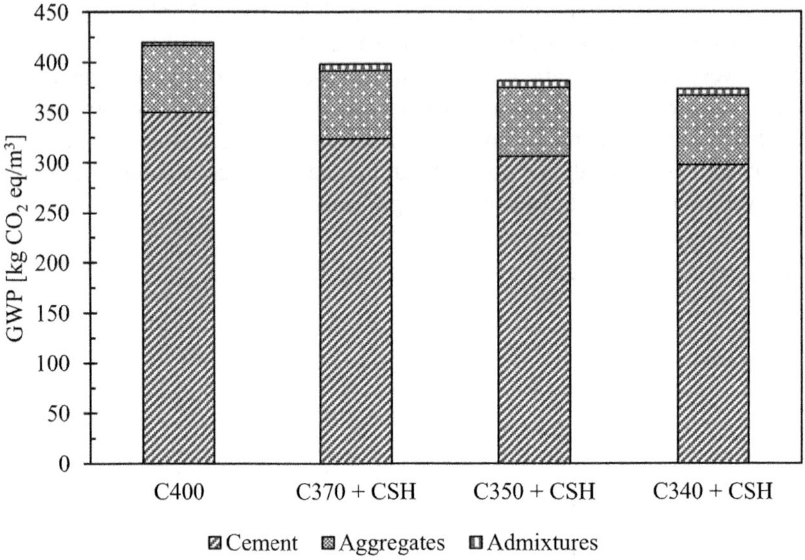

Fig. 4. Global warming potential estimation

The environmental sustainability of the optimized mixtures is further supported by their economic viability, as shown in the comparative production cost analysis presented in Fig. 5.

The relative cost, normalized with respect to the market price for producing 1 m^3 of the reference mixture (C400), is divided into contributions from cement, aggregates and admixtures. All optimized mixtures showed production costs comparable to or slightly lower than the reference. This result demonstrates that the economic impact of the C-S-H-based admixture is effectively balanced by the cost savings associated with lower cement content.

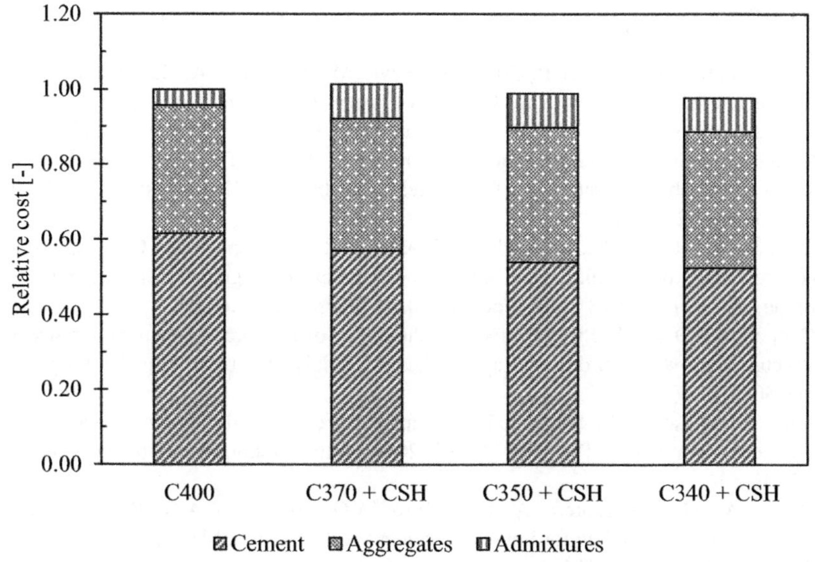

Fig. 5. Relative production costs estimation

4 Conclusions

The experimental investigation presented in this study aimed to evaluate the feasibility of large-scale implementation of nanotechnology, specifically C-S-H nucleation, as a viable strategy to reduce the environmental impact of precast concrete production. A reference mixture, representative of a typical industrial precast concrete and proportioned to meet both early and long-term compressive strength requirements, was gradually optimized reducing the cement content by up to 60 kg/m^3 at fixed water/cement ratio. To compensate for the slower hydration kinetics, typically associated with lower cement dosages, a commercial C-S-H-based seeding admixture was incorporated.

Considering the results obtained, the following conclusions can be drawn:

- The addition of a C-S-H nucleator seeding admixture enabled a reduction in cement content by up to 50 kg/m^3, with negligible influence on the hydration kinetics. Notably, both early-age and 28-days compressive strengths were improved compared to the reference mixture;
- The reduction in cement content led to a decrease in the global warming potential (GWP) per cubic meter of precast concrete by up to 11%, proving the environmental benefits of the proposed optimization strategy;
- The use of a C-S-H-based seeding admixture in the optimized mixtures did not result in increased production costs, demonstrating that the environmental gains can be achieved without economic penalties.

These results support the adoption of nanotechnology in precast concrete as a scalable strategy to reduce the environmental impact, without compromising economic feasibility, manufacturing process efficiency or long-term durability.

References

1. Renne, N., Kara De Maeijer, P., Craeye, B., Buyle, M., Audenaert, A.: Sustainable assessment of concrete repairs through Life Cycle Assessment (LCA) and Life Cycle Cost Analysis (LCCA). Infrastructures 7 (2022). https://doi.org/10.3390/infrastructures7100128
2. Coffetti, D., Crotti, E., Gazzaniga, G., Carrara, M., Pastore, T., Coppola, L.: Pathways towards sustainable concrete. Cem. Concr. Res. **154**, 106718 (2022). https://doi.org/10.1016/j.cemconres.2022.106718
3. Shamass, R., Rispoli, O., Limbachiya, V., Kovacs, R.: Mechanical and GWP assessment of concrete using blast furnace slag, silica fume and recycled aggregate. Case Stud. Construct. Mater. **18**, e02164 (2023). https://doi.org/10.1016/j.cscm.2023.e02164
4. Coppola, L., et al.: The improvement of durability of reinforced concretes for sustainable structures: a review on different approaches. Materials **15** (2022). https://doi.org/10.3390/ma15082728
5. John, E., Matschei, T., Stephan, D.: Nucleation seeding with calcium silicate hydrate – a review. Cem. Concr. Res. **113**, 74–85 (2018). https://doi.org/10.1016/j.cemconres.2018.07.003
6. Cuesta, A., Morales-Cantero, A., De la Torre, A.G., Aranda, M.A.G.: Recent advances in C-S-H nucleation seeding for improving cement performances. Materials **16** (2023). https://doi.org/10.3390/ma16041462
7. Golewski, G.L., Szostak, B.: Application of the c-s-h phase nucleating agents to improve the performance of sustainable concrete composites containing fly ash for use in the pre-cast concrete industry Materials **14** (2021). https://doi.org/10.3390/ma14216514
8. Feng, X., Zhuo, C., Yin, S.: The application of C-S–H accelerators in the precast concrete industry: early-age properties and CO2 footprint analysis. J. Cleaner Prod. **435**, 140558 (2024). https://doi.org/10.1016/j.jclepro.2024.140558
9. Zeyad, A.M., et al.: Review on effect of steam curing on behavior of concrete, Cleaner Mater. **3**, 100042 (2022). https://doi.org/10.1016/j.clema.2022.100042
10. Min, T.B., Cho, I.S., Park, W.J., Choi, H.K., Lee, H.S.: Experimental study on the development of compressive strength of early concrete age using calcium-based hardening accelerator and high early strength cement. Constr. Build. Mater. **64**, 208–214 (2014). https://doi.org/10.1016/j.conbuildmat.2014.04.053
11. Wang, Y., et al.: Accelerators for normal concrete: a critical review on hydration, microstructure and properties of cement-based materials. Cement Concr. Compos. **134**, 104762 (2022). https://doi.org/10.1016/j.cemconcomp.2022.104762
12. Rixom, R., Mailvaganam, N.: Chemical Admixtures for Concrete. 3rd edn. (2002). https://doi.org/10.4324/9780203017241
13. Aïtcin, P.C., Flatt, R.J.: Science and technology of concrete admixtures (2015). https://doi.org/10.1016/C2015-0-00150-2
14. Wang, F., Kong, X., Jiang, L., Wang, D.: The acceleration mechanism of nano-C-S-H particles on OPC hydration. Constr. Build. Mater. **249**, 118734 (2020). https://doi.org/10.1016/j.conbuildmat.2020.118734

Performance and Evolution of Materials for 3D Construction Printing: An Overview from Conventional to Extreme Environments

V. Sangiorgio[1](✉), M. Valeri[1], C. Cantagallo[1], C. Guth[2], and F. Salese[1,2]

[1] Department of Engineering and Geology (INGEO), D'Annunzio University of Chieti –Pescara, Viale Pindaro, 42, 65127 Pescara, Italy
valentino.sangiorgio@unich.it

[2] International Research School of Planetary Sciences, Via Antonio Lo Feudo 1, 65127 Pescara, Italy

Abstract. Additive manufacturing is revolutionizing the construction sector, from traditional residential buildings to emergency shelters and even extreme environments, by simplifying on-site operations.

A critical challenge in this field is the development of sustainable materials for additive manufacturing. The housing crisis in low-income countries demands high-performance materials with simple, low-cost production cycles and locally available resources. Furthermore, 3D printing applications in extreme environments, such as deserts, polar regions, and planetary habitats, including those on the Moon or Mars, require innovative and printable material solutions.

Several reviews exist on specific 3D printing material typologies, but a comprehensive overview focusing on material performance and production processes is still lacking. Accordingly, this study pursues two main objectives:

(1) to deliver a critical review and comprehensive overview of 3D printable construction materials, analyzing their historical evolution, production cycles, mechanical, physical, and thermal properties, as well as sustainability and compatibility with 3D printing;
(2) to integrate insights from the ArchiSpace project, funded by the European Union (MSCA Staff Exchange), which focuses on designing and constructing habitats in terrestrial analogs of planetary environments.

A systematic literature review was conducted using an integrated approach combining qualitative analysis and bibliometric methods, offering a comprehensive overview of the state of the art in the field. As a result, a major gap identified is the understanding of geopolymer matrix properties, particularly regarding mix design, performances, and 3D printing processes. This dual focus on terrestrial and extraterrestrial applications highlights the transformative potential of 3D printing and provides a roadmap for future research and implementations in traditional and extreme environments.

Keywords: 3D construction printing · Geopolymers · Regolith-simulants · Sustainable construction materials · In situ resource utilization (ISRU)

M. Posani and S. Singh (Eds.): RYS 2025, RILEM Bookseries 67, pp. 305–319, 2026.
https://doi.org/10.1007/978-3-032-14652-6_24

1 Introduction

In recent years, additive manufacturing has emerged as a disruptive force in the construction industry, with 3D Construction Printing (3DCP) offering innovative solutions to the structural, logistical, and environmental limitations of traditional building methods. By enabling process automation, reducing material waste and construc-tion time, and allowing for high geometric flexibility and customization, 3DCP ad-dresses critical global challenges such as urban overpopulation, labor shortages, and the urgent need for sustainable and resilient buildings [1]. A prominent application of 3DCP is in the development of emergency housing for regions affected by natural disasters, armed conflicts, or humanitarian crises. Here, the use of locally available materials significantly reduces logistical and environmental costs, enhancing community resilience. Similarly, in extreme environments—deserts, polar regions, and extraterrestrial surfaces (Moon and Mars)—3D printing enables rapid deployment and habitat construction despite severe constraints like temperature extremes and supply limitations [2]. A cornerstone of this approach is In-Situ Resource Utilization (ISRU), which promotes the use of indigenous materials for on-site construction, minimizing reliance on Earth-based logistics and maximizing sustainability [3]. Key to the success of 3DCP in such settings is the development of high-performance, sustainable materials that exhibit optimal mechanical strength, durability, rheological behavior, and environmental compatibility across diverse conditions [4]. Despite the rapid growth of the field, scientific literature re-mains fragmented, often focusing on isolated properties like compressive strength or interlayer adhesion [5]. Additionally, most research still centers around Portland cement [6], with limited attention to low-impact alternatives such as geopolymers, earth-based composites, natural fibers, and simulated regolith mixtures [7]. Recent studies underscore the promise of geopolymers for space and emergency applications due to their ability to set in oxygen-free environments and at low temperatures, combined with the use of industrial or natural waste-based precursors, resulting in signifi-cantly reduced environmental impact [8]. Materials like fly ash [9], slag [10], and metakaolin [11] also show considerable potential for enhancing the mechanical strength, durability, and sustainability of 3D-printable construction materials. Parallel advancements in simulated regolith-based composites, often combined with sulfur, polymers, or epoxy resins, enable the fabrication of extraterrestrial structures via 3D printing or laser sintering technologies [12]. This context reveals the urgent need for a transdisciplinary and systemic approach to materials design in additive construction. While several reviews address specific 3D printing material typologies, a comprehensive synthesis that considers a wide spectrum and comparison of materials (including both material performance and production processes) remains absent. This study presents a comprehensive overview of the evolution of materials for 3D construction printing, analyzing current performance trends and innovations. Emphasis is placed on both terrestrial and extraterrestrial applications, proposing a research roadmap for sustainable, scalable, and resilient construction. The work is enriched by insights from the European MSCA Staff Exchange project ArchiSpace, which investigates habitat construction in planetary analog environments, providing crucial knowledge for developing self-sufficient housing in extreme contexts.

2 Methods

This work adopts a methodological framework structured into four main phases: (1) definition of the review framework, (2) systematic collection and selection of sources, (3) critical and interpretive analysis of the contents, and (4) structured writing of the results. The entire process was developed in accordance with the four guiding principles for scientific reviews outlined by Snyder (2019) [13].

2.1 Design of the Review

This review defines a research strategy aimed at analyzing the materials used in 3D printing for construction, with a particular focus on applications in critical contexts such as emergency housing and extreme environments. The goal is to provide an updated overview of the advancements in materials for 3D Construction Printing, highlighting trends, innovations, and opportunities with a sustainable perspective. The research is motivated by the growing importance of sustainability in the construction sector and integrates contributions from the European project ArchiSpace, which focuses on the design of habitats in planetary analog environments using local resources, low-impact techniques, and energy-efficient systems.

Each of these contains a set of fundamental research questions.

1) What are the main raw materials used in the production of geopolymers?
2) How can the raw materials used be classified?
3) Into how many and which categories can the geopolymers developed so far be devided?
4) What are the sustainability prospects of traditional materials, geopolymers, and regolith in the context of 3D printing for construction?

2.2 Conduction of the Review

The review combined bibliometric and qualitative analyses to thoroughly assess the state of the art on sustainable materials, geopolymers, and 3D printing applications in construction, focusing on emergency housing and extreme environments. Using targeted keywords such as "3D printing of constructions," "emergency housing," "extreme environments," "sustainable materials," and "geopolymers," searches were conducted in academic databases including Google Scholar, ScienceDirect, and Wiley Online Library. An initial pool of 80 articles was screened by abstract to exclude irrelevant studies. A subsequent full-text qualitative review applied criteria based on methodological rigor, publication within the last 15 years, and relevance to sustainability and the use of geopolymers in 3D printing. Ultimately, 38 high-impact studies were selected, enabling the identification of key trends, challenges, and opportunities that form a solid foundation for further research phases.

2.3 Analysis

The analysis systematically collects and critically processes data on the evolution of materials and technologies in construction. Using advanced visualization tools like charts and comparative tables, it ensures clear and objective presentation. The study focuses on three key innovation areas: 3D printing, geopolymers, and extraterrestrial materials, lunar and Martian regolith, for space habitat construction (Fig. 1).

1) 3D Construction printing
   - A qualitative analysis was conducted to explore the historical evolution and development of traditional construction materials in the context of 3D printing technologies.
2) Geopolymers
3) This section is structured into three analytical stages:
   - A qualitative analysis of the historical background and development of geopolymeric materials.
   - A bibliometric analysis aimed at identifying the most commonly utilized geopolymers in current literature and industrial applications.
   - A qualitative in-depth analysis focusing on the characteristics and performance of the most relevant geopolymeric systems.
4) Regolith-Based Materials
   - A qualitative analysis was performed to investigate the potential of regolith as a novel and promising material for future construction technologies, particularly in extraterrestrial environments.

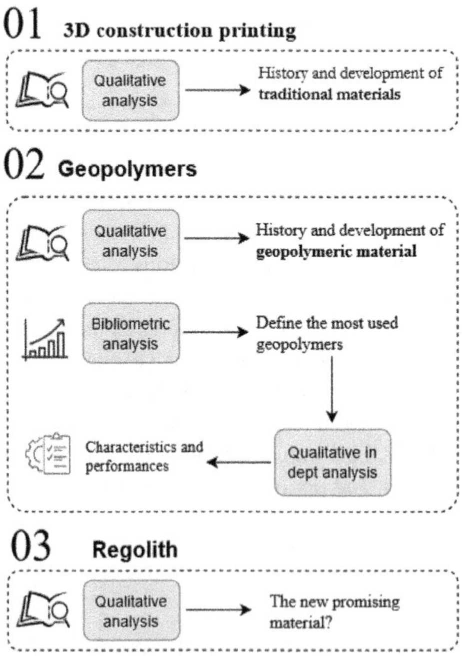

Fig. 1. Overview of the proposed framework for bibliographic analysis and bibliometric analysis.

2.4 Writing up the Review

The final phase involves drafting the review, highlighting the importance of the historical-critical analysis of sustainable materials, with particular attention to the use of geopolymers and materials from the Moon and Mars, as a foundation for the development of advanced, high-performance solutions, ideal for construction in extreme environments. Additionally, the potential of 3D printing as an emerging technology is examined, capable of transforming the construction market and steering it towards more sustainable, innovative, and highly performing future scenarios.

3 Results and Discussion

In this section, the results of the qualitative analysis are presented first, followed by those from the bibliometric analysis for each innovation sector.

3.1 3D Construction Printing: A Technological Revolution for the Future of Building

3D printing in the construction sector is rooted in pioneering technological innovations, beginning in 1941 with Urschel's "Wall Building Machine," which laid the foundation for automated building. A crucial breakthrough occurred in 1996 with Khoshnevis's Contour Crafting, introducing layer-by-layer cement extrusion that enabled complex and customized structures. In 2005, Dini's D-Shape technology expanded 3D printing capabilities through binder jetting on sand, facilitating advanced architectural components [14].

The year 2014 marked global visibility for the technology, with its adoption in residential and infrastructure projects demonstrating significant reductions in time and cost, while enabling more sustainable solutions [15]. In 2015, ICON built the first 3D-printed house in the USA, confirming the feasibility of producing functional, affordable homes rapidly [14]. In 2016, Apis Cor and Winsun showcased large-scale 3D printing potential by printing houses in 24 h and entire residential complexes, respectively [15].

Simultaneously, the integration of industrial robots, mobile robots, and collaborative robots (cobots) has enhanced the precision, efficiency, and versatility of automated construction, expanding 3D printing to diverse materials and geometries while lowering operational costs. The Yhnova project in France exemplifies the successful use of robotic platforms to print entire buildings [16].

These developments represent key milestones in the evolution of 3D printing in construction, with the potential to radically transform the industry by offering innovative, sustainable, and cost-effective solutions to meet current housing and environmental challenges [15]. Figure 2 shows the Historical Evolution of Traditional Materials in 3D Printing Technologies.

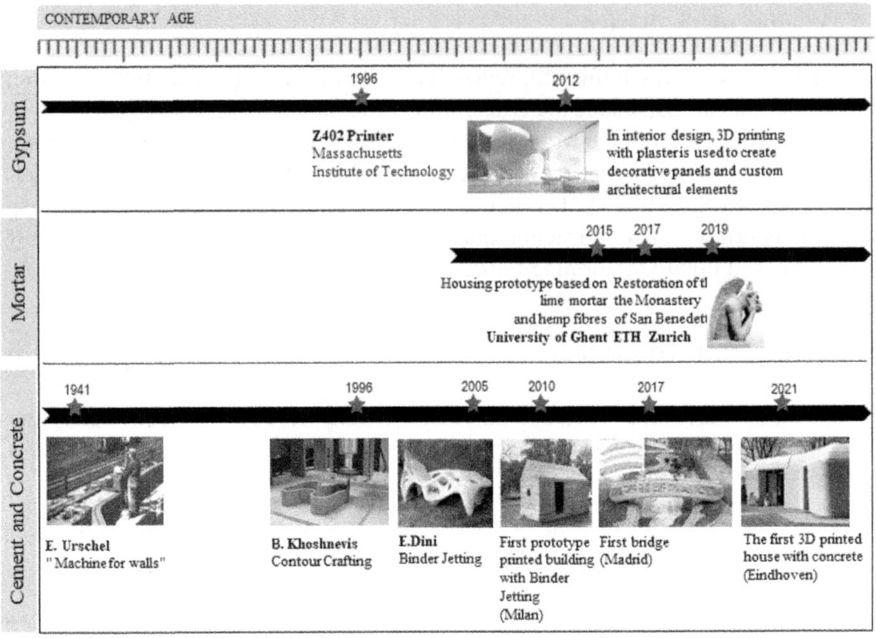

Fig. 2. Historical Evolution of Traditional Materials in 3D Printing Technologies.

3.2 Geopolymers, Historical Evolution and Perspectives in the Sustainable Landscape

History.

Geopolymers are emerging as a cutting-edge class of inorganic binders synthesized from aluminosilicate materials in highly alkaline environments. Defined by a low-temperature polycondensation process, they present a sustainable and low-carbon alternative to conventional binders like Portland cement. The term "geopolymer" was coined in 1978 by French chemist Joseph Davidovits, who identified parallels between these synthetic materials and natural geological processes.

In the 1970s, Davidovits proposed a revolutionary hypothesis: that ancient structures such as the Egyptian pyramids may have been constructed using early forms of synthetic stone produced through low-temperature chemical reactions. This led to the establishment of the Institut Géopolymère in 1981, a pioneering center for geopolymer research and industrial development. By the 1990s, the focus expanded to the use of industrial by-products, with seminal work by Van Jaarsveld, Van Deventer, and Lukey demonstrating the viability of fly ash and blast furnace slag as geopolymer precursors—advancing the concept of circular resource valorization.

Between 1998 and 2002, the Geopolymer Institute, under Davidovits' leadership, convened the first international conferences dedicated to geopolymers, fostering a global, multidisciplinary scientific network. This culminated in the 2005 Geopolymer World Congress, confirming Davidovits' foundational role in the field.

A major milestone was the 2009 publication of *Geopolymers: Structures, Processing, Properties, and Industrial Applications* by J.L. Provis and J.S.J. Van Deventer, which defined the scientific landscape and set the agenda for future research.

The evolution of geopolymers—from theoretical concept to industrial innovation—marks a paradigm shift in materials science, centered on sustainability, waste reuse, and reduced environmental impact. Today, geopolymers are recognized as key enablers of low-carbon technologies and are at the forefront of developing resilient, future-ready construction solutions [17]. Figure 3 shows the historical evolution of geopolymers.

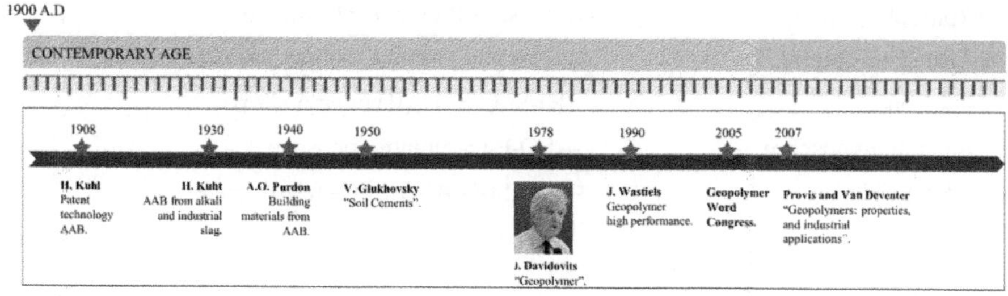

Fig. 3. Historical evolution of geopolymers.

Structure, Synthesis and Raw Materials.

Geopolymers represent a new generation of inorganic binders, characterized by a three-dimensional amorphous or semicrystalline network composed of interconnected mineral units via strong covalent bonds. The term "geopolymer" is derived from the combination of the prefix "geo", which refers to the analogy with natural geological processes, and the suffix "polymer", highlighting the macromolecular nature of the material, resulting from polycondensation reactions in an alkaline environment [17].

Geopolymers are composed of the molecular units (or chemical groups) as presented in Table 1.

The structural and performance characteristics of geopolymers critically depend on the chemical composition of the precursors and the specific synthesis conditions.

Among the key parameters that influence the formation and evolution of the geopolymeric matrix are the solid-to-liquid ratio, the concentration and type of alkaline activators (such as sodium hydroxide and sodium silicate), the pH of the system, the activation temperature, and the polymerization time. Fine regulation of these variables allows for control over the reaction kinetics, the microstructure of the final product, and consequently, its mechanical, chemical, and physical properties.

One of the most notable features of geopolymers is their remarkable versatility in the selection of raw materials. The precursors that can be used include both natural sources—such as metakaolins, volcanic ashes, and activated clays—as well as industrial waste materials, such as fly ash, blast furnace slag, and recycled glass residues. Figure 4 shows an overview of both natural and artificial sourced geopolymers.

Geopolymers are currently developed and applied in six main classes of materials (Table 2).

Table 1. Classification of chemical groups.

Geopolymeric network	Units molecular
poly(siloxo)	-Si-O-Si-O- siloxo
poly(sialate)	-Si-O-Al-O- sialate
poly(sialate-siloxo)	-Si-O-Al-O-Si-O-sialate-siloxo
poly(sialate-disiloxo)	-Si-O-Al-O-Si-O-Si-O-sialate-disiloxo
poly(phosphate)	-P-O-P-O- phosphate
poly(phospho-siloxo)	-P-O-Si-O-P-O- phospho-siloxo
poly(phospho-sialate)	-P-O-Si-O-Al-O-P-O- phospho-sialate
poly-silicone	-(R)-Si-O-Si-O-(R) organo-siloxo
poly(alumino-phospho)	-Al-O-P-O- alumino-phospho
poly(ferro-sialate)	-Fe-O-Si-O-Al-O-Si-O- ferro-sialate

Table 2. Class of geopolymers.

	Type	Composed	Si:Al
1	Waterglass-based geopolymer	poly(siloxonate)	1:0
2	Kaolinite / Hydrosodalite-based geopolymer	poly(sialate)	1:1
3	Metakaolin MK-750-based geopolymer	poly(sialate-siloxo)	2:1
4	Calcium-based geopolymer	(Ca, K, Na)-sialate	1,2,3
5	Rock-based geopolymer	poly (sialate-multisiloxo)	$1 < Si:Al < 5$
6	Silica-based geopolymer	poly(siloxonate)	$Si:Al > 5$

To better understand the diffusion and use of materials in 3D construction printing, a bibliometric analysis was carried out using the Scopus database. Figure 5a shows a keyword map highlighting frequently used terms and their co-occurrence, while Fig. 5b (Top 50 Keywords Plus) confirms key themes such as metakaolin, fly ash, and slags, followed by compressive strength and geopolymer. The dataset focuses on commonly used raw materials in geopolymer matrices, analyzed from physical-mechanical and thermal perspectives. Data processing was performed using R, and the results were visualized with VOSviewer, offering a clear overview of emerging trends and research networks in the construction sector.

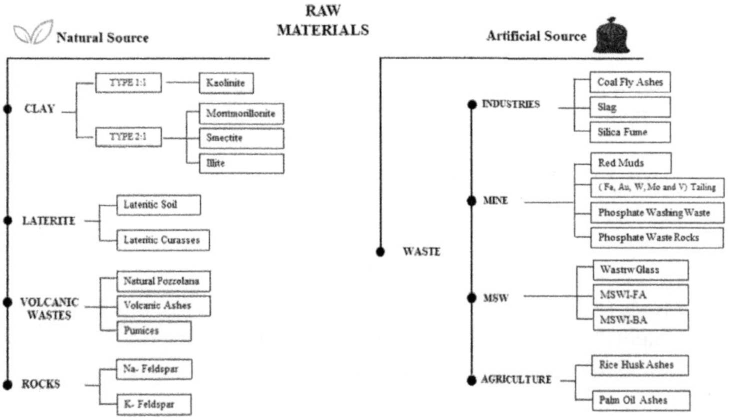

Fig. 4. Classification raw material.

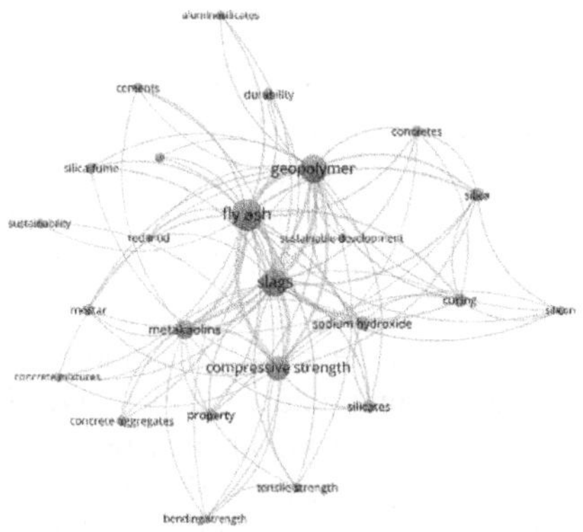

Fig. 5. (a). Co-occurrence keywords network. Software: Vos Viewer.

Fig. 5. (b). Top50 Keywords Plus on construction materials. Software:Bibliometrix.

This bibliometric analysis highlights that the most studied geopolymers are those based on Fly Ash, Slag, and Metakaolin. Consequently, these three precursors and their corresponding geopolymers were examined in greater detail, focusing on activation conditions, curing, compressive strength (MPa), tensile strength (MPa), and shrinkage (%). The results of this in-depth investigation are presented in Table 3 based to studies of Zhuand et al. [9], Zhang et al. [18] and He et al. [19].

Table 3. Characteristics of the most widely used geopolymer-based raw materials.

Precursor	Activation conditions	Curing	Compressive strength (MPa)	Tensile Strength (MPa)	Shrinkage (%)
Fly Ash	NaOH 10–14 M + NaSiO3 (ratio 0.7–2.5)	24-48h at 60–80 °C or ambient (28 days)	30–70 15–35	2.5–5.0	0.01–0.05
Slag	NaOH 8–12 M + NaSiO3 (ratio 0.6–1.5)	24-72h at 60–85 °C or ambient (28 days)	50–80 30–60	3.0–6.0	
Metakaolin	NaOH 6–10 M ± NaSiO3 or Ca $(OH)_2$	Ambient temperature	40–100	4.0–7.5	

Sustainability.
The application of geopolymers in the construction industry offers a transformative alternative to conventional binders, combining environmental sustainability with industrial efficiency. These amorphous, inorganic materials enable a substantial reduction in natural resource consumption and CO_2 emissions, aligning with circular economy principles. Unlike ordinary Portland cement (OPC), which emits approximately 0.87 tons of CO_2 per ton produced, geopolymer systems—particularly those based on activated fly ash—can lower emissions to just 0.16 tons. This footprint can be reduced even further through the use of waste-derived alkaline activators, such as calcium hydroxide, and by minimizing or replacing sodium silicate with recycled alternatives from glass or alumina production. A key advantage of geopolymers lies in their ability to valorize industrial by-products, including fly ash, blast furnace slag, and even sewage sludge, thus closing material loops in high-impact sectors. To enable large-scale adoption, recent research has focused on the development of "ready-mix" dry geopolymer formulations that require only water for activation—mirroring the practicality of conventional cements. Innovative strategies involve thermally treating blends of carbonates (e.g., $CaCO_3$, $MgCO_3$) and aluminosilicate powders at 400–500 °C to yield geopolymeric powders suitable for immediate, field-ready use. The performance and competitiveness of these materials depend on a deep understanding of their chemical and structural behavior. In particular, the dissolution of silica and alumina and the subsequent formation of key binding phases—such as N-A-S-H and N-C-A-S-H gels—are critical to achieving desired mechanical and durability properties.

Advancing research in these domains is essential for optimizing material design, reducing production costs, and positioning geopolymers as a truly sustainable and high-performance solution for next-generation construction [20].

To sum up, the Fig. 6 presents a comparative analysis of the production cycles of traditional and innovative materials.

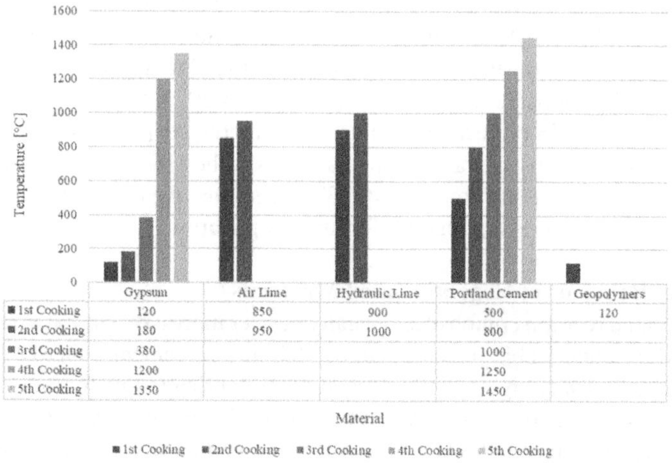

Fig. 6. Comparative Analysis of the Production Cycle of Traditional and Innovative Materials.

3D Printing.
3D-printed geopolymers represent an innovative combination of eco-friendly, high-performance materials integrated with 3D printing technology, leading to a true revolution in the construction sector. The adoption of 3D-printed geopolymers has involved numerous research institutes and universities internationally, facilitating significant progress in various industrial and application fields. Between 2015 and 2020, 3D printing with geopolymers saw notable developments, thanks to the refinement of printing technologies and the adaptation of inorganic materials for building applications [21]. A pioneering project in this field was the "Sustainable Housing Project" at the University of Melbourne, completed in 2015 [14]. In 2020, the American company ICON completed projects in collaboration with Habitat for Humanity, using geopolymer-based materials for housing construction. These projects are of particular significance as they address housing emergencies, with a specific focus on areas affected by natural disasters. The geopolymers used in these interventions are made from industrial waste, such as fly ash, activated by alkaline agents to produce durable and sustainable structures [22]. In 2023, the company Renca (Resilient Environmental and Natural Construction Architecture) built the world's first 3D-printed house using geopolymer mortar, located in the western United States desert, in collaboration with Strong Print 3D and Geopolymer International [23]. These projects, developed in various global regions, highlight how 3D-printable geopolymers represent an emerging technology with significant potential to promote sustainability, efficiency, and innovation in the construction industry.

3.3 Regolith-Based Construction Materials

The interest in geopolymers and alternative materials is not limited to more sustainable construction processes on Earth, but also extends to Martian and Lunar construction, where some recent studies have investigated the use of biopolymers to bind mineral materials [24]. Among alternative materials, Lunar regolith has been studied as a construction material since Apollo missions returned samples in 1969. Due to scarcity of actual lunar soil, terrestrial simulants such as JSC-1A, LHS-1, and LMS-1 have been developed to replicate its properties [25, 26]. In 2023, the Italian Space Agency announced the GLAMS project developing 3D-printable geopolymers using lunar simulants. Given the high cost of transporting materials to the Moon ($130,000/kg), regolith enables cost-effective in-situ construction [27]. Missions like GLAMS and ICON show growing interest in leveraging regolith for lunar application.

Mechanical Behavior of Regolith-Based Materials
Geopolymerized regolith composites typically have compressive strengths ranging from 25 to 40 MPa, comparable to the strength of Portland cement concrete [27]. These strength properties vary significantly depending on fabrication methods, including sintering, geopolymerization, or epoxy reinforcement. Laser-sintered regolith bricks can achieve strengths of up to 45 MPa [28] which are sufficient for lunar construction in reduced gravity. Epoxy-regolith composites exceed 100 MPa in strenght, but since they require epoxy transported from Earth, they are not fully compatible with ISRU approaches. The lunar environment itself can reduce material strength by between 8% and 70%. This trend points to the need for specialized environmental testing to ensure material reliability. Solar UV radiation causes lunar regolith to become electrostatically charged, making particles adhere strongly to each other. This chemical interaction is difficult to manage, but it might offer benefits in lunar construction prove useful in lunar construction. Methods such as solar, microwave, and laser sintering aim to convert regolith into solid forms without added binders [25].

The GLAMS project employed 3D printing using the WASP 40100 LDM platform to create geopolymer blends with foaming agents, improving both printability and strength. Recent advancements also include autonomous additive manufacturing systems like ATHLETE and REGOLIGHT. Direct ink writing (DIW) with viscosity-tuned regolith paste offers high-resolution control of shape and density. The relationship between yield stress, plastic viscosity, and printability is well demonstrated in DIW regolith-epoxy composites [29]. However, current challenges remain. Lunar regolith is different from place to place; feldspathic highlands are richer in aluminum- and plagioclase, while maria regions contain more iron- and pyroxene [30]. Bonding layers tend to adhere poorly, and maintaining stable regolith structures is difficult due to the absence of water and air. Low gravity and vacuum conditions impair heat transfer during sintering and curing, requiring process modifications. The absence of standardized simulants identical chemical and mineral compositions complicates comparisons across different studies [31, 32].

3.4 Roadmap and Future Direction

In summary, a roadmap of the most significant developments related to new materials and 3D printing is proposed (Fig. 7). The analysis of the evolution of 3D printing in the construction sector highlights how technological advancements are closely linked to the development and adoption of innovative materials.

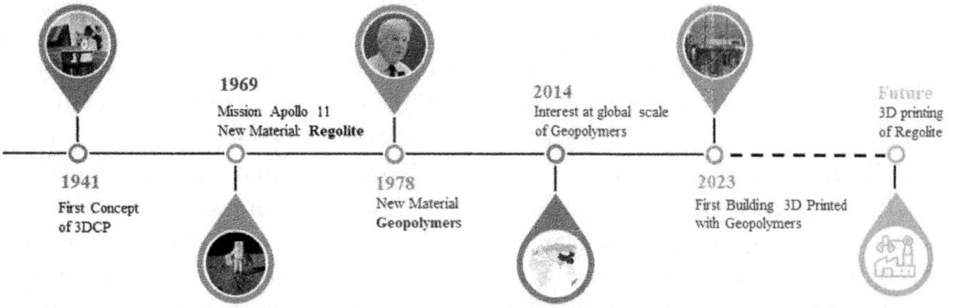

Fig. 7. Roadmap and future direction of innovative materials in 3d printing.

The roadmap highlights key historical and technological milestones in 3D construction printing:

- **Year 1941**: William E. Urschel patented one of the first extrusion machines for automated wall construction, paving the way for modern 3D printing in construction.
- **Year 1969**: During the Apollo 11 mission, lunar regolith samples were collected, showing its potential for extraterrestrial construction due to its properties.
- **Year 1978**: Joseph Davidovits introduced the term "geopolymer," identifying it as a sustainable alternative to Portland cement with thermal and chemical stability.
- **Year 2014**: Growing global interest in geopolymers for 3D printing due to environmental benefits and suitability for extreme environments.
- **Year 2023**: Renca, with Strong Print 3D and Geopolymer International, built the first 3D-printed house using geopolymer mortar in a U.S. desert, demonstrating practical use in harsh conditions.
- **Future Challenge.** The key open question is whether highly sustainable materials like regolith 3D can be effectively printed, crucial for in-situ habitat construction on the Moon and Mars to optimize local resources and reduce material transport during space missions

4 Conclusions

This study investigates the current status and future prospects of innovative materials and 3D printing technologies in construction, with a focus on applications in remote and emergency settings. It combines quantitative bibliometric analysis—using keyword co-occurrence networks and Keyword Plus methods—to identify the most researched geopolymers, and qualitative analysis to trace the development of 3D printing technologies alongside the evolution of geopolymers. The research highlights fly ash, slag, and metakaolin-based geopolymers as the primary materials studied, detailing their compressive strengths and suitability for structural use: fly ash mixtures activated with NaOH and sodium silicate show strengths of 30–70 MPa, slag-based mixtures achieve 50–80 MPa, and metakaolin-based geopolymers demonstrate the highest strengths of 40–100 MPa. The study also examines challenges related to these materials in additive manufacturing, such as variability in raw material precursors that affects mechanical and rheological consistency. Looking ahead, the research points to the need for deeper investigation into geopolymer performance to develop optimized mixtures tailored for 3D printing, with special emphasis on utilizing local in-situ resources. This approach is crucial not only for extreme terrestrial environments but also for off-Earth construction applications on the Moon and Mars, where local resource availability is essential. The use of lunar and Martian regolith emerges as a promising opportunity to build emergency housing, drastically reducing the dependency on materials shipped from Earth and enhancing self-sufficiency in long-duration space missions. Ultimately, harnessing local resources can foster the creation of resilient habitats both in space and in vulnerable areas on Earth affected by natural disasters. The proposed work has the limitation of not providing an in-depth analysis of a single material, but rather offering a broad overview intended to provide researchers and practitioners with a reference framework. Future research will build on this overview to select the most performant and sustainable materials for specific construction applications. Ultimately, a new 3D-printable construction material will be developed to be sustainable, printable, and both thermally and mechanically efficient.

Acknowledgements. The Digital Fabrication Lab of the INGEO Department at the D'Annunzio University of Chieti-Pescara is acknowledged for providing the space, equipment, and support necessary during the implementation phases of this work. This work has received financial and institutional support from the European Union through the Next Generation EU program, funded under the National Recovery and Resilience Plan (PNRR), as well as from the Ministry of Universities and Research, the funds were obtained through co-financing between the University of 'G. D'Annunzio' (Ud'A) DM 639 (24/04/2024) and the FABLAB Bitonto Association. F. Salese, C. Guth, V. Sangiorgio, C. Cantagallo are supported by MSCA Staff Exchange - ArchiSpace - funded by the European Commission, Grant no. 101183089.

References

1. Ngo, T., Kashani, A., Imbalzano, G., Nguyen, K., Hui, D.: Additive manufacturing (3D printing): a review of materials, methods, applications and challenges. Compos. B Eng. **143**, 172–196 (2018)
2. Feng, Y., Cui, S.: A review of emergency response in disasters: present and future perspectives. Nat. Hazards **105**, 1109–1138 (2021)
3. Ducan, B.P., Smith, M.A.: The practical application of additive manufacturing for extreme environments. MRS Commun. **15**, 81–89 (2025)
4. Morel, J.C., Mesbah, A., Oggero, M., Walker, P.: Building houses with local materials means to drastically reduce the environmental impact of construction. Build. Environ. **36**(10), 1119–1126 (2001)
5. Bazli, M., Ashrafi, H., Rajabipour, A., Kutay, C.: 3D printing for remote housing: benefits and challenges. Autom. Constr. **148**, 104772 (2023)
6. Buswell, R.A., De Silva, L., Jones, S.Z., Dirrenberger, J.: 3D printing using concrete extrusion: a roadmap for research. Cem. Concr. Res. **112**, 37–49 (2018)
7. Lee, S., Van Riessen, A.: A review on geopolymer technology for lunar base construction. Materials **15**(13), 4516 (2022)
8. Mendoza-Cachú, D., López-Martínez, A.A., Franco-Urquiza, E.A.: Geopolymers for space applications part II: synthesis and physical characterization. Ceramics **7**(4), 1584–1599 (2024)
9. Zhuang, Y.X., et al.: Fly ash-based geopolymer: clean production, properties and applications. J. Cleaner Prod. **125**, 253–267 (2016)

10. Niu, F.S., Zhou, S.S., Liu, S.X., Zhang, J.X.: Study of manufacture process and properties of tailings and slag based geopolymers. Adv. Mater. Res. **156–157**, 803–807 (2010)
11. Katsiki, A., Hertel, T., Tysmans, T., Pontikes, Y., Rahier, H.: Metakaolinite phosphate cementitious matrix: inorganic polymer obtained by acidic activation. Materials **12**(3), 442 (2019)
12. Cesaretti, G., Dini, E., De Kestelier, X., Colla, V., Pambaguian, L.: Building components for an outpost on the Lunar soil by means of a novel 3D printing technology. Acta Astronaut. **93**, 430–450 (2014)
13. Snyder, H.: Literature review as a research methodology: an overview and guidelines. J. Bus. Res. **104**(C), 333–339 (2019)
14. Chen, K., Liu, Q., Chen, B., Zhang, S., Ferrara, L., Li, W.: Effect of raw materials on the performance of 3D printing geopolymer a review. J. Build. Eng. **84**(May), 108501 (2024)
15. Volpe, S., Sangiorgio, V., Fiorito, F., Varum, H.: Overview of 3D construction printing and future perspectives: a review of technology, companies and research progression. Archit. Sci. Rev. **67**(1), 1–22 (2022)
16. Gul, J.Z., et al.: 3D printing for soft robotics – a review. Sci. Technol. Adv. Mater. **19**(1), 243–262 (2018)
17. Davidovits, J.: Geopolymer: inorganic polymeric new materials. J. Ther. Anal. **37**, 1633–1656 (1991)
18. Zhang, P., Gao, Z., Wang, J., Guo, J., Hu, S., Ling, Y.: Properties of fresh and hardened fly ash/slag based geopolymer concrete: a review. J. Cleaner Prod. **270**, 122389 (2020)
19. He, P., et al.: Effects of Si/Al ratio on the structure and properties of metakaolin based geopolymer. Ceram. Int. **42**(13), 14416–14422 (2016)
20. Provis, J.L., Van Deventer, J.S.J.: Geopolymers: Structures, Progessing, Proprieties and Industrial Applications. Edit Woodhead (2009)
21. Leonelli, C., Romagnoli, M.: Chemically Activated Inorganic Polymer Geopolymers. 2nd edn. Icers, Bologna (2013)
22. https://msd.unimelb.edu.au. Accessed 28 Mar 2025
23. Biggerstaff, A.O., Lepech, M.D., Loftus, D.J.: Evaluation of a biopolymer-bound soil composite for 3D printing on the lunar surface. In: Earth and Space 2021, pp. 36–51 (2021)
24. García-Alvarado, R., Moroni-Orellana, G., Perez, P.: Architectural evaluation of 3D-printed buildings. Rev. Build. **11**(6), 254 (2021)
25. https://www.3dnatives.com/it. Accessed 28 Mar 2025
26. Shazwan Suhaizan, M., Tran, P., Exner, A., Falzon, B.G.: Regolith sintering and 3D printing for lunar construction: an extensive review on recent progress. Progress Addit. Manuf. **9**, 1715–1736 (2023)
27. Zheng, X., Zhao, C., Sun, X., Dong, W.: Lunar regolith geopolymer concrete for in-situ construction of lunar bases: a review. Polymers **16**, 1582 (2024)
28. https://www.voxelmatters.com/glams-project-advances-3d-printed-lunar-regolith-structures/. Accessed 20 May 2025
29. Gupta, N., Dawara, V., Kumar, A., Viswanathan, K.: Synthetic space bricks from lunar and martian regolith via sintering (2023)
30. Bao, C., Wang, Y., Pearce, G., Zhao, P., Liu, M., Mushtag, R.T.: 3D printing of regolith-based epoxy composites with excellent temperature resistance and mechanical strength. Acta Astronaut. **229**, 787–803 (2025)
31. Zhou, C., et al.: Properties and characteristics of regolith-based materials for extraterrestrial construction. Eng. **37**, 159–181 (2024)
32. Farries, K.W., Visintin, P., Smith, S.T., Eyk, P.V.: Sintered or melted regolith for lunar construction: state-of-the-art review and future research directions. Constr. Build. Mater. **296**, 123627 (2021)

Performance of LC_3 Cement and Other Low-Clinker Cements: Impact of Normative Duality

Luis Rolando González Aguila(✉), José Fernando Martirena Hernández, and Yosvany Diaz Cárdenas

Center for Research and Development of Structures and Materials (CIDEM), Universidad Central Marta Abreu de las Villas, Santa Clara, Cuba
lgaguila@uclv.cu, fmartirena@lc3.ch

Abstract. The following study focuses on achieving harmonization of cement standards at the international level, with a trend towards the use of sustainable materials that increasingly lead to a Net Zero society. Several cements with different clinker additions were analyzed in order to evaluate their performance by compressive strength as well as the influence of fineness and workability parameters might have on them, following the procedures established in the international standards ASTM and CEN. It was evident that LC_3 cement, despite the differences found in the testing procedures, maintained always excellent quality. This study allows us to assess whether the methodological differences identified between both standards influence the physical-mechanical parameters of the cement and promotes the need for the application of this cement in international projects.

Keywords: LC_3 cement · sustainability · standards

1 Introduction

The construction industry, a key pillar of economic growth and development, faces the challenge of reducing its environmental impact, while meeting the growing demand for infrastructure. Portland cement, as the main component of concrete, is one of the most utilized materials worldwide, however, its manufacture is highly energy consuming and a major source of CO_2 emissions, accounting for 6–8% of global anthropogenic emissions. These emissions come mainly from the decarbonation of limestone and the use of fossil fuels at temperatures above 1450 °C for clinker production. The projected increase in construction needs by 2050, which may lead to an expansion in cement output, further underlines the urgency of decarbonation, in line with the Paris Agreement [1].

Reducing the cement industry's carbon footprint by 2050 calls for integrating new technological trends, particularly the partial substitution of clinker with supplementary cementitious materials (SCMs) such as fly ash, blast furnace slag, and calcined clays. The inclusion of SCMs leads to a direct reduction in clinker usage and associated CO_2 emissions and may enhance concrete's durability and mechanical performance in the long term [2].

M. Posani and S. Singh (Eds.): RYS 2025, RILEM Bookseries 67, pp. 320–333, 2026.
https://doi.org/10.1007/978-3-032-14652-6_25

Among SCMs, calcined clays and especially metakaolin-rich clays stand out for their high reactivity and abundant global availability. Their effectiveness is rooted in the pozzolanic reaction between the silica(SiO_2) and alumina (Al_2O_3) in calcined clay (activated between 700–800 °C) and the calcium hydroxide ($Ca(OH)_2$) originating from clinker hydration, creating additional phases like hydrated calcium silicates (C-S-H)and hydrated calcium aluminates (C-A-H), that densify the microstructure, lower porosity which boost mechanical strength, and enhance durability [3, 4].

The ternary Limestone Calcined Clay Cement (LC_3) is a low-clinker cement that harnesses the synergy of calcined clay (30%), limestone (15%), and a reduced amount of clinker (50%), enabling a reduction of nearly 40% in CO_2 emissions without sacrificing strength with similar or superior performances to traditional Portland cement [5],the chemistry of LC_3 involves complex interactions: metakaolin reacts with portlandite and carbonate to form additional hydrated silicates and stable carboaluminates, enhancing the properties of mortars and concretes.

Despite their clear benefits, the regulatory adoption of such cements remains slow. Standards, though essential for ensuring quality and safety, often lag behind technological innovations, especially when prescriptive approaches mandate specific compositions or testing methods. This creates a regulatory inertia that hinders adoption [6].

Recently, there has been a push toward performance-based standards, which set requirements for performance properties rather than fixed proportions, facilitating innovation and allowing the use of low-clinker cements, such as LC_3.

1.1 Global Standardization and Barriers to the Adoption of New Standards

Standardization underpins quality, safety, and international interoperability in the cement sector [7]. Yet, there is a persistent need for greater flexibility such as that afforded by ASTM C1157 [8], which focuses on performance rather than composition. While prescriptive standards like ASTM C595 [9] (for blended hydraulic cements, ASTM C618 [10] (for coal fly ash and raw or calcined natural pozzolan for use in concrete), and traditional ASTM C150 [11] (for Portland cement), specify permissible raw materials and formulations. Hooton [12] highlights that despite their introduction, and the existing historical record which corroborates their performance with ASTM C595 [9] and C150 [11] specifications. Usage of performance specifications is limited by user/owner concerns over covering all long-term durability aspects, suggesting that resistance to change is more cultural and economic than technical [13, 14].

The European standards (EN 197-1 [15] (for common cements) and EN 197-5 [16] (for Portland-composite cements CEM II/C-M and composite cements CEM VI)) have gradually expanded to include a greater diversity of cements but stablish rigid compositional limits. This duality between prescriptive and performance-based approaches particularly between ASTM and EN frameworks creates challenges for the acceptance and diffusion of innovative, more sustainable cements on a global scale [17].

Performance-based standards specify targets to be achieved or functions to be provided, leaving the means to achieve them to the market. In other words, the composition will not be prescribed, but it will specify in generic terms how a cement should behave when exposed to certain conditions of use over its lifetime to meet the needs of different users. This allows manufacturers to innovate in composition as long as the final product meets the established performance criteria. On the one hand, prescriptive specification of cement constituents has the advantage of simplifying the process, including quality control, as the composition is known and predictable. On the other hand, the introduction of innovations in cement composition nowadays requires modification of prescriptive standards, a process that demands a long period of time, consensus among multiple stakeholders (manufacturers, regulators, users), and ultimately delays the effective introduction of new products into the market [12, 18]. This gap between technological innovation and regulatory adaptation is a significant barrier to decarbonation.

Internationally, there are significant differences in Portland cement standards, influenced by factors such as raw material availability, regional climatic conditions, and technical aspects of each region. Examples cited by John et al. [17] include discrepancies in the maximum amount of limestone allowed as a filler in standards (e.g., some regions allow a higher percentage of limestone than others), as well as strict limits on constituent substitution in a Portland cement mix. Furthermore, obsolete test methods must be replaced by new methods, such as calorimetry (which allows monitoring cement hydration kinetics) or accelerated carbonation (which evaluates long-term durability), although their adoption is challenging due to the investment in equipment and necessary training. The inclusion of criteria such as carbon capture would also facilitate the use of calcined clay, despite its demonstrated technical success over the years, by recognizing its contribution to emission reduction.

As expressed by Hooton [12]at the ACI concrete convention, there is an urgent need to learn to use performance-based testing. Although ASTM C1157 was adopted in 1992 and even included in the ACI 318 code (Building Code Requirements for Structural Concrete), only a small number of North American states have adopted it. Company owners not only want compliance with performance nevertheless usually determine the constituents of the material based on traditional practices, which limits the flexibility to incorporate innovations, because regulatory restrictions can lead to an unnecessary increase in CO2 emissions harmful to the environment and society, by not allowing the use of more sustainable cements. Therefore, modifying standard specifications to include new blends is a complex process, which may be unjustified due to the existing variety of cements, but the development of new performance-based standards could represent an opportunity to change the governance of the cement sector by updating test methods, harmonizing global standards, and integrating climate policies into regulations.[19].

Currently, the use of low-carbon or low-clinker cements in construction is hampered by the lack of coverage in the proper standards, which can lead to certain prohibitions or the need for special approvals. Although some standards such as ACI 318-19 allow the use of SCMs, the requirements are based on data from Portland cement and traditional SCMs, which makes the implementation of novel materials difficult. However, most current standards such as ASTM C1157 [8]and EN 197-5 [16] allow the use of more traditional cementitious materials such as fly ash, slag, and pozzolans, which facilitates their application and supports the use of LC3 as a ternary CEM II/C-M cement in EN 196-6 [20] with limitations on clinker content (50–64%).

This paper compares the differences established in the ASTM and CEN test methods related to performance-based parameters such as compressive strength and fineness, with the aim of evaluating the future behavior of low-clinker cements such as LC3 in the face of this normative duality and proposing ways for greater harmonization.

2 Materials and Methods

The experimental program was designed to evaluate the physical and mechanical properties of various cements, considering performance specifications such as compressive strength and fineness of grinding of the different binders used. Five types of cement, representative of current formulations and innovations in the sector, were selected for a comparative analysis:

Portland Cement (CEM I 42.5): This cement, with a strength class of 42.5 MPa, was used as a reference material due to its widespread use and its role as the basis for most existing standards. Various curing ages were evaluated at the facilities of the Carlos Marx Cement Factory, which allowed obtaining performance data under real production conditions.

Pozzolanic Portland Cement (CEM IV/A 42.5): This cement with a strength class of 42,5 contains natural pozzolans and clinker content less than 85%.

Limestone Portland Cement (CEM II A/LL 32.5) This cement with a strength class of 32,5 contains limestone additions, Limestone can act as an inert filler or participate in reactions with clinker aluminates to form carboaluminates, contributing to microstructure densification and early strength.

Limestone Portland Cement (CEM II B/L) This cement with a strength class of 25 MPa according to the producer contains higher limestone additions, reaching 26% of limestone additions.

LC3 Formulations: These formulations represent the forefront in low-clinker cements, highlighting the potential of calcined clays, even low-grade clays. For both LC3 formulations, kaolinitic clay with an approximate 37.5% kaolin equivalent content [21] was used. Kaolin is a clay mineral that, when calcined at specific temperatures, transforms into metakaolin, a highly reactive pozzolan and the suitability of kaolinitic clays as SCMs is established for clays that have more than 18%. [22], the clay used for each formulation has 19.68% see Table 1.

LC3-50: This structural cement, with 50% clinker, 30% calcined clay and 15% limestone, plus gypsum was designed to offer performance comparable to traditional Portland cement (CEM I type) in terms of strength. All supplementary materials (were used as partial replacements for Portland clinker by mass).

LC3-35: A cement oriented towards masonry applications, with a higher calcined clay content and lower clinker content, makes it suitable for non-structural or low-strength elements, maximizing CO2 emission reduction.

The average values of each series of fracture tests performed in the mortar press were recorded and used for statistical analysis, ensuring data reliability.

Table 1. Chemical composition of the raw clay by XRF

%SiO2	%Al	%Fe	%Ca	%Mg	PPI
34.48	19.53	8.9	8.93	0	16.23
34.23	20.14	8.9	9.84	0	16.58
34.26	20.66	8.9	9.2	0	16.11
35.18	17.84	8.14	9.28	0	18.46
34.13	20.17	8.08	9.28	0	17.09

2.1 Determination of Cement Fineness

Fineness is a crucial parameter that directly influences the kinetics of cement hydration, its workability, and, consequently, its early strength and long-term property development. Higher fineness generally implies a larger specific surface area available for reaction with water, which accelerates the hydration process. For this research, fineness was determined using two standardized testing methodologies, and the same size of mesh was used as a reference slightly from the standard, and following a direct comparison of results under different normative criteria:

ASTM Method (ASTM C1891 [23]): This method, titled "Test Method for Fineness of Hydraulic Cement by Air Jet Sieving at 45-μm (No. 325)", based on the principle of air-jet sieving. An EOLO FTBLA series vacuum air jet sieve, equipped with a 90 μm mesh, was used. 10 g of cement was placed on the sieve, and the test was carried out until the mass retained on the sieve remained constant, indicating that no more material passed through the mesh. This method is particularly effective for separating fine particles and preventing agglomeration, which can occur in mechanical sieving. Air pressure and test time were strictly controlled to ensure reproducibility of results.

EN Method (EN 196-6 [20]]): This method, part of the European standard "Methods of testing cement - Part 6: Determination of fineness; EN 196-6:2018", also uses an air-jet sieve, but with a 90 μm mesh. In this case, 25 g of cement was weighed. After 5 min of continuous sieving, the mass of the residue retained on the sieve was carefully determined with an accuracy of 0.01 g. This method is more common in Europe and is often associated with a larger mesh opening, which may result in slightly different fineness values compared to the ASTM method, even for the same material. The difference in sample mass and test time are also key factors distinguishing both procedures.

Comparing the results obtained by both methods is crucial to understanding how normative variations can influence the classification and acceptance of cements internationally.

2.2 Determination of Cement Compressive Strength

Compressive strength is one of the fundamental and most widely accepted parameters for determining the quality and performance of cements. This test provides a direct measure of the mortar's or concrete's ability to withstand loads, which is essential for structural design. For this study, two compressive strength testing methodologies were applied, allowing a comparative evaluation under different normative frameworks:

ASTM Method (ASTM C109 [24]): This method, "Standard Test Method for Compressive Strength of Hydraulic Cement Mortars (Using 2-in. or [50-mm] Cube Specimens)", uses 50 mm cube specimens. For mortar preparation, standard graded sand meeting the requirements of specification C778 [25] was used, ensuring uniform particle size distribution and composition. Laboratory environmental conditions were strictly controlled: temperature at (23 ± 3) °C and relative humidity above 50%, to minimize variability in cement hydration. The mix consisted of 2.75 parts sand to one part cement, with a water-cement ratio of 0.485 for pure Portland cement. For other cements, the w/c ratio was adjusted according to a pre-established consistency table, based on the normal consistency test, to ensure uniform workability of the paste. The molding and compaction process was manual, performed in mortar layers approximately 25 mm thick (approximately half the mold depth) in all compartments of the molds. In each compartment, tamping was performed 32 times in approximately 10 s, distributed in 4 rounds, each at right angles to the other and consisting of eight contiguous passes over the sample surface. Once the tamping of the first layer in all cube compartments was complete, the compartments were filled with the remaining mortar and then tamped in the same manner. Specimens were cured in lime-saturated water until testing ages. Compressive strength was determined at 24 h, 48 h, 3 days, 7 days, and 28 days, providing a strength development curve over time.

ASTM C348 [26] ASTM C349 [27] are methods cover by ASTM to evaluate the flexural and compressive strength of hydraulic-cement mortars using portions of prisms broken in flexure. First, under method C 348, a standard mortar (1 part cement and 2.75 parts sand by mass, with a specific flow) is prepared, and 4 × 4 × 16 cm prismatic specimens are molded in triple-gang molds, compacted in two layers, and cured for one day in the molds before demolding. Flexural strength is determined by applying a load to the center of the specimen until fracture. Immediately after the flexural break, or within a specific time (10 min for 24 h specimens, 30 min for others), both broken portions of each prism are used for the compressive strength test according to method C 349. For this, the portions are placed in an aligning device with specific bearing plates (40.0 ± 0.1 mm by 50.8 mm) in a hydraulic testing machine, applying the load until failure. The compressive strength values obtained by this method are for reference purposes only and do not substitute cement acceptance values obtained with 50 mm cubes according to method C 109/C 109M [24].

EN Method (EN 196-1 [28]): This method, "Methods of testing cement - Part 1: Determination of strength", uses 4 × 4 × 16 cm prismatic specimens. These specimens are first tested for flexural strength, and the two resulting halves are then used for the compression test. The mix consisted of three parts standard sand with 450 g of cement for a series of three standard mortar specimens. The mixing operation was performed automatically using a standard mortar mixer, which ensures greater homogeneity. Once mixing was complete, molding was carried out using a shock or vibrating table that was raised vertically and fell freely 60 times in 60 s, compacting the mortar uniformly. First, a layer of mortar was placed and then compacted. At the end of the setting time (generally 24 h), a second layer was placed and the previous process was repeated. The surface was then leveled, and the mold was placed in a curing chamber at a uniform temperature and controlled humidity. After 24 h, the specimens were demolded and left to cure in water until the testing age. The Cuban standard NC 506 uses the same method for the production of 4 × 4 × 16 cm prismatic specimens as described in EN 196-1. However, unlike EN 197-1, NC 1340 [23] establishes that the testing times must be 3, 7, and 28 days. Therefore, a combination of the testing ages established in both standards (48 h, 3, 7, and 28 days) was used to obtain a more complete view of mortar behavior over time and facilitate comparison.

Differences in specimen geometry (cubes vs. prisms), compaction methods (manual vs. vibrating table), and sand specifications can significantly influence compressive strength results, making direct comparison between the two normative systems a challenge and a key point of this study.

3 Results and Discussion

3.1 Fineness of Cements

The evaluation of cement fineness, performed using both testing methodologies (ASTM C1891 and EN 196-6), revealed interesting patterns and subtle differences (Fig. 1). It was observed that LC3 cements, in general, exhibited slightly higher fineness than traditional Portland cement, link to the grinding process. This finding is consistent with the nature of LC3 components; calcined clay and limestone typically require finer grinding to maximize their reactivity and effectively contribute to strength development. Despite this higher fineness, all evaluated cements, including LC3, remained within the permissible range established by relevant standards for their respective classes.

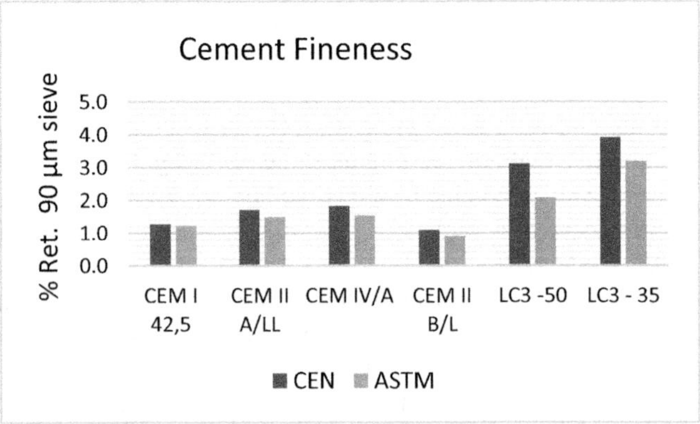

Fig. 1. Fineness of cements, % Retained sieve.

The variations observed between the fineness values determined by both standards were not statistically significant in most cases. However, a consistent trend was identified: the fineness obtained by the ASTM method was slightly lower than that obtained by the EN method, the difference in the amount of material introduced into the sieve (10 g for ASTM vs. 25 g for EN) and the testing principle (air-jet sieving, which prevents agglomeration of fine particles more efficiently) can also influence the results.

Higher fineness, regardless of the method, generally implies a larger specific surface area of the cement, which can lead to faster strength gain at early ages, as well as improved workability of the cement paste. The consistency in LC3 fineness, despite methodological differences, suggests acceptable quality in the grinding process but improvable.

3.2 Compressive Strength

The experimental study of mortar compressive strength, using both ASTM and EN methods, provided valuable information about the performance of different cement types and the influence of normative duality. An initial analysis, which compared the performance of three different batches of pure Portland cement (CEM I 42.5), with their corresponding standards, demonstrated a significant difference between the strength requirements of both standards, as shown in Fig. 2a), 2b) and 2c).

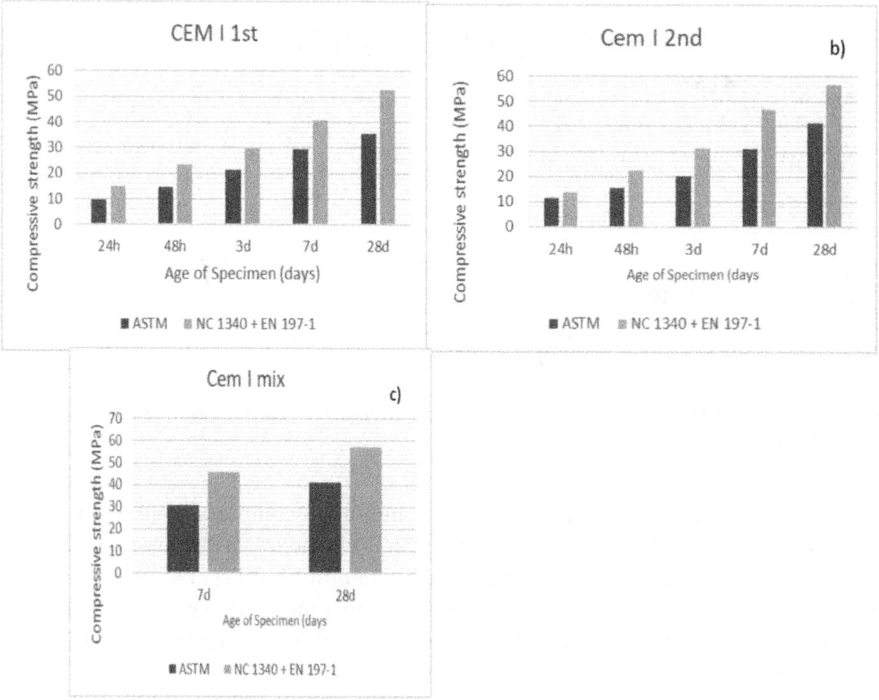

Fig. 2. Compressive strength at different ages using ASTM and EN methods a) Cement I (first batch), b) Cement I (second batch), c) Composite cement I.

The results showed that all specimens Fig. 3a) b), regardless of cement type or testing method, met the minimum compressive strength requirements established by both standards. This is a crucial finding, appreciable in Fig. 4 as it validates the ability of low-clinker cements, such as LC3, to meet the performance levels required by current regulations.

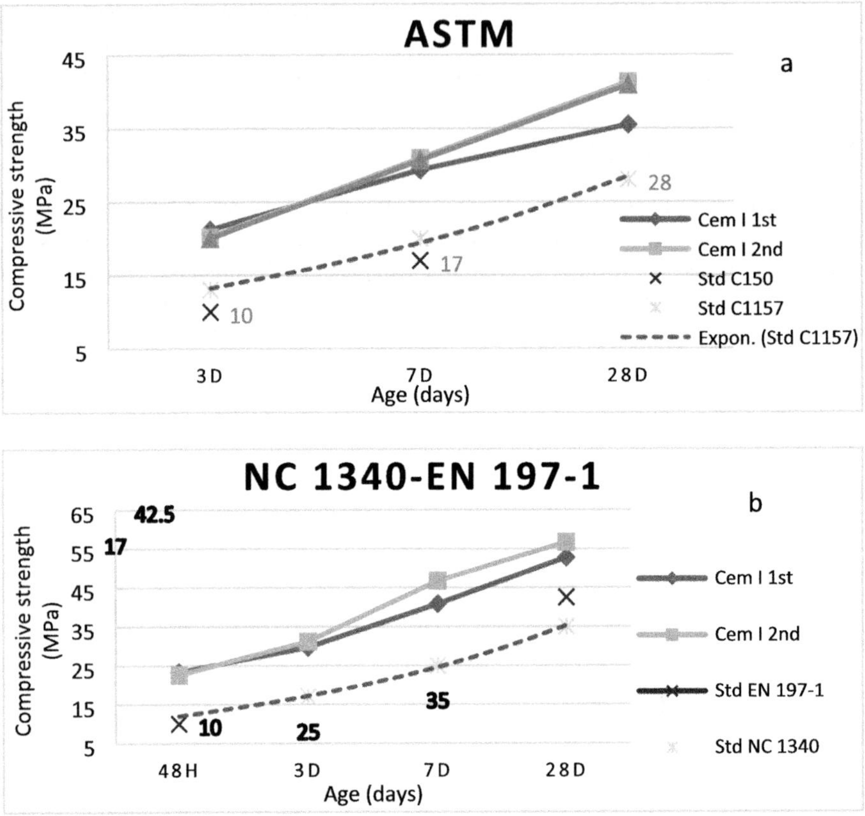

Fig. 3. a) Comparison of compressive strength requirements according to methods described in a) ASTM and b) NC1340 and EN 197-1.

A more detailed analysis revealed that the minimum compressive strength requirements are lower in ASTM standards compared to EN standards, which is consistent with other studies [29]. This difference in acceptable minimum values can have significant implications for the classification and use of cements internationally. The reasons for this disparity are multifaceted and are likely related to methodological differences in the preparation of specimens and the constituent materials:

- Influence of sand: ASTM and EN standards specify standardized sands with different particle size distributions and compositions. The sand used in the ASTM C109 method (standard graded sand) may have characteristics that influence the packing density and the formation of the mortar matrix differently from the sand used in EN 196-1. These differences can affect the final strength of the mortar.

- Compaction method: The ASTM C109 method uses manual tamping, while the EN 196-1 method uses a jolting or vibrating table for compaction. Manual compaction can introduce greater variability and a lower compaction density compared to mechanical compaction, which could result in lower strengths. More efficient compaction reduces the porosity of the mortar, which leads to higher strength.
- Specimen geometry: The compression tests on cubes (ASTM) and prisms (EN) also introduce differences. 50 mm cubes tend to show slightly lower compressive strengths than 4 × 4 × 16 cm prisms (tested in halves after flexure) due to confinement effects and the height/width ratio.

Despite these methodological differences, a strong correlation was observed between the compressive strength results obtained by the ASTM and EN methods, with a correlation coefficient $r = 0.95$. This high value indicates a robust linear relationship between the results of both normative systems: a cement that shows high strength with one method will generally also show high strength with the other. This suggests that, although the absolute values may differ, the relative ability of cements to develop strength is consistent across both standards.

However, the quantitative relationship revealed that the compressive strength values obtained with the ASTM method represented an average of 77.6 ± 9.3% of the values reported under the EN method. This percentage difference is significant and demonstrates the prevalence of the European standard's strength requirements. This implies that a cement that barely meets the ASTM minimums might not meet those of EN, or that a cement designed to meet EN will far exceed the requirements of ASTM. This disparity hampers the adoption of cements specifically developed to meet ASTM specifications in markets operating under EN regulations, and vice versa, creating a technical and commercial barrier.

A final comparison of several cements applying ASTM C349 vs EN 196-1 standards show similar tendencies Fig. 4. As in the previous case, Portland cement and Pozzolanic Portland cement exceeded the minimum compressive strength requirements. In the case of LC3 cements, the two formulations showed a promise performance:

- LC3-50: Fabricated as a structural cement with 50% clinker, showed strengths comparable to the reference Portland cement, which validates its potential as a sustainable alternative for high strength applications. The contribution of calcined clay and limestone to the densification of the microstructure and the formation of additional hydrated phases explains this performance.
- LC3-35: Designed for masonry applications with a lower clinker content and a higher clay content, this cement also met the requirements for masonry, demonstrating its viability for lower-strength applications where emission reduction is a priority.

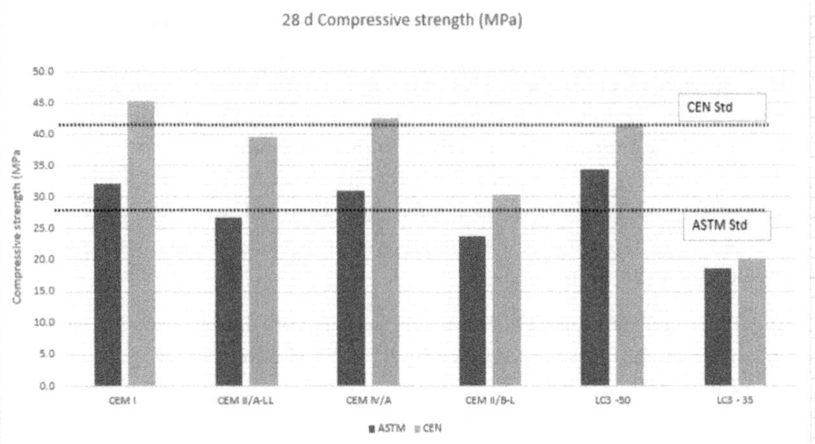

Fig. 4. Comparison of compressive strength of cements using 40x40x160 mm samples.

The kaolin content in the calcined clays used in the LC3 formulations was consistent with the strength values obtained. A higher reactivity of metakaolin contributes to a greater formation of C-S-H and C-A-H, which translates directly into higher strength. The consistency in the performance of LC3, despite the normative duality, highlights its potential to be a globally applicable cementitious material.

3.3 Implications of Normative Duality

The coexistence of different normative systems (ASTM and EN) and the significant variations in requirements and test methods pose considerable challenges for the global cement and construction industry. The 22.4% difference in average strength values between ASTM and EN is not trivial and has important practical implications:

- International Trade: A cement produced and certified under one standard (e.g., ASTM) may not be automatically accepted in a market that requires certification under another standard (e.g., EN), even if its intrinsic performance is similar or superior. This forces manufacturers to perform duplicate tests or to produce specific formulations for different markets, which increases costs and supply chain complexity.
- Innovation and Adoption: The introduction of new materials such as LC3 is hampered by the need to demonstrate conformity with multiple sets of standards, each with its own specificities. This slows down the research, development, and commercialization process, delaying the adoption of more sustainable solutions.
- Structural Design: Engineers and designers must be aware of the differences in strength requirements between standards when specifying materials. A design based on a cement that complies with ASTM standards might require adjustments if the project is executed in a region with stricter EN standards, or vice versa.
- Quality Perception: The superiority of the European standard's (EN) strength requirements can create a perception of higher quality for cements certified under this standard, which could influence purchasing decisions and market confidence. However,

this does not necessarily imply that ASTM cements are of lower quality, but that the evaluation criteria are different.

To overcome this duality, a concerted effort towards international harmonization of standards is required. This does not mean eliminating one standard in favor of another, but identifying common performance criteria and accepting equivalent test methods. Fostering performance-based standards, which focus on the final properties of the material rather than its specific composition, is a fundamental step in this direction. This would allow manufacturers to innovate and develop more sustainable cements while ensuring that the final products meet the required levels of safety and durability anywhere in the world.

4 Conclusions

This study has provided a preliminary evaluation of the performance of several cements, including innovative LC3 formulations, under the standard ASTM and EN procedures, revealing differences between both normative frameworks.

1. The production of ternary cements formulated under current conditions from calcined clays and limestone allows for a substantial reduction in clinker content and, consequently, in CO_2 emissions. These cements effectively contribute to environmental sustainability, while offering competent performance requirements that make them viable for construction applications.
2. Despite achieving slightly superior fineness, LC3 cements demonstrated properties equivalent to Portland cement, regardless of the standard used for evaluation. It is expected that if fineness equivalent to Portland cement and a kaolin content greater than 40% were achieved, strengths even superior to those achieved in Portland cement mortars could be obtained, opening new possibilities for optimizing LC3 performance.
3. A preliminary performance evaluation of several cements carried out using the standard ASTM and EN procedures, showing a strong linear correlation with a coefficient $r = 0.95$. This high coefficient indicates that a cement with high compressive strength values according to the ASTM methodology tend to remain high in the EN standards, underscoring the need for greater normative harmonization to facilitate the global implementation of more sustainable cements.

Acknowledgements. The authors wish to express their sincere gratitude to the specialists and quality personnel of the cement laboratory at the Karl Marx Cement Factory for their invaluable attention and assistance in carrying out a part of the experimental studies for this work. Their collaboration was fundamental to obtaining the data presented.

References

1. Verschaeve, J.: Moving to performance-based cement and concrete standards in Europe. Brussel (2024)

2. Beushausen, H., Torrent, R., Alexander, M.G.: Performance-based approaches for concrete durability: state of the art and future research needs. Cem. Concr. Res. **119**, 11–20 (2019). https://doi.org/10.1016/j.cemconres.2019.01.003
3. Taylor, H.F.W.: Cement Chemistry 2nd edition. Thomas Telford Publishing, Thomas Telford Services Ltd., 1 Heron Quay, London E144JD (1997)
4. Cancio Díaz, Y., et al.: Limestone calcined clay cement as a low-carbon solution to meet expanding cement demand in emerging economies. Dev. Eng. **2**, 82–91 (2017). https://doi.org/10.1016/j.deveng.2017.06.001
5. LC3 Project: https://lc3.ch/
6. Scrivener, K., Izoret, L.: Cement performance-based standards
7. European Standards - CEN-CENELEC 2. https://www.cencenelec.eu/european-standardization/european-standards/. Accessed 24 May 2025
8. Performance specification for hydraulic cement. ASTM International, West Conshohocken, PA (2023). https://doi.org/10.1520/C1157_C1157M-23
9. ASTM: Specification for blended hydraulic cements. ASTM International, West Conshohocken, PA (2014). https://doi.org/10.1520/C0595_C0595M-14
10. Specification for coal fly ash and raw or calcined natural Pozzolan for use in concrete. ASTM International, West Conshohocken, PA (2019). https://doi.org/10.1520/C0618-19
11. ASTM, C150 standard specification for Portland cement 1. www.astm.or
12. Hooton, R.D.: Changes to ASTM cement and concrete standards to facilitate sustainable initiatives (2023)
13. Sutter, L.L., Hooton, R.D.: Progress towards sustainability through performance-based standards and specifications. Cem. Concr. Res. **174**, 107303 (2023). https://doi.org/10.1016/J.CEMCONRES.2023.107303
14. Hooton, R.D.: Bridging the gap between research and standards. Cem. Concr. Res. **38**(2), 247–258 (2008). https://doi.org/10.1016/j.cemconres.2007.09.012
15. Cement – Part 1: Composition, specifications and conformity criteria for common cements; German version EN 197-1:2011 (2021)
16. Cement – Part 5: Portland-composite cement CEM II/C-M and Composite cement CEM VI; German version EN 197-5:2021 (2021). www.din.de
17. John, V.M., Quattrone, M., Abrão, P.C.R.A., Cardoso, F.A.: Rethinking cement standards: opportunities for a better future (2019). https://doi.org/10.1016/j.cemconres.2019.105832
18. Douglas Hooton, R.: Current developments and future needs in standards for cementitious materials (2015). https://doi.org/10.1016/j.cemconres.2015.05.022
19. Kanavaris, F., et al.: Standardization of low clinker cements containing calcined clay and limestone: a review by RILEM TC-282 CCL. Mater. Struct. Materiaux et Construct. **56**(9) (2023). https://doi.org/10.1617/s11527-023-02257-y
20. Methods of testing cement – Part 6: Determination of fineness; German version EN 196-6:2018 (2018)
21. Avet, F., Scrivener, K.: Simple and reliable quantification of kaolinite in clay using an oven and a balance. RILEM Bookseries (2020). https://doi.org/10.1007/978-981-15-2806-4_17
22. Díaz, A.A., Almenares Reyes, R.S., Carratalá, F.A., Martirena Hernández, J.F.: Proposal of a methodology for the preliminary assessment of kaolinitic clay deposits as a source of SCMs. In: RILEM Bookseries, Springer Netherlands, pp. 29–34 (2018). https://doi.org/10.1007/978-94-024-1207-9_5
23. Test method for fineness of hydraulic cement by air jet sieving at 45-m (No. 325), Jan. 15, 2020, ASTM International, West Conshohocken, PA. https://doi.org/10.1520/C1891-20
24. Standard test method for compressive strength of hydraulic cement mortars (using 2-in. or [50-mm] cube specimens) 1. www.astm.org,
25. Standard specification for standard sand C778. www.astm.org

26. c348 standard test method for flexural strength of hydraulic-cement mortars. www.astm.org
27. C349 standard test method for compressive strength of hydraulic-cement mortars (Using Portions of Prisms Broken in Flexure) 1. www.astm.org
28. DIN-Normenausschuss Bauwesen (NABau) Datum/Uhrzeit des Ausdrucks: Methods of testing cement – Part 1: Determination of strength; EN 196-1:2016 (2016)
29. Siddique, M., Islam, M., Halder, U.: A comparative study on physical properties of cement: ASTM and BDS EN standards, p. 2016. https://consensus.app/papers/a-comparative-study-on-physical-properties-of-cement-astm-siddique-islam/9f7646a3d2a9588790f9df396d350a0d/

Predictive Modeling of Consistency Class Using Machine Learning Techniques on Industrial Concrete Datasets

Abdelhamid Hafidi[1,2,3], Benoit Hilloulin[1], Ilhame Harbouz[4(✉)], Ahmed Loukili[1,2], and Ammar Yahia[3,5]

[1] Nantes Université, Ecole Centrale Nantes, CNRS, GeM, UMR 6183, F-44000 Nantes, France
[2] Edycem, 3 Rue Nicolas Sadi Carnot, 85600 Montaigu-Vendée, France
[3] Department of Building and Civil Engineering, Université de Sherbrooke, Sherbrooke J1K 2R1, Québec, Canada
[4] Green Tech Institute (GTI), Mohammed VI Polytechnic University (UM6P), Benguerir 43150, Morocco
Ilhame.harbouz@um6p.ma
[5] Department of Architecture, Canadian Dubai University, Dubai, UAE

Abstract. In this study, we applied machine learning techniques to classify concrete into consistency classes, using real-world data provided by an industrial partner in France. The dataset includes over 480 different mixtures and more than 1,515 concrete samples, most of which incorporate supplementary cementitious materials (SCMs) such as slag and fillers. There are twelve input features and one output representing each of the five consistency classes. The machine learning models demonstrated strong performance, with the Light Gradient Boosting Machine emerging as the best performer, with an F1-score exceeding 88%. Using Shapley Additive Explanations, we identified the water-to-binder ratio, admixture dosage, and binder content as the most influential factors affecting the slump. Overall, this research underscores the potential of machine learning to enhance concrete mix design and promote more sustainable construction practices, though future work is needed to validate the models across diverse regions, raw materials, and mix designs to ensure broader applicability and robustness.

Keywords: Machine Learning · Consistency Class · Supplementary Cementitious Materials · Shapley Additive Explanations

1 Introduction

Concrete is a complex material composed of coarse and fine aggregates, water, cement, and various admixtures [1–3]. The incorporation of supplementary cementitious materials (SCMs) such as fly ash, silica fume, and mineral additives enhances both the performance and strength of concrete [4–8]. Many of these SCMs are industrial byproducts, which present significant environmental challenges if not properly used [9–11].

Understanding fresh-state properties, such as slump and slump flow, is essential for achieving desired performance in concrete applications. Accurate prediction of these

M. Posani and S. Singh (Eds.): RYS 2025, RILEM Bookseries 67, pp. 334–347, 2026.
https://doi.org/10.1007/978-3-032-14652-6_26

properties is crucial to meeting design specifications. In this context, machine learning (ML) has emerged as a powerful tool.

ML, a subset of Artificial Intelligence, develops algorithms that learn and improve from historical data without requiring explicit programming [12, 13]. This makes ML particularly effective for predicting concrete properties and identifying structural issues—areas where traditional algorithms often fall short [14, 15].

Among these properties, slump is one of the most widely used indicators for assessing workability due to its strong correlation with other rheological characteristics of concrete [16, 17]. Initially, slump prediction relied on analytical, empirical, and semi-empirical models to estimate workability [18–20]. However, large discrepancies between model predictions and experimental results were often observed.

With the use of advanced models such as Artificial Neural Networks (ANN) and tree-based algorithms, the slump prediction has significantly improved, addressing the complex multi-variable nature of concrete mixtures. Chen et al. [21] found that ANN models outperformed traditional Multiple Linear Regression (MLR), yielding lower prediction errors. Yeh [22] further highlighted the nonlinear relationships between concrete components and slump, which can be effectively analyzed through response trace plots.

Recent studies have continued to explore and validate ML applications in slump prediction. Zhang et al. [23] compared seven ML models and identified XGBoost and Random Forest as the most accurate for industrial concrete mixtures. Similarly, Nadimalla et al. [24] proposed a ML model for predicting both strength and workability, with Decision Tree models outperforming Multi-Layer Regression in slump prediction. However, Multi-Layer Regression still performed well in estimating the compaction factor, a key indicator of concrete consistency [25]. These advancements affirm that machine learning provides a robust framework for improving the predictive accuracy of concrete fresh properties, offering greater reliability and precision than traditional methods.

The primary distinction between this study and the referenced literature lies in the nature and the length of the dataset. This research relies on one year (2023) of industrial data, which includes a broad range of concrete formulations incorporating various cement types, concrete strength classes, and two supplementary cementitious materials (SCMs): slag and fillers. A significant challenge arises from the variability introduced by the uncontrolled conditions typical of on-site testing. Nevertheless, the detailed documentation provided by the industrial partner ensured that all relevant information was available, enabling accurate selection and analysis of the concrete formulations used in this study.

2 Methodology

The methodology for predicting concrete consistency using ML is presented in Fig. 1 and structured into four main steps:

(I) **Database Creation:** A comprehensive dataset comprising 1,515 concrete samples was compiled. The input features represent various mixture parameters, while the output corresponds to the classified concrete consistency.

(II) **Data Preprocessing:** The dataset was cleaned (by removing the missing values) and standardized to ensure consistent formatting and reliable feature scaling. Categorical variables were encoded using the LabelEncoder from scikit-learn.

(III) **Modelling:** Four ML algorithms - Random Forest (RF), Decision Tree (DT), eXtreme Gradient Boosting (XGB), and LightGBM (LGBM) - were trained on 80% of the dataset, with the remaining 20% used for testing. Model performance was evaluated using the weighted F1-score, precision, and recall to address class imbalance. The model with the highest performance across these metrics was selected for further analysis.

(IV) **Analysis:** The predictive performance of the models was visualized using heatmaps. Model interpretability was achieved through SHAP (SHapley Additive exPlanations), which identified the most influential features affecting the consistency class predictions.

2.1 Database Description

Table 1 provides a statistical description of the dataset's input and output parameters. The dataset includes twelve variables: eleven input features related to mix design, and one output variable representing concrete consistency.

Cement content ranges from 58 to 450 kg/m^3, with the lower values reflecting high substitution rates of cement with supplementary cementitious materials (SCMs). The average binder content (cement + SCMs) is 313.2 kg/m^3.

The dataset includes two types of SCMs: slag and fillers, used individually or in combination. These SCMs are encoded as categorical variables, and their usage varies significantly, with substitution rates ranging from 0% to 84.94%, highlighting the wide variability in SCM utilization.

The water-to-binder ratio (w/b) ranges from 0.33 to 0.81, with a mean of 0.56. Aggregate content also shows substantial variation: fine aggregates range from 280 to 1360 kg/m^3, and coarse aggregates from 640 to 1615 kg/m^3. The admixture dosage varies from 0 to 8.46 l.

The dataset includes two additional categorical features: concrete strength class (ranging from C16/20 to C50/60) and cement type (from CEM I to CEM V) which are also encoded for use in machine learning models.

The target variable—concrete consistency—is classified into five categories according to standard specifications [26]. Fig. 2 illustrates the distribution of the inputs, highlighting the diversity of the mixtures. This heterogeneity reinforces the relevance of using ML to develop robust and generalizable predictive models.

As shown in Fig. 3, the S3 consistency class is the most prevalent, which aligns with industrial norms where S3 is often the preferred and validated consistency class for production. To address this class imbalance, class weights were applied during model training and evaluation. Additionally, stratified data splitting was employed to maintain proportional class representation in both training and test datasets.

Figure 4 presents the correlation matrix, offering a visual representation of the degree of correlation among all the parameters. Linear regression is employed to adjust one variable relative to another based on an affine relationship [27]. The strongest correlation is observed between the SCM percentage and cement content, which is expected since

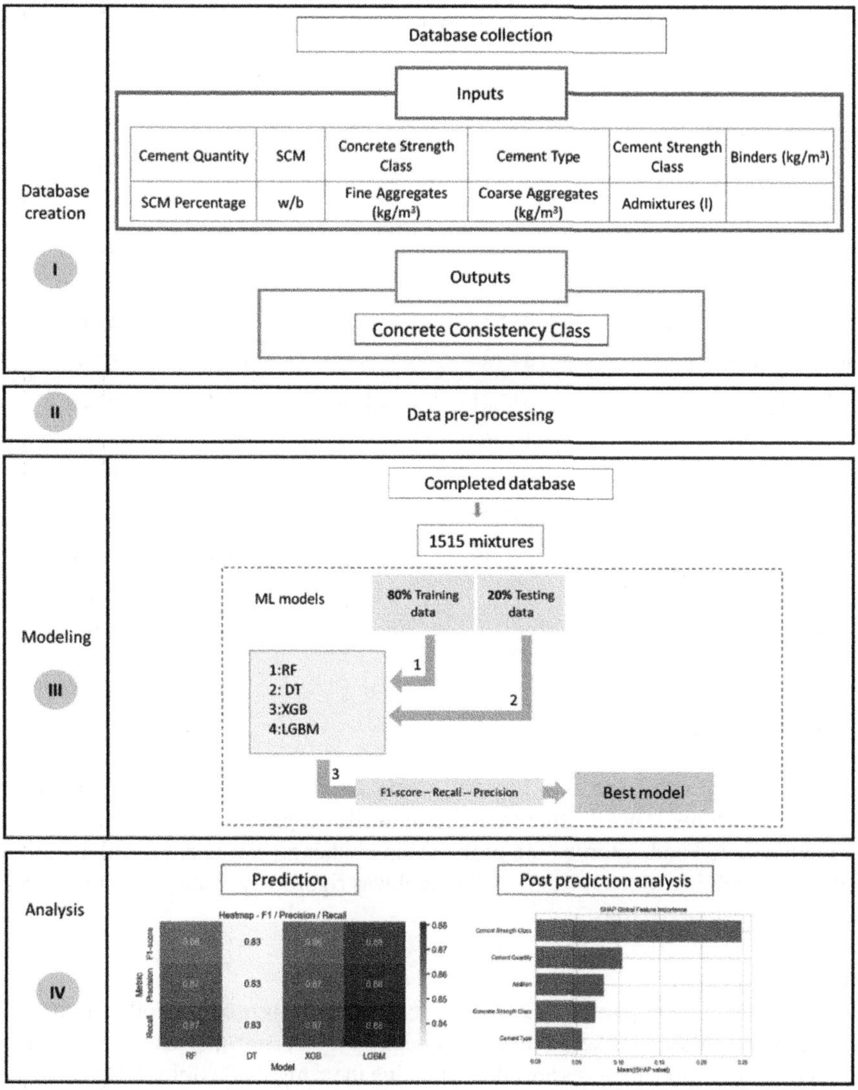

Fig. 1 Flowchart Methodology

the amount of supplementary cementitious materials (SCMs) is directly influenced by the type and cement content. Additionally, a significant correlation is noted between the water-to-binder ratio and the binder content, which is expected due to the proportional relationship between water, cement, and SCMs. Overall, the correlation matrix highlights weak interdependence among most parameters, particularly between inputs and outputs, ensuring that the analysis remains unbiased and independent across all features.

Table 1. Statistical description of the database features

Variable	Mean	Std	Min	Max
Encoded Variables				
Concrete Strength Class			0.0	7.0
SCM				
Cement Type			0.0	4.0
Numerical Variables				
Cement Content (kg/m^3)	262.4	60.07	58.0	450.0
Cement Strength Class	43.18	2.35	42.5	52.5
Binders (kg/m^3)	313.21	44.34	204.0	562.0
SCM Percentage (%)	16.26	14.08	0.0	84.94
w/b Ratio	0.56	0.08	0.33	0.81
Fine Aggregates (kg/m^3)	847.23	50.42	280.0	1360.0
Coarse Aggregates (kg/m^3)	990.18	59.83	640.0	1615.0
Admixtures (l)	1.44	0.89	0.0	8.46
Concrete Consistency			0.0	4.0

2.2 Machine Learning Models

This study incorporates a total of four ML models, including a mix of both simple and complex algorithms. This diverse selection is well-recognized for its effectiveness in solving prediction and optimization problems in civil engineering applications [28]. To ensure the reliability of the results, each model was rigorously evaluated using separate training and testing datasets.

Decision Tree (DT)

Decision Trees are non-parametric machine learning models used for both classification and regression tasks. Unlike parametric models, they do not increase the number of parameters when more features are added, making them highly flexible. A Decision Tree consists of two main elements: nodes and branches. During the training phase, each node evaluates a specific data feature to divide the observations, while during prediction, it directs a single data point along a specific path based on its features, ultimately leading to a final prediction.

Random Forest (RF)

The RF algorithm, initially introduced by Breiman [29], is a well-known supervised learning technique used for both classification and regression tasks. RF combines the principles of bootstrap aggregation (bagging) and random subspace selection to create an ensemble of independent decision trees. Each tree is built through a random sampling process, where a subset of input features is selected for each bootstrap sample. The final

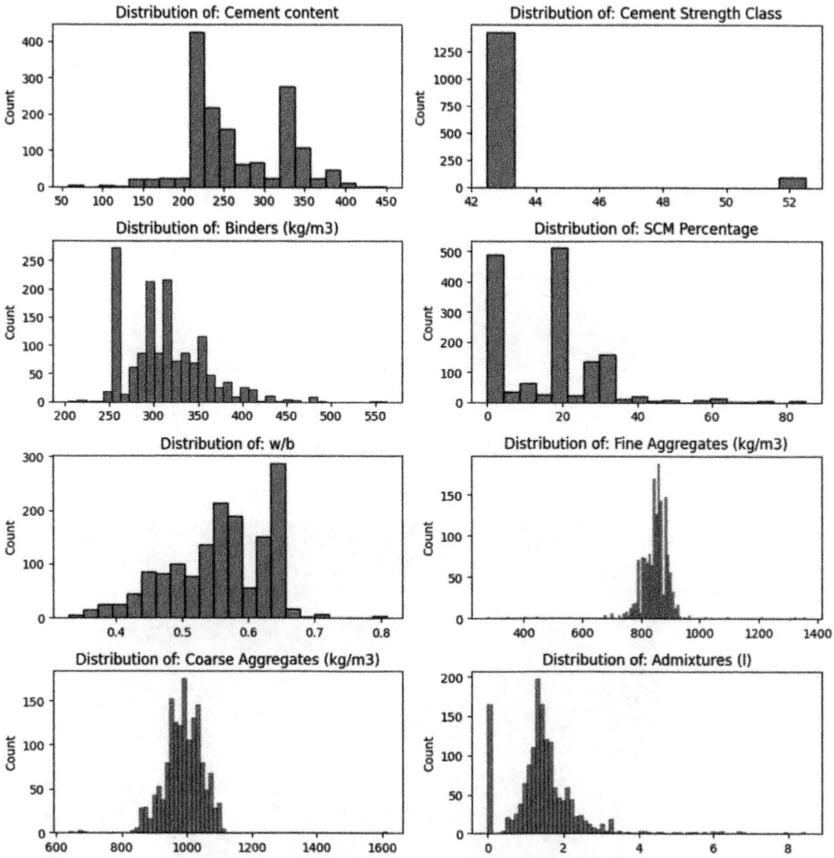

Fig. 2 Inputs distributions

prediction of the model is obtained by aggregating the outputs of all trees, typically through a majority vote (for classification) or averaging (for regression).

Extreme Gradient Boosting (XGB)

XGB is an advanced implementation of the gradient boosting framework, designed to enhance efficiency and reduce overfitting. XGB refines the classical gradient boosting method by introducing regularization and optimized tree-building strategies. The predictions are calculated through the aggregation of multiple additive functions, expressed mathematically as follows:

$$Y_i^k = Y_i^{k-1} + \mu f_k \tag{1}$$

where Y_i^k represents the prediction at iteration k, Y_i^{k-1} is the previous prediction, μ is the learning rate, and f_k is the new estimator. The model improves progressively with each iteration by introducing a new estimator f_k that minimizes the following objective

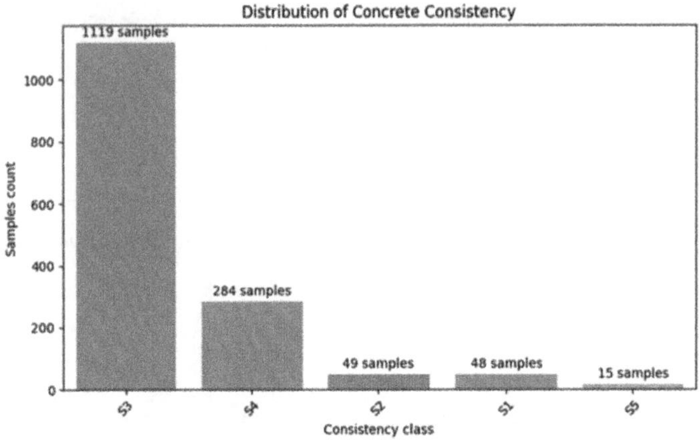

Fig. 3 Distribution of consistency classes

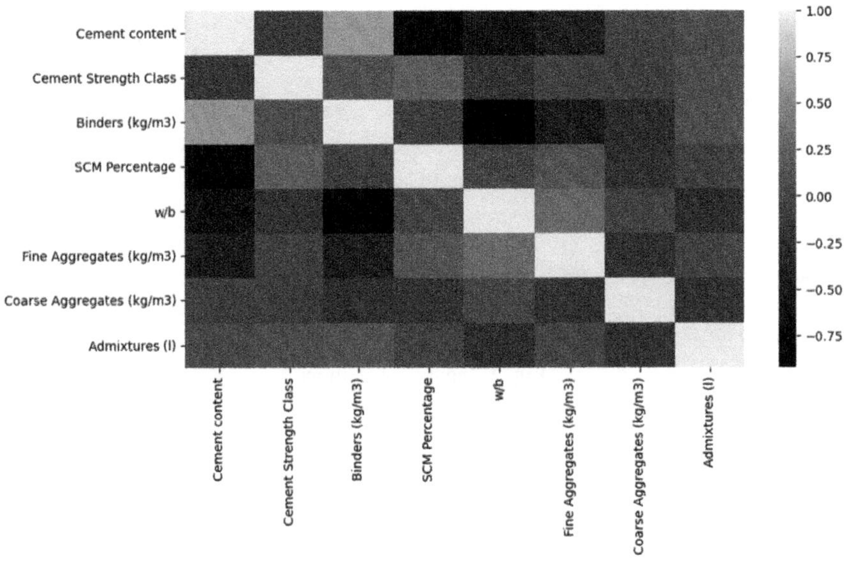

Fig. 4 Correlation matrix

function:

$$\text{objective} = \lambda T + \sum_{j=0}^{n} \left[E_j w + \frac{1}{2}(F_j + \gamma) w_j^2 \right] \quad (2)$$

where T is the number of leaves in the k-th tree, λ and γ are regularization parameters, E_j, F_j are the first and second gradients of the loss function, respectively, and w_j is the prediction value (or leaf score) for the j-th leaf of the tree.

Light Gradient Boosting Machine (LGBM)
LGBM, developed by Microsoft [30], is an optimized variant of the traditional Gradient Boosted Decision Trees (GBDT). Unlike typical level-wise growth in decision trees, LGBM employs a leaf-wise growth strategy, which expands the leaf that minimizes the loss the most. This approach significantly improves prediction accuracy and reduces overfitting by constraining tree depth. However, LGBM requires careful hyperparameter tuning due to its complexity [31].

2.3 Model Evaluation Metrics

To evaluate the accuracy and reliability of the classification models, three key metrics are employed: Precision, Recall, and F1-score. These metrics are fundamentally based on the concepts of True Positive (TP), False Positive (FP), True Negative (TN), and False Negative (FN), which are defined as follows:

- **True Positive (TP)**: The model correctly predicts the positive class (class 1 is predicted and is actually 1).
- **False Positive (FP)**: The model predicts the positive class (1), but the actual class is negative (0).
- **True Negative (TN)**: The model correctly predicts the negative class (class 0 is predicted and is actually 0).
- **False Negative (FN)**: The model predicts the negative class (0), but the actual class is positive (1).

The three-evaluation metrics are defined as follows:

Precision
Precision measures the proportion of true positive predictions among all instances classified as positive. It is represented as:

$$\text{Precision} = \frac{TP}{TP+FP} \tag{3}$$

Recall
Recall, also known as sensitivity or true positive rate, quantifies the model's ability to identify all relevant instances. It is defined as:

$$\text{Recall} = \frac{TP}{TP+FN} \tag{4}$$

F1-Score
The F1-score is the harmonic mean of precision and recall, offering a balanced measure that accounts for both false positives and false negatives:

$$\text{F1} - \text{score} = 2 \times \frac{\text{Precision} \times \text{Recall}}{\text{Precision} + \text{Recall}} \tag{5}$$

These metrics provide a comprehensive understanding of model performance, particularly in scenarios where class distributions are imbalanced.

3 Results and Discussion

3.1 Models Results

Comparison of Models

Figure. 5 presents a comparative evaluation of four machine learning models: RF, DT, XGB, and LGBM, based on F1-score, Precision, and Recall.

Among these models, LGBM consistently outperforms the others across all three metrics, achieving an F1-score of 0.88, which highlights its strong predictive capability. XGB follows closely, with balanced Precision and Recall values of 0.87, highlighting its solid generalization capabilities. RF also performs well, with an average score of 0.86 across the three metrics

In contrast, the DT model shows the lowest performance, reflecting its limitations in modeling the complex relationships between mixture parameters and consistency classes. This underperformance may be attributed to a potential class imbalance, particularly with the dominance of class S3, despite the implementation of calibrated splitting across all classes within the dataset.

Overall, this analysis underscores the robustness of gradient boosting algorithms (XGB and LGBM) for predicting concrete consistency classes. Their enhanced learning mechanisms and built-in regularization help reduce overfitting, making them well-suited for handling complex and imbalanced datasets.

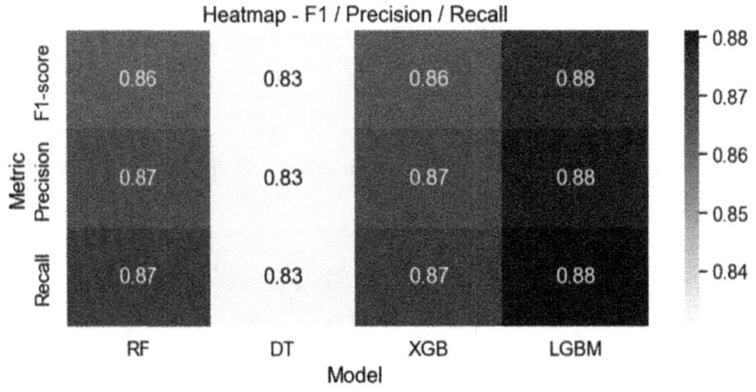

Fig. 5 Heatmap of F1-score, Precision, and Recall for the evaluated ML models.

Confusion Matrix

Figure 6 presents the confusion matrix for the LGBM model on the test set, identified as the best-performing model, providing a detailed visualization of its classification performance across all five classes. The matrix displays the actual classes on the y-axis and the predicted classes on the x-axis, with color intensity indicating the number of samples per category.

The diagonal elements represent correct predictions, where the predicted class matches the true class. Notably, class 2 (S3 consistency) shows the highest number of accurate predictions, with 214 samples correctly classified. This high precision is primarily attributed to its dominance in the dataset, associated with the robustness of the model, highlighting LGBM's strong ability to predict this major class accurately.

Interestingly, class 4 (S5) exhibits perfect accuracy, with all 3 samples correctly classified. Although the sample size is small, the model's ability to precisely detect this class is noteworthy.

Off-diagonal elements indicate misclassifications. For instance, 21 samples from class 3 (S4) were incorrectly predicted as class 2. Similarly, a few misclassifications occurred in classes 0 and 1, though these errors are minimal.

Overall, the confusion matrix demonstrates LGBM's high predictive accuracy for the dominant class (S3) and acceptable performance **for the remaining classes**, with limited misclassification. This analysis supports the model's robustness in classifying the different consistency levels of concrete mixtures.

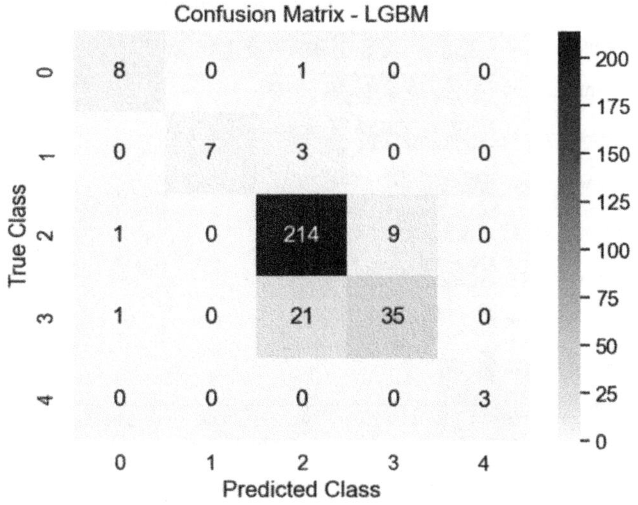

Fig. 6 Models results

3.2 Features Analysis

Figure 7 presents the ranking of the most influential features affecting the model's predictions, as determined by SHAP analysis. The features are displayed in descending order based on their average absolute SHAP values, with the most impactful features at the top and the least impactful at the bottom. The analysis indicates that Binders (kg/m^3), water-to-binder ratio (w/b) and Admixtures (l) are the most influential factors, each contributing on an average SHAP value of 0.12 and 0.07 respectively. This result is consistent with experimental findings and qualitatively confirms the robustness of the best-performing model.

This indicates that variations in these components significantly affect the model's predictions, highlighting their critical role in optimizing concrete properties.

Cement Content and Fine Aggregates (kg/m^3) are ranked next, with SHAP values of 0.04 and 0.03 respectively, indicating their notable effect on the consistency class. The concrete strength class, Coarse Aggregates (kg), and Cement Type show moderate contributions to the model's predictions, with mean SHAP values gradually decreasing. Additionally, SCM Type and Cement Strength Class are identified as the least influential features, suggesting a relatively limited impact on the final predictions. This analysis highlights the critical importance of Binders, Admixtures, and the water-to-binder ratio in driving the properties of the concrete mixture, aligning well with findings in existing concrete research.

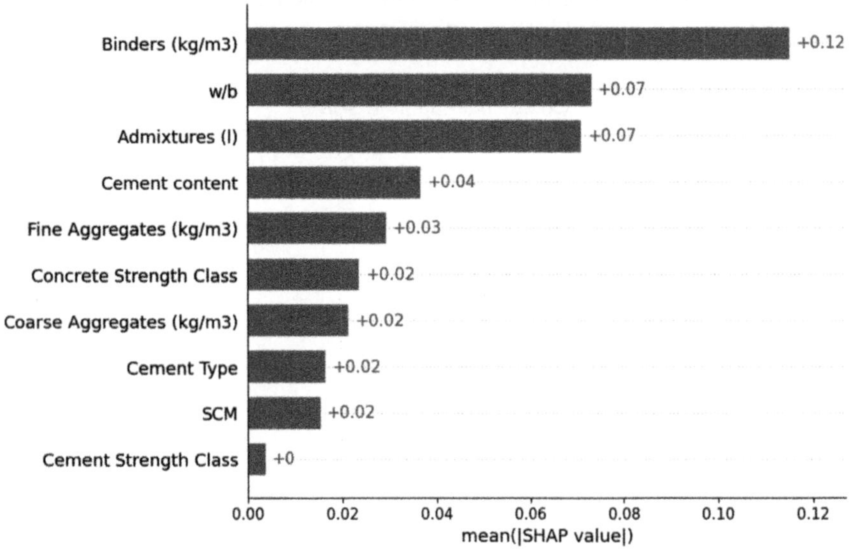

Fig. 7 Features Importance

4 Conclusions and Outlooks

This study evaluates the performance of ensemble learning classification techniques in predicting the slump class of concrete. Four ML algorithms were employed: DT, RF, XGB, and LGBM. The analysis was conducted using a dataset of 1,515 concrete samples collected from industrial partners. Input features included mixture proportions and the use of SCMs. Based on the results of this analysis, the following conclusions can be drawn:

- All machine learning models demonstrated high precision in predicting concrete slump classes, highlighting the robustness of these techniques.
- The tested ML models yielded comparable results, confirming both the consistency and the reliability of these ensemble learning methods.

- Among the tested models, LGBM achieved the best performance, accurately predicting 267 out of 303 specimens in the test set.
- SHAP analysis was used to analyze feature importance for each prediction. According to this analysis, binder content, admixture dosage, water-to-binder ratio (w/b), and fine aggregates quantity were identified as the most influential factors.

Despite the inherent complexity of fresh-state concrete properties, primarily due to the interactions between mix components, this study paves the way for further research into the application of machine learning in concrete technology. These approaches hold promises for adoption by concrete producers, not only to optimize fresh-state performance but also to support predictions of strength and long-term durability.

References

1. Nili, M., Afroughsabet, V.: Combined effect of silica fume and steel fibers on the impact resistance and mechanical properties of concrete. Int. J. Impact Eng. **37**(8), 879–886 (2010). https://doi.org/10.1016/j.ijimpeng.2010.03.004
2. Nataraja, M.C., Dhang, N., Gupta, A.P.: Statistical variations in impact resistance of steel fiber-reinforced concrete subjected to drop weight test. Cem. Concr. Res. **29**(7), 989–995 (1999). https://doi.org/10.1016/S0008-8846(99)00052-6
3. Song, P.S., Wu, J.C., Hwang, S., Sheu, B.C.: Assessment of statistical variations in impact resistance of high-strength concrete and high-strength steel fiber-reinforced concrete. Cem. Concr. Res. **35**(2), 393–399 (2005). https://doi.org/10.1016/j.cemconres.2004.07.021
4. Khan, M., Cao, M., Xie, C., Ali, M.: Effectiveness of hybrid steel-basalt fiber reinforced concrete under compression. Case Stud. Constr. Mater. **16**, e00941 (2022). https://doi.org/10.1016/j.cscm.2022.e00941
5. Khan, M., Cao, M., Chaopeng, X., Ali, M.: Experimental and analytical study of hybrid fiber reinforced concrete prepared with basalt fiber under high temperature. Fire Mater. **46**(1), 205–226 (2022). https://doi.org/10.1002/fam.2968
6. Khan, M., Ali, M.: Improvement in concrete behavior with fly ash, silica-fume and coconut fibres. Constr. Build. Mater. **203**, 174–187 (2019). https://doi.org/10.1016/j.conbuildmat.2019.01.103
7. Ahmad, A., Farooq, F., Ostrowski, K.A., Śliwa-Wieczorek, K., Czarnecki, S.: Application of novel machine learning techniques for predicting the surface chloride concentration in concrete containing waste material. Materials. **14**(9), 2297 (2021). https://doi.org/10.3390/ma14092297
8. Thomas, B.S., et al.: Sugarcane bagasse ash as supplementary cementitious material in concrete – a review. Mater. Today Sustain. **15**, 100086 (2021). https://doi.org/10.1016/j.mtsust.2021.100086
9. Ahmad, W., Ahmad, A., Ostrowski, K.A., Aslam, F., Joyklad, P.: A scientometric review of waste material utilization in concrete for sustainable construction. Case Stud. Constr. Mater. **15**, e00683 (2021). https://doi.org/10.1016/j.cscm.2021.e00683
10. Tomczak, K., Jakubowski, J., Kotwica, Ł.: Key factors determining the self-healing ability of cement-based composites with mineral additives. Materials. **14**(15), 15 (2021). https://doi.org/10.3390/ma14154211
11. Nafees, A., et al.: Modeling of mechanical properties of silica fume-based green concrete using machine learning techniques. Polymers. **14**(1), 1 (2022). https://doi.org/10.3390/polym14010030

12. Michalski, R.S., Carbonell, J.G., Mitchell, T.M. (eds.): Machine Learning. Springer Berlin Heidelberg, Berlin, Heidelberg (1983). https://doi.org/10.1007/978-3-662-12405-5
13. Bonaccorso, G.: Machine Learning Algorithms. Packt Publishing Ltd (2017)
14. Ghahramani, Z.: Probabilistic machine learning and artificial intelligence. Nature. **521**(7553), 7553 (2015). https://doi.org/10.1038/nature14541
15. Taffese, W.Z., Sistonen, E.: Machine learning for durability and service-life assessment of reinforced concrete structures: recent advances and future directions. Autom. Constr. **77**, 1–14 (2017). https://doi.org/10.1016/j.autcon.2017.01.016
16. C. F. Ferraris' F. D. Larrard, Testing and Modelling of Fresh Concrete Rheology, National Institute of Standards and Technology, Gaithersburg, MD, NIST. IR 6094, 1998. doi:https://doi.org/10.6028/NIST.IR.6094.
17. Murata, J.: Flow and deformation of fresh concrete. Mater. Constr. **17**(2), 117–129 (1984). https://doi.org/10.1007/BF02473663
18. Ferraris, C.F., Obla, K.H., Hill, R.: The influence of mineral admixtures on the rheology of cement paste and concrete. Cem. Concr. Res. **31**(2), 245–255 (2001). https://doi.org/10.1016/S0008-8846(00)00454-3
19. Schowalter, W.R., Christensen, G.: Toward a rationalization of the slump test for fresh concrete: comparisons of calculations and experiments. J. Rheol. **42**(4), 865–870 (1998). https://doi.org/10.1122/1.550905
20. G. Christensen, Modelling the Flow of Fresh Concrete: the Slump Test. 1991. Consulté le: 6 novembre (2023). [En ligne]. Disponible sur: http://gateway.proquest.com/openurl?url_ver=Z39.88-2004&rft_val_fmt=info:ofi/fmt:kev:mtx:dissertation&res_dat=xri:pqm&rft_dat=xri:pqdiss:9135849
21. (PDF) Modeling slump of concrete using the group method data handling algorithm (2010) | Li Chen | 5 Citations. Consulté le: 22 novembre (2023). [En ligne]. Disponible sur: https://typeset.io/papers/modeling-slump-of-concrete-using-the-group-method-data-3qunhsovxq
22. Yeh: Exploring concrete slump model using artificial neural networks. J. Comput. Civ. Eng. **20**(3), 217–221 (2006). https://doi.org/10.1061/(ASCE)0887-3801
23. Zhang, X., Akber, M.Z., Zheng, W.: Predicting the slump of industrially produced concrete using machine learning: a multiclass classification approach. J. Build. Eng. **58**, 104997 (2022). https://doi.org/10.1016/j.jobe.2022.104997
24. Nadimalla, A.M., Masjuki, S.A., Saad, S.A., Ali, M.: Machine learning model to predict slump, VEBE and compaction factor of M sand and shredded pet bottles concrete. IOP Conf. Ser.: Mater. Sci. Eng. **1244**(1), 012023 (2022). https://doi.org/10.1088/1757-899X/1244/1/012023
25. Mahajan, L., Mahadik, S., Bhagat, S.R.: Investigation of Fly ash concrete by slump cone and compaction factor test. IOP Conf. Ser.: Mater. Sci. Eng. **970**(1), 012011 (2020). https://doi.org/10.1088/1757-899X/970/1/012011
26. NF EN 206, Afnor EDITIONS. Consulté le: 16 juin (2023). [En ligne]. Disponible sur: https://www.boutique.afnor.org/fr-fr/norme/nf-en-206/beton-specification-performances-production-et-conformite/fa171582/1465
27. Determination of the Coefficient of Correlation | Science. Consulté le: 29 septembre (2023). [En ligne]. Disponible sur: https://www.science.org/doi/10.1126/science.30.757.23
28. Liang, M., Chang, Z., Wan, Z., Gan, Y., Schlangen, E., Šavija, B.: Interpretable ensemble-machine-learning models for predicting creep behavior of concrete. Cem. Concr. Compos. **125**, 104295 (2022). https://doi.org/10.1016/j.cemconcomp.2021.104295
29. Breiman, L.: Random forests. Mach. Learn. **45**(1), 5–32 (2001). https://doi.org/10.1023/A:1010933404324

30. Kopitar, L., Kocbek, P., Cilar, L., Sheikh, A., Stiglic, G.: Early detection of type 2 diabetes mellitus using machine learning-based prediction models. Sci. Rep. **10**(1), 1 (2020). https://doi.org/10.1038/s41598-020-68771-z
31. G. Ke et al.: LightGBM: A Highly Efficient Gradient Boosting Decision Tree, in Advances in Neural Information Processing Systems, Curran Associates, Inc., (2017). Consulté le: 19 décembre 2024. [En ligne]. Disponible sur: https://proceedings.neurips.cc/paper_files/paper/2017/hash/6449f44a102fde848669bdd9eb6b76fa-Abstract.html

Screening Filamentous Fungi for the Development of Sustainable Mycelial Composites

Giuseppe Ciaramella Moita[1(✉)], Marcelle Lisbôa de Sousa[1], Evanildo Francisco de Souza Júnior[2], Romildo Dias Toledo Filho[1], and Eliana Flavia Camporese Servulo[1]

[1] Federal University of Rio de Janeiro, Rio de Janeiro, Brazil
giuseppe.moita@coc.ufrj.br
[2] Chinese Academy of Sciences, Dalian Institute of Chemical Physics, Dalian, China

Abstract. Mycelium-based composites (MBCs) are sustainable biomaterials gaining increasing attention, particularly in construction, due to their recyclability, biodegradability, and thermal and acoustic insulation properties. This study investigated the influence of different lignocellulosic biomasses on the mycelial growth of filamentous fungi and their laccase (Lac) production, aiming to identify optimal conditions for MBC development. Seven fungal species—five ascomycetes (*Trichoderma longibrachiatum, Talaromyces amestolkiae, Aspergillus flavus, Fusarium oxysporum, Aspergillus niger*), one mucoromycete (*Rhizopus oryzae*), and one basidiomycete (*Pycnoporus sanguineus*)—were tested for Lac activity, a key enzyme for lignin degradation and biomass colonization. Bamboo, rice husk, and wood sawdust were selected as substrates, previously characterized for bulk density, water absorption, and particle size. Mycelial growth was assessed under two conditions: substrates moistened with distilled water or supplemented with potato dextrose broth (PDB). Fungi with higher Lac activity exhibited greater colonization capacity. PDB supplementation improved growth in species with limited development in water-only conditions. Bamboo showed the highest compatibility among the substrates, likely due to its finer particle size, which may enhance mass transfer and nutrient diffusion. The only strain to produce a cohesive, macroscopically visible composite was *P. sanguineus* when cultivated on bamboo. These findings highlight the relevance of selecting compatible fungal-biomass pairs and optimizing environmental conditions to advance the development of MBCs as eco-friendly construction materials.

Keywords: Mycelium-based composites · Lignocellulosic biomass · Laccase activity

1 Introduction

Mycelium-based composites (MBCs) are emerging as a sustainable and innovative solution within the construction sector. Mycelium is the vegetative structure of fungi, composed of a network of filaments known as hyphae. These filaments grow and intertwine,

M. Posani and S. Singh (Eds.): RYS 2025, RILEM Bookseries 67, pp. 348–360, 2026.
https://doi.org/10.1007/978-3-032-14652-6_27

forming a dense matrix that can be used to produce composites when combined with agricultural residues or other organic substrates [1, 2].

Among the enzymes secreted by fungi during this process, laccase plays a central role in lignin degradation. Lignin acts as a protective barrier around cellulose and hemicellulose, limiting microbial access to these polysaccharides. By breaking down lignin, laccase facilitates the exposure of fermentable sugars, enabling fungi to access energy sources and more effectively decompose the substrate. Therefore, fungi with higher laccase production tend to have a competitive advantage in colonizing lignocellulosic materials, contributing to faster and more uniform matrix formation in MBCs [3, 4]. In this sense, the ability of mycelium to break down organic matter and convert waste into valuable resources makes it a key component in the development of environmentally friendly and renewable building materials [5].

MBCs offer a range of advantageous properties and construction benefits. They contribute to waste reduction in two ways: by using organic waste as raw material and by being fully biodegradable, thus aligning with circular economy principles [6]. The production process of MBCs also requires significantly less energy and fewer resources than conventional materials used in non-structural applications, such as insulation and partitioning [7]. These composites can serve as sustainable alternatives to materials like polystyrene [8], offering excellent thermal and acoustic performance due to their cellular structure and contributing to greater energy efficiency in buildings [6] and improving fire safety [9].

In the context of scaling up MBC production for use in construction, studying fungal growth kinetics becomes essential, as biological manufacturing is inherently slower than conventional methods. Growth rate directly influences production time and poses a major constraint for large-scale applications [10]. Given the wide range of possible variable combinations (fungal species, substrate types, etc.), a better understanding of fungal growth dynamics is key to improving process efficiency [11, 12].

Considering the importance of rapid colonization for the efficient fabrication of MBCs, this study investigates the capacity of fungi to produce laccase, the most attractive ligninolytic enzyme, and the growth kinetics of selected filamentous fungi on different lignocellulosic biomasses. The fungal strains evaluated include *Aspergillus flavus*, *A. niger*, *Fusarium oxysporum*, *Pycnoporus sanguineus*, *Rhizopus oryzae*, *Trichoderma longibrachiatum*, and *Talaromyces amestolkiae*. Bamboo residue, rice husk, and wood sawdust were chosen as growth substrates. The aim is to identify fungi-substrate combinations with faster growth and greater composite-forming capacity, establishing a foundation for future development of MBCs.

2 Material and Methods

2.1 Biomasses

Three lignocellulosic biomasses, differing in origin, chemical composition, and granulometric profile, were selected as substrates for this study (Fig. 1). The first was a finely milled bamboo residue, obtained as a byproduct from an industrial laminated bamboo processing unit. The second biomass was rice husks, collected from a rice milling facility. The third was wood sawdust composed of a heterogeneous blend of species, including

Cedrela fissilis, *Hymenolobium petraeum*, *Erisma uncinatum* Warm., *Manilkara salzmanni*, and *Pinus elliottii*. These materials were characterized by their water absorption, particle size distribution, and bulk density. For the water absorption analysis, the method used was the "tea bag method," as previously described by Tinoco [13]. The test was performed in triplicate for each biomass, as well as for the tea bag itself, to correct for water absorbed by the bag. The particle size distribution was determined according to the NBR NM 248 standard [14], which applies to both fine and coarse aggregates. An adapted version of NBR 16916 [15], originally developed for fine aggregates, was used to determine the bulk density.

(a) (b) (c)

Fig. 1. Visual aspects of the biomasses used: (a) bamboo; (b) wood; and (c) rice husk.

2.2 Fungi

Seven different fungi were used in this study, designated as *Laboratory Batch* (LB) strains: *Trichoderma longibrachiatum* (LB1), *Pycnoporus sanguineus* (LB2), *Talaromyces amestolkiae* (LB3), *Aspergillus flavus* (LB4), *Fusarium oxysporum* (LB5), *Aspergillus niger* (LB6), and *Rhizopus oryzae* (LB7). Six strains were provided by the National Institute of Technology (NIT-Rio), except for LB2, which was obtained through a donation from the Culture Collection of Basidiomycete Fungi (CCIBt-SP) from the Sao Paulo Institute of Botanic.

Laccase (Lac) is the main enzyme microorganism used to degrade lignin in biomasses, playing a key role in forming mycelium-based composites. Therefore, a preliminary qualitative colorimetric assay was conducted to evaluate the Lac production capacity of the selected fungal strains. The method applied was based on the protocol described by Djokić et al. [16]. The procedure began with the preparation of a specific culture medium for Lac detection, composed of the following components per liter of distilled water: 1.644 g of ABTS (2,2′-azino-bis (3-ethylbenzothiazoline-6-sulfonic acid)), 39 g of PDA (Potato Dextrose Agar), and 0.160 g of copper sulfate ($CuSO_4$). The pH of the medium was adjusted to 4.5 using 1.0 M HCl before sterilization in an autoclave to improve laccase production [17]. Inoculated plates were then incubated at 25 ± 1 °C for 7 days. The appearance of a blue-green halo around the fungal colony was considered a positive indication of Lac activity (Fig. 2).

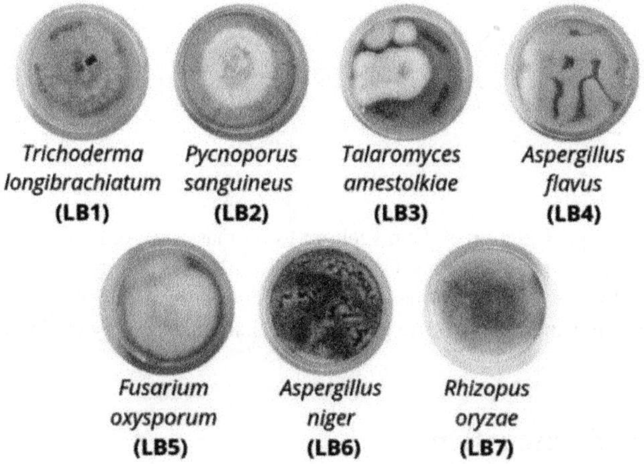

Fig. 2. Representative macroscopic morphology of filamentous fungi cultured on solid medium.

2.3 Mycelial Growth on Biomasses

Mycelial growth was studied in Petri dishes (90 mm), in which 5 g of biomass was placed in closed plates and sterilized in an autoclave. The samples were then dried in an oven at 100 ± 5 °C for 24 h to remove any remaining moisture. Subsequently, each sample was supplemented with 5 g of liquid, either distilled water or potato dextrose medium (200 g/L potato infusion and 20 g/L glucose), with the aim of assessing whether the fungus requires external support to initiate or enhance its growth. The fungal inoculum was prepared from a mycelial plug taken from a culture previously grown for 7 days on PDA (200 g/L potato infusion, 20 g/L glucose, and 17 g/L agar).

Mycelial growth on the biomass was monitored through photographic records and quantified using ImageJ software. Growth was calculated based on the mycelium-to-biomass surface ratio (MSR), as defined in Eq. (1), where A_b represents the total area of the biomass present in the Petri dish, and A_m corresponds to the area covered by the mycelium. A schematic illustration is provided in Fig. 3 to facilitate understanding of the MSR concept.

$$MSR = \frac{A_m}{A_b} \tag{1}$$

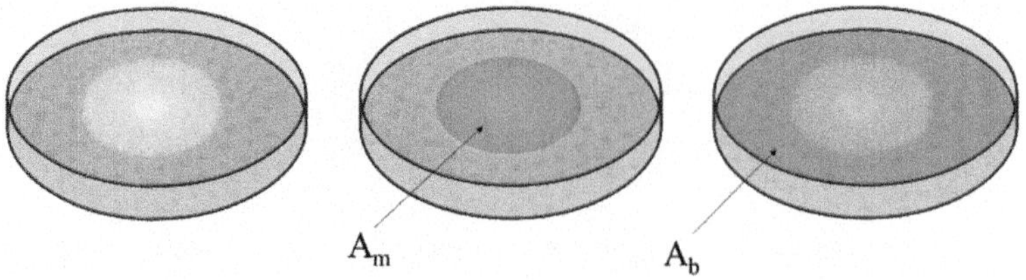

Fig. 3. Schematic representation of the mycelium-to-biomass surface ratio (MSR) used to assess fungal growth. The left panel shows a Petri dish with a fungal colony growing radially over a lignocellulosic substrate. The central panel highlights in red the area occupied by the mycelium (A_m), while the right panel highlights in green the total area covered by the biomass (A_b) within the Petri dish.

3 Results and Discussion

3.1 Biomasses Characterization

The characterization of the biomasses is presented in Table 1, which provides bulk density and water absorption values, and in Fig. 4, which shows the particle size distribution of the materials. The results are consistent with the visual characteristics of the substrates: bamboo exhibits the smallest particle size, lowest density, and highest water absorption. Wood has a coarser particle size, higher density, and lower water absorption. Finally, rice husk is the densest among all materials, with a uniform particle size distribution, as expected, and a water absorption rate similar to wood's. Previous studies support these results [18], although factors such as biomass origin, processing, and others may affect the physical properties of the material.

Table 1. Physical characterization of the biomasses

Biomass	Apparent bulk density (g/cm^3)	Water absorption (%)
Bamboo	0,26	375,33
Wood	0,48	222,61
Rice husk	0,60	229,81

These physical characteristics of the biomasses are important in the study of MBCs, as they can directly influence mycelial growth. A finer biomass, which results in lower particle size and apparent bulk density, may lead to a greater surface area for contact with the mycelium and, consequently, increased access to nutrients by the microorganism, thereby facilitating the composite formation process. On the other hand, a material with larger particle size and/or higher density may hinder this access and delay the composite formation process.

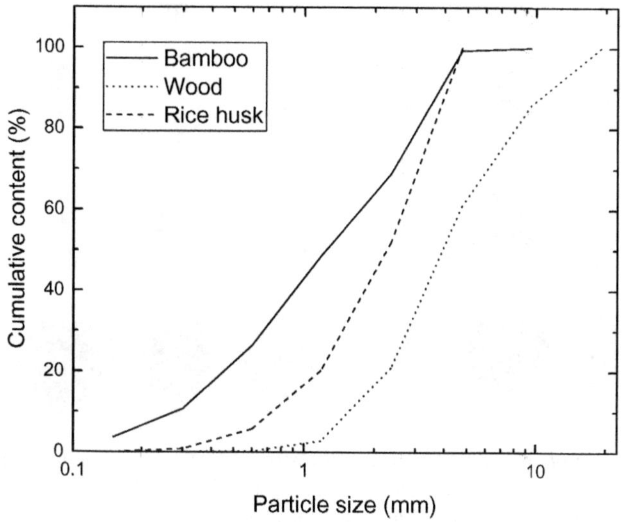

Fig. 4. Particle size distribution of the biomasses

3.2 Laccase Analysis

In the colorimetric assay used to assess laccase production by the selected fungi, enzyme activity is indicated not only by the area stained blue but also by the intensity of the coloration. The greater the area and/or the more intense the blue color, the higher the production of the target enzyme. From Fig. 5, it can be observed that *P. sanguineus* (LB2) is the highest producer of Lac enzyme, as it presents the most intense coloration across its growth area. In contrast, *F. oxysporum* (LB5) shows a darker coloration only in the central region of the colony, despite covering almost the entire Petri dish, while *A. niger* (LB6) occupies the largest area overall but displays a much lighter coloration, which may indicate lower laccase production despite its wider spread. This suggests that LB2 is the most efficient in lignin degradation. It should be noted, however, that this is a predominantly qualitative analysis, as no image analysis software or quantitative technique was applied.

The partial depolymerization of lignin mediated by Lac plays a pivotal role in the initial stages of lignocellulosic degradation, as it facilitates microbial access to structurally embedded polysaccharides such as cellulose and hemicellulose, components that are more readily metabolized. This enzymatic mechanism reflects a key ecological function of certain fungi in wood decay and underscores their biotechnological relevance. In the context of MBCs production, fungal strains exhibiting elevated Lac activity may represent more effective candidates, given their enhanced capacity to initiate substrate colonization, promote matrix interpenetration, and possibly influence the structural integrity and cohesion of the resulting bio-composites (Fig. 5).

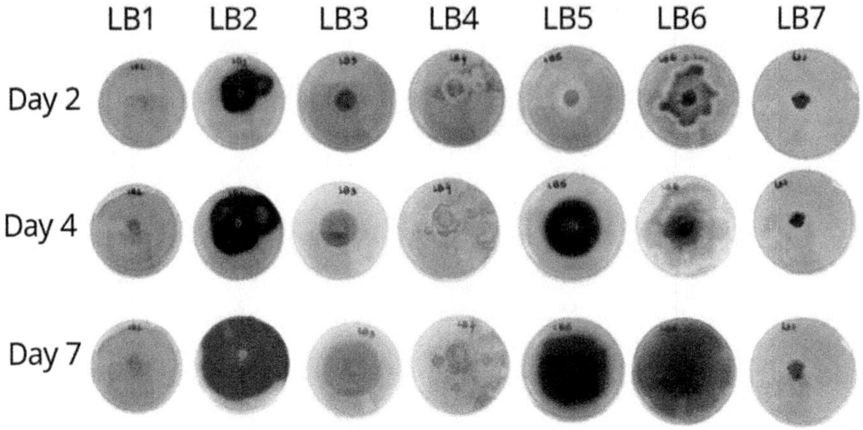

Fig. 5. Time-course progression of pigment diffusion and halo formation by fungal strains cultured on chromogenic agar medium. Laccase activity is evidenced by the development of a blue coloration around the inoculum, resulting from the enzymatic oxidation of the ABTS substrate. *Trichoderma longibrachiatum* (LB1), *Pycnoporus sanguineus* (LB2), *Talaromyces amestolkiae* (LB3), *Aspergillus flavus* (LB4), *Fusarium oxysporum* (LB5), *Aspergillus niger* (LB6), and *Rhizopus oryzae* (LB7).

3.3 Mycelial Growth on Biomasses

The analysis of fungal growth on different types of biomass revealed interesting results, with the majority of strains exhibiting some degree of colonization. Fig. 6 illustrates examples of distinct growth patterns observed for *Pycnoporus sanguineus*, *Trichoderma longibrachiatum*, and *Aspergillus niger* on bamboo biomass, using either distilled water (H_2O) or potato dextrose broth (PDB) as the moistening medium.

In this figure, *P. sanguineus* demonstrates robust initial colonization of the biomass, followed by the development of an extensive hyphal network that eventually interconnected the entire biomass surface. Such mycelial growth is essential for the future formation of composites, as it ensures the physical binding of bamboo particles into a cohesive structure.In contrast, *A. niger* exhibited a faster colonization rate than P. sanguineus, but without producing a dense hyphal network capable of interlinking the bamboo particles. Although the fungal growth was vigorous, the absence of extensive hyphal binding could limit its effectiveness in composite formation. Finally, for some species, such as *T. longibrachiatum*, even superficial growth was insufficient to cover the entire biomass layer within the study period. This suggests a limited colonization capacity for certain fungi, both in depth and in surface coverage, which could restrict their application in large-scale composite production.

A summary of the mycelial growth patterns observed in this study is presented in Fig. 7, which shows the mycelium-to-biomass surface ratio (MSR) for each fungal strain across the different biomasses and experimental conditions evaluated.

As shown in Fig. 7, mycelial growth, expressed as the mycelium-to-biomass surface ratio (MSR), was notably higher when lignocellulosic substrates were supplemented with PDB rather than moistened with water alone. This observation aligns with expectations,

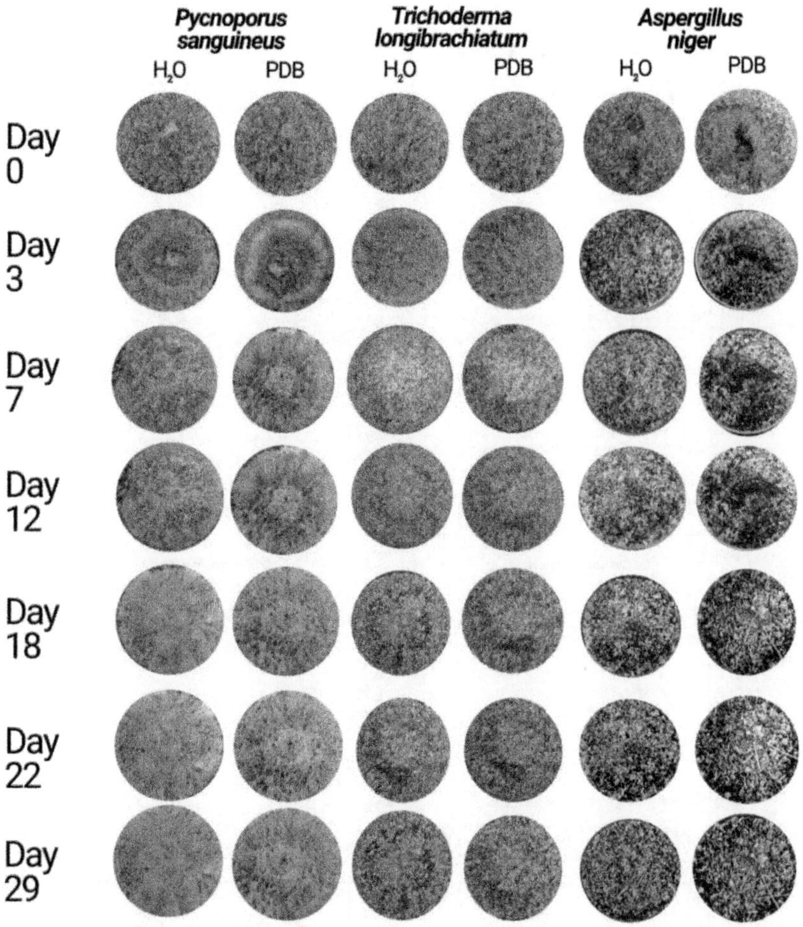

Fig. 6. Time-lapse photographic record of mycelial growth of *Pycnoporus sanguineus*, *Trichoderma longibrachiatum*, and *Aspergillus niger* on bamboo under two moisture conditions: distilled water (H_2O) and potato dextrose broth (PDB) to compare growth dynamics and colonization patterns.

as the nutrient-rich medium provides additional resources that accelerate colonization. Across all tested biomasses, fungi reached peak development within 3 to 5 days under PDB conditions, while in water-moistened substrates growth typically stabilized only after 5 to 7 days.

Bamboo consistently emerged as the most favorable matrix for fungal development among the substrates evaluated. In addition to its finer particle size (Fig. 4), bamboo presented a much lower apparent bulk density (Table 1) compared to rice husk and wood, resulting in a higher apparent surface area available for colonization. This combination likely enhanced aeration and mass transfer within the substrate, enabling deeper and more uniform hyphal penetration. It is possible that this superior performance is directly related to the physical properties analyzed in this study, as previously discussed. In

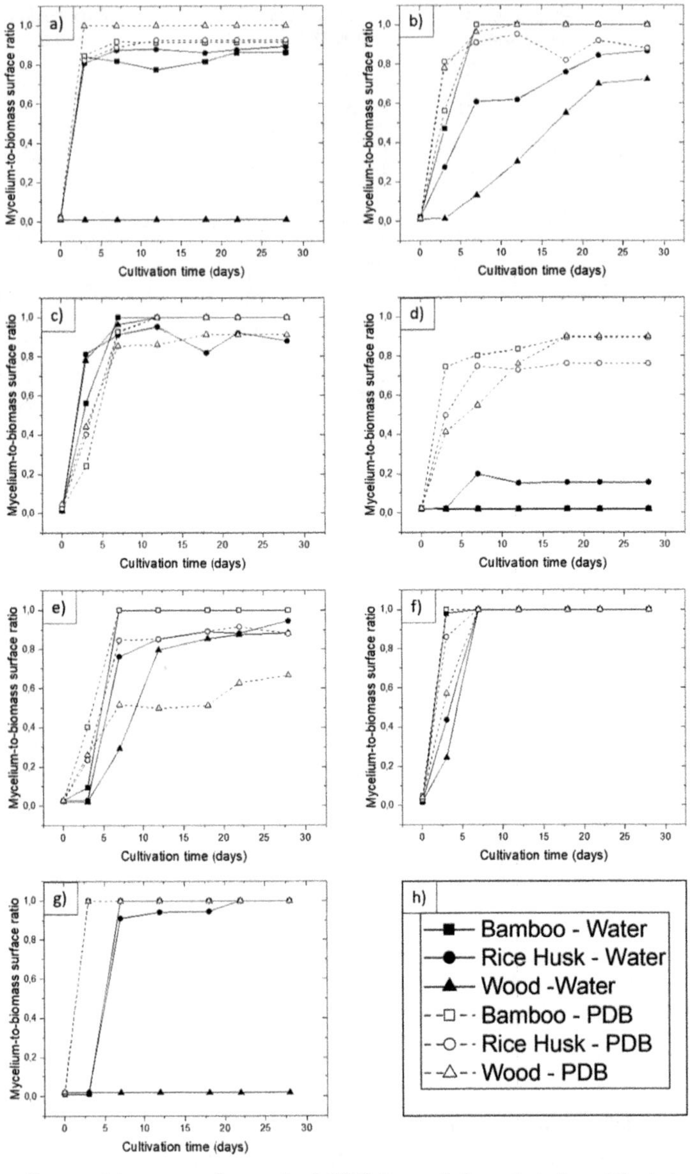

Fig. 7. Mycelium-to-biomass surface ratio (MSR) for each fungal strain cultivated on three different lignocellulosic biomasses—bamboo, rice husk, and wood sawdust—under two moisture conditions: distilled water and PDB. The MSR was calculated based on the proportion of biomass surface colonized by mycelium over the incubation period, allowing quantitative comparison of colonization efficiency between strains, substrates, and growth conditions. (a) LB1 – *Trichoderma longibrachiatum*; (b) LB2 – *Pycnoporus sanguineus*; (c) LB3 – *Talaromyces amestolkiae*; (d) LB4 – *Aspergillus flavus*; (e) LB5 – *Fusarium oxysporum*; (f) LB6 – *Aspergillus niger*; and (g) LB7 – *Rhizopus oryzae*. (h) Growth scale legend used for visual assessment.

contrast, although permissive to fungal colonization, rice husk and wood—particularly under PDB supplementation—did not support the formation of cohesive MBCs. For future studies, it would be valuable to investigate the chemical composition of these three biomasses, particularly their lignin and cellulose content, to better understand the factors influencing fungal colonization and composite formation.

For the two biomasses other than bamboo, fungal growth was largely restricted to superficial spreading, with limited structural integration or binding of the particulate matrix. This observation underscores a critical distinction between mere colonization and effective composite formation, the latter requiring both metabolic activity and mechanical cohesion mediated by hyphal networks. A more in-depth investigation is needed to elucidate the reasons for this outcome, especially considering that various types of biomasses have been reported as suitable for MBC production [19] including all those studied here [20–24].

A qualitative–quantitative divergence was observed in fungal performance. While several strains exhibited robust surface growth reaching a MSR of 1, only *P. sanguineus*, *F. oxysporum*, and *R. oryzae* displayed behaviors partially aligned with the structural objectives of MBC development. This indicates that several fungi were able to grow superficially enough to cover the entire biomass surface, often aided by PDB, but were unable to form dense hyphal networks that physically bind substrate particles, ensuring cohesion beyond mere surface coverage. *P. sanguineus* was the most successful, achieving substantial colonization and effective particle binding on bamboo substrates. Notably, it was able to form dense, cohesive composites even without nutritional supplementation, indicating a strong innate ability to degrade lignin and infiltrate the substrate matrix (Fig. 8). This was an expected outcome, as this species has been previously reported for similar applications [25, 26]. *F. oxysporum*, while also effective, was less successful than *P. sanguineus* and showed marked improvement only under PDB supplementation. In contrast, *R. oryzae*, despite its high growth rate, failed to establish any structural integration with the substrate, instead producing a superficial, loose mycelial layer resembling a biofilm, lacking the depth and cohesion required for true composite formation. These differences highlight the varying capacities of fungal species to not only colonize but also mechanically bind lignocellulosic and directly connect with their enzymatic activity profiles.

These biological behaviors align with the Lac activity profiles observed for the respective strains (Fig. 5). Two of the three fungi capable of partial or full composite formation were also the highest Lac producers. This correlation highlights the relevance of oxidative enzyme activity as a functional biomarker for effective MBC candidates. Lac-mediated oxidative depolymerization of lignin not only facilitates structural loosening of the biomass but also promotes fungal access to cellulosic fractions, which are more readily metabolized and likely play a role in the anchorage and growth of hyphal structures within the substrate. Despite the facilitating role of PDB, its necessity appears substrate- and species-dependent. In the most favorable scenario (*P. sanguineus* on bamboo), the contribution of PDB was marginal, suggesting that intrinsic enzymatic capabilities and physical substrate compatibility are more critical determinants of composite viability than exogenous nutrient enrichment. This has important implications for the scalability and sustainability of MBC production, where reducing reliance on costly or non-renewable supplements is desirable.

Fig. 8. Mycelium-based composite produced by *Pycnoporus sanguineus* (LB2) grown on bamboo substrate moistened with distilled water. The image shows the cohesive structure formed through dense hyphal networks binding the bamboo particles, illustrating the capacity of LB2 to achieve particle integration even without nutrient supplementation.

Future investigations should focus on strategies to enhance Lac production in less efficient fungi, potentially through co-culture approaches, metal ion supplementation (e.g., Cu^{2+}), or genetic engineering. Moreover, gaining deeper insight into the dynamics of mycelial infiltration beyond the surface and into the internal structure of the biomass will be crucial for improving the mechanical properties and structural integrity of the resulting composites. Ultimately, integrating biochemical, physical, and structural criteria will be essential to enable the rational selection of fungal strains and substrates, leading to the efficient, cost-effective, and structurally robust production of mycelium-based composites.

4 Conclusions

This study highlights the biotechnological potential of filamentous fungi for producing mycelium-based composites (MBCs) and underscores the importance of selecting both high-performing fungal strains and compatible lignocellulosic substrates. Among the species evaluated, *P. sanguineus* and *F. oxysporum* yielded the most promising results, particularly on bamboo, which proved to be the most favorable substrate for effective colonization and structural integration. The physical characteristics of the biomasses, together with their laccase activity profiles, were decisive in promoting mycelial infiltration and particle cohesion. While PDB supplementation accelerated initial growth in several strains, its impact was minimal in the most efficient fungus–substrate combinations, indicating that intrinsic metabolic capacity and substrate compatibility may be more critical than external nutrient enrichment. These findings provide valuable insights into the parameters that govern MBC development and point to future research needs, including strategies to enhance enzymatic performance, optimize nutrient inputs, and engineer substrates to improve composite integrity. Advancing in these directions will be key to scaling up the sustainable production of bio-based materials with improved structural and functional properties.

References

1. Sydor, M., Cofta, G., Doczekalska, B., Bonenberg, A.: Fungi in mycelium-based composites: usage and recommendations. Materials. **15**, 6283 (2022)
2. Majib, N.M., Yaacob, N.D., Ting, S.S., et al.: Fungal mycelium-based biofoam composite: a review in growth, properties and application. Prog. Rubber Plast. Recycl. Technol. **41**, 91 (2024)
3. Xie, N., Chapeland-Leclerc, F., Silar, P., Ruprich-Robert, G.: Systematic gene deletions evidences that laccases are involved in several stages of wood degradation in the filamentous fungus Podospora anserina. Environ. Microbiol. **16**, 141–161 (2014). https://doi.org/10.1111/1462-2920.12253
4. Tiwari, A., Chen, C.-W., Haldar, D., et al.: Laccase in biorefinery of lignocellulosic biomass. Appl. Sci. **13**, 4673 (2023). https://doi.org/10.3390/app13084673
5. Irbe, I., Kirpluks, M., Kampuss, M., et al.: Assessing the conformity of mycelium biocomposites for ecological insulation solutions. Materials. **17** (2024). https://doi.org/10.3390/ma17246111
6. Bonga, K.B., Bertolacci, L., Contardi, M., et al.: Mycelium Agrowaste-bound biocomposites as thermal and acoustic insulation materials in building construction. Macromol. Mater. Eng. **309**, 2300449 (2024). https://doi.org/10.1002/mame.202300449
7. Zhang, X., Hu, J., Fan, X., Yu, X.: Naturally grown mycelium-composite as sustainable building insulation materials. J. Clean. Prod. **342**, 130784 (2022). https://doi.org/10.1016/j.jclepro.2022.130784
8. Aranda-Calipuy, M.M., Roncal-Lázaro, A., Quezada-Alvarez, M.A., et al.: Pleurotus ostreatus mycelium and sugarcane bagasse as substitute environment-friendly material for polystyrene foam. Sustainability. **15**, 9157 (2023). https://doi.org/10.3390/su15129157
9. Jones, M., Bhat, T., Huynh, T., et al.: Waste-derived low-cost mycelium composite construction materials with improved fire safety. Fire Mater. **42**, 816–825 (2018). https://doi.org/10.1002/fam.2637
10. Jones, M., Huynh, T., John, S.: Inherent species characteristic influence and growth performance assessment for mycelium composite applications. Adv. Mater. Lett. **9**, 71–80 (2018). https://doi.org/10.5185/amlett.2018.1977
11. Sydor, M., Bonenberg, A., Doczekalska, B., Cofta, G.: Mycelium-based composites in art, architecture, and interior design: a review. Polymers (Basel). **14**, 145 (2021). https://doi.org/10.3390/polym14010145
12. Majib, N.M., Yaacob, N.D., Ting, S.S., et al.: Fungal mycelium-based biofoam composite: a review in growth, properties and application. Prog. Rubber, Plast. Recycl. Technol. **41**, 91–123 (2025). https://doi.org/10.1177/14777606241252702
13. Pimentel Tinoco, M., Gouvêa, L., de Cássia Magalhães Martins, K., et al.: The use of rice husk particles to adjust the rheological properties of 3D printable cementitious composites through water sorption. Constr. Build. Mater. **365**, 130046 (2023). https://doi.org/10.1016/j.conbuildmat.2022.130046
14. Associação Brasileira de Normas Técnicas (ABNT) (2003) NBR NM 248 Agregados - Determinação da composição granulométrica. ABTN, Rio de Janeiro

15. Associação Brasileira de Normas Técnicas (ABNT) (2021) NBR 16916 Agregado miúdo-Determinação da absorção de água. ABNT, Rio de Janeiro
16. Djokić, I., Knežević, A., Savković, Ž., et al.: Characterization of Culturable Mycobiome of newly excavated ancient wooden vessels from the Archeological site of Viminacium Serbia. J. Fungi. **10**, 345 (2024). https://doi.org/10.3390/jof10050343
17. Díaz, R., Téllez-Téllez, M., Bibbins-Martínez, M.D., et al.: Influence of initial pH of the growing medium on the activity, production and expression profiles of laccases produced by Pleurotus ostreatus in submerged fermentation. Electron. J. Biotechnol. **16**, 6 (2013). https://doi.org/10.2225/vol16-issue4-fulltext-6
18. Rocha, M., Gomes, B., Aguiar, A., et al.: Fire reaction properties of bio-based aggregates used in thermally insulated building components. Fire Mater. **48**, 62–78 (2024). https://doi.org/10.1002/fam.3166
19. Huang, Z., Wei, Y., Hadigheh, S.A.: Variations in the properties of engineered mycelium-bound composites (MBCs) under different manufacturing conditions. Buildings. **14** (2024)
20. Mbabali, H., Lubwama, M., Yiga, V.A., et al.: Development of Rice husk and sawdust mycelium-based bio-composites: optimization of mechanical, physical and thermal properties. J. Inst. Eng. (India): Ser. **105**, 97–117 (2024). https://doi.org/10.1007/s40033-023-00458-x
21. Sun, W., Tajvidi, M., Hunt, C.G., et al.: Fully bio-based hybrid composites made of wood, fungal mycelium and cellulose Nanofibrils. Sci. Rep. **9**, 3766 (2019). https://doi.org/10.1038/s41598-019-40442-8
22. Saez, D., Grizmann, D., Trautz, M., Werner, A.: Exploring the binding capacity of mycelium and wood-based composites for use in construction. Biomimetics. **7**, 78 (2022). https://doi.org/10.3390/biomimetics7020078
23. Soh, E., Chew, Z.Y., Saeidi, N., et al.: Development of an extrudable paste to build mycelium-bound composites. Mater. Des. **195**, 109058 (2020). https://doi.org/10.1016/j.matdes.2020.109058
24. Gan, J.K., Soh, E., Saeidi, N., et al.: Temporal characterization of biocycles of mycelium-bound composites made from bamboo and Pleurotus ostreatus for indoor usage. Sci. Rep. **12**, 19362 (2022). https://doi.org/10.1038/s41598-022-24070-3
25. Bruscato, C., Malvessi, E., Brandalise, R.N., Camassola, M.: High performance of macrofungi in the production of mycelium-based biofoams using sawdust — sustainable technology for waste reduction. J. Clean. Prod. **234**, 225–232 (2019). https://doi.org/10.1016/j.jclepro.2019.06.150
26. Santos, I.S., Nascimento, B.L., Marino, R.H., et al.: Influence of drying heat treatments on the mechanical behavior and physico-chemical properties of mycelial biocomposite. Compos. Part B Eng. **217**, 108870 (2021). https://doi.org/10.1016/j.compositesb.2021.108870

Study on the Mechanical Properties of Permeable Concrete with Ground Granulated Blast Furnace Slag

Monika Verma[1(✉)] and Geeta Mehta[2]

[1] Department of Engineering, University of Naples Parthenope, Naples, Italy
monika.verma001@studenti.uniparthenope.it
[2] School of Civil Engineering, Lovely Professional University, Phagwara, Punjab, India
mehtaanu6@gmail.com

Abstract. Like regular concrete, permeable concrete is a composite material made of particles joined by a cementitious binder. However, the material has a network of spaces that let water through since it includes less fine aggregate. Effective water management is the main goal of permeable concrete, especially in urban settings where stormwater management is crucial. It lessens the strain on drainage systems, replenishes groundwater, and lessens the impact of the heat island. This study examines the effects of using ground granulated blast furnace slag (GGBS) in permeable concrete mixtures in place of the cement as a partial replacement. GGBS was used in different percentages (10%, 20%, 30%, 40%, 50%, and 60%) to replace the cement. At 7, 14, 28 and 56 days, mechanical characteristics such as compressive strength, split tensile strength, and flexural strength alongside durability properties such as porosity, permeability, and water absorption properties were assessed. According to the findings, adding GGBS to permeable concrete mixtures greatly increases their strength, especially when replacement amounts are between 30% and 50% due to pozzolanic activity and microstructural densification. Porosity decreased from 26.0% to 17.4%, water absorption dropped to 10.9%, and permeability remained within functional limits, confirming the mix's hydraulic efficiency. When GGBS is added to concrete, its compressive, tensile, and flexural strengths are higher than those of normal concrete, improving the concrete's mechanical performance. Nevertheless, by lessening the material's environmental impact, GGBS substitution improves the sustainability of permeable concrete. GGBS is a more environmentally friendly alternative to typical cement manufacture since it helps reduce the large carbon emissions linked to it. The findings highlight a trade-off between enhanced short-term mechanical properties and potential limitations in the long-term performance of permeable concrete with GGBS, emphasizing the need for further research into optimizing GGBS content for balanced performance and sustainability.

Keywords: Permeable concrete · GGBS · compressive strength · permeability · porosity · water absorption

M. Posani and S. Singh (Eds.): RYS 2025, RILEM Bookseries 67, pp. 361–385, 2026.
https://doi.org/10.1007/978-3-032-14652-6_28

1 Introduction

1.1 General

Concrete remains one of the most widely used construction materials, forming the backbone of modern infrastructure (Nilimaa, 2023). However, because of the substantial carbon dioxide (CO_2) emissions linked to cement production, the increasing demand for concrete presents serious environmental challenges (Dobrucali, 2024). In response to the environmental challenges posed by conventional concrete production, researchers and industry professionals are actively exploring sustainable alternatives. Permeable concrete has emerged as a sustainable construction material, offering dual benefits of reduced environmental footprint and enhanced urban hydrological performance (Xiao et al., 2025). Its interconnected pore structure—typically achieving permeability rates ranging from 2 to 20 mm/s—facilitates efficient infiltration of stormwater, thereby promoting groundwater recharge and significantly mitigating surface runoff. These hydraulic properties make it particularly suitable for applications in urban pavements, parking lots, and low-traffic roads where conventional impervious surfaces exacerbate flooding and water quality issues (Antunes, Ghisi, & Severis, 2020; Azad, Sheikh, & Hai, 2024). Because of its high porosity, permeable concrete helps water seep in and lessens surface runoff, which helps with problems like groundwater depletion and urban flooding (Adresi et al., 2023). In the pursuit of sustainable concrete, ground granulated blast furnace slag (GGBS), a byproduct of the steel industry, has shown promise as a supplementary cementitious material (SCM) (Couvidat et al., 2020). GGBS, which is high in calcium and amorphous silica, improves the durability and mechanical qualities of concrete while also significantly lowering CO2 emissions.(Ahmad et al., 2022). Replacing Ordinary Portland Cement (OPC) with Ground Granulated Blast Furnace Slag (GGBS) reduces carbon emissions and enables the beneficial reuse of industrial by-products, contributing to both sustainable construction and effective waste management (Imbabi, Carrigan, & McKenna, 2012). Beyond its sustainability benefits, incorporating GGBS in permeable concrete enhances mechanical performance through pozzolanic reactions that generate additional C–S–H, leading to improved compressive strength and abrasion resistance—key factors for long-term durability in porous pavement systems (Couvidat et al., 2021) (Dash & Kar, 2018). It is perfect for applications like parking lots, driveways, walkways, and low-traffic roads because it mitigates heat island effects, replenishes groundwater, and reduces surface runoff, all of which address environmental issues (Althoey et al., 2023a). Though its porous nature makes it more vulnerable to freeze-thaw cycles, its compressive strength and durability are adequate for certain applications even though it lacks the load-bearing capability of traditional concrete. Its structural components include aggregates, a Transition Zone (TZ), and Hardened Cement Paste (HCP) (Zhou et al., 2024). Characterized by reduced fine aggregate content, permeable concrete features interconnected voids that enable water infiltration. Its performance is defined more by porosity and permeability than compressive strength, supporting sustainable construction through pollutant reduction, waste utilization, and lower emissions, while also offering effective stormwater management and a reduced environmental footprint (Singh, Sampath, & Biligiri, 2020).

Permeable concrete keeps its open structure for water flow while requiring controlled water-to-cement ratios and using supplementary cementitious materials (SCMs) less frequently than HPC. Although it doesn't directly reduce CO2 emissions like HPC (Verma et al., 2025), its sustainability is found in stormwater management and urban cooling, which promote environmentally friendly growth (Knight, Cunningham, & Miller, 2023). Moreover, its main function is different from that of high-early strength concrete, permeable concrete can use comparable materials and methods to satisfy certain needs for quick construction, longevity, and environmental advantages (Kováč & Sičáková, 2018). Permeable concrete is becoming more and more popular in infrastructure projects where porosity and permeability are important because of its reputation for controlling rainwater, lowering surface runoff, and fostering sustainability. However, some applications, including parking lots and sidewalks, could also need early load-bearing capacity, which can be achieved with customized methods (Drake, Bradford, & Marsalek, 2013). High-early strength cement, for example, can retain permeability while offering the required early performance. While attaining sufficient binding for early strength, a low water-to-cement ratio guarantees the integrity of the void structure. For expedited projects, chemical admixtures, like accelerators, can maximize workability and setting time without sacrificing permeability (Ghafoori et al., 2025).

Furthermore, adding mineral admixtures (Verma et al., 2024) such as silica fume and ground granulated blast furnace slag (GGBS) improves workability, durability, and environmental sustainability by lowering carbon emissions and increasing resilience to stressors like freeze-thaw cycles (Gupta, Raj, & Kumar Sahu, 2022) (Xu et al., 2023). Chemical admixtures and Ground Granulated Blast Furnace Slag (GGBS) greatly improve the performance and sustainability of permeable concrete. In line with the environmental objectives of permeable concrete, GGBS lowers CO_2 emissions and helps conserve energy because it takes less energy to make than Portland cement (Silva et al., 2023). GGBS enhances the material's qualities while encouraging resource conservation by using industrial by-products. It increases strength and durability without sacrificing permeability and improves workability and compaction, guaranteeing uniform void formation (Onsongo et al., 2025). Additionally, GGBS enhances resistance to alkali-silica reactions, chloride penetration, and sulphate attacks—all of which are critical for long-term durability under challenging conditions (Xu et al., 2022). The optimization of permeable concrete also heavily relies on chemical admixtures. Accelerators speed up setting and curing for applications that need early strength, retarders delay setting in hot conditions to facilitate placement and finishing, and water reducers increase workability while preserving structural integrity (Pacheco, 2024). In order to ensure that permeable concrete can satisfy a variety of infrastructure objectives, the efficacy of these materials is dependent on variables such as aggregate size, cement characteristics, and needed permeability (Kia, Wong, & Cheeseman, 2017). In conclusion, even though the objectives of permeable concrete are different from those of high-early strength concrete, the careful application of GGBS and admixtures enables it to strike a balance between environmental stewardship, speed of building, and durability (Zhang et al., 2024).

Since their inception in the 1960s, high-range water reducers, also known as superplasticizers, have been essential to contemporary concrete technology (Verma et al., 2025). By increasing placement and compaction while preserving the void structure necessary for water permeability, they show great promise in the creation of permeable concrete, even if their primary application is to improve the workability and flowability of conventional concrete (Shah, Khan, & Khan, 2013). Sulfonated Melamine-Formaldehyde Condensates (SMF), Sulfonated Naphthalene-Formaldehyde Condensates (SNF), Modified Lignosulfonates (MLS), and Polycarboxylate derivatives are common types that improve flowability without adding water by dispersing cement particles and releasing bound water. By maintaining the interconnected void network for water infiltration and guaranteeing adequate flowability for uniform placement, even in difficult-to-reach places, superplasticizers improve the performance of permeable concrete (Zhong & Wille, 2016). They make it possible to lower the water content, which aids in regulating overall permeability and preserving sufficient strength. Superplasticizers create long-lasting permeable concrete that efficiently controls stormwater while maintaining structural integrity in a variety of environmental circumstances by maximizing the water-to-cement ratio. Superplasticizers are an excellent addition to permeable concrete compositions because of their combined advantages of workability and endurance (Volder, Watson, & Viswanathan, 2009).

1.2 Need of the Study

The purpose of this study is to assess the mechanical characteristics of permeable concrete that contains different amounts of GGBS in place of cement. In order to comprehend how GGBS affects the entire performance of permeable concrete, the study examines the changed mixes' compressive strength, flexural strength, and permeability characteristics. The results of this study are intended to give a thorough grasp of GGBS's feasibility in permeable concrete applications. This study advances green construction technology and encourages the use of environmentally friendly building materials by showcasing its potential to enhance sustainability and mechanical performance.

The main difficulty with permeable concrete is striking a balance between strength, permeability, and longevity. Cement is a basic building material that is well-known for its stability and strength. Because concrete's permeability greatly affects its durability and factors like carbonation, chloride assault, and sulfate attack can erode it, this balance is especially important in severe conditions (Chindaprasirt et al., 2008). Fillers and pozzolanic materials can be added to permeable concrete to enhance its performance and longevity. When derived from industrial waste, these materials not only improve strength but also preserve the necessary permeability. They also help the environment by lowering CO_2 emissions and solving waste disposal issues (Marathe, Durga, & and Sadowski, 2024). Permeable concrete, which puts permeability and long-term durability above quick strength growth, is less affected by blended concrete's generally low early-age strength, particularly when slag cement is employed (Liu et al., 2018). This problem still exists, though, when early strength is necessary for quick construction. Superplasticizers and blended materials are important tools for improving the performance of concrete, and current research is still focused on finding the ideal balance between strength, durability, and permeability (Cavusoglu, 2024). Because of its environmental benefits and capacity to satisfy strict building specifications, permeable concrete continues to be an essential material as the need for sustainable infrastructure increases. In conclusion, permeable concrete may be made more workable, long-lasting, and environmentally friendly by using blended ingredients and superplasticizers, guaranteeing that it satisfies the unique requirements of sustainable building projects (Althoey et al., 2023b).

2 Materials and Mix Design

2.1 OPC 53-Grade Cement

Because of its advantageous qualities, such as its low chloride content, which improves corrosion resistance and endurance, Ordinary Portland Cement 53-Grade Cement was selected for permeable concrete. Its strong early strength is essential for pavements' initial load-bearing capability, and its high fineness enhances workability while preserving the water-to-cement ratio (Cai et al., 2023). By lowering the heat of hydration, the lower C3A fraction minimizes thermal cracking and increases durability in the face of temperature changes. Applications needing early strength and consistent quality, such high-traffic areas and large-scale infrastructure, are best suited for OPC 53 (Nasir, Al-Amoudi, & Maslehuddin, 2017). Table 1 and 2 represents the chemical composition and physical properties of OPC 53 grade.

Table 1. Chemical composition of OPC 53

Oxides	CaO	SiO_2	Al_2O_3	Fe_2O3	MgO	K_2O
Percentage	63.28	19.92	6.44	1.16	0.63	0.31

Table 2. Physical Properties of OPC 53

Test	Units	Test Result	Permitted Range	As per
Specific Gravity	-	3.10	-	IS4031(1988)–Part1
Fineness	m^2/kg	410	Min 370m^2/kg	
Normal Consistency	%	30	-	IS4031(1988)–Part4
Initial Setting Time	minutes	85	Min 60 min	IS4031(1988)–Part5
Final Setting Time	minutes	340	Max 600 min	IS4031(1988)–Part5
Soundness	mm	1.0	<10 mm	

2.2 Ground Granulated Blast Furnace Slag (GGBS)

Ground Granulated Blast-Furnace Slag (GGBS) was selected due to its substantial environmental benefits and ability to enhance the performance of permeable concrete. GGBS is a by-product of the efficient use of slag that would otherwise be waste in the production of iron. Because GGBS requires less energy during manufacture than one-fifth that of Portland cement and generates significantly less carbon dioxide, it is a sustainable choice for permeable concrete (Divsholi, Lim, & Teng, 2014). In addition to reducing the carbon footprint of construction, GGBS in permeable concrete increases durability and decreases permeability, strengthening the concrete's resistance to hostile environments (Alkhawaldeh et al., 2024). GGBS helps create ecologically friendly permeable concrete, which is crucial for modern sustainable building methods, by conserving energy and natural resources. Calcium silicates and aluminates, along with additional bases, make up the majority of the non-metallic substance known as GGBS (P, Nagarajan, & Shashikala, 2019). This material's numerous chemical and physical characteristics were thoroughly examined. Table 3 and 4 represents chemical and physical characteristics of Ground Granulated Blast Furnace Slag.

Table 3. Chemical composition of GGBS

Compound	CaO	SiO_2	Al_2O_3	Fe_2O_3	MgO	SO_3	CaO/SiO_2	Loss on ignition
Actual Value	41.0	33.2	13.3	1.3	11.6	1.0	1.23	0.5
Range as per BS:6699–1992	32–45	32–42	7–16	0.1–1.5	5–15	2.5 max	1.4 max	3 max

Table 4. Physical Properties of GGBS

Test	Units	Test Result
Specific Gravity	-	2.91
Fineness	m^2/kg	450
Glass Content	%	93
Bulk Density	Kg/m^3	1100
Color	-	Dull White

2.3 Superplasticizer

Superplasticizers are useful for producing high-strength, low-permeability concretes by reducing water requirements by 12–25% without compromising workability (Pereira, Evangelista, & de Brito, 2012). The superplasticizer utilized in this investigation is Master Glenium ACE 30, which was purchased commercially from Hindustan Construction Chemicals in Visakhapatnam. The second-generation polycarboxylic ether polymer is the basis for the new admixture Master Glenium ACE 30, which is renowned for its capacity to produce significant early strength gains. Table 5 lists all of Master Glenium ACE 30's characteristics.

Table 5. Properties of Super plasticizer Master Glenium®ACE30

Test	Units
Form or state	Liquid
Color	Reddish Brown
Specific gravity	1.06 to 1.09
Chloride content	Nil
Air entrainment	Approximately1% additional air is entrapped
pH	≥6
Principal constituent	Polycarboxylate Ether

2.4 Coarse Aggregate

Machine-crushed coarse aggregate of two different sizes, 20 mm and 10 mm was used in this research. The coarse aggregate had a uniform shape, free from any flaky particles. The results of the sieve analysis and the physical properties of the coarse aggregate are presented in Tables 6 and 7.

Table 6. Physical Properties of Coarse Aggregates

Test	Units	Test Result
Specific Gravity	-	2.74
Fineness	-	7.17
Water absorption	%	0.62
Impact strength	%	10.3

Table 7. Sieve Analysis of Coarse Aggregates

IS Sieve Size	Weight Retained (Gm)	Percentage Weight Retained	Cumulative Retained Percentage	Percentage Passing
20 mm	877	877	17.54	82.46
10 mm	4085	4962	99.24	0.76
4.75 mm	38	5000	100	0
2.36 mm	0	5000	100	0
1.18 mm	0	5000	100	0
600	0	5000	100	0
300	0	5000	100	0

(continued)

Table 7. (*continued*)

IS Sieve Size	Weight Retained (Gm)	Percentage Weight Retained	Cumulative Retained Percentage	Percentage Passing
150	0	5000	100	0
Total			716.78	

2.5 Water

Potable water is used in this study, which complied with IS: 3025-1986. The water's potability is confirmed by the analysis results, which are shown in Tables 8.

Table 8. Characteristics of water

Parameter	Value
pH	6.8
Taste	agreeable
Odour	agreeable
Turbidity (Nt units)	4

2.6 Mix Design

Mix design, which determines the optimal proportions of cement, sand, and aggregates to yield the required strength, porosity, and permeability, is an essential stage in the production of permeable concrete. When designing a mix, strength and porosity must be balanced. This is particularly important for permeable concrete structures that require both effective water penetration and structural stability. With a few adjustments, the mix design for M40 grade permeable concrete was completed in compliance with IS: 10262-2019 to guarantee appropriate permeability.

Table 9. Mix proportions for M40 grade concrete

Cement (binder)	Coarse aggregate	Water
390 kg/m^3	1242.43 kg/m^3	156.0 kg/m^3
1	5.5	0.4

The final mix proportions for M40 grade permeable concrete are designed to ensure adequate permeability while maintaining structural integrity and durability, as detailed in the mix proportions Table 9. These proportions aim to create a permeable concrete mix that balances the need for water infiltration with the required load-bearing capacity, ensuring both environmental performance and long-term durability.

3 Experimental Programme

3.1 Batching and Mixing

The first step in making permeable concrete is batching, which involves guaranteeing the right ratios (Tan et al., 2024). Due of its higher accuracy and consistency in ingredient proportioning, weigh batching is recommended over volume batching. Batching is followed by mixing, where the components are fully mixed by hand or by machine, depending on the needs and size of the project (Kolapo et al., 2012). Weigh batching and machine mixing were used in this work to guarantee accurate measurement and consistent mixing, both of which are essential for obtaining the desired qualities of permeable concrete (Table 10).

Table 10. The mix proportions of blended concrete

Mix Designation	% Replacement of permeable cement with GGBS	OPC-53 content kg/m^3	GGBS content kg/m^3	Coarse aggregate content kg/m^3	Water kg/m^3
BC0	0%	390	0	1242.43	156
BC10	10%	351	39	1242.43	156
BC20	20%	312	78	1242.43	156
BC30	30%	273	117	1242.43	156
BC40	40%	234	156	1242.43	156
BC50	50%	195	195	1242.43	156
BC60	60%	156	234	1242.43	156

In the production of permeable concrete, weigh batching is preferred over volume batching due to its superior accuracy and consistency. This method involves measuring all constituents by mass using calibrated scales or automated batching systems, ensuring precise control over mix proportions. Such precision is essential for achieving the desired mechanical strength, interconnected void structure, and permeability characteristics of pervious concrete. Most concrete specifications mandate batching by mass, as powder materials, sand, and coarse aggregates can only be accurately measured by weight. While water and liquid admixtures may be batched by either volume or weight, mass-based measurement remains the standard for solid constituents (Yuan et al., 2021). Additionally, the widespread use of ready-mixed concrete plants has led to batching processes that are predominantly automatic or semi-automatic, further enhancing consistency and quality control. In this study, weight-based replacement of cement with GGBS was adopted in accordance with IS:10262–2019 guidelines.

3.2 Casting and Curing of Specimens

Molds are cleaned, greased, and securely fastened before use to guarantee correct dimensions and stop slurry leaks. After cleaning, the molds are set on a vibrating platform and filled with three layers of permeable concrete. The molds are then vibrated until the cement paste visibly coated the surface after vibration. In accordance with IS: 516-1969 regulations, excess concrete is removed, and the top is completed smooth. To ensure sufficient hydration and the desired strength and durability of the permeable concrete, the cubes are demolded after 24 h and then immersed in water to cure at room temperature.

3.3 Experimental Testing

The experimental program for permeable concrete is structured into two parts. The first part focuses on workability and the second part involves strength and durability tests such as compressive strength, flexural strength, split tensile strength, and porosity, permeability and water absorption tests. Specimens are casted and prepared accordingly to conduct these tests (Fig. 1, 2).

Fig. 1. Casting of cylinders with different proportions

Fig. 2. Pervious Concrete Cubes

3.4 Workability of Permeable Concrete

The workability of the permeable concrete was assessed by varying the proportion of GGBS in the mixes through a number of experiments. The results gathered from these experiments are shown in Table 11. The workability of the permeable concrete is impacted by the proportion of GGBS. As per the performed tests of slump-cone and compaction factor, maximum workability is observed BC40. Workability of permeable concrete mix tends to increase with replacement of cement with GGBS up to 40%, whereas it shows a decrease for BC50 and BC60.

Table 11. Workability behavior of concrete mix with Super plasticizer 0.7%

Binder	w/b ratio	Slump	Compaction
BC0	0.40	45	0.82
BC10	0.40	58	0.86
BC20	0.40	62	0.87
BC30	0.40	66	0.89
BC40	0.40	67	0.89
BC50	0.40	58	0.86
BC60	0.40	40	0.81

The elevated slump values observed in certain mixes reaching up to 67 mm for the BC40 blend are primarily attributed to the incorporation of a high-range water-reducing admixture (Master Glenium ACE 30) and the fine particle size of GGBS, both of which enhance the workability of the mix. While conventional permeable concrete typically exhibits low workability to preserve its open pore structure, this study aimed to improve placement uniformity and compaction efficiency without compromising permeability. Accordingly, vibration-based compaction was employed during specimen preparation, which is more appropriate for mixes with moderate slump values (Yu et al., 2023).

3.5 Compressive Strength of Permeable Concrete

The mixes chosen to assess permeable concrete's compressive strength are listed in Fig. 3. The results for different blends of permeable concrete at varying ages—7, 14, 28, 56 days—are displayed in Fig. 3. The compressive strength was measured using 150 mm cubes. The compressive strength of each mix was determined as the average value obtained from three specimens. It was calculated using Equation (1):

$$f_c = F/A$$

where

f_c (N/mm^2) represents the compressive strength, F (N) is the maximum applied load at failure, and A (mm^2) is the cross-sectional area of the specimen (Bai et al., 2024).

The variation in compressive strength with the percentage replacement of cement with GGBS blends outperform control (BC0: 49.06 N/mm^2) for BC10 to BC40, peaking at 54.21 N/mm^2 for BC40. Optimal GGBS range: 30–40% — the pozzolanic reaction and filler effect improve strength beyond the plain cement mix. Strength continues to rise, especially for GGBS mixes. BC40 achieves 55.60 N/mm^2 — the highest — showing the long-term strength benefits of GGBS. Even BC60, though lower (45.06 N/mm^2), shows substantial improvement compared to its 7-day strength, proving GGBS's long-term pozzolanic activity. Partial replacement of cement with GGBS improves long-term strength due to ongoing pozzolanic reactions. For structural applications where early strength isn't critical, 30–40% GGBS is ideal.

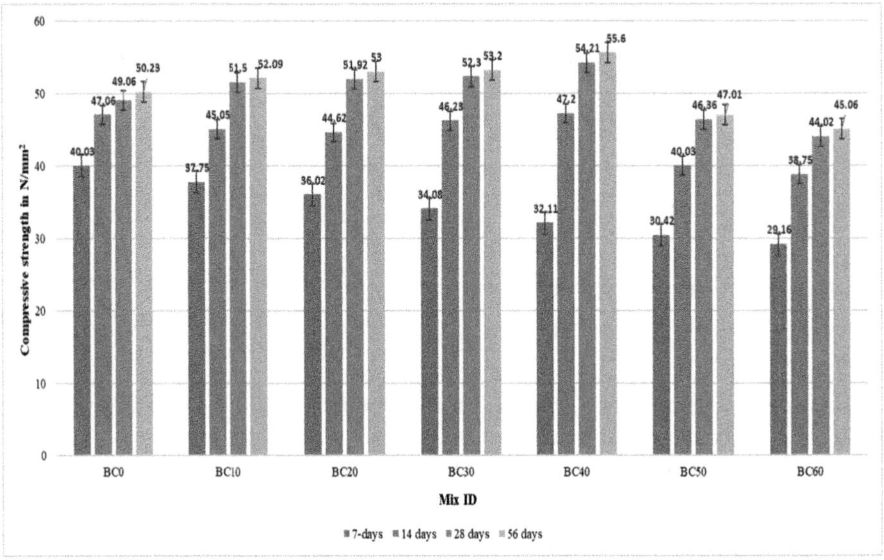

Fig. 3. Compressive strength after curing days in N/mm^2

At 7 days, mixes with higher GGBS content (e.g., BC40–BC60) exhibit lower compressive strength compared to the control mix (BC0). This reduction is primarily due to the slow reactivity of GGBS, which does not contribute significantly to early hydration. GGBS lacks the clinker phases present in OPC that drive rapid strength gain, resulting in a dilution effect where the reduced OPC content limits the formation of early calcium silicate hydrate (C–S–H). By 14 days, strength begins to recover, particularly in mixes with 30–40% GGBS replacement. This improvement is attributed to the initiation of pozzolanic reactions, wherein GGBS reacts with calcium hydroxide ($Ca(OH)_2$) released during OPC hydration. This reaction produces additional C–S–H gel, which enhances the microstructure and contributes to strength development (Gao et al., 2005; Ma et al., 2025). At 28 and 56 days, mixes with 30–40% GGBS (BC30 and BC40) outperform the control mix, indicating the long-term benefits of pozzolanic activity. The continued formation of secondary C–S–H gel leads to microstructural densification, reduced porosity, and improved mechanical performance (Ou & Kishi, 2025). The optimal strength observed in BC40 (54.21 N/mm^2 at 28 days and 55.60 N/mm^2 at 56 days) suggests that this replacement level achieves a favourable balance between OPC hydration and GGBS reactivity. Beyond 40% replacement, strength begins to decline again. This is likely due to insufficient calcium hydroxide availability to sustain pozzolanic reactions and excessive dilution of OPC, which compromises the binder's early and long-term strength development (Gao et al., 2019). The strength trends confirm that pozzolanic activity of GGBS plays a pivotal role in long-term strength enhancement, particularly at moderate replacement levels (30–40%). These findings align with established behaviour of GGBS in concrete systems and underscore its suitability for sustainable permeable concrete applications when optimized appropriately.

3.6 Split Tensile Strength of Permeable Concrete

All mixes split tensile strength is determined after seven and twenty-eight days. The outcomes are displayed in Figure 4.

Tensile strength increases with age for all mixes, aligning with typical concrete hydration behaviour. Moderate substitution levels (particularly 30% and 40%) seem to perform best, approaching or even exceeding control mix strength at later ages. Excessive substitution (beyond 50%) likely hampers both early and long-term strength. The BC40 mix shows promising long-term performance, possibly indicating improved microstructural development or pozzolanic activity at this replacement level. BC30 and BC40 display promising strength development. By 56 days, BC40 (5.1 N/mm^2) even surpasses BC0 (4.7 N/mm^2), suggesting potential long-term benefits from the replacement of cement with GGBS.

A gradual decline in tensile strength is observed with increasing GGBS content, from 4.80 N/mm^2 (BC0) to 3.59 N/mm^2 (BC60). This reduction is attributed to the low early reactivity of GGBS, which delays the formation of binding phases such as C–S–H. The dilution of OPC at higher replacement levels reduces the availability of early hydration products, weakening the matrix at this stage. A significant strength gain is observed in mixes BC30 and BC40, with BC40 achieving the highest tensile strength at 56 days (5.80 N/mm^2). This improvement is due to the pozzolanic reaction between GGBS and calcium hydroxide, which produces additional C–S–H gel, enhancing the bond between aggregates and the cement matrix (Wang, Yang, & Wu, 2024). The refined pore structure and improved interfacial transition zone (ITZ) contribute to better crack resistance and tensile performance. Beyond 40% GGBS, tensile strength declines again, indicating that excessive replacement may lead to insufficient calcium hydroxide for sustained pozzolanic activity and a weakened binder phase (He et al., 2022).

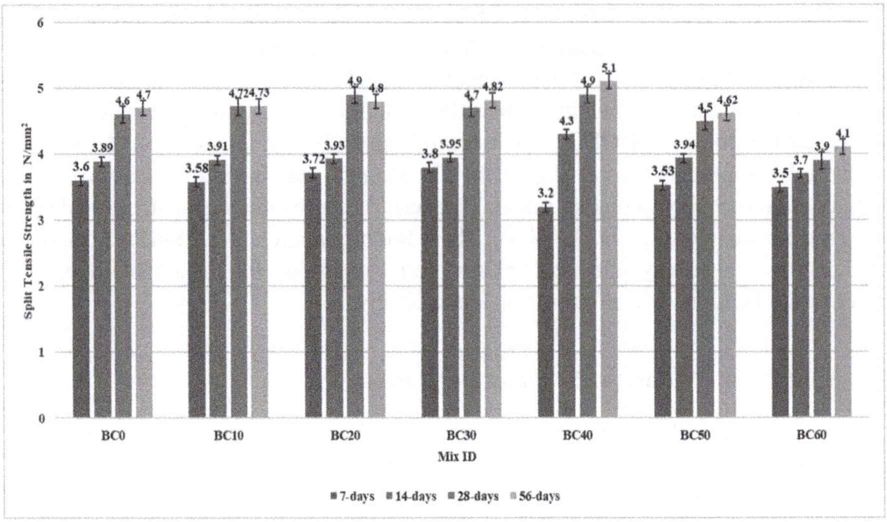

Fig. 4. Split Tensile strength of concrete at different days for different % replacement by GGBS in N/mm^2

These results show that although adding GGBS to permeable concrete generally increases its split tensile strength, the best results are obtained when the replacement level is 40%. In permeable concrete applications, there appears to be an ideal GGBS content for striking a balance between strength and permeability, as the split tensile strength tends to decline beyond this point.

3.7 Flexural Strength of Permeable Concrete

At 7 days, mixes with higher GGBS content (e.g., BC10–BC20) exhibit lower flexural strength compared to the control mix (BC0: 4.1 N/mm^2). This reduction is attributed to the limited early hydration of GGBS, which delays the formation of binding phases and results in a weaker matrix at early curing stages. By 28 and 56 days, mixes with 30–40% GGBS replacement (e.g., BC30 and BC40) show improved flexural strength, surpassing the control mix. This enhancement is due to the pozzolanic reaction between GGBS and calcium hydroxide, forming additional C–S–H gel that strengthens the matrix and improves crack resistance. The filler effect of GGBS also contributes to better particle packing and a refined microstructure, which enhances flexural performance (Karadumpa & Pancharathi, 2021; Wang, Yang, & Wu, 2024). Beyond 40% replacement, flexural strength begins to decline again, indicating that excessive GGBS may lead to insufficient early hydration and reduced binder efficiency. The lower strength at these levels suggests a threshold beyond which the benefits of pozzolanic activity are outweighed by the dilution of OPC.

28 Days Observation:

- Clear trend: Flexural strength improves significantly as GGBS increases up to 40%, reaching 4.68 N/mm^2 (BC40) — outperforming the control (BC0: 4.62 N/mm^2).

- Beyond 40%, strength starts dropping (BC50: 4.2 N/mm^2, BC60: 3.88 N/mm^2), showing that 40% GGBS offers an optimal blend of ductility and strength.

56 Days Observation:

- Long-term gains are visible, especially in BC30 and BC40, which peak at 4.76 N/mm^2 (BC40).
- High GGBS mixes (BC50 & BC60) stay lower at 4.31 N/mm^2 and 4.05 N/mm^2, reflecting slow reactivity at higher replacements and potential porosity retention.
- Optimum Flexural Strength is observed at 30–40% GGBS replacement.
- Initial Dip at 14 Days suggests possible curing sensitivity or mix water redistribution affecting flexural stiffness short-term.
- Long-Term Strength Benefit: GGBS improves flexural strength over time, especially in the 30–40% range, confirming its role in enhancing durability and crack resistance.
- High GGBS (>50%) leads to decreased flexural strength — likely due to lower early hydration and microstructural changes (Fig. 5).

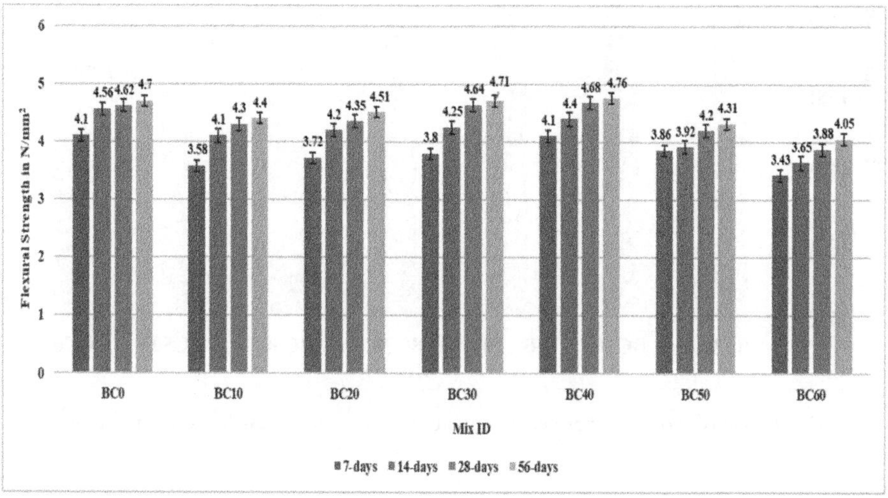

Fig. 5. Flexural strength of concrete at 7, 14, 28 and 56 days for different % replacement by GGBS in N/mm^2

3.8 Density of Permeable Concrete with Varying GGBS Content

The density of both fresh and hardened permeable concrete demonstrates a consistent decreasing trend with increasing replacement levels of Ordinary Portland Cement (OPC) by Ground Granulated Blast Furnace Slag (GGBS). As shown in the results, the density decreases progressively from 1791.6 kg/m^3 for the control mix (BC0) to 1779 kg/m^3 for the mix with 60% GGBS replacement (BC60). This represents an approximate reduction of 2–3 kg/m^3 for every 10% increment in GGBS content. This trend is primarily attributed to the lower specific gravity of GGBS (2.91) compared to OPC (3.10). When GGBS is used as a mass-based replacement, the lighter material occupies a greater volume within the binder matrix, thereby increasing the paste volume while maintaining the same total mass. This volumetric expansion leads to a marginal reduction in overall concrete density (Ahmad et al., 2022). Additionally, the finer particle size of GGBS contributes to improved particle packing, which may slightly influence the internal structure but does not offset the reduction in mass-based density (Yahyaee & Elize, 2024). While the change in density is relatively small, it is systematic and relevant for applications where weight-sensitive design or thermal performance is critical. Moreover, this observation reinforces the importance of considering volumetric effects and material-specific properties when designing sustainable concrete mixes with supplementary cementitious materials (A & O.M, 2024; Chen, Wu, & Liang, 2024) (Fig. 6).

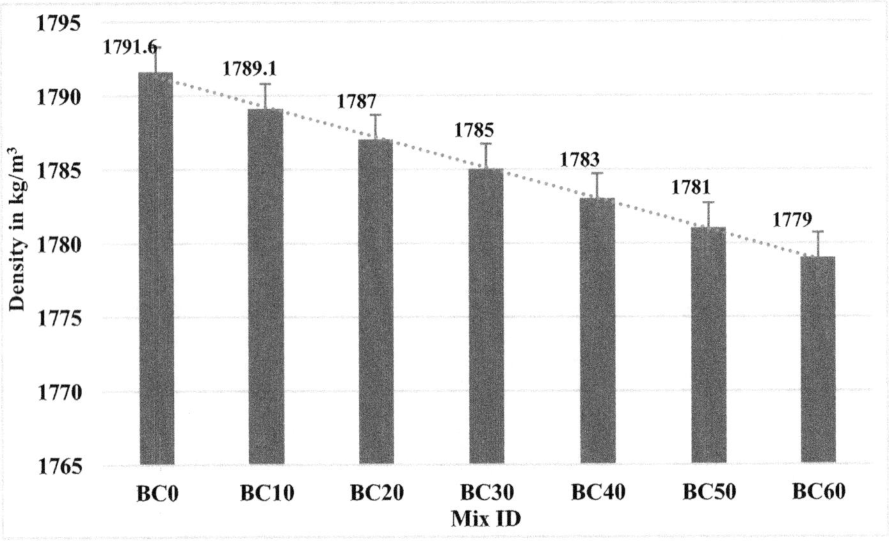

Fig. 6. Density of fresh permeable concrete with varying GGBS content in kg/m^3

3.9 Porosity of Hardened Concrete with Varying GGBS Content

The porosity of permeable concrete exhibits a clear and consistent decreasing trend with increasing replacement levels of Ordinary Portland Cement (OPC) by Ground Granulated Blast Furnace Slag (GGBS). The porosity (W) of the concrete specimens was calculated using Eq. (3):

$$W = [1\text{--}[(M_2 - M_1)/V_0 P_w]] \times 100\%$$

Where W (%) is the porosity, M_1 (kg) is the dry mass of the specimen, M_2 (kg) is the saturated mass, V_0 (m^3) is the measured volume of the specimen, and P_w (kg/m^3) is the density of water. This method quantifies the volume of voids within the concrete matrix relative to its total volume, which is critical for assessing the permeability and durability of permeable concrete (Bai et al., 2024).

As shown in the results, porosity decreases from 26.0% in the control mix (BC0) to 17.4% in the mix with 60% GGBS replacement (BC60). This represents a total reduction of 8.6 percentage points, with an average decrease of approximately 1.4% for every 10% increment in GGBS content.

This reduction is primarily attributed to the filler effect and pozzolanic reactivity of GGBS. The finer particle size of GGBS enhances the packing density of the binder matrix, effectively filling voids between cement grains and aggregates (Gholampour & Ozbakkaloglu, 2017). Additionally, the pozzolanic reaction between GGBS and calcium hydroxide leads to the formation of additional calcium silicate hydrate (C–S–H) gel, which further densifies the microstructure and reduces capillary porosity (Dom et al., 2022). These combined effects result in a more compact and less permeable matrix, even within the inherently porous structure of permeable concrete (Saini, Saikia, & Rajput, 2025). The observed trend highlights the potential of GGBS not only to improve mechanical performance but also to enhance the durability and long-term performance of permeable concrete by reducing its susceptibility to water ingress and associated deterioration mechanisms (Dom et al., 2022) (Fig. 7).

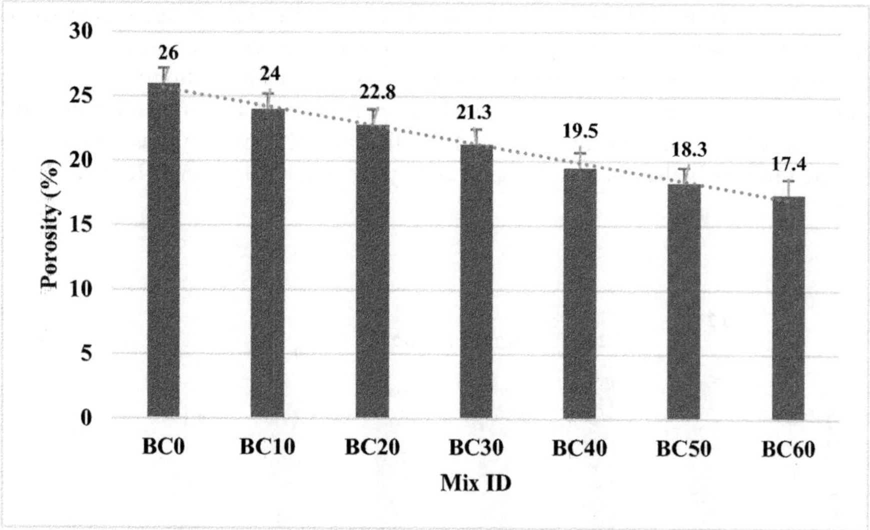

Fig. 7. Porosity of permeable concrete with varying %age of GGBS content at 28 days

3.10 Permeability of Hardened Concrete with Varying GGBS Content

The permeability of permeable concrete mixes shown in Fig. 8 represents a distinct decreasing trend with increasing replacement levels of Ordinary Portland Cement (OPC) by Ground Granulated Blast Furnace Slag (GGBS). The permeability coefficient (K) of the concrete specimens was calculated using Equation (2):

$$K = (V \times L)/(A \times \Delta h \times t)$$

Where K (mm/s) is the permeability coefficient, V (L) is the volume of water passing through the specimen, L (mm) is the height of the specimen, A (mm^2) is the cross-sectional area, Δh (mm) is the constant hydraulic head, and t (s) is the duration of the test. This method allows for quantifying the rate at which water permeates through the porous structure, which is critical for evaluating the functional performance of permeable concrete (Bai et al., 2024).

As presented in the results, the permeability coefficient declines from 0.294 cm/s for the control mix (BC0) to 0.192 cm/s for the mix with 60% GGBS replacement (BC60). This represents a total reduction of approximately 35%, with the most significant drop occurring beyond the 40% replacement level. This reduction in permeability is primarily attributed to the microstructural densification induced by GGBS (Munjal, Hau, & Hon Arthur, 2021). The finer particle size of GGBS enhances the packing density of the binder matrix, while the pozzolanic reaction with calcium hydroxide leads to the formation of additional calcium silicate hydrate (C–S–H) gel (He et al.). These effects collectively reduce the size and connectivity of capillary pores, thereby limiting the ease with which water can pass through the concrete (Jwaida et al., 2025). While a moderate reduction in permeability is beneficial for improving durability and resistance to aggressive agents, excessive reduction—particularly beyond 40% GGBS—may compromise the functional performance of permeable concrete, which relies on sufficient infiltration capacity (Saini, Saikia, & Rajput, 2025). These findings underscore the importance of optimizing GGBS content to balance mechanical strength, durability, and permeability, ensuring that the concrete remains both structurally sound and hydraulically functional (Kadhum et al., 2025).

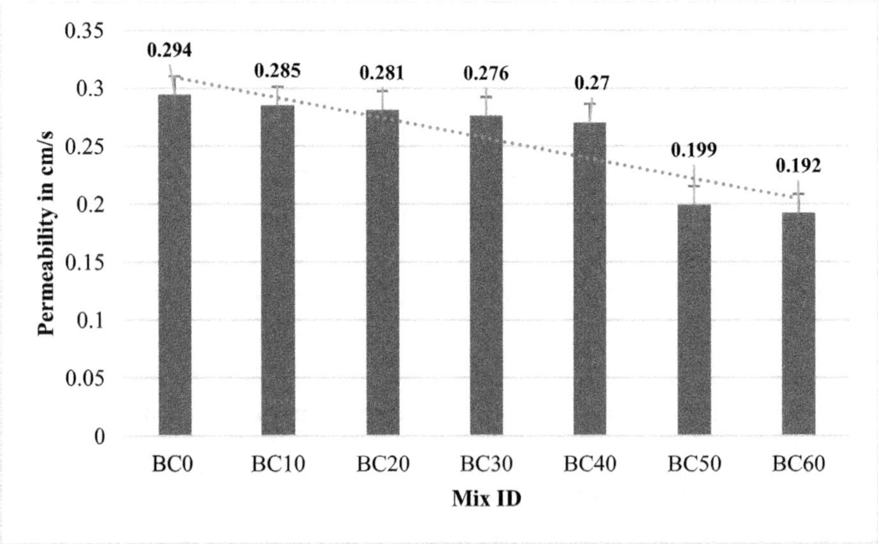

Fig. 8. Permeability of permeable concrete with varying %age of GGBS content at 28 days

3.11 Water Absorption

The water absorption behavior of permeable concrete mixes incorporating varying proportions of Ground Granulated Blast Furnace Slag (GGBS) demonstrates a clear influence on pore structure and durability (Al-Otaibi, 2008). The control mix (BC0), containing no GGBS, exhibited the highest water absorption at 13.2%, indicative of a more porous matrix with limited refinement of capillary voids as represented in Fig. 9. As GGBS content increased, water absorption progressively decreased, reaching a minimum of 10.9% at 40% replacement (BC40). This reduction is attributed to the pozzolanic reaction between GGBS and calcium hydroxide, which forms additional calcium silicate hydrate (C–S–H) (Li et al., 2025), thereby densifying the microstructure and reducing permeability to water (Aghaeipour & Madhkhan, 2017). Notably, mixes BC10 through BC40 showed consistent improvement in water resistance, with BC30 and BC40 offering the most favorable balance between permeability and durability. However, beyond 40% GGBS, water absorption began to rise again, with BC60 recording 13.5%, surpassing even the control mix. This reversal suggests that excessive GGBS may hinder early-age hydration and lead to incomplete matrix development, resulting in increased porosity (Azimi & Toufigh, 2023). Overall, the data indicate that partial replacement of OPC with GGBS up to 40% enhances the durability of permeable concrete without compromising its functional porosity, making BC30–BC40 the optimal range for sustainable and resilient pavement applications.

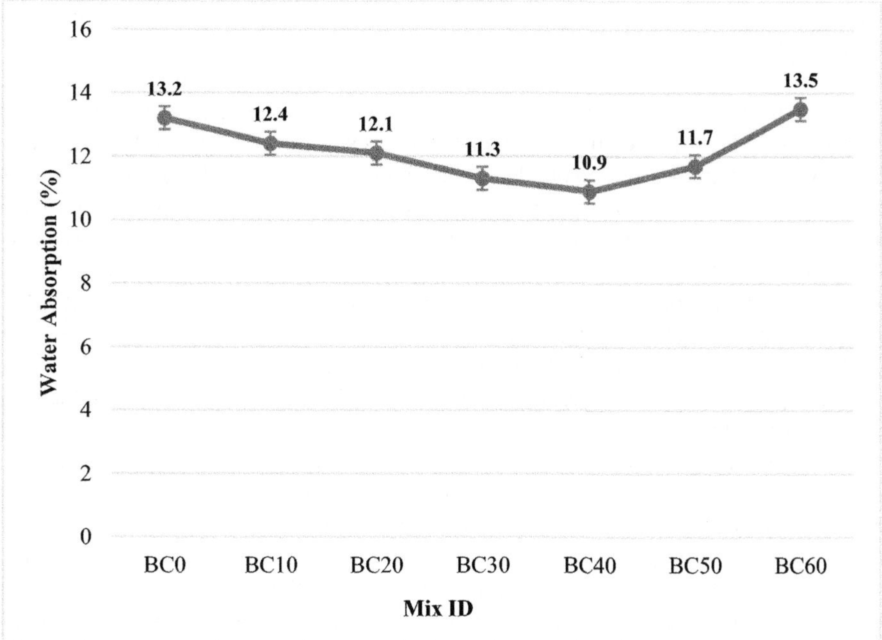

Fig. 9. Water Absorption of permeable concrete with varying %age of GGBS content at 28 days

4 Conclusion

This study comprehensively evaluated the mechanical and durability performance of permeable concrete incorporating varying proportions of Ground Granulated Blast Furnace Slag (GGBS) as a partial replacement for Ordinary Portland Cement (OPC). The results demonstrate that GGBS significantly influences the fresh and hardened properties of permeable concrete, with optimal performance observed at 30–40% replacement levels.

Workability improved consistently with increasing GGBS content up to 40%, as evidenced by higher slump and compaction factor values. This enhancement is attributed to the finer particle size and smooth texture of GGBS, which improves paste flow and reduces internal friction. However, beyond 40%, workability declined, likely due to excessive dilution of cementitious content and reduced cohesiveness.

Compressive strength results revealed that early-age strength (7 days) decreased with higher GGBS content due to its slower hydration kinetics. However, at later ages (28 and 56 days), mixes with 30–40% GGBS (particularly BC40) outperformed the control mix (BC0), achieving a peak strength of 55.60 MPa at 56 days. This improvement is attributed to the pozzolanic reaction between GGBS and calcium hydroxide, which forms additional calcium silicate hydrate (C–S–H), densifying the matrix and enhancing strength. Split tensile and flexural strength followed similar trends. BC40 exhibited the highest tensile strength (5.80 MPa at 56 days) and flexural strength (4.76 MPa at 56 days), surpassing the control mix. These gains are linked to improved microstructural development, better aggregate–paste bonding, and a refined interfacial transition zone due to GGBS's filler and pozzolanic effects.

Density of the concrete decreased slightly with increasing GGBS content, from 1791.6 kg/m^3 (BC0) to 1779 kg/m^3 (BC60), due to the lower specific gravity of GGBS. Despite this reduction, the mixes maintained sufficient mass for structural applications. Porosity and water absorption both decreased significantly up to 40% GGBS, indicating improved durability. Porosity dropped from 26.0% (BC0) to 17.4% (BC60), while water absorption reached its lowest at 10.9% for BC40. These reductions are due to the densification of the matrix and improved particle packing from GGBS. Permeability, a critical property for permeable concrete, showed a controlled decline with increasing GGBS. The permeability coefficient decreased from 0.294 cm/s (BC0) to 0.192 cm/s (BC60). While this indicates reduced infiltration capacity, mixes up to 40% GGBS maintained sufficient permeability for functional performance, balancing hydraulic conductivity with enhanced durability.

From an environmental perspective, the use of GGBS significantly reduces the carbon footprint associated with cement production and promotes the beneficial reuse of industrial by-products. This contributes to sustainable construction practices by lowering greenhouse gas emissions, conserving natural resources, and addressing waste management challenges. Future research should explore the long-term performance of GGBS-based permeable concrete under various environmental conditions, including freeze-thaw cycles and chemical exposure. Additionally, the integration of other supplementary cementitious materials (SCMs), optimization of aggregate gradation, and the use of nano-additives could further enhance the performance and sustainability of permeable concrete systems along with microstructure analysis using SEM, XRD and TGA technologies. Overall, the findings suggest that 30–40% GGBS replacement offers the most favorable balance between mechanical strength, permeability, and durability. This range optimizes the pozzolanic activity of GGBS without excessively compromising early-age performance or infiltration capacity, making it highly suitable for environmentally conscious infrastructure applications.

References

Tamilarasan, A., Suganya, O.M.: Investigation on the properties of sustainable alkali-activated GGBS mortar with nano silica and dredged marine sand: mechanical, microstructural and durability aspects. Case Stud. Constr. Mater. **20**, e03148 (2024). https://doi.org/10.1016/j.cscm.2024.e03148

Adresi, M., Yamani, A., Karimaei Tabarestani, M., Rooholamini, H.: A comprehensive review on pervious concrete. Constr. Build. Mater. **407**, 133308 (2023). https://doi.org/10.1016/j.conbuildmat.2023.133308

Aghaeipour, A., Madhkhan, M.: Effect of ground granulated blast furnace slag (GGBFS) on RCCP durability. Constr. Build. Mater. **141**, 533–541 (2017). https://doi.org/10.1016/j.conbuildmat.2017.03.019

Ahmad, J., et al.: A comprehensive review on the ground granulated blast furnace slag (GGBS) in concrete production. Sustainability. **14**(14), 8783 (2022) https://www.mdpi.com/2071-1050/14/14/8783

Al-Otaibi, S.: Durability of concrete incorporating GGBS activated by water-glass. Constr. Build. Mater. **22**(10), 2059–2067 (2008). https://doi.org/10.1016/j.conbuildmat.2007.07.023

Alkhawaldeh, A.A., Judah, H.I., Shammout, D.Z., Almomani, O.A., Alkhawaldeh, M.A.: Sustainability evaluation and life cycle assessment of concretes including pozzolanic by-products and alkali-activated binders. Results Eng. **23**, 102569 (2024). https://doi.org/10.1016/j.rineng.2024.102569

Althoey, F., Ansari, W.S., Sufian, M., Deifalla, A.: Advancements in low-carbon concrete as a construction material for the sustainable built environment. Dev. Built Environ. **16**, 100284 (2023). https://doi.org/10.1016/j.dibe.2023.100284

Antunes, L.N., Ghisi, E., Severis, R.M.: Environmental assessment of a permeable pavement system used to harvest stormwater for non-potable water uses in a building. Sci. Total Environ. **746**, 141087 (2020). https://doi.org/10.1016/j.scitotenv.2020.141087

Azad, A., Sheikh, M.N., Hai, F.I.: A critical review of the mechanisms, factors, and performance of pervious concrete to remove contaminants from stormwater runoff. Water Res. **251**, 121101 (2024). https://doi.org/10.1016/j.watres.2024.121101

Azimi, Z., Toufigh, V.: Influence of blast furnace slag on pore structure and transport characteristics in low-calcium Fly-ash-based Geopolymer concrete. Sustainability. **15**(18), 13348 (2023) https://www.mdpi.com/2071-1050/15/18/13348

Bai, X., Zhou, H., Bian, X., Chen, X., Ren, C.: Compressive strength, permeability, and abrasion resistance of pervious concrete incorporating recycled aggregate. Sustainability. **16**(10), 4063 (2024) https://www.mdpi.com/2071-1050/16/10/4063

Cai, X., Yang, D., Zhang, D., Cui, J., Wang, W., Liu, L.: Development of high-early-strength low-carbon engineered cementitious composites with calcium sulfoaluminate cement incorporating high-volume fly ash. Case Stud. Constr. Mater. **18**, e01959 (2023). https://doi.org/10.1016/j.cscm.2023.e01959

Cavusoglu, I.: Superplasticizer dosage effect on strength, microstructure and permeability enhancement of cementitious paste fills. Minerals. **14**, 1242 (2024). https://doi.org/10.3390/min14121242

Chen, W., Wu, M., Liang, Y.: Effect of SF and GGBS on pore structure and transport properties of concrete. Materials. **17**, 1365 (2024). https://doi.org/10.3390/ma17061365

Chindaprasirt, P., Hatanaka, S., Chareerat, T., Mishima, N., Yuasa, Y.: Cement paste characteristics and porous concrete properties. Constr. Build. Mater. **22**, 894–901 (2008). https://doi.org/10.1016/j.conbuildmat.2006.12.007

Couvidat, J., et al.: Greening effect of slag cement-based concrete: environmental and ecotoxicological impact. Environ. Technol. Innovation. **22**, 101467 (2021). https://doi.org/10.1016/j.eti.2021.101467

Couvidat, J., Diliberto, C., Meux, E., Izoret, L., Lecomte, A.: Greening effect of concrete containing granulated blast-furnace slag composite cement: is there an environmental impact? Cem. Concr. Compos. **113**, 103711 (2020). https://doi.org/10.1016/j.cemconcomp.2020.103711

Dash, S., Kar, B.: Environment friendly pervious concrete for sustainable construction. IOP Conf. Ser. Mater. Sci. Eng. **410**, 012005 (2018). https://doi.org/10.1088/1757-899X/410/1/012005

Divsholi, B.S., Lim, T.Y.D., Teng, S.: Durability properties and microstructure of ground granulated blast furnace slag cement concrete. Int. J. Concr. Struct. Mater. **8**(2), 157–164 (2014). https://doi.org/10.1007/s40069-013-0063-y

Dobrucali, E.: Relationship between CO2 emissions from concrete production and economic growth in 20 OECD countries. Buildings. **14**(9), 2709 (2024) https://www.mdpi.com/2075-5309/14/9/2709

Dom, A., Jamaluddin, N., Abdul Hamid, N.A., Hoon, C.: A review: GGBS as a cement replacement in concrete. IOP Conf. Ser. Earth Environ. Sci. **1022**, 012044 (2022). https://doi.org/10.1088/1755-1315/1022/1/012044

Drake, J., Bradford, A., Marsalek, J.: Review of environmental performance of permeable pavement systems: state of the knowledge. Water Qual. Res. J. Can. **48**, 2013 (2013). https://doi.org/10.2166/wqrjc2013.055

Gao, D., Meng, Y., Yang, L., Tang, J., Lv, M.: Effect of ground granulated blast furnace slag on the properties of calcium sulfoaluminate cement. Constr. Build. Mater. **227**, 116665 (2019). https://doi.org/10.1016/j.conbuildmat.2019.08.046

Gao, J.M., Qian, C.X., Liu, H.F., Wang, B., Li, L.: ITZ microstructure of concrete containing GGBS. Cem. Concr. Res. **35**(7), 1299–1304 (2005). https://doi.org/10.1016/j.cemconres.2004.06.042

Ghafoori, N., Maler, M.O., Najimi, M., Hasnat, A., Gbadamosi, A.: Role of cement type on properties of high early-strength concrete. J. Compos. Sci. **9**(1), 3 (2025) https://www.mdpi.com/2504-477X/9/1/3

Gholampour, A., Ozbakkaloglu, T.: Performance of sustainable concretes containing very high volume class-F Fly ash and ground granulated blast furnace slag. J. Clean. Prod. **162**, 1407–1417 (2017). https://doi.org/10.1016/j.jclepro.2017.06.087

Gupta, M., Raj, D.R., Kumar Sahu, D.A.: Effect of Rice husk ash, silica fume & GGBFS on compressive strength of performance based concrete. Mater. Today Proc. **55**, 234–239 (2022). https://doi.org/10.1016/j.matpr.2021.06.343

He, J., Liu, T., Wu, W., Ling, X., Yu, Q.: Mechanism of SCMs on the hydration, pore structure, and durability of UHPFRC. J. Sustain. Cem.-Based Mater., 1–18. https://doi.org/10.1080/21650373.2025.2499063

He, T., Xu, Z., Li, Z., Zhao, X., Zhao, S., Liu, Y.: Study on the relationship between the particle size distribution characteristics of ground granulated blast furnace slag and its mortar properties. Frontiers Mater. **9**, 962279 (2022). https://doi.org/10.3389/fmats.2022.962279

Imbabi, M.S., Carrigan, C., McKenna, S.: Trends and developments in green cement and concrete technology. Int. J. Sustain. Built Environ. **1**(2), 194–216 (2012). https://doi.org/10.1016/j.ijsbe.2013.05.001

Jwaida, Z., et al.: Investigating the properties of composite cement-based mortar containing high volumes of GGBS and CCR. J. Compos. Sci. **9**(6), 301 (2025) https://www.mdpi.com/2504-477X/9/6/301

Kadhum, A., Al-Ameeri, A., Edrees, S., Rafiq, M.: High-performance eco-friendly concrete with improved strength, chloride ion penetration, and corrosion resistance by the high volume of ground granulate blast furnace slag (GGBS). Zastita Materijala. **66**, 845 (2025). https://doi.org/10.62638/ZasMat1325

Karadumpa, C., Pancharathi, R.: Influence of Particle Packing Theories on Strength and Microstructure Properties of Composite Cement–Based Mortars. J. Mater. Civ. Eng. **33**, 04021267 (2021). https://doi.org/10.1061/(ASCE)MT.1943-5533.0003848

Kia, A., Wong, H.S., Cheeseman, C.R.: Clogging in permeable concrete: a review. J. Environ. Manag. **193**, 221–233 (2017). https://doi.org/10.1016/j.jenvman.2017.02.018

Knight, K.A., Cunningham, P.R., Miller, S.A.: Optimizing supplementary cementitious material replacement to minimize the environmental impacts of concrete. Cem. Concr. Compos. **139**, 105049 (2023). https://doi.org/10.1016/j.cemconcomp.2023.105049

Kolapo, O., Babafemi, A.J., Umoh, A., Olawuyi, B.: Effect of batching methods on the fresh and hardened properties of concrete. Int. J. Res. Rev. Appl. Sci. **13**, 773–779 (2012)

Kováč, M., Sičáková, A.: Pervious concrete as an environmental solution for pavements: focus on key properties. Environments. **5**(1), 11 (2018) https://www.mdpi.com/2076-3298/5/1/11

Li, M., Wang, L., Wang, J., Li, X., Xu, F., Liu, S.: Hydration properties of cement paste containing high volume ground granulated blast furnace slag (GGBFS). J. Therm. Anal. Calorim. **150**, 13099 (2025). https://doi.org/10.1007/s10973-025-14563-4

Liu, H., Luo, G., Wei, H., Yu, H.: Strength, permeability, and freeze-thaw durability of pervious concrete with different aggregate sizes, porosities, and water-binder ratios. Appl. Sci. **8**, 1217 (2018). https://doi.org/10.3390/app8081217

Ma, N., Ye, H., Choi, C.E., Zhang, J.: Strength enhancement of ca(OH)2 activated ground granulated blast furnace slag-stabilized dredged sediments using Na2CO3. Acta Geotech. **20**(2), 945–964 (2025). https://doi.org/10.1007/s11440-024-02470-1

Marathe, S., Durga, P.L., Sadowski, Ł.: Engineering of alkali-activated permeable pavement composites with agro-industrial wastes. Int. J. Pavement Eng. **25**(1), 2431600 (2024). https://doi.org/10.1080/10298436.2024.2431600

Munjal, P., Hau, K.K., Hon Arthur, C.C.: Effect of GGBS and curing conditions on strength and microstructure properties of oil well cement slurry. J. Build. Eng. **40**, 102331 (2021). https://doi.org/10.1016/j.jobe.2021.102331

Nasir, M., Al-Amoudi, O., Maslehuddin, M.: Effect of placement temperature and curing method on plastic shrinkage of plain and pozzolanic cement concretes under hot weather. Constr. Build. Mater. **152**, 943–953 (2017). https://doi.org/10.1016/j.conbuildmat.2017.07.068

Nilimaa, J.: Smart materials and technologies for sustainable concrete construction. Dev. Built Environ. **15**, 100177 (2023). https://doi.org/10.1016/j.dibe.2023.100177

Onsongo, S.K., Olukuru, J., Munyao, O.M., Mwabonje, O.: The role of agricultural ashes (rice husk ash, coffee husk ash, sugarcane bagasse ash, palm oil fuel ash) in cement production for sustainable development in Africa. Discov. Sustain. **6**(1), 62 (2025). https://doi.org/10.1007/s43621-025-00841-6

Ou, G., Kishi, T.: A numerical model for the densification process of C-S-H in microstructural development of Portland cement paste. Adv. Cem. Res., 1–15 (2025). https://doi.org/10.1680/jadcr.24.00157

Saranya, P., Nagarajan, P., Shashikala, A.P.: Ecofriendly GGBS concrete: a state-of-the-art review. IOP Conf. Ser. Mater. Sci. Eng. **330**, 012057 (2019). https://doi.org/10.1088/1757-899X/330/1/012057

Pacheco, R. T.: Unlocking Concrete's Potential: a Comprehensive Guide to Admixtures and their Impact on Construction (2024)

Pereira, P., Evangelista, L., de Brito, J.: The effect of superplasticisers on the workability and compressive strength of concrete made with fine recycled concrete aggregates. Constr. Build. Mater. **28**(1), 722–729 (2012). https://doi.org/10.1016/j.conbuildmat.2011.10.050

Saini, K., Saikia, S.K., Rajput, A.S.: Durability assessment of ground granulated blast furnace slag-based CO2 sequestered concrete. J. Sustain. Cem.-Based Mater. **14**(7), 1427–1443 (2025). https://doi.org/10.1080/21650373.2025.2509033

Shah, A., Khan, S., Khan, R.: Effect of high range water reducers (HRWR) on the properties and strength development characteristics of fresh and hardened concrete. IJST, Trans. Civ. Eng. **37**, 513–517 (2013)

Silva, L.H.P., et al.: Use of blast furnace slag in cementitious materials for pavements - systematic literature review and eco-efficiency. Sustain. Chem. Pharm. **33**, 101030 (2023). https://doi.org/10.1016/j.scp.2023.101030

Singh, A., Sampath, P.V., Biligiri, K.P.: A review of sustainable pervious concrete systems: emphasis on clogging, material characterization, and environmental aspects. Constr. Build. Mater. **261**, 120491 (2020). https://doi.org/10.1016/j.conbuildmat.2020.120491

Tan, K., et al.: Innovative high-strength, high-permeability concrete for large-scale applications in permeable subgrade of highway tunnel. Case Stud. Constr. Mater. **20**, e02977 (2024). https://doi.org/10.1016/j.cscm.2024.e02977

Verma, M., Sharma, A., Chaturvedi, I., Imtiyaz, W., Bharti, S., Kumar, A.: Optimization of production variables for metakaolin and rice husk ash-based geopolymer cement. IOP Conf. Ser. Earth Environ. Sci. **1327**(1), 012002 (2024). https://doi.org/10.1088/1755-1315/1327/1/012002

Verma, M., Sharma, S., Colangelo, F., Farina, I.: Experimental study on durability properties of densified small particles based concrete. In: 18th International Brick and Block Masonry Conference. springer, Cham (2025)

Volder, A., Watson, W., Viswanathan, B.: Potential use of pervious concrete for maintaining existing mature trees during and after urban development. Urban For. Urban Green. **8**, 249–256 (2009). https://doi.org/10.1016/j.ufug.2009.08.006

Wang, Z., Yang, J., Wu, Y.: Particle packing optimization for CCR-GGBS-FA binder stone waste Pavement Base material. Appl. Sci. **14**(17), 7979 (2024) https://www.mdpi.com/2076-3417/14/17/7979

Xiao, Q., Cai, L., Liu, Y., Zhang, M., Hu, X.: Impact of pore geometry on mechanical properties and anti-clogging efficiency in porous drainage pavement structures. Case Stud. Constr. Mater. **23**, e04929 (2025). https://doi.org/10.1016/j.cscm.2025.e04929

Xu, D., et al.: The influence of curing regimes on hydration, microstructure and compressive strength of ultra-high performance concrete: a review. J. Build. Eng. **76**, 107401 (2023). https://doi.org/10.1016/j.jobe.2023.107401

Xu, Z., et al.: Promoting utilization rate of ground granulated blast furnace slag (GGBS): incorporation of nanosilica to improve the properties of blended cement containing high volume GGBS. J. Clean. Prod. **332**, 130096 (2022). https://doi.org/10.1016/j.jclepro.2021.130096

Yahyaee, T., Elize, H.S.: A comprehensive study on mechanical properties, durability, and environmental impact of fiber-reinforced concrete incorporating ground granulated blast furnace slag. Case Stud. Constr. Mater. **20**, e03190 (2024). https://doi.org/10.1016/j.cscm.2024.e03190

Yu, F., Guo, J., Liu, J., Cai, H., Huang, Y.: A review of the pore structure of pervious concrete: Analyzing method, characterization parameters and the effect on performance. Constr. Build. Mater. **365**, 129971 (2023). https://doi.org/10.1016/j.conbuildmat.2022.129971

Yuan, Q., Liu, Z., Zheng, K., Ma, C.: Chapter 3 - Portland cement concrete. In: Yuan, Q., Liu, Z., Zheng, K., Ma, C. (eds.) Civil Engineering Materials, pp. 59–204. Elsevier (2021). https://doi.org/10.1016/B978-0-12-822865-4.00003-9

Zhang, Y., et al.: Early-age cracking of Fly ash and GGBFS concrete due to shrinkage, creep, and thermal effects: a review. Materials. **17**, 2288 (2024). https://doi.org/10.3390/ma17102288

Zhong, R., Wille, K.: Compression response of normal and high strength pervious concrete. Constr. Build. Mater. **109**, 177–187 (2016). https://doi.org/10.1016/j.conbuildmat.2016.01.051

Zhou, D., Liu, J., Cheng, L., An, S., Huo, L.: Pore structures and deterioration mechanism of concrete after cryogenic freeze-thaw cycles: effects of moisture contents and aggregates. J. Build. Eng. **96**, 110620 (2024). https://doi.org/10.1016/j.jobe.2024.110620

Transforming Industrial by-Products into Sustainable Cementitious Solutions for CO_2 Reduction

Nabil Khatib[1(✉)], Mohamed El Habib Hitar[2], El Mostapha Lotfi[2], El Houssaine Ouacha[1,3], Hicham Banouni[1], and Bouazza Faiz[1]

[1] Laboratory of Metrology and Information Processing, Ibn Zohr University, Faculty of Science, B.P, 8106 Agadir, Morocco
nabil.khatib@edu.uiz.ac.ma

[2] Laboratory of Spectroscopy, Molecular Modeling, Materials, Nanomaterials, Water and Environment, Materials For Environment Team, ENSAM, Mohammed V University in Rabat, Rabat, Morocco

[3] Laboratory of Science and Technology Research (LRST), ESEF, Ibn Zohr University, Agadir, Morocco

Abstract. The increasing demand for sustainable construction materials has propelled research into innovative solutions that minimize environmental impact. My current work focuses on exploring the potential of supplementary cementitious materials (SCMs) derived from industrial by-products, particularly mining waste, as a sustainable alternative to conventional cement. This approach aligns with the global imperative to reduce CO_2 emissions associated with cement production, one of the largest industrial contributors to greenhouse gas emissions. By leveraging advanced characterization techniques and a deep understanding of material science, this research aims to transform inert waste into active components that enhance the performance and sustainability of cementitious systems. Through this, the study aspires to contribute to a circular economy model while addressing critical challenges in waste management and carbon footprint reduction in the construction industry.

Keywords: Mine Tailings · Supplementary Cementitious Materials (SCMs) · CO_2 Reduction · Waste Valorization

1 Introduction

Globally, mining generates vast quantities of waste, with mine tailings being among the largest streams, reaching nearly 4 billion tonnes [1, 2]. These fine-grained residues, consisting of crushed rock, process water, and chemical additives, are commonly stored in dams. Poorly managed disposal has resulted in serious environmental hazards, such as acid mine drainage [3], airborne dust [4], structural failures [5], and contamination of soils and ecosystems [6, 7].

M. Posani and S. Singh (Eds.): RYS 2025, RILEM Bookseries 67, pp. 386–399, 2026.
https://doi.org/10.1007/978-3-032-14652-6_29

Recent advancements in sustainable construction have moved beyond viewing mine tailings as mere inert fillers or backfill material. Today, tailings are being reengineered into active components of cementitious systems. For example, [8] demonstrated how gold mining tailings could be transformed into bricks that meet strength and absorption standards, offering both environmental and socioeconomic benefits in remote regions. Similarly, [9] showed that Pb–Zn tailings from the Touissit-Bou Beker mining district in Morocco could be effectively reprocessed by froth flotation to reduce environmental contaminants, producing a cleaner material suitable for reuse in construction. This was further supported by [10], who confirmed the value of these tailings through detailed geometallurgical and environmental assessments. Adding to this evidence, [11] demonstrated that sulfidic tailings can actively participate in cement hydration reactions and immobilize hazardous elements when blended into various cement systems.

In a broader context, [12] highlighted that mine tailings, along with other industrial residues, contain essential oxides (e.g., CaO, SiO_2, Al_2O_3, Fe_2O_3) suitable for clinker formation, though the presence of impurities such as alkalis and heavy metals necessitates careful handling during processing.

The application of mine tailings has also been explored in more technically demanding concrete formulations. [13] evaluated molybdenum tailings in self-compacting concrete, finding that up to 20% replacement of cement maintains both flowability and strength. Similarly, [14] studied copper mine tailings and slag, concluding that their performance – though limited in early reactivity – can be comparable to traditional supplementary cementitious materials (SCMs) at low dosages. Extending this scope, [15] showed that a combination of mine tailings powder and sand can significantly improve the mechanical and durability properties of ultra-high-performance fiber-reinforced concrete, even under sulfate exposure and shrinkage conditions.

In parallel, innovation in binder technologies is shifting toward entirely waste-derived systems [16]. introduced a waste-based green cement (WGC) activated with alkaline compounds sourced solely from industrial waste. Their binder reached up to 42 MPa in compressive strength, representing a major step forward in reducing reliance on Portland cement and commercial activators. Supporting this direction, [17] developed mechanical and thermal activation protocols to enhance the reactivity of graphite mine tailings, showing promising results in solubility and hydration behavior.

These contributions point to a growing consensus: mine tailings – regardless of origin – can be transformed into performance-enhancing components of modern, sustainable construction materials. Their reactivity, stability, and environmental safety are being increasingly validated, provided appropriate treatment and mix design strategies are applied.

Building on this momentum, the cement industry has also embraced the use of industrial by-products like fly ash, silica fume, and slag. These materials not only enhance the durability of concrete and lower production costs but also reduce the environmental impact of cement manufacturing, which currently accounts for 5–8% of global anthropogenic CO_2 emissions [18 – 20]. Consequently, replacing Portland cement—even partially—with reactive or inert waste materials has become a recognized climate mitigation strategy.

Yet, despite these innovations, research specifically targeting Zn–Pb mine tailings remains sparse. Most existing studies are limited to individual case analyses, lacking comparative insight into different tailings types and their interaction with cement chemistry. Our study addresses this gap by evaluating Zn–Pb tailings from Morocco as potential SCMs. The aim is to reduce the cement content in construction materials, thereby cutting CO_2 emissions and aligning with policy mechanisms such as the EU's Mécanisme d'Ajustement Carbone aux Frontières (MACF), which highlights cement as a priority sector for decarbonization (European Commission, 2023 [21]). This work also supports the implementation of Best Available Techniques (BAT) by minimizing waste generation and promoting its reuse—contributing to a circular economy in both mining and construction industries.

2 Results and Discussion

2.1 Sampling and Pre-Treatment Workflow

Sample collection was meticulously planned and executed through drilling at multiple predefined points across the tailings dam, following a well-distributed mesh pattern to ensure representative sampling (Fig. 1). The samples were then subjected to pre-treatment steps before laboratory testing. These included drying at 105 °C to eliminate moisture and grinding to achieve a target particle fineness of approximately 100 μm. This process aimed to enhance the material's reactivity by increasing its specific surface area and reducing particle size. The entire sampling and pre-treatment process, conducted within the mine's onsite laboratory, is detailed in the flowchart presented in Fig. 2.

The pre-treatment was essential due to the inherently poor fineness of raw mine tailings, which typically exhibit relatively large particle sizes ($D_{50} > 100$ μm) and low specific surface area (SSA). Such characteristics reduce the tailings' reactivity as SCMs, as larger particles provide less surface area for chemical interaction. Specifically, a lower SSA results in fewer contact points between the particles and the surrounding hydration products in the cementitious matrix, slowing down the pozzolanic reaction and compromising mechanical performance.

For this study, the tested specimen is a homogenized composite of all field-collected samples, hereafter referred to as **Sample AD**.

2.2 Mineralogical Fingerprint from XRD

The diffractogram (Fig. 3) of the tested sample is dominated by sharp, well-resolved reflections attributable to quartz and aluminosilicate frameworks (plagioclase-type calcium–aluminum silicates and minor Mg–Al silicates), with occasional iron-oxide peaks. The overall background is flat and the broad diffuse halo in the 20–35° 2θ region (Cu Kα) is weak, indicating that the glassy/amorphous fraction is limited. In crystalline tailings, such patterns are typical of thermodynamically stable silicates produced by slow geological or high-temperature industrial formation, rather than quenched glasses. This by itself argues against intrinsic hydraulicity (self-setting in water) because hydraulic behavior in SCMs is commonly associated with Ca-rich, metastable, and/or amorphous phases that hydrate rapidly [22 – 24].

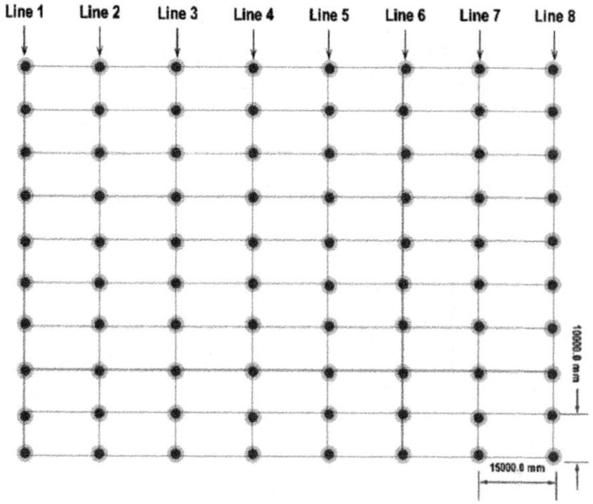

Fig. 1. Well-Distributed Mesh Pattern for Representative Sampling of Tailings at former Dams

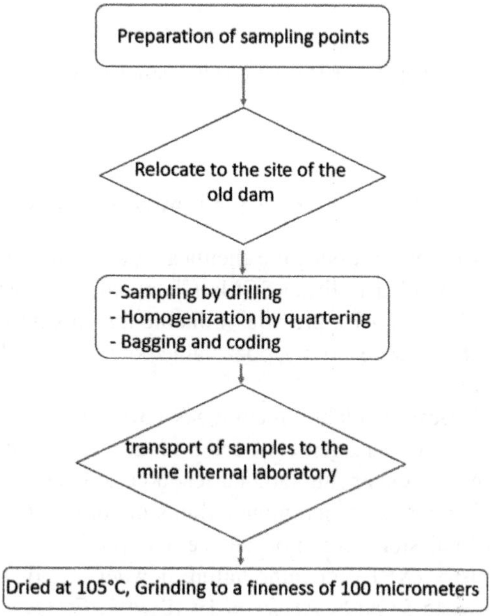

Fig. 2. Flowchart of Onsite Sampling and Pre-Treatment Process Conducted in the Mine Laboratory

Methodologically, quantitative confirmation should be obtained by Rietveld refinement with an internal standard (e.g., 10 wt% corundum) to determine the amorphous content and to apportion crystalline phases; without an internal standard, amorphous fractions are systematically underestimated [22, 26]. For defensible phase balances, we can further use finely ground, back-loaded mounts, a step size $\leq 0.02° 2\theta$, and a counting time sufficient to resolve minor peaks; also applying micro-absorption corrections if heavy oxides (Mn/Ti/Zr) are abundant [22, 26].

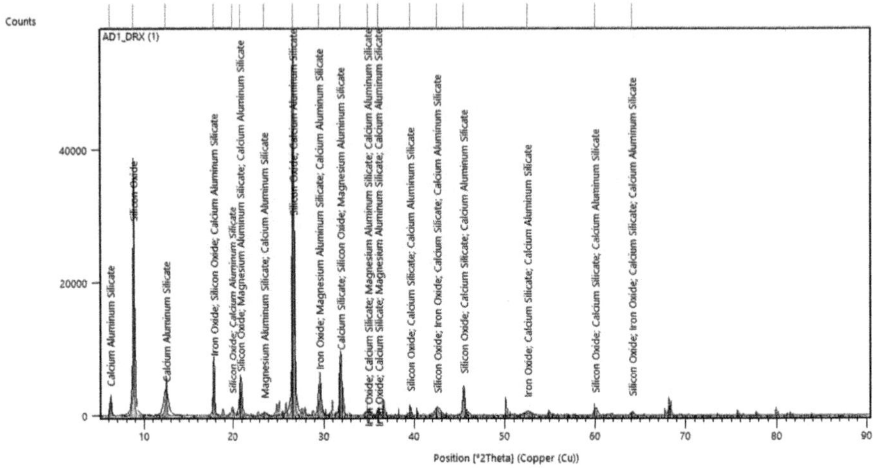

Fig. 3. XRD pattern with phase labels.

2.3 Chemistry from XRF: What Drives (or Limits) Reactivity

Beyond physical preparation, assessing the chemical composition of the mine tailings is essential for evaluating their feasibility as SCMs. Chemical characterization, particularly through X-ray fluorescence (XRF), provides valuable insights into the oxide composition, which directly influences reactivity, pozzolanic potential, and long-term durability in cementitious systems.

In alignment with Best Available Techniques (BAT) guidelines for environmentally responsible mine waste management, such evaluations contribute to both reducing environmental burdens and enhancing circular resource utilization. From a sustainability perspective, valorizing silicate-rich mine tailings in construction aligns with global efforts to reduce CO_2 emissions from Portland cement production.

Bulk oxide chemistry (XRF) for mine tailing powder tested AD, as mentioned in Table 1, shows $SiO_2 = 45.53$ wt%, $Al_2O_3 = 19.40$ wt%, $Fe_2O_3 = 0.77$ wt%, $CaO = 3.77$ wt%, $K_2O = 4.20$ wt%, $SO_3 = 2.73$ wt%, and a notably high $MnO = 15.53$ wt% (with $ZnO < 0.1$ wt% and PbO: not detected), as summarized in Table 1. The sum $(SiO_2 + Al_2O_3 + Fe_2O_3) = 65.70$ wt% is typical of aluminosilicate SCM candidates, while the low CaO (3.77 wt%) argues against stand-alone hydraulic setting. This chemistry positions the tested mine tailing AD primarily as a latent pozzolan: in high-pH pore solutions, the aluminosilicate phases can partially dissolve, consume portlandite [$Ca(OH)_2$], and form secondary C-(A)-S-H/C-S-H and related hydrates when AD is blended with Portland cement [23 – 25]. High SiO_2 and Al_2O_3 contents are especially desirable because they drive these pozzolanic reactions: dissolved silicate and aluminate species recombine with $Ca(OH)_2$ (a by-product of clinker hydration) to generate binding gels that improve strength and durability. In parallel, silicate-rich SCMs can reduce ASR susceptibility by binding part of the alkalis into hydrates and lowering transport, thereby limiting the alkali inventory available for deleterious reactions with reactive aggregates.

Two compositional features require targeted consideration. (i) Alkalis: the measured $K_2O = 4.20$ wt% (with Na_2O typically minor) can accelerate aluminosilicate dissolution—beneficial to early pozzolanic kinetics—but may also elevate pore-solution alkalinity, which can exacerbate ASR depending on aggregate reactivity and overall mix design; therefore, pore-solution measurements and accelerated ASR screening (e.g., AMBT/CSA) are recommended, particularly at higher replacement levels [24, 25]. (ii) Heavy oxides: the elevated $MnO = 15.53$ wt% (and trace Ti/Zr where present) generally occurs as dense, largely inert oxides under cementitious conditions; these phases increase true density and may influence color but are not expected to contribute significantly to binding phases. If any Mn-bearing particles are finely dispersed and redox-active, they may subtly alter early hydration kinetics—a possibility best assessed via isothermal calorimetry on blended pastes [24].

From a durability standpoint, maintaining a high pore-solution pH (≥ 12.5) remains essential to preserve the passive layer on reinforcing steel. Portland cement typically ensures this alkalinity, and the tested sample AD is intended as a partial replacement; nonetheless, chemical evaluation and mix optimization are prudent to verify that the passive condition of steel is preserved and that any pH effects are offset by the alkalinity supplied by the cement and the hydroxide balance of the blend. Taken together, the measured XRF values in Table 1—coupled with microstructural/mineralogical evidence and calorimetry—support using the tested Mine tailing powder AD as a supplementary cementitious material that reacts pozzolanically when blended with Portland cement, provided ASR risk is managed through testing and the binder design maintains the necessary alkalinity for reinforcement protection [23 – 25].

Table 1. Average Oxide Composition of Pretreated Mine Tailing Samples Determined by XRF Analysis

Oxide	SiO_2	Al_2O_3	Fe_2O_3	CaO	SO_3	K_2O	MnO	TiO_2 / ZrO_2	MgO	ZnO / PbO	SrO
Value (%)	45.53	19.40	0.77	3.77	2.73	4.20	15.53	1.75 / 1.57	1.57	<0.1 / ND	3.17

2.4 Carbonate Content (Calcimetry) and its Implications

Acid calcimetry (CO_2-evolution), using Bernard's calcimeter, has given a result of $CaCO_3 = 6.67973$ wt% (≈ 6.68 wt%). Converting stoichiometrically, this corresponds to $CaO_{eq} \approx 0.56 \times 6.67973 = 3.74$ wt%, which is essentially identical to the CaO measured by XRF. Therefore, most of the calcium occurs as carbonates rather than as hydraulic calcium silicates. This is important: carbonates in SCMs act mainly via filler/nucleation effects and through carboaluminate formation with aluminates in blended systems, not through self-hydration [24]. In Ca-poor, Si–Al-rich materials like the tested mine tailing powder AD, this carbonate fraction can still be beneficial by accelerating early reactions of the cement paste and stabilizing hemicarboaluminate/monocarboaluminate, especially in limestone-containing cements [23 – 25].

2.5 Microstructure and Local Chemistry from SEM/EDX

Backscattered-electron imaging (Figs. 4, 5 and 6) shows a fine-grained silicate matrix with angular particles and a heterogeneous pore distribution, consistent with crushed aluminosilicate tailings. EDX point/area analyses (Fig. 7) confirm an O–Si–Al-rich matrix with Fe and K as minors and only limited Ca, matching the XRF-based Ca deficiency. Importantly, trace heavy-element carriers (e.g., Zn, Pb) can appear locally in EDX despite being low/undetected in bulk XRF, reflecting sparse, micron-scale grains (sulfides/oxides or encapsulated inclusions). This micro-heterogeneity has two practical implications: (i) heavy elements are likely to be immobilized by cement hydration products once blended (encapsulation plus sorption on C-(A)-S-H and AFm phases), but (ii) leaching assessments should be run on both the raw powder and hydrated composites to verify environmental safety under the intended exposure classes [27, 28].

From a reactivity standpoint, the SEM evidence of tight, crystalline grains implies diffusion-limited dissolution unless the material is mechanically activated. Increasing fineness (Blaine SSA or laser-diffraction d50 reduction) increases the density of defects and reactive surface sites, which can significantly boost pozzolanic reaction rates even when the long-range order remains crystalline [22, 24].

Fig. 4. SEM micrograph at 10 μ magnification

Fig. 5. SEM micrograph at 5 μ magnification

Fig. 6. SEM micrograph at 2 μ magnification

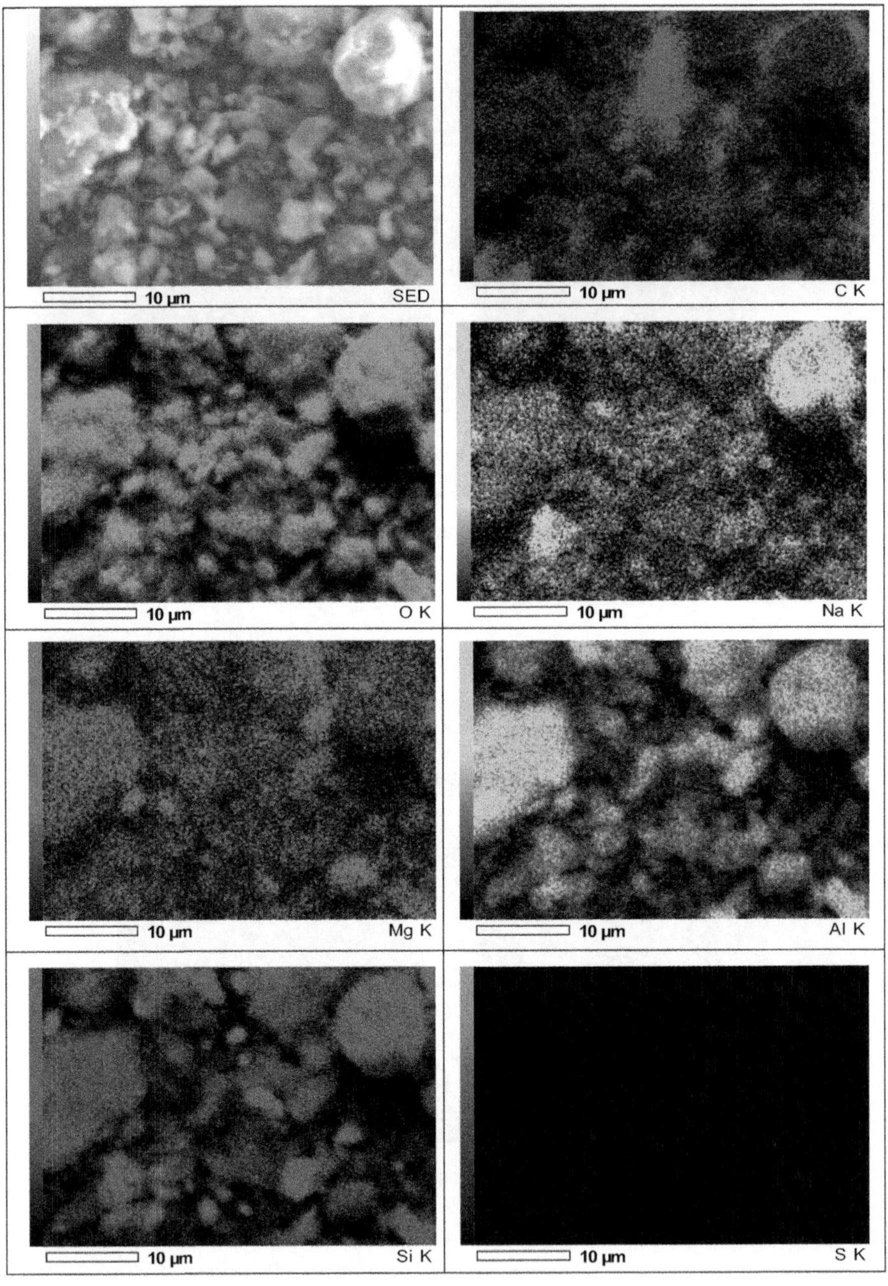

Fig. 7. EDX elemental maps for tested mine tailing powder (C, O, Na, Mg, Al, Si, S, K, Ca, Fe, Cu, Zn, Pb).

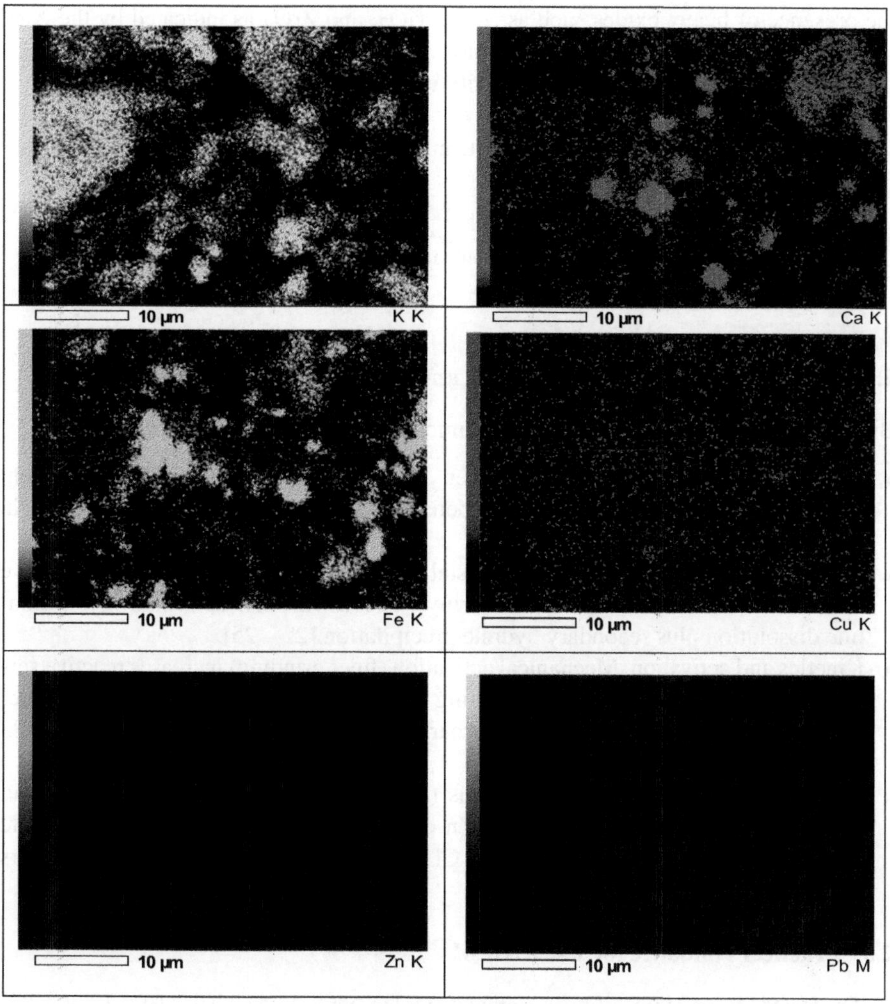

Fig. 7. (*continued*)

2.6 Physical Characterization via Pycnometry

In addition to chemical composition, the physical characteristics of mine tailings must be evaluated to assess their behavior in cementitious systems. One of the key parameters is true density, which provides insight into the compactness and mineralogical nature of the solid phase. It is a useful indicator for calculating specific surface area (when combined with particle size distribution data) and understanding the material's bulk behavior.

To determine the true density of the sample, pycnometry was conducted using Diethyl phthalate (purity 99.50%, Aldrich Chemistry) as the displacement fluid. This method is described in details by our previous study [30]. The resulting density was measured at 3.103 g/cm^3, as mentioned also in Table 2. This relatively high value is consistent with the presence of heavy oxides such as MnO, TiO_2, and ZrO_2 as indicated by the XRF results. It also suggests that the tailings have potential as a dense, compact filler or as a pozolanic additive, provided their reactivity is validated through further testing.

Table 2. True Density of Mine Tailings Determined by Pycnometry Using Diethyl Phthalate

Sample	Density (g/cm^3)	Liquid Used	Purity	Supplier
AD	3.103	Diethyl phthalate	99.50%	Aldrich Chemistry

2.7 Integrated Appraisal of Reactivity and Performance Implications

Bringing together chemistry, mineralogy, microstructure, and calorimetry:

- Calcium speciation. Agreement between XRF CaO and calcimetry CaO-equivalent confirms that Ca is predominantly carbonate, corroborating the absence of Ca-rich, metastable hydraulic glass.
- Structure vs. reactivity. XRD shows crystalline aluminosilicates with low glass; hence intrinsic hydraulicity is unlikely. Reactivity proceeds via pozzolanic pathways—alkaline dissolution plus secondary hydrate precipitation [22 – 25].
- Kinetics and activation. Mechanical activation (finer grinding) increases reactive surface area and defect density, accelerating dissolution and C-(A)-S-H growth; limestone in the binder can further promote carboaluminate formation and nucleation [23 – 25].
- Durability and safety. Moderate alkalis (notably K_2O) require ASR management at the mix level [24, 25]. Sparse Pb/Zn carriers observed by SEM/EDX motivate leaching tests (EN 12457–2, US-EPA TCLP) on the raw tailing and hydrated binders to verify immobilization [27, 28].

2.8 Practical Guidance for Use as SCM

- The tested Pb-Zn Mine tailing is a latent pozzolan: SiO_2 45.53 wt%, Al_2O_3 19.40 wt%, CaO 3.77 wt%, $CaCO_3$ 6.68 wt% ($CaO_{e\varphi} \approx 3.74$ wt%), MnO 15.53 wt%.
- XRD: predominantly crystalline with limited glass $\rightarrow$ low intrinsic hydraulicity.
- SEM/EDX: fine, angular aluminosilicate grains; Ca limited; trace heavy-element inclusions localized.
- In alkaline pore solution, the tested mine tailing powder AD consumes portlandite and forms C-(A)-S-H/carboaluminates (pozzolanic pathway).
- Use level: 15–30 wt% cement replacement to reduce clinker and CO_2 while maintaining performance.
- Fineness target: Blaine $>400\ m^2\cdot kg^{-1}$ or $d_{50} \leq 10\ \mu m$ to overcome diffusion-limited dissolution.
- Verify reactivity: isothermal calorimetry (extra heat vs. control), TGA/DTG (CH$\downarrow$, bound water$\uparrow$), QXRD on hydrated pastes.
- Durability/mechanics: to include strength index and transport tests; manage ASR risk ($K_2O \approx$ 4–5 wt%) via mix design and aggregate choice.

- Environmental compliance: to run EN 12457–2 / US-EPA 1311 (TCLP) to confirm immobilization of trace carriers.
- Recommend ultrasound NDT: monitoring pulse velocity/attenuation/reflection coefficient [30 – 34] to track setting, CH consumption, and microstructure densification non-destructively.

3 Conclusion

The Zn–Pb mine tailing tested is a Si–Al-rich, Ca-poor mine tailing (SiO_2 45.53 wt%, Al_2O_3 19.40 wt%, CaO 3.77 wt%; $CaCO_3$ 6.68 wt% → $CaO_{e\varphi} \approx 3.74$ wt%) whose XRD pattern is predominantly crystalline with little glass, while SEM/EDX shows fine, angular aluminosilicate grains—altogether indicating low intrinsic hydraulicity but credible latent pozzolanic behavior. Used as a secondary binder at 15–30 wt% replacement, Sample AD should consume portlandite in alkaline pore solution and form secondary C-(A)-S-H/carboaluminates, enabling clinker reduction and CO_2 savings without compromising performance when fineness is high (Blaine >400 $m^2 \cdot kg^{-1}$ or $d_{50} \leq 10$ μm). Verification should couple isothermal calorimetry (additional heat vs. control), TGA/DTG (CH↓, bound water↑), and QXRD on hydrated pastes, complemented by mechanical and transport indices. Alkalis ($K_2O \approx 4.20$–5 wt%) must be managed to control ASR via aggregate selection, SCM synergy, and, where needed, pore-solution checks; the high MnO (15.53 wt%) is largely inert but warrants monitoring for subtle effects on early hydration. Environmental safety should be documented with EN 12457–2 / US-EPA 1311 (TCLP) leaching tests. As a rapid, non-destructive process monitor, integrate ultrasound NDT (pulse velocity/attenuation) to track setting and microstructure densification in Mine tailing sample tested as blended systems. Overall, the evidence supports the tested mine tailing as a viable SCM for durable, lower-carbon binders aligned with circular-economy goals.

References

1. Lottermoser, B.G.: Mine Wastes: Characterization, Treatment and Environmental Impacts. Mine Wastes (Third Edition): Characterization, Treatment and Environmental Impacts (2010). https://doi.org/10.1007/978-3-642-12419-8
2. Hannah, R., Rosado, P.: Global cement production has plateaued over the last decade. In: Our World in Data (2025)
3. Dold, B.: Evolution of acid mine drainage formation in sulphidic mine tailings. Minerals. **4**(3), 621–641 (2014). https://doi.org/10.3390/min4030621
4. Dudka, S., Adriano, D.: Environmental impacts of metal ore mining and processing: a review. J. Environ. Qual. **26**(3), 590–602 (1997). https://doi.org/10.2134/jeq1997.00472425002600030003x
5. Macklin, M.G., et al.: The long term fate and environmental significance of contaminant metals released by the January and march 2000 mining tailings dam failures in Maramures County, upper Tisa Basin. Romania. Appl. Geochem. **18**(2), 241–257 (2003). https://doi.org/10.1016/S0883-2927(02)00123-3
6. Riget, F., Johansen, P., Asmund, G.: Uptake and release of lead and zinc by blue mussels. Experience from transplantation experiments in Greenland. Mar. Pollut. Bull. **34**(10), 805–815 (1997). https://doi.org/10.1016/S0025-326X(97)00028-3

7. Perner, K., et al.: Contamination of arctic fjord sediments by Pb-Zn mining at Maarmorilik in central West Greenland. Mar. Pollut. Bull. **60**(7), 1065–1073 (2010). https://doi.org/10.1016/j.marpolbul.2010.01.019
8. Silva, E.M., Vriesde, E., Druiventak, A., Veiga, M., De Tomi, G.: Production of cement-tailings bricks with artisanal gold mining waste. Cleaner Waste Syst. **11**, 100269., ISSN 2772-9125 (2025). https://doi.org/10.1016/j.clwas.2025.100269
9. Khalil, A., et al.: Sustainable reprocessing of Pb–Zn mine tailings through froth flotation for resource recovery and environmental remediation in abandoned mining regions. Miner. Eng. **222**, 109132., ISSN 0892-6875 (2025). https://doi.org/10.1016/j.mineng.2024.109132
10. El Aallaoui, A., et al.: Investigating the reprocessing potential of abandoned zinc-lead tailings ponds: a comprehensive study using physicochemical, mineralogical, and 3D geometallurgical assessments. Miner. Eng. **209**, 108634., ISSN 0892-6875 (2024). https://doi.org/10.1016/j.mineng.2024.108634
11. Martins, N.P., Helser, J., Plötze, M., et al.: Effect of sulfidic mine tailings used as mineral admixtures on the hydration of common and alternative cements. Mater. Struct. **57**, 19 (2024). https://doi.org/10.1617/s11527-023-02289-4
12. Mache, E., Rajczakowska, M., Cwirzen, A.: Process residues in cement clinker production: a review. Waste Manag. Bull. **3**(3), 100205., ISSN 2949-7507 (2025). https://doi.org/10.1016/j.wmb.2025.100205
13. Jeong, J.G., Kim, J.Y., Kim, J.U., et al.: Self-compacting concrete using molybdenum mineral tailings: materials and mechanical property. Discov Civ Eng. **1**, 108 (2024). https://doi.org/10.1007/s44290-024-00109-4
14. Zeng, H., Ramanathan, S., Kim, H.J.: Effect of copper mine tailings and copper slag on the hydration, microstructure, and mechanical property in Portland cement. Results Eng. **26**, 104890., ISSN 2590-1230 (2025). https://doi.org/10.1016/j.rineng.2025.104890
15. Zaid, O., Ahmed, M., Yosri, A.M., et al.: Evaluating the impact of mine tailings wastes on the development of sustainable ultra high performance Fiber reinforced concrete. Sci. Rep. **15**, 6285 (2025). https://doi.org/10.1038/s41598-025-88683-0
16. Perera, M.A.G.P., Ranjith, P.G.: Greener horizons: revolutionizing construction materials with waste-based innovations - an experimental study. J. Build. Eng. **103**, 112211., ISSN 2352-7102 (2025). https://doi.org/10.1016/j.jobe.2025.112211
17. Maruthupandian, S., Chaliasou, N.A., Akram, S., Chrysanthou, A., Kanellopoulos, A.: Characterisation and activation of graphite mine tailings for use in cementitious binders. Constr. Build. Mater. **473**, 140923., ISSN 0950-0618 (2025). https://doi.org/10.1016/j.conbuildmat.2025.140923
18. Assi, L., Carter, K., Deaver, E., Anay, R., Ziehl, P.: Sustainable concrete: building a greener future. J. Clean. Prod. **198**, 1641–1651 (2018). https://doi.org/10.1016/j.jclepro.2018.07.123
19. Chen, C., Habert, G., Bouzidi, Y., Jullien, A.: Environmental impact of cement production: detail of the different processes and cement plant variability evaluation. J. Clean. Prod. **18**(5), 478–485 (2010). https://doi.org/10.1016/j.jclepro.2009.12.014
20. Siddique, R.: Utilization of industrial by-products in concrete. Proc. Eng. **95**(Scescm), 335–347 (2014). https://doi.org/10.1016/j.proeng.2014.12.192
21. European Commission. (2023). Decarbonisation options for the cement industry. Publications Office of the European Union. https://publications.jrc.ec.europa.eu/repository/handle/JRC131246
22. Scrivener, K., Snellings, R., Lothenbach, B. (eds.): A Practical Guide to Microstructural Analysis of Cementitious Materials, 1st edn. CRC Press (2016). https://doi.org/10.1201/b19074
23. Taylor, H.F.W.: Cement Chemistry, 2nd edn, p. 361. Thomas Telford Publishing, London (1997). https://doi.org/10.1680/cc.25929

24. Snellings, R., Mertens, G.: Jan Elsen; supplementary cementitious materials. Rev. Mineral. Geochem. **74**(1), 211–278 (2012). https://doi.org/10.2138/rmg.2012.74.6
25. Thomas, M.: The effect of supplementary cementing materials on alkali-silica reaction: a review. Cem. Concr. Res. **41**(12), 1224–1231., ISSN 0008-8846, (2011). https://doi.org/10.1016/j.cemconres.2010.11.003
26. Stutzman, P.E.: Guide for X-Ray Powder Diffraction Analysis of Portland Cement and Clinker. National Institute of Standards and Technology, Gaithersburg, MD., NISTIR 5755, (1996)
27. EN 12457–2. Characterization of waste—Leaching—Compliance test for leaching of granular waste materials and sludges.
28. US-EPA Method 1311 (TCLP). Toxicity Characteristic Leaching Procedure.
29. Londono-Zuluaga, D., Gholizadeh-Vayghan, A., Winnefeld, F., et al.: Report of RILEM TC 267-TRM phase 3: validation of the R3 reactivity test across a wide range of materials. Mater. Struct. **55**, 142 (2022). https://doi.org/10.1617/s11527-022-01947-3
30. Khatib, N., et al.: Monitoring early age behavior of mortar by ultrasound technique for environmental issues: effects of sand grading and salinity of mixing water. Constr. Build. Mater. **306**, 124852., ISSN 0950–0618 (2021). https://doi.org/10.1016/j.conbuildmat.2021.124852
31. Khatib, N., Ouacha, E.H., Banouni, H., et al.: Application of a novel ultrasound technique for real-time characterization of alkali accelerating admixture effects on cement mortar hydration. J Build Rehabil. **10**, 145 (2025). https://doi.org/10.1007/s41024-025-00659-9
32. Khatib, N., et al.: Effects of temperature and clay on hydration behavior of mortar: a study through ultrasound technique. Instrum. Mesure Métrologie. **23**(3), 243–251 (2024). https://doi.org/10.18280/i2m.230308
33. Khatib, N., Ouacha, E.H., Faiz, B., Ezzaidi, M., Banouni, H.: Analysis of the attenuative behaviour of accelerated cement based materials through a series of ultrasound pulse Echo measurements. Eng. Solid Mech. **7**, 109–120 (2019). https://doi.org/10.5267/j.esm.2019.4.002
34. Khatib, N., Ouacha, E.H., Faiz, B., Ezzaidi, M., Banouni, H.: Ultrasound study of setting and hardening behaviour of mortar using Portland composite cements CPJ45 and CPJ35 and different dosages of an alkali-free setting accelerator for shotcrete. J. Eng. Appl. Sci. **13**, 7150–7157 (2018)

Earth-Based Materials

A Combined Laboratory and Field Study to Assess the Durability of CEBs Against Water-Mediated Weathering

Rafail Panagiotou(✉) and Ioannis Ioannou(✉)

Department of Civil and Environmental Engineering, University of Cyprus, Nicosia, Cyprus
panagiotou.rafail@ucy.ac.cy

Abstract. Water-mediated weathering is probably the main deterioration mechanism affecting compressed earth blocks (CEBs). Among the water-mediated mechanisms affecting the performance of CEBs in use, wind-driven rain, rising damp and alternating wetting and drying cycles are the most prominent ones. In this study, spray erosion and drip tests were first conducted in the lab to assess the effect of wind-driven rain on CEBs. Furthermore, capillary absorption experiments and wet/dry cycles were carried out to investigate their response against rising damp and alternating wetting and drying, respectively. The results of the laboratory tests showed that the CEBs under investigation were not severely affected in spray erosion and drip tests, while they presented a comparatively low capillary absorption coefficient. Furthermore, they were resistant to alternating wetting and drying cycles. Following the laboratory tests, wallettes using the same CEBs were built and exposed to real environmental conditions in a rural environment. The wallettes were built on a stone foundation to prevent them from being eroded due to direct contact with liquid water and the ground. The wallettes showed negligible erosion after an entire year of natural exposure. More specifically, no degradation was observed due to wind-driven rain, though localized erosion was evident in areas where surface run-off of water occurred. The results of this study show that there is an agreement between the laboratory tests and the natural exposure of CEB wallettes, thus suggesting that the CEBs under investigation may be used locally to build contemporary structures, without being in severe danger of water-mediated erosion.

Keywords: Compressed Earth Blocks · Natural Exposure · Spray Test · Drip Test

1 Introduction

Compressed earth blocks (CEBs) are vulnerable to direct contact with liquid water. This is highlighted in several normative documents that deal with earth-based building materials [1–6]. According to the New Zealand Standards [4], an earthen structure should present a minimum service life of 50 years to be considered durable. Considering CEBs' vulnerability to direct contact with liquid water, special design should be applied

M. Posani and S. Singh (Eds.): RYS 2025, RILEM Bookseries 67, pp. 403–414, 2026.
https://doi.org/10.1007/978-3-032-14652-6_30

to earthen structures, e.g., large roof overhangs, water-resistant foundation or external plastering, to protect them against weathering [2, 3, 5, 6]. However, attention should also be given to the performance of CEBs themselves against water action.

Different water-mediated weathering phenomena, e.g., rising damp, wind-driven rain, wetting/drying cycles, deteriorate CEBs [7]. The most destructive one is probably wind-driven rain [8]. The effect of wind-driven rain is usually assessed in the laboratory by the NZS 4298 [5] spray erosion test. In general, stabilized CEBs pass the test [9–11], while non-stabilized ones fail [12, 13]. However, some researchers argue that the test is too harsh and does not predict the actual performance of CEBs [14, 15].

Another way to assess the effect of rainfall is the drip test. Bruno et al. [16] argue that cement and biological stabilization enhance the resistance of CEBs against drip erosion. The latter is further confirmed by Aguilar et al. [17], who observed increased erosion resistance by using chitosan biopolymer, either as an admixture or as a surface coating. They also observed that the effect of chitosan as a surface coating was much stronger compared to its addition as an admixture. Furthermore, Danso [18] and Panagiotou et al. [7] argue that non-stabilized CEBs also pass this test and can, thus, be considered suitable for construction.

Capillary absorption experiments are usually used to assess the behavior of CEBs against rising damp. However, direct contact with water usually results in disintegration and alteration of the microstructure of the CEBs [19]. Hall and Djerbib [20] proposed placing the sample on a saturated porous medium with negligible hydraulic resistance, e.g., a wick, to protect it from disintegration. Microstructural alteration, however, may only be avoided by using inert wetting liquids, e.g., organic liquids, instead of water [21].

As stated above, laboratory tests may be used to assess the expected performance of CEBs in use; however, they cannot determine with certainty their real durability. Natural exposure can give valuable information about the durability of CEBs; however, such information is usually limited to the specific exposure area and needs a lot of time to be derived. Guettala et al. [22] tried to correlate the results of laboratory testing on CEBs with the results from 4 years of natural exposure of CEB wallettes. They concluded that laboratory tests are harsher than outdoor conditions, giving higher values of erosion for CEBs. Seco et al. [23] exposed stabilized cylindrical samples of CEBs to outdoor conditions for approximately 18 months and then tried to correlate the results with various tests; they concluded that only the total absorption test agrees with the erosion pattern observed. Furthermore, Bui et al. [24] exposed rammed earth walls to outdoor conditions for 20 years and observed that walls with stabilized rammed earth presented erosion of about 2 mm in thickness, while walls built with non-stabilized rammed earth exhibited 6.4 mm thickness erosion. Thus, they concluded that, even unrendered non-stabilized rammed earth can be used in their region or other similar regions.

This study aims to assess the durability performance of non-stabilized CEBs produced in Cyprus against water-mediated weathering. Firstly, extensive laboratory testing, including spray erosion, drip, capillary absorption and wet/dry cycles, was carried out. Following the laboratory tests, CEB wallettes were built and exposed to real environmental conditions. The combination of laboratory testing and natural exposure is anticipated to help in assessing the performance of CEBs against water-mediated weathering.

2 Materials and Methods

In the current study, a soil-sand mixture was used to produce CEBs. The mixture consisted of 75% soil and 25% quarried sand. The mixing percentages were determined after extensive laboratory characterization of the raw materials [7] to obtain an optimized granular composition satisfying the criteria of CRATerre-EAG [3] for the production of CEBs.

2.1 Production of CEBs

The soil mixture was mixed with 16% w/w water, which was the optimum moisture content determined by proctor compaction tests [7]. Subsequently, an AECT IMPACT 2001A commercial hydraulic press, which applies a pressure of ca. 2200 psi, was used to produce 300 mm x 150 mm x 100 mm CEBs. Following their production, the CEBs were stacked on pallets and stored in the laboratory (T = 25 °C ± 5 °C, RH = 55% ± 10%) for at least 3 months.

2.2 Laboratory Testing

Bulk Density

The bulk density of the CEBs was calculated according to EN 772-13 [25] on 6 full CEB blocks. The specimens were dried in an oven at 105 °C until constant mass and then they were weighed. The external dimensions of the blocks were measured using a vernier caliper and the volume of each block was calculated. The density of the blocks was the quotient of their mass and volume.

Spray Erosion Test

The spray erosion test was conducted on 6 full CEB blocks. According to NZS 4298 [5], the specimen was placed 470 mm away from a nozzle, which sprayed at a pressure of 50 kPa for 1 h. At the end of the test, the erosion depth was measured using a rod with a 10 mm diameter.

Drip Test

The drip test was conducted according to NZS 4298 [5] on 6 prismatic specimens with dimensions 100 mm × 100 mm x 50 mm. A wick was placed 400 mm above an inclined (~27°) specimen and 100 mL of water was left to fall in 20–60 min. The depth of erosion was measured at the end of the test using a rod with a 3.15 mm diameter.

Capillary Absorption

Capillary absorption experiments were conducted on 6 cubic specimens with 70 mm edge, following EN 1925 [26]. The specimens were placed in an oven at 110 °C until constant mass. A non-reactive liquid, i.e., n-heptane, was used in the test to enable the calculation of the capillary absorption coefficient. The tests were carried out in a controlled environment at 20 °C.

Wet/Dry Appraisal Test

The NZS 4298 [5] wet/dry appraisal test was conducted on 6 cubic specimens with 70 mm edge. The specimens were partially immersed in 1 cm deep water for 4 min. The immersion time was determined following the results of the spray erosion test and the recommendation of the wet/dry appraisal test. Subsequently, the specimens were placed in an oven at 65 °C for 24 h to dry. In total, 6 wet/dry cycles were carried out and the specimens were examined after each cycle to check whether they satisfied the criteria of NZS 4298 [5].

2.3 Natural Weathering

Construction of Wallettes

Four CEB wallettes were built in a rural environment in Cyprus. The wallettes were built in pairs on a stone foundation, following the traditional practice used in Cyprus for the construction of earthen structures. Two (2) different locally quarried stones were used; Prastio and Troodos stone. A lower sorptivity characterizes the latter, which is an igneous stone, compared to the former, which is a limestone [27]; this enables the study of the effect of the foundation properties on the CEB wallettes performance. Each pair of wallettes was oriented in different directions, i.e., South-North and East-West, to assess the effect of orientation on the ultimate erosion. All the wallettes were covered with roofing tiles and left unrendered.

Assessment Methodology

Assessment of the weathering of the wallettes was complicated due to the coupled effect of various phenomena, e.g., rising damp, wind-driven rain and wetting/drying cycles. Monitoring started in November 2023 and lasted for a full year. Visual observation was carried out at least once a month, focusing on identifying problems caused by wind-driven rain or alternating wetting and drying cycles. Furthermore, every 2 months, high-resolution images were taken from various angles, with sufficient overlap to prepare 3D models of the wallettes using photogrammetry. The images were inserted into Agisoft Metashape software and they were automatically aligned. Subsequently, a dense point cloud was generated, based on the common points identified by the software. Finally, a textured 3D mesh was reconstructed based on the original photos to enhance the realism of the acquired 3D model. The effect of wind-driven rain was assessed based on the pit depth caused by the water droplets. Furthermore, the effect of alternating wetting/drying cycles was assessed based on the observance of various problems, such as fretting, cracking and efflorescence.

3 Results and Discussion

3.1 Laboratory Testing

The results of the bulk density, drip and spray erosion tests are summarized in Table 1. The value of the bulk density calculated is quite high and this may be attributed to the high compaction pressure applied during the production of the CEBs and the granular composition of the soil mixture used. The CEBs showed great performance in both erosion tests. In the drip test, they presented a pit depth of only 2.45 mm, corresponding to a limiting erodibility index (LEI) of 2, according to NZS 4298 [5]. This value is considered acceptable, as it is < 15 mm, which is the upper acceptability limit for this test. In the spray erosion test, the CEBs showed an erosion rate of 0.16 mm/min, which is linked to a LEI of 1. The LEI value assigned is the optimum, while the rate of erosion is much lower than the upper acceptability limit for this test, which is 2 mm/min. It is worth noting that the difference in the LEI assigned by the 2 tests is because a LEI = 1 can only be assigned using the spray erosion test, according to NZS 4298 [5]; the lowest LEI that can be assigned using the drip test is LEI = 2. Furthermore, a large coefficient of variation (COV) was observed in both tests, which is attributed to the inhomogeneity of the CEBs. However, the results of both tests showed that the CEBs under study can withstand water erosion.

Table 1. Bulk density, drip and spray erosion tests results.

Bulk Density		Drip test			Spray erosion test		
Bulk density, kg/m^3	COV, %	Pit depth, mm	COV, %	LEI	Rate of erosion, mm/min	COV, %	LEI
2070	1	2.45	35	2	0.16	26	1

The studied CEBs successfully passed the NZS 4298 [5] wet/dry appraisal test. Figure 1 shows the condition of a test specimen at the end of the test. The test specimen remained intact after the 6 cycles, showing material loss only in the water-submerged portion. However, according to the standard, a minimum of 50 mm fragment should be lost from the material to fail the test.

Fig. 1. Characteristic image of a CEB at the end of wet/dry appraisal test.

The results of the capillary absorption experiments indicated that the studied CEBs exhibited linear behavior, in line with the square-root of time law (Fig. 2). Furthermore, a relatively low capillary absorption coefficient was calculated, i.e., 84 $g/m^2sec^{1/2}$, with a COV = 10%. This capillary absorption coefficient is lower than respective ones reported in other studies dealing with earth-based building materials [19, 28]; the latter may be attributed to the increased density of the CEBs resulting from the high compaction pressure applied during their production. The low capillary absorption coefficient, together with the results of the drip and spray erosion tests, suggest that the studied non-stabilized CEBs can withstand water erosion and migration, likely due to their dense microstructure.

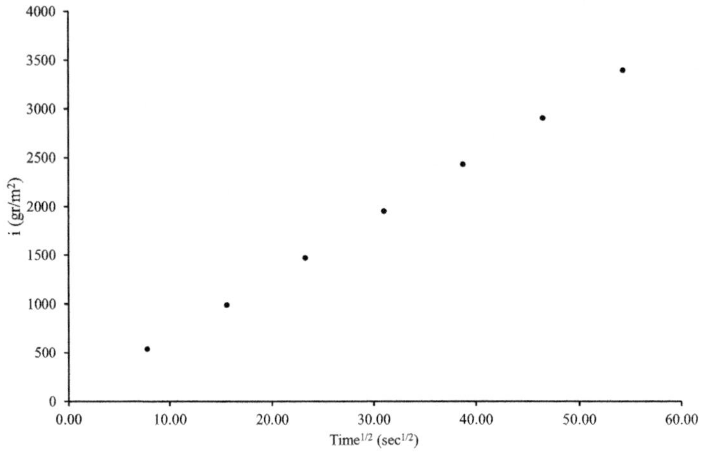

Fig. 2. Characteristic i vs $t^{1/2}$ curve for the CEBs under study.

3.2 Natural Weathering

The results of the natural weathering are shown in Figs. 3, 4 and 5. Figure 3 shows images exported from textured 3D models of the CEB wallettes with Prastio stone foundation. Generally, the CEBs remained unaltered after a full year of outdoor exposure. However, localized erosion was observed in some areas of the wallettes; this was attributed to surface run-off originating from the roofing tiles. Severe cracking was also observed in the north facade, especially where localized erosion occurred. This was much more severe compared to the other facades, due to the higher moisture retained in winter and the accelerated drying in summer. Surface cracking was observed in the western façade, which also did not receive sufficient direct sunlight in winter. In contrast, cracking was not observed in the eastern and southern facades, due to gentler drying conditions, i.e., less intense drying during winter and spring. However, in these 2 facades, and especially in the southern one, white efflorescence was observed on the lower part of the CEB wallette, i.e., on the first course of CEBs (Fig. 4). This was attributed to rising damp, as the efflorescence formed a straight line parallel to the foundation.

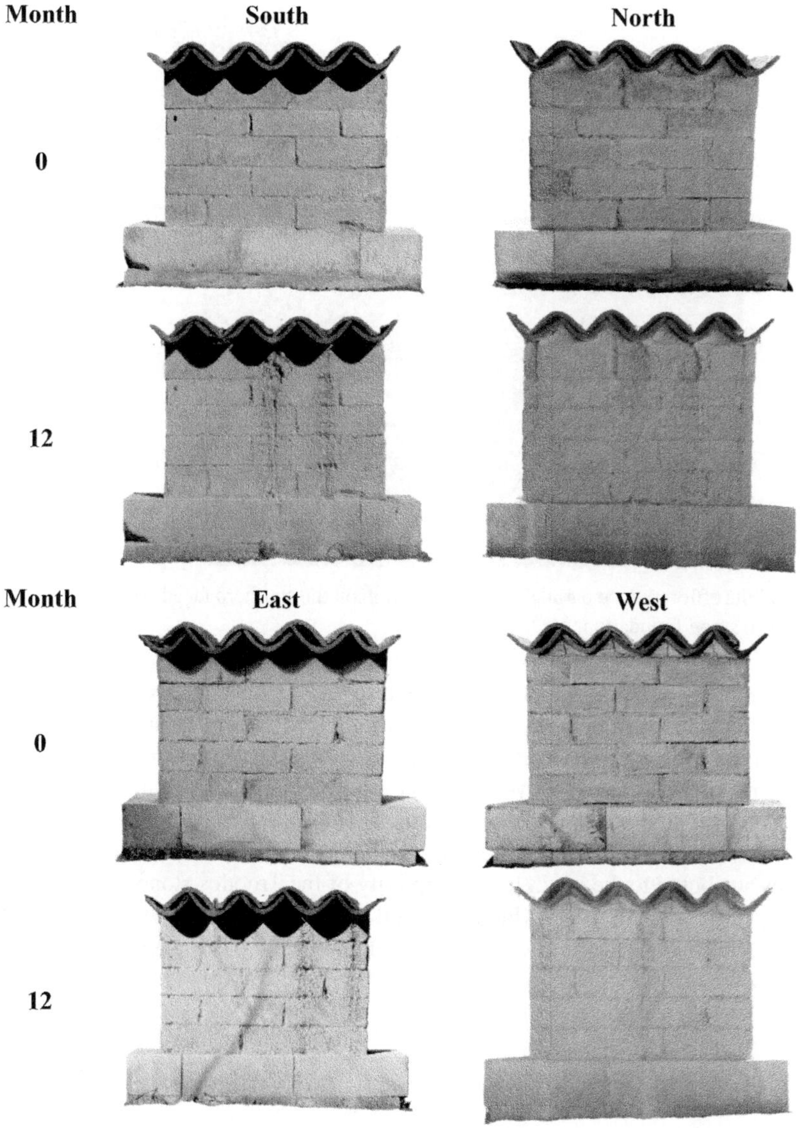

Fig. 3. Images from textured 3D models of the CEB wallettes with Prastio stone foundation.

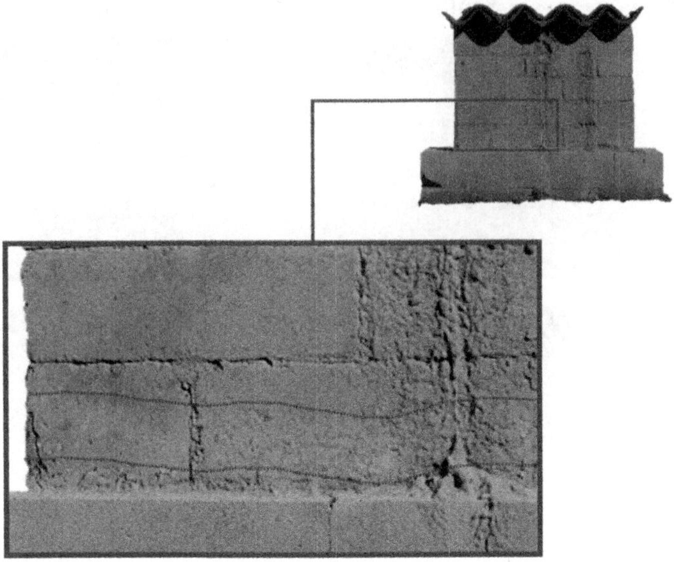

Fig. 4. White efflorescence parallel to the foundation on the southern façade of the CEB wallettes with Prastio stone foundation.

Figure 5 shows images exported from textured 3D models of the CEB wallettes with Troodos stone foundation. Again, the CEBs remained generally unaltered. However, surface run-off due to the poor placement of the roofing tiles produced localized erosion. Surface cracking was also observed, as in the case of the wallettes built on Prastio stone foundation, especially on the northern façade. However, efflorescence in the lower brick courses was not observed in the case of the wallettes built on Troodos stone foundation. This was attributed to the much lower sorptivity of the Troodos stone, compared to the sorptivity of the Prastio stone, which inhibited ground moisture from rising in the CEB wallettes. This highlights the effect of the wallette foundation in the protection of earthen structures.

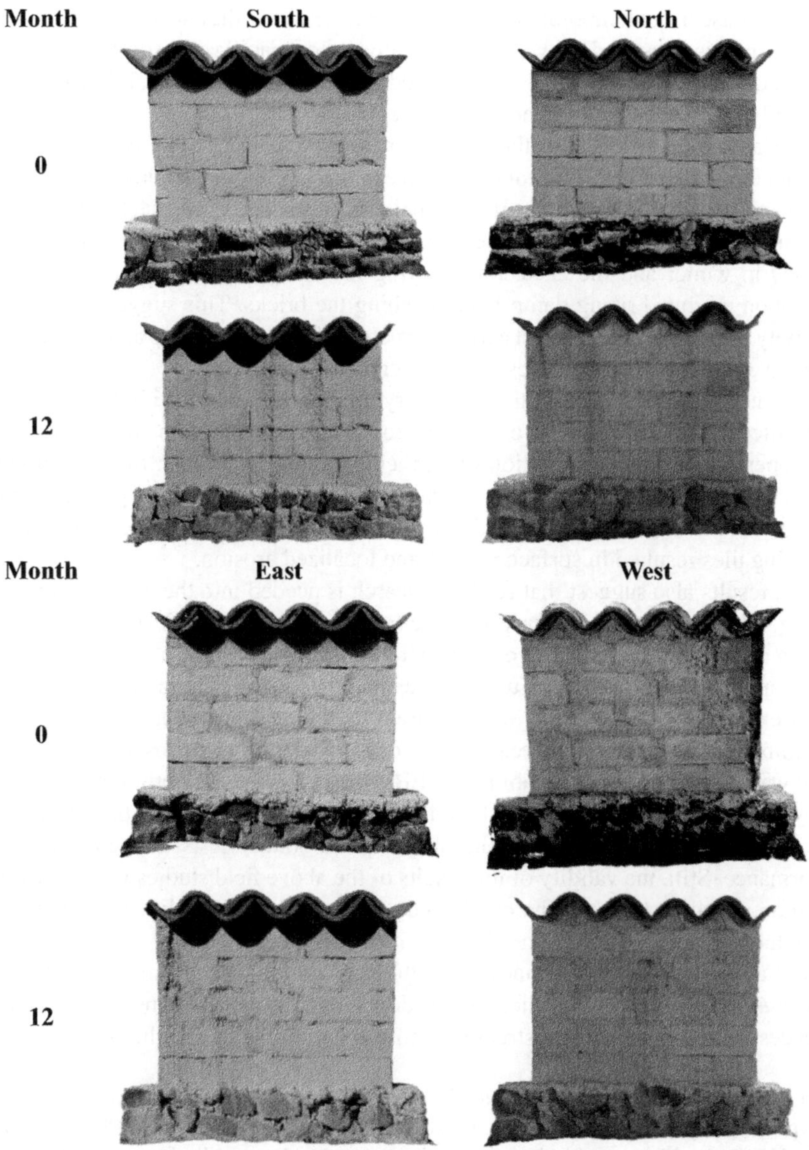

Fig. 5. Images from textured 3D models of the CEB wallettes with Troodos stone foundation.

4 Conclusions

In the current study, a combined laboratory and field study of the durability of non-stabilized CEBs against water-mediated weathering was conducted.

The results of the laboratory investigation showed that the non-stabilized CEBs under study were very durable. In particular, they were assigned the lowest limiting erodibility index, i.e., LEI = 1, following the spray erosion test. Thus, they are considered suitable for use in contemporary construction. Furthermore, they successfully passed the wet/dry appraisal test and showed a comparatively low capillary absorption coefficient. In conclusion, the laboratory results showed that the studied non-stabilized CEBs are expected to withstand water attacks in the field.

The latter was confirmed by the results of natural weathering of CEB wallettes. These were not eroded after one full year of natural exposure. Although some localized erosion was observed due to surface run-off, the majority of the bricks were left intact. Surface cracking was observed on the northern and western facades due to the higher moisture retained in winter and the accelerated drying in summer. The lower sorptivity stone foundation inhibited rising damp from reaching the bricks. This suggests that a lower sorptivity foundation may protect earthen structures from rising damp and, subsequently, from the effects of wet/dry cycles, e.g., salt crystallization.

In conclusion, the results of the laboratory investigation agree with those of the field study after the first full year of exposure. However, great attention should be given to the proper design and construction of earthen structures to protect them from natural elements. Improper design and/or construction might yield durability problems during their service life, as was observed in the case of this field study, where the poor placement of roofing tiles resulted in surface run-off and localized erosion.

The results also suggest that further research is needed into the erosion of CEBs in service. Although it was observed that the laboratory results replicate the CEB actual performance, this was confirmed only for a limited amount of time in a rural environment alone, and it, therefore, needs further investigation to obtain reliable results. Studies for an extended period of time in both rural/urban environments may also enable the determination of a quantitative weathering index for the CEB wallettes. Furthermore, this study was conducted on non-stabilized CEBs that presented the optimum performance set by the standards. Field studies should also be conducted on CEBs that show poorer performance in the lab to assess the ability of laboratory tests to predict in-service performance. Still, the validity of the results of the above field studies will be restricted to the areas where they will be conducted, due to the specific climatic conditions in which the wallettes will be exposed.

The above will help better understand the degradation mechanisms that affect CEB structures and will enable stakeholders to adopt better design and construction practices to successfully protect earthen structures from water-mediated weathering.

Acknowledgements. The authors would like to acknowledge financial support from the Republic of Cyprus, through the Cyprus Research and Innovation Foundation (Project PHD IN INDUSTRY/1222/0081). They would also like to thank Mr. M. Tapakoudis from Gigantas Antaios Touvlopiio Ltd for his help with the production of CEBs.

References

1. ASTM International: Standard guide for design of earthen wall building systems. ASTM E2392, ASTM International, West Conshohocken, PA, USA (2010)
2. Volhard, F., Röhlen, U.: Rules governing building with earth: Edition 2007, 3rd edn. Vieweg Verlag, Wiesbaden (2007)

3. CRAterre-EAG: Compressed earth blocks standards. CDI and CRAterre-EAG Publications, Belgium (1998)
4. Standards New Zealand: Engineering design of earth buildings. NZS 4297, Standards New Zealand, Wellington, New Zealand (1998)
5. Standards New Zealand: Materials and workmanship for earth buildings. NZS 4298, Standards New Zealand, Wellington, New Zealand (1998)
6. Standards New Zealand: Earth building not requiring specific design. NZS 4299, Standards New Zealand, Wellington, New Zealand (1998)
7. Panagiotou, R., Kyriakides, M.A., Illambas, R., Ioannou, I.: An experimental approach for the investigation of the performance of non-stabilized Compressed Earth Blocks (CEBs) against water-mediated weathering. J. Cult. Herit. **57**, 184–193 (2022)
8. Medvey, B., Dobszay, G.: Durability of stabilized earthen constructions: a review. Geotech. Geol. Eng. **38**, 2403–2425 (2020)
9. Sumanasena, V., Rajabipour, A., Bazli, M., Kutay, C., Guo, D.: Strength and erosion resistance of mudbrick as an alternative local material for Australia's Northern Territory remote housing. Case Stud. Constr. Mater. **16**, e01023 (2022)
10. Nshimiyimana, P., Messan, A.: Courard L (2021) Hydric and durability performances of compressed earth blocks stabilized with industrial and agro by-product binders: calcium carbide residue and rice Husk Ash. J. Mater. Civ. Eng. **33**(6), 04021121 (2021)
11. Danso, H., Adu, S.: Characterization of compressed earth blocks stabilized with clay pozzolana. J. Civ. Environ. Eng. **9**, 331 (2019)
12. Mahdad, M., Benidir, A.: Hydro-mechanical properties and durability of earth blocks: influence of different stabilisers and compaction levels. Int. J. Sustain. Build. Technol. Urban Dev. **9**(2), 44–60 (2018)
13. Eires, R., Camoes, A., Jalali, S.: Enhancing water resistance of earthen buildings with quicklime and oil. J. Clean. Prod. **142**, 3281–3292 (2017)
14. Beckett, C.T.S., Jaquin, P.A., Morel, J.C.: Weathering the storm: a framework to assess the resistance of earthen structures to water damage. Constr. Build. Mater. **242**, 118098 (2020)
15. Meek, A.H., Beckett, C.T.S., Elchalakani, M.: Alternative stabilised rammed earth materials incorporating recycled waste and industrial by-products: Durability with and without water repellent. Constr. Build. Mater. **265**, 120629 (2020)
16. Bruno, A.W., Scott, B., D'Offay-Mancienne, Y., Perlot, C.: Recyclability, durability and water vapour adsorption of unstabilised and stabilised compressed earth bricks. Mater. Struct. **53**, 149 (2020)
17. Aguilar, R., et al.: The potential use of chitosan as a biopolymer additive for enhanced mechanical properties and water resistance of earthen construction. Constr. Build. Mater. **114**, 625–637 (2016)
18. Danso, H., Martinson, D.B., Ali, M., Williams, J.B.: Physical, mechanical and durability properties of soil building blocks reinforced with natural fibres. Constr. Build. Mater. **101**, 797–809 (2015)
19. Subramaniaprasad, C.K., Abraham, B.M., Nambiar, E.K.K.: Sorption characteristics of stabilised soil blocks embedded with waste plastic fibres. Constr. Build. Mater. **63**, 25–32 (2014)
20. Hall, M., Djerbib, Y.: Moisture ingress in rammed earth: Part 1—the effect of soil particle-size distribution on the rate of capillary suction. Constr. Build. Mater. **18**, 269–280 (2004)
21. Hall, C., Hoff, W.D.: Water transport in brick, stone and concrete. CRC Press, Oxon, 3rd edn. (2022)
22. Guettala, A., Abibsi, A., Houari, H.: Durability study of stabilized earth concrete under both laboratory and climatic conditions exposure. Constr. Build. Mater. **20**, 119–127 (2006)

23. Seco, A., Urmeneta, P., Prieto, E., Marcelino, S., Garcia, B., Miqueleiz, L.: Estimated and real durability of unfired clay bricks: determining factors and representativeness of the laboratory tests. Constr. Build. Mater. **131**, 600–605 (2017)
24. Bui, Q.B., Morel, J.C., Reddy, B.V.V., Ghayad, W.: Durability of rammed earth walls exposed for 20 years to natural weathering. Build. Environ. **44**, 912–919 (2009)
25. European Committee for Standardization (CEN) (2000) Methods of test for masonry units – Part 13: Determination of net and gross dry density of masonry units (except for natural stone).EN 772-13, CEN, Brussels, Belgium
26. European Committee for Standardization (CEN): Natural stone test methods – Determination of water absorption coefficient by capillarity. EN 1925, CEN, Brussels, Belgium (1999)
27. Modestou, S., Theodoridou, M., Fournari, R., Ioannou, I.: Physico-mechanical properties and durability performance of natural building and decorative carbonate stones from Cyprus. Geological Society, London, Special Publications **416**, 145–162 (2016)
28. Indekeu, M.L., Feng, C., Janssen, H., Woloszyn, M.: Experimental study on the capillary absorption characteristics of rammed earth. Constr. Build. Mater. **283**, 122689 (2021)

Bio-Based Fluidification of Earth Mortar Using Tannin Addition

S. Auger[1(✉)], J.-E. Aubert[1], L. Walter[2], J.-C. Garrigues[2], and E. Perez[3]

[1] LMDC, UPS - INSA Toulouse, Toulouse, France
s_auger@insa-toulouse.fr
[2] UMR EcoFog, Université de Guyane, Kourou, Guiana, France
[3] SOFTMAT, UPS - CNRS Toulouse, Toulouse, France

Abstract. The construction sector accounts for approximately 40% of energy consumption and 20% of greenhouse gas emissions in France, primarily due to the use of cement. This could be mitigated by replacing cement with raw earth in many cases. However, performance and processing of raw earth need significant improvements to enable industrial-scale production.

It is therefore essential to reduce water content while maintaining consistency. This study investigates the combination of two types of tannins with three different types of soils in both unmodified and alkaline environments, as alternatives to environmentally harmful chemical fluidizers. The pH of the mixing solutions was adjusted using sodium hydroxide. Consistency was quantified using a penetrometer.

Results indicated that the addition of both tannins reduced the water content of all soils while maintaining consistency (-17% in average). In an alkaline environment, this effect was enhanced for kaolinite (-42%), whereas it diminished for smectite (-10%). For the one rich in calcite, no significant change in behavior was observed between neutral and high pH levels (-28%). In the presence of too much tannin, a surface-active effect took place, aerating the mortar.

This study validates the use of tannins to fluidify kaolinite, however it highlights that various soils react differently. Further investigations, such as zeta potential measurements, are needed to confirm the adsorption behavior of tannins, particularly for kaolinite in a basic environment, and to assess whether the basic conditions inhibit adsorption in the case of smectites.

Keywords: Earth-based construction · clay material · dispersant · smectite · density

1 Introduction

Concrete is, by volume, the second most consumed materials in the world after water [1]. Its typical composition includes water, aggregates, additives, and cement. The latter has an estimated global production of 4.2 billion cubic meters annually [2], cement manufacturing alone contributes to about 5–8% of global greenhouse gas emissions [3, 4] mainly due to the clinkerization process. In this context, replacing concrete in applications where its high compressive strength is not essential is critical to reducing the environmental footprint of the construction sector.

M. Posani and S. Singh (Eds.): RYS 2025, RILEM Bookseries 67, pp. 415–427, 2026.
https://doi.org/10.1007/978-3-032-14652-6_31

Among low-carbon alternatives, raw earth stands out for its local availability and simple process of mixing inorganic soil, water, and possibly additives, followed by drying. It is fully recyclable if not incorporating cement [5], positively impacts indoor comfort through improved thermal inertia [6] and hygrothermal regulation [7]. Despite its many advantages and demonstrated durability through world-wide century-old earthen building, raw earth construction faces legal, cultural, and technical challenges. Among the most critical technical issues are extended drying times, which slow down construction, and significant shrinkage, which can lead to cracking or structural instability. To mitigate these problems, dispersing agents such as sodium hexametaphophate or carboxymethylated xylan [8–12] are commonly used in the ceramic industry. They adsorb on the sides of clay platelet, creating repulsion forces, thus liberating water molecules and enhancing fluidification. However, their environmental impact makes them unsuitable for large-scale use in sustainable construction.

Tannins, a plant-derived polyphenolic compounds traditionally used for leather tanning, dyeing, and ink production, represent a promising bio-based alternative. It is used traditionally in earth-based recipe [13], and it has been studied for its fluidification properties mainly since 2020. Most studies focused on the dispersing mechanisms of tannins in soils rich in kaolinite clay [14–17] leaving soils composed of illite or smectite clays largely underexplored [18]. While hydrolysable tannin have been the primary focus, condensed tannins were not entirely overlooked [19, 20]. The mechanism related to the molecular structure of hydrolysable tannins involves the adsorption of molecules on the sides of clay platelets, resulting in electro-steric repulsion. When sodium hydroxide is added to create a basic environment (tannin alone is acid), hydrolysable tannin breaks into glucose and negatively charged molecules. The first creates more steric hindrance while the latter adsorbs. Research on condensed tannin suggest that the same mechanism happens, but due to their different structures, condensed tannin is less effective. In addition to their rheological effects, tannins have also been associated with improved mechanical performance, not only due to enhanced packing density but also through the formation of ferric tannate complexes when soils are particularly rich in iron oxides.

This study aims to deepen the understanding of tannin-based dispersion mechanisms in clay mortars, by comparing the performance of hydrolysable and condensed tannins, both with and without NaOH, across different clay mineralogy. The used criterion was to reduce water content while staying at constant workable consistency.

2 Materials and Procedures

2.1 Materials

Three soils, referred to as B, N, and M, were used in this study. Soils B and N were collected from brickworks in southwestern France, while soil M was obtained in northeastern France. Soils were preserved at 8% water content. Standardized sand was used (NF EN 196-1). The grain size distribution and plasticity of the three soils were measured and are very similar. Therefore, they are not used in this article as a a differentiating factor.

Two commercial tannins were chosen. Both were bought at Green'ing, a French retailer of natural products. Chestnut, a hydrolysable tannin, derived from the Castanea sativa tree, was produced in southern France. Acacia, a condensed tannin, extracted from the Acacia mearnsii tree, was produced in South Africa. Both extracts were previously analyzed [19]. Since the products are commercial, and that after comparison, no significative change in the infrared spectra were found, no further analysis was done.

Tap water and sodium hydroxide from Solvay Chemicals International pure at 99.99% was used. In solution, chestnut tannin is more acidic than acacia: pH of chestnut alone was 3.18 ± 0.04 and pH of acacia was 4.39 ± 0.03. Preliminary adjustments of NaOH/tannin ratio were made, and the adapted amounts of NaOH needed to reach pH = 8 and pH = 10 are presented in Table 1.

Table 1. NaOH/tannin ratio used

pH	8	10
NaOH/chestnut ratio	0.095	0.160
NaOH/acacia ratio	0.030	0.140

2.2 Material Characterization

Chemical Composition of Soils

Major oxide composition was evaluated on the basis of macro elemental analysis performed by Inductively Coupled Plasma-Atomic Emission Spectrometry (ICP-AES) on crushed samples smaller than 80 µm.

XRD Measurements

Minerals were identified by X-ray diffraction with a Bruker D8 Advance X-ray diffractometer, equipped with a SOLLER monochromator having a copper anticathode (λ = 1.54 Å) on crushed samples smaller than 80 µm.

PH Measurement

PH of the mixing solution was measured before being added to the mortar. A GLP sensION + PH31 pH meter was used with a resolution of 0.01.

2.3 Mortar Preparation

The percentage of sand was chosen in order to have linear shrinkage small enough (<2,5%) according to the DIN 18946 standard. This criterion led to use 70% of sand for soil N and B, and 80% for soil M. Water content was adjusted systematically to guarantee a fixed consistency using the procedure described in the next section. NaOH was added first, in the required amount according to Table 1, then tannin, gradually, while the solution was mixed by magnetic stirring until all tannin was dissolved. The pH of the solution was measured for each mortar preparation.

The mortar was prepared according to the NF EN 196–1 standard. Three consistency tests were done with a penetrometer, if the right depth range was reached, then samples were cast and water content measured. If not, one or two percent of water (at the same pH of the adjusted pH) were added, the mixing procedure was started again, and a new consistency test was carried.

2.4 Consistency Measurement

For consistency measurement, a semi-automatic cone penetrometer Proviteq was used to measure yield stress. This test consisted of filling a 45 mm high – 70 mm diameter mold with fresh mortar, and to drop a cone (80g – 30° – 35 mm) in it. The cone was free to sink during five seconds before being held by the penetrometer, thus the penetration depth could be measured. This method is widely used to measure liquid limits of soils (NF EN ISO 17892-12).

A target interval of penetration range for a suitable mortar was needed. A control mortar was made, with consistency measured with flow table test (standards DIN 18946 and NF EN 1015–3). The consistency of the same mortar was then measured thanks to penetrometer, and a target range of penetration was set with proportionality rules (Table 2).

Table 2. Target range of penetration depth

Soil	Diameter (flow table) [mm]	Depth (penetrometer) [mm]
B	170 to 180	33.6 to 38.6
N	170 to 180	30.7 to 34.4
M	170 to 180	33.3 to 37.1

The penetration test was chosen over the flow table test because it's reliable, effective, efficient and can be linked to the yield stress in further studies [21].

2.5 Mechanical Tests

Once the consistency was right, the samples were cast in steel mold, according to EN 196–1 standard. To accelerate drying without changing any properties, they were put in ovens at 40 °C. At this temperature, it has been reported that no change in the chemical structure of biobased additive such as tannin happen [22]. After 24h, they were taken out of the molds, and after 72h, out of the oven, after reaching constant mass (mass variation $< 0.2\%$ in 24h). Then samples were put in a room at 20 °C, 50% $\pm$ 10% RH for a week so they could stabilize, until reaching constant mass. Compression tests were carried out with press IGM and a 50kN sensor, after measuring mass and dimensions with a caliper to get density.

3 Results

3.1 Material Characterization

The chemical composition of the three soils is presented in Table 3.

Table 3. Chemical composition of soils M, B and N in %

	SiO_2	Al_2O_3	Fe_2O_3	MnO	MgO	CaO	Na_2O	K_2O	TiO_2	P_2O_5	LOI	Total
B	58.1	14.3	4.8	0.1	2.2	5.6	0.7	2.8	0.6	0.1	10.5	99.8
N	61.0	16.3	5.4	0.1	1.4	2.6	0.2	3.7	0.7	0.1	8.3	99.7
M	45.4	13.5	4.4	0.1	1.4	14.2	0.4	2.0	0.7	0.1	17.7	99.8

The XRD patterns of the three soils are shown in Fig. 1.

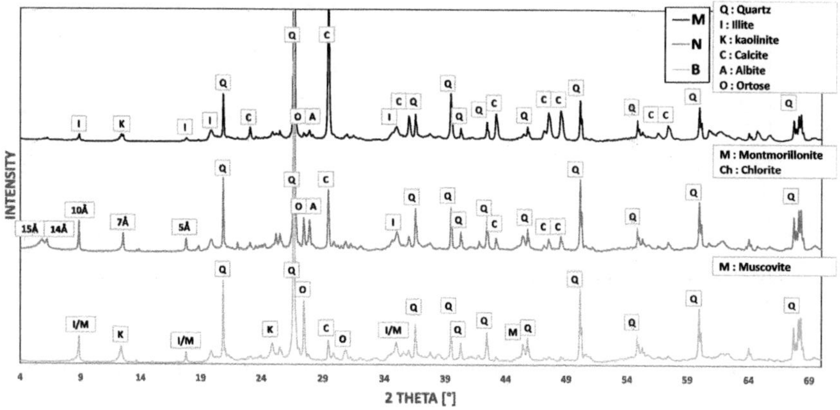

Fig. 1. X-Ray diffraction patterns of soils M, B and N

The three chosen soils were very different in their composition, as it can be seen on the diffractogram (Fig. 1), and from the chemical composition (Table 3). Concerning types of clay minerals, B and M were mainly composed of kaolinite and illite whereas N was composed of montmorillonite and illite. Concerning other minerals, M contained much more calcite than the others. Moreover, B contained muscovite (this was shown by the presence of flakes during grinding). Finally, N had higher albite content than the others. These soils came from the south of France; therefore, they had a low amount of iron oxides.

3.2 Influence of Tannin Content with Soil B

Two different percentages of tannin were tested at natural pH and elevated pH (pH = 8) with soil B. The results concerning water reduction, shrinkage, density and compressive strength (C S) are presented in Fig. 2. For the various properties, standard deviation can be read on the graphs, based on six samples for compressive strength and on three for linear shrinkage.

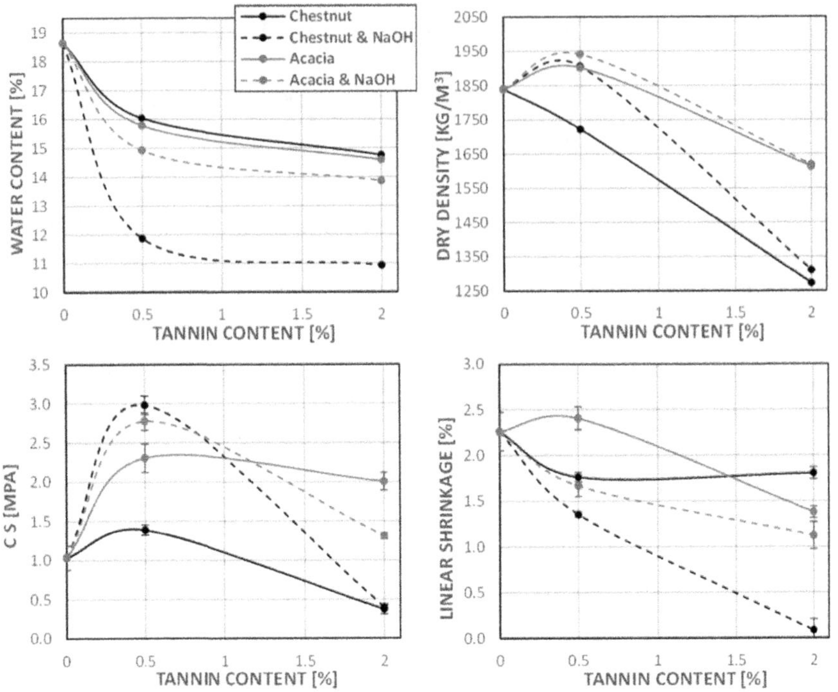

Fig. 2. Water content, compressive strength, density and shrinkage of clay B versus tannin content

The use of tannin at natural or elevated pH consistently reduced the water content by 3% to 6%. The greatest reduction was systematically observed at an elevated pH of 8, with approximately 63% reduction, compared to 20–40% at natural pH, corroborating previous findings [14, 18, 19]. However, increasing the tannin concentration from 0.5% to 2% did not lead to any significant additional reduction, suggesting that a 0.5% dosage is sufficient to disperse the soil B.

Concerning compression strength and density, they both reached a maximum before 0.5% tannin content, then decreasing. The drop in compressive strength at 2% tannin content was more marked for mixtures with NaOH than for mixtures with tannin alone, whereas for density it depended on the tannin. At 2% the addition of NaOH did not influenced density.

As far as shrinkage is concerned, the trend was not usual. Shrinkage decreased as density decreased too at very similar water content (±1% water content). It meant that with the same amount of water that evaporated, mixture with 2% of tannin did not shrink as much as mixture with 0.5% tannin content. The effect was more pronounced in the presence of NaOH.

Two additional qualitative effects had to be underlined. First a coat appeared, darker than the inside, and thicker on the free surface when drying. It was thinner for sample with 0.5% tannin than samples with 2% tannin, and thicker for acacia than for chestnut. No effect on mechanical properties has been found. Secondly, air was drowned in mixture with 2% of tannin. Quantity of drown air or thickness of the coat were not measured because not of interest in the present study of water content reduction, but both effects can be seen on Fig. 3.

Fig. 3. Soil B with 2% of acacia tannin

3.3 Fluidification in an Alkaline Environment

Figure 4 shows the effect of tannins on mortar fluidification with and without an alkaline environment; water content is measured before molding. For each tannin three points were measured: the first one at low pH is tannin without NaOH, the others were around pH = 8 and pH = 10 reached in the mixing solution thanks to NaOH addition.

Two different control tests without tannin were done for each soil. One with tap water (pH = 8) and the second with water with NaOH (adjusted to reach pH = 10). This allowed checking that most of the fluidification was not due to increasing the concentration of OH^-. This could influence the fixation of proton on the sides of clay sheet and induce deflocculation [23].

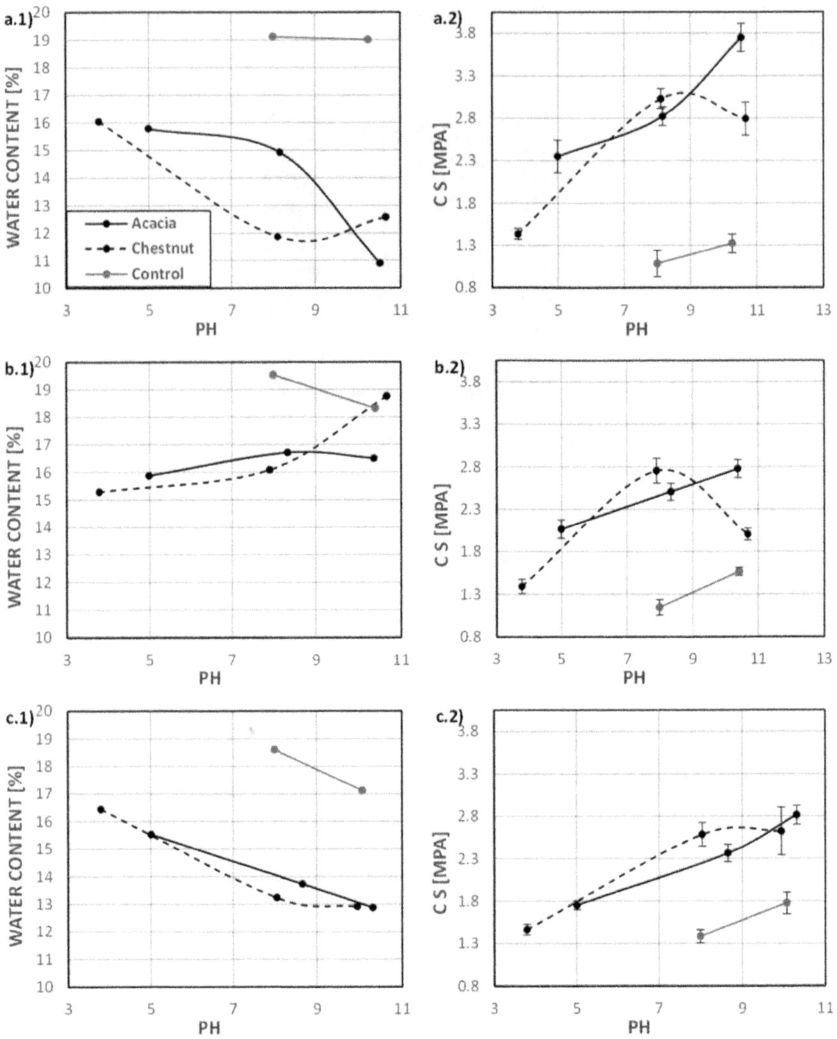

Fig. 4. Mortar water content on the left, compression strength (C S) on the right for a) Soil B, b) Soil N, c) Soil M

Results were similar for soil B and M: both tannins had a fluidification effect. At pH = 8 and pH = 10, the fluidification couldn't be attributed to NaOH, since when added alone the effect was much smaller (curve in grey on Fig. 4). When the water content reduction was enhanced with addition of NaOH for condensed tannin, it didn't show much difference for hydrolysable tannin. The most important water reduction happened for soil B with acacia tannin: the reduction reached 8% in comparison with control samples while it decreased only by 6.5% with chestnut. For soil M, the increase in pH from 8 to 10 didn't have a significant effect: around 1% less of water.

Soil N reacted differently: if tannin alone enabled to keep the same consistency with less water content, this effect was reduced by the presence of NaOH, and even cancelled for chestnut at pH = 10. Overall, both tannins had a similar effect, which didn't depend on the soil when tannin is alone. The reduction was effective but limited: 2% to 5% less water is needed. When NaOH was added to reach a basic pH, different soil had different reactions: water reduction was enhanced for B, is slightly enhanced for M and almost eliminated for N.

Concerning compressive strength, B and M follow the same trend, which is the opposite to the water content. The less water there is in the mortar, the higher the compressive strength is, as well as the density (Fig. 4), and the smaller is the shrinkage (not represented here, but the trend is exactly the opposite of compressive strength). Compressive strength with acacia increased as the pH increased while for chestnut there seemed to be an optimum before reaching a plateau. For chestnut results with pH = 8 and pH = 10 were the same.

Concerning soil N, tannin alone brought little improvement. At pH = 8 and pH = 10, water content was higher than with tannin alone, but the compressive strength was still better. For chestnut at pH = 10, the water content was the same than for control sample (around 18.5%) but compressive strength increased from 1.5 MPa to 2 MPa. It meant that new bond formed in the mortar, density was not the only criterion.

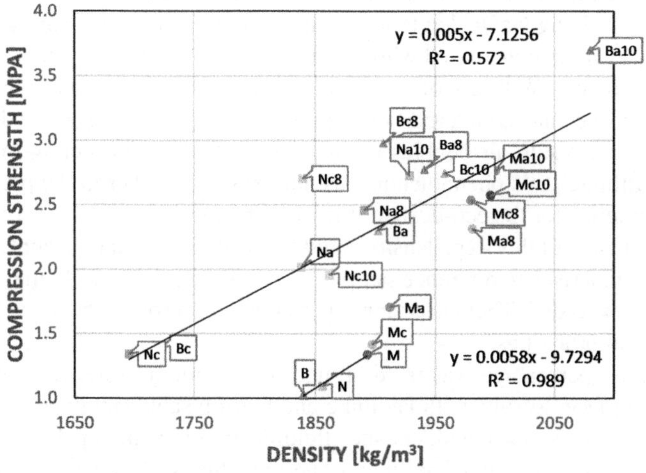

Fig. 5. Compressive strength versus density

Figure 5 shows compressive strength in relation to density. For each point, the first letter is the soil (B: triangles, N: squares or M: circle), the second the tannin (a for acacia, c for chestnut) and the number, if present, stands for the pH of the mixing solution when NaOH was added.

If a straight line is drawn through points B, N, and M (samples without tannin), it becomes clear that all other samples lie above this line. Moreover, a second line representing the average trend of the tannin-containing samples appears nearly parallel to the first one. This suggests that soils with and without tannin follow different density–strength relationships. The upward shift of the tannin samples implies that the addition of tannins not only caused a denser microstructure but also introduced additional bonding mechanisms. It is also worth noting that, for mixtures at pH = 10, the visual 'coating' on particles appeared lighter compared to other conditions, which may indicate a difference in tannin interaction or distribution.

4 Discussion

This study aimed to reduce water content of soil mortar thanks to tannin addition at constant consistency. The first step was to analyze materials, then to optimize tannin percentage and finally to test the effect of condensed tannin on different types of soil.

The second step of the study showed that there was an optimal tannin content around 0.5%. However, in contrast to other studies, no coagulation or diminution of the fluidification process were shown when this percentage was exceeded [14, 18]. A decrease in density with no significative decrease in water content (at constant consistency) followed. Air trapped in the mortar during mixing was found responsible of this density loss. As visible on Fig. 3, with 2% of tannin, the sample had plenty of holes, due to bubble present in the mortar. The ability of tannin to decrease the interfacial tension has been reported in the literature [24, 25]. Therefore, the main hypothesis was that in the mortar, it decreased the interfacial tension, and when overdosed, air was needed for the mix to remain stable. Adding NaOH led to a decrease in both density and shrinkage. It meant that the sample volume increased while its mass remained relatively constant, resulting in higher air content. With 2% of tannin, rather than promoting water removal, NaOH appeared to have enhanced interparticle repulsion—likely due to surface deprotonation at high pH. As a result, air bubbles became more stable, contributing to the reduced shrinkage. Although this phenomenon was not explored further in the present study, it could be of interest for the eco-design of mineral foams.

As a conclusion of this step, density could be a criterion for tannin content. If density decrease in comparison to reference sample, as it was the case for [19], the optimum tannin content is exceeded. Then, less tannin is a better compromise between fluidification and mechanical properties.

Once tannin percentage was fixed around the optimum, water reduction for other soils can take place. Results with kaolinite and hydrolysable tannin confirms what has been found in other studies, with enhance fluidification in at high pH thanks to electro-steric repulsion [14]. Still with kaolinite, water reduction with condensed tannin is more influenced by the pH change from 8 to 10 compared to hydrolysable tannin. The chemical structures of hydrolysable and condensed tannins differ significantly, which influences their reactivity and behavior in alkaline environments. Hydrolysable tannins are composed of a central polyol core-typically glucose-esterified with phenolic acids such as gallic or ellagic acid. These ester bonds are susceptible to hydrolysis under acidic or basic conditions, releasing the phenolic units and sugar moiety [14]. In contrast, condensed tannins, also known as proanthocyanidins, are oligomeric or polymeric chains such as catechin and epicatechin. These structures are more chemically stable and less prone to hydrolysis. As a result, condensed tannins require stronger alkaline conditions for full ionization of the molecule.

Soil M reacted similarly with both tannins. There was an improvement of the water content reduction with the addition of NaOH, but the rise from pH = 8 to pH = 10 didn't influence much water reduction. It questioned the ability of tannin to adsorb better on illite at high pH and experiment such as zeta potential measurement could answer this question. The experiment also demonstrated that calcite wasn't important in the fluidification process of clay. It was reported in the literature that small amounts of tannins could adsorb on calcite [26], however it does not impact the fluidification process, neither the compression resistance.

Soil N reacted differently, fluidified by tannin alone, but the addition of NaOH lowered the water reduction. Montmorillonite is known to be a reactive clay, with high Cation Exchange Capacity (CEC) [6]. The results suggested that in an alkaline environment, tannins - particularly hydrolysable ones - may form stable complexes with interfoliar cations rather than adsorbing onto the clay sheets, thereby inhibiting their fluidizing effect. Experiment such as zeta potential measurement, or at least tests with a weaker base, as done by [19] would give precious results to better understand this mechanism.

From the mechanical point of view, there was a linear relationship between density and resistance. This relation was already demonstrated for earth mortar [27]. However, with tannin addition, compressive strength was higher at same density (Fig. 5), meaning that the bonds in the material were different. This difference was due to the formation of ferric tannate, a complex already identified in the literature [28, 29], which enhanced mechanical properties, and oxidate on the surface of the sample, creating a "coat". Acacia is found to form more complex than chestnut when alone and at pH = 10. This was justified by its better compressive strength and the formation of a bigger coat on the samples.

5 Conclusion

The variation in tannin content suggests that density is an important criterion for detecting whether or not the optimum tannin content has been exceeded.

By comparing hydrolysable and condensed tannins, this study reveals not only a difference in effectiveness according to pH and soil type, but also a distinction in the chemical mechanisms involved. While hydrolysable tannins favor electro-steric repulsion at moderate pH, condensed tannins show their full potential at higher pH, suggesting a slower but potentially more structuring mechanism. These differences influence not only fluidification, but also mechanical performances, with condensed tannins inducing higher strengths, probably via the formation of more stable ferric complexes. Both tannins are interesting for fluidification of clay mortar.

The results highlight that variability of clay minerals forbid to draw general conclusion. What is effective for some is not for other, as soil N, composed of montmorillonite, demonstrates in this study. Before being used at an industrial scale, compatibility test must be done between soil and additive.

Further experiments with more parameters, such as weaker base or measuring other data with a zettameter or adsorption rate thanks to organic carbon measurement would bring more results about chemical reactions and deepen the understanding of the mechanisms involved when tannins are added to earth materials.

References

1. Scrivener, K.L., John, V.M., Gartner, E.M.: Eco-efficient cements: potential economically viable solutions for a low-CO2 cement-based materials industry. Cement Concrete Res. **114**, 2–26 (2018)
2. Global cement and concrete association. The GCCA 2050 cement and concrete industry roadmap for net zero concrete
3. Barcelo, L., Kline, J., Walenta, G., Gartner, E.: Cement and carbon emissions. Mater Struct. juin **47**(6), 1055–1065 (2014)
4. Benhelal, E., Zahedi, G., Shamsaei, E., Bahadori, A.: Global strategies and potentials to curb CO2 emissions in cement industry. J. Clean. Prod. **51**, 142–161 (2013)
5. Melià, P., Ruggieri, G., Sabbadini, S., Dotelli, G.: Environmental impacts of natural and conventional building materials: a case study on earth plasters. J. Clean. Prod. **80**, 179–186 (2014)
6. Lagouin, M., Aubert, J.E., Laborel-Préneron, A., Magniont, C.: Influence of chemical, mineralogical and geotechnical characteristics of soil on earthen plaster properties. Constr. Build. Mater. **304**, 124339 (2021)
7. Darling, E.K., Cros, C.J., Wargocki, P., Kolarik, J., Morrison, G.C., Corsi, R.L.: Impacts of a clay plaster on indoor air quality assessed using chemical and sensory measurements. Build. Environ. **57**, 370–376 (2012)
8. Konduri, M.K.R., Fatehi, P.: Dispersion of kaolin particles with carboxymethylated xylan. Appl. Clay Sci. **137**, 183–191 (2017)
9. Landrou, G., Brumaud, C., Winnefeld, F., Flatt, R., Habert, G.: Lime as an anti-plasticizer for self-compacting clay concrete. Materials **9**(5), 330 (2016)
10. Landrou, G., Brumaud, C., Plötze, M.L., Winnefeld, F., Habert, G.: A fresh look at dense clay paste: Deflocculation and thixotropy mechanisms. Colloids and Surfaces A: Physicochemical and Engineering Aspects. Févr **539,** 252-60 (2018)
11. Perrot, A., Rangeard, D., Levigneur, A.: Linking rheological and geotechnical properties of kaolinite materials for earthen construction. Mater. Struct. **49**(11), 4647–4655 (2016)
12. Perrot, A., Rangeard, D., Menasria, F., Guihéneuf, S.: Strategies for optimizing the mechanical strengths of raw earth-based mortars. Constr. Build. Mater. **167**, 496–504 (2018)
13. Vissac, A., Bourgès, A., Gandreau, D., Anger, R., Fontaine, L. : argiles & biopolymères - les stabilisants naturels pour la construction en terre (2017)
14. Du, Y., Brumaud, C., Winnefeld, F., Lai, Y.H., Habert, G.: Mechanisms for efficient clay dispersing effect with tannins and sodium hydroxide. Colloids and Surfaces A: Physicochemical and Engineering Aspects. **630**, 127589 (2021)
15. Lovage, C., Prud'homme, E., Jorand, Y.: Viscosity control of kaolinites dispersion via addition of tannin and ferric chloride. In: Beckett, C., Bras, A., Fabbri, A., Keita, E., Perlot, C., Perrot, A., éditeurs. Second RILEM International Conference on Earthen Construction [Internet]. Cham: Springer Nature Switzerland; 2024 [cité 22 avr 2025], pp. 91-8. (RILEM Bookseries; vol. 52). Disponible sur: https://doi.org/10.1007/978-3-031-62690-6_10
16. Walter, L., Estevez, Y., Medjigbodo, G., Linguet, L., Nait, O.: Toward Sustainable Poured Earth Construction Using Lateritic Soil and Local Extract of Tannin
17. Walter, L., Medjigbodo, G., Estevez, Y., Linguet, L., Nait-Rabah, O.: Tannin with sodium carbonate: a single additive for poured earth concrete with tropical soils. Results Eng. **25**, 103981 (2025)
18. Clausell, J.R., Signes, C.H., Solà, G.B., Lanzarote, B.S.: Improvement in the rheological and mechanical properties of clay mortar after adding Ceratonia Siliqua L. extracts. Construction and Building Materials **237**, 117747 (2020)

19. Walter, L., Estevez, Y., Medjigbodo, G., Aubert, J.E., Linguet, L., Nait-Rabah, O.: Influence of eco-friendly dispersants on the properties of a lateritic soil-based mortar. Constr. Build. Mater. **411**, 134307 (2024)
20. Yi, D.: A new Low Carbon Clay BasedConcrete with Improved Water Resistance. [ETH Zurich] (2023)
21. Perrot, A., Rangeard, D., Lecompte, T.: Field-oriented tests to evaluate the workability of cob and adobe. Mater Struct. avr **51**(2), 54 (2018)
22. Keita, I., Sorgho, B., Dembele, C., Plea, M., Zerbo, L., Guel, B., et al.: Ageing of clay and clay–tannin geomaterials for building. Construction and Building Materials. **61**, 114–119 (2014)
23. Anger, R. : Approche granulaire et colloïdale du matériau terre pour la construction
24. Dai, H., Chen, Y., Zhang, S., Feng, X., Cui, B., Ma, L., et al.: Enhanced interface properties and stability of lignocellulose nanocrystals stabilized pickering emulsions: the leading role of tannic acid. J. Agric. Food Chem. **69**(48), 14650-61 (2021)
25. Wang, Y.H., Lin, Y., Yang, X.Q.: Foaming properties and air–water interfacial behavior of corn protein hydrolyzate–tannic acid complexes. J. Food Sci. Technol. **56**(2), 905–913 (2019)
26. Tangarfa, M., Semlali Aouragh Hassani, N., Alaoui, A.: Behavior and Mechanism of Tannic Acid Adsorption on the Calcite Surface: Isothermal, Kinetic, and Thermodynamic Studies. ACS Omega **4**(22), 19647-54 (2019)
27. Ardant, D., Brumaud, C., Perrot, A., Habert, G.: Robust clay binder for earth-based concrete. Cem. Concr. Res. **172**, 107207 (Oct 2023)
28. Du, Y., Brumaud, C., Habert, G.: Water Stabilization of Clay Bricks with Improved Tannin and Iron Mixes. In: Barcelona, Spain; 2022 [cité 12 juin 2024], pp. 377–81. Disponible sur: https://www.scientific.net/CTA.1.377
29. Du, Y., Habert, G., Brumaud, C.: Influence of tannin and iron ions on the water resistance of clay materials. Constr. Build. Mater. **323**, 126571 (2022)

Bio-Stabilised Earthen Materials: A Perspective on the Potential Contribution to Climate Change Adaptation and Mitigation

Magda Posani[1(✉)], Guillaume Habert[2], Yannick Igor Fogue Djombou[3], Yi Du[2,4], Evy Vereecken[5,6], Pierre Esteve-Bourrel[7,8], Léo Pinchard[9], Alessia Emanuela Losini[10], Guilherme Barreto Arez Coelho[11], Lola Ben-Alon[12], Olga Beatrice Carcassi[12], Mansoure Dormohmadi[13,14], Alejandro Jiménez Rios[15], Muralidhar Kamath[16], Michele Libralato[17], Joana Maia[18], Christina Makoundou[19], Bruno Malet-Damour[3], Alessandra Ranesi[9], Céline Perlot[20,21], Agostino Walter Bruno[22], Snežana Vučetić[23], and Ana Margarida Armada Brás[24]

[1] Department of Civil Engineering, Aalto University, 02150 Espoo, Finland
magda.posani@aalto.fi

[2] Chair of Sustainable Construction, ETH Zurich, 8093 Zurich, Switzerland

[3] PIMENT Laboratory, University of La Réunion, 97430 Le Tampon, France

[4] Laboratory for Concrete & Asphalt - Empa, 8600 Dübendorf, Switzerland

[5] Building Physics and Sustainable Design, KU Leuven, 3000 Leuven, Belgium

[6] Monuments and Monumental Decoration Lab - KIK-IRPA, 1000 Brussels, Belgium

[7] Instit. for Infrastruct. and Environment, University of Edinburgh, Edinburgh EH9 3JL, UK

[8] Versailles School of Architecture, Paris-Saclay University, 78000 Versailles, France

[9] CERIS, Dept. of Civil Eng. Arch. and Georesources - IST, 1049-001 Lisbon, Portugal

[10] LOCIE - Université Savoie Mont Blanc, 73000 Chambéry, France

[11] Dept. of Civil Eng, Universidade NOVA de Lisboa, 2829-516 Caparica, Portugal

[12] Graduate School of Architecture, Planning and Preservation, Columbia University, New York 10027, USA

[13] DBI – The Danish Institute of Fire and Security Technology, 2650 Hvidovre, Denmark

[14] Department of the Built Environment, Aalborg University, 2450 Copenhagen, Denmark

[15] Department of Architecture & Civil Engineering, University of Bath, Bath BA27AY, UK

[16] Technical Services - Apple Chemie India Private Limited, Nagpur 440001, India

[17] Department of Engineering and Architecture, University of Udine, 33100 Udine, Italy

[18] CONSTRUCT-LFC, University of Porto, 4200-465 Porto, Portugal

[19] Department of Construction Engineering, University of Antwerp, 2020 Antwerp, Belgium

[20] E2S UPPA, SIAME - Université de Pau et des Pays de l'Adour, 64600 Anglet, France

[21] Institut Universitaire de France, 75231 Paris, France

[22] Dipartimento di Ingegneria Civile, Chimica e Ambientale, Università di Genova, 16145 Genoa, Italy

[23] Faculty of Technology, University of Novi Sad, 21000 Novi Sad, Serbia

[24] School of Civil Engineering and Built Environment, Liverpool John Moores University, Liverpool L3 3AF, UK

M. Posani and S. Singh (Eds.): RYS 2025, RILEM Bookseries 67, pp. 428–449, 2026.
https://doi.org/10.1007/978-3-032-14652-6_32

Abstract. With its potential for low embodied carbon, bio-stabilised earth offers a sustainable alternative to traditional carbon-intensive building materials. Moreover, unlike lime and cement stabilisation, bio-stabilisation methods hold the promise to improve the durability of earthen materials while retaining their moisture buffering capacity and recyclability.

Despite the promising characteristics of bio-stabilised earth, research on the topic remains limited. The RILEM TC BEC aims to address this gap by advancing the understanding of bio-stabilisation. Its objectives include developing a classification system for bio-stabilisation techniques, examining their effects on the microstructure and chemical composition of earthen materials, and evaluating their macroscopic impact on mechanical and hygrothermal properties. The TC also aims to develop and promote a performance-based design approach for diverse applications of bio-stabilised earthen materials. Finally, the TC intends to evaluate the role of bio-stabilised earth constructions in climate change adaptation and mitigation strategies.

This paper introduces the work of the RILEM TC BEC and presents findings from its literature review, focusing on the potential of bio-stabilised earth for climate change adaptation and mitigation.

Keywords: Polymers · Fibers · Availability · Comfort · Durability · End of life

1 Introduction

Global efforts to reduce carbon emissions under the Paris Agreement have fallen short, with current policies predicting a 2.8 °C temperature rise by the end of the century, far exceeding the target of keeping it below 2 °C, ideally 1.5 °C [1]. To achieve substantial reductions in greenhouse gas emissions by 2030, a system-wide transformation, including the building sector, is essential. The United Nations (UN) urges moving towards a zero-carbon building stock, emphasizing both operational and embodied carbon perspectives. While many governments focus on energy-efficient buildings and cleaner energy systems [2], thus targeting operational emissions, less attention is given to the embodied emissions in building materials [3, 4], which come with significant carbon footprints. This highlights the urgent need for low-carbon building materials to help meet Paris Agreement goals.

Earth-based materials, with near-zero embodied carbon and potential for full recyclability, are a promising solution for sustainable construction [5]. However, due to their relatively weak mechanical performance and susceptibility to water and erosion, mineral stabilisers like lime and cement are often added to the material. This mineral stabilisation tends to solve the problems mentioned above, but it negatively impacts the recyclability and moisture buffering capacity of the earthen materials [6–13]. Responding to the need for alternative solutions, the RILEM technical committee (TC) BEC, formed in 2022, focuses on *"Bio-stabilised Earth-Based Construction"*. The committee's primary objectives include creating a clear classification of bio-stabilisation methods and exploring their effects on material performance and durability.

Bio-stabilization of earthen materials is hereby defined as the use of a product of biological origin or of biological processes [14] by the addition of materials which can help "binding the earth particles", typically with the aim to reduce volume change, or improve strength or durability against liquid water [15].

The work of TC BEC has classified bio-stabilisation methods into biopolymers, bio-fibres, and bacteria/enzymatic-based processes and is working on a review of the existing literature on the topic. The review considers the effect of bio-stabilisation on the microstructure and chemical composition of earthen materials, as well as evaluating their macroscopic impact on mechanical and hygrothermal properties. Furthermore, the TC is also working to develop and promote a performance-based design approach for diverse applications of bio-stabilised earthen materials. Finally, the TC intends to evaluate the role of bio-stabilised earth constructions in climate change adaptation and mitigation strategies. This paper summarises the committee's efforts and findings in this area.

2 Methods and Scope

Based on a literature review, this paper aims to provide an overview on the potential of bio-stabilised earthen construction for climate change adaptation and mitigation. Climate change, driven primarily by human activities since the 1800s, has caused significant alterations in global temperatures and weather patterns, leading to extreme climate events. Addressing these challenges in the construction sector requires mitigation and adaptation strategies [16]. Mitigation reduces greenhouse gas emissions through sustainable design practices. Adaptation ensures durability, comfort, and low energy needs under changing weather conditions. Bio-stabilised earthen materials can be advantageous in both respects. Ideally, they aim to be low-carbon materials and support circularity [11, 17, 18], thus contributing to mitigation strategies. Additionally, these materials can enhance climate adaptation through their thermal inertia and moisture buffering ability, which are rarely found in modern, conventional constructions [19–21]. Moreover, bio-stabilisation can increase the durability of earthen materials against water erosion. These factors are promising for improving the resilience of buildings to extreme weather events, such as heatwaves and heavy rains.

In this paper, the potential of bio-stabilised earthen construction for climate change adaptation and mitigation is evaluated from a Life Cycle perspective, as schematically represented in Fig. 1. The research starts with considerations on the availability of raw materials, moves on to the environmental impact of material production (processing and manufacturing), evaluates effectiveness during the building's operational phase, and concludes with discussions on the end-of-life stage of the materials. All relevant points of the life cycle are considered, except for transportation. Transportation specifics depend on the location, and this work presumes that soil and raw materials for stabilisation are locally sourced to minimise the environmental impact of transportation. All the points addressed are pertinent to climate change mitigation. The discussion on the operational phase of bio-stabilised earthen buildings is also relevant to adaptation strategies.

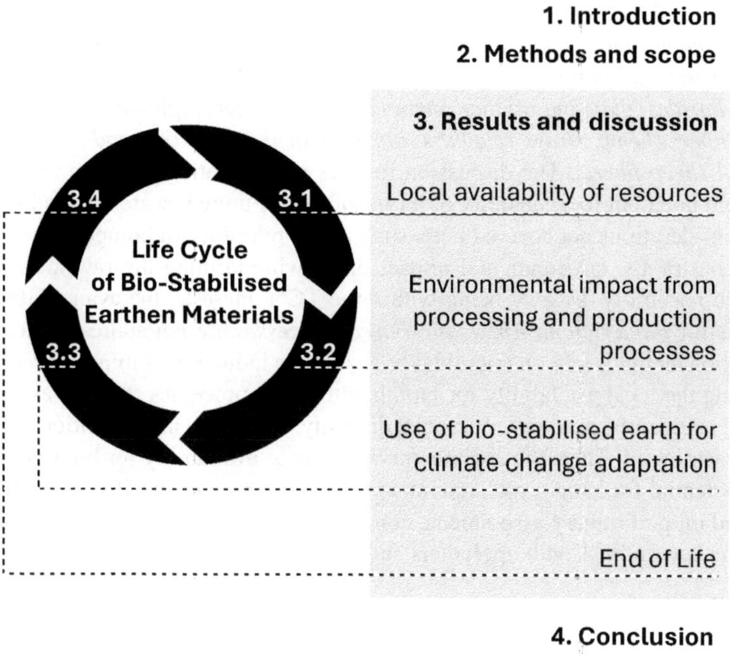

Fig. 1. Schematic overview of the structure of the paper.

3 Results and Discussion

3.1 Local Availability of Resources

Earth is one of the most accessible and sustainable construction materials. Often treated as waste, excavated earth from urban expansion and infrastructure projects presents a sustainable alternative to conventional quarry extraction. Studies show that reusing this excavated soil in construction can significantly reduce environmental impacts, such as habitat destruction, soil erosion, and landscape alteration [22, 23]. Research offers valuable examples highlighting the widespread availability of earth and its potential for reuse in construction. The Brittany region of France is reported to produce 2.8 million tonnes of excavated soil annually, of which 0.7 million tonnes is suitable for earthen construction—enough to support 52% of local housing needs [24]. Similarly, Switzerland generates tens of millions of tonnes of excavation waste annually [25], while Sweden and Portugal contribute millions and hundreds of thousands of tonnes, respectively [22]. In China, subway construction alone produces over a million tonnes of soil yearly [26].

Compared to earth, bioproducts are much more susceptible to local production diversity and can be more or less abundant, depending strongly on the local agricultural production and environmental factors, resulting in a rich regional variability. This subsection explores the local availability of resources for the bio-stabilisation of raw earth within a global context. While the outcomes presented are not exhaustive, they aim to offer an overview, particularly emphasizing the biopolymers and bio-fibers deemed of interest for the work of RILEM TC BEC, schematically represented in Fig. 2. This includes *complex molecules (lignin and tannin), proteins (casein and sodium caseinate, albumin, keratin), lipids (linseed, kapok, and grape seed oils, as well as carnauba wax), polysaccharides (guar and xanthan gums, carrageenan, and alginate), fungi (mycelium), and bio-fibers (hemp, straw, rice husk, bamboo, arundo donax, wood particles, typha, and wool sheep fibers).* The discussion focuses on estimating the geographical availability and potential for scalability of bio-stabilisers. Future research will address additional considerations not covered here, such as the relevance of competitive industries, ethical constraints, environmental impact of growing/farming the raw materials, and economic feasibility. Despite being within the TC's interests, the availability of local resources for *Bacteria and Enzymatic-Based Processes* are not addressed here. These stabilisation methods rely on microbial or enzymatic induction of mineral precipitation. Evaluating the local availability for biomineralization processes requires assessing the presence of essential nutrients, microbial diversity, environmental conditions, and available infrastructure. Thus, the assessment of resource availability for biomineralization and its potential for large-scale implementation is deferred to specific research efforts.

Based on preliminary assessments conducted by TC BEC, the following summary highlights key findings on biopolymers and bio-fibers.

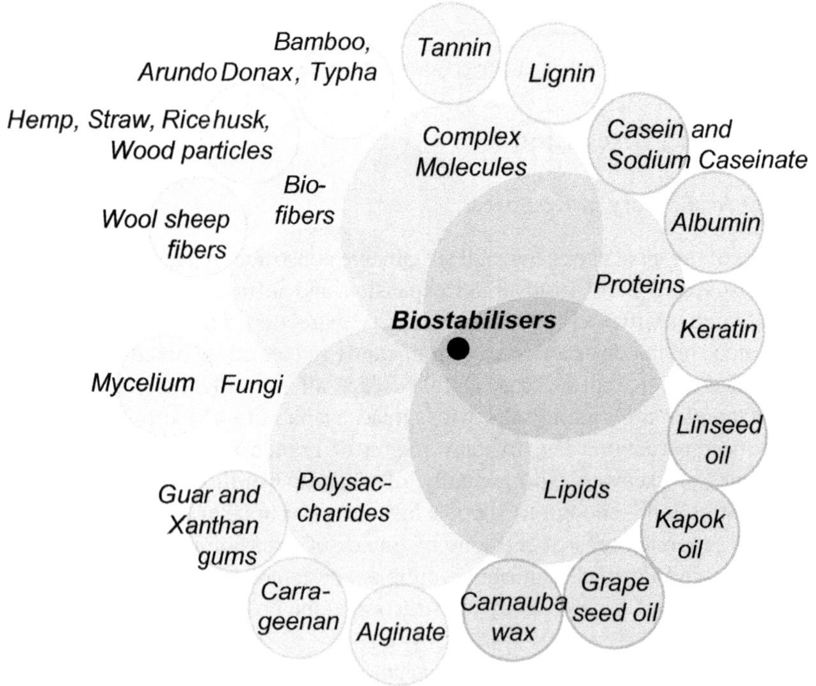

Fig. 2. Schematic overview of biopolymers and bio-fibers deemed of interest for the work of RILEM TC BEC.

Biopolymers. *(p1) Lignin.* Lignin, the most abundant aromatic polymer in nature, constitutes 10–35% of lignocellulosic raw materials [27]. An estimated 1.5 billion tonnes of lignin are produced globally each year, with 60 million tonnes originating as a by-product of the pulp and paper industry. However, around 95% of this lignin is under-utilised, primarily burned as low-value biomass fuel [28]. Major producers include Borregaard (Norway), Tembac (Canada), Nippon Paper Industries (Japan), ALMT (India), and Domjo AB (Sweden) [29]. *(p2) Tannin.* Tannins, abundant phenolic metabolites in plants, are the second most plentiful natural aromatic macromolecules after lignin. Extracted primarily from species like acacia, quebracho, and chestnut using hot water, they are commercially available as powders or concentrated solutions. The annual global production is approximately 200,000 tonnes, 90% of which are condensed tannins [30, 31], and it serves diverse industries. Major production regions include Europe, the Americas, Africa, and Asia [32].

(p3) Casein and Sodium Caseinate. Casein, a protein-based biopolymer, constitutes 80% of the proteins in cow's milk [33]. It is extracted by adjusting milk pH and separating water. Sodium caseinate is a soluble form that can be obtained by neutralising acid casein with NaOH [34]. Both casein and sodium caseinate salt biopolymers can be used as soil stabilisers [33]. Milk is the main raw material necessary to produce these biopolymers, with the annual availability of milk globally being more than 950 million tonnes [35]. India leads the production, providing 25% of the global supply, followed by the European Union (17%), the USA (11%), Pakistan (7%), and China and Brazil (4% each) [35]. *(p4) Albumin.* Albumin, a highly abundant negatively charged protein [36], is found in both animal (e.g., cow, chicken, salmon) and plant sources (e.g., peanuts, sunflower). For instance, in animal serum, albumin constitutes 35–50% of total protein, and it can be separated from its source through several methods, including chromatography, solvent extraction, electrophoresis, and adsorption [37]. Market-available albumins include egg white ovalbumin and bovine serum albumin (BSA) [38]. Ovalbumin, which makes up over 50% of chicken egg white protein, offers potential for earth stabilisation [39–41]. With a chicken population of over 33 billion globally in 2020, and commercial chickens laying over 300 eggs yearly per hen on average, an estimated 0.4 billion tonnes of albumin could be potentially available every year (disregarding the competition with the food industry) [35]. China leads global egg production (40%), followed by the USA (8%) and India (5%) [35]. BSA is a slaughterhouse by-product, it is less accessible due to waste management challenges (potentially dangerous pathogens involved), while still holding potential. On this matter, global cattle production is relevant, with Brazil and India leading at 15% and 12%, respectively. Other relevant contributors are the USA, Ethiopia, China, Argentina, and Pakistan [35]. Cultural factors may limit the availability of bovine blood in some regions, especially India. *(p5) Keratin.* Keratin, a durable fibrous protein, is essential for structural protection in body parts like hair, wool, claws, and feathers. Widely used in biomaterials, agriculture, and textiles, keratin is predominantly extracted from wool, which is composed of 95–98% proteins, primarily α-keratins [42]. Wool processing for keratin can be done in several ways, including removing lipids using detergent, sodium hypochlorite, or organic solvents [43]. Global wool production in 2023 was approximately 1.9 million tonnes [44], with a consequent estimate of potentially available keratin of more than 1.5 million annually. As per FAO data on sheep farming

in 2024, China leads global production (15%), followed by India (6%) and Australia (5%), with relevant contributions from countries like Iran, Nigeria, Türkiye, and the UK (3–4%) [35].

(p6) Linseed oil. Linseed, or flaxseed, comes from the flax plant (*Linum usitatissimum*) and is cultivated for its oil-rich seeds used in dietary, industrial, and nutritional applications. Global linseed oil production in 2021 was approximately 830,000 tonnes, with leading producers including China, Belgium, the USA, Germany, Russia, India, Ethiopia, Kazakhstan, Canada, and Poland [35]. *(p7) Kapok oil.* The kapok tree (*Ceiba pentandra*) is native to tropical regions of the Americas, West Africa, and northern South America [45]. The global production of kapok oil was led by Indonesia and Thailand, as per 2021 data, with an annual production of 201,000 and 84,000 tonnes, respectively [35]. *(p8) Grapeseed oil.* Grapeseed, a key component of grape pomace—a winemaking by-product—comprises about 25% of the pomace [46]. Grapeseed oil, valued for its similarity to cooking oils like olive oil, is highly sought after for its versatility in both food and non-food industries. As per FAO data in 2023 [35], major producers are China (19% of global production), Italy (9%), France (9%), the USA (7%), Spain (7%), India (5%), and Türkiye (5%). Additional contributions come from Chile, South Africa, and Egypt (3% each). *(p9) Carnauba Wax.* Carnauba wax, also known as Brazilian wax or palm wax, is a hard wax derived from the leaves of the Copernicia prunifera palm, native to the northeastern area of Brazil [47]. The wax is obtained by collecting, drying, and beating the leaves to release the wax, followed by refining and bleaching [47]. Known for its versatility and durability, this wax is of great importance in the chemical, electronic, cosmetic, food and pharmaceutical industries [48]. Brazil is considered the leading producer, with a reported annual production of carnauba powder of around 17,000 tonnes in 2012 [48].

(p10) Guar Gum. Guar gum, a natural polysaccharide, is extracted from the seeds of *Cyamopsis tetragonolobus* and *Cyamopsis psoraloides* [49]. It thrives in hot, dry climates and sandy soil [50]. India dominates global production, contributing 90% of the world's supply, with 70% grown in Rajasthan and neighbouring regions [51]. In 2010, India was estimated to have over 3 million hectares of guar crops, with production of 644 kg/ha, suggesting an annual output of 1.9 million tonnes [52]. *(p11) Xanthan Gum.* Xanthan gum, produced by *Xanthomonas* bacteria, is the first industrially produced biopolymer [53], known for high productivity. China (100,000 tonnes), the USA (50,000 tonnes), and Italy (18,000 tonnes), dominate global exports, contributing 81.4% of the 200 million kilograms traded in 2016 [54]. *(p12) Carrageenan.* Carrageenans, derived from red algae like *Chondrus crispus* and *Euchema spinosum* [55], are widely used in food and industrial applications. These red seaweeds are mostly encountered along the Atlantic coasts of North America, Europe, and the western Pacific coasts of Korea and Japan [56]. In 2001, the main producers of seaweed that could be used as carrageenan source were ranked as follows [57]. The Philippines (78%), Indonesia (17%), and Tanzania (5%) are the main producers of *Kappaphycus alvarezii and Eucheuma denticulatum, collectively producing a total of 120,000 dry tonnes* annually. Approximately 14,000 dry tonnes of *Gigartina skottsbergii* are harvested annually in Chile, while around 1,500 dry tonnes come from Mexico, Morocco, and Peru. In contrast, global production of *Chondrus crispus* is significantly lower, at about 4,000 dry tonnes

per year, with Canada contributing 51%, Europe (France, Spain, and Portugal) 36%, and the Republic of Korea 13%. Nowadays, Indonesia leads global production of carrageenan seaweed, producing 8 million tonnes, followed by the Philippines producing 2 million tonnes [35]). *(p13) Alginate.* Alginate is a biopolymer recognised for its biocompatibility, non-toxicity, non-immunogenicity, and biodegradability [58]. Alginate is found in the outer cell wall of brown algae, such as kelps. The principal component of alginate is alginic acid, while sodium alginate is the sodium salt of alginic acid [58]. Sodium alginate is non-toxic, environmentally stable, and exhibits strong gelation, film-forming, and complexing properties. The gel formed by sodium alginate is soft and soluble in alkaline solutions. An irreversible chemical reaction with polyvalent cations (excluding magnesium) forms crosslinking bonds, resulting in a thermo-irreversible gel. Alginate is widely used to encapsulate chemical and biological compounds, and its applications span across various fields, including agriculture, food technology, pharmaceuticals, cosmetics, chemical engineering, environmental engineering, and the paper and textile industries [58]. Research on global mass flow of cultivated seaweed in 2019 [59], shows an annual production of more than 16 million wet tonnes, with almost 5 million tonnes used in Alginate production. As of 2009, Alginate production was estimated to be distributed across the following regions [60]: 44% in Asia and the Pacific (notably China and Japan), 29% in Europe (with significant manufacturing in Norway), and 3% in the Americas (notably Chile and Peru).

(p14) Mycelium. Mycelium is a thin, filament-like structure that forms the roots (*hyphae*) of fungi, providing them with food, water, and nutrients. Mycelium is plentiful and grows quickly, making it suitable for various applications, including construction. It can naturally grow as a network of fine threads or be cultivated using agricultural waste [61]. Optimal growth conditions for mycelium only include appropriate nutrients (commonly provided by/available in agricultural byproducts), and specific ranges of relative humidity, temperature, and airflow for inoculation [62], enabling its production in manufacturing spaces without strong regional constraints. Mycelium is considered to hold potential for the implementation of bio-stabilised earth applications [63].

Bio-fibers. *(f1) Hemp Fibers.* Hemp, derived from *Cannabis sativa*, yields two primary co-products of interest for the construction industry: bast fibers and hurds/shives (the woody core of the hemp stalk) [64]. Hurds find application in thermal insulation materials such as hempcrete [65]. Hemp fibers, among the many applications such as textiles, papers, and maritime ropes and sails [66], can be used for earth stabilisation [67]. Native to India and Persia, hemp is now grown in nearly all temperate and tropical regions worldwide [68]. Hemp fiber production was estimated at approximately 300 tonnes per year in 2021 [69], with major producers including France, China, North Korea, Poland, and the United States. Other notable hemp-producing countries include the Netherlands, Australia, Austria, Chile, Italy, Lithuania, Russia, Romania, and Ukraine [35] *(f2) Straw.* Straw, a by-product of the production of cereals, is widely available globally. Rice straw is predominantly produced in countries such as China, India, Bangladesh, Indonesia, Vietnam, Thailand, Myanmar, the Philippines, Cambodia, Pakistan, Brazil, Japan, Nigeria, and the USA. These nations collectively produced more than 907 million tonnes of rice in 2022 [35], which equates to over 1.3 billion tonnes of rice straw, given an approximate rice grain-to-straw ratio of 1:1.5 [70]. Wheat straw is also significant, with global wheat production estimated at around 770 million tonnes in 2020–2021 [35]. Given that the straw-to-wheat grain ratios range from 0.9 to 2.4, global straw production is estimated to be between 700 and 1,600 million [71]. Leading wheat straw producers include China, India, Russia, the USA, France, Ukraine, Australia, Pakistan, Canada, Germany, Turkey, Argentina, and the United Kingdom [35]. Additionally, straw from other crops like maize and barley is common. *(f3) Rice Husk.* Rice husks, constituting 22% of milled rice weight [72], are abundant in Asia, the Americas, and Africa. Global annual rice production reached 787 million tonnes in 2021 [35] Asia is the global leader in rice production, with China, Thailand, Vietnam, and Bangladesh being other top producers [35]. These are followed by the Americas, primarily through the contribution of Brazil and the USA, and then Africa, with Egypt, Nigeria, and Madagascar being notable contributors [35]. The reuse of rice husks in construction is limited, one factor being their significant utilization within the energy sector [73].

(f4) Arundo Donax. Arundo donax, commonly known as giant reed or giant cane, is an herbaceous, perennial, invasive plant [74]. Originally from East Asia, it grows globally in moist environments like riverbanks, streams, and ditches [75]. Due to its rapid growth, it is primarily cultivated for biofuel production, especially in Europe and the United States [76]. It is particularly prevalent in southern Europe [77]. Studies on annual crop yields for *Arundo donax* plantations report values ranging from 13.6 to 51.4 tonnes of dry matter per hectare [78–80]. *(f5) Bamboo.* Bamboo is a non-wood forest product and wood substitute, found globally, particularly in tropical, subtropical, and temperate regions [81]. The major bamboo-growing regions, covering approximately 35,000 hectares, include Africa (4,648 ha), Asia (24,877 ha), and the Americas (5,514 ha) [82]. *(f6) Wood Particles. Forestry* residues, including bark, dead trees, off-cuts, and small-diameter trees, are often utilised for low-cost goods [83]. On a large scale, this biomass is primarily used as fuel, accounting for 63%, while recycled wood particles constitute only 15% of global waste [84]. According to a report on 2023 data [85], Africa was the largest wood fuel-producing region, contributing 38% - 741 million m3 - of global removals. The Asia-Pacific region followed with a 36% share (701 million m3), then Latin America and the Caribbean (14%), Europe (9%), and Northern America (3%). These regions are proportionally expected to be the main contributors to the production of wood particles. *(f7) Typha.* Typha, or cattails, are a global genus of freshwater plants crucial for ecological balance [86]. Found primarily in the Northern Hemisphere, species from the Typhaceae family thrive in wetlands worldwide, including notable growth in regions in North America, Europe, Asia, and Australia [87] Typha's adaptability to various wetland ecosystems, ability to flourish in disturbed environments due to human activities, rapid growth rates, and aggressive clonal propagation highlight its potential as a fast-growing bio-resource [87].

(f8) Wool Sheep Fibers. Wool, a natural fiber derived from the fleece of animals such as sheep, goats, and llamas, finds it primary use in the textile industry [90]. An average sheep produces 4.5 kg of wool annually and must be shorn regularly for health reasons [88]. In 2022, the global sheep population was 1.296 billion, distributed mainly across China, India, and Australia, resulting in a global production of clean wool of over 1 million tonnes [94]. However, wool production generates significant waste, with 75%

of wool from European farms being unsuitable for the textile industry and discarded as special waste [91, 95], indicating potential for alternative applications.

The literature consulted on the local availability of bio-stabilisers suggests the following observations, ordered intuitively from the least to the most location-specific. *Mycelium* can be produced worldwide. *Casein and sodium caseinate* are significant in milk-producing regions, which includes India, the EU, the USA, Pakistan, and China and Brazil. *Albumin* is critical in egg-producing areas (Ovalbumin), including China, the US and India. *Lignin* is especially pertinent for regions with wood, paper pulp, or other lignocellulosic material-related industries. Key producers include Norway, Canada, Japan, India, and Sweden. Similarly, *wood particles* are relevant in regions with forestry-related industries. *Keratin* is significant for regions that have relevant wool production, such as China, India and Australia, and to a minor extent also Iran, Nigeria, Türkiye, and the UK. Similar considerations apply to *wool sheep fibers*. *Hemp fibres, straw, and rice husk* are widely available and closely linked to agricultural production, particularly relevant in regions that grow hemp, cereals, and rice. *Bamboo* is significant for tropical, subtropical, and temperate regions, including parts of Africa, Asia, and the Americas. *Tannin and Typha* are found in most continents, with tannin being particularly significant in regions with acacia, quebracho, and chestnut trees, while typha is relevant in locations with wetlands. *Linseed oil* holds potential in regions having flax cultivations, such as China, Belgium, the US, Germany, Russia, India, Ethiopia, Kazakhstan, Canada, and Poland. *Carrageenan and alginates* are of relevance in areas having access to seaweed. Specifically, carrageenan is relevant in Indonesia and the Philippines, and to a lesser extent in Tanzania, Chile, Canada, and parts of Europe (France, Spain, and Portugal). Alginates are significant in China and Japan, Norway, Chile, and Peru. *Arundo donax* grows in warm, temperate, and subtropical climates, with Southern Europe and the US currently having relevant amounts. *Grapeseed oil* holds potential in winemaking regions. At the national level, it can be particularly relevant for China, Italy, France, the US, and Spain. *Xanthan gum* is predominantly produced in China, the USA, and Italy. *Guar gum* is primarily produced in India, but plantations could be suitable for other countries with hot, dry climates and sandy soil. Finally, *carnauba wax* production has great potential in Brazil.

3.2 Environmental Impact from Processing and Production Processes

The production of bio-stabilised earth begins with sourcing raw materials, primarily earth and bio-stabilisers. Earth is commonly sourced from excavation waste from construction sites, causing minor embodied impact. On the contrary, sourcing strategies for bio-stabilisers can significantly affect the environmental footprint of the final material [10]. Bio-stabilisers can be often derived from agricultural byproducts or waste, offering a sustainable alternative to virgin materials. After sourcing, primary materials often require processing before they can be used as bio-stabilisers in earthen materials. Efficient extraction methods have been developed for various plants and animals, depending on the specific type of bio-stabiliser considered. For example, hemp shives are a by-product of hemp fiber production. After cultivation and harvesting of hemp straw, decortication process separates the fibers from the shives, then the products are cleaned to remove impurities [89].

Biopolymers generally require a more complex production process than bio-fibres. For instance, the production of guar gum, involves processes like roasting, differential attrition, sieving, and polishing. The Guar seeds are broken to separate the germ from the endosperm, producing unhusked guar splits. These splits are then polished to remove the husk, yielding refined guar splits, which are further processed into guar gum powder with different grades based on colour, mesh size, viscosity, and hydration rate [51]. Despite the relative complexity of the processing, two studies conducted in the context of Southwestern US [90, 91] highlighted that the environmental impact of guar gum production is majorly due to the farming stage, rather than material processing. The two studies considered a cradle-to-gate approach (excl. Biogenic carbon removal) consisting of farming, transport, and processing. Ref. [90] estimates the greenhouse gas emissions for the production of 1 kg of guar gum to 1.87 kg $_{CO2\text{-}eq}$, 82% of which due to farming (with irrigation and land preparation being the main contributors). Ref. [91] similarly estimated a Global Warming Potential (GWP) of 2.67 kg $_{CO2\text{-}eq}$ per kg of Guar gum. The environmental impact was found to be mainly (86%-97%) due to farming, in all the 10 environmental categories assessed. Irrigation and land preparation were found to be the two largest contributors to most impact categories (e.g., 55% and 14% for GWP, respectively). Furthermore, a literature review on the life cycle assessment of lignin highlights the environmental impact of processing of this biopolymer [92]. The study reports a cradle-to-gate impact of 1 kg of lignin range from 0.1 to 2.5 kg CO_2-eq, depending on production process (kraft, organosolv, soda, and sulfite methods). The previous examples show how Life Cycle Assessment (LCA) can be used to analyse the cradle-to-gate environmental impact of bio-stabilisers, useful to compare the various locally available stabilisers for an informed environmental choice. Depending on the stages of the life cycle considered, cradle-to gate or cradle to site approaches can be used, or more comprehensive ones such as cradle-to-grave (from resource extraction to disposal) or cradle-to-cradle (from resource extraction to recycling) can be employed. Depending on the goal of the assessment, the evaluation can move from bio-stabilisers alone to bio-stabilised earth products (e.g., compressed earth blocks) or components (e.g., walls).

For instance, Ref. [93] examined earth blocks containing straw or sawdust in Cyprus, using a cradle-to-site approach. The study compared three scenarios: on-site production with locally available soil and transported straw or sawdust, on-site production with transported raw materials, and factory production followed by transportation to the site. The results showed a GWP (excluding biogenic carbon removal) ranging between 0.0018 and 0.013 kg $_{CO2\text{-}eq}$ per kilogram of stabilised earth blocks. The findings confirmed that minimising transportation and utilising locally available resources can significantly reduce the environmental footprint of the blocks. In Ref. [94], the scale is increased, considering 1 m^2 of wall assembly as functional unit, with a cradle to gate approach. The study showcased that walls based on bio-fiber-earth composites have lower environmental impact than conventional systems based on concrete masonry units, in the four environmental categories assessed.

Current studies suggest that resourcing bio-stabilisation such as bio-fibers could relevantly lower the environmental impact of earthen materials, compared to cement stabilisation [95]. Given the importance of this topic, future studies on the environmental impact of bio-stabilisation methods and bio-stabilised earthen products, including end-of-life assessments, would significantly contribute to the existing literature. Over the coming years, TC BEC will include this objective in its research agenda.

3.3 Use of Bio-Stabilised Earth for Climate Change Adaptation

This section discusses the potential of bio-stabilised earth construction for adaptation strategies. It starts with considerations on the potential impact of future weather on the built environment and continues by discussing assessment tools useful to evaluate the performance and resilience of bio-stabilised earthen constructions under future climatic conditions.

Climate change adaptation encompasses all actions that help reduce the vulnerability of individuals and communities to the adverse effects of climate change [96, 97]. The construction industry is increasingly acknowledging the urgent need for adaptation strategies. As global temperatures rise and extreme weather events become more frequent, the sector must define and embrace practices to ensure the resilience of buildings. These strategies can include climate-adaptive design principles and constructive systems that mitigate discomfort, and energy demands due to extreme climatic conditions, while enhancing durability against extreme weather events.

Changes in Rain and Wind Patterns. The Intergovernmental Panel on Climate Change [98] reports that the coming decades will bring more intense and frequent extreme rainfall and stronger tropical cyclones to most regions worldwide, potentially leading to heightened wind-driven rain impacts on building facades. Liquid water and moisture are primary drivers of material degradation, especially for earthen constructions where water can cause erosion and collapses [99]. Stabilising earthen materials and repairing existing structures are critical for mitigating moisture-driven damages. *Changes in Temperatures.* Air temperature fluctuations significantly impact building degradation processes. Particularly, they are of concern in those cold climates where it leads to increased freeze-thaw cycles [100]. Additionally, global warming will increase mean temperatures in most regions, resulting in hot extremes and heatwaves, exacerbating overheating and discomfort inside buildings [101], and increasing cooling demands. These temperature trends will affect degradation and indoor comfort, posing challenges for durability of constructions and occupants well-being.

Potential of Bio-Stabilised Earthen Architecture for Climate Adaptation. Bio-stabilised earthen materials offer significant opportunities for climate change adaptation. Their potential for creating thermally heavy constructions can mitigate overheating extremes through thermal inertia, passively cooling interiors during hot daytime and releasing heat during cooler nights [102–104]. Earthen materials also excel in moisture buffering, passively contributing to regulate indoor relative humidity (RH) and reducing daily and seasonal peaks [21]. This improves the occupants' comfort and health, as extreme (very low and very high) indoor RH levels can cause respiratory problems and promote allergen growth [105]. By stabilising RH levels, energy demands for cooling and humidity control are reduced [105–107]. Unlike conventional mineral stabilisers, bio-stabilisers hold the potential to preserve the good moisture buffering ability of earthen materials [108, 109]. Finally, bio-stabilisers enhance resistance to water erosion and reduce water absorption rates, increasing durability during extreme rain events and floods compared to traditional unstabilised earthen materials [110, 111]. This dual benefit of preserving the passive indoor RH regulation and improving durability, creates a great potential for climate resilient buildings. Quantitative examples can be found in the literature. For instance, a recent study[109] observed starch to improve moisture buffering capacity of earth samples, with an increase of up to 27% (using a hydrogel starch solution with 2.5% starch content) in moisture uptake after 12h of exposure to high RH. Additionally, in 2023, good durability performance against liquid water erosion were observed [111] in starch-stabilized earth (using an aqueous solution containing 5% w/v starch, with a water-to-clay ratio of 30 wt%), with mass loss of 3.3% after 180 min of immersion in water, versus 100% mass loss after only 30 min in unstabilised earth. Few studies similarly observed that lignin can significantly reduce erosion in silty-sand soils [112, 113]. Remarkably, adding just 0.6 wt% lignosulfonate led to 85-fold reduction in the erosion coefficient. Molasse-stabilisation was also observed to reduce liquid water and erosion in earth composites, e.g. a 10 vol% addition in earth plasters was found to decrease the capillary water absorption coefficient by almost 80% and increase the time of resistance to erosion (during immersion in water) from 25 to 48 min [114].

Methods for assessing climate change adaptation potential. *Experimental Methods.* Experimental methods that assess bio-stabilised earth performance, focusing on factors like erosion, frost susceptibility, and hygrothermal properties can be applied to evaluate the potential bio-stabilised earth materials for climate resilient construction. Additionally, long-term field monitoring can be implemented to evaluate real-world durability and performance. In field tests, particular attention can be given to periods that effectively represent anticipated future extremes, such as monitoring during heatwaves and heavy rain showers. Accelerated aging tests that simulate extreme climatic conditions can also help to predict the future durability of bio-stabilised earthen materials. *Numerical Methods.* Numerical modelling complements experimental methods, enabling evaluations under projected climate scenarios. Of relevance are dynamic hygrothermal models, which can be used to predict the effect of future climate on indoor comfort, energy consumption, and moisture-related degradation risks for earthen constructions. At the component level, tools like WUFI Pro [115], DELPHIN [116], and Comsol Multiphysics [117] are of use in simulating the coupled heat and moisture transfer through building components, providing insights into the behaviour of earthen materials under realistic climatic conditions. At the whole-building level, dynamic models, including WUFI Plus [115] and EnergyPlus [118] (when using the combined heat and moisture finite element algorithm), can be used to integrate the building envelope's hygrothermal performance with energy demand and indoor comfort evaluations. Since the outdoor weather is an input of the simulations, all aforementioned tools enable researchers to investigate the hygrothermal behaviour of earthen constructions under the stress imposed by future climate. Outdoor weather data is typically available in the form of climatic files representing current climatic patterns. However, to account for the effects of climate change, future-projected climate data files can be used.

Future weather files (for dynamic hygrothermal/energy simulations). Future weather files are developed through methodologies that incorporate projections based on IPCC climate scenarios, such as Representative Concentration Pathways (RCPs) and

Shared Socioeconomic Pathways (SSPs). One common approach is to download regional climate projection data e.g., those available from the CORDEX (Coordinated Regional Downscaling Experiment) database [119], which provides high-resolution climate projections. Methods like those outlined by Machard et al. [120] can be used to downscale such daily climate data to hourly time series, allowing to create hourly climatic files for numerical simulations [121]. Alternatively, software like Meteonorm [122] offers pre-assembled and ready-to-use future weather files, enabling faster integration into simulations. This approach eliminates the need for downscaling and manipulation of data, resulting in a simplified workflow compared to the CORDEX-based approach. While Meteonorm datasets can be used for heat and temperature-related investigations, there are doubts in the scientific community about their reliability concerning wind-driven rain and humidity data. One last approach relies on the morphing method, which mathematically transforms present-day weather data to match projected variables of a climate change scenario. Typically, a morphing tool is fed with current local climatic datasets, such as the Test Reference Year (TRY) files, and modifies them to match the projected IPCC climate scenarios. As explained in Ref. [123], the morphing method is known for preserving the local weather characteristics, which is particularly important when designing or analysing the performance of buildings. Morphing tools such 'The Morph EPW App' [124, 125], 'CCWorldWeatherGen'[126] and the 'Future Weather Generator' [123] can be used for the scope. Other morphing tools are also available, each with its specific limitations [123].

For earthen constructions, future climate considerations are critical, particularly for durability against extreme wind-driven rain and potential for reducing indoor discomfort during heatwaves. A combination of experimental and numerical methods can be instrumental in evaluating the potential of bio-stabilised earthen constructions under future climate conditions, to assess its potential for implementing climate-adapted buildings. This approach will be incorporated into the research plans of the TC BEC.

3.4 End of Life

The transition to a circular economy has become a prominent policy objective worldwide. The construction sector is a critical focus in this context, consuming approximately 50% of all extracted raw materials and generating 25–30% of the total waste in the European Union [127]. This issue is mirrored globally, with China and the United States producing significant quantities of construction and demolition waste (CDW), contributing to a global environmental challenge [128]. Earthen materials offer promising opportunities for circularity, as their end-of-life options include safe landfilling or being recycled without significant performance loss [5]. However, the choice of stabilisers significantly influences their circularity potential.

Circularity Performance of Bio-stabilised Earthen Materials. Unstabilised earthen materials are inherently recyclable due to their capacity for re-plasticisation, enabling reuse without significant loss of performance [12, 129, 130]. However, stabilisers like cement and lime reduce this potential by altering the material's chemistry and requiring energy-intensive processes for downcycling [11]. In contrast, bio-stabilisers such as guar gum or xanthan gum may allow easier recycling, or even composting in some cases, depending on the specific biopolymer used. The primary recycling methods for bio-stabilised earth materials involve crushing, washing, and drying, followed by reformulation with or without additives. Bruno et al. [6] demonstrated that guar gum-stabilised compressed earth bricks retained comparable compressive strength after recycling, while cement-stabilised bricks exhibited reduced performance. Muguda et al. [131] observed changes in particle size distribution and Atterberg limits after recycling, with stabiliser type influencing the degree of these changes. For instance, xanthan gum-stabilised tiles showed significant agglomeration of fines, whereas guar gum stabilisation had minimal impact.

The recyclability of bio-stabilised earth materials remains an area of active research, with studies highlighting variability in outcomes depending on the stabiliser adopted. Although full restoration of properties is rarely achieved, preliminary results indicate that bio-stabilised earth has a promising potential to be effectively recycled. More research is needed on the topic, and TC BEC aims to contribute constructing a growing body of knowledge on the matter.

4 Conclusion

This paper offers a brief introduction to the current state of the art on bio-stabilised earth construction, focusing on resource availability, environmental impact of production processes and material sourcing, potential for climate change adaptation, and recyclability. Bio-stabilised earth is an expanding field of research with considerable potential for sustainable and climate-resilient construction. Nonetheless, the performance of bio-based additives in stabilising earthen materials remains underexplored, especially when compared to the well-established literature on cement- and lime-based stabilisation solutions. Given the promising results presented, the TC BEC is committed to advancing the scientific and practical knowledge of bio-stabilised earthen construction and encourages wider engagement from the academic and professional communities. In particular, WP5 will focus on establishing the environmental impact of different bio-stabilisation methods, as well as their ability to improve climate-change resilience of constructions, and their effect on the end-of-life recyclability of earthen materials.

References

1. United Nations Emissions Gap Report 2022. Accessed 12 Aug 2024
2. Galimshina, A., Moustapha, M., Hollberg, A., et al.: Strategies for robust renovation of residential buildings in Switzerland. Nat. Commun. **15**, 2227 (2024)
3. Priore YD, Jusselme T, Habert G (2022) Exploring long-term building stock strategies in Switzerland in line with IPCC carbon budgets. In: IOP Conference Series: Earth and Environmental Science. IOP Publishing, p 12023
4. Röck, M., Saade, M.R.M., Balouktsi, M., et al.: Embodied GHG emissions of buildings – The hidden challenge for effective climate change mitigation. Appl. Energy **258** (2020). https://doi.org/10.1016/j.apenergy.2019.114107
5. Brumaud, C., Du, Y., Ardant, D., Habert, G.: Earth, the new liquid stone: development and perspectives. Mater Today Commun. **39**, 108959 (2024)

6. Bruno, A.W., Scott, B., D'Offay-Mancienne, Y., Perlot, C.: Recyclability, durability and water vapour adsorption of unstabilised and stabilised compressed earth bricks. Mater. Struct. **53**, 1–15 (2020)
7. Arrigoni, A., Grillet, A.-C., Pelosato, R., et al.: Reduction of rammed earth's hygroscopic performance under stabilisation: an experimental investigation. Build. Environ. **115**, 358–367 (2017). https://doi.org/10.1016/j.buildenv.2017.01.034
8. Saidi, M., Cherif, A.S., Zeghmati, B., Sediki, E.: Stabilization effects on the thermal conductivity and sorption behavior of earth bricks. Constr. Build. Mater. **167**, 566–577 (2018). https://doi.org/10.1016/J.CONBUILDMAT.2018.02.063
9. Gentile, V., Velasquez, J.D.V., Fantucci, S., et al.: 3D-printed clay components with high surface area for passive indoor moisture buffering. J. Build. Eng., 109631 (2024)
10. Koul, B., Yakoob, M., Shah, M.P.: Agricultural waste management strategies for environmental sustainability. Environ. Res. **206** (2022). https://doi.org/10.1016/j.envres.2021.112285
11. Losini, A.E., Grillet, A.C., Bellotto, M., et al.: Natural additives and biopolymers for raw earth construction stabilization – a review. Constr. Build. Mater 304 (2021)
12. Reddy, B.V.V., Kumar, P.P.: Embodied energy in cement stabilised rammed earth walls. Energy Build **42**, 380–385 (2010)
13. Pacheco-Torgal, F., Jalali, S.: Earth construction: lessons from the past for future eco-efficient construction. Constr. Build. Mater. **29**, 512–519 (2012)
14. Ilman, B., Balkis, A.P.: Sustainable biopolymer stabilized earthen: Utilization of chitosan biopolymer on mechanical, durability, and microstructural properties. J. Build. Eng. **76** (2023). https://doi.org/10.1016/j.jobe.2023.107220
15. Zealand, S.N.: NZS 4298 (1998): Materials and workmanship for earth buildings (1998)
16. Posani, M., Veiga, M.D.R., De Freitas, V.P.: Historic buildings resilience: a view over envelope energy retrofit possibilities (2018)
17. Perlot, C., Bruno, A.W., Posani, M., et al.: Challenges for bio-stabilised earth-based construction. In: International Conference on Earthen Construction. Springer, pp. 101–111 (2024)
18. Bruno, A.W., Lalicata, L.M., Abdallah, R., et al.: Synergic effect of hydrated lime and guar gum stabilisation on the mechanical, thermal and hygroscopic behaviour of a Ligurian earth material. Constr. Build. Mater. **439**, 137258 (2024)
19. Ben-Alon, L., Rempel, A.R.: Thermal comfort and passive survivability in earthen buildings. Build Environ. **238** (2023). https://doi.org/10.1016/j.buildenv.2023.110339
20. Zhang, L., Sang, G., Han, W.: Effect of hygrothermal behaviour of earth brick on indoor environment in a desert climate. Sustain. Cities Soc. **55** (2020). https://doi.org/10.1016/j.scs.2020.102070
21. McGregor, F., Heath, A., Shea, A., Lawrence, M.: The moisture buffering capacity of unfired clay masonry. Build Environ. **82** (2014). https://doi.org/10.1016/j.buildenv.2014.09.027
22. Hale, S.E., Roque, A.J., Okkenhaug, G., et al.: The reuse of excavated soils from construction and demolition projects: Limitations and possibilities. Sustainability (Switzerland) **13** (2021). https://doi.org/10.3390/su13116083
23. Agharroud, K., Bounab, A., El Maftouhi, T., et al.: Assessment of the sustainability of aggregate quarrying practices in northern Morocco: a case regarding the eastern provinces of the Tangier Peninsula. Bull. Eng. Geology Environ. **82** (2023). https://doi.org/10.1007/s10064-023-03252-x
24. Hamard, E., Lemercier, B., Cazacliu, B., et al.: A new methodology to identify and quantify material resource at a large scale for earth construction – Application to cob in Brittany. Constr Build Mater 170 (2018). https://doi.org/10.1016/j.conbuildmat.2018.03.097
25. Ardant, D., Brumaud, C., Perrot, A., Habert, G.: Robust clay binder for earth-based concrete. Cem. Concr. Res. (2023) (2023). https://doi.org/10.1016/j.cemconres.2023.107207

26. Zhang, N., Duan, H., Sun, P., et al.: Characterizing the generation and environmental impacts of subway-related excavated soil and rock in China. J. Clean. Prod. **248**,(2020). https://doi.org/10.1016/j.jclepro.2019.119242
27. Jędrzejczak, P., Collins, M.N., Jesionowski, T., Klapiszewski, Ł.: The role of lignin and lignin-based materials in sustainable construction – A comprehensive review. Int. J. Biol. Macromol. **187** (2021)
28. Tribot, A., Amer, G., Alio, M.A., et al.: Wood-lignin: Supply, extraction processes and use as bio-based material. Eur. Polym. J. **112**, 228–240 (2019)
29. Mandlekar, N., Cayla, A., Rault, F., et al.: An overview on the use of lignin and its derivatives in fire retardant polymer systems. In: Lignin - Trends and Applications (2018)
30. Du, Y., Habert, G., Brumaud, C.: Design of tannin-based poured earth material via deflocculation-coagulation control: additive selection and tannin variation. ACS Sustain. Chem. Eng. **10**, 14495–14502 (2022)
31. Pizzi, A.: Tannins: Prospectives and actual industrial applications. Biomolecules 9 (2019)
32. Pagliaro, M., Albanese, L., Scurria, A., et al.: Tannin: a new insight into a key product for the bioeconomy in forest regions. Biofuels, Bioproducts Biorefining 15 (2021). https://doi.org/10.1002/bbb.2217
33. Fatehi, H., Abtahi, S.M., Hashemolhosseini, H., Hejazi, S.M.: A novel study on using protein based biopolymers in soil strengthening. Constr. Build. Mater. **167**, 813–821 (2018)
34. Gagnon, R.D., Bazinet, L., Mikhaylin, S.: Functional Properties of Casein and Caseinate Produced by Electrodialysis with Bipolar Membrane Coupled to an Ultrafiltration Module. Membranes (Basel) **12** (2022). https://doi.org/10.3390/membranes12030270
35. FAO (2024) FAOSTAT Data. https://www.fao.org/faostat/en/#data. Accessed 12 Aug 2024
36. Moman, R.N., Gupta, N., Varacallo, M.: Physiology, Albumin (2023)
37. Hutapea, T.P.H., Madurani, K.A., Syahputra, M.Y., et al.: Albumin: Source, preparation, determination, applications, and prospects. J. Sci. Adv. Mater. Devices **8** (2023)
38. Doppalapudi, S., Katiyar, S., Domb, A.J., Khan, W.: Biodegradable natural polymers. Advanced polymers in medicine 33–66 (2015)
39. Vissac, A., Bourgès, A., Gandreau, D., et al.: argiles & biopolymères-les stabilisants naturels pour la construction en terre (2017)
40. Winkler, E.M.: The effect of blood on clays. Soil Sci. **82**, 157–172 (1956)
41. Abdelhadi, M., Gotoh, K.: Efficiency of albumin and eggshell powder in soil stabilization. Soils Foundations **38** (1998). https://doi.org/10.3208/sandf.38.3_109
42. Gong, H., Zhou, H., Forrest, R.H.J., et al.: Wool keratin-associated protein genes in sheep—A review. Genes (Basel) **7** (2016)
43. Giteru, S.G., Ramsey, D.H., Hou, Y., et al.: Wool keratin as a novel alternative protein: A comprehensive review of extraction, purification, nutrition, safety, and food applications. Compr. Rev. Food Sci. Food Saf. **22** (2023). https://doi.org/10.1111/1541-4337.13087
44. Textile Exchange: Materials Market Report 2024. In: (2024). https://textileexchange.org/app/uploads/2024/09/Materials-Market-Report-2024.pdf
45. Kozłowski, R.M., Mackiewicz-Talarczyk, M., Barriga-Bedoya, J.: New emerging natural fibres and relevant sources of information. In: Handbook of Natural Fibres: Second Edition (2020)
46. Yang, C., Shang, K., Lin, C., et al.: Processing technologies, phytochemical constituents, and biological activities of grape seed oil (GSO): a review. Trends Food Sci. Technol. 116 (2021)
47. Cargo handbook Carnauba Wax. In: https://www.cargohandbook.com/Carnauba_Wax
48. de Freitas, C.A.S., de Sousa, P.H.M., Soares, D.J., et al.: Carnauba wax uses in food – A review. Food Chem **291** (2019)

49. Tahmouzi, S., Meftahizadeh, H., Eyshi, S., et al.: Application of guar (Cyamopsis tetragonoloba L.) gum in food technologies: A review of properties and mechanisms of action. Food Sci. Nutr. **11** (2023)
50. Nishinari, K., Takemasa, M., Zhang, H., Takahashi, R.: Storage plant polysaccharides: xyloglucans, galactomannans, glucomannans. In: Comprehensive Glycoscience: From Chemistry to Systems Biology (2007)
51. Mudgil, D., Barak, S., Khatkar, B.S.: Guar gum: processing, properties and food applications - a review. J. Food Sci. Technol. **51** (2014)
52. Chauhan, J.S., Singh, K.H., Choudhury, P.R., et al.: Growth, varietal development and cultivar diversity in seed production chain of cluster bean (Cyamopsis tetragonoloba) in India. Indian J. Agric. Sci. **93** (2023)
53. Rosalam, S., England, R.: Review of xanthan gum production from unmodified starches by Xanthomonas comprestris sp. Enzyme Microb. Technol. **39**, 197–207 (2006)
54. UN (2023) UN Comtrade Database. https://comtradeplus.un.org/. Accessed 14 Aug 2024
55. Rupert, R., Rodrigues, K.F., Thien, V.Y., Yong, W.T.L.: Carrageenan From Kappaphycus alvarezii (Rhodophyta, Solieriaceae): Metabolism, Structure, Production, and Application. Front Plant Sci. **13** (2022)
56. Necas, J., Bartosikova, L.: Carrageenan: a review. Vet. Med. (Praha) **58**, 187–205 (2013)
57. FAO: Seaweeds used as a Source of Carrageenan (2024). https://www.fao.org/4/y4765e/y4765e09.htm. Accessed 14 Aug 2024
58. Yang, J.S., Xie, Y.J., He, W.: Research progress on chemical modification of alginate: a review. Carbohydr. Polym. **84** (2011)
59. Janke, L.: Mapping the global mass flow of seaweed: cultivation to industry application. J. Ind. Ecol. **28**, 1256–1269 (2024)
60. Bixler, H.J., Porse, H.: A decade of change in the seaweed hydrocolloids industry. J. Appl. Phycol. **23**,(2011). https://doi.org/10.1007/s10811-010-9529-3
61. Gaff, M., Hosseini, S.B., Paulasová, D., et al.: Functionality of production processes in mycelium-based composites. Compos. Struct., 118309 (2024)
62. Jones, M., Huynh, T., Dekiwadia, C., et al.: Mycelium composites: a review of engineering characteristics and growth kinetics. J. Bionanosci. **11** (2017)
63. Bhurtel, A., Salifu, E., Siddiqua, S.: Composite biomediated engineering approaches for improving problematic soils: Potentials and opportunities. Sci. Total Environ. **914** (2024)
64. Arnaud, L., Boyeux, B., Hustache, Y.: Hemp and the construction industry. In: Hemp: Industrial Production and Uses. CABI Wallingford UK, pp 239–259 (2013)
65. Evrard, A., De Herde, A.: Hygrothermal performance of Lime-Hemp wall assemblies. J. Build. Phys. **34** (2010). https://doi.org/10.1177/1744259109355730
66. Crini, G., Lichtfouse, E., Chanet, G., Morin-Crini, N.: Traditional and New Applications of Hemp (2020)
67. Faria, P., Beckett, C.T.S., Fabbri, A., et al.: RILEM contribution to earthen building. In: International Conference on Earthen Construction. Springer, pp 194–205 (2024)
68. Manaia, J.P., Manaia, A.T., Rodriges, L.: Industrial hemp fibers: an overview. Fibers **7** (2019). https://doi.org/10.3390/fib7120106
69. Textile Exchange: Growing Hemp for the Future (2023). https://textileexchange.org/app/uploads/2023/04/Growing-Hemp-for-the-Future-1.pdf. Accessed 14 Aug 2024
70. Singh, L., Brar, B.S.: A review on rice straw management strategies. Nature Environ. Pollution Technol. **20** (2021)
71. Engel, R.E., Long, D.S., Carlson, G.R.: Predicting straw yield of hard red spring wheat. In: Agronomy Journal (2003)
72. Njogu, P., Kinyua, R., Muthoni, P., Nemoto, Y.: Thermal Gasification of Rice Husks from Rice Growing Areas in Mwea, Embu County, Kenya. Smart Grid and Renewable Energy 06 (2015). https://doi.org/10.4236/sgre.2015.65010

73. Jyothsna, G., Bahurudeen, A., K Sahu, P.: Sustainable utilisation of rice husk for cleaner energy: a circular economy between agricultural, energy and construction sectors. Mater. Today Sustainability **25** (2024). https://doi.org/10.1016/j.mtsust.2024.100667
74. Cintura, E., Faria, P., Molari, L., et al.: Characterization of an Arundo donax-based composite: a solution to improve indoor comfort. Ind. Crops Prod. **208**, 117756 (2024)
75. Lansdown, R.: Arundo donax, Giant Reed. In: The IUCN Red List of Threatened Species (2013). https://doi.org/10.2305/IUCN.UK.2013-1.RLTS.T164340A1043245.en. Accessed 13 Aug 2024
76. Heaton, E., Voigt, T., Long, S.P.: A quantitative review comparing the yields of two candidate C4 perennial biomass crops in relation to nitrogen, temperature and water. Biomass Bioenergy **27** (2004). https://doi.org/10.1016/j.biombioe.2003.10.005
77. Molari, L., Coppolino, F.S., García, J.J.: Arundo donax: A widespread plant with great potential as sustainable structural material. Constr. Build Mater. 268 (2021). https://doi.org/10.1016/j.conbuildmat.2020.121143
78. Ceotto, E., Vasmara, C., Marchetti, R., et al.: Biomass and methane yield of giant reed (Arundo donax L.) as affected by single and double annual harvest. GCB Bioenergy **13** (2021). https://doi.org/10.1111/gcbb.12790
79. Alexopoulou, E., Zanetti, F., Scordia, D., et al.: Long-term yields of switchgrass, giant reed, and miscanthus in the mediterranean basin. Bioenergy Res **8** (2015). https://doi.org/10.1007/s12155-015-9687-x
80. Danelli, T., Sepulcri, A., Masetti, G., et al.: Arundo donax l. Biomass production in a polluted area: effects of two harvest timings on heavy metals uptake. Appl. Sci. (Switzerland) **11** (2021). https://doi.org/10.3390/app11031147
81. Maxim, L., Paudel, S., Piazza, M., et al.: World bamboo resources A thematic study prepared in the framework of the Global Forest Resources Assessment 2005 (2005). https://www.fao.org/3/a1243e/a1243e00.pdf
82. FAO: Global Forest Resources Assessment - 2020 Main report (2020). https://www.fao.org/3/ca9825en/ca9825en.pdf. Accessed 13 Aug 2023
83. Coudert, L., Blais, J.F., Mercier, G., et al.: Remediation processes for wood treated with organic and/or inorganic preservatives. In: Handbook of Recycled Concrete and Demolition Waste (2013)
84. Dobrzanski, E., Ferreira, E.S., Tiwary, P., et al.: Size-structure-property relationship of wood particles in aqueous and dry insulative foams. Carbohydr. Polym. **335**, 122077 (2024)
85. FAO: Global forest products facts and figures 2023. Rome (2024)
86. Svedarsky, D., Grosshans, R., Venema, H., et al.: Integrated management of invasive cattails (Typha spp.) for wetland habitat and biofuel in the Northern Great Plains of the United States and Canada: A review. Mires and Peat **25** (2019). https://doi.org/10.19189/MaP.2018.APG.367
87. Li, X., Newnham, R., Vandergoes, M.J., et al.: Insights into the natural and cultural history of Typha orientalis (Raupō) in Aotearoa New Zealand. PLOS Water **3**, e0000240 (2024)
88. IWTO International Wool Textile Organisation. In: 2024. https://iwto.org/. Accessed 12 Aug 2024
89. Chen, P., Han, Y., Jia, F., et al.: Investigation of the mechanism of aerodynamic separation of rice husks from brown rice following paddy hulling by coupled CFD-DEM. Biosyst. Eng. **218** (2022). https://doi.org/10.1016/j.biosystemseng.2022.03.015
90. Antonanzas, J., Idowu, O.J., Ray, D.T., et al.: Assessment of guar gum production cost and environmental impact in the US Southwest. Ind Crops Prod 203 (2023). https://doi.org/10.1016/j.indcrop.2023.117116
91. Summers, H.M., Sproul, E., Seavert, C., et al.: Economic and environmental analyses of incorporating guar into the American southwest. Agric. Syst. **191**,(2021). https://doi.org/10.1016/j.agsy.2021.103146

92. Moretti, C., Corona, B., Hoefnagels, R., et al.: Review of life cycle assessments of lignin and derived products: Lessons learned. Science of the Total Environment **770** (2021)
93. Christoforou, E., Kylili, A., Fokaides, P.A., Ioannou, I.: Cradle to site Life Cycle Assessment (LCA) of adobe bricks. J. Clean. Prod. **112** (2016). https://doi.org/10.1016/j.jclepro.2015.09.016
94. Ben-Alon, L., Loftness, V., Harries, K.A., Cochran Hameen, E.: Life cycle assessment (LCA) of natural vs conventional building assemblies. Renewable and Sustainable Energy Reviews 144 (2021). https://doi.org/10.1016/j.rser.2021.110951
95. Ramezannia, A., Gocer, O., Bashirzadeh Tabrizi, T.: The life cycle assessment of stabilized rammed earth reinforced with natural fibers in the context of Australia. Constr. Build. Mater. **416**,(2024). https://doi.org/10.1016/j.conbuildmat.2024.135034
96. UN: What is climate change adaptation and why is it crucial? (2024). https://climatepromise.undp.org/news-and-stories/what-climate-change-adaptation-and-why-it-crucial. Accessed 14 Aug 2024
97. UN: Climate Adaptation (2024). https://www.un.org/en/climatechange/climate-adaptation. Accessed 14 Aug 2024
98. IPCC: Climate change 2023: Synthesis report. Contribution of working groups I, II and III to the sixth assessment report of the Intergovernmental panel on climate change. European University Institute (2023)
99. Hart, S., Raymond, K., Williams, C.J., et al.: Modeling earthen treatments for climate change effects. Heritage **6**,(2023). https://doi.org/10.3390/heritage6050222
100. Richards, J., Guo, Q., Viles, H., et al.: Moisture content and material density affects severity of frost damage in earthen heritage. Sci. Total. Environ. **819**,(2022). https://doi.org/10.1016/j.scitotenv.2022.153047
101. Attia, S., Benzidane, C., Rahif, R., et al.: Overheating calculation methods, criteria, and indicators in European regulation for residential buildings. Energy Build **292**,(2023). https://doi.org/10.1016/j.enbuild.2023.113170
102. Brambilla, A., Jusselme, T.: Preventing overheating in offices through thermal inertial properties of compressed earth bricks: a study on a real scale prototype. Energy Build **156**, 281–292 (2017). https://doi.org/10.1016/J.ENBUILD.2017.09.070
103. Fernandes, J., Mateus, R., Gervásio, H., et al.: Passive strategies used in Southern Portugal vernacular rammed earth buildings and their influence in thermal performance. Renew. Energy **142**,(2019). https://doi.org/10.1016/j.renene.2019.04.098
104. Posani, M., Veiga, R., Freitas, V.: Post-Insulating Traditional Massive Walls in Southern Europe: A moderate thermal resistance can be more effective than you think. Energy Build (2023). https://doi.org/10.1016/j.enbuild.2023.113299
105. Posani, M., Voney, V., Odaglia, P., et al.: Low-carbon indoor humidity regulation via 3D-printed superhygroscopic building components. Nat. Commun. **16**, 425 (2025)
106. Osanyintola, O.F., Simonson, C.J.: Moisture buffering capacity of hygroscopic building materials: experimental facilities and energy impact. Energy Build **38**, 1270–1282 (2006). https://doi.org/10.1016/j.enbuild.2006.03.026
107. Zhang, M., Qin, M., Rode, C., Chen, Z.: Moisture buffering phenomenon and its impact on building energy consumption. Appl. Therm. Eng. **124**,(2017). https://doi.org/10.1016/j.applthermaleng.2017.05.173
108. Guihéneuf, S., Rangeard, D., Perrot, A., et al.: Effect of bio-stabilizers on capillary absorption and water vapour transfer into raw earth. Mater. Struct./Materiaux et Constructions **53** (2020). https://doi.org/10.1617/s11527-020-01571-z
109. Trambitski, Y., Kizinievič, O., Kizinievič, V.: Effect of retrograded starch hydrogel on the hygroscopic and durability properties of clay composites. Materials **15**,(2022). https://doi.org/10.3390/ma15248815

110. Achenza, M., Fenu, L.: On earth stabilization with natural polymers for earth masonry construction. Mater. Struct./Materiaux et Constructions **39**,(2006). https://doi.org/10.1617/s11527-005-9000-0
111. Zavaleta, D., Nanez, R., Silva, G., et al.: Evaluation of Chitosan and Potato Starch as Stabilizers to Improve the Mechanical and Water Durability Properties of Printable Earth-Based Matrices Reinforced with Sisal Fibers. In: Materials Science Forum (2023)
112. Indraratna, B., Muttuvel, T., Khabbaz, H.: Modelling the erosion rate of chemically stabilized soil incorporating tensile force - deformation characteristics. Can. Geotech. J. **46**,(2009). https://doi.org/10.1139/T08-103
113. Indraratna, B., Muttuvel, T., Khabbaz, H., Armstrong, R.: Predicting the erosion rate of chemically treated soil using a process simulation apparatus for internal crack erosion. J. Geotech. Geoenviron. Eng. **134**,(2008). https://doi.org/10.1061/(asce)1090-0241(2008)134:6(837)
114. Pedergnana, M., Elias-Ozkan, S.T.: Impact of Local Additives on Properties of Earthen Plasters. Terra 2016 (2016)
115. Fraunhofer Institute for Building Physics IBP. WUFI, Product overview. https://wufi.de/en/software/product-overview/. Accessed 29 Mar 2018
116. Bauklimatik-Dresden: DELPHIN (2024). https://bauklimatik-dresden.de/delphin/index.php. Accessed 16 Aug 2024
117. COMSOL (2024) COMSOL Multiphysics Simulation Software. https://www.comsol.com/comsol-multiphysics. Accessed 16 Aug 2024
118. U.S. Department of Energy (2024) EnergyPlus. https://energyplus.net/. Accessed 16 Aug 2024
119. CORDEX (2019) Coordinated Regional Climate Downscalling Experiment. http://www.cordex.org/. Accessed 16 Aug 2024
120. Machard, A., Inard, C., Alessandrini, J.-M., et al.: A methodology for assembling future weather files including heatwaves for building thermal simulations from the European coordinated regional downscaling experiment (EURO-CORDEX) climate data. Energies (Basel) **13**, 3424 (2020)
121. Coelho, G.B.A., Henriques, F.M.A.: Performance of passive retrofit measures for historic buildings that house artefacts viable for future conditions. Sustain. Cities Soc. **71**,(2021). https://doi.org/10.1016/j.scs.2021.102982
122. METEOTEST Meteonorm - VErsion 6.0. Meteotest, Bern, Switzerland
123. Rodrigues, E., Fernandes, M.S., Carvalho, D.: Future weather generator for building performance research: an open-source morphing tool and an application. Build. Environ. **233**,(2023). https://doi.org/10.1016/j.buildenv.2023.110104
124. The Morph EPW App. In: https://www.morphepw.app/
125. Belcher, S.E., Hacker, J.N., Powell, D.S.: Constructing design weather data for future climates. Build. Serv. Eng. Res. Technol. **26**,(2005). https://doi.org/10.1191/0143624405bt112oa
126. Jentsch, M.F., James, P.A.B., Bourikas, L., Bahaj, A.B.S.: Transforming existing weather data for worldwide locations to enable energy and building performance simulation under future climates. Renew. Energy **55**,(2013). https://doi.org/10.1016/j.renene.2012.12.049
127. European Commission (2021) European Commission Construction and Demolition Waste (CDW). http://ec.europa.eu/environment/waste/construction_demolition.htm
128. Menegaki, M., Damigos, D.: A review on current situation and challenges of construction and demolition waste management. Curr. Opin. Green Sustain. Chem. **13** (2018)

129. Bui, Q.B., Morel, J.C., Venkatarama Reddy, B.V., Ghayad, W.: Durability of rammed earth walls exposed for 20 years to natural weathering. Build. Environ. **44** (2009). https://doi.org/10.1016/j.buildenv.2008.07.001
130. Pelicaen, E., Janssens, B., Knapen, E.: Circular building with raw earth: A qualitative assessment of two cases in Belgium. In: IOP Conference Series: Earth and Environmental Science (2021)
131. Muguda, S., Hughes, P.N., Augarde, C.E., et al.: Geotechnical characterisation of recycled biopolymer-stabilised earthen materials. In: 17th European Conference on Soil Mechanics and Geotechnical Engineering, ECSMGE 2019 – Proceedings (2019)

Exploring Alternative Stabilization for Compressed Earth Blocks: A Systematic Literature Review

Ricardo Cruz[1(✉)], José Alexandre Bogas[1], Alessandra Ranesi[2], and Paulina Faria[2]

[1] CERIS—Civil Engineering Research and Innovation for Sustainability, Instituto Superior Técnico, Universidade de Lisboa, Av. Rovisco Pais, 1, 1049-001 Lisboa, Portugal
ricardojtcruz@tecnico.ulisboa.pt

[2] CERIS NOVA—Civil Engineering Research and Innovation for Sustainability, NOVA School of Science and Technology, NOVA University Lisbon, 2829-516 Caparica, Portugal

Abstract. Compressed earth blocks (CEB) are gaining attention as a sustainable alternative to traditional masonry units due to their low embodied energy, reduced carbon emissions and use of locally sourced materials. However, their vulnerability to water, particularly in regions with heavy rainfall and flooding, remains a major challenge. To address this, Portland cement is often used as a stabilizer, but its significant environmental impact has driven researchers to seek more sustainable alternatives. In response, recent studies have explored partial replacement of Portland cement with industrial by-products and wastes such as fly ash and ground granulated blast furnace slag. While some of these substitutions showed promising results, they still raise concerns.

This systematic review evaluates the potential of stabilizing CEB solely with alternative reactive materials with low embodied energy. Using the PRISMA framework, relevant studies from the Web of Science database were analyzed. Findings indicate that both organic and inorganic alternative binders can substantially enhance the mechanical strength of unstabilized CEB. By highlighting eco-efficient stabilization strategies, this review contributes to advancing sustainable practices in earth-based construction.

Keywords: Earth construction · alternative stabilization · mechanical strength · sustainability

1 Introduction

In response to the urgent need for sustainable and low-carbon alternatives in the construction industry, compressed earth blocks (CEB) have gained renewed interest as an environmentally friendly masonry unit [1]. Traditionally composed of locally sourced soil, CEB are compacted into blocks without the need for firing, which significantly reduces energy consumption and carbon emissions compared to conventional masonry

Presenter of the paper during the Conference: Ricardo Cruz (#48687)

M. Posani and S. Singh (Eds.): RYS 2025, RILEM Bookseries 67, pp. 450–469, 2026.
https://doi.org/10.1007/978-3-032-14652-6_33

units [2]. However, to improve structural integrity and durability, particularly in load-bearing applications or in areas with high moisture exposure, hydraulic binders are commonly added [3]. While cement and hydraulic lime have been widely used for stabilization, their high environmental impact has prompted a shift toward alternative binders derived from industrial by-products, agricultural residues, and bio-based materials with low embodied energy and CO_2 emissions.

The partial replacement of cement and hydraulic lime for CEB stabilization with greener products represents a pragmatic strategy to reduce environmental impacts. Numerous studies reviewed in the literature have explored this hybrid stabilization approach, where materials with low embodied energy are blended with cement or lime to achieve synergistic effects. While this strategy has been widely promoted [4–7], it still presents several limitations when assessed from an environmental performance perspective. Although it reduces the volume of high embodied energy and carbon intensive binders, it continues to rely on cement and lime undermining the full potential transition to low-impact construction. Additionally, for example fly ash (FA), a by-product of pulverized coal combustion in thermal power plants [8], faces uncertain long-term availability as global energy systems shift away from coal-based power generation.

This literature review systematically examines the existing research on the use of low embodied-energy binders for the stabilization of CEB. A meta-analysis was conducted using data from 29 studies selected based on defined inclusion and exclusion criteria. The analysis focuses on the testing methods employed for physical and mechanical characterization, as well as the reported results. Additionally, the review explores correlations between key properties – such as porosity and bulk density – and how these relationships may be influenced by different stabilization approaches.

2 Methodology

A systematic literature review was conducted in accordance with the PRISMA (Preferred Reporting Items for Systematic Reviews and Meta-Analyses) methodology, recognized for its applicability across diverse research domains.

The literature search was performed using the Web of Science (WoS) database. Only peer-reviewed journal articles were considered, to ensure the inclusion of studies subjected to rigorous peer-review. Given the diversity in terminology across the literature, the search strategy employed a range of semantic variations, including synonyms (e.g., "block" and "brick"), plural forms, and both American and British spellings (e.g., "stabilized" and "stabilised"). The review was restricted to studies published between January 2014 and December 2024.

An initial set of 441 records was retrieved. Literature reviews or studies that employed geopolymer stabilization were excluded at the outset. Although geo-polymerization is a viable stabilization technique, it was excluded from this review due to its high energy demands – primarily from the calcination of aluminosilicate precursors and the need for elevated curing temperatures – as well as the substantial costs of alkaline activators like sodium silicate [9]. These factors pose significant barriers to the widespread adoption of geopolymer technology in CEB. Subsequently, the titles, keywords, and abstracts were screened, resulting in the exclusion of 190 of these studies. At this stage, studies that focused on stabilizing CEB solely with Portland cement or hydraulic lime or on the incorporation of fibers or aggregates that would solely modify the earth particle size distribution (physical stabilization) were removed. Physical stabilization would expand the volume of literature, making the review less focused and harder to manage within reasonable limits given the paper's length limitations. The remaining articles underwent full-text review. In this stage, articles that just partially replaced Portland cement or hydraulic lime, without fully addressing the scope of this review, were excluded. Following this rigorous process, 29 scientific articles were deemed eligible. These were organized alphabetically by the first author's name, and relevant data were extracted manually or using image and graph digitization software (WebPlotDigitizer). The extracted data were compiled in Microsoft Excel (version 2502) and included the publication data and the results of various characterization tests for both control samples and those incorporating alternative binders. Control samples (unstabilized CEB, UCEB) served as performance benchmarks in each study. Any variations in performance relative to the control samples were identified for each measured property. In instances where data were unclear or not reported, the corresponding fields were not considered.

3 Publication Data Analysis

This chapter offers a critical overview of recent research trends related to the stabilization of CEB with low embodied-energy binders.

In terms of publication trends, the number of studies remained relatively the same between 2014 and 2019, accounting for only 14% of the total publications (5 occurrences in 6 years). A marked increase occurred in 2020 (+2 occurrences); however, this growth stabilized in subsequent years, with publications in 2021, 2022 and 2023 (4 occurrences each year). A second notable increase was observed in 2024, which comprised 34% of the total publications (10 occurrences). Figure 1 presents the publication and citation distribution over the years. The number of citations were obtained using WoS database. By the end of 2024, the number of citations from 29 articles published between 2014 and 2024, was 445. The number of journals identified were 17, being mainly Q1 in Building and Construction, Civil and Structural Engineering or Materials Science. In fact, of the 29 selected studies, approximately 28% were published by *Construction and Building Materials* (Elsevier) and 14% by *Journal of Building Engineering* (Elsevier). The remaining 58% was released by other journals such as *Buildings* (MDPI) and *Journal of Materials in Civil Engineering* (ASCE).

A total of 24 different products for CEB for stabilization were found and classified into two macro-groups: organic and inorganic stabilizers.

Inorganic stabilizers were used in 69% of the studies included in the review. Wastes from different sectors were used as stabilizers, namely: mining and quarry waste (phosphate washing sludge, PWS; and phosphogypsum, PG); polymeric waste (glass powder, GP) and industrial waste and by-products such as ashes derived from coal; municipal solid waste, and other industrial combustion processes, including fly ashes (FA), municipal solid waste incinerator bottom ash (MSWIBA), medical waste incinerator bottom ash (MWIBA), sugarcane bagasse ash (SBA), cotton waste ash (CWA), rice husk ash (RHA), and wood biomass ash (WBA). Additional industrial wastes and by-products

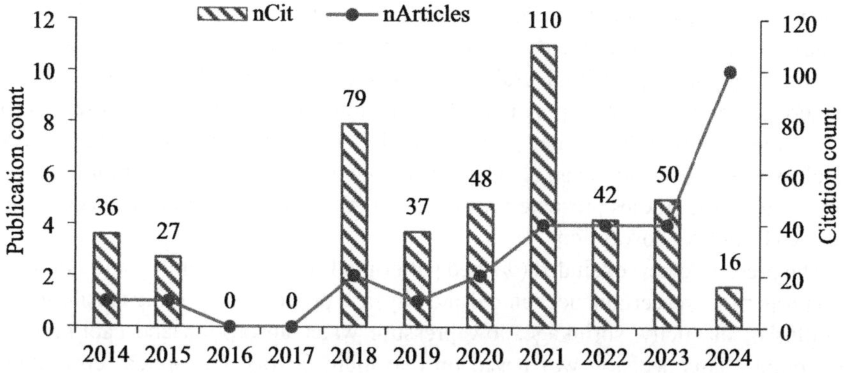

Fig. 1. Number of citations (nCit) and publications (nArticles) per year.

came from the metal industry (granulated ground blast furnace slag, GGBFS; and silica fume, SF), acetylene gas production (calcium carbide residue, CCR), construction and demolition waste industry (recycled cement, RC) and cement industry (cement kiln dust, CKD).

Organic binders were used in 31% of the studies analyzed and were mainly vegetal or animal based. Among vegetal bio-based stabilizers (24% of the studies) were found sugarcane molasses (SCM), cow dung (grass fibers processed in digestion), palm kernel oil residue (PKO residue), alginate, Xanthan gum, guar gum, carob gum. The remaining 7% (of the total) of the studies reported animal bio-based stabilizers such as animal glue and ovalbumin. The different categories and their percentages of occurrence can be seen in Fig. 2.

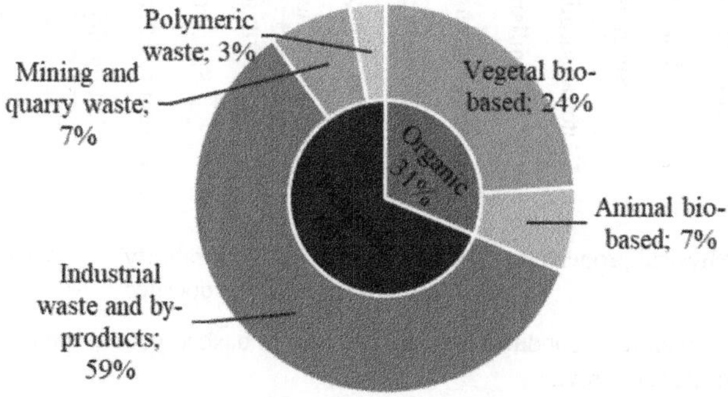

Fig. 2. Distribution (%) of CEB stabilizers included in the review.

The tests were initially grouped into four main broad categories: i) physical properties; ii) mechanical properties; iii) durability properties; iv) microstructural analysis. Approximately 93% of the articles included in the review presented the mechanical characterization of the CEB, assessing compressive strength, bending strength, splitting strength and modulus of elasticity. The compressive strength was addressed in all the studies presenting the CEB mechanical characterization, as it is considered the most critical performance indicator. Physical characterization was run in about 76% of the studies, involving measurements of bulk density, porosity, shrinkage, water absorption by immersion and capillarity. Durability tests, such as wetting-drying cycles, drip test, accelerated erosion testing (spray test) and abrasion resistance were less observed (only 41% of the studies) although being an essential factor in earth-based construction. Nevertheless, water absorption can also be considered a key indicator of durability. Finally, about 41% of the studies presented results from microstructural analysis, such as SEM, MIP, XRD and N_2-adsorption.

The present review excluded (dashed pattern in Fig. 3) all durability and microstructural properties and certain mechanical and physical properties (splitting tensile strength, modulus of elasticity, shrinkage, low pressure water absorption and capillary water absorption). This decision was based on two main factors: (i) limited representation of these properties in the reviewed literature (Fig. 3), which would hinder meaningful comparative analysis; and (ii) their inclusion would have made the review difficult to manage within reasonable limits given the paper's length limitations. Therefore, bulk density, porosity, compressive strength, and bending strength were prioritized. In addition, the microstructural mechanisms responsible for the variations in these properties will also be explored. Focusing exclusively on these parameters ensures consistency and strengthens the reliability of the conclusions drawn from the reviewed studies.

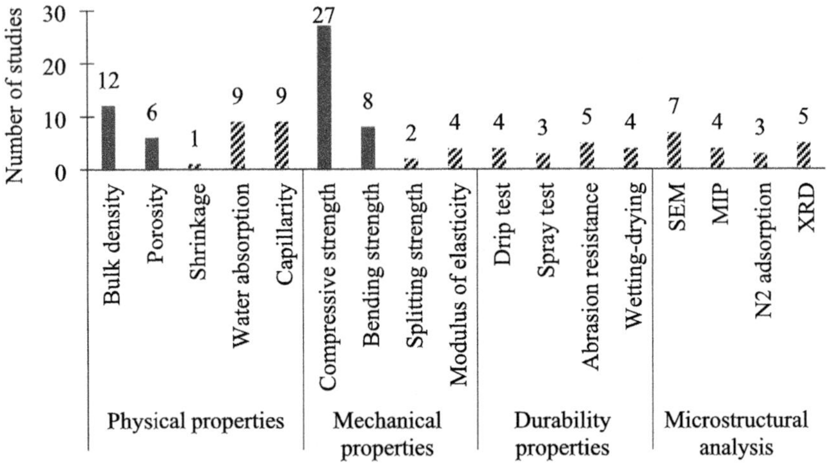

Fig. 3. Most common tests conducted by the authors. In dashed pattern the properties excluded from the present literature review.

4 Testing Methods

Accurate and consistent testing methods are essential for evaluating the performance and suitability of CEB in construction applications. To this end, a variety of standardized and non-standardized testing techniques have been developed to assess the key properties of CEB. This subchapter presents a comprehensive overview of the testing methods employed in the reviewed literature, to identify gaps or inconsistencies in current testing practices.

4.1 Bulk Density

Bulk density is a key physical property, representing the mass per unit volume, including both solid material and pore spaces. It is critical for evaluating the quality, strength, and thermal performance of CEB and serves as an indirect indicator of compaction and structural integrity. Accurate measurement is therefore essential. However, methods for determining bulk density differed across studies, with standards such as EN 772-13 [10], ASTM C134-95 [11] or NF P18-459 [12] applied in 58.3%, 16.7% and 8.3% of cases, respectively. Differences were noted, for example, in the preconditioning of specimens. Some studies [13–18] oven-dried the blocks until constant mass at 105 ± 5 °C, others [19, 20] at 60 ± 5 °C, while some others [3, 15, 16, 21, 22] reported values for air-dried blocks at ambient laboratory conditions (typically 23 ± 3 °C and 55–75% relative humidity). These differing drying methods lead to variations in moisture content, which directly affect the measured mass and, consequently, bulk density. In addition, most studies (11 out of 12 studies) determined volume using geometric measurements but one study [23] employed hydrostatic weighing for stabilized CEB, a more precise method that accounts for surface irregularities but cannot be used on UCEB due to their susceptibility to water.

4.2 Total Porosity

Total porosity (Φ) represents the proportion of the material's volume that is occupied by voids or pore spaces. It plays a crucial role in CEB performance and serves as an indirect indicator of compaction and the internal structure of the material. Again, the methods used for the determination of total porosity vary significantly from study to study. Nshimiyimana *et al.* [23, 24] determined Φ as being the ratio between bulk density and skeletal density (ρ_s) obtained through helium pycnometer testing. Other researchers [3, 15–18] considered Φ (Eq. 1) as the sum of the volume of voids after compaction (V_V) with the volume of voids from free hydration water (V_{H2O}), determined (Eq. 2) as the difference between the volume of mixing water (M_w) and the volume occupied by hydration products. The latter was calculated as the product of: i) α_H the long-term degree of hydration; ii) w_b the bound water (by weight of binder) at full hydration, and iii) M_b the mass of binder.

$$\Phi = V_V + V_{H_2O} \tag{1}$$

$$V_{H_2O} = \frac{M_W - \alpha_H * w_b * 0.746M_b}{\rho_w} \tag{2}$$

Other studies [3, 17, 18, 20] estimated the porosity through MIP and/or N_2 adsorption. However, these techniques are more suitable for analyzing pore size distribution and connectivity in the accessible pore network rather than determining the total porosity of CEB. For example, MIP only measures pores that are accessible to mercury. Closed pores cannot be intruded by mercury and therefore go undetected.

4.3 Compressive Strength

Compressive strength is one of the most critical mechanical properties used to evaluate the structural performance and load-bearing capacity of CEB. The determination of compressive strength involves subjecting a block or a representative specimen to a compressive load until failure, using a compression testing machine. To ensure consistent and reliable results, standardized procedures or national specifications are followed for sample preparation, curing, and testing. Among the standards procedures reported in the reviewed literature, examples include EN 772–1 [25], HB-195 [26], XP P13–901 [27] or IS 3495 [28], which were applied in 25.0%, 25.0%, 17.8% and 3.6% of cases, respectively.

Four different approaches were detected (Fig. 4). Approximately 15% of the studies [22, 23, 29, 30] followed the RILEM TC 164 recommendations, cutting blocks in half and stacking them with earth mortar to increase the slenderness ratio and approximate the results to unconfined compressive strength. However, Olivier *et al.* [31] found that this method yielded lower compressive strength values compared to cylindrical samples with a slenderness ratio of 1.5. This underestimation occurs regardless of the earth mortar's composition, as its strength and stiffness are generally lower than those of CEB [32, 33]. Other researchers measured compressive strength directly on whole [14–18, 34–40], half blocks [3, 13, 19, 20, 41, 42] or cylinders [21, 43–45] (46%, 23% and 15%, respectively). When using half blocks, the sawing process introduces local stresses that damage the sample, as reported by Aranda-Jiménez *et al.* [46], who observed a 21.7% reduction in compressive strength and higher variability compared to whole blocks. It is important to note that most researchers did not account for confinement effects in their tests on whole or halved blocks. The remaining studies [3, 15, 16] applied correction factors based on sample slenderness, referencing those found in HB-195 [26], which are derived from Krefeld's work [47] on fired bricks. However, Aubert *et al.* [32] argued that the correction factors proposed by Heathcote & Jankulovski [48] are more suited for CEB.

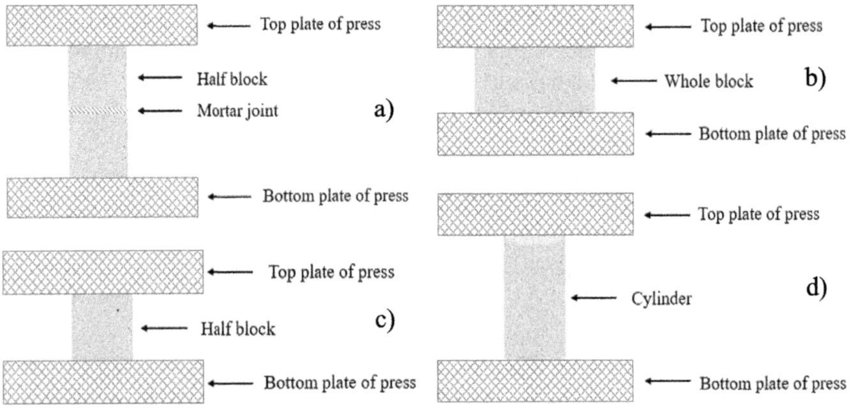

Fig. 4. Compressive strength testing scheme: a) half-block settled with mortar on half-block – RILEM TC 164 recommendation; b) whole block; c) half block; d) cylinder.

Another important factor influencing the compressive strength of CEB is their moisture content. Testing blocks in wet or saturated conditions often results in significantly lower strength values compared to those measured in oven-dried or air-dried specimens.

To determine the dry compressive strength, samples are typically oven-dried until their mass become stable. However, drying temperatures reported in the literature vary widely, ranging from 45 ± 5 °C to 105 ± 5 °C. According to Fabbri *et al.* [49], drying at the higher end of this range (105 ± 5 °C) can alter the material's properties, potentially producing misleading strength values. Conversely, lower temperatures, such as 45 ± 5 °C, may not fully eliminate moisture and can lead to inconsistent results due to fluctuations in the oven's relative humidity.

When assessing wet or saturated compressive strength, the duration of water immersion becomes a critical parameter. Studies have employed immersion times ranging from as little as 10 min to over 48 h. Short durations are generally used to evaluate the sample's initial water sensitivity, whereas longer immersion periods are intended to simulate full saturation, reflecting a worst-case exposure scenario.

4.4 Flexural Strength

Among the reviewed studies, the most widely used method was the three-point bending test. In this approach, a specimen is supported at both ends and subjected to a centrally applied load until failure occurs (Fig. 5), in accordance with standards such as EN 772-6 [50], ASTM C78-09 [51] or SNI 03-6458 [52]. The flexural strength is then calculated based on the maximum load applied, the span length, and the dimensions of the specimen.

Among the analysed studies, only one used the four-point bending test, applying the load at two equally spaced points until failure (Fig. 5). According to Paul and Gnanendran [53], the four-point bending configuration is more suitable when failure is predicted within the middle third of the span, as it provides a more even stress distribution.

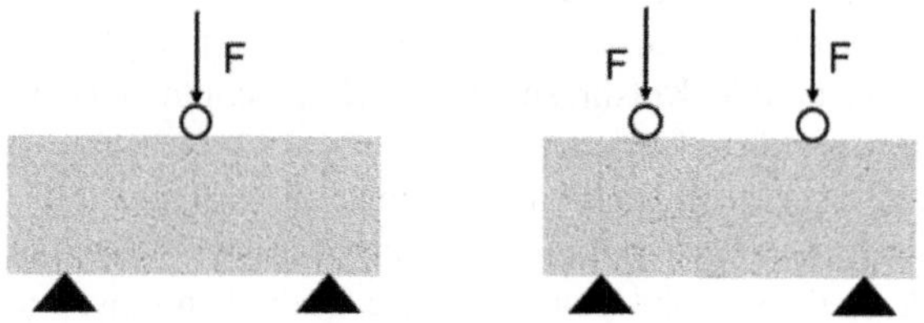

Fig. 5. Flexural strength testing scheme: three-point (on the left) and four-point (on the right) bending test.

Moreover, within the included literature, the blocks were preconditioned at different moisture conditions: air-dried [15, 19, 20, 29, 36, 41, 42], oven-dried [13, 37] or saturated [13, 37]. Another key factor is the loading rate. In general, higher loading rates tend to result in greater apparent flexural strengths. In fact, a rapid load gives shorter time for microcracks to grow and propagate within the material, increasing the material's resistance to failure. Conversely, slower loading rates give cracks more time to develop and allow for greater stress redistribution, which often leads to lower measured strength. In the reviewed literature, a loading rate of 5 mm/min was used for displacement-controlled tests [37] and from 40 to 400 N/s for force-driven tests [13, 15, 19, 20, 29, 36, 41, 42].

More recently, the RILEM Technical Committee TC 274-TCE [54] published guidelines and best practices for determining the flexural strength of raw-earth bricks using the three-point bending method. The authors recommended to precondition the specimens at 20 ± 3 °C and 50 ± 10% relative humidity and to apply a loading rate below 1.5 mm/min for displacement-controlled tests and equal to 50 N/s for force-controlled tests.

5 Results Overview

5.1 Bulk Density and Total Porosity

Despite the differences in procedures, the general trend is that the incorporation of organic or inorganic binders decreased the bulk density of CEB compared to the control samples (Table 1, in brackets). On one hand, for instance, the use of animal glue (1 wt%) and xanthan gum (1 wt%) led to a reduction in ρ of approximately 4% compared to unstabilized blocks, dropping from 2199 kg/m^3 to around 2113 and 2108 kg/m^3, respectively (Table 1). Similarly, CEB stabilized with 10 wt% CWA showed a 4% decrease in ρ (from 1847 to 1769 kg/m3). The most pronounced reduction was observed for CEB stabilized with CCR, with reductions of bulk density by 5% to 18% across different replacement levels (5–25 wt%). The combination of CCR and RHA in different proportions also resulted in a decreasing trend in bulk density. This behavior is primarily attributed to the physical characteristics of the binders. Both CCR and RHA have significantly lower specific density compared to the lateritic soil [23, 24]. As the proportion of these binders increases, the block becomes lighter, leading to a consistent decrease in bulk density. On the other hand, some materials demonstrated the tendency to increase ρ of CEB. Notably, PG at higher contents (up to 80%) increased ρ by up to 25%, likely due to its higher packing capacity. Likewise, the use of cow dung at 15–30 wt% replacement increased ρ by 3% to 9%, indicating potential for enhancing the compactness when well-processed. For example, Kulshreshtha *et al.* [55] recommends the removal of fibers of fresh cow-dung, what is not practical.

When reported, Φ generally exhibited the inverse trend than ρ (Fig. 6 and Table 1). The relationship shown in Fig. 6 holds consistently regardless of the type and amount of alternative binder used and is independent of the method employed to determine ρ and Φ, with the exception of one study [20]. Rivéra-Gómez *et al.* [20] incorporated 3 wt% alginate into three different clayey soils to produce CEB and estimated Φ using MIP. As discussed in Sect. 4.1, MIP is known to underestimate the total porosity of CEB and is therefore not considered a reliable method for this measurement. For example, Cruz & Bogas [3] found that MIP porosity accounted for only about 75–79% of Φ. Based on this correction, the adjusted Φ values for the samples studied by Rivéra-Gómez *et al.* [20] would be approximately 47.0, 46.6, and 42.6% for CEB produced with Soil I, Soil II and Soil III, respectively, instead of 36.2, 35.9, 32.8% found by the authors. As with bulk density, Φ decreased in alternative stabilized CEB compared to UCEB (Table 1). However, stabilization can alter the pore size distribution and pore tortuosity. For example, a coarser pore structure was observed in UCEB compared to RC-stabilized CEB [3, 17], despite lower Φ.

Table 1. Physical properties of CEB stabilized with alternative binders and unstabilized CEB (in brackets).

Ref	Binder (wt %)	ρ (kg/m^3)	Φ (%)
[3]	RC (4–8%)	1960–1980 (2000)	28.8–29.8 (26.2)
[13]	MWIFA (10–40%)	1649–1533[a] (1653[a])	-
[14]	Cow-dung (15–30%)	1797–1910[a] (1748[a])	-
[15–18]	RC (5–10%)	1718–1729 (1815)	39.3–39.0 (34.3)
	RC (10%)	1640[a] (1745[a])	
[17, 18]	RC (8%)	1778[a] (1835[a])	32.8 (30.4)
[19]	CWA (10%)	1769[a] (1847[a])	-
[20]	Alginate (3%) – Soil I	1510[a]	36.2
	Alginate (3%) – Soil II	1450[a]	35.9
	Alginate (3%) – Soil III	1390[a]	32.8
[21]	Animal glue (1%)	2113 (2199)	-
	Xantham gum (1%)	2108 (2199)	-
[22]	PG (20–80%)	1792–2166 (1739)	-
[23, 24]	CCR (5–25%)	1711–1477[a] (1801[a])	38–45 (35)
	CCR:RHA (20:0–12:8)	1522–1578[a] (1801[a])	44–39 (35)

Notation: RC – recycled cement; MWIFA – medical waste incinerator fly ash; CWA – cotton waste ash; PG – phosphogypsum; CCR – calcium carbide residue; RHA: rice husk ash; ρ – bulk density; Φ – porosity; [a] – oven dried

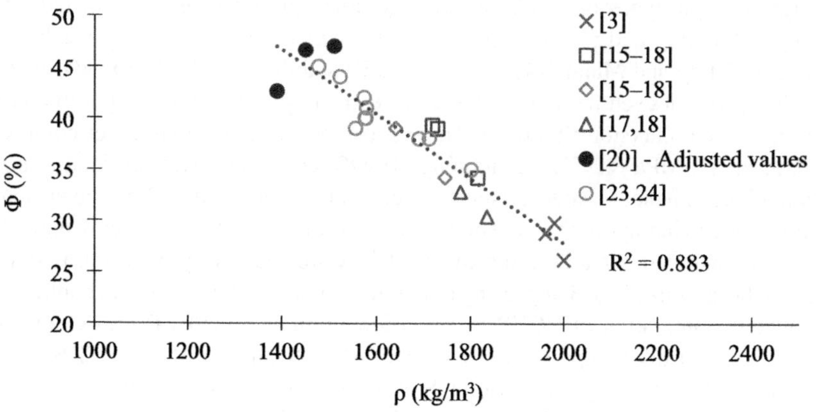

Fig. 6. Total porosity (Φ) as a function of bulk density (ρ).

5.2 Compressive Strength

Vegetal or animal bio-based binders showed improvement of compressive strength in some cases higher than others (Table 2). For instance, guar gum (2–4% wt.) offered just a slight improvement in strength relative to control sample, whereas SCM presented outstanding performance, with 2.3–3.4 times increase in dry strength for 5–10% wt dosage. Surprisingly, the XRD of CEB containing SCM revealed that no new phases were generated [35]. However, the SEM micrograph revealed a sticking behaviour of SCM that glued the earth particles, possibly responsible for the increased mechanical performance [35]. The same conclusions were obtained by Ouedraogo *et al.* [43] for CEB stabilized with 2–4% wt. of ovalbumin. The authors obtained 2.0–2.3 or 2.8–3.8 times increase in compressive strength depending on the earth used (Soil I and Soil II, respectively). The variation in strength development between formulations was attributed to the differing mineralogical compositions. In addition to quartz, feldspars, and illite, Soil I and Soil II contained montmorillonite and kaolinite, respectively [43]. The presence of expansive clays led to drying cracks, which negatively impacted strength. Interestingly, ovalbumin improved the water resistance of Soil I more than that of Soil II (Table 2), likely due to stronger molecular interactions with the clay particles [43]. In other study, Nouaouria & Nouaouria [45] reported a 2–time increase in compressive strength for CEB stabilized with 1 wt% carob gum, which further increased to 2.8 times at a 2 wt% dosage. The researchers attributed this strength improvement with the same type of gelling and entanglement mechanisms found for SCM by Chantit *et al.* [35], mentioned above. For the same reason, Ma *et al.* [21] reported an approximate 2–time increase in compressive strength after incorporating 1% wt of animal glue or xanthan gum. In addition, the researchers also observed the formation of hydrogen, covalent, and ionic bonds between the biomolecules and clay. The slight difference in compressive strength between the stabilization of CEB with animal glue and Xanthan gum (Table 2) was attributed to their bonding mechanisms: animal glue formed a more cohesive and durable matrix through multiple types of bonds, whereas Xanthan gum primarily interacted through hydrogen bonding and physical entanglement, which led to a slightly weaker structure [21]. Yalley and Manu [14] developed CEB stabilized with 15–30 wt% cow dung and observed that even the lowest dosage effectively prevented disintegration during a 10-min water immersion (Table 2). They also reported a maximum wet compressive strength of 2.8 MPa for CEB containing 20 wt% cow dung. Kulshreshtha *et al.* [55] attributed the enhanced water resistance of cow dung-stabilized CEB to the presence of naturally occurring hydrophobic, small-sized microbial aggregates in cow dung. Their study demonstrated a marked improvement in water durability, with the first visible crack in the stabilized CEB appearing only after 6 h of water exposure, in contrast to the complete disintegration of UCEB within a few minutes. Finally, Rivéra-Gomez *et al.* [20] reported air-dry compressive strengths of 3.9, 3.8, and 1.7 MPa for CEB produced with three different clayey soils—Soil I, Soil II, and Soil III, respectively—and stabilized with 3% alginate (Table 2). Based on the Φ values of the CEB made with these soils (Table 1), a higher compressive strength was expected for the block produced with Soil III due to the lower porosity. However, the CEB made with Soil III exhibited a poorer pore distribution, with an average diameter of 129.9 nm – approximately 2.2 times larger than that of the CEB produced using Soil I.

Table 2. Compressive strength of organic binder-stabilized CEB and unstabilized CEB (in brackets).

Ref	Binder (wt %)	Compressive strength (MPa)		
		Dry	Air-dry	Wet
[14]	Cow-dung (15–30%)	4.6–5.8 (4.6)	-	0.9–2.8
[20]	Alginate (3%) – Soil I	-	3.9	-
	Alginate (3%) – Soil II	-	3.8	-
	Alginate (3%) – Soil III	-	1.7	-
[21]	Xanthan gum (1%)	-	6.8 (3.6)	-
	Animal glue (1%)	-	7.2 (3.6)	-
[34]	Guar gum (2–4%)	-	2.8–2.9 (2.6)	-
[35]	SCM (5–10%)	7.5–11.1 (3.3)	-	6.7–9.9
[40]	PKO residue (2–8%)	4.0–4.3 (4.0)	-	-
[43]	Ovalbumin (2–4%) – Soil I	10.9–12.9 (5.5)	-	2.9–5.7
	Ovalbumin (2–4%) – Soil II	8.6–11.8 (3.1)	-	4% – 2.6
[45]	Carob gum (1–2%)	-	1.6–2.2 (0.8)	-

Notation: SCM – sugarcane bagasse molasse; PKO – palm kern oil

For the inorganic stabilization the highest improvements were observed for CEB stabilized with 10 wt% CWA (1.6 times), 10 wt% SCA (1.7 times), 20 wt% WBA (Type B – 2 times), 30 wt% MWIFA (2.4 times), 8 wt% RC (2.7 times), 25 wt% CKD (Soil I – 3.1 times; Soil II – 3.3 times), 15 wt% CCR (4.2 times) and 16:4 wt% CCR:RHA (6.4 times). According to Mebarkia *et al.* [42], the chemical analysis of CKD revealed a high tricalcium silicate (C_3S) content, which is associated with enhancing the durability of clayey soils and a reduction in their swelling behavior. Accordingly, XRD analysis of CEB produced with either Soil I or Soil II and containing 25% CKD revealed the presence of C-S-H despite its low crystallinity. For CEB stabilized with WBA (Type A and Type B), the researchers observed the formation of C-S-H and C-A-H [29]. However, the type B WBA presented a higher content of MgO, characterized by its hydrophobicity [29]. In addition, the reaction between SiO_2, Al_2O_3 and MgO produced magnesium silicate hydrate (M-S-H) leading to a more compact microstructure [29] and, consequently, higher strength values. For CEB stabilized with CCR or CCR:RHA blends, the XRD patterns revealed the presence of C-S-H and C-A-H in both types of CEB [56]. However, the SEM micrographs of CCR CEB revealed uncemented pores visible in the inter-particle regions, while CCR:RHA CEB exhibited a more cohesive microstructure, with inter-particle zones well cemented. This difference in the microstructure helps explain the greater improvement in compressive strength observed in the CCR:RHA samples. For RC, the improvement in strength can be mainly attributed to the C-S-H covering the earth particles surface and filling the inter-particle regions [3]. In addition, as mentioned in Sect. 5.1, RC stabilization altered the pore size distribution and pore tortuosity of UCEB, promoting the formation of a less coarse pore structure. In the case of MWIFA, its high CaO content (63.9%) promoted the occurrence of pozzolanic reactions with the aluminosilicates of the earth [13]. Salim *et al.* [39] also suggested the occurrence of pozzolanic reactions when SCA was added. However, the researchers reported a low content of CaO in SCA (2.8%), which makes the strength improvement more likely to be due to a filling effect. Similar conclusions can be drawn from the findings of Salem *et al.* [44] and Elahi *et al.* [37], who reported modest strength improvements – 1.1 to 1.3 times for 4–10 wt% SF and 1.3 to 1.5 times for 10–30 wt% FA, respectively (Table 3) – suggesting limited pozzolanic activity, corroborating the chemical compositions of the materials. In fact, CEB stabilized with FA crumbled upon water immersion regardless of the dosage (Table 3). The remaining studies are those of Latha *et al.* [36], Sekhar *et al.* [38], Benaicha *et al.* [41] and Oubaha *et al.* [22], who stabilized CEB with 5–25 wt% MSWIBA, 25 wt% GGFBS, 1.5–4.5 wt% GP and 20–80 wt% PG, respectively (Table 3). Again, the slight improvements in compressive strength were attributed to the occurrence of pozzolanic reactions. However, in the works of Latha *et al.* [36] and Oubaha *et al.* [22] there was a strength reduction above a certain binder content, suggesting the presence of unreacted particles.

The ratio between wet and dry compressive strengths (wet/dry) is a critical indicator of durability. This ratio was the highest for CEB stabilized with 5–10 wt% SCM (0.89), indicating excellent water resistance (Fig. 7). However, the blocks were only immersed for 2h, evaluating solely the sample's initial water sensitivity. As mentioned in Sect. 4.3, the immersion times and oven-drying temperatures are not standardized across the reviewed literature, which makes the comparison of the wet/dry ratio between studies impossible to perform. Nevertheless, for all reviewed studies the optimal composition presented a wet/dry ratio higher than 0.33 (Fig. 7), considered to be the minimum for CEB to have good durability in outdoor environments [3].

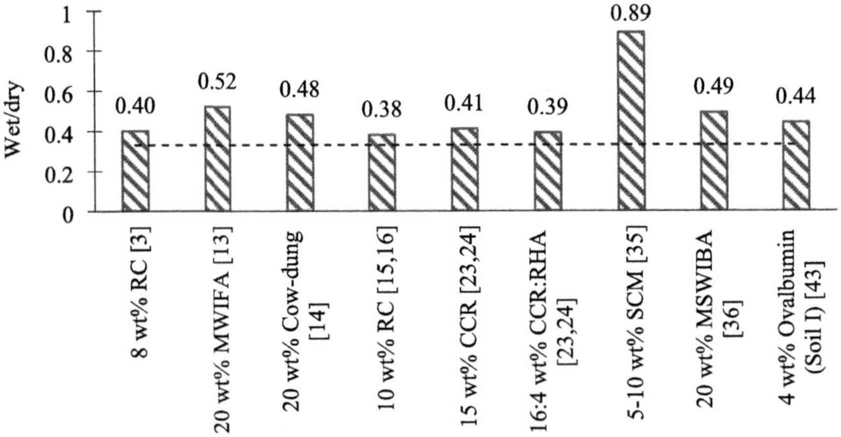

Fig. 7. Optimal ratio between wet and dry compressive strengths (wet/dry) reported by the authors [3, 13–16, 23, 24, 35, 36, 43] and (dashed) the value proposed by [3].

Table 3. Compressive strength of inorganic binder-stabilized CEB and unstabilized CEB (in brackets).

Ref	Binder (wt %)	Compressive strength (MPa)		
		Dry	Air-dry	Wet
[3]	RC (4–8%)	8% – 12.0 (7.4)	6.0–9.8 (3.6)	8% – 4.8
[13]	MWIFA (10–40%)	2.1–2.9 (1.2)	-	0.7–1.5
[15–18]	RC (5–10%)	10% – 6.5 (4.3)	2.5–4.4 (2.3)	10% – 2.5
[17, 18]	RC (8%)	-	5.5 (2.1)	-
[19]	CWA (10%)	-	3.0 (1.9)	-
[22]	PG (20–80%)	-	1.5–1.0 (1.3)	-
[23, 24]	CCR (5–25%)	3.0–4.6 (1.1)	-	1.6–2.7
	CCR:RHA (20:0–12:8)	4.4–7.0 (1.1)	-	1.8–2.7
[29]	WBA Type A (5–20%)	-	13.0–14.3 (12.5)	-
	WBA Type B (5–20%)	-	13.1–25.9 (12.5)	-
[30]	PWS (60%)	-	4.5	-
[36]	MSWIBA (5–25%)	3.6–4.0 (3.5)	-	1.7–2.0
[37]	FA (10–30%)	0.8–0.9 (0.6)	-	X
[38]	GGBFS (25%)	2.1 (1.5)	-	X
[39]	SCA (3–10%)	2.5–3.8 (2.3)	-	-
[41]	GP (1.5–4.5%)	-	5.8–7.1 (5.0)	-
[42]	CKD (5–25%) – Soil I	-	20.3–26.5 (8.6)	-
	CKD (5–25%) – Soil II	-	15.1–24.7 (7.4)	-
[44]	SF (4–10%)	-	1.8–2.0 (1.6)	-

Notation: RC – recycled cement; MWIFA – medical waste incinerator fly ash; CWA – cotton waste ash; PG – phosphogypsum; CCR – calcium carbide residue; RHA – rice husk ash; WBA – wood biomass ash; PWS – phosphate washing sludge; MSWIBA – municipal solid waste incinerator bottom ash; FA – fly ash; GGBFS – ground granulated blast-furnace slag; SCA – sugarcane bagasse ash; GP – glass powder; CKD – cement kiln dust; SF – silica fume

5.3 Flexural Strength

As observed with compressive strength, incorporating inorganic alternative stabilizers also enhanced CEB flexural strength. Compared to the control samples, the flexural strength increased by maximum of 1.2 times with CKD (Soil I) [42], 1.3 times with 20 wt% MSWIBA [36] and 20 wt% FA [37], 2.3 times with 10 wt% RC [15], 2.4 times with 10 wt% CWA [19], 2.5 times with 20% WBA type A [29], 2.8 times with 25 wt% CKD (Soil II) [42], 3.3 times with 20 wt% WBA type B [29], and 3.9 times with 30 wt% MWIFA [13] (Table 4). Once again, these improvements can be attributed to hydration and pozzolanic reactions, whose resulting products densify the earth matrix and enhance the flexural strength of the CEB. Meanwhile, Benaicha *et al.* [41] observed a decrease in flexural strength with the addition of 1.5–4.5 wt% GP, despite the previously discussed improvement in compressive strength. The researchers, however, did not provide any explanation for this reduction. Within the studies included in the review, only Rivera-Gómez *et al.* [20] studied the flexural strength of CEB stabilized with organic binders. The authors reported a flexural strength of 0.91, 0.86 and 0.39 MPa for CEB produced with Soil I, II and III, respectively and stabilized with 3 wt% alginate. However, no UCEB was reported as reference. Nevertheless, it is anticipated that the gelling and entanglement mechanisms previously discussed would significantly enhance flexural strength.

Although flexural strength and compressive strength reflect distinct mechanical behaviors, studies [15, 57] have demonstrated a clear correlation between the two. Analysis of the data from the reviewed literature – specifically the reported values of compressive and three-point flexural strengths – reveals a consistent linear relationship, irrespective of the type and content of alternative binders used (Fig. 8). However, the nature of this relationship is influenced by the method employed to determine compressive strength. Notably, differences arise depending on whether the RILEM TC 164 recommendation is followed or if the test is conducted directly on full-sized or halved blocks.

Table 4. Flexural strength of alternative binder-stabilized CEB and unstabilized CEB (in brackets).

Ref	Binder (wt %)	Flexural strength (MPa)
[13]	MWIFA (10–40%)	0.30–0.49 (0.10)
[15]	RC (5–10%)	0.52–0.93 (0.41)
[19]	CWA (10%)	0.74 (0.31)
[20]	Alginate (3%) – Soil I	0.91
	Alginate (3%) – Soil II	0.86
	Alginate (3%) – Soil III	0.39
[29]	WBA Type A (5–20%)	0.90–1.43 (0.77)
	WBA Type B (5–20%)	1.32–2.50 (0.77)
[36]	MSWIBA (5–25%)	0.40–0.49 (0.38)
[37]	FA (10–30%)	0.21–0.24 (0.18)
[42]	CKD (5–25%) – Soil I	6.33–7.39 (6.08)
	CKD (5–25%) – Soil II	3.20–6.39 (2.32)

Notation: MWIFA – medical waste incinerator fly ash; RC – recycled cement; CWA – cotton waste ash; WBA – wood biomass ash; MSWIBA – municipal solid waste incinerator bottom ash; FA – fly ash; CKD – cement kiln dust

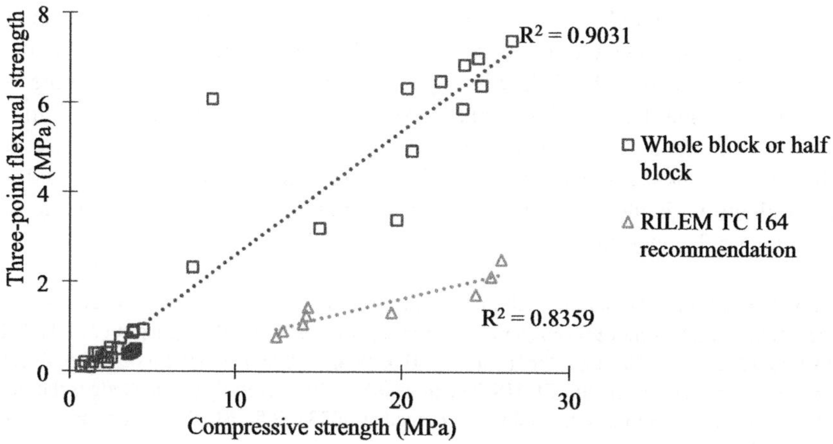

Fig. 8. Compressive strength as a function of the three-point flexural strength.

6 Conclusions

This systematic literature review explored the potential of low-embodied energy alternative binders to stabilize CEB. From 29 peer-reviewed studies published between 2014 and 2024, this review highlights a growing interest in the use of organic and inorganic waste-derived materials to improve the properties of UCEB while keeping their environmental footprint low.

Both organic and inorganic binders contributed to improved performance, despite it being through different mechanisms. Organic materials such as sugarcane molasses, ovalbumin, alginate, and carob gum enhanced mechanical strength by forming adhesive matrices through gelling and binding effects. These improvements were often achieved without the formation of new mineralogical phases, indicating that physical and chemical interactions between the binders and soil particles played a critical role. In contrast, inorganic binders such as recycled cement, calcium carbide residue, wood biomass ash and cement kiln dust typically improved mechanical performance through pozzolanic and hydration reactions. These reactions led to the development of strengthening phases such as calcium silicate hydrate, calcium aluminate hydrate, and magnesium silicate hydrate, which helped densify the earth matrix and enhance overall block integrity.

In terms of physical properties, the inclusion of alternative binders generally resulted in a reduction in bulk density and a corresponding increase in porosity. However, the extent of this effect varied according to the binder's characteristics and dosage. In addition, compared to UCEB, the compressive strength showed marked improvements, increasing by as much as 6.4 times for CEB stabilized with 16:4 wt% blends of calcium carbide residue and rice husk ash.

Despite these promising outcomes, the review identified the application of different testing procedures for the same property in different studies. This lack of standardization complicates direct comparisons between studies and highlights the need for harmonized testing protocols. Nevertheless, several organic stabilizers, such as sugarcane molasses and cow dung, introduced encouraging enhancements in CEB durability, with wet/dry strength ratios exceeding the minimum threshold considered suitable for exterior use.

In summary, the findings of this review support the technical viability and environmental benefits of using alternative, low-carbon binders for stabilizing CEB. However, further advancements in this field will require durability characterization and the establishment of standardized testing frameworks. Moreover, once the effects of different stabilization methods on mechanical performance and durability are fully assessed, a multi-criteria analysis, considering cost-effectiveness, environmental impact, and stabilizer availability, should guide the informed selection process. These steps are essential for scaling up the adoption of sustainable earth-based construction practices.

Acknowledgements. The authors acknowledge the Portuguese Foundation for Science and Technology (FCT, https://ror.org/00snfqn58) for funding this research through project PTDC/ECI-CON/0704/2021, and through the project UIDB/06438/2025 (https://doi.org/10.54499/UID/06438/2025) of the research unit CERIS. The first author also gratefully acknowledges the support of FCT through the doctoral scholarship referenced as 2023.05651.BD (https://doi.org/10.54499/2023.05651.BD).

References

Sinha, S., Sudarsan, J.S.: Building a greener future: how earth blocks are reshaping sustainability and circular economy in construction. Architecture **5**, 25 (2025). https://doi.org/10.3390/architecture5020025

Turco, C., Abedi, M., Teixeira, E.R., Mateus, R.: Thermophysical properties of compressed earth blocks incorporating natural materials. Energies (Basel) **17**, 2070 (2024). https://doi.org/10.3390/en17092070

Cruz, R., Bogas, J.A.: Durability of compressed earth blocks stabilised with recycled cement from concrete waste and incorporating construction and demolition waste. Constr. Build. Mater. **450**, 130673 (2024). https://doi.org/10.1016/j.conbuildmat.2024.138673

Islam, M.S., Elahi, T.E., Shahriar, A.R., Mumtaz, N.: Effectiveness of fly ash and cement for compressed stabilized earth block construction. Constr. Build. Mater. **255**, 119392 (2020). https://doi.org/10.1016/j.conbuildmat.2020.119392

Cabrera, S., Elert, K., Guilarducci, A., Margasin, A.: The effect of local pozzolans and lime additions on the mineralogical, physical and mechanical properties of compressed earth blocks in Argentina. Revista de la construcción **21**, 248–263 (2022). https://doi.org/10.7764/rdlc.21.2.248

Ouma, J., Ongwen, N., Ogam, E., et al.: Acoustical properties of compressed earth blocks: Effect of compaction pressure, water hyacinth ash and lime. Case Stud. Const. Mater. **18**, e01828 (2023). https://doi.org/10.1016/J.CSCM.2023.E01828

Ojerinde, A., Adekunle, A., Fatai, O., et al.: The partial replacement of ordinary Portland cement with rice husk ash to stabilize compressed earth blocks for affordable building Materials. In: PLEA 2018 - 34th International Conference on Passive and Low Energy Architecture. Hong Kong, pp. 321–325 (2018)

Singh, N.B.: Fly ash in the construction industry. In: Handbook of Fly Ash, 1st edn., pp. 565–610. Elsevier, London (2022)

Pol Segura, I., Ranjbar, N., Juul Damø, A., et al.: A review: Alkali-activated cement and concrete production technologies available in the industry. Heliyon **9**, e15718 (2023). https://doi.org/10.1016/J.HELIYON.2023.E15718

EN 772-13: Methods of test for masonry units - Part 13: Determination of net and gross dry density of masonry units (except for natural stone). European Committee for Standardization (CEN) (2000)

ASTM C134-95: Standard Test Methods for Size, Dimensional Measurements, and Bulk Density of Refractory Brick and Insulating Firebrick. American Society for Testing & Materials (ASTM) (2016)

NF P18-459: Concrete - Testing hardened concrete - Testing porosity and density. Association Francaise de Normalisation (2010)

Al Biruni, M.T., Sarker, H.S., Charu, W.U.S., Ahmed, T.: Sustainable use of medical waste incineration fly ash in compressed stabilized earth blocks. Constr. Build. Mater. **452**, 138886 (2024). https://doi.org/10.1016/j.conbuildmat.2024.138886

Paa, P., Yalley, K., Manu, D.: Strength and durability properties of cow dung stabilised earth brick. Civil Environ. Res. **3**, 117–125 (2013)

Bogas, J.A., Real, S., Cruz, R., Azevedo, B.: Mechanical performance and shrinkage of compressed earth blocks stabilised with thermoactivated recycled cement. J. Build. Eng. **79**, 107892 (2023). https://doi.org/10.1016/j.jobe.2023.107892

Cruz, R., Bogas, J.A., Balboa, A., Faria, P.: Water resistance of compressed earth blocks stabilised with thermoactivated recycled cement. Materials **17**, 5617 (2024). https://doi.org/10.3390/ma17225617

Real, S., Bogas, J.A., Cruz, R., Gomes, M.G.: Eco-recycled cement's effect on the microstructure and hygroscopic behaviour of compressed stabilised earth blocks. J. Build. Eng. **95**, 110227 (2024). https://doi.org/10.1016/j.jobe.2024.110227

Real, S., Bogas, J.A., Cruz, R., et al.: Thermal performance of compressed earth blocks stabilised with thermoactivated recycled cement. Energy Build **314**, 114288 (2024). https://doi.org/10.1016/j.enbuild.2024.114288

Imar, M.M., Diakhate, M., Moungam, L.M.B.A., et al.: Valorization of cotton waste ash in the formulation of compressed earth blocks. Mech. Adv. Mater. Struct. **31**, 5220–5233 (2024). https://doi.org/10.1080/15376494.2023.2213240

Rivera-Gómez, C., Galán-Marín, C., López-Cabeza, V.P., Diz-Mellado, E.: Sample key features affecting mechanical, acoustic and thermal properties of a natural-stabilised earthen material. Constr. Build. Mater. **271**, 121569 (2021). https://doi.org/10.1016/j.conbuildmat.2020.121569

Ma, J., Abdelaal, A., Zhang, H., et al.: Ultra-compressed earth block stabilized by bio-binder for sustainable building construction. Case Stud. Constr. Mater. **21**, e03523 (2024). https://doi.org/10.1016/j.cscm.2024.e03523

Oubaha, S., Hakkou, R., Taha, Y., et al.: Elaboration of compressed earth blocks based on phosphogypsum and phosphate mining by-products. J. Build. Eng. **62**, 105423 (2022). https://doi.org/10.1016/j.jobe.2022.105423

Nshimiyimana, P., Messan, A., Courard, L.: Physico-mechanical and hygro-thermal properties of compressed earth blocks stabilized with industrial and agro by-product binders. Materials **13**, 3769 (2020). https://doi.org/10.3390/ma13173769

Nshimiyimana, P., Messan, A., Courard, L.: Hydric and durability performances of compressed earth blocks stabilized with industrial and agro by-product cinders: calcium carbide residue and rice husk ash. J. Mater. Civ. Eng. **33**, 4021121 (2021). https://doi.org/10.1061/(ASCE)MT.1943-5533.0003745

EN 772-1: Methods of test for masonry units - Part 1: Determination of compressive strength. European Committee for Standardization (CEN) (2011)

HB 195: The Australian Earth Building Handbook. Standards Australia, Sydney, Australia (2002)

XP P13-901 Compressed Earth Blocks for Walls and Partitions. Definitions. Specifications. Test Methods. Delivery Acceptance Conditions. Association Française de Normalisation (AFNOR), Paris, France (2001)

IS 3495: Methods of tests of burn clay building blocks - Part 1: Determination of compressive strength. Indian Standards (1992)

Nadia, B., Fatma, K., Nasser, C.: Mechanical, thermal and durability investigation of compressed earth bricks stabilized with wood biomass ash. Constr. Build. Mater. **364**, 129874 (2023). https://doi.org/10.1016/j.conbuildmat.2022.129874

Inabi, O., Khalil, A., Zouine, A., et al.: Investigation of the innovative combined reuse of phosphate mine waste rock and phosphate washing sludge to produce eco-friendly bricks. Buildings **14**, 2600 (2024). https://doi.org/10.3390/buildings14092600

Olivier, M., Mesbah, M.A., Gharbi, Z., Morel, J.-C.: Test method for strength tests on blocks of compressed earth. Materials and Structures/Materiaux et Constructions **30**, 515–517 (1997)

Aubert, J.E., Maillard, P., Morel, J.C., Al Rafii, M.: Towards a simple compressive strength test for earth bricks? Mater. Struct. **49**, 1641–1654 (2016). https://doi.org/10.1617/s11527-015-0601-y

Lan, G., Chao, S., Wang, Y., Cui, Y.: Methods to test the compressive strength of earth blocks. Adv. Mater. Sci. Eng. **2021**, 1767238 (2021). https://doi.org/10.1155/2021/1767238

Bruno, A.W., Scott, B., D'Offay-Mancienne, Y., Perlot, C.: Recyclability, durability and water vapour adsorption of unstabilised and stabilised compressed earth bricks. Mater. Struct./Materiaux et Constructions **53**, 149 (2020). https://doi.org/10.1617/s11527-020-015 85-7

Chantit, F., El, A.F.E., Kchikach, A.: Performance of compressed earth bricks reinforced with sugar industrial by-products bagasse and molasse: mechanical, physical and durability properties. J. Build. Eng. **90**, 109403 (2024). https://doi.org/10.1016/j.jobe.2024.109403

Thennarasan Latha, A., Murugesan, B., Thomas, B.S.: Compressed stabilized earth block incorporating municipal solid waste incinerator bottom ssh as a partial replacement for fine aggregates. Buildings **13**, 1114 (2023). https://doi.org/10.3390/buildings13051114

Elahi, T.E., Shahriar, A.R., Islam, M.S.: Engineering characteristics of compressed earth blocks stabilized with cement and fly ash. Constr. Build. Mater. **277**, 122367 (2021). https://doi.org/10.1016/j.conbuildmat.2021.122367

C. Sekhar, D., Nayak, S.: Utilization of granulated blast furnace slag and cement in the manufacture of compressed stabilized earth blocks. Constr. Build. Mater. **166**, 531–536 (2018). https://doi.org/10.1016/j.conbuildmat.2018.01.125

Salim, R.W., Ndambuki, J.M., Adedokun, D.A.: Improving the bearing strength of sandy loam soil compressed earth block bricks using sugercane bagasse ash. Sustainability **6**, 3686–3696 (2014). https://doi.org/10.3390/su6063686

Yalley, P.P., Badu, E.: Stabilising earth brick with palm kernel oil residue for construction of low cost housing. Advancements in Civil Engineering & Technology **1**, 1 (2018). https://doi.org/10.31031/acet.2018.01.000504

Benaicha, A.C., Fourar, A., Mansouri, T., Fawaz, M.: Valorization of sediment extracted from the dam in construction works. Model Earth Syst. Environ. **8**, 4093–4102 (2022). https://doi.org/10.1007/s40808-021-01326-x

Mebarkia, R., Bouzeroura, M., Chelouah, N.: Study of the effect of cement kiln dust on the mechanical, thermal and durability properties of compressed earth blocks. Constr. Build. Mater. **349**, 128707 (2022). https://doi.org/10.1016/j.conbuildmat.2022.128707

Ouedraogo, K.A.J., Aubert, J.E., Tribout, C., et al.: Ovalbumin as natural organic binder for stabilizing unfired earth bricks: Understanding vernacular techniques to inspire modern constructions. J. Cult. Herit. **50**, 139–149 (2021). https://doi.org/10.1016/j.culher.2021.05.004

Ait Salem, A.A., Benchouk, A., Belayachi, N.: Influence of silica fume on the mechanical, thermal and water absorption behavior of compressed earth bricks. MRS Adv. **8**, 976–981 (2023). https://doi.org/10.1557/s43580-023-00615-0

Nouaouria, A., Nouaouria, M.S.: Mechanical behaviour and durability of compressed earth blocks treated with bio-binder. Indian Geotech. J. **54**, 1187–1197 (2024). https://doi.org/10.1007/s40098-023-00794-5

Aranda-Jiménez, Y., Zuñiga-Leal, C., Moreno-Chimely, L., Robles-Aranda, M.E.: Compressed earth blocks (CEB) compression tested under two earth standards. Cogent Eng. **10**, 2163116 (2023). https://doi.org/10.1080/23311916.2022.2163116

Krefeld, W.J.: Effect of shape of specimen on the apparent compressive strength of brick masonry. In: Proceedings of the American Society of Materials. Philadelphia, pp. 363–372 (1938)

Heathcote, K., Jankulovski, E.: Aspect ratio correction factors for soilcrete blocks. Australian Civil Engineering Transactions, Institution of Engineers Australia **34**, 309–312 (1992)

Fabbri A, Morel J-C, Gallipoli D.: Assessing the performance of earth building materials: a review of recent developments. RILEM Techn. Lett. **3**, 46–58 (2018). https://doi.org/10.21809/rilemtechlett.2018.71

EN 772-6 (2001) Methods of test for masonry units - Part 6: Determination of bending tensile strength of aggregate concrete masonry units. European Committee for Standardization (CEN)

ASTM C78-09: Standard Test Method for Flexural Strength of Concrete (Using Simple Beam with Third-Point Loading). American Society for Testing & Materials (ASTM) (2010)

SNI 03-6458: Methods for Flexural Strength of Soil-cement Using Simple Beam with Third-point Loading. National Standardization Agency of Indonesia (2000)

Paul, D., Gnanendran, C.: Characterisation of lightly stabilised granular base materials by flexural beam testing and effects of loading rate. Geotech. Test. J. **35** (2012). https://doi.org/10.1520/GTJ104183

Perlot, C., Prime, N., Aubert, J.-E., et al.: Recommendation of TC RILEM TC 274-TCE: 3-point bending test procedure for earthen bricks—quality control of earth bricks for structural masonry by flexural strength. Mater. Struct. **58** (2025). https://doi.org/10.1617/s11527-025-02702-0

Kulshreshtha, Y., Vardon, P.J., Meesters, G., et al.: What makes cow-dung stabilised earthen block water-resistant. Constr. Technol. Archit. **1**, 540–548 (2022). https://doi.org/10.4028/www.scientific.net/cta.1.540

Nshimiyimana, P., Messan, A., Zhao, Z., Courard, L.: Chemico-microstructural changes in earthen building materials containing calcium carbide residue and rice husk ash. Constr. Build. Mater. **216**, 622–631 (2019). https://doi.org/10.1016/j.conbuildmat.2019.05.037

Morel, J.C., Pkla, A., Walker, P.: Compressive strength testing of compressed earth blocks. Constr. Build. Mater. **21**, 303–309 (2007). https://doi.org/10.1016/j.conbuildmat.2005.08.021

Fire Resistance and Insulation Properties of Earth Plaster on Corncob-Based Thermal Insulation Boards

Selina Vaculik[1(✉)], Lea Magdalena Scholz[2], Laurens Bekemans[2], and Thomas Matschei[1]

[1] Institute of Building Materials Research, RWTH Aachen University, 52062 Aachen, Germany
vaculik@ibac.rwth-aachen.de

[2] Construction and Design, RWTH Aachen University, 52062 Aachen, Germany

Abstract. This paper investigates the fire resistance and insulation properties of earth plaster applied to corncob thermal insulation boards. The corncob boards were produced using a hydraulic heat press with a water-starch ratio of 10%, providing a unique composite material for thermal insulation. The study focuses on the fire resistance of the earth plaster when applied to corncob boards and the thermal performance of the boards themselves. The plastered corncob boards were subjected to flame exposure tests, in which the impact of fire on the plaster layer and the underlying insulation was analysed. The thermal degradation, material strength, and heat transfer properties were carefully monitored to understand how the earth plaster influences the fire behaviour of the corncob-based insulation.

In addition to fire resistance, the insulation properties of the corncob boards were evaluated. The thermal conductivity and insulation efficiency were measured under controlled conditions, comparing the performance of corncob boards both with and without the earth plaster layer.

The findings demonstrate that even a thin layer of earth plaster can significantly enhance fire resistance by delaying or preventing ignition and reducing heat transfer. The compressed corncob boards and the loose corncob exhibited enhanced thermal insulation performance, making them promising options for circular building materials. This research demonstrates the potential of utilising agricultural by-products, such as corncob and combined earth plaster, to develop sustainable, fire-resistant insulation solutions for the construction industry.

Keywords: fire resistance · earth plaster · corn cob · thermal insulation

1 Introduction

The building sector increasingly incorporates bio-based insulation materials derived from renewable resources to reduce embodied carbon and construction-related waste [1, 2]. Agricultural residues, such as cereal straw, hemp hurd, and corn residues, are considered low-impact alternatives to conventional insulation. Corncob, the lignified core of maize, represents a particularly suitable raw material due to its availability, independence from food production, and potential for use in lightweight insulation systems [2].

M. Posani and S. Singh (Eds.): RYS 2025, RILEM Bookseries 67, pp. 470–482, 2026.
https://doi.org/10.1007/978-3-032-14652-6_34

In one study, it was demonstrated that corncob samples withstood several minutes of direct flame exposure before undergoing complete combustion, whereas conventional polymer foams failed within seconds under comparable conditions [2]. Based on these results, moderate fire resistance was attributed to untreated corncob. A 2024 review further highlighted the promising fire performance and material durability of corncob-based composites used in natural construction systems [3].

Despite such findings, the high flammability of plant-based insulations remains a critical limitation. In their uncoated state, most of these materials exhibit rapid ignition, flame spread, and structural degradation, resulting in low fire classifications (Euroclass E or lower) [4]. To ensure compliance with building regulations, biogenic insulation systems must be protected by noncombustible claddings. For example, wall assemblies in residential construction are required to meet classification K110 or EI30 [4]. Experimental investigations confirm this need for protection. Uncoated straw and hemp panels undergo rapid charring and disintegration upon exposure to flames, whereas assemblies finished with mineral plaster exhibit improved stability [4, 5]. It was demonstrated in [5] that applying clay or lime plaster to straw bale walls significantly enhanced their fire resistance. In applied construction testing, thick mineral coatings enabled biofiller wall systems to achieve 30-min integrity and insulation performance under thermal load [4].

Earthen plasters provide a material-compatible and sustainable solution for enhancing fire resistance in biogenic wall systems. Clay and loam-based plasters are essentially incombustible and capable of delaying thermal penetration due to their high thermal mass and low thermal conductivity [6]. Their function as passive fire barriers is well established in historical earth-building traditions and contemporary laboratory research [4, 5].

Although corncob-based insulation has shown favourable properties in single-material tests, its behaviour as part of a composite with clay plaster remains largely unexplored. A deeper understanding of this combination is necessary to evaluate its potential as a safe and sustainable construction material. The present study investigates the fire resistance of compressed corncob boards with one or two layers of clay plaster applied as a surface finish. The flame exposure method used here is based on a small-scale propane burner setup previously introduced in [7], which allows for reproducible fire testing under controlled conditions. The experimental analysis focuses on ignition behaviour, temperature development, and structural stability during flame contact.

2 Materials and Methods

2.1 Raw Materials and Board Manufacturing

The test specimens were produced using untreated corncob granulate, a chopped by-product from maize harvesting (supplier: Haxthausen, origin: Italy). The corn cobs, initially marketed as barbecue fuel, were shredded using a standard shredder to a maximum particle size of 30 mm and then sieved to obtain a particle size range of 0 to 20 mm. Any fraction larger than 20 mm was reprocessed and re-screened until it passed 20 mm. The resulting grading was close to a Fuller type curve. A modified potato starch binder (supplier: Alfa Aesar) was added at 10 wt%, and water was sprayed evenly until the dry starch was visibly moistened. The moistened mixture was homogenised in a tumbler and filled into a 150 × 150 mm aluminium mould lined with anti-adhesive foil. Using a laboratory hot press (Collin), the specimens were compressed at 150 °C and 3 bar. Pressing was carried out in multiple steps with intermediate venting to allow moisture escape. After demoulding, the specimens were dried at 40 °C to mass consistency (see Acknowledgements).

A total of 12 boards were produced with identical composition and processing parameters. The average density of the boards was 324 kg/m^3, with a variation coefficient of 3.77%, ensuring internal consistency within the sample series. The average thickness of the boards is 21.62 ± 4.7 mm for the overall production. All boards used later in the fire experiments were first measured for thermal conductivity.

2.2 Clay Plaster Application

Six of the specimens were coated with a clay-based finishing plaster. A commercial clay plaster with rounded sand 0 to 1.4 mm was mixed with 14% water in a forced mortar mixer. Plaster was applied to the corncob boards in two layers of 3 mm each (see Fig. 1). One type of board received a single plaster layer. In comparison, the other was finished with two successive layers, totalling approximately 6 mm in thickness. The plaster shows about 2% drying shrinkage. Some isolated microcracks in the single-layer system could be observed.

Fig. 1. Structure of corncob boards with unprocessed surface and plaster layers

The second plaster layer was modified with chopped flax fibres at 1% by mass with a fibre length of 5 mm to reduce shrinkage and improve surface cohesion. The fibre-reinforced second layer reduced visible crack widths. The plaster was applied by hand, using a trowel, directly onto the pre-wetted rough surface of the corncob boards (see Fig. 1). Adhesion between plaster and corncob board was evaluated qualitatively. The plaster mechanically interlocked with the rough surface of the corncob board. After drying, specimens could be handled and inverted without loss of plaster. No delamination was observed during handling or after fire exposure (see Fig. 11). All coated specimens were air-dried for at least seven days and then stored at 40 °C until mass consistency before flame exposure to ensure dimensional stability and avoid moisture-related failure during testing.

2.3 Specimen Overview and Classification

The following three specimen types were tested:

Specimen 1: uncoated board (reference),
Specimen 2: board with single plaster layer (≈3 mm),
Specimen 3: board with double plaster layer (≈6 mm, fibre-reinforced).

2.4 Thermal Conductivity Measurement

Thermal conductivity was measured for all specimens used in the fire resistance tests to assess the composite material's insulating capacity. The measurements were conducted using a guarded hot plate apparatus (Lambda EP500e, Lambda Meßtechnik GmbH, Dresden) according to ISO 8302 [8]. Specimens were dried at 40 °C to mass constancy before testing because the conductivity of bio-based boards can increase with moisture. The water content of the samples was therefore negligible. The testing temperature was varied between 10 °C, 23 °C, and 38 °C, with the reported values corresponding to the standard 23 °C condition.

2.5 Flame Exposure Testing

Two types of flame exposure tests were conducted to assess the fire resistance of the corncob-based insulation materials. In total, nine specimens were used for each setup:

(a) Open-Air Propane Burner Test
The primary flame exposure test was conducted outdoors using a high-energy propane burner and the setup shown in Fig. 2. The specimens were mounted vertically on a metal rack, and the flame was applied in two sequential positions at a distance of 35 cm and 70 cm from the burner nozzle, respectively. Each flame application lasted 30 s, followed by a brief cooling phase. The flame was directed at the lower edge of the specimen, simulating a realistic ignition scenario from below. A thermal imaging camera (Testo 872) was positioned at the back of the board to continuously monitor the backside temperature development during and after flame contact. The test enabled the evaluation of ignition behaviour, flame spread, thermal shielding, and structural degradation of both coated and uncoated corncob boards under high-heat flux conditions.

(b) Enclosed Chamber Test (Methane Combustion):
Complementary flame tests were conducted in a closed steel fire chamber with a volume of approximately 1 m^3. A methane burner was used to generate a steady flame in a low-ventilation environment. Specimens were placed on a ceramic support and exposed

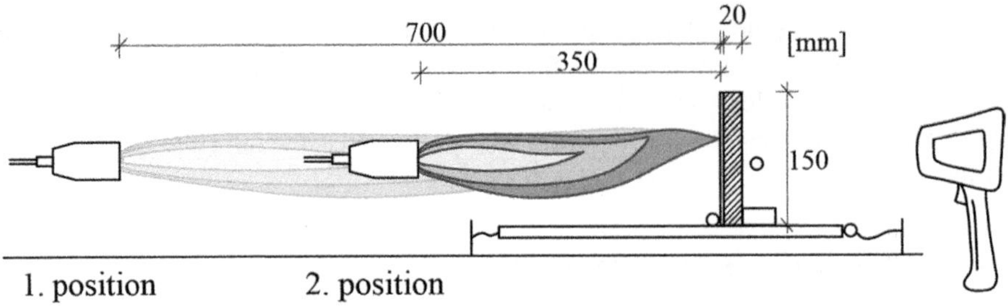

Fig. 2. Test setup for propane burner exposure with measurement configuration

to flame in two orientations: from below and from the side. This approach was used to simulate different fire impact scenarios.

Both raw materials (loose corncob granulate) and compressed board specimens were tested under identical boundary conditions. To adapt to the tool design, up to four sub-samples with maximum dimensions of 75x75 were taken from each 150 mm plate and tested. The focus of the chamber test was on ignition resistance, charring behaviour, smoke production, and structural changes under oxygen-restricted combustion settings. The specimens were positioned horizontally on a ceramic grate above the burner (see Fig. 3). The angle of the burner can be adjusted to 45° and 90°. Flame exposure lasted up to 3 min, and the temperature development on both the exposed and unexposed sides was monitored using infrared thermography. Smoke generation, visual degradation, and the structural integrity of the plaster and board composite were qualitatively assessed after the material had cooled.

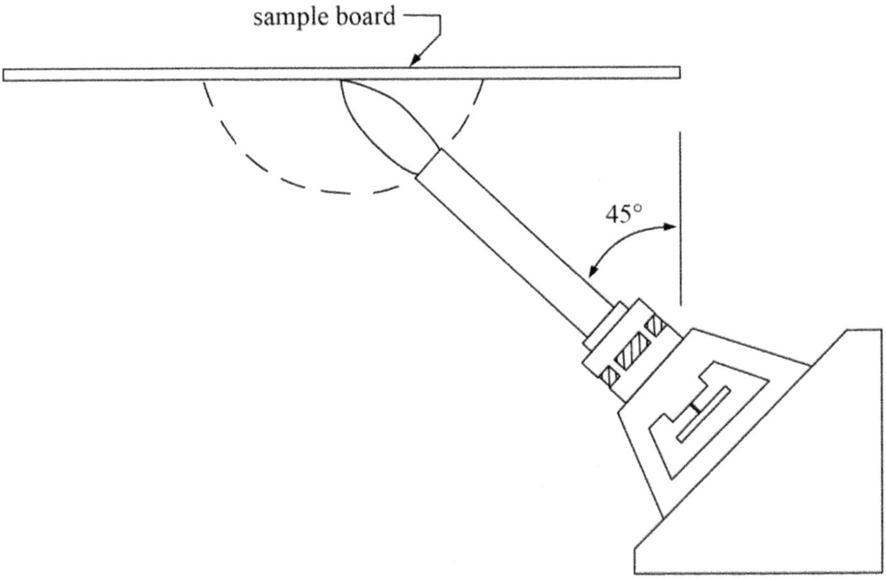

Fig. 3. Vertical burning test in adaptation to UL94 [9]

This dual-testing approach enabled the evaluation of fire resistance under both well-ventilated (open flame) and enclosed fire conditions. Differences in flame penetration, surface charring, and temperature distribution were used to assess the effectiveness of the plaster layers as passive fire protection measures.

3 Results and Discussion

3.1 Thermal Conductivity

The thermal conductivity of the uncoated specimen (Specimen 1) was determined to be 71.1 mW/(m·K). The specimen with a single clay plaster layer (Specimen 2) exhibited a slightly higher value of 73.3 mW/(m·K), and the double-coated specimen (Specimen 3) reached 75.4 mW/(m·K). These results and their variability are shown in Fig. 4.

Despite this effect, all values remain well below 100 mW/(m·K) [10], confirming the suitability of the material for thermal insulation applications in interior wall systems. The thermal conductivity values are comparable to those of other natural insulation materials such as hemp-lime or straw panels (50-60 mW/(m·K) [11, 12].

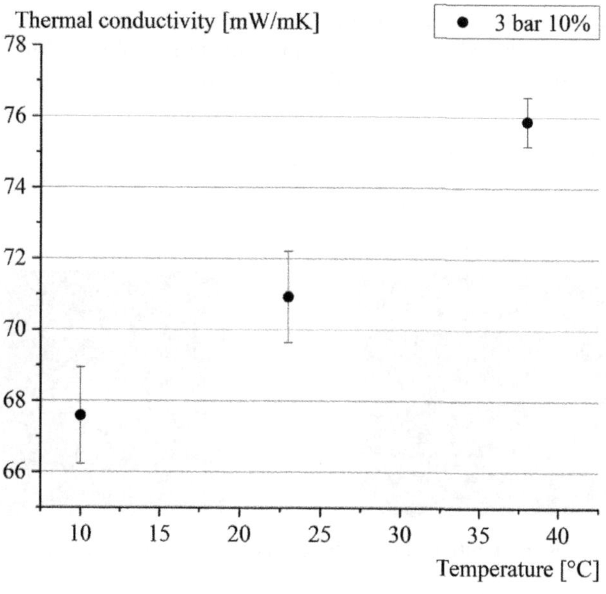

Fig. 4. Thermal conductivity at different temperatures (n = 4)

These findings confirm that corncob-based boards can offer a favourable balance between thermal insulation and passive fire protection, making them a viable option for sustainable and safe construction systems. These results confirm that the plaster coatings marginally increase the overall thermal conductivity due to their higher material density and mineral content. Nonetheless, all values remain within a range suitable for interior insulation applications, particularly when considering the composite's fire safety potential.

3.2 Fire Performance: Ignition and Flame Spread

The uncoated reference specimens (Specimen 1) ignited within seconds after flame contact with the propane gas test. The flame spread rapidly across the exposed surface, and continuous combustion was observed after the burner was turned off. This behaviour is consistent with fire class E as defined by EN 13501-1 [13], where ignition and flame propagation occur without delay and self-extinguishing characteristics. The board was primarily consumed in the central exposure area, and charring extended visibly to the edges.

In addition to the controlled flame exposure using propane, supplementary experiments were conducted in a closed combustion chamber fueled with methane to simulate a confined fire scenario. In the methane-fired enclosure, loose corncob bulk material ignited readily and sustained combustion, confirming its high flammability in an unbound state (see Fig. 5). However, when pressed into board form, the material exhibited only superficial charring, as shown in Fig. 6, without sustained ignition. This suggests that densification effectively reduces oxygen access and changes the combustion dynamics toward carring rather than open flame. The sample was then separated to determine the depth of the destroyed part within it.

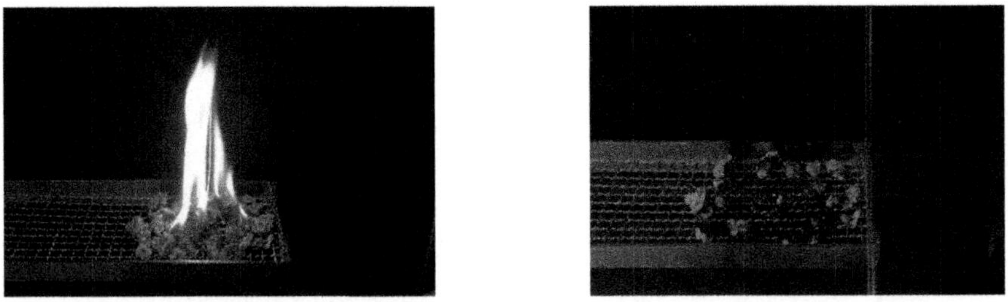

Fig. 5. Ignition of the loose corn cob and observed damage

In contrast, the plastered specimens (Specimens 2 and 3) showed no visible ignition or open flames during either of the flame application cycles (see Fig. 7). The surface darkened slightly but did not combust. The single-layer plaster exhibited minor cracking and partial detachment after cooling, but no combustion of the underlying material occurred. The double-layer system remained intact throughout the entire test, showing no signs of ignition or surface degradation. These observations confirm that even a relatively thin layer of earth plaster can prevent surface ignition and act as an effective flame retardant.

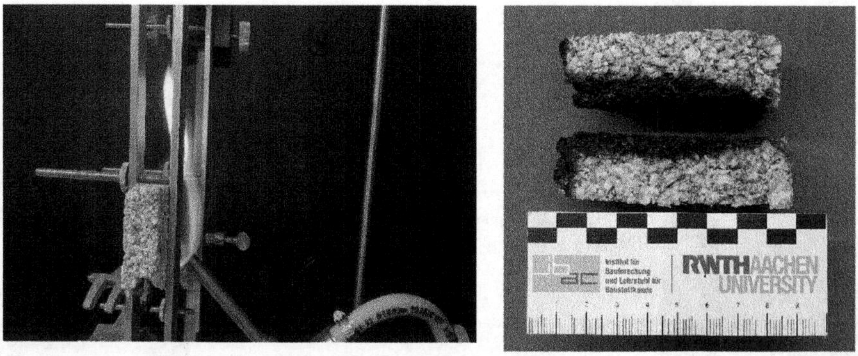

Fig. 6. Board testing and charring

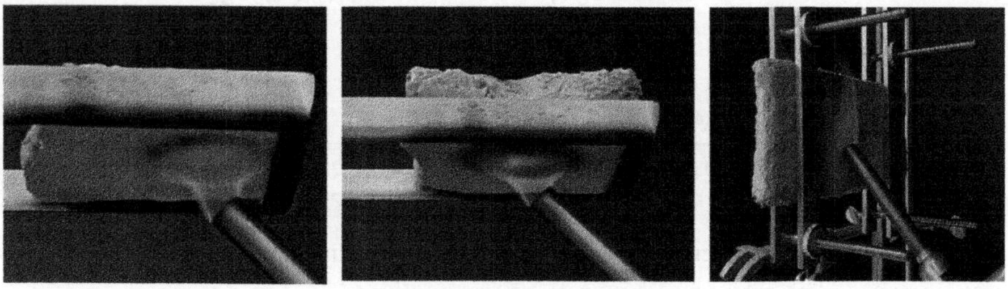

Fig. 7. Coated board flame tests

3.3 Heat Transmission and Backside Temperature

Thermographic imaging of the uncoated board showed a rapid temperature rise at the back surface. The thermal camera's ceiling of 130 °C was exceeded within less than one minute, indicating that heat was transmitted through the entire board thickness almost immediately after exposure (see Fig. 8). This behaviour is consistent with the low thermal mass and high porosity of the untreated corncob matrix.

In contrast, the plastered specimens demonstrated a significantly delayed temperature increase. The single-layer board reached the 130 °C limit only during the second flame cycle, while the double-layer system remained below this threshold for the entire duration of the test. The clay plaster functions as a thermal barrier, slowing the propagation of heat through the composite. This effect, in combination with the board's intrinsic insulation properties, suggests that a properly dimensioned plaster layer can maintain the structure's integrity even under direct flame impact.

In addition to the propane burner trials, thermographic measurements were also conducted during the enclosed methane chamber tests (see Fig. 9). A thermal imaging camera was used to monitor the temperature development at the rear surface of the specimens under low-ventilation fire exposure (see Fig. 10). Similar to the open-air test, uncoated boards reached high backside temperatures within a short time. In contrast, plastered specimens demonstrated a notably slower temperature rise.

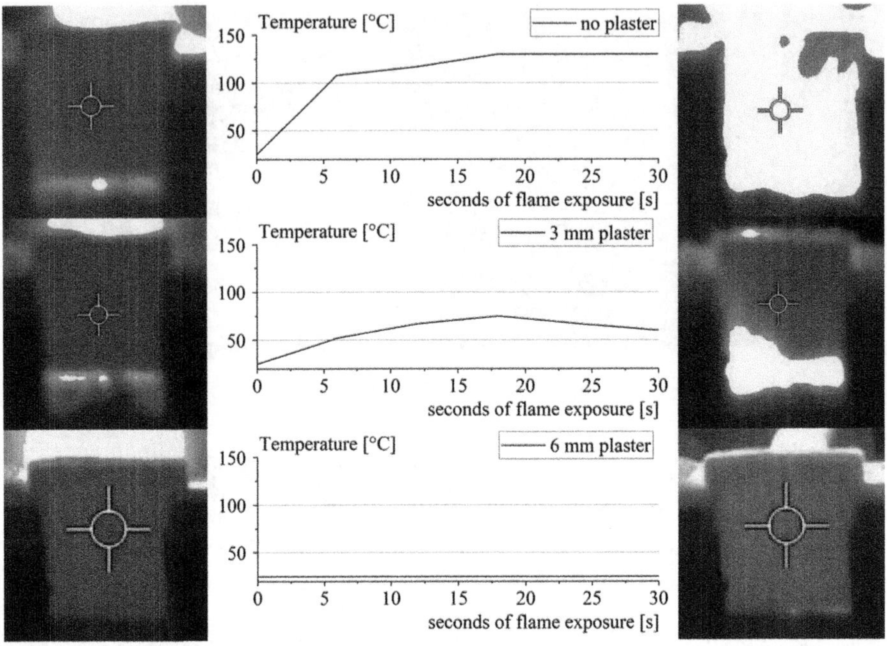

Fig. 8. Comparison of the temperature curve with plaster thicknesses of none, 3 mm, and 6 mm

Fig. 9. Thermal images while fire testing

To further investigate the effects of heat retention, additional thermographic recordings were taken immediately after removing the specimens from the fire chamber (see Fig. 11). These measurements revealed a continued temperature increase on the backside of uncoated boards, indicating internal heat storage and delayed surface emission. In contrast, the plastered specimens exhibited more stable temperature profiles during the cooling phase, confirming that the clay layer serves not only as a heat barrier during flame contact but also as a moderator of residual thermal load. Although the plaster slightly increases thermal conductivity (see Fig. 4), the plastered surface acts as a thermal barrier that delays heat transfer to the back face.

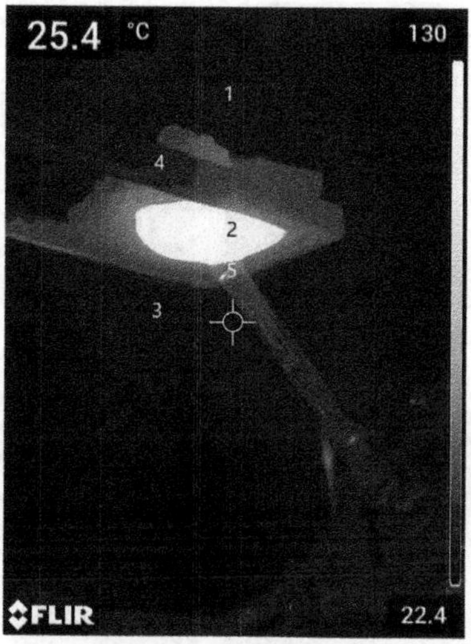

1 – top side
2 – source of fire
3 – bottom side
4 – holder
5 – flame

Fig. 10. Thermal imaging camera during flame exposure

These findings underscore the significance of post-exposure thermal behaviour in assessing fire safety. Clay plaster not only enhances resistance to flame impact but also mitigates afterheat effects that could compromise structural integrity or fire containment.

3.4 Smoke Generation and Combustion Residues

The uncoated specimen produced dense grey smoke (see Fig. 12) shortly after ignition, accompanied by glowing combustion and audible crackling. This is characteristic of uncontrolled pyrolysis in lignocellulosic materials. Afterburning persisted for more than one minute post-exposure, and the specimen showed complete degradation in the exposed area.

In contrast, the plastered boards generated minimal smoke. No afterburning or smouldering was observed following the removal of the flame source. The minor shrinkage cracks in the plaster of Specimen 2 did not lead to fire penetration, and no gas-phase combustion was observed. The double-layer specimen (Specimen 3) remained smoke-free and structurally intact. These results suggest that the clay plaster coating not only prevents ignition but also inhibits the formation of volatile gases and oxygen access to the bio-based core.

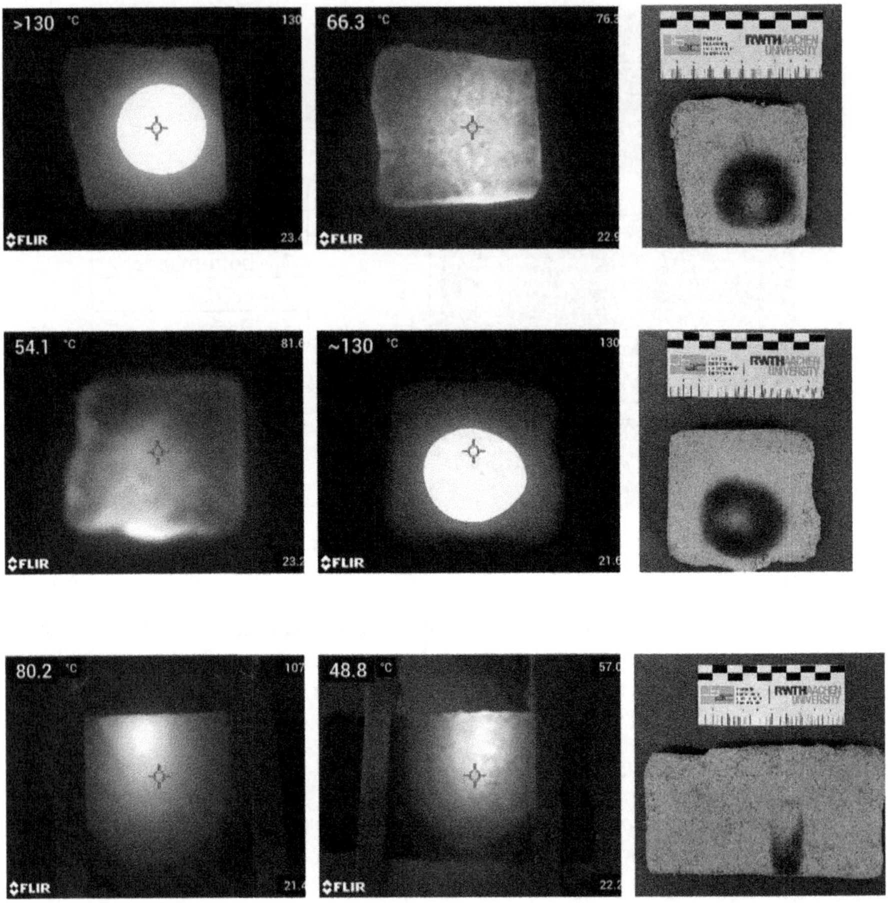

Fig. 11. Temperature measurements after fire tests

Additional observations were made during the enclosed methane chamber tests. The loose corncob granulates exhibited visible smoke formation during combustion, consistent with its high surface area and low thermal stability. Pressed boards without plaster showed only mild smoke release, while plastered specimens remained free of visible smoke throughout the test. These differences underline the importance of both densification and mineral surface treatment in reducing the release of combustion gases during fire exposure.

Fig. 12. Afterburning and smoke production of the tested sample

4 Conclusion

The study confirms the inherent flammability of biogenic insulation materials, such as corncob-based particleboards, and highlights the critical role of earthen plaster coatings in enhancing fire resistance. While the uncoated reference specimen ignited rapidly and transmitted heat without delay, even a single layer of clay plaster significantly delayed ignition and reduced heat transfer. The application of a second, fibre-reinforced plaster layer further enhanced structural integrity and thermal shielding, entirely preventing ignition and visible damage. Supplementary fire tests conducted in a closed methane chamber demonstrated that the uncoated loose corncob material poses a high fire risk, with rapid ignition and pronounced smoke formation. In contrast, compressed boards showed reduced combustibility, and plastered specimens exhibited no visible ignition, even under multidirectional flame exposure. These tests confirmed that the clay coating remains effective not only under open-flame conditions but also in low-ventilation fire scenarios.

Thermographic post-exposure analysis revealed a measurable afterheat effect in uncoated boards, with surface temperatures continuing to rise briefly after the flame was removed. Plastered specimens, however, remained thermally stable, indicating that the clay layer contributes not only to fire resistance but also to post-fire heat dissipation. Furthermore, the study highlights significant differences in smoke generation: while loose bulk material produced dense smoke, pressed boards emitted only light smoke, and plastered specimens remained virtually smoke-free. These characteristics are essential for safe application in building interiors. Thermal conductivity measurements revealed that the addition of clay plaster results in only moderate increases in λ-values. All tested variants remained within a range suitable for interior insulation purposes. When balanced against the substantial improvement in fire behaviour, the increase in thermal conductivity is acceptable and does not compromise the material's usability. Overall, the results demonstrate that corncob insulation systems can be made fire-compatible through simple, mineral-based surface treatments. Earth plaster proves to be a technically effective and ecologically consistent solution for enhancing the safety of biogenic insulation. Future research should address long-term durability, moisture buffering, and the integration of prefabricated application methods for scalable use in construction.

5 Acknowledgments

The authors would like to thank Univ.-Prof. Dr. rer. nat. Oliver Weichold, ibac, and his team for the support and assistance during board pressing and for the opportunity to use the heat press and burning chamber.

References

1. Lu, Z., et al.: Climate mitigation potential of biobased insulation materials: a comprehensive review and categorisation. J. Clean. Prod. (2024): 143356. Cmadeubi.wordpress.com. Corncob insulation research and fire testing. Accessed Apr 2025.
2. Pinto, J., et al.: Characterization of corn cob as a possible raw building material. Constr. Build. Mater. **34**, 28–33 (2012)
3. Yang, Y., Haurie, L., Wang, D.-Y.: Bio-based materials for fire-retardant application in construction products: a review. J. Therm. Anal. Calorim., 1–20 (2022)
4. Kjer-Hansen, C.: Test of biogenic fire protection paves the way for sustainable construction. DBI. (2024)
5. Garas, G., Allam, M., Mamdouh, K.: Straw bale fire test on cement plaster mixes. WIT Trans. Built Environ. **108**, 51–59 (2009)
6. Volhard, F.: Wood needs mass–improving timber construction with earth building materials. LEHM (2020)
7. Vaculik, Selina, and Thomas Matschei. "Fire Resistance of Wood Fiber Reinforced Concrete." In: International Conference on Bio-Based Building Materials. Cham: Springer Nature Switzerland (2023).
8. ISO 8302:1991; Thermal insulation — Determination of steady-state thermal resistance and related properties — Guarded hot plate apparatus
9. UL 94 Tests for Flammability of Plastic Materials for Parts in Devices and Appliances;
10. Ali, A., Issa, A., Elshaer, A.: A comprehensive review and recent trends in thermal insulation materials for energy conservation in buildings. Sustainability. **16**(20), 8782 (2024)
11. Laborel-Préneron, A., Magniont, C., Aubert, J.-E.: Characterization of barley straw, hemp shiv and corn cob as resources for bioaggregate based building materials. Waste Biomass Valorization. **9**(7), 1095–1112 (2018)
12. Nguyen, S.T., et al.: Modeling thermal conductivity of hemp insulation material: a multi-scale homogenization approach. Build. Environ. **107**, 127–134 (2016)
13. DIN EN 13501; Klassifizierung von Bauprodukten und Bauarten zu ihrem Brandverhalten – Teil 1: Klassifizierung mit den Ergebnissen aus den Prüfungen zum Brandverhalten von Bauprodukten Deutsche Fassung EN 13501-1:2018

Additive Manufacturing with Biomaterials: Insights from a Meta-Analysis of Key Parameters

Melany Nicole Medina Pérez[1], Diego Griffith[1], Khristel María Rodríguez[1], Rafael Duque[1], Luis Castillero[1], and Maria De Los Angeles Ortega Del Rosario[1,2](✉)

[1] Research Group in Design, Manufacturing and Materials (DM+M), School of Mechanical Engineering, Universidad Tecnológica de Panamá, Panama City 0819-07289, Panama
{maria.ortega,maria.ortega}@utp.ac.pa
[2] Sistema Nacional de Investigación (SNI), Clayton, City of Knowledge Biulding 205, Panama City 0816-02852, Panama

Abstract. The integration of biomaterials into additive manufacturing (AM) offers a promising approach to sustainable construction; however, their evaluation remains fragmented and methodologically inconsistent. This study presents a meta-analysis of scientific literature published between 2020 and 2025, aiming to identify viable biomaterials for 3D printing in terms of mechanical performance, environmental impact, and economic feasibility. A hybrid multi-criteria decision-making (MCDM) framework, combining Fuzzy Analytic Hierarchy Process (Fuzzy AHP), COPRAS, and VIKOR, was applied to assess and rank six alternative materials, including mycelium, clay-based composites, geopolymers, and biochar-enhanced mixes. Quantitative data on compressive strength, CO_2 emissions, and unit cost were extracted and normalized to enable cross-comparison. Results showed substantial variation in material rankings across individual methods, highlighting the sensitivity of outcomes to the evaluation approach. To address this, a Borda Count aggregation was employed, revealing geopolymers and biochar-based composites as top-performing alternatives overall. Among bio-based materials, lime-sisal fiber and mycelium demonstrated moderate performance with relatively consistent rankings. This multi-criteria evaluation provides a structured framework for material selection in AM, supporting informed decision-making for the development of resilient and low-carbon construction solutions.

Keywords: Additive manufacturing · biomaterials · sustainable construction · meta-analysis · process parameters · multi-criteria decision-making (MCDM)

1 Introduction

The architecture, engineering, and construction (AEC) industry is among the sectors with the highest environmental impact, accounting for approximately 38% of global greenhouse gas emissions and 40% of natural resource consumption [1]. This context

M. Posani and S. Singh (Eds.): RYS 2025, RILEM Bookseries 67, pp. 483–499, 2026.
https://doi.org/10.1007/978-3-032-14652-6_35

increasingly demands innovative models and technologies integrating sustainability, circular economy principles, and environmental resilience [2–5]. One emerging technology gaining increasing attention and application in addressing this challenge is additive manufacturing (AM), commonly known as 3D printing. AM can produce complex geometries without traditional molds, enabling individualized and collaborative customization. Furthermore, it significantly reduces material waste, optimizes the use of raw materials, and incorporates materials directly into the manufacturing process, contributing to a progressive reduction in carbon footprint [6–8]. These capabilities underscore the need for further exploration of alternative materials, particularly biomaterials, which offer promising pathways to address both environmental concerns and the structural demands of contemporary construction [6, 9, 10].

However, despite the advantages of AM, its widespread reliance on conventional materials, such as concrete and calcium phosphate cements, continues to generate high carbon footprints and exacerbates the depletion of virgin resources [11]. Moreover, the processing required for printable concrete often results in a higher carbon footprint per cubic meter than that of traditional concrete [12]. This reinforces the importance of identifying new material alternatives that, when combined with AM technologies, can offer lower-impact solutions aligned with the principles of sustainable construction.

When examining the development of AM with biomaterials, it becomes clear that scientific literature has evolved in a fragmented and heterogeneous manner. Numerous studies lack a systematic framework for characterizing or classifying results, making it difficult to identify patterns, critical variables, and comparable conditions across different approaches. Thus, meta-analysis emerges as a valuable method for synthesizing existing findings and quantitatively evaluating the influence of key parameters, such as printing process, geometry, efficiency, and the structural performance of components [13, 14].

The growing concern for developing more sustainable, less polluting, and even regenerative urban environments is generating new opportunities for sustainable and regenerative design, including the integration of renewable resources and agro-industrial by-products that are still commonly treated as waste [15, 16]. Materials currently regarded as waste may become valuable raw materials, capable not only of meeting ecological criteria but also providing additional functional properties to structures, such as bioreceptivity [17, 18], thermal insulation [19–22], acoustic insulation [15, 23] and hygroscopic regulation, including moisture absorption capacity [24–26].

In the context of sustainable construction, biomaterials for AM offer potential to reduce environmental impact and enable innovative design. However, existing literature is fragmented and lacks an integrated analysis of key mechanical, thermal, and environmental performance parameters. While numerous studies explore individual biomaterials in AM, few provide quantitative comparisons that jointly assess environmental, mechanical, and economic aspects. This study addresses that gap through a meta-analysis of scientific publications from 2020 to 2025 on biomaterials used in AM for sustainable construction.

Guided by the question: Which biomaterials used in AM offer the best technical and functional performance for sustainable construction? the study applies multi-criteria decision-making methods to compare materials based on mechanical, thermal, and environmental criteria. The aim is to identify those with the greatest potential for practical implementation in sustainable building applications.

2 Methodology

A systematic review was conducted in Scopus using keywords such as additive manufacturing, biomaterials, and sustainable construction. A total of 78 articles were identified. Of these, 38 provided consistent quantitative data and were included in the results analysis. The remaining 40 were excluded from the quantitative section due to insufficient or non-standardized data but were used to support the theoretical and methodological framework. Studies unrelated to 3D printing in construction or focused on conventional materials were excluded during initial screening. Figure 1 presents the research framework and selection criteria for the bibliometric analysis. Due to the limited availability of integrative studies, multi-criteria decision-making (MCDM) was applied to evaluate mechanical, economic, and environmental performance. Quantitative data on mechanical properties, cost, and environmental impact were extracted from the selected articles. Given the variety of criteria reported, MCDM methods were used to integrate these variables and generate a performance-based ranking of materials. The following section outlines the methodology.

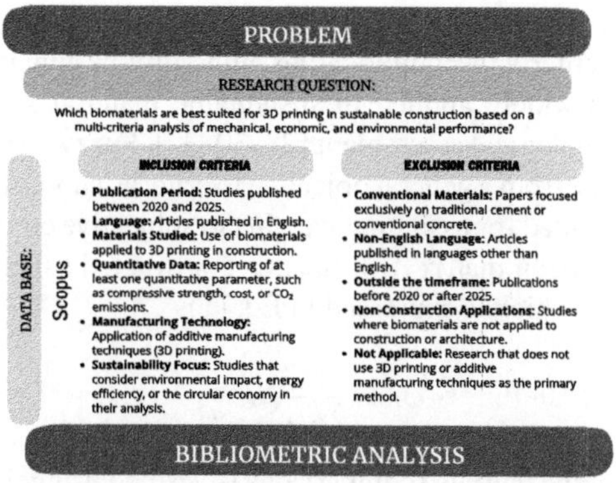

Fig. 1. Research framework and selection criteria for the bibliometric analysis of biomaterials used in 3D printing for sustainable construction.

Although similar rankings could be obtained using a single MCDM method, integrating Fuzzy AHP, COPRAS, and VIKOR allowed for cross-validation and robustness of results in a field where the literature is fragmented and highly heterogeneous. This integration was selected to validate consistency and enhance robustness in ranking biomaterials. This study employs MCDM to systematically compare and rank key parameters identified in the literature. These methods are particularly suitable for addressing complex decision-making scenarios and offer a structured analytical framework that is particularly valuable in contexts characterized by uncertainty and methodological heterogeneity, such as the evaluation of biomaterials and their mechanical and environmental properties.

The general workflow of the MCDM methodology used in this study is illustrated in Fig. 2, including the problem structuring (defining the problem, alternatives, and criteria) and the problem analysis phase (quantitative and qualitative evaluation leading to decision-making).

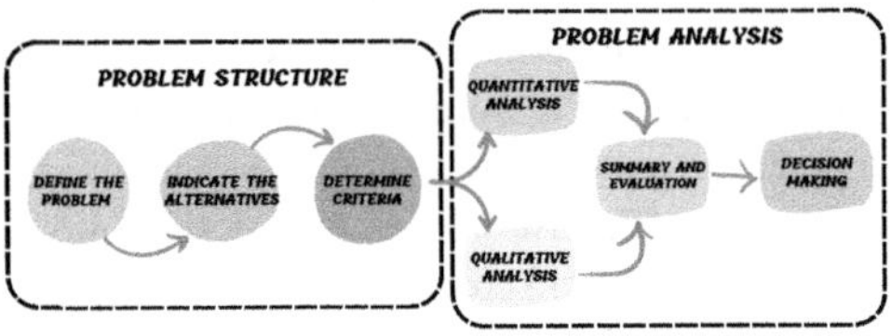

Fig. 2. Flowchart of the methodology used for the MCDM, adapted from [27].

2.1 Description of the Methods

Fuzzy AHP (Analytic Hierarchy Process). It extends the classical AHP method by Saaty (1980) to support complex decision-making involving multiple criteria under uncertainty. By using triangular fuzzy numbers (TFNs), it allows more flexible pairwise comparisons of criteria, accommodating ambiguity in human judgment [27–29]. The method has been widely applied in sustainable material selection, AM process optimization, and technological assessment [28, 30–32]. Fuzzy AHP structures decision problems hierarchically, from a general objective to criteria, sub-criteria, and alternatives [33]. It is particularly suited for assigning weights to subjective criteria through pairwise comparisons. A comparison matrix $A = \left(a_{ij}\right)$, is constructed using Saaty's scale (1–9), and the priority vector, which is expressed in Eq. (1):

$$\omega_i = \frac{\sum_{j=1}^{n} a_{ij}}{\sum_{i=1}^{n} \sum_{j=1}^{n} a_{ij}} \tag{1}$$

The consistency of the pairwise comparisons is verified using the Consistency Index (CI) and the Consistency Ratio (CR). The CI measures the coherence of judgments and is calculated as Eq. (2):

$$CI = \frac{\lambda_{max} - n}{n-1} \tag{2}$$

where λ_{max} is the largest eigenvalue of the comparison matrix, and n is the number of criteria. The CR assesses the acceptability of the inconsistency (valid when $CR < 0.1$)

$$CR = \frac{CI}{RI} \tag{3}$$

Previous studies have shown that this method yields reliable weighting of criteria, even in contexts with high variability [14, 34].

Hodonou [30] applied AHP to select optimal combinations of materials, processes, and designs in additive manufacturing. In the present study, AHP served as the initial step in a hybrid MCDM framework, providing the weights used in COPRAS and VIKOR analyses.

COPRAS (Complex Proportional Assessment). Developed by Zavadskas, it is a multi-criteria method designed to rank alternatives by simultaneously considering benefit and cost criteria. It evaluates the proportional contribution of each alternative toward achieving the overall objective, enabling transparent and robust decision-making by identifying the most influential variables [35]. It is valued for its computational simplicity and ability to handle quantitative and heterogeneous criteria, making it particularly suitable for AM, where mechanical, thermal, environmental, and economic factors often interact. Its primary goal is to determine the utility degree of each alternative, based on criteria weights, commonly obtained through AHP or Fuzzy AHP.

The process begins with the construction of a decision matrix $X = \left[x_{ij}\right]$, where each row represents an alternative and each column corresponds to a quantitative criterion [35]. To enable comparability among the different criteria, benefit-type criteria are normalized using Eq. (4):

$$r_{ij} = \frac{x_{ij}}{\sum_{i=1}^{m} x_{ij}} \quad (4)$$

Meanwhile, for cost-type criteria, normalization is carried out using Eq. (5):

$$r_{ij} = \frac{min(x_i)}{x_{ij}} \quad (5)$$

The weighted matrix is calculated by multiplying each normalized value by its corresponding weight, w_j, which was previously obtained through Fuzzy AHP (Eq. (6)):

$$q_{ij} = r_{ij} \cdot w_j \quad (6)$$

Partial sums are calculated based on the type of criterion. For each alternative i, the sums S_i^+ (benefit criteria) and S_i^- (cost criteria), are obtained using the expressions: $S_i^+ = \sum_{j \epsilon B} q_{ij}$ y $S_i^- = \sum_{j \epsilon C} q_{ij}$, respectively. Finally, the relative utility of each alternative is determined using the equation: $Q_i = S_i^+ + \frac{minS^-}{S_i^-} \cdot \sum_{j \epsilon C} q_{ij}$, which allows for the establishment of a performance ranking, with the alternative having the highest Q_i value being the most preferred.

VIKOR (VIšekriterijumsko KOmpromisno Rangiranje). Introduced by Opricovic and Tzeng [36], it is a MCDM approach designed to select the best alternative under conflicting criteria, allowing the identification of compromise solutions that balance efficiency, stability, and equilibrium. Its main strength lies in identifying the absolute optimal alternative and determining the one closest to the theoretical ideal, considering both total utility and maximum dissatisfaction (compromise ranking). This feature is particularly valuable in the context of AM with biomaterials, where materials often involve trade-offs among mechanical, thermal, and environmental properties. VIKOR has been successfully applied to evaluate Fused Deposition Modeling (FDM) parameter

[37], select sustainable material combinations [38], and compare formulations in 3D printing under ecological criteria [39, 40]. Here, VIKOR is employed as the third stage of the meta-analysis to evaluate the relative efficiency of each biomaterial, considering simultaneously mechanical, thermal, and sustainability criteria that were previously weighted using AHP.

It begins with the construction of a matrix including the performance of each alternative for each criterion. The ideal values f_j^+ (most desirable) y f_j^- (least desirable) are identified for each criterion, which establishes a reference framework for evaluation [36]. Next, for each alternative i the aggregated utility function S_i is calculated, as defined in Eq. (7):

$$S_i = \sum_{j=1}^{n} w_j \cdot \frac{f_j^+ - f_{ij}}{f_j^+ - f_j^-} \tag{7}$$

Likewise, the maximum regret function R_i, is determined, which accounts for the criterion with the most significant deviation from the ideal value, as shown in Eq. (8):

$$R_i = max_j = \left[w_j \cdot \frac{f_j^+ - f_{ij}}{f_j^+ - f_j^-} \right] \tag{8}$$

Subsequently, the compromise index Q_i, is calculated, which combines S_i y R_i sing a balance parameter v (commonly v = 0.5), as expressed in Eq. (9):

$$Q_i = v \cdot \frac{S_i - S^+}{S^- - S^+} + (1 - v) \cdot \frac{R_i - R^+}{R^- - R^+} \tag{9}$$

where S^+, S^-, R^+, R^-, represent the maximum and minimum values among all alternatives. Finally, the alternatives are ranked according to Q_i, and the best solution is determined based on compromise acceptance conditions, such as acceptable advantage and decision stability [36–38]. Where S^+, S^-, R^+, R^-, represent the maximum and minimum values among all alternatives.

These methods were selected for their capability to integrate multiple criteria with different units and scales, enabling a balanced assessment of mechanical performance, environmental impact, and economic feasibility. In this process, the data were normalized, weights were assigned to the criteria through a Saaty matrix, and the overall scores for each alternative were calculated for each method. The following diagram illustrates how these three methods are integrated into a multi-criteria decision-making process *(see* Fig. 3*):*

Borda Count. To integrate the rankings from Fuzzy AHP, COPRAS, and VIKOR, a final ranking was generated using the Borda Count method, which assigns points based on each material's position in the individual rankings. By summing these points, the method produces an average ranking that reflects overall performance across all evaluations, reducing bias from any single method and providing a robust and balanced material prioritization.

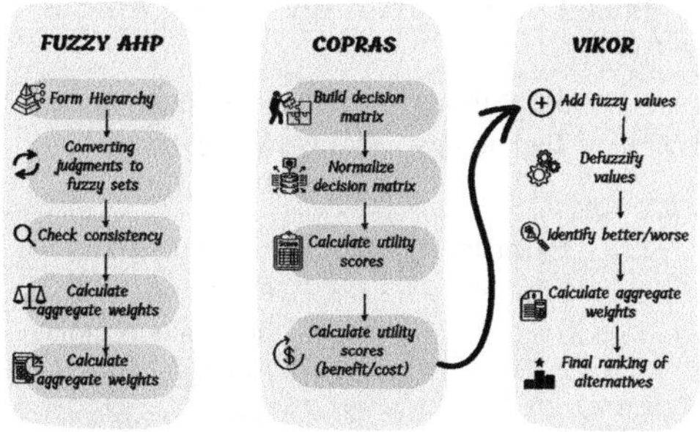

Fig. 3. Integration of the three multicriteria analysis methods [41].

2.2 Qualitative Analysis of the Articles

To identify trends, approaches, and benefits associated with biomaterials in additive manufacturing, a comparative qualitative analysis was conducted on selected scientific articles. A synthesis table was prepared including nine key variables: article title, authors, biomaterial type, composition, AM technology, mechanical properties, reported values, additional tests, and highlighted benefits.

This analysis helped reveal relationships between material types and functional performance, as well as trends in technologies, reinforcements, and applications (see Table 1). The most frequently studied materials were clay-based composites, stabilized earth, and bio-based reinforcements such as natural fibers (hemp, straw, rice husk) and biogenic materials (mycelium, algae, biochar). Their popularity is linked to local availability, printability (mainly via extrusion), and low embodied carbon. Clay is valued for its abundance and rheological properties; natural fibers improve strength, reduce shrinkage, and enhance insulation. More innovative applications explore biochar and mycelium for their potential in circularity and the integration of waste or living systems into construction.

Notably, most studies were published in the last five years, highlighting a rapidly growing field. Rather than reiterating established knowledge, the analysis uncovers emerging material combinations that merge vernacular construction with digital fabrication, reinforcing the need for structured multi-criteria frameworks to guide material selection based on performance and sustainability.

Since methods like COPRAS and VIKOR are particularly sensitive to extreme values due to their reliance on best and worst-case comparisons, the Borda aggregation mitigates the influence of outliers by emphasizing relative positions rather than absolute scores. This approach provides a robust final ranking, reducing the risk of bias from any single method and offering a more reliable comparison across materials.

Table 1. Qualitative Analysis of the Articles.

Authors	Type of Biomaterial	3D Printing Technology	Evaluated Mechanical Properties	Benefits
[42]	Cobcrete, Hempcrete, Bamboo	Extrusion-based 3D printing	Compression, tensile strength, flexural strength (modulus of rupture), split tensile strength, and plasticity	Natural, low-carbon materials with high insulation, strength, and formwork-free printability
[43]	Stabilized earth with plant fibers	KUKA KR 20–3 robotic arm - Turbosol Poli-T pump (large-scale extrusion)	Compressive strength, shear strength (yield strength)	Low-impact, extrudable mix enabling stable robotic printing and modular scalability
[44]	Earth, plant fibers, biopolymers, mycelium, algae, and bacteria	Mainly Material Extrusion without heat, also Binder Jetting and printing with living organisms	Compressive strength, tensile strength, elastic modulus, plasticity, surface roughness, interlayer adhesion	Low-carbon, printable, and circular materials with digital-traditional integration
[45]	Cement – biochar	Direct Ink Writing (DIW) with modified extrusion	Compressive strength	Biochar-based, stable-printing composite that reduces cement use and promotes circularity
[46]	Clay with sawdust and natural fibers	Direct Ink Writing (DIW)	Compressive strength, shrinkage, rheology	Sustainability, reduces cracks, improves rheological behavior
[47]	Cob (clay, straw, sand)	Extrusion-based 3D printing	Compressive and tensile strength	Low carbon, reusable materials, vernacular construction method adapted for 3DP
[48]	3D printed cob (earth, straw)	Extrusion-based 3D printing	Compressive strength, elastic modulus	Resource efficiency, reduces material use, suitable for low-rise buildings

(*continued*)

Table 1. *(continued)*

Authors	Type of Biomaterial	3D Printing Technology	Evaluated Mechanical Properties	Benefits
[49]	Clay with natural fibers, rice husk, marble dust, etc	Extrusion-based 3D printing	Compressive and flexural strength, shrinkage	Low embodied carbon, cost-effective, printable with natural additives
[44]	Earth with wheat straw, hemp, banana, and kenaf fibers	Extrusion-based 3D printing	Compressive strength, thermal conductivity	High fiber content, improved thermal and mechanical performance, low carbon

3 Results

The meta-analysis included 38 studies reporting experimental data on biomaterials for additive manufacturing. While many of these studies addressed mechanical properties, such as compressive strength, flexural strength, and elastic modulus, not all provided complete and comparable numerical data. Only those offering consistent, quantitative results were selected for the multi-criteria decision-making (MCDM) analysis (Table 2). Environmental indicators, including carbon footprints, were inconsistently reported across the literature, and comprehensive life-cycle assessments remained scarce. Economic aspects were also considered in a limited number of studies, often lacking detailed or standardized cost data.

Although five key criteria were initially defined for the MCDM analysis: tensile strength, compressive strength, thermal resistance, water absorption, and sustainability, the literature review revealed that compressive strength and sustainability were the most frequently and consistently reported. Data on tensile strength, thermal behavior, and water absorption were either scarce or inconsistent. Consequently, the MCDM analysis was based on the criteria with sufficient quantitative data, ensuring objectivity and comparability [31, 50].Although five key criteria were initially defined for the MCDM analysis: tensile strength, compressive strength, thermal resistance, water absorption, and sustainability, the literature review revealed that compressive strength and sustainability were the most frequently and consistently reported. Data on tensile strength, thermal behavior, and water absorption were either scarce or inconsistent. Consequently, the MCDM analysis was based on the criteria with sufficient quantitative data, ensuring objectivity and comparability [31, 50].

Hierarchical weights were calculated using the Fuzzy AHP method [31], while COPRAS [19] and VIKOR [36] techniques were applied to generate integrated rankings based on the mechanical (1), environmental (3), and economic performance (5) of the selected biomaterials. This combined approach enabled the identification of materials with the most significant potential for sustainable construction through additive manufacturing [38, 39]. The results are presented from the prioritization of criteria to the final ranking of alternatives, allowing a direct comparison between methods based on proportional utility (COPRAS) and compromise solutions (VIKOR), while capturing trade-offs among structural performance, environmental impact, and material cost. Compressive strength, unit cost, and CO_2 emissions were identified as key parameters to support a multidimensional assessment across performance domains. The corresponding quantitative data are compiled in (see Hierarchical weights were calculated using the Fuzzy AHP method [31], while COPRAS [19] and VIKOR [36] techniques were applied to generate integrated rankings based on the mechanical (1), environmental (3), and economic performance (5) of the selected biomaterials. This combined approach enabled the identification of materials with the most significant potential for sustainable construction through additive manufacturing [38, 39]. The results are presented from the prioritization of criteria to the final ranking of alternatives, allowing a direct comparison between methods based on proportional utility (COPRAS) and compromise solutions (VIKOR), while capturing trade-offs among structural performance, environmental impact, and material cost. Compressive strength, unit cost, and CO_2 emissions were identified as key parameters to support a multidimensional assessment across performance domains. The corresponding quantitative data are compiled in (see Table 2).

After compiling quantitative data on compressive strength, cost, and CO_2 emissions, it was found that tensile and buckling properties were either absent or inconsistently reported across the reviewed studies. Although a few works address these aspects, such as the evaluation of tensile strength in cob mixtures [52] and the assessment of structural performance under compressive loading [53], most of the research has focused primarily on compressive strength. Given its frequent inclusion in literature and its fundamental role in the structural behavior of 3D-printed bio-based materials, compressive strength was selected as the primary mechanical benchmark for comparative analysis.

In cases where studies reported value ranges for specific criteria, the arithmetic mean was used to represent each material in the decision matrix, as the fuzzy AHP, COPRAS, and VIKOR methods require a single representative value per criterion. This approach enabled consistent comparison across all materials while minimizing bias from outlier values.

Therefore, tensile and buckling properties were not considered in the present study. Several of the evaluated biomaterials were also compared with widely recognized low-carbon alternatives, such as geopolymers, magnesia-based cements, and kaolinite-based binders, which are known for their low CO_2 emissions and reduced environmental impact. This comparison provided a reference framework for interpreting the sustainability and relative performance of the proposed materials.

Fuzzy AHP, COPRAS, and VIKOR were implemented using a Python-based framework. Input data were normalized, and criteria weights were derived from the Saaty matrix to calculate aggregate scores for each material across all methods. Results were visualized in graphs and tables to compare material performance across evaluation frameworks *(see* Fig. 4*)*. This comparison revealed significant variation in rankings: for instance, VIKOR ranked hempcrete highest, while Fuzzy AHP and COPRAS favored kaolinite-based cement and cement–biochar composites, respectively. These

Table 2. Quantitative data extracted from the literature included compressive strength (MPa), material cost (USD·kg^{-1}), and carbon dioxide emissions (kg CO_2), enabling a comparative evaluation of the mechanical performance, economic feasibility, and environment.

Authors	Materials	Compressive strength (MPa)	Cost ($•kg^{-1})	CO_2 (kg)
[42, 51–53]	Hempcrete	0.2–0.5	0.21–0.24	1.5–2.1
[49]	Lime - sisal fiber	1.26	0.151	0.05239
[24, 44, 54–62]	Mycelium (pure fungal mycelium grown in mold)	0.07–0.3	0.07–0.17	0.263
[45, 63, 64]	Cement–biochar	3.2 - 6	0.081–0.103	0.03125
[65, 66]	Biomass-fungi composites (hemp hurd, sawdust colonized by fungi)	0.003–0.09	0.07–0.17	0.263
[67–70]	Earth-fiber composites (straw, hemp, banana, kenaf)	0.4–1.35	0.25–0.8	0.39
[1, 71, 72]	Kaolinite – based Cement	35.9	0.07–0.33	0.15–0.20
[42, 43, 73, 74]	Magnesia – Based cement	0.84–1.16	0.10–0.15	0.3766
[8, 75, 76]	Geopolymers	32.6	0.366	0.103–0.120

discrepancies highlight how sensitive rankings can be to the decision-making method applied.

To address this variability, a Borda Count aggregation was applied. The final ranking identified geopolymers as the top-performing material overall, followed by cement–biochar composites and kaolinite-based cement. Among the biomaterials, lime-sisal fiber and mycelium achieved moderate rankings due to their relatively stable performance, while materials such as hempcrete and earth-fiber composites, despite strong results in specific methods, were less consistent and ranked lower overall.

Together, these results illustrate both the strengths and limitations of single-method evaluations. While individual methods can highlight specific material advantages, aggregated approaches such as Borda Count mitigate bias and offer a more balanced perspective. This multi-criteria evaluation strategy underscores the importance of combining diverse decision-making frameworks for more informed and robust material selection in sustainable construction. While the analysis confirms trends already reported in the literature, such as the sustainability potential of mycelium, this study contributes by integrating fragmented data into a coherent decision-making framework, enabling clearer comparisons and practical material prioritization.

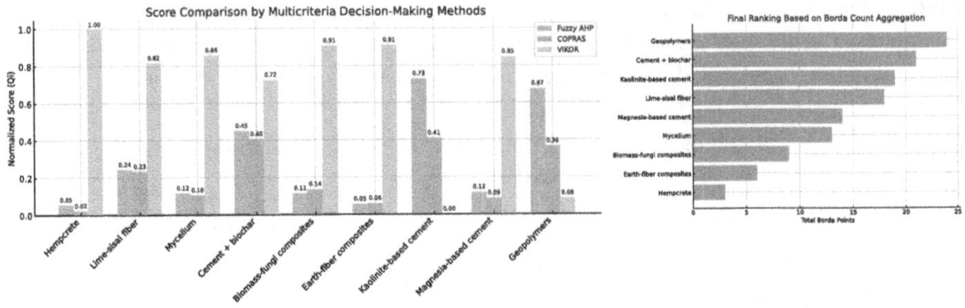

Fig. 4. Normalized score comparison (Q_i) of sustainable materials using Fuzzy AHP, COPRAS, and VIKOR decision-making methods (left) and Final ranking of sustainable materials using Borda Count aggregation (right).

4 Conclusion

This study conducted a structured multi-criteria meta-analysis to compare nine biomaterials for additive manufacturing in sustainable construction. Geopolymers and cement–biochar composites emerged as the highest-performing materials, offering a strong balance between mechanical strength, low carbon footprint, and economic feasibility. Among bio-based alternatives, lime-sisal fiber and mycelium showed consistent, moderate performance, while others, such as hempcrete and earth-fiber composites, performed well in specific methods but lacked overall consistency.

This research provides a methodological foundation to objectively assess bio-based materials under defined criteria. While the results reaffirm trends already noted in the literature, such as the environmental advantages of biochar and the structural promise of geopolymers, the key contribution lies in applying a hybrid MCDM framework that highlights the relative strengths and limitations of each material in context. The use of Borda Count aggregation allowed for a more balanced final ranking, mitigating the influence of extreme scores from individual methods.

The findings emphasize the need for further empirical validation, particularly for bio-based materials like mycelium, whose strong lab-scale performance must be confirmed in real-world applications. Overall, this work contributes to the development of decision-support tools for early-stage material selection and supports the transition toward biomaterials that are not only innovative but also measurable, scalable, and aligned with sustainable construction goals.

References

1. Abu-Ennab, L., Dixit, M.K., Birgisson, B., Pradeep Kumar, P.: Comparative life cycle assessment of large-scale 3D printing utilizing kaolinite-based calcium sulfoaluminate cement concrete and conventional construction. Clean. Environ. Syst. **5** (2022). https://doi.org/10.1016/j.cesys.2022.100078
2. Tohidi, A., Gomaa, M., Haeusler, M.H., Shiel, J.: 3D printing self-shading wall structure with earth: enhancing thermal properties in Earthen Architecture through Computational tool path design, inspired by nature & vernacular architecture. Presented at the Proceedings of the International Conference on Computer-Aided Architectural Design Research in Asia (2024)

3. Uddin, Md.A., et al.: Sustainable construction practices in urban areas: innovative materials, technologies, and policies to address environmental challenges. Energy Build. **341**, 115831 (2025). https://doi.org/10.1016/j.enbuild.2025.115831
4. Fonseca, M., Matos, A.M.: 3D construction printing standing for sustainability and circularity: material-level opportunities. Materials **16** (2023). https://doi.org/10.3390/ma16062458
5. Thiel, C., Hechtl, C.M., Gehlen, C., Kränkel, T.: Sustainability potential of additive manufactured concrete structures – studies on the life cycle assessment and circularity of an extruded exterior wall. In: RILEM Bookseries, pp. 13–21 (2024). https://doi.org/10.1007/978-3-031-70031-6_2
6. Tavares, T.M., Ganga, G.M.D., Godinho Filho, M., Rodrigues, V.P.: The benefits and barriers of additive manufacturing for circular economy: a framework proposal. Sustain. Prod. Consump. **37**, 369–388 (2023). https://doi.org/10.1016/j.spc.2023.03.006
7. Ortega Del Rosario, M.D.L.Á., Beermann, K., Chen Austin, M.: Environmentally responsive materials for building envelopes: a review on manufacturing and biomimicry-based approaches. Biomimetics **8**, 52 (2023). https://doi.org/10.3390/biomimetics8010052
8. Bong, S.H., Du, H.: Sustainable additive manufacturing of concrete with low-carbon materials. In: Sustainable Concrete Materials and Structures, pp. 317–341 (2024). https://doi.org/10.1016/B978-0-443-15672-4.00011-5
9. Ortega Del Rosario, M.D.L.A., Castaño, C., Chen Austin, M.: Biodesign as a tool to achieve sustainable construction through additive manufacturing. In: Arruda, A.J.V., Palombini, F.L. (eds.) Biomimetics, Biodesign and Bionics: Technological Advances Toward Sustainable Development, pp. 251–282. Springer, Cham (2024). https://doi.org/10.1007/978-3-031-51311-4_10
10. Solano, T., et al.: Human-centered and regenerative design: leveraging biomaterials for climate-responsive built environment. IntechOpen (2025). https://doi.org/10.5772/intechopen.1009900
11. Arash, M., Mohammad Abu Hasan, K., Golam, K.: Assessing the environmental impact of building houses in remote areas: 3D printing vs. traditional construction techniques. J. Build. Eng. **102**, 111968 (2025). https://doi.org/10.1016/j.jobe.2025.111968
12. Perrot, A., et al.: Snapshot on 3D printing with alternative binders and materials: earth, geopolymers, gypsum and low carbon concrete. Cem. Concr. Res. **185**, 107651 (2024). https://doi.org/10.1016/j.cemconres.2024.107651
13. Rudin, R., Zboinska, M.A., Sämfors, S., Gatenholm, P.: RePrint. In: Presented at the Hybrids and Haecceities - Proceedings of the 42nd Annual Conference of the Association for Computer Aided Design in Architecture, ACADIA 2022 (2023)
14. Ransikarbum, K., Pitakaso, R., Kim, N.: A decision-support model for additive manufacturing scheduling using an integrative analytic hierarchy process and multi-objective optimization. Appl. Sci. **10**, 5159 (2020). https://doi.org/10.3390/app10155159
15. Paramatti, M., Romani, A., Pugliese, G., Levi, M.: PLA feedstock filled with spent coffee grounds for new product applications with large-format material extrusion additive manufacturing. ACS Omega **9**, 6423–6431 (2024). https://doi.org/10.1021/acsomega.3c05669
16. Alonso Madrid, J., Sotorrío Ortega, G., Gorostiza Carabaño, J., Olsson, N.O.E., Tenorio Ríos, J.A.: 3D claying: 3D printing and recycling clay. Crystals **13** (2023). https://doi.org/10.3390/cryst13030375
17. Sochůrkovà, P., Devyatkina, S., Kordovà, S., Vaško, I., Tsikoliya, S.: Bioreceptive parameters for additive manufacturing of clay based composites. In: Kontovourkis, O., Phocas, M.C., Wurzer, G. (eds.) Proceedings International Conference Education Research Computing Aided Architectural Design in Europe, pp. 45–54. Education and research in Computer Aided Architectural Design in Europe (2024). https://doi.org/10.52842/conf.ecaade.2024.1.045

18. Stohl, L., Manninger, T., von Werder, J., Dehn, F., Gorbushina, A., Meng, B.: Bioreceptivity of concrete: a review. J. Build. Eng. **76**, 107201 (2023). https://doi.org/10.1016/j.jobe.2023.107201
19. de Rubeis, T., Ciccozzi, A., Giusti, L., Ambrosini, D.: On the use of 3D printing to enhance the thermal performance of building envelope – a review. J. Build. Eng. **95** (2024). https://doi.org/10.1016/j.jobe.2024.110284
20. Blanco, I., Siracusa, V.: The use of thermal techniques in the characterization of bio-sourced polymers. Materials. **14** (2021). https://doi.org/10.3390/ma14071686
21. de Rubeis, T., Ciccozzi, A., Pasqualoni, G., Paoletti, D., Ambrosini, D.: On the use of waste materials for thermal improvement of 3D-printed block—an experimental comparison. Buildings **13** (2023). https://doi.org/10.3390/buildings13051136
22. Robertson, O., Høgdal, F., McKay, L., Lenau, T.: Fungal future: a review of mycelium bio-composites as an ecological alternative insulation material. In: Presented at the Proceedings of the NordDesign 2020 Conference, NordDesign 2020 (2020)
23. Rezaieyan, E., et al.: Acoustic properties of natural fiber reinforced composite micro-perforated panel (NFRC-MPP) made from cork fiber and polylactic acid (PLA) using 3D printing (2024). https://www.scopus.com/inward/record.uri?eid=2-s2.0-85182870065&doi=10.1016%2fj.jobe.2024.108491&partnerID=40&md5=80aa2163998f138b6a86d9cd5adc45e5. https://doi.org/10.1016/j.jobe.2024.108491
24. Cintura, E., et al.: Hygroscopicity and morphology of bio-based boards—the influence of the formulation. Appl. Sci. **14**, 873 (2024). https://doi.org/10.3390/app14020873
25. Claude, V., Nguyen, E., Delhaye, A., Mayeux, A., Charron, S.: Hygroscopic and thermal inertia impact of biobased insulation in a wood frame wall. In: Amziane, S., Merta, I., and Page, J. (eds.) Bio-Based Building Materials, pp. 355–372. Springer, Cham (2023). https://doi.org/10.1007/978-3-031-33465-8_28
26. Bhuvaneswari, V., et al.: A critical review on hygrothermal and sound absorption behavior of natural-fiber-reinforced polymer composites. Polymers **14**, 4727 (2022). https://doi.org/10.3390/polym14214727
27. Hurtado, G.B.T.: El Proceso de análisis jerárquico (AHP) como herramienta para la toma de decisiones en la selección de proveedores : aplicación en la selección del proveedor para la Empresa Gráfica Comercial MyE S.R.L. https://sisbib.unmsm.edu.pe/bibvirtual/monografias/basic/toskano_hg/contenido.htm. Accessed 07 May 2025
28. Alyamani, R., Long, S.: The application of fuzzy analytic hierarchy process in sustainable project selection. Sustainability **12**, 8314 (2020). https://doi.org/10.3390/su12208314
29. Saaty, T.L.: The Analytic Hierarchy Process: Planning, Priority Setting, Resource Allocation. New York; London (1980)
30. Hodonou, C., Balazinski, M., Brochu, M., Mascle, C.: Material-design-process selection methodology for aircraft structural components: application to additive vs subtractive manufacturing processes. Int. J. Adv. Manuf. Technol. **103**, 1509–1517 (2019). https://doi.org/10.1007/s00170-019-03613-5
31. Subramani, R., Vijayakumar, P., Rusho, M.A., Kumar, A., Shankar, K.V., Thirugnanasambandam, A.K.: Selection and optimization of carbon-reinforced polyether ether ketone process parameters in 3D printing—a rotating component application. Polymers **16**, 1443 (2024). https://doi.org/10.3390/polym16101443
32. Zeynali, M., Aghdaie, M.H., Rezaeiniya, N., Zolfani, S.H.: A hybrid fuzzy multiple criteria decision making (MCDM) approach to combination of materials selection. AJBM. **6**, 11171–11178 (2012). https://doi.org/10.5897/AJBM11.2022
33. Saaty, T.L.: The analytic hierarchy process: planning, priority setting, resource allocation. McGraw-Hill International Book Co., New York; London (1980)

34. Skrzek, K., Mazgajczyk, E., Dybała, B.: Application of fuzzy logic-based expert advisory systems in optimizing the decision-making process for material selection in additive manufacturing. Materials. **18**, 324 (2025). https://doi.org/10.3390/ma18020324
35. Zavadskas, E.K., Kaklauskas, A., Turskis, Z., Tamošaitien, J.: Multi-attribute decision-making model by applying grey numbers. Informatica **20**, 305–320 (2009). https://doi.org/10.15388/Informatica.2009.252
36. Opricovic, S., Tzeng, G.-H.: Compromise solution by MCDM methods: a comparative analysis of VIKOR and TOPSIS. Eur. J. Oper. Res. **156**, 445–455 (2004). https://doi.org/10.1016/S0377-2217(03)00020-1
37. Raykar, S.J., D'Addona, D.M.: Selection of best printing parameters of fused deposition modeling using VIKOR. Mater Today Proc. **27**, 344–347 (2020). https://doi.org/10.1016/j.matpr.2019.11.104
38. Agrawal, R.: Sustainable material selection for additive manufacturing technologies: a critical analysis of rank reversal approach. J. Clean. Prod. **296**, 126500 (2021). https://doi.org/10.1016/j.jclepro.2021.126500
39. Elloumi, A., Jerbi, A., Ben Amor, R., Souissi, S.: A Comparative study of gray relational analysis and VlseKriterijumska Optimizacija I Kompromisno Resenje approaches for enhancing mechanical properties and productivity in 3D-Printed copper-filled PLA parts. J. Elastomers Plast. **56**, 675–692 (2024). https://doi.org/10.1177/00952443241254939
40. Barad, S., Jagadish, Adapa, S.K., Yanda, S.: Experimental optimization of green FDM process parameters: an integrated MCDM approach. In: Dikshit, M.K., Soni, A., Davim, J.P. (eds.) Advances in Manufacturing Engineering, pp. 233–242. Springer, Singapore (2023). https://doi.org/10.1007/978-981-19-4208-2_17
41. Salimi, A.H., et al.: Exploring the role of advertising types on improving the water consumption behavior: an application of integrated fuzzy AHP and fuzzy VIKOR method. Sustainability. **12**, 1232 (2020). https://doi.org/10.3390/su12031232
42. Binega Yemesegen, E., Memari, A.M.: A review of experimental studies on Cob, Hempcrete, and bamboo components and the call for transition towards sustainable home building with 3D printing. Constr. Build. Mater. **399** (2023). https://doi.org/10.1016/j.conbuildmat.2023.132603
43. Zavaleta, D., et al.: 3D-printing of a basic housing unit prototype using earthen-based matrices stabilized with rice husk fibers. J. Build. Eng. **103** (2025). https://doi.org/10.1016/j.jobe.2025.112111
44. Carcassi, O.B., Ben-Alon, L.: Additive manufacturing of natural materials. Autom. Constr. **167** (2024). https://doi.org/10.1016/j.autcon.2024.105703
45. Vergara, L.A., Perez, J.F., Colorado, H.A.: 3D printing of ordinary Portland cement with waste wood derived biochar obtained from gasification. Case Stud. Constr. Mater. **18** (2023). https://doi.org/10.1016/j.cscm.2023.e02117
46. Gyawali, B., Haghnazar, R., Akula, P., Alba, K., Nasir, V.: A review on 3D printing with clay and sawdust/natural fibers: printability, rheology, properties, and applications. Resul. Eng. **24**, 103024 (2024). https://doi.org/10.1016/j.rineng.2024.103024
47. Alqenaee, A., Memari, A.: Experimental study of 3D printable cob mixtures. Constr. Build. Mater. **324**, 126574 (2022). https://doi.org/10.1016/j.conbuildmat.2022.126574
48. Gomaa, M., Vaculik, J., Soebarto, V., Griffith, M., Jabi, W.: Feasibility of 3DP cob walls under compression loads in low-rise construction. Constr. Build. Mater. **301**, 124079 (2021). https://doi.org/10.1016/j.conbuildmat.2021.124079
49. Faleschini, F., Trento, D., Masoomi, M., Pellegrino, C., Zanini, M.A.: Sustainable mixes for 3D printing of earth-based constructions. Constr. Build. Mater. **398**, 132496 (2023). https://doi.org/10.1016/j.conbuildmat.2023.132496

50. Gangwar, S., Saxena, P., Virmani, N., Biermann, T., Steinnagel, C., Lachmayer, R.: Selection of a suitable additive manufacturing process for soft robotics application using three-way decision-making. Int. J. Adv. Manuf. Technol. **132**, 2003–2015 (2024). https://doi.org/10.1007/s00170-024-13398-x
51. Sinka, M., Spurina, E., Korjakins, A., Bajare, D.: Hempcrete – CO2 neutral wall solutions for 3D printing. Environ. Clim. Technol. **26**, 742–753 (2022). https://doi.org/10.2478/rtuect-2022-0057
52. Asghari, N., Memari, A.M.: State of the art review of attributes and mechanical properties of hempcrete. Biomass. **4**, 65–91 (2024). https://doi.org/10.3390/biomass4010004
53. Fekete, L.M.: Analysis of concrete with chaff. Acta Mater. Transylvan. **6**, 90–94 (2023). https://doi.org/10.33924/amt-2023-02-05
54. Aiduang, W., et al.: A review delving into the factors influencing mycelium-based green composites (MBCs) production and their properties for long-term sustainability targets. Biomimetics. **9**, 337 (2024). https://doi.org/10.3390/biomimetics9060337
55. Appels, F.V.W., et al.: Fabrication factors influencing mechanical, moisture- and water-related properties of mycelium-based composites. Mater. Des. **161**, 64–71 (2019). https://doi.org/10.1016/j.matdes.2018.11.027
56. Bitting, S., Derme, T., Lee, J., Van Mele, T., Dillenburger, B., Block, P.: Challenges and opportunities in scaling up architectural applications of mycelium-based materials with digital fabrication. Biomimetics. **7**, 44 (2022). https://doi.org/10.3390/biomimetics7020044
57. Dessi-Olive, J.: Strategies for growing large-scale mycelium structures. Biomimetics. **7**, 129 (2022). https://doi.org/10.3390/biomimetics7030129
58. Ghazvinian, A., Roha Viera, F., Farrokhsiar, P.: (PDF) Mycelium-based bio-composites for architecture: assessing the effects of cultivation factors on compressive strength. In: ResearchGate (2025). https://doi.org/10.5151/proceedings-ecaadesigradi2019_465
59. Müller, C., Klemm, S., Fleck, C.: Bracket fungi, natural lightweight construction materials: hierarchical microstructure and compressive behavior of Fomes fomentarius fruit bodies. Appl. Phys. A **127**, 178 (2021). https://doi.org/10.1007/s00339-020-04270-2
60. Etinosa, P.O., et al.: Self-organized mycelium biocomposites: Effects of geometry and laterite composition on compressive behavior. J. Mech. Behav. Biomed. Mat. **142**, 105831 (2023). https://doi.org/10.1016/j.jmbbm.2023.105831
61. Früchtl, M., et al.: Sustainable pultruded sandwich profiles with mycelium core. Polymers **15**, 3205 (2023). https://doi.org/10.3390/polym15153205
62. Livne, A., Wösten, H.A.B., Pearlmutter, D., Gal, E.: Fungal Mycelium bio-composite acts as a CO2-sink building material with low embodied energy. ACS Sustainable Chem. Eng. **10**, 12099–12106 (2022). https://doi.org/10.1021/acssuschemeng.2c01314
63. Rose, D., Shirzad, S.: Innovations in green concrete: combining Metakaolin and Arundo grass biochar for enhanced sustainability. Sustainability. **16**, 11219 (2024). https://doi.org/10.3390/su162411219
64. Mensah, R.A., et al.: Biochar-added cementitious materials—a review on mechanical, thermal, and environmental properties. Sustainability. **13**, 9336 (2021). https://doi.org/10.3390/su13169336
65. Pohl, C., et al.: Establishment of the basidiomycete fomes fomentarius for the production of composite materials. Fungal Biol. Biotechnol. **9**, 4 (2022). https://doi.org/10.1186/s40694-022-00133-y
66. Rahman, A.M., et al.: Characterization of 3D printed samples from biomass-fungi composite materials. Manufacturing Letters. **41**, 1136–1142 (2024). https://doi.org/10.1016/j.mfglet.2024.09.139

67. Carcassi, O.B., Maierdan, Y., Akemah, T., Kawashima, S., Ben-Alon, L.: Maximizing fiber content in 3D-printed earth materials: printability, mechanical, thermal and environmental assessments. Constr. Build. Mater. **425**, 135891 (2024). https://doi.org/10.1016/j.conbuildmat.2024.135891
68. Madhusudhana, H.K., et al.: Analysis of the effect of parameters on fracture toughness of hemp fiber reinforced hybrid composites using the ANOVA method. Polymers **13**, 3013 (2021). https://doi.org/10.3390/polym13173013
69. Shin, E., Carcassi, O.B., Maierdan, Y., Kawashima, S., Ben-Alon, L.: Developing 3D-printed natural fiber-rich earth materials in construction. In: Beckett, C., Bras, A., Fabbri, A., Keita, E., Perlot, C., Perrot, A. (eds.) Second RILEM International Conference on Earthen Construction. pp. 32–41. Springer, Cham (2024). https://doi.org/10.1007/978-3-031-62690-6_4
70. Kamaruddin, N.: (PDF) Quantifying of farmers' acceptance and perception in developing kenaf, Hibiscus cannabinus, industry in Malaysia. ResearchGate. (2012). https://doi.org/10.1504/IJGE.2012.051491
71. Elhoweris, A., Galan, I., Glasser, F.P.: Stabilisation of α' dicalcium silicate in calcium sulfoaluminate clinker. Adv. Cem. Res. **32**, 112–124 (2020). https://doi.org/10.1680/jadcr.18.00050
72. Nadhifah, N., et al.: Mesoporous aluminosilicate from metakaolin with natural surfactants for waste cooking oil conversion to hydrocarbon fuels via low-temperature pyrolysis. Case Stud. Chem. Environ. Eng. **11**, 101111 (2025). https://doi.org/10.1016/j.cscee.2025.101111
73. Bernard, E., et al.: MgO-based cements – current status and opportunities. RILEM Technical Letters. **8**, 65–78 (2023). https://doi.org/10.21809/rilemtechlett.2023.177
74. Xu, K., Gao, H.: Preparation, performance optimization, and environmental benefits of precisely carbonated reactive magnesia cement. Sci. J. Technol. **7**, 25–32 (2025). https://doi.org/10.54691/7ayf8v47
75. Cavalieri, M., Ferrara, P.L., Finocchiaro, C., Martorana, M.F.: An economic analysis of the use of local natural waste: volcanic ash of MT. Etna volcano (Italy) for geopolymer production. Sustainability **16**, 740 (2024). https://doi.org/10.3390/su16020740
76. Amari, S., Millar, G.J., Darestani, M., Samali, B.: (PDF) Engineering and Life Cycle Assessment (LCA) of sustainable zeolite-based geopolymer incorporating blast furnace slag. ResearchGate. (2025). https://doi.org/10.3390/su16010440
77. Witherspoon, R., Wang, H., Aravinthan, T., Omar, T.: SSEE 2009: Solutions for a Sustainable Planet, vol. 1, pp. 311–329, Engineers Australia, Barton, A.C.T (2009). https://search.informit.org/doi/abs/10.3316/INFORMIT.767090402496926

Multicriteria Analysis for Material Selection in Artificial Reefs with Additive Manufacturing

Edgar Mauricio Vásquez Trejos[1(✉)], Maria De Los Ángeles Ortega Del Rosario[1], and Shannon Weeks[2]

[1] Research Group in Design, Manufacturing and Materials (DM+M), Universidad Tecnológica de Panamá, Panamá City 0819-07289, Panama
{edgar.vasquez4,maria.ortega}@utp.ac.pa

[2] Autoridad de los Recursos Acuáticos de Panamá, Panamá City, Panama

Abstract. The increasing use of artificial reefs (ARs) to restore marine ecosystems, particularly amid climate change and coral reef degradation, underscores the need to understand how morphology and materials influence their environmental integration and bioreceptivity. Traditionally made from waste materials, ARs have evolved into prefabricated concrete structures to address concerns about material ecotoxicity. More recently, additive manufacturing (AM) has enabled the fabrication of complex, porous structures that better mimic natural reefs, incorporating bioreceptive materials like clay-based mixtures and marine biomass remnants to enhance ecosystem integration. This research proposes a systematic methodology for selecting materials for AM-based ARs, considering mechanical performance, biocompatibility, cost, environmental impact, and carbon footprint. The selection process integrates three multicriteria decision-making (MCDM) methods: Analytic Hierarchy Process (AHP) for weighting criteria, and TOPSIS and VIKOR for ranking materials based on ideal and compromise solutions. The Borda count technique is used to balance rankings from the TOPSIS and VIKOR methods. Data collection includes laboratory experiments, literature reviews, and database analysis. By optimizing material selection, this methodology enhances understanding of the interactions between key parameters and improves design strategies for AM-based ARs, ultimately supporting more effective habitat restoration.

Keywords: Artificial Reef · Additive Manufacturing · Material Selection · Multicriteria Decision Making

1 Introduction and Background

The application of artificial reefs (ARs) as a marine ecosystem restoration technique has gained significant attention over the last two decades. This increasing interest stems from the urgent need to mitigate the degradation of coral reefs, which has been primarily driven by anthropogenic activities and climate-related threats. Factors such as ocean acidification, rising sea temperatures, land-based runoff contamination, and excessive solar radiation exposure have collectively contributed to the decline of these critical ecosystems. The degradation of coral reefs poses a severe threat not only to marine biodiversity,

M. Posani and S. Singh (Eds.): RYS 2025, RILEM Bookseries 67, pp. 500–512, 2026.
https://doi.org/10.1007/978-3-032-14652-6_36

living biomass, and habitat complexity but also to human activities that depend on healthy reef systems. Industries such as tourism, fisheries, and coastal protection services rely on the ecological stability of coral reefs, making their conservation and restoration an essential priority for both environmental sustainability and socio-economic development [1–4].

The effectiveness of artificial reefs in the restoration process is deeply linked to their morphological characteristics. The ability of ARs to replicate the structural complexity and intricate features of natural reefs determines their level of integration within the marine environment and their capacity to support marine fauna. These complex morphologies enhance biodiversity by providing shelter, breeding grounds, and protective habitats for a wide range of marine species. Additionally, the structural complexity of artificial reefs increases their effectiveness as substrates for coral settlement, algal colonization, and overall reef development. As a result, the more closely an artificial reef mimics natural reef functions, the greater its potential to contribute to long-term ecological stability and resilience [5, 6].

Beyond morphological design, the materials used in artificial reef construction play a crucial role in determining their bioreceptivity and overall effectiveness. The selection of appropriate materials influences the extent to which ARs can support marine life and integrate into the surrounding ecosystem. Using non-toxic, non-acidifying materials is fundamental to ensuring that artificial reefs do not introduce harmful substances into the marine environment. The development of ARs must prioritize materials that promote biological colonization while avoiding negative impacts on water chemistry and reef-associated organisms. Ensuring that artificial reefs are constructed from environmentally sustainable and biologically receptive materials is therefore a key factor in their success as a restoration tool, as it directly affects their ability to function as long-term habitats for marine species and facilitate natural reef growth [7, 8].

Over the last few decades, the progressive understanding of key characteristics of artificial reefs (ARs) for restoration has enabled significant advancements in their design and manufacturing techniques. Early artificial reefs were constructed using discarded materials, such as old ships, tires, and concrete blocks, which were dropped into the sea. While these structures provided some level of habitat, their ability to effectively mimic the complexity and ecological functions of natural reefs was limited. Their irregular shapes, lack of surface roughness, and insufficient shelter spaces restricted their potential for marine colonization and biodiversity enhancement [9].

As the demand for more functional and ecologically effective artificial reefs grew, the approach shifted towards designing and manufacturing prefabricated modules with more controlled structures. These new artificial reef designs focused on replicating the intricate and heterogeneous characteristics of natural reefs, allowing for better integration into marine environments. The use of mold-prefabricated modules made from cementitious and earth-based materials became a widely adopted method due to its ability to provide more stable and durable reef structures (See Fig. 1). These prefabricated ARs offered improved surface properties, allowing for better settlement of marine organisms such as corals, algae, and invertebrates, which contribute to the development of a self-sustaining ecosystem [9].

The rapid expansion of additive manufacturing (AM) over the past decade has transformed artificial reef fabrication, providing an innovative alternative to conventional construction methods. Unlike conventional molding and assembly techniques, AM enables the precise fabrication of highly intricate, bioinspired structures that closely replicate the morphological complexity of natural reefs. This capability enables the creation of artificial reefs with optimized surface textures, porosity, and microstructures, thereby enhancing their capacity to support marine life. These features not only provide an ideal substrate for coral settlement but also improve water flow dynamics, reduce sediment accumulation, and create refuge spaces for various marine species. Another key advantage of AM technology is its versatility in material selection. The ability to fabricate artificial reefs using a wide range of materials, including sustainable, non-toxic, and biocompatible options, has significantly improved their environmental integration. Modern AM techniques enable the incorporation of eco-friendly cementitious blends, geopolymers, and natural aggregates, ensuring that artificial reefs do not introduce harmful substances or contribute to ocean acidification. This material flexibility further enhances the long-term success of artificial reef installations by promoting biological colonization and reducing negative environmental impacts.

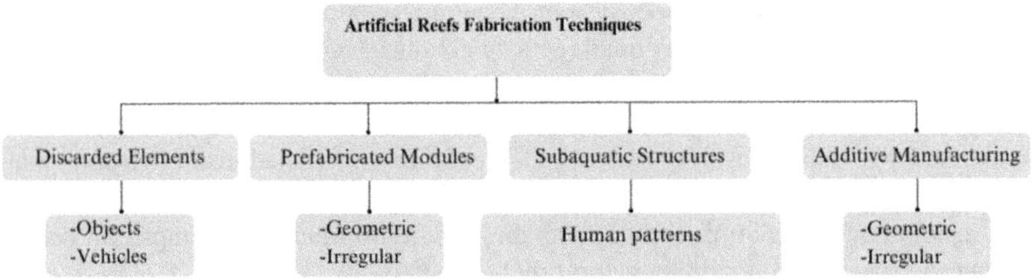

Fig. 1. Main construction techniques used for Artificial Reef fabrication.

With its capacity for precision, customization, and scalability, additive manufacturing has emerged as one of the most promising fabrication methods for artificial reefs. Its ability to produce complex, site-specific reef structures tailored to different environmental conditions makes it a game-changer in marine habitat restoration. By enabling the design of reefs that closely mimic the structural and functional properties of natural ecosystems, AM holds great potential for advancing the effectiveness of artificial reefs in biodiversity conservation and marine ecosystem rehabilitation [10, 11].

The material selection for 3D-printed structures presents a challenge, as it involves a crucial decision that affects multiple phases of the design and fabrication process. Selecting a material that has ideal mechanical performance, printability, biocompatibility, degradability, porosity, cost, environmental impact, and carbon footprint constitutes a Multi-Criteria Decision-Making (MCDM) problem, as it requires evaluating various factors simultaneously to determine the best alternative [10, 11].

MCDM is a mathematical framework designed to support decision-making in complex scenarios involving multiple alternatives and evaluation criteria. This method is widely used in contexts that demand optimized solutions to intricate problems by systematically comparing and ranking alternatives based on predefined criteria [12].

Several MCDM tools, such as the Analytical Network Process (ANP), Analytical Hierarchy Process (AHP), Preference Selection Index (PSI), Multi-Attribute Utility Theory (MAUT), Vlse Kriterijumska Optimizacija Kompromisno Resenje (VIKOR), and the Technique for Order of Preference by Similarity to the Ideal Solution (TOPSIS), among others, are extensively used in the material selection process within the manufacturing industry. These tools provide a structured approach to evaluating different materials, ensuring that the selected option meets technical, economic, and environmental requirements [12–14].

In additive manufacturing (AM), Multi-Criteria Decision-Making (MCDM) methods are widely used to support both technology and material selection, given the diversity of available techniques and material options. These decisions require evaluating numerous parameters—such as accuracy, speed, material compatibility, printability, mechanical performance, environmental impact, and cost—which influence the overall quality and feasibility of the process [15–18]. By systematically comparing these factors, MCDM tools provide a structured and effective approach to identify optimal solutions tailored to the specific requirements of each application [19–21].

2 Mathematical Methodology

This study employs an MCDM framework that integrates AHP, TOPSIS, and VIKOR methods. To ensure consistency among the rankings produced by the different methods, the Borda count method is applied as a consensus-based aggregation technique.

2.1 Analytic Hierarchy Process for Weighting Criteria

The Analytic Hierarchy Process (AHP) technique is based on the pair-wise comparison of different criteria and sub-criteria and how important the relative and subjective weight is between them [22]. This relative relationship between the criteria $C_1, C_2, \ldots, C_N$ can be expressed through a pairwise $N \times N$ size matrix A which is shown in Eq. (1) [23]:

$$A = \begin{bmatrix} a_{11} & \cdots & a_{1j} \\ \vdots & \ddots & \vdots \\ a_{i1} & \cdots & a_{ij} \end{bmatrix} \tag{1}$$

These comparisons are made using the judgment scale shown in Table 1.

Where a_{ij} represents a comparison of criteria C_i respect to C_j. In the AHP method, the pair-wise matrix is normalized by dividing every value of the column by the total sum of the column. The priority vector that shows the relative importance between the criteria is calculated by dividing the normalized matrix by the number of elements in the columns, as shown in Eq. (2) [23].

$$w = \frac{1}{n}\sum_{j=1}^{n} \frac{a_{ij}}{\sum_{i=1}^{n} a_{ij}}, \quad i, j = 1, 2, \cdots, n \tag{2}$$

Table 1. Judgment scale for criteria comparison.

Intensity of preference	Definition
1	Equally preferred
2	Slightly preferred
3	Moderately preferred
4	Moderately to strongly preferred
5	Strongly preferred
6	Strongly to very strongly preferred
7	Very strongly preferred
8	Very strongly to extremely preferred
9	Extremely preferred

The Analytic Hierarchy Process requires confirming the consistency, coherence, and accuracy of the comparison by calculating the consistency ratio (*CR*). The *CR* is calculated using Eq. (3), and its value must be less than 0.1 [23].

$$CR = \frac{CI}{RI} \tag{3}$$

where *RI* is a random index, it was experimentally established by Saaty [22]. *CI* is a consistency index calculated using the expression of Eq. (4) using the consistency eigenvector and the number of rows in the pair-wise matrix [23].

$$CI = \frac{\lambda_{\max} - n}{n - 1} \tag{4}$$

2.2 TOPSIS and VIKOR Process for Ranking Criteria

In this paper, TOPSIS and VIKOR methods are employed. Both methods require the construction of a decision matrix D, as shown in Eq. (5) [24].

$$D_{mn} = \begin{bmatrix} x_{11} \; x_{12} \cdots x_{1j} \\ x_{21} \; x_{22} \cdots x_{2j} \\ \vdots \quad \vdots \quad \ddots \quad \vdots \\ x_{i1} \; x_{i2} \cdots x_{ij} \end{bmatrix} \tag{5}$$

where x_{ij} is the value of the *i* criterion in the *j* alternative, *m* is the alternative, and *n* is the criterion.

The TOPSIS method enables the selection of the best alternative based on its shortest or farthest Euclidean distance from the ideal solution, respectively. Every element of the matrix is normalized using Eq. (6) [24].

$$r_{ij} = \frac{x_{ij}}{\sqrt{\sum_{i=1}^{m} x_{ij}^2}} \tag{6}$$

This normalized matrix is multiplied by the weight obtained by the AHP method. Every value of this weighted matrix is compared with ideal and non-ideal solutions using the expression of Eq. (7) to get the Euclidean distance between the best and worst alternatives [24].

$$S_i^+ = \sqrt{\sum_{j=1}^{n} \left(v_{ij} - v_j^+\right)^2} \tag{7}$$

The Euclidean distance for the best and worst alternatives is used to calculate the relative proximity with the TOPSIS index, as shown in Eq. (8) [24]. A higher value of the TOPSIS index indicates a more optimal alternative.

$$C_i = \frac{S_i^-}{S_i^+ + S_i^-} \tag{8}$$

The VIKOR method is based on the idea of a compromise solution. This method evaluates the degree of closeness of alternatives to the ideal solution. The conflicts between other options are ranked with conflicting factors. The method implements two weighing functions, utility and regret measures, instead of the Euclidean distance used in the TOPSIS technique [25]. The utility measures the global performance of every alternative considering the weighing addition of its difference with the ideal solution; the expression is shown in Eq. (9) [26].

$$S_i = \sum_{j=1}^{n} w_j \cdot \frac{f_j^+ - x_{ij}}{f_j^+ - f_j^-} \tag{9}$$

where f_j^+ and f_j^- are the best and worst values, and w_j is the weight obtained in AHP. Regret measures the global dissatisfaction of every criterion for each alternative and is shown in Eq. (10) [26].

$$R_i = \max_j \left(w_j \cdot \frac{f_j^+ - x_{ij}}{f_j^+ - f_j^-} \right) \tag{10}$$

Utility and Regret measures are used to calculate the VIKOR index shown in Eq. (11) [26]. This index is used to rank the alternatives and select the most optimum option.

$$Q_i = v \cdot \frac{S_i - S^-}{S^+ - S^-} + (1 - v) \cdot \frac{R_i - R^-}{R^+ - R^-} \tag{11}$$

2.3 Borda Count for Ranking Balance

The Borda count is a group decision method used for combining different techniques and obtaining a final consolidated ranking. In this method, every alternative is weighted according to its position in each ranking, as shown in Eq. (12).

$$P_{ik} = m - r_{ik} + 1 \tag{12}$$

where Pik are the points assigned to alternative i in method k, m is the number of alternatives and r_{ik} is the position in the k ranking. These points are added for every alternative according to Eq. (13) [27].

$$P_i^{\text{total}} = \sum_{k=1}^{K} P_{ik} \tag{13}$$

The best alternative is the one that has the highest value of P_i^{total}. The final mathematical methodology framework for this research is shown in Fig. 2.

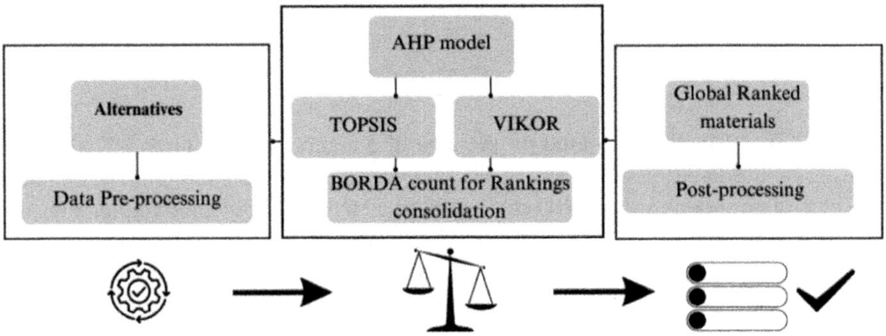

Fig. 2. Proposed mathematical methodology framework for multicriteria decision making.

3 Results, Findings, and Discussion

The main parameters selected for this research, along with their respective pairwise matrix and result weights from the AHP, are shown in Table 2. The relative weights between each parameter are based on the importance of each parameter with the others. Mechanical performance is prioritized over other parameters due to the importance of structural integrity under aquatic conditions. Biocompatibility is determined based on its potential marine eco-toxicity, and its weight is the second most important factor in the matrix. The cost, environmental impact, and carbon footprint are equally important factors. All parameters, except mechanical performance, must be minimized. The values of the consistency index (CI) and the consistency ratio are 0.028 and 0.025, respectively. Therefore, the AHP matrix is consistent. Nonetheless, it should be noted that the assignment of relative weights to the criteria is inherently subjective, as it relies on expert judgment and context-specific priorities. This subjectivity may influence the final material rankings, particularly when the performance of alternatives is similar. Therefore, sensitivity analysis or consensus-building methods are recommended in future work to assess the robustness of the results and reduce potential bias in the weighting process.

Table 2. Pairwise matrix for AHP and resulting weights.

	MP	Bio	Cost	Env	CF	AHP
Mechanical Performance (MPa)	1	2	4	4	5	0.43
Biocompatibility	0.5	1	3	3	4	0.28
Cost (USD)	0.25	0.33	1	2	2	0.13
Environmental impact	0.25	0.33	0.5	1	2	0.1
Carbon footprint	0.2	0.25	0.5	0.5	1	0.06

The nine main materials used in artificial reef fabrication are considered as alternatives for multi-criteria decision-making (MCDM) analysis [10, 11]. These materials and their values for every parameter are extracted from the literature and technical data available online presented in Table 3. Their values are presented in Table 4.

Table 3. List of materials, abbreviations, and data sources used in the evaluation.

Material Description	Ref	Bio
Polylactic Acid	PLA	[28] Moretti et al., 2021
Portland Cement	PC	[29] Sangiorgio et al., 2025
Cement with Seashell Aggregates	C.S	[10, 30] Yoris-Nobile et al., 2023
Cement with Limestone Aggregates	C.L	[10, 30] Yoris-Nobile et al., 2023
Cement with Sandglass Aggregates	C.G	[10, 30] Yoris-Nobile et al., 2023
Geopolymer with Seashell Aggregates	G.S	[10, 30] Yoris-Nobile et al., 2023
Geopolymer with Limestone Aggregates	G.L	[10, 30] Yoris-Nobile et al., 2023
Geopolymer with Sandglass Aggregates	G.G	[10, 30] Yoris-Nobile et al., 2023
Clay	Clay	[31] Marcelino-Sadaba et al., 2017

Table 4. Values of material parameters for MCDM.

Criteria	PLA	PC	C.S	C.L	C.G	G.S	G.L	G.G	Clay
MP	60	40	54	69	75	32	49	40	55
Bio	0.01	2240.00	168.00	92.80	237.00	290.00	248.00	238.00	0.02
Cost	15.00	0.10	0.05	0.05	0.04	0.21	0.20	0.20	0.52
Env	134.00	411.00	10.40	6.18	15.30	1.73	14.80	15.60	0.02
CF	17.00	119.00	3.11	2.49	6.30	3.65	2.67	5.44	0.01

Table 4 presents the raw values used for each material and criterion in the MCDM analysis. These values were collected from published scientific articles, environmental databases, and technical datasheets. Each entry corresponds to a specific property such as mechanical strength (in MPa), biocompatibility (toxicity-related indicators), economic cost (USD/kg), environmental impact (eco-toxicity or impact score), and carbon footprint (kg CO_2-eq/kg material). Although some simplification was necessary to harmonize heterogeneous data sources, the selection process aimed to preserve the relative differences between materials.

The results of the TOPSIS and VIKOR methods are presented in Tables 5 and 6, respectively. These two methods are weighed and combined in a single ranking using the Borda count; the results are shown in Table 7. These rankings align with biological performance reported in the literature. For example, Berman et al. [11] observed a 40% increase in coral larval settlement on 3D-printed clay-based substrates, and Oren et al. [5] found higher fish aggregation on complex bioinspired geometries produced via AM. Such findings underscore the relevance of including biocompatibility and surface features in the material selection process for AM-based artificial reefs.

Table 5. Ranking obtained of TOPSIS method.

Alternative	Score	Rank
C.G	0.943521	1
C.L	0.915666	2
Clay	0.866111	3
C.S	0.839395	4
G.L	0.797626	5
G.G	0.753285	6
G.S	0.710184	7
PLA	0.678403	8
PC	0.297746	9

Table 6. Ranking obtained of VIKOR method.

Alternative	S	R	Q	Rank
C.G	0.036517	0.029624	0	1
C.L	0.074379	0.06	0.063043	2
Clay	0.204132	0.2	0.323926	3
PLA	0.321169	0.15	0.3391	4
C.S	0.235109	0.21	0.356958	5
G.L	0.297276	0.26	0.460625	6
G.G	0.387649	0.35	0.632952	7
G.S	0.469905	0.43	0.787407	8
PC	0.790478	0.35	0.900094	9

Table 7. Consolidated ranking using Borda count.

Alternative	VIKOR Rank	TOPSIS Rank	Borda Score	Final Rank
C.G	1	1	2	1
C.L	2	2	4	2
Clay	3	3	6	3
C.S	5	4	9	4
G.L	6	5	11	5
G.G	7	6	13	6
G.S	8	7	15	7
PLA	4	8	12	8
PC	9	9	18	9

The results obtained by applying TOPSIS, VIKOR, and Borda count methods show consistency in identifying the top-performing alternatives. Cement with Limestone and Glass Sand aggregates were ranked as the best options across all methods, with scores close to the ideal solution. In the case of the VIKOR method, these alternatives also present the lowest regret value (R), meaning that even their worst-performing criterion remains close to the ideal, more so than in any other alternative. These findings are consistent with previous research in the literature, where only the TOPSIS method was applied [10, 30]. The agreement across methods suggests a robust decision-making framework.

Clay was ranked equally across all three methods, highlighting its stability and balanced performance compared to other alternatives. Its performance could be further improved if the raw material is sourced locally and processed with low industrial impact, which would reduce its environmental footprint. The results for each method and alternative are illustrated in Fig. 3.

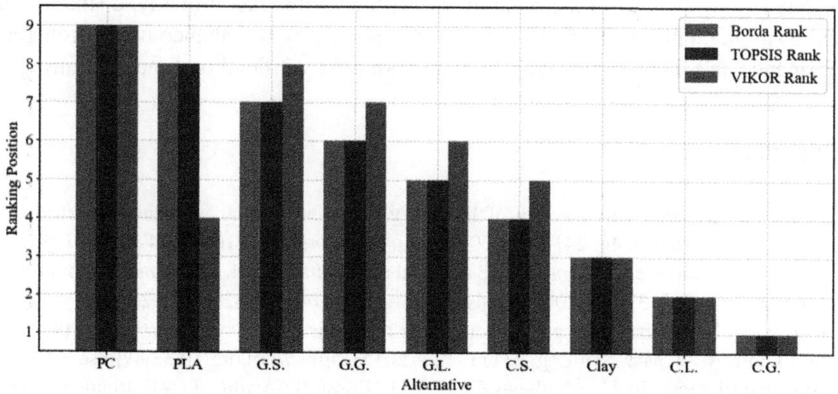

Fig. 3. Comparison between TOPSIS, VIKOR, and Borda count rankings for each material.

Cement with Seashells showed a moderate performance, consistently ranking between 4th and 5th position across the methods. In contrast, geopolymer-based materials showed lower performance in all three methods compared to their cement-based counterparts.

PLA is the alternative that exhibited the least consistency among the methods. It was ranked 3rd in TOPSIS, but 8th in VIKOR. This discrepancy can be attributed to PLA's high biocompatibility, which is favored by VIKOR's prioritization of minimizing the worst-performing criterion (regret). However, in the final Borda count, PLA still ranked 8th, suggesting that although it has strong attributes, they do not compensate for its overall weaknesses across other criteria.

Finally, Portland Cement consistently ranked as the least favorable option. It obtained the lowest score in TOPSIS, the highest Q value in VIKOR, and the worst rank in Borda count. This is primarily due to its high environmental impact, large carbon footprint, and poor biocompatibility.

4 Conclusion and Future Work

This work presents a robust methodology for multicriteria decision-making in the material selection for additive manufacturing artificial reefs. In this research, we integrate the AHP, TOPSIS, VIKOR, and Borda count methods for an integrated consideration of mechanical, ecological, economic, and environmental aspects. The main findings of these methods reveal that the cementitious mixes with limestone and glass aggregate are the most balanced and efficient options, combining high structural performance with low environmental impact and carbon footprint.

The consistent performance of clay in each method demonstrates its potential as a sustainable option, particularly when its origin and local processing are optimized. In contrast, other materials, such as PLA, exhibit less stable behavior during evaluation. On the other hand, Portland cement was systematically identified as the worst alternative due to its high environmental impact and low biocompatibility.

For future work, expanding the experimental data through field and biocolonization tests is proposed to validate the model's predictions. Additionally, a study is proposed to implement locally collected and processed clay to enhance its environmental performance and increase its viability as an option for artificial reef manufacturing.

References

1. Baine, M.: Artificial reefs: a review of their design, application, management and performance. Ocean Coast. Manag. **44**, 241–259 (2001). https://doi.org/10.1016/S0964-5691(01)00048-5
2. Ammar, M.: Coral reef restoration and artificial reef management, future and economic. Open Environ. Eng. J. **2**, 37–49 (2009). https://doi.org/10.2174/1874829500902010037
3. Rinkevich, B.: Climate change and active reef restoration—ways of constructing the "Reefs of Tomorrow." J. Mar. Sci. Eng. **3**, 111–127 (2015). https://doi.org/10.3390/jmse3010111
4. Bracho-Villavicencio, C., Matthews-Cascon, H., Rossi, S.: Artificial reefs around the world: a review of the state of the art and a meta-analysis of its effectiveness for the restoration of marine ecosystems. Environments **10** (2023). https://doi.org/10.3390/environments10070121

5. Oren, A., et al.: Three-dimensional-printed coral-like structures as a habitat for reef fish. J. Mar. Sci. Eng. 11 (2023). https://doi.org/10.3390/jmse11040882
6. Riera, E., Mauroy, B., Francour, P., Hubas, C.: Establishing complexity targets to enhance artificial reef designs. Sci. Rep. **14**, 22060 (2024). https://doi.org/10.1038/s41598-024-72227-z
7. Vivier, B., et al.: Marine artificial reefs, a meta-analysis of their design, objectives and effectiveness. Glob. Ecol. Conserv. **27**, e01538 (2021). https://doi.org/10.1016/j.gecco.2021.e01538
8. Higgins, E., Metaxas, A., Scheibling, R.: A systematic review of artificial reefs as platforms for coral reef research and conservation. PLoS ONE **17**, e0261964 (2022). https://doi.org/10.1371/journal.pone.0261964
9. Matus, I., Góis, J., Vaz-Pires, P., Rocha, A., Alves, J.: Artificial reefs through additive manufacturing: a review of their design, purposes and fabrication process for marine restoration and management. Rapid Prototyp. J. **30**, 87–122 (2024). https://doi.org/10.1108/RPJ-07-2023-0222
10. Yoris-Nobile, A.I., et al.: Artificial reefs built by 3D printing: systematisation in the design, material selection and fabrication. Constr. Build. Mater. 362 (2023). https://doi.org/10.1016/j.conbuildmat.2022.129766
11. Berman, O., et al.: Design and application of a novel 3D printing method for bio-inspired artificial reefs. Ecol. Eng. **188**, 106892 (2023). https://doi.org/10.1016/j.ecoleng.2023.106892
12. Sahoo, S.K., Goswami, S.: A comprehensive review of multiple criteria decision-making (MCDM) methods: advancements, applications, and future directions. Decis. Mak. Adv. **1**, 25–48 (2023). https://doi.org/10.31181/dma1120237
13. Noryani, M., Sapuan, S., M.T., M.: Multi-criteria decision-making tools for material selection of natural fibre composites: A review. J. Mech. Eng. Sci. 12, 3330–3353 (2018). https://doi.org/10.15282/jmes.12.1.2018.5.0299
14. Schuhmann, D., Rupp, M., Merkel, M., Harrison, D.K.: Additive vs. conventional manufacturing of metal components: selection of the manufacturing process using the AHP method. Processes **10** (2022). https://doi.org/10.3390/pr10081617
15. Huang, M., Chen, L., Zhong, Y., Qin, Y.: A generic method for multi-criterion decision-making problems in design for additive manufacturing. Int. J. Adv. Manuf. Technol. **115**, 2083–2095 (2021). https://doi.org/10.1007/s00170-021-06832-x
16. Ren, D., Choi, J.-K., Schneider, K.: A multicriteria decision-making method for additive manufacturing process selection. Rapid Prototyp. J. **28**, 77–91 (2022). https://doi.org/10.1108/RPJ-11-2021-0302
17. Subramani, R., et al.: A recent trend on additive manufacturing sustainability with supply chain management concept, multicriteria decision making techniques. Adv. Mater. Sci. Eng. (2022). https://doi.org/10.1155/2022/9151839
18. Falkonakis, I., Lotfian, S., Yeter, B.: Multi-criteria decision analysis of an innovative additive manufacturing technique for onboard maintenance. Sustain. Switz. **16** (2024). https://doi.org/10.3390/su16093763
19. Zaman, U.K.U., Rivette, M., Siadat, A., Mousavi, S.M.: Integrated product-process design: Material and manufacturing process selection for additive manufacturing using multi-criteria decision making. Robot. Comput.-Integr. Manuf. **51**, 169–180 (2018). https://doi.org/10.1016/j.rcim.2017.12.005
20. Zhang, H., et al.: Materials selection of 3D-printed continuous carbon fiber reinforced composites considering multiple criteria. Mater. Des. **196** (2020). https://doi.org/10.1016/j.matdes.2020.109140
21. Zakeri, S., Chatterjee, P., Konstantas, D., Ecer, F.: A decision analysis model for material selection using simple ranking process. Sci. Rep. **13**, 8631 (2023). https://doi.org/10.1038/s41598-023-35405-z

22. Peko, I., Gjeldum, N., Bilić, B.: Application of AHP, fuzzy AHP and PROMETHEE method in solving additive manufacturing process selection problem. Teh. Vjesn. **25**, 453–461 (2018). https://doi.org/10.17559/TV-20170124092906
23. Saaty, T., Vargas, L., St, C.: The analytic hierarchy process (2022)
24. Madanchian, M., Taherdoost, H.: A comprehensive guide to the TOPSIS method for multi-criteria decision making. Sustain. Soc. Dev. **1** (2023). https://doi.org/10.54517/ssd.v1i1.2220
25. Khan, M., Kumam, P., Kumam, W.: Theoretical justifications for the empirically successful VIKOR approach to multi-criteria decision making. Soft Comput. **25** (2021). https://doi.org/10.1007/s00500-020-05548-6
26. Mardani, A., Zavadskas, E.K., Govindan, K., Amat Senin, A., Jusoh, A.: VIKOR technique: a systematic review of the state of the art literature on methodologies and applications. Sustainability **8** (2016). https://doi.org/10.3390/su8010037
27. Saari, D.G.: Selecting a voting method: the case for the Borda count. Const. Polit. Econ. **34**, 357–366 (2023). https://doi.org/10.1007/s10602-022-09380-y
28. Moretti, C., et al.: Cradle-to-grave life cycle assessment of single-use cups made from PLA. PP and PET. Resour. Conserv. Recycl. **169**, 105508 (2021). https://doi.org/10.1016/j.resconrec.2021.105508
29. Sangiorgio, V., Bianchi, I., Forcellese, A.: Advancing decarbonization through 3D printed concrete formworks: life cycle analysis of technologies, materials, and processes. Energy Build. **332**, 115444 (2025). https://doi.org/10.1016/j.enbuild.2025.115444
30. Yoris-Nobile, A.I., et al.: Life cycle assessment (LCA) and multi-criteria decision-making (MCDM) analysis to determine the performance of 3D printed cement mortars and geopolymers. J. Sustain. Cem.-Based Mater. **12**, 609–626 (2023). https://doi.org/10.1080/21650373.2022.2099479
31. Marcelino-Sadaba, S., Kinuthia, J., Oti, J., Seco Meneses, A.: Challenges in Life Cycle Assessment (LCA) of stabilised clay-based construction materials. Appl. Clay Sci. **144**, 121–130 (2017). https://doi.org/10.1016/j.clay.2017.05.012

Potential of Treating War-Polluted Soils for Earth-Based Construction in Ukraine

Anna Pomazanna[1(✉)], Mykhailo Shevchenko[1], and Oleksandr Bonchkovskyi[2]

[1] Kharkiv School of Architecture, Lviv, Ukraine
a.pomazanna@khsa.edu.ua

[2] Institute of Geography of the National Academy of Sciences of Ukraine, Kyiv, Ukraine

Abstract. The study investigates the feasibility of repurposing soils contaminated by military activity in Kharkiv Oblast, Ukraine, as sustainable feedstock for post-conflict reconstruction. While research in Ukraine has centered on agricultural and ecological impacts of soil contamination, its role in post-war reconstruction remains largely unexamined. In this study nine soil samples from two soil profiles were analysed for heavy-metal content. After conducting particle-size distribution analysis and performing field tests for earth-based materials on the loess sublayer from Kharkiv Oblast, two remediation strategies were piloted. The dilution strategy blends contaminated soils with secondary aggregates to reduce heavy-metal concentrations below regulatory thresholds while producing prototypes of compressed earth blocks (CEB). The encapsulation strategy encloses the contaminated core within protective insulation and clay or lime plaster layers, immobilizing residual metals through sorption and pH buffering. Both approaches demonstrated technical viability but face regulatory and methodological hurdles. Local resource constraints and the urgent need for standardized long-term leachate and dust-release protocols are also highlighted. These findings indicate that war-impacted soils can be transformed into safe, locally sourced building materials, thereby reducing landfill burdens and supporting low-carbon reconstruction in Ukraine.

Keywords: Soil Contamination · Post-war Reconstruction · Earth-based Construction · Ukraine · Circular Building Materials

1 Introduction

In recent years, the full-scale Russian invasion of Ukraine has inflicted not only profound human suffering but also unprecedented damage to both built and natural environments. Preliminary assessments by the Initiative on GHG Accounting of War estimate that over the first three years of the full-scale war (February 2022–February 2025), direct warfare and necessary reconstruction activities have generated approximately 230 Mt CO_2e (emissions comparable to the combined annual output of several mid-sized European nations) and nearly 62.2 Mt CO_2e in "grey energy" associated with rebuilding infrastructure, of which about 40% stems from building reconstruction alone [1]. The climate and material footprints of reconstruction underscore the urgent need for sustainable, locally

M. Posani and S. Singh (Eds.): RYS 2025, RILEM Bookseries 67, pp. 513–526, 2026.
https://doi.org/10.1007/978-3-032-14652-6_37

sourced building solutions that can accelerate recovery while minimizing environmental harm. For most war-damaged regions in eastern and southern Ukraine, using local earth as a natural, circular building material can be a feasible alternative due to local traditions, readily available resources, and existing industrial infrastructure [2].

Across Ukraine's war-torn territories, the destruction of military and civilian infrastructure has led to extensive soil disturbances, with high levels of heavy-metal contamination. Already since 2014 military actions in Luhansk and Donetsk Oblast have introduced a broad spectrum of chemical pollutants, especially heavy metals such as lead, copper, chromium, manganese, vanadium, cobalt, etc. [3] Major overviews of the most recent state of soils in the areas of an active warfare include the comprehensive assessment by the Ukrainian Researchers' Society [4], which draws on extensive field surveys conducted throughout the year, and Ecoaction's 2023 report, which employs a combination of remote sensing and targeted field observations to map war-induced soil damage in Kharkiv and Donetsk oblasts [5]. Earlier field investigations document acute heavy-metal contamination directly resulting from military operations [6]. These findings highlight the dual challenge of ecological degradation and public-health risk posed by war-affected soils.

Although soil contamination in Ukraine has primarily been studied through the lens of agricultural productivity and ecosystem health, its implications for post-conflict reconstruction remain underexplored. In many small towns and villages, polluted earth is routinely mixed with demolition debris, while in urban centers pre-war excavated soils already harbor elevated heavy-metal burdens [7]. The resulting materials are often deemed hazardous waste and destined for landfills, exacerbating disposal pressures and squandering potentially valuable local resources. To date, no comprehensive methodology has assessed whether these contaminated substrates can be rehabilitated and repurposed safely for building applications.

The study carried out in terms of the Bauhaus Earth Experimental Fellowship 2024–2025 addresses two intertwined questions: (1) can warfare-impacted soils from Kharkiv Oblast be treated to serve as locally sourced feedstock for reconstruction, and (2) can their war-related contaminants be effectively immobilized to ensure long-term safety for occupants and the environment? To answer these questions, we assessed structural potential of the soils in the region, applied ICP-MS and ICP-AES to quantify heavy metal concentrations in nine soil samples collected at two profiles in Kharkiv Oblast, and piloted both dilution and encapsulation strategies for contaminant immobilization.

By reframing contaminated soils as raw materials for circular economy building systems, our work extends remediation approaches developed for agricultural soils into the realm of sustainable construction. In doing so, we contribute a novel perspective on valorizing war-damaged soils, offering pathways to reduce landfill burdens, conserve virgin resources, and support Ukraine's resilient, low-carbon reconstruction.

The remainder of this paper is organized as follows. In Section two we describe sampling protocols, analytical procedures, provide an overview of the soil remediation strategies and explain the design approach towards the dilution and encapsulation experiments. Section three presents results on contaminant profiles, and the efficacy of our pilot remediation strategies, accompanied by a discussion of human-health and environmental implications. Finally, the concluding section synthesizes our findings, outlines limitations, and proposes future research directions and policy recommendations.

2 Materials and Methods

2.1 Soil Sampling and Contamination Assessment in War-Affected Areas of Kharkiv Oblast

Soil samples for this study were collected in the Kharkiv region (Kharkiv Oblast), Ukraine, in November 2024. Two sampling sites were selected based on the field research conducted by the Ukrainian Researchers Society in the summer of 2023. Both locations are situated in areas previously occupied by the Russian military, where direct sources of military-related pollution could be identified. The selection of these sites was also influenced by safety considerations, particularly the presence of landmines, which restricted access to certain zones.

The first site, located near the village of Fedorivka, is situated on agricultural land that was used as a launch site for the "Grad" multiple rocket launcher system in 2022. A soil survey conducted in 2023 recorded elevated levels of lead contamination in this area. The second site, in the village of Mala Rohan near Kharkiv city, was the site of a helicopter crash in 2022. Soil samples taken from this location in the same year showed significant cadmium contamination [4].

The focus of the soil investigation was the soil profile. Although the parent material relevant for earthen construction (predominantly loess and loess-like loam) is located at depths of approximately 80 cm, upper soil horizons were also sampled to assess the vertical migration of contaminants. At the Fedorivka site four distinct soil horizons were sampled (FED-01 – FED-04): the chernic (A) horizon, two transitional (B) horizons, and the loess layer. Notably, the soil in this area had been compacted by heavy military machinery. In Mala Rohan, five horizons were identified and sampled (MRO-01 – MRO-05): the chernic (Ah) horizon, a humic-transitional (A) horizon, two further transitional (Bwh and Bw) horizons, and the loess. The soil at the Fedorivka site is attributed to Calcic Chernozem, whereas at the Mala Rohan it is classified as Haplic Chernozem.

All soil samples were analyzed for 12 heavy metals and metaloids: mercury (Hg), cadmium (Cd), lead (Pb), arsenic (As), copper (Cu), cobalt (Co), selenium (Se), zinc (Zn), nickel (Ni), chromium (Cr), strontium (Sr), and vanadium (V). Three samples (MRO-02, MRO-05, and FED-04) were cross-tested in three independent laboratories to ensure analytical precision. Particular attention was given to samples from the loess layer. The remaining samples were analyzed in only one laboratory. Concentrations of heavy metals were determined using atomic emission spectrometry in Ukrainian laboratories and mass spectrometry in a German laboratory. However, atomic emission spectrometry can produce inaccuracies due to spectral overlaps among certain elements. These inaccuracies may be mitigated by expert interpretation of spectra and precise instrument calibration.

The test results were compared with data from the 2023 survey to assess contamination dynamics over time. Laboratories 2 and 3 produced results within the permissible margin of error [8], while Lab 1 reported greater variability. Based on consistency with the previous year's survey, the results from Laboratory 3 are considered the most reliable. A summary of the findings for five elements is provided in Table 1.

Ukrainian environmental legislation [11] sets the threshold for arsenic at 2 mg/kg and for cadmium at 1,5 mg/kg. This is currently the only legal document in Ukraine that defines limit values for heavy metals in dry matter, without specific provisions for construction materials. We have also studied relevant German regulations since a significant part of the research was conducted in Germany. The German Substitute Building Materials Ordinance [9], which governs the use of recycled materials in construction, sets limits of 2 mg/kg for cadmium and 40 mg/kg for arsenic. The Technical Guidelines for the Recycling of Mineral Waste [10], which provide application-specific criteria for mineral waste reuse, prescribe stricter thresholds of 1 mg/kg for cadmium and 15 mg/kg for arsenic.

Table 1. Heavy metals concentration in the soil samples, mg/kg

Code	Mercury (Hg)			Cadmium (Cd)			Lead (Pb)			Arsenic (As)			Copper (Cu)		
	Lab 1	*Lab 2*	*Lab 3*	*Lab 1*	*Lab 2*	*Lab 3*	*Lab 1*	*Lab 2*	*Lab 3*	*Lab 1*	*Lab 2*	*Lab 3*	*Lab 1*	*Lab 2*	*Lab 3*
MRO-01	0,052			4,8			4,4			(<0,2)			17,3		
MRO-02	0,044	(<0,1)		3,6	0,62		3,7	15,2		(<0,2)	7,1		14,2	22,3	
MRO-03	0,047			3,7			3,6			(<0,2)			13,8		
MRO-04	0,044			3,5			3,5			(<0,2)			10,2		
MRO-05	0,046	(<0,1)	(<0,05)	3,6	0,59	**4**	3,1	11,1	10	(<0,2)	5,8	**8**	10,4	17,2	25
MRO-04 2023					41,23			9,34						64,98	
Background					**0,87**			8,15						8,15	
FED-01	0,05			4,2			5,1			(<0,2)			21		
FED-02	0,049			3,9			4,8			(<0,2)			18,5		
FED-03	0,051			4,2			3,4			(<0,2)			15,4		
FED-04	0,048	(<0,1)	(<0,05)	4,1	0,73	0,2	3,3	17,5	12	(<0,2)	6,2	**10**	13,4	25,4	21
FED-05 2023					(<0,1)			8,75						5,19	
Background		(<0,1)			0.6			12.5			**6,5**			6.1	
MPC Ukraine		2,1			**1,5**			32			**2**				
***ErsatzbaustoffV* Germany**		0,6			**2**			140			40			80	

At both sites, arsenic concentrations exceeded the Ukrainian legal limit. In addition, cadmium levels in the Mala Rohan samples also surpassed the Ukrainian threshold. These concentrations are also above permissible levels set by German regulations for cadmium, while arsenic concentrations fall within the German limits but still raise concerns in the local context.

Further testing of the samples included analysis for polycyclic aromatic hydrocarbons (PAHs) and per- and polyfluoroalkyl substances (PFAS). Low concentrations of PAHs were detected exclusively in the chernic (humic) horizons at both sites. Given the project's focus on subsoil layers for use as earthen construction material, topsoil contamination with PAHs was not considered critical for further analysis. PFAS testing, limited to the loess layers (parent material), yielded no detectable concentrations.

Based on these findings, the study focuses on the implications of heavy metal contamination, particularly cadmium and arsenic, for the use of local earth as a sustainable building material. Treatment strategies for contaminated loess are discussed in the following sections.

2.2 Remediation Strategies

Soil remediation techniques have evolved across multiple disciplines (particularly soil science, environmental engineering, and agronomy) and were originally developed to rehabilitate agricultural lands contaminated by pesticides, heavy metals, or industrial effluents. These approaches can, however, be adapted to the context of sustainable post-war reconstruction, where the objective shifts from restoring arable productivity to ensuring the safe reuse of excavated earth as building materials. reviewing established strategies thus provides a foundation for selecting methods that both immobilize harmful contaminants and align with sustainable construction practices. In terms of this research project multiple remediation strategies were assessed in order to choose the most efficient context-specific approach.

Chemical stabilization employs sorptive amendments to immobilize heavy metals and metalloids via adsorption, precipitation, or complexation reactions. Biochar in particular has garnered interest for its high surface area and tunable properties [12]. Physical remediation methods, most notably soil washing, use aqueous or chelating solutions to detach pollutants from coarser fractions, thereby recovering sand and gravel for reuse and concentrating contaminants in clay- and silt-rich sludges [13]. However, effectiveness of the chelate-assisted leaching can be highly dependent on soil pH, competing ion concentrations, and the proportion of fines in the soil matrix [14]. In regions such as Kharkiv Oblast, where native soils contain less than 20% sand, this technique offers limited resource recovery and generates substantial volumes of secondary waste that require secure disposal. In-situ soil venting has similarly limited applicability: while this treatment can volatilize certain organic pollutants, it offers almost no benefit for heavy-metals-polluted soils and entails high energy and infrastructural demands [15]. In contrast, thermal treatment can permanently remove certain heavy metals from soil; however, similar to soil washing, it irreversibly alters soil properties, requires substantial energy and infrastructural investment [16], and may be impractical in the Ukrainian context, where energy infrastructure has been severely damaged.

Biological remediation, encompassing bioaugmentation with pollutant-degrading microorganisms and phytoremediation using hyperaccumulator plants, presents an environmentally friendly alternative for removing contaminants [17]. Phytoremediation has the added advantage that harvested biomass may be valorized: either through energy recovery via combustion or incorporation into biogenic construction composites [18]. However, the time scales required to achieve regulatory compliance often span multiple growing seasons, making these strategies less feasible for the urgent reconstruction needs of war-affected communities.

Containment (or encapsulation) aims to physically immobilize pollutants within a geotechnical or construction structure, thereby preventing leaching and human exposure. Clay minerals, with their high cation-exchange capacity and swelling properties, can trap heavy metals within lattice structures [19]. When used as binders in earth blocks or rammed earth construction these qualities of clay can be used to immobilize heavy metals inside the built structure. Although containment does not reduce total contaminant mass, it effectively binds metals *in situ* and transforms contaminated soil into structural elements. In the context of reconstruction, encapsulation leverages existing earth-construction traditions and local materials while ensuring long-term environmental safety.

The method of dilution is relatively novel and suggests mixing contaminated soil with "clean" aggregates such as reclaimed concrete rubble, calcareous gravel, or local sand, lowers the concentration of pollutants below threshold limits prescribed by building-materials regulations [20]. Furthermore, dilution seamlessly integrates with circular-economy frameworks, promoting the reuse of demolition debris and minimizing the need for virgin materials.

Given the low sand fraction of Kharkiv soils, the urgency of post-war reconstruction, and the imperative to repurpose on-site earth resources, this study prioritizes dilution, in order to achieve compliance through strategic blending, and containment, that can immobilize residual contaminants within earth blocks. These dual strategies maximize material reuse, minimize waste generation, and uphold environmental safety, thereby supporting resilient and sustainable rebuilding efforts.

2.3 Dilution Strategy

In order to evaluate the dilution strategy for contaminated loess from Kharkiv Oblast, we adapted the protocol established under project upMIN100 ("Upcycling of Mineral Construction and Demolition Waste") commissioned by the German Federal Institute for Research on Building, Urban Affairs and Spatial Development (BBSR). In that study, researchers blended recycled aggregates (loam, clay, and silt laden with heavy metals) at varying ratios with uncontaminated primary materials to manufacture earth blocks and mortars that simultaneously met German Substitute Building Materials Ordinance *(Ersatzbaustoffverordnung)* thresholds and achieved required compressive strengths [20]. Drawing on their methodology, our work sought to reproduce and extend these findings using soils endemic to Kharkiv.

The Kharkiv samples present distinctive granulometric characteristics: sand fractions of approximately 6% at Fedorivka and 15% at Mala Rohan. Such low coarse-fraction content precludes direct use as CEB feedstock and necessitates augmentation with higher–sand-content aggregates. To this end, we sourced washed sand from a soil-washing facility in Berlin, where the project was conducted. The reclaimed sand from the washing facility met all regulatory limits for heavy metals; however, its mercury content was higher than that of the original Kharkiv soil samples. Consequently, blending this sand with contaminated loess effectively dilutes heavy-metal concentrations in both components of the mixture.

Mixture optimization began by targeting pollutant dilution rather than contaminant removal, incorporating exclusively secondary raw materials to avoid introducing new pollutants. We prepared three primary blend ratios: 50% sand / 50% silt and clay, 60% sand / 40% silt and clay, and 70% sand / 30% silt and clay. Each mixture was thoroughly homogenized, molded into standard CEB specimens. Specimens underwent heavy-metal analysis (ICP-MS).

2.4 Encapsulation Strategy

In this study, the encapsulation strategy remains in the conceptual and prototyping phase. The primary objective is to immobilize heavy metals (HMs) within contaminated earthen materials by enclosing them with protective layers composed of uncontaminated materials. This approach aims to prevent direct contact and minimize the potential for leaching of HMs into the environment.

Encapsulation techniques have been previously employed, notably in the management of lead-based paints, where covering layers are applied to prevent exposure when removal is impractical due to financial or technical constraints. However, the long-term effectiveness of such methods can be compromised by the degradation of the covering material, leading to the re-exposure of the underlying contaminants [21].

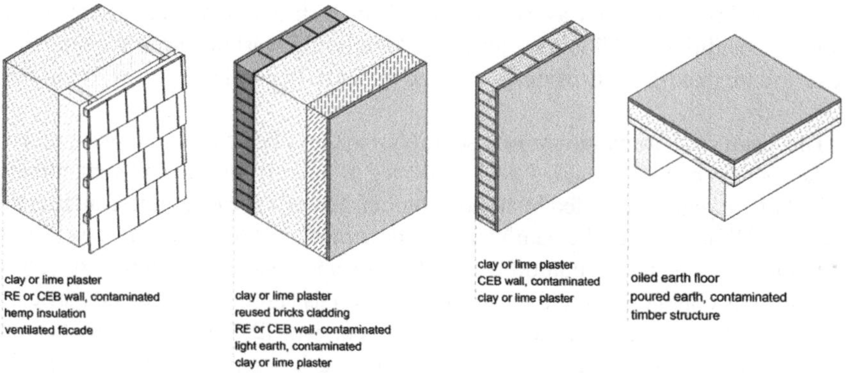

Fig. 1. Details designed to encapsulate earthen material contaminated with HMs (in red). From left to right: exterior wall with ventilated facade, exterior wall plastered, interior wall plastered, floor slab. Made with compressed earth blocks (CEB) or rammed earth (RE).

In the context of earthen construction materials, the inherent properties of clay minerals offer a promising avenue for HM immobilization. Recent studies on historical earthen structures have demonstrated the capacity of earth-based materials to adsorb heavy metals within the matrix, with minimal leaching observed beyond the surface layer of 0–3 cm [22]. Clay particles possess high cation exchange capacities and surface areas, facilitating the adsorption and complexation of metal ions, thereby reducing their mobility [23]. Studies have demonstrated that the mobility of HMs is significantly influenced by soil pH, with lower pH levels increasing metal solubility and mobility. Therefore, maintaining a stable, neutral to slightly alkaline pH within the encapsulated material is crucial to prevent HM mobilization. [24, 25].

To address these considerations, the project proposes the development of earth-based construction elements wherein the load-bearing components are fabricated from HM-contaminated soils, encapsulated within multiple protective layers. Specifically, a 3–5 cm layer of clay or lime plaster is applied to both the interior and exterior surfaces of the structural elements. Protective elements can also include different bio-based insulation material, however, the study on the potential migration of HMs from clay to plant fibers has to be studied additionally. This design aims to physically isolate the contaminated core, mitigate dust generation, and buffer against environmental factors that could alter the pH and promote HM leaching.

Prototypical building details have been designed (Fig. 1) to illustrate the practical implementation of this strategy. These designs prioritize user safety and environmental protection by preventing the dissemination of contaminants from the encapsulated materials.

3 Results and Discussion

3.1 Discussion of Arsenic and Cadmium Contamination in War-Affected Soils

Given the project's focus on researching contaminated soil as a resource for reconstruction in Kharkiv Oblast, it was essential to distinguish pollutants directly attributable to recent hostilities on Ukrainian territory. Identifying warfare-specific contaminants allows to anticipate their environmental behavior and tailor scalable mitigation strategies accordingly.

The measured concentrations of arsenic (As) in the collected soil samples exceed the regulatory thresholds. Arsenic is a key indicator of military-related soil contamination, particularly in regions affected by armed conflict. Historical case studies, such as those from World War I and the Vietnam War, have documented elevated arsenic levels in soils due to the use of chemical munitions and other arsenic-containing compounds [26, 27].

Arsenic is commonly found in explosive materials and metallic alloys, particularly as an additive in copper-based ammunition components. [28] In addition to its anthropogenic sources, arsenic can also occur naturally, as a component of certain geological formations. Furthermore, it is frequently present in agrochemicals such as insecticides, pesticides, and herbicides, leading to its accumulation in soils of agricultural and especially viticultural areas [29].

Despite its known environmental and health hazards, arsenic's behavior in soils (especially those with typical Ukrainian pedogenic and geological characteristics) remains insufficiently studied. Research conducted in Canada, however, has demonstrated that arsenic tends to exhibit limited vertical and lateral mobility in soil profiles [30].

In Ukraine, systematic studies on arsenic contamination in war-affected soils are scarce. One contributing factor is the lack of widespread use of inductively coupled plasma (ICP) instrumentation in state laboratories. Instead, arsenic concentrations are often determined using atomic absorption spectroscopy (AAS), which, while effective, may lack the sensitivity and multi-elemental capabilities required for complex contamination assessments involving metalloids such as arsenic and selenium (Se).

Cadmium (Cd), in contrast, is a widespread contaminant in urban environments. While not typically associated with explosive materials, cadmium is present in propellants used in rocket fuel and has been reported in the composition of armored vehicle plating. High concentrations of cadmium have been recorded in post-conflict regions such as the Persian Gulf following military operations in the early 1990s. These levels are often attributed to hydrocarbon combustion near oil and gas extraction sites, where cadmium may be released as a byproduct [31, 32].

Numerous studies mentioned in the Introduction section of this paper have documented the extent of military-related soil contamination in Ukraine by combining traditional field sampling with satellite remote sensing and GIS-based spatial analysis to map degraded areas and prioritize sampling efforts. However, these investigations have predominantly focused on assessing environmental and agronomic risks rather than exploring the reuse potential of contaminated soils. Reframing excavated polluted soils as feedstock for circular building materials can mitigate hazards and divert soil waste streams into productive reuse.

3.2 Environmental Dispersion and Toxicological Risks of Heavy Metals

To assess the potential human health hazards posed by heavy metals in earth-based construction materials, it is first necessary to understand their pathways of exposure. Heavy metals may enter the body via inhalation of contaminated dust, ingestion of tainted food or water. Although many earthen and masonry products appear inert, weathering and mechanical wear can liberate metal-bearing particulates, fumes, or leachates that contaminate indoor air, groundwater, and adjacent soils. [33] Chronic exposure to cadmium, lead, and mercury has been conclusively linked to carcinogenesis, while other elements, such as arsenic, chromium, and nickel, are recognized for their mutagenic and endocrine-disrupting properties, with several acting as so-called "metal–estrogens" that may elevate risks for hormone-dependent cancers [34].

When building materials are applied directly to or in contact with soil, rainwater percolation can mobilize soluble metal species into the subsurface environment. The composition and volume of leachate are governed by factors including material mineralogy, pH, and ionic strength [35]. Laboratory leaching tests such as those prescribed by the German *Ersatzbaustoffverordnung* and the Dutch Soil Quality Decree quantify release rates under standardized hydraulic and chemical conditions. These norms predict field performance and classify building materials for different types of landfills at the end of their lifecycle [9, 36].

A substantial portion of landfilled solid waste originates from the construction and demolition (C&D) sector, which accounts for 30% of total global waste production. On average, over 35% of C&D waste is disposed of in landfills annually [37]. Ukraine currently faces a critical waste management issue, with over 90% of waste being landfilled [38]. Since solid waste is disposed of in both regulated and unregulated landfills (either dumped on the soil surface or buried) monitoring maximum permissible concentrations (MPCs) of toxic substances in soil is crucial. Currently, Ukraine has approximately 300 hazardous waste storage facilities, many of which were constructed without adequate technical safeguards, posing significant regional environmental risks [39].

Therefore, the proposed strategies seek to divert material that would otherwise be classified as hazardous waste and destined for landfills, where it poses significant long-term risks to human health and ecosystems, by safely repurposing it into construction elements. This approach prevents further dispersal of contaminants and transforms waste into valuable local materials for reconstruction projects.

3.3 Dilution and Encapsulation Strategies

Application of the dilution strategy to Kharkiv-region loess blended with Berlin-washed sand demonstrated clear benefits in reducing heavy-metal concentrations. In all mixtures, cadmium levels in the loess component fell below both German and Ukrainian legal thresholds, while arsenic concentrations were reduced by at least 50%. Simultaneously, the comparatively higher mercury content of the washed sand was effectively balanced by the lower–mercury loess, yielding final mixtures within permissible limits. Even the 50/50 blend achieved a marked reduction in cadmium and arsenic concentrations compared to untreated loess, and the 60/40 blend emerged as optimal, striking a balance between pollutant dilution and mechanical performance (Table 2).

Table 2. Heavy metals concentration in the soil/sand mixtures, mg/kg

Code	Mercury	Cadmium	Lead	Arsenic	Copper	Zinc	Nickel	Chromium
FED100	**(<0,05)**	**0,2**	**12**	**10**	**21**	**52**	**38**	**47**
FED5050	(<0,05)	0,1	36	7	27	47	23	24
FED6040	0,053	(<0,1)	22	4	20	35	14	16
FED7030	0,072	(<0,1)	33	4	29	44	14	14
MRO100	**(<0,05)**	**4**	**10**	**8**	**25**	**46**	**31**	**37**
MRO5050	(<0,05)	2,1	33	7	29	55	23	27
MRO6040	0,052	0,4	80	5	47	60	16	15
MRO7030	0,052	0,4	37	4	47	60	16	15
SAND100	**0,07**	**(<0,1)**	**40**	**(<2)**	**26**	**40**	**4**	**3**
MPC Ukraine	2,1	1,5	32	2				
ErsatzbaustoffV Germany	0,6	2	140	40	80	300	100	

Securing suitable dilution agents is critical. It is recommended to source secondary aggregates such as crushed demolition rubble from war-damaged structures, which, once validated by leachate and dry-matter tests, can substitute for washed sand. Once validated by leachate and dry-matter tests they can substitute for washed sand used for the experiment. Ukraine currently lacks soil-washing facilities, and moreover, the fine-textured, clay- and silt-rich soils of Kharkiv would render such facilities inefficient. Thus, local procurement of demolition waste aligns with both material availability and circular-economy principles. Nonetheless, rigorous testing protocols must be developed to certify these secondary materials for safe use.

The encapsulation strategy, although still at the design and prototyping stage, directly addresses the principal hazard of heavy-metal leaching into soil and groundwater. By enclosing contaminated earth cores, it aims to immobilize metals through sorption and pH buffering. This method offers high energy efficiency by avoiding large-scale material transport or thermal treatments, and it leverages the inherent sequestration capacity of clay minerals.

A comprehensive life-cycle assessment (LCA) of this strategy is necessary to evaluate environmental trade-offs and health-risk reductions. The possible direction for utilizing contaminated material is implementing material passports that facilitate monitoring and reuse of the materials throughout the life cycle [40]. Another option is to include contamination records in the Environmental Product Declarations (EPDs) which are already part of the discussion on the environmental safety of the building materials in the context of the post-war recovery in Ukraine [41]. These documents could include the contaminant profile and encapsulation details of each element, facilitating safe future deconstruction (Fig. 2).

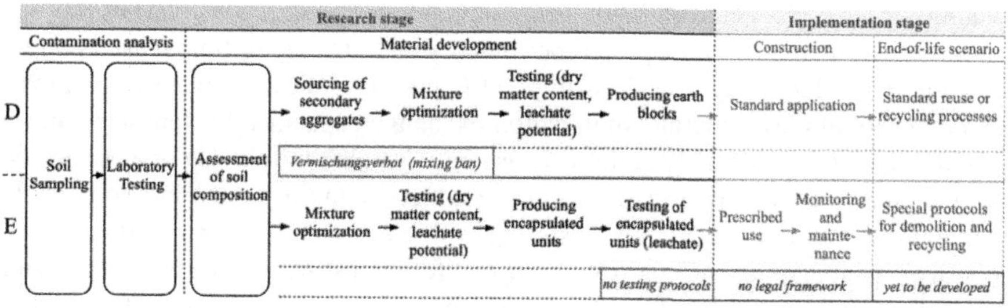

Fig. 2. Diagram comparing the development stages of the dilution (D) and encapsulation (E) strategies, highlighting key steps and associated challenges at each phase.

4 Conclusions

The research set out to determine whether war-impacted soils from Kharkiv Oblast can be safely repurposed as local construction materials by rendering heavy-metal contaminants immobile. Focusing on Kharkiv's loess, we analyzed clay, silt, and sand fractions to ensure our methods were both context-specific and practically achievable within the six-month study period.

Despite the study's contributions, several key limitations should be acknowledged. Spatially, only two soil profiles (nine discrete depth samples) were collected in Kharkiv Oblast, a density that may not capture small-scale heterogeneity in heavy-metal distributions; optimal monitoring designs often require larger sample sizes or more intensive spatial coverage to achieve reliable estimates of pollutant variability [42]. We partially mitigated these constraints by integrating published regional data from the Ukrainian Researchers Society [4]. Analytically, we employed ICP-MS for quantifying quasi-total heavy-metal concentrations in dry matter. The same method was used to verify contaminant levels in compressed earth block (CEB) mixtures for the dilution strategy. Notably, this project did not include leachate-generation tests, which is necessary for understanding long-term contaminant release under simulated rainfall or soil-contact scenarios. This is especially pertinent for encapsulation approaches, where dust-generation resistance and leachate performance govern human and ecological safety over decades.

This study demonstrated that heavy-metal–contaminated soils from Kharkiv Oblast can be transformed into viable construction materials if appropriate immobilization strategies are employed. By blending loess rich in cadmium and arsenic with secondary aggregates (initially washed sand in Berlin trials) we achieved contaminant concentrations below both Ukrainian maximum permissible concentrations and the German *Ersatzbaustoffverordnung* thresholds. This dilution approach offers a practicable route to valorize soils that would otherwise be classified as hazardous waste. However, regulatory frameworks existing inside the EU such as Germany's mixing ban *(Vermischungsverbot)* prohibit the intentional mixing of contaminated and uncontaminated materials, presenting a notable barrier to implementation despite technical efficacy [43]. Besides, in the Ukrainian context, it is recommended to source locally available secondary aggregates, particularly crushed demolition rubble from war-damaged structures, and certify it through standardized leachate testing. Prioritizing such recycled feedstocks conserves local sand deposits and minimizes the energy and water footprints associated with soil washing [44], a critical consideration for sustainable post-war reconstruction.

The encapsulation strategy prototypes employ 3–5 cm clay or lime plasters optionally with an insulation layer made with plant fibers to physically isolate contaminated cores and exploit clay minerals' high cation-exchange capacities for long-term heavy-metal immobilization [45]. Maintaining near-neutral to alkaline pH within these layers is essential, as acidic conditions markedly increase metal solubility and migration [24, 25]. By integrating periodic surface re-plastering with clean materials or introducing fresh secondary aggregates during controlled end-of-life deconstruction, the project envisions a dynamic hybrid pathway of the two proposed strategies that incrementally dilutes residual contaminants over a structure's lifespan.

Looking forward, the absence of recognized protocols to assess long-term leachate release and dust-generation resistance in encapsulated earthen units remains a critical knowledge gap [22]. Establishing "living-lab" pilot facilities under controlled but realistic conditions will be important for validating performance, informing policy development, and refining life-cycle assessments across technical, environmental, and health dimensions [46]. Underpinned by rigorous testing and circular economy principles, these combined dilution and encapsulation strategies can help to repurpose war-impacted soils safely, reducing landfill burdens and supplying urgently needed building materials for Ukraine's sustainable post-conflict recovery.

References

1. de Klerk, L.: Climate damage caused by Russia's war in Ukraine, 24 February 2022–23 February 2025: Preliminary assessment by the Initiative on GHG Accounting of War (2025)
2. Pomazanna, A., Huthöfer, S.: Das Potential industriell gefertigter Lehmsteine für den Wiederaufbau der Ukraine. ce/papers **8**, 93–107 (2025)
3. Kravchenko, O. (ed.): Воєнні дії на сході України – цивілізаційні виклики людству. ЕПЛ, Львів (2015)
4. Bonchkovskyi, O., et al.: War-induced soil disturbances in north-eastern Ukraine (Kharkiv region): Physical disturbances, soil contamination and land use change. Sci. Total Environ. **964**, 178594 (2025)
5. Holubtsov, O., Sorokina, L., Splodytel, A., Chumachenko, S.: Вплив війни росії проти України на стан українських ґрунтів: Результати аналізу. Центр екологічних ініціатив «Екодія», Київ (2023)
6. Baliuk, S.A., et al.: Вплив збройної агресії та воєнних дій на сучасний стан ґрунтового покриву, оцінка шкоди та збитків, заходи з відновлення. ФОП Бровін О. В., Харків (2022)
7. Karmazynenko, S.P., et al.: Важкі метали у компонентах навколишнього середовища м. Маріуполь (еколого-геохімічні аспекти). Інтерсервіс, Київ (2014)

8. International Organization for Standardization. EN ISO 17294-2:2016 — Water quality — Application of inductively coupled plasma mass spectrometry (ICP-MS) — Part 2: Determination of selected elements including uranium isotopes. ISO/CEN, Geneva (2016)
9. Umweltbundesamt. Verordnung über Anforderungen an den Einbau von mineralischen Ersatzbaustoffen in technische Bauwerke (2012)
10. LAGA. TR Boden, M20: Technische Richtlinie zur Verwertung von mineralischen Abfällen (2002)
11. Ministry of Health of Ukraine. Про затвердження Гігієнічних регламентів допустимого вмісту хімічних речовин у ґрунті (2020)
12. Ghandali, M.V., et al.: Heavy metals immobilization and bioavailability in multi-metal contaminated soil under ryegrass cultivation as affected by ZnO and MnO_2 nanoparticle-modified biochar. Sci. Rep. **14**, 10684 (2024)
13. Thaler, P., Walter, B.: EX-SITU Behandlung von kontaminierten Böden: Anlagen in Österreich und angewandte Praxis. Umweltbundesamt GmbH, Wien (2012)
14. Khare, S., et al.: Bio-chelate assisted leaching for enhanced heavy metal remediation in municipal solid waste compost. Sci. Rep. **14**, 14238 (2024)
15. Nadim, F., et al.: Detection and remediation of soil and aquifer systems contaminated with petroleum products: an overview. J. Pet. Sci. Eng. **26**(1–4), 169–178 (2000)
16. Li, J., et al.: Removal and immobilization of heavy metals in contaminated soils by chlorination and thermal treatment on an industrial scale. Chem. Eng. J. **359**, 385–392 (2019)
17. Sales da Silva, I.G., et al.: Soil bioremediation: overview of technologies and trends. Energies **13**(18), 4664 (2020)
18. Cozma, P., et al.: Coupling phytoremediation with plant biomass valorisation and metal recovery: an overview. In: 2021 International Conference on e-Health and Bioengineering (EHB), pp. 1–4. IEEE, Iasi (2021)
19. Yuan, G.D., Theng, B.K.G., Churchman, G.J., Gates, W.P.: Clays and clay minerals for pollution control. In: Bergaya, F., Lagaly, G. (eds.) Developments in Clay Science, vol. 5, pp. 587–644. Elsevier, Amsterdam (2013)
20. Klinge, A., et al.: Upcycling mineralischer Bau- und Abbruchabfälle (upMIN 100): 100-prozentige Substitution natürlicher Gesteinskörnungen und Bindemittel in wiederverwendbaren Lehmbaustoffen. BBSR-Online-Publ. 81/2024, Bonn (2024)
21. Mears, A., Ruth, J.: Innovative practices for healthier homes: a case study. Parsons School of Design, Healthy Materials Lab, New York (2020)
22. Kalbe, U., Schröder, S.A., Ziegert, C.: Vorstudie zur Identifikation und Einordnung von historischen Lehmbauteilen für die Wiederverwendung hinsichtlich der Anreicherung von Schadstoffen und bauschädlichen Salzen. BAM/GOLEHM, Berlin (2024)
23. Bradl, H.B.: Adsorption of heavy metal ions on soils and soil constituents. J. Colloid Interface Sci. **277**(1), 1–18 (2004)
24. Alloway, B.J.: Heavy Metals in Soils. Springer, Dordrecht (1995)
25. Scheffer, F., Schachtschabel, P.: Lehrbuch der Bodenkunde, vol. 16. Enke, Stuttgart (2010)
26. Bausinger, T., Bonnaire, E., Preuß, J.: Exposure assessment of a burning ground for chemical ammunition on the Great War battlefields of Verdun. Sci. Total. Environ. **382**(2–3), 259–271 (2007)
27. Radke, B., et al.: Arsenic-based warfare agents: production, use, and destruction. Crit. Rev. Environ. Sci. Technol. **44**(14), 1525–1576 (2014)
28. Vidosavljević, D., et al.: Soil contamination as a possible long-term consequence of war in Croatia. Acta Agriculturae Scandinavica, Section B – Soil Plant Sci. **63**(4), 322–329 (2013)
29. Singh, R., et al.: Arsenic contamination, consequences and remediation techniques: a review. Ecotoxicol. Environ. Saf. **112**, 247–270 (2015)
30. Ursitti, F., et al.: Assessing and managing exposure from arsenic in CCA-treated wood play structures. Can. J. Public Health **95**(6), 429–433 (2004)

31. Sadiq, M., AlThagafi, K.M., Mian, A.A.: Preliminary evaluation of metal contamination of soils from the Gulf War activities. Bull. Environ. Contam. Toxicol. **49**, 633–639 (1992)
32. Fowler, S., Price, A.: Contaminant enrichment and toxicity of the Gulf's sea surface microlayer: Baseline for determining long-term persistence. Mar. Pollut. Bull. **174**, 113–122 (2022)
33. Mears, A., et al.: Color × Health: Affordable Housing Starts with Healthier Spaces. Parsons Sch. of Design, New York (2020)
34. Campanale, C., et al.: A detailed review study on potential effects of microplastics and additives of concern on human health. Int. J. Environ. Res. Public Health **17**(4), 1212 (2020)
35. Negash, A., Verschoor, A.: Critical emission limit values for building materials: technical background, interpretation and reconstruction. RIVM Letter Rep 2022-0112 (2022)
36. The Netherlands, Ministry of Housing, Spatial Planning and the Environment & State Secretary for Transport, Public Works and Water Management (2007) Decree No. 469 containing rules relative to quality of soil
37. Purchase, C.K., et al.: Circular economy of construction and demolition waste: a literature review on lessons, challenges, and benefits. Mater (Basel) **15**(1), 76 (2021)
38. Vitenko, T., Marynenko, N., Kramar, I.: European experience in waste management. Environ. Sci. Proc. **9**, 17 (2021)
39. DLF Webpage, Ukrainian National Waste Management Strategy until 2030. https://dlf.ua/en/ukrainian-national-waste-management-strategy-until-2030-approved/. Accessed 15 May 2025
40. Costa, A.R., Hoolahan, R.: Materials Passports: Accelerating Material Reuse in Construction. Lancaster Univ./Orms, London (2024)
41. UNIDO & Ministry of Economy of Ukraine. Impact assessment of policy options related to the guiding framework of standards and technical regulations on the green reconstruction of Ukraine (2024)
42. Molla, A., et al.: Evaluating sample sizes and design for monitoring and characterizing the spatial variations of potentially toxic elements in soil. Sci. Total. Environ. **847**, 157489 (2022)
43. German Bundestag. Act reorganising the law on closed cycle management and waste (Gesetz zur Neuordnung des Kreislaufwirtschafts- und Abfallrechts) of 24 February 2012 (2012)
44. Kim, D.-H., et al.: Environmental assessment on a soil washing process of a PB-contaminated shooting range site: a case study. Environ. Sci. Pollut. Res. Int. **20**, 4563–4574 (2013)
45. Otunola, B.O., Ololade, O.O.: A review on the application of clay minerals as heavy-metal adsorbents for remediation purposes. Environ. Technol. Innov. **18**, 100692 (2020)
46. ENOLL. Policy recommendations for integrating living labs in Framework Programme 10 (F10). European Network of Living Labs (2025). https://enoll.org/wp-content/uploads/2025/01/Policy-Recommendations-for-Integrating-Living-Labs-in-FP10.pdf. Accessed 15 May 2025

Recent Advancements on the Effect of Bio-Stabilisation on the Mechanical and Hygrothermal Properties of Earthen Building Materials

Alessia Emanuela Losini[1,18], Agostino Walter Bruno[2,18](✉), Magda Posani[3,18], Samuel J. Armistead[4,18], Adewumi John Babafemi[5,18], Paulina Faria[6,18], Simon Guihéneuf[7,18], Ioannis Ioannou[8,18], Alejandro Jiménez Rios[9,18], Muralidhar Kamath[10,18], Derrick Mwebaza[11,18], Rafail Panagiotou[8,18], Alessandra Ranesi[12,18], Magdalini Theodoridou[11,18], Selina Vaculik[13,18], Snežana Vucetic[14,18], Guillaume Habert[15,18], Céline Perlot[16,18], and Ana Margarida Armada Brás[17,18]

[1] LOCIE, CNRS, Université Savoie Mont Blanc, 73000 Chambery, France
[2] Dip. Di Ingegneria Civile, Chimica E Ambientale, Università Di Genova, Genoa, Italy
agostinowalter.bruno@unige.it
[3] Department of Civil Engineering, Aalto University, 02150 Espoo, Finland
[4] Department of Architecture and Civil Engineering, Chalmers University of Technology, Gothenburg, Sweden
[5] Department of Civil Engineering, Stellenbosch University, Stellenbosch, South Africa
[6] CERIS, Department of Civil Engineering, NOVA University Lisbon, 2829-516 Caparica, Portugal
[7] Univ. Bretagne Sud, IRDL, UMR CNRS 6027, 56100 Lorient, France
[8] Department of Civil and Environmental Engineering, University of Cyprus, Nicosia, Cyprus
[9] Department of Architecture and Civil Engineering, University of Bath, Bath, UK
[10] General Manager Technical Services, Apple Chemie India Private Limited, Nagpur, India
[11] Hub for Biotechnology in the Built Environment, School of Engineering, Newcastle University, Newcastle Upon Tyne, UK
[12] Department of Civil Engineering, Architecture and Georesources, Instituto Superior Técnico, University of Lisbon, CERIS, Av. Rovisco Pais, 1049-001 Lisbon, Portugal
[13] Institute of Building Materials Research, RWTH Aachen University, 52062 Aachen, Germany
[14] Faculty of Technology, University of Novi Sad, Novi Sad, Serbia
[15] Chair of Sustainable Construction, ETH Zurich, Zurich, Switzerland
[16] Université de Pau Et Des Pays de L'Adour, SIAME, 2S UPPA Anglet, France
[17] Institut Universitaire de France, Paris, France
[18] Department of Civil Engineering and Built Environment, Liverpool John Moores University, Liverpool L3 3AF, UK

Abstract. Raw earth is a viable alternative to traditional energy intensive building materials like concrete and fired clayey materials. It can be locally sourced and serves as an excellent hygrothermal buffer, regulating indoor temperature and relative humidity. Unlike mainstream building materials, earth is also fully recyclable

M. Posani and S. Singh (Eds.): RYS 2025, RILEM Bookseries 67, pp. 527–544, 2026.
https://doi.org/10.1007/978-3-032-14652-6_38

at the end of its life cycle. However, stabilisation – the addition of supplementary material to enhance mixture stability – is often required to enhance the durability of earthen building products and their resistance to liquid water erosion, typically increasing their environmental impact.

To mitigate this impact, various bio based stabilisation methods have recently been explored by the scientific community. These involve the addition of biopolymers, bio-fibres or the triggering of bio-mineralisation processes. Still, the effects of these novel bio-stabilisation methods on the mechanical and hygrothermal behaviour of earthen building composites/products require further investigation. This paper provides an overview of existing experimental methods for measuring the mechanical, hygroscopic and thermal properties of bio-stabilised earthen composites and products, along with key findings from the literature. Insights from this review can help guide future research on bio-stabilised earthen products.

Keywords: Bio-stabiliser · Earth-based materials · Mechanical behaviour · Moisture buffering · Sorption-desorption isotherm · Thermal properties

Abbreviations: MBV (Moisture Buffering Value), RH (Relative Humidity), EMPD (Effective Moisture Penetration Depth), WVP (Water Vapour Permeability), UCS (Unconfined Compressive Strength), OT (Oat), TF Thypa), LIG (lignin sulfonate stabilised samples), WOOL(stabilised samples with wool sheep fibres, TAN (tannins stabilised samples), SC (urease from liquid soybeans extract), SP (fine soybeans powder) and SCP (mixture of both liquid and fine powder soybeans) stabilised compressed earth samples using enzyme-induced carbonate precipitation, wt% (% by weight), vol% (% by volume).

1 Introduction

Earthen buildings have long been recognised for their technical efficiency, sustainable and environmentally friendly properties, making them a viable alternative to conventional construction materials. Their use has been revitalised in modern construction due to increasing concerns about energy efficiency, indoor air quality, and the reduction of carbon footprints [1]. The hygroscopic, thermal, and mechanical properties of earth-based building materials are widely examined in the literature, as they can play a key role in regulating indoor environments without jeopardizing the structural performance [2, 3].

One of the most significant advantages of earthen construction is their ability to regulate indoor relative humidity fluctuations through moisture exchange with the surrounding air [4]. This hygroscopic behaviour can enhance occupant comfort, reduce energy demand for climate control, and prevent issues related to excessive indoor humidity, such as condensation and mould growth. The incorporation of bio-stabilisers, such as natural polymers, plant fibres, and bio-compounds, can influence this moisture-regulating ability [5]. Thus, a deeper understanding of the impact of bio-stabilisers on the water vapour permeability and sorption behaviour of earth-based materials is required. Additionally, bio-stabilisation can help reduce the capillary water absorption of earth-based materials, potentially increasing their durability. Although standardised testing protocols exist for evaluating these properties, several methodological differences were observed in the literature on earthen composites often undermining their reproducibility and comparability [6].

In addition to moisture regulation, the thermal performance of earth-based building products plays a crucial role in energy efficiency and indoor comfort [7]. Thermal conductivity and specific heat capacity determine a material's ability to transfer and store heat, influencing the overall thermal performance of buildings. The thermal properties of earth-based building material including bio-stabilised ones, commonly vary depending on moisture content, density, and composition. As in the moisture related properties, there is also lack of standardised experimental methods to measure thermal properties and in the literature, inconsistencies are often found between different testing approaches. Indeed, steady-state and transient techniques are commonly used to assess thermal properties, but the absence of standardised transient testing procedures, for instance, further complicates the interpretation of thermal results when testing bio-stabilised earthen specimens.

Mechanical behaviour is another critical aspect influencing the applicability of bio-stabilised earthen construction. Strength and stiffness assessments, including unconfined compressive strength (UCS) tests, modulus of elasticity measurements, and flexural strength evaluations, are essential for determining structural feasibility. Furthermore, those properties can change with moisture content [8]. The integration of bio-stabilisers can significantly alter mechanical properties by modifying bonding mechanisms, reducing brittleness, and enhancing ductility. However, the limited standardisation in testing approaches for bio-stabilised earthen materials highlights the need to establish reliable methodologies that ensure reproducibility and applicability in construction practice [9].

Despite their many advantages, bio-stabilised earthen materials remain underexplored in terms of their long-term performance, durability, and compatibility with modern construction standards. While various experimental techniques were employed to assess their hygroscopic, thermal, and mechanical properties, inconsistencies in testing procedures and lack of standardisation have hindered direct comparisons and broader implementation. This study aims to provide a comprehensive analysis of experimental methods used to evaluate bio-stabilised earthen materials, highlighting key challenges and advancements in the field. The work hereby presented was carried out within the RILEM Technical Committee TC BEC, specifically under Work Package 3 (WP3) – *Impact of bio-additives and bio-stabilisation methods on the short- and long-term hygrothermal and mechanical performance.*

2 Experimental Methods

There are test methods specifically developed for some building materials and products, such as cementitious ones, that are hardly adapted to earthen building products. Sometimes they are defined as fitting all building products, although some more specific products can hardly use them. That is why different research teams use different test methods to assess earthen building products and particularly biostabilised ones. This review tries to contribute to reverse this situation.

2.1 Hygroscopic Behaviour

This section briefly reviews the main aspects of experimental procedures commonly adopted to study the hygroscopic behaviour of earthen building materials, focusing on moisture adsorption-desorption tests and quantification of water vapour permeability. Liquid transfer is also considered, focusing on the measurement of capillary water absorption.

Moisture Adsorption-Desorption and Moisture Buffering

The moisture adsorption-desorption of bio-stabilised earth materials is typically assessed by measuring moisture content under steady-state conditions or during cyclic humidity variations at a constant temperature. Sorption isotherms are determined at successive relative humidity (RH) levels, ensuring steady-state and measuring the equilibrium moisture content, both in adsorption and desorption process [10]. When the moisture adsorbed is not fully released during desorption, a variation in the equilibrium moisture content for the same RH level is observed in the material, commonly referred to as hysteresis. Hygroscopic materials happen to show hysteresis in desorption possibly related to capillary condensation and for some binders (i.e., lime, gypsum) to the trigger of new chemical reactions (hydration, carbonation) [11].

Experimental setups used for testing the sorption isotherm include climatic chambers, desiccators, and dynamic vapour sorption devices. Differences in methodology and material heterogeneity in terms of size and composition might affect results comparability [6, 12]. For example, the use of a few grams of earth powder instead of larger bulk samples might bring some differences in sorption isotherm results. Similarly, the drying methodology setting in temperature, RH level and equilibrium criterion to prepare the samples for the test might impact the results.

Cyclic tests simulate indoor RH conditions by exposing materials to daily cycles alternating between high and low RH levels (exposure duration and RH levels depending on the selected method). Standardised methods such as ISO 24353, and the NORDTEST protocol [14] define specific RH levels and durations to evaluate moisture buffering capacity. The MBV (Moisture Buffering Value), as defined in the NORDTEST, quantifies the material's ability to adsorb and release moisture under dynamic conditions. MBV is therefore defined as the mass change (Δm, kg) normalised by the exposed surface area (A, m2) and the RH variation (ΔRH, %), expressed in kg/(m2·RH). Sample dimensions, surface of exposure, and air velocity play a critical role in ensuring reliable measurements. The calculation of the effective moisture penetration depth (EMPD) allows to estimate the depth of specimens involved in the moisture buffering and to set the minimum thickness for the test. The EMPD of earthen materials is in the range of 4–30 mm, in accordance with data from the literature [3, 13, 15–17]. A thickness greater than 20 mm is recommended in absence of EMPD data [18].

Water Vapour Permeability

The WVP (Water Vapour Permeability) quantifies the ability of a material to let moisture pass through under a constant vapour pressure gradient. Its evaluation is crucial for understanding how bio-stabilisation can affect the moisture buffering performance of earthen materials. WVP can be assessed through the cup method standardised by ISO 12572 [19], under isothermal conditions. The test prescribes sealing a sample over a cup while controlling both internal and external RH levels and tracking mass variations until steady-state condition is achieved. The dry cup test (typically (9/50% RH) assesses inward moisture flow, while the wet cup test (typically 93/50% RH) evaluates outward flux from the cup. The WVP of the specimen, δ (kg/(m·s·Pa)), is then calculated by means of Eq. 1 [20]:

$$\delta = G/\left(\left(A \cdot \Delta p_v\right) \cdot d\right) \tag{1}$$

where G is the steady-state vapour flow (kg·s^{-1}), A is the exposed surface area (m^2), Δp_v is the water vapour pressure difference across the specimen (Pa) and d is the specimen thickness (m). It can also be expressed by the equivalent thickness of air S_d (m) or the water vapour resistance factor μ (-).

Experimental accuracy in WVP assessment depends on sample properties, test setup, and environmental control. Achieving 0% RH inside dry cups is particularly challenging for highly permeable samples. Pazera and Salonvaara [21] observed that the 0% RH condition is unrealistic when dealing with earthen samples, with RH measured values reaching up to 10% RH in the first 24h of test. However, using saturated salt solutions instead of silica gel is considered to potentially improve the test reliability, better maintaining low RH levels inside the cup [2]. Wet and dry cup results are often non-comparable for porous, hygroscopic materials like earth-based ones, since the WVP increases at higher RH due to enhanced liquid transport [22, 23]. The contributions of vapour and liquid transfer remain difficult to separate, requiring careful specification of test conditions for meaningful cross-laboratory comparisons. Research has shown that sample thickness significantly affects WVP measurements, with thinner samples more influenced by surface film resistance [2].

Capillary Water Absorption

Capillary water absorption testing is commonly employed to determine the capillary water absorption coefficient A_{value}, (kg/m^{2}s$^{1/2}$) or the sorptivity S (mm/min$^{1/2}$) [7, 24]. Standard testing procedures, such as EN 15801 [12] and EN 1015–18 [13], involve partial immersion of specimens in water, but for unstabilised earth materials, this often leads to erosion and inaccurate results. To mitigate this, alternative methods involve placing samples on saturated porous media, such as sponges or sand, ensuring consistent absorption while minimising material loss [25]. Refinements, such as filter paper placement and polyurethane foam supports, have further improved unidirectional water absorption while preventing direct contact with liquid water. The standard test involves periodical weighing of specimens by manually moving the sample from the test setup to a balance, with consequent risk of particle loss. A basket can be used to move the samples, but the implementation of an automated measurement method could represent a significant advancement in improving the reliability and replicability of the test [26].

Despite various methodological advancements, significant discrepancies exist in experimental setups and testing durations across studies, complicating direct comparisons. Some approaches, such as those outlined in EN 15801 [27], adapt existing standards to earth-based materials, while others employ novel techniques, including measuring water absorption height or mass loss over time. Additionally, DIN 18945 [28] introduces a qualitative suction test, categorising earthen materials based on their resistance to swelling, cracking, and spalling over specified durations. These methodological variations highlight the urgent need for standardised testing protocols to ensure the reproducibility and comparability of capillary water absorption measurements in earth building products.

The liquid transport coefficient for suction D_ω (m^2/s) is a crucial parameter for hygrothermal simulations. However, due to the high sensitivity of earthen materials to liquid water, direct experimental determination can be challenging. As an alternative, an indirect approach, using the capillary water absorption coefficient A_{value} ($kg/m^2s^{1/2}$), has been proposed, establishing a mathematical relationship with D_ω (Eq. 2):

$$D_\omega = 3.8 \cdot (A_{value}/\omega_f)^2 \cdot 1000^{[(\omega/\omega f)-1]} \tag{2}$$

where ω and ωf (kg/m^3) are the current and saturated water mass per unit volume.

2.2 Thermal Behaviour

Thermal conductivity is a material property that quantifies the rate at which heat is conducted through a material. It quantifies the ability of a material to transfer heat in response to a temperature gradient, under steady-state conditions. It is often expressed in watts per meter-kelvin (W/m·K), with lower values indicating better thermal insulation ability. Experimental techniques for measuring thermal conductivity are classified into steady-state and transient methods [29]. Steady-state methods, such as heat flow meters and guarded hot plate systems, impose a constant temperature difference across the specimen. Once thermal equilibrium is reached, the resulting heat flux is measured, and thermal conductivity λ is calculated using Fourier's law (Eq. 3):

$$\lambda = (q \cdot d)/\Delta T \tag{3}$$

where q (W/m2) is the heat flux, d (m) is the specimen thickness, and ΔT (K) is the temperature difference across the specimen.

Transient methods, including transient line source, hot wire, and hot disktechniques, involve applying a heat pulse to the sample and monitoring the temperature response over time. These methods provide rapid and simultaneous measurements of thermal conductivity, heat capacity, and thermal diffusivity. They are considered to be particularly suitable for tests when investigating moisture-dependent thermal properties: due to the quick measurement procedure (few minutes), the samples conditioned at specific temperature and relative humidity can be rapidly tested in a different environment, having little time to change their moisture content. For instance, Giuffrida et al. [30] measured the thermal conductivity of earth samples equalised in a climatic chamber at T = 20 °C and RH levels between 15% and 70%. However, transient methods often exhibit greater variability, requiring multiple tests to ensure reliable results.

While transient methods offer efficiency, steady-state methods provide higher accuracy despite requiring longer duration and the need to maintain different RH levels over extended periods. The absence of standardised transient testing procedures further complicates their adoption in building material assessments. Comparative studies highlight inconsistencies between techniques; Colinart et al. [31] found that hot disk measurements yielded higher thermal conductivity values, with discrepancies increasing as the material density increased. Similarly, Gomes et al. [32] reported up to 30% lower thermal conductivity values in steady-state methods compared to transient techniques. Given these variations, it is essential to specify both the methodology (e.g., test equipment reference, setting for the instrumentation) and the testing conditions in experimental studies (e.g., measurement, conditioning and pre-conditioning RH and temperature conditions). Interpretation of thermal conductivity results should be done comparatively rather than as absolute values, accounting for the adopted test procedures and potential measurement uncertainties.

Another fundamental thermal property is the specific heat capacity C_p (J/kg·K), which quantifies the heat stored per unit mass of a material as its temperature rises by 1 K. Despite its significance, data on bio-stabilised earth materials remain scarce, prompting the adaptation of methods used for other masonry materials. Differential scanning calorimetry has been widely employed, as demonstrated by Losini et al. [33] for biopolymer-stabilised rammed earth and Alassaad et al. [34] for lightweight earth products containing phase change materials. The latter calculated C_p using a mixing law based on the mass fractions of individual components. Other indirect techniques involve thermal effusivity measurements, which relate C_p to thermal conductivity and density [35] or guarded hot plate apparatuses that integrate heat flow to determine internal energy storage [36]. Moussa et al. [37] measured multiple thermal properties of compressed earth blocks using the DEsProTherm system, ensuring unidirectional heat flow. The adiabatic calorimeter technique, reviewed by Giuffrida et al. [38], assesses C_p by immersing pre-conditioned samples in water and tracking temperature variations. However, its applicability to bio-stabilised earth is questionable due to liquid water exposure. Another approach involves dynamic thermal measurement devices (e.g., ISOMET), which estimate thermal conductivity and volumetric heat capacity dynamically over short testing durations [39]. While various methods exist, the absence of standardised procedures for bio-stabilised earthen materials underscores the need for careful selection and adaptation of testing protocols.

2.3 Mechanical Behaviour

This section briefly summarises the experimental procedures commonly used to determine the stiffness and strength of bio-stabilised earth-based materials. These properties are commonly assessed through laboratory tests, with unconfined compressive strength (UCS) tests being a standard approach. The modulus of elasticity E is typically derived from the slope of the initial tangent or a secant to the compressive stress-axial strain curve, often taken between 20% and 30% of the peak compressive strength to balance accuracy and avoid plastic deformation effects [40]. Alternatively, unloading paths during cyclic loading can be used, though stress cycles are typically limited to one-third of the peak strength to minimise plastic strain [41]. Various studies have evaluated the impact of biopolymer stabilisation on stiffness and strength, employing flexural strength [5]. Parracha et al. [42] measured the dynamic modulus of elasticity using a zeus resonance meter in line with EN 14146 [43].

Few studies have incorporated displacement sensors in UCS tests to measure both axial and radial strain, allowing for the determination of Poisson's ratio (v), which describes the material's tendency to expand or contract perpendicularly to an applied load. Combined with *E*, Poisson's ratio enables the calculation of the shear modulus (*G*) using the relation $G = E/2(1 + v)$. Alternatively, the shear modulus can be estimated through tensile strength tests on bowtie samples or three-point bending tests using prismatic specimens [44, 45]. These methods provide valuable insights into the mechanical properties of earth products, though the standardisation of such approaches for bio-stabilised earth remains an area for further development. The moisture content when testing earth-based samples should always be quantified.

3 Experimental Findings

3.1 Moisture-Related Behaviour and Capillary Water Absorption

Sorption isotherm measurements have been conducted on various earth-based materials under steady-state conditions using different methods, including saturated salt solutions, climate chambers and dynamic vapor sorption device.

Lima et al. [46] did not find much influence on the water vapor adsorption and desorption curves, using DIN 18947 [43] samples and a similar test method, when adding oat (OF) and typha (TF) fibres to an earth mortar, most probably because of the already high hygroscopicity of the reference mortar (Fig. 1).

Research on bio-stabilised rammed earth [33] observed that 1 wt% of tannins or lignin sulfonate, as well as 0.25 wt% of sheep wool fibres, had a marginal impact on the sorption isotherm of the material. Tannin was found to slightly increase equilibrium moisture contents in the range of 20–80 RH%, while lignin sulfonate slightly increased it at 97 RH%. Another recent study [48] found that light-earth composites (with 35 wt% of reed fibres) displayed greater hygroscopicity than cob (with 2.5 wt% of flax straw fibres), with mass variations reaching up to 9% at 90% RH. This outcome is likely due to the higher hygroscopicity of the fibres compared to the earth used in the study. It is important to remember that clay mineralogy plays an important role on the hygroscopic behaviour [26] therefore, the moisture related properties of bio-stabilised earthen materials is the result of the combination of clay mineralogy with the type and amount of bio-stabiliser.

Under cyclic conditions, studies have explored the impact of bio-fibres and biopolymers on the MBV of earth-based products. In earth blocks stabilised with 20 vol% of gypsum and 10 vol% of lime, together with 15–30 vol% of rice husk, the material containing 30 vol% of rice husk was found to have a higher MBV than that with 15 vol% of rice husk [49]. A study on biopolymer-stabilised earth [50] observed that guar gum and xanthan gum increased the MBV of earthen materials significantly. While earth-based materials are highly hygroscopic, further research is needed to confirm the long-term effects of bio-stabilisers on their moisture buffering performance.

Under cyclic conditions, studies have explored the impact of bio-fibres and biopolymers on the MBV of earth-based products. In earth blocks stabilised with 20 vol% of gypsum and 10 vol% of lime, together with 15–30 vol% of rice husk, the material containing 30 vol% of rice husk was found to have a higher MBV than that with 15 vol% of rice husk [49]. A study on biopolymer-stabilised earth [50] observed that guar gum and xanthan gum increased the MBV of earthen materials significantly. While earth-based materials are highly hygroscopic, further research is needed to confirm the long-term effects of bio-stabilisers on their moisture buffering performance.

Several studies have investigated the WVP of bio-stabilised earthen materials. Research on earth blocks [51], rammed earth [33] and plastering mortars [46] stabilised

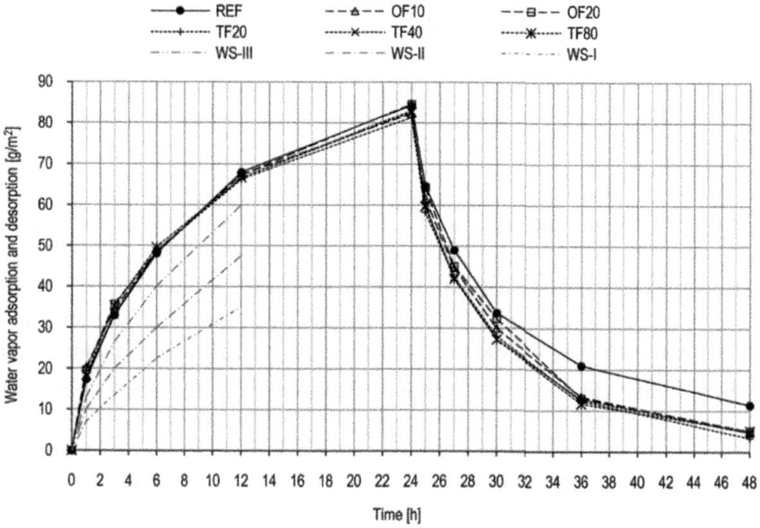

Fig. 1. Water vapor sorption/desorption curves of earth mortars with different additions of oat (OF) and typha (TF) fibres [46], and compared to DIN 18947 limits of sorption classes [47].

with linseed oil, xanthan gum, tannins, and lignin sulfonate has shown a wide range of WVP values, from 7 to 20 ng/m·s·Pa, equivalent to a 20 to 10 water vapour resistance factor, depending on the stabiliser type and dosage. In earth blocks, 1–2 wt% of linseed oil was found to reduce permeability by about 20%, and 0–5-1 wt% of XG enhanced it by about 10%. In rammed earth, the use of 1 wt% of tannin and 0.25 wt% of sheep wool fibres were observed to maintain the material's WVP, whereas 1 wt% of lignin sulfonate decreased it by about 20–25%. Fibre-stabilised mortars containing up to 20 vol% of OF and TF fibres were found to have comparable WVP to the reference unstabilised earth mortar. The latter result suggests that a moderate volume addition of bio-fibres may have only a marginal effect on permeability.

Despite these findings, the influence of bio-stabilisation on WVP remains underexplored, particularly for bio-mineralised earth products. Existing studies indicate that the impact of stabilisation depends not only on the specific biopolymer, other bio-compounds or bio-fibre used, but also on the earth material itself. While some stabilisers preserve or improve permeability, a comprehensive characterisation of bio-stabilised earth products is still lacking, highlighting the need for further research in this area.

The impact of fibre type and content on the water absorption of adobes has been widely studied, with varying conclusions. Several studies have reported an increase in water absorption with higher fibre content, as observed for oil palm empty fruit bunch fibres [52], chicken feather and sugarcane bagasse fibres [53], and straw and hemp [54]. Conversely, other studies found that certain fibres reduce water absorption, such as cassava peel [55] and wood fibres with high lignin content [56].

The differences in reported trends are likely due to fibre type, size, and content, as well as their impact on the material's porosity. For instance, sawdust adobes tend to have higher capillary absorption due to finer pores, whereas straw-stabilised adobes exhibit lower absorption, attributed to the formation of larger pores [57]. These findings highlight the need for further investigation to better understand the complex relationship between fibre inclusion and water absorption in earth-based materials.

3.2 Thermal Behaviour

Studies on the thermal conductivity of bio-stabilised earthen materials mainly focus on the effects of bio-fibres, biopolymers and other organic additives. While bio-fibres generally reduce the thermal conductivity, thus improving insulating performance [32, 46], biopolymers like linseed oil or lignin sulfonate tend to increase it. Some unconventional stabilisers, such as wood biomass ashes and E. coli cells, were also found to lower the thermal conductivity [42]. Most studies use transient measurement techniques, with ISOMET 2104 and KD2 Thermal Properties Analyser being common choices.

Several studies on earth-based materials highlight the role of moisture content and curing conditions on thermal properties, yet inconsistencies in defining "dry" samples and the absence of unstabilised reference measurements in some cases remain an issue [6]. Moreover, the presence of fibre and biopolymer in the matrix raises concerns about the maximal temperature that biobased materials can be exposed to without undergoing degradation.

Losini et al. [33] performed hot disk measurements and showed increased conductivity for lignin sulfonate stabilised samples (LIG) compared to the reference, unstabilised rammed earth (MIX). Stabilised samples with wool fibres (WOOL) and with tannins addition (TAN) showed negligible differences, as illustrated in Fig. 2. Steady-state methods, such as heat flux meters, revealed a strong correlation between moisture content and conductivity in cob and light earth [58], as shown in Fig. 3.

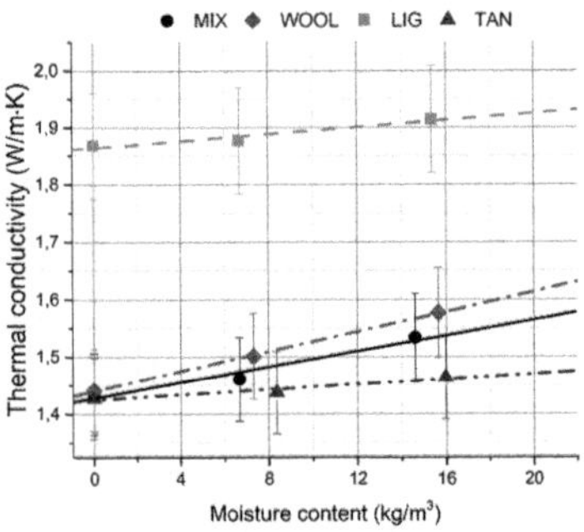

Fig. 2. Thermal conductivity of reference unstabilised rammed earth (MIX) and stabilised mixtures WOOL (wool 0.25 wt%), TAN (tannins 1 wt%) and LIG (lignin sulfonate 1 wt%) at 0%, 55% and 84% RH. Vertical bars represent standard error [33].

However, data on specific heat capacity remain scarce. Losini et al. [33] and Abbas et al. [35] provided some insights, showing variations between 600 and 2200 J/(kg·K), yet further studies are needed for a comprehensive understanding of bio-stabiliser effects on thermal properties.

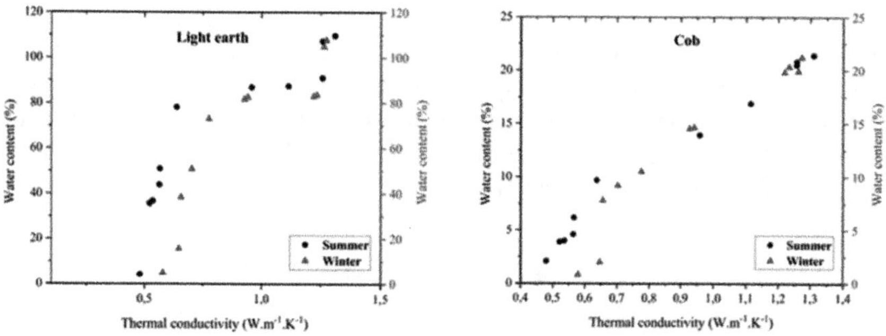

Fig. 3. Thermal conductivity evolution in 'summer' and 'winter' drying conditions for light earth and cob [58].

3.3 Mechanical Behaviour

Lima et al. [46] tested the effect of OF and TF fibres on an earth mortar, showing that flexural, compressive and adhesive strengths varied with the type and content of the fibres (Fig. 4).

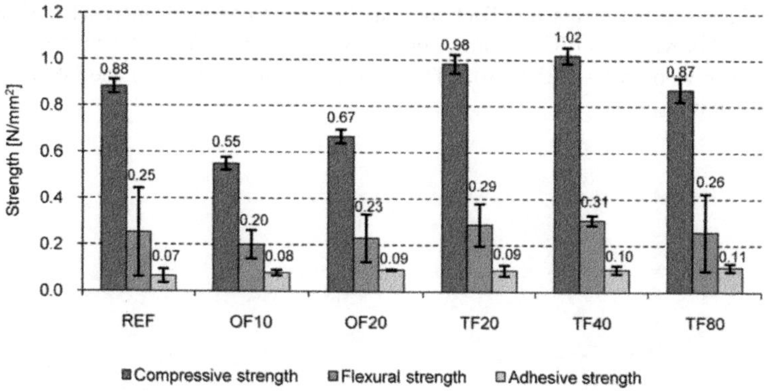

Fig. 4. Flexural, compressive and adhesive strengths of earth mortars with different contents of OT (oat) and TF (thypha) fibres [46].

Losini et al. [59] investigated the stiffness of rammed earth materials stabilised with biopolymers and bio-fibres, finding that lignin sulfonate significantly increased the secant modulus of elasticity for the stabilised samples (LIG), while other stabilisers, including tannin and grape-seed flour (TAN and GRA), slightly reduced the stiffness of the specimens (Fig. 5). Similarly, Parracha et al. [42] observed that adding E. coli cells to earth plaster reduced the dynamic modulus of elasticity from 3527 $\pm$ 45 MPa to 1958 $\pm$ 142 MPa. In contrast, Muguda et al. [45] demonstrated that engineered earth stabilised with xanthan and guar gum exhibited increasing stiffness and strength with higher biopolymer content, reaching optimal performance at 3% (Fig. 6).

The influence of bio-fibres on elastic behaviour was further explored by Laborel-Préneron et al. [5] who reported that increasing the fibre content from 3% to 6% reduced the Young modulus due to the compressibility of fibres. Additionally, Cuccurullo et al. [60] tested enzyme-induced carbonate precipitation (EICP) in compressed earth blocks, showing that while different stabilisation methods (i.e. urease from liquid soybeans extract SC, fine soybeans powder SP or a mixture of both liquid and fine powder soybeans SCP) resulted in varying stiffness, all EICP-treated samples exhibited reduced stiffness compared to unstabilised specimens (Fig. 7).

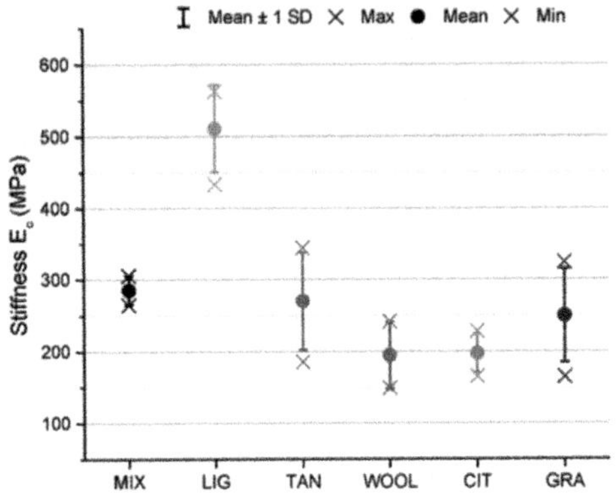

Fig. 5. Stiffness of unstabilised (MIX) and stabilised rammed earth materials[59]. Stabilized samples are abbreviated as LIG (lignin sulfonate 1 wt% addition), TAN (tannin 1 wt% addition), WOOL (sheep wool fibres 0.25 wt%), CIT (citrus peel 1 wt% addition) and GRA (grape-seed flower 1% addition).

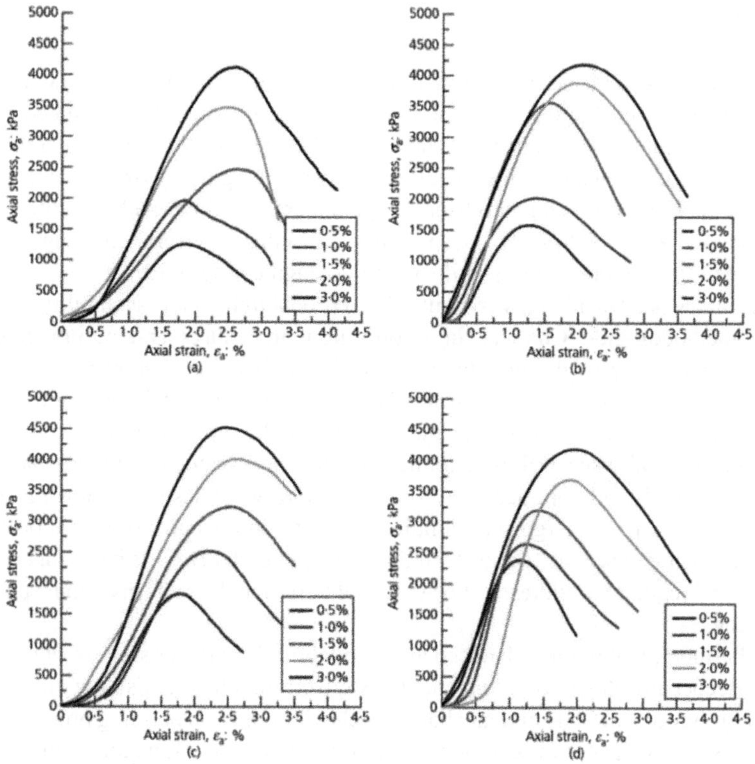

Fig. 6. Stress–strain behaviour in UC tests for different stabiliser percentage varying from 0.5 to 3%: (a) guar gum, 7 days; (b) guar gum, 28 days; (c) xanthan gum, 7 days; (d) xanthan gum, 28 days [45].

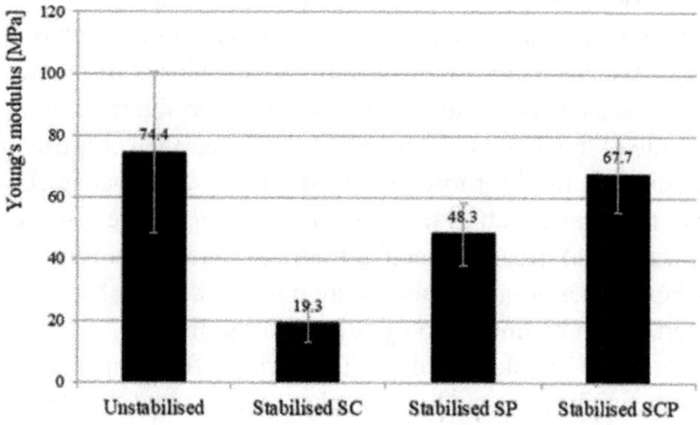

Fig. 7. Young's modulus of unstabilised and SC (urease from liquid soybeans extract); SP (fine soybeans powder) and SCP (mixture of both liquid and fine powder soybeans) stabilised compressed earth samples using enzyme-induced carbonate precipitation [60].

4 Conclusions

The study of bio-stabilised earthen materials represents a promising pathway towards more sustainable and resilient construction practices. The use of bio-mediated and bio-inspired techniques has demonstrated the potential to enhance the mechanical properties and durability of earthen building materials while reducing the environmental footprint associated with conventional stabilisers (e.g., cement and lime) and presenting higher reuse potential at the end-of-life. Despite these advantages, several challenges must be addressed to fully integrate bio-stabilisation into engineering practice.

One of the key challenges remains the reproducibility and scalability of experimental methodologies. Changes in stabilisation procedures and curing conditions can significantly impact the effectiveness of bio-stabilisation. Future research should focus on developing standardised protocols for laboratory and field testing to facilitate comparisons across different studies and ensure reliable implementation.

Another critical aspect is the economic feasibility and practicality of bio-stabilisation techniques in real-world applications. While laboratory studies have shown promising results, the transition to large-scale implementation requires cost-effective production methods, and practical construction techniques. Collaborative efforts between academia, industry, and regulatory bodies will be essential to bridge the gap between research and application.

Furthermore, integrating advanced and standardised characterisation techniques can provide deeper insights into the mechanisms governing bio-stabilisation and its effect on the hygroscopic, thermal and mechanical behaviour of earth building products. These approaches can help optimise bio-stabilisation conditions, ultimately enhancing the overall performance of bio-stabilised earth products.

Current research trends in biostabilised earthen materials both investigate the use of ancestral use of biopolymers in traditional earthen construction and the possibility of using new waste and by-products for biostabilisation. The most promising supplementary materials are not so easy to classify, since the criteria are several (e.g. mechanical, hygrothermal, durability, sustainability) and they might not match all of them [1]. In the present study, we presented some of the most promising biostabilisers, which are mainly still at trial and lab scale state of advancement. However, the traditional use of biostabilisers makes some of them already currently used for earthen construction techniques, such are straw, currently used for torchis, adobe, plasters, and bauge construction techniques [2]. Other materials, such as tannins and cow dung, were also traditionally used for protective plasters in African vernacular traditions and are currently studied for their properties. Further studies about durability are now the center of interest to understand how these properties can be improved by using biostabilisation.

In conclusion, while bio-stabilised earth materials offer a sustainable alternative to conventional earth stabilisation methods, further research and development are needed to overcome existing limitations. By addressing challenges related to reproducibility, durability, scalability, and economic feasibility, bio-stabilisation can emerge as a viable solution for environmentally friendly construction projects. Continued interdisciplinary collaboration and innovation will be crucial in advancing this field and unlocking its full potential.

References

1. Pacheco-Torgal, F., Jalali, S.: Earth construction: lessons from the past for future eco-efficient construction. Constr. Build. Mater. **29**, 512–519 (2012)
2. Fabbri, A., Aubert, J.-E., Bras, A.A., et al.: Hygrothermal and acoustic assessment of earthen materials. Test Characterisation Earth-based Build Mater Elem State-of-the-Art Rep RILEM TC 274-TCE 83–126 (2022)
3. McGregor, F., Heath, A., Shea, A., Lawrence, M.: The moisture buffering capacity of unfired clay masonry. Build. Environ. **82**, 599–607 (2014). https://doi.org/10.1016/j.buildenv.2014.09.027
4. Soudani, L., Woloszyn, M., Fabbri, A., et al.: Energy evaluation of rammed earth walls using long term in-situ measurements. Sol. Energy **141**, 70–80 (2017). https://doi.org/10.1016/j.solener.2016.11.002
5. Laborel-Préneron, A., Aubert, J.-E., Magniont, C., et al.: Effect of plant aggregates on mechanical properties of earth bricks. J. Mater. Civ. Eng. **29**, 4017244 (2017)
6. Colinart, T., et al.: Preliminary results of interlaboratory tests on hygrothermal properties of CEB. In: Springer (ed) Second RILEM International Conference on Earthen Construction (2024)
7. Hall, M., Allinson, D.: Analysis of the hygrothermal functional properties of stabilised rammed earth materials. Build. Environ. **44**, 1935–1942 (2009). https://doi.org/10.1016/j.buildenv.2009.01.007
8. Chauhan, P., El Hajjar, A., Prime, N., Plé, O.: Unsaturated behavior of rammed earth: Experimentation towards numerical modelling. Constr. Build. Mater. **227**, 116646 (2019). https://doi.org/10.1016/j.conbuildmat.2019.08.027
9. Fabbri, A., Morel, J.C., Aubert, J.E., Bui, Q.B., Gallipoli D.R.B.V.: Testing and characterisation of earth-based building materials and elements. RILEM state-of-the-art reports. In: Codes and Standards on Earth Construction, pp 243–259 (2022)
10. ISO 12571:2013(E). International Standard Hygrothermal performance of building materials and products — Determination of hygroscopic sorption properties (2013)
11. Ranesi, A., Faria, P., Veiga, M do R.: Traditional and modern plasters for built heritage: suitability and contribution for passive relative humidity regulation. Heritage **4**, 2337–2355 (2021)
12. Losini, A.E., Woloszyn, M., Chitimbo, T., et al.: Extended hygrothermal characterization of unstabilized rammed earth for modern construction. Constr. Build Mater. **409** (2023). https://doi.org/10.1016/j.conbuildmat.2023.133904
13. McGregor, F., Heath, A., Fodde, E., Shea, A.: Conditions affecting the moisture buffering measurement performed on compressed earth blocks. Build. Environ. **75**, 11–18 (2014). https://doi.org/10.1016/j.buildenv.2014.01.009
14. Rode, C.: Moisture Buffering of Building Materials Department of Civil Engineering Technical University of Denmark (2005)
15. McGregor, F., Heath, A., Maskell, D., et al.: A review on the buffering capacity of earth building materials. Proc. Inst. Civ. Eng. Constr. Mater **169**, 241–251 (2016)
16. McGregor, F., Fabbri, A., Ferreira, J., et al.: Procedure to determine the impact of the surface film resistance on the hygric properties of composite clay/fibre plasters. Mater. Struct. **50**, 1–13 (2017)
17. Losini, A.E.: Rammed earth stabilization with waste or recycled materials and natural additives : characterizations and simulations. Thesis at Universite Savoie Mont Blanc, Génie Civil et Sciences de l'Habitat (2021)
18. Arrigoni, A., Grillet, A.C., Pelosato, R., et al.: Reduction of rammed earth's hygroscopic performance under stabilisation: an experimental investigation. Build. Environ. **115**, 358–367 (2017). https://doi.org/10.1016/j.buildenv.2017.01.034

19. NSAI (2001) EN ISO 12572 - Hygrothermal performance of building materials and products - Determination of water vapour transmission properties, pp. 1–35
20. Künzel, H.M.: Simultaneous Heat and Moisture Transport in Building Components One- and two-dimensional calculation using simple parameters (1995)
21. Pazera, M., Salonvaara, M.: Examination of stability of boundary conditions in water vapor transmission tests. J. Build. Phys. **33**, 45–64 (2009)
22. Colinart, T., Glouannec, P.: Accuracy of water vapor permeability of building materials reassessed by measuring cup's inner relative humidity. Build. Environ. **217**, 109038 (2022)
23. Ranesi, A., Veiga, M.R., Faria, P.: Laboratory characterization of relative humidity dependent properties for plasters: a systematic review. Constr. Build. Mater. **304**, 124595 (2021)
24. Fabbri, A., Soudani, L., Mcgregor, F., Morel, J.: Analysis of the water absorption test to assess the intrinsic permeability of earthen materials. Constr. Build. Mater. **199**, 154–162 (2019). https://doi.org/10.1016/j.conbuildmat.2018.12.014
25. Du, Y., Habert, G., Brumaud, C.: Influence of tannin and iron ions on the water resistance of clay materials. Constr. Build. Mater. **323**, 126571 (2022)
26. Lima, J., Faria, P., Santos Silva, A.: Earth plasters: the influence of clay mineralogy in the plasters' properties. Int J Archit Herit **14**, 948–963 (2020)
27. (2009) EN 15801 Conservation of cultural property - Test methods - Determination of water absorption by capillarity
28. (2024) DIN 18945:2013–08 Earth blocks - Terms and definitions, requirements, test methods
29. Posani, M., Veiga, R., de Freitas, V.P.: Thermal mortar-based insulation solutions for historic walls: an extensive hygrothermal characterization of materials and systems. Constr. Build. Mater. **315**, 125640 (2022)
30. Giuffrida, G., Costanzo, V., Nocera, F., et al.: Natural and recycled stabilizers for rammed earth material optimization. In: International Conference on Sustainability in Energy and Buildings, pp 164–174. Springer (2023)
31. Colinart, T., Pajeot, M., Vinceslas, T., et al.: Thermal conductivity of biobased insulation building materials measured by hot disk: possibilities and recommendation. J Build Eng **43**, 102858 (2021)
32. Gomes, M.G., Flores-Colen, I., Da Silva, F., Pedroso, M.: Thermal conductivity measurement of thermal insulating mortars with EPS and silica aerogel by steady-state and transient methods. Constr. Build. Mater. **172**, 696–705 (2018)
33. Losini, A.E., Grillet, A.-C., Vo, L., et al.: Biopolymers impact on hygrothermal properties of rammed earth: from material to building scale. Bulging Environ **233**, 110087 (2023). https://doi.org/10.1016/j.buildenv.2023.110087
34. Alassaad, F., Touati, K., Levacher, D., Sebaibi, N.: Effect of latent heat storage on thermal comfort and energy consumption in lightweight earth-based housings. Build. Environ. **229**, 109915 (2023)
35. Abbas, M.S., McGregor, F., Fabbri, A., et al.: Effect of moisture content on hygrothermal properties: comparison between pith and hemp shiv composites and other construction materials. Constr. Build. Mater. **340**, 127731 (2022)
36. El Fgaier, F., Lafhaj, Z., Antczak, E., Chapiseau, C.: Dynamic thermal performance of three types of unfired earth bricks. Appl. Therm. Eng. **93**, 377–383 (2016)
37. Ouedraogo, M., Dao, K., Millogo, Y., et al.: Physical, thermal and mechanical properties of adobes stabilized with fonio (Digitaria exilis) straw. J. Build. Eng. **23**, 250–258 (2019). https://doi.org/10.1016/j.jobe.2019.02.005
38. Giuffrida, G., Caponetto, R., Nocera, F.: Hygrothermal properties of raw earth materials: a literature review. Sustainability **11** (2019)
39. Theodoridou, M., Kyriakou, L., Ioannou, I.: PCM-enhanced lime plasters for vernacular and contemporary architecture. Energy Procedia **97**, 539–545 (2016)

40. Rozzi, V., Bruno, A.W., Fabbri, A., et al.: Stabilising compressed earth materials with untreated and thermally treated recycled concrete: a multi-scale investigation. J. Clean. Prod. **423**, 138614 (2023)
41. Bui, Q.-B., Morel, J.-C.: Assessing the anisotropy of rammed earth. Constr. Build. Mater. **23**, 3005–3011 (2009)
42. Parracha, J.L., Pereira, A.S., Velez da Silva, R., et al.: Effect of innovative bioproducts on the performance of bioformulated earthen plasters. Constr. Build. Mater. **277**, 122261 (2021). https://doi.org/10.1016/j.conbuildmat.2021.122261
43. EN 14146:2004 (Main) Natural stone test methods - Determination of the dynamic modulus of elasticity (by measuring the fundamental resonance frequency)
44. Morel, J.C., Pkla, A.: A model to measure compressive strength of compressed earth blocks with the '3 points bending test.' Constr. Build. Mater. **16**, 303–310 (2002)
45. Muguda, S., Booth, S.J., Hughes, P.N., et al.: Mechanical properties of biopolymer-stabilised soil-based construction materials. Géotechnique Lett. **7**, 309–314 (2017)
46. Lima, J., Faria, P.: Eco-efficient earthen plasters: The influence of the addition of natural fibers. RILEM Bookseries **12**, 315–327 (2016). https://doi.org/10.1007/978-94-017-7515-1_24
47. (2024) DIN 18947 Earth plasters - Requirements, test and labelling
48. Azil, A., Touati, K., Sebaibi, N., et al.: Monitoring of drying kinetics evolution and hygrothermal properties of new earth-based materials using climatic chamber simulation. Case Stud. Constr. Mater. **18**, e01798 (2023). https://doi.org/10.1016/j.cscm.2022.e01798
49. Antunes, A., Faria, P., Silva, V., Brás, A.: Rice husk-earth based composites: a novel bio-based panel for buildings refurbishment. Constr. Build. Mater. **221**, 99–108 (2019). https://doi.org/10.1016/j.conbuildmat.2019.06.074
50. Muguda, S., Lucas, G., Hughes, P.N., et al.: Durability and hygroscopic behaviour of biopolymer stabilised earthen construction materials. Constr. Build Mater. **259**, 119725 (2020). https://doi.org/10.1016/j.conbuildmat.2020.119725
51. Guihéneuf, S., Perrot, A., Rangeard, D.: Linseed oil and xanthan gum: promising stabilisers for earthen building materials. Constr. Technol. Archit. **1**, 245–254 (2022)
52. Ismail, S., Yaacob, Z.: Properties of laterite brick reinforced with oil palm empty fruit bunch fibres. Pertanika J Sci Technol **19**, 33–43 (2011)
53. Salih, M.M., Osofero, A.I., Imbabi, M.S.: Critical review of recent development in fiber reinforced adobe bricks for sustainable construction. Front. Struct. Civ. Eng. **14**, 839–854 (2020)
54. Laborel-Préneron, A., Magniont, C., Aubert, J.-E.: Hygrothermal properties of unfired earth bricks: effect of barley straw, hemp shiv and corn cob addition. Energy Build **178**, 265–278 (2018)
55. Villamizar, M.C.N., Araque, V.S., Reyes, C.A.R., Silva, R.S.: Effect of the addition of coal-ash and cassava peels on the engineering properties of compressed earth blocks. Constr. Build. Mater. **36**, 276–286 (2012)
56. Millogo, Y., Morel, J.-C., Aubert, J.-E., Ghavami, K.: Experimental analysis of Pressed Adobe Blocks reinforced with Hibiscus cannabinus fibers. Constr. Build. Mater. **52**, 71–78 (2014)
57. Costi de Castrillo, M., Ioannou, I., Philokyprou, M.: Reproduction of traditional adobes using varying percentage contents of straw and sawdust. Constr. Build. Mater. **294**, 123516 (2021). https://doi.org/10.1016/j.conbuildmat.2021.123516
58. Azil, A., Touati, K., Sebaibi, N., et al.: Monitoring of drying kinetics evolution and hygrothermal properties of new earth-based materials using climatic chamber simulation. Case Stud. Constr. Mater. **18**, e01798 (2023). https://doi.org/10.1016/j.cscm.2022.e01798

59. Losini, A.E., Grillet, A.-C., Woloszyn, M., et al.: Mechanical and Microstructural Characterization of Rammed Earth Stabilized with Five Biopolymers. Mater **15** (2022). https://doi.org/10.3390/ma15093136
60. Cuccurullo, A., Gallipoli, D., Bruno, A.W., et al.: Earth stabilisation via carbonate precipitation by plant-derived urease for building applications. Geomech. Energy Environ. **30**, 100230 (2022)

Recyclability of Soils Stabilised with Natural Hydraulic Lime for Earthen Constructions

Chiara Turco(✉), Elisabete Teixeira, and Ricardo Mateus

ISISE, ARISE, Department of Civil Engineering, University of Minho, Guimarães, Portugal
chiaraturco@civil.uminho.pt

Abstract. The construction sector has a significant impact on the environment through greenhouse gas emissions, energy consumption, resource depletion and waste generation. In response, earth is increasingly recognised as a sustainable construction material due to its abundance, local availability, non-toxicity, and excellent hygroscopic properties. While research on earthen materials has underscored its environmental and thermo-hygrometric benefits, the recyclability of earth, particularly when stabilised with mineral binders, remains poorly understood. Stabilised soils are frequently regarded as less recyclable, yet their actual reuse potential has been scarcely investigated. This research investigates the recyclability of lime-stabilised soils, focusing on two soil types with different grain sizes. Two recycling methods were evaluated: recycling the stabilised soil without further modification and applying additional stabilisation in each cycle. These methods were tested for three recycling cycles to evaluate the behaviour of the material and performance over time. The change in aggregate size after each cycle, compressive strength and water absorption by capillarity were studied. The study found that additional stabilisation must be applied for recycling without downcycling previously stabilised soils. The results show that if the same amount of lime is applied at each cycle, the strength increases exponentially. Due to the irreversible nature of mineral stabilisation, it is recommended to use cementitious stabilisers only when necessary.

Keywords: Recyclability · Stabilised Soil · Natural Hydraulic Lime · Earthen Constructions · Sustainable Construction Materials

1 Introduction

The construction industry plays an important role in environmental challenges, contributing to greenhouse gas emissions, depletion of natural resources, high energy consumption, and waste generation [1–3]. To tackle these problems, earthen materials are gaining attention as an environmentally friendly alternative. Their wide availability, local sourcing potential, non-toxic nature and superior ability to regulate internal moisture make them an ideal and sustainable choice for building applications [4]. The recyclability of earth as a construction material is frequently highlighted as a key advantage driving its resurgence in sustainable construction and circular economy strategies. The reversibility of clayey soil structures when interacting with water [5], i.e. the ability to return

M. Posani and S. Singh (Eds.): RYS 2025, RILEM Bookseries 67, pp. 545–559, 2026.
https://doi.org/10.1007/978-3-032-14652-6_39

to its initial state with minimal energy input, positions earth as a uniquely sustainable material. However, the introduction of stabilisers such as cement and lime complicates this narrative, sparking a debate on environmental trade-offs and the recyclability of the earth at the end of its life.

In stabilised soils, crystallisation structures are formed when cementing materials are deposited in contact with particles. These structures are characterised by high rigidity, which makes the material more resistant and less deformable [6]. However, unlike clay alone, they are irreversible when in contact with water [5]. A central critique comes from the latest RILEM state-of-the-art report [7], which states that even low-percentage stabilisation of earthen walls can significantly increase energy consumption and limit recyclability. This view aligns with Van Damme and Houben [8], who argue against cement stabilisation due to its unfavourable mechanical benefit-to-environmental cost ratio. While cement stabilisers provide structural improvements, these gains often do not justify the environmental costs, particularly when considering earth's potential as a low-energy, reusable material. Contrary perspectives, such as those presented by Pacheco-Torgal and Jalali [9], suggest that even stabilised soils retain a degree of recyclability. They assert that lime- and cement-stabilised earth can still be reused in construction, mitigating concerns about waste generation. Subsequent discussions, including those by Marsh et al. [10] and Morel et al. [11], emphasised the need for a case-by-case evaluation, considering factors like climatic conditions, soil type, and stabilisation method. These discussions highlight the complex interplay between mechanical performance, environmental impact, and recyclability, with some studies advocating for minimal stabilisation or alternative methods such as biopolymers to balance these priorities [12–15]. From a life-cycle perspective, Fernandes et al. [16] considered adding a small percentage of hydraulic lime (one per cent in volume) to recycled soils to produce new earth blocks. The mentioned amount of lime was supplied by a Portuguese manufacturer based on its experience in recycling defective dried blocks or dismantled walls, showing a viable path for the reuse of stabilised earth. However, experimental studies exploring the functional performance of recycled stabilised earth remain scarce.

To the best of the authors' knowledge, Bruno et al. [15] provided the first analysis on the recyclability of stabilised soils, revealing that non-stabilised soils and soils stabilised with biopolymers retained their mechanical properties after recycling, while soils stabilised with cement showed significant reductions in stiffness and strength. Other studies followed, but all focused on the recyclability of biopolymer-modified soils [13, 14]. Yet, most of the earthen building stock and commercially available solutions make use of cementitious binders. What happens at the end of their life cycle? What properties does previously stabilised soil have?

This paper seeks to address this gap by analysing the recyclability potential of stabilised soils for earthen construction, i.e. the reuse of recycled soils from compressed earth blocks (CEBs) for the same structural function without recycling in non-structural applications. The experimental design involves three recycling cycles of soil derived from CEBs stabilised with 5% by volume of natural hydraulic lime (NHL), produced by a Portuguese company, of two different types, one fine-grained and one coarse-grained. The campaign consists of two experiments: (i) Recycling of stabilised soils without further modification (Scenario 1); (ii) Recycling of stabilised soils by applying additional stabilisation in each cycle (Scenario 2). The recycling potential is assessed by analysing the size distribution of aggregates, compressive strength tests and evaluating the capillarity coefficient over three cycles. The study aims to clarify the end-of-life structural performance of stabilised soils and contribute to a better understanding of their role in sustainable and circular construction. By providing new experimental data, this study informs the ongoing debate and supports the development of more realistic life cycle assessment (LCA) scenarios for earthen materials.

2 Materials and Methods

2.1 Recycled Soil Characteristics

This study evaluates the recycling behaviour of soils stabilised with NHL. The soils were obtained from CEBs produced by a Portuguese company and previously characterised [17]. The original soil, extracted in Beja (Alentejo, southern Portugal), was sieved to two maximum particle sizes following the company's specifications: ≤2 mm (R1) and ≤12.5 mm (R2). Both soils were stabilised with 5% NHL by volume and compacted according to the company's practice. After testing, part of the blocks was preserved, and between 20 and 22 months later, the remaining material was gently crushed with a metal rolling pin to recover the recycled soils.

The fine fraction presented a liquid limit of 29.5% and a plasticity index of 11% (NP 143 [18]), classifying the soil as an inorganic clay of medium plasticity according to the Unified Soil Classification System. The particle density was 2.71 g/cm^3 (NP 83 [19]). Mineralogical analysis confirmed the presence of quartz and non-swelling kaolinitic clay minerals, such as muscovite and clinochlore [20]. The optimum water content, determined by the Proctor test (LNEC E 197 [21]) on the original soil, was 12.2% for R1 and 12% for R2. The Proctor test was not repeated on the recycled soils due to limited material availability.

2.2 Natural Hydraulic Lime and Stabilisation Mechanisms

NHL is a naturally occurring mixture of calcium carbonate (limestone), clay and other silicate/aluminate minerals. For this study, a commercial NHL5, i.e. an eminently hydraulic lime, was used to apply additional stabilisation to part of the analysed samples. The same NHL was used by the manufacturer for the stabilisation of the CEBs from which the soil was recycled.

Stabilisation with NHL improves the workability of soil and the rigidity of compacted mixtures and water absorption [6, 22]. The development of strength is dependent upon the formation of crystalline inter- and intra-aggregate bonds, and the secondary contribution of desiccated gelatinous and amorphous products [6]. After long-term curing, adequate stabilisation of soil with lime makes the material conform to elastic principles, evidencing a near-linear stress-strain relationship with brittle failures. The stabilisation mechanism is based upon hydration and carbonation reactions. As soon as lime and water are mixed with soil, free calcium ions are released (cation exchange), leading to an increase in the pH that encourages the solubility of clay minerals, mainly silica. The hydration of these compounds produces calcium silicate hydrates (C–S–H) or calcium aluminate hydrates (C–A–H). Alongside hydraulic components, large $Ca(OH)_2$ (portlandite) crystals are forming simultaneously, which will eventually react with available CO_2 to form calcite, $CaCO_3$.

When lime is used for stabilisation, reactive clay is attacked, and the natural cohesion is lowered by the addition of water. A change in texture is also displayed in the soil, tending to approach that of coarse silt [6]. Structural arrangements of reaction products and soil elements contribute to the rigidity of stabilised earthen mixtures. Reduction in permeability and porosity is due to infilling of cracks and voids with precipitated reaction products. Prompt and adequate compaction is essential for establishing points of contact between structural units and for early strength development.

2.3 Design of the Mixtures

Four mixtures were designed for the experiments. M1 and M2 with the finer-grained soil (R1), M3 and M4 with the coarser-grained soil (R2). While M1 and M3 involve soil reuse without further modification, M2 and M4 involve applying additional stabilisation in each cycle (5v.% NHL). A target density, reproducing that of the original CEBs (R1 = 1720 kg/m^3, and R2 = 1840 kg/m^3), was used for the new specimens. The material quantities (expressed in g) are shown in Table 1 and refer to the quantity required to produce a cylindrical specimen of 7 cm in diameter and 14 cm in height.

Table 1. Design of the mixtures.

	Target density [g/cm^3]	Soil		Water	Water	Water	NHL	
M1	1.72	100%	927 g	R1 type	12.2%	113 g	–	
M2							5v.%	35 g
M3	1.84	100%	991 g	R2 type	12%	119 g	–	
M4							5v.%	34 g

2.4 Experimental Protocol

Before use, the recycled soils were dried in the oven at 105 °C. After cooling, they were gently crushed with a heavy metallic rolling pin to avoid lumps. The aggregate size distribution was recorded before each recycling cycle to control granulometry modifications. After that, every soil was again mixed and prepared for the sample making.

According to the designed mixtures presented in Table 1, when lime was used, the ingredients were mixed dry first. To limit the loss of mass, all operations were conducted in a tray, so that all the material involved could be collected. Water was added gradually, and the ingredients were thoroughly mixed. The mixing process was executed manually. The material was poured into a cylindrical mould and compacted until the desired density was reached using a hydraulic jack within a steel frame. Eight cylindrical samples with a diameter of 7 cm and a height of 14 cm were prepared for each mixture. After demoulding, the specimens were cured at room temperature (lab conditions, approximately 20 °C and 50% relative humidity) for 12 days and then dried in the oven at 40 °C for 48 h to remove any moisture excess. Dry samples were weighed to measure the bulk density. Five of them were used for compression tests. The remaining three specimens were used for capillarity tests. All the material is saved after every test, dried again in the oven for 24 h, stored in plastic bags for one week, and used again for another cycle.

Aggregate Size Distribution. Analysis of the aggregate size distribution of the recycled soils was performed before each cycle by dry sieve analysis to monitor anomalies due to material crushing after every cycle. For each mixture, 1.5 kg of recycled soil was passed through nine sieves using a machine that vibrates them for five minutes. Sieves with the following apertures were used: 12.5 mm, 8 mm, 4 mm, 2 mm, 1 mm, 0.5 mm, 0.25 mm, 0.125 mm, 0.063 mm. The material that passed through the latter sieve was collected in the final closed container. The amount of material retained by each sieve was recorded. With these data, the aggregate size curves were plotted.

For this study, which is practical, the overall analysis of the fraction of gravel (comprising particles of size > 2 mm), sand (comprising particles of size between 0.063 mm and 2 mm), and fines (comprising all particles < 0.063 mm) is also included and is shown using bar graphs.

Compressive Strength Test. The compressive strength test was conducted on dry cylindrical specimens with an MTS machine. The test was performed under displacement control at a speed of 0.5 mm/min. For the calculation of the stress-strain curves, the axial deformation was determined as the change in specimen height from the initial value. It was therefore assumed that the change in height of the specimen is equal to the displacement recorded by the transducer installed on the loading system. From the test, the compressive strength at the peak and the apparent elastic modulus (tangent stiffness) are deduced (Fig. 1).

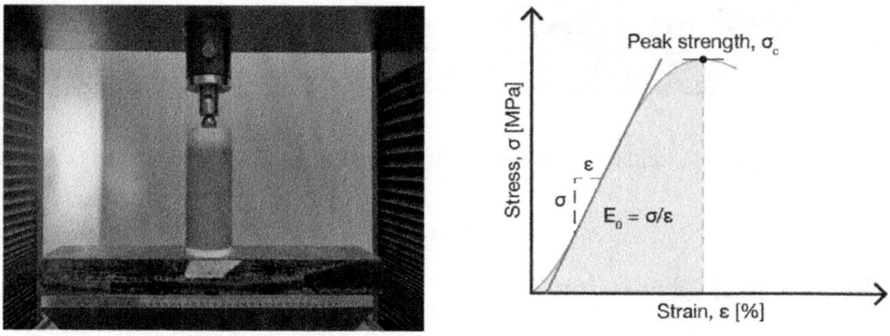

(a) (b)

Fig. 1. Compression test setup: (a) a specimen during testing, and (b) parameters calculation.

Peak compressive strength σ_c, expressed in MPa, was calculated by dividing the peak load F (N) by the contact area A (mm^2). The apparent elastic modulus (or tangent stiffness) E_0, expressed in MPa, was calculated as the slope of the straight line of the force-displacement curves over the stress range 0.1–0.2 MPa for un-restabilised M1 and

M3 samples, and 0.4–0.5 MPa for re-stabilised M2 and M4 samples. The stress limits were established to eliminate initial settlement effects and to reduce the occurrence of plastic deformations [23]. Due to the assumptions being made, the data on apparent elastic modulus should be considered accurate in relative rather than absolute terms [23]. Five samples per mixture were analysed. The mean value and standard error are presented in the results.

Statistical analysis (t-test) was conducted to verify the significance of the observed variation across cycles (Cycle I vs. Cycle II, Cycle II vs. Cycle III), considering, for all tests, $\alpha = 0.05$.

Capillarity Test. The capillarity test was conducted by adapting the experimental protocol from LNEC E 393 [24]. The dry specimens were placed in a transparent plastic box, on a rubber perforated sheet, in contact with a thin layer of water (5 ± 1 mm). The capillary rise of water was observed at regular time intervals between 5 min and 24 h. The centimetres of capillary water rise and the change in mass of the specimens were recorded at these intervals. To obtain the capillarity curves, the mass of water absorbed by the absorption area is plotted against the square root of time. The average of the curves of the three tested samples is presented.

The capillarity coefficient C_b is then calculated as the slope of the straight line of each curve between 10 min and 5 h (Fig. 2). When this was not possible, it was adjusted to the last reliable measurement. The mean value and standard error are presented.

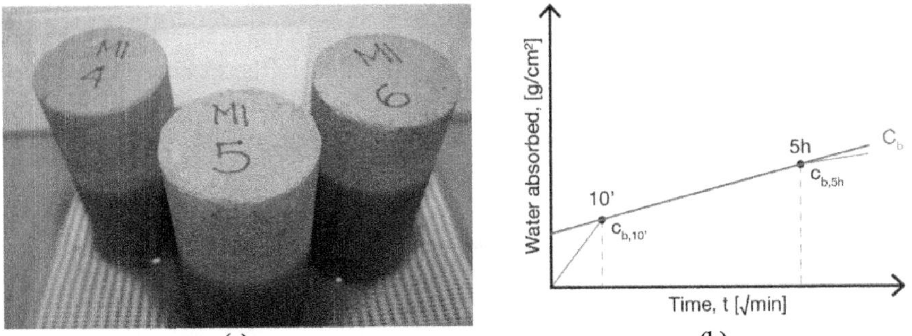

(a) (b)

Fig. 2. Capillarity test setup: (a) some specimens during testing, and (b) parameters calculation.

3 Results and Discussion

3.1 Aggregate Size Distribution

The results of the dry sieve analysis are shown in Fig. 3.

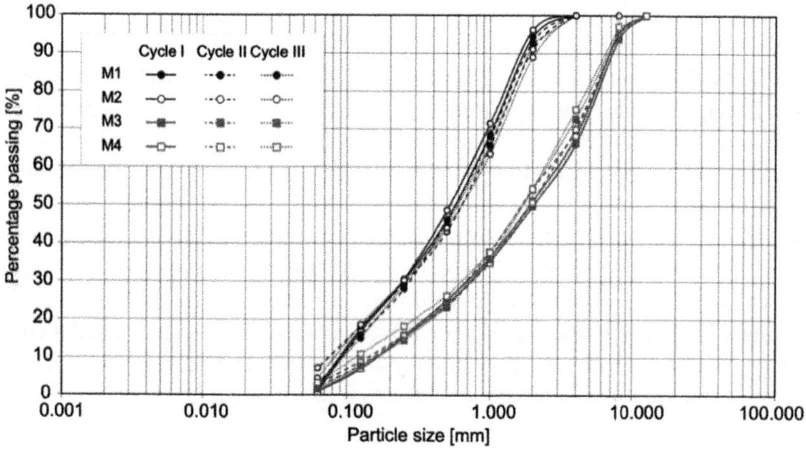

Fig. 3. Granulometric curves of the soils before every recycling cycle.

After three cycles, in the case of both soils, the curves are still well graded, and no anomalies are evident. However, the analysis of the distribution of the soil-aggregate fractions (gravel-size, sand-size and fines) reveals a certain tendency for the non-re-stabilised mixtures (M1 and M3) to increase the sandy fraction, while the re-stabilised mixtures (M2 and M4) tend to increase the fine fraction over the cycles (Fig. 4).

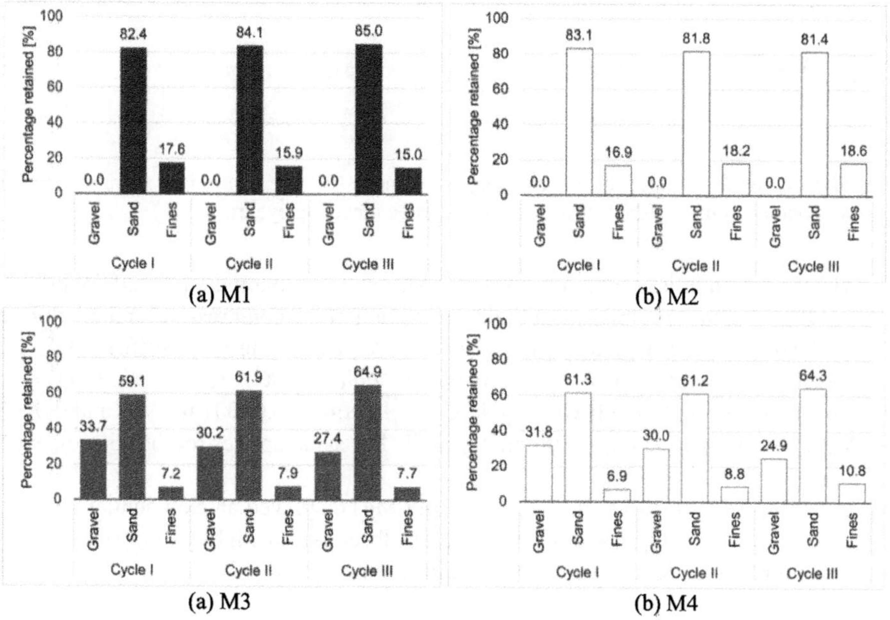

Fig. 4. Distribution of soil-aggregate fractions across the three recycling cycles for (a) M1, (b) M2, (c) M3, and (d) M4 mixtures. The following fractions were considered: gravel >2 mm, sand 2–0.063 mm, fines <0.063 mm.

In the non-re-stabilised materials (M1 and M3), the tendency to increase the sandy fraction is attributed to particle aggregation (Fig. 4(a)) and, in the case of more coarse soils (M3 and M4) also to the breaking or disintegration of gravel-sized aggregates into smaller aggregates (Fig. 4(c) e (d)). In the re-stabilised materials (M2 and M4), the observed increase in the fine fraction is instead related to pulverisation [6], due to the breaking of brittle cementitious bonds and the possible presence of unreacted lime (Fig. 4(b) and (d)).

This trend contrasts with the results of studies on soils stabilised with biopolymers, where fines generally decrease due to particle agglomeration caused by the presence of the binder [14]. Cementitious bonds are different from the hydrogel bonds created by biopolymers [25]. The behaviour observed for lime-stabilised soils is attributed to the breaking of the brittle hydraulic lime bonds under mechanical action. However, it is not possible to compare this result with any literature due to the lack of studies.

3.2 Compression Test

Figure 5 illustrates the results of the compression tests expressed in terms of mean peak compressive strength and apparent elastic modulus.

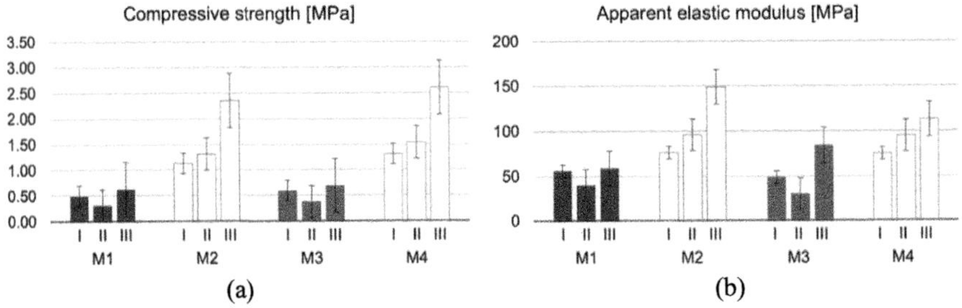

Fig. 5. Compression test results: (a) average peak compressive strength and (b) average apparent elastic modulus of the materials tested across the three recycling cycles.

The non-re-stabilised materials (M1 and M3) show an anomalous trend. With both R1 and R2 soils, the specimens exhibited a reduction of the compressive strength during Cycle II with respect to Cycle I, followed by a recovery of strength and stiffness in Cycle III. The reduction was deemed significant with $p\text{-}value = 0.0001$ for M1, and 0.0017 for M3 when comparing Cycle II with Cycle I, and $p\text{-}value = 0.00001$ for M1, and 0.0003 for M3 when comparing Cycle III with Cycle II. The same applies for the stiffness with $p\text{-}values << 0.5$.

In contrast, re-stabilised materials (M2 and M4) displayed an exponential increase in the compressive strength and stiffness over the three cycles, highlighting a cumulative strengthening effect due to progressive stabilisation (Fig. 6).

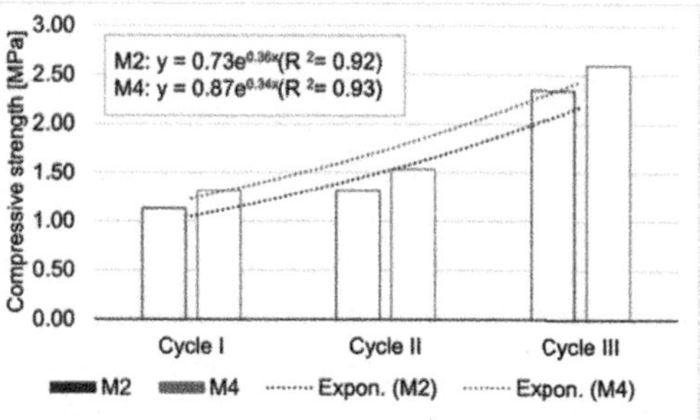

Fig. 6. Exponential increase of the compressive strength across three recycling cycles when additional stabilisation is applied.

Experimental observations revealed the presence of cracks in the samples for Cycle II (Fig. 7), which may explain the observed decrease in mechanical performance.

(a) (b)

Fig. 7. (a) M1 samples before starting the Cycle II tests, and (b) details on one of the highlighted fractures.

The possibility of these defects being related to the manufacturing process was excluded, both because the process was executed by the same person and in the same conditions, but also because the problem consistently occurred only during Cycle II and for both soils involved.

The results are not comparable with adequate literature. In Bruno et al. [15], the recycled soil is stabilised with cement. With each cycle (without re-stabilisation), the mechanical properties of the test specimens gradually deteriorate due to the breaking of the cement bond [15]. In the present study, the stabilising agent used is NHL, and there is a recovery that is not easy to interpret since a consistent reduction in both compressive strength and stiffness was also expected.

Some considerations can be made about the nature of the bonds. The primary difference between cement and NHL is that, despite similar hydration mechanisms and C–S–H bond formation, the lime reacts more slowly at the early stage, but strength develops over time, also because of carbonation [22]. In [22], it is reported that curing times between 7 and 60 weeks might be required for lime-stabilisation to reach maturity. The same does not apply to cement, as its reactivity stops after initial hydration. This is indeed demonstrated by the poor strength, stiffness and failure energy of cement-stabilised recycled materials presented in Bruno et al. [15]. Therefore, the strength and stiffness recovery observed in this experiment during Cycle II may be due to the long-term lime effect [6].

As for the second scenario, to the best of the author's knowledge, the progressive stabilisation at each cycle has never been practised. In absolute terms, materials with added stabiliser demonstrate markedly higher capacity compared to their non-re-stabilised counterparts, as detailed in the Supplementary Material at the end of this paper, in which the experimental stress-strain curves are provided. Results suggest that this is the only way to recycle stabilised soil without the occurrence of unforeseeable situations. The way consists of applying additional stabilisation at each cycle. Yet, environmental concerns are claimed.

3.3 Capillarity Test

The graphs of Fig. 8 illustrate the results of the capillarity tests.

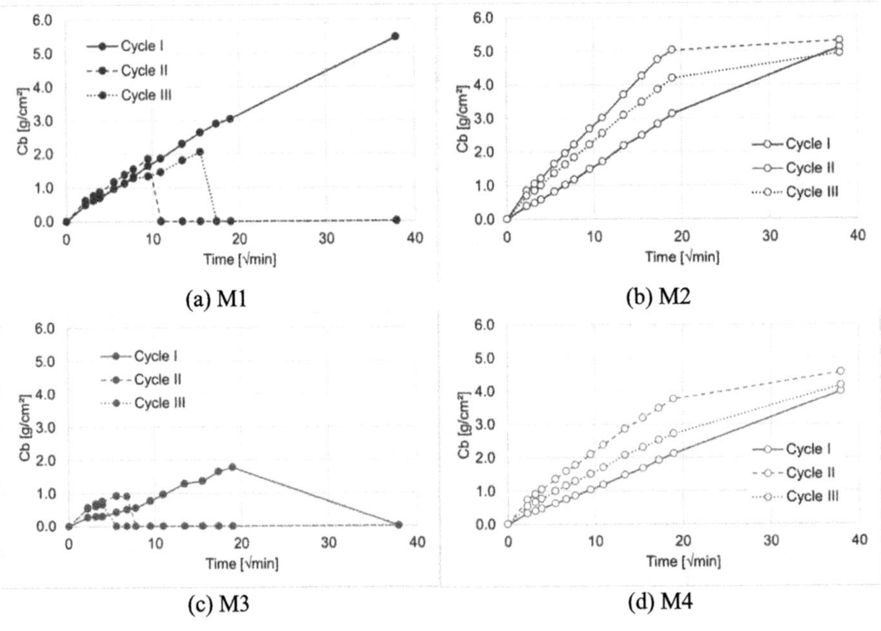

Fig. 8. Capillarity curves across the three recycling cycles for (a) M1, (b) M2, (c) M3, and (d) M4 mixtures.

The results indicate that non-re-stabilised materials (M1 and M3) are highly susceptible to water. In contrast, re-stabilised materials (M2 and M4) exhibit robust water resistance across all cycles. While this second result was expected, given the water stability of the crystalline structures of the cementitious bonds [5], the first result allows considerations to be made about the state of the soil at the time of the first recycling cycle.

The M1 and M3 curves show that non-re-stabilised materials withstand the test, yet with some difficulties, only during the first reuse. M1 samples from Cycle I resist water for 24 h, but subsequent cycles show marked vulnerability (Fig. 8(a)). M3 samples perform worse, failing within 6 h in Cycle I and progressively deteriorating in subsequent cycles (Fig. 8(c)). When additional stabilisation is applied to recycled soils, the materials exhibit no discomfort and provide consistent and reliable test results. However, it should be noted that for both M2 and M4, the average curves show the same trend: in both cases, the highest rate of water uptake occurs during Cycle II and the lowest during Cycle I, as suggested by the slope of the curves after 10 min (Fig. 8(b) and (d)). The first 10 min are usually excluded from the discussion, as the beginning of the test, the rate of water absorption is always more pronounced [26].

The image in Fig. 9 shows the capillarity coefficients calculated as the slope of the straight line in the time interval from 10 min to 5 h.

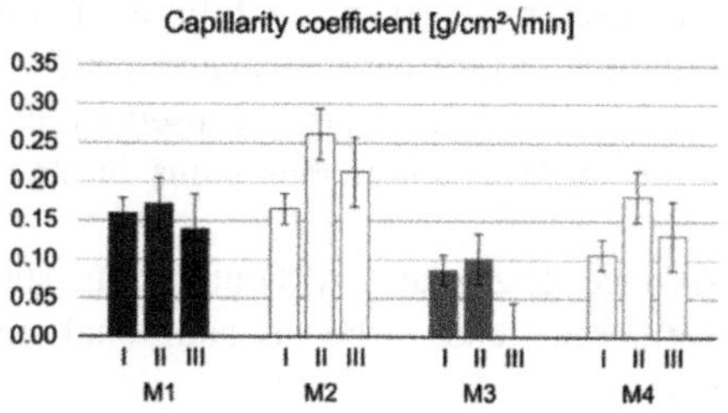

Fig. 9. Capillarity coefficients and capillarity water rise velocity.

The results allow for a more immediate interpretation and verify the hypothesis that finer-grained soil materials (M1 and M2) tend to retain more water than coarse-grained soils (M3 and M4) [15]. Tough, when no further stabilisation is applied, coarser soil samples (M3) were more vulnerable to water than those made from finer soil (M1).

4 Conclusions

Although the environmental and thermo-hygrometric advantages of earthen materials are well established, the recyclability of soils stabilised with mineral binders remains insufficiently understood. Stabilised soils are frequently regarded as less recyclable, yet their actual reuse potential has been scarcely investigated.

This study explores the recyclability of stabilised soils obtained from compressed earth blocks (CEBs) stabilised with natural hydraulic lime (NHL). The blocks, produced by a Portuguese company for a previous study on the influence of particle size, provided two soil types with distinct maximum grain sizes (fine and coarse). After drying, crushing, and rewetting, two recycling strategies were assessed: (i) Recycling of stabilised soils without further modification (Scenario 1); (ii) Recycling of stabilised soils by applying additional stabilisation in each cycle (Scenario 2). Both processes were repeated over three consecutive recycling cycles to evaluate the evolution of material performance. For each cycle, grain size distribution, compressive strength, and capillary water absorption were determined.

The first scenario revealed significant limitations. For both recycled soils (finer R1 and coarser R2), a transient reduction in peak compressive strength and stiffness was observed, primarily attributed to defects such as cracks identified during specimen characterisation. The deterioration of stabilised materials without additional stabilisation was further confirmed by capillarity tests, which indicated a marked loss of water resistance. In contrast, the second scenario produced more consistent results across both soils. The progressive addition of lime in each cycle led to an exponential increase in peak compressive strength and stiffness, while the specimens remained stable under water exposure. Overall, the main outcomes can be summarised as follows:

- Recycling lime-stabilised soils promotes particle aggregation, reflected in an increase in the sand fraction. When additional stabilisation is applied, however, the fine fraction increases at each cycle, suggesting bond breakage and disaggregation induced by the enhanced brittleness of the material. In all recycling cycles, the coarse fraction consistently decreases (notably in cases M3 and M4 of R2 soil).
- Without additional stabiliser, recycled soils exhibit temporary or permanent reductions in compressive strength and stiffness, attributed to defects generated by mechanical disaggregation and irreversible changes caused by cementation. In contrast, progressive stabilisation leads to continuous gains in strength and stiffness over successive cycles.
- Similarly, water resistance deteriorates without further stabilisation, in line with the mechanical performance trends, whereas additional stabilisation preserves or improves this property.

These findings indicate that previously stabilised soils cannot be reliably recycled without further treatment, as their behaviour is unpredictable. Conversely, the application of additional stabilisation at each cycle proved effective, enabling the material to be reused for the same purpose without down-cycling. Nonetheless, the environmental implications of repeated stabilisation must be carefully considered. The results, therefore, underline the importance of evaluating the necessity of stabilisation in construction practice, given its irreversible impact on soil recyclability.

4.1 Limitations and Future Studies

In this study, the same moulding water content (12–12.2%) was used for all cycles, without adjustments for hydration demand or granulometry changes. This simplification may have affected the achieved dry density and partly contributed to variability in the mechanical performance. Future studies should correct the moisture setting at each cycle to more accurately capture the compaction state of the material.

The formation of hydrated and carbonated phases, as well as the potential reactivation of residual lime and resumption of hydration upon re-compaction, were not assessed. These mechanisms could influence the mechanical and durability properties of recycled materials but may also underlie the differences between lime and other stabilisers reported in the literature. Future work should therefore include detailed microstructural and mineralogical characterisation (e.g., XRD, TGA, SEM) to clarify the role of these processes in the recyclability of stabilised soils.

5 Supplementary Material

5.1 Experimental Stress-Strain Curves from Compression Tests of M1–4 Mixtures Across the Three Recycling Cycles

The following figure shows the stress-strain curves of the compression tests performed on the materials analysed in this study.

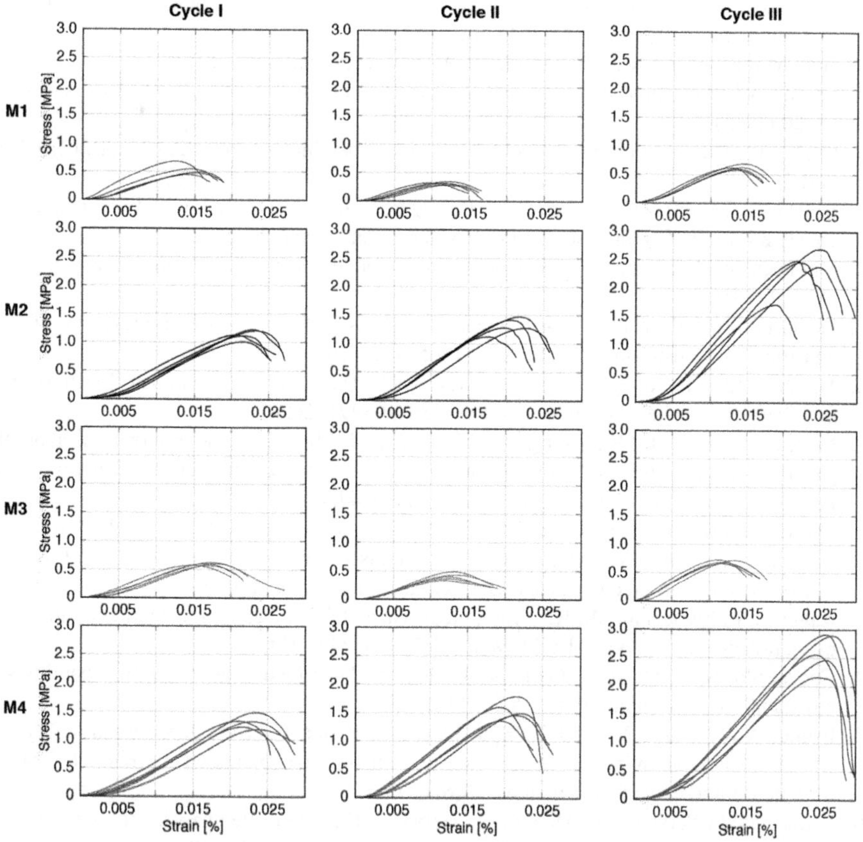

The curves illustrate the progressive improvement in strength and stiffness of the re-stabilised materials (M2 and M4) across successive recycling cycles compared to the non-re-stabilised ones (M1–M3).

Acknowledgments. This work was financed by national funds through FCT – Foundation for Science and Technology, under grant agreement UIBD/150874/2021 attributed to the first author. The work was also partly financed by FCT/MCTES through national funds (PIDDAC) under the R&D Unit Institute for Sustainability and Innovation in Structural Engineering (ISISE), under reference UIDB/04029/2020 (https://doi.org/10.54499/UIDB/04029/2020), and under the Associate Laboratory Advanced Production and Intelligent Systems ARISE under reference LA/P/0112/2020.

A special thanks to César Gonçalves for the valuable support during the lab work.

References

1. Huang, L., Krigsvoll, G., Johansen, F., Liu, Y., Zhang, X.: Carbon emission of global construction sector. Renew. Sustain. Energy Rev. **81**, 1906–1916 (2018)
2. Al-Numan, B.S.O.: Construction industry role in natural resources depletion and how to reduce it. In: *Natural Resources Deterioration in MENA Region: Land Degradation, Soil Erosion, and Desertification*. Cham: Springer International Publishing; 2024. p. 93–109
3. European Environment Agency. Buildings and construction. https://www.eea.europa.eu/en/topics/in-depth/buildings-and-construction
4. Hall, M.R., Lindsay, R., Krayenhoff, M.: Modern earth buildings (2012)
5. Osipov, V.I.: Structural bonds of and properties of clays. Bull. Int. Assoc. Eng. Geol. **12**, 13–20 (1975)
6. Croft, J.B.: The structures of soils stabilized with cementitious agents. Eng. Geol. **2**(2), 63–80 (1967)
7. Fabbri, A., Morel, J.-C., Aubert, J.E., Bui, Q.B., Gallipoli, D., Reddy, B.V.V.: Testing and characterisation of earth-based building materials and elements (2022)
8. Van Damme, H., Houben, H.: Earth concrete. Stabilization revisited. Cem. Concr. Res. **114**, 90–102 (2018)
9. Pacheco-Torgal, F., Jalali, S.: Earth construction: Lessons from the past for future eco-efficient construction. Constr. Build. Mater. **29**, 512–519 (2012)
10. Marsh, A.T.M., Heath, A., Walker, P., Reddy, B.V.V., Habert, G.: Discussion of "Earth concrete Stabilization revisited." Cem. Concr. Res. **130**, 105991 (2020)
11. Morel, J.-C., Aubert, J.E., Millogo, Y., Hamard, E., Fabbri, A.: Some observations about the paper "Earth construction: lessons from the past for future eco-efficient construction" by F. Pacheco-Torgal and S. Jalali. Constr. Build. Mater. **44**, 419–421 (2013)
12. Muguda, S., Hughes, P.N., Augarde, C.E., Perlot, C., Gallipoli, D., Bruno, A.W.: Geotechnical characterisation of recycled biopolymer-stabilised earthen materials. In: XVII European Conference on Soil Mechanics and Geotechnical Engineering (2019)
13. Muguda, S., Wyndham, M.: Recycling biopolymer stabilised earthen construction materials using β-mannanase. Constr. Build. Mater. **457**, 139334 (2024)
14. Cuccurullo, A., Muguda, S., Gallipoli, D., Perlot, C. Evaluation of recycling capabilities of bio-stabilised earth building materials (2022)
15. Bruno, A.W., Scott, B., D'Offay-Mancienne, Y., Perlot, C.: Recyclability, durability and water vapour adsorption of unstabilised and stabilised compressed earth bricks. Mater. Structur./Materiaux et Constructions. 53(6) (2020)
16. Fernandes, J., Peixoto, M., Mateus, R., Gervásio, H.: Life cycle analysis of environmental impacts of earthen materials in the Portuguese context: rammed earth and compressed earth blocks. J. Clean. Prod. (2019)
17. Turco, C., Paula Junior, A., Jacinto, C., Fernandes, J., Teixeira, E., Mateus, R.: Influence of particle size on compressed earth blocks properties and strategies for enhanced performance. Appl. Sci. (Switzerland) **14**(5) (2024)
18. Instituto Português da Qualidade (IPQ). NP 143: Determinação dos Limites de Consistência (1969)
19. Instituto Português da Qualidade (IPQ). NP 83: Determinação Da Densidade das Partículas (1965)
20. Turco, C., Abedi, M., Teixeira, E., Mateus, R.: Thermophysical properties of compressed earth blocks incorporating natural materials. Energies (Basel) **17**(9) (2024)
21. Laboratório Nacional de Engenharia Civil (LNEC). LNEC E 197: Solo – Ensaio de compactação (1966)

22. Ciancio, D., Beckett, C.T.S., Carraro, J.A.H.: Optimum lime content identification for lime-stabilised rammed earth. Constr. Build. Mater. **53**, 59–65 (2014)
23. Bruno, A.W., et al.: Synergic effect of hydrated lime and guar gum stabilisation on the mechanical, thermal and hygroscopic behaviour of a Ligurian earth material. Constr. Build. Mater. **439**, 137258 (2024)
24. Laboratório Nacional de Engenharia Civil (LNEC). LNEC E 393: Betões – Determinação da absorção de água por capilaridade (1993)
25. Muguda, S., et al.: Mechanical properties of biopolymer-stabilised soil-based construction materials. Geotech. Lett. **7**(4), 309–314 (2017)
26. Boulmaali, M., Belhamri, A.: Investigation of thermal inertia and hydric properties of an eco-building material: compressed stabilized earth blocks. Heat Mass Transfer / Wärme- und Stoffübertragung. **59**(4), 713–727 (2022)

Regenerative Retrofitting of Built Environment Assets via Ecologically Active Soils

Alejandro Jiménez Rios[1(✉)], Juliana Calabria-Holley[1], and Francisco Javier Castilla Pascual[2]

[1] Department of Architecture and Civil Engineering and Centre for Regenerative Design and Engineering for a Net Positive World (RENEW), University of Bath, Bath, UK
ajr225@bath.ac.uk

[2] Department of Applied Mechanics and Project Engineering, University of Castilla – La Mancha, Albacete, Spain

Abstract. It is estimated that 80% of the buildings that will exist in 2050 have already been constructed. However, many of these buildings perform poorly in terms of energy efficiency and fail to comply with net-positive solutions. Natural soil is a low-embodied energy, recyclable, and healthy material that has been used for millennia in our buildings, following vernacular traditions and sustainable practices. When enhanced to be ecologically active, soil can support plant growth, offering a path to retrofitting buildings that enhance thermal insulation, expand green urban surfaces, improve air quality, and elevate both aesthetics and occupant well-being. Soils are considered ecologically active if they can provide the necessary humidity level and nutrition conditions for plant germination and long-term viability. This paper presents the rationale and the objectives of the Regenerative Retrofitting Via Ecologically Active Soil Structures (Reeco-Soil) project, which aims to establish a comprehensive understanding of the state-of-the-art and regenerative potential of retrofitting built environment assets using ecologically active soil structures. The study employs a methodology based on PRISMA guidelines. Preliminary results are presented according to keywords clusters and in terms of year publication distribution. Approximately 70% published within the last decade, indicating a strong and current evidence base. Ultimately, this project aims to generate a pathway to enable a flourishing relationship between humans and the living world by introducing a regenerative and innovative approach to building retrofitting.

Keywords: Regenerative Design · Ecologically Active Soils · Regenerative Building Retrofitting · Green Urban Surfaces · Aerial Additive Manufacturing

1 Introduction

1.1 Rationale

The global construction sector faces a critical challenge: how to mitigate the environmental footprint of our built environment while adapting to climate change and avoiding biodiversity loss. It is estimated that approximately 80% of the buildings that will exist

M. Posani and S. Singh (Eds.): RYS 2025, RILEM Bookseries 67, pp. 560–570, 2026.
https://doi.org/10.1007/978-3-032-14652-6_40

in 2050 have already been constructed [1]. Unfortunately, most of these buildings perform poorly in terms of energy efficiency and environmental integration, and they fall short of delivering net-positive outcomes. Net-positive systems actively contribute to the regeneration of ecosystems and human well-being, rather than merely preventing harm, ultimately renewing humans unity with nature [2]. This implies that the window of opportunity for shaping the future performance of the built environment lies not predominantly in new builds, but in the adaptation, retrofitting, and reuse of existing assets. In comparison with the demolition and construction of a new building, retrofitting and reusing an existing one could lead to up to 70% reduction of the upfront embodied carbon in construction [3].

Conventional retrofitting strategies often rely on energy-intensive materials and processes, raising concerns about their long-term sustainability [4]. There is an urgent need to develop regenerative retrofitting approaches. Such approaches need to be holistic and go beyond carbon neutrality to actively restore ecological systems and support human health and comfort, allowing humans and nature to survive, thrive, and co-evolve [5].

In this context, soil emerges as a powerful yet underexplored ally. Natural soil is a low-embodied energy, healthy, and recyclable material that has been used for millennia in architecture and construction across cultures [6–8]. Rooted in vernacular traditions, earthen construction techniques embody principles of circularity and sustainability [9]. Recent innovations suggest that soil, when designed to be ecologically active, can be reimagined as a dynamic material capable of supporting plant growth [10]. Ecologically active soils which are engineered to provide optimal humidity and nutrient conditions, can form living building surfaces that sequester carbon, offer thermal insulation, foster biodiversity, and enhance aesthetics and occupant well-being.

Despite growing interest in green infrastructure [11] and nature-based solutions [12], a central challenge persists: how can we integrate green areas into existing buildings in a regenerative and innovative way? Recent advances on aerial additive manufacturing [13] and 3D printing [14], could foster the potential of activated soil-based retrofitting interventions. A systematic, interdisciplinary understanding of how ecologically active soils can be applied to the retrofitting of the built environment is currently lacking.

1.2 Objectives

This paper introduces and frames the objectives of the Regenerative Retrofitting Via Ecologically Active Soil Structures (Reeco-Soil) project, conducted in collaboration between the University of Bath, UK, and Universidad Castilla La Mancha, Spain. The primary goal of this project is to investigate the state-of-the-art and regenerative potential of using ecologically active soil to retrofit existing buildings. Specifically, the study addresses the following key question:

- What are the current materials (soil types), methods (application techniques), and biological components (plant species) explored in the literature for integrating ecologically active soils into building retrofitting strategies?

To answer this question, the article presents the first stages of a systematic literature review (SLR) conducted in accordance with PRISMA guidelines [15]. By synthesising existing research, the Reeco-Soil project aims to identify knowledge gaps, technological potentials, and future directions for advancing regenerative retrofitting strategies of built environment assets.

2 Methods

This systematic literature review follows the Preferred Reporting Items for Systematic reviews and Meta-Analyses 2020 guidelines [15] and is reported in alignment with the PRISMA-P [16] and PRISMA-S [17] checklists. The review focuses on peer-reviewed academic literature and relevant grey literature describing real-world applications or prototypes. The Reeco-Soil project has been registered in the Open Science Framework website [18] and the protocol adopted for this SLR added to the OSF Registries [19].

Two electronic databases were searched using carefully constructed Boolean search strings: Scopus (Elsevier) and Web of Science Core Collection (Clarivate). The search strategy included keywords and wildcards related to soil types (e.g., "bioactiv* soil", "earthen material*", "engineer* soil"), green infrastructure functions (e.g., "plant* growth", "urban green*", "green infrastructur*"), and fabrication techniques (e.g., "3D print*", "extrus*", "spray*", "additive manufactur*", "digital fabricat*"). These queries were refined using filters for publication years (2000–2025), document types (article, conference paper, book chapter), and subject areas (Engineering, Materials Science, Environmental Sciences, and Energy). Only English-language studies were included. Additional real-world project examples will be sourced through design and construction websites such as Dezeen, Designboom, The Architects' Journal, and New Civil Engineering, where case studies of soil-based digital fabrication methods were browsed manually. The search was conducted on 30/04/2025. The full search strategy followed is detailed in the accompanying PRISMA-S checklist [20].

Studies were included if they focused on the application of soil or earthen materials in building retrofitting, the use of soil as a plant-supporting or thermal-regulating substrate, or the development of techniques (such as extrusion or printing) to integrate soil into the building envelope. The review considered both experimental and conceptual works that align with regenerative, sustainable, or ecological design principles. Exclusion criteria included studies unrelated to the built environment, those focused solely on traditional earthen architecture without a retrofit or innovation focus, or those lacking relevance to either materials, vegetation, or application methods. All identified records were exported to EndNote for data management. Duplicate entries were removed using a combination of filters matching on title and manual validation. The PRISMA protocol checklist developed for this SLR can be found elsewhere [21].

The following stages of the review selection process will involve a two-step screening procedure. First, titles, abstracts, and keywords will be reviewed against the eligibility criteria. Second, full texts will be assessed in full for their final inclusion or rejection. Screening will be performed by two independent screeners, with regular quality checks for consistency. Disagreements on the selection process would be resolved by a third reviewer who will ultimately decide on the selection or rejection of the record. Data will then be extracted manually into a structured spreadsheet, capturing variables such as soil type, application method, plant species, region, environmental context, and any reported outcomes related to building performance or ecological impact. Any significant amendments to the protocol will be reported transparently in the final publication of the SLR. Therefore, in this paper we present the methodology adopted to conduct the SLR as well as the initial identification of records and bibliographic mapping. Full results will be presented in a further publication.

3 Results and Discussion

3.1 Identification of Records

The total number of records found were 648 and 332 from Scopus and Web of Science respectively. A Findable, Accessible, Interoperable and Reusable [22] bibliographic database has been created with all the records identified at this stage [23]. After the deduplication was conducted, 240 records were removed and 720 remained (see Fig. 1).

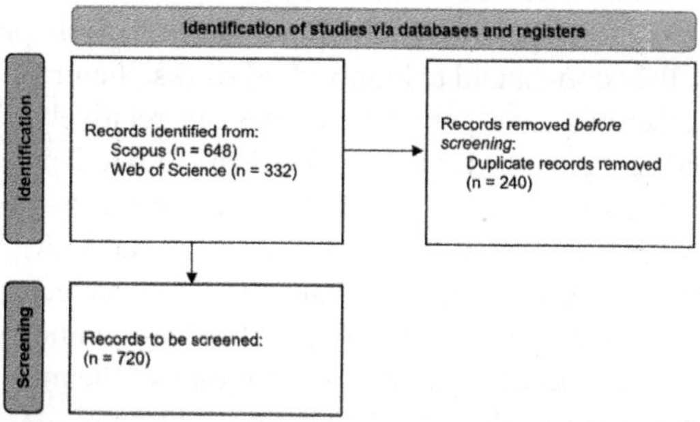

Fig. 1. PRISMA 2020 flow diagram [15] for the first stages of the Reeco-Soil SLR presenting the number of records that were originally identified and screened.

The bar chart presented in Fig. 2 displays the number of records identified per year from 2000 to 2025 following the implementation of the Reeco-Soil SLR search strategy. The volume of publications shows a clear upward trend over time, with minimal output before 2010 and a notable acceleration from 2015 onward. Peak publication years include 2020 (71 records), 2022 (68), and 2023 (64), reflecting increasing scholarly attention to topics related to soil-based retrofitting, additive manufacturing, and ecological materials in the built environment. 37.5% of the identified records after deduplication were published between 2021 and 2025, and 32.6% between 2016 and 2020. This indicates that approximately 70% of the records were published within the last decade, ensuring that the review will reflect the current state of research and recent technological developments in the field. This fact strengthens the relevance and timeliness of the findings expected from the next stages of the review process.

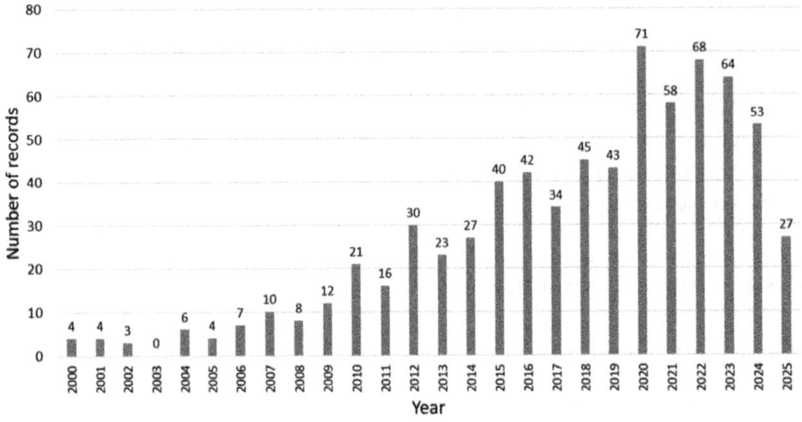

Fig. 2. Number of records found per year after deduplication with zero records found from 2003 and 71 from 2020, the year with more records.

Table 1 presents the most cited articles found after the deduplication of records was conducted. A quick content analysis on the titles of the records presented in Table 1 shows that although they correspond to highly cited works, their relevance to the current SLR is limited. Furthermore, the publication dates are relatively old, being 2017 the latest one and extending up to 2011. On the other hand, Table 2 presents the most cited papers in the last 5 years (2020 until 2025). These more recent works' titles result to be of more relevance to answer the question posed by our SLR. This highlights the worth of the work presented, which reflects a higher interest in recent years in the sector related to the use of natural soil, vegetation in urban environments, and regenerative design approaches for the retrofitting of existing buildings. The most cited authors were Ingham J. M. (701 citations), Dizhur D. (524 citations), Prota A. (482 citations), Lignola G. P. (477 citations), and Griffith M. C. (371 citations).

3.2 Bibliographic Mapping

Figure 3 presents the network visualization of keywords co-occurrence across the records identified after deduplication for the Reeco-Soil SLR generated using VOSviewer. The yellow cluster is centred around terms such as soil, infiltration, stormwater management, and engineering geology, which are directly relevant to the project's focus on ecologically active soil retrofitting and are expected to yield high-value sources. The green cluster includes keywords such as building envelopes, energy conservation, and environmental impact, also of high relevance given the project's emphasis on improving the thermal and ecological performance of buildings. The red cluster represents the largest group of keyword co-occurrences, primarily focusing on retrofitting, fibre-reinforced polymers, strengthening, and coatings. While some of these records (especially those related to surface treatments or materials compatible with soil systems) may be relevant, others focused on concrete construction and structural design may fall outside the scope of the review. Lastly, the blue cluster groups terms related to seismic retrofitting, failure mechanisms, and structural response, which are generally less aligned with the ecological and environmental objectives of the Reeco-Soil project. Despite this, the distribution of keywords suggests a strong representation of highly relevant topics, supporting the expectation that a sufficient number of records will remain after the screening process ensuring the SLR's relevance.

Table 1. Articles with the highest number of citations found after the deduplication stage was completed.

Reference	Title	Year	# Citations
[24]	Externally bonded grids as strengthening and seismic retrofitting materials of masonry panels	2011	256
[25]	Seismic behaviour of ordinary masonry buildings during the 2016 central Italy earthquakes	2016	233
[26]	Performance assessment of basalt FRCM for retrofit applications on masonry	2017	220
[27]	A qualification method for externally bonded Fibre Reinforced Cementitious Matrix (FRCM) strengthening systems	2015	214
[28]	Experimental testing and numerical modelling of masonry wall solution with PCM incorporation: A passive construction solution	2012	194
[29]	Thermal energy storage characteristics of bentonite-based composite PCMs with enhanced thermal conductivity as novel thermal storage building materials	2016	171
[30]	URM walls strengthened with fabric-reinforced cementitious matrix composite subjected to diagonal compression	2014	166
[31]	Experimental evaluation of insulation materials for walls and roofs and their impact on indoor thermal comfort under composite climate	2013	140

The Reeco-Soil project is expected to generate valuable insights into the use of ecologically active soils for the regenerative retrofitting of existing buildings, offering potential benefits such as enhanced thermal performance, increased urban greenery, and improved environmental quality. The information presented in this conference paper, including the systematic review protocol, detailed search strategy, and initial keyword co-occurrence analysis, demonstrates methodological transparency and provides a solid foundation for the work to be done in upcoming stages. By documenting the review process and presenting preliminary findings, the paper supports reproducibility, helps refine future research directions, and contributes to the knowledge base necessary for advancing sustainable and low-carbon building retrofitting solutions.

Table 2. Articles from 2020 until 2025 with the highest number of citations found after the deduplication stage was completed.

Reference	Title	Year	# Citations
[32]	Numerical analysis of heavy cob walls' hygrothermal behavior	2025	98
[33]	Out-of-Plane Flexure of Masonry Panels with External Thermal Insulation	2021	93
[34]	Influence of the rain on algae growth on building facades. A predictive model based on neural networks	2023	91
[35]	Block masonry equation-based model updating of a masonry minaret and seismic performance evaluation	2023	91
[36]	Architectural Heritage and Archetypal Landscape Approaches Facing Environmental Hazards	2024	87
[37]	A structural wall incorporating biosourced earth for summer thermal comfort improvement: Hygrothermal characterization and building simulation using calibrated PMV-PPD model	2022	82
[38]	Cyclic two-way out-of-plane testing of unreinforced masonry walls retrofitted using composite materials	2020	81
[39]	Seismic performance enhancement of masonry building models strengthened with the cost-effective materials under bi-directional excitation	2021	76

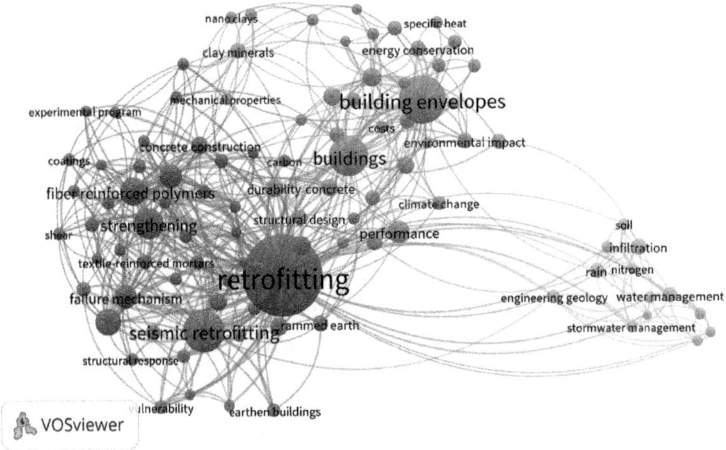

Fig. 3. Network visualization of keywords co-occurrence created with VOSviewer [40] where four distinct cluster can be identified by the color of their components. The lines connect the keywords that appear together in the same manuscripts.

4 Conclusion

The Reeco-Soil project investigates the potential of using biologically activated soils for the regenerative retrofitting of existing buildings. Its primary aim is to identify soil compositions and application methods that support plant growth on building surfaces, contributing to improved thermal insulation, enhanced urban biodiversity, and reduced environmental impact. By integrating ecological principles with low-carbon construction strategies, the project's goal is to transform underperforming building envelopes into ecologically active surfaces. The use of natural and low-embodied energy materials such as un-stabilized soils offers a promising path to meet net-positive performance targets while revitalizing the existing built environment. Nevertheless, a number of technical challenges (i.e., extra mass added to the structure, efficient management of the water irrigation systems, unforeseen decay of existing structural materials, etc.) would need to be overcome before the use of living soils can be successfully adopted for the retrofitting of existing buildings.

The systematic literature review undertaken as part of the project and partially presented in this paper (methodology and preliminary results), was designed following PRISMA guidelines and applied a robust search strategy where relevant records were found using Scopus and Web of Science databases. A total of 720 unique records were identified after deduplication, with approximately 70% published within the last decade, indicating a strong and current evidence base. A co-occurrence network analysis of keywords revealed four primary thematic clusters. Two of these, related to soils, infiltration, stormwater management (yellow), and building envelopes, energy conservation, and environmental impact (green), were found to be highly relevant to the project's objectives.

The next stages of the project will involve the detailed screening of the 720 identified records to determine their eligibility for inclusion in the final synthesis. This process will follow predefined inclusion and exclusion criteria and aim to extract quantitative and qualitative data relevant to soil types, application techniques, environmental performance, and biological viability. The findings will form the basis for a peer-reviewed systematic literature review article and support the development of new experimental work and a future funding proposal. Ultimately, the outcomes of this review will inform the design and implementation of nature-based retrofitting strategies and contribute to the advancement of regenerative design approaches in architecture and engineering.

References

1. Parish, E.: Energy efficiency measures will lead the way to net zero buildings (2022). https://www.theclimategroup.org/our-work/news/energy-efficiency-measures-will-lead-way-net-zero-buildings#:~:text=New%20buildings%20are%20more%20energy,emissions%20for%20decades%20to%20come. Accessed 08 Apr 2025
2. RENEW. RENEW A Manifesto for Regenerative Design and Engineering. University of Bath (2025). https://www.bath.ac.uk/publications/centre-for-regenerative-design-engineering-for-a-net-positive-world-renew-manifesto/attachments/regenerative-design-engineering-for-a-net-positive-world-centre-manifesto-2025.pdf

3. Norman, J., et al.: The Future of Structural Design. The Institution of Structural Engineers (2025). https://www.istructe.org/resources/guidance/the-future-of-structural-design/
4. Fořt, J., Černý, R.: Limited interdisciplinary knowledge transfer as a missing link for sustainable building retrofits in the residential sector. J. Clean. Prod. **343**, 131079 (2022). https://doi.org/10.1016/j.jclepro.2022.131079
5. Broadbent, O., Norman, J.: The Regenerative Structural Engineer. The Institution of Structural Engineering (2024). https://www.istructe.org/resources/guidance/the-regenerative-structural-engineer/
6. Jiménez Rios, A., O'Dwyer, D.: Earthen buildings in Ireland. In: 6th International Congress on Construction History. CRC Press, Brussels, Belgium (2018). https://zenodo.org/records/7494070#.Y66eldWZM2w
7. Zalamea-León, E., et al.: Housing development through the BIM methodology to reach the powerhouse standard by applying rammed-earth techniques and solar energy. J. Archit. Eng. **30**(1), 04024002 (2024). https://doi.org/10.1061/JAEIED.AEENG-1647
8. Alfaro Malatesta, S., Yuste Miguel, B., Palme, M.: Los Cielos de paja y barro aligerado: respuesta adaptativa al clima extremo en contexto andino: straw skies and lightened mud: adaptive response to extreme weather in an Andean context. Maskana **14**(2), 59–73 (2024). https://doi.org/10.18537/mskn.14.02.06
9. Jiménez Rios, A.: Cob vernacular architecture conservation and its contribution towards a sustainable built environment. In: Hanaor, D. (ed.) Sustainable Materials for the Built Environment. Springer, In Press
10. Barnes, S., et al.: 3D printing of ecologically active soil structures. Addit. Manuf. **52**, 102670 (2022). https://doi.org/10.1016/j.addma.2022.102670
11. Piesik, S.: Habitat: Vernacular Architecture for a Changing Planet. Thames & Hudson Ltd., UK (2017)
12. Nehren, U., et al.: Towards a typology of nature-based solutions for disaster risk reduction. Nature-Based Solut. **3**, 100057 (2023). https://doi.org/10.1016/j.nbsj.2023.100057
13. Zhang, K., et al.: Aerial additive manufacturing with multiple autonomous robots. Nature **609**(7928), 709–717 (2022). https://doi.org/10.1038/s41586-022-04988-4
14. Zavaleta, D., et al.: 3D-printing of a basic housing unit prototype using earthen-based matrices stabilized with rice husk fibers. J. Build. Eng. **103**, 112111 (2025). https://doi.org/10.1016/j.jobe.2025.112111
15. Page, M.J., et al.: The PRISMA 2020 statement: an updated guideline for reporting systematic reviews. BMJ **372**, n71 (2021). https://doi.org/10.1136/bmj.n71
16. Moher, D., et al.: Preferred reporting items for systematic review and meta-analysis protocols (PRISMA-P) 2015 statement. Syst. Rev. **4**(1), 1 (2015). https://doi.org/10.1186/2046-4053-4-1
17. Rethlefsen, M.L., et al.: PRISMA-S: an extension to the PRISMA statement for reporting literature searches in systematic reviews. Syst. Rev. **10**(1), 39 (2021). https://doi.org/10.1186/s13643-020-01542-z
18. Jiménez Rios, A., Calabria-Holley, J., Castilla Pascual, F.J.: Regenerative Retrofitting Via Ecologically Active Soil Structures (Reeco-Soil) Project. OSF (2025). https://doi.org/10.17605/OSF.IO/FGBYW
19. Jiménez Rios, A., Calabria-Holley, J., Castilla Pascual, F.J.: Regenerative Retrofitting Via Ecologically Active Soil Structures (Reeco-Soil) Registry. OSF Registries (2025). https://doi.org/10.17605/OSF.IO/9PGZ4
20. Jiménez Rios, A., Calabria-Holley, J., Castilla Pascual, F.J.: REECO-SOIL Systematic Literature Review Search Strategy. Zenodo (2025). https://doi.org/10.5281/zenodo.15356620
21. Jiménez Rios, A., Calabria-Holley, J., Castilla Pascual, F.J.: REECO-SOIL Systematic Literature Review Protocol. Zenodo (2025). https://doi.org/10.5281/zenodo.15356806

22. Wilkinson, M.D., et al.: The FAIR Guiding Principles for scientific data management and stewardship. Sci. Data **3**(1), 160018 (2016). https://doi.org/10.1038/sdata.2016.18
23. Jiménez Rios, A., Calabria-Holley, J., Castilla Pascual, F.J.: Bibliographic Dataset for the Regenerative Retrofitting Via Ecologically Active Soil Structures (Reeco-Soil) Research Project. Zenodo (2025). https://doi.org/10.5281/zenodo.15318914
24. Papanicolaou, C., Triantafillou, T., Lekka, M.: Externally bonded grids as strengthening and seismic retrofitting materials of masonry panels. Constr. Build. Mater. **25**(2), 504–514 (2011). https://doi.org/10.1016/j.conbuildmat.2010.07.018
25. Sorrentino, L., et al.: Seismic behaviour of ordinary masonry buildings during the 2016 central Italy earthquakes. Bull. Earthq. Eng. **17**(10), 5583–5607 (2019). https://doi.org/10.1007/s10518-018-0370-4
26. Lignola, G.P., et al.: Performance assessment of basalt FRCM for retrofit applications on masonry. Compos. B Eng. **128**, 1–18 (2017). https://doi.org/10.1016/j.compositesb.2017.05.003
27. Ascione, L., de Felice, G., De Santis, S.: A qualification method for externally bonded Fibre Reinforced Cementitious Matrix (FRCM) strengthening systems. Compos. B Eng. **78**, 497–506 (2015). https://doi.org/10.1016/j.compositesb.2015.03.079
28. Silva, T., et al.: Experimental testing and numerical modelling of masonry wall solution with PCM incorporation: a passive construction solution. Energy Build. **49**, 235–245 (2012). https://doi.org/10.1016/j.enbuild.2012.02.010
29. Sarı, A.: Thermal energy storage characteristics of bentonite-based composite PCMs with enhanced thermal conductivity as novel thermal storage building materials. Energy Convers. Manage. **117**, 132–141 (2016). https://doi.org/10.1016/j.enconman.2016.02.078
30. Babaeidarabad, S., De Caso, F., Nanni, A.: URM walls strengthened with fabric-reinforced cementitious matrix composite subjected to diagonal compression. J. Compos. Constr. **18**(2), 04013045 (2014). https://doi.org/10.1061/(ASCE)CC.1943-5614.0000441
31. Kumar, A., Suman, B.M.: Experimental evaluation of insulation materials for walls and roofs and their impact on indoor thermal comfort under composite climate. Build. Environ. **59**, 635–643 (2013). https://doi.org/10.1016/j.buildenv.2012.09.023
32. Kabore, A., Bendouma, M., Ouellet-Plamondon, C.: Numerical analysis of heavy cob walls' hygrothermal behavior. Energy Build. **331**, 115351 (2025). https://doi.org/10.1016/j.enbuild.2025.115351
33. Manos, G.C., Melidis, L., Katakalos, K.: Out-of-plane flexure of external Thermal-insulation attached to Non-load-bearing masonry walls. Structures **53**, 779–796 (2023). https://doi.org/10.1016/j.istruc.2023.04.098
34. D'Orazio, M., Quagliarini, E., Gianangeli, A.: Influence of the rain on algae growth on building facades. A predictive model based on neural networks. Build. Environ. **246**, 110990 (2023). https://doi.org/10.1016/j.buildenv.2023.110990
35. Onat, O., Toy, A.T., Özdemir, E.: Block masonry equation-based model updating of a masonry minaret and seismic performance evaluation. J. Civ. Struct. Heal. Monit. **13**(6), 1221–1241 (2023). https://doi.org/10.1007/s13349-023-00703-7
36. Bostenaru Dan, M., et al.: Architectural heritage and archetypal landscape approaches facing environmental hazards. Sustainability **16** (2024). https://doi.org/10.3390/su16041505
37. Charai, M., Mezrhab, A., Moga, L.: A structural wall incorporating biosourced earth for summer thermal comfort improvement: hygrothermal characterization and building simulation using calibrated PMV-PPD model. Build. Environ. **212**, 108842 (2022). https://doi.org/10.1016/j.buildenv.2022.108842
38. Padalu, P.K.V.R., Singh, Y., Das, S.: Cyclic two-way out-of-plane testing of unreinforced masonry walls retrofitted using composite materials. Constr. Build. Mater. **238**, 117784 (2020). https://doi.org/10.1016/j.conbuildmat.2019.117784

39. Banerjee, S., Nayak, S., Das, S.: Seismic performance enhancement of masonry building models strengthened with the cost-effective materials under bi-directional excitation. Eng. Struct. **242**, 112516 (2021). https://doi.org/10.1016/j.engstruct.2021.112516
40. van Eck, N.J., Waltman, L.: VOSviewer Manual. Manual for VOSviewer version 1.6.18, **1**(0) (2022)

Ultralight Earth: A Preliminary Look into Thermal Performance of Locally Sourced Fibre-Earth Composites

Johanna Hyrkäs[1], Matti Kuittinen[1], and Magda Posani[2](✉)

[1] Department of Architecture, Aalto University, Espoo, Finland
[2] Department of Civil Engineering, Aalto University, Espoo, Finland
magda.posani@aalto.fi

Abstract. This paper presents a preliminary investigation into ultralight earth (ULE) insulation materials developed using locally sourced clay and bio-fibers from Finnish agriculture and industrial side streams. The study responds to the urgent need for construction materials that have low carbon footprint and high carbon storage. This would support Finland's legally binding 2035 carbon neutrality target and broader bioeconomy objectives.

Four ULE composites, based on rye straw, hemp shives, reed chaff, and recycled wood, were developed in collaboration with a local clay craftsperson and tested for dry density and thermal conductivity. The resulting materials exhibited dry densities below 450 kg/m3, qualifying as ultralight, and achieved thermal conductivity values as low as 0.047 W/(m·K). These results are competitive with conventional insulation materials, such as mineral wool and expanded polystyrene. Notably, the straw-, reed-, and hemp-based ULEs met the thermal insulation threshold of 0.065 W/(m·K), as defined in European guidelines for ETICS, while all formulations qualified as thermal insulating mortars under EN 998-1:2017.

These findings demonstrate the feasibility of ULE materials as sustainable alternatives for thermal insulation in subarctic climates. The study also lays the groundwork for future optimisation of mix designs to enhance performance and durability, paving the way for wider adoption in both new and renovated buildings. This work contributes to closing the knowledge gap on ULE applications in Nordic countries and supports a circular, low-carbon construction sector in Finland.

Keywords: Ultralight Earth · Bio-based Insulation · Clay Composites · Thermal Conductivity · Cold Climate Construction · Bioeconomy

1 Introduction

1.1 Potential of Light Earth Implementation in Finland

In the European Union (EU), buildings are responsible for nearly 40% of total greenhouse gas (GHG) emissions [1], deriving from both operational energy use, such as heating and cooling, and the embodied emissions of construction materials [2, 3]. Recognising this

M. Posani and S. Singh (Eds.): RYS 2025, RILEM Bookseries 67, pp. 571–584, 2026.
https://doi.org/10.1007/978-3-032-14652-6_41

impact, the Intergovernmental Panel on Climate Change (IPCC) identified the building sector as holding the greatest potential for emission reductions within the EU, estimating that an 80–90% cut in emissions by 2050 is both achievable and necessary [4].

To support this decarbonisation trajectory, renovation of the existing building stock is widely regarded as preferable to demolition and reconstruction, due to its significantly lower carbon footprint [5]. Among renovation strategies, deep energy retrofits offer particularly promising results [6, 7], by drastically reducing operational emissions and enabling the use of low-carbon materials to move towards climate-neutral buildings. At the same time, when demolition and reconstruction are deemed necessary, low-carbon strategies should be implemented throughout the process, with strong consideration given to low carbon material solutions. This approach is strongly supported by the EU, with the 2024 recast of the Energy Performance of Buildings Directive (EPBD, Directive 2024/1275 [8]) marking a shift towards whole life-cycle carbon accounting and setting a clear path towards a nearly zero-energy, net zero-emission building stock.

Nordic countries are setting even more ambitious climate targets than those of the EU. Finland, for example, has committed to achieving carbon neutrality by 2035, fifteen years ahead of the EU goal [9], under the legally binding Finnish Climate Change Act, enacted in 2022 [10]. This makes Finland's target the third most ambitious in the world, following Norway and Uruguay [11]. However, the transition to carbon neutrality is complex and further complicated by a decline in the ability of Finland's land-use sector to act as carbon sink, due to unrealistic estimates of available biomass leading to increased logging rates, and as a result of the methodology used to calculate the carbon balance of the LULUCF (Land Use, Land-Use Change, and Forestry) sector [12].

Furthermore, the IPCC has called for developing new CO_2 removals (CDRs) to complement those provided by ecosystem services. By 2060, they estimate that 7,1 Gt of novel CDRs be needed [13]. In this context, the building sector may offer a viable solution, as there are multiple approaches for storing carbon in building products [14, 15]. Local forestry and agricultural by-products can be processed into bio-based building materials and thereby store biogenic carbon. These materials hold the potential to have high contents of biogenic carbon, and effectively increase carbon storages in new and renovated buildings [16–18]. Furthermore, their use aligns with Finland's broader aim of developing a stronger bioeconomy, promoting not only climate action but also sustainable regional development [19].

This paper explores the potential of combining local clay and biofibers from agricultural and industrial side streams to create thermal insulation materials with high thermal performance and low carbon footprints. Substantial amount of four local fibres (rye straw, reed, hemp shives and crushed demolition wood) is implemented in the mix design in order to produce very light earth materials with insulating properties comparable to those of conventional petrochemical and mineral insulation solutions. This step is essential to enable usability in the Finnish context, which is characterised by very cold winters. Clay is characterised by a near-zero carbon footprint, as well as high potential for circularity and recyclability [20–22]. Furthermore, clay is a widely available material in Finland and can be sourced from excavation waste. Clay combined with biofibres from locally available side streams could pave a way for new insulation materials that can support both the Finnish transition to a strong local bioeconomy and mitigation of emissions from the construction sector.

1.2 Use of Fibres in Traditional Earthen Construction

The use of earth as a building material, often combined with plant fibres, dates back thousands of years and is rooted in several ancient civilizations. Fibres such as straw, grass, wood and other cellulose-based fibres were historically incorporated into earth matrices to enhance the performance of earth building materials like adobe, cob, and occasionally rammed earth. These natural additives improved mechanical properties, reduced shrinkage cracking, and increased durability.

Rammed earth is made from a mixture of soils (clay, silt, sand, and gravel) compacted in layers. Compressed earth blocks (CEB) are technically moulded rammed earth. While rammed earth traditionally did not include fibres, [23, 24] experiments with adding fibres to improve CEB's properties are plenty. Low percentage of natural fibres (0.125 – 0.25 wt%) added to compressed earth mix have been in some cases found to improve compressive strength [25, 26], but more often added fibre reduced the compressive and tensile strength of the material [27–29]. However, moderate amount of fibres (0.80 – 10 wt%) has been found to clearly improve some other qualities like sound insulation and damping [29], moisture buffering capacities [30, 31] and durability [25, 32]. Adobe bricks, which can technically be conceived as moulded cob, are traditionally made from a mixture of soil, water, and straw. In classical adobe and cob common fibre, usually straw, content is around 20–40% by volume [33, 34] or of 0.2–3.3% by weight [35, 36]. The addition of straw fibres is a key feature, as it improves the mechanical properties and durability of the bricks.

While aforementioned earthen building techniques produce relatively heavy materials, with densities ranging from 1200 to 2400 kg/m^3 [37, 38], light earth mixes have lower density, typically in the range of 200 to 1200 kg/m^3 [38, 39], and thus different mechanical and hygrothermal properties. The most defining characteristic of light earth is its high thermal insulation performance, with observed thermal conductivity ranging from 0.05 to 0.2 W/(m.K)] [39]. As so, light earth is particularly promising for construction in Nordic regions with cold climates like Finland, where yearly mean temperature, although on the rise, is currently around 2 °C [40].

1.3 Modern Light Earth (LE) and Ultralight Earth (ULE) Insulation

Light earth (LE) is typically defined as a mix of earth and biofibres with a density lower than 1200 kg/m^3 [41]. However, this definition covers a wide variety of materials, with very different hygrothermal and mechanical properties. Thus, it is beneficial to establish a category for very light earth mixes, considered particularly promising for application as thermal insulation in Nordic regions. According to Labat et al. [42], only mixes with densities lower than 450 kg/m^3 would be suitable for use as thermal insulation material in modern construction. Although other suggested definitions of light-earth suitable for insulation purposes exist (below 500 kg/m^3 [43]), in this work, the authors propose setting the limit to 450 kg/m^3 and calling such composites *ultralight earth* (ULE).

Frank Volhard presented current version of light earth in the first version of his seminal book "Light Earth" in 1983, with latest edition re-published 2016 [38]. According to Volhard, LE is a mix of earth and lightweight materials, which constitute majority of the volume. In the mix, the role of earth is to bind the fibres together. The technique is based on traditional straw-clay (1200–1700 kg/m^3) approach and, to the best knowledge of the authors, it was first mentioned in German "Lehmbauordnung" from 1944 [41] as an earth mixed with lightweight aggregate with a bulk density lower than 1 200 kg/m^3. The most common fibre used in traditional light earth is wheat straw, thus in many languages light earth is referred to as "straw clay" (*stroh lehm* in German, *argile paille* in French). More recently, the inner part of hemp stalks, hemp shives, have gained popularity in LE, most probably due to French innovation hempcrete, which is a mix of hemp shives and lime.

According to scientific literature, hemp shives and straw are still the most commonly used fibres in ULEs, as reported in Table 1. Other cellulose-based fibres like rape straw, sunflower bark and pith, reed, sawdust and Typha Australis have also been experimented with in ULE mixes. Most of the research utilised local natural earths previously used by ceramists or clay building professionals. The lightest mixes (around 200 kg/m^3) were obtained using hemp shives, but comparably light ones were achieved using other fibres like straw. The lowest thermal conductivity recorded is 0.050 W/(m.K) [44], achieved with the use of wheat straw. Sunflower pith and hemp shives were observed to produce almost equally low readings, in the order of 0.055–0.060 W/(m.K) [45–47].

Table 1. Density and thermal conductivity in ULE, according to the literature.

Earth	Fibres	Density [kg/m^3]	Thermal conductivity [W/(m.K)]	Ref
Raw natural earth	Rape straw, sunflower bark and pith	235–714	0.055–0.158	[45]
Raw natural earths	Hemp shives	196–345	0.060–0.100	[39]
Raw natural earths	Reed, hemp shives	399–703	0.120–0.200	[48]
Basecoat plaster	Straw	343–369	0.060–0.083	[49]
Raw natural earth	Wheat straw, sawdust	264–729	0.050–0.150	[44]
Raw natural earths	Straw	241–531	0.071–0.120	[41]
Raw natural earth	Typha Australis	304–586	0.065–0.112	[50]
Raw natural earths and mining side streams	Hemp shives	195–310	0.060–0.097	[47]

2 Materials

2.1 Raw Materials

The earth used for the mixes was the only commercially available, dried and powdered Finnish clay from a major brick producer. Particle size distribution was determined according to ISO 17892-4 and the results showed clay content of 56% with the rest of the particles being silt. Liquid limit was determined with drop cone test according to ISO 17892-12:2018 and plastic limit according to ISO 17892-12:2018. Liquid limit was 51.9% and plastic limit 27.4% returning the plasticity index of 24.5%. Organic contents, determined by loss on ignition method, was low (1%). The material can be thus classified using Unified Soil Classification System (USCS) as inorganic clay of high plasticity.

The fibres used in the study are rye straw, reed, hemp shives and crushed demolition wood. They were chosen according to local availability and specific properties. Rye has traditionally been cultivated in Finland, hundred years ago it constituted one fourth of all grassland crops. However, its cultivation has since declined to current 3% of all arable land [51]. Compared to other popular cereal crops, rye straw has higher lignin content [52]. Lignin is considered to provide water resistance as well as strength to the composite material, which makes rye straw interesting alternative to more commonly used wheat and barley straws. **Reed,** *Phragmites australis, is a* common herbaceous plant growing on the banks of water bodies. Reeds have benefited from the eutrophication of water bodies, and removing their growth helps remove nutrients from the waters. The reed chaff used in the study originates from a private foundation's project focused on improving the condition of the Baltic Sea, where methods for winter harvesting of common reed and utilizing the produced material are being developed [53]. **Hemp shive** is the soft inner part of the hemp stalk that is produced as a byproduct of fibre hemp cultivation. It has enjoyed high popularity in LE research due to its lightness, good thermal insulation properties, and adhesive-promoting surface microstructure. The cultivation of fibre hemp has been limited in Finland, but a company that provided the material for the research has been developing methods, creating network of farmers and investing significantly in starting local production. Traditionally, **wood** has been used in Finland especially for the construction of single-family houses. Significant waste and side streams are generated from the demolition of buildings and the woodworking industry: in 2022, 2.2 million tons of wood waste was produced, which corresponds to total amount of all household and mixed waste in Finland [54]. Woodchips produced from garden and forestry residue by cutting have been previously mixed with earth by traditional clay builders in Finland. Demolition wood is crushed instead of cut, which according to preliminary tests seems to improve the adhesion to clay. In addition the wood is dry, which reduces the initial moisture content of the material and thus decreases the risk of microbial growth.

2.2 Fibres Processing

The rye straw was delivered in small bales by a local farmer and was cut to maximum 6 cm long pieces. It is notable that the fraction contained a lot of spikes, and they were kept in the mixture. Reed chaff was mainly utilized in the form processed by the harvesting and crushing machine. Occasionally, segments longer than 6cm were found, and they were shortened to a length of maximum 6 cm. Hemp shives were used as received from the producer. The demolition wood was divided into two fractions: the smaller fraction passed through a 4 mm sieve, and a larger fraction, from which pieces exceeding approximate dimensions of 1×6 cm were removed. Since the smallest wood particles easily settle at the bottom, the final mix was adjusted to include 20 wt% of fine material and 80 wt% of coarser fraction.

2.3 Sample Preparation

One clay-fibre mix was developed for each type of fibre. The design and manufacturing of the mixes was implemented by a local craftsperson according to clay building conventions. The aim was to achieve lightweight composites with minimal binder content. The mix designs are reported in Table 2.

Each batch of the mix contained the same amount of water and earth, only the amount of fibres varied. In this mix, clay acts as the binder, and to ensure proper cohesion, all fibre surfaces must be coated with clay slip. The fibre quantity in each mix was therefore determined by the coating capacity of the slip: a fixed volume of clay slip was prepared, and fibres were gradually added and mixed in. Once the slip could no longer fully coat the fibres, evidenced by the appearance of dry, uncoated areas, the maximum fibre content was considered to be reached. This process was carried out by a local craftsperson to ensure the necessary expertise for accurately estimating the maximum fibre content in the mix design.

Samples were manufactured separately in order to minimise the risk of sedimentation compromising the uniformity of samples. Powdered clay was added to water and mixed at medium velocity in small industrial mixer for 10 min. The created clay slip was poured on correct amount of fibres and mixed by hand to achieve steady consistency where all surfaces of fibres are covered with clay.

The composite mix obtained was then moved into a plywood mould (20 × 20 × 4 cm) in two layers, each of which was compacted after deposition using a small wooden rammer. The compacted composite was pushed out of the mould immediately and left to dry for 4 days in room temperature (22 °C ± 1 °C) and relative humidity (RH 52 ± 5%). The sample preparation steps are illustrated in Fig. 1.

The samples were left to dry on a perforated metal sheet for the first two days, then turned upside down for the following two days to ensure effective drying through both main planar surfaces. The drying process was completed in an oven set to 60 °C, where the samples were dried until reaching equilibrium in three successive weighing (change of weight below 1% in weighing performed 24 h apart).

Table 2. Final mix designs.

Fibre	Water [wt%]	Earth [wt%]	Fibre [wt%]
Straw	40.0	40.0	20.0
Hemp	38.0	38.0	24.0
Reed	38.0	38.0	24.0
Wood	33.0	33.0	34.0

Fig. 1. Samples preparation: (a) mixing water and earth, (b) pouring of clay slip on fibres, (c) ULE composite after manual mixing, (d) setting of ULE mix in the mould, (e) compaction, (f, g) extraction from mould into the perforated metal sheet, (h) extracted wet sample, and (i) drying at room temperature.

3 Laboratory Tests

All tests were performed on three samples. Results are expressed as the average of the three measurements and standard deviation is provided as an indication of the variability in the measured properties.

3.1 Dry Bulk Density

The dry density of the samples was determined using the dry weight of the samples and volume determined through dimensional measurement. Specimens were dried in a ventilated oven at 60 °C until constant mass was achieved, defined as a mass change of less than 0.1% over 24 h. After drying, the mass (m) of the specimen was recorded using a precision balance (accuracy ± 0.01 g). The volume (V) was calculated based on length, width, and thickness measurements taken using a digital calliper (accuracy ± 0.01 mm). The dry bulk density (ρ) was then calculated using the standard formulation as mass divided by volume.

3.2 Thermal Conductivity Test

Thermal conductivity was measured using a Heat Flow Meter (HFM) apparatus developed by NETZSCH in compliance with ASTM-C518 and ISO 8301, specifically a HFM 446 Lambda Eco-Line Small [55].

The apparatus is a stand-alone system for measuring thermal conductivity in the range of 0.007 to 2.0 W/(m·K), with an accuracy of ±2%. It consists of independently temperature-controlled upper and lower plates enclosed in a protective case to minimise heat losses [56]. The plates operate within a temperature range of –20 °C to 90 °C, making the system suitable for assessing insulation materials under diverse environmental conditions. Each plate has a surface area of approximately 20.5 × 20.5 cm, with the effective measurement area restricted to the central 10 × 10 cm to reduce edge effects. Accurate surface temperature measurements are achieved using thermocouples (accuracy ± 0.01 °C), affixed directly to the top and bottom surfaces of the sample using metal tape.

Test specimens (20 × 20 × 4 cm) were oven-dried at 60 °C until mass stabilisation was confirmed by a change of less than 0.1% over 24 h. During testing, each sample was placed between the two plates of the HFM, the protective case of the HFM was closed to ensure minimum interference of outdoor environment with the measurement, and the defined temperature gradient was imposed with plate temperatures set to –10 °C and + 10 °C, resulting in an average testing temperature of 0 °C for thermal conductivity evaluation.

Heat flux through the sample was measured by two calibrated heat flux transducers. The measurements taken after reaching thermal equilibrium, with data recorded at 1-min intervals, were used to compute the final thermal conductivity value. At least 30 readings were considered in each equilibrium measurement quantification, and they were used to calculate the thermal conductivity of the sample. This was done automatically by the HFM software, based on the measured heat flux, temperature gradient, and sample thickness.

4 Results and Discussion

Results from preliminary tests are reported in Table 3. Results indicate relevant differences in thermal conductivity among the mixes utilizing alternative fibres. As the thermal conductivity is linearly dependent of the density, most probably the differences derive from the distinct characteristics, especially the dry bulk density, of the of various fibres. ULE made from rye straw and reed emerge as the most promising options, with thermal conductivity in the order of 0.05 W/(m.K). This is in line with best-achieved performance in the literature with ULE, and comparable to conventional insulation materials like mineral wool and expanded polystyrene (EPS), which are normally considered to have a thermal conductivity in the order of 0.04 W/(m.K) [57].

Table 3. Density and thermal conductivity of the mixes.

ULE mix	Density [kg/m^3]	Thermal conductivity [W/(m.K)]
Straw	280 ± 6	0.047 ± 0.001
Hemp	315 ± 7	0.064 ± 0.003
Reed	312 ± 12	0.047 ± 0.001
Wood	388 ± 12	0.072 ± 0.003

The preliminary results hereby observed are considered very promising, particularly when compared to a variety of insulation materials and standards, as shown in Fig. 2. According to guidelines ETAG 004:2013 [58] for external thermal insulation composite systems (ETICS), all ULE materials hereby presented can be ranked as thermal insulation, except for the wood-based composite (thermal conductivity above 0.065 W/(m.K)). According to standard EN 998-1:2017: "Specification for mortar for masonry—Part 1: Rendering and plastering mortar" [59], all solutions can be ranked as thermal insulating mortars. However, it is worth noting that both standards prescribe additional requirements for insulation materials. For example, EN 998-1 specifies that thermal insulating mortars must have compressive strengths in the range of 0.4 to 5.0 N/mm2. ETAG 004 includes further requirements; for instance, when insulation materials are susceptible to degradation from humidity, tensile strength must be tested after exposure to heat–moisture conditions at 70 °C and 95% relative humidity. These and other durability-related requirements are highly relevant for the future development of ULE insulation materials. Future studies should investigate the ability of ULE to meet existing standards for mechanical performance and durability and assess the suitability of these standards for non-conventional bio-based and clay-based insulation materials.

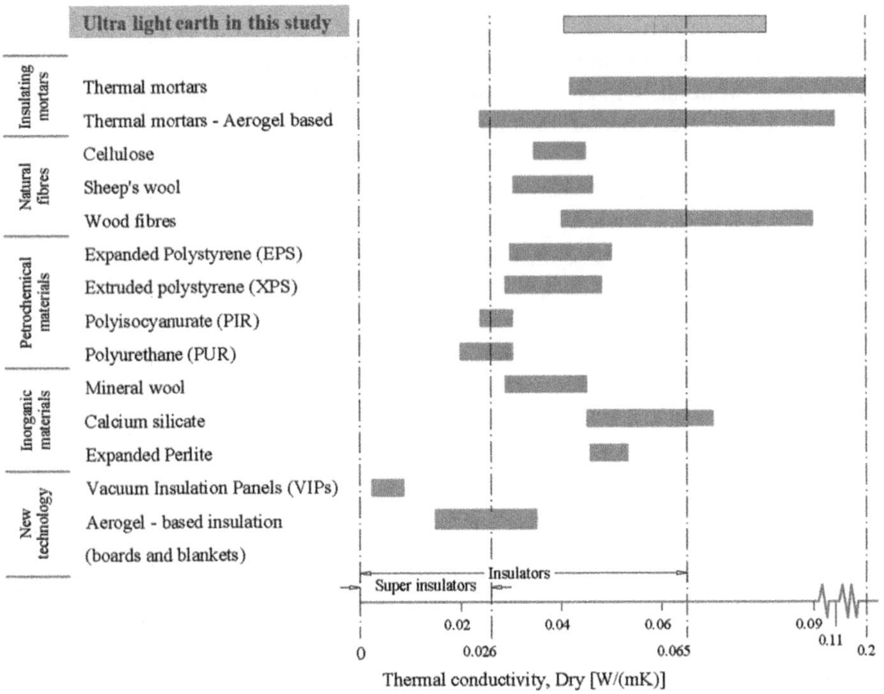

Fig. 2. Thermal conductivity of common thermal insulation materials and reference values from guidelines and standards: 0.065 W/(m·K) threshold for themal insulation materials [58] and 0.2 W/(m·K) threshold for of thermal mortars [59]. The scheme is adapted from Ref. [60], with copyrights permission from the authors.

5 Conclusion

This paper presents a preliminary study on ultralight earth insulation materials produced using locally sourced raw materials from agricultural and industrial side streams in Finland. By combining clay with bio-byproducts, the study demonstrates the potential of the composites to produce sustainable, carbon-storing insulation products that also contribute to value-added utilisation of local biomass, thereby supporting the Finnish bioeconomy.

The preliminary material development and testing presented showcase encouraging results: all composites qualify as thermal insulating mortars, and most meet the criteria for thermal insulation materials, with thermal conductivity values comparable to conventional mineral and petrochemical alternatives. These findings lay the groundwork for more detailed future investigations aimed at optimising the mix design to further reduce thermal conductivity and enhance the materials' competitiveness with conventional insulation products.

Given the limited research on ultralight earth materials for use in cold climates, this study provides a valuable foundation for future work tailored to the Finnish context. Subsequent research should focus on ULE application for improving thermal performance in both renovation and new construction interventions in subarctic conditions.

As a bio-based material, ULE also presents certain challenges, particularly concerning its susceptibility to microbial growth, especially during initial drying period. Since the binder is unstabilised clay, issues related to durability and performance after natural and accelerated ageing, including exposure to liquid water infiltration, should be addressed in future studies. Further research is needed to evaluate these aspects and ensure the practical applicability of ULE in real-world interventions. Additional work could also explore optimal multilayer insulation system designs that protect the ULE layer and enhance long-term durability.

References

1. Habert, G.: The Routledge Handbook of Embodied Carbon in the Built Environment (2023)
2. Priore, Y.D., Jusselme, T., Habert, G.: Embodied net-zero compatible buildings? They already exist!. J. Phys. Conf. Ser. **2600**(15) (2023)
3. Röck, M., et al.: Embodied GHG emissions of buildings – the hidden challenge for effective climate change mitigation. Appl. Energy **258**, 114107 (2020)
4. Shukla, A.R.: Summary for policymakers in climate change 2022: mitigation of climate change. In: Contribution of Working Group III to the Sixth Assessment Report of the IPCC (2022)
5. Leichter, M., Piccardo, C.: Assessing life cycle sustainability of building renovation and reconstruction: a comprehensive review of case studies and methods. Build. Environ. **262**, 111817 (2024)
6. Galimshina, A., Moustapha, M., Hollberg, A., Lasvaux, S., Sudret, B., Habert, G.: Strategies for robust renovation of residential buildings in Switzerland. Nat. Commun. **15**(1), 1–12 (2024)
7. Galimshina, A., et al.: Bio-based materials as a robust solution for building renovation: a case study. Appl. Energy **316**, 119102 (2022)
8. EU. DIRECTIVE (EU) 2024/1275 OF THE EUROPEAN PARLIAMENT AND OF THE COUNCIL on the energy performance of buildings. Official Journal of the European Union (2024)
9. Kuittinen, M., Häkkinen, T.: Reduced carbon footprints of buildings: new finnish standards and assessments. Build. Cities **1**(1), 182–197 (2020)
10. Finnish government. Inclusive and Competent Finland — a Socially, Economically and Ecologically Sustainable Society. Helsinki: Programme of Prime Minister Sanna Marin's Government 2019 (2019)
11. UNEP. Emissions Gap (2019)
12. Majava, A., Vadén, T., Toivanen, T., Järvensivu, P., Lähde, V., Eronen, J.T.: Sectoral low-carbon roadmaps and the role of forest biomass in Finland's carbon neutrality 2035 target. Energy Strateg. Rev. **41**, 100836 (2022)
13. Pörtner, H.-O., et al.: IPCC, 2022: Summary for Policymakers. Cambridge University Press, Cambridge, UK and New York, NY, USA (2022)
14. Kuittinen, M., Zernicke, C., Slabik, S., Hafner, A.: How can carbon be stored in the built environment? A review of potential options (2021)
15. Van Roijen, E., Miller, S.A., Davis, S.J.: Building materials could store more than 16 billion tonnes of CO2 annually. Science **387**(6730), 176–182 (2025)
16. Göswein, V., Arehart, J., Phan-Huy, C., Pomponi, F., Habert, G.: Barriers and opportunities of fast-growing biobased material use in buildings. Build. Cities **3**(1), 745–755 (2022)
17. Göswein, V., Reichmann, J., Habert, G., Pittau, F.: Land availability in Europe for a radical shift toward bio-based construction. Sustain. Cities Soc. **70**, 102929 (2021)

18. Habert, G., Iannaccone, G., Pittau, F.: The renovation of the building stock in Europe: an essential opportunity to store carbon in buildings. Riv. Tema **05**(02) (2021)
19. Finnish Government. The Finnish Bioeconomy Strategy: Sustainably towards higher value added. Publ. Finnish Gov., no. 2022:5 (2022)
20. Ben-Alon, L., Loftness, V., Harries, K.A., DiPietro, G., Hameen, E.C.: Cradle to site Life Cycle Assessment (LCA) of natural vs conventional building materials: a case study on cob earthen material. Build. Environ. **160** (2019)
21. Perlot, C., Bruno, A.W., Posani, M., Habert, G., Vučetić, S.: Challenges for bio-stabilised earth-based construction. RILEM Bookseries **52**, 101–111 (2024)
22. Pelé-Peltier, A., Charef, R., Morel, J.-C.: Factors affecting the use of earth material in mainstream construction: a critical review (2022)
23. Parracha, J.L., Silva, A.S., Cotrim, M., Faria, P.: Mineralogical and microstructural characterisation of rammed earth and earthen mortars from 12th century Paderne Castle. J. Cult. Herit. **42**, 226–239 (2020)
24. Chen, L., et al.: Practice and reflection on rammed earth architecture: the case study of tiles hill–xiangshan campus reception centre in China. Buildings **14**(12) (2024)
25. Danso, H., Martinson, D.B., Ali, M., Williams, J.B.: Physical, mechanical and durability properties of soil building blocks reinforced with natural fibres. Constr. Build. Mater. **101**, 797–809 (2015)
26. Saleem, M.A., Abbas, S., Haider, M.: Jute fiber reinforced compressed earth bricks (FR-CEB)-a sustainable solution. Pak. J. Engg. Appl. Sci. **19**(2005), 83–90 (2016)
27. Charai, M., Salhi, M., Horma, O., Mezrhab, A., Karkri, M., Amraqui, S.: Thermal and mechanical characterization of adobes bio-sourced with Pennisetum setaceum fibers and an application for modern buildings (2022)
28. Li Piani, T., et al.: Dynamic behaviour of adobe bricks in compression: the role of fibres and water content at various loading rates. Constr. Build. Mater. **230** (2020)
29. Sadouri, R., Chantit, F.: Physical, mechanical, and durability properties of compressed earth blocks filled by juncus plant fibers. **42**(1), 194–205 (2024)
30. Charai, M., Mezrhab, A., Moga, L.: A structural wall incorporating biosourced earth for summer thermal comfort improvement: hygrothermal characterization and building simulation using calibrated PMV-PPD model. Build. Environ. **212**, 108842 (2022)
31. Palumbo, M., McGregor, F., Heath, A., Walker, P.: The influence of two crop by-products on the hygrothermal properties of earth plasters. Build. Environ. **105**, 245–252 (2016)
32. Laborel-Préneron, A., Giroudon, M., Aubert, J.E., Magniont, C., Faria, P.: Experimental assessment of bio-based earth bricks durability. IOP Conf. Ser. Mater. Sci. Eng. **660**(1), 1–10 (2019)
33. Costi de Castrillo, M., Ioannou, I., Philokyprou, M.: Reproduction of traditional adobes using varying percentage contents of straw and sawdust. Constr. Build. Mater. **294**, 123516 (2021)
34. Calatan, G., Hegyi, A., Dico, C., Mircea, C.: Determining the optimum addition of vegetable materials in adobe bricks. Procedia Technol. **22**, 259–265 (2016)
35. Haddad, K., Latif, E., Lannon, S.: The state of the art of cob construction: a comprehensive review of the optimal mixtures and testing methods. RILEM Bookseries **45**, 212–231 (2023)
36. Costi de Castrillo, M., Philokyprou, M., Ioannou, I.: Comparison of adobes from pre-history to-date. J. Archaeol. Sci. Reports **12**, 437–448 (2017)
37. Sabbà, M.F., Tesoro, M., Falcicchio, C., Foti, D.: Rammed earth with straw fibers and earth mortar: Mix design and mechanical characteristics determination. Fibers **9**(5) (2021)
38. Volhard, F.: Light Earth Building: a handbook for building with wood and earth. Basel, Switzerland: Birkhäuser (2016)
39. Vinceslas, T., Colinart, T., Lenormand, H., de Menibus, A.H., Hamard, E., Lecompte, T.: Hygrothermal properties of light earth insulation materials: evaluation of uncertainties and consequences. Acad. J. Civ. Eng. **37**(2), 198–203 (2019)

40. Tietäväinen, H., Tuomenvirta, H., Venäläinen, A.: Annual and seasonal mean temperatures in Finland during the last 160 years based on gridded temperature data. Int. J. Climatol. **30**(15), 2247–2256 (2010)
41. Lehmbauordnung: Verordnung über Lehmbauten vom 4. Okt. 1944. Deutscher Heimat-Verlag (1947)
42. Labat, M., Magniont, C., Oudhof, N., Aubert, J.-E.E.: From the experimental characterization of the hygrothermal properties of straw-clay mixtures to the numerical assessment of their buffering potential. Build. Environ. **97**, 69–81 (2016)
43. Verron, L., et al.: Estimating and mapping the availability of earth resource for light earth building using a soil geodatabase in Brittany (France). Resour. Conserv. Recycl. **184** (2022)
44. Jové-Sandoval, F., García-Baños, E.M., Barbero-Barrera, M.M.: Characterisation and thermal improvement of adobe walls from earth-straw lightweight panels. MRS Adv. **9**(2), 71–77 (2024)
45. Brouard, Y., Belayachi, N., Hoxha, D., Ranganathan, N., Méo, S.: Mechanical and hygrothermal behavior of clay – Sunflower (Helianthus annuus) and rape straw (Brassica napus) plaster bio-composites for building insulation. Constr. Build. Mater. **161**, 196–207 (2018)
46. Colinart, T., Vinceslas, T., Lenormand, H., De Menibus, A.H., Hamard, E., Lecompte, T.: Hygrothermal properties of light-earth building materials. J. Build. Eng. **29**, 101134 (2020)
47. Vinceslas, T., Colinart, T., Hamard, E., de Ménibus, A.H., Lecompte, T., Lenormand, H.: Light earth performances for thermal insulation: application to earth–hemp. In: Earthen Dwellings and Structures: Current Status in their Adoption, 1st ed., K. Gourav and S. N. Ullas, Eds. Springer Singapore, pp. 357–367 (2019)
48. Goodhew, S., Boutouil, M., Streiff, F., Le Guern, M., Carfrae, J., Fox, M.: Improving the thermal performance of earthen walls to satisfy current building regulations (2021)
49. Holzhueter, K., Itonaga, K.: The potential for light straw clay construction in Japan: an examination of the building method and thermal performance. J. Asian Archit. Build. Eng. **16**(1), 209–213 (2017)
50. Niang, I., et al.: Hygrothermal performance of various typha–clay composite. J. Build. Phys. **42**(3), 316–335 (2018)
51. Luke Natural Resource Institute Finland. Use of arable land area by Year and Species (2021). https://statdb.luke.fi/PxWeb/pxweb/en/LUKE/LUKE__02/Maatalous__04/Tuotanto__22/Kaytossa/oleva/maatalousmaa/03_Peltoala_1910_ja_1920-.px/table/tableViewLayout2/?loadedQueryId=f4852cd7-ccfe-4f68-a240-845791a3f283&timeType=item. Accessed 24 May 2025
52. Xu, F.: Structure, Ultrastructure, and Chemical Composition. Fundamentals and Applications, p. 33,1st ed. Elsevier (2010)
53. Coastal Reed Project - John Nurminen Foundation. https://johnnurmisensaatio.fi/en/our-work/projects/coastal-reed-project/. Accessed 05 June 2025
54. Statistics Finland's. Waste generation by industry by Year (2022). https://pxdata.stat.fi/PxWeb/pxweb/en/StatFin/StatFin__jate/statfin_jate_pxt_12qw.px/table/tableViewLayout1/. Accessed 25 May 2025
55. HFM 446 Lambda Small Eco-Line - NETZSCH Analyzing & Testing. https://analyzing-testing.netzsch.com/en/products/heat-flow-meter-hfm-and-guarded-heat-flow-meter-ghfm/hfm-446-lambda-eco-line-1. Accessed 05 June 2025
56. El Assaad, M., Plantec, Y., Colinart, T., Lecompte, T.: Influence of moisture transfer on thermal conductivity measurement by HFM: measurement accuracy on insulation materials and consequences on building energy assessments. Energy Build. **320**, 114635 (2024)
57. Posani, M., Veiga, R., de Freitas, V.P.: Thermal renders for traditional and historic masonry walls: comparative study and recommendations for hygric compatibility. Build. Environ. **228**, 109737 (2023)

58. EOTA (European Organisation for Technical Approvals). ETAG 004: External thermal insulation composite systems (ETICS) with rendering. Guideline for European Technical Approval, pp. 1–229 (2013)
59. British Standards Institution. CEN. EN 998-1:2017—Specification for Mortar for Masonry—Part 1: Rendering and Plastering Mortar (2017)
60. Posani, M., Veiga, R., De Freitas, V.P.: Retrofitting historic walls: feasibility of thermal insulation and suitability of thermal mortars. Heritage **4**(3), 2009–2022 (2021)

GPSR Compliance

The European Union's (EU) General Product Safety Regulation (GPSR) is a set of rules that requires consumer products to be safe and our obligations to ensure this.

If you have any concerns about our products, you can contact us on ProductSafety@springernature.com

In case Publisher is established outside the EU, the EU authorized representative is:

Springer Nature Customer Service Center GmbH
Europaplatz 3
69115 Heidelberg, Germany

Batch number: 10755248

Printed by Printforce, the Netherlands

RILEM Youth Symposium 2025 – RYS2025

RILEM Bookseries

Volume 67

RILEM, The International Union of Laboratories and Experts in Construction Materials, Systems and Structures, founded in 1947, is a non-governmental scientific association whose goal is to contribute to progress in the construction sciences, techniques and industries, essentially by means of the communication it fosters between research and practice. RILEM's focus is on construction materials and their use in building and civil engineering structures, covering all phases of the building process from manufacture to use and recycling of materials. More information on RILEM and its previous publications can be found on www.RILEM.net.

Indexed in SCOPUS, Google Scholar and SpringerLink.

Magda Posani · Surender Singh
Editors

RILEM Youth Symposium 2025 – RYS2025

1st RILEM Youth Symposium on Building Materials and Construction

Set 2

Springer

Editors
Magda Posani
Department of Civil Engineering
Aalto University
Espoo, Finland

Surender Singh
Department of Civil Engineering
IIT Madras
Chennai, Tamil Nadu, India

ISSN 2211-0844 ISSN 2211-0852 (electronic)
RILEM Bookseries
ISBN 978-3-032-14651-9 ISBN 978-3-032-14652-6 (eBook)
https://doi.org/10.1007/978-3-032-14652-6

This Springer imprint is published by the registered company Springer Nature Switzerland AG
The registered company address is: Gewerbestrasse 11, 6330 Cham, Switzerland

Contents

Cementitious Materials

Innovative Low-Carbon Materials

Service Life and Durability

Hygrothermal Behaviour

Cultural Heritage

Environmental Impact Assessment

RILEM Publications

The following list is presenting the global offer of RILEM Publications, sorted by series. Each publication is available in printed version and/or in online version.

RILEM Proceedings (PRO)

PRO 1: Durability of High Performance Concrete (ISBN: 2-912143-03-9; e-ISBN: 2-351580-12-5; e-ISBN: 2351580125); *Ed. H. Sommer*

PRO 2: Chloride Penetration into Concrete (ISBN: 2-912143-00-04; e-ISBN: 2912143454); *Eds. L.-O. Nilsson and J.-P. Ollivier*

PRO 3: Evaluation and Strengthening of Existing Masonry Structures (ISBN: 2-912143-02-0; e-ISBN: 2351580141); *Eds. L. Binda and C. Modena*

PRO 4: Concrete: From Material to Structure (ISBN: 2-912143-04-7; e-ISBN: 2351580206); *Eds. J.-P. Bournazel and Y. Malier*

PRO 5: The Role of Admixtures in High Performance Concrete (ISBN: 2-912143-05-5; e-ISBN: 2351580214); *Eds. J. G. Cabrera and R. Rivera-Villarreal*

PRO 6: High Performance Fiber Reinforced Cement Composites - HPFRCC 3 (ISBN: 2-912143-06-3; e-ISBN: 2351580222); *Eds. H. W. Reinhardt and A. E. Naaman*

PRO 7: 1st International RILEM Symposium on Self-Compacting Concrete (ISBN: 2-912143-09-8; e-ISBN: 2912143721); *Eds. Å. Skarendahl and Ö. Petersson*

PRO 8: International RILEM Symposium on Timber Engineering (ISBN: 2-912143-10-1; e-ISBN: 2351580230); *Ed. L. Boström*

PRO 9: 2nd International RILEM Symposium on Adhesion between Polymers and Concrete ISAP '99 (ISBN: 2-912143-11-X; e-ISBN: 2351580249); *Eds. Y. Ohama and M. Puterman*

PRO 10: 3rd International RILEM Symposium on Durability of Building and Construction Sealants (ISBN: 2-912143-13-6; e-ISBN: 2351580257); *Ed. A. T. Wolf*

PRO 11: 4th International RILEM Conference on Reflective Cracking in Pavements (ISBN: 2-912143-14-4; e-ISBN: 2351580265); *Eds. A. O. Abd El Halim, D. A. Taylor and El H. H. Mohamed*

PRO 12: International RILEM Workshop on Historic Mortars: Characteristics and Tests (ISBN: 2-912143-15-2; e-ISBN: 2351580273); *Eds. P. Bartos, C. Groot and J. J. Hughes*

PRO 13: 2nd International RILEM Symposium on Hydration and Setting (ISBN: 2-912143-16-0; e-ISBN: 2351580281); *Ed. A. Nonat*

PRO 14: Integrated Life-Cycle Design of Materials and Structures - ILCDES 2000 (ISBN: 951-758-408-3; e-ISBN: 235158029X); (ISSN: 0356-9403); *Ed. S. Sarja*

PRO 15: Fifth RILEM Symposium on Fibre-Reinforced Concretes (FRC) - BEFIB'2000 (ISBN: 2-912143-18-7; e-ISBN: 291214373X); *Eds. P. Rossi and G. Chanvillard*

PRO 16: Life Prediction and Management of Concrete Structures (ISBN: 2-912143-19-5; e-ISBN: 2351580303); *Ed. D. Naus*

PRO 17: Shrinkage of Concrete – Shrinkage 2000 (ISBN: 2-912143-20-9; e-ISBN: 2351580311); *Eds. V. Baroghel-Bouny and P.-C. Aïtcin*

PRO 18: Measurement and Interpretation of the On-Site Corrosion Rate (ISBN: 2-912143-21-7; e-ISBN: 235158032X); *Eds. C. Andrade, C. Alonso, J. Fullea, J. Polimon and J. Rodriguez*

PRO 19: Testing and Modelling the Chloride Ingress into Concrete (ISBN: 2-912143-22-5; e-ISBN: 2351580338); *Eds. C. Andrade and J. Kropp*

PRO 20: 1st International RILEM Workshop on Microbial Impacts on Building Materials (CD 02) (e-ISBN 978-2-35158-013-4); *Ed. M. Ribas Silva*

PRO 21: International RILEM Symposium on Connections between Steel and Concrete (ISBN: 2-912143-25-X; e-ISBN: 2351580346); *Ed. R. Eligehausen*

PRO 22: International RILEM Symposium on Joints in Timber Structures (ISBN: 2-912143-28-4; e-ISBN: 2351580354); *Eds. S. Aicher and H.-W. Reinhardt*

PRO 23: International RILEM Conference on Early Age Cracking in Cementitious Systems (ISBN: 2-912143-29-2; e-ISBN: 2351580362); *Eds. K. Kovler and A. Bentur*

PRO 24: 2nd International RILEM Workshop on Frost Resistance of Concrete (ISBN: 2-912143-30-6; e-ISBN: 2351580370); *Eds. M. J. Setzer, R. Auberg and H.-J. Keck*

PRO 25: International RILEM Workshop on Frost Damage in Concrete (ISBN: 2-912143-31-4; e-ISBN: 2351580389); *Eds. D. J. Janssen, M. J. Setzer and M. B. Snyder*

PRO 26: International RILEM Workshop on On-Site Control and Evaluation of Masonry Structures (ISBN: 2-912143-34-9; e-ISBN: 2351580141); *Eds. L. Binda and R. C. de Vekey*

PRO 27: International RILEM Symposium on Building Joint Sealants (CD03; e-ISBN: 235158015X); *Ed. A. T. Wolf*

PRO 28: 6th International RILEM Symposium on Performance Testing and Evaluation of Bituminous Materials - PTEBM'03 (ISBN: 2-912143-35-7; e-ISBN: 978-2-912143-77-8); *Ed. M. N. Partl*

PRO 29: 2nd International RILEM Workshop on Life Prediction and Ageing Management of Concrete Structures (ISBN: 2-912143-36-5; e-ISBN: 2912143780); *Ed. D. J. Naus*

PRO 30: 4th International RILEM Workshop on High Performance Fiber Reinforced Cement Composites - HPFRCC 4 (ISBN: 2-912143-37-3; e-ISBN: 2912143799); *Eds. A. E. Naaman and H. W. Reinhardt*

PRO 31: International RILEM Workshop on Test and Design Methods for Steel Fibre Reinforced Concrete: Background and Experiences (ISBN: 2-912143-38-1; e-ISBN: 2351580168); *Eds. B. Schnütgen and L. Vandewalle*

PRO 32: International Conference on Advances in Concrete and Structures 2 vol. (ISBN (set): 2-912143-41-1; e-ISBN: 2351580176); *Eds. Ying-shu Yuan, Surendra P. Shah and Heng-lin Lü*

PRO 33: 3rd International Symposium on Self-Compacting Concrete (ISBN: 2-912143-42-X; e-ISBN: 2912143713); *Eds. Ó. Wallevik and I. Níelsson*

PRO 34: International RILEM Conference on Microbial Impact on Building Materials (ISBN: 2-912143-43-8; e-ISBN: 2351580184); *Ed. M. Ribas Silva*

PRO 35: International RILEM TC 186-ISA on Internal Sulfate Attack and Delayed Ettringite Formation (ISBN: 2-912143-44-6; e-ISBN: 2912143802); *Eds. K. Scrivener and J. Skalny*

PRO 36: International RILEM Symposium on Concrete Science and Engineering – A Tribute to Arnon Bentur (ISBN: 2-912143-46-2; e-ISBN: 2912143586); *Eds. K. Kovler, J. Marchand, S. Mindess and J. Weiss*

PRO 37: 5th International RILEM Conference on Cracking in Pavements – Mitigation, Risk Assessment and Prevention (ISBN: 2-912143-47-0; e-ISBN: 2912143764); *Eds. C. Petit, I. Al-Qadi and A. Millien*

PRO 38: 3rd International RILEM Workshop on Testing and Modelling the Chloride Ingress into Concrete (ISBN: 2-912143-48-9; e-ISBN: 2912143578); *Eds. C. Andrade and J. Kropp*

PRO 39: 6th International RILEM Symposium on Fibre-Reinforced Concretes - BEFIB 2004 (ISBN: 2-912143-51-9; e-ISBN: 2912143748); *Eds. M. Di Prisco, R. Felicetti and G. A. Plizzari*

PRO 40: International RILEM Conference on the Use of Recycled Materials in Buildings and Structures (ISBN: 2-912143-52-7; e-ISBN: 2912143756); *Eds. E. Vázquez, Ch. F. Hendriks and G. M. T. Janssen*

PRO 41: RILEM International Symposium on Environment-Conscious Materials and Systems for Sustainable Development (ISBN: 2-912143-55-1; e-ISBN: 2912143640); *Eds. N. Kashino and Y. Ohama*

PRO 42: SCC'2005 - China: 1st International Symposium on Design, Performance and Use of Self-Consolidating Concrete (ISBN: 2-912143-61-6; e-ISBN: 2912143624); *Eds. Zhiwu Yu, Caijun Shi, Kamal Henri Khayat and Youjun Xie*

PRO 43: International RILEM Workshop on Bonded Concrete Overlays (e-ISBN: 2-912143-83-7); *Eds. J. L. Granju and J. Silfwerbrand*

PRO 44: 2nd International RILEM Workshop on Microbial Impacts on Building Materials (CD11) (e-ISBN: 2-912143-84-5); *Ed. M. Ribas Silva*

PRO 45: 2nd International Symposium on Nanotechnology in Construction, Bilbao (ISBN: 2-912143-87-X; e-ISBN: 2912143888); *Eds. Peter J. M. Bartos, Yolanda de Miguel and Antonio Porro*

PRO 46: ConcreteLife'06 - International RILEM-JCI Seminar on Concrete Durability and Service Life Planning: Curing, Crack Control, Performance in Harsh Environments (ISBN: 2-912143-89-6; e-ISBN: 291214390X); *Ed. K. Kovler*

PRO 47: International RILEM Workshop on Performance Based Evaluation and Indicators for Concrete Durability (ISBN: 978-2-912143-95-2; e-ISBN: 9782912143969); *Eds. V. Baroghel-Bouny, C. Andrade, R. Torrent and K. Scrivener*

PRO 48: 1st International RILEM Symposium on Advances in Concrete through Science and Engineering (e-ISBN: 2-912143-92-6); *Eds. J. Weiss, K. Kovler, J. Marchand and S. Mindess*

PRO 49: International RILEM Workshop on High Performance Fiber Reinforced Cementitious Composites in Structural Applications (ISBN: 2-912143-93-4; e-ISBN: 2912143942); *Eds. G. Fischer and V. C. Li*

PRO 50: 1st International RILEM Symposium on Textile Reinforced Concrete (ISBN: 2-912143-97-7; e-ISBN: 2351580087); *Eds. Josef Hegger, Wolfgang Brameshuber and Norbert Will*

PRO 51: 2nd International Symposium on Advances in Concrete through Science and Engineering (ISBN: 2-35158-003-6; e-ISBN: 2-35158-002-8); *Eds. J. Marchand, B. Bissonnette, R. Gagné, M. Jolin and F. Paradis*

PRO 52: Volume Changes of Hardening Concrete: Testing and Mitigation (ISBN: 2-35158-004-4; e-ISBN: 2-35158-005-2); *Eds. O. M. Jensen, P. Lura and K. Kovler*

PRO 53: High Performance Fiber Reinforced Cement Composites - HPFRCC5 (ISBN: 978-2-35158-046-2; e-ISBN: 978-2-35158-089-9); *Eds. H. W. Reinhardt and A. E. Naaman*

PRO 54: 5th International RILEM Symposium on Self-Compacting Concrete (ISBN: 978-2-35158-047-9; e-ISBN: 978-2-35158-088-2); *Eds. G. De Schutter and V. Boel*

PRO 55: International RILEM Symposium Photocatalysis, Environment and Construction Materials (ISBN: 978-2-35158-056-1; e-ISBN: 978-2-35158-057-8); *Eds. P. Baglioni and L. Cassar*

PRO 56: International RILEM Workshop on Integral Service Life Modelling of Concrete Structures (ISBN 978-2-35158-058-5; e-ISBN: 978-2-35158-090-5); *Eds. R. M. Ferreira, J. Gulikers and C. Andrade*

PRO 57: RILEM Workshop on Performance of cement-based materials in aggressive aqueous environments (e-ISBN: 978-2-35158-059-2); *Ed. N. De Belie*

PRO 58: International RILEM Symposium on Concrete Modelling - CONMOD'08 (ISBN: 978-2-35158-060-8; e-ISBN: 978-2-35158-076-9); *Eds. E. Schlangen and G. De Schutter*

PRO 59: International RILEM Conference on On Site Assessment of Concrete, Masonry and Timber Structures - SACoMaTiS 2008 (ISBN set: 978-2-35158-061-5; e-ISBN: 978-2-35158-075-2); *Eds. L. Binda, M. di Prisco and R. Felicetti*

PRO 60: Seventh RILEM International Symposium on Fibre Reinforced Concrete: Design and Applications - BEFIB 2008 (ISBN: 978-2-35158-064-6; e-ISBN: 978-2-35158-086-8); *Ed. R. Gettu*

PRO 61: 1st International Conference on Microstructure Related Durability of Cementitious Composites 2 vol., (ISBN: 978-2-35158-065-3; e-ISBN: 978-2-35158-084-4); *Eds. W. Sun, K. van Breugel, C. Miao, G. Ye and H. Chen*

PRO 62: NSF/ RILEM Workshop: In-situ Evaluation of Historic Wood and Masonry Structures (e-ISBN: 978-2-35158-068-4); *Eds. B. Kasal, R. Anthony and M. Drdácký*

PRO 63: Concrete in Aggressive Aqueous Environments: Performance, Testing and Modelling, 2 vol., (ISBN: 978-2-35158-071-4; e-ISBN: 978-2-35158-082-0); *Eds. M. G. Alexander and A. Bertron*

PRO 64: Long Term Performance of Cementitious Barriers and Reinforced Concrete in Nuclear Power Plants and Waste Management - NUCPERF 2009 (ISBN: 978-2-35158-072-1; e-ISBN: 978-2-35158-087-5); *Eds. V. L'Hostis, R. Gens and C. Gallé*

PRO 65: Design Performance and Use of Self-consolidating Concrete - SCC'2009 (ISBN: 978-2-35158-073-8; e-ISBN: 978-2-35158-093-6); *Eds. C. Shi, Z. Yu, K. H. Khayat and P. Yan*

PRO 66: 2nd International RILEM Workshop on Concrete Durability and Service Life Planning - ConcreteLife'09 (ISBN: 978-2-35158-074-5; ISBN: 978-2-35158-074-5); *Ed. K. Kovler*

PRO 67: Repairs Mortars for Historic Masonry (e-ISBN: 978-2-35158-083-7); *Ed. C. Groot*

PRO 68: Proceedings of the 3rd International RILEM Symposium on 'Rheology of Cement Suspensions such as Fresh Concrete (ISBN 978-2-35158-091-2; e-ISBN: 978-2-35158-092-9); *Eds. O. H. Wallevik, S. Kubens and S. Oesterheld*

PRO 69: 3rd International PhD Student Workshop on 'Modelling the Durability of Reinforced Concrete (ISBN: 978-2-35158-095-0); *Eds. R. M. Ferreira, J. Gulikers and C. Andrade*

PRO 70: 2nd International Conference on 'Service Life Design for Infrastructure' (ISBN set: 978-2-35158-096-7, e-ISBN: 978-2-35158-097-4); *Eds. K. van Breugel, G. Ye and Y. Yuan*

PRO 71: Advances in Civil Engineering Materials - The 50-year Teaching Anniversary of Prof. Sun Wei' (ISBN: 978-2-35158-098-1; e-ISBN: 978-2-35158-099-8); *Eds. C. Miao, G. Ye and H. Chen*

PRO 72: First International Conference on 'Advances in Chemically-Activated Materials – CAM'2010' (2010), 264 pp, ISBN: 978-2-35158-101-8; e-ISBN: 978-2-35158-115-5, *Eds. Caijun Shi and Xiaodong Shen*

PRO 73: 2nd International Conference on 'Waste Engineering and Management - ICWEM 2010' (2010), 894 pp, ISBN: 978-2-35158-102-5; e-ISBN: 978-2-35158-103-2, *Eds. J. Zh. Xiao, Y. Zhang, M. S. Cheung and R. Chu*

PRO 74: International RILEM Conference on 'Use of Superabsorbent Polymers and Other New Additives in Concrete' (2010) 374 pp., ISBN: 978-2-35158-104-9; e-ISBN: 978-2-35158-105-6; *Eds. O. M. Jensen, M. T. Hasholt and S. Laustsen*

PRO 75: International Conference on 'Material Science - 2nd ICTRC - Textile Reinforced Concrete - Theme 1' (2010) 436 pp., ISBN: 978-2-35158-106-3; e-ISBN: 978-2-35158-107-0; *Ed. W. Brameshuber*

PRO 76: International Conference on 'Material Science - HetMat - Modelling of Heterogeneous Materials - Theme 2' (2010) 255 pp., ISBN: 978-2-35158-108-7; e-ISBN: 978-2-35158-109-4; *Ed. W. Brameshuber*

PRO 77: International Conference on 'Material Science - AdIPoC - Additions Improving Properties of Concrete - Theme 3' (2010) 459 pp., ISBN: 978-2-35158-110-0; e-ISBN: 978-2-35158-111-7; *Ed. W. Brameshuber*

PRO 78: 2nd Historic Mortars Conference and RILEM TC 203-RHM Final Workshop – HMC2010 (2010) 1416 pp., e-ISBN: 978-2-35158-112-4; *Eds. J. Válek, C. Groot and J. J. Hughes*

PRO 79: International RILEM Conference on Advances in Construction Materials Through Science and Engineering (2011) 213 pp., ISBN: 978-2-35158-116-2, e-ISBN: 978-2-35158-117-9; *Eds. Christopher Leung and K. T. Wan*

PRO 80: 2nd International RILEM Conference on Concrete Spalling due to Fire Exposure (2011) 453 pp., ISBN: 978-2-35158-118-6, e-ISBN: 978-2-35158-119-3; *Eds. E. A. B. Koenders and F. Dehn*

PRO 81: 2nd International RILEM Conference on Strain Hardening Cementitious Composites (SHCC2-Rio) (2011) 451 pp., ISBN: 978-2-35158-120-9, e-ISBN: 978-2-35158-121-6; *Eds. R. D. Toledo Filho, F. A. Silva, E. A. B. Koenders and E. M. R. Fairbairn*

PRO 82: 2nd International RILEM Conference on Progress of Recycling in the Built Environment (2011) 507 pp., e-ISBN: 978-2-35158-122-3; *Eds. V. M. John, E. Vazquez, S. C. Angulo and C. Ulsen*

PRO 83: 2nd International Conference on Microstructural-related Durability of Cementitious Composites (2012) 250 pp., ISBN: 978-2-35158-129-2; e-ISBN: 978-2-35158-123-0; *Eds. G. Ye, K. van Breugel, W. Sun and C. Miao*

PRO 84: CONSEC13 - Seventh International Conference on Concrete under Severe Conditions – Environment and Loading (2013) 1930 pp., ISBN: 978-2-35158-124-7; e-ISBN: 978-2-35158-134-6; *Eds. Z. J. Li, W. Sun, C. W. Miao, K. Sakai, O. E. Gjorv and N. Banthia*

PRO 85: RILEM-JCI International Workshop on Crack Control of Mass Concrete and Related issues concerning Early-Age of Concrete Structures – ConCrack 3 – Control of Cracking in Concrete Structures 3 (2012) 237 pp., ISBN: 978-2-35158-125-4; e-ISBN: 978-2-35158-126-1; *Eds. F. Toutlemonde and J.-M. Torrenti*

PRO 86: International Symposium on Life Cycle Assessment and Construction (2012) 414 pp., ISBN: 978-2-35158-127-8, e-ISBN: 978-2-35158-128-5; *Eds. A. Ventura and C. de la Roche*

PRO 87: UHPFRC 2013 – RILEM-fib-AFGC International Symposium on Ultra-High Performance Fibre-Reinforced Concrete (2013), ISBN: 978-2-35158-130-8, e-ISBN: 978-2-35158-131-5; *Ed. F. Toutlemonde*

PRO 88: 8th RILEM International Symposium on Fibre Reinforced Concrete (2012) 344 pp., ISBN: 978-2-35158-132-2, e-ISBN: 978-2-35158-133-9; *Ed. Joaquim A. O. Barros*

PRO 89: RILEM International workshop on performance-based specification and control of concrete durability (2014) 678 pp, ISBN: 978-2-35158-135-3, e-ISBN: 978-2-35158-136-0; *Eds. D. Bjegović, H. Beushausen and M. Serdar*

PRO 90: 7th RILEM International Conference on Self-Compacting Concrete and of the 1st RILEM International Conference on Rheology and Processing of Construction Materials (2013) 396 pp, ISBN: 978-2-35158-137-7, e-ISBN: 978-2-35158-138-4; *Eds. Nicolas Roussel and Hela Bessaies-Bey*

PRO 91: CONMOD 2014 - RILEM International Symposium on Concrete Modelling (2014), ISBN: 978-2-35158-139-1; e-ISBN: 978-2-35158-140-7; *Eds. Kefei Li, Peiyu Yan and Rongwei Yang*

PRO 92: CAM 2014 - 2nd International Conference on advances in chemically-activated materials (2014) 392 pp., ISBN: 978-2-35158-141-4; e-ISBN: 978-2-35158-142-1; *Eds. Caijun Shi and Xiadong Shen*

PRO 93: SCC 2014 - 3rd International Symposium on Design, Performance and Use of Self-Consolidating Concrete (2014) 438 pp., ISBN: 978-2-35158-143-8; e-ISBN: 978-2-35158-144-5; *Eds. Caijun Shi, Zhihua Ou and Kamal H. Khayat*

PRO 94 (online version): HPFRCC-7 - 7th RILEM conference on High performance fiber reinforced cement composites (2015), e-ISBN: 978-2-35158-146-9; *Eds. H. W. Reinhardt, G. J. Parra-Montesinos and H. Garrecht*

PRO 95: International RILEM Conference on Application of superabsorbent polymers and other new admixtures in concrete construction (2014), ISBN: 978-2-35158-147-6; e-ISBN: 978-2-35158-148-3; *Eds. Viktor Mechtcherine and Christof Schroefl*

PRO 96 (online version): XIII DBMC: XIII International Conference on Durability of Building Materials and Components (2015), e-ISBN: 978-2-35158-149-0; *Eds. M. Quattrone and V. M. John*

PRO 97: SHCC3 – 3rd International RILEM Conference on Strain Hardening Cementitious Composites (2014), ISBN: 978-2-35158-150-6; e-ISBN: 978-2-35158-151-3; *Eds. E. Schlangen, M. G. Sierra Beltran, M. Lukovic and G. Ye*

PRO 98: FERRO-11 – 11th International Symposium on Ferrocement and 3rd ICTRC - International Conference on Textile Reinforced Concrete (2015), ISBN: 978-2-35158-152-0; e-ISBN: 978-2-35158-153-7; *Ed. W. Brameshuber*

PRO 99 (online version): ICBBM 2015 - 1st International Conference on Bio-Based Building Materials (2015), e-ISBN: 978-2-35158-154-4; *Eds. S. Amziane and M. Sonebi*

PRO 100: SCC16 - RILEM Self-Consolidating Concrete Conference (2016), ISBN: 978-2-35158-156-8; e-ISBN: 978-2-35158-157-5; *Ed. Kamal H. Kayat*

PRO 101 (online version): III Progress of Recycling in the Built Environment (2015), e-ISBN: 978-2-35158-158-2; *Eds. I. Martins, C. Ulsen and S. C. Angulo*

PRO 102 (online version): RILEM Conference on Microorganisms-Cementitious Materials Interactions (2016), e-ISBN: 978-2-35158-160-5; *Eds. Alexandra Bertron, Henk Jonkers and Virginie Wiktor*

PRO 103 (online version): ACESC'16 - Advances in Civil Engineering and Sustainable Construction (2016), e-ISBN: 978-2-35158-161-2; *Eds. T. Ch. Madhavi, G. Prabhakar, Santhosh Ram and P. M. Rameshwaran*

PRO 104 (online version): SSCS'2015 - Numerical Modeling - Strategies for Sustainable Concrete Structures (2015), e-ISBN: 978-2-35158-162-9

PRO 105: 1st International Conference on UHPC Materials and Structures (2016), ISBN: 978-2-35158-164-3, e-ISBN: 978-2-35158-165-0

PRO 106: AFGC-ACI-fib-RILEM International Conference on Ultra-High-Performance Fibre-Reinforced Concrete – UHPFRC 2017 (2017), ISBN: 978-2-35158-166-7, e-ISBN: 978-2-35158-167-4; *Eds. François Toutlemonde and Jacques Resplendino*

PRO 107 (online version): XIV DBMC – 14th International Conference on Durability of Building Materials and Components (2017), e-ISBN: 978-2-35158-159-9; *Eds. Geert De Schutter, Nele De Belie, Arnold Janssens and Nathan Van Den Bossche*

PRO 108: MSSCE 2016 - Innovation of Teaching in Materials and Structures (2016), ISBN: 978-2-35158-178-0, e-ISBN: 978-2-35158-179-7; *Ed. Per Goltermann*

PRO 109 (2 volumes): MSSCE 2016 - Service Life of Cement-Based Materials and Structures (2016), ISBN Vol. 1: 978-2-35158-170-4, Vol. 2: 978-2-35158-171-4, Set Vol. 1&2: 978-2-35158-172-8, e-ISBN: 978-2-35158-173-5; *Eds. Miguel Azenha, Ivan Gabrijel, Dirk Schlicke, Terje Kanstad and Ole Mejlhede Jensen*

PRO 110: MSSCE 2016 - Historical Masonry (2016), ISBN: 978-2-35158-178-0, e-ISBN: 978-2-35158-179-7; *Eds. Inge Rörig-Dalgaard and Ioannis Ioannou*

PRO 111: MSSCE 2016 - Electrochemistry in Civil Engineering (2016), ISBN: 978-2-35158-176-6, e-ISBN: 978-2-35158-177-3; *Ed. Lisbeth M. Ottosen*

PRO 112: MSSCE 2016 - Moisture in Materials and Structures (2016), ISBN: 978-2-35158-178-0, e-ISBN: 978-2-35158-179-7; *Eds. Kurt Kielsgaard Hansen, Carsten Rode and Lars-Olof Nilsson*

PRO 113: MSSCE 2016 - Concrete with Supplementary Cementitious Materials (2016), ISBN: 978-2-35158-178-0, e-ISBN: 978-2-35158-179-7; *Eds. Ole Mejlhede Jensen, Konstantin Kovler and Nele De Belie*

PRO 114: MSSCE 2016 - Frost Action in Concrete (2016), ISBN: 978-2-35158-182-7, e-ISBN: 978-2-35158-183-4; *Eds. Marianne Tange Hasholt, Katja Fridh and R. Doug Hooton*

PRO 115: MSSCE 2016 - Fresh Concrete (2016), ISBN: 978-2-35158-184-1, e-ISBN: 978-2-35158-185-8; *Eds. Lars N. Thrane, Claus Pade, Oldrich Svec and Nicolas Roussel*

PRO 116: BEFIB 2016 – 9th RILEM International Symposium on Fiber Reinforced Concrete (2016), ISBN: 978-2-35158-187-2, e-ISBN: 978-2-35158-186-5; *Eds. N. Banthia, M. di Prisco and S. Soleimani-Dashtaki*

PRO 117: 3rd International RILEM Conference on Microstructure Related Durability of Cementitious Composites (2016), ISBN: 978-2-35158-188-9, e-ISBN: 978-2-35158-189-6; *Eds. Changwen Miao, Wei Sun, Jiaping Liu, Huisu Chen, Guang Ye and Klaas van Breugel*

PRO 118 (4 volumes): International Conference on Advances in Construction Materials and Systems (2017), ISBN Set: 978-2-35158-190-2, Vol. 1: 978-2-35158-193-3, Vol. 2: 978-2-35158-194-0, Vol. 3: ISBN:978-2-35158-195-7, Vol. 4: ISBN:978-2-35158-196-4, e-ISBN: 978-2-35158-191-9; *Eds. Manu Santhanam, Ravindra Gettu, Radhakrishna G. Pillai and Sunitha K. Nayar*

PRO 119 (online version): ICBBM 2017 - Second International RILEM Conference on Bio-based Building Materials, (2017), e-ISBN: 978-2-35158-192-6; *Eds. Sofiane Amziane and Mohammed Sonebi*

PRO 120 (2 volumes): EAC-02 - 2nd International RILEM/COST Conference on Early Age Cracking and Serviceability in Cement-based Materials and Structures, (2017), Vol.

1: 978-2-35158-199-5, Vol. 2: 978-2-35158-200-8, Set: 978-2-35158-197-1, e-ISBN: 978-2-35158-198-8; *Eds. Stéphanie Staquet and Dimitrios Aggelis*

PRO 121 (2 volumes): SynerCrete18: Interdisciplinary Approaches for Cement-based Materials and Structural Concrete: Synergizing Expertise and Bridging Scales of Space and Time, (2018), Set: 978-2-35158-202-2, Vol.1: 978-2-35158-211-4, Vol. 2: 978-2-35158-212-1, e-ISBN: 978-2-35158-203-9; *Eds. Miguel Azenha, Dirk Schlicke, Farid Benboudjema and Agnieszka Knoppik*

PRO 122: SCC'2018 China - Fourth International Symposium on Design, Performance and Use of Self-Consolidating Concrete, (2018), ISBN: 978-2-35158-204-6, e-ISBN: 978-2-35158-205-3; *Eds. C. Shi, Z. Zhang and K. H. Khayat*

PRO 123: Final Conference of RILEM TC 253-MCI: Microorganisms-Cementitious Materials Interactions (2018), Set: 978-2-35158-207-7, Vol.1: 978-2-35158-209-1, Vol.2: 978-2-35158-210-7, e-ISBN: 978-2-35158-206-0; *Ed. Alexandra Bertron*

PRO 124 (online version): Fourth International Conference Progress of Recycling in the Built Environment (2018), e-ISBN: 978-2-35158-208-4; *Eds. Isabel M. Martins, Carina Ulsen and Yury Villagran*

PRO 125 (online version): SLD4 - 4th International Conference on Service Life Design for Infrastructures (2018), e-ISBN: 978-2-35158-213-8; *Eds. Guang Ye, Yong Yuan, Claudia Romero Rodriguez, Hongzhi Zhang and Branko Savija*

PRO 126: Workshop on Concrete Modelling and Material Behaviour in honor of Professor Klaas van Breugel (2018), ISBN: 978-2-35158-214-5, e-ISBN: 978-2-35158-215-2; *Ed. Guang Ye*

PRO 127 (online version): CONMOD2018 - Symposium on Concrete Modelling (2018), e-ISBN: 978-2-35158-216-9; *Eds. Erik Schlangen, Geert de Schutter, Branko Savija, Hongzhi Zhang and Claudia Romero Rodriguez*

PRO 128: SMSS2019 - International Conference on Sustainable Materials, Systems and Structures (2019), ISBN: 978-2-35158-217-6, e-ISBN: 978-2-35158-218-3

PRO 129: 2nd International Conference on UHPC Materials and Structures (UHPC2018-China), ISBN: 978-2-35158-219-0, e-ISBN: 978-2-35158-220-6;

PRO 130: 5th Historic Mortars Conference (2019), ISBN: 978-2-35158-221-3, e-ISBN: 978-2-35158-222-0; *Eds. José Ignacio Álvarez, José María Fernández, Íñigo Navarro, Adrián Durán and Rafael Sirera*

PRO 131 (online version): 3rd International Conference on Bio-Based Building Materials (ICBBM2019), e-ISBN: 978-2-35158-229-9; *Eds. Mohammed Sonebi, Sofiane Amziane and Jonathan Page*

PRO 132: IRWRMC'18 - International RILEM Workshop on Rheological Measurements of Cement-based Materials (2018), ISBN: 978-2-35158-230-5, e-ISBN: 978-2-35158-231-2; *Eds. Chafika Djelal and Yannick Vanhove*

PRO 133 (online version): CO2STO2019 - International Workshop CO2 Storage in Concrete (2019), e-ISBN: 978-2-35158-232-9; *Eds. Assia Djerbi, Othman Omikrine-Metalssi and Teddy Fen-Chong*

PRO 134: 3rd ACF/HNU International Conference on UHPC Materials and Structures - UHPC'2020, ISBN: 978-2-35158-233-6, e-ISBN: 978-2-35158-234-3; *Eds. Caijun Shi and Jiaping Liu*

PRO 135: Fourth International Conference on Chemically Activated Materials (CAM2021), ISBN: 978-2-35158-235-0, e-ISBN: 978-2-35158-236-7; *Eds. Caijun Shi and Xiang Hu*

PRO 136: Lime Based Materials for Repairing Historic Structures (2022), ISBN: 978-2-35158-240-4, e-ISBN: 978-2-35158-237-4; *Eds. Ioanna Papayianni and Jan Válek*

PRO 137 (online version): V International Conference Progress of Recycling in the Built Environment (2023), e-ISBN: 978-2-35158-238-1; *Eds. Ines Döring, Yury Villagrán Zaccardi and Anette Müller*

RILEM Reports (REP)

Report 19: Considerations for Use in Managing the Aging of Nuclear Power Plant Concrete Structures (ISBN: 2-912143-07-1); *Ed. D. J. Naus*

Report 20: Engineering and Transport Properties of the Interfacial Transition Zone in Cementitious Composites (ISBN: 2-912143-08-X); *Eds. M. G. Alexander, G. Arliguie, G. Ballivy, A. Bentur and J. Marchand*

Report 21: Durability of Building Sealants (ISBN: 2-912143-12-8); *Ed. A. T. Wolf*

Report 22: Sustainable Raw Materials - Construction and Demolition Waste (ISBN: 2-912143-17-9); *Eds. C. F. Hendriks and H. S. Pietersen*

Report 23: Self-Compacting Concrete state-of-the-art report (ISBN: 2-912143-23-3); *Eds. Å. Skarendahl and Ö. Petersson*

Report 24: Workability and Rheology of Fresh Concrete: Compendium of Tests (ISBN: 2-912143-32-2); *Eds. P. J. M. Bartos, M. Sonebi and A. K. Tamimi*

Report 25: Early Age Cracking in Cementitious Systems (ISBN: 2-912143-33-0); *Ed. A. Bentur*

Report 26: Towards Sustainable Roofing (Joint Committee CIB/RILEM) (CD 07) (e-ISBN 978-2-912143-65-5); *Eds. Thomas W. Hutchinson and Keith Roberts*

Report 27: Condition Assessment of Roofs (Joint Committee CIB/RILEM) (CD 08) (e-ISBN 978-2-912143-66-2); *Ed. CIB W 83/RILEM TC166-RMS*

Report 28: Final report of RILEM TC 167-COM 'Characterisation of Old Mortars with Respect to Their Repair (ISBN: 978-2-912143-56-3); *Eds. C. Groot, G. Ashall and J. Hughes*

Report 29: Pavement Performance Prediction and Evaluation (PPPE): Interlaboratory Tests (e-ISBN: 2-912143-68-3); *Eds. M. Partl and H. Piber*

Report 30: Final Report of RILEM TC 198-URM 'Use of Recycled Materials' (ISBN: 2-912143-82-9; e-ISBN: 2-912143-69-1); *Eds. Ch. F. Hendriks, G. M. T. Janssen and E. Vázquez*

Report 31: Final Report of RILEM TC 185-ATC 'Advanced testing of cement-based materials during setting and hardening' (ISBN: 2-912143-81-0; e-ISBN: 2-912143-70-5); *Eds. H. W. Reinhardt and C. U. Grosse*

Report 32: Probabilistic Assessment of Existing Structures. A JCSS publication (ISBN 2-912143-24-1); *Ed. D. Diamantidis*

Report 33: State-of-the-Art Report of RILEM Technical Committee TC 184-IFE 'Industrial Floors' (ISBN 2-35158-006-0); *Ed. P. Seidler*

Report 34: Report of RILEM Technical Committee TC 147-FMB 'Fracture mechanics applications to anchorage and bond' Tension of Reinforced Concrete Prisms – Round Robin Analysis and Tests on Bond (e-ISBN 2-912143-91-8); *Eds. L. Elfgren and K. Noghabai*

Report 35: Final Report of RILEM Technical Committee TC 188-CSC 'Casting of Self Compacting Concrete' (ISBN 2-35158-001-X; e-ISBN: 2-912143-98-5); *Eds. Å. Skarendahl and P. Billberg*

Report 36: State-of-the-Art Report of RILEM Technical Committee TC 201-TRC 'Textile Reinforced Concrete' (ISBN 2-912143-99-3); *Ed. W. Brameshuber*

Report 37: State-of-the-Art Report of RILEM Technical Committee TC 192-ECM 'Environment-conscious construction materials and systems' (ISBN: 978-2-35158-053-0); *Eds. N. Kashino, D. Van Gemert and K. Imamoto*

Report 38: State-of-the-Art Report of RILEM Technical Committee TC 205-DSC 'Durability of Self-Compacting Concrete' (ISBN: 978-2-35158-048-6); *Eds. G. De Schutter and K. Audenaert*

Report 39: Final Report of RILEM Technical Committee TC 187-SOC 'Experimental determination of the stress-crack opening curve for concrete in tension' (ISBN 978-2-35158-049-3); *Ed. J. Planas*

Report 40: State-of-the-Art Report of RILEM Technical Committee TC 189-NEC 'Non-Destructive Evaluation of the Penetrability and Thickness of the Concrete Cover' (ISBN 978-2-35158-054-7); *Eds. R. Torrent and L. Fernández Luco*

Report 41: State-of-the-Art Report of RILEM Technical Committee TC 196-ICC 'Internal Curing of Concrete' (ISBN 978-2-35158-009-7); *Eds. K. Kovler and O. M. Jensen*

Report 42: 'Acoustic Emission and Related Non-destructive Evaluation Techniques for Crack Detection and Damage Evaluation in Concrete' - Final Report of RILEM Technical Committee 212-ACD (e-ISBN: 978-2-35158-100-1); *Ed. M. Ohtsu*

Report 45: Repair Mortars for Historic Masonry - State-of-the-Art Report of RILEM Technical Committee TC 203-RHM (e-ISBN: 978-2-35158-163-6); *Eds. Paul Maurenbrecher and Caspar Groot*

Report 46: *S*urface delamination of concrete industrial floors and other durability related aspects guide - Report of RILEM Technical Committee TC 268-SIF (e-ISBN: 978-2-35158-201-5); *Ed. Valerie Pollet*

Bio-based Materials

Enhancing Urban Biodiversity Through Bioreceptive Materials: Insights on Design, Properties, and Additive Manufacturing Applications

Rafael Andres Castillo-Watts[1](✉), Maria De Los Ángeles Ortega Del Rosario[1], and Miguel Chen Austin[2]

[1] Research Group in Design, Manufacturing and Materials (DM+M), School of Mechanical Engineering, Universidad Tecnológica de Panamá, Panama City 0819-07289, Panama
maria.ortega@utp.ac.pa

[2] Research Group in Energy and Comfort in Bioclimatic Buildings (ECEB), Faculty of Mechanical Engineering, Universidad Tecnológica de Panamá, Ciudad de Panamá, Panama

Abstract. Bioreceptivity refers to a material's capacity to support biological colonization without undergoing biodeterioration, a concept of growing relevance in engineering design. Current challenges in the architecture, engineering, and construction industry, like waste generation, CO_2 emissions, water footprints, climate change, and increasing urbanization, have driven the search for strategies to mitigate these impacts. One promising approach involves promoting urban biodiversity by developing bioreceptive surfaces and facades, which have garnered recent attention. This work synthesizes findings from various studies on bioreceptive materials, emphasizing the intrinsic and extrinsic properties most relevant for effective design considerations.

Key intrinsic factors influencing bioreceptivity include surface roughness, porosity, and chemical composition, though water retention remains the primary factor for biological growth. Bioreceptivity is also highly dependent on extrinsic environmental conditions such as humidity, air quality, and the availability of organisms in the surroundings, which must be considered during the design phase. In this sense, additive manufacturing (AM) emerges as a valuable tool for developing bioreceptive structures, as it can fabricate complex morphologies that support spore adhesion and biological colonization. Furthermore, experimental strategies, such as using organic waste and living materials, explore novel ways to enhance bioreceptivity while promoting ecological sustainability.

This analysis emphasizes the integration of materials science, ecological principles, and innovative manufacturing techniques to enhance bioreceptive performance. These materials offer promising pathways for improving urban biodiversity and supporting ecological restoration efforts. This work aims to develop a comprehensive decision-making framework for engineers, designers, and architects, highlighting key parameters for promoting bioreceptive structures and fostering bio-inclusive urban spaces.

Keywords: Additive manufacturing · bioreceptivity · ecological restoration · sustainability · urban biodiversity

M. Posani and S. Singh (Eds.): RYS 2025, RILEM Bookseries 67, pp. 587–602, 2026.
https://doi.org/10.1007/978-3-032-14652-6_42

1 Introduction

Urbanization and climate change are major drivers of biodiversity loss in urban environments, altering land use, habitat continuity, and ecological balance. Recent estimates by the World Bank indicate that more than half of the global population—over 4 billion people—currently resides in urban areas. This figure is projected to exceed 70% by 2050, implying a continued expansion of urban land and increased anthropogenic pressures on natural systems embedded within cities (World Bank, 2024) [1]. As the world continues to urbanize, sustainable development increasingly depends on the successful management of urban growth, especially in low- and lower-middle-income countries, where the urbanization rate is expected to be faster [2–6].

This uncontrolled urban expansion has significantly reduced green spaces and exacerbated the "Urban Heat Island" effect, a phenomenon where heat accumulates around built environments due to the retention of solar energy in impervious surfaces and the lack of vegetation [7–9]. In this context, it is essential to adopt architectural design approaches that integrate both ecological and functional considerations.

One of these approaches is bioreceptivity, where materials are selected to facilitate the passive colonization of organisms. Instead of vegetated systems, bioreceptive design focuses on construction materials that can passively support biological community establishment. This approach meets the need for compact design which harmonizes urban development with the need for nature, thus achieving conservation goals for urban development and biodiversity amenities, while nurturing the relation of architecture with nature and the ecological processes that are bound to alter.

The implementation of bioreceptive materials represents a highly valuable strategy for integrating biodiversity into cities, allowing built surfaces to host living organisms without compromising their integrity [10, 11]. Unlike traditional green walls, which require high maintenance costs and mechanical irrigation systems, bioreceptive solutions operate passively by leveraging material properties such as roughness, porosity, and capillary water retention [12]. In this way, the formation of microhabitats is promoted, contributing to both environmental quality and ecological regeneration.

Sustainable materials are recognized not only for reducing negative environmental impacts but also for enhancing occupant health and lowering carbon emissions, as acknowledged by international certification systems such as LEED v4.1 [13]. Alongside, the study of bioreceptivity in concrete and natural materials has emerged as a promising strategy to mitigate global warming and restore ecological functions in highly urbanized areas [14].

Simultaneously, additive manufacturing (AM) is emerging as a central enabler to implement innovative bioreceptive solutions [15, 16]. This is due to its capacity to construct intricate geometries, tailor surface textures, and minimize material usage. These are all factors that make it possible to design architectural structures suited for biological colonization [17]. Also, when these structures are integrated with the sustainability principles of biomimicry, AM transforms into a driver for the development of built ecosystems that coexist and thrive alongside organisms [18].

Given the growing relevance of these approaches, this article provides a comprehensive review of the key factors influencing the promotion of bioreceptivity and ecological design, as well as the role of additive manufacturing in creating resilient, sustainable, and bioinclusive urban environments.

2 Methodology of the Literature Review

The review aimed to analyze the scientific principles and existing work on the use of bioreceptive materials for augmenting urban biodiversity through additive manufacturing. The articles included in the review focused on bioreceptivity, sustainable materials, additive manufacturing, and bio-integrated architectural design related to clay and cement composites.

The literature search was conducted using two primary academic databases: Scopus, selected for its comprehensive coverage of peer-reviewed scientific publications, and Google Scholar, which offered additional access to interdisciplinary and emerging research. Search queries combined relevant keywords such as bioreceptive, additive manufacturing, 3D printing, clay, cement, biodiversity, colonization, simulation, and ecological design.

The selected studies were then categorized into thematic focus areas, including material formulation, rheological behavior, surface texture optimization, and environmental performance in the urban setting. This thematic classification enabled a structured analysis of the key factors influencing the development and implementation of bioreceptive materials in urban environments.

3 Bioreceptivity: Concepts and Influencing Factors

3.1 Definition

Bioreceptivity is defined as the inherent capacity of a material to support the colonization of living organisms without compromising its structural integrity [10]. In the last couple of years, this term has been pivotal in reconsidering urban design from an ecological perspective. While traditionally associated with microalgae, lichens, mosses, and bacteria, the concept can also be extended to structures that indirectly support larger organisms such as insects, birds, or small mammals, especially when integrated into broader ecological design strategies [12, 15]. Thus, bioreceptivity can be understood not only as a passive surface condition, but also as an active enabler of multispecies cohabitation.

3.2 Types of Bioreceptivity

In recent years, the concept of bioreceptivity has gained prominence in rethinking urban design through an ecological lens. As shown in Fig. 1, bioreceptivity can be classified into four levels based on how materials interact with living organisms. By enabling the natural colonization of organisms such as algae, mosses, lichens, and bacteria, bioreceptive materials contribute to promoting local biodiversity, purifying air, and restoring microhabitats on building surfaces [10].

- Primary bioreceptivity. The initial ability of a material to be colonized by organisms immediately after it has been manufactured.

- Secondary bioreceptivity. The capacity to be colonized when the material has undergone natural weathering and modification of the environment.
- Tertiary bioreceptivity. Occurs when human activity modifies the surface of the material to promote or reduce colonization.
- Quaternary bioreceptivity. Refers not to the material but to the accumulation of external factors on its surface, which indirectly promote biological growth.

This classification serves as a framework for evaluating how material characteristics and environmental conditions jointly determine the potential for bio-colonization, an essential consideration in ecological architectural design.

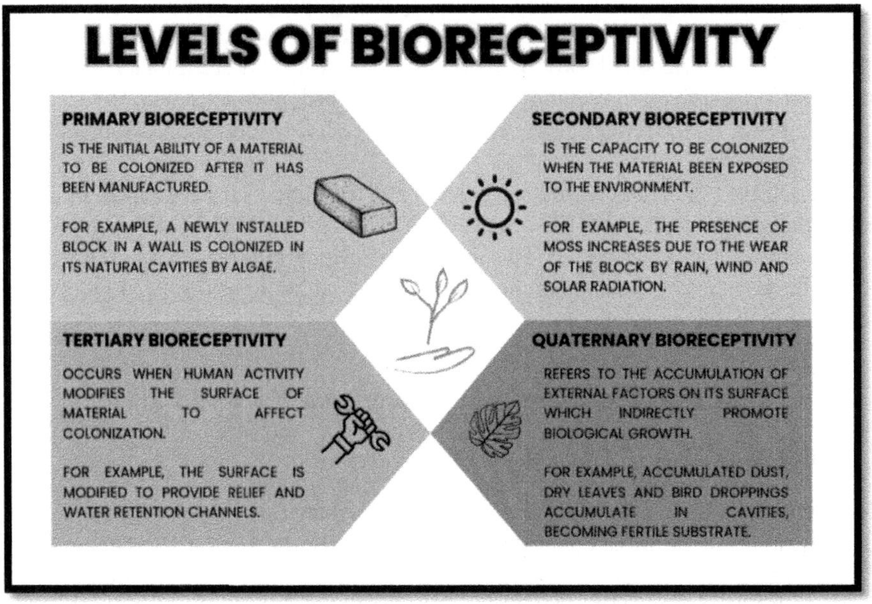

Fig. 1. Schematic representation of the four levels of bioreceptivity in materials. The illustration uses the example of a clay block progressing through these stages under varying environmental and design conditions.

3.3 Intrinsic Factors

Intrinsic factors relate to the physical and chemical characteristics of a material that determine its inherent bioreceptivity. Among them, one can define:

Surface Roughness and Texture. Surface roughness plays a significant role in the primary adhesion of microbes. Sochůrková et al. [19] demonstrated that even minor roughness on a ceramic surface significantly improves bioreceptivity. This roughness acts as a micro-refuge, promoting biofilm formation and water retention. Likewise, Lim and Lharchi [5] demonstrated that the incorporation of rough or grooved morphologies in 3D-printed concrete triggered cyanobacterial growth by creating zones of shading and niches with high water-retention capacities, which significantly prevent water evaporation. The results highlight the potential of design-controlled roughness as a method for regulating biological colonization.

Porosity and Capillary Water Retention. An open, porous texture favors capillary movement and the storage of moisture, which is essential for microbial life. For instance, Sochůrková et al. [15] incorporated sawdust and sediment into clay, resulting in fired ceramics with different porosity features. Sawdust was used as an organic pore-former, which burnt out during firing, thus creating a network of voids that increased water absorption. The greater water retention capacity was linked with high surface humidity over time, which supported microbial colonization. However, this increase in porosity also reduced mechanical stability, highlighting a trade-off between ecological function and structural integrity. Moreover, Vázquez-Nion et al. [20] emphasized that physical characteristics of the material, such as porosity and surface roughness, play a more decisive role in promoting bioreceptivity than chemical composition alone.

Chemical Composition and Nutrient Enrichment. The incorporation of nutrients within the matrix plays a key role in promoting biological colonization. Veeger et al. [21] examined concrete formulations enriched with bone meal, a phosphorus- and calcium-rich additive. Their findings highlighted that phosphorus enhanced the colonization of algal organisms by serving as an immediately accessible nutrient source at the material surface. The pattern of nutrient distribution within the matrix affected the density of growth, i.e., even dispersion is required for uniform bioreceptive effectiveness. These findings validate the argument that bioreceptivity can be improved not just by surface treatment but also by bulk composition changes.

Surface Chemistry and pH Level. While it is commonly held that the optimal pH for biological colonization is generally below 10, Veeger et al. [21] demonstrated that biological growth is still possible at high levels of alkalinity, as long as other conditions (such as nutrient availability, surface roughness, and moisture retention) are adequate. Throughout the study, the pH was not significantly influenced by the addition of bone meal; however, algal colonization was observed due to a synergistic interaction among surface roughness, nutrient availability, and sufficient porosity. It contradicts the widespread assumption that pH is a primary limiting factor but instead demonstrates intricate interrelationships among various intrinsic variables that influence bioreceptive outcomes.

These insights provide a foundational understanding of the material parameters that influence bioreceptivity. Those parameters can potentially be fine-tuned through emerging fabrication methods, including additive manufacturing.

3.4 Extrinsic Environmental Factors

Extrinsic factors refer to the environmental conditions that affect colonization after the material has been installed. Unlike intrinsic factors, which are linked to the composition and properties of the material, these factors depend on the environment in which the structure is located and are essential to ensuring proper bioreceptivity.

Ambient Humidity. Ambient humidity is the most critical bioreceptivity parameter since it provides climatic conditions for the growth, development, and survival of biological communities [22, 23].

In the study conducted by Prieto et al., the surface and subsurface moisture content of the material was found to be directly correlated with the degree of biological colonization. Using a portable, non-destructive microwave-based measurement device, the authors demonstrated that areas with higher water retention in internal layers supported a greater density of organisms [22]. These results highlight the significance of deep moisture as a stable water reservoir, enabling species to persist even during dry periods. This insight is particularly relevant for designing materials and structures intended to support long-term ecological functions in variable climates.

Sunlight and Shading. Sunlight affects surface temperature and evaporation rates, thereby influencing the persistence of humidity and the microclimatic conditions essential for colonization. In their experiments, Lim and Lharchi [24] printed samples with geometrical variations, finding that recessed or shaded areas-maintained surface moisture longer and enabled the establishment of cyanobacteria.

These results suggest that incorporating micro-shading elements into printed structures can be a passive approach to enhance bioreceptivity in sun-exposed environments.

Ambient Temperature. It is a factor that regulates both the metabolism of colonizing microorganisms and the physical and chemical conditions of bioreceptive surfaces. The findings by Quaglarini et al. [25] showed that at temperatures above 35 °C, the growth of certain green algae, such as *Chlorella mirabilis*, is inhibited. In accelerated growth trials on fired bricks, it was observed that at a moderate temperature of 27.5 °C, biofouling was significantly promoted compared to cold temperatures of 10 °C, where the populated area was reduced by up to 75%.

For effective bioreceptivity to develop in materials exposed to the environment, it is not enough for a single extrinsic factor to be in optimal condition; a synergistic combination of multiple environmental variables, especially temperature and relative humidity, is essential. Several studies have shown that the interaction between these factors can enhance or limit biological colonization, even when substrate properties are favorable.Materials for bioreceptive applications.

4 Materials for Bioreceptive Applications

Understanding how various construction materials interact with biological agents is crucial for designing surfaces that support ecological colonization. Each material exhibits distinct physical and chemical characteristics that directly affect its bioreceptive potential. Over the past decade, experimental research has played a crucial role in identifying the conditions that optimize bioreceptivity across different substrates. These studies provide empirical evidence that helps define how some variables, such as firing, temperature, mix design, additive content, or surface morphology, influence colonization success [11, 15, 16, 19, 21, 26, 27].

This section synthesizes the findings of such studies by categorizing materials into two main groups: bioreceptive concrete and ceramic materials. Each group is analyzed based on the methodologies used to enhance their bioreceptive performance and their adaptability to additive manufacturing techniques.

4.1 Bioreceptive Concrete

Concrete is one of the most widely used materials in the construction industry, making it a strategic candidate for bioreceptive applications. Numerous experimental studies have focused on optimizing concrete formulations to enhance their bioreceptive performance.

Veeger et al. [21] identified high surface roughness, open porosity, capillary water content, and abrasion pH below 10 as key factors for enhancing bioreceptivity. Bone ash improved phosphorus availability, promoting growth, while surface retarders increased wettability by removing the hydrophobic cement layer. Crushed expanded clay aggregates also boosted colonization, whereas increasing the water-to-cement ratio had little effect. They concluded that pH reduction below 10 is unnecessary, as biofilm growth occurred even at higher pH levels.

Karmoker et al. [14] investigated a further design incorporating brick chips, bone ash sieved to 150 μm, and vinegar as an acidifying agent to lower the pH. They incorporated these mixes into hexagonal concrete structures, mimicking beehive cells, which created microclimates that capture rain, dew, and dust moisture. Additionally, a biofilm paste composed of lab-grown Sphagnum moss and yogurt was applied to the surface, yielding substantial moss growth after five weeks.

Bone et al. [26] The study compared three types of mortar (ordinary Portland cement, Portland cement with industrial waste, and a low-carbon natural cement) in coastal environments over one year. The results showed that the low-carbon natural cement had greater species diversity. In contrast, the ordinary Portland cement showed a larger colonization area, suggesting that surface complexity and roughness, which were greater in this mortar, may significantly influence bioreceptivity.

Finally, Jakubovskis et al. [28] designed laminar concrete wall panels composed of two layers: a high porosity, lightweight bioreceptive layer, and a structural fiber-reinforced concrete core. The panels promoted colonization by natural, non-vascular plant species in the climatic conditions of Northern Europe and sustained biological growth for two years following installation without nutrient supplementation. A new bio-reinforcement composite consisting of recycled paper pulp and topsoil also improved the system's environmental integration and long-term sustainability, demonstrating concrete's potential not only for colonization but also for energy conservation and urban air filtration.

4.2 Ceramic Materials

Ceramics and clay-based materials have been a focus of interest in bioreceptive studies due to their inherent porosity, textural variability, and compatibility with additive manufacturing. Their mineral composition, microstructure, and environmental responsiveness make them ideal for facilitating microbial colonization and ecological integration in building applications.

Numerous studies have explored improving fired clay bioreceptivity through material modification and surface design. Sochůrková et al. [15] investigated clay composites with organic additives (sawdust, excavation sediments) produced via additive manufacturing and fired at 1000 °C, 1100 °C, and 1200 °C. Higher sawdust content increased water absorption due to porous structures formed by combustion of organic matter, while clay–sediment composites showed better mechanical stability and lower moisture fluctuation, favoring long-term colonization.

In another study conducted by Sochůrková et al. [19], confirmed that both material composition and environmental conditions strongly affect colonization. Biofilms and additives such as agar-agar, xanthan gum, sawdust, or river sediment enhanced growth. Firing at 1100 °C offered the best porosity–stability balance. Surface micro-irregularities and specific geometries improve moss attachment, especially in shaded, humid areas. Overall, microtopography, formulation, and environmental exposure are key factors for optimizing clay-based bioreceptive systems.

Lim and Lharchi [24] combined parametric design with 3D printing to enhance water diffusion in clay structures, identifying water availability as the most critical factor for biocolonization. Using varying layer heights, nozzle widths, curvatures, and channel geometries, they found that thicker nozzles (8 mm) and higher layer heights improved diffusion and moss growth by increasing porosity. Steeper curves aided edge diffusion, while concave interiors retained moisture longer. Wider and deeper channels optimized water delivery and minimized overflow, whereas smooth surfaces showed limited colonization, underscoring the role of surface roughness.

Lastly, Rotondi et al. [27] focused on a circular design strategy, incorporating organic waste into clay to create porous channels upon firing. They discovered that, by mixing organic waste with clay, the organic waste disappears after the firing process, leaving gaps, porosities, and heterogeneities that, managed through design, allow for biocolonization patterns to be created for the desired aesthetic and functional purposes.

These findings reveal that the bioreceptivity of ceramic and clay-based materials is not only determined by their intrinsic porosity and composition but also by their processing, forming, and intersection with ecological principles. In conjunction with additive manufacturing, they offer a flexible and sustainable platform for creating biologically active surfaces.

4.3 Comparative Analysis of Bioreceptive Materials

While each material type presents distinct advantages for bioreceptive applications, their performance must be evaluated comparatively to inform material selection in real-world applications. Although experimental studies have demonstrated that ceramic and concrete-based substrates can exhibit bioreceptive potential under certain conditions, their environmental impact, particularly in terms of costs, environmental impact and spatial restrictions, must also be considered [29]. As can be seen in Table 1, earthen and ceramic materials such as rammed earth and clay plaster exhibit considerably lower Global Warming Potential (GWP) compared to the concrete and lime-cement counterparts [18].

Table 1. Global warming potential based on life cycle stage (in kg CO_2/m^3 equivalents) extracted from Bühler et al. [18]. The original data were retrieved from the ÖKOBAUDAT platform, a standardized database for ecological evaluations of buildings maintained by the Federal Ministry for Housing, Urban Development, and Building of Germany.

Material	Production	Demolition	Transport	Disposal	Recycling Potential	Total Global Warming Potential
1. Rammed earth	9.3	1.6	6.0	6.8	−2.9	20.8
2. Clay plaster	93.2	–	2.5	2.8	−3.9	94.6
3. Compressed earth brick	93.6	–	6.3	4.1	−1.8	99.6
4. Burnt clay brick	138.3	0.3	3.2	0.3	−7.0	135.2
5. Concrete	197.0	3.1	12.0	6.0	−21.40	196.7
6. Concrete brick	242.40	1.3	5.1	13.5	4.1	258.2

As shown in Table 1, the Global Warming Potential of various building materials exhibits notable differences in terms of environmental impact throughout their life cycle. Materials such as earth and ceramics have considerably lower carbon emissions compared to conventional alternatives such as concrete. Based on this analysis, ceramic and earth materials not only contribute to bioreceptivity but also align with the objectives of low-carbon building.

In addition to the experimental advances discussed, several broader environmental, functional, and cultural benefits have been associated with the use of earth-based materials in architecture. These materials not only support biological colonization but also contribute to sustainable construction practices, thermal regulation, and the integration of local ecosystems. Table 2 summarizes fifteen key reasons why earth-based materials are considered a valuable alternative to conventional construction materials [18].

Table 2. Reasons to build with earth-based materials as an alternative to conventional materials extracted from different sources [7, 13, 18, 21, 24, 27, 28, 30].

Reason	Description	References
1. Lowers resource depletion	Utilizes locally sourced materials, reducing the need for imported construction materials and minimizing the environmental impact	[18, 27]
2. Reduces carbon footprint	Earthen materials require minimal processing, significantly reducing CO_2 emissions	[13]

(continued)

Table 2. (*continued*)

Reason	Description	References
3. Enhance indoor air quality	Regulate humidity and absorb toxins to promote a healthier environment	[13]
4. Reduces waste	Generates minimal waste due to the ability to reuse and recycle materials directly on-site	[18, 27]
5. Lowers energy consumption	The buildings require less energy for heating and cooling, reducing the overall energy demand and associated emissions	[18, 28]
6. Offer thermal mass benefits	The high thermal mass of earth walls buffers interior spaces from extreme outdoor temperatures, reducing the need for heating and cooling	[18, 28]
7. Promotes biodiversity	By using non-toxic, natural materials, local ecosystems and biodiversity are supported	[19, 21, 24, 27]
8. Sustainable practices	Encourages the adoption of more sustainable, low-impact construction practices	[18, 27, 31]
9. Aesthetic and cultural value	Earthen buildings can be moulded into unique shapes, offering aesthetic diversity and preserving cultural heritage through traditional construction	[18, 30]
10. Sustainable material lifecycle	The materials can be returned to earth at the end of a building´s cycle	[18, 27]
11. Provides a renewable resource	Earth, as a material, can be considered a potentially renewable resource if managed sustainably	[18]
12. Acoustic insulation properties	The density of the walls provides excellent sound insulation, creating quieter and more serene indoor environments	[7, 28]
13. Conserves water	Earthen materials require less water than conventional construction methods	[24, 27]
14. Reduces the heat island effect	Can help to reduce the urban heat island effect, contributing to cooler urban environments	[7, 28]

(*continued*)

Table 2. (*continued*)

Reason	Description	References
15. Flexibility in design and use	Offers versatility in design, allowing for creative and adaptative uses that can evolve with minimal environmental impact	[17, 24, 30]

5 Additive Manufacturing and Design Strategies for Bioreceptive Systems

As urban areas strive to reconnect with natural processes, bioreceptive surfaces have emerged as a passive method for enhancing biodiversity in constructed environments. Material selection is key, but it is through design and fabrication technologies that the full potential of bioreceptivity can be realized. In this context, additive manufacturing has become a transformative tool, enabling the production of complex geometries that can actively support biological colonization.

Additive manufacturing emerged in the mid-1990s as a digital fabrication technique that utilizes mathematical modeling and computational assistance to fabricate three-dimensional objects layer by layer [32]. In bioreceptive applications, additive manufacturing enables designers to control geometry on multiple scales from surface micro-texture to overall form, manipulating how moisture, shade, and biological organisms interact with a structure [24].

5.1 Case Studies and Architectural Applications

Moreover, the integration of natural and porous materials into additive manufacturing workflows expands the ecological possibilities of printed forms. As a case study, we can mention The Vertical Food Farm by LAMÁQUINA. They demonstrate the potential of additive manufacturing, combining urban agriculture and ecological design. Printed clay modules were stacked vertically to create a façade-integrated farming system. These elements host edible plants while simultaneously supporting microbial life and vegetative cover [33].

The startup Urban Reef, located in the Netherlands, has focused on using additive manufacturing with natural materials to create structures capable of hosting ecosystems that are easy to integrate into existing buildings [34]. Thanks to additive manufacturing, they have achieved good results in designing complex geometries using a combination of ceramic and organic materials.

Urban Reef has carried out various projects focused on bioreceptivity, but among the most notable is the project in collaboration with Woonzorg Nederland [35], which created a porous structure inspired by natural forms. This "urban reef" was made in the garden of a residential complex for seniors to develop greener and more pleasant spaces, attracting a diverse range of organisms and encouraging interaction between people and nature.

They have also developed the "Rain Reef" project [36], developed in collaboration with Citylab010. It explores rainwater capture to improve urban microclimates and biodiversity. Using additive manufacturing and ceramics, collectors are designed to integrate with drainage systems, overflowing when full. These low-maintenance devices create cooler, more humid microclimates, with interior temperatures averaging 2 °C lower than the surroundings, helping mitigate the heat island effect.

While most studies in bioreceptivity focus on microorganisms and non-vascular plants, the concept also includes faunal interactions. In urban ecosystems, many animals face severe habitat fragmentation, and built structures offer and opportunity for creating artificial microhabitats that respond to their shelter, nesting, and climate needs. Additive manufacturing allows for the integration of faunal-responsive geometries such as shaded cavities, modular perches, overhangs, or thermally buffered voids, which can be adapted to specifics species through parametric design.

Artist Raphaël Emine [37] created a project that involves creating additively printed ceramic structures specifically designed to house insects in urban environments. These interventions, beyond their artistic dimension, exemplify the application of bioreceptive design principles oriented not only toward vegetal colonization but also toward the facilitation of faunal occupation. In this way, they highlight the potential of additive manufacturing to integrate functional microhabitats within the built environment while simultaneously fostering ecological awareness. Figure 2 shows the case studies mentioned above.

Fig. 2. Case studies illustrating the use of additive manufacturing to enhance biodiversity in urban environments. A) 3D-printed vertical farming structure made from natural materials, designed to support biodiversity in indoor environments [21]. Reproduced with permission of the author via Copyright Clearance Center (License Number: 6033241014844). B) Woonzorg project by Urban Reef: 3D-printed urban reefs fostering bioreceptivity in residential areas. Image courtesy of Max Latour, Urban Reef [32]. C) Rain Reef by Urban Reef: ceramic 3D-printed structures that collect rainwater and promote micro-ecosystems. Image courtesy of Max Latour, Urban Reef [33]. D) 3D-printed ceramic sculptures that serve as insect habitats while providing aesthetic value. Image courtesy of 3D WASP and Jérôme Valton [34].

5.2 Challenges and Future Perspectives

Although additive manufacturing offers great potential for bioreceptive architectural applications, key technical and conceptual challenges remain. A major issue is the absence of standardized criteria for assessing printability, with current methods relying largely on empirical trial-and-error, limiting reproducibility and scalability across different conditions [33].

Another critical barrier is the complex rheology and drying behavior of natural material mixtures, where achieving a suitable compromise between extrudability, structural stability, and environmental durability remains unresolved.

Future research should establish quantitative metrics for bioreceptivity, optimize bio-based binders and additives, and use computational tools to simulate colonization and long-term environmental responses. Designs should also incorporate niches, recesses, and shaded spaces to support birds, bats, and small mammals, promoting multispecies habitats aligned with inclusive biodesign.

6 Conclusion

This article provides a comprehensive overview of the key parameters influencing bioreceptive system development, emphasizing the role of material properties and environmental conditions. Bioreceptivity is defined as the capacity of a material to support biological colonization without degradation, and it depends on both intrinsic (physical and chemical properties) and extrinsic (climatic and environmental) factors. Understanding their interaction is essential for the effective design of bioreceptive surfaces.

The article analyzed two material groups for bioreceptive applications (concrete and ceramics) finding that optimized compositions and surface features support both microbial and plant growth. Additive manufacturing was identified as a key enabler, offering precise control of geometry, porosity, and texture to create structures that retain moisture, form microclimates, and promote colonization.

In conclusion, this study highlights that the future of bioreceptive architecture depends not only on material selection but also on integrating environmental intelligence and fabrication technologies. Clay-based materials and concrete, when combined with digitally guided design and ecological strategies, provide a promising pathway for developing living infrastructure that enhances urban resilience, biodiversity, and climate adaptation.

References

1. Urban Development - Overview. In: World Bank. https://www.worldbank.org/en/topic/urbandevelopment/overview
2. Dutta, U.P., Hazarika, N.: Urbanization and carbon emissions: panel evidence from 68 low-income and lower-middle-income countries. J. Urban Aff. **47**, 269–288 (2023). https://doi.org/10.1080/07352166.2023.2174871
3. Szabó, B., Korányi, D., Gallé, R., et al.: Urbanization decreases species richness, and increases abundance in dry climates whereas decreases in wet climates: a global meta-analysis. Sci. Total. Environ. **859**, 160145 (2023). https://doi.org/10.1016/j.scitotenv.2022.160145
4. Buzási, A., Csizovszky, A.: Urban sustainability and resilience: what the literature tells us about "lock-ins"? Ambio **52**, 616–630 (2023). https://doi.org/10.1007/s13280-022-01817-w
5. Wang, D., Dong, L., Di, S.: Data-driven comparison of urban sustainability towards sustainable urban development under sustainable development goals (SDGs): a review based on bibliometric analysis. Front Energy Res. **11** (2023). https://doi.org/10.3389/fenrg.2023.1168126
6. Uddin, Md.A., Shahabuddin, M., Jameel, M., et al.: Sustainable construction practices in urban areas: innovative materials, technologies, and policies to address environmental challenges. Energy Build. **341**, 115831 (2025). https://doi.org/10.1016/j.enbuild.2025.115831
7. Shishegar, N.: The Impacts of Green Areas on Mitigating Urban Heat Island Effect: A Review. In: ResearchGate (2014). https://www.researchgate.net/publication/271206461_The_Impacts_of_Green_Areas_on_Mitigating_Urban_Heat_Island_Effect_A_Review_Published_in_2014. Accessed 9 Apr 2025
8. Bonilla-Rodríguez, M., Arroyo-Lambaer, D., Castillo, A., et al.: Urban ecological restoration: setting priorities for restoring native vegetation in lava field remnants in Mexico City. Front Sustain. Cities **3** (2021). https://doi.org/10.3389/frsc.2021.709357

9. Xhexhi, K.: Climate Parameters, Heat Islands, and the Role of Vegetation in the City. In: Xhexhi, K. (ed.) Ecovillages and Ecocities: Bioclimatic Applications from Tirana, Albania, pp. 149–170. Springer International Publishing, Cham (2023)
10. Guillitte, O.: Bioreceptivity: a new concept for building ecology studies. Sci. Total. Environ. **167**, 215–220 (1995). https://doi.org/10.1016/0048-9697(95)04582-L
11. Stohl, L., Manninger, T., von Werder, J., et al.: Bioreceptivity of concrete: a review. J. Build. Eng. **76**, 107201 (2023). https://doi.org/10.1016/j.jobe.2023.107201
12. Cruz, M.: Beckett R BIORECEPTIVE DESIGN A novel approach towards bio-digital materiality
13. LEED v4.1 | U.S. Green Building Council. https://www.usgbc.org/leed/v41
14. Karmoker, U., Paul, S.J.K., Ray, P.D., Chakrabartty, M.: A More Bio-Receptive Concrete Façade Design (2023)
15. Sochůrková, P., Devyatkina, S., Kordová, S., et al.: Bioreceptive parameters for additive manufacturing of clay based composites. Nicosia 45–54 (2024)
16. Abdallah, Y.K., Estévez, A.T.: 3D-printed bioreceptive tiles of reaction-diffusion (Gierer–Meinhardt model) for multi-scale algal strains' passive immobilization. Buildings **13**, 1972 (2023). https://doi.org/10.3390/buildings13081972
17. Ortega Del Rosario, M.D.L.A., Castaño, C., Chen Austin, M.: Biodesign as a tool to achieve sustainable construction through additive manufacturing. In: Biomimetics, Biodesign and Bionics. Springer, Cham, pp 251–282 (2024)
18. Bühler, M., Hollenbach, P., Köhler, L., Armstrong, R.: Unlocking resilience and sustainability with earth-based materials: a principled framework for urban transformation. Front Built. Environ. **10**, 1385116 (2024). https://doi.org/10.3389/fbuil.2024.1385116
19. Sochůrková, P., Vaško, I., Tsikolia, S., et al.: Bioreceptivity as a Factor of Additive Fabrication: Material Research on 3D Printed Structures affecting Urban Spaces, pp. 115–124 (2023)
20. Vázquez-Nion, D., Silva, B., Prieto, B.: Influence of the properties of granitic rocks on their bioreceptivity to subaerial phototrophic biofilms. Sci. Total. Environ. **610–611**, 44–54 (2018). https://doi.org/10.1016/j.scitotenv.2017.08.015
21. Veeger, M., Ottelé, M., Prieto, A.: Making bioreceptive concrete: formulation and testing of bioreceptive concrete mixtures. J. Build. Eng. **44**, 102545 (2021). https://doi.org/10.1016/j.jobe.2021.102545
22. Prieto, B., Young, M.E., Turmel, A., Fuentes, E.: Role of masonry fabric subsurface moisture on biocolonisation. A case study. Build. Environ. **210**, 108690 (2022). https://doi.org/10.1016/j.buildenv.2021.108690
23. Crispim, C.A., Gaylarde, P.M., Gaylarde, C.C.: Algal and cyanobacterial biofilms on calcareous historic buildings. Curr. Microbiol. **46**, 0079–0082 (2003). https://doi.org/10.1007/s00284-002-3815-5
24. Lim, A.C.S., Lharchi, A.: Parameters for Bio-receptivity in 3D Printing. In: Faircloth, B., Pedersen Zari, M., Thomsen, M.R., Tamke, M. (eds.) Design for Climate Adaptation, pp. 701–715. Springer International Publishing, Cham (2023)
25. Quagliarini, E., Gianangeli, A., D'Orazio, M., et al.: Effect of temperature and relative humidity on algae biofouling on different fired brick surfaces. Constr. Build. Mater. **199**, 396–405 (2019). https://doi.org/10.1016/j.conbuildmat.2018.12.023
26. Bone, J.R., Hall, A.E., Stafford, R., Herbert, R.J.H.: Inconsistent bioreceptivity of three mortar mixes in subtidal sites. Ecol. Eng. **204**, 107265 (2024). https://doi.org/10.1016/j.ecoleng.2024.107265
27. Rotondi, C., Gironi, C., Ciufo, D., et al.: Bioreceptive ceramic surfaces: material experimentations for responsible research and design innovation in circular economy transition and "ecological augmentation." Sustainability **16**, 3208 (2024). https://doi.org/10.3390/su16083208

28. Jakubovskis, R., Malaiškienė, J., Gribniak, V.: Bio-colonization layered concrete panel for greening vertical surfaces: a field study. Case Stud. Constr. Mater. **19**, e02394 (2023). https://doi.org/10.1016/j.cscm.2023.e02394
29. Veeger, M., Nabbe, A., Jonkers, H., Ottelé, M.: Bioreceptive concrete: State of the art and potential benefits
30. Marchese, K.: The "urban reef" concept aims at making cities more biodiverse. In: DesignWanted (2021). https://designwanted.com/urban-reef-biodiverse/
31. Ortega Del Rosario, M.D.L.Á., Beermann, K., Chen Austin, M.: Environmentally responsive materials for building envelopes: a review on manufacturing and biomimicry-based approaches. Biomimetics **8**, 52 (2023). https://doi.org/10.3390/biomimetics8010052
32. Zhao, Q., Gao, C., Zhang, Y., Zhang, Y.: Advances and application potential in the research of silicate mineral-based 3D printing materials. Prog. Mater. Sci. **152**, 101450 (2025). https://doi.org/10.1016/j.pmatsci.2025.101450
33. Gyawali, B., Haghnazar, R., Akula, P., et al.: A review on 3D printing with clay and sawdust/natural fibers: printability, rheology, properties, and applications. Results Eng. **24**, 103024 (2024). https://doi.org/10.1016/j.rineng.2024.103024
34. Urban Reef. Innovative 3D-Printed Habitats for Urban Biodiversity (2024)
35. Welkom bij Woonzorg Nederland|Woonzorg Nederland - betaalbare woningen voor senioren. https://www.woonzorg.nl/nieuws/feestelijke-onthulling-3d-koraalrif-in-rhoon. Accessed 10 May 2025
36. Matser D Enhancing the bioreceptivity of ceramic RainReefs
37. Severi, A.: 3D Printed Ceramic Sculptures to House Insects|Raphaël Emine. In: 3D Printers|WASP (2025). https://www.3dwasp.com/en/3d-printed-ceramic-sculptures-to-house-insects/. Accessed 10 May 2025

3D Printing of Wood Particle Reinforced Clay Composite in Direct Ink Writing

Biva Gyawali[1](✉), Kai Bentley[2], Pavan Akula[3], Kamran Alba[4], and Vahid Nasir[1]

[1] Department of Wood Science and Engineering, Oregon State University, Corvallis, USA
gyawalb@oregonstate.edu

[2] School of Mechanical, Industrial and Manufacturing Engineering, Oregon State University, Corvallis, USA

[3] School of Civil and Construction Engineering, Oregon State University, Corvallis, USA

[4] Cullen College of Engineering, University of Houston, Texas, USA

Abstract. Additive manufacturing technology offers huge potential for practicing the circular economy, automation in construction, and mass customization in production. It offers freedom in the design, fabrication, and development of novel structures in construction. Environmentally sustainable construction can be promoted by utilizing forest-based products and earthen materials using this technique. There are several studies on the 3D printing of earthen-based material, such as clay, using natural fibers. However, the study on clay 3D printing using forest-based materials is limited. This study proposes optimizing formulations for 3D printing based on clay, wood particles, and water content. Furthermore, this research delves into the impact of printing parameters and materials (wood particles, water content) on the buildability of 3D-printed structures. The result shows that an increase in nozzle size, i.e., extrusion rate, shows a positive impact on buildability, while higher printing speed has a negative impact on buildability. Similarly, wood particles have shown a reinforcing effect, whereas higher water content has reduced the buildability height. The result shows that the average buildability of the mixture ranges from a minimum of 17 mm (mix design: clay, 5 *wt.%* wood particle, 0.5 *wt. %* xanthan gum, 70 *wt.%* water content printed at 90 *mm/s* using 3 *mm* nozzle diameter) to a maximum of 236 *mm* (mix design: clay, 10 *wt.%* wood particle, 0.5 *wt.%* xanthan gum, 50 *wt.%* water content printed at 30 *mm/s* using 6 *mm* nozzle diameter). Furthermore, this study briefly discusses the opportunities for future research in this field.

Keywords: Buildability · Additive Manufacturing · Printing Speed · Earthen-based Material · Forest-based Materials

1 Introduction

The growing attention to automation and robotics in construction is offering unprecedented opportunities to tackle inefficiency and low productivity [1] while reducing labor costs, offering design freedom, and improving quality [2]. Additive manufacturing processes, including 3D printing, support innovative product development, freedom in material selection [3], opportunities for mass customization, and manufacturing products with complex geometries [4].

M. Posani and S. Singh (Eds.): RYS 2025, RILEM Bookseries 67, pp. 603–611, 2026.
https://doi.org/10.1007/978-3-032-14652-6_43

Additive manufacturing supports cleaner production through providing opportunities for using the waste or byproducts of renewable materials, leading to a reduced carbon footprint of the production [5]. Environmentally friendly construction can be promoted using sustainable options like earthen-based materials combined with renewable materials, including agricultural waste or forest-based bioproducts [6, 7]. Due to their local availability, low cost, and reusability, earthen materials are gaining popularity for affordable housing construction [8]. The byproducts of the wood industry often end up in landfill sites or are used in biomass or pulp and paper applications [9]. In literature, the byproducts, such as sawdust or wooden chips, have been mixed with soil or clay to produce several products, including fired clay bricks [10], ceramic pot filters [11], briquettes [12] etc. Natural fibers like wheat straw [13], rice husk [14], and hemp fiber [15] have been used as reinforcing agents in the 3D printing of soil or clay. Biopolymers such as xanthan gum and geopolymers such as lime and cement have been used as binding agents in the slurry. These studies have reported that the incorporation of natural fibers assists in reducing the crack formation during drying shrinkage. However, there is limited research on the 3D printing area using earthen-based materials and forest-based products.

Evaluating the Printability of material, including studying the flowability, extrudability, and buildability during the 3D printing process [6, 16] are of critical importance. Buildability is one of the important printability criteria reflecting the ability of the slurry to preserve its desired shape under the weight of successive printed layers before the occurrence of collapse [17]. Buildability height of a slurry could be impacted by printing parameters such as nozzle diameter and printing speed [18]. Geometries such as a wall [19], cone [20], or hollow cylinder [19] have been used in literature to conduct a buildability assessment of 3D printable slurry.

Despite the significance of buildability on the assessment of printability of natural fibers reinforced soil-based composites, there is a gap in the literature on the influence of slurry composition, printing speed, and factors affecting extrusion rate during printing (i.e., diameter of nozzle and layer height of filament) [6]. Furthermore, literature lacks studies on the interaction of the slurry's composition and the printing parameters on the buildability height. This study focuses on using statistical tools to analyze the printing height of slurry during the 3D printing of wood particles reinforced earthen slurry. This will provide insight into the optimal design of the mixture and printing parameters to achieve optimal printing height.

2 Methodology

2.1 Materials

Douglas fir (*Pseudotsuga menziesii)* sawdust residue was collected from Arauco company in Albany, Oregon. The moisture content of the sawdust was 50%, was dried for 75 min using a dryer, reducing its moisture content below 15%. The sawdust was grinded using a Wiley mill, and the output of the mill was sieved in a sieve shaker. The wood particles passing through 250 *microns* were collected and used in this experiment. The wood particles were stored in the environmental conditioning chamber with a temperature of 20 °C and 65% relative humidity, maintaining the equilibrium moisture content of

12% in the wood particles. The clay used in this work was composed of following minerals: Kaolinite (10–25 *wt.%*), feldspar (10–25 *wt.%*), crystalline silica- quartz (10–25 *wt.%*), kaolin (10–25 *wt.%*), talc- steatite (<2 *wt.%*), titanium dioxide (< 2 *wt.%*), mica group (< 2 *wt.%*), dolomite (< 2 *wt.%*). The supplier (*Laguna Clay Co.*) also reports a 'Kaolin' fraction, which corresponds to a bulk clay fraction of commercial clay, mostly consisting of Kaolinite minerals. Xanthan gum (XG) was used as a binding agent in the slurry.

2.2 Preparation of Slurry

At the beginning, the quantity of wood particles was selected based on the literature review [21, 22], while making the mixture (i.e., 0, 5, 10 *wt.%* of total clay and wood particles mixture). The fixed quantity of xanthan gum was used in each formulation, i.e., 0.5 *wt.%* of the total mixture of clay and wood particles. Water was added at five different levels in the mixture (i.e., 50, 55, 60, 65, 70 *wt.%* of total clay, xanthan gum, and wood particles mixture). To select a suitable range of water content, a preliminary investigation was carried out. It was observed that slurry prepared with no wood particles and water content higher than 70 wt.% experienced an excessive slumping issue during 3D printing. Also, when 10 wt.% of wood particles were used in the mixture with less than 50 wt.% of water content, the slurry had very poor extrudability, causing difficulty in smooth printing. The required quantity of xanthan gum was first mixed with the desired amount of water using a spatula for one minute and then further mixed in a Hobart Planetary mixer N50 at low speed (136 *rpm*) for 2 min. The manually mixed clay and wood particles were added to this solution. Homogeneous mixing was carried out in the Hobart mixer for 5 min at the same speed (136 *rpm*). The prepared slurry was manually loaded into the tube of the 3D printer.

2.3 Experimental Design

3D Potterbot 10XL with 3600 *ml* extruder with movement of printing head in z-axis and the build plate in x- and y-axis was used in this study. To study the impact of printing parameters on the printing height, extrusion rate (which can be adjusted by nozzle diameter and layer height), and printing speed were selected. In this study, the printing height is defined as the height at which the 3D printed hollow cylinder can resist the weight of successive printed layers, and it is recorded at the time of collapse of the structure, as shown in Fig. 1. Two different nozzle diameters (3 and 6 *mm*) and printing speeds of 30, 60, and 90 *mm/s* were selected. A model of a hollow cylinder of diameter 100 *mm* and height of 300 *mm* was designed in Fusion 360. Ultimaker Cura was used to develop a G-code script, which was selected for printing. The layer heights of 1.5 *mm* and 3 *mm* were used for 3 *mm* and 6 *mm* nozzle diameters, respectively, to improve the quality of print. The extrusion rate (E) changed when two different nozzle diameters were used. The change was due to the combined effect of nozzle diameter and layer height. In the case of a 6 *mm* nozzle diameter, the extrusion rate was E = 16.161 mm^3/s, and in the case of a 3 *mm* nozzle diameter, E = 3.865 mm^3/s. A total of 90 different printing conditions were tested, based on the combination of three different mixtures of clay and wood particles, two extrusion rates, five water contents, and three printing speeds. Each printing condition was repeated three times to ensure consistency in results. Figure 1 shows the procedure for 3D printing of wood reinforced clay slurry adopted in this work.

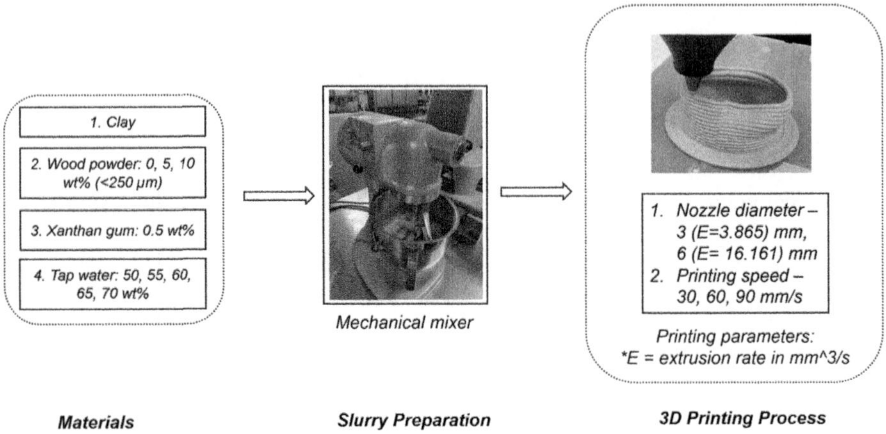

Fig. 1. Procedure for 3D printing of wood particle reinforced clay slurry

3 Results and Observations

The effect of wood particles, water content, extrusion rate, and printing speed was studied using Minitab statistical analysis. The main effect and interaction of the parameters on the printing height were studied using Analysis of Variance (ANOVA) at a 0.05 significance level. The experimental observation showed that the average buildability of the mixture ranges from a minimum of 17 *mm* (mix design: clay, 5 *wt.%* wood particle, 0.5 *wt.%* xanthan gum, 70 *wt.%* water content printed at 90 *mm/s* using 3 *mm* nozzle diameter) to a maximum of 236 *mm* (mix design: clay, 10 *wt.%* wood particle, 0.5 *wt.%* xanthan gum, 50 *wt.%* water content printed at 30 *mm/s* using 6 *mm* nozzle diameter).

3.1 Effect of Parameters on Buildability Height

ANOVA analysis and the interaction between individual parameters were studied. The analysis showed the significance of all the factors on the printing height, with the extrusion rate showing the highest F-value (1557.92), followed by water content (1396.04), wood particles (246.20), and printing speed (139.42), and the P-values for all these four parameters were 0.000. Figure 2 shows that increasing the printing speed from 30 *mm/s* to 90 *mm/s* reduces the printing height of the cylinder. A similar observation was reported in [18]. The authors conducted 3D printing of geopolymer-based concrete structures and explained that the lower printing speed allows the material to properly solidify deposited layers, helping them to resist the loads of upper layers.

The extrusion rate (nozzle diameter) is an important parameter impacting the quality of print. Figure 2 shows that the printing height is highly improved when the extrusion rate increases from 3.856 mm^3/s to 16.161 mm^3/s. This observed improvement can be attributed to the increased layer height and higher layer stability. Water content is another critical factor that governs the rheological properties of slurry. Figure 2 shows that an increase in water content reduces the print height. However, in the case of wood particles, it showed the non-linear effect on the height of the print. At 5% wood particles, the average print height was reduced; however, at 10% wood particles, the buildability height was enhanced.

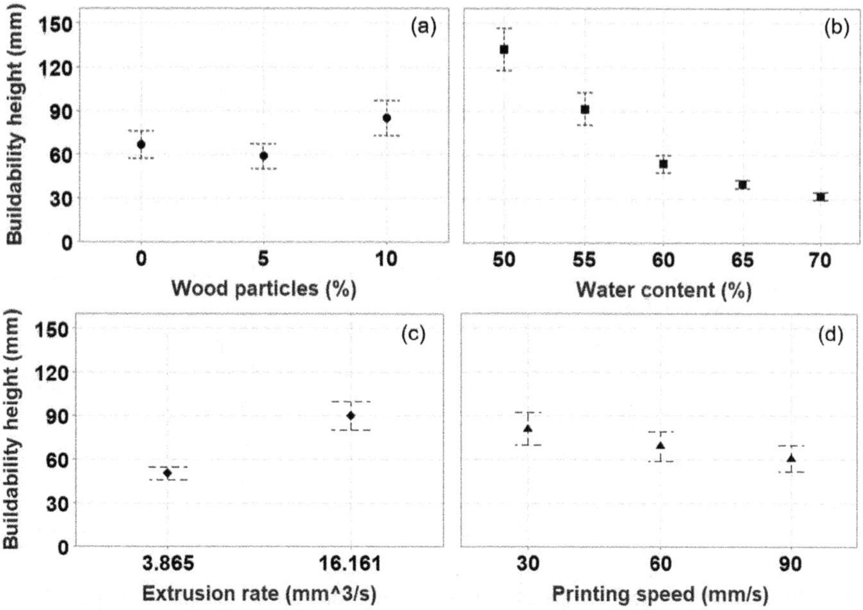

Fig. 2. Interval plot of parameters: (a) wood particles, (b) water content, (c) extrusion rate, (d) printing speed to study their effect on the buildability height

ANOVA analysis also revealed that the interactions between some factors were statistically significant. The most important interaction was observed between the water content and extrusion rate (F-value = 223.87, P-value = 0.000). Additionally, the statistically significant interaction between parameters was wood particles and extrusion rate (F-value = 37.26, P-value = 0.000), wood particles and water content (F-value = 25.24, P-value = 0.000), and water content and printing speed (F-value = 16.24, P-value = 0.000). However, in the case of printing speed with wood particles and extrusion rate, the interaction was not statistically significant (P-value = 0.058 and 0.609, respectively). The third-order interaction revealed that the interaction between wood particles, water content, and extrusion rate (F-value = 15.87) was also statistically significant (P-value = 0.000). Figure 3 shows the summary of interaction plots between the parameters studied. It was observed that the larger extrusion rate increases buildability height, but this is more prominent at lower water content (i.e., 50%). Another interesting observation is the interaction between water content and wood particles. At 5% wood particle, the printing height was reduced at 50, 55, and 60 *wt.%* water content. However, the printing height slightly increased at 65 and 70 *wt.%* water content when the clay slurry was reinforced by 5 *wt.%* wood particles. The complex behavior of the interaction between wood particles and water content can be better understood by conducting rheological tests of the slurry for various mix designs and linking the results to the criteria of printability [6]. Furthermore, the interaction between the wood particle content and extrusion rate shows that the positive impact of wood particles is significant when a larger nozzle diameter is used. Thus, the interaction plot highlights the complex role of printing parameters and composition of slurry on the height of the buildability of 3D printed structures.

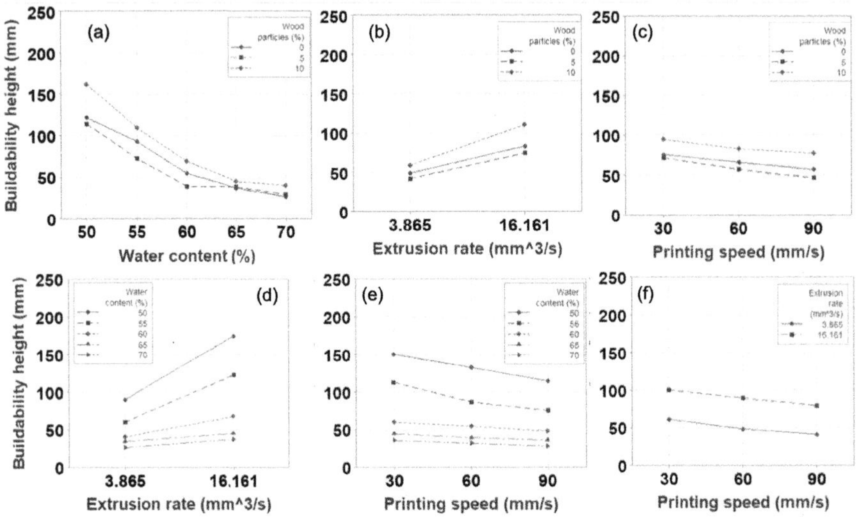

Fig. 3. Interaction plot between parameters estimating the print height (a) wood particles and water content, (b) wood particles and extrusion rate, (c) wood particles and printing speed, (d) water content and extrusion rate, (e) water content and printing speed, and (f) extrusion rate and printing speed.

4 Discussion and Conclusion

This research studies the role of mixture composition (wood particle and water content) and printing parameters (nozzle diameter/extrusion rate and printing speed) on the printing height of the wood particle reinforced clay composites. Statistical analysis has been used to analyze the printing height, and the results show that the printing parameters and slurry composition play a significant role in obtaining the optimal printing height. Mixture compositions, including wood particles and water content, and printing parameters, significantly impact the buildability of the clay slurry. Wood particles show a nonlinear role and highly interact with the extrusion rate and water content. Understanding the interaction between the parameters is also critical for optimal printing. Hence, this study reveals that ink composition, including water content, quantity of wood particles, and printing parameters, including nozzle diameter and printing speed, are critical for the constructability of the 3D printable wood particle reinforced clay composites, which are useful for the optimization of the 3D printable mixtures.

The limitations of this study and the potential future research directions, along with proposed methodologies and obstacles, are summarized below:

a. Post printing characterization:

- This study mainly focused on the buildability height of the fresh slurry. Future work could focus on evaluating the mechanical properties, such as compressive strength, following the ASTM C109 standard. Mechanical strength of 3D printed specimens developed from different mix designs could provide further insights into the effect of material composition on the strength of 3D printed composites. However, cracks and shrinkage issues might be observed.
- Long-term durability of the 3D printed specimens could be assessed through various durability tests such as wetting-drying cycles, freeze-thaw cycles, or accelerated weathering to simulate outdoor exposure (e.g., direct sunlight, rainfall, and humidity). Wood particles might improve the resistance of the specimens to water-induced damage. The role of wood particles could be investigated to study the cracks and dimensional stability of 3D printed specimens. However, microcracks and volumetric swelling or shrinkage might be observed depending on the test procedure adopted.

b. Binder optimization:

This study doesn't consider the role of a binder such as xanthan gum in the buildability of the 3D printable slurry. Identifying an optimal range and type of binders, such as biopolymers and geopolymers, can result in a slurry having good extrudability and printability characteristics and can mitigate issues like slumping or clogging during extrusion. However, increasing the binder content beyond a threshold may lead to poor extrudability. One of the main challenges of using biopolymers is their low water resistance. Therefore, the weather resistance of slurry containing biopolymer and its outdoor application should be further studied.

c. Rheological characterization:

Future studies could be focused on oscillatory and flow rheology tests to characterize the fresh properties of slurry, which could be linked to the printability criteria. Addition of shear-thinning agents and plasticizers will improve the rheological properties and flowability of the slurry, and an optimal mix design can be developed by incorporating different rheological modifiers and binders. However, measurement of highly viscous slurry might be challenging in the rheometer.

By focusing on these areas, future studies could aim to develop an optimal mix design for the 3D printing of wood particles reinforced clay composites with improved printability, strength, and durability.

Acknowledgement. This research was supported by TDRS Fund from the Division of Research and Innovation - Office for Research Advancement and Center for the Future of Forests and Society (CFFS) at Oregon State University.

References

1. Davila Delgado, J.M., et al.: Robotics and automated systems in construction: Understanding industry-specific challenges for adoption. J. Build. Eng. **26**, 100868 (2019). https://doi.org/10.1016/j.jobe.2019.100868
2. Huang, P.Y., Sakurai, M.: Factor automation: the Japanese experience. IEEE Trans. Eng. Manage. **37**(2), 102–108 (1990). https://doi.org/10.1109/17.53712
3. Paoletti, I.: Mass customization with additive manufacturing: new perspectives for multi performative building components in architecture. In: Procedia Engineering (2017). https://doi.org/10.1016/j.proeng.2017.04.275
4. Hasani, N., et al.: Outlook on human-centred design in industry 5.0: towards mass customisation, personalisation, co-creation, and co-production. Int. J. Sustain. Eng. **18**(1), 2486343 (2025). https://doi.org/10.1080/19397038.2025.2486343
5. Majeed, A., et al.: A big data-driven framework for sustainable and smart additive manufacturing. Robot. Comput. Integr. Manufact. **67**, 102026 (2021). https://doi.org/10.1016/j.rcim.2020.102026
6. Gyawali, B., Haghnazar, R., Akula, P., Alba, K., Nasir, V.: A review on 3D printing with clay and sawdust/natural fibers: printability, rheology, properties, and applications. Results Eng. **24**, 103024 (2024). https://doi.org/10.1016/j.rineng.2024.103024
7. Gyawali, B.: Evaluating the mechanical properties of 3D printable earthen material reinforced by forest-based products. Presented at the FPS-SWST Joint Convention, Fort Collins, Colorado, June 15-20. https://doi.org/10.13140/RG.2.2.27307.55845
8. Kulshreshtha, Y., et al.: The potential and current status of earthen material for low-cost housing in rural India. Constr. Build. Mater. **247** (2020). https://doi.org/10.1016/j.conbuildmat.2020.118615
9. Worku, L.A., Bachheti, A., Bachheti, R.K., Rodrigues Reis, C.E., Chandel, A.K.: Agricultural residues as raw materials for pulp and paper production: overview and applications on membrane fabrication. Membranes **13**(2), 2 (2023). https://doi.org/10.3390/membranes13020228
10. Alabduljabbar, H., Benjeddou, O., Soussi, C., Khadimallah, M.A., Alyousef, R.: Effects of incorporating wood sawdust on the firing program and the physical and mechanical properties of fired clay bricks. J. Build. Eng. **35** (2021). https://doi.org/10.1016/j.jobe.2020.102106
11. Ekpunobi, U.E., Agbo, S.U., Ajiwe, V.I.E.: Evaluation of the mixtures of clay, diatomite, and sawdust for production of ceramic pot filters for water treatment interventions using locally sourced materials. J. Environ. Chem. Eng. **7**(1) (2019). https://doi.org/10.1016/j.jece.2018.11.036
12. Chukwuneke, J.L., Umeji, A.C., Obika, E.N., Fakiyesi, O.B.: Optimisation of composite briquette made from sawdust /rice husk using starch and clay binder. Int. J. Integr. Eng. **13**(4) (2021). https://doi.org/10.30880/ijie.2021.13.04.019
13. Carcassi, O.B., Maierdan, Y., Akemah, T., Kawashima, S., Ben-Alon, L.: Maximizing fiber content in 3D-printed earth materials: printability, mechanical, thermal and environmental assessments. Constr. Build. Mater. **425**(135891) (2024)
14. Ferretti, E., Moretti, M., Chiusoli, A., Naldoni, L., de Fabritiis, F., Visonà, M.: Rice-husk shredding as a means of increasing the long-term mechanical properties of earthen mixtures for 3D printing. Materials **15**(3) (2022). https://doi.org/10.3390/ma15030743
15. Yang, H.Q., Klug, C., Schmitz, T.H.: Fiber-reinforced clay: an exploratory study on automated thread insertion for enhanced structural integrity in LDM. Ceramics **6**(3) (2023). https://doi.org/10.3390/ceramics6030084
16. Rückrich, S., Agranati, G., Grobman, Y.J.: Earth-based additive manufacturing: a field-oriented methodology for evaluating material printability. Architectural Sci. Rev. **66**(2) (2023). https://doi.org/10.1080/00038628.2022.2154739

17. Malaeb, Z., Hachem, H., Tourbah, A., Maalouf, T., El Zarwi, N., Hamzeh, F.: 3D concrete printing: machine and mix design. Int. J. Civil Eng. Technol. **6** (2015)
18. Khan, S.A., et al.: The impact of nozzle diameter and printing speed on geopolymer-based 3D-Printed concrete structures: Numerical modeling and experimental validation. Results Eng. **21**, 101864 (2024). https://doi.org/10.1016/j.rineng.2024.101864
19. Bryson, Z., Srubar, W., S. łdots, K., łdots, N.: Towards 3D printed earth-and bio-based insulation materials: a case study on light straw clay. In: 8th International Conference on Non-conventional Materials and Technologies (2022)
20. Bhusal, S., Sedghi, R., Hojati, M.: Evaluating the printability and rheological and mechanical properties of 3D-printed earthen mixes for carbon-neutral buildings. Sustainability **15**(21) (2023). https://doi.org/10.3390/su152115617
21. Kudahettige-Nilsson, R.L., Ullsten, H., Henriksson, G.: Plastic composites made from glycerol, citric acid, and forest components. BioResources **13**(3) (2018). https://doi.org/10.15376/BIORES.13.3.6600-6612
22. Chawla, K., Singh, J., Singh, R.: On chemical assisted blending of secondary recycled ABS with bakelite and wood dust for fused filament fabrication. Adv. Mater. Process. Technol. **9**(1), 116–137 (2023). https://doi.org/10.1080/2374068X.2022.2085955

Bio-Encapsulation of Phase Change Materials via Coacervation and Spray Drying for Integration into Building Materials

Chiara Pezzoli[1,2(✉)], Francesca Zanoni[2], Saulo Rocha Ferreira[1,3], and Antonio Caggiano[1,4]

[1] University of Genoa, Genova, Italia
chiara.pezzoli@edu.unige.it
[2] Sphera Encapsulation s.r.l., Verona, Italia
[3] University of Buenos Aires, Ciudad Autonoma de Buenos Aires, Buenos Aires, Argentina
[4] Federal University of Lavras (UFLA), Lavras, Brazil

Abstract. With the continuous rise of the global population, energy consumption has significantly increased, leading to a pressing need for more efficient and sustainable solutions in the building sector. Phase Change Materials (PCMs) have emerged as a promising alternative for thermal energy storage (TES) due to their ability to store and release heat during phase transitions. However, their integration into construction materials remains a challenge, primarily due to their leakage and low thermal conductivity. This study explores the microencapsulation of PCMs through the complex coacervation technique coupled with spray drying using carboxymethylcellulose (CMC) and potato proteins (PP) as shell materials, with the deposition of an additional shell of calcium carbonate. Various CMC-PP ratios were tested, with 1:2 and 1:2.5 yielding the most promising results. Encapsulation efficiency of PCM (EE %) reached 88.41 ± 2.01% and 66.84 ± 3.84% for these ratios, respectively, while the maximum yield of process (PY %) obtained in the spray drying phase was 38.71 ± 3.45%. Despite process optimizations, the yield remained relatively low, indicating the need for further improvements in microcapsules robustness. To evaluate better the performance in building materials for the use as TES system, additional studies on calorimetry, mechanical resistance, and compatibility will be performed in future.

Keywords: PCM · Microencapsulation · Thermal Energy Storage · Coacervation · Spray dry · Carbon-neutral buildings

1 Introduction

The human population is constantly growing, and with it, the consumption of energy which has been provoking in recent times a severe shortage of fossil fuels. Energy efficiency of buildings plays a preponderant role in this phenomenon, and it has been the core of multiple research projects around the globe. As an example, in Europe. Around the 40% of the total energy consumption is related to buildings, which are guilty of producing 36% of the total greenhouse emission of CO_2 in Europe [1]. This phenomenon is

M. Posani and S. Singh (Eds.): RYS 2025, RILEM Bookseries 67, pp. 612–623, 2026.
https://doi.org/10.1007/978-3-032-14652-6_44

mostly registered in developed countries, caused by the increasing living standards and the necessity to ensure indoor thermal comfort. Undoubtedly, this model is not any more sustainable and a solution, allowing for the improvement in energy conservation and optimized consumption is strongly needed. The employment of phase change materials (PCM), being one of the most popular thermal energy storage (TES) systems, is increasing. This roots in the fact that PCMs, when subjected to a certain temperature, change the phase from solid to liquid and during this phase transitions they can absorb heat from the surroundings. On the contrary, during night when the temperature decreases, they return to their solid form, liberating in a passive way the accumulated energy in the form of heat [2]. PCMs are therefore the most promising candidates to store energy, since they show high energy storage density, good isothermal operating 2characteristics, and finally a higher sustainability. Among them, organic PCM (e.g. paraffin and fatty acids/esters) appear to be more chemically stable, not corrosive, recyclable, and moreover, show no supercooling effect when compared to their inorganic counterpart (e.g., inorganic salt such as potassium nitrate) [2]. Anyway, the main challenge to overcome to take advantage of these materials is to find a way to increase their storage efficacy and thermal conductivity and to subsequently integrate this type of system into buildings materials.

The most common strategies implemented for the inclusion of PCM in buildings are: the direct incorporation of PCM in construction materials (e.g., cement paste) where the material is immerged directly in PCM baths; The stabilization in pre-formed shapes (e.g., plastic or metal container) and finally, the encapsulation of PCM (e.g., macro-encapsulation, micro-encapsulation and nano-encapsulation depending on the size of the produced particles).

Among the discussed options for PCM inclusion, microencapsulation, seems to be the most promising approach to allow their compartmentalization in building materials, and seen from researchers as a more effective route to achieve a higher stability, storage capacity and better thermal functionality of PCMs (Liu). Microencapsulation corresponds to a process in which single droplets or particles are surrounded by a continuous film or a protective membrane composed of a polymeric material (shell) [3]. Among the advantages offered by encapsulation applied to PCMs are the reduced rate of evaporation, the lower reactivity toward the outside environment, the decrease of the diffusion rate and leakage during phase transition, since the PCM encapsulates can tolerate multiple volume changes and the presence of the shell can increase their heat transfer area [4].

There are three different types of methods that can be used to obtain micro-encapsulated PCMs (MPCMs), named chemical, physical, and the combination of the two. Among the first type, the main are: suspension polymerization [5], dispersion-polymerization, emulsion-polymerization, and in-situ polymerization [2]. These are all methods that are currently well known and already established for the encapsulation of PCMs, but some concerns are raised from the environmental point of view, since the polymers that are used to create the external shell of the particles are mainly composed of plastic-based materials (e.g., polystyrene (PS), poly(ethyl methacrylate) (PEMA) and polyvinyl acetate (PVA)) [6]. Moreover, in most cases, the formation of the shell is obtained through the addition of formaldehyde (e.g., in the case of melamine-formaldehyde systems), which acts as a crosslinker, raising concern about health issues related to the presence of toxic residues in the final product [7].

In this work, the combination of complex coacervation and spray drying, will be explored as a strategy to produce encapsulated bio-based phase change materials (MPCMs).

Complex coacervation is a physicochemical encapsulation method that involves three main steps. In the initial step, two immiscible phases, one containing the core material (PCM), and a second containing the encapsulating polymers and a surfactant are produced and mixed in the form of an emulsion. In the second step,

the change of the pH induces the deposition of the polymer onto the dispersed droplets [8]. The last step involves the crosslinking of the droplets by using a cross-linking agent such as formaldehyde or a natural cross-linker (e.g., tannic acid) to form a rigid and stable shell around the PCM core [8].

One of the main advantages of complex coacervation is its ability to produce microcapsules with a very thin shell, allowing for a higher PCM payload (e.g., up to 70–80%), which is essential to obtain an efficien2t energy storage system [2]. Furthermore, in the present study an additional mineralized shell will be added to the formed coacervate to enhance their thermal conductivity, as well as their mechanical and chemical stability, particularly under the extreme alkaline pH conditions typical of construction materials (e.g., pH 12) [9].

In the present work, the developed MPCMs were dried by using a spray drying process. Spray drying is a physical method that involves the use of high temperature to promote the immediate evaporation of water from a solution to form a fine powder [5].

This study represents a one-of-a-kind attempt, never reported in literature, to combine different encapsulation approaches (e.g., coacervation and spray dry) along with shell functionalization applied to a biobased source of PCM.

2 Materials

Potato protein isolate (PP) SOLANIC 300, was purchased from Avebe (NL). Sodium carboxymethylcellulose (CMC) was purchased from Farmalabor (IT), phase change material (PCM) was purchased from CRODA (UK) commercial name CrodaTherm 24W (Melting temperature 23 °C, Heat capacity of 210 J/g as reported from the producer).

Tween 80, Span 80, tannic acid, $CaCl_2$ and Na_2CO_3 were all sourced from Sigma Aldrich (USA). SDS was purchased from Chem-Lab (BG). Glycerol was obtained from ACEF (IT) and the maltodextrin DE 19 (MD) from Avebe (NL).

3 Methods

3.1 Microencapsulation of PCM Through Complex Coacervation and Shell Mineralization

The PCM microparticles (MPCMs) were produced via complex coacervation process. In detail, a solution of CMC 3% w/v, SDS 0.6% w/v, Tween 80 1.5% w/v and Span 80 0.7% w/v was prepared in distilled water, separately a solution of PP 7% w/v was prepared by dissolving the powder in distilled water. The two different solutions were left dissolving at 40 °C under continuous stirring for 30 min.

The PCM (e.g., 64% w/v) was added to the CMC and surfactants solution and emulsified by using an Ultra-Turrax homogenizer (IKA-Werke, DE) at 15.000 rpm for 4 min. The emulsion was later subjected to sonication to create a finer emulsion, by using an ultrasonic generator (Bandelin, DE) applied for 4 min at 40% amplitude operating at 20 kHz, with a 10 s on/10 s off pulse cycle. The resulting emulsion was added to the protein solution and kept under stirring at 500 rpm for 30 min.

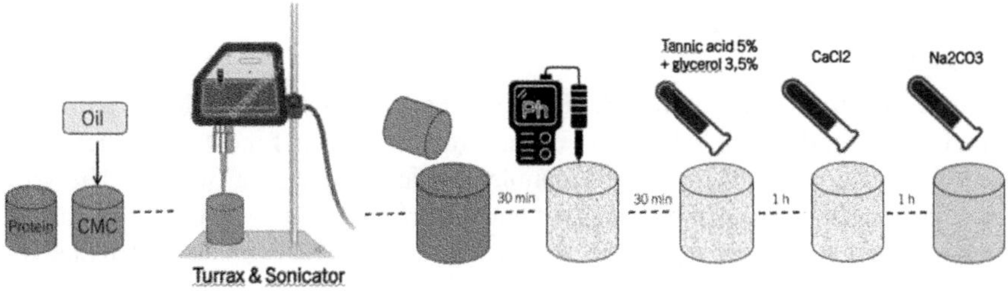

Fig. 1. MPCMs production by complex coacervation and shell mineralization process.

After 30 min, the pH of the solution was adjusted to 4 ± 0.5 by using a solution of HCl 0.5 M, the solution was stirred for 30 additional minutes after which a solution containing 5% w/v tannic acid with 3.5% w/v glycerol was added to crosslink the coacervates.

After 1 h, the creation of a mineralized shell around the formed coacervate was obtained by adding a 4% w/v $CaCl_2$ solution, followed 1 h by the addition of a 6% w/v Na_2CO_3 solution. Finally, the MPCMs were left at room temperature overnight. The process scheme is represented in Fig. 1.

3.2 MPCMs Preliminary Resistance upon Drying

To evaluate whether the produced MPCMs could withstand the drying process (necessary for the spray drying step), a preliminary test was performed, where a drop of the dispersion of MPCMs was placed on a microscope slide (without a coverslip), and allowed to air-dry until complete water evaporation to observe their structural integrity using an optical microscope OPTECH BM 60 (USA).

3.3 Spray Drying

The prepared MPCMs with the addition of 20% w/v maltodextrin DE 19 was then dried using a Mini spray dryer B-290 BUCHI (CH) with the following equipment set up (inlet temperature: 120 °C, outlet temperature: 75 °C, air flow: 40 L/min, aspirator: 100%, pump: 25%).

3.4 MPCMs Characterization

Particle Size and Zeta-Potential
The particle size distribution of the MPCMs was analysed using a Mastersizer 3000 (Malvern, UK). The sample was diluted upon the right obscuration value (e.g., $< 20\%$) and the Refractive Index (RI) of the material was 1.59. The superficial charge of the MPCMs were evaluated by using a Nano Zetasizer (Malvern, UK), where the sample was diluted 200 times in demineralized water and the following parameters were used: RI of 1.45, adsorption of 0.001 and attenuator value equal to 5.

Encapsulation Efficiency
Total PCM

For the determination of the total PCM content in the dried MPCMs, 1 g of powder was added to 5 mL of HCl 4M and left on a rotating stirrer for 30 min. Then, 15 mL of ethyl acetate was added, and the sample was left stirring for 30 min. The sample was centrifuged at 5000 rpm for 5 min and the supernatant recovered. The obtained supernatant solution was evaporated, and the amount of PCM extracted calculated using an analytical balance.

Superficial PCM

To determine the amount of superficial PCM non included in the encapsulates or exposed on the surface, 1 g of MPCMs powder was manually shaken with 10 mL of ethyl acetate for 30 s. The sample was centrifuged at 5000 rpm for 5 min, the supernatant recovered and the amount of PCM was weighted through an analytical balance.

The encapsulation efficiency (EE%) of dried MPCMs was evaluated by using Eq. 1:

$$EE\% = \frac{(total\ PCM\ -\ surface\ PCM)}{Theoretical\ PCM} \times 100 \tag{1}$$

where the theoretical PCM is equivalent to the starting PCM amount in g used for the encapsulation.

Process Yield
The process yield (PY %) was used to assess the efficiency of the spray-drying process. To evaluate this parameter, the powder collected from the spray dry was weighed by using an analytical balance, and the yield was calculated with the following equation (Eq. 2)

$$PY\ \% = \frac{weight\ of\ the\ obteined\ powder}{total\ starting\ solid\ material} \times 100 \tag{2}$$

MPCMs Surface Evaluation
Environmental scanning electron microscopy (ESEM) was applied to investigate the morphology of the obtained MPCMs. The images were obtained by an ESEM EVO LS25 (ZEISS, DE). The samples were placed on double-sided adhesive carbon tabs mounted on ESEM stubs and then analysed.

4 Statistical Analysis

All measurements were performed in triplicate, and results were reported as mean value followed by its standard deviation. Statistical analysis was performed using one-way analysis of variance (ANOVA). The significance level (alpha) was set to 0.05.

5 Results and Discussion

5.1 Microencapsulation of PCM

Preparation of MPCMs Through Complex Coacervation Followed by Shell Mineralization

To find the right ratio between CMC and PP multiple tests (e.g., 1:1, 1:1.2, 1:2 and 1:2.5) were conducted. The coacervation process was conducted at pH 4 $\pm$ 0.05, since this was found to be the pH value where the two polymers were more reactive (data not shown). At this pH, indeed, the amino groups of lysine residues present in the PP, are positively charged [10] and at the same, the carboxylic groups of CMC are highly negatively charged, since they are present in their dissociated form (e.g., pKa $\approx$ 3.5) [11]. The presence in the same solution of chargeable polymers is the basis for the complex coacervation process, and to form resistant particles it is of fundamental importance to identify the concentration at which electrostatic interactions between them are favoured [8].

The obtained results (Fig. 2), indicated that by using a ratio of 1:1 (CMC-PP), the obtained capsules were not able to withstand the drying process on the glass slide causing their rupture and leakage of the contained PCM core as visible in Fig. 2 A In the image, indeed, typically coacervate structure are not visible, that are instead observable in the case of the MPCMs obtained by using the ratio 1:1.2, 1:2 and 1:2.5 (CMC-PP). We hypothesize that in the case of the 1:1 (CMC-PP) ratio the amount of polymers used were not enough to create a stable shell around the PCM droplets. Ferrera et al. also highlight the importance of finding the appropriate ratio between the materials used to form microcapsules via complex coacervation. In their study, involving gelatin and Arabic gum, they observed that when the amount of Arabic gum is insufficient, microcapsule formation does not occur [12]. As can be observed in Fig. 2 B, C and E the formed MPCMs particles were found spherical in shape, with a tendency to aggregate upon drying. This can be an effect correlated to the tendency of particles to come close to each other upon the evaporation of water, due to capillary forces that arise when the liquid–air interface recedes and overcomes the electrostatic repulsion between particles [13].

The size of the MPCMs obtained were similar for all the tested ratio, ranging from 13.2 $\pm$ 0.17 μm in the case of the ratio 1:1.2 (CMC-PP), 9.85 $\pm$ 0.05 when the tested ratio was 1:2 (CMC-PP) and 9.76 $\pm$ 0.09 μm whe2n the ratio was 1:2.5 (CMC-PP). As can be observed, smaller particles size was obtained when the concentration of solids was higher, this can be a consequence of the presence of more proteins in the preparation solution that are well known emulsifier, enabling the creation of smaller particles [2, 14]. Several studies had already underlined the importance of encapsulates size when it comes to the encapsulation of PCM, linked to the fact that smaller particles

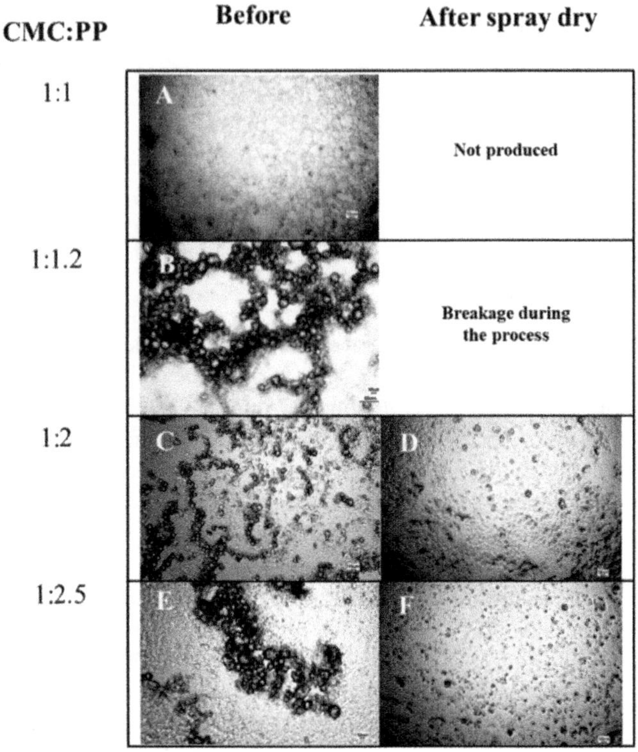

Fig. 2. MPCMs particles observed by an optical microscope OPTECH BM 60 (USA) (magnification 20x, bright field) with different CMC:PP ratio dried on a glass slide 2before and after spraying\dry (scale bar 10 μm).

showed a higher mechanical resistance [2, 15]. Meanwhile the higher payload of PCM is strongly requested from an application point of view [2]. The obtained size range is similar with those reported in the literature. For example, in the article of Bayés-García et al., (2010), where MPCMs were synthesized using complex coacervation, the average particle sizes was around 12 μm [16]. In other studies, where the production of MPCMs was carried out via interfacial or in situ polymerization, the resulting microparticle sizes were comparable. For instance, Jin et al. (2023) reported MPCMs with particle sizes ranging from 6.36 to 98.21 μm, while Parameshwaran et al. (2021) obtained sizes between 2 and 10 μm [17, 18].

Another important aspect of microparticles relies on their superficial charge, since this characteristic can be used to understand better the possible interaction of the microparticles with the external environment. The detected superficial charge in the case of the developed MPCMs were highly negative for all the produced samples, ranging between -36.7 ± 0.11 mV, -36.69 ± 0.23 mV and -39.62 ± 0.39 mV, in the case of the CMC-PP ratio 1:1.2, 1:2 and 1:2,5 respectively. The high negative surface charge can be attributed to the addition of the mineralized shell that increased the pH value of the solution up to 7.5, where the amino groups of the PP become highly negative, contributing to the overall negative charge given by the presence of the CMC [11].

This characteristic is particularly relevant, as Uchikawa et al. reported that cement particles exhibit a positive surface charge in the early stages of hydration, primarily due to the adsorption of Ca^{2+} ions. Consequently, the introduction of negatively charged microparticles could enhance the stability of the system through electrostatic interactions, promoting better dispersion and reducing early-stage aggregation [19].

Due to the observed results and the successful MPCMs formation the 1:1.2, 1:2 and 1:2.5 CMC-PP samples were selected to proceed with the spray drying process.

Spray Drying of MPCMs

Preliminary tests were conducted to understand the feasibility of the drying process, and to define the setup of the equipment. As an example, different amounts of MD (e.g., 5, 10 and 20% w/v) were added to the MPCMs dispersion to identify the appropriate amount required to improve the drying process. In literature, it is reported that MD can improve the spray dry process, due to their high glass temperature and role in preventing the adhesion of microcapsules on the dryer walls [20]. In this preliminary test (data not shown), it was observed that the most promising results were obtained by adding 20% w/v of and in the case of the 1:1.2 (CMC-PP) sample, even if the encapsulates were formed and stable (Fig. 2) the spray drying process was not possible, indeed the MPCMs were broken inside the main chamber losing their core.

Table 1. MPCMs characterization after spray drying process

	1:2 20MD	1:2,5 20MD
Payload (%)	37.95	37.56
Solid content formulation (%)	33.70	33.92
EE %	88.41 ± 2.01	66.84 ± 3.84
PY %	36.1 ± 0.82	38.71 ± 3.45

Meanwhile, the MPCMs prepared by using the 1:2 and 1:2.5 (CMC-PP) ratio was able to endure such conditions.

In Table 1 the characteristics of the obtained MPCMs after spray dry are reported, among them the PCM content w/w expressed as Payload (%), the total solids content in the formulation, the encapsulation efficiency (EE %), and process yield (PY %).

As can be observed, the formulation with the highest EE % was the one produced with a CMC-PP ratio of 1:2, that registered a value of 88.41 ± 2.01%. The obtained values were in line with those reported in the literature and, in some cases, even higher than those observed for encapsulation via coacervation of PCM but is worth noting that these results were achieved without the use of spray drying and limited to the coacervation part. For instance, Chang et al., (2016) [21], which used canola protein isolate and chitosan reported an EE % of 78%, while Mancer et al., (2018) [22] EE % values up to 84% using soybean protein isolate and pectin.

Regarding the PY %, the best performance was observed in the case of the sample produced with the 1:2.5 CMC-PP, reaching a value of 38.71 ± 3.45% with a significant difference when compared to the ratio 1:2 CMC-PP sample, that was found to achieve a PY % of 36.1 ± 0.82 (p value = 0.27). Although relatively low, this value lacks references in the literature for the yield of spray-dried MPCM particles produced by complex coacervation.

MPCMs Morphology Characterization

The eSEM was used to evaluate the morphology of the best MPCMs samples (e.g., CMC-PP 1:2 and 1:2.5 ratio). In Fig. 3 it can be observed that both the samples were very simil8ar (Fig. 3).

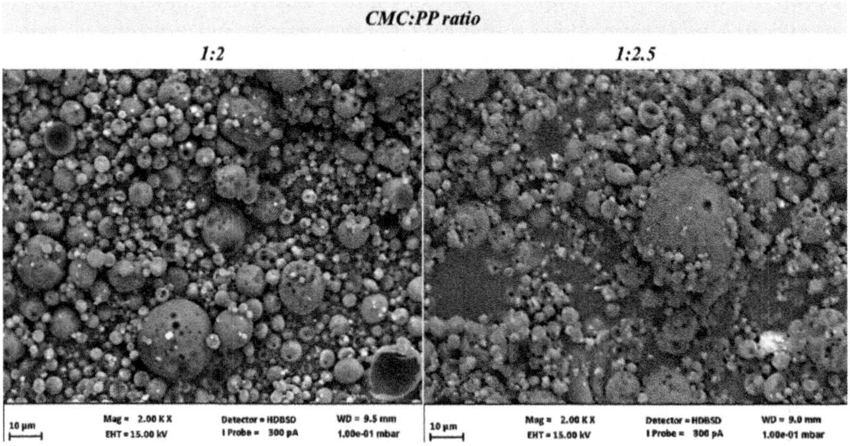

Fig. 3. eSEM images of spray- dried powders

Indeed, the MPCMs were mainly characterized by spherical shape particles with a heterogeneous distribution, with most of them showing a particle size smaller than 10 µm, although larger particles with a diameter of around 30 and 40 µm were also present. In both cases, the surface of the MPCMs particles seems to be smooth, with the absence of wrinkled particles. Additionally, it was observed the presence of darker round spot on most of the surface of the particles, that was already described in the case of the work of Ogrodowska et al. (2017), relative to a rich in oil emulsion encapsulated by spray drying [23] confirming the successful PCM inclusion in the particles structure. It was hypothesized that the MPCMs exhibit a core–shell structure, in which an oil core is encapsulated by a matrix of proteins and carboxymethyl cellulose (CMC), crosslinked by glycerol and tannic acid. An additional outer layer of calcium carbonate and maltodextrins was proposed to further encapsulate the structure, enhancing the mechanical stability of the particles.

6 Conclusions

Thanks to the use of CMC and proteins as encapsulant materials along with the combination of different encapsulation techniques, it was possible to efficiently microencapsulate PCM. The best-performing samples were found to be those sample produced with a CMC-to-protein ratio of 1:2 and 1:2.5 with the addition of 20% w/v of MD to improve the drying process, achieving an encapsulation efficiency of 88.41 $\pm$ 2.01% and 66,84 $\pm$ 3.48%, respectively. These formulations enable us to achieve a payload of PCM of 37.95% and 37.56% respectively. The obtained values of payload are still lower if compared to the one present in the market (e.g.,78% for the Micronal DS 5008 X from BASF, or 70% for the mPCMs produced by Microtek Laboratories).Despite the difference in the amount of encapsulated PCM, it is important to highlight that commercially available products typically use capsules made from low-sustainability materials, such as polymerized acrylates (MMA/PETRA) or melamine-formaldehyde resins, which, unlike bio-based capsules, enable the formation of more rigid shells and allow for a higher oil loading.

The process yield obtained with the spray dry was 38.71 $\pm$ 3.45% in the sample with a 1:2.5 CMC-PP ratio indicating a promising process that can be further improved through additional studies aiming at evaluating the influence of parameters such as the inlet temperature, the sprayed volume on the drying process but also focusing particularly on improving MPCMs robustness and reducing their size to avoid breakage in the main chamber of the spray drier.

To better understand the possibility of their usage in the building sector, further characterization will be required in terms of calorimetry and mechanical resistance but also the evaluation of their affinity with the mix used to produce building materials such as bricks and cement paste. This will allow to obtain a complete evaluation of the possibility to use this innovative encapsulation approach to create a smart energy storage unit for buildings.

Acknowledgment. The authors gratefully acknowledge the support received from the following research projects:

- This work was partially supported by the European Union's Horizon 2020 research and innovation programme under the Marie Skłodowska-Curie Staff Exchange (MSCA-SE) action (GA 101086440 - Project acronym: BEST - Bio-based Energy-efficient materials and Structures for Tomorrow).

- Further support was provided by the Marie Skłodowska-Curie Postdoctoral Fellowship (GA 101153320 – Project acronym: SMART-NRG – Solutions for constructions and building materials based on Energy Absorption).

- The first author also acknowledge the co-funded PhD scholarship initiative supported by the Italian Ministry of University and Research under Ministerial Decree No. 630 (April 24, 2024), in collaboration with Sphera Encapsulation Srl and the Department of Civil, Chemical and Environmental Engineering (DICCA) of the University of Genoa.

- Finally, this work was developed within the framework of the PRIN 2022 project OPTIMA (OPTImized use of phase change MAterials for energy and seismic retrofitting of buildings), funded by the Italian Ministry of University and Research (MUR) for the period 2025–2027.

References

1. Soares, N., Costa, J., Gaspar, A., Santos, P.: Review of passive PCM latent heat thermal energy storage systems towards buildings' energy efficiency. Energy Build. **59**, 82–103 (2013). https://doi.org/10.1016/j.enbuild.2012.12.042
2. Giro-Paloma, J., Martínez, M., Cabeza, L.F., Fernández, A.I.: Types, methods, techniques, and applications for microencapsulated phase change materials (MPCM): a review. Renew. Sustain. Energy Rev. **53**, 1059–1075 (2015). https://doi.org/10.1016/j.rser.2015.09.040
3. Tyagi, V., Buddhi, D., Kothari, R., Tyagi, S.: Phase change material (PCM) based thermal management system for cool energy storage application in building: an experimental study. Energy Build. **51**, 248–254 (2012). https://doi.org/10.1016/j.enbuild.2012.05.023
4. Sarı, A., Alkan, C., Bilgin, C.: Micro/nano encapsulation of some paraffin eutectic mixtures with poly(methyl methacrylate) shell: preparation, characterization and latent heat thermal energy storage properties. Appl. Energy **136**, 217–227 (2014). https://doi.org/10.1016/j.apenergy.2014.09.047
5. Borreguero, A., Valverde, J., Rodríguez, J., Barber, A., Cubillo, J., Carmona, M.: Synthesis and characterization of microcapsules containing Rubitherm®RT27 obtained by spray drying. Chem. Eng. J. **166**(1), 384–390 (2010). https://doi.org/10.1016/j.cej.2010.10.055
6. Sarkar, S., Mestry, S., Mhaske, S.: Developments in phase change material (PCM) doped energy efficient polyurethane (PU) foam for perishable food cold-storage applications: a review. J. Energy Storage **50**, 104620 (2022). https://doi.org/10.1016/j.est.2022.104620
7. Tillet, G., Boutevin, B., Ameduri, B.: Chemical reactions of polymer crosslinking and post-crosslinking at room and medium temperature. Prog. Polym. Sci. **36**(2), 191–217 (2010). https://doi.org/10.1016/j.progpolymsci.2010.08.003
8. Hawlader, M.N.A., Uddin, M.S., Zhu, H.J.: Encapsulated phase change materials for thermal energy storage: experiments and simulation. Int. J. Energy Res. **26**(2), 159–171 (2002). https://doi.org/10.1002/er.773
9. Wang, T., Wang, S., Luo, R., Zhu, C., Akiyama, T., Zhang, Z.: Microencapsulation of phase change materials with binary cores and calcium carbonate shell for thermal energy storage. Appl. Energy **171**, 113–119 (2016). https://doi.org/10.1016/j.apenergy.2016.03.037
10. Bhutto, R.A., et al.: Potato protein as an emerging high-quality: source, extraction, purification, properties (functional, nutritional, physicochemical, and processing), applications, and challenges using potato protein. Food Hydrocolloids **157**, 110415 (2024). https://doi.org/10.1016/j.foodhyd.2024.110415
11. Koupantsis, T., Pavlidou, E., Paraskevopoulou, A.: Flavour encapsulation in milk proteins – CMC coacervate-type complexes. Food Hydrocolloids **37**, 134–142 (2013). https://doi.org/10.1016/j.foodhyd.2013.10.031
12. Ferreira, S., Nicoletti, V.R.: Microencapsulation of ginger oil by complex coacervation using atomization: effects of polymer ratio and wall material concentration. J. Food Eng. **291**, 110214 (2020). https://doi.org/10.1016/j.jfoodeng.2020.110214
13. Thill, A., Spalla, O.: Aggregation due to capillary forces during drying of particle submonolayers. Colloids Surf. A **217**(1–3), 143–151 (2003). https://doi.org/10.1016/s0927-7757(02)00569-1
14. Zhang, M., Fan, L., Liu, Y., Huang, S., Li, J.: Effects of proteins on emulsion stability: the role of proteins at the oil–water interface. Food Chem. **397**, 133726 (2022). https://doi.org/10.1016/j.foodchem.2022.133726
15. Liu, C., Rao, Z., Zhao, J., Huo, Y., Li, Y.: Review on nanoencapsulated phase change materials: preparation, characterization and heat transfer enhancement. Nano Energy **13**, 814–826 (2015). https://doi.org/10.1016/j.nanoen.2015.02.016

16. Bayés-García, L., Ventolà, L., Cordobilla, R., Benages, R., Calvet, T., Cuevas-Diarte, M.: Phase change materials (PCM) microcapsules with different shell compositions: preparation, characterization and thermal stability. Sol. Energy Mater. Sol. Cells **94**(7), 1235–1240 (2010). https://doi.org/10.1016/j.solmat.2010.03.014
17. Jin, X., Haider, M.Z., Ahn, J., Fang, G., Hu, J.W.: Development of nanomodified eco-friendly thermal energy storing cementitious composite using PCM microencapsulated in biosourced encapsulation shell. Case Stud. Constr. Mater. **19**, e02447 (2023)
18. Parameshwaran, R., Naresh, R., Ram, V.V., Srinivas, P.: Microencapsulated bio-based phase change material-micro concrete composite for thermal energy storage. J. Build. Eng. **39**, 102247 (2021). https://doi.org/10.1016/j.jobe.2021.102247
19. Uchikawa, H., Hanehara, S., Sawaki, D.: The role of steric repulsive force in the dispersion of cement particles in fresh paste prepared with organic admixture. Cem. Concr. Res. **27**(1), 37–50 (1997). https://doi.org/10.1016/s0008-8846(96)00207-4
20. Caliskan, G., Dirim, S.N.: The effects of the different drying conditions and the amounts of maltodextrin addition during spray drying of sumac extract. Food Bioprod. Process. **91**(4), 539–548 (2013). https://doi.org/10.1016/j.fbp.2013.06.004
21. Chang, P.G., Gupta, R., Timilsena, Y.P., Adhikari, B.: Optimisation of the complex coacervation between canola protein isolate and chitosan. J. Food Eng. **191**, 58–66 (2016). https://doi.org/10.1016/j.jfoodeng.2016.07.008
22. Mancer, D., Allemann, E., Daoud, K.: Metformin hydrochloride microencapsulation by complex coacervation: study of size distribution and encapsulation yield using response surface methodology. J. Drug Deliv. Sci. Technol. **45**, 184–195 (2018). https://doi.org/10.1016/j.jddst.2018.03.015
23. Ogrodowska, D., Tańska, M., Brandt, W.: The influence of drying process conditions on the physical properties, bioactive compounds and stability of encapsulated pumpkin seed oil. Food Bioprocess Technol. **10**(7), 1265–1280 (2017). https://doi.org/10.1007/s11947-017-1898-z

Enhancing Olive Husk Properties Through Alkali, Thermal, and Biopolymer Treatments for Sustainable Building Applications

Halima Belhadad[1,2](✉), Bellel Nadir[1], and Ana Bras[2]

[1] Energy Physics Laboratory (EPL), Faculty of Exact Sciences, University Constantine 1, Route Ain El Bey, 25000 Constantine, Algeria
belhadadhalima@outlook.fr

[2] Built Environment and Sustainable Technologies (BEST) Research Institute, School of Civil Engineering and Built Environment, Liverpool John Moores University, Liverpool L3 3AF, UK

Abstract. Olive husk (OH), a plentiful by-product of olive oil production, offers promising potential as a renewable aggregate in sustainable construction. This study investigates three modification pathways, alkali treatment with 6% NaOH (ATOH), controlled thermal treatment (TTOH), and biopolymer coating with gum arabic (GAOH), to enhance OH's compatibility with cementitious and other mineral binders. The treatments were designed to reduce residual oil content and improve interfacial bonding while preserving OH's inherent dual structure, comprising compacted olive stones and a fibrous fraction. Performance was assessed against key application-oriented specifications: reduced hydrophobicity for improved binder adhesion, retention of particle integrity to maintain structural duality, low thermal conductivity for insulation efficiency, and optimised density/porosity balance to ensure lightweight performance without compromising durability. Characterisation, following RILEM TC 236-BBM protocols, included bulk density, porosity, thermal conductivity, water absorption, and particle size distribution, complemented by thermogravimetric/differential scanning calorimetry (TGA/DSC) and Fourier-transform infrared spectroscopy (FTIR). Thermal treatment achieved the lowest density (450.7 $kg{\cdot}m^{-3}$), highest porosity (54.0%), and best insulation performance ($\lambda = 0.0838$ $W{\cdot}m^{-1}{\cdot}K-1$). Alkali treatment markedly reduced oil content, improving microstructural integration despite higher water uptake (62.1%), while GAOH yielded the densest structure (592.1 $kg{\cdot}m^{-3}$, 41.6% porosity) with reduced water absorption (53.8%) but higher conductivity ($\lambda = 0.1024$ $W{\cdot}m^{-1}{\cdot}K^{-1}$). Particle size analysis revealed fibre disaggregation for ATOH and TTOH, enhancing dispersion, whereas GAOH induced controlled agglomeration. Thermal and chemical analyses confirmed treatment-specific modifications, including lignin/hemicellulose reduction (ATOH), partial cellulose degradation (TTOH), and gum arabic driven surface functionalisation (GAOH). These findings demonstrate that targeted treatments can tailor OH properties for integration into high-performance, lightweight, and thermally efficient bio-based composites.

M. Posani and S. Singh (Eds.): RYS 2025, RILEM Bookseries 67, pp. 624–641, 2026.
https://doi.org/10.1007/978-3-032-14652-6_45

Graphical Abstract

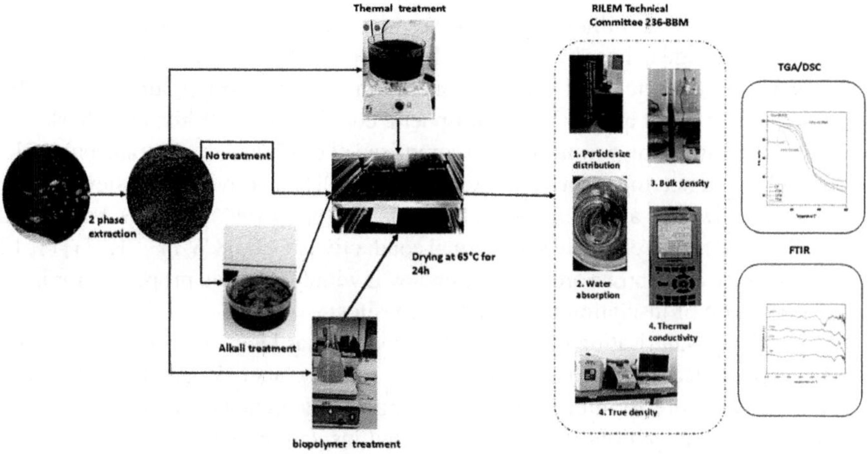

Keywords: olive husk · thermal treatment · alkali treatment · thermal conductivity · TGA/DSC · FTIR

1 Introduction

The construction sector remains one of the most resource-intensive hu-man activities, responsible for nearly 40% of global energy consumption and natural resource depletion [1, 2]. In the context of ongoing climate challenges and the imperative to reduce greenhouse gas emissions, the development of sustainable building materials has become a research priority. Building products not only influence the embodied energy of construction systems, but also affect long-term performance in terms of hygrothermal regulation, indoor air quality, and material durability [3]. Consequently, integrating renewable and locally available resources into construction practice is a crucial step toward low-impact, resilient buildings.

One promising strategy involves the valorisation of agricultural residues as bio-based aggregates for insulation and lightweight composite materials. These bio-wastes are produced in large quantities, particularly in agri-food sectors, and their use in building materials offers a dual environmental benefit: reducing landfilling needs and lowering the environmental footprint of conventional materials [4]. Recent studies have investigated a wide range of such residues, including rice husks, wheat straw, sun-flower stalks, and olive pruning waste, for applications in plasters, light-weight concrete, insulation boards, and earth-based systems.

Raw olive husk (OH), a lignocellulosic by-product of olive oil extraction, constitutes a largely underutilized yet abundant agricultural residue with promising potential for thermal insulation applications. In olive oil production, both traditional two- and three-phase extraction processes generate substantial quantities of by-products: approximately 400–800 kg of wet pomace (35–60% moisture) per tonne of processed olives, alongside 200 kg of oil [5]. With global olive oil production exceeding 2.56 million metric tons in 2023/24 and projected to reach 3.37 million metric tons in 2024/25 [6], the Mediterranean basin, led by Spain, Italy, Greece, Turkey, Tunisia, Morocco, and Portugal, produces several million tonnes of these residual streams annually [7, 8]. When inadequately managed, these by-products alter soil pH, release phenolic compounds that contaminate groundwater, and emit methane during decomposition, affecting soil biodiversity, surface water quality, and atmospheric conditions [9]. While olive husk (OH) denotes the fibrous, semi-porous fraction comprised of crushed pits, skin, and pulp. Olive pomace (OP), is a heterogeneous mixture containing OH and olive mill wastewater. Raw OH is characterized by a high lignocellulosic content (up to 49%) [10], low bulk density (420–640 kg/m^3), and relatively low thermal conductivity (<0.08 W·m^{-1}·K^{-1}) [11, 12]. Its high porosity and fibrous architecture endow it with favourable properties for its use in the production of insulation bio-based composites.

Although the valorisation of olive oil by-products has been extensively explored in construction applications such as concretes [13, 14], mortars [15], and fired clay bricks [16], primarily as pore-forming agents or mostly as supplementary cementitious material as ash [17]. The integration of OH into dedicated insulation materials remains notably limited. This underutilization stems largely from concerns over the residual oil content in olive husk (OH), which typically ranges between 4.5% and 9% [5]. The residual oil creates a hydrophobic layer on OH particles, reducing adhesion with hydrophilic binders (e.g., PVA) and compromising mechanical stability.

To address these challenges, surface and chemical modification treatments are frequently employed to enhance the performance of lignocellulosic aggregates. Methods such as alkali treatment, thermal degradation, and biopolymer coatings have shown potential to improve the hydrophilicity, roughness, and surface energy of bio-based particles, thereby improving matrix compatibility and mechanical integrity [18, 19]. How-ever, existing research is limited with regard to systematic comparisons of such treatments applied specifically to olive husk.

In this context, the present study investigates the impact of three targeted treatment methods, namely alkali treatment (6% NaOH), thermal boiling, and gum arabic biopolymer coating, on the physical, thermal, and surface characteristics of olive husk. The goal is to evaluate how each treatment enhances the material's potential for incorporation into cementitious and earth-based construction systems. Key properties, including bulk density, porosity, thermal conductivity, water absorption, and chemical composition, are analysed using standard protocols such as those recommended by RILEM Technical Committee 236-BBM. Additionally, Thermogravimetric Analysis (TGA/DSC) and Fourier Transform Infrared Spectroscopy (FTIR) are employed to understand the thermal degradation behaviour and chemical transformations of the treated materials.

The work further explores the practical implications of these treatments by testing the integration of treated and untreated OH in lightweight cement mortars, offering insights into workability and mix stability. The findings aim to inform the design of bio-based composite materials that combine thermal efficiency, mechanical stability, and environmental sustainability, thereby contributing to a circular economy approach in construction.

2 Materials and Treatment

The olive husk (OH) used in this study is a lignocellulosic agricultural by-product derived from the two-phase mechanical extraction process of olive oil. Raw OH was sourced directly from a local olive oil mill in Gualma, algeria during the 2022/2023 harvest season. As received, the material comprised a heterogeneous mixture of crushed olive pits, skin, pulp fragments, and residual oil (≈8.3% oil content by dry mass, as reported in prior characterization[13]). To align with sustainability goals, OH was used in its unrefined form to minimize processing energy. The following pre-processing steps were applied (Fig. 1):

Sieving: Separation using a mechanical sieve shaker to isolate the 0.5–2.0 mm particle fraction (ISO 3310–1 standard sieves).
Drying: Oven-drying at 65 ± 5 °C for 24 h to constant mass (<0.5% moisture; Memmert UF110 oven).
Storage: Dried OH was stored in airtight containers at 25 °C and 40% RH until use.

Fig. 1. Olive husk used in this study.

2.1 Treatments

The treatments were designed to reduce residual oil content and improve interfacial compatibility with cementitious and other mineral binders, while preserving OH's dual structure of compacted olive stones and fibrous pulp. Performance was assessed against key application-oriented specifications: reduced hydrophobicity for enhanced binder adhesion, adequate particle integrity to retain structural duality, low thermal conductivity for insulation efficiency, and controlled density/porosity to balance strength and weight.

Alkali treatment

The alkali treatment was designed to eliminate surface impurities, reduce oil content, and increase the surface reactivity of OH by partial delignification. The process involved (Fig. 2):

- immersing the air-dried OH in a 6% (w/w) sodium hydroxide (NaOH) aqueous solution for a duration of 8 h at ambient temperature.
- After 8h, the samples were thoroughly rinsed with distilled water until a neutral pH was attained,

The treated OH was then oven-dried at 65 °C for 48 h to reach a constant mass.

Fig. 2. Photographic documentation of the alkali treatment process, including stages: (a) material preparation,(b) immersion, (c)rinsing, and (d) drying.

Thermal Treatment

The thermal treatment was selected for its ability to reduce the oil residues in the bio-aggregate and simultaneously preserve its intrinsic fibrous and structural characteristics. The procedure was:

boiling the OH for 4 h in distilled water, using a mass ratio of 0.1 OH-to-water.
the material was oven-dried at 65 °C for 48 h.

This treatment aimed to reduce residual oil, microbial load, and improve dimensional stability without inducing significant degradation of the bio-aggregate matrix (Fig. 3).

Biopolymer Treatment (Gum Arabic Coating)

The third treatment involved the application of a natural hydrophilic biopolymer, *gum arabic* (GA), selected for its notable oil holding capacity (OHC) of 0.87g/g [20] and its potential to enhance the interfacial behaviour of bio-aggregates. The process followed:

The olive husk particles were soaked in an 18% GA aqueous solution for 24 h at ambient temperature to ensure sufficient diffusion and coating.

Fig. 3. Photographic documentation of the thermal treatment process, including stages: (a) material preparation, (b) boiling, and (c) drying.

Following the soaking phase, the samples were oven-dried at 60 °C for 48 h until a constant mass was achieved.

This biopolymer treatment was specifically intended to entrap residual oil within the OH structure, thereby mitigating oil migration and improving compatibility with both cementitious and earthen binders. Moreover, the gum arabic coating is expected to contribute to enhanced water resistance and better binder-aggregate adhesion by forming a hydrophilic yet cohesive interface layer around the bio-aggregate.

The gum Arabic used in this work was supplied by Sigma-Aldrich under product Number: G9752 with an oil holding capacity (OHC) of 0.87 g/g, indicating its ability to absorb oil. The GA properties provided by supplier are presented in Table 1 (Fig. 4).

Table 1. Physico-chemical properties of Gum Arabic (Sigma-Aldrich, Product No. G9752)

Colour	White to Faint Beige
Synonyms	Acacia gum
Form	Powder
Moisture Content	$< 15\%$
Insoluble Residue	$< 0.20\%$
Density	1.4 g/cm^3
specific mass	13.5 KN / m3
Molecular weight	$\approx 0.25 \times 106 \approx 0.25 \times 106$
pH	4.5
Oil holding capacity (OHC)	0.87 g/g
ash	3.3%

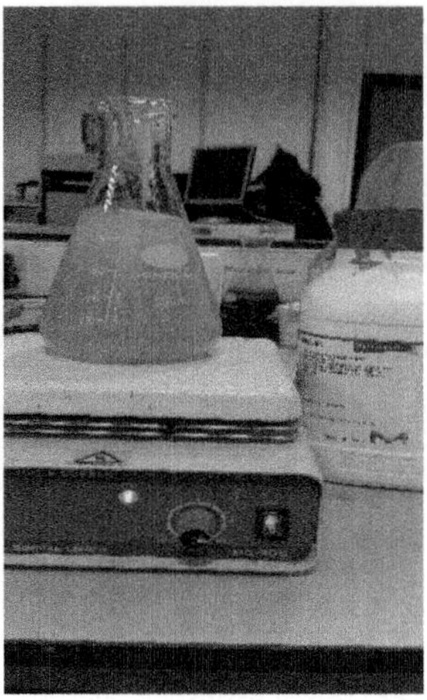

Fig. 4. Sigma-Aldrich (Product No. G9752) gum Arabic, and a 18% GA solution

3 Experimental Methodology

The physical and physicochemical characterisation of untreated and pre-treated olive husk (OH) was conducted in accordance with Liuzzi et al. [22] and the testing recommendations of RILEM Technical Committee 236-BBM [21], originally validated for hemp shiv but applicable to lignocellulosic residues such as OH due to their analogous anatomical structure, porosity, and moisture interaction behaviour. All measurements were performed in triplicate, with results reported as mean values. A coefficient of variation (CV) below 5% was adopted as the acceptance threshold to ensure statistical reliability.

3.1 Sample Preparation and Preconditioning

Representative sampling was achieved using the coning and quartering method to prevent segregation between coarse and fine particles. Preconditioning was carried out in a ventilated oven at 60 °C until constant mass (variation $< 0.1\%$ over 24 h), followed by stabilisation in a stable lab condition (20 $\pm$ 2 °C; 45 $\pm$ 5% RH). This step ensured removal of free moisture and equilibration to standard laboratory conditions, thereby minimising variability in hygroscopic measurements.

3.2 Particle Size Distribution

Particle size distribution of olive husk (OH) was determined according to the recommendations of RILEM TC 236-BBM, which identifies two principal techniques for characterising plant-based aggregates: mechanical sieving and image analysis. In this work, mechanical sieving was selected, as it is well suited for aggregates with approximately spheroidal or irregular–rounded morphologies, such as OH particles, corn cob granules, and other lignocellulosic by-products.

The test was conducted using a stack of standardised square-mesh sieves. The sieve apertures used were 5.00 mm, 2.50 mm, 1.25 mm, 0.80 mm, 0.50 mm, 0.25 mm, and 0.08 mm, with a solid pan to collect the fines. Approximately 100 g of oven-dried OH was placed on the top sieve and agitated in a mechanical sieve shaker for 15 min. This procedure determines the equivalent particle width by identifying the smallest aperture through which each particle can pass when oriented along its length.

The retained mass on each sieve was measured with an accuracy of ± 0.01 g, and the proportion of each particle size class was expressed as a percentage of the total sample mass. Each test was carried out in triplicate, and mean values were reported when the coefficient of variation was below 5%.

3.3 Bulk, True Particle Density and Porosity

Bulk density was determined following the RILEM TC 236-BBM procedure. A glass cylindrical container of (diameter = 4 cm height = 7 cm) was filled with OH without compaction, and the mass was measured using a precision balance (±0.01 g). And bulk density (ρ_{bulk}) was calculated as:

$$P_{bulk} = m/V \tag{1}$$

where m is the sample mass and V is the container volume. Each test was repeated three times, and the mean value was retained when the coefficient of variation (CoV) was <5%.

True particle density (ρ_{true}) was evaluated using helium pycnometry inspired by Liuzzi et al. [22], which ensures high accuracy for irregular and porous particles by quantifying the solid-phase volume through gas displacement. In laboratories lacking a pycnometer, the ethanol displacement method was applied as a validated alternative for porous bio-aggregates, as ethanol penetrates open pores without swelling lignocellulosic matter.

Open porosity (Φ) was then calculated using Eq. (2):

$$\Phi = 1 - \rho_{bulk}/\rho_{true} \tag{2}$$

where Φ: is the total porosity in %, ρ_{bulk}: is the bulk density in kg/m^3 and ρ_{true}: is the true density in kg/m^3.

3.4 Thermal Conductivity

Thermal conductivity was measured using a transient plane source device (ISOMET 2104, Applied Precision, Slovakia) with a 60 mm contact probe, in compliance with the RILEM TC 236-BBM protocols. OH was loosely compacted into a glass cell (diameter = 20 cm; height = 6 cm) to achieve a specimen thickness > 4 cm, ensuring compliance with thermal insulation products requirements. Measurements were performed under lab conditions (20 ± 2 °C; RH = 45 ± 5%) at three points per sample. Thermal conductivity is the primary determinant of OH's potential as a bio-based filler candidate; measurement under controlled environmental conditions ensures that results are comparable with other biobased aggregates and fibres.

3.5 Water Absorption Capacity

Water absorption was assessed according to RILEM TC 236-BBM. Oven-dried samples were immersed in distilled water at ambient temperature, and their mass was recorded after immersion durations of 1 min, 15 min, 4 h, 24 h, and 48 h. After removal from water, the samples were gently surface-dried with blotting paper prior to weighing.

Each recorded mass value was corrected for the initial oven-dry mass, and the water absorption at time W(t), was calculated according to Eq. (3):

$$W(t) = [M(t) - M(0)]/M0 \tag{3}$$

where M(t) is the mass at time t and M0 is the oven-dry mass. For OH samples, equilibrium water uptake was generally reached within 24 h, with negligible variation ($< 1\%$) between 24 h and 48 h.

3.6 Chemical and Thermal Analysis

TGA/DSC was performed using a NETZSCH LTTS instrument on a sample mass of approximately (mean mass $\approx$51.77 mg). The samples were heated from 20 °C to 600 °C with a heating rate of 10 °C/min. These analyses quantified volatile matter, thermal stability, and ash yield, parameters that influence fire resistance and long-term thermal performance.

FTIR spectra were obtained using a JASCO FT/IR 4100 spectrophotometer by directly scanning crushed samples. The spectra were recorded in transmission mode over a frequency range from 500–4000 cm^{-1} with a resolution of 2 cm^{-1}. This analysis was employed to detect functional group modifications induced by chemical or thermal treatments, providing insight into treatment effectiveness and potential impacts on OH's hydrophilicity and bonding with binders.

4 Results and Discussion

The particle size distribution (PSD) of olive husk (OH) and its treated variants demonstrates the significant influence of surface condition and inter-particle cohesion on granulometric behaviour. As seen in Fig. 5, in the untreated OH, 66% passed the 5 mm sieve, but only ~30% was below 2 mm. This apparent coarseness is partly attributed to oil-rich pulp and fibrous residues promoting agglomeration, which masks the true distribution of finer particles. Both alkali-treated (ATOH) and thermally treated (TTOH) samples exhibited a substantial shift toward finer fractions, with 42.3% and 46.2% passing 2 mm, respectively. These treatments reduced oil content and disrupted fibre–pulp bonding, leading to disaggregation and a more representative PSD. The thermal process in particular enhanced friability, with nearly 8% $<$ 0.5 mm. Conversely, the gum arabic-coated OH (GAOH) showed a coarser profile, with only 22.5% $<$ 2 mm. The biopolymer coating likely induced surface agglomeration and particle bonding, reducing the mobility of fines during sieving.

The physical and hydrothermal properties of olive husk (OH) and its treated derivatives reveal the substantial effect of pre-treatment methods on aggregate performance.

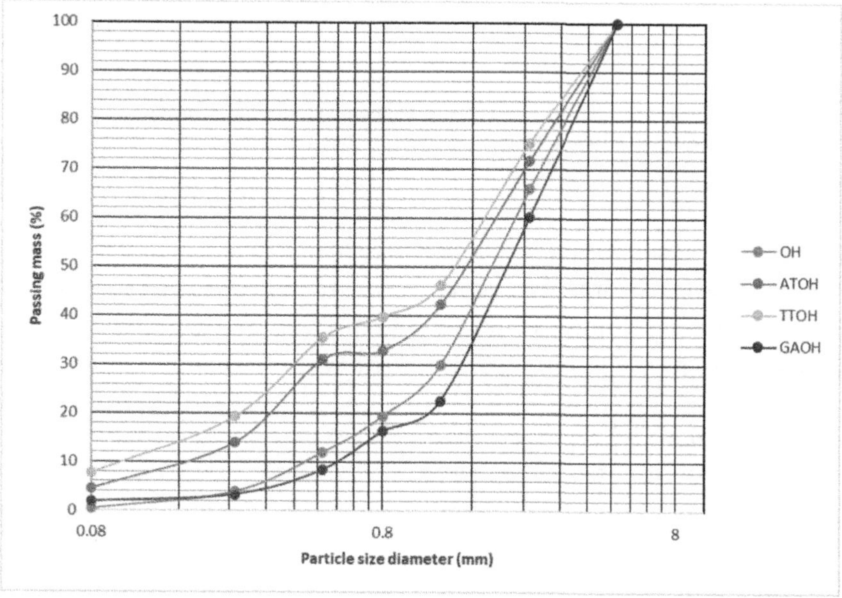

Fig. 5. Particle size distribution of olive husk OH, TTOH,ATOH, and GAOH

The physical, hygrothermal, and thermal properties of untreated olive husk (OH) and its modified derivatives reveal a strong influence of pre-treatment type on aggregate performance (Table 2). The untreated OH exhibited a bulk density of 549.3 kg·m¯3 and a true density of 1292.3 kg·m¯3, corresponding to an open porosity of 56.8%. This intermediate porosity reflects its heterogeneous structure, composed of dense woody endocarp fragments embedded in an oil-rich mesocarp, which provides void space without the extreme friability of highly porous pith materials.

Treatment type substantially modified OH performance. Alkali treatment (ATOH) produced a slight reduction in bulk density (531.2 kg·m¯3) but a pronounced decrease in true density (958 kg·m¯3), leading to a reduction in porosity to 44.5%. This densification at the microstructural level is attributed to the removal of oils and extractives, combined with partial dissolution of hemicellulose and amorphous lignin phases, which induces particle contraction. Thermal treatment (TTOH) generated the lowest bulk density (450.7 kg·m¯3) while maintaining moderate porosity (54.0%), consistent with thermal softening, microcracking, and partial fibre degradation. Gum arabic treatment (GAOH), in contrast, resulted in the highest bulk density (592.1 kg·m¯3) and lowest porosity (41.6%), indicating significant pore filling and surface smoothing due to the biopolymer coating.

Thermal conductivity (λ) remained low for all OH types, ranging from 0.0833 $W{\cdot}m^{-1}{\cdot}K^{-1}$ (OH) to 0.1024 $W{\cdot}m^{-1}{\cdot}K^{-1}$ (GAOH). While the gum arabic coating slightly increased λ due to improved particle contact and reduced porosity, all values remained within the typical range for lightweight insulating aggregates, such as corn cob (0.096 $W{\cdot}m^{-1}{\cdot}K^{-1}$) [4]. Notably, although OH conductivity values are higher than those of cork (≈0.04–0.05 $W{\cdot}m^{-1}{\cdot}K^{-1}$) [4] and rice husk (0.059–0.079 $W{\cdot}m^{-1}{\cdot}K^{-1}$)

Table 2. Physico-thermal parameters of olive husk and other plant-based aggregates

Bio aggregates	Bulk density (Kg/m3)	True density (Kg/m3)	Porosity (%)	Thermal conductivity W/ (m K)	Final water absorption (w_{∞})	Reference
Olive husk (OH)	549,33	1292.3	56,84	0.0833	50,44	This study
ATOH	531.24	958	44.55	0.0871	62.1	This study
TTOH	450.7	979	53.96	0.0838	58.72	This study
GAOH	592.1	1014.13	41.6	0.1024	53.82	This study
sunflower pith	15.9	-	97.3	0.036	917	[4]
Corn cob	497	-	72.05	0.096	125	[4]
Rice husk	103–112	780	91.2	0.059–0.079	124	[4]
Hemp shives	114	1465	82.5	0.49–0.057	406	[4]

[4], they are comparable to hemp shiv [16] and still suitable for use in thermal insulation composites when combined with low-λ binders.

Water absorption capacity displayed divergent responses to treatment. ATOH reached the highest absorption (62.1 wt%), reflecting increased accessibility of capillary pores following delignification and oil removal. TTOH also showed higher absorption (58.7 wt%) than raw OH (50.4 wt%), likely due to fragmentation-induced surface area gains. GAOH exhibited only a modest increase (53.8 wt%), suggesting that the gum arabic layer partially mitigates capillary transport, despite its hydrophilic chemistry.

When positioned relative to other plant-based aggregates, OH occupies an intermediate density–porosity–absorption space. Materials such as sunflower pith and rice husk, with porosities above 90%, display water uptakes several times higher (≈900 wt% and ≈124 wt%, respectively [4]) but offer superior insulation potential due to their extremely low densities. Conversely, OH's denser and more mechanically stable structure offers advantages for composite durability and dimensional stability, particularly in humid environments. This balance between insulation capacity and structural integrity positions OH as a promising bio-aggregate for cementitious and earthen composites where both low thermal conductivity and controlled water affinity are desired.

From a materials engineering perspective, the treatments applied here offer pathways for property tuning: alkali treatment for enhanced water uptake in moisture-buffering applications, thermal treatment for lightweight aggregates with improved friability, and gum arabic coating for reduced porosity and greater dimensional stability. Such versatility supports OH's potential as a regionally abundant, sustainable alternative to conventional aggregates, with the additional benefit of valorising an agro-industrial by-product.

Thermogravimetric and differential scanning calorimetry (TGA/DSC) measurements were performed using a NETZSCH LTTS instrument. For each treatment, three independent replicates (three different specimens) were analysed to assess repeatability. Approximately 50 mg of finely crushed material was used per run (mean mass ≈ 51.8 mg). Samples were heated from 20 °C to 600 °C at 10 °C·min^{-1} under a nitrogen atmosphere. Reported thermal metrics (T_5%, and residue at 600 °C) correspond to the mean values of the three replicates; the associated standard deviation (SD) or relative standard deviation (RSD) is given where relevant. Representative TGA/DSC traces are shown in Fig. 6.

The temperature at 5% mass loss (T_5%) was used as an indicator of thermal stability. Raw olive husk (OH) exhibited the highest T_5% at 232.35° ± 0.5C, reflecting greater initial resistance to thermal decomposition, likely due to the presence of residual oils and extractives. This relatively high initial degradation temperature; compared to other studies on different types of olive husk, which degrade around 200 °C [23], suggests a favourable resistance to early thermal decomposition.

Alkali-thermally treated olive husk (ATOH) showed a reduced T_5% of 196.24 ± 1.3 °C, while thermally treated OH (TTOH) and gum arabic-coated OH (GAOH) exhibited significantly lower T_5% values of 121.52 ± 0.8 °C and 114.80 ± 1.7 °C, respectively. The early degradation onset in TTOH and GAOH is attributed to the partial removal of stabilizing components during treatment (TTOH) and the thermal sensitivity of the gum arabic coating (GAOH).

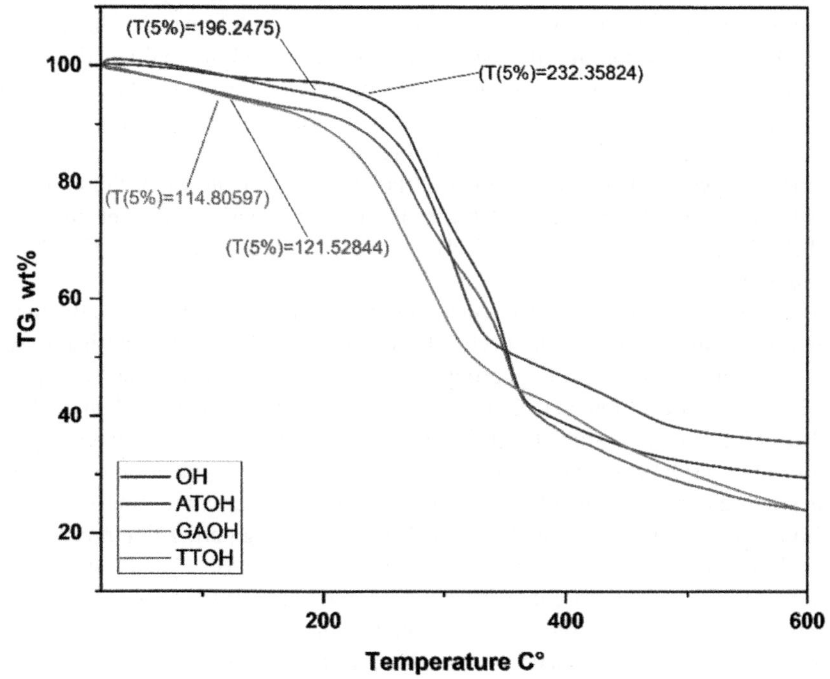

Fig. 6. TGA plots of the different Olive Husk: OH, ATOH, TTOH, and GAOH

The main mass loss occurred between 200 °C and 400 °C, associated with the degradation of hemicellulose and cellulose. OH exhibited the highest mass loss in this range (45.81 ± 2.2%%), followed by ATOH (43.53 ± 3.5%) and TTOH (41 ± 1.9%). These values reflect the progressive removal of thermolabile fractions and increased thermal maturity of the treated materials. GAOH, despite its early onset of degradation, displayed a similar mass loss pattern, indicating that the internal structure of the lignocellulosic core remains largely intact under the coating. The residue at 600 °C provides an estimate of the inorganic content and char stability. ATOH showed the highest ash yield (35.37 ± 0.9%), suggesting a concentration of thermally stable mineral phases due to the removal of volatiles during alkali treatment. OH exhibited a moderate ash content (29.40 ± 1.1%), while TTOH and GAOH had lower residues of 23.80 ± 2.7% and 24.15 ± 1.3%, respectively. These lower values imply more complete volatilization of organic components and reduced inorganic retention, which may affect char stability and fire resistance performance. In summary, ATOH demonstrated improved thermal degradation behaviour with higher residual mass, indicating better fire endurance potential. TTOH offered a balance between thermal resistance and reduced ash formation, while GAOH, despite offering better surface compatibility, showed limited thermal stability due to early decomposition of the gum coating.

The DSC curves (Fig. 7) highlight key thermal transitions in raw and treated olive husk samples. An initial endothermic peak below 120 °C, more pronounced in OH and GAOH, corresponds to moisture evaporation. Between 200–400 °C, all samples exhibit exothermic reactions linked to the decomposition of hemicellulose and cellulose. TTOH shows the highest exothermic intensity, indicating increased thermal reactivity due to structural modification during thermal treatment. ATOH presents a sharp exothermic peak near 320 °C, likely reflecting alkali-induced structural changes. In contrast, OH exhibits a more gradual decomposition profile. GAOH displays the lowest exothermic response, suggesting a stabilising effect of the gum arabic matrix that limits oxidative degradation. These differences underscore the impact of each treatment on the thermal behaviour of olive husk, with TTOH being the most reactive and GAOH the most thermally stable.

The FTIR spectra (Fig. 8) reveal the main functional groups present in the raw and treated olive husk samples. All spectra display a broad O–H stretching band around 3300 cm^{-1}, indicative of hydroxyl groups from lignocellulosic components [24]. Peaks near 2910 cm^{-1} are attributed to C–H stretching vibrations in aliphatic chains [25]. The band at ~1730 cm^{-1}, associated with C = O stretching in hemicellulose or ester groups, is reduced in ATOH and nearly absent in TTOH, reflecting degradation or removal of hemicellulose during treatment. Peaks around 1600 cm^{-1} (aromatic ring vibrations of lignin) are more pronounced in OH and GAOH, suggesting partial lignin preservation [24, 25]. Significant changes in the 1000–1200 cm^{-1} region (C–O and C–C stretching of cellulose and hemicellulose) are observed in ATOH and TTOH, confirming structural alterations. GAOH shows broader and more intense bands in this region, likely due to the presence of gum arabic polysaccharides. These spectral differences confirm the effectiveness of chemical and thermal treatments in modifying the lignocellulosic structure, particularly by reducing hemicellulose and enhancing cellulose accessibility.

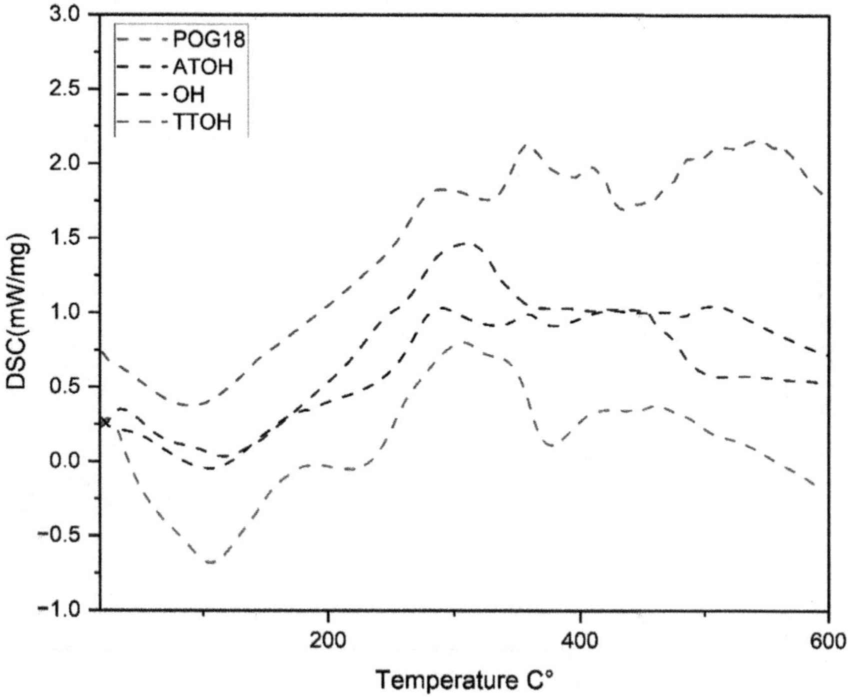

Fig. 7. DSC curve of Olive Husk: OH, ATOH, TTOH, and GAOH

From a life cycle standpoint, olive husk (OH) offers compelling environmental potential as a bio-aggregate. A comprehensive review by Grazieschi et al.[26] benchmarked conventional insulation materials against a functional unit of 1 m2 achieving 1 m2·K/W over 50 years. For example, glass wool requires 16–31 MJ/FU and emits 0.6–1.2 kg CO_2-eq/FU, while EPS demands 44–78 MJ/FU and emits 1.9–3.5 kg CO_2-eq/FU. In contrast, raw or minimally processed lignocellulosic bio-aggregates like OH likely fall within the lower end of embodied energy ranges (e.g., < 10 MJ/FU) due to minimal industrial processing. Consequently, integrating OH into building composites could dramatically reduce both embodied energy and carbon emissions compared to traditional insulations, particularly when OH replaces high-impact mineral fillers in Mediterranean region applications.

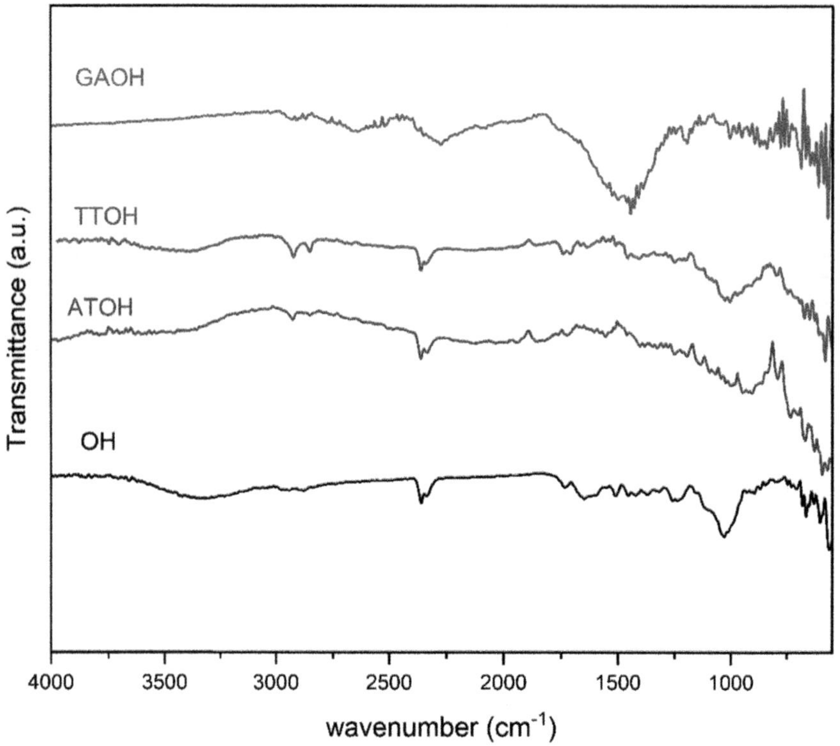

Fig. 8. FTIR spectra of the Olive Husk: OH, ATOH, TTOH, and GAOH

5 Conclusion

This study examined the influence of three pre-treatment strategies, alkali treatment (ATOH), thermal treatment (TTOH), and gum arabic coating (GAOH), on the particle size distribution, physical, thermal, and hygric properties of olive husk (OH), a readily available bio-waste in the Euro-Mediterranean region. Granulometric analysis revealed that ATOH and TTOH markedly increased the fine fraction (<2 mm) by disrupting oil-rich agglomerates and fibre–pulp bonds, whereas GAOH produced a coarser distribution due to surface-induced agglomeration. Bulk and true densities were strongly influenced by treatment: ATOH promoted densification through extractive removal, TTOH produced the lowest bulk density indicative of internal friability, and GAOH increased both densities via polymer film deposition.

Porosity trends mirrored these density changes, with reductions to 44.5% (ATOH) and 41.6% (GAOH), while TTOH maintained a higher porosity (54.0%) due to partial thermal degradation of fibrous components. Thermal conductivity values for all samples (0.0833–0.1024 $W{\cdot}m^{-1}{\cdot}K^{-1}$) remained within the range of lightweight insulating materials, reinforcing the potential of OH, particularly in treated forms, for insulation applications. Water absorption capacity was highest in ATOH (62.1%) owing to greater pore accessibility, followed by TTOH (58.7%), while GAOH exhibited reduced uptake (53.8%) as a result of hydrophilic polymer surface sealing.

Thermal analyses (TGA/DSC) demonstrated enhanced stability in ATOH and TTOH, with lower volatile mass losses and more distinct decomposition stages; ash contents ranged from 23.8–35.4%, reflecting treatment-specific compositional shifts. FTIR results corroborated these findings, showing lignin and oil reduction in ATOH and characteristic C–O and O–H bands from gum arabic in GAOH, confirming successful chemical surface modification.

From a life-cycle perspective, OH offers a significant environmental advantage over conventional insulation materials. Minimally processed OH is expected to remain well below conventional materials benchmarks (often <10 MJ/FU), particularly when sourced from local agro-industrial by-products, thereby reducing both embodied energy and carbon footprint while contributing to waste valorisation.

Collectively, these results indicate that targeted pre-treatments can tailor OH properties to meet specific performance requirements, thereby enhancing its integration into eco-efficient construction materials. Given its hygrothermal regulation potential, low environmental impact, and regional abundance, OH is particularly promising for Mediterranean climates where both insulation and moisture buffering are critical. Future research will focus on incorporating optimised OH into earthen and cementitious mortars and concretes to improve thermal efficiency, workability, and long-term durability while quantifying full life-cycle performance.

References

1. Ramesh, T., Prakash, R., Shukla, K.K.: Life cycle energy analysis of buildings: an overview. Energy Build. **42**, 1592–1600 (2010). https://doi.org/10.1016/j.enbuild.2010.05.007
2. Cabeza, L.F., Rincón, L., Vilariño, V., Pérez, G., Castell, A.: Life cycle assessment (LCA) and life cycle energy analysis (LCEA) of buildings and the building sector: a review. Renew. Sustain. Energy Rev. **29**, 394–416 (2014). https://doi.org/10.1016/j.rser.2013.08.037
3. British Practice for Low Carbon Buildings. For low carbon & buildings, May 2012 (2012)
4. Bakkour, A., Ouldboukhitine, S.E., Biwole, P., Amziane, S.: A review of multi-scale hygrothermal characteristics of plant-based building materials. Constr. Build. Mater. **412**, 134850 (2024). https://doi.org/10.1016/j.conbuildmat.2023.134850
5. Medouni-Haroune, L., Zaidi, F., Medouni-Adrar, S., Kecha, M.: Olive pomace: from an olive mill waste to a resource, an overview of the new treatments. J. Crit. Rev. **5**(6), 1–6 (2018). https://doi.org/10.22159/jcr.2018v5i5.28840
6. International Olive Council. International Olive Council Newsletter, No. 187, December 2023. [Newsletter] (2023)
7. García Martín, J.F., et al.: Energetic valorisation of olive biomass: olive-tree pruning, olive stones and pomaces. Processes **8**(5), 511 (2020). https://doi.org/10.3390/pr8050511
8. Cintura, E., Nunes, L., Esteves, B., Faria, P.: Agro-industrial wastes as building insulation materials: a review and challenges for Euro-Mediterranean countries. Ind. Crops Prod. **171**, 113833 (2021). https://doi.org/10.1016/j.indcrop.2021.113833
9. Enaime, G., Dababat, S., Wichern, M., Lübken, M.: Olive mill wastes: from wastes to resources. Environ. Sci. Pollut. Res. Int. **31**(14), 20853–20880 (2024). https://doi.org/10.1007/s11356-024-32468-x

10. Louaar, N., Lebouazda, H.: Caractérisations physico-chimiques de grignon d'olive. Mémoire, Université des Frères Mentouri Constantine 1. [Thesis] (2019)
11. Kuruc, M., Štefunková, Z.: Properties of olive stones with a view to their use as lightweight aggregate in construction mortars. Slovak J. Civ. Eng. **32**(1), 52–57 (2024). https://doi.org/10.2478/sjce-2024-0007
12. Cintura, E., Nunes, L., Faria, P.: Characterization of agro-wastes to be used as aggregates for eco-efficient insulation boards. In: Proceedings of the International Conference Construction, Energy Environment & Sustainability, Coimbra, Portugal, 12–15 October 2021
13. Belhadad, H., Bellel, N., Bras, A.: Exploring the dual nature of olive husk: fiber/aggregate in lightweight bio-concrete for enhanced hygrothermal, mechanical, and microstructural properties. Buildings **15**(1950), 1–26 (2025). https://doi.org/10.3390/buildings15111950
14. Belhadad, H., Bellel, N., Bras, A.: Influence of alkali treatment of olive husk on the physical, mechanical, and hygrothermal performance of a new bio-lightweight concrete. In: Amziane, S.J., Toledo Filho, R.D., da Gloria, M.Y.R., Page, Ed., (eds.) Bio-Based Building Materials - Proceedings of ICBBM 2025.ICBBM 2025. RILEM Bookseries, vol 61. Springer, Cham (2025). https://doi.org/10.1007/978-3-031-92874-1_61
15. Merroun, O., Maalouf, C., Bogard, F.: Mechanical performance of cement mortar with olive pomace aggregates and olive mill wastewater: an experimental investigation (2023). https://doi.org/10.1080/23311916.2023.2212522
16. La Rubia-García, M.D., et al.: Assessment of olive mill solid residue (pomace) as an additive in lightweight brick production. Constr. Build. Mater. **36**, 495–500 (2012). https://doi.org/10.1016/j.conbuildmat.2012.06.009
17. Tayeh, B.A., Nyarko, M.H., Zeyad, A.M., Al Harazin, S.Z.: Properties and durability of concrete with olive waste ash as a partial cement replacement Properties and durability of concrete with olive waste ash as a partial cement replacement (2021). https://doi.org/10.12989/acc.2021.11.1.000
18. Gallala, W., Khater, H.M., Gaied, M.E.: Production of low cost biocomposite made of palm fibres waste. RICA (May), [Article number] (2020). https://doi.org/10.20937/RICA.53541
19. Chandrasekar, M., Ishak, M.R., Sapuan, S.M., Leman, Z., Jawaid, M.: A review on the characterisation of natural fibres and their composites after alkali treatment and water absorption. Plast. Rubber Compos. **46**(3), 119–136 (2017). https://doi.org/10.1080/14658011.2017.1298550
20. Bashir, M., Haripriya, S.: Assessment of physical and structural characteristics of almond gum. Int. J. Biol. Macromol. **93**, 476–482 (2016). https://doi.org/10.1016/j.ijbiomac.2016.09.009
21. Amziane, S., Collet, F., Lawrence, M., Magniont, C., Picandet, V., Sonebi, M.: Recommendation of the RILEM TC 236-BBM: characterisation testing of hemp shiv to determine the initial water content, water absorption, dry density, particle size distribution and thermal conductivity. Mater. Struct. Constr. **50**(3), 1–11 (2017). https://doi.org/10.1617/s11527-017-1029-3
22. Liuzzi, S., Rubino, C., Martellotta, F., Stefanizzi, P., Casavola, C., Pappalettera, G.: Characterization of biomass based materials for building applications: the case of straw and olive tree waste. Ind. Crops Prod. **147**, 112229 (2020). https://doi.org/10.1016/j.indcrop.2020.112229
23. Wzorek, M., Junga, R., Yilmaz, E., Bozhenko, B.: Thermal decomposition of olive-mill byproducts: a TG-FTIR approach. Energies **14**, 4123 (2021). https://doi.org/10.3390/en14144123
24. Romano, A., Grammatikos, S., Riley, M., Bras, A.: Physicochemical characterisation of bio-based insulation to explain their hygrothermal behaviour. Constr. Build. Mater. **258**, 120163 (2020). https://doi.org/10.1016/j.conbuildmat.2020.120163

25. Mashodi, N., Rahim, N.Y., Muhammad, N., Asman, S.: Evaluation of extra virgin olive oil adulteration with edible oils using ATR- FTIR spectroscopy. Malaysian J. Appl. Sci. **5**(1), 35–44 (2020)
26. Grazieschi, G., Asdrubali, F., Thomas, G.: Embodied energy and carbon of building insulating materials: a critical review. Cleaner Environ. Syst. **2**, 100032 (2021). https://doi.org/10.1016/j.cesys.2021.100032

Modelling Transport Phenomena of Mortars Made with bioPCM-Recycled Wood Aggregates

Hala Salhab[1,2](✉), Sergio Nardini[2], Bernardo Buonomo[2], Vincenzo Bianco[3], Ignacio Peralta[4], and Antonio Caggiano[1,5]

[1] DICCA, Università degli studi di Genova, Genova, Italy
hala.salhab@unicampania.it
[2] Engineering Department, Università della Campania L. Vanvitelli, Aversa, Italy
[3] Engineering Department, Università degli studi di Napoli Parthenope, Napoli, Italy
[4] Laboratorio de Flujometría (FLOW), FRSF-UTN, Lavaisse 610, Argentina
[5] LMNI, FIUBA, University of Buenos Aires, Buenos Aires, Argentina

Abstract. This study investigates the thermal behavior and moisture transport phenomena by capillary action of sustainable cementitious composites made with recycled wood aggregates filled with bio-based phase change materials, aiming to enhance energy efficiency and durability in building applications. Two numerical models are developed and validated: an enthalpy-based thermal model using the Apparent Calorific Capacity Method, and a moisture diffusion model based on Richards' equation and extended Darcy's law. Experimental data were obtained from three mortar types: i. ordinary Portland cement, ii. Wood mortar, and iii. a third type fabricated with PCM–RWA aggregates labeled as "NRG-WOOD"—and used to benchmark the simulations. The thermal model successfully captured latent heat storage and release during heating and cooling cycles, while the hygric model accurately predicted capillary water absorption across all mixes. Notably, the NRG-WOOD mortar exhibited significantly reduced water uptake (<5 kg/m^2), attributed to modified pore structures induced by PCM encapsulation, as reflected in elevated Raleigh-Ritz parameters. Both models showed strong agreement with experimental results without requiring post-calibration, underscoring their robustness and predictive reliability. This dual functionality—enhanced thermal regulation and reduced moisture ingress—highlights the potential of PCM–RWA composites as sustainable, high-performance building materials.

Keywords: Phase Change Materials (PCMs) · Recycled Wood Aggregates (RWAs) · Thermal Energy Storage (TES) · Apparent Calorific Capacity Method (ACCM) · Transport Phenomena · Capillary Water Absorption

1 Introduction

In recent years, growing concerns over energy sustainability and global climate change have driven increased interest in innovative technologies aimed at improving the energy efficiency of buildings [1]. The building sector alone accounts for more than one-third

M. Posani and S. Singh (Eds.): RYS 2025, RILEM Bookseries 67, pp. 642–656, 2026.
https://doi.org/10.1007/978-3-032-14652-6_46

of global energy consumption [2]. This figure continues to rise in tandem with population growth and the escalating demand for indoor thermal comfort. As a result, there is a pressing need to develop advanced materials and solutions that can maintain indoor temperatures within comfortable ranges while simultaneously reducing the energy required to achieve these conditions [3, 4]. One promising strategy to enhance the thermal performance of buildings involves the integration of thermal energy storage (TES) systems, particularly using Phase Change Materials (PCMs), which offer a practical and applicable solution for incorporation into various building components [5, 6].

Numerous experimental studies have explored the integration of PCMs into cementitious composites to enhance their thermal and structural performance. These investigations, conducted across multiple scales, have consistently shown that PCMs significantly improve the thermal energy storage capacity of concrete and mortar systems [7, 8]. Some contributions reviewed the effect of the PCMs on the resulting mechanical capacities of concrete, with results varying according to the incorporation method and material type [9, 10]. Notably, PCMs have also been shown to reduce heat hydration during early curing, helping to mitigate thermal cracking in significant structural elements [11]. Moreover, their use has been studied under extreme conditions, including high temperatures, freeze–thaw cycles, and chemically aggressive environments, as well as in geopolymer systems [12–15].

Numerical modeling has become a fundamental tool for analyzing the coupled thermal and moisture behavior of cementitious composites enhanced with phase change materials. In the context of thermal energy storage, particularly under conditions replicated by Dynamic Kernel Calorimetry (DKK), most simulation approaches are grounded in extensions of the classical Stefan problem [16], which governs heat transfer involving phase transitions. Solutions to this problem are generally classified into three categories: (i) the fixed grid method [17], (ii) the deformed grid method [18], and (iii) hybrid combinations of the two. Within the fixed grid framework [19], the enthalpy-based method (EM) has shown high efficacy in capturing latent heat effects, with the Apparent Calorific Capacity Method (ACCM) [20] and the Heat Source Method (HSM) [21] representing two main alternatives used for solving the EM.

Beyond thermal behavior, simulating the capillary water absorption behavior, which is essential for evaluating the durability of PCM-containing composites, requires a different class of models tailored to moisture transport phenomena. While many traditional approaches treat the material as a homogeneous medium at the macroscopic scale [22], recent research has increasingly focused on mesoscale modeling to capture the heterogeneous nature of concrete better. For example, Schlangen et al. [23] developed a lattice-based mesoscale model to investigate how varying water content influences drying shrinkage, cracking, and mechanical properties, including tensile strength and ductility. Similarly, Segura and Carol [24] proposed a zero-thickness interface formulation to simulate moisture diffusion across discontinuities, significantly enhancing resolution in heterogeneous media. Other notable contributions include the works of Caggiano et al. [25], who introduced mesoscale models specifically focused on capillary water penetration in concrete.

In this context, the present study aims to numerically investigate the thermal behavior and moisture transport in cementitious mortars incorporating bio-based Phase Change

Materials (bio-PCMs) and Recycled Wood Aggregates (RWAs). Two coupled phenomenological models are developed: (1) a thermal model based on the Apparent Calorific Capacity Method (ACCM) to simulate heat transfer with phase change, and (2) a moisture transport model governed by Richards' equation and extended Darcy's law to capture capillary water absorption. While no direct modeling of durability is performed, moisture ingress is used as an indicator of material performance related to durability. Heat and moisture transport are treated independently in this work; potential coupling effects are not included, and the implications of this simplification are discussed. The paper is organized as follows: Sect. 2 presents the materials, experimental methods, and modeling approaches; Sect. 3 reports numerical and experimental comparisons. Finally, Sect. 4 concludes with a summary of the main findings and recommendations.

2 Materials and Methods

This section summarizes the experimental results considered as a benchmark in this study and presents a comparison of these results with the corresponding simulations.

2.1 Outline of Experimental Test Setup

Experimental tests were conducted on three different mortar mixtures, prepared and tested according to EN 196-1 (2005) [26] at the University of Genova. The selection of these three mortar types (OPC, Wood, and NRG-Wood) was made to enable a functionally relevant comparison that isolates the effects of incorporating RWAs and bio-PCMs. Specifically, this allows the evaluation of (i) the baseline behavior of Ordinary Portland Cement mortar (OPC), (ii) the influence of wood-based aggregates on moisture transport and porosity (Wood mortar), and (iii) the combined effect of PCMs and RWAs on both thermal regulation and moisture resistance (NRG-Wood mortar). The mixtures were cast using limestone Portland cement (CEM II/B-LL 32.5 R). In addition, PureTemp 25 (supplied by Sphera Encapsulation S.r.l., Verona, Italy) was used in this study, and its thermal and physical properties are summarized in [27]. Table 1 shows the mix proportions of the three different mortar mixtures. More specifically, the first row indicates the labels assigned to each mixture type, providing essential details about the composition of each mortar. For instance, the label "OPC" refers to the reference mixture, which contains neither PCMs nor RWAs. The "Wood-mortar" label indicates a partial replacement approach, where 40% of the sand is substituted with recycled wood aggregates. The label "NRG-WOOD mortar" refers to a similar partial replacement, where 40% of the sand is replaced with NRG-WOOD aggregates. More details about the production process of the NRG-WOOD aggregates, which consist of wood impregnated with PCM and coated with two protective layers of epoxy resin and sand powder to prevent PCM leakage, can be found in [28]. After one day, the specimens were demolded and stored underwater for 28 days to ensure proper curing at a temperature of 21 °C. Following this period, the thermal and physical properties of the specimens were evaluated.

Further details of these experimental results are available in [28] and are omitted from this paper for the sake of brevity.

Table 1. Mix design overview of the three mortars studied.

Label	Density (Kg/m^3)	OPC Mortar (kg/m^3)	Wood-Mortar (kg/m^3)	NRG-WOOD Mortar (kg/m^3)
Cement	3810	559.94	559.94	530.59
Sand	2670	1521.04	940.29	891.00
Wood	1000	–	287.18	–
NRG-WOOD	1300	–	–	412.29
Water	1970	250.32	238.43	284.73
S-P	1000	3.92	5.69	5.39
Ca-(Cl)$_2$	2500	0.00	16.80	–
w/c ratio %	–	0.45	0.43	0.54

2.2 Thermal Modeling Approach

The basic equations, employed for predicting phase transformation phenomena in PCM cement-based systems, are described in this section.

Enthalpy-Based and Apparent Calorific Capacity Method:
The basic equation describing a heat conduction problem can be written as follows:

$$\frac{\partial Q}{\partial t} = \nabla.(\lambda \nabla T) + q_v \qquad \forall x \in \Omega \tag{1}$$

where Q is the heat of the system [J], t the time [s], λ the thermal conductivity of the material [W/m × K] (depending on temperature T [K]and position vector $\boldsymbol{x}$ of the considered body Ω), $\dot{q}_v$ is the possible source term [W/m^3] while ∇. and ∇ are the divergence and gradient tensorial operators, respectively.

In thermodynamics, a small amount of heat added to a system (dQ) is defined by employing the first law of thermodynamics as:

$$dU = \delta Q - pdV \tag{2}$$

where dU is a variation of the internal energy of the system [J], δQ is the incremental heat input [J] and $p.dV$ the rate of work spent [Pa × m^3], indicated as dW (under the simplified hypothesis that $dW = pdV$).

By introducing the definition of the enthalpy of a homogeneous system, $H = U + pdV$, and by combining Eqs. (1) and (2), and adopting the hypothesis of a constant pressure process, the following enthalpy-based equation can be reached:

$$\frac{\partial H}{\partial t} = \nabla.(\lambda \nabla T) + \dot{q}_v \forall \boldsymbol{x} \in \Omega \tag{3}$$

which is the most used equation for solving phase changes in construction and building material applications, and it is commonly addressed as the enthalpy-based method.

The ACCM allows for describing the enthalpy evolution of a system in terms of an apparent (or sometimes called effective) heat capacity during the thermal phase change. The approach is based on the following chain rule:

$$\frac{\partial H}{\partial t} = \frac{\partial H}{\partial T}\frac{\partial T}{\partial t} \tag{4}$$

Then, by introducing the so-called temperature-dependent apparent (effective) heat capacity, defined as follows:

$$\frac{\partial H}{\partial T} = \rho C_{eff}(T) \tag{5}$$

where ρ is the density [Kg/m^3], C_{eff} (T) is the apparent heat capacity [J/Kg.K], a function of temperature that includes the latent effect.

Equation (3) is modified into the following non-linear transient heat equation:

$$\rho C_{eff}(T)\frac{dT}{dt} = \nabla.(\lambda \nabla T) + \dot{q}_v \qquad \forall \boldsymbol{x} \in \Omega \tag{6}$$

To complete the above problem statement of the ACCM approach outlined in Eq. (6), Initial Conditions (ICs) and Boundary Conditions (BCs) need to be employed.

2.3 Moisture Diffusion Approach

This section presents the basic equations of the moisture diffusion approach proposed by Caggiano et al. [25], which are considered in this work for simulating the capillary water absorption for the three studied mortars.

Governing equations ruling the drying/wetting behavior of three types of mortar assume a strong nonlinearity of material parameters depending on the water content and the local porosity. Richards' equation, establishing the transport of water for unsaturated materials, in the unsteady state form, is employed as follows:

$$\frac{\partial \theta}{\partial \mathrm{t}} = \nabla.\left(\mathrm{K}_{\theta}.\nabla_p\right) \tag{7}$$

where θ represents the water content [%], K_θ is the unsaturated water permeability [m^4 N^{-1} s^{-1}] and p is the pore pressure [N/m^2]; ∇· and ∇ are the divergence and gradient tensor operators, respectively. Equation (7) represents the so-called "mixed form" of Richards' equation, containing both pressure and water content. When $p(\theta)$ is known, a saturation-based expression can be formulated as follows:

$$\frac{\partial \theta}{\partial \mathrm{t}} = \nabla.(D_\theta.\nabla_\theta) \tag{8}$$

being D_θ the capillary diffusivity [m^2 s^{-1}] [29]:

$$D_\theta = K_\theta.\frac{\partial \theta}{\partial \mathrm{t}} \tag{9}$$

The capillary (pore) pressure takes the form [30] is a key input to Eqs. (7, 8) and is related to the pore structure of the material. It is computed from the Laplace-Young equation:

$$p = \frac{2CB\gamma_{1v}\cos\theta_c}{\ln(1 - \frac{\theta}{\phi})} \tag{10}$$

where B is the Raleigh-Ritz pore size distribution constant [m^{-1}], γ_{lv} [N/m^1] is the liquid-vapor interfacial energy (surface tension), θ_c is the equilibrium contact angle between the liquid and the solid [deg], C is an empirical constant [—] and ϕ the open porosity [%], the values and justification for these parameters are further discussed in Sect. 3.2.

This formulation assumes idealized pore geometries at the Representative Elementary Volume (REV) scale. While this is suitable for OPC mortar, its application to compose systems with encapsulated PCMs like NRG-WOOD mortar introduces simplifications. The heterogeneity introduced by the epoxy-shelled PCM capsules and irregular wood aggregates may locally violate REV assumptions. Nonetheless, the model remains valid at the mesoscale by averaging these heterogeneities over larger volumes. This limitation is acknowledged, and its influence is partially addressed through calibration of the pore size distribution parameter B and porosity ϕ, both of which effectively capture the altered capillary behavior in the PCM-containing mortar.

Zhou [31] proposal for the unsaturated water permeability was adopted in this study:

$$D_\theta = \frac{C\gamma_{1v}\cos\theta_c}{25B\mu}\frac{\phi^2}{(\phi - \theta)ln^2\left(1 - \frac{\theta}{\phi}\right)}[1 - [1 - \ln\left(1 - \frac{\theta}{\phi}\right) * \left(1 - \frac{\theta}{\phi}\right)]^2 \tag{11}$$

where μ is the dynamic viscosity of water (1.002×10^{-3} N $\times$ s/m^2) taken as in [25], θ_s is the saturated water content, approximated as ϕ, and θ_r is the residual water content estimated as 0.05%.

Further details and the extension of the model for a mesoscale approach are available in Caggiano et al. [25]. While Eq. (11) was originally formulated for more homogeneous porous media, it is applied here under the assumption that mesoscale averaging over representative domains allows effective parameterization. The strong agreement between simulated and experimental absorption curves supports the validity of this approach. However, the model's limitations in capturing local discontinuities, such as at the PCM–matrix interface, must be acknowledged and motivate future work toward fully coupled, multiscale frameworks.

3 Results and Discussion

This section reports the description of the numerical results and shows their comparisons against the experimental data.

3.1 Thermal-Energy Storage DKK Tests in Spherical Samples

The numerical simulations are based on the assumption that the NRG-Wood mortar can be considered as a continuum and a homogeneous medium. In this context, a homogenization technique was employed for evaluating the effective specific heat capacity C_{eff}

of the NRG-Wood mortars. It is based on the mixture theory by using the volume percentages of each component, such as RWAs, cement paste, PCMs, epoxy resin, and sand powder. More precisely, the model smears out the specific heat capacity of the RWAs $C_{RWA}(T)$, cement paste $C_{paste}(T)$, epoxy resin $C_{epoxy}(T)$, sand powder $C_{sand}(T)$ and the apparent specific heat capacity of the PCM $C_{app,PCM}(T)$, through adopting the volume fraction of each component as the smeared out (weighting) factor χ.

The specific heat capacities of each component were experimentally determined with DSC measurements and calorimetry tests. RWAs, cement pastes, epoxy resin and sand powder have only the sensible heat storage part (Fig. 1 a), while the PCMs have an apparent specific heat capacity $C_{app,PCM}(T)$ that incorporates the additional latent behavior within the temperature range of the phase change (during melting and solidification, as shown in Fig. 1 b). More detailed thermal and physical properties of the PCM (PureTemp 25) can be found in Reference [27].

Based on these values, the effective heat capacity C_{eff} of the composite mortars was calculated through a homogenization approach using volume-weighted averaging in two steps:

PCM-RWAs Aggregates:

$$C_{eff,PCM-RWA}(T) = \chi_{RWA} \times C_{RWA}(T) + \chi_{epoxy} \times C_{epoxy}(T) + \chi_{sand} \times C_{sand}(T) + \chi_{PCM} \times C_{app,PCM}(T) \tag{12}$$

where χ_{RWA}, χ_{epoxy}, χ_{sand}, χ_{PCM} are the volume fraction of the recycled woods, epoxy resin, sand powder and the filled PCMs, respectively.

Then, at a composite level (i.e., PCM-RWA mortar), a homogenized overall system of the effective specific heat capacity, C_{eff}, was determined by weighting the heat capacities $C_{eff,PCM\text{-}RWA}(\theta)$, evaluated through Eq. (26), and $C_{paste}(\theta)$ of the individual material components by their volume fractions ψ:

PCM-RWA mixtures:

$$C_{eff}(T) = \psi_{paste} \times C_{paste}(T) + \psi_{PCM-RWA} \times C_{eff,PCM-RWA}(T) \tag{13}$$

where $\psi_{PCM-RWAs}$, ψ_{paste} are the volume fractions of the PCM-RWAs and the paste, respectively. The product $\rho Ceff$, shown in Fig. 2, represents the volumetric heat capacity of the mortar systems and is directly related to their thermal inertia (ρ = 2070 kg/m^3 for OPC mortar, 2010 kg/m^3 for Wood mortar, and 1840 kg/m^3 for NRG-Wood mortar). A higher ρC_{eff} indicates greater ability to resist rapid temperature changes, which is beneficial for thermal regulation in building materials under transient heat loading.

By implementing the ACCM procedure described in Sects. 2, temperature evolutions were simulated and compared with the experimental data reported in [28]. More in detail, the spherical samples made of RWA mortars (with and without PCMs) were simulated with the proposed heat flow model. All employed input parameters in the simulations are summarized in Table 2. The thermal conductivity λ values were obtained experimentally, while the convective heat transfer coefficients for heating and cooling (*h* heating and *h* cooling) were assumed and adjusted by fitting to the experimental temperature profiles.

The numerical simulations, based on the previously defined input parameters, were compared with the experimental data. Figure 3 presents three graphs illustrating the temperature evolution for the control mixtures, obtained using the DKK method, a transient

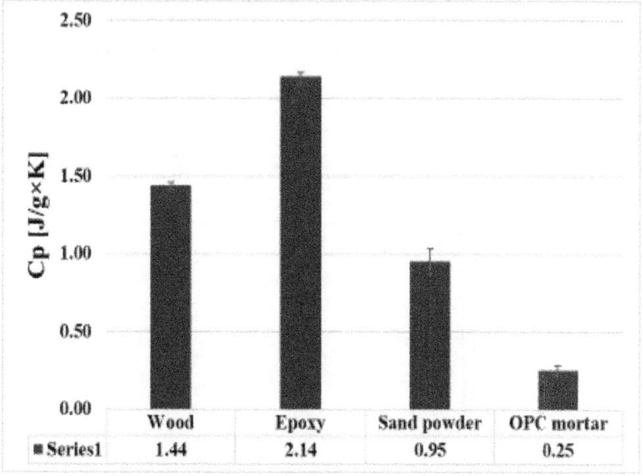

(a)

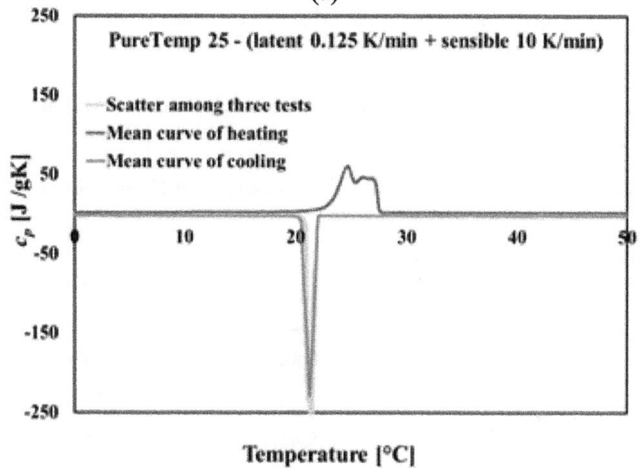

(b)

Fig. 1. (a) C_p results of mortar components using calorimetry test; (b) C_p results of PureTemp 25 PCM showing latent and sensible heat during heating and cooling cycle using DSC measurements.

thermal test that uses a spherical sample with a diameter of 80 mm to determine the thermal behavior of mortars (for more details about the DKK test, see [28]). Overall, the modeling approach successfully captured the experimental temperature trends with high accuracy. In particular, the simulation results for the reference mixtures, shown in Fig. 3a, exhibit excellent agreement with the corresponding experimental measurements. The heating and cooling curves overlap closely in both cycles, confirming the suitability of the model for simulating heat conduction in conventional cementitious matrices without PCM or wood aggregates.

Similarly, the simulations for the wood mortar demonstrate good alignment with the experimental data, accurately reproducing the observed temperature evolution and

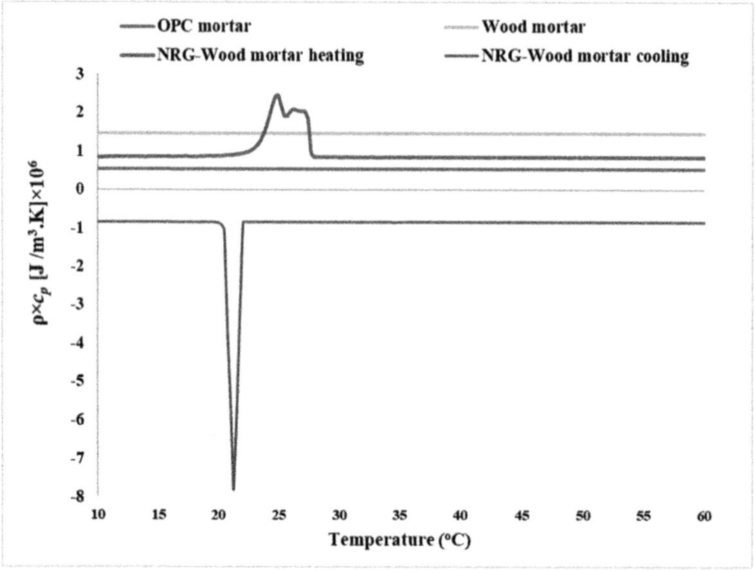

Fig. 2. $\rho \times C_{eff}$ of three mortar systems with and without PCM and RWAs for heating and cooling.

Table 2. Thermal parameters used in the numerical model for the Thermal-Energy Storage (DKK) test.

Numerical parameters	OPC	Wood mortar	NRG-Wood mortar
C_{eff} (J.m^{-3}K^{-1})		Figure 2	
λ (W/m.K)	1.1105	0.8459	0.7191
$h_{heating}$ (W.m^{-2} K^{-1})	2.65	9.25	5.00
$h_{cooling}$ (W.m^{-2} K^{-1})	1.8	5.85	3.75

trends. It is worth noting that, although the numerical simulation of the NRG-Wood mortar demonstrates the expected thermal behavior of the PCM, such as latent heat storage during heating and delayed temperature drop during cooling, this effect appears less pronounced in the experimental results. This discrepancy may be attributed to several factors. First, incomplete PCM activation can occur during experiments due to non-uniform temperature distribution, inadequate thermal contact, or insufficient heating/cooling rates that do not fully engage the phase transition. Second, PCM leakage, degradation, or imperfect encapsulation within the recycled wood aggregates can reduce the effective latent heat contribution. Additionally, thermal inertia of the surrounding matrix and measurement sensitivity limits may further obscure the PCM's thermal response during testing. Despite this, the simulation remains a valuable tool for understanding the theoretical performance potential of PCM-enhanced systems under idealized conditions.

In this context, it may be important to highlight that nearly all input parameters were derived directly from the characterization of the thermal tests, as well as from the

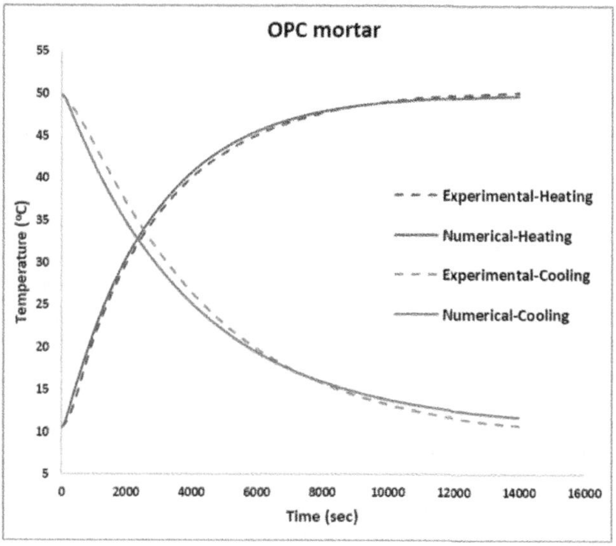

(a)

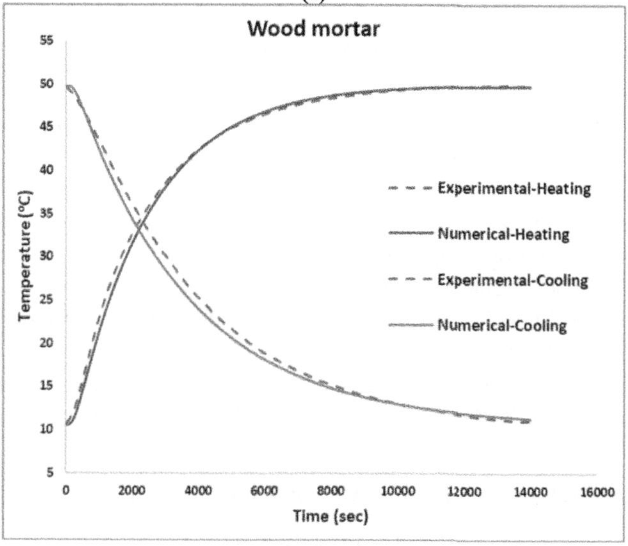

(b)

Fig. 3. Comparison between experimental (markers) and numerical (solid) T evolution during heating/cooling cycles for: (a) OPC, (b) Wood mortar, and (c) NRG-Wood mortar.

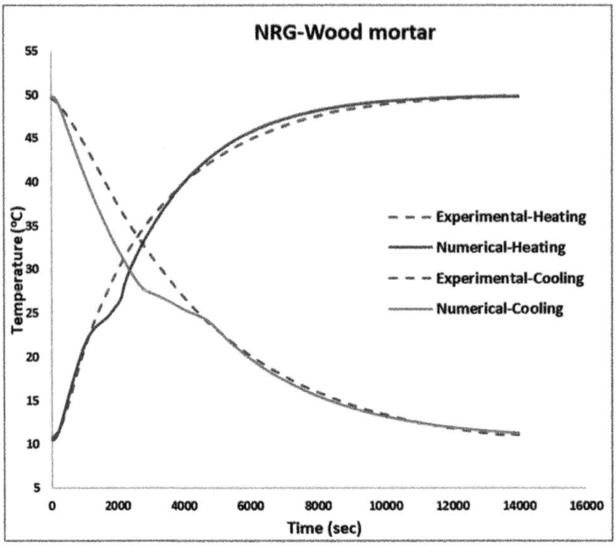

(c)

Fig. 3. (*continued*)

homogenized effective heat capacity (C_{eff}) model, which was constructed by weighing the experimentally determined heat capacities of the individual material components according to their respective volume fractions. Using this set of input parameters, accurate numerical predictions for all three mortar systems were achieved without the need for additional recalibration or optimization procedures. Furthermore, the results demonstrate that a single heat transfer coefficient, represented by the h (Robin) boundary condition parameter in Table 2, was sufficient to ensure an excellent agreement between the simulated and experimental data across all mixtures. These findings support the validity of the assumptions made in the model, namely the adoption of spherical symmetry and the treatment of the PCM-RWA composite mortars as a homogeneous medium.

3.2 Capillary Water Absorption in Cementitious Mortar

In this section, the predictions of the diffusive model are compared against the results of the specimens subjected to capillary transport tests. Figure 3 compares the behavior in terms of water absorption versus time. Two specimens per mixture were tested, with data available at several time intervals: 10 min, 20 min, 30 min, 60 min, 90 min, 120 min, 180 min, 240 min, and 420 min, as well as 24 h.

Material properties were chosen according to [25]. Particularly, the adopted parameters were:

$$\lambda = 7.2 \times 10^{-5}\text{N.mm}^{-1}, \mu = 1.002 \times 10^{-3}\text{N.s.mm}^{-2}, cos\theta c = 0.94, \ \text{C} = 2.15.$$

The open porosity for each type of mortar, ϕ [%] and the calibrated Raleigh-Ritz pore size distribution B (see Table 3) is listed below. These parameters were assumed and calibrated by fitting the numerical model to the experimental capillary absorption data:

Table 3. Moisture-transport parameters used in the numerical model for the capillary water absorption test.

	OPC Mortar	Wood mortar	NRG-Wood mortar
ϕ	0.43	0.20	0.026
B	19.5×10^7	2.3×10^7	1×10^6

It can be generally observed that the absorption responses were very well captured by the nonlinear diffusive model. The model predicted capillary water absorption accurately across all mortar mixtures, confirming the robustness of the numerical approach. Figure 4 specifically illustrates a comparative analysis of three mortar types: Wood mortar, OPC mortar, and NRG-Wood mortar, under both experimental and numerical conditions.

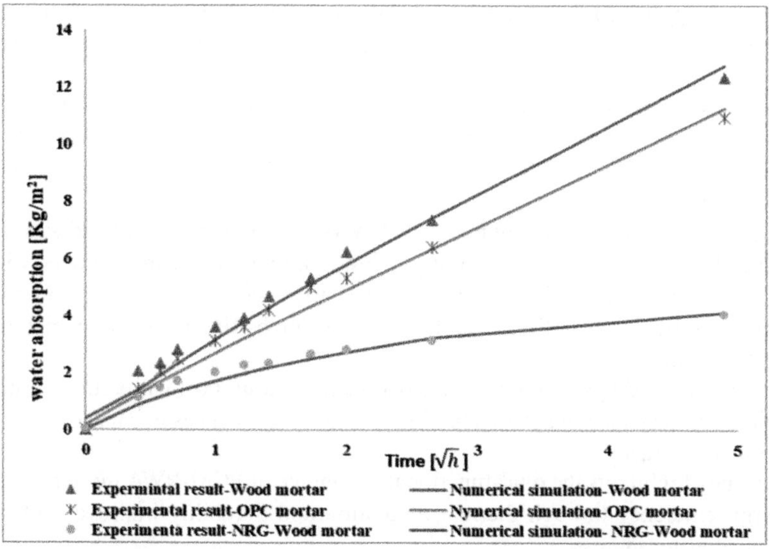

Fig. 4. Comparison of experimental and simulated capillary water absorption for OPC, Wood, and NRG-Wood mortars.

From Fig. 4, Wood mortar demonstrated the highest water absorption rate, exceeding 13 kg/m^2 by the end of the test. This is due to the porous nature of wood, which promotes greater capillary action and moisture uptake. OPC mortar followed with a moderately lower absorption level, while NRG-Wood mortar showed a significantly reduced capillary uptake, remaining under 5 kg/m^2 throughout the testing duration, which can be attributed to the incorporation of PCMs, which alter the microstructure of the mortar matrix. As discussed in Guardia et al. [32], the addition of PCMs increases the Raleigh-Ritz pore size distribution parameter B, indicating a shift toward finer pore structures and lower permeability. Notably, even though some PCM-modified mortars possess relatively high open porosity, the value of *B* becomes a more dominant factor in controlling water ingress.

Moreover, the excellent agreement between experimental and numerical results across all three mixtures supports the validity of the used FEM model, which is based on an extended Darcy's law and the nonlinear formulation of Richards' equation. The model's accuracy reinforces the importance of incorporating pore structure parameters such as B and porosity ϕ into transport simulations to reliably predict moisture diffusion in cementitious systems.

4 Conclusions

This work aimed at investigating the effect of PCMs mixed into several cement-based mortars on thermal energy storage and capillary water absorption. To achieve this, two numerical models were developed and validated to simulate the thermal and moisture behavior of mortars incorporating RWAs and bio-based PCMs.

The following main points can be highlighted:

- The enthalpy-based thermal model, implemented through the ACCM, successfully captured the thermal response of PCM-modified mortars, including latent heat effects, under both heating and cooling cycles.
- The moisture transport model, based on Richards' equation and an extended Darcy's law, accurately predicted capillary water absorption across three mortar types: OPC, Wood mortar, and NRG-Wood mortar, demonstrating strong agreement with experimental results.
- Mortars containing PCM-impregnated RWAs (NRG-Wood) exhibited significantly reduced water absorption, remaining below 5 kg/ml^2, compared to over 13 kg/m^2 for wood-only mortars. This improvement is attributed to the modified pore structure introduced by PCMs, reflected in the higher Raleigh-Ritz parameter, which governs capillary suction.
- The thermal and hygric numerical models required no post-calibration, confirming their robustness and predictive reliability based solely on measured material properties and volume fractions.
- The study highlights the dual-functional advantage of PCM-RWA mortars, enhanced energy efficiency through thermal regulation and durability improvement through reduced water ingress.
- Future work should focus on evaluating the long-term performance of PCM-RWA mortars under repeated thermal cycling, aging effects, and environmental exposure, to better assess their real-world durability and reliability. Additionally, the integration of multi-scale modeling approaches could help capture the inherent heterogeneity of PCM-impregnated composite mortars, particularly at the meso- and microstructural levels. Finally, the current work treats heat and moisture transport independently; however, coupled hygrothermal modeling may provide more accurate predictions for applications where simultaneous heat and moisture gradients are present. This represents a valuable direction for future model development.

Acknowledgment. Part of this study was developed within the framework of PRIN 2022 project OPTIMA (OPTImized use of phase change MAterials for energy and seismic retrofitting of buildings), funded by Italian MUR 2025–2027. Support was provided by EU funding: this work was partially financed by the Europe-an Union's Horizon 2020 research and innovation programme under the Marie Skłodowska-Curie Staff Exchange action (GA 101086440 – Project acro-nym: BEST, Bio-based Energy-efficient materials and Structures for Tomor-row), and the Marie Skłodowska-Curie Postdoctoral Fellowship (GA 101153320 – Project acronym: SMART-NRG, Solutions for constructions and building materials based on Energy Absorption).

References

1. Hembade, L., Neithalath, N., Rajan, S.: Understanding the energy implications of phase-change materials in Concrete Walls through finite-element analysis. J. Energy Eng. **140**(1), 04013009 (2014)
2. IEA (International Energy Agency). Energy balances of OECD countriesd2010 edition. 2010. Paris, France
3. Kheradmand, M., Azenha, M., de Aguiar, J.L.B., Castro-Gomes, J.: Experimental and numerical studies of hybrid PCM embedded in plastering mortar for enhanced thermal behaviour of buildings. Energy **94**, 250–261 (2016)
4. Ricklefs, A., Thiele, A.M., Falzone, G., Sant, G., Pilon, L.: Thermal conductivity of cementitious composites containing microencapsulated phase change materials. Int. J. Heat Mass Transf. **104**, 71–82 (2017)
5. Zhou, D., Zhao, C.-Y., Tian, Y.: Review on thermal energy storage with phase change materials (PCMs) in building applications. Appl. Energy **92**, 593–605 (2012)
6. Devaux, P., Farid, M.M.: Benefits of PCM underfloor heating with PCM wallboards for space heating in winter. Appl. Energy **191**, 593–602 (2017)
7. Entrop, A.G., Brouwers, H.J.H.: Angelina HME Reinders. "Experimental research on the use of micro-encapsulated Phase Change Materials to store solar energy in concrete floors and to save energy in Dutch houses". Solar Energy **85**(5), 1007–1020 (2011)
8. Cabeza, L.F., Castellón, C., Nogués, M., Medrano, M., Leppers, R., Zubillaga, O.: Use of microencapsulated PCM in concrete walls for energy savings. Energy Build. **39**(2), 113–119 (2007)
9. Cao, L., Di, Su., Tang, Y., Fang, G., Tang, F.: Properties evaluation and applications of thermal energystorage materials in buildings. Renew. Sustain. Energy Rev. **48**, 500–522 (2015)
10. Padala, S.K., Deshpande, S.J., Bhattacharjee, B.: Assessment of setting characteristics, water absorption, thermal performance and compressive strength of energy-efficient phase change material (PCM)–ashcrete blocks. Sādhanā **46**(2), 103 (2021)
11. Eddhahak, A., Drissi, S., Colin, J., Caré, S., Neji, J.: Effect of phase change materials on the hydration reaction and kinetic of PCM-mortars. J. Therm. Anal. Calorim. **117**, 537–545 (2014)
12. Banu, D., Feldman, D., Haghighat, F., Paris, J., Hawes, D.: Energy-storing wallboard: flammability tests. J. Mater. Civ. Eng. **10**(2), 98–105 (1998)
13. Sakulich, A.R., Bentz, D.P.: Increasing the service life of bridge decks by incorporating phase-change materials to reduce freeze-thaw cycles. J. Mater. Civ. Eng. **24**, 1034–1042 (2012)
14. Hunger, M., Entrop, A.G., Mandilaras, I., Brouwers, H.J.H., Founti, M.: The behavior of self-compacting concrete containing microencapsulated Phase Change Materials. Cement Concr. Compos. **31**, 731–743 (2009)

15. Shadnia, R., Zhang, L., Li, P.: Experimental study of geopolymer mortar with incorporated PCM. Constr. Build. Mater. **84**, 95–102 (2015)
16. Crank, J.: Free and Moving Boundary Problems. Clarendon, Oxford (1984)
17. Eyres, N.R., Hartree, D. R., Ingham, J., Jackson, R., Sarjant, R.J., Wagstaff, J.B.: The calculation of variable heat flow in solids. Philosop. Trans. Roy. Soc. Lond. Ser. A, Math. Phys. Sci. **240**(813), 1–57 (1946)
18. Lynch, D.R., O'Neill, K.: Continuously deforming finite elements for the solution of parabolic problems, with and without phase change. Int. J. Numer. Meth. Eng. **17**(1), 81–96 (1981)
19. Udaykumar, H.S., Mittal, R., Shyy, W.: Computation of solid–liquid phase fronts in the sharp interface limit on fixed grids. J. Comput. Phys. **153**(2), 535–574 (1999)
20. Šavija, B., Schlangen, E.: Use of phase change materials (PCMs) to mitigate early age thermal cracking in concrete: theoretical considerations. Constr. Build. Mater. **126**, 332–344 (2016)
21. Rostamizadeh, M., Khanlarkhani, M., Sadrameli, S.M.: Simulation of energy storage system with phase change material (PCM). Energy Build. **49**, 419–422 (2012)
22. Martín-Pérez, B., Pantazopoulou, S.J., Thomas, M.D.A.: Numerical solution of mass transport equations in concrete structures. Comput. Struct. **79**, 1251–1264 (2001)
23. Schlangen, E., Koenders, E.A.B., van Breugel, K.: Influence of internal dilation on the fracture behaviour of multi-phase materials. Eng. Fract. Mech. **74**, 18–33 (2007)
24. Segura, J.M., Carol, I.: On zero-thickness interface elements for diffusion problems. Int. J. Numer. Anal. Meth. Geomech. **28**, 947–962 (2004)
25. Caggiano, A., Schicchi, D.S., Mankel, C., Ukrainczyk, N., Koenders, E.A.B.: A mesoscale approach for modeling capillary water absorption and transport phenomena in cementitious materials. Comput. Struct. **200**, 1–10 (2018)
26. EN 196-1, Methods of Testing Cement – Part 1: Determination of Strength, 26 (2005)
27. PureTemp, PureTemp 25 Technical Data Sheet (2021). https://puretemp.com/wpcontent/uploads/2021/06/PureTemp25TechnicalDataSheet.pdf. Accessed 2 Dec 2024
28. Salhab, H., Zanjani, M., Nardini, S., et al.: Experimental characterization of sustainable cementitious composites: thermal energy storage with recycled wood aggregates and bio-based phase change materials. Int. J. Thermophys. **46**, 87 (2025). https://doi.org/10.1007/s10765-025-03548-y
29. Hall, C.: Anomalous diffusion in unsaturated flow: fact or fiction? Cem. Concr. Res. **37**, 378–385 (2007)
30. Aligizaki, K.: Pore Structure of Cement-Based Materials. CRC Press, London (2005)
31. Zhou, J.: Performance of engineered cementitious composites for concrete repairs. Ph.D. Thesis, Delft University of Technology, the Netherlands (2011)
32. Guardia, C., Schicchi, D.S., Caggiano, A., et al.: On the capillary water absorption of cement-lime mortars containing phase change materials: experiments and simulations. Build. Simul. **13**, 19–31 (2020). https://doi.org/10.1007/s12273-019-0556-y

Polylaminate Waste Valorisation into Mycelium-Based Composites for Sustainable Insulation Applications

Joni Wildman[1(✉)], Valeria Cascione[2], Pete Walker[1], Daniel Henk[3], and Andrew Shea[1]

[1] Department of Civil Engineering and Architecture, University of Bath, Bath, UK
jlw89@bath.ac.uk
[2] Institute of Sustainability and Climate Change, University of Bath, Bath, UK
[3] Milner Centre for Evolution, University of Bath, Bath, UK

Abstract. This study addresses two critical challenges in advancing a sustain-able and circular economy. The construction industry must transition to low-impact materials, as many conventional insulation materials deplete finite resources and have high embodied carbon. Sustainable, effective alternatives are essential to reduce both operational and embodied carbon of buildings. Concurrently, the world faces an increasing volume of waste; recycling options are often preferable to landfill and incineration but can impose environmental burdens. One problematic material is waste polylaminate carton (e.g. Tetra-Pak) which cannot be easily recycled alongside paper products, with the recycling process complex and region-dependent.

Mycelium-based composites (MBCs) have emerged as a promising bio-based material capable of transforming lignocellulosic waste into effective insulation products. In their manufacture, fungal mycelium grows on a substrate, forming a natural adhesive as it colonises the material. Once fully colonised, the material is dried to kill the fungus and produce an inert material. While agricultural byproducts, such as straw and hemp-shiv, are common substrate choices, fungi can also grow on more complex waste streams, including polylaminate materials.

In this study, the successful growth and binding of polylaminate waste with mycelium to produce MBCs is demonstrated. MBCs were produced using Lentinus tigrinus mycelium with hemp-polylaminate blends in ratios (by mass of hemp:polylaminate) of 1:0, 2:1, 1:2, and 0:1. Thermal conductivity (λ) measurements were conducted, with the measured thermal conductivity of the pure polylaminate MBCs $\lambda = 0.035 \pm 0.002$ W/mK. Furthermore, increasing the proportion of polylaminate in the substrate blend did not significantly affect λ. Water absorption properties were also evaluated, showing that higher Tetra-Pak content led to reduced water absorption. Additionally, a life cycle assessment (LCA) was conducted, revealing that these materials exhibit a low global warming potential (GWP), suggesting they could be low-carbon alternative to conventional insulation. These findings highlight the potential of utilising complex waste streams from the food and beverage industry to produce mycelium-based insulation materials. This approach contributes to reducing environmental burdens associated with waste while promoting circular and sustainable practices within the construction sector.

M. Posani and S. Singh (Eds.): RYS 2025, RILEM Bookseries 67, pp. 657–673, 2026.
https://doi.org/10.1007/978-3-032-14652-6_47

Keywords: Mycelium-based composite · Bio-based insulation · Waste valorisation · Tetra-Pak · Circular economy

1 Introduction

Non-circular practices in material production and waste management have led to significant greenhouse gas (GHG) emissions and the proliferation of landfill waste. These issues span across many industrial sectors, presenting an urgent need for sustainable solutions [1]. Among the most responsible sectors is the construction industry, which accounts for approximately 39% of global carbon emissions [2], and in the UK accounts for 62% of waste production [3]. When assessing the carbon emissions of the built environment, both operational carbon and embodied carbon need to be considered. A significant proportion of operational carbon emissions come from heating and cooling [4]; the use of effective insulation materials can reduce this burden and represent an important solution [5]. Insulation materials can be characterised by their thermal conductivity (λ), which is defined according to Fourier's law of conduction [6]. Materials can be considered to be insulators if they have $\lambda < 0.1$ W/m.K [7]. Typical insulation materials include EPS, XPS, and mineral wool; these are effective insulators ($\lambda = 0.03$–0.05 W/m.K) [8]. However, these traditional insulation materials can have high embodied carbon footprints; they are energy intensive to produce, deplete finite resources, and contribute to the growing waste problem as they are challenging to recycle [9, 10].

To address the challenge of reducing both operational and embodied carbon, bio-based solutions have emerged as promising insulation options [11]. Among these, mycelium-based composites (MBCs), which harness the natural binding and growth properties of fungal mycelium, have gained attention for their potential to produce lightweight, durable, and biodegradable materials [12, 13]. Mycelium is the vegetative part of a fungus, comprised of a network of thread-like filaments called hyphae [14]. Mycelium-based composites are typically produced by cultivating fungal mycelium on agricultural waste substrates, which provide nutrients for the fungi and serve as the structural matrix for the final material [15]. While agricultural byproducts like straw or sawdust are commonly used, they also serve as feedstocks for other bio-based applications, such as hempcrete or biofuels [16]. Within MBC production, there is a possibility to explore alternative substrates, particularly complex waste streams that are otherwise difficult to recycle.

One such waste stream is polylaminated materials, such as Tetra-Pak waste. Tetra-Pak products are multilayered packaging items commonly used in the food and beverage industry. They are comprised of six layers of paperboard, plastic (polyethylene), and aluminium and are designed to provide durable and shelf-stable containers that protect their contents from light, air, and contamination [17, 18]. In 2024 179 billion Tetra-Pak cartons were produced world-wide, an increase of 5% on the previous year [19]. However, recycling rates of Tetra-Pak were only 25% thus approximately 132 billion cartons ended up in landfill or incinerated [20].

Despite some recycling processes being technically feasible, the recycling of polylaminates remains energy-intensive, costly, and limited to specific regions due to the challenges of separating its composite materials. Consequently, much of this waste ends up in landfills or incinerators, contributing further to environmental degradation [21].

Previous studies have explored the valorisation of Tetra-Pak waste for construction materials, highlighting its potential as an alternative raw material for sustainable and eco-friendly applications. Bonocore et al. ([22]) demonstrated that discarded polylaminate containers can be used as insulating panels by layering discarded sheets and using a heat treatment method, however did not explicitly calculate thermal conductivitymoisture ab. Considering other applications in construction, studies have explored the incorporation of Tetra-Pak waste in composite panels, polymer concrete, and prefabricated building components. Research by Bekhta et al. (2016) on composite panels indicates that Tetra-Pak particles, combined with polyethylene-based waste films, can be used to manufacture binderless boards that meet European particleboard standards, although mechanical properties are influenced by the ratio of waste components [23]. In polymer concrete, Martínez-López et al. (2015) investigated the use of Tetra-Pak waste as a partial replacement for silica sand, finding that while mechanical strength decreases with higher Tetra-Pak content, gamma irradiation can enhance certain properties [24]. These studies collectively underscore the feasibility of repurposing Tetra-Pak waste in construction while also addressing environmental concerns associated with its disposal.

This paper explores the valorisation of Tetra-Pak waste as a substrate for mycelium-based composites, converting a challenging waste stream into sustainable insulation materials. Thermal performance is assessed through thermal conductivity measurements. Liquid water absorption in full immersion conditions is also evaluated, as it influences durability, performance, and mould susceptibility. Additionally, a life cycle assessment (LCA) of the lab-scale production quantifies the environmental impact, with a focus on global warming potential (GWP) to evaluate sustainability. The goals of the study are as follows:

- To determine if fungal mycelium successfully grow on and bind Tetra-Pak waste to form a mycelium-based composite
- To measure the thermal conductivity of Tetra-Pak based mbcs and compare with conventional insulation and MBCs made using hemp
- To measure the water absorption properties of MBCs made with increasing ratios of Tetra-Tak
- To conduct an LCA of the lab scale production of the Tetra-Pak MBCs and compare with traditional insulation materials

These aims collectively make progress towards assessing the suitability and feasibility of using Tetra-Pak as a substrate in MBC production. Through this, the research aligns with broader sustainability goals by addressing both waste valorisation and carbon footprint reduction in the built environment.

By leveraging the digestive capabilities of fungi to bind and process complex substrates, this approach aims to address two critical challenges: reducing polylaminate waste and producing environmentally friendly insulation materials for the construction industry. This study explores the material properties of these composites, the feasibility of utilising Tetra-Pak waste as a mycelium substrate, and the broader implications for waste management and circular economy practices.

2 Materials and Methods

2.1 Mycelium

A liquid culture inoculum of *Lentinus tigrinus* was obtained from kingdomLCs. The identity of the species was verified based on sequencing of the ITS4 region. The inoculum was subsequently propagated, maintained, and stored on malt extract agar with yeast extract (MEYA).

Mycelium growing on MEYA was then used to inoculate hydrated and sterilised brown rice. The fungi was allowed to colonise the rice for 5 days in an incubator at 23 °C. The grain colonised by mycelium was subsequently used as grain spawn to form the composites.

2.2 Tetra-Pak

The waste Tetra-Pak material used in this study was sourced from post-consumer drink cartons, primarily consisting of empty milk and juice containers collected over several weeks. The collected cartons were thoroughly cleaned prior to use. The collected cartons were flattened and cut into rectangular sections with dimensions of 20 mm × 10 mm. Hemp-shiv was obtained without additives and of dimensions 10 mm x 5 mm x 2 mm. Mixtures of cut Tetra-Pak and hemp shiv were prepared with the following mass ratios of hemp shiv to Tetra-Pak: 1:0, 2:1, 1:2, and 0:1 (by mass of hemp:polylaminate). Each mixture was placed in a separate bucket and soaked in water for 24 h. Following soaking, excess water was drained from the mixtures and mixed with plain flour (10% flour by weight). Each mixture was then transferred to a separate autoclavable bags and sterilised in an autoclave ready to be used in the composite production step.

2.3 MBC Production

In a biosafety cabinet the substrate and grain spawn were combined in a ratio of 10:1 with the grain spawn broken up into mycelium-colonised individual grains. A large bowl and spoon, cleaned with 70% isopropanol, were used to mix the spawn and substrate. The mixture was then packed in to cylindrical moulds (height 10 mm, diameter 90 mm) and rectangular moulds (150 mm x 150 mm x 20 mm) cleaned with 70% isopropanol. Once the moulds were filled they were covered with a sheet of parafilm and secured with tape to allow for air exchange, whilst minimising contamination during incubation. The specimens were then incubated at 23 °C until fully colonised (5 days for cylindrical moulds, 7 days for rectangular moulds). The specimens were then removed from the moulds and weighed. They were then dried in an oven at 50 °C for 48 h. The specimens were weighed after 40 h and after 48 h to ensure that no further mass was lost upon 8 h additional drying, therefore indicating that the specimens were not retaining moisture. For each of the four mixtures (hemp-shiv- Tetra-Pak ratios 1:0, 2:1, 1:2, 0:1) five cylindrical specimens were made and three rectangular specimens were made.

2.4 Thermal Characterisation

The thermal conductivity of each rectangular specimen was determined using a Heat Flow Meter according to ASTM-C518 [25]. A Heat Flow Meter is a steady state method of measuring thermal conductivity. It consists of two plates held at two fixed temperatures, with the material being measured placed between these two plates such that heat flows through the material from the hotter plate to the colder plate. A steady state temperature gradient is reached before a measurement is made. Thermal conductivity is determined by the Heat Flow Meter using Fourier's Law of heat conduction: $q = -\lambda \nabla T$ q is the heat flux (W/m^2), λ is the thermal conductivity of the material (W/m.K), and ∇T is the temperature gradient. Five thermal analysis tests were performed on each of the 12 specimens (60 tests in total) using a Thermtest HFM-25. Each specimen was stored in the laboratory at room conditions for 1 week (approx. 20 °C and 50% RH) prior to testing. Specimen density was calculated according to ASTM C303 prior to testing [26]. The test was conducted at a mean temperature of 20 °C with a 20 °C temperature difference between the plates.

2.5 Immersion Test

The liquid water absorption of the MBCs produced in this study was tested in accordance with ASTM C7163 *Water Absorption by Immersion of Thermal Insulation Materials* [27]. Four specimens were tested for each composition ratio (as described in sect. 2.3). The specimens were weighed immediately prior to full immersion. Specimens were fully submerged in water at 20 °C, with a thin wire rack placed on top to prevent floating and ensure complete immersion. The test was carried out at room temperature (20–23 °C). After 2 h the specimens were removed from the water and allowed to drain for 10 min. At that point, any remaining surface water was removed by blotting with paper towels and the specimens were weighed. Subsequently, the specimens were re-immersed in the water for an additional 22 h to give a total immersion time of 24 h. After the additional 22 h, the specimens were removed from the water and allowed to drain for 10 min before any surface water was again removed by blotting. The specimens were then weighed. The percentage water absorbed by weight was calculated based on the difference in mass following immersion and the original mass.

2.6 LCA

A life cycle assessment (LCA) of the pure Tetra-Pak substrate MBC produced at a lab-scale was conducted. The LCA was completed in alignment with EN 15804:2012 + A2:2019 [28]. The goal of this study is to evaluate the environmental impacts of Tetra-Pak based mycelium-based composites used for insulation applications and compare these to LCAs conducted in the literature on conventional insulation materials and MBCs based on lignocellulosic substrates, with a focus on global warming potential (GWP) indicators.

System Description: The system boundaries and processes considered for evaluating the production of the MBCs are depicted in Fig. 1. The processing of Tetra-Pak involved manual cutting of the cartons, of which did not involve any inventory input. Here, the Tetra-Pak is considered a waste within the LCA framework, and therefore the impacts associated with Tetra-Pak production are not included in the LCA. In an attributional LCA waste materials are treated as having zero upstream impact, and the Tetra-Pak used here was sourced from post-consumer waste [29]. Once the substrate is prepared, the stages of MBC production include substrate soaking, substrate sterilisation, grain spawn production, mixing of substrate and grain spawn, incubation, and drying. Collectively, these steps correspond to the stages A1-A3 of the life cycle (cradle to gate). To determine the steps in the Tetra-Pak MBC production process that contribute most to its environmental impact, the Global Warming Potential of each stage was assessed. These stages include substrate preparation and soaking, substrate sterilisation, grain spawn production, mixing, incubation, and drying, with the associated inputs detailed in Fig. 1. This analysis aims to identify opportunities for process improvements, hence why the impacts were categorised by individual production processes rather than on A1-A3 stages.

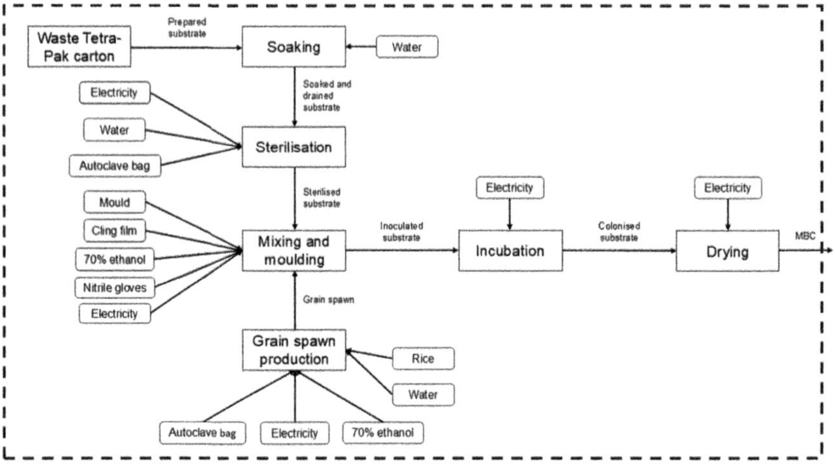

Fig. 1. Figure showing the system boundaries adopted in the LCA of mycelium-based composites produced using Tetra-Pak and *Lentinus tigrinus* grain spawn. The boundaries include the substrate preparation and the mycelium-based composite production (A1-A3 stages). The different stages of the mycelium-based composite production process as shown in square boxes, with the inputs at each stage shown in rounded boxes. **Functional Unit:** The functional unit (FU) of this LCA study is defined as the mass (kg) of MBC insulation required to achieve a thermal resistance (R-value) of 1 m^2K/W over an area of 1 m^2. The relationship between these is expressed by the equation:

$$FU = R\lambda\rho A$$

where: R is the thermal resistance (equal to 1 (m^2K/W)), λ is the thermal conductivity in W/(m.K), ρ is the density in kg/m^3, and A is the area (equal to 1 m^2) [30].

This FU choice quantifies the volume of insulation material needed to achieve the specified thermal resistance. It is assumed that λ and ρ are the same for composites made with A equal to 1 m^2 as for those produced in this study.

Inventory: The life cycle inventory (LCI) was primarily based on laboratory measurements. Where direct measurements were not possible, the data were supplemented using literature sources and technical data sheets. Background data were sourced from the Ecoinvent V3.6 database [31] and OpenLCA [32] was used for the LCA analysis.

Biogenic carbon was considered by applying the static −1/+1 approach, following the current version of the EN 15804 standard (CEN 2019). The carbon content of the Tetra-Pak was calculated based on the material composition provided by Tetra-Pak, stating that their cartons are composed of approximately 70% paperboard by weight [18]. Given that paper-based materials typically contain about 44–50% carbon by weight, the carbon content attributable to the paper component of the Tetra-Pak can be estimated accordingly [33].

To calculate the amount of CO_2 sequestered in these organic substrates, the molar masses of carbon (C) and carbon dioxide (CO_2) are considered. CO_2 has a molar mass of 44 g per mole, while C has a molar mass of 12 g per mole, resulting in a mass ratio of CO_2 to C of 44/12 = 3.67 [34]. Using this ratio, the fact Tetra-Pak is 70% paper, and taking the carbon content of paper as 44%, the mass of CO_2 assumed to be sequestered per kg of Tetra-Pak was 1.13 kg.

3 Results and Discussion

3.1 MBC Production

Figure 2 shows the resulting mycelium-based composites made using increasing proportions of Tetra-Pak. MBCs were successfully produced in all cases, with three replicates made for each. This showed that mycelium can successfully bind a pure Tetra-Pak substrate, with good structural integrity and homogeneity observed.

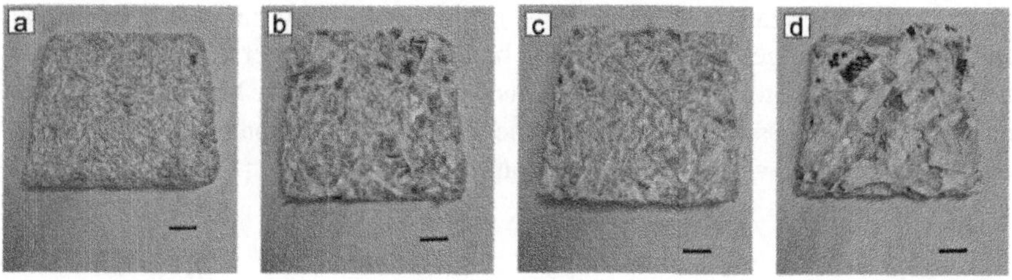

Fig. 2. Mycelium-based composites made with different compositions of hemp-shiv and Tetra-Pak with the following ratios of hemp-shiv: Tetra-Pak of a) 1:0, b) 2:1, c) 1:2), and d) 0:1.

3.2 Thermal Performance

Figure 3 shows the thermal conductivity (λ) measured for each of the different substrate mixes used in the study, and Fig. 3 shows the density (ρ). All specimens had a λ measurement of less than 0.1 W/m.K and therefore demonstrated good insulative capabilities.

The lowest average λ was for the pure hemp-shiv substrate (0.0342 ± 0.001 W/m.K) and the greatest for the 33% Tetra-Pak substrate (0.0364 ± 0.001 W/m.K). A one-way ANOVA test was conducted to assess the statistical significance of these differences. The results indicated that there was no statistically significant difference in λ between the groups ($p = 0.4 > 0.05$) therefore showing that substituting the hemp-shiv substrate for increasing amounts of the Tetra-Pak waste did not adversely affect the good thermal performance of the materials. Similarly, although the 100% Tetra-Pak substrates had the highest average density (175 ± 11 kg/m^3), the difference between average density of the substrate mixes was not statistically significant ($p = 0.561 > 0.05$). Furthermore, there was no significant correlation between λ and ρ when considering all specimens ($r = -0.02$, $p > 0.05$). Correlations between λ and ρ may be observed if specimen were produced with a wider range of densities (e.g. by packing the substrate into the mould at different densities), and this may suggest that other factors are dominant (e.g. homogeneity and anisotropy).

3.3 Immersion Test

Figure 4 shows the results of the water immersion test for the MBCs made using different proportions of hemp-shiv and Tetra-Pak. The results suggest that increasing the proportion of Tetra-Pak reduced the water absorbed after 24 h, however, after 2 h the difference in water absorption between the different compositions was similar. This could be because between 2 h and 24 h of soaking the MBCs with a higher proportion of Tetra-Pak become saturated sooner than those with more hemp-shiv, as hemp-shiv is more porous. This is reflected in the higher absorption of raw hemp-shiv compared to raw chopped Tetra-Pak, as shown in the right hand columns of Fig. 4. Although the reduction in water absorption after 24 h with increasing Tetra-Pak content is advantageous when considering insulation materials, the water absorption of these materials remains high. This is observed in other studies of bio-based construction materials and is a wider problem with these materials that requires further research to mitigate. The insulation materials should not get wet (and will not be immersed in water during use), however could still be exposed to water ingress events, and therefore these liquid water absorbing properties may pose issues. Efforts to reduce the water absorption (such as hydrophobic coatings) will need to align with the environmental benefits of these materials.

3.4 LCA

In this section, the results of the life cycle assessment (LCA) of the MBCs produced with Tetra-Pak are presented and discussed. The environmental impacts associated with the declared FU of MBC are analysed, with a particular focus on comparing GWP indicators. This focus is due to the construction industry's significant contribution to global carbon emissions and the increasing emphasis on reducing embodied carbon in building materials.

Based on the FU equation detailed in sect. 2.6, the mass of MBC corresponding to the chose FU in this LCA was 6.04 kg. This was calculated using the measured average thermal conductivity and density presented in sect. 3.2 for the pure Tetra-Pak MBC.

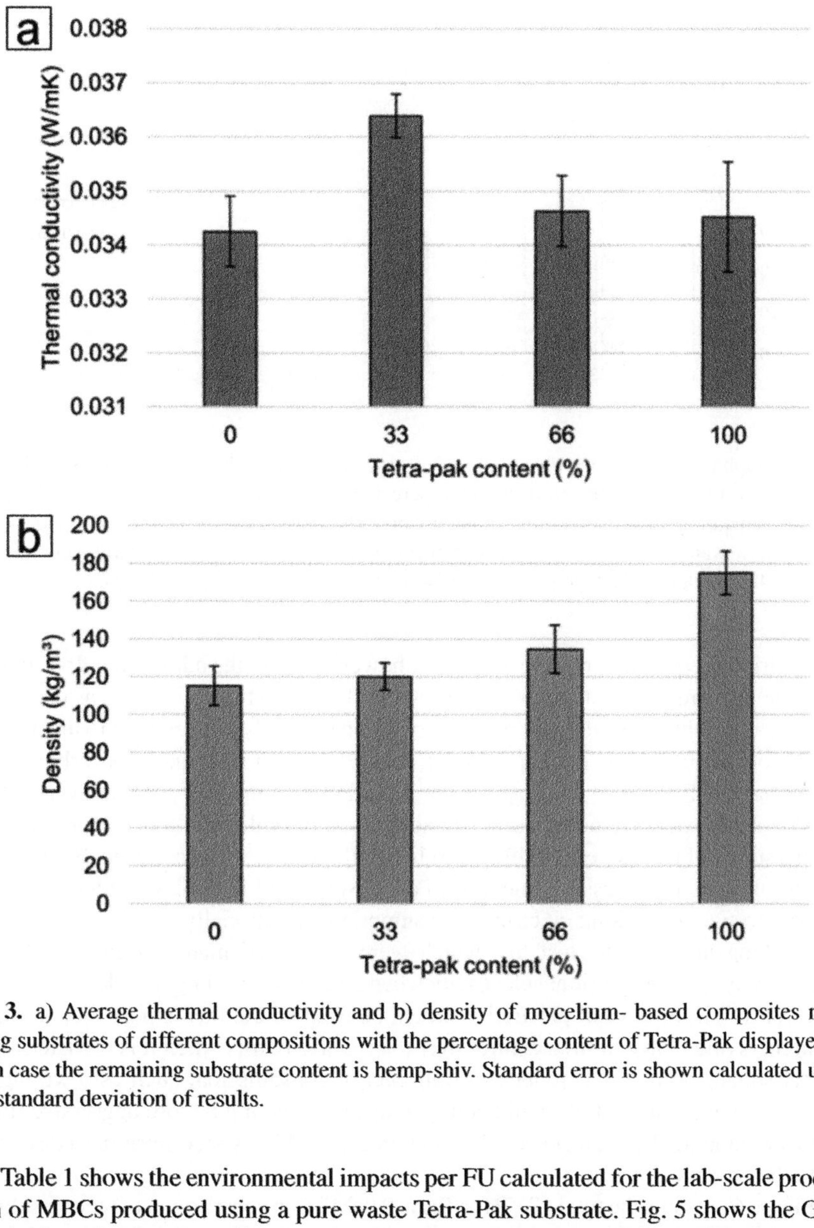

Fig. 3. a) Average thermal conductivity and b) density of mycelium- based composites made using substrates of different compositions with the percentage content of Tetra-Pak displayed. In each case the remaining substrate content is hemp-shiv. Standard error is shown calculated using the standard deviation of results.

Table 1 shows the environmental impacts per FU calculated for the lab-scale production of MBCs produced using a pure waste Tetra-Pak substrate. Fig. 5 shows the GWP (total, biogenic, fossil, and land-use-and-land-use-change) associated with each step of the production process. The steps of the process that contributed the most to GWP total were the incubation and drying stages, primarly due to their GWP fossil contribution which is due to the electricity demands of running an incubator and an oven respectively. The grain spawn production had the next highest contribution to GWP total- this is due to the GWP fossil contribution of the incubation of the grain spawn as the mycelium is colonising the grain. The substrate sterilisation stage contributed to the GWP fossil due

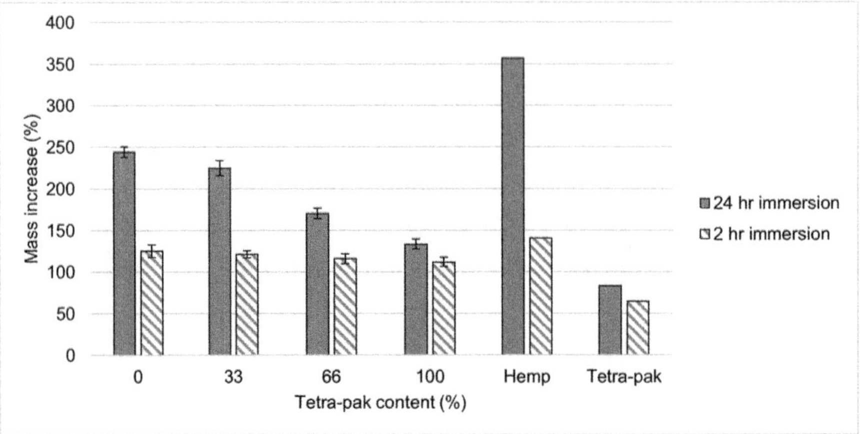

Fig. 4. Graph to show the percentage mass increase of mycelium-based composites made using substrates of different compositions with the percentage content of Tetra-Pak displayed. In each case the remaining substrate content is hemp-shiv. The mass increase is upon 2 h and 24 h of soaking in a water immersion test. The percentage mass increase of raw hemp-shiv and raw chopped Tetra-Pak are displayed for comparison. Standard errors bars are shown.

the electricity required to run the autoclave, however due to the addition of flour into the substrate mixture at this stage the GWP total is slightly net negative here as the flour is biogenic in origin. The substrate preparation stage is largely net negative to the negative GWP biogenic impact as the paper in the Tetra-Pak substrate sequesters carbon. There is minimal GWP fossil in this stage as the Tetra-Pak was manually cut and washed.

This study was conducted at the lab scale as a feasibility demonstration, therefore various stages of the process are likely to change if the process is scaled up. For instance, instead of manual processing, the Tetra-Pak waste would require shredders to cut the material to size, increasing electricity consumption. Additionally, steps such as mixing and packing into mouthe test lds, which were performed manually in the lab-scale process, may require additional electricity when automated at a larger scale. To minimise the GWP fossil, non-fossil electricity should be used where this is possible. However, researchers often have limited control over the electricity mix when it is predetermined. Efforts should be made to reduce overall energy consumption, such as lowering the temperature and duration of incubation, perhaps by selecting a more aggressive fungal strain, or reducing the oven-powered drying time, possibly by incorporating a desiccator to aid the process.

Figure 6 presents the total GWP for Tetra-Pak MBC in comparison to other insulation materials, with the GWP provided for each material per functional unit (FU) as specified in this study. GWP values for EPS, ICB, PUR, and XPS were obtained from Pargana et al. (2014) [30] and data for glass wool and rock wool were sourced from Ricciardi et al. (2014) [35]. The GWP values for the sawdust-based MBC, cellulose-based MBC, and the bamboo MBC were obtained from Alaux et al. (2023) [36], Livne et al. (2022) [37], and Carcassi et al. (2022) [38] respectively. With a value of −1.35 kg CO_2 eq, the Tetra-Pak MBC has a considerably lower total GWP than conventional insulation materials

Table 1. Environmental impact indicators for the conducted LCA, for the mass (kg) of Tetra-Pak MBC insulation required to achieve a thermal resistance (R-value) of 1 (m^2K/W) over an area of 1 m^2, in accordance with EN-15804. This analysis was conducted for Tetra-Pak MBCs produced on a laboratory. Acronyms: GWP-biogenic (climate change, biogenic), GWP-fossil (climate change, fossil), GWP-LULUC (climate change, land use and land use change), GWP-total (climate change, total), ODP (ozone depletion), AP (acidification), EP f (eutrophication, freshwater), EP m (eutrophication, marine), EP t (eutrophication, terrestrial), POCP (photochemical ozone formation), ADPE (resource use, minerals and metals), and ADPF (resource use, fossils).

Impact category	Unit	Result
GWP- Biogenic	kg CO_2 eq	−9.78
GWP- Fossil	kg CO_2 eq	8.34
GWP- LULUC	kg CO_2 eq	0.09
GWP- Total	kg CO_2 eq	−1.35
ODP	kg CFC_{11} eq	7.70E-07
AP	molc H^+ eq	0.06
EP f	kg P eq	0.003
EP m	kg N eq	0.03
EP t	molc N eq	0.19
POCP	kg NMVOC eq	0.02
ADPE	kg Sb eq	8.27E-05
ADPF	MJ	136.89
WDP	m^3	751.80

such as expanded polystyrene (EPS), polyurethane (PUR), and extruded polystyrene (XPS) with total GWP per FU of 99 kg CO_2 eq, 99.9 kg CO_2 eq, and 100.1 kg CO_2 eq respectively. Glass wool and rock wool insulation have lower GWP than EPS, PUR, and XPS, however are still higher in terms of GWP impact than the Tetra-Pak MBCs, with total GWP of 10 kg CO_2 eq and 2.5 kg CO_2 eq for glass wool and rock wool respectively.

The high GWP of various traditional insulation materials can be attributed to their reliance on fossil-based raw materials and energy-intensive manufacturing processes. EPS, PUR, and XPS are derived from petrochemical sources, and the blowing agents used in their manufacture can be potent green-house gases. Additionally, the fossil GWP is not offset by a negative biogenic GWP as they are non-biogenic. In comparison, MBCs, including Tetra-Pak MBC, use substrates that are biogenic in origin (e.g. paper and agricultural waste).

When comparing these materials, it is noted that the GWP values for traditional insulation materials are taken from EPDs and represent industrial-scale production, while the GWP values for Tetra-Pak MBCs and other MBCs are based on lab-scale studies. As such, the reported values for MBCs may be subject to change if scaled up, though this could potentially lead to further efficiencies and reductions in environmental impact through process optimisation, material sourcing improvements, and economies of scale.

When compared to other MBCs, Tetra-Pak MBC performs similarly, though slight variations exist due to differences in substrate composition and processing. The Cellulose & Rapeseed MBC demonstrates a slightly better environmental performance of −1.58 kg CO_2 eq, while Sawdust MBC and Bamboo MBC exhibit slightly higher emissions of 0.56 kg CO_2 eq and 7.39 kg CO_2 eq. However, direct comparison between these MBCs materials is complicated as the lab processes adopted in each of these studies is subject to variation and uncertainties

The low GWP reported for all MBCs in Fig. 6 reflects the consistent finding that MBCs could be a sustainable alternative to traditional insulation materials. Future research should directly compare MBCs made within the same study using the same/similar protocol, and include consideration of regional availability and other material properties beyond thermal conductivity. Beyond environmental and thermal performance of different MBCs, the choice of substrate could be influenced by local waste sources, desired material properties, and aesthetic requirements. For instance, substrates such as Tetra-Pak waste could be preferable in urban environments where composite packaging waste is abundant, while agricultural waste-based MBCs may be more suitable in rural regions with access to biomass byproducts.

Overall, the results support the environmental advantages of MBCs in terms of reducing GWP of insulating homes. Furthermore, the use of complex waste streams like Tetra-Pak, instead of conventional agricultural waste, to produce MCS, aligns with previous findings on the sustainability potential of MBCs. By repurposing post-consumer composite waste, Tetra-Pak MBC could reduce both waste generation and carbon emissions.

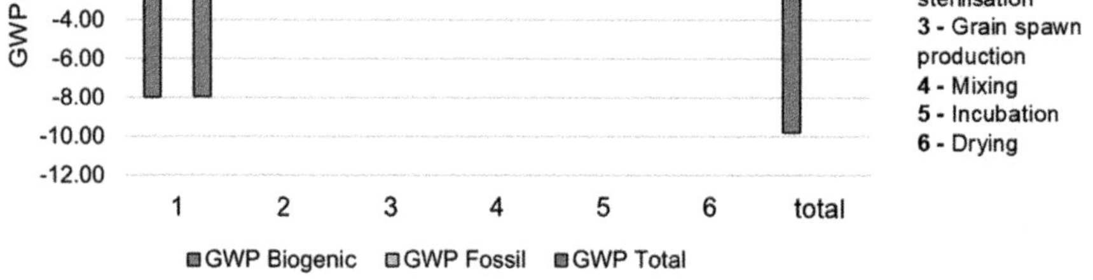

Fig. 5. Graph showing the LCA environmental indicators of global warming potential (GWP) (GWP total, GWP biogenic, GWP fossil, for Tetra-Pak mycelium-based composites produced at a laboratory using *Lentinus tigrinus* mycelium. The functional unit (FU) of this LCA study is defined as the mass (kg) of MBC insulation required to achieve a thermal resistance (R-value) of 1 ($m^2 \cdot K/W$) over an area of 1 m^2. Indicators are shown for the key stages in the MBC production process.

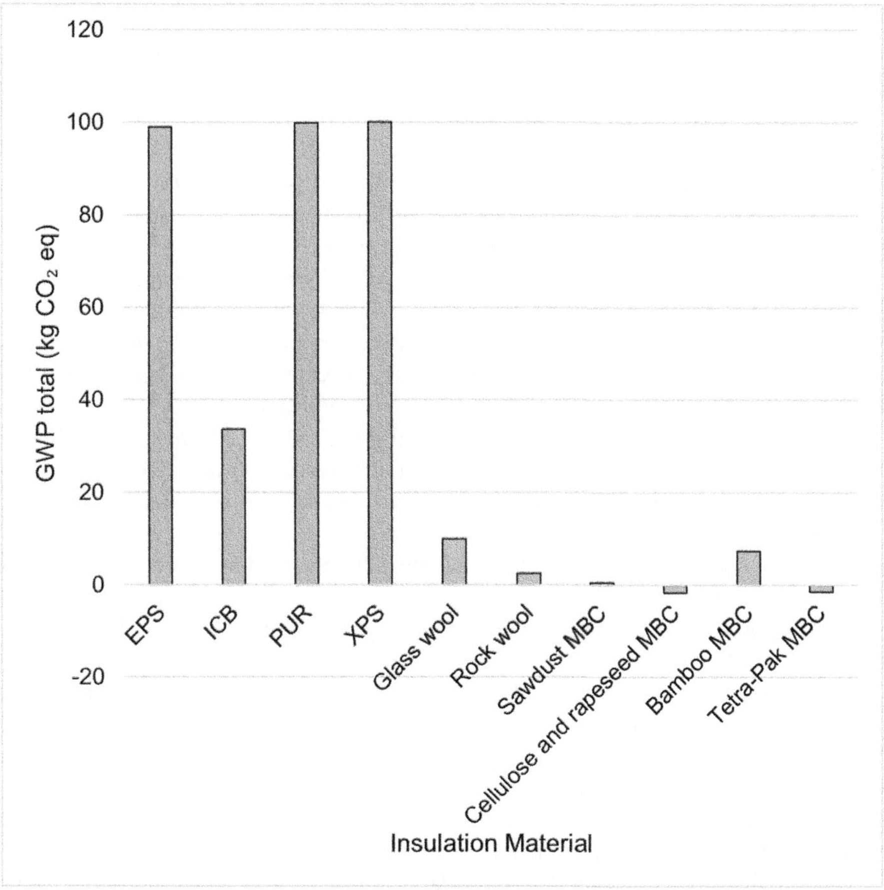

Fig. 6. Graph to show the GWP total for the A1-A3 stages of the production of different insulation materials. Values are given for a FU defined as the mass (kg) of MBC insulation required to achieve a thermal resistance (R-value) of 1 ($m^2 \cdot K/W$) over an area of 1 m^2. GWP values for EPS, ICB, PUR, and XPS were obtained from Pargana et al. (2014) [30] and data for glass wool and rock wool were sourced from Ricciardi et al. (2014) [35]. The GWP values for the sawdust-based MBC, cellulose-based MBC, and the bamboo MBC were obtained from Alaux et al. (2023) [36], Livne et al. (2022) [37], and Carcassi et al. (2022) [38] respectively.

4 Conclusion

This study addressed two key sustainability challenges; that of reducing polylaminate waste, particularly post-consumer Tetra-Pak, and developing sustainable insulation materials for the construction industry. Conventional insulation materials, while effective at reducing operational energy demands, contribute significantly to embodied carbon and pose recycling challenges. Simultaneously, Tetra-Pak waste can be difficult to recycle due to its complex multi-layered structure, leading to high landfill and incineration rates. By investigating the feasibility of Tetra-Pak derived mycelium-based composite, this research aimed to explore an innovative valorisation pathway that diverts waste from disposal whilst contributing to the development of bio-based, low-carbon insulation materials.

This study successfully demonstrated the feasibility of using Tetra-Pak waste as a substrate in MBCs for insulation applications. MBCs were produced using *Lentinus tigrinus* mycelium with four different hemp-shiv to Tetra-Pak ratios (1:0, 2:1, 1:2, 0:1); all mixtures resulted in well-colonised and structurally intact materials. Thermal conductivity (λ) measurements showed that all samples exhibited good insulation properties, with values ranging from $\lambda = 0.0342 \pm 0.001$ W/mK (pure hemp-shiv) to $\lambda = 0.0364 \pm 0.001$ W/mK (33% Tetra-Pak). Importantly, it was shown that increasing the proportion of Tetra-Pak waste did not significantly impact λ. Density (ρ) measurements revealed a slight increase with higher Tetra-Pak content (175 ± 11 kg/m^3 for 100% Tetra-Pak), but this difference was not statistically significant.

Water absorption tests indicated that while all MBCs absorbed water, increasing the proportion of Tetra-Pak waste reduced the overall water uptake after 24 h, potentially benefitting the moisture resistance of bio-based materials in insulation applications. However, the high overall water absorption highlighted the need for further research into hydrophobic treatments or protective coatings for practical use- a common problem when considering bio-based materials.

A life cycle assessment (LCA) of this lab-scale production of Tetra-Pak-based MBCs revealed a negative total GWP of -1.35 kg CO_2 eq per functional unit, which is lower than conventional insulation materials such as expanded polystyrene, polyurethane, and extruded polystyrene. The most carbon-intensive production stages were incubation and drying, primarily due to electricity consumption, while the biogenic carbon content of the Tetra-Pak material contributed to a net-negative GWP impact. These results underscore the potential of using Tetra-Pak waste to produce bio-based insulation materials with promising environmental and thermal performance.

Overall, this study indicates that Tetra-Pak waste could be valorised into mycelium-based composites for sustainable insulation applications. The findings demonstrate that mycelium effectively binds polylaminate waste, producing materials with thermal conductivity values comparable to conventional insulation, and shows that MBCs can effectively incorporate complex waste streams beyond agricultural waste. The reduced water absorption with higher Tetra-Pak content suggests potential durability benefits, while the life cycle assessment underscores the lower environmental impact of MBCs compared to fossil-based alternatives. While the study confirms the feasibility of using Tetra-Pak waste as a mycelium substrate, further research is needed to optimise production at scale, enhance moisture resistance, and evaluate long-term durability in real-world conditions. By integrating waste valorisation with bio-based material innovation, this work supports the transition toward circular economy practices in construction, demonstrating a viable pathway for both waste reduction and low-carbon building materials.

References

1. Hailemariam, A., Erdiaw-Kwasie, M.O.: Towards a circular economy: Implications for emission reduction and environmental sustainability. en. In: Business Strategy and the Environment 32.4 (May 2023), pp. 1951–1965. issn:0964–4733, 1099–0836. https://onlinelibrary.wiley.com/doi/10.1002/bse.3229
2. World Green Building Council: Bringing Embodied Carbon Upfront. World Green Building Council (2019) url: https://worldgbc.org/climateaction/embodied-carbon/
3. SP Group Global Ltd. Waste Management in the UK–in Numbers. Accessed 09 Apr 2025. (2024). https://www.s-pgroup.com/blog/blog8/waste-management-in-the-uk-in-numbers-4.
4. International Energy Agency. Buildings. Accessed 09 Apr 2025. (2025). https://www.iea.org/energy-system/buildings.
5. Aditya, L., et al.: A review on insulation materials for energy conservation in buildings. In: Renewable and Sustainable Energy Reviews 73 (June 2017), pp. 1352–1365. issn: 13640321. https://doi.org/10.1016/j.rser.2017.02.034
6. Yuksel, N.: The review of some commonly used methods and techniques to measure the thermal conductivity of insulation materials. En. In: Insulation Materials in Context of Sustainability. Ed. by Amjad Almusaed and Asaad Almssad. InTech (Aug. 2016). https://doi.org/10.5772/64157. isbn:978-953-51-2624-9
7. DIN41082. doi: 10.31030/1929159. https://www.beuth.de/de/-/-/167922321.1
8. Anh, L.D.H., Pasztory, Z.: "An overview of factors influencing thermal conductivity of building insulation materials". en. In: Journal of Building Engineering 44 (Dec. 2021), p. 102604. issn: 23527102. https://doi.org/10.1016/j.jobe.2021.102604
9. Dickson, T., Pavıa, S.: Energy performance, environmental impact and cost of a range of insulation materials. en. In: Renewable and Sustainable Energy Reviews 140 (Apr. 2021), p. 110752. issn: 13640321. https://doi.org/10.1016/j.rser.2021.110752 doi:10.1016/j.rser.2021.110752.
10. Grazieschi, G., Asdrubali, F., Thomas, G.: Embodied energy and carbon of building insulating materials: A critical review. en. In: Cleaner Environmental Systems 2 (June 2021), p. 100032. issn: 26667894. https://doi.org/10.1016/j.cesys.2021.100032doi:10.1016/j.cesys.2021.100032.
11. Raja, P., et al.: A review of sustainable bio-based insulation materials for energy-efficient buildings. en. In: Macromolecular Materials and Engineering 308.10 (Oct. 2023), p. 2300086. issn: 1438–7492, 1439–2054. https://doi.org/10.1002/mame.202300086doi:.
12. Camilleri, E., et al.: Mycelium-based composites: An updated comprehensive overview. en. In: Biotechnology Advances 79 (Mar. 2025), p. 108517. issn: 07349750. https://doi.org/10.1016/j.biotechadv.2025.108517
13. Van Wylick, A., et al.: Mycelium composites and their biodegradability: an exploration on the disintegration of mycelium-based materials in soil. In: pp. 652–659, Barcelona, Spain (Jan. 2022). https://doi.org/10.4028/www.scientific.net/CTA.1.652.url. https://www.scientific.net/CTA.1.652
14. Moore, D., Robson, G.D., Trinci, A.P.J.: Twenty first century guidebook to fungi. Eng. 2nd ed. Cambridge, United Kingdom: Cambridge university press (2020). isbn: 978-1-108-74568-0.
15. Yang, L., Park, D., Qin, Z.: Material function of mycelium-based bio-composite: a review. Front. Materials. **8**, 737377. issn: 2296-8016 (Sept. 2021). https://doi.org/10.3389/fmats.2021.737377
16. Hoang, T.-D., et al.: "Valorization of agriculture residues into value added products: a comprehensive review of recent studies". En. Chem. Rec. **24**(8), e202300333. issn: 1527-8999, 1528-0691 (2024). https://doi.org/10.1002/tcr.202300333

17. Zawadiak, J.: Tetra Pak Recycling–Current Trends and New Developments. en. Am. J. Chem. Eng. **5**(3), 37 (2017). issn: 2330–8605. doi:10.11648/j.ajche.20170503.12.
18. Tetra Pak. Tetra Pak Official Website. Accessed 09 Apr 2025. https://www.tetrapak.com/.
19. Tetra Laval Group. Tetra Laval Annual Report 2023–2024: Sustainability. Accessed 09 Apr 2025 (2024). https://publications.inpublix.com/tetralaval-annualreport-2023-2024/tetra-pak/sustainability/.2
20. Rachal, M.: Tetra Pak reports plant-based plastics demand growing in North America. In: Packaging Dive (June 2024). Accessed 09 Apr 2025. https://www.packagingdive.com/news/tetra-pak-sustainability-report-fy23-packaging-recycling/718034/.
21. Dong, H., et al.: "Life cycle environmental and economic assessment of Tetra Pak recycling technologies". en. In: Resources, Conservation and Recycling 202 (Mar. 2024), p. 107355. issn: 09213449. https://doi.org/10.1016/j.resconrec.2023.107355
22. Bonocore, G., De Luca, P.: "Preparation and characterization of insulating panels from recycled Polylaminate (tetra Pak) materials". En. Sustainability. **14**(11), 6858. issn: 2071–1050 (2022). https://doi.org/10.3390/su14116858
23. Bekhta, P., et al.: "Properties of composite panels made from tetra-Pak and polyethylene waste material". En. J. Polym. Environ. **24**(2), 159–165. issn: 1566-2543, 1572-8900 (2016). https://doi.org/10.1007/s10924-016-0758-7
24. Martınez-Lopez, M., et al.: "Waste materials from tetra Pak packages as reinforcement of polymer concrete". En. Int. J. Polymer Sci. **2015**, 1–8. issn: 1687-9422, 1687-9430 (2015). https://doi.org/10.1155/2015/763917
25. ASTM International: Standard Test Method for Steady-State Thermal Transmission Properties by Means of the Heat Flow Meter Apparatus. https://www.astm.org/c0518-21.html. ASTM International (2021)
26. ASTM International. ASTM C303–10: Standard Test Method for Dimensions and Density of Preformed Block and Board-Type Thermal Insulation. West Conshohocken, PA, 2010. doi:https://doi.org/10.1520/C0303-10. url: https://www.astm.org/c0303-10.html.
27. ASTM International: Standard Terminology for Sustainability Relative to the Performance of Buildings. https://www.astm.org/c7163-18.html. ASTM International (2018)
28. European Committee for Standardization (CEN). EN 15804:2012+A2:2019- Sustainability of Construction Works: Environmental Product Declarations-Core Rules for the Product Category of Construction Products. Brussels, Belgium, 2019. url: https://www.cen.eu.
29. Daniela Dominguez Aldama et al. "Allocation methods in life cycle assessments (LCAs) of agri-food co-products and food waste valorization systems: Systematic review and recommendations". en. In: Journal of Cleaner Production 421 (Oct. 2023), p. 138488. issn: 09596526. https://doi.org/10.1016/j.jclepro.2023.138488
30. Pargana, N., et al.: "Comparative environmental life cycle assessment of thermal insulation materials of buildings". en. In: Energy and Buildings 82 (Oct. 2014), pp. 466–481. issn: 03787788. https://doi.org/10.1016/j.enbuild.2014.05.057.3
31. Centre, E.: Ecoinvent Database v3.6, Zurich, Switzerland (2019). https://www.ecoinvent.org
32. OpenLCA (GreenDelta GmbH).: Open Source Life Cycle Assessment Software, Berlin, Germany (2024) url: https://www.openlca.org
33. Confederation of European Paper Industries (CITPA). CITPA Guidelines for Calculating Carbon Footprints for Paper-Based Packaging. Accessed: 2025-04-09. 2018. url: https://www.citpa-europe.org/sites/default/files/CITPA%20CO2%20Guidelines%202017%20FINAL%20with%20cover%20page.pdf.
34. Wood and wood-based products-calculation of the biogenic carbon content of wood and conversion to carbon dioxide. European Committee for Standardization, 2014. url: https://www.en-standard.eu/une-en16449-2014-wood-and-wood-based-products-calculation-of-thebiogenic-carbon-content-of-wood-and-conversion-to-carbondioxide/.

35. Ricciardi, P., Belloni, E., Cotana, F.: Innovative panels with recycled materials: Thermal and acoustic performance and Life Cycle Assessment. en. Appl. Energy **134**, 150–162 (Dec. 2014) . issn: 03062619. https://doi.org/10.1016/j.apenergy.2014.07.112
36. Alaux, N., et al.: Environmental potential of fungal insulation: a prospective life cycle assessment of mycelium-based composites. en. Int. J. Life Cycle Assess. **29**(2), 255–272 (Feb. 2024). issn: 0948–3349, 1614–7502. https://doi.org/10.1007/s11367-023-02243-0
37. Livne, A., et al.: "Fungal Mycelium bio-composite acts as a CO2-sink building material with low embodied energy". en. ACS Sustain. Chem. Eng. **10**(37), 12099–12106 (Sept. 2022), issn: 2168–0485, 2168–0485. https://doi.org/10.1021/acssuschemeng.2c01314
38. Carcassi, O.B., et al.: Carbon Footprint Assessment of a Novel Bio-Based Composite for Building Insulation. en. Sustainability **14**(3), 1384 (Jan. 2022) . issn: 2071–1050. https://doi.org/10.3390/su14031384

Bituminous Materials and Polymers

Determination of Complex Modulus from a Back-Analysis Process of Frequency Response Functions for Asphalt Mixtures Using FEM Modeling

Thamires Ximenes Cavalcante[1](✉), Ana Karoliny Lemos Bezerra[2], Mauro José da Silva Filho[3], Lucas Feitosa de Albuquerque Lima Babadopulos[3], and Jean-Claude Carret[1]

[1] École de Technologie Supérieure, Montreal, Canada
thamires.ximenes-cavalcante.2@ens.etsmtl.ca
[2] ENTPE, Ecole Centrale de Lyon, CNRS, LTDS, UMR5513, 69518 Vaulx-en-Velin, France
[3] Federal University of Ceará, Fortaleza, Brazil

Abstract. This paper presents a method to determine the complex modulus of asphalt mixtures (AM) from Frequency Response Functions (FRFs) using a back-analysis approach. The main objective was to optimize the process for obtaining the linear viscoelastic (LVE) properties of AM based on non-destructive impact resonance test (IRT) data. The methodology combines an optimization code developed in MATLAB with numerical simulations in COMSOL, using the Finite Element Method (FEM) in 2D and the 2S2P1D rheological model to simulate the LVE behavior. The combined use of MATLAB and COMSOL allowed for efficient integration between numerical simulation and the optimization process, resulting in a more robust and automated methodology for analyzing viscoelastic materials. The results obtained from the IRT were compared with those from the Tension-Compression test (TCT). A good agreement was observed between the results of both tests, demonstrating the effectiveness of the proposed method. The expected impact of this research is to contribute to the advancement of computational tools for material characterization in engineering, with potential applications in road infrastructure projects and other related fields.

Keywords: Asphalt mixture · Optimization · Back-analysis · Impact resonance tests (IRT) · Frequency Response Function (FRF)

1 Introduction

The linear viscoelastic (LVE) properties of asphalt mixtures have been investigated through non-destructive techniques (NDT) in recent years. Among these, ultrasonic testing and impact resonance test (IRT) are frequently employed to analyze the mechanical material's response, in terms of complex modulus [1–5]. In IRT, characterization is performed either by evaluating the fundamental resonance frequency or by measuring

M. Posani and S. Singh (Eds.): RYS 2025, RILEM Bookseries 67, pp. 677–690, 2026.
https://doi.org/10.1007/978-3-032-14652-6_48

the Frequency Response Function (FRF) [6–13]. The results of recent research have shown that it is possible to carry out these non-destructive tests more quickly, economically and easily when compared to traditional tests carried out with robust hydraulic press. One of the challenges of using NDT is analyzing the results, because they do not allow for the direct determination of the LVE properties of asphalt mixtures. To obtain these properties, a back-analysis process is required [10–13]. In the context of pavement materials, inverse analysis combined with the Finite Element Method (FEM) is a well-established approach, particularly in applications such as Falling Weight Deflectometer (FWD) tests, where FEM simulations are used to determine the stiffness of different pavement layers under dynamic loading conditions [14, 15].

For the LVE characterization of asphalt mixtures, the complexity comes from the material's modulus dependence on loading frequency and temperature. This becomes particularly challenging in IRT tests, as an infinite spectrum of frequencies is simultaneously excited. Conversely, for purely elastic materials such as cement-based composites or steel, the analysis of IRT outcomes is considerably simpler, relying on the direct application of analytical equations that link resonant frequency, geometric parameters, and material stiffness [4, 9, 16, 17].

For viscoelastic materials, such as asphalt mixtures, a more complex process is required. Modelling is used to optimize the FRF based on variations in the parameters of an LVE model, such as the 2S2P1D model [18–21]. Through an iterative procedure, the viscoelastic parameters are adjusted to represent the material's stiffness behavior across different frequencies and temperatures. This optimization process is guided by minimizing the error between the experimental and the modeled FRF responses, refining the material parameters iteratively [10–13].

Carret et al. [10–13] introduced five distinct inverse analysis methods for this purpose, of which three demonstrated satisfactory and reliable outcomes. These methodologies are generally structured into two primary phases. In the initial phase, an optimization is conducted at each temperature, extracting the LVE parameters for that specific condition and contributing to the construction of a segment of the master curve, typically represented by the norm of the complex modulus.

In the subsequent phase, the overall LVE behavior is determined by synthesizing the master curve segments obtained from each tested temperature. The main differences among the proposed methods lie in the selection of the viscoelastic model variables to be optimized versus those that remain constant, as well as whether only a single peak or multiple resonance peaks are considered. All the details, step-by-step procedures and methodological distinctions for each approach can be found in [10–13].

In this context, this article develops an optimization code based on the inverse analysis method proposed in [12, 13, 22], with the aim of initiating its validation and opening perspectives for the optimization process to be extended to different materials. The objective of this work was to contribute to the validation of methods for analyzing IRT test results that allow for the complete and adequate linear viscoelastic characterization of asphalt mixtures, taking into account robust rheological models. For that, the results of the IRT and Tension-Compression tests (TCT) were compared for two different asphalt mixture specimens. This research is part of RILEM's MWP Technical Committee - Mechanical Wave Propagation to Characterize Bituminous Mixtures, which officially started in 2024. The committee's main objective is to standardize ultrasonic resonance and impact methods for assessing the stiffness and viscoelastic properties of asphalt materials. As highlighted in its opening letter [23], one of the main challenges in this area is the lack of standardized test procedures and guidelines for interpreting the results for non-destructive methods. This study contributes directly to the committee's objectives by offering data analysis tools that can facilitate the development of reliable protocols for laboratory and field applications.

2 Materials and Methods

2.1 Materials

The materials used were an asphalt mix with 20% of Reclaimed Asphalt Pavement (RAP), bitumen type PG58S-28 and bitumen content of 5.75% of the total mass of the mix. The asphalt mixtures were designed based on a reference ESG-10 mixture, a dense-graded asphalt concrete commonly used in Quebec for surface courses. Two specimens were manufactured and compacted using the Superpave Gyratory Compactor (SGC), in 150 mm diameter molds, following the Quebec Ministry of Transport (MTQ) [24] standard. After compaction, the specimens were cored transversely to extract samples with a diameter of around 60 mm. The test specimens were compacted so as to have similar air void content. Full details of the mixing, compaction and drilling process can be found in [25]. The main characteristics of the two specimens are shown in Table 1.

Table 1. Main characteristics of the tested specimens.

Specimen	Mass (kg)	Height (m)	Diameter (m)	Air Void (%)
RAP20-I	0.7423	0.11842	0.057	1.478
RAP20-II	0.7466	0.11803	0.057	1.269

2.2 Methods

Specimen RAP20-I was tested only with IRT and the RAP20-II was tested only with TCT. It is worth noting that the TCT was performed for the purpose of serving as a benchmark for comparison and validation of the results obtained with the IRT and the code developed, since the TCT is a traditional test that is already standardized and widely accepted.

Impact Resonance Test

The Impact Resonance Test (IRT) is a dynamic testing method based on the phenomenon of resonance of mechanical compression waves. In this test, an excitation is typically applied at the center of one end of the specimen using a hammer, while the amplified longitudinal response is captured by a piezoelectric sensor, an uniaxial accelerometer, positioned at the center of the opposite end. Fig. 1 shows the IRT configuration for the longitudinal mode.

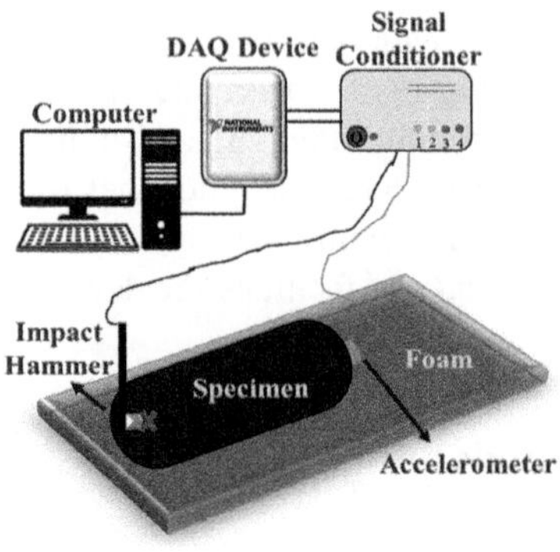

Fig. 1. Configuration of the IRT [26].

The RAP20-I sample was tested at 4 temperatures (−20, 0, 10 and 20 °C), with five impacts applied at each temperature condition. All tests were conducted within 2 min of removing the specimens from the climatic chamber. Therefore, temperature changes within the specimens during this short period were considered negligible and unlikely to have influenced the characterization results. The data recorded by the hammer load cell and the accelerometer are acquired in the time domain, as shown in Fig. 2 for a temperature of −20 °C. These signals are then transformed into the frequency domain by applying the Fast Fourier Transform (FFT), with a resolution of 1 Hz. The FFT was performed using a fixed time window up to 1.0×10^{-3} s, corresponding to the duration of the impact force signal. Fig. 3 shows the frequency-domain transform of the signal shown in Fig. 2.

The FRF provides characterization over a wide frequency range and is obtained as a ratio between the output frequency-domain signal (from the accelerometer) and the input signal (from the hammer) of the IRT. An example of the experimental RAP20-I FRF obtained in the test carried out in this work is shown in Fig. 4a. In addition, Fig. 4b shows the coherence function (CF), an indication of the quality of the measured signals, where it can be seen that in the resonant frequency peak range, between 15000 Hz and 20000 Hz, the CF is approximately 1, indicating a good quality signal.

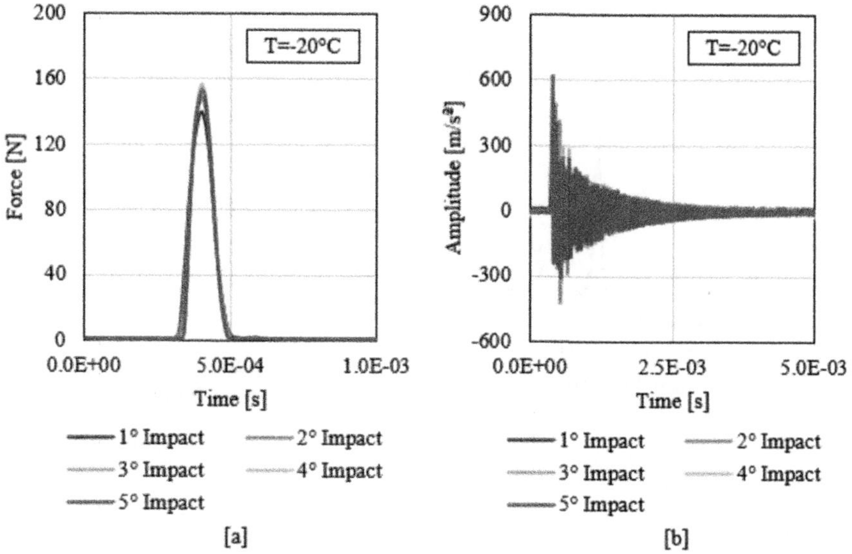

Fig. 2. Dynamic test experimental data for specimen RAP20-I at −20 °C (5 Impacts): [a] force in time domain; and [b] acceleration in time domain.

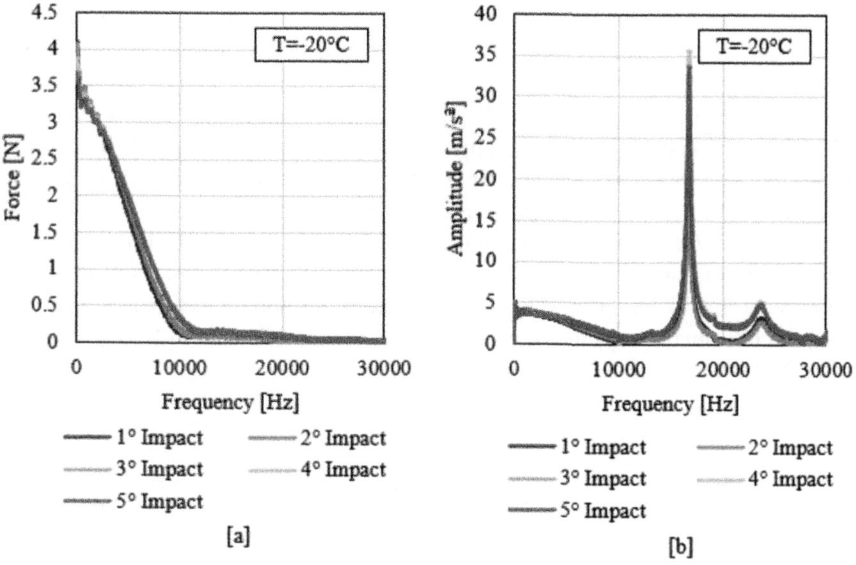

Fig. 3. Dynamic test experimental data for specimen RAP20-I at −20 °C (5 Impacts): [a] force in frequency domain; and [b] acceleration in frequency domain.

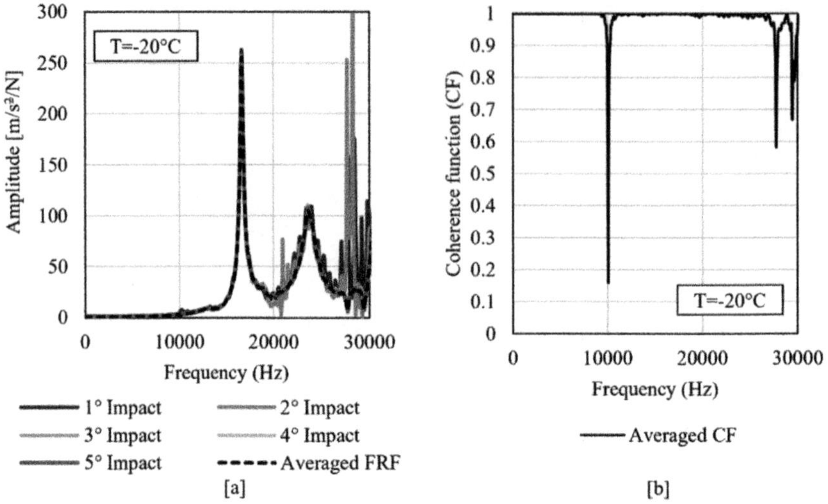

Fig. 4. Dynamic test experimental data for specimen RAP20-I at −20 °C (5 Impacts): (a) FRFs and averaged FRF; and (b) Coherence functions.

Tension-Compression Test (TCT)

The traditional test to determine complex modulus used was the cyclic tension-compression (TC) test on cylindrical specimens, which is a homogenous test. In this type of test, a hydraulic press is used in strain controlled mode to apply a homogeneous axial loading. The targeted axial loading strain amplitude was about 50 μm/m. In this case, the LVE behavior is considered. The axial load is measured with a load cell of 25kN. The specimen is glued to the upper and lower aluminum caps. The axial strain is measured with three extensometers placed at 120° from each other. Tests were performed inside the thermal chamber at eight different temperatures (−35, −25, −15, −5, 5, 15, 25 and 35 °C) and seven frequencies (0.01, 0.1, 0.3, 1, 3 and 10 Hz). The complex modulus and Poisson's ratio values are deduced from the measurements of the sensors and the according to Eq. (1) and Eq. (2) of the behavior homogeneous isotropic. More details and information on the execution of this test can be found at [27, 28].

$$E^*(\omega) = \frac{\sigma_1^*}{\varepsilon_1^*} = \left|E^*(\omega)\right|e^{i\varphi_E} \quad (1)$$

$$V^*(\omega) = \frac{\varepsilon_1^*}{\varepsilon_1^*} = \left|V^*(\omega)\right|e^{i\varphi_v} \quad (2)$$

where ω represents the angular frequency ($\omega = 2\pi f$), where f is the frequency. σ_1^* is axial stress, ε_1^* is axial strain, ε_2^* is radial strain, φ_E and φ_v are the phases of the complex modulus and of the complex Poisson's ratio respectively.

Linear Viscoelastic Modeling with 2S2P1D Model

Using experimental data from TC tests or IRT tests it is possible to perform the modeling using a viscoelastic linear model. For that, the 2S2P1D continuous spectrum model, developed at ENTPE [18, 19], was adopted to represent the linear viscoelastic behavior of asphalt materials tested in this study. The 2S2P1D model is represented by a spring in parallel with a series association of a spring, two parabolic elements, and one dashpot. In the three-dimensional [20, 21] case the expressions for the complex modulus and the complex Poisson's ratio, for isotropic behavior, are shown in Eq. (3) and Eq. (4), respectively.

$$E^*_{2S2P1D}(\omega) = E_{00} + \frac{E_0 - E_{00}}{1 + \delta(jw\tau)^{-k} + (jw\tau)^{-h} + (jw\beta\tau)^{-1}} \quad (3)$$

$$V^*_{2S2P1D}(\omega) = V_{00} + \frac{V_0 - V_{00}}{1 + \delta(jw\tau_V)^{-k} + (jw\tau_v)^{-h} + (jw\beta\tau_v)^{-1}} \quad (4)$$

where ω represents the angular frequency ($\omega = 2\pi f$), where f is the frequency. E_0 and E_{00} are the modulus at high and low frequencies, respectively, and ν_0 and ν_∞ are the Poisson's ratios at high and low frequencies. The parameters k and h are dimensionless constants with values between 0 and 1. δ is another dimensionless parameter, and η relates to the Newtonian viscosity. The characteristic times of the complex modulus (τ_e) and Poisson's ratio (τ_v) vary only with temperature.

The Time-Temperature Superposition Principle (TTSP) is applicable to asphalt mixtures in linear domains [29, 30]. Therefore, the characteristic times at any temperature T can be determined by multiplying the shift factor by the characteristic time of the reference temperature. The shift factor (a_T) is calculated using the Williams–Landel–Ferry (WLF) [22], presented in Eq. (5).

$$\log\,(a_T) = \frac{C_1\left(T - T_{Ref}\right)}{C_2 + T - T_{Ref}} \quad (5)$$

where C_1 and C_2 are the WLF constants, and T_{ref} is the reference temperature.

Optimization Process for IRT Data with Code Developed

Based on the experimental FRFs (see Fig. 4 and Fig. 5), 11 frequency points were selected along the first resonance peak. These points were identified using a computer algorithm that selects 5 points on each side of the peak within a 70% amplitude threshold, in addition to the peak point itself. Fig. 5 presents the graphical output generated by the code, illustrating the selected points for the IRT test conducted at a temperature of 10 °C.

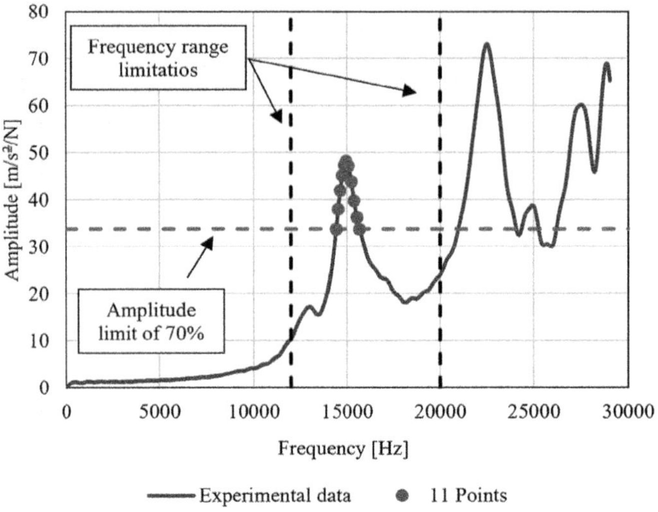

Fig. 5. Selection of the 11 frequency points along the first resonance peak for a temperature of 10 °C.

The values of the 11 selected points, along with the experimental curve and numerical model, are used as input data for the optimization code. The optimization process involves comparing the 11 experimental points with those obtained from the numerical simulation, based on initial parameters defined by the user.

The optimization was carried out in MATLAB using the "*fminsearch*" function, following the methodology proposed by Carret et al. [13]. The optimization code is directly integrated with the COMSOL numerical simulator. The optimization process returns to the user the four variables of the 2S2P1D model (E_0, k, δ and τ_E), which are varied in the process of back-calculating the complex modulus at the resonance frequencies of the peaks. The selection of parameters was based on the parametric analysis conducted by Carret et al. [11, 13], who identified the four parameters with the greatest influence on the FRFs. The remaining parameters of the 2S2P1D model were fixed based on reference material, as reported by Carret et al. [11, 13]. The initial values for the four optimized 2S2P1D parameters were chosen randomly, taking values within a range of +/− 15% around the average values of the reference material. This iterative process continues until the variation in error between two successive iterations falls below 1%. The error function to be minimized was defined by Eq. (6),

$$Error = \sum_{i=1}^{11}\left(\left|\frac{|H_{Exp_i}| - |H_{Comp_i}|}{|H_{Exp_i}|}\right|\right) \tag{6}$$

where, H_{Exp_i} is the experimental FRF, H_{Comp_i} is the computational FRF.

The numerical FRFs are calculated using the finite element method (FEM) by solving Eq. (7) in the frequency domain,

$$-\rho\omega^2 \boldsymbol{u} - \nabla.\sigma = \boldsymbol{F}_p^{i\Phi} \tag{7}$$

where ρ is the apparent density of the material, ω is the angular frequency, $\boldsymbol{u}$ is the displacement vector, ∇ is the gradient tensor, and σ is the Cauchy stress tensor, i is a complex number such as $i^2 = -1$, $\boldsymbol{F_p}$ is the unity load equal to 1 N and Φ is the phase of the cyclic load. In solving Eq. (7), free boundary conditions are assumed throughout the model, except at the impact point, where a point load is applied.

3 Results

3.1 Optimization Process

Table 2 shows the combination of four parameter (E_0, k, δ and τ_E) values, that resulted in the lowest error for each temperature, as well as the value of the error function and the number of iterations required for the process to converge. Fig. 6 shows the graphical result obtained at the end of the optimization, in which we can compare the curve generated from the initial parameters (start), the optimized curve and the experimental curve obtained in the IRT test for the RAP20-I specimen, at a temperature of −20 °C. It can be seen that the optimized curve is in good agreement with the experimental curve.

Table 2. Result of the optimization process for the IRT test for specimen RAP20-I.

Temp. (°C)	$E_0(MPa)$	$\tau_E(s)$	δ	k	Error	N° of iterations
−20	40419	0.101	1.463	0.152	1.227	163
0	41459	103.195	1.943	0.106	1.072	205
10	35807	0.109	2.384	0.248	0.815	199
20	38314	0.008	0.668	0.143	1.341	222

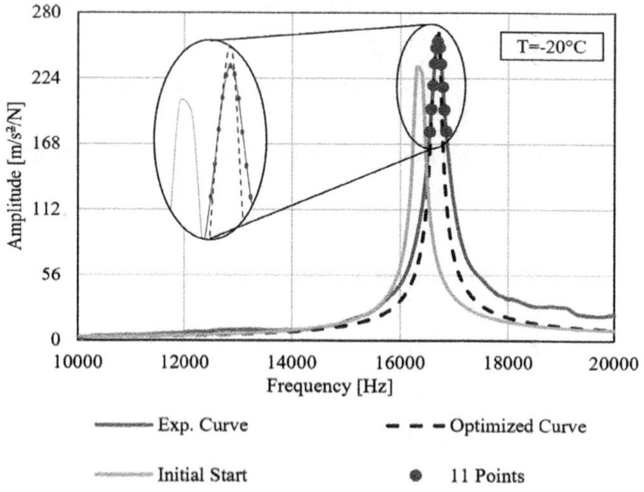

Fig. 6. Graphical result of the optimization process for the IRT test for specimen RAP20-I for a temperature of −20 °C.

3.2 Comparison Between TC and IRT Tests

In order to compare the results obtained with the two tests carried out in this study, known as the TC and the IRT tests, Cole-Cole graphs and master curves of the complex modulus and phase angle were plotted. Fig. 7 shows the Cole-Cole plots, which represents the relationship between the real and imaginary parts of the complex modulus. The curves form a typical arc, indicating the viscoelastic behavior of the material, where energy dissipation (imaginary component) decreases as stiffness (real component) increases.

In the Fig. 9 is possible to see that the data points formed a single curve and, in this case, the TTSP is verified, suggesting that master curves can be constructed for any reference temperature. For that, the shift factors were calculated using the WLF equation, shown in the Fig. 8. The shift factors are the horizontal translation factors of the isotherms. The master curve of complex modulus and phase angle are represented in Fig. 9 and Fig. 10, respectively, for the reference temperature of 10 °C. The master curves allow us to evaluate the linear viscoelastic behavior of the material over a wide frequency range.

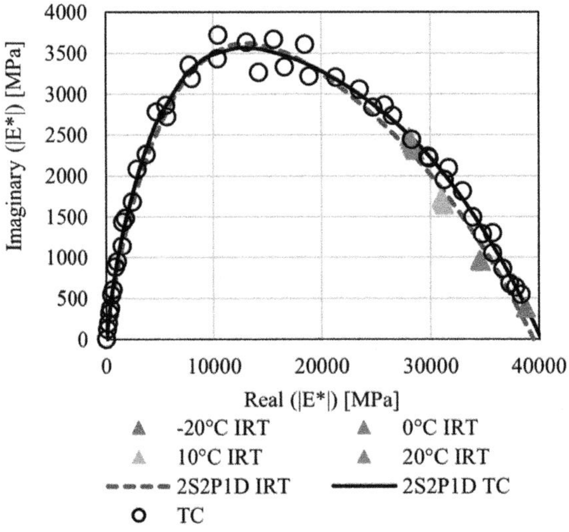

Fig. 7. Cole-Cole diagram obtained from the dynamic tests (IRT) and from the tension–compression tests (TCTs).

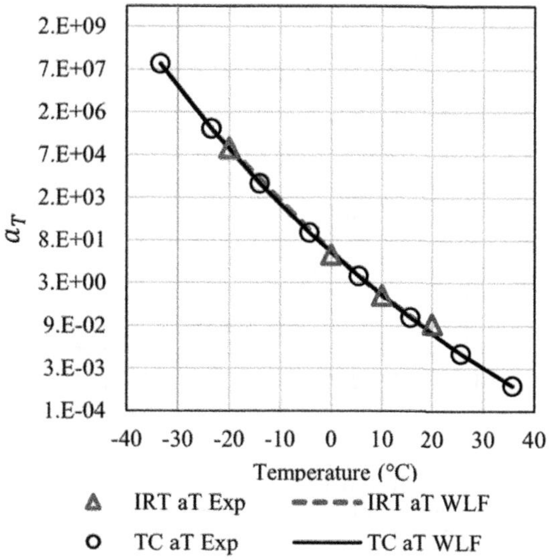

Fig. 8. Shift factors of the WLF equation (Eq. (5)) obtained from the dynamic tests (IRT) and from the tension–compression tests (TCT).

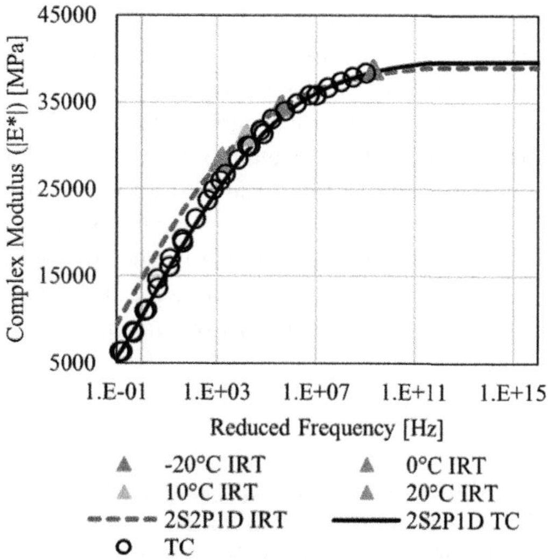

Fig. 9. Master curve for the norm of the complex modulus at 10 °C obtained from the dynamic tests (IRT) and from the tension–compression tests (TCT).

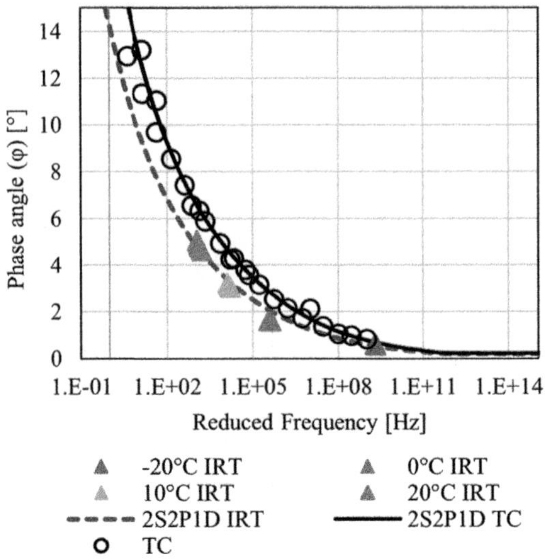

Fig. 10. Master curve of the phase angle at 10 °C obtained from the dynamic tests (IRT) and from the tension–compression tests (TCT).

Observing the graphs presented above, it is possible to see a good agreement between the results of TCT and IRT, using the optimization code in the inverse analysis process developed in this study. Similar results have been found in previous studies [12, 13], where greater differences were also observed at reduced low frequencies, or higher temperatures. But, in general the differences are small and can be justified by the different levels of deformation at which the two tests are carried out. The traditional tension compression test is carried out at 50 μm/m and the IRT at around 0.1 to 0.6 μm/m. This difference causes the phenomenon of non-linearity observed in asphalt materials [31], where the norm of the complex modulus increases and the phase angle decreases when the deformation level decreases. Another possible reason for the differences could be that the results are from different specimens, where the TCT was carried out on the RAP20-II specimen and the IRT on the RAP20-I specimen.

4 Conclusions

This study developed and demonstrated the application of an integrated optimization methodology using MATLAB and COMSOL to determine the complex modulus of asphalt mixtures through inverse analysis of Frequency Response Functions (FRFs) obtained from Impact Resonance Tests (IRT). The use of the 2S2P1D model in finite element simulations enabled accurate numerical representation of the mechanical behavior of asphalt mixtures. The comparison between the IRT and traditional Tension-Compression tests showed good agreement, especially in the characterization of the complex modulus and phase angle across a wide frequency range. Although small discrepancies were observed at lower frequencies in the master curves, these were attributed to differences in levels of deformation at which the two tests are carried out and specimen variability. Overall, the results confirm the reliability and effectiveness of the proposed methodology, reinforcing its potential for use in the non-destructive characterization of viscoelastic materials. This approach contributes to the advancement of computational tools in pavement engineering, especially when it comes to the future standardization of IRT, and offers a promising alternative for efficient and precise material evaluation, which is one of the expectations of RILEM's MWP Technical Committee.

References

1. Di Benedetto, H., Sauzéat, C., Sohm, J.: Stiffness of bituminous mixtures using ultrasonic wave propagation. Road Mater. Pavement Des. **10**(4), 789–814 (2009)
2. Mounier, D., Di Benedetto, H., Sauzéat, C.: Determination of bituminous mixtures linear properties using ultrasonic wave propagation. Constr. Build. Mater. **36**, 638–647 (2012)
3. Norambuena-Contreras, J., Castro-Fresno, D., Vega-Zamanillo, A., Celaya, M., Lombillo-Vozmediano, I.: Dynamic modulus of asphalt mixture by ultrasonic direct test. NDT & E International. **43**(7), 629–634 (2010)
4. Whitmoyer, S.L., Kim, Y.R.: Determining asphalt concrete properties via the impact resonant method. J. Test. Eval. **22**(2), 139–148 (1994)
5. Lacroix, A., Kim, Y.R., Far, M.S.S.: Constructing the dynamic modulus mastercurve using impact resonance testing. Journal of the Association of Asphalt Paving Technologists. **78**, 67–102 (2009)
6. Rupitsch, S.J., Ilg, J., Sutor, A., Lerch, R., Döllinger, M.: Simulation based estimation of dynamic mechanical properties for viscoelastic materials used for vocal fold models. J. Sound Vib. **330**(18–19), 4447–4459 (2011)
7. Gudmarsson, A., Ryden, N., Di Benedetto, H., Sauzéat, C.: Complex modulus and complex Poisson's ratio from cyclic and dynamic modal testing of asphalt concrete. Constr. Build. Mater. **88**, 20–31 (2015)
8. Gudmarsson, A., Ryden, N., Birgisson, B.: Application of resonant acoustic spectroscopy to asphalt concrete beams for determination of the dynamic modulus. Mater. Struct. **45**(12), 1903–1913 (2012)
9. Bezerra, A.K.L., Melo, A.R.S., Freitas, I.L.B., Babadopulos, L.F.A.L., Carret, J.C., Soares, J.B.: Determination of modulus of elasticity and Poisson's ratio of cementitious materials using S-wave measurements to get consistent results between static, ultrasonic and resonant testing. Constr. Build. Mater. **398**, 132456 (2023)
10. Carret, J.C., Di Benedetto, H., Sauzéat, C.: Multi-modal dynamic linear viscoelastic back analysis for asphalt mixes. J. Nondestruct. Eval. **37**, 35 (2018)
11. Carret, J.C., Di Benedetto, H., Sauzéat, C.: Linear viscoelastic behavior of asphalt mixes from dynamic frequency response functions. Int. J. Numer. Anal. Methods Geomech. **44**(7), 1019–1031 (2020)
12. Carret, J.C., Pedraza, A., Di Benedetto, H., Sauzéat, C.: Comparison of the 3-dim linear viscoelastic behavior of asphalt mixes determined with tension-compression and dynamic tests. Constr. Build. Mater. **174**, 529–536 (2018a)
13. Carret, J.C., Di Benedetto, H., Sauzéat, C.: Characterization of asphalt mixes behaviour from dynamic testing and comparison with conventional complex modulus tests. Appl. Sci. **8**(11), 1–15 (2018c)
14. Li, M., Wang, H., Xu, G., Xie, P.: Finite element modeling and parametric analysis of viscoelastic and nonlinear pavement responses under dynamic FWD loading. Constr. Build. Mater. **141**, 23–35 (2017)

15. Liu, P., Wang, D., Oeser, M.: Application of semi-analytical finite element method to analyze asphalt pavement response under heavy traffic loads. J. Traffic Transport. Eng. (English edition). **4**(2), 206–214 (2017)
16. Martinček, I.: The determination of Poisson's ratio and the dynamic modulus of elasticity from the frequencies of natural vibration in thick circular plates. J. Sound Vib. **2**(3), 257–270 (1965)
17. Boz, I., Solaimanian, M.: Investigating the effect of rejuvenators on low-temperature properties of recycled asphalt using impact resonance test. Int. J. Pavement Eng. **19**(11), 1007–1016 (2016)
18. Di Benedetto, H., Olard, F., Sauzéat, C., Delaporte, B.: Linear viscoelastic behavior of bituminous materials: from binders to mixtures. Road Mater. Pavement Des. **5**(2), 163–202 (2004)
19. Olard, F., Di Benedetto, H.: General 2S2P1D model and relation between the linear viscoelastic behaviors of bituminous binders and mixes. Road Mater. Pavement Des. **4**(2), 185–224 (2003)
20. Tiouajni, S., Di Benedetto, H., Sauzéat, C., Pouget, S.: Approximation of linear viscoelastic model in the 3 dimensional case with mechanical analogues of finite size—application to bituminous materials. Road Mater. Pavement Des. **12**(4), 897–930 (2011)
21. Di Benedetto, H., Delaporte, B., Sauzéat, C.: Three dimensional linear behavior of bituminous materials: experiments and modeling. ASCE Int. J. Geomech. **7**(2), 149–157 (2007)
22. Ferry, J.D.: Viscoelastic Properties of Polymers, 3rd edn. John Wiley & Sons, New York (1980)
23. Carret, J.C., Babadopulos, L.F.A.L.: Opening letter of RILEM TC MWP: mechanical wave propagation to characterize bituminous mixtures. RILEM Tech Lett. **10**, 15–21 (2025)
24. Bezerra, A.K.L., Carret, J.-C., Babadopulos, L.F.A.L., Soares, J.B.: Determination of the stiffness of ecological asphalt mixtures with non-destructive impact resonance tests. Mater. Today Proc. **84**(Part 1), 163–170 (2023)
25. LC 26–004. Formulation des enrobés à l'aide de la presse à cisaillement giratoire selon la méthode du laboratoire des chaussées, Ministry of Transport of Quebec (MTQ), 2020.
26. Gayte, P., Di Benedetto, H., Sauzéat, C., Nguyen, Q.: Influence of transient effects for analysis of complex modulus tests on bituminous mixtures. Road Mater. Pavement Des. **17**(1), 271–289 (2015)
27. Perraton, D., Di Benedetto, H., Sauzéat, C., Hofko, N., Graziani, A., Nguyen, Q.: 3 dim experimental investigation of linear viscoelastic properties of bituminous mixtures. Mater. Struct. **49**(1), 4813–4829 (2016)
28. Nguyen, H.M., Pouget, S., Di Benedetto, H., Sauzéat, C.: Time-temperature superposition principle for bituminous mixtures. Eur. J. Environ. Civ. Eng. **13**(10), 1095–1107 (2009)
29. Nguyen, Q.T., Di Benedetto, H., Sauzéat, C., Tapsoba, N.: Time-temperature superposition principle validation for bituminous mixes in the linear and nonlinear domain. ASCE J. Mater. Civ. Eng. **25**(9), 1181–1188 (2013)
30. Mangiafico, S., Babadopulos, L.F.A.L., Sauzéat, C., Di Benedetto, H.: Nonlinearity of bituminous mixtures. Mech. Time-Depend Mater. **22**(1), 29–49 (2018)
31. Filho MJS, Babadopulos LFAL, Carret JC, Soares JB (2024) Influence of test configuration parameters on the results of impact resonance tests for stiffness determination. In: 38th Congresso de Pesquisa e Ensino em Transportes, Florianópolis, Proceedings, Galoá.

Cement and Concrete

Durability Performance of Ternary Blended Self-consolidating Concrete: A Comparative Study with Portland Limestone Cement-Based SCC

Timothy Kofi Ametefe(✉) and Mark Bediako

CSIR – Building and Road Research Institute, Kumasi, Ghana
tkametefe@csir.brri.org

Abstract. The construction industry increasingly emphasizes sustainability and durability, astmespecially in concrete technology, due to environmental concerns associated with cement production and the need for resilient infrastructure. A broader study which adopted the simplex centroid mixture design approach developed a self-consolidating concrete (SCC) mix with a ternary mixed binder, optimizing the content of waste glass powder (GP) and low-grade calcined clay (CC). The goal was to develop a more sustainable and durable SCC by reducing the reliance on traditional cement and divert glass waste from landfills. For this study, two types of concrete mixes were prepared: a control SCC mix (CM) with Portland limestone cement (PLC) as the only binder and another with the ternary mixed binder (TM). The ternary blend constituted of 73% PLC, 21% GP and 6% CC. The durability performance including water absorption, sorptivity, resistance to acid attack and rapid chloride permeability were evaluated over curing periods of 28, 56 and 90 days and compared with the control mix. Results indicate that TM had consistently lower water absorption (up to 13%), reduced initial sorptivity (up to 19%), and significantly better resistance to chloride ingress and acid attack compared to CM. The ternary blended SCC mix was found to be a more durable alternative to conventional PLC-based SCC, with reduced permeability and improved resistance to chemical attacks.

Keywords: self-consolidating concrete · ternary blend · calcined clay · waste glass powder · durability · sustainability

1 Introduction

The construction industry is undergoing a significant transformation as it seeks to address the dual challenges of environmental sustainability and the demand for durable, resilient infrastructure. Concrete, the most widely used construction material, plays a pivotal role in this transition. However, the production of Portland cement, a primary component of conventional concrete, is a major contributor to global carbon dioxide emissions, accounting for 8–10% of the world's total emissions [1]. This environmental impact, coupled with the increasing volume of industrial and post-consumer waste, has driven

M. Posani and S. Singh (Eds.): RYS 2025, RILEM Bookseries 67, pp. 693–706, 2026.
https://doi.org/10.1007/978-3-032-14652-6_49

the search for alternative materials and innovative concrete technologies that can reduce reliance on traditional cement while enhancing the durability performance of concrete structures.

Self-consolidating concrete (SCC), a highly flowable and non-segregating concrete, has gained prominence in modern construction due to its ability to fill complex forms and densely reinforced structures without the need for mechanical vibration [2]. While SCC provides significant benefits in workability and construction efficiency, its high cement content raises concerns about sustainability and long-term durability, making these key areas worth considering for research. Recent studies have investigated the impact of incorporating various supplementary cementitious materials (SCMs), including fly ash (FA), ground granulated blast furnace slag (GGBFS), and silica fume (SF), as partial cement replacements in SCC, focusing on their effects on its durability. Leung et al. [3] reported reduced surface water absorption and sorptivity when they partially replaced OPC with FA and SF. Similarly, Sai et al. [4] found that incorporating FA and SF decreased chloride ingress in concrete. Ali-Boucetta et al. [5] studied the effects of GGBFS and glass powder (GP) in SCC. They reported lower sorptivity, reduced gas permeability coefficient and enhanced resistance to chloride ion penetration. El Mir et al. [6] evaluated the water absorption and chloride ion migration of SCC when they partially replaced limestone filler with metakaolin and silica fume. The authors reported that due to the reduced porosity of the interfacial transition zone (ITZ) on account of pozzolanic reactions and capillary pore enhancement, the addition of metakaolin and silica fume reduces the water absorption and chloride ion migration of SCC.

Extensive research has demonstrated that SCMs enhance the microstructure of concrete by promoting the formation of secondary hydration products and exerting a filler effect. These mechanisms contribute to the development of a denser, more tortuous, and less interconnected pore network, thereby significantly improving the durability properties of the material [7–10]. However, the availability of conventional SCMs, such as FA, GGBFS and SF, are expected to become increasingly constrained in the near future due to the ongoing global transition toward cleaner energy production and decarbonized steel manufacturing processes. Consequently, there is a growing imperative to explore alternative SCMs that have received comparatively less attention, including waste glass powder and calcined clays (CC), to ensure the continued development of sustainable cementitious systems [11, 12]. Additionally, waste glass, a non-biodegradable material, poses significant environmental challenges when disposed of in landfills [13]. According to Ali-Boucetta et al. [5], the global recycling rate for waste glass stood at just 32%, in 2018. Thus, the majority of discarded glass ultimately finds its way to landfill sites, where it occupies precious land space indefinitely. According to Tariq et al. [14], the worldwide availability of waste glass constitutes roughly around 3% of the overall cement consumption. Although this represents a relatively small fraction, it presents a viable alternative for diverting waste away from landfills.

This study builds upon a broader research effort that employed the simplex centroid mixture design approach to develop an optimized ternary blended SCC mix incorporating 21% GP and 6% CC as partial replacements for Portland limestone cement (PLC). The objective of this study was to evaluate the durability performance of this ternary blended SCC mix (TM) compared to a control SCC mix (CM) with only PLC as the binder. Key durability indicators, including water absorption, sorptivity, resistance to acid attack, and rapid chloride permeability, were examined over curing periods of 28, 56, and 90 days.

1.1 Materials

The concrete mix consisted of CEM II/B-L Portland limestone cement (42.5R), calcined clay (CC) (70% $<$ 75 µm), waste glass powder (GP) (70% $<$ 75 µm), well-graded pit sand, 14 mm graded coarse aggregates, potable water, and a polycarboxylate-based high-range water reducer (HRWR). The specific gravities of PLC, CC, and GP were 3.097, 2.653, and 2.503, respectively, with surface areas of 402 m^2/kg, 420 m^2/kg, and 497.5 m^2/kg. Table 1 presents the chemical composition of the binder constituents, while Fig. 1 illustrates the particle size distribution of the aggregates.

Table 1. Chemical composition of binder constituents

Chemical composition (%)	PLC	GP	CC
Al_2O_3	7.03	1.79	21.9
Fe_2O_3	3.36	0.276	4.33
MgO	3.12	1.6	0.23
SiO_2	30	70.7	70.9
CaO	50.4	9.67	0.114
K_2O	1.09	0.273	1.06
Na_2O	–	15.1	–
SO_3	3.95	0.145	0.06

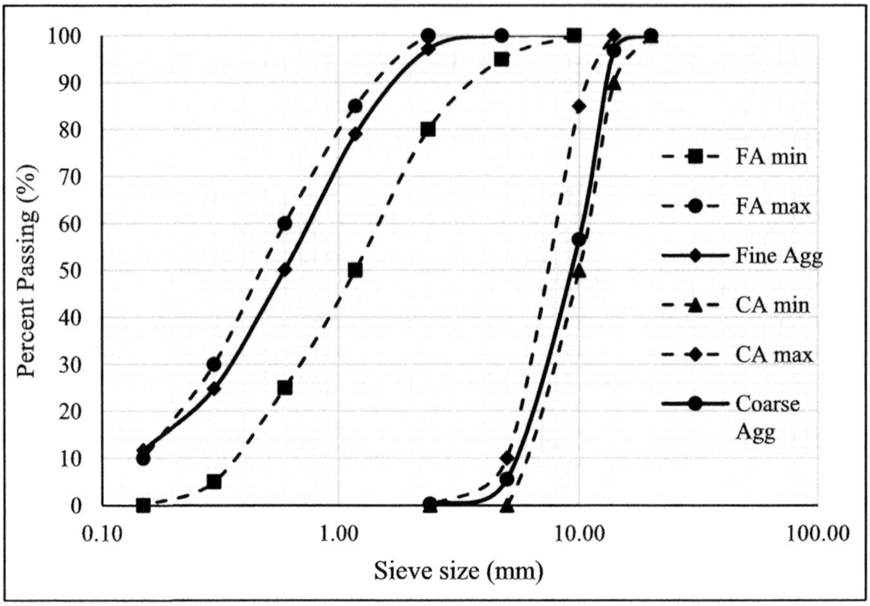

Fig. 1. Particle size distribution of aggregates

2 Methods

2.1 SCC Mix Design, Fresh Properties and Sample Preparation

The ternary mix (TM) represents the optimized proportions of PLC, GP, and CC, determined using the simplex centroid mixture design approach from a broader study, whereas the control mix (CM) contained only PLC as binder. The input factors—X_1 (PLC), X_2 (CC), and X_3 (GP)—were optimized based on multiple responses, including slump flow diameter, slump flow time (T_{500}), V-funnel time, L-box ratio, blocking step, HRWR content, compressive strength, flexural strength, ultrasonic pulse velocity, water absorption, porosity, sorption, and rapid chloride permeability. Figure 2 presents a ternary plot of the experimental matrix used to derive the proportions for seven SCC mixes in the optimization process. Given the non-hydraulic nature of CC and GP, concrete mixtures were not viable across the entire simplex centroid region. Therefore, a feasible sub-region was defined within the complete simplex to ensure realistic mix proportions. Multi-objective optimization was performed using the Derringer and Suich desirability function, assigning equal weight to each parameter to balance all performance properties without compromising any critical aspect.

The SCC mix design and proportioning followed the guidelines specified in ACI 237R [15] and EFNARC [16] specifications. Table 2 presents the mix proportions per cubic meter for both mixes. The SCC mixtures were designed to achieve SF2 slump flow (660–750 mm), viscosity classifications of VS2 (>2 s, based on T_{500}) and VF2 (9–25 s, based on V-funnel time), and a passing ability of PA2 (L-box ratio of 0.8–1.0) with a visual stability index (VSI) of 0. A minimum compressive strength of 30 MPa at 28 days was targeted.

The mix proportions of both SCC mixes were measured and mixed using a concrete mixer. The fresh properties, including filling ability, passing ability and resistance to segregation were analysed, ensuring they achieved the specified design parameters outlined above. 100 mm cubic metallic moulds and cylindrical moulds of 100 mm diameter and 50 mm height were then filled without any external vibration. The fresh concrete samples were covered with a plastic sheet and wet burlap, maintaining a temperature of 23 ± 2 °C for 24 h. After this initial curing period, the concrete specimens were demoulded and immersed in lime-saturated water until they were due for testing at the designated ages.

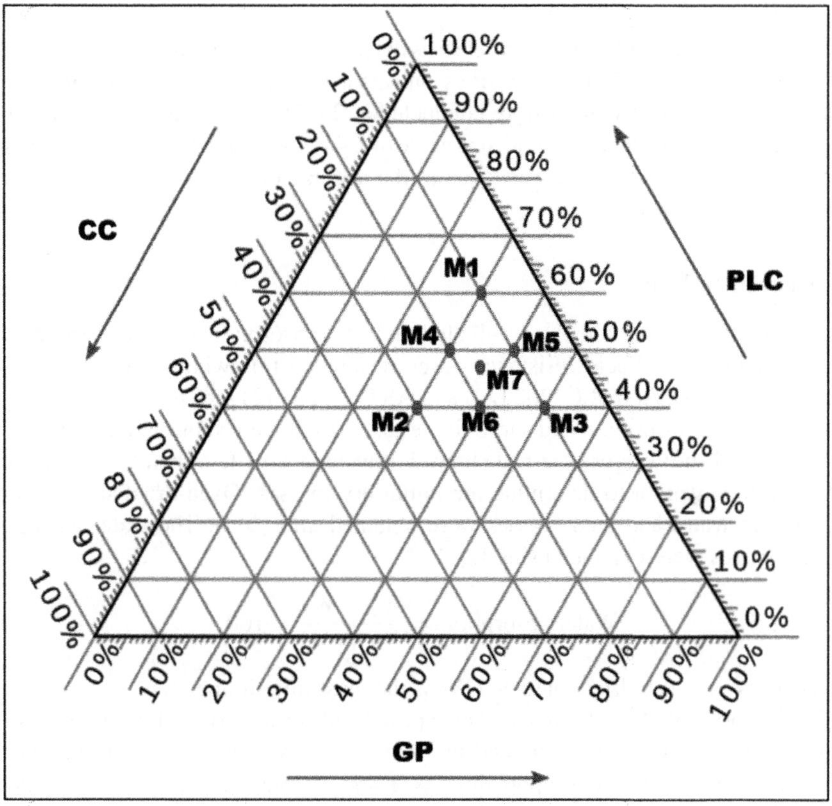

Fig. 2. ternary plot of mixture design points

Table 2. Mix proportions of control and ternary mixes per cubic meter

Mix ID	PLC (kg)	CC (kg)	GP (kg)	Water (kg)	Coarse Aggregates (kg)	Fine Aggregates (kg)	HRWR (%)
CM	496	–	–	199	810	849	1.2
TM	362	30	104	199	810	849	1.398

2.2 Compressive Strength

The compressive strength of the concrete was determined in accordance with EN 12390–3 [17] using a Matest Servo Cyber-Plus universal testing machine with a maximum capacity of 1500 kN. The tests were conducted at 28, 56 and 90 days. Concrete cubes were placed in the machine and loaded at a rate of 2.4 kN/s until failure. The compressive strength for each concrete age was taken as the average of the maximum compressive force of three samples.

2.3 Durability Tests

To assess durability, key indicators—including water absorption, sorptivity, rapid chloride permeability, and acid resistance—were evaluated following ASTM C642 [18], ASTM C1585 [19], ASTM C1202 [20] and ASTM C267 [21], respectively.

For the water absorption and sorptivity tests, samples were subjected to oven drying at 105 °C until the weight obtained showed a difference of less than 0.2% compared to the previous weight to determine the initial dry masses. Oven-dried samples were immersed in water and their masses were obtained after 24 h. The water absorption values of mixes were obtained using Eq. 1.

$$\text{Water absorption} = \frac{W_2 - W_1}{W_1} \times 100 \tag{1}$$

Where W_1 is oven-dried mass and W_2 is the mass after immersion.

For the sorptivity test, the side and unexposed end surfaces of each oven-dried specimen were sealed with duct tape, and the initial mass was measured. During testing, specimens were placed on support devices inside a water container, ensuring about 2 mm water level above the support. Mass measurements were recorded at 60 s, 5, 10, 20, 30, and 60 min, followed by hourly measurements up to 6 h. Additional measurements were taken once daily for 3 days and at three intervals between days 4 and 7, each spaced at least 24 h apart, with a final measurement on day 8. Water absorption at each time interval was determined using Eq. 2.

$$\mathbf{I} = \frac{\mathbf{m_t}}{\mathbf{a} \times \boldsymbol{\rho}} \tag{2}$$

where m_t is the mass change at time t (g), a is the exposed surface area (mm^2), and ρ is the density of water (g/mm^3).

The initial rate of water absorption was obtained from the slope of the absorption vs. square root of time plot between 1 min and 6 h, ensuring a correlation coefficient ≥ 0.95. The secondary rate of absorption was determined from the slope of the plot for measurements taken from day 1 onwards.

For the rapid chloride permeability test, samples were subjected to vacuum saturation using an intelligent concrete vacuum saturated machine for 22 h (3 h of dry saturation, 1 h of wet saturation and 18 h of static saturation). Samples were immediately placed in the applied voltage cells with the positive terminal of the test host connected to the cell containing a 0.3 N sodium hydroxide solution and the negative terminal connected to the cell containing a 3% sodium chloride solution. A constant voltage of 60 V was passed through the concrete specimens for six hours and the total charge that passed through each specimen was recorded.

The evaluation of the acid resistance of SCC samples cured for 28, 56 and 90 days was conducted by subjecting them to a 5% sulphuric acid (H_2SO_4) solution for 28 days, following the ASTM C267 standard [21]. Prior to immersing the cured samples in the acidic solution, their initial weights were measured and recorded. Subsequently, the weights of the specimens were determined and recorded at 7, 14, 21, and 28 days, during the immersion period and the corresponding weight losses were calculate. The compressive strengths of the specimens were determined after the 28 days immersion and the Compressive Strength Change Factor (CSCF, %) was obtained using Eq. 3.

$$CSCF = \frac{fcu_{x\ days} - fcu_{immersed\ samples}}{fcu_{x\ days}} \times 100 \tag{3}$$

Where,

CSCF, % is the compressive strength change factor;

$fcu_{x\ days}$ is the compressive strength obtained after 28, 56 or 90 days curing.

$fcu_{immersed\ samples}$ is the compressive strength obtained after immersion of samples in acidic solution.

3 Results and Discussion

3.1 Fresh Properties and Compressive Strength of Mixes

The fresh properties and compressive strength of both concrete mixes are presented in Table 3. From the table, both mixes achieved all the specified fresh properties. However, TM exhibited a slightly higher deformability and potentially better passing ability and stability in complex formwork or densely reinforced structures due to a marginally higher L-Box ratio and higher T_{500} and V-funnel times. The improved deformability of TM can be attributed to the synergic interaction between the spherical shaped and smooth textured glass powder and the calcined clay particles, which decreased the interparticle friction in the mix [22]. The prolonged flow times in TM can be linked to the increased surface area of CC and GP, which elevated their water absorption potential. This resulted in a depletion of free water for lubrication, thereby increasing mixture viscosity [23]. The increase in mixture viscosity facilitates the effective transportation of aggregates as concrete flows. This explains the improved passing ability of TM [22]. The high stability of both mixes can be attributed to the high paste content, which enhances cohesion among the mix components, thereby enhancing resistance to segregation [24].

CM exhibited significantly higher compressive strength than TM at early ages: 34% greater at 3 days, 48% at 7 days, 13% at 28 days, and 3% at 56 days. However, by 90 days, the strength for TM slightly exceeded that of CM by approximately 1%. This indicates that the early-age strength gain in CM was more rapid, likely due to faster PLC hydration [25]. Thus, the high CaO content of PLC accelerates C-S-H formation from C_3A through nucleation effect, contributing substantially to the early strength development in CM [26]. On the other hand, strength development in TM, driven by the pozzolanic activation of GP and CC, was slower initially but ultimately surpassed that of CM at later age (90 days). This suggests that the impact of GP and CC on strength becomes more pronounced over time, compensating for the slower early reaction rate and ultimately resulting in improved long-term performance [14].

Table 3. Fresh properties and compressive strength of mixes.

Fresh Property	CM	TM	Criteria
Slump flow diameter (mm)	707.5	747.5	SF2 (660–750 mm)
Slump flow time (T_{500}) (s)	4	5.6	VS2 (> 2 s)
L-box ratio	0.968	0.989	PA2 (0.8–1.0)
V-funnel time (s)	9.2	9.7	VF2 (9–25 s)
Visual stability index	0	0	0
Compressive strength (MPa)			> 30 MPa
28 days	43.3	37.8	
56 days	43.7	42.4	
90 days	43.2	43.7	

3.2 Durability Properties

Transport Properties. Figure 3 presents the results obtained for the water absorption test and RCPT of CM and TM. The results demonstrate a consistent reduction in both water absorption and chloride ingress at all curing ages with the inclusion of GP and CC in TM. Compared to CM, TM showed a progressive decrease in water absorption by 3%, 7% and 13% at 28, 56 and 90 days, respectively. CM exhibited moderate chloride ingress, with total charge values ranging between 2000 and 4000 coulombs across all curing ages. In contrast, TM demonstrated significantly reduce chloride permeability, with total charges of 1000 and 1073 coulombs at 28 and 56 days, respectively, falling within the low chloride ingress range (1000–2000 coulombs). At 90 days, TM exhibited very low chloride ingress, with a total charge in the range of 100–1000 coulombs.

The observed reduction in both water absorption and chloride ingress in TM compared to CM highlights the effectiveness of GP and CC in enhancing the durability properties of concrete. The progressive decline in water absorption suggests that the pozzolanic reactions of GP and CC contribute to pore refinement over time, leading to a denser microstructure. This aligns with established literature where SCMs reduce capillary porosity, thereby limiting water penetration [5, 14, 24, 27]. Similarly, the lower chloride penetrability in TM reinforces this improvement in impermeability which can be attributed to the reduced interconnectivity of pores [28].

Table 4 presents the initial and secondary rate of absorption for both mixes whereas Fig. 4 shows a sample plot of water absorption against the square root of time (CM at 28 days). TM, juxtaposed against CM, consistently exhibited lower initial and secondary sorptivity at all ages, which suggests that the continued pozzolanic reaction of GP and CC, improved pore refinement and reduced interconnectivity of pores reduce water transport into the deeper concrete layers [29–31].

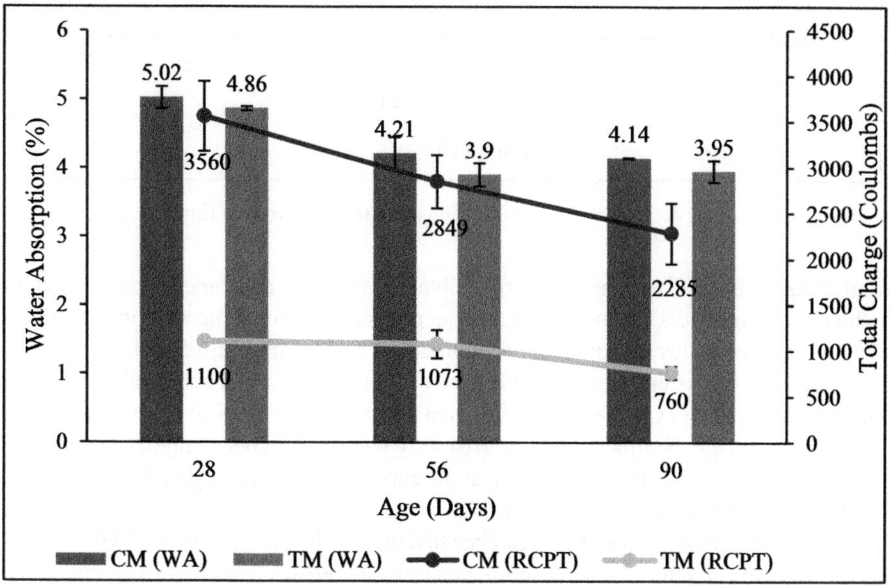

Fig. 3. water absorption test and RCPT results

Table 4. initial and secondary sorptivity of concrete mixes.

Mix	Age (days)	Initial sorptivity, mm/$\sqrt{s}$ (R^2)	Secondary sorptivity, mm/$\sqrt{s}$ (R^2)
CM	28	0.017 (0.9995)	0.008 (0.9693)
	56	0.024 (0.9971)	0.010 (0.9536)
	90	0.026 (0.9959)	0.008 (0.9581)
TM	28	0.025 (0.9975)	0.006 (0.9656)
	56	0.022 (0.9891)	0.004 (0.9183)
	90	0.021 (0.9871)	0.004 (0.9909)

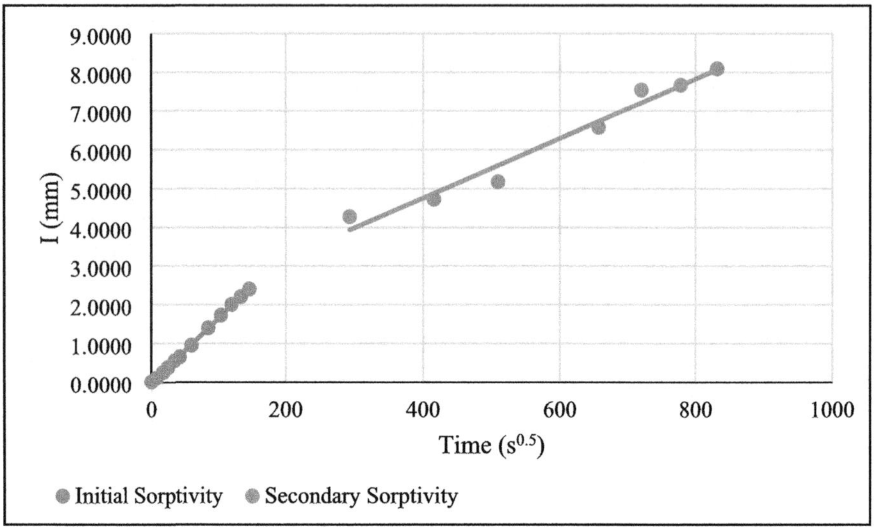

Fig. 4. plot of water absorption against square root of time.

Acid Attack. Table 5 outlines the results obtained from the resistance of acid attack on both concrete mixes. The immersion of concrete samples in acidic solution led to the dissolution of hydrates and peeling off aggregates, as shown in Fig. 5 and Fig. 6. From the Table, CM consistently exhibited higher weight loss than TM across all immersion durations and curing ages, demonstrating that TM possesses significantly greatly resistance to acid attack. Similarly, CM consistently recorded higher compressive strength change factor values compared to TM at all curing ages. This further confirms TM's superior resistance to acid degradation.

The enhanced durability of TM can be attributed to the incorporation of GP and CC, which refine the concrete microstructure and reduce porosity. This improvement results from the formation of secondary calcium silicate hydrate (C-S-H) and calcium aluminosilicate hydrate (C-A-S-H) phases, which strengthen the matrix while simultaneously reducing the amount of calcium hydroxide, a highly acid-susceptible component [32, 33]. Furthermore, the addition of GP and CC lowers the alkalinity of the pore solution, reducing the solubility and dissolution rate of calcium-based compounds in acidic environments [34]. This contributes to an improved buffering capacity and enhanced acid neutralization potential, thereby delaying both the onset and progression of acid-induced deterioration [35, 36].

Table 5. resistance to acid attack on SCC mixes

Mix	Age (days)	weight loss, % (standard deviation)				Compressive strength after immersion (MPa)	CSCF (%)
		7 days	14 days	21 days	28 days		
CM	28	7.30 (0.29)	11.35 (0.61)	14.26 (0.59)	17.06 (0.38)	10.8 (0.52)	75.0
	56	4.21 (0.13)	8.66 (0.95)	10.94 (1.06)	14.31 (1.19)	11.4 (1.10)	73.9
	90	3.59 (0.20)	4.12 (0.10)	4..94 (0.30)	6.07 (0.24)	19.0 (0.37)	56.1
TM	28	−1.62 (0.13)	0.67 (0.02)	3.62 (0.30)	6.58 (0.94)	15.9 (0.45)	58.0
	56	−1.26 (0.01)	1.07 (0.02)	1.52 (0.12)	8.74 (0.31)	18.8 (1.29)	55.8
	90	0.21 (0.21)	0.28 (0.01)	0.45 (0.06)	0.50 (0.06)	22.7 (1.35)	48.1

Fig. 5. visual observation of CM samples at various ages in acidic solution

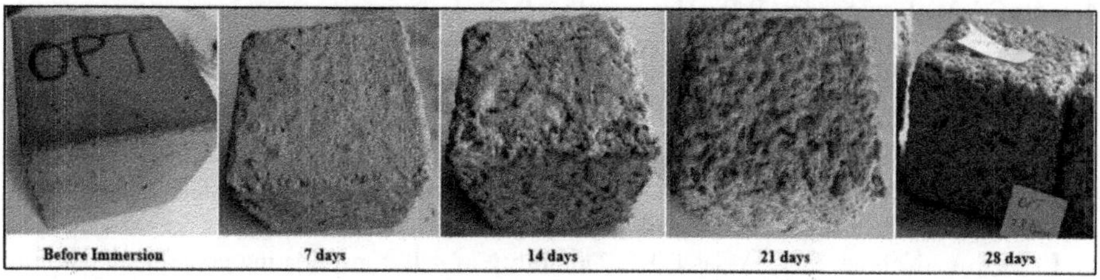

Fig. 6. visual observation of TM samples at various ages in acidic solution

4 Conclusions and Recommendation

This study assessed the durability performance of a ternary blended SCC incorporating glass powder and calcined clay, with the aim of developing a sustainable and durable alternative to conventional SCC made solely with Portland limestone cement. Transport properties including water absorption, sorptivity and rapid chloride permeability test as well as resistance to acid attack were evaluated. Based on the findings, the following conclusions and recommendation can be made.

1. The addition of glass powder and calcined clay improves the filling ability, passing ability and stability of SCC. No adverse effects were observed on the fresh properties of SCC.
2. Reductions in water absorption, chloride ingress and sorptivity indicate improved transport properties with the inclusion of glass powder and calcined clay owing to a denser and more impermeable microstructure.
3. The incorporation of glass powder and calcined clay in SCC enhances its resistance to acid attack and compressive strength deterioration in acidic solutions.
4. Ternary blended SCC system incorporating glass powder and calcined clay can achieve durability enhancements without sacrificing fresh and mechanical performance. It is therefore recommended for structural members exposed to aggressive and harsh environmental conditions.

References

1. Bediako, M., Valentini, L.: Strength performance and life cycle assessment of high-volume low-grade kaolin clay pozzolan concrete: a Ghanaian scenario. Case Studies in Construction Materials. **17**, e01679 (2022). https://doi.org/10.1016/j.cscm.2022.e01679
2. Güneyisi, E., Gesoğlu, M., Algin, Z.: Performance of self-compacting concrete (SCC) with high-volume supplementary cementitious materials (SCMs). In: Eco-Efficient Concrete, pp. 198–217. Elsevier Ltd. (2013). https://doi.org/10.1533/9780857098993.2.198
3. Leung, H.Y., Kim, J., Nadeem, A., Jaganathan, J., Anwar, M.P.: Sorptivity of self-compacting concrete containing fly ash and silica fume. Constr. Build. Mater. **113**, 369–375 (2016). https://doi.org/10.1016/j.conbuildmat.2016.03.071
4. Sai, C.V.V., Kameswara, R.B., Hanumantha, R.C.: Investigation on chloride penetration in concrete mixes of different cement replacement percentages with fly ash and silica fume. In: Materials Today: Proceedings, vol. 33, pp. 820–827. Elsevier Ltd. (2020). https://doi.org/10.1016/j.matpr.2020.06.270
5. Ali-Boucetta, T., Behim, M., Cassagnabere, F., Mouret, M., Ayat, A., Laifa, W.: Durability of self-compacting concrete containing waste bottle glass and granulated slag. Constr. Build. Mater. **270** (2021). https://doi.org/10.1016/j.conbuildmat.2020.121133
6. El Mir, A., Nehme, S.G., Assaad, J.J.: Durability of self-consolidating concrete containing natural waste perlite powders. Heliyon. **6**(1), e03165 (2020). https://doi.org/10.1016/j.heliyon.2020.e03165
7. Shakouri, M., Exstrom, C.L., Ramanathan, S., Suraneni, P.: Hydration, strength, and durability of cementitious materials incorporating untreated corn cob ash. Constr. Build. Mater. **243**, 118171 (2020). https://doi.org/10.1016/j.conbuildmat.2020.118171

8. Ndahirwa, D., Zmamou, H., Lenormand, H., Leblanc, N.: The Role of Supplementary Cementitious Materials in Hydration, Durability and Shrinkage of Cement-Based Materials, their Environmental and Economic Benefits: a Review Cleaner Materials. Elsevier Ltd. (2022, September 1). https://doi.org/10.1016/j.clema.2022.100123
9. Justnes, H.: How SCMs improve concrete durability-a fundamental view (2016). https://doi.org/10.18552/2016/scmt4kn4
10. Pang, L., Liu, Z., Wang, D., An, M.: Review on the application of supplementary cementitious materials in self-compacting concrete. Crystals. **12**(2) (2022). https://doi.org/10.3390/cryst12020180
11. GCCA: The GCCA 2050 Cement and Concrete Industry Roadmap for Net Zero Concrete, London (2021)
12. Schmidt, W., Commeh, M., Olonade, K., Schiewer, G.L., Dodoo-Arhin, D., Dauda, R., et al.: Sustainable circular value chains: from rural waste to feasible urban construction materials solutions. Dev. Built Environ. **6** (2021). https://doi.org/10.1016/j.dibe.2021.100047
13. Zhang, Q., Wang, R., Shen, Y., Zhan, L., Xu, Z.: An ignored potential microplastic contamination of a typical waste glass recycling base. J. Hazard. Mater. **422** (2022). https://doi.org/10.1016/j.jhazmat.2021.126854
14. Tariq, S., Scott, A.N., Mackechnie, J.R., Shah, V.: Durability of high volume glass powder self-compacting concrete. Appl. Sci. (Switz.). **10**(22), 1–20 (2020). https://doi.org/10.3390/app10228058
15. ACI 237R : Self-Consolidating Concrete. American Concrete Institute (2007).
16. EFNARC The European Guidelines for Self-Compacting Concrete Specification, Production and Use 'The European Guidelines for Self Compacting Concrete' (2005). Retrieved from www.efnarc.org
17. EN 12390-3: Testing Hardened Concrete. Part 3: Compressive Strength of Test Specimens. British Standards Institution (2019).
18. ASTM C642 Standard Test Method for Density, Absorption, and Voids in Hardened Concrete (2006). Retrieved from www.astm.org
19. ASTM C1585 Standard test method for measurement of rate of absorption of water by hydraulic-cement concretes. ASTM International (2004). Retrieved from www.astm.org,
20. ASTM C1202 Standard test method for electrical indication of concrete's ability to resist chloride ion penetration 1. ASTM International (2012). https://doi.org/10.1520/C1202-12
21. ASTM C267-01 Standard Test Methods for Chemical Resistance of Mortars, Grouts, and Monolithic Surfacings an d Polymer Concretes. ASTM International (2012)..
22. Promsawat, P., Chatveera, B., Sua-iam, G., Makul, N.: Properties of self-compacting concrete prepared with ternary Portland cement-high volume fly ash-calcium carbonate blends. Case Studies in Construction Materials. **13**, e00426 (2020). https://doi.org/10.1016/j.cscm.2020.e00426
23. Bediako, M., Atiemo, E.: Influence of higher volumes of clay Pozzolana replacement levels on some technical properties of cement pastes and mortars. Journal of Scientific Research and Reports. **3**(23), 3018–3030 (2014). https://doi.org/10.9734/JSRR/2014/9046
24. Matos, A.M., Ramos, T., Nunes, S., Sousa-Coutinho, J.: Durability enhancement of scc with waste glass powder. Mater. Res. **19**(1), 67–74 (2016). https://doi.org/10.1590/1980-5373-MR-2015-0288
25. Barrett, T., Sun, H., Weiss, W.J.: Performance of Portland Limestone Cements: Cements Designed to be more Sustainable that Include up to 15% Limestone Addition (2013). https://doi.org/10.5703/1288284315335
26. Lea, F. M. (Frederick M., Hewlett, P. C., & Liska, M. (2019). Lea's Chemistry of Cement and Concrete. (P. C. Hewlett & M. Liska, Eds.) (5th ed.). Oxford, United Kingdom: Butterworth-Heinemann.

27. Ghafoori, N., Najimi, M., Sharbaf, M.: Properties of self-consolidating concrete containing natural pozzolan. Retrieved from. (2016) https://www.researchgate.net/publication/305118254
28. Zhang, Y., Yang, Z., Ye, G.: Dependence of unsaturated chloride diffusion on the pore structure in cementitious materials. Cem. Concr. Res. **127** (2020). https://doi.org/10.1016/j.cemconres.2019.105919
29. Du, H., Tan, K.H.: Transport properties of concrete with glass powder as supplementary cementitious material. ACI Mater. J. **112**(3), 429–437 (2015) https://doi.org/10.14359/51687363
30. Omrane, M., Rabehi, M.: Effect of natural pozzolan and recycled concrete aggregates on thermal and physico-mechanical characteristics of self-compacting concrete. Constr. Build. Mater. **247** (2020). https://doi.org/10.1016/j.conbuildmat.2020.118576
31. Kumar, R., Mohd Yaseen, A.Y.B., Shafiq, N., Jalal, A.: Influence of Metakaolin, Fly ash and Nano silica on mechanical and durability properties of concrete. Key Eng. Mater. **744**, 8–14 (2017). https://doi.org/10.4028/www.scientific.net/KEM.744.8
32. Koushkbaghi, M., Kazemi, M.J., Mosavi, H., Mohseni, E.: Acid resistance and durability properties of steel fiber-reinforced concrete incorporating rice husk ash and recycled aggregate. Constr. Build. Mater. **202**, 266–275 (2019). https://doi.org/10.1016/j.conbuildmat.2018.12.224
33. Bustos, F., Martinez, P., Videla, C., Lopez, M.: Reducing concrete permeability by using natural pozzolans and reduced aggregate-to-pasteratio. J. Civ. Eng. Manag. **21**(2), 165–176 (2015). https://doi.org/10.3846/13923730.2013.802719
34. Albert, C., Isgor, O.B., Angst, U.: Exploring machine learning to predict the pore solution composition of hardened cementitious systems. Cem. Concr. Res. **162** (2022). https://doi.org/10.1016/j.cemconres.2022.107001
35. Cyr, M.: Influence of supplementary cementitious materials (SCMs) on concrete durability. In: Eco-Efficient Concrete, pp. 153–197. Elsevier Ltd. (2013). https://doi.org/10.1533/9780857098993.2.153
36. Mlinárik, L., Kopecskó, K.: The influence of combined application of two SCMs on the corrosion and acid attack durability of mortars. Periodica Polytechnica Civil Engineering. **61**(2), 313–321 (2017). https://doi.org/10.3311/PPci.9352

Effect of Silica Fume Variants on the Mechanical and Permeation Characteristics of Ultra High Performance Concrete

Lakshmi Thotakura(✉), Ganesh Babu Kodeboyina, and Deepti Avirneni

Ecole Centrale School of Engineering, Mahindra University, Hyderabad 500043, India
Lakshmi20pcie001@mahindrauniversity.edu.in

Abstract. Ultra-High Performance Concrete (UHPC) with its superior mechanical strength, durability has a wide applicability in engineering. This study explores the effects of three different variants of silica fume like grey, white, and fumed silica on the mechanical and permeation characteristics of UHPC. Experimental investigations were carried out to assess the mechanical properties of compressive and tensile strength, and durability performance through permeation. The results reveal that the type of silica fume significantly influences UHPC's performance and provides insights into optimizing sustainable UHPC mix designs using locally available silica fumes. The study will also help in using these concretes for the different requirements of structural and architectural applications appropriately.

Keywords: Silica Fume · Ultra High Performance Concrete · Permeation · Efficiency · Sustainable

1 Introduction

Ultra high performance concrete (UHPC) is a next generation high strength composite with possibilities both in structural and architectural applications, due to its high mechanical strength, impermeability and durability characteristics. Most of the earlier efforts were focused on the strength characteristics of UHPC and not many studies were available on durability or permeation. Ultra high performance concrete (UHPC) are high strength cementitious composites with compressive strengths in excess of 130 MPa [1]. The strength of the concrete is directly related to the porosity of the composite and by reducing porosity a very high strength can be produced. These high-strength concretes need high cement contents, and the higher the cement content results in higher carbon emissions affecting the environment. Reducing the carbon footprint by minimizing the cement content and the incorporation of supplementary cementitious materials containing highly reactive silica, that reacts with calcium hydroxide released during hydration of cement is proposed [2]. These secondary cementitious compounds also help in achieving additional strength as well as better performance.

Silica fume (SF) is an industrial byproduct of the silicon or the ferrosilicon industry, primarily with a high amorphous content and a particle size bordering the nano range.

M. Posani and S. Singh (Eds.): RYS 2025, RILEM Bookseries 67, pp. 707–717, 2026.
https://doi.org/10.1007/978-3-032-14652-6_50

Due to this high SiO_2 content of about 97%, SF is highly reactive and is utilized as a high efficiency pozzolan in concrete [3]. Abbas et al. achieved a UHPC with a compressive strength of 151 MPa at 28 days with 20% SF [4]. The mixture contained SF with 95% SiO_2, 30% quartz powder (QP), and quartz sand (QS), and fibrous additions at a water content of 0.23. A split tensile strength of 9.4 MPa with no fibers was also reported. Also, the initial and secondary sorptivity coefficients were observed to be 0.0631 mm/$\sqrt{}$sec and 0.0432 mm/$\sqrt{}$sec, and the permeable pore space (PPS) was 3.52%. A UHPC with a strength of 166 MPa was reported in a cement matrix containing about 18% SF in combination with quartz powder, sand, and 2% fibers [5].

UHPC of 122.2 MPa was also developed with a QS and waste foundry sand (WSF) combination in a matrix containing 670.5 kg/m^3 of cement and 10% SF with basalt aggregate (BA) [6]. The split tensile and flexural strengths achieved were 7.5 MPa, and 9.5 MPa respectively at 28 days. Tafraoui et al. investigated a mixture of 25% of SF with a cement content of 691 kg/m^3 along with 138 kg/m^3 of metal fibers, QS and QP achieving a compressive strength of 155 MPa at 28 days [7]. Sohail et al. investigated the performance of UHPC of 161 MPa incorporating 23% SF combined with fly ash (FA) [8]. The corresponding flexural strength was 11 MPa and the initial sorptivity rate was 0.00113 mm/$\sqrt{}$sec. It also exhibited a porosity of 1.43%. Hasan et al. studied UHPC containing 25% silica fume reporting 28-day compressive strengths of 88.3 MPa for plain and 113.9 MPa for fiber-reinforced mixes [9]. The corresponding split tensile strengths were 7.6 MPa and 8.85 MPa, respectively and the permeable porosities (PPS) were 9.9% and 16.7%. Alsalman et al. investigated UHPC with a binder content of 890 kg/m^3 having silica fume at 5, 10, 15, and 20%, achieving compressive strengths in the range of about 125 at 90 days [10].

Most of these investigations on UHPCs were all done by using the silica fume with about 98% silica content. The aim of the present effort is to understand the effectiveness of a few locally available silica fumes in UHPC compositions. Locally available silica fumes of three verities, grey silica fume (SG) that is commonly used but with only 90% SiO_2 content, white silica fume (SW) of almost the same fineness as the gray with 99% SiO_2 and a super fine fumed silica (FS) with about 98% SiO_2 in combination with the gray silica fume were studied. The performance of these three different types of silica fumes was evaluated through workability, compressive and tensile strength parameters. The durability and other characteristics affecting their environmental stability are studied through the permeability to water and chlorides, apart from the pH levels to understand the corrosion probabilities.

2 Materials and Methodology

2.1 Materials

After a close look at the literature along with some preliminary studies a silica fume content of 25% was chosen for both grey and white varieties, while 0.01% of fumed silica was added to 25% gray SF mixture in the third composition. While most earlier attempts on UHPCs used high strength cement or even a proprietary premix composition, the presently used cement was an ordinary Portland cement of Euro C 32.5 grade with a fineness of 0.32 m^2/gm. Quartz powder (QP) passing through 2000 mesh and fine quartz sand (QS) of ASTM standard sand grading [11] formed the remaining components. The water binder ratios (w/b) were kept at around 0.25 and to make the concrete workable, a polycarboxylic ether (PCE) based water reducer from SIKA was used. The chemical compositions of these materials are given in Table 1.

Table 1. Chemical characteristics of materials

Chemical Composition	C	SG	SW	FS	QP	QS
CaO (%)	61.85	0.75	0.00	—	0.03	0.01
SiO_2 (%)	20.07	90.20	99.92	98.61	99.72	99.02
Al_2O_3 (%)	5.32	1.10	0.031	—	0.13	0.77
Fe_2O_3 (%)	4.62	1.38	0.012	—	0.05	0.22
MgO (%)	2.54	0.40	0.00	—	0.00	—
SO_3 (%)	2.34	0.50	—	—	0.03	—
LoI (%)	2.41	2.50	0.001	3.96	0.09	0.8
Fineness (m^2/g)	0.32	20	22	184	2000 (mesh)	ASTM Sand
Specific Gravity	2.83	2.65	2.64	2.20	2.65	2.68

(C- cement; SG- silica fume-grey; SW- silica fume-white; FS- fumed silica; QP– quartz powder; QS– quartz sand; SP– superplasticizer; LoI- Loss on ignition).

2.2 Concrete Mix Details and Experimental Investigations

The UHPC compositions studied contain three different varieties of silica fume and were compared with a control concrete without silica fume (US0) and ASTM standard mortar (AS0) [12]. To make a homogenous mixture, these cementitious materials were mixed in a Hobart-type mixer for 15 to 25 min until they resulted in a thixotropic matrix and were compacted with a handheld vibrator. Immediately, after casting the moulds were covered with plastic sheets to prevent evaporation. The detailed mix proportions of these UHPC concretes with the different silica fume varieties are given in Table 2.

Table 2. Ultra-high performance concrete matrix compositions

Mix details	C	SG	SW	FS	QP	QS	SP (%)	w/b
AS0	1.00	0.00			0.00	2.75	0	0.485
US0	1.00	0.00			0.38	1.07	2	0.35
SG25	1.00	0.25			0.25	1.05	2	0.24
SW25	1.00		0.25		0.25	1.05	2	0.22
SGF25	1.00	0.25		0.01	0.25	1.05	2	0.24

These compositions were studied for their performance in terms of workability, strength and permeation characteristics apart from the XRD analysis.

2.3 Workability and Mechanical Characteristics

The flowability measured by using ASTM C 230 method [13] for all the concretes (Fig. 1). The concrete was cast into 40 × 40 × 160 mm moulds as shown in Fig. 1. They were demoulded the next day and the samples were cured in normal tap water until testing. A loading rate of 0.05 mm/sec was found to be the most appropriate for the compressive strength test, which is in line with that suggested by the earlier researchers [14]. The split tensile and the flexural tensile strengths of these samples were also evaluated. The ultrasonic pulse velocities were also measured at a frequency of 50 kHz according to ASTM C 597 to understand the quality of the concrete [15].

Fig. 1. Flow of UHPC and their samples preparation

2.4 Permeation Characteristics

The strength of concrete directly depends on the porosity of the matrix which also allows the ingress of aggressive environmental forces into the pore system affecting the durability of the cementitious composites. The permeation characteristics were studied through the permeable pore space, water absorption, capillary absorption and chloride penetration.

Studies on water absorption and permeable pore space (PPS) of the UHPC matrix were conducted in accordance with ASTM C 642 [16]. The samples were dried in a hot air oven at a temperature of 105 °C for about 72 h to obtain the oven dry mass (M_O). The specimens are then cooled in a desiccator till they are immersed in water and saturated mass (M_S) was measured at intervals to understand the water absorption characteristics. After about 72 h the apparent submerged mass (M_A) of the specimen is established. The absorption rate and permeable pore space are then obtained by using Eq. (1).

$$\text{PPS } (\%) = \frac{M_s - M_O}{M_s - M_A} \tag{1}$$

Another way to understand permeation characteristics is capillary absorption, or surface absorption studied through the rate of capillary permeation as per ASTM C 1585 [17]. To start with, the oven dry samples were taken, and the sides of the samples were coated with epoxy. After the epoxy is fully dry the bottom cut surface of the specimen is exposed to water. The capillary absorption is measured continuously by using Eq. (2), to establish the initial rate of absorption (until 6 h) and the secondary rate of absorption (until 72 h).

$$\text{Rate of capillary absorption } (\%)\, \mathrm{I} = \frac{M_t}{a * \rho} \tag{2}$$

where, I' is the rate of capillary absorption, 'M_t' is the mass change at a given time, and 'a' is the exposed cross-sectional area while 'ρ' is the density of water.

The chloride ion penetration characteristics of the concretes were investigated through the NT Build 443 test [18]. Initially a solution of 165 gms. of NaCl per dm^3 is prepared and then the cut surface of a concrete sample is exposed to this chloride solution while all the remaining surfaces are sealed with epoxy coating. The specimens have to be exposed for at least 35 days, but considering the very high impermeability of the UHPC the exposure period was increased to 90 days. The samples were then split into two and the chloride penetration depths were measured by spraying 0.1 molar $AgNO_3$ solution. The penetration is characterized by a whitish or greyish white surface precipitate on the samples.

2.5 XRD Analysis

After the compression testing the samples were powdered, sieved through a 200 micron mesh, and pre-dried in a vacuum desiccator for subjecting them to XRD evaluation. The XRD measurements were performed using a Rigaku SmartLab diffractometer with CuKα radiation (40 kV, 50 mA), scanning from 3° to 90° of 2θ° at a step size of 0.01° and a rate of 3°/min. Phase identification was carried out using the ICDD database, following the approach of Lee et al. [19].

3 Results and Discussions

3.1 Workability and Mechanical Characteristics

The flowability measurements show that the workability was very good for all the UHPC compositions (Fig. 1). The flow diameters of the different cementitious compositions indicating their workability was presented in Table. 3. All the UHPC mixes with all three silica fumes show flow value beyond 200 mm, while the control (US0) and ASTM mortar (AS0) exhibit very little workability [4]. The compressive strengths at 28 and 90 days are presented in Fig. 2. It can be seen that the composition containing silica fume grey (SG25) even with only 90% SiO_2 content exhibited the highest compressive strength of 144 MPa, while that with silica fume white (SW25) even with its 98% SiO_2 resulted in a marginally lower strength of 134 MPa.

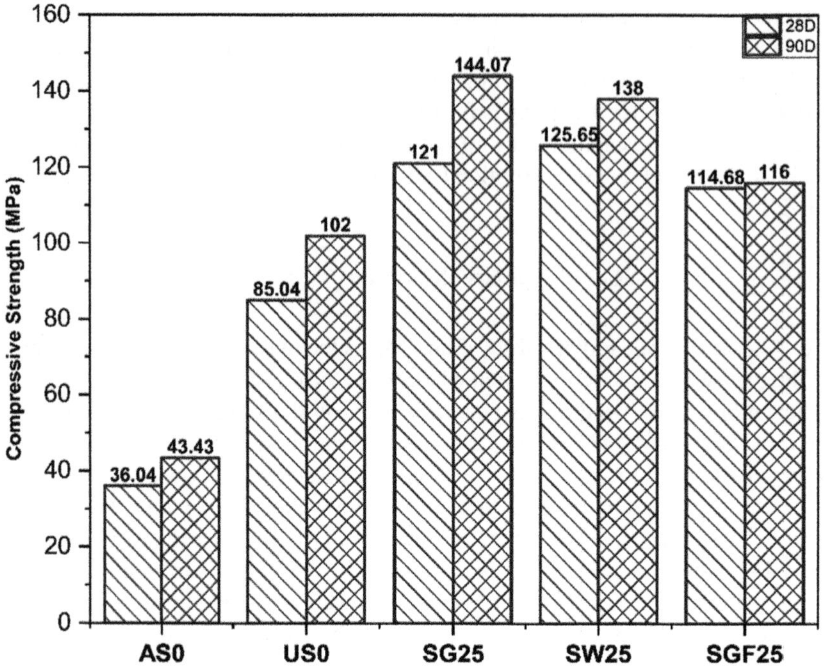

Fig. 2. Compressive Strength of UHPC at 28 and 90 days

Compared to these two composites the third containing silica fume grey along with 0.01% fumed silica with (SGF25) resulted in much lower strength of only 116 MPa. The ultrasonic pulse velocities of all compositions studied, including the control and the ASTM mortar, were found to be above 4.5 km/sec, indicating that all of them belong to the excellent concrete quality as per ASTM [15]. Table 3 presents the split tensile and flexural tensile strengths of these concrete mixtures, which appear to be proportional to the corresponding compressive strengths. Table 3 also contains the pH values of the different concretes studied and it can be seen that most of them, but for US0, are all above 12, indicating no major depletion of the alkalinity even with 25% silica fume in the compositions.

Table 3. Workability and mechanical strength characteristics

Mix details	Flow (mm)	Compressive strength (MPa)	Split tensile strength (MPa)	Flexural tensile Strength (MPa)	Pulse velocity (km/sec)	pH
AS0	100	43	6.60	2.40	4.76	12.31
US0	131	102	9.49	5.59	4.89	11.54
SG25	229	144	9.85	12.54	5.06	12.34
SW25	220	138	8.50	7.95	4.76	12.37

(continued)

Table 3. (*continued*)

Mix details	Flow (mm)	Compressive strength (MPa)	Split tensile strength (MPa)	Flexural tensile Strength (MPa)	Pulse velocity (km/sec)	pH
SGF25	205	116	—	11.29	5.40	12.23

3.2 Permeation Characteristics

The permeable pore space and its size distribution which allows the environmental ingress (ingress of chlorides, sulphates and carbonation etc.) has a profound influence on the concrete durability. The details of the different permeation characteristics of the compositions in terms of the water absorption characteristics, permeable pore space, rate of capillary absorption and their initial and secondary sorptivity coefficients as well as the chloride permeation depths studied were all presented in Table 4.

Table 4. Permeation characteristics of UHPC concretes

Mix details	30 min absorption	72 h absorption	Permeable pore space	Capillary absorption	Initial rate of absorption	Secondary rate of absorption	Chloride Depth
units	(wt. %)	(wt. %)	(vol. %)	(ml/m^2/s)	(mm/$\sqrt{}$sec)	(mm/$\sqrt{}$sec)	(mm)
AS0	4.44	14.29	6.81	0.46	0.0225	0.0001	40
US0	2.94	9.04	4.34	0.30	0.0103	0.0012	7
SG25	1.38	5.97	2.54	0.10	0.0021	0.0005	3
SW25	1.49	5.84	3.25	0.07	0.0025	0.0007	4
SGF25	2.60	8.67	4.45	0.11	0.0033	0.0007	4

The capillary pore structure formed during the excess water escaping the matrix during compaction forming the connected capillary pore channels as altered by the hydration activity later allows the movement of fluids and environment. This will affect concrete alkalinity, remove the passivating layer on the reinforcement and lead to corrosion or deterioration of the concrete. According to CEB recommendations, a 30 min water absorption of concrete below 3% represents concrete of good quality [20]. The absorption characteristics, permeable pore space and the rate of capillary absorption appear to be inversely proportional to the compressive strength.

In fact, these silica fume concretes show a 30 min absorption below 3% with SG25 showing the lowest absorption. The permeable pore space (PPS) also follows a similar trend, with SG25 and SW25 which have the highest strength resulted in about 6% porosity. However, SGF25, even with an additional 0.01% fumed silica even with such a high specific surface showed a high permeation. Similarly, the total absorption (at 72 h) also appears to follow a similar trend.

The 30 min capillary absorption studied according to BS 1881 [21], resulted in values below the limit of 0.35 for concretes of poor quality, except for AS0 the ASTM plain mortar. All the UHPC compositions resulted in concretes of good quality with values below 0.17 [20]. The initial and secondary absorptions were also below 0.01 mm/√sec for all SF concretes, indicating a very low sorptivity for the UHPC mixes. Also, with the observed low porosity the chloride permeation depth observed in all the UHPC compositions was below 5 mm for UHPC even after the increased exposure period of 90 days compared to the 35 days specified by the test NT Build 443 [18].

3.3 XRD Analysis

The analysis of the concrete compositions was done by using XRD and the obtained data refinement was done through Rietveld analysis. The phase composition was established using ICDD database. The XRD patterns of different constituents of the concretes were given in Fig. 4. It can be seen that a broad peak at around a 2θ° of 22 to 24 degrees was present in silica fume-gray indicating the presence of amorphous silica, while there was no definite indication of such diffused peak in silica fume-white. Fumed silica also exhibited a broad peak similar to silica fume gray. May be this vary high reactivity of the fumed silica was instrumental in consuming the nascent calcium hydroxide in the cement hydration reaction which inhibited the later strength development resulting in the lower strength observed in SGF25 composition. This stifling effect can also be seen in the fact that there is no significant improvement in strength after 28 days, unlike an increase of about 20% in all other compositions.

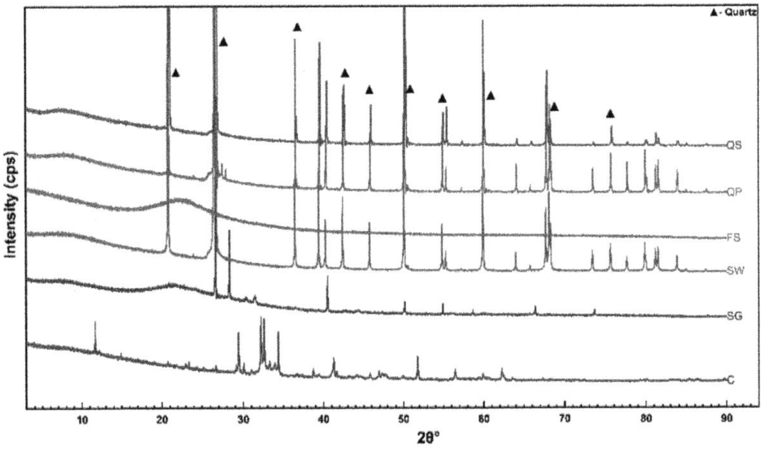

Fig. 4. XRD of the cementitious materials

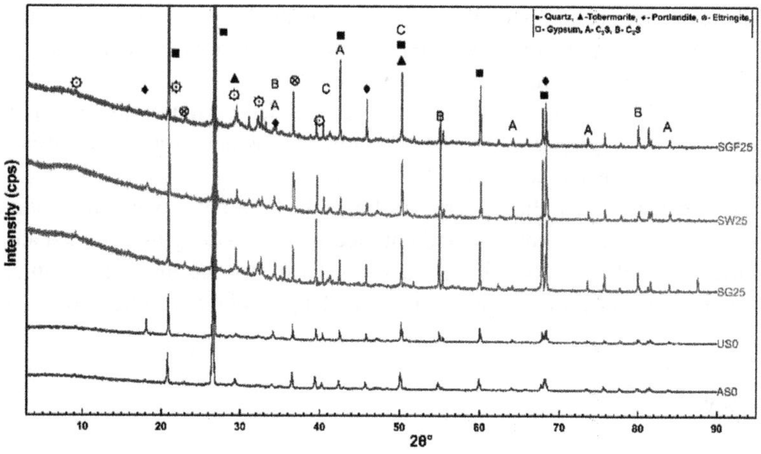

Fig. 5. XRD of the UHPC concrete with silica fume variants

The XRD patterns of the different fully cured concrete compositions at 90 days were also ascertained to understand the final composition characteristics and are presented in Fig. 5. After refinement, the hydration products of alite, belite, quartz, tobermorite, portlandite, ettringite and gypsum were identified in these UHPC and control concretes shown in Fig. 5. The evaluation of the individual peaks that represent the different phase compositions are not being attempted at this stage. However, it can be seen that all the XRD patterns of the composition containing the different types of silica fume appear to be very similar, though the individual peak strength may vary.

4 Conclusions

The present investigation was an attempt to look at the viability of using three different varieties of locally available silica fumes to arrive at strengths reaching up to the UHPC levels while having high performance capabilities expected. The following are the major conclusions in this regard.

- The effectiveness of the different locally available silica fume variants in the development of UHPC concrete results in concretes of strengths ranging from around 115 to 145 MPa.
- It was also seen that the addition of a very small quantity of 0.01% fumed silica to the composition resulting in 145 MPa with gray silica fume and lowering of the strength to 115 MPa, maybe due to the reduced flowability and the corresponding compaction problems associated with it.
- The composition containing silica fume-grey exhibited a better performance in terms of both the mechanical strengths and permeation characteristics among all variants, followed by silica fume-white and the composition containing fumed silica in line with the compressive strengths.
- However, the ultrasonic pulse velocities of more than 4.5 km/sec appear to indicate that all the compositions including the ASTM mortar mix are of a very good concrete quality.

- The different permeation and sorptivities established through several different test methods indicate that the resulting UHPC compositions are of a very good quality of concrete as per the CEB regulations.
- The depth of chloride ingress, even with the extended period of almost three times, was very small showing that the UHPC compositions have significant resistance to chloride penetration and are only below 4 mm.
- The XRD patterns also indicate very similar peaks for all the composition containing the different silica fume variants and coupled with the strengths obtained show that all of them exhibit good pozzolanic reactivity. However, the addition of fumed silica of very high specific surface appears to stifle the pozzolanic reaction considerably.

References

1. NF P 18-470. Concrete - Ultra-high performance fibre-reinforced concretes -Specifications, performance, production and conformity. AFNOR, La Plaine Saint-Denis (2016)
2. Davis, R.E.: A review of pozzolanic materials and their use in concretes. ASTM International, West Conehohocken (1950)
3. SFA homepage. https://www.silicafume.org/general-silicafume.html
4. Abbas, S., Soliman, A.M., Nehdi, M.L.: Exploring mechanical and durability properties of ultra-high performance concrete incorporating various steel fiber lengths and dosages. Constr. Build. Mater. **75**, 429–441 (2015)
5. Randl, N., Steiner, T., Ofner, S., Baumgartner, E., Mészöly, T.: Development of UHPC mixtures from an ecological point of view. Constr. Build. Mater. **67**, 373–378 (2014)
6. Smarzewski, P.: Mechanical properties of ultra-high performance concrete with partial utilization of waste foundry sand. Buildings **10**(1), 11 (2020)
7. Tafraoui, A., Escadeillas, G., Vidal, T.: Durability of the ultra high performances concrete containing metakaolin. Constr. Build. Mater. **112**, 980–987 (2016)
8. Sohail, M.G., Kahraman, R., Al Nuaimi, N., Gencturk, B., Alnahhal, W.: Durability characteristics of high and ultra-high performance concretes. J. Build. Eng. **33**, 101669 (2021)
9. Hasan, T.M., Gilbert, L., Allena, S., Owusu-Danquah, J., Torres, A.: Development of non-proprietary ultra-high performance concrete mixtures. Buildings **12**(11), 1865 (2022)
10. Alsalman, A., Dang, C.N., Hale, W.M.: Development of ultra-high performance concrete with locally available materials. Constr. Build. Mater. **133**, 135–145 (2017)
11. ASTM C 778. Standard specification for standard sand. American Society for Testing and Materials, USA (2006)
12. ASTM C 109. Standard test method for compressive strength of hydraulic cement mortars. American Society for Testing and Materials, USA (2008)
13. ASTM C 230. Standard Specification for Flow Table for Use in Tests of Hydraulic Cement. American Society for Testing and Materials, USA (2014)
14. Zhong, R., Wille, K., Viegas, R.: Material efficiency in the design of UHPC paste from a life cycle point of view. Constr. Build. Mater. **160**, 505–513 (2018)
15. ASTM C 597. Standard test method for pulse velocity through concrete. American Society for Testing and Materials, USA (2009)
16. ASTM C 642. Standard test method for density, absorption, and voids in hardened concrete. American Society for Testing and Materials, USA (1997)
17. ASTM C 1585. Standard test method for measurement of rate of absorption of water by hydraulic-cement concretes. American Society for Testing and Materials, USA (2004)

18. NT Build 443. Concrete hardened: accelerated chloride penetration. Nord test, Espoo, Finland (1995)
19. Lee, N.K., Koh, K.T., Kim, M.O., Ryu, G.S.: Uncovering the role of micro silica in hydration of ultra-high performance concrete (UHPC). Cem. Concr. Res. **104**, 68–79 (2018)
20. CEB. Diagnosis and Assessment of concrete Structures: State-of-Art Report. Bull.d' Information n.192 (1989)
21. BS 1881-208. Testing concrete - Recommendations for the determination of the initial surface absorption of concrete, BSI. London (1996)

Influence of Magneto-Rheological Response on the Early Hydration and Microstructure of Cement Paste

Feng Hu and Dengwu Jiao(✉)

Department of Architecture and Civil Engineering, City University of Hong Kong, Kowloon Tong, Hong Kong SAR, China
dengwu.jiao@cityu.edu.hk

Abstract. The rheological properties of concrete are crucial for its workability and mechanical performance, but the requirements may conflict at various stages of construction. For example, 3D concrete printing demands low yield stress and viscosity for pumping and extruding, while requiring higher static yield stress post-extrusion to prevent collapse. Therefore, it is necessary to regulate the rheological properties to balance these conflicting requirements. Magneto-rheological control is an innovative approach that involves adding magnetic particles to cementitious materials and applying an external magnetic field. This approach offers new possibilities for the active regulation of the rheological properties of cementitious materials. However, the effects of magnetic fields and magnetic particle additions on early microstructure and hydration behavior are not fully understood. To address this gap, this study investigates the impact of magneto-rheological response on the early hydration and microstructure in cement paste. The results show that the magnetic field induces a stronger magneto-rheological response with higher water to cement ratios and increased particle content. While it does not alter early hydration products, it negatively affects early microstructural compactness by creating weak zones between clustered magnetic particles. These findings contribute to a better understanding of magneto-rheological control and its potential in construction applications.

Keywords: Active rheology control (ARC) · Early hydration · Microstructure · Cement paste · Magnetic field

1 Introduction

The rheological properties of concrete are crucial for its workability, placement, and mechanical performance. In 3D concrete printing, for instance, low yield stress ensures smooth extrusion, while high static yield stress enhances buildability. Effective control of these properties is essential to balance conflicting demands across various construction stages [1, 2]. Traditional rheology control methods, such as optimizing mix proportions and adding chemical admixtures, are limited to the mixing phase and cannot actively adjust the rheology of post-mixing concrete. In contrast, active rheology control (ARC) enables real-time adjustment, including magnetic fields, vibration, microwaves, and temperature [3].

M. Posani and S. Singh (Eds.): RYS 2025, RILEM Bookseries 67, pp. 718–724, 2026.
https://doi.org/10.1007/978-3-032-14652-6_51

Among these, magneto-rheological control has gained significant attention due to its reversibility and responsiveness [4, 5]. By incorporating magnetic particles (e.g., Fe_3O_4, carbonyl iron), cement paste exhibits tunable stiffness under a magnetic field, as particle aggregation enhances structural build-up. Pulsed magnetic fields can also induce a micro-stirring effect, which momentarily reduces stiffness. Beyond rheology, magnetic particles influence hydration and microstructure evolution. For example, nano-Fe_3O_4 can act as nucleation sites, accelerating hydration and improving early strength [6, 7]. Additionally, magnetic fields alone can also influence hydration kinetics, microstructure, and strength development, with studies showing their role in refining C-S-H gel and enhancing early-age strength [8].

While magneto-rheological control presents a promising approach for rheological regulation, challenges remain in understanding the synergistic effects of magnetic particles and external fields on hydration and microstructure in cement paste. This study investigates the influence of magneto-rheological response on the early hydration and microstructure of cement paste incorporating Fe_3O_4 particles. The findings contribute to advancing ARC for applications such as pumping, formwork casting, and 3D concrete printing.

2 Materials and Methods

2.1 Materials

The cement paste was prepared using CEM I 42.5 N ordinary Portland cement, micron-sized Fe_3O_4 particles, and deionized water. The chemical composition of the cement is provided in Table 1. The Fe_3O_4 particles have an average size of 35 μm, with a density of 5.18 g/cm^3. To achieve appropriate flowability, water-to-cement (w/c) ratios of 0.4 and 0.5 were selected, and magnetic particles were added at mass fractions of 5% and 10% based on the total weight of water and cement.

Table 1. Main chemical composition of cement (in wt%).

	CaO	SiO_2	Al_2O_3	Fe_2O_3	MgO	SO_3	K_2O	P_2O_5
Cement	64.10	18.70	4.42	4.86	3.33	3.19	0.76	0.11

2.2 Testing Methods

The magneto-rheological response was evaluated through fluidity tests conducted under different magnetic field conditions. Fresh cement paste was poured into an aluminum mold, and after excess paste was removed, the mold was lifted vertically to allow free spreading. The spread diameters were measured in two perpendicular directions, averaged, and recorded at 15-min intervals for one hour after mixing. For magnetic field exposure, a square-shaped NdFeB magnet (20 mm thick) was positioned horizontally beneath the plate, while the reference group was tested without a magnetic field. To analyze the early hydration behavior, fresh cement paste was cast into plastic molds and placed in a container with or without a magnetic field. A magnetic field was generated by two 150 mm × 50 mm × 20 mm NdFeB magnets positioned 50 mm apart. After exposure, hydration was rapidly halted by immersion in liquid nitrogen (>99% purity) for 5 min, followed by vacuum drying at 20 ± 2 °C for 6 h. The dried samples were ground to < 75 μm powder and scanned via X-ray diffraction (XRD) over 5–70° (Cu target) at 5°/min. To further examine the microstructure of the cement paste and the clustering behavior of magnetic particles, scanning electron microscopy (SEM) was employed. In addition, early compressive strength of the samples was measured at 1-day to assess the effect of magnetic field exposure on early strength.

3 Results and Discussion

3.1 Time-Dependent Flowability Behavior

The flowability behavior of cement paste over time is shown in Fig. 1. All mixtures exhibit a consistent decrease in fluidity over time, primarily due to early hydration. In the absence of a magnetic field, the addition of magnetic particles reduces the fluidity of the paste compared to the reference groups, with a more pronounced reduction as the particle content increases. This is because the additional magnetic particles consume free water to coat themselves, thereby reducing the thickness of the water film between particles [9]. The reference groups (0.4_R and 0.5_R) without magnetic particles show minimal fluidity differences under different magnetic field conditions, indicating an insignificant magneto-rheological response. However, pastes containing magnetic particles display significantly lower fluidity under magnetic field (MF), demonstrating a sustained magneto-rheological effect. This is due to magnetic particles becoming magnetized under the magnetic field, forming dipoles that align into chain-like or clustered structures, which increase the paste's stiffness [10]. Moreover, this effect is more pronounced with higher magnetic particle content.

For the mixture with w/c of 0.5 and 5%MP, the fluidity reductions under MF compared to NMF are 7.85%, 8.35%, 14.77%, and 12.65% at 15, 30, 45, and 60 min, respectively. For the mixture with w/c of 0.5 and 10%MP, the reductions are 27.75%, 26.75%, 25.32%, and 29.37% at the same intervals. On the other hand, reducing the w/c ratio decreases the magneto-rheological response. Specifically, as shown in Fig. 1, at w/c of 0.4 and 10%MP, the fluidity reductions are 18.18%, 17.67%, 21.21%, and 19.91% at 15, 30, 45, and 60 min, respectively, which are significantly lower than those observed for w/c = 0.5 under the same magnetic field conditions. This is because low-w/c paste exerts higher viscous resistance on the magnetic particles, making it more difficult to move to form chain-like or clustered structures [11].

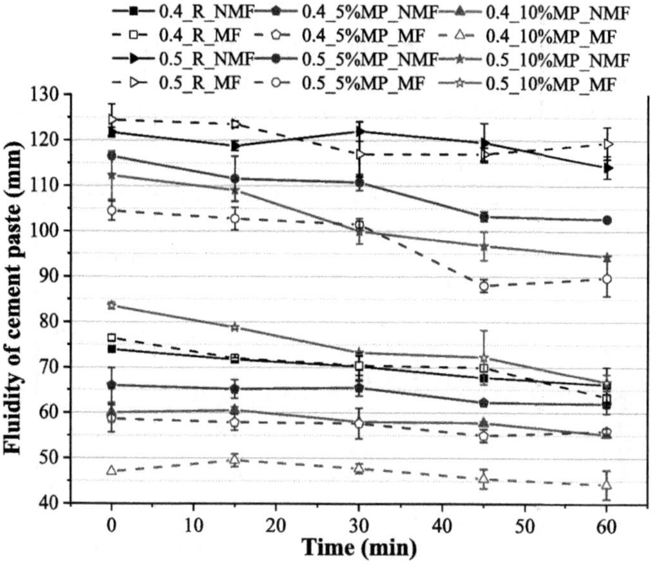

Fig. 1. Effect of magnetic field on the time-dependent fluidity of cement pastes.

3.2 Early Hydration Behavior and Microstructure

The XRD results of cement paste under varying magnetic field conditions are presented in Fig. 2. The incorporation of magnetic particles does not significantly alter the primary hydration products of the cement paste, which include unhydrated tricalcium silicate (C_3S), dicalcium silicate (C_2S), portlandite, ettringite, and calcite [12]. Notably, the addition of magnetic particles introduces a distinct peak for magnetite at approximately 35.5°, confirming their well distribution within the paste. Ettringite is a key component of early hydration products, but rapid freezing with liquid nitrogen may remove its bound water, affecting its microstructure and resulting in weak peaks in the XRD results. Apart from this, the primary early hydration products show no significant differences from the reference group. Previous studies have shown that nano-Fe_3O_4 can partially replace Al to form Fe-ettringite [6], but this process is difficult to achieve in early hydration, explaining the absence of Fe-ettringite peaks in the XRD results. Furthermore, the application of a magnetic field does not lead to the formation of new crystalline phases, as no significant changes in the types of primary hydration products are observed.

Although the magneto-rheological response does not alter the primary hydration products of cement paste, its impact on the early microstructure can't be overlooked. Figure 3 illustrates the microstructural morphology of cement paste with w/c of 0.45 and 5%MP after 30 min of magnetic field exposure. Under the influence of the magnetic field, these particles tend to form chain-like or clustered structures. However, their irregular shapes lead to the formation of weak zones between particles, where early hydration products fail to effectively fill the gaps. This suggests that the magneto-rheological response can negatively affect the compactness of the microstructure, which is reflected in the 1-day compressive strength of the pastes with 5%MP. The pastes without magnetic field (NMF) exhibits a strength of 17.1 ± 0.2 MPa, whereas the samples exposed to a magnetic field show a reduction in strength after 60 min of exposure (Table 2). Moreover, the directional orientation of magnetic particles leads to anisotropic strength development, with a more pronounced decline in the perpendicular direction observed after 30 min of magnetic field exposure, which further intensifies at 60 min.

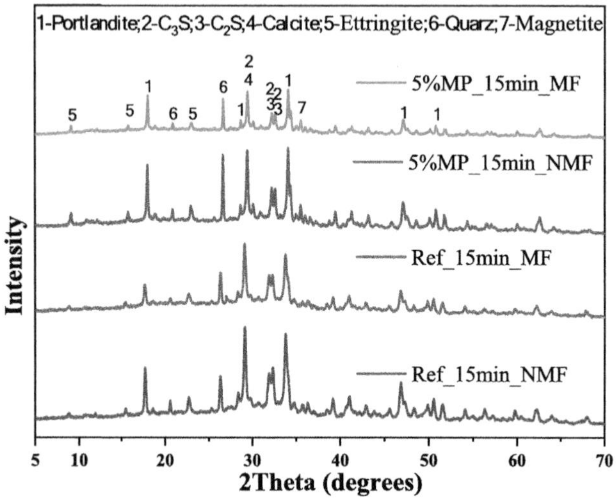

Fig. 2. XRD patterns of cement paste under different magnetic field conditions (w/c = 0.45).

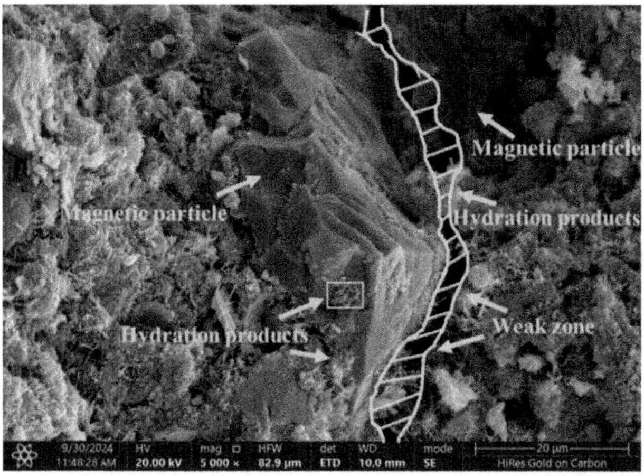

Fig. 3. Morphology of cement paste after 30min magnetic field exposure.

Table 2. 1-day compressive strength of cement paste with 5%MP under different magnetic field exposure durations and measured along different directions (w/c = 0.45).

Exposure time (min)	MF_Parallel	MF_Perpendicular
30	17.4 ± 0.35	16.4 ± 0.61
60	17.0 ± 0.14	15.9 ± 0.49

4 Conclusions

The effects of magneto-rheological response on early hydration and microstructure in cement paste containing magnetic particles were studied. The results showed that the magneto-rheological response is more pronounced at lower-stiffness paste (i.e., paste with higher water content) and higher magnetic particle content. The magneto-rheological response does not alter the primary hydration products of the cement paste. However, it induces weak zones within the paste, compromising the compactness of the early microstructure and leading to anisotropic behavior in early strength.

Acknowledgement. Financial supports from National Natural Science Foundation of China (52408295), Guangdong Basic and Applied Basic Research Foundation (2025A1515012809), Shenzhen Science and Technology Program (JCYJ20230807115000001), and Early Career Scheme (ECS) from Research Grants Council of Hong Kong SAR (21211024) are greatly appreciated.

References

1. Rahul, A.V., Mohan, M.K., De Schutter, G., Van Tittelboom, K.: 3D printable concrete with natural and recycled coarse aggregates: rheological, mechanical and shrinkage behaviour. Cem. Concr. Compos. **125**, 104311 (2022)
2. Kanagasuntharam, S., Ramakrishnan, S., Muthukrishnan, S., Sanjayan, J.: Effect of Magnetorheological additives on the buildability of 3D concrete printing. J. Build. Eng. **74**, 106814 (2023)
3. De Schutter, G., et al.: Active control of properties of fresh and hardening concrete. RILEM Techn. Lett. **8**, 198–203 (2023)
4. Jiao, D.: Active rheology control of cementitious materials using magnetic field. Ghent University (2021)
5. Chibulu, C., Jiao, D., Yardimci, M., De Schutter, G.: Magneto-rheology control of cement paste containing Fe3O4 nanoparticles in view of reducing or preventing formwork leakage. Cem. Concr. Compos. **142**, 105176 (2023)
6. Amin, M., El-Gamal, S., Hashem, F.: Effect of addition of nano-magnetite on the hydration characteristics of hardened Portland cement and high slag cement pastes. J. Therm. Anal. Calorim. **112**, 1253–1259 (2013)
7. He, Y., Lu, L., Sun, K., Wang, F., Hu, S.: Electromagnetic wave absorbing cement-based composite using Nano-Fe3O4 magnetic fluid as absorber. Cem. Concr. Compos. **92**, 1–6 (2018)

8. Soto-Bernal, J.J., Gonzalez-Mota, R., Rosales-Candelas, I., Ortiz-Lozano, J.A.: Effects of static magnetic fields on the physical, mechanical, and microstructural properties of cement pastes. Adv. Mater. Sci. Eng. **2015**(1), 934195 (2015)
9. Jiao, D., Lesage, K., Yardimci, M.Y., El Cheikh, K., Shi, C., De Schutter, G.: Structural evolution of cement paste with nano-Fe3O4 under magnetic field-Effect of concentration and particle size of nano-Fe3O4. Cem. Concr. Compos. **120**, 104036 (2021)
10. Jiao, D., Lesage, K., Yardimci, M.Y., Shi, C., De Schutter, G.: Flow behavior of cementitious-like suspension with nano-Fe3O4 particles under external magnetic field. Mater. Struct. **54**(6), 209 (2021)
11. Jiao, D., Lesage, K., Yardimci, M.Y., El Cheikh, K., Shi, C., De Schutter, G.: Rheological behavior of cement paste with nano-Fe3O4 under magnetic field: magneto-rheological responses and conceptual calculations. Cem. Concr. Compos. **120**, 104035 (2021)
12. Ylmén, R., Jäglid, U., Steenari, B.-M., Panas, I.: Early hydration and setting of Portland cement monitored by IR, SEM and Vicat techniques. Cem. Concr. Res. **39**(5), 433–439 (2009)

Micromechanical Modeling of Spatially Heterogeneous Interfacial Transition Zone (ITZ) in Recycled Aggregate Concrete

Ahmed O. Shittu[1](✉), Feng Li[2], Elske Linß[2], Florian Kleiner[3], Christiane Rößler[3], and Luise Göbel[1]

[1] Chair of Mechanics of Engineering Materials, F. A. Finger-Institute for Building Material Science and Institute for Structural Mechanics, Bauhaus-Universität Weimar, Weimar, Germany
ahmed.olanrewaju.shittu@uni-weimar.de

[2] Research Group "Sensor Technology for Products and Processes", Materials Research and Testing Institute Weimar, Weimar, Germany

[3] Chair of Building Materials, F. A. Finger-Institute for Building Material Science, Bauhaus-Universität Weimar, Weimar, Germany

Abstract. The use of recycled aggregates (RA) in concrete has gained increasing attention due to its potential to enhance sustainability in construction. However, the varying properties of different RAs can significantly impact concrete performance, both at the microstructural and macroscopic levels. A key factor influencing mechanical behavior of concrete is the interfacial transition zone (ITZ), where aggregates interact with cement paste. The ITZ plays a crucial role in determining concrete's strength, durability, and overall performance. Despite its importance, many existing models oversimplify the ITZ as a uniform layer, ignoring its actual variability.

This study examines the impact of spatially varying ITZ properties in concrete containing two different RAs, including abraded concrete aggregate (RCa) and recycled concrete aggregate (RCu). A micromechanical multiscale model is applied to represent the heterogeneous nature of concrete, spanning from the micrometer-scale of hydration products to the centimeter-scale of the bulk material. In addition to existing approaches, the present model incorporates the non-uniform characteristics of the ITZ.

Using insights from nanoindentation and SEM-EDX analyses, it was observed that ITZ thickness, porosity, and mechanical properties vary depending on the type of RA and its interaction with the new cement paste. These variations are integrated into the micromechanical model, resulting in a more accurate analytical representation of the elastic and compressive behavior of concrete with recycled aggregates. This approach also enhances understanding of how different RAs influence the ITZ structure and, consequently, the overall material performance.

Keywords: Recycled Aggregates · Mechanical Properties · Porosity · Multiscale Model · Micromechanics · Interfacial Transition Zone

M. Posani and S. Singh (Eds.): RYS 2025, RILEM Bookseries 67, pp. 725–738, 2026.
https://doi.org/10.1007/978-3-032-14652-6_52

1 Introduction

The construction industry's pursuit of sustainability has significantly propelled the use of recycled aggregates (RA) in concrete. While offering considerable environmental benefits, the incorporation of RA introduces complexities due to their inherent variability, which can distinctly influence concrete performance, with downstream effects at the microstructural level due to its interaction with cement paste. [1]. A pivotal element governing the mechanical integrity and durability of concrete is the interfacial transition zone (ITZ) which is a micro-region mediating the interaction between aggregates and the cement paste matrix [2, 3]. The characteristics of this zone are critical determinants of the material's strength, transport properties, and overall macroscopic behavior.

Despite its acknowledged importance, many conventional modeling approaches simplify the ITZ as a homogenous layer with uniform properties throughout the concrete volume [e.g., with a thickness assumed to be 10μm]. This simplification often fails to capture the nuanced reality, especially in RA concrete where the ITZ's nature can be profoundly affected by the type of recycled material (e.g., porous recycled brick versus denser recycled concrete aggregate) and the pre-existing conditions of the RA itself, such as adhered old mortar [4, 5]. Consequently, the ITZ properties like thickness, porosity, and local stiffness are not constant but exhibit significant spatial variability, differing not only between RA types but also potentially around a single RA particle.

This study addresses this critical gap by investigating the implications of considering spatially varying ITZ properties in concrete incorporating two different recycled aggregate types, namely abraded concrete aggregate (RCa) and recycled concrete aggregate (RCu). An advanced micromechanical multiscale model is employed, capable of bridging observations from the micrometer scale of cement hydration products to the centimeter scale of the concrete composite [6, 7]. Unlike models assuming a uniform ITZ, the present approach explicitly accounts for its heterogeneous nature. Drawing insights from experimental characterizations like nanoindentation and SEM-EDX analysis, which reveal that ITZ features are influenced by RA type and the interplay between old and new cementitious phases [8–11], these variations are integrated into the model.

Various homogenization schemes in the context of micromechanical modeling, such as the Mori-Tanaka scheme [12], have been employed to estimate the effective properties of concrete [4, 6]. However, these models face challenges in accurately representing the graded nature of the ITZ. The ITZ exhibits non-uniform gradients in porosity and stiffness, with its Young's modulus often reported to be 20% to 80% of the bulk cement paste [13–16]. Other modeling concepts such as hard core-soft shell formulation [14] or power-law distributions for ITZ porosity [13] have attempted to address this complexity. However, many models still rely on assumed ITZ properties, which may not capture the actual variations, especially when incorporating RAs.

The primary objective is to develop a sophisticated computational representation for predicting both the elastic stiffness and compressive strength of RA concrete. This refined modeling approach aims to provide enhanced insights into how different RA types interact with the surrounding cement paste, modify the ITZ structure, and ultimately dictate the composite's performance. Such understanding is paramount for fostering more reliable and efficient utilization of recycled materials in sustainable concrete production, moving beyond empirical design towards a fundamentally grounded predictive capability.

2 Micromechanical Framework for Recycled Aggregate Concrete

2.1 Hierarchical Material Representation and Scale Transitions

Concrete is recognized as a heterogeneous material with a complex microstructure evolving across multiple length scales, from nanometers (e.g., C-S-H gel) to decimeters (e.g., coarse aggregates) [6]. Its primary constituents include aggregates (natural and/or recycled), cement paste, and the ITZs enveloping the aggregates. To model this complexity, a continuum micromechanics approach is adopted, treating different phases as quasi-homogeneous subdomains with effective material properties.

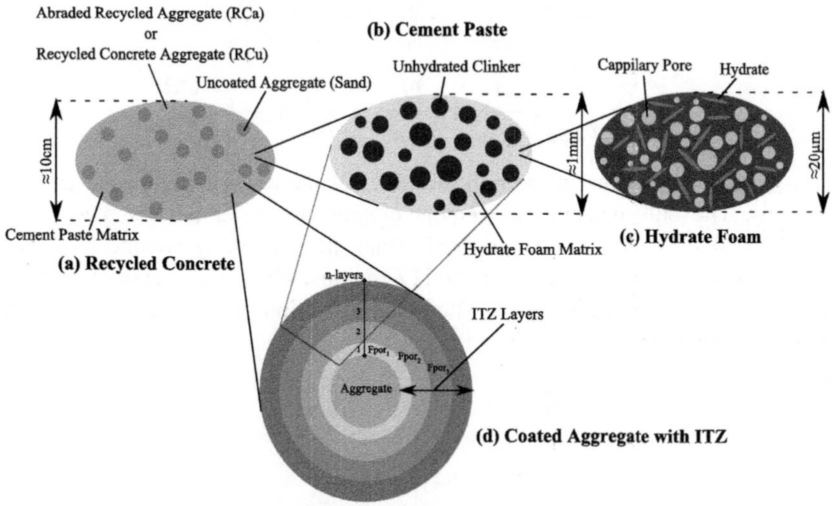

Fig. 1. Multiscale representation of recycled concrete and its constituent phases: (a) Recycled Concrete, (b) Cement Paste, (c) Hydrate Foam, (d) Coated Aggregate with multi-layered ITZ.). The 2D sketch refers to 3D volume elements.

This facilitates a detailed analysis of recycled concrete, which comprises various aggregate sizes and types (including RA with their specific micro (stress) field interactions), embedded within a cement paste matrix. At the macroscopic level (Fig. 1a), recycled concrete is envisioned as a composite of (recycled) aggregates which are coated by a multi-layered ITZ, and uncoated sand particles, all embedded in a bulk cement paste matrix. The cement paste itself is a hierarchical material (Fig. 1b, 1c), comprising unhydrated clinker grains within a hydrate foam matrix. This matrix, in turn, consists of capillary pores and hydrate gel, which is distinctively modeled as infinitely long, needle-shaped inclusions in contrast to the other quasi-spherical phases, as detailed in [7, 17].

2.2 Characterization-Driven Phase Definition and Properties

The predictive power of any micromechanical model hinges critically on an accurate and as-realistic-as-possible representation of its constituent phases. This involves not only identifying the distinct materials present (e.g., aggregates, cement paste, interfacial zones) and their shapes but also precisely quantifying their volumetric contributions and intrinsic mechanical properties. These elements collectively form the fundamental input definition and parameterization for the model.

Bulk Material Phases (Aggregates and Cement Paste): At the concrete scale, the total aggregate content is determined by the mix proportions, see Table 1. Given the recycled aggregate-to-cement mass ratio (ra/c) and sand-to-cement ratio (s/c) in the mix, the volume fractions of coarse recycled aggregate and sand can be estimated by converting these mass ratios using the respective bulk densities. The volume fractions of the aggregates can thus be estimated as:

$$f_{ra}^{conc} = \frac{{}^{ra}\!/_{c}}{\rho_{ra}}\left(\frac{1}{\rho_{clin}} + \frac{{}^{w}\!/_{c}}{\rho_{water}} + \frac{{}^{ra}\!/_{c}}{\rho_{ra}} + \frac{{}^{s}\!/_{c}}{\rho_{sand}}\right)^{-1}, \tag{1}$$

$$f_{sand}^{conc} = \frac{{}^{s}\!/_{c}}{\rho_{sand}}\left(\frac{1}{\rho_{clin}} + \frac{{}^{w}\!/_{c}}{\rho_{water}} + \frac{{}^{ra}\!/_{c}}{\rho_{ra}} + \frac{{}^{s}\!/_{c}}{\rho_{sand}}\right)^{-1} \tag{2}$$

where ρ_{ra} and ρ_{sand} represent the bulk densities of the recycled aggregate and sand, respectively. The bulk densities of cement clinker and water are given as $\rho_{clin} = 3.15g/cm^3$ and $\rho_{water} = 1.00g/cm^3$ [18]. Then, the remaining volume in the RVE is occupied by cement paste (including the ITZ regions, which are essentially part of the paste phase surrounding aggregates). The volume fraction of cement paste (including ITZ) can be written as:

$$f_{cp}^{conc} = 1 - f_{ra}^{conc} - f_{sand}^{conc} \tag{3}$$

Table 1. Mix design parameters

Mass Ratio	Value (–)
Sand to Cement	3.52
Water to Cement	0.55
RCa to Cement	1.37
RCu to Cement	1.29

At the level of the cement paste (the "matrix" phase of the concrete), the internal composition evolves with hydration, which relates the time-dependent amounts of anhydrous and hydrated phases in cement paste [1]. Herein, the Powers' model [19] is employed to estimate the volume fractions of unhydrated clinker, hydration products, and pores in the bulk paste based on the initial mix parameters (w/c ratio) and the degree of hydration (ξ). The volume fractions are determined according to:

$$f_{hyd}^{cp} = \frac{43.15\xi}{20 + 63w/c} \geq 0, \tag{4}$$

$$f_{clin}^{cp} = \frac{20(1 - \xi)}{20 + 63w/c} \geq 0. \tag{5}$$

Within the hydrate foam phase, the volume fractions of hydrate gel f_{hyd}^{hf} and capillary pores f_{pore}^{hf} are computed as:

$$f_{pore}^{hf} = \frac{63(w/c - 0.367\xi)}{20\xi + 63w/c} \geq 0, f_{hyd}^{hf} = 1 - f_{pore}^{hf} \tag{6}$$

Modeling the Spatially Graded ITZ: With the known aggregates volume fraction the volume fraction of the ITZ phase can be determined. In this work, the statistical geometric model of Lu and Torquato [20] is employed to account for the overlap of ITZ layers in a random aggregate packing. This model expresses the ITZ volume fraction (f_{ITZ}^{conc}) as a function of the aggregate volume fraction (V_a), the average ITZ thickness (t_{ITZ}), the number density of aggregate particles (N_V), which is related to particle size distribution, and c, d, g aggregate size gradation parameters (e.g. coefficients for Fuller or equal volume fraction distributions [21].

$$f_{ITZ}^{conc} = 1 - V_a - (1 - V_a)\exp\Big[-\pi N_V\Big(ct_{ITZ} + dt_{ITZ}^2 + gt_{ITZ}^3\Big)\Big]. \tag{7}$$

To quantify the inhomogeneities within the ITZ, the porosity increase factor (F_{por}) proposed by Königsberger and Staquet's [4] is used. This factor accounts for the experimentally observed increase in ITZ porosity while ensuring that hydrate and clinker proportions remain consistent with those in the bulk paste. The clinker and hydrate foam volume fractions in the ITZ are given by:

$$f_{clin}^{ITZ-cp} = \frac{1 - f_{pore}^{hf} f_{hf}^{cp} F_{por}}{1 - f_{pore}^{hf} f_{hf}^{cp}} f_{clin}^{cp}, f_{hf}^{ITZ-cp} = 1 - f_{clin}^{ITZ-cp} \tag{8}$$

$$f_{pore}^{ITZ-hf} = \frac{1 - f_{pore}^{hf} f_{hf}^{cp} F_{por}}{f_{hf}^{ITZ-cp}}, f_{hyd}^{ITZ-hf} = 1 - f_{pore}^{ITZ-hf} \tag{9}$$

2.3 Elastic Properties of Material Phases

Each material phase defined in the current model is characterized by its elastic properties, (assumed linear elastic behavior for small strains, see Table 2). All phases are considered as isotropic. The elastic stiffness tensor $\mathbb{C}_i$ is calculated by combining the elastic bulk modulus (k_i) and shear modulus (μ_i) of the respective material phase i. The equation is as follows:

$$\mathbb{C}_i = 3k_i\mathbb{I}^{vol} + 2\mu_i\mathbb{I}^{dev}, i = \{hyd, pore, clin, RCaorRCu, sand\}, \tag{10}$$

where $\mathbb{I}^{vol}$ and $\mathbb{I}^{dev}$ denote the volumetric and deviatoric parts of the symmetric fourth-order identity tensor [22].

Table 2. Stiffness parameters of different phases

Phases	Bulk Modulus k [GPa]	Shear Modulus μ [GPa]	Bulk Density ρ [gcm^{-3}]	Source
Clinker	116.58	53.81	3.15	[4]
Hydrates	18.69	11.76	2.07	[4]
Sand	35.35	29.91	2.24	[4]
RCa	21.58	18.48	2.28	Homogenization of normal concrete
RCu	21.58	18.48	2.27	Homogenization of normal concrete

3 Homogenization and Upscaling of Elasticity and Strength

3.1 Upscaling Elastic Stiffness of Recycled Aggregates Concrete

Based on the morphological representation of the material, the effective elastic properties can be derived using homogenization techniques within the framework of continuum micromechanics. Accordingly, a macroscopically homogeneous strain field is applied across the boundary of an RVE. The local response within the heterogeneous microstructure is related to the macroscopic (effective) strain through the strain localization tensor, denoted by $\mathbb{A}$.. Specifically, the local strain tensor ε(x) at a given point x within the RVE is linearly linked to the macroscopic strain tensor E through the relation:

$$\varepsilon(\mathbf{x}) = \mathbb{A}(\mathbf{x}) : \mathbf{E}. \tag{11}$$

To establish a link between microstructural behavior and the macroscopic response of concrete, the effective strain and stress tensors are defined as volume averages of their respective local fields within the RVE. The effective (macroscopic) strain tensor **E** is expressed as:

$$\mathbf{E} = \frac{1}{V}\int_V \varepsilon(\mathbf{x})d\mathrm{V}. \tag{12}$$

Similarly, the macroscopic stress tensor $\mathbf{\Sigma}$ is given by the volume average of the local stress field $\sigma(\mathbf{x})$

$$\mathbf{\Sigma} = \frac{1}{V}\int_V \sigma(\mathbf{x})d\mathrm{V}. \tag{13}$$

At each point within the heterogeneous domain, a linear-elastic constitutive relationship governs the local stress–strain behavior, defined as:

$$\sigma(\mathbf{x}) = \mathbb{C}(\mathbf{x}) : \varepsilon(\mathbf{x}), \text{ and macroscopically as } \mathbf{\Sigma} = \mathbb{C}_{hom} : \mathbf{E}, \tag{14}$$

where $\mathbb{C}_{hom}$ denotes the homogenized elasticity tensor. For that, each material phase (both coated and uncoated aggregate) is considered embedded in an infinite medium of homogenized bulk cement paste. The overall stiffness tensor for recycled concrete ($\mathbb{C}_{Rcon}$) is then expressed as:

$$\begin{aligned} \mathbb{C}_{Rcon} &= \sum_i f_i \mathbb{C}_i : \mathbb{A}_{\infty,i} : \left(\sum_i f_i \mathbb{A}_{\infty,i} \right)^{-1} \\ \forall i &\in \{sand, RA, ITZlayer1, 2, \ldots.., n\}. \end{aligned} \tag{15}$$

where $\mathbb{C}_i$ denotes the fourth-order stiffness matrix of the ith phase, f_i is the equivalent weighted volume fraction, $\mathbb{A}_{\infty,i}$ is the strain concentration tensor for each phase, and n is the number of concentric ITZ layers.

The strain concentration tensor for uncoated aggregates can be determined using the Mori–Tanaka scheme:

$$\mathbb{A}_{\infty,i} = \left[\mathbb{I} + \mathbb{P}^0_{sph} : \left(\mathbb{C}_i - \mathbb{C}^0 \right) \right]^{-1} \forall i \in \{sand\} \tag{16}$$

where $\mathbb{P}^0_{sph}$ represents the fourth-order Hill tensor for spherical inclusions, and $\mathbb{C}^0$ is the stiffness matrix of the surrounding matrix phase.

For coated aggregates, the strain concentration tensor is more complex due to the variation in stiffness across multiple ITZ layers. The Herve and Zaoui model [23] is applied to determine the strain concentration tensor at each ITZ, accounting for the spatial variations. Detailed formulation can be found in [24].

3.2 Compressive Strength Prediction

To estimate the compressive strength, the multiscale model is further exploited by introducing a failure criterion, as previously reported [7, 17]. Accordingly, macroscopic failure is initiated when a critical stress state is reached in the most vulnerable microstructural component of the concrete. Local stresses in each phase are then estimated by applying a macroscopic uniaxial stress and using homogenization framework in reverse downscaling of the compressive loading to the hydrates within the ITZ phase [4]. The failure-inducing hydrate stress invariants for hydrates located in one of the ITZ layers are defined separately for the volumetric and the deviatoric parts of the microscopic stresses. It is formulated for the hydrate phase oriented in (φ,ϑ)-direction, where φ is the azimuthal angle and ϑ is the polar angle in a spherical coordinate system defining the orientation of the hydrate needles. The second order moment of the stresses are given as:

$$\begin{aligned} \overline{\overline{\sigma}}^{dev}_{hyd;\varphi,\vartheta} &= \sqrt{\frac{2\mu^2_{hyd}}{f_{hyd;\varphi,\vartheta}} \varepsilon_i : \frac{\partial \mathbb{C}_i}{\partial \mu_{hyd;\varphi,\vartheta}} : \varepsilon_i}, \\ \forall i &\in \{CP, ITZ\} \\ \overline{\overline{\sigma}}^{vol}_{hyd;\varphi,\vartheta} &= \sqrt{\frac{3k^2_{hyd}}{f_{hyd;\varphi,\vartheta}} \varepsilon_i : \frac{\partial \mathbb{C}_i}{\partial k_{hyd;\varphi,\vartheta}} : \varepsilon_i}, \end{aligned} \tag{17}$$

where $f_{hyd;\varphi,\vartheta}$ denotes the bulk cement paste or ITZ phase-specific volume fractions of the oriented hydrates $k_{hyd;\varphi,\vartheta}$ and $\mu_{hyd;\varphi,\vartheta}$ denote their bulk and shear moduli, respectively.

The hydrate needle failure is modeled by means of a Drucker–Prager criterion, given as:

$$\mathfrak{f}\left(\overline{\overline{\sigma}}^{vol}_{hyd}, \overline{\overline{\sigma}}^{dev}_{hyd}\right) = \frac{\overline{\overline{\sigma}}^{dev}_{hyd}}{\sqrt{2}} - k_{hyd} + \alpha_{hyd}\frac{\overline{\overline{\sigma}}^{vol}_{hyd}}{\sqrt{3}} \leq 0, \quad (18)$$

whereby k_{hyd} and α_{hyd} denote the Drucker–Prager constants of the hydrates. The values for the Drucker-Prager constants k_{hyd} and α_{hyd}, , as well as the assumed acicular (needle-like) geometry and orientation distribution of the hydrates, are adopted from the well-established model by Pichler et al. [7], which has been validated for cementitious systems. This representation is crucial as the needle-like shape effectively models the stress concentrations within the hydrate phase that precipitate failure under compressive load.

3.3 Experimental Determination of Model Input Parameters

The micromechanical model presented herein is validated against experimental data obtained from an associated laboratory test. This experimental procedure, detailed further in companion studies [10, 11], involved casting concrete specimens with RCa and RCu aggregates using the mix designs outlined in Table 1. For microstructural characterization, samples were cut from these specimens, embedded in epoxy, and polished to a fine finish. SEM-BSE was then used to acquire high-resolution images of the ITZ. Porosity along defined straight lines was quantified from these images using a grayscale segmentation technique in ImageJ, as described in [10, 11]. Macroscopic mechanical properties, specifically the Young's modulus and compressive strength, were determined at 2, 7, 14, 28, 56, and 91 days according to DIN EN 12390-13:2012.

4 Results and Discussion

4.1 ITZ Microstructure: Porosity Distribution

Figure 2, below show the distinctive porosity around the recycled aggregates relating the average value at each data point.

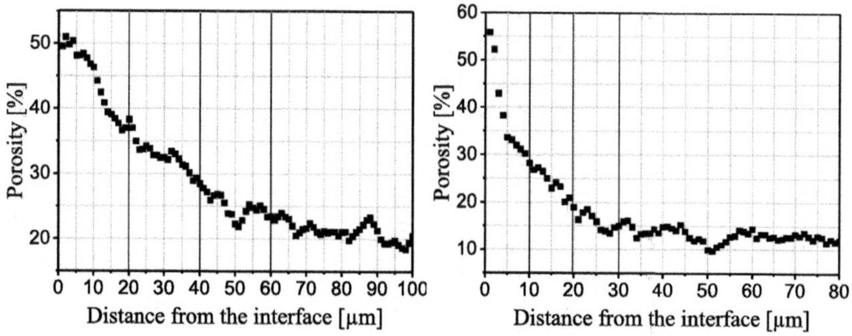

Fig. 2. Distribution of average porosity around recycled aggregate (a) RCu, (b) RCa.

Porosity decreases with increasing distance from the aggregate, reflecting the gradual variation within the ITZ before stabilizing in the bulk cement paste. The ITZ surrounding RCu aggregate extends up to 60 μm, while that of RCa aggregates is thinner and amounts to 30 μm. This difference arises from the presence of adhered mortar around RCu, which increases porosity, while abrasion of RCA reduces the adhered mortar and consequently the ITZ thickness. To model these variations, the ITZ was divided into three layers, with a 20 μm thickness for RCu and 10 μm for RCA. This porosity distribution is used to determine the F_{por} values given in the table below (Table 3).

Table 3. Value of porosity ratio at different ITZ layers

Aggregates	F_{por} (–)		
	Layer 1	Layer 2	Layer 3
RCu	1.6	1.4	1.1
RCa	1.4	1.2	1.1

4.2 Elastic Modulus and Compressive Strength Evolution

Finally, the model's predictions for elastic modulus and compressive strength are now compared against the experimental data. As described in Sect. 3.3, these tests were conducted at various ages, which were then related to their degrees of hydration using the loss of ignition method [25].

The comparison of the modeled stiffness evolutions with their experimental counterparts are summarized in Figs. 3. In addition, the model response of a similar model not accounting explicitly for the ITZ is included. Accordingly, the model predicts the stiffness evolution satisfactory well, though the significance of the ITZ is negligible in this case. This relatively minor influence on the overall elastic modulus can be attributed to stiffness being a rather volume-averaged property, which is less sensitive to the localized characteristics of the ITZ.

Quantitative comparison at a hydration degree of 0.92 indicates that the model including the ITZ provides slightly improved predictions for both aggregate types. The percentage differences between the modeled elastic modulus (with ITZ) and experimental data were only minor, about 1.5% for RCu and 1.2% for RCa (Fig. 4).

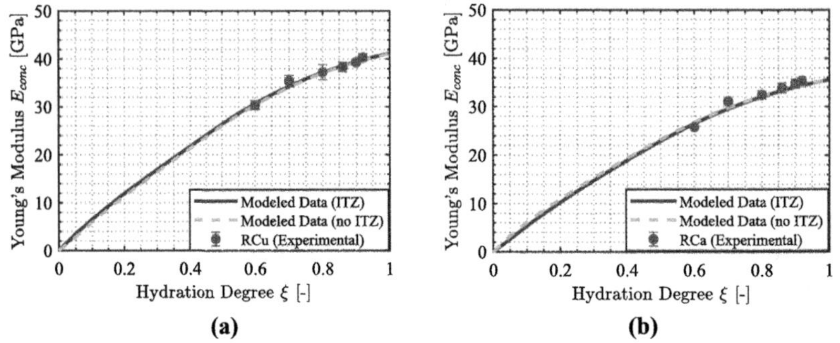

Fig. 3. Experimentally determined and modeled young's modulus of recycled concrete: (a) RCu, (b) RCa.

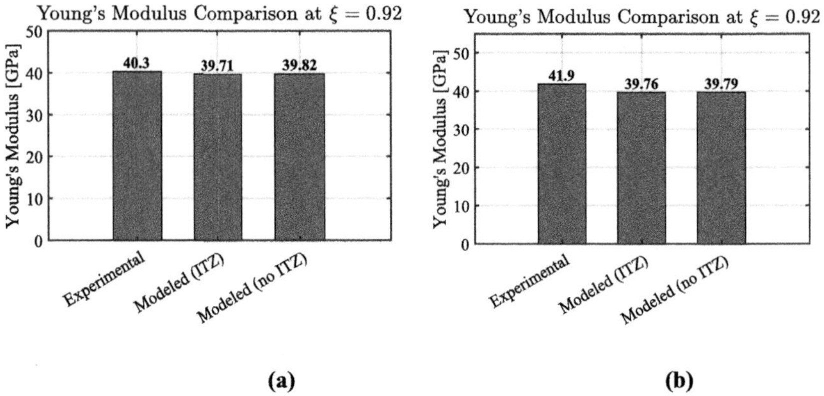

Fig. 4. Elastic modulus comparison of the model with and without ITZ with the experimental data: (a) RCu, (b) RCa.

Experimental data and the model responses regarding the evolution of compressive strength as a function of hydration degree are contrasted in Fig. 5. Again, a satisfactory agreement is observed. However, the influence of the ITZ is more pronounced for estimating the compressive strength than the elastic stiffness. For a given degree of hydration of 0.92, the model considering the spatially varying ITZ provided even improved estimates of compressive strength compared to the model neglecting the ITZ, especially for RCa. The percentage differences between the modeled strength (with ITZ) and experimental data amount to 5.1% for RCu and to 2.1% for RCa. In contrast, the model without considering an ITZ exhibits larger deviations when compared to experimental values. This highlights the critical role of accurately representing the heterogeneous ITZ, specifically its weaker, more porous layers, in predicting the ultimate compressive strength, which is often governed by failure initiation in these zones (Fig. 6).

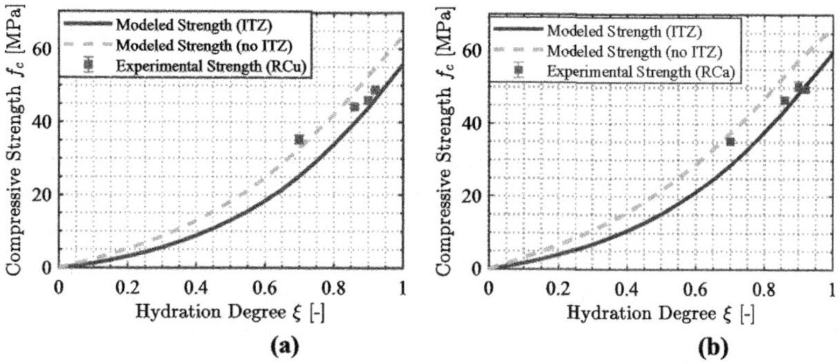

Fig. 5. Compressive strength validation of recycled concrete with experimental data: (a) RCu (b) RCa.

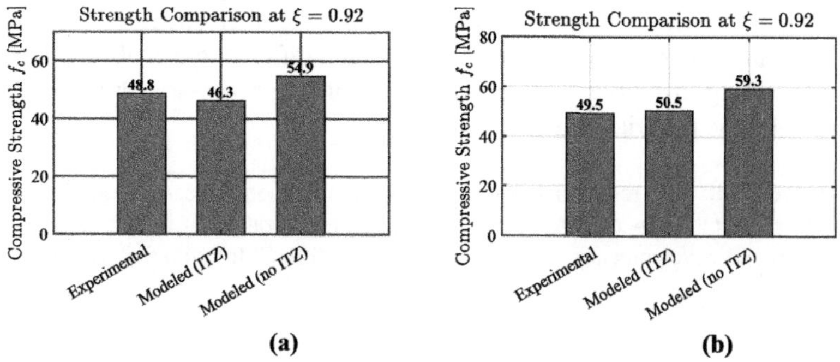

Fig. 6. Compressive strength comparison of the model with and without ITZ with the experimental data: (a) RCu (b) RCa.

4.3 Discussion on Model Assumptions and Sensitivity

It is important to acknowledge that the predictive accuracy of the model is dependent on several key input parameters and assumptions. The ITZ, for instance, was discretized into three layers herein. This choice was made to balance the accurate representation of the porosity gradient observed experimentally (Fig. 2) with computational efficiency. Using fewer layers might fail to capture the sharp decrease in porosity near the aggregate, potentially underestimating the ITZ's impact on strength. Conversely, more layers would increase computational cost without significantly altering the overall homogenized properties.

The elastic moduli for the recycled aggregates (RCa and RCu) were determined through homogenization of normal concrete. While this is a reasonable starting point, these values can vary depending on the source concrete. A higher or lower aggregate stiffness would proportionally shift the predicted composite Young's modulus up or down, though its influence on strength prediction, which is governed by failure in the weakest phase (ITZ hydrates), would be less direct. Current studies focus more intensively on the relevance of the considered model input parameters by means of probabilistic studies.

5 Conclusion

This study successfully applied a micromechanical multiscale model to investigate the impact of spatially heterogeneous Interfacial Transition Zones (ITZ) on the mechanical properties of recycled aggregate concrete. The experimentally determined differences of the ITZ thickness and porosity distribution around two different recycled aggregates types, namely recycled concrete aggregate (RCu) and abraded concrete aggregate (RCa), could be directly linked to the presence or removal of adhered mortar. By incorporating these spatially varying ITZ properties into the model, estimations for both elastic modulus and compressive strength evolutions were improved when compared to experimentally determined data. While the effect on the elastic stiffness can be expectedly considered to be relatively minor, the accurate representation of the ITZ's heterogeneous nature proved crucial for predicting compressive strength, demonstrating the ITZ as a critical factor governing the failure behavior of recycled aggregate concrete. This refined modeling approach provides enhanced insights into the microstructural influences on the macroscopic performance of concrete containing different types of recycled aggregates, contributing to a more fundamental understanding necessary for the sustainable utilization of these materials in construction.

Acknowledgement. This research is supported by the German Research Foundation (DFG) through the project Micro- and macro-mechanical characterization of contact zones in recycled concretes and development of a multiscale model (MiMaCMo) with the grant number GO 3270/4-1.

References

1. Göbel, L., Lahmer, T., Osburg, A.: Uncertainty analysis in multiscale modeling of concrete based on continuum micromechanics. Eur. J. Mech. A. Solids **65**, 14–29 (2017). https://doi.org/10.1016/j.euromechsol.2017.02.008
2. Scrivener, K.L., Crumbie, A.K., Laugesen, P.: The interfacial transition zone (ITZ) between cement paste and aggregate in concrete. Interface Sci. **12**(4), 411–421 (2004). https://doi.org/10.1023/B:INTS.0000042339.92990.4c
3. Maso, J.C.: Interfacial Transition Zone in Concrete. CRC Press, Boca Raton (1996).https://doi.org/10.1201/9781482271560
4. Königsberger, M., Staquet, S.: Micromechanical multiscale modeling of ITZ-driven failure of recycled concrete: effects of composition and maturity on the material strength. Appl. Sci. **8**(6), 976 (2018). https://doi.org/10.3390/app8060976
5. Djerbi, A.: Effect of recycled coarse aggregate on the new interfacial transition zone concrete. Constr. Build. Mater. **190**, 1023–1033 (2018). https://doi.org/10.1016/j.conbuildmat.2018.09.180

6. Bernard, O., Ulm, F.-J., Lemarchand, E.: A multiscale micromechanics-hydration model for the early-age elastic properties of cement-based materials. Cem. Concr. Res. **33**(9), 1293–1309 (2003). https://doi.org/10.1016/S0008-8846(03)00039-5
7. Pichler, B., Hellmich, C., Eberhardsteiner, J.: Spherical and acicular representation of hydrates in a micromechanical model for cement paste: prediction of early-age elasticity and strength. Acta Mech. **203**(3–4), 137–162 (2009). https://doi.org/10.1007/s00707-008-0007-9
8. Yang, C.C.: Effect of the transition zone on the elastic moduli of mortar. Cem. Concr. Res. **28**(5), 727–736 (1998). https://doi.org/10.1016/S0008-8846(98)00035-0
9. Medina, C., Zhu, W., Howind, T., Sánchez de Rojas, M.I., Frías, M.: Influence of interfacial transition zone on engineering properties of the concrete manufactured with recycled ceramic aggregate. J. Civ. Eng. Manag. **21**(1), 83–93 (2014). https://doi.org/10.3846/13923730.2013.802727
10. Li, F., Kleiner, F., Rößler, C., Göbel, L., Linß, E.: Micromechanical characterization of the interfacial transition zone in recycled brick concrete by means of combined nanoindentation and SEM, In: Proceedings of the 12th International Conference on Fracture Mechanics of Concrete and Concrete Structures (FraMCoS-12) (2025)
11. Li, F., Göbel, L., Forbrig, G., Rößler, C., Kleiner, F., Linß, E.: Microstructural and micromechanical study of the interfacial transition zone of recycled concrete by combined nanoindentation and SEM-EDX technique. In: 1st International Conference on Concrete and Circular Economy (CCE2025) (2025)
12. Mori, T., Tanaka, K.: Average stress in matrix and average elastic energy of materials with misfitting inclusions. Acta Metall. **21**(5), 571–574 (1973). https://doi.org/10.1016/0001-6160(73)90064-3
13. Lutz, M.P., Zimmerman, R.W.: Effect of an inhomogeneous interphase zone on the bulk modulus and conductivity of a particulate composite. Int. J. Solids Struct. **42**(2), 429–437 (2005). https://doi.org/10.1016/j.ijsolstr.2004.06.046
14. Garboczi, E.J., Schwartz, L.M., Bentz, D.P.: Modeling the influence of the interfacial zone on the DC electrical conductivity of mortar. Adv. Cem. Based Mater. **2**(5), 169–181 (1995). https://doi.org/10.1016/1065-7355(95)90001-2
15. Hashin, Z., Monteiro, P.: An inverse method to determine the elastic properties of the interphase between the aggregate and the cement paste. Cem. Concr. Res. **32**(8), 1291–1300 (2002). https://doi.org/10.1016/S0008-8846(02)00792-5
16. Ramesh, G., Sotelino, E.D., Chen, W.F.: Effect of transition zone on elastic moduli of concrete materials. Cem. Concr. Res. **26**(4), 611–622 (1996). https://doi.org/10.1016/0008-8846(96)00016-6
17. Pichler, B., Hellmich, C.: Upscaling quasi-brittle strength of cement paste and mortar: a multiscale engineering mechanics model. Cem. Concr. Res. **41**(5), 467–476 (2011). https://doi.org/10.1016/j.cemconres.2011.01.010
18. Göbel, L., Königsberger, M., Osburg, A., Pichler, B.: Viscoelastic behavior of polymer-modified cement pastes: insight from downscaling short-term macroscopic creep tests by means of multiscale modeling. Appl. Sci. **8**(4), 487 (2018). https://doi.org/10.3390/app8040487
19. Powers, T.C., Brownyard, T.L.: Studies of the physical properties of hardened portland cement paste. J. Proc. **43**(9), 549–602 (1947). https://doi.org/10.14359/15303
20. Lu, B., Torquato, S.: Nearest-surface distribution functions for polydispersed particle systems. Phys. Rev. A Atomic Molec. Opt. Phys. **45**(8), 5530–5544 (1992). https://doi.org/10.1103/PhysRevA.45.5530
21. Sun, G., Sun, W., Zhang, Y., Liu, Z.: Numerical calculation and influencing factors of the volume fraction of interfacial transition zone in concrete. Sci. China Technol. Sci. **55**(6), 1515–1522 (2012). https://doi.org/10.1007/s11431-011-4737-x

22. Königsberger, M., Hlobil, M., Delsaute, B., Staquet, S., Hellmich, C., Pichler, B.: Hydrate failure in ITZ governs concrete strength: a micro-to-macro validated engineering mechanics model. Cem. Concr. Res. **103**, 77–94 (2018). https://doi.org/10.1016/j.cemconres.2017.10.002
23. Herve, E., Zaoui, A.: Inclusion-based micromechanical modelling. Int. J. Eng. Sci. **31**(1), 1–10 (1993). https://doi.org/10.1016/0020-7225(93)90059-4
24. Nežerka, V., Zeman, J.: A micromechanics-based model for stiffness and strength estimation of cocciopesto mortars. Acta Polytechnica (2012)
25. Göbel, L., Bos, C., Schwaiger, R., Flohr, A., Osburg, A.: Micromechanics-based investigation of the elastic properties of polymer-modified cementitious materials using nanoindentation and semi-analytical modeling. Cem. Concr. Compos. **88**, 100–114 (2018). https://doi.org/10.1016/j.cemconcomp.2018.01.010

Performance of Magnesium-Silicate-Hydrate Engineered Cementitious Composites (MSH-ECC) at Sub-Elevated Temperatures (up to 200 °C)

Dhanendra Kumar[1,2](✉), Nabodyuti Das[2], Jitendra Patel[3], and En-Hua Yang[2]

[1] Indian Institute of Technology (IIT) Jammu, Jammu and Kashmir 181221, India
dhanendra.kumar@iitjammu.ac.in
[2] Nanyang Technological University (NTU), Singapore 639798, Singapore
[3] University of Bologna, via Terracini 28, 40131 Bologna, Italy

Abstract. Engineered Cementitious Composites (ECC) have shown better resistance at high temperatures than conventional fiber-reinforced concretes (FRC). The enhanced high-temperature performance results from a combination of the following features: presence of polymer fibers, lack of coarse aggregates, and high-thermal resistance cementitious matrix. Since the thermal stability of the cementitious matrix significantly affects the high-temperature performance, this study aims to study the influence of magnesium-silicate-hydrate (MSH) based cementitious matrix on high-temperature residual behavior. The MSH cementitious matrix utilizes reactive magnesia cement (MgO-based binder) and silica fume as a reactive silica source. This study experimentally investigates the mechanical performance of MSH-based ECC at sub-elevated temperatures (up to 200 °C). The highest investigated temperature was limited to 200 °C as the polymer fibers [polyethylene (PE)] melt around 160–170 °C. The mechanical performance was characterized under uniaxial tension (tensile strength and ductility) and compression under residual testing conditions. The study showed a linear reduction in residual compressive strength and first cracking strength of the MSH-ECC. The tensile ductility was completely lost after the melting of PE fiber above 150 °C, showing a remarkable drop in both tensile strength and strain capacity at 200 °C. The XRD characterization, however, showed no noticeable changes in the chemical composition of the cementitious matrix, and observed damages were primarily attributed to physical causes.

Keywords: MSH-ECC · SHCC · High temperature · Tensile ductility · Residual behavior

1 Introduction

Engineered Cementitious Composites (ECC) or Strain Hardening Cementitious Composites (SHCC) belong to a class of ductile fiber-reinforced concrete. ECC exhibits tensile ductility similar to metals when loaded under direct tension, in contrast to

M. Posani and S. Singh (Eds.): RYS 2025, RILEM Bookseries 67, pp. 739–750, 2026.
https://doi.org/10.1007/978-3-032-14652-6_53

brittle/quasi-brittle behavior in conventional concrete [1, 2]. The tensile strain capacity is generally greater than 1%, and strain capacities greater than 10% have also been achieved. The tensile strain-hardening in ECC is a pseudo-hardening behavior in the sense that instead of elongation in the composite, the saturated multiple cracks are bridged by randomly distributed polymeric microfibers, providing crack-bridging capacity until failure [3]. The tensile ductility in ECC is achieved through judicious tailoring of the cementitious matrix, fiber selection, and fiber-matrix interface, guided by micromechanics principles [4]. The strain-hardening behavior under tension imparts significant energy-absorbing capacity and damage tolerance in ECC, making them promising for seismic, blast, impact, and high-durability infrastructure [2, 5].

In recent years, investigations have also been conducted into the high-temperature performance of ECC [6–10]. The studies demonstrated significant improvement in the residual compressive strength over conventional concrete at high temperatures up to 600 °C [8, 9]. This enhancement is attributed to several key factors: the presence of polymer fibers that bridge microcracks and create open channels for vapor pressure release, the absence of coarse aggregates which minimizes thermal incompatibility at the aggregate-paste interface, and the use of a thermally stable cementitious matrix enriched with supplementary cementitious materials (SCMs) [9, 11, 12]. On the contrary, the notable tensile properties of ECC suffer significantly once the polymer fibers melt. Polymer fibers commonly used in ECC, such as polyethylene (PE) and polyvinyl alcohol (PVA), melt within the temperature range of 160–230 °C, depending on the fiber type. Once fibers are lost, the matrix becomes highly brittle and susceptible to brittle failure, with notable loss in both tensile strength and ductility. Similar to conventional concrete, the high-temperature performance of ECC depends on the mixture composition.

The ECC typically includes a high amount of SCMs to reduce the ordinary Portland cement (OPC) content due to the lack of filler materials such as coarse aggregates [13]. Sahmaran et al. [12] have demonstrated improved high-temperature residual performance in ECC with high fly ash content, which is primarily attributed to secondary hydration reactions forming more CSH at elevated temperatures (around 200 °C). However, Kumar et al. [14] later found that such enhancement is observed only at early ages (28 days) and later (150 days); such benefits are lost as no secondary hydration reactions happen when the composites mature. The slag and calcium carbonate-based ECC have shown significantly better residual compressive behavior up to 600 °C [9]. The improved response is due to the more thermally stable cementitious matrix comprising calcium carbonate-based filler replacing siliceous fillers, minimizing portlandite content, and forming refractory compounds. A detailed discussion on the high-temperature behavior of ECC is provided in [10].

Recently, Magnesium-Silicate-Hydrate (MSH) cementitious system-based ECC has been developed by the authors of this paper, which presents benefits in utilizing magnesium oxide (MgO) based sustainable binder and low fiber volume fraction [15–20]. The MSH binder system offers a reduced carbon footprint compared to OPC, primarily due to the lower calcination temperature required to produce reactive MgO (~700–1000 °C vs. ~ 1400–1450 °C for OPC) and its greater potential for CO_2 sequestration through carbonation [20]. The MSH cementitious system utilizes MgO powder and a reactive silica source such as silica fume. The system gains strength via hydration reaction, forming brucite [$Mg(OH)_2$] and MSH gel, the major binding materials [18, 19]. The details of the MSH cementitious system are given in the following references [20–23]. The MSH cementitious matrix is a low pH system, which makes it appealing for waste encapsulation and nuclear waste storage structures. On the other hand, low pH and highly brittle behavior of the MSH system compared to the OPC-based system could compromise the tensile properties and durability. Authors have recently developed various versions of MSH-ECC investigating the influence of various mixture constituents and their content on the mechanical performance at room temperatures [15, 16, 24–26]. A limited number of studies have also studied the durability of MSH-based conventional concrete [20, 27]. However, to the best of our knowledge, the influence of high temperatures on MSH-ECC has not been investigated yet.

The development of MSH cementitious system-based composite is still in the early stages, with studies exploring its performance under various mechanical and environmental loads. To build confidence and promote wider application, it is also necessary to investigate the behavior at high temperatures. This study thus aims to investigate MSH-ECC behavior at sub-elevated temperatures up to 200 °C. The present study focuses on the performance of MSH-ECC under sub-elevated temperatures (up to 200 °C), which are relevant in emerging applications like industrial facilities, tunnels, etc., and other environments where moderate thermal exposure is common. Also, the highest temperature was limited to 200 °C because the polyethylene fibers in the MSH-ECC melt around 160–165 °C, rendering the complete loss of tensile ductility. The study experimentally investigates the residual mechanical behavior under compression and uniaxial tension at room temperature (30 °C), 100 °C, 150 °C, and 200 °C. The influence of temperature on the chemical composition of the MSH matrix was also studied. The details of the experimental investigation and discussion of the results are presented in the following sections.

2 Materials and Methods

2.1 Materials

The cementitious binder used in this study consists of magnesium oxide (MgO) powder and silica fume (SF). The MgO was procured from Liaoning Metals and Minerals Enterprise Co. Ltd., China, while the silica fume was supplied by Elkem Pvt. Ltd. The chemical composition of the binder, determined through X-ray fluorescence (XRF) analysis, is presented in Table 1. The mixture composition of the MSH-ECC is provided in Table 2. Potable water was used in the mixture, and Sika ViscoCrete-2192 was incorporated as a superplasticizer (SP) to enhance the workability of the MSH-ECC. Polyethylene (PE) fiber with a length of 12 mm, a diameter of 23 μm, elastic modulus of 113 GPa, a nominal tensile strength of 3250 MPa, density of 960 kg/m^3 (specifications provided by the manufacturer), and melting point typically in the range of 160–165 °C, was used in the mixture. The specific gravity of the MgO and SF was 3.58 and 2.20, respectively. PE fibers were incorporated at a low volume fraction of 1.2% relative to the total volume of the composite.

Table 1. Chemical composition of MgO and silica fume (in wt. %).

Material	MgO	CaO	SiO_2	Al_2O_3	Fe_2O_3
MgO	89.9	2.8	5.1	0.7	1.0
SF	0.7	1.2	92.9	0.4	1.4

Table 2. Mixture composition (w.r.t. weight of MgO).

MgO	Silica Fume	Water	SP	PE fiber (% vol. Fraction)
1.0	1.0	0.80	0.05	1.2

2.2 Specimen Preparation

The MSH-ECC was prepared using a Hobart planetary mixer. The dry binder components MgO powder and SF were blended at low speed for 3 min. Subsequently, water and approximately 70–80% of the SP were added to the dry mixture, and mixing continued at low speed for 5–6 min until the desired consistency and flowability were achieved. Next, the mixing speed was increased to medium for one minute to ensure complete homogenization of the cementitious matrix. PE fibers were then gradually introduced while maintaining a low mixing speed to minimize the risk of fiber agglomeration. After adding all fibers, the mix was blended at medium speed for 2 min. The resulting mixture was manually inspected for fiber clumping. If clumps were observed, the mixer was operated at high speed for an additional 30 s to apply high shear and disperse the fibers effectively. The remaining SP was added to achieve the required workability and fiber distribution. Finally, the fresh MSH-ECC mix was cast into molds and covered for one day. After demolding, the specimens were cured in water for 56 days at 27 ± 2 °C. A longer curing duration was preferred to achieve a higher degree of hydration and stabilize the mechanical properties.

For each target temperature (room temperature/30 °C, 100 °C, 150 °C, and 200 °C), three cubes and four dogbone specimens were prepared to evaluate the uniaxial compressive strength and tensile properties (strength and ductility) of the MSH-ECC. The cubic specimens were 50 mm on each side. The dogbone specimens had a length of 350 mm, with tensile deformations recorded within a central gauge region of approximately 125 ± 3 mm. This gauge section featured a reduced cross-sectional area (width: 35 mm; thickness: 15 mm) to ensure that most cracks developed within the intended region during tensile testing.

2.3 Testing Procedures

The mechanical properties were characterized in residual conditions, i.e., the specimens were cooled down to room temperature after exposure to the target high temperature and then subjected to mechanical loading under room temperature conditions. The mechanical testing was done after 56 days of casting. All MSH-ECC specimens, except the control group [tested at room temperature (around 30 °C)], were subjected to thermal exposure in a temperature-controlled muffle furnace. The specimens were heated to the target temperatures of 100 °C, 150 °C, and 200 °C at a controlled rate of 10 °C per minute. Once the target temperature was reached, the specimens were maintained at that temperature for 1 h to ensure a steady-state thermal condition. As there are no standards for high-temperature residual characterization, the adopted heating protocol has been commonly used for the characterization of ECC high-temperature residual behavior [6, 8, 9]. Cooling was performed passively since the furnace had no active cooling system. Depending on the exposure temperature, the specimens were left to cool inside the furnace for 1 to 2 days. The cooling duration was higher in the case of 150 °C and 200 °C due to the higher thermal inertia of the system. The specimens were taken out of the furnace when the specimen temperature was below 40 °C to prevent any thermal shock to the specimens.

After high-temperature exposure, the weight loss was recorded for both cube and dogbone specimens. It was calculated as the percentage reduction in specimen weight before and after heating, reflecting moisture loss and thermal degradation caused by elevated temperatures.

The XRD characterization of the cementitious matrix after exposure to elevated temperatures was conducted using Bruker D8 Advance to determine the changes in the chemical composition. The parameters of X-ray characterizations were: 2θ scan step – 0.04°, 2θ scanning range – 5° to 80°, CuKα radiation of 1.54 Å wavelength, accelerating current – 10 mA, and voltage – 30 kV.

Compression testing was conducted in load-controlled mode on cube specimens at a 55 kN/min loading rate using a Recherches & Réalisations REMY testing machine (France). The thermal effects on the tensile properties of MSH-ECC were evaluated through uniaxial tensile tests on dogbone specimens. The tensile test was conducted using a 250 kN Instron 5569 universal testing machine (UTM). A constant loading rate of 0.5 mm/min was applied during the test. Tensile deformation over the gauge length was recorded using two linear variable displacement transducers (LVDTs) mounted on either side of the dogbone specimen. The average of the two LVDT readings was used to calculate strain for the uniaxial tensile stress-strain response of the composites.

3 Results and Discussion

3.1 Weight Loss

The weight loss with temperature in cube (compression) and dogbone (tension) specimens is shown in Fig. 1. The rate of weight loss was higher up to 100 °C and then slowed down from 100 °C to 200 °C. The initial weight loss till 100 °C is due to the loss of free water or moisture in the composite's pores. The further weight loss is due to the adsorbed water in the MSH gels and brucite, the two common hydration products in the MSH cementitious system. The additional weight loss from 150 °C to 200 °C also accounts for the loss of PE fibers, which melt around 165 °C. The magnitude of weight loss was similar for both cube and dogbone specimens, suggesting a similar degree of thermal exposure and that the adopted heating protocol is sufficient to make the core of the specimen reach the target temperature. The weight loss in MSH-ECC is higher than that of OPC-based ECC [6, 8, 9, 12], which is attributed to the higher binder content and, thus, greater water content.

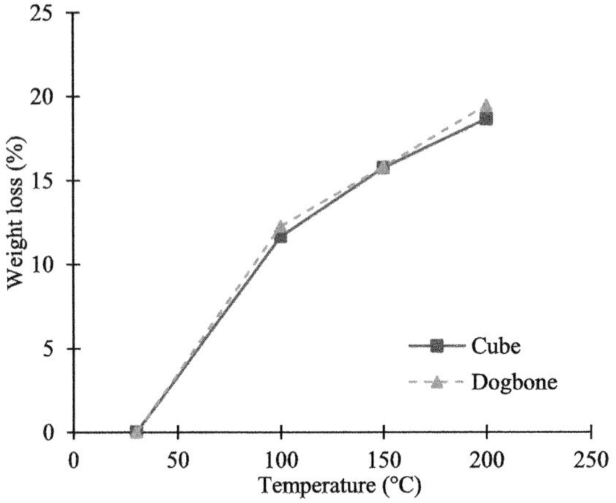

Fig. 1. Weight loss in specimens with temperature.

3.2 XRD Characterization

The XRD pattern of the MSH cementitious matrix after being subjected to elevated temperatures up to 200 °C is shown in Fig. 2. The major phases found were brucite and M-S-H gel as the hydration products, and Periclase and Magnesite from the unreacted MgO powder in the mixture. The M-S-H is observed as broad humps in the XRD spectra as it is amorphous/poorly crystalline [28, 29]. It can be seen that no significant changes in the chemical composition of the cementitious matrix were observed till 200 °C. There might be some loss of physically adsorbed water from brucite and M-S-H, but no noticeable decomposition or formation of new compounds. The dehydration of brucite could be seen from the reduction in the intensity peaks with increased temperature. The XRD characterization shows that the changes observed in the mechanical properties of the MSH-ECC result from the physical processes, such as loss of moisture and associated cracking within the cementitious matrix, rather than any substantial phase decomposition or new compound formation.

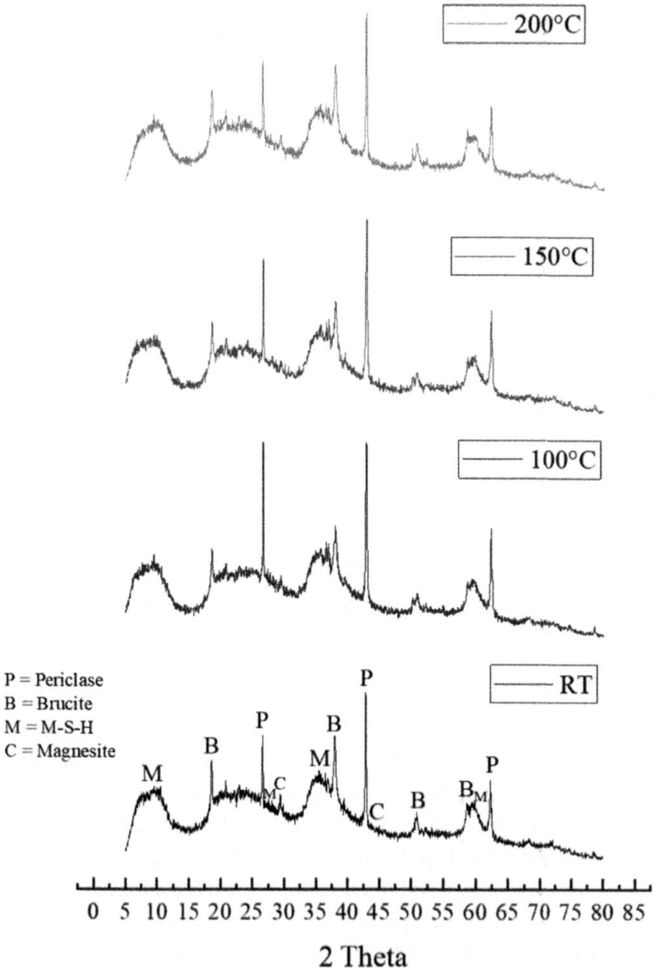

Fig. 2. XRD patterns of MSH matrix at various temperatures.

3.3 Compressive Strength

The variation in the residual compressive strength and normalized residual compressive strength of MSH-ECC with temperatures up to 200 °C is shown in Fig. 3. The residual compressive strength showed a monotonous reduction with an increase in temperature, unlike OPC-based ECC [14, 30]. The reduction in compressive strength follows higher water loss observations (reported in Sect. 3.1). A significant loss in moisture after exposure to high temperatures was observed. It is expected that notable weight loss (> 10%) at a low temperature of 100 °C might have caused shrinkage of the matrix and induced microcracks, which are at large bridged by PE fibers but compromised the strength of the composite. The reduction in the first cracking of the cementitious matrix (which correlates with compressive strength) observed from the tensile stress-strain response shown in FIG also suggests the same. Overall, the MSH-ECC showed poorer residual performance than reported in literature [10] for OPC-based ECC at elevated temperatures up to 200 °C.

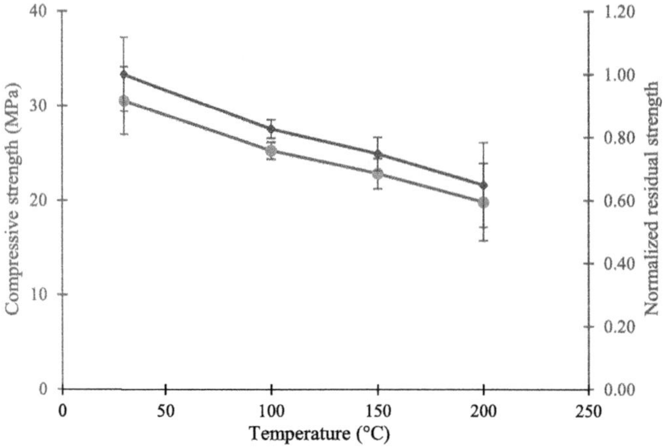

Fig. 3. Variation in compressive strength with temperature.

3.4 Tensile Strength

The tensile stress-strain plots of MSH-ECC at various temperatures up to 200 °C are shown in Fig. 4. The different colors represent individual specimens tested at the same temperature, included to illustrate variability and cover the standard deviation. The strain-hardening response was observed up to 150 °C and lost thereafter. In spite of tensile strain-hardening behavior up to 150 °C, the stress-strain response was different. The drop in tensile stress, which indicates a sudden drop in the stiffness due to the crack formation, was minimal or absent at higher temperatures (100 °C and 150 °C), suggesting the fiber engages the matrix as soon as the crack is formed in the composite. A detailed discussion of the ultimate tensile strength and tensile strain capacity is provided in the subsequent paragraphs.

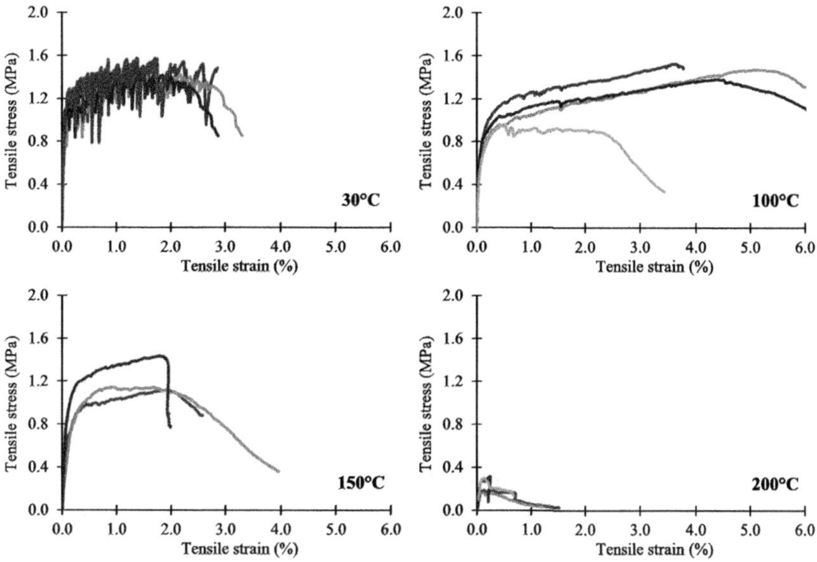

Fig. 4. Tensile stress-strain response at various temperatures.

The variation in the ultimate tensile strength and tensile strain capacity of MSH-ECC at various temperatures up to 200 °C is shown in Fig. 5. It can be observed that ultimate tensile strength shows an initial monotonic drop up to 150 °C, similar to compressive strength. However, the strength reduces significantly from 150 °C to 200 °C due to the loss of polymeric PE fibers. Such sharp reduction in tensile strength has been observed in polymer fiber-based ECC. The ultimate tensile strength reaches below 0.3 MPa at 200 °C, making the composite unusable.

The tensile strain capacity of MSH-ECC at room temperature was around 2.5%, significantly lower than reported for MSH-ECC. The difference in the tensile strength could have been due to the change in the properties of MgO powder, such as fineness, calcination temperature, and chemical composition. The compressive strength was also noticeably lower than reported earlier. Also, a slightly higher fiber volume fraction of 1.2% than required (i.e., the optimum fiber volume fraction of 0.5% [15]) for pure paste-based MSH-ECC also lowered the tensile strain capacity. The tensile strain capacity increases sharply at 100 °C to around 4%. This sudden rise is attributed to the reduction in the first cracking strength of the matrix with almost similar fiber bridging capacity, thus allowing the formation of more saturated multiple cracking. The tensile strain capacity drops again at 150 °C to around 2%. The tensile strain-hardening behavior is completely lost above 150 °C, which is attributed to the loss of microfibers in the matrix.

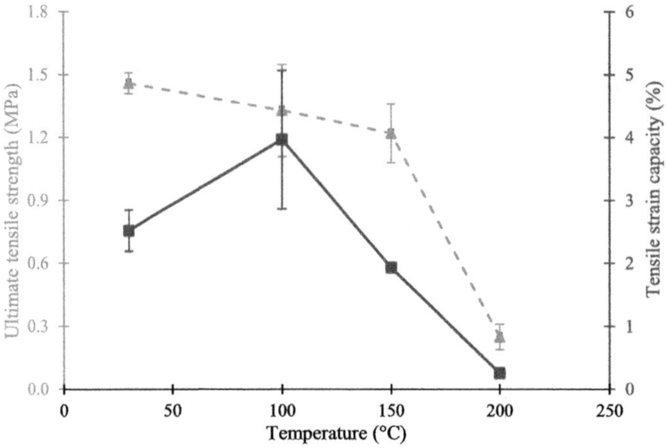

Fig. 5. Variation in tensile properties with temperature.

4 Conclusions

This study reports the behavior of Magnesium-Silicate-Hydrate (MSH) cementitious matrix-based Engineered cementitious Composites (ECC), referred to as MSH-ECC, at sub-elevated temperatures up to 200 °C. The behavior of MSH-ECC was characterized by compression and uniaxial tension under residual conditions after exposure to 100 °C, 150 °C, and 200 °C.

The key findings of the experimental investigation are summarized below:

- The weight loss in MSH-ECC is significantly higher than that reported for OPC-based ECC. The weight loss was similar for cube and dogbone specimens.
- The XRD characterization showed no noticeable changes in the chemical composition of the MSH matrix up to 200 °C. The minimal changes observed were related to the dehydration of brucite (due to loss of physically adsorbed water).
- The compressive strength showed monotonic reduction up to 200 °C, with residual compressive strength reducing to 65% of the room temperature value at 200 °C.
- The reduction in the first cracking strength of the MSH-ECC follows the compressive strength reduction pattern. The ultimate strength reduces marginally up to 150 °C and then drops suddenly at 200 °C, along with the complete loss of tensile strain-hardening due to the melting of PE fibers. The tensile strain capacity showed initial improvement at 100 °C and then dropped again at 150 °C, followed by complete loss of ductility at 200 °C.

This study reports the findings of the preliminary study investigating the performance of MSH cementitious matrix-based ECC and understanding the interaction between MSH systems and fiber at elevated temperatures. This study will be extended to understand how the MSH cementitious matrix degrades with further increases in temperatures, such as 400 °C and 600 °C. Measures to improve the tensile properties after the loss of PE fibers, such as fiber-hybridization, will be explored. All the results will share more insights into the high-temperature performance of MSH-ECC.

Acknowledgements. This study was conducted at the School of Civil and Environmental Engineering (CEE), Nanyang Technological University (NTU), Singapore. Financial support from the Ministry of National Development, Singapore (CoT-V1–2020-1) is appreciated. The assistance of the Protective Engineering (PE Lab) and Construction Technology (CT Lab) laboratory staff with the experimental investigation is duly acknowledged.

References

1. Li, V.C.: Engineered cementitious composites (ECC)-tailored composites through micromechanical modeling, in: N. Banthia, A. Bentur, a.A.M. (Eds.), Fiber Reinforced Concrete: Present and the Future. Canadian Society of Civil Engineers (1997)
2. Li, V.C.: On engineered cementitious composites (ECC): a review of the material and its applications. J. Adv. Concr. Technol. **1**(3), 215–230 (2003)
3. Li, V.C., Mishra, D.K., Wu, H.-C.: Matrix design for pseudo-strain-hardening fibre reinforced cementitious composites. Mater. Struct. **28**, 586–595 (1995)
4. Kanda, T., Li, V.C.: Practical design criteria for saturated pseudo strain hardening behavior in ECC. J. Adv. Concr. Technol. **4**(1), 59–72 (2006)
5. Li, V.C., Kanda, T.: Engineered cementitious composites for structural applications. J. Mater. Civ. Eng. **10**(2), 66–69 (1998)
6. Sahmaran, M., Lachemi, M., Li, V.C.: Assessing mechanical properties and microstructure of fire-damaged engineered cementitious composites. ACI Mater. J. **107**(3) (2010)
7. Bhat, P.S., Chang, V., Li, M.: Effect of elevated temperature on strain-hardening engineered cementitious composites. Constr. Build. Mater. **69**, 370–380 (2014)
8. Deshpande, A.A., Kumar, D., Ranade, R.: Influence of high temperatures on the residual mechanical properties of a hybrid fiber-reinforced strain-hardening cementitious composite. Constr. Build. Mater. **208**, 283–295 (2019)
9. Kumar, D., Ranade, R.: Influence of matrix-modification and fiber-hybridization on high-temperature residual mechanical performance of strain-hardening cementitious composites. Constr. Build. Mater. **302**, 124157 (2021)
10. Kumar, D.: Strain Hardening Cementitious Composites for Fire Resilient Infrastructure. Ph.D. Thesis, State University of New York at Buffalo, New York, USA (2021)
11. Liu, J.-C., Tan, K.H.: Fire resistance of strain hardening cementitious composite with hybrid PVA and steel fibers. Constr. Build. Mater. **135**, 600–611 (2017)
12. Şahmaran, M., Özbay, E., Yücel, H.E., Lachemi, M., Li, V.C.: Effect of fly ash and PVA fiber on microstructural damage and residual properties of engineered cementitious composites exposed to high temperatures. J. Mater. Civ. Eng. **23**(12), 1735–1745 (2011)
13. Yang, E.-H., Yang, Y., Li, V.C.: Use of high volumes of fly ash to improve ECC mechanical properties and material greenness. ACI Mater. J. **104**(6), 620 (2007)
14. Kumar, D., Soliman, A.A., Ranade, R.: Effects of fly ash content and curing age on high temperature residual compressive strength of strain-hardening cementitious composites. In: RILEM-fib International Symposium on Fibre Reinforced Concrete, pp. 3–12. Springer, Heidelberg (2021)
15. Sonat, C., Kumar, D., Koh, W.C., Li, J., Unluer, C., Yang, E.-H.: Strain hardening magnesium-silicate-hydrate composites with extremely low fiber dosage of 0.5% by volume. Cem. Concr. Compos. **142**, 105200 (2023)
16. Kumar, D., Mi, T., Yang, E.H.: Improving the material sustainability of strain-hardening magnesium-silicate-hydrate composite by incorporating aggregates. Constr. Build. Mater. **407**, 133576 (2023)

17. Sonat, C., He, S., Li, J., Unluer, C., Yang, E.-H.: Strain hardening magnesium-silicate-hydrate composites (SHMSHC) reinforced with short and randomly oriented polyvinyl alcohol microfibers. Cem. Concr. Res. **142**, 106354 (2021)
18. Zhang, T., Cheeseman, C.R., Vandeperre, L.J.: Development of low pH cement systems forming magnesium silicate hydrate (MSH). Cem. Concr. Res. **41**(4), 439–442 (2011)
19. Zhang, T., Vandeperre, L.J., Cheeseman, C.R.: Formation of magnesium silicate hydrate (MSH) cement pastes using sodium hexametaphosphate. Cem. Concr. Res. **65**, 8–14 (2014)
20. Tran, H.: Development of magnesium silicate hydrate binder systems. Ph.D. Thesis, University of Canterbury, New Zealand (2019)
21. Li, Z., et al.: Characterization of reaction products and reaction process of MgO–SiO2–H2O system at room temperature. Constr. Build. Mater. **61**, 252–259 (2014)
22. Jia, Y., et al.: Role of sodium hexametaphosphate in MgO/SiO_2 cement pastes. Cem. Concr. Res. **89**, 63–71 (2016)
23. Sonat, C., Dung, N., Unluer, C.: Performance and microstructural development of MgO–SiO2 binders under different curing conditions. Constr. Build. Mater. **154**, 945–955 (2017)
24. Kumar, D., Yang, E.-H.: Mechanical performance of engineered cementitious composites (ECC) utilizing magnesium-silicate-hydrate ($MgO\text{-}SiO_2$) binders. In: RILEM-fib International Symposium on Fibre Reinforced Concrete pp. 91–98. Springer, Heidelberg (2024)
25. Kumar, D., Sanalkumar, K.U., Yang, E.-H.: Strain-hardening magnesium-silicate-hydrate composites (SHMSHC) utilizing reactive magnesia cement (MgO) and calcined clay (metakaolin). Cement Concr. Compos. **152**, 105656 (2024)
26. Mi, T., Kumar, D., Faris, M.S., Yang, E.-H.: Sustainable strain-hardening magnesium silicate hydrate composite. Maga. Concr. Res. **77** (21–22), 1223–1233 (2025)
27. Tran, H., Scott, A., Dhakal, R.: Mechanical and durability properties of magnesium silicate hydrate binder concrete. Mag. Concr. Res. **72**(13), 693–702 (2020)
28. Zhang, T., Zou, J., Wang, B., Wu, Z., Jia, Y., Cheeseman, C.R.: Characterization of magnesium silicate hydrate (MSH) gel formed by reacting MgO and silica fume. Materials **11**(6), 909 (2018)
29. Bernard, E., et al.: Characterization of magnesium silicate hydrate (MSH). Cem. Concr. Res. **116**, 309–330 (2019)
30. Kumar, D., Deshpande, A.A., Ranade, R.: Influence of fiber length on the mechanical behavior of steel-PVA hybrid fiber-reinforced strain-hardening cementitious composites at high temperatures. Indian Concr. J. **93**(12), 30–38 (2019)

Sustainable Concrete Production Using Waste Foundry Sand as Recycled Fine Aggregate

Joseph Pugh[1,2](✉), Diane Gardner[1], and Riccardo Maddalena[1]

[1] School of Engineering, Cardiff University, Cardiff CF24 3AA, UK
pughj7@cardiff.ac.uk

[2] Knights Brown, 160 Christchurch Road, Ringwood BH24 3AR, Hampshire, UK

Abstract. As engineers look to resolve the significant environmental issues associated with both the overextraction of natural resources, and the recycling of 'hard-to-abate' industrial material, the use of Waste Foundry Sand (WFS) within concrete production may provide a robust solution to both, as we transition to a more circular economy.

WFS is a non-hazardous industrial byproduct arising from the cast metal industry, produced in quantities exceeding 100 million tonnes annually. Currently, there is no accepted solution for WFS reuse, with only 15% being recycled and the rest immediately landfilled. Primarily a high-quality silica sand, WFS may be used as direct replacement for fine aggregate within concrete production; this practice therefore has dual benefit, namely resolving significant waste management issues within foundries, whilst also limiting raw material content within concrete production.

This research investigates the raw material properties of WFS (silica and chromite) with a perspective on suitability as recycled aggregate, as well as relative impact on physical, mechanical and durability properties of subsequent concrete with its inclusion. In contrast to other work, here WFS was successfully implemented up to 100% fine aggregate replacement; yielding sufficiently workable and durable, structural grade concrete without the aid of additions or durability specific design.

At 100% fine aggregate replacement, compressive strength exceeding 54 and 44 MPa were achieved with chromite and silica WFS respectively. All WFS concrete displayed suitably resilient mixes in terms of chloride penetration, with all silica WFS mixes exhibiting improvements in bulk electrical resistivity up to a maximum of 5.8 kΩ.cm at 50% replacement. Highly comparable properties in terms of density, open porosity and water absorption were measured relative to control samples, further demonstrating the suitability of WFS as an alternative aggregate. The results of this collaborative research contribute towards alignment of construction and manufacturing industries for sustainable concrete production.

Keywords: Waste Foundry Sand · sustainable concrete · recycled aggregate · industrial waste · resource efficiency · circular economy

M. Posani and S. Singh (Eds.): RYS 2025, RILEM Bookseries 67, pp. 751–768, 2026.
https://doi.org/10.1007/978-3-032-14652-6_54

1 Introduction

The global foundry industry accounts for the annual production of more than 100 million tons of cast metal [1], with the demand for such products continuing to grow [2]. Such production however comes at a cost, not only with process embodied carbon and raw material usage, but also significant waste production, one form of which is Waste Foundry Sand (WFS).

As a means of manufacturing moulding media for cast metal production, a high-quality virgin silica sand is mixed with binder material yielding a hardened mould [3]. Such products can produce high grade, precise metallic parts for the automotive industry and beyond. Yet, despite the high quality of the moulding sand facilitating multiple reuses [4], the chemical and physical properties of their subsequent components however are short lived, resulting in eventual degradation and disposal. At this point in the life cycle the byproduct is deemed WFS, of which around 100 million tonnes is produced annually [5] with only 15% recycled [6].

The built environment and construction sector are also notoriously unsustainable, accounting for approximately 38% of global carbon emissions [7], much of which can be attributed to the concrete industry. Despite the many active fields of research in producing more sustainable concrete, the concept of raw material usage is often overlooked despite being an immediate threat to both the environment, and the future success of the industry. A recent United Nations report has suggested 40–50 billion tonnes of aggregate is extracted from the natural environment annually [8], of which it is estimated over half is consumed by the concrete industry [9]. Such extraction has had catastrophic consequences on the natural environment; the prevalence of which is exacerbated by the fact that global building sand usage is projected to increase by around 45% by 2060 [10]. The need for alternative aggregate sources is clearly now more important than ever.

A potential solution that simultaneously contributes to solving both the overextraction of natural sand resources, and the WFS management issue may therefore be to repurpose such waste as a fine aggregate replacement within concrete. In general, there are two main WFS variations, distinguished by binder type; greensand which uses clay as a binder material, and chemically bound sand which uses a polymer activated by a catalyst. Chemically bound sand is becoming increasingly common [11] and typically consists of phenolic-urethanes, epoxy-resins, furfuryl alcohol, and sodium silicates [12]. An additional subcategory of foundry sand waste may be produced during fettling or surface enhancement [13], although this volume is marginal in scale compared to the substantial quantities of waste foundry sand created during mould processing, and there is currently no evidence of fettling sand reuse within concrete within published literature.

Typically, the inclusion of WFS within concrete, whether as a partial or complete replacement of natural sand, tends to reduce the workability of the mixture [14–24]. Such a result is often attributed to the increased fineness of WFS [25], impurity presence and increased water demand [26] and can lead to poor compaction and more numerous voids within the hardened mixture, having a subsequent impact on physical and mechanical properties. The impact of WFS inclusion on compressive strength is highly variable, but generally displays relative decreases as compared to control samples yet is deemed tolerable up to a given replacement percentage [15, 16, 25–28]. Inferior strength properties with WFS inclusion have been explained by increased surface area of the WFS causing a reduction in the amount water-cement gel in the matrix, hence leading to poor binding of coarse and fine aggregates [20, 24, 26, 29–32], as well as increased voids due to poor compaction [24]. There is evidence however, of improvement in compressive strength with WFS inclusion [14, 17, 18, 20–22, 29, 32–36] attributed to the fine nature of WFS resulting in a more dense matrix [14, 18, 20, 21, 24, 29, 32, 37], the impact of increased silica content [18, 29, 38], as well as better regularity of particles [29, 38].

The durability properties of WFS concrete prove far less investigated than those of mechanical and physical properties, and perhaps even more variable. There is evidence of contrary results in terms of resistance to carbonation [14, 24, 25, 38], sulphate [25, 39, 40] and cyclic weathering [21, 40–43] where the impact of WFS may be deemed inconclusive. It may generally be concluded however, that WFS improves resistance to both corrosion [21, 22, 33] and abrasion [20, 22, 33]. The optimum fine aggregate replacement level outlined within literature is generally in the region of 15–30% by weight for standard concrete [4–6].

Despite the significant amount of WFS concrete research, limited publications have assessed the impact of full fine aggregate replacement with WFS [19, 27, 38]; where typically significant negative impacts on physical, mechanical and durability properties are displayed when doing so. There is strong literary evidence of similar generic trends with reductions in concrete performance beyond 50% replacement, justifying the limited investigation considering full replacement. Research conducted on 100% WFS substitution showed more than 50% reduction in compressive strength as compared to control samples at all curing periods, with equivalent magnitudes of increase in water absorption [27]. Another study has also shown similar evidence with comparable reductions in the same properties at full replacement [19]. There is evidence however, albeit limited, of successful full fine aggregate replacement with chemically bound WFS [38]; the work presented in this paper is both the most promising, and perhaps only example so far evidencing suitability of such for potential scalability. Increases in compressive strength were shown relative to control samples at all replacement levels by 28-days curing, achieving up to 48.5 MPa at 100% replacement; alongside adequate workability, water absorption, carbonation, and tensile properties with uniform calcium silicate hydrate (C-S-H) phases' distribution. Given the significant variability of WFS and its ensuing concrete that have previously been outlined, such research has concluded that the usually recommended replacement threshold of 30% is to be challenged on a case-by-case basis, as a means of optimising both environmental and economic benefits. The production, therefore, of a structural grade, durable and sustainable concrete product replacing 100% of the virgin fine aggregate can be a real possibility when considering a high quality WFS product. The research in the present paper offers supporting evidence to prove suitability of chemically bound WFS for full replacement in sustainable concrete production.

2 Materials and Methods

Two types of WFS obtained from Weir Minerals were used in the testing, namely chromite (C-WFS) and silica (S-WFS). S-WFS is a chemically bound WFS initially consisting of furfuryl alcohol, formaldehyde, and phenol binder, which degrade during the casting process leaving a layer of melted residue coating grains. C-WFS is a more refined waste, used as an additive to produce better surface finish to cast metal products.

Particle size analysis was conducted as per BS EN 933-1:2012 [44], with density of aggregate material determined in accordance with BS EN 1097-3:1998 [45] and BS EN 1097-6:2022 [46]. X-Ray Diffraction (XRD) analyses were conducted using a Philips PW1710 Powder Diffractometer. Powdered samples were run at Cu Kα radiation of 35 kV and 40 mA, scanning between 5 and 75° 2θ at a scan speed of 0.02° 2θ/s. Scanning Electron Microscopy (SEM) was conducted using a Zeiss Sigma HD Electron Microscope.

A preliminary mix design was conducted using a BRE design methodology [47] targeting 30 MPa compressive strength and 60–180 mm slump. This design was subsequently amended to increase the overall fine aggregate content and retain a fine to coarse aggregate ratio close to one as a means of facilitating suitable workability. The final amended mix design ratio used for testing was 1: 1.5: 2: 0.38 (cement: fine aggregate: coarse aggregate: water). Five series were cast; two incorporating C-WFS and two incorporating S-WFS each at both 50% and 100% fine aggregate substitution, alongside a control mix with 100% Virgin Fine Aggregate (VFA). MasterGlenium ACE 499 superplasticizer was used where required to reach target slump. Mix details and references are outlined in Table 1.

Table 1. Mix proportions and fresh properties of amended BRE designed WFS mixes.

Mix reference	Cement (kg/m^3)	Virgin fine aggregate (kg/m^3)	WFS (kg/m^3)	Coarse aggregate (kg/m^3)	Water (kg/m^3)	Superplasticizer content (%CEM)	Slump (mm)	Vebe (seconds)
CTRL	492	738	0	984	187	0.00	35	5.66
C-WFS50	492	369	369	984	187	0.00	110	0.55
C-WFS100	492	0	738	984	187	0.00	130	0.10
S-WFS50	492	369	369	984	187	0.00	35	6.55
S-WFS100	492	0	494	984	187	0.33	170	1.68

The VFA used was a 0–4 mm quartz sand with limestone blend. The densities of each of the fine aggregate materials are displayed in Table 2. The virgin coarse aggregate used was a 4–10 mm limestone aggregate of particle density 2845 kg/m^3. The cement used was CEM II/A-L 32,5, a limestone blend Portland cement of density 3090 kg/m^3. The proportioning of mixes by mass-based replacement is acknowledged as having a potential impact on comparability of results related to C-WFS mixes. Consideration of such when assessing results, however, ensures findings remain valid in terms of proving feasibility of WFS material in concrete.

Workability tests were conducted on fresh concrete mixes as per BS EN 12350-2:2019 [48] and BS EN 12350-3:2019 [49]. Compressive and tensile strength testing were conducted as per BS EN 12390-3:2019 [50] and BS EN 12390-6:2019 [51] respectively, using a Controls Automax testing machine. Density and open porosity of concrete

Table 2. Density values for fine aggregate material.

Material	Loose bulk density (kg/m^3)	Particle density (kg/m^3)
VFA	1575	2719
C-WFS	2409	4133
S-WFS	1456	2611

samples were determined through a combination of BS EN 12390-7:2019 [52] and HAMSTAD WP1 [53]. Assessment of water absorption rate was conducted in accordance with ASTM C1585-20 [54]. Bulk electrical resistivity was conducted using a Giatec RCON with device and method compliant with ASTM C1876-19 [55]. All hardened testing was conducted following 7- and 28-day curing within a water tank at 20 °C.

3 Results and Discussion

Grading analysis of fine aggregates and relevant blends are displayed in Fig. 1. Both of the WFS products generally show a low particle size variability as compared to VFA, with a significant sample mass represented by the 0.1–0.6 mm segment. Interestingly, and in contrast to other publications [56–60], the portion of 'fine' material is less significant, and even lower than that displayed by VFA. C-WFS is slightly finer than S-WFS with average fineness moduli of 1.58 and 1.66 respectively, as compared to 2.43 for VFA.

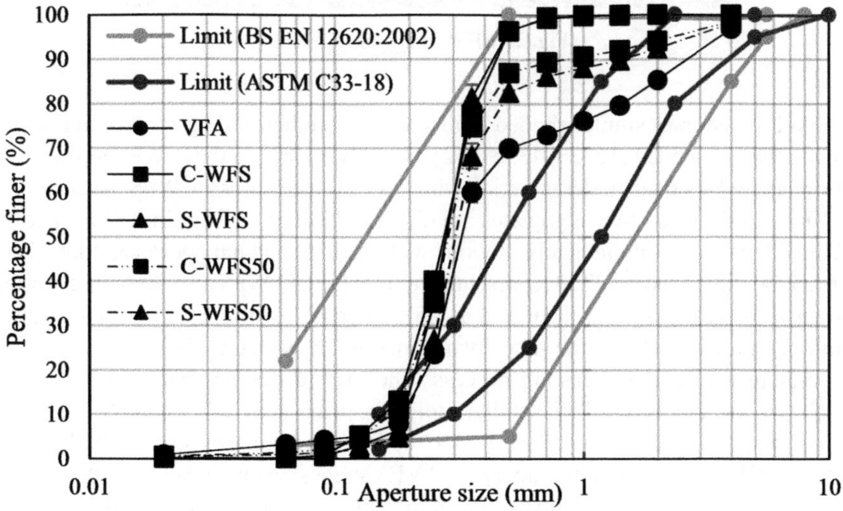

Fig. 1. Grading analysis of VFA, C-WFS and S-WFS.

The WFS curves however are near identical, so much so that they are difficult to distinguish on the graph, which may be anticipated given their dual application in foundry engineering; both are described as a 'fine graded fine aggregate' within BS EN 12620:2002 + A1:2008, with the VFA also falling within this category. A key objective of the present study is to evaluate the feasibility of high-volume replacement of VFA with WFS. Indirect assessment of the potential impact of partial and full replacement on particle packing characteristics, with adjustments to aggregate grading is therefore monitored as a potentially contributing variable. SEM images of C-WFS, and S-WFS are displayed in Figs. 2 and 3. The different types of particles within C-WFS are clearly visible with the denser metallic particles, identifiable by both the lighter shade of grey and the angular particle shape, as well as silica grains similar to that of S-WFS.

Fig. 2. SEM image of C-WFS.

S-WFS possesses similar shaped grains to that of other findings within literature of sub-angular to rounded [16, 38, 43, 61, 62]. Notably visible on the S-WFS grain surface is the degraded residual binder material, which is responsible for the WFS's black colour.

XRD results of all fine aggregates are displayed in Fig. 4. The similarities between VFA and S-WFS are immediately obvious with distinct matching peaks at 20.7°, 26.6°, 36.6°, 50.4°, 55.6° and 60.2° 2θ representing quartz [63]. C-WFS also shows the presence of quartz. S-WFS has been identified as 95% quartz, 3% iron and 2% cristobalite by semi-quantitative analysis; the latter two minerals likely due to the metallic and high temperature nature of the casting process. The main difference between S-WFS and VFA lies in the identified (albeit smaller) dolomite peaks at 31°, 45.1° and 51.2° 2θ [64], and calcite peaks at 29.4°, 47.7° and 48.8° 2θ [65]; these correspond to the limestone blend of VFA. Semi-quantitative analysis of VFA suggests a mineral composition of 75% quartz, 15% dolomite, 9% calcite and 1% albite/calcian. As the name suggests, alongside quartz the other main constituent of C-WFS is chromite; clear peaks corresponding to this mineral were identified at 18.2°, 30.2°, 35.5°, 53.8°, and 63.6° 2θ [66].

The effect of WFS on workability results was variable, C-WFS clearly displayed marked improvements in workability with higher replacement showing increased slump

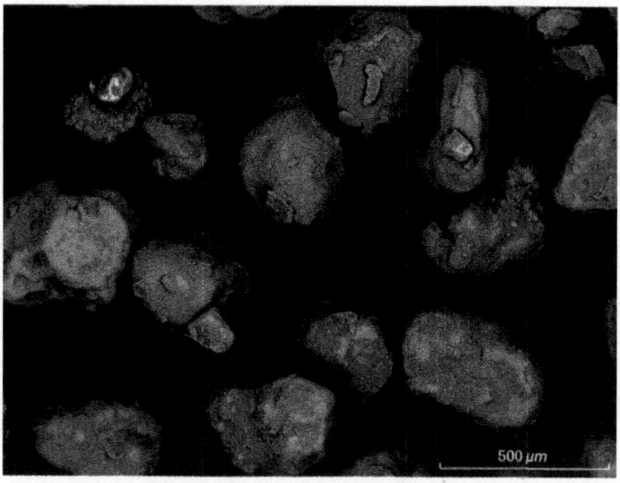

Fig. 3. SEM image of S-WFS.

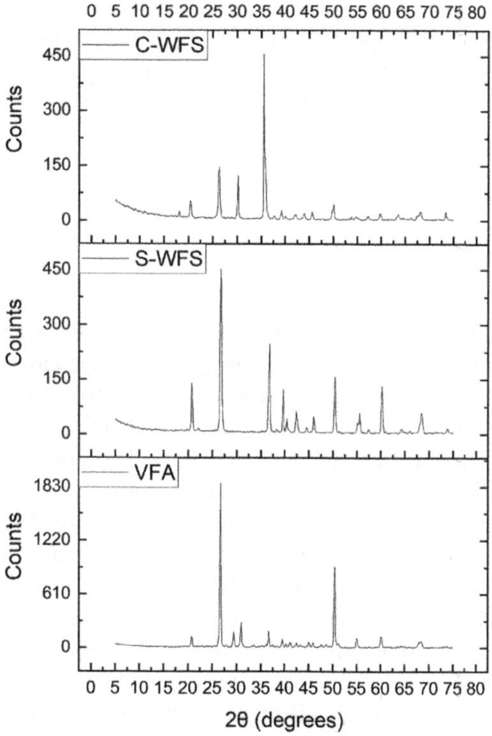

Fig. 4. XRD results for virgin fine aggregate, silica and chromite WFS.

and decreased Vebe time. It is possible that chromite particles absorbed less water than silica grains, leaving more free water in the mix for workability retention. Additionally, both the effect of aggregate grading variation by inclusion of increased portion of finer graded WFS, and mass-based proportioning with high density aggregates must be considered as having an influence on workability. The S-WFS however has limited impact on workability up to 50% replacement, with the mix behaving similarly to that of the control, but a marked reduction in workability is displayed beyond this point, hence the required addition of superplasticizer to render comparable.

The reduced workability with incorporation of S-WFS may be explained by increased fineness [25], impurity presence and increased water demand [26] of the WFS itself. Nonetheless, such adequate and even improved workability with WFS incorporation is in stark contrast to many other findings within literature that outline inferior workability properties [14–24]. This highlights the potential for high-volume replacement despite previous concerns regarding the performance of WFS in fresh concrete. Results for 28-day bulk and matrix density are displayed in Table 3, with an open porosity comparison displayed in Fig. 5.

Table 3. Average 28-day bulk and matrix densities for WFS mixes with standard deviation.

Mix reference	Average bulk density (kg/m^3)	Average matrix density (kg/m^3)
CTRL	2402 ± 7	2536 ± 11
C-WFS50	2490 ± 3	2651 ± 3
C-WFS100	2690 ± 3	2840 ± 6
S-WFS50	2329 ± 4	2526 ± 18
S-WFS100	2279 ± 2	2476 ± 25

All mixes are within the expected density range for a standard concrete, aside from the C-WFS samples which are slightly higher given the increased particle density of this material. S-WFS mixes show decreased bulk and matrix densities as compared to control samples consistent with other research [26, 40, 67], the latter due to the more significant number of inaccessible internal pores as well as reduced particle density, and the former due to both this and an increased open porosity.

Inferior particle packing characteristics induced via a reduced volume of particles in the 2–4 mm range at full WFS replacement could be considered as potentially contributing towards increased open porosity. Given the consistent grading of both C-WFS and S-WFS however, it is clear there is another active porosity inducing mechanism within S-WFS mixes. Such increased open porosity within S-WFS samples is likely formed via impurities within the binder material impacting cement hydration reactions. Such phenomena are less notable within C-WFS mixes but clearly have a significant impact on the concrete matrix when S-WFS is incorporated given the 48% and 51% increases as compared to control samples. Other research has also suggested that WFS impurities may impact hydration processes [15, 26].

It does appear however, that the S-WFS100 is undergoing more enhanced curing from 7-to 28-day than that of control with a 25% reduction in porosity as compared to 17%

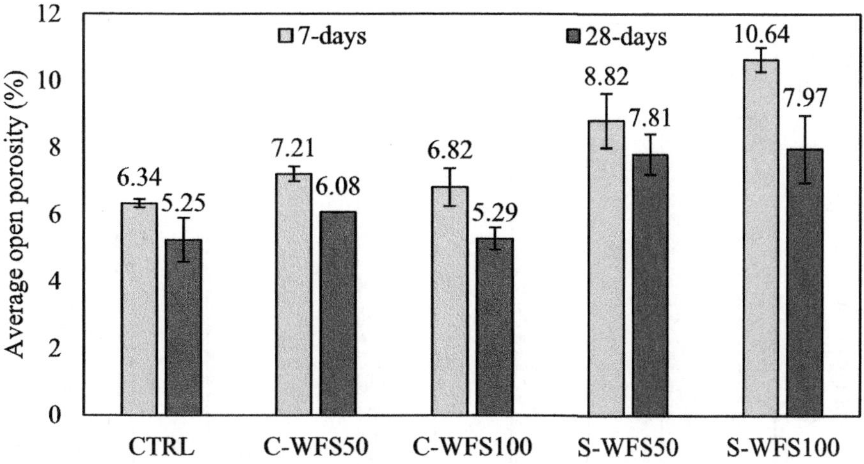

Fig. 5. Open porosity results for 7- and 28-day specimens. Error bars indicate the standard deviation.

for control. Such a result could possibly be described by the nature of the high-quality S-WFS product having higher quartz content than VFA, which when heated to extremely high temperatures within the foundry partially formed cristobalite as detected in XRD analysis. There is evidence of waste sand containing quartz and cristobalite contributing to pozzolanic reactions and long-term strength [68], as well as other crystalline quartz-based sands also presenting as partially pozzolanic [69, 70]. The likelihood of further improved curing and hence reduced porosity in the longer term is therefore promising. Nonetheless, at this stage, given such high open porosity values further investigation into likely fluid penetration properties would be useful given the increased susceptibility to such action. Compressive and tensile strength results are displayed in Figs. 6 and 7.

There is a clear trend in terms of reduction in compressive strength with increased proportion of WFS replacement, yet all series evidenced concrete suitable for structural grade applications at 28-days [71]. S-WFS100 samples showed the most significant reductions, with a 28.7% drop as compared to the control specimens; C-WFS100 samples showed reductions of just 12.6% over the same time scale. The extent of replacement with C-WFS appears to have limited impact in terms of further reducing compressive strength, with 28-day results of 54.90 MPa and 54.78 MPa at 50% and 100% substitution, respectively. The same is true for S-WFS with a 5.1% reduction from 50% to 100% replacement. Highly comparative compressive strength is typically hard to achieve at such high-volume replacement, with most research identifying significant performance impacts beyond 15–30% replacement [4–6, 72–78]. The superior performance of C-WFS concrete compared to that of S-WFS is likely due to the greater suitability of C-WFS physical properties; namely increased particle angularity and density, evidenced in Fig. 2 and Table 3 respectively. Increased particle angularity of fine aggregate has been evidenced as improving compressive strength [79], as has the use of high density fine aggregate [80]. Additionally, the reduced surface contamination of C-WFS grains when compared to S-WFS grains in terms of residual binder likely also contributed to

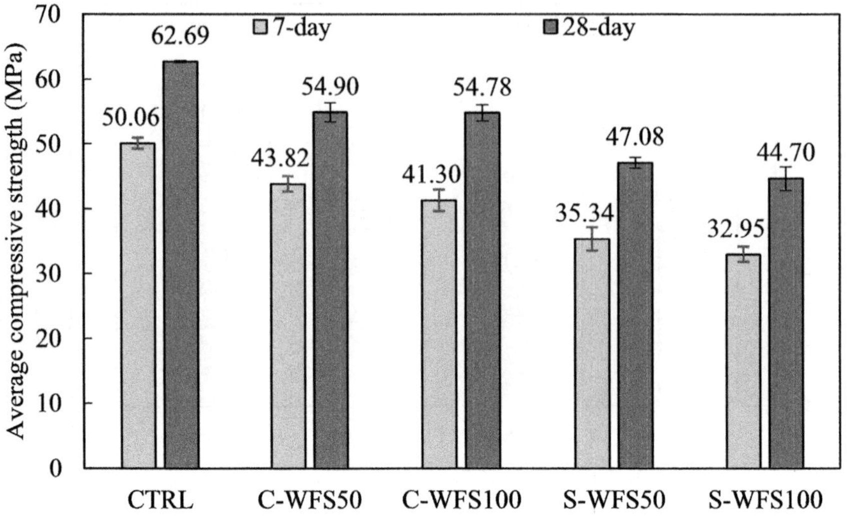

Fig. 6. Compressive strength results for 7- and 28-day specimens. Error bars indicate the standard deviation.

improved bonding due to limited interference with cement hydration, having previously been identified as having the potential to impact the hydration process [15, 26].

As anticipated, all samples show increases in compressive strength from 7 to 28-days, yet the most significant improvements were displayed in S-WFS50 and S-WFS100 samples with 33.2% and 35.6% respectively. Given control samples showed a 25.2% increase, such results could be further evidence of S-WFS's contribution towards improved curing, and aligns with other research where WFS samples tend towards control values with longer curing period [26, 29]. Such a phenomena is common in previously assessed porosity results and aligns with more significant open porosity reductions. Tensile strength properties follow near identical trends to those of compressive strength, although less significant increases are shown with curing period, and a drop in ultimate strength for S-WFS50 is displayed. Although unusual, such a minor reduction is not of concern given the highly variable nature of concrete tensile properties. The results are reflective of a direct correlation between increased porosity and reduced mechanical properties. S-WFS mixes have previously been identified as having comparatively higher open porosity, which may have had a direct impact on reducing overall strength as is consistent with both standard concrete and many other materials [81].

Water absorption rate values at 7- and 28-days are displayed in Fig. 8. C-WFS samples have the most significant impact on increased water absorption rate at 7-days, with S-WFS appearing to have limited effect with largely similar results to the control series. Such results suggest the increased open porosity of S-WFS samples did not contribute to increasing susceptibility to water penetration by capillary action at any replacement level or curing period. Such behaviour may be described by the fact that the water absorption rate, is dependent on microscopic properties such as pore distribution, size and structure [82, 83], not only overall open porosity.

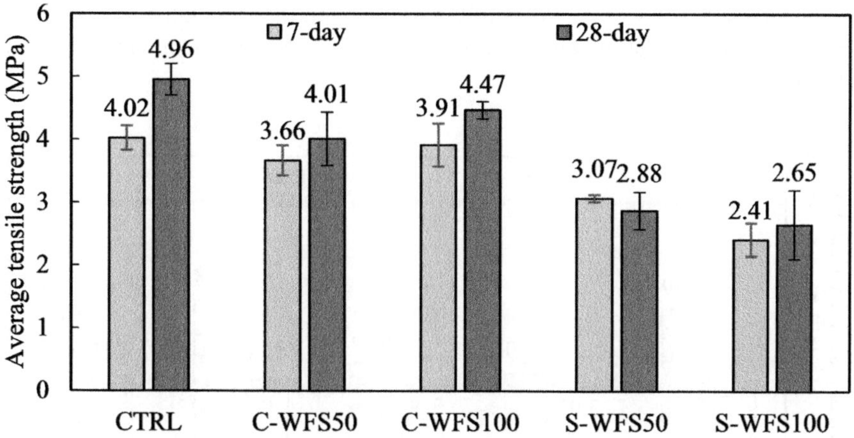

Fig. 7. Tensile strength results for 7- and 28-day specimens. Error bars indicate the standard deviation.

The pore structure of C-WFS50 proved most suitable for capillary transport in this case, to which the mass-based proportioning approach, and hence reduced mix fine content may have contributed. C-WFS mixes, at 28-days retained largely similar sorptivity values to those of S-WFS samples with relative improvements at 100% substitution. All values at 28-days are considered acceptable for concrete of this nature [84].

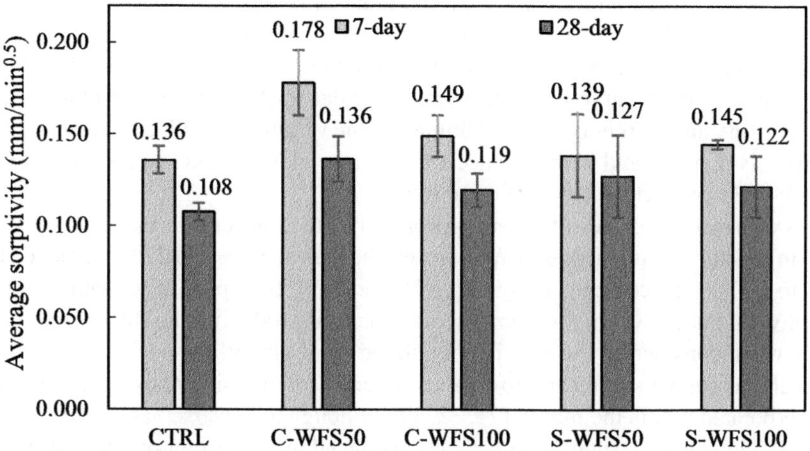

Fig. 8. Sorptivity results for samples at 7- and 28-day specimens. Error bars indicate the standard deviation.

Bulk electrical resistivity results are shown in Fig. 9. 28-day cured samples all displayed moderate durability properties in terms of likely chloride penetration, despite their increased porosities. In fact, S-WFS samples showed improvements as compared to control samples suggesting superior durability. There is also potential however, that the binder material coating WFS particles may have contributed to increased resistivity;

it is known that S-WFS binder content is phenolic, with such moulding material known to be electrically insulating with dielectric strength of 250–400 V/mm and dissipation factor of 0.03–0.3 [85].

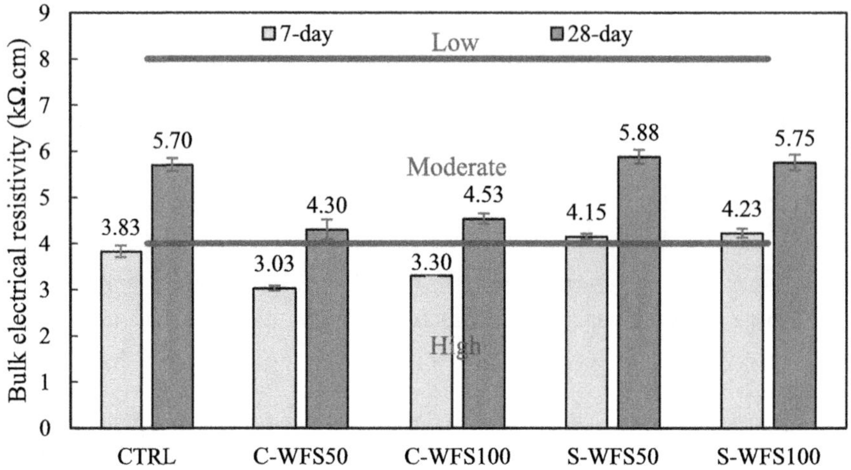

Fig. 9. Bulk electrical resistivity results for 7- and 28-day specimens with outlined likely chloride penetration limits in red. Error bars indicate the standard deviation.

The impact of chromite content on the reduced resistivity of C-WFS mixes must also be considered; there is potential that the conductive nature of chromite as a mineral [86] contributed less resistance to current flow, and hence such results are possibly not reflective of inferior durability properties, particularly when considered alongside low open porosity and sorptivity values. Ultimately, the variation in bulk electrical resistivity of mixes is minimal and within an acceptable range, nonetheless suggesting suitability of both WFS products in terms of durability.

As with mechanical results, there appears to be little impact in terms of further reductions in resistivity with increased WFS, given changes of 5.3% and 2.2% displayed from 50% to 100% replacement for both C-WFS and S-WFS respectively. Such results are reflective of successful high-volume replacement in producing durable mixes; particularly when considering the mix is not tailored for a durability specific design. Other research has shown similar improvements in terms of corrosion resistance when WFS is incorporated within the matrix [21, 32, 39], although the reverse has also been shown [24, 25]. Nonetheless, such improvements displayed in the current research at full substitution are reflective of a high quality WFS product with the potential to fully replace virgin fine aggregates.

4 Conclusion

The inclusion of WFS within concrete is evidently both feasible and, in many ways, beneficial given the production of durable, structural grade concretes at full fine aggregate substitution. S-WFS clearly has a significant impact in terms open porosity, displaying a 51% increase at full substitution as compared to control samples; yet this increase has limited effect on the physical and durability properties in terms of both water and rapid chloride penetration where either minimal change or marked improvements were displayed. Reductions in strength are displayed in all samples incorporating WFS as compared to control samples, albeit all less than 30% and within an acceptable range. Structural grade concrete of 54.78 MPa and 44.70 MPa was produced at 100% replacement for both C-WFS and S-WFS samples respectively, suggesting high volume replacement of VFA with this high-quality waste product is largely feasible in terms of mechanical properties. The impact of inferior particle packing due to both mass-based proportioning of C-WFS mixes, and reduced proportion of 2-4mm aggregate by full WFS substitution is acknowledged, but in this case proven not to degrade concrete properties to an extent whereby mixes became unfeasible.

Further investigation into the interaction between binder material and cement is required in order to optimise the curing process that may be impacted by this combination; it is very likely that the increased open porosity is a direct consequence of this chemical reaction. In doing so, an understanding of the binder chemistry and environmental impacts associated with such would also be beneficial to work towards a sustainable removal strategy. There is potential to further improve mix properties, and perhaps counter any negative impacts associated with the inclusion of WFS by addition of supplementary cementitious material. Further investigation into this topic is required, alongside long-term durability investigation. Both a life cycle, and economic assessment of the ensuing concrete product and its constituents are necessary in order to fully understand the environmental impacts associated with producing this material, and its feasibility at scale.

In terms of engineering performance, however, it is evident that there is clear potential for WFS concrete application within the sector. This study presents evidence that full replacement of VFA with WFS can produce a sustainable concrete exhibiting physical characteristics comparable to those of conventional concrete. The resulting material demonstrates both sufficient durability and robust mechanical performance, without the use of an application or performance-optimised mix design. The trialling of WFS concrete within a site scenario is paramount to proving such materials are not only feasible under lab conditions, but transferrable to the construction industry. Further work to explore the latter via a long term site trial alongside Knights Brown Construction Ltd. is already underway.

Acknowledgements. The EPSRC-IAA project WASTEREBUILT (524170) is acknowledged. The authors acknowledge Weir Group Plc, Knights Brown and Carbon Upcycling Technologies for supporting the PhD studentship. The authors acknowledge the assistance of Mr Carl Wadsworth in casting and preparing samples.

References

1. Casting, M.: Census of World Casting Production - Fewer Castings Made in 2020. American Foundry Society, Modern Casting (2021)
2. EFF: The European Foundry Industry at a Glance. . In: European Foundry Federation (EFF) (2024). https://eff-eu.org/statistics/. Accessed 27 Mar 2025
3. FIRST: Foundry Sand Facts for Civil Engineers (FHWA-IF-04-004). Butterworth-Heinemann, Fall River, MA (2004)
4. Siddique, R., Noumowe, A.: Utilization of spent foundry sand in controlled low-strength materials and concrete. Resour. Conserv. Recycl. **53**, 27–35 (2008). https://doi.org/10.1016/J.RESCONREC.2008.09.007
5. Ahmad, J., Zhou, Z., Martínez-García, R., et al.: Waste foundry sand in concrete production instead of natural river sand: a review. Materials **15** (2022). https://doi.org/10.3390/ma15072365
6. Tittarelli, F.: Waste foundry sand. Waste and Supplementary Cementitious Materials in Concrete: Characterisation, Properties and Applications, pp. 121–147 (2018). https://doi.org/10.1016/B978-0-08-102156-9.00004-3
7. Construction Industry Council: Construction Industry Council: Policy and Public Affairs - Climate Change. In: CIC (2021). https://www.cic.org.uk/policy-and-public-affairs/climate-change#:~:text=The%20built%20environment%20and%20construction,size%20of%20Paris%20every%20week. Accessed 19 Jan 2024
8. United Nations Environment Programme: Sand and Sustainability: Finding New Solutions for Environmental Governance of Global Sand Resources (2019)
9. United Nations Environment Programme U: Sand, Rarer than One Thinks: UNEP Global Environmental Alert Service (GEAS) (2014)
10. Zhong, X., Deetman, S., Tukker, A., Behrens, P.: Increasing material efficiencies of buildings to address the global sand crisis. Nat. Sustain. **5**, 389–392 (2022). https://doi.org/10.1038/s41893-022-00857-0
11. Cast Metals Federation: UK Casting Industry Census 2017 (2017)
12. Siddique, R., Singh, G.: Utilization of waste foundry sand (WFS) in concrete manufacturing. Resour. Conserv. Recycl. **55**, 885–892 (2011). https://doi.org/10.1016/J.RESCONREC.2011.05.001
13. Hall, H.T.: Fettling. In: Castings, pp. 117–123. Macmillan Education UK, London (1971)
14. Siddique, R., Aggarwal, Y., Aggarwal, P., et al.: Strength, durability, and micro-structural properties of concrete made with used-foundry sand (UFS). Constr. Build. Mater. **25**, 1916–1925 (2011). https://doi.org/10.1016/J.CONBUILDMAT.2010.11.065
15. Ganesh Prabhu, G., Hyun, J.H., Kim, Y.Y.: Effects of foundry sand as a fine aggregate in concrete production. Constr. Build. Mater. **70**, 514–521 (2014). https://doi.org/10.1016/J.CONBUILDMAT.2014.07.070
16. Mushtaq, S.M., Siddique, R., Goyal, S., Kaur, K.: Experimental studies and drying shrinkage prediction model for concrete containing waste foundry sand. Clean. Eng. Technol. **2**, 100071 (2021). https://doi.org/10.1016/J.CLET.2021.100071
17. Kavitha, O.R., Shyamala, G., Akshana, V.: Study of sustainable concrete property containing waste foundry sand. Mater. Today Proc. **39**, 855–860 (2021). https://doi.org/10.1016/J.MATPR.2020.10.359
18. Siddique, R., de Schutter, G., Noumowe, A.: Effect of used-foundry sand on the mechanical properties of concrete. Constr. Build. Mater. **23**, 976–980 (2009). https://doi.org/10.1016/J.CONBUILDMAT.2008.05.005
19. Coppio, G.J.L., de Lima, M.G., Lencioni, J.W., et al.: Surface electrical resistivity and compressive strength of concrete with the use of waste foundry sand as aggregate. Constr. Build. Mater. **212**, 514–521 (2019). https://doi.org/10.1016/J.CONBUILDMAT.2019.03.297

20. Singh, G., Siddique, R.: Abrasion resistance and strength properties of concrete containing waste foundry sand (WFS). Constr. Build. Mater. **28**, 421–426 (2012). https://doi.org/10.1016/J.CONBUILDMAT.2011.08.087
21. Siddique, R., Singh, G., Singh, M.: Recycle option for metallurgical by-product (Spent Foundry Sand) in green concrete for sustainable construction. J. Clean. Prod. **172**, 1111–1120 (2018). https://doi.org/10.1016/J.JCLEPRO.2017.10.255
22. Manoharan, T., Laksmanan, D., Mylsamy, K., et al.: Engineering properties of concrete with partial utilization of used foundry sand. Waste Manage. **71**, 454–460 (2018). https://doi.org/10.1016/J.WASMAN.2017.10.022
23. Saha, S., Rajasekaran, C., More, A.P.: Use of foundry sand as partial replacement of natural fine aggregate for the production of concrete. In: Lecture Notes in Civil Engineering, pp. 61–71. Springer (2019)
24. Ahmad, J., Aslam, F., Zaid, O., et al.: Mechanical and durability characteristics of sustainable concrete modified with partial substitution of waste foundry sand. Struct. Concr. **22**, 2775–2790 (2021). https://doi.org/10.1002/suco.202000830
25. Ganesh Prabhu, G., Bang, J.W., Lee, B.J., et al.: Mechanical and durability properties of concrete made with used foundry sand as fine aggregate. Adv. Mater. Sci. Eng. **2015**, 1–11 (2015). https://doi.org/10.1155/2015/161753
26. Basar, H.M., Deveci Aksoy, N.: The effect of waste foundry sand (WFS) as partial replacement of sand on the mechanical, leaching and micro-structural characteristics of ready-mixed concrete. Constr. Build. Mater. **35**, 508–515 (2012). https://doi.org/10.1016/J.CONBUILDMAT.2012.04.078
27. Khatib, J.M., Herki, B.A., Kenai, S.: Capillarity of concrete incorporating waste foundry sand. Constr. Build. Mater. **47**, 867–871 (2013). https://doi.org/10.1016/j.conbuildmat.2013.05.013
28. dos Santos, L.F., Magalhães, R.S., Barreto, S.S., et al.: Characterization and reuse of spent foundry sand in the production of concrete for interlocking pavement. J. Build. Eng. **36**, 102098 (2021). https://doi.org/10.1016/j.jobe.2020.102098
29. Siddique, R., Singh, G., Belarbi, R., et al.: Comparative investigation on the influence of spent foundry sand as partial replacement of fine aggregates on the properties of two grades of concrete. Constr. Build. Mater. **83**, 216–222 (2015). https://doi.org/10.1016/J.CONBUILDMAT.2015.03.011
30. Naik, T.R., Patel, V.M., Parikh, D.M., Tharaniyil, M.P.: Utilization of used foundry sand in concrete. J. Mater. Civ. Eng. **6**, 254–263 (1994). https://doi.org/10.1061/(ASCE)0899-1561(1994)6:2(254)
31. Monosi, S., Tittarelli, F., Giosuè, C., Ruello, M.L.: Effect of two different sources and washing treatment on the properties of UFS by-products for mortar and concrete production. Constr. Build. Mater. **44**, 260–266 (2013). https://doi.org/10.1016/J.CONBUILDMAT.2013.02.029
32. Singh, G., Siddique, R.: Effect of waste foundry sand (WFS) as partial replacement of sand on the strength, ultrasonic pulse velocity and permeability of concrete. Constr. Build. Mater. **26**, 416–422 (2012). https://doi.org/10.1016/J.CONBUILDMAT.2011.06.041
33. Thiruvenkitam, M., Pandian, S., Santra, M., Subramanian, D.: Use of waste foundry sand as a partial replacement to produce green concrete: mechanical properties, durability attributes and its economical assessment. Environ. Technol. Innov. **19**, 101022 (2020). https://doi.org/10.1016/J.ETI.2020.101022
34. Priyadarshini, M., Giri, J.P.: Use of recycled foundry sand for the development of green concrete and its quantification. J. Build. Eng. **52**, 104474 (2022). https://doi.org/10.1016/J.JOBE.2022.104474
35. de Martins, M.A.B., Barros, R.M., Silva, G., dos Santos, I.F.S.: Study on waste foundry exhaust sand, WFES, as a partial substitute of fine aggregates in conventional concrete. Sustain. Cities Soc. **45**, 187–196 (2019). https://doi.org/10.1016/J.SCS.2018.11.017

36. Bilal, H., Yaqub, M., Rehman, S.K.U., et al.: Performance of foundry sand concrete under ambient and elevated temperatures. Materials **12**, 2645 (2019). https://doi.org/10.3390/ma12162645
37. Noronha Marques, E., Bergmann, C.P., Masuero, A.B.: Analysis of the technical feasibility of sustainable concrete production using waste foundry sand as a fine aggregate. ACS Omega **8**, 46406–46413 (2023). https://doi.org/10.1021/acsomega.3c02998
38. Mavroulidou, M., Lawrence, D.: Can waste foundry sand fully replace structural concrete sand? J. Mater. Cycles Waste Manag. **21**, 594–605 (2019). https://doi.org/10.1007/s10163-018-00821-1
39. Gurumoorthy, N., Arunachalam, K.: Durability studies on concrete containing treated used foundry sand. Constr. Build. Mater. **201**, 651–661 (2019). https://doi.org/10.1016/J.CONBUILDMAT.2019.01.014
40. Smarzewski, P., Barnat-Hunek, D.: Mechanical and durability related properties of high performance concrete made with coal cinder and waste foundry sand. Constr. Build. Mater. **121**, 9–17 (2016). https://doi.org/10.1016/j.conbuildmat.2016.05.148
41. Naik, T.R., Kraus, R.N., Ramme, B.W., Canpolat, F.: Effects of fly ash and foundry sand on performance of architectural precast concrete. J. Mater. Civ. Eng. **24**, 851–859 (2012). https://doi.org/10.1061/(ASCE)MT.1943-5533.0000432
42. Naik, T.R., Kraus, R.N., Chun, Y., et al.: Properties of field manufactured cast-concrete products utilizing recycled materials. J. Mater. Civ. Eng. **15**, 400–407 (2003). https://doi.org/10.1061/(ASCE)0899-1561(2003)15:4(400)
43. Guney, Y., Sari, Y.D., Yalcin, M., et al.: Re-usage of waste foundry sand in high-strength concrete. Waste Manage. **30**, 1705–1713 (2010). https://doi.org/10.1016/J.WASMAN.2010.02.018
44. BSI: Tests for geometrical properties of aggregates. Part 1: Determination of particle size distribution. Sieving method - BS EN 933-1:2012. British Standards Institution, London (2012)
45. BSI: Tests for mechanical and physical properties of aggregates. Part 3: Determination of loose bulk density and voids - BS EN 1097-3:1998. British Standards Institution, London (1998)
46. BSI: Tests for mechanical and physical properties of aggregates. Part 6: Determination of particle density and water absorption - BS EN 1097-6:2022. British Standards Institution, London (2022)
47. Teychenné, D., Franklin, R., Erntroy, H.: Design of Normal Concrete Mixes, 2nd edn. Building Research Establishment (BRE), London (1997)
48. BSI: Testing Fresh Concrete. Part 2: Slump test - BS EN 12350-2:2019. British Standards Institution, London (2019)
49. BSI: Testing fresh concrete. Part 3: Vebe test - BS EN 12350-3:2019. British Standards Institution, London (2019)
50. BSI: Testing hardened concrete. Part 3: Compressive strength of test specimens - BS EN 12390-3:2019. British Standards Institution, London (2019)
51. BSI: Testing hardened concrete. Part 6: Tensile splitting strength of test specimens - BS EN 12390-6:2023. British Standards Institution, London (2019)
52. BSI: Testing hardened concrete. Part 7: Density of hardened concrete - BS EN 12390-7:2019. British Standards Institution, London (2019)
53. Roels, S., Carmeliet, J., Hens, H.: WP1: Round Robin Experimental Work - Work program, instruction document and interlaboratory comparison of the results (2002)
54. ASTM: Standard Test Method for Measurement of Rate of Absorption of Water by Hydraulic-Cement Concretes - ASTM C1585 – 20. ASTM International, West Conshohocken, PA (2020)

55. ASTM: Standard Test Method for Bulk Electrical Resistivity or Bulk Conductivity of Concrete - C1876 - 19. West Conshohocken, PA (2019)
56. Liu, S., Zheng, W., Wu, F.: Preparation of ultra-high performance concrete containing waste foundry sand and its application in structures. Structures **58**, 105472 (2023). https://doi.org/10.1016/j.istruc.2023.105472
57. Liu, S., Zheng, W., Wang, Y.: Utilization of waste foundry sand and fly ash in the production of steel fibre reinforced concrete. J. Clean. Prod. **433**, 139872 (2023). https://doi.org/10.1016/j.jclepro.2023.139872
58. Bochare, R., Dagliya, M., Paliwal, N., et al.: Sustainable concrete production using toxic foundry sand and its subsequent effect on water contamination. Sci. Total. Environ. **923**, 171551 (2024). https://doi.org/10.1016/j.scitotenv.2024.171551
59. Anand Kumar, B.G., Basutkar, S.M., Renukadevi, M.V., et al.: Use of Reclaimed Foundry Sand (RFS) as Fine Aggregate in Mortars and Concrete, pp 153–163 (2024)
60. Del Angel, G., Sainz-Aja, J.A., Tamayo, P., et al.: Effect of recycled foundry sand on the workability and mechanical properties of mortar. Appl. Sci. **13**, 3436 (2023). https://doi.org/10.3390/app13063436
61. Paul, P., Belhaj, E., Diliberto, C., et al.: Comprehensive characterization of spent chemical foundry sand for use in concrete. Sustainability **13**, 12881 (2021). https://doi.org/10.3390/su132212881
62. Deng, A., Tikalsky, P.J.: Geotechnical and leaching properties of flowable fill incorporating waste foundry sand. Waste Manage. **28**, 2161–2170 (2008). https://doi.org/10.1016/J.WASMAN.2007.09.018
63. RRUFF Quartz R040031. In: RRUFF Database. https://rruff.info/quartz/display=default/R040031. Accessed 3 Apr 2025
64. RRUFF Dolomite R040030. In: RRUFF Database. https://rruff.info/dolomite/display=default/R040030. Accessed 3 Apr 2025
65. RRUFF Calcite R040070. In: RRUFF Database. https://rruff.info/calcite/display=default/R040070. Accessed 3 Apr 2025
66. RRUFF Chromite R110059. In: RRUFF Database. https://rruff.info/chromite/display=default/R110059. Accessed 3 Apr 2025
67. Gholampour, A., Zheng, J., Ozbakkaloglu, T.: Development of waste-based concretes containing foundry sand, recycled fine aggregate, ground granulated blast furnace slag and fly ash. Constr. Build. Mater. **267**, 121004 (2021). https://doi.org/10.1016/j.conbuildmat.2020.121004
68. Furlani, E., Zanocco, M., Tubaro, E., et al.: Waste silica sand as a possible pozzolanic filler to produce cement mortars: experimental investigation. J. Mater. Cycles Waste Manag. **26**, 1795–1803 (2024). https://doi.org/10.1007/s10163-024-01939-1
69. Guettala, S., Mezghiche, B.: Compressive strength and hydration with age of cement pastes containing dune sand powder. Constr. Build. Mater. **25**, 1263–1269 (2011). https://doi.org/10.1016/j.conbuildmat.2010.09.026
70. Liu, M., Liu, E., Hao, J.L., et al.: Hydration and material properties of blended cement with ground desert sand. Constr. Build. Mater. **389**, 131624 (2023). https://doi.org/10.1016/j.conbuildmat.2023.131624
71. BSI: Concrete - Complementary British Standard to BS EN 206. Part 1: Method of specifying and guidance for the specifier - BS8500-1:2023. British Standards Institution, London (2023)
72. Singh, J., Dhiman, V.K.: Impact of used foundry sand on concrete's characteristics—an overview, pp. 659–676 (2021)
73. Bhardwaj, A., Kumar, P., Siddique, S., Shukla, A.: Comprehensive review on utilization of waste foundry sand in concrete. Eur. J. Environ. Civ. Eng. **27**, 1056–1087 (2023). https://doi.org/10.1080/19648189.2022.2070778

74. Del Angel, G.G., Thomas, C.: The use of foundry sand for recycled aggregate concrete. In: The Structural Integrity of Recycled Aggregate Concrete Produced With Fillers and Pozzolans, pp. 3–24 (2022). https://doi.org/10.1016/B978-0-12-824105-9.00014-7
75. Bhardwaj, B., Kumar, P.: Waste foundry sand in concrete: a review. Constr. Build. Mater. **156**, 661–674 (2017). https://doi.org/10.1016/J.CONBUILDMAT.2017.09.010
76. Kumar, S., Silori, R., Kumar Sethy, S.: Insight into the perspectives of waste foundry sand as a partial or full replacement of fine aggregate in concrete. Total Environ. Res. Themes **6**, 100048 (2023). https://doi.org/10.1016/J.TOTERT.2023.100048
77. Mehta, V.: Sustainable approaches in concrete production: an in-depth review of waste foundry sand utilization and environmental considerations. Environ. Sci. Pollut. Res. **31**, 23435–23461 (2024). https://doi.org/10.1007/s11356-024-32785-1
78. García, G., Cabrera, R., Rolón, J., et al.: Systematic review on the use of waste foundry sand as a partial replacement of natural sand in concrete. Constr. Build. Mater. **430**, 136460 (2024). https://doi.org/10.1016/j.conbuildmat.2024.136460
79. Burgmann, S., Breit, W.: Impact of crushed natural and recycled fine aggregates on fresh and hardened mortar properties. Constr. Mater. **4**, 37–57 (2023). https://doi.org/10.3390/constrmater4010003
80. Ouda, A.S.: Development of high-performance heavy density concrete using different aggregates for gamma-ray shielding. HBRC J. **11**, 328–338 (2015). https://doi.org/10.1016/j.hbrcj.2014.06.010
81. Neville, A.M.: Properties of Concrete, 5th edn. Pearson, Essex (2011)
82. Hall, C.: Water movement in porous building materials—I. Unsaturated flow theory and its applications. Build. Environ. **12**, 117–125 (1977). https://doi.org/10.1016/0360-1323(77)90040-3
83. Claisse, P.A.: The transport properties of concrete and the equations that describe them. In: Transport Properties of Concrete, pp. 1–16. Elsevier (2014)
84. Hall, C., Raymond Yau, M.H.: Water movement in porous building materials—IX. The water absorption and sorptivity of concretes. Build. Environ. **22**, 77–82 (1987). https://doi.org/10.1016/0360-1323(87)90044-8
85. Ibeh, C.C.: Phenol-formaldehyde resins. In: Handbook of Thermoset Plastics, pp. 23–71. Elsevier (1998)
86. Sun, W., Jiang, J., Dai, L., et al.: Electrical properties of dry polycrystalline olivine mixed with various chromite contents: Implications for the high conductivity anomalies in subduction zones. Geosci. Front. **12**, 101178 (2021). https://doi.org/10.1016/j.gsf.2021.101178

Structural Performance and Design

3D Finite Element Analysis on the Effects of Boundary Elements Confinement on Cyclic Performance of 3D Printed Concrete Wall

Syed Bustan Fatima Warsi[1(✉)], Biranchi Panda[1,2], and Pankaj Biswas[1,2]

[1] Center for Intelligent Cyber Physical Systems, Indian Institute of Technology Guwahati, Guwahati, India
w.syed@iitg.ac.in

[2] Sustainable Resources for Additive Manufacturing (SreAM) Lab, Department of Mechanical Engineering, Indian Institute of Technology Guwahati, Guwahati, India

Abstract. 3D concrete printing offers substantial benefits to the construction industry, including the fabrication of customized components, elimination of formwork, and reduction in material waste. However, achieving structural integrity and ductility through effective reinforcement integration remains a significant challenge, particularly under cyclic loading conditions. To address this, the present study investigates the influence of reinforcement detailing on the cyclic performance of 3D printed concrete (3DPC) walls. Finite element (FE) simulations are conducted on 3DPC walls with confined boundary elements to evaluate their lateral load resistance and deformation capacity. The walls are designed in accordance with codal provisions and assessed based on ductility, energy dissipation, and lateral load resistance properties. The FE model incorporates advanced material modeling to capture nonlinear cyclic behavior and provides insights into load carrying capacity, stress distribution and failure mechanisms. The effects of key design parameters such as configuration of confined boundary elements, reinforcement placement, and material anisotropy arising from the layer-by-layer printing process are considered through the micro modelling approach.

The results demonstrate that confined boundary elements significantly enhance the cyclic performance of 3DPC walls by improving ductility and delaying the catastrophic failure. Additionally, the study identifies that such reinforcement strategies are compatible with the 3D printing process, ensuring both practicality and structural effectiveness. The outcomes have important implications on bridging the gap in codified design approaches for 3DPC, while offering a framework for designing seismic-resistant 3D-printed structures and advancing their application in dynamic and large-scale structural systems.

Keywords: 3D Concrete Printing · Finite Element Analysis · Confined Boundary Elements · Structural Design · Cyclic Loading · Seismic Resistance

M. Posani and S. Singh (Eds.): RYS 2025, RILEM Bookseries 67, pp. 771–783, 2026.
https://doi.org/10.1007/978-3-032-14652-6_55

1 Introduction

3D concrete printing, an emerging digital construction technology, has gained significant attention in recent years for its potential to revolutionize the construction industry by automating the fabrication of building components, reducing formwork, and minimizing material waste [1]. This additive manufacturing technique facilitates the fabrication of geometrically intricate and customized structures with enhanced efficiency and minimal labor requirements [2]. However, while advancements in material development and printing technology have propelled 3DPC into practical applications, structural performance under dynamic and seismic loading conditions remains a significant concern—primarily due to challenges in reinforcement integration and anisotropic material behavior induced by the layer-by-layer deposition process [3].

Subsequent One of the critical limitations in structural 3DPC applications is the inability to incorporate conventional reinforcement strategies without interrupting the printing process. This shortcoming can drastically affect the load-carrying capacity, ductility, and energy dissipation properties of printed structural members when subjected to cyclic or seismic loading. Several studies have attempted to address reinforcement compatibility by investigating techniques such as post-tensioning, embedment of steel bars, or incorporation of fibers [4]. However, these methods often compromise either the automation aspect of 3DPC or fail to meet the structural standards required for seismic zones.

To overcome these challenges, the concept of confined boundary elements—a well-established strategy in conventional masonry and reinforced concrete (RC) design for enhancing seismic performance—can be revisited in the context of 3DPC [5–7]. Notably, the integration of confined boundary reinforcement can be achieved through hybrid construction techniques, wherein reinforcement cages can be post-installed or embedded during the printing process. This approach not only enhances structural integrity and ductility but also maintains the automation and efficiency benefits of 3D concrete printing. Boundary confinement has been shown to significantly improve lateral load resistance and delay the onset of critical damage by enhancing the energy absorption and ductility of structural walls. While its effectiveness is well-documented in cast-in-place RC shear walls, limited research exists on its application within 3D printed systems, particularly under cyclic loading conditions.

In this context, it becomes essential to ensure that proposed reinforcement solutions are not only effective in enhancing cyclic performance but are also consistent with codal provisions for structural walls. Confined boundary elements offer such a pathway, as they can be designed within the framework of established seismic design codes while still being adapted for integration in 3DPC through hybrid or post-installed reinforcement approaches.

Furthermore, numerical modeling techniques such as finite element analysis (FEA) have emerged as indispensable tools to explore the nonlinear cyclic behavior of model materials and structural configurations without the constraints of full-scale experimental setups. Recent studies have attempted to capture the structural response of 3DPC components using various numerical approaches, ranging from homogenized isotropic models to meso-scale and discrete element techniques [8–10]. Homogenized isotropic models are computationally efficient and widely used; however, they overlook the intrinsic anisotropy induced by layer-by-layer deposition, resulting in limited accuracy in predicting localized cracking and interfacial weaknesses [8]. Meso-scale models, which explicitly represent filaments or aggregate distributions, provide detailed insight into microstructural behavior but demand high computational cost and are often restricted to small-scale specimens [9]. Similarly, discrete element methods are effective in describing fracture processes but struggle with scaling to structural dimensions [10]. In light of these limitations, the present study advances a finite element method (FEM)-based strategy that explicitly integrates anisotropy and interlayer separation into the modeling framework. This approach balances computational tractability with physical realism, enabling the accurate capture of directional stiffness variation, crack initiation, and failure modes in 3DPC elements. By contextualizing the model against existing methods, the originality of this work lies in demonstrating how anisotropic, layered representations enhance predictive accuracy and provide a more reliable tool for assessing the mechanical behavior of hardened 3DPC components in structural applications.

Micro-modelling approaches that account for the anisotropic properties of 3DPC, such as interlayer bonding and directional strength variation, have enabled more accurate predictions of crack propagation, failure mechanisms, and stress redistribution [11].

Accordingly, this study numerically investigates the cyclic behavior of 3DPC walls incorporating confined boundary elements, with a focus on the effects of reinforcement detailing and boundary configurations using micro-modelling simulations. The objective is to offer practical insights for the seismic design of 3D printed concrete structures and to address existing limitations within current seismic design frameworks.

2 Materials and Methods

2.1 Finite Element Model Development

A comprehensive 3D finite element (FE) model is developed using ABAQUS/Standard to simulate the cyclic behavior of 3DPC walls with and without confined boundary elements. The geometry of the wall specimens are modeled based on typical configurations conforming to ACI 318 (2019) [12] and IS 456 (2000) [13] provisions for seismic-resistant reinforced concrete walls. To replicate the layer-by-layer anisotropic behavior of 3DPC, a micro-modeling approach is adopted where each printed layer is individually modelled, meshed, and material properties were directionally defined through the cohesive zone modelling with surface to surface interaction (Fig. 1). To further ensure realistic representation, the model accounts for the directional anisotropy of 3DPC arising from the layer-by-layer printing process. This inclusion allows the simulation to capture interlayer bond effects, directional stiffness variations, and localized crack initiation, thereby strengthening the reliability of the cyclic performance predictions.

In 3D finite element composite interface model, the linear and non-linear behaviors of the printed layers and interfaces are explicitly defined. The model employs 3D hexahedral-shaped eight-node linear brick elements with reduced integration and hourglass control (type C3D8R) to represent the cast top and bottom beams and 3DPC layers. Steel reinforcing bars are represented using the wire feature, associated with three-dimensional truss elements with two nodes (T3D2).

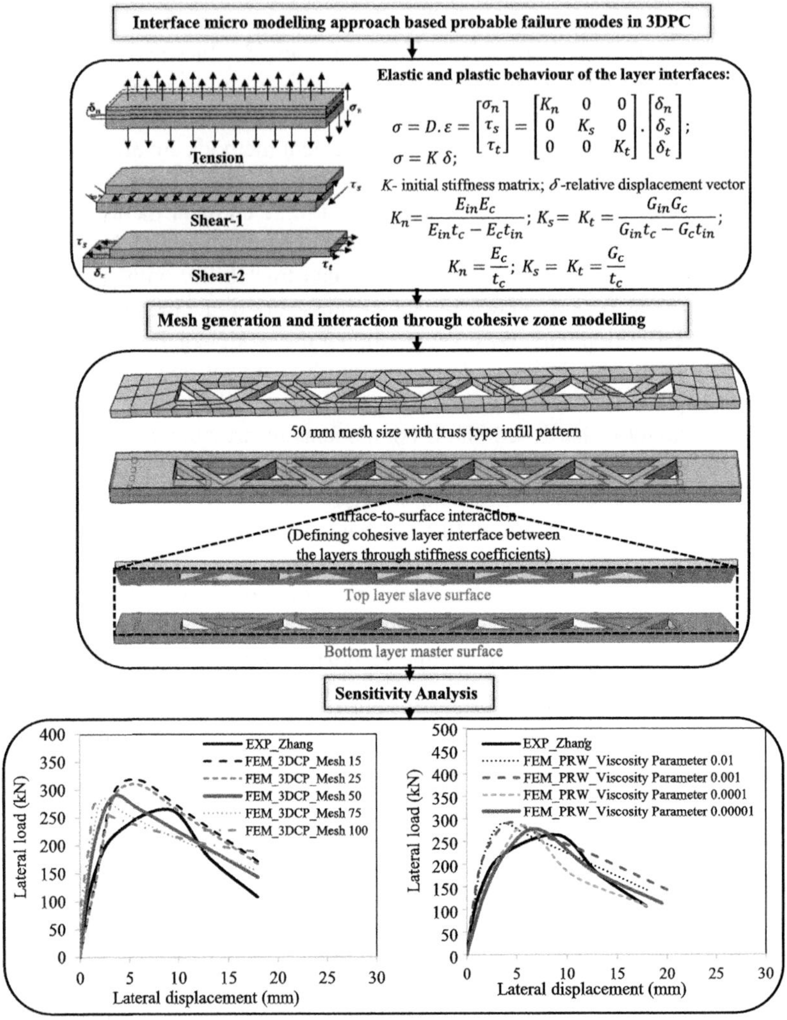

Fig. 1. Methodological approach for analyzing 3DPC wall under cyclic loading though interface micro modelling (reproduced from [11]).

For the interfaces between printed layers, a surface-to-surface discretization method is applied, incorporating cohesive interaction in the interaction property manager and utilizing a finite sliding formulation (Fig. 1b). Embedded body constraints are employed to facilitate load transfer and ensure proper bonding between the steel reinforcement and the 3DPC matrix. This constraint ensures that the reinforcement nodes exhibit identical displacement to the surrounding concrete nodes. Before selecting the mesh size for discretization purpose, a sensitivity analysis is carried out with varying mesh sizes and viscosity parametrs and 50 mm mesh with 0.00001 viscosity parameter is selected (Fig. 1).

These selections ensured accurate representation of the physical behavior while maintaining feasible simulation times. The total computational time for the micro-modelling simulations of the 3DPC walls is approximately 1528 min, which is considered reasonable for the scale of analysis conducted. This demonstrates that the adopted modeling strategy can provide reliable results without imposing impractical computational demands, thereby supporting its potential applicability in design practice.

2.2 Material Properties and Constitutive Modeling

Considering that concrete in the wall experiences both compressive and tensile stresses, the concrete damage plasticity (CDP) model is employed incorporating damage characteristics to capture the nonlinear cyclic response of the 3DPC material. The input parameters were calibrated using published experimental data on 3DPC behavior under cyclic loading [11]. Reinforcement bars were modeled as embedded truss elements using a bilinear isotropic hardening law. Table 1, serves as input in ABAQUS to accurately capture the behaviour of 3DPC wall under quasi-static cyclic loading.

Table 1. Input parameters used for the numerical simulation [11].

Interface-based model		Concrete Damage Plasticity model	
Young modulus of 3DPC, Ec (N/mm^2)	24584	Ultimate compressive and tensile stress of 3DPC, f_{cm}, f_{ctm} (N/mm^2)	27.35, 3.27
Poisson ratio, υ	0.2	Strain at peak stress, ε_{c1}	0.0022
Shear modulus of 3DPC, G_c (N/mm^2)	10243.33	k	2.32
layer thickness, t_c (mm)	15	Compressive yield stress of 3DPC, σ_{co} ($0.4 f_{cm}$) (N/mm^2)	14.4
Normal interface elastic stiffness, K_n (N/mm^3)	1639	Density (N/mm^3)	21000
Tangential interface elastic stiffness, $Ks = Kt$ (N/mm^3)	683, 683	Dilation Angle (degree)	35°
Specific fracture toughness for Mode I, G_{IC} (N/mm)	0.29	Flow potential eccentricity	0.1
Specific fracture toughness for Mode II and III, G_{IIC}, G_{IIIC} (N/mm)	0.6, 0.6	Ratio of biaxial compressive yield stress to uniaxial compressive yield stress (f_{b0}/f_{c0})	1.16
Friction coefficient	0.75	Ratio between second stress invariant on the tensile meridian and that on the compressive meridian, K	0.67
		Viscosity parameter	0.00001

2.3 Boundary Element Configurations and Loading Protocol

The analysis is conducted using the Abaqus dynamic-implicit solver in three steps: initial, gravity and pre-compression, and in-plane cyclic loading. The initial step initializes the boundary conditions with the bottom surface of the foundation beam fixed using the encastre boundary condition, which are mandatory and default for all analyses. In the subsequent step, gravity load and pre-compression load are applied (Fig. 2). Finally, in-plane cyclic displacement-controlled loading is applied at the cross-sectional center of the top beam (Fig. 2b).

Low-cycle reversed quasi-static lateral loading is applied at the top of the wall with fixed boundary conditions at the base. The loading protocol followed the guidelines of FEMA 461 (2007) [14] to simulate earthquake-induced cyclic displacement histories. Axial load representing gravity effects are also applied as a constant compressive load throughout the simulation.

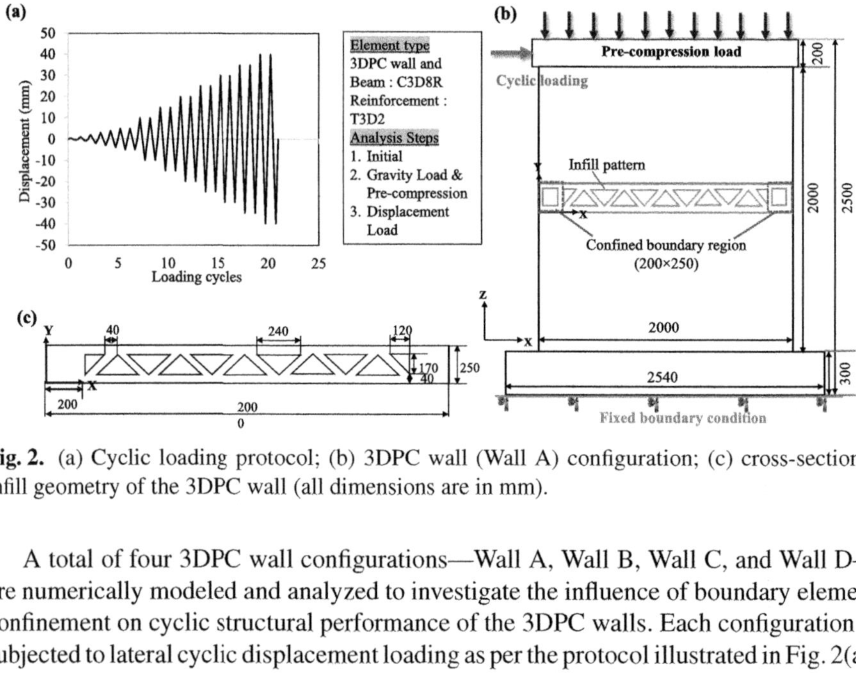

Fig. 2. (a) Cyclic loading protocol; (b) 3DPC wall (Wall A) configuration; (c) cross-sectional infill geometry of the 3DPC wall (all dimensions are in mm).

A total of four 3DPC wall configurations—Wall A, Wall B, Wall C, and Wall D—are numerically modeled and analyzed to investigate the influence of boundary element confinement on cyclic structural performance of the 3DPC walls. Each configuration is subjected to lateral cyclic displacement loading as per the protocol illustrated in Fig. 2(a), with the loading steps and boundary conditions as defined in Fig. 2(b).

- **Wall A** is modeled without any solid boundary elements, serving as the control specimen to understand the baseline behavior of a plain 3DPC infill wall.
- **Wall B** incorporates solid boundary elements without reinforcement, simulating the effect of added mass and stiffness without reinforcement.
- **Wall C** introduces vertical reinforcement within the solid boundary columns, representing an unconfined reinforced boundary system.

- **Wall D** includes both vertical reinforcement and lateral confinement in the form of transverse ties (stirrups), simulating confined boundary elements as per seismic design recommendation.

In practical terms, such a configuration (Wall D) can be achieved through a hybrid approach, wherein an infill printing pattern can be designed to create cavities in the boundary zones. These cavities allow subsequent placement of column reinforcement along with stirrups, which can then be conventionally cast with concrete. This ensures structural feasibility of the reinforcement detailing while maintaining compatibility with 3D printing processes.

2.4 Reinforcement Detailing of Designed 3DPC Walls

The reinforcement detailing of the structural 3DPC walls are illustrated in Fig. 3, which presents the configurations developed in accordance with the provisions outlined in ACI 318:19 [12]. The reinforcement layouts were finalized following a comprehensive design procedure aimed at evaluating the structural behavior under lateral cyclic loading. The detailed design procedure is described in [15].

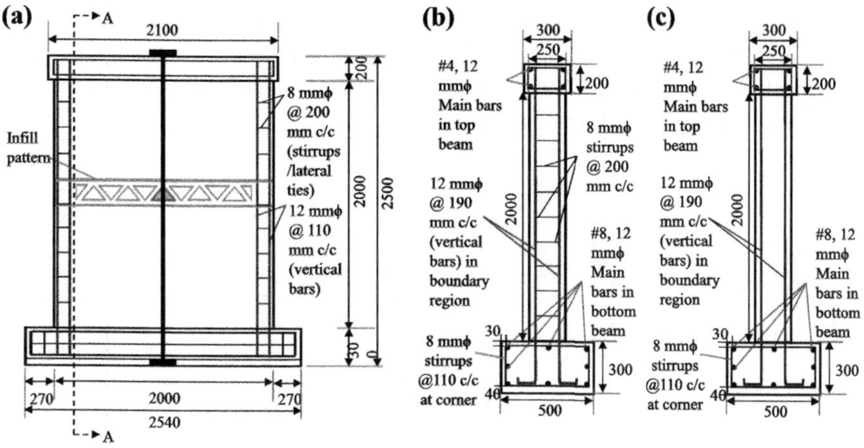

Fig. 3. Reinforcement details of the structural 3DPC wall (a) elevation of Wall D; (b) cross-section detailing of (b) Wall D along A-A; (c) Wall C along A-A.

To ensure a controlled comparison, key design parameters—concrete compressive strength, reinforcement bar strength and diameter, axial load, aspect ratio, slenderness ratio, and all other geometric parameters are kept constant across all wall models. This approach isolates the effect of boundary confinement detailing on the cyclic performance of the 3DPC walls.

3 Results and Discussion

This section evaluates the lateral performance of 3DPC walls under cyclic loading. Key parameters include peak lateral load, corresponding displacement, stress distribution, and ductility index. The results of load-displacement curves are shown in Fig. 4.

3.1 Effect of Boundary Elements on Lateral Displacement Capacity

The lateral displacement capacity increases progressively with enhancements in boundary element detailing. Wall A, which lacks any solid or reinforced boundary elements, records the lowest displacement (13.85 mm), indicating a limited ability to deform plastically (Fig. 4 and Fig. 5a). With the addition of solid boundary columns in Wall B, displacement capacity increases by 61.67%, as these columns provide improved anchorage and resistance to lateral forces. Wall C, which incorporates vertical reinforcement into the boundary elements, further increases displacement by 79.31% over Wall A. This improvement reflects the vertical bars' ability to resist tensile stresses and delay crack formation. Wall D, which combines vertical and transverse confinement reinforcement, exhibits the highest displacement (27.34 mm), marking a 97.35% increase from Wall A and a 10.06% improvement over Wall C. The additional confinement enhances crack control and post-yield deformation, promoting a highly ductile response.

These trends demonstrate that confined reinforcement improves strain compatibility and restrains lateral crack widening, thus facilitating enhanced lateral displacement without significant degradation in structural integrity.

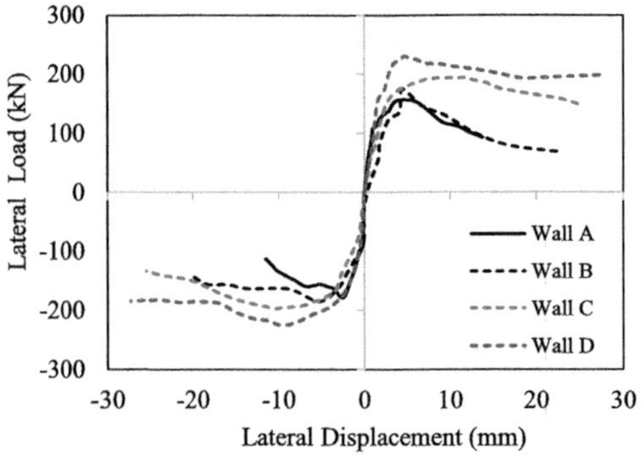

Fig. 4. Effect of boundary element on cyclic in-plane loading capacity of 3DPC wall.

3.2 Impact on Load-Carrying Capacity

A similar trend appears in the peak lateral load capacity (Fig. 4). Wall A, exhibiting early brittle failure due to the absence of confinement, carries the lowest load (155.99 kN) (Fig. 5a). The introduction of solid boundary columns in Wall B improves load resistance by 11.85%, due to better load transfer and suppression of premature shear cracking, but showing almost similar failure profile (Fig. 5b). Wall C, with vertical reinforcement, sustains a further increase of 23.67% compared to Wall A, owing to enhanced flexural resistance and energy absorption capacity. Wall D achieves the highest peak load of 229.14 kN, which corresponds to a 46.89% increase from Wall A and a 31.33% gain over Wall B. The difference between Wall C and Wall D (18.78%) underscores the additional strength imparted by transverse confinement, which limits concrete spalling and bar buckling under cyclic loads which ultimately restricted wall topling and concrete crushing (Fig. 5d). Mechanically, these load enhancements result from improved confinement effectiveness, increased axial load capacity in boundary zones, and a delayed onset of material degradation.

3.3 Stress Distribution and Crack Propagation

The stress contour plots reveal a clear correlation between boundary detailing and failure mode. Wall A shows localized stress concentration at the base, leading to premature shear cracking and rapid stiffness loss and resulted in overturning of the wall (Fig. 5a). Wall B redirects stress paths upward due to improved confinement at the edges, resulting in delayed overturning and minor concrete crushing (Fig. 5b). In Wall C, vertical reinforcement spreads the stress along the wall height, promoting a more ductile flexural failure pattern without overturning (Fig. 5c). Wall D exhibits the most favorable distribution: cracks form in a stable and well-distributed manner, with confinement reinforcement preventing excessive damage at critical regions such as toe crushing and also restricted overturning. This balanced stress path maintains structural performance even after yielding (Fig. 5d).

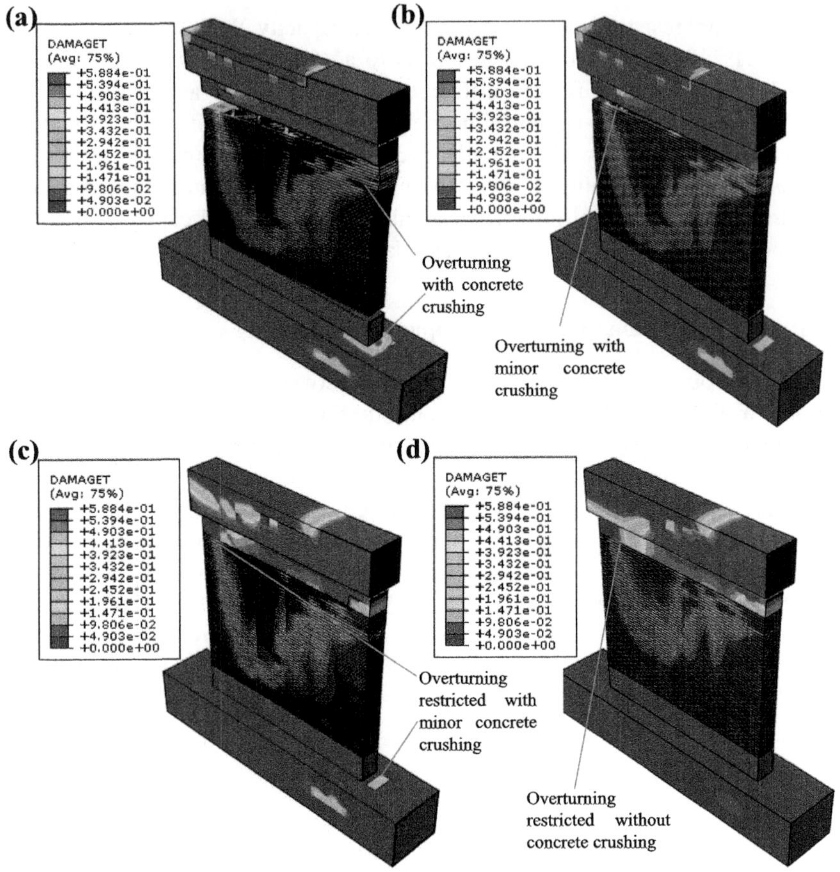

Fig. 5. Failure mode of (a) Wall A; (c) Wall B; (c) Wall C; (d) Wall D.

3.4 Ductility Index

The ductility index (μ), defined as the ratio of ultimate displacement (d_{ult}) to the yield displacement (d_{yield}), provides insight into the deformation capacity of the 3DPC wall beyond its elastic limit (Fig. 6). As shown in Table 2, Wall D exhibits the highest ductility index, reaching a value of 11.74, while Wall A displays the lowest index of 5.88, indicating a relatively brittle response. Wall B, with only solid boundary columns, achieves a ductility index of 8.66, indicating moderate improvement over Wall A. The presence of these solid elements allows the wall to sustain larger displacements before failure, though without the additional reinforcement, its post-peak behavior remains less controlled.

Wall C, with vertical reinforcement in the boundary elements, demonstrates a significant improvement, achieving a ductility index of 10.07. This reflects the vertical bars' ability to engage more effectively after yielding and contribute to strain redistribution along the height of the wall. However, the highest ductility, again, is observed in Wall D, where both vertical and lateral reinforcement confine the boundary regions, promoting enhanced post-yield performance and stable hysteretic behavior under cyclic loading.

This supports the argument for codified design practices incorporating such confinement strategies for printed structures in seismic zones. The method employed to evaluate the ductility of the simulated 3DPC walls is illustrated in Fig. 6. The load–displacement curve is idealized as a bilinear representation (elastic-perfectly-plastic) [15].

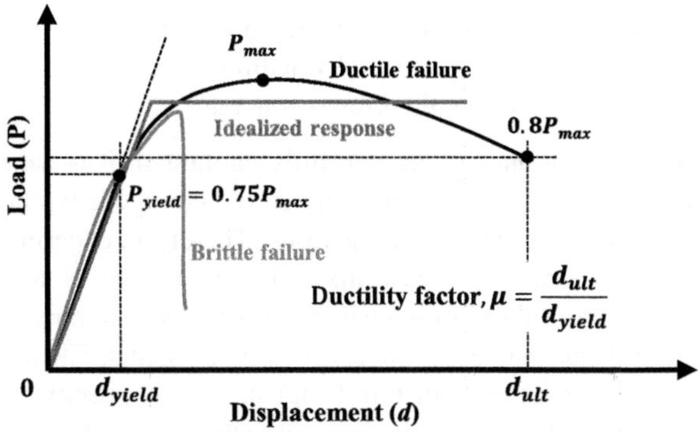

Fig. 6. Ductility assessment through idealised load–displacement curve [15].

Table 2. In plane displacement derived from the load displacement response to calculate the ductility Index.

Wall type	Loading direction	d_{yield}	d_{ult}	μ
Wall A	Push (+)	+ 2.63	+ 13.85	5.88
	Pull (−)	−1.78	−11.56	
Wall B	Push (+)	+ 2.69	+ 22.39	8.66
	Pull (−)	−2.23	−20.05	
Wall C	Push (+)	+ 2.72	+ 24.84	10.07
	Pull (−)	−2.33	−25.65	
Wall D	Push (+)	+ 2.63	+ 27.94	11.74
	Pull (−)	−2.14	−27.53	

4 Conclusions

This numerical investigation demonstrates the significant role of confined boundary elements in enhancing the cyclic performance of 3DPC walls. The integration of confined boundary reinforced elements enhances lateral load-bearing capacity, increases deformability, delays failure progression, and promotes a more ductile structural response under seismic loading conditions. The key conclusions drawn from the analysis are:

- Wall A, lacking any solid boundary and reinforcement, exhibits the lowest load-bearing capacity, displacement, and ductility among all configurations.
- Incorporation of solid boundary elements in Wall B leads to moderate improvements in lateral performance, with noticeable gains in both load and displacement capacities.
- Vertical reinforcement in Wall C significantly enhances both the strength and deformability, as evidenced by a higher ductility index and improved energy absorption.
- Confined boundary elements in Wall D provide improved stress distribution and control crack propagation, promoting a ductile failure mode.
- Proper reinforcement placement, especially within boundary zones, is essential to counteract material anisotropy in 3D printed concrete.

While the modelling parameters are carefully calibrated, it is acknowledged that numerical simulations inherently simplify material heterogeneity, interfacial bonding effects, and construction-related imperfections. Therefore, experimental validation remains indispensable to confirm these results under real loading conditions.

These findings contribute to the growing body of knowledge aiming to integrate practical reinforcement strategies into 3D printed structural elements. The outcomes serve as a framework for developing codified seismic design provisions specific to 3DPC systems and support the broader adoption of additive manufacturing in structural applications.

Acknowledgement. This work is supported by IITG Technology Innovation and Development Foundation (IITGTI&DF) at IIT Guwahati through grant number DST/NMICPS/TIH12/IITG/2020. The corresponding author acknowledges the support received from DST (file no: DST/IC/IC-IMPACTS/ 2022/P-6) in the present work.

References

1. Buswell, R.A., Leal de Silva, W.R., Jones, S.Z., Dirrenberger, J.: 3D printing using concrete extrusion: a roadmap for research. Cem. Concr. Res. **112**, 37–49 (2018)
2. Panda, B., Lim, J.H., Tan, M.J.: Mechanical properties and deformation behaviour of early age concrete in the context of digital construction. Compos. B Eng. **165**, 563–571 (2018)
3. Wolfs, R.J.M., Bos, F.P., Salet, T.A.M.: Early age mechanical behaviour of 3D printed concrete: numerical modelling and experimental testing. Cem. Concr. Res. **106**, 103–116 (2019)
4. Warsi, S.B.F., Panda, B., Biswas, P.: Exploring fibre addition methods and mechanical properties of fibre-reinforced 3D printed concrete: a review. Dev. Built Environ. **16**, 100295 (2023)
5. Borah, B., Kaushik, H.B., Singhal, V.: Analysis and design of confined masonry structures: review and future research directions. Buildings **13**, 1282 (2023)
6. Hidalgo, P.A., Ledezma, C.A., Jordan, R.M.: Seismic behavior of squat reinforced concrete shear walls. Earthq. Spectra **18**(2), 287–308 (2002)
7. Zhao, J., Shen, F., Si, C., Sun, Y., Yin, L.: Experimental investigation on seismic resistance of RC shear walls with CFRP bars in boundary elements. Int. J. Concr. Struct. Mater. **14**, 1–20 (2020)
8. Van den Heever, M., Bester, F., Kruger, J., van Zijl, G.: Numerical modelling strategies for reinforced 3D concrete printed elements. Addit. Manuf. **50**, 102569 (2022)

9. Lourenço, P.B., Rots, J.G.: Multisurface interface model for analysis of masonry structures. J. Eng. Mech. **123**(7), 660–668 (1997)
10. Huang, P., Pan, X., Niu, Y., Du, L.: Concrete failure simulation method based on discrete element method. Eng. Fail. Anal. **139**, 106505 (2022)
11. Warsi, S.B.F., Panda, B., Biswas, P.: Structural analysis of 3D-printed concrete walls under quasi-static cyclic loading using composite micro-model. Prog. Addit. Manuf., 1–21 (2024)
12. ACI Committee, Building code requirements for structural concrete, 318, American Concrete Institute (2019)
13. IS 456-2000: Plain and Reinforced Concrete- Code of Practice (Fourth Edition), Bureau of Indian Standards (2000)
14. FEMA 461: Interim Protocols for Determining Seismic Performance. Characteristics of Structural and Nonstructural Components Through Laboratory Testing. Washington, DC: Federal Emergency Management Agency (2007)
15. Warsi, S.B.F., Panda, B., Biswas, P.: Design of earthquake-resistant 3D printed concrete wall based on ACI 318-19: Analytical investigation and numerical modelling. Structures, p. 109371 (2025)

Assessment of Design Codification for Cast-In Anchors in Fibre Reinforced Concrete

Mahsa Asadollahi(✉), Marina Bock, and Nikolaos I. Tziavos

Department of Civil Engineering, Aston University, Birmingham, UK
masad24@aston.ac.uk

Abstract. Cast-in anchors are embedded into concrete substrates to provide sufficient anchorage for both structural and non-structural components. In recent years, the use of higher strength cementitious mixtures (>50 MPa) often with the inclusion of steel fibres have been gaining popularity due to superior mechanical properties and durability. However, current design codification has been developed for normal strength concrete substrates which may lead to overly conservative or even unsafe designs for cast-in anchors in Ultra High-Performance Fibre Reinforced Concrete (UHPFRC). This paper presents an assessment of currently employed design standards and analytical models from the literature. Experimental results from 247 tests have been collated and compared to design predictions. The analysis of the experimental data base focused on the following parameters: concrete mixture type and compressive strength, bearing pressure under anchor heads, and embedment depth. Design predictions significantly underestimate concrete breakout capacity in UHPFRC slabs. In some cases where cast-in fasteners are used in thin slabs, design models were found to be unsafe, with breakout capacity being overestimated by much as 28% due to specimen dimensions having an effect on failure mode. This highlights the inadequacy of current design approaches for high-strength, fibre-reinforced concretes, as they often neglect critical parameters such as anchor and specimen geometry, as well as fibre content.

Keywords: Cast-in anchors · UHPFRC · Experimental testing · Pull-out test · Concrete Capacity Design (CCD) · Design standards

1 Introduction

Fasteners embedded in concrete are widely used in various civil engineering applications. In precast construction, serving as lifting inserts [1] and employed to enable composite action – whether in sandwich panels [2] or, in bridge engineering, between steel girders and concrete decks [3]. In power plants they have been used to fix mechanical equipment or machinery [4]. Fasteners are broadly categorised by installation type into cast-in (preplaced) and post-installed [5]. Typical cast-in fastenings include single headed studs, L- or J-bolts, anchor channels and embedded plates with headed studs amongst others. In contrast, post-installed systems comprise mechanical expansion, undercut or epoxy-bonded anchors. Cast-in systems allow for more accurate positioning during installation and can be placed closer to concrete edges when appropriate reinforcement is provided [6] and as a result, they are often the preferred system for structural applications.

M. Posani and S. Singh (Eds.): RYS 2025, RILEM Bookseries 67, pp. 784–795, 2026.
https://doi.org/10.1007/978-3-032-14652-6_56

In recent years, cast-in fasteners have also been employed in façade engineering, driven by the rise of complex architectural designs and the superior strength and durability of Ultra-High-Performance Fibre-Reinforced Concrete (UHPFRC) [7]. UHPFRC also allows for notable weight reduction, making it ideal for curved façades. Figure 1 shows examples where fibre-reinforced concrete supports modern building envelopes. However, current design codes do not specifically address cast-in fasteners in UHPFRC, and there is a lack of a dedicated analytical approach. Several experimental studies on cast-in place fastening systems in fibre-reinforced substrates has been recently published. This paper presents an in-depth assessment of the commonly employed CCD method to high strength fibre-reinforced substrates to provide a further understanding on the applicability of current codification.

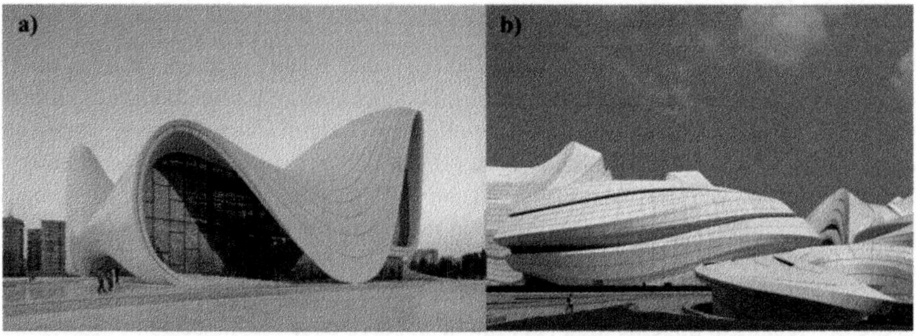

Fig. 1. Modern applications of fibre-reinforced substrates in building engineering: a) Heydar Aliyev Center, Baku [8] and b) Changsha Meixihu International Culture and Arts Centre, China [9].

2 Cast-In Fasteners for Load-Transfer Connections

Depending on their applications and load transfer mechanisms, cast-in anchors are subjected to tensile, shear, or a combined loading regime. Additionally, if eccentricities are present, they may also be subject to bending and torsion [10]. Failure modes are influenced by type of loading, material properties, as well as geometric parameters. Under tensile loading, typical failure modes include steel failure, pull-out failure, concrete breakout (CB) (see Fig. 2), side-face blowout, concrete splitting, and bond failure [11]. This study focuses on single cast-in anchors under tensile loads without edge effects or eccentricities. For investigations of connection behaviour under shear loads, the reader is referred to [12, 13]. The concrete cone breakout is illustrated in Fig. 2 and forms the foundation for the Concrete Capacity Design (CCD) method adopted in design standards [11, 14].

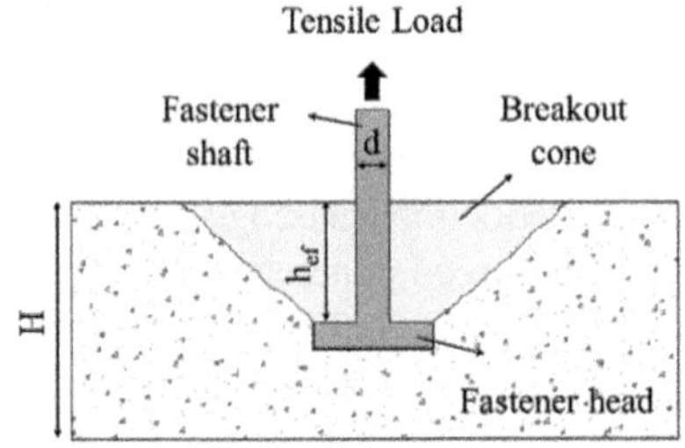

Fig. 2. Concrete cone breakout failure mode under tensile load, where H is the depth of the slab, d is the fastener shaft diameter, and h_{ef} is the effective embedment depth.

2.1 Design Codification

The design of cast-in anchors in concrete has advanced notably in recent decades, driven by the need for safer, more reliable structural connections. Early methods were mostly empirical, relying on manufacturer data with little standardisation. A key breakthrough came with the CCD method by Fuchs et al. [15], which introduced fracture mechanics to better predict concrete cone failure under tension. This method was later adopted in major codes like ACI [11], marking a shift toward more reliable and safer design. In Europe, the evolution of anchor design codes culminated in the publication of EN 1992-4 (EC2) [14], a key milestone that consolidated European Technical Approval Guidelines [16] and CEN Technical Specification [17] into a standard code for the design of fastenings in concrete.

Concrete cone breakout failure is the predominant tensile failure mode in anchorage design, especially in shallow members [18]. This failure is characterised by a conical fracture surface that extends from the anchor head to the free surface. Early models prior to the 1990s idealised this failure as a 45° cone with uniform stress distribution without accounting for the size effect, inherent in concrete fracture behaviour [19]. ACI [11] and EC2 [14] account for the quasi-brittle nature of concrete failure, recognizing fracture propagation at approximately 35° from the anchor head. The nominal breakout strength in current existing codes primarily depends on two factors: i) the effective embedment depth and ii) the concrete's compressive strength. It increases proportionally to the embedment depth raised to the 1.5 power and is further influenced by the square root of the mean concrete compressive strength [20]. For a detailed account on code development the reader is referred to earlier studies [5, 21].

According to ACI [11] the concrete breakout capacity of an isolated cast-in anchor, without edge effects and adjacent anchors, can be estimated using the general expression:

$$N_u = af_c^{0.5}h_{ef}^{\beta} \tag{1}$$

where N_u is the concrete breakout capacity under tensile load, h_{ef} is the effective embedment depth and f_c is the mean concrete compressive strength. For uncracked concrete, parameters α and β are 16.8 and 1.5 respectively. In EC2 [14] this capacity is estimated by:

$$N_u = k_1 f_{ck}^{0.5}h_{ef}^{1.5} \tag{2}$$

where k_1 is a constant parameter that varies depending on whether the anchor is cast-in (12.7) or post-installed (11) and f_{ck} is the nominal characteristic compressive cylinder strength.

2.2 Analytical Models for UHPFRC

Choi et al. [13] extended the CCD model for single cast-in anchors to estimate concrete breakout strength in UHPFRC under tensile loading, assuming no edge effects or overlapping failure cones. The proposed analytical model was calibrated using 17 tensile tests, introducing a modified constant to reflect the higher tensile strength of UHPFRC and its enhanced breakout capacity, as expressed in Eq. (3):

$$N_u = 26.1 f_c^{0.5} h_{ef}^{1.5} \tag{3}$$

Nevertheless, the abovementioned modification does not account for the effect of member thickness and anchor head size directly which can influence the tensile capacity of the connections significantly especially for large headed anchors [21, 22]. Nilforoush [22] proposed an analytical model incorporating modification factors to account for the effects of member thickness (Ψ_H), anchor head size (Ψ_{AH}) and surface reinforcement (Ψ_{Sr}). The modified Equations and modification factors are expressed as follows:

$$N_{u,m} = 6.7 f_c^{0.5} h_{ef}^{5/3} \tag{4}$$

$$N_u = N_{u,m}\, \Psi_H\, \Psi_{AH}\, \Psi_{Sr} \tag{5}$$

where $N_{u,m}$ refers to the mean concrete cone breakout failure load for a single anchor, assuming it is not influenced by adjacent anchors or the edges of the concrete member.

3 Design Assessment

3.1 Experimental Data

For this study a total of 247 unconfined pull-out tests [23] on cast-in anchors within concrete substrates were collated. The test data were categorised based on the compressive strength as follows:

- Normal strength: $f_c \leq 50$ MPa
- High Performance: $50 < f_c < 90$ MPa
- Ultra-High-Performance: $f_c \geq 90$ MPa

The resulting notation is: i) Normal strength Concrete (NC), ii) Normal strength Reinforced Concrete (NRC), iii) Normal strength Fibre Reinforced Concrete (NFRC), iii) High Performance Concrete (HPC), iv) High Performance-Fibre Reinforced Concrete (HPFRC) and v) UHPFRC. An overview of the collated experimental results is shown in Table 1. The investigations incorporate various fibre reinforcements, such as hooked end, straight steel fibres (SF), polyoxymethylene (POM), and hybrid combinations.

Table 1. Summary of collated experimental data

Source	Nr of tests	**fc (MPa)	Concrete type	Fibre type	Fibre content (%)	***Failure mode
Choi et al. [13]	1	33	NC	-	-	CB
Nilforoush [22]	13	34	NC	-	-	CB or CB & S
*Elfgren and Nilsson [24]	30	31.1	NRC	-	-	-
*Eligehausen et al. [25]	34	27.1, 26.4, 26.6	NRC	-	-	-
*Zhao [26]	43	16.2,18.9, 19.3, 20.1, 23.5	NRC	-	-	-
*Lee et al. [27]	3	37.2	NRC	-	-	-
Nilforoush [22]	6	34	NRC	-	-	CB
	4	34	NFRC	SF	1	CB & RC
Li et al. [28]	18	71.1	HPC	-	-	CB or StF
Li et al. [3]	6	71.3	HPC	-	-	CB
Nilforoush [22]	2	68	HPC	-	-	CB
	4	68	HPFRC	SF	1	CB & RC
Lu et al. [29]	4	83.8	HPFRC	SF	2	CB & S or S
	10	126.7	UHPFRC	SF	2	CB or CB & S
Li et al. [3]	6	93	UHPFRC	SF	1	CB or StF
	12	97.9	UHPFRC	Hybrid	2	CB or StF
Choi et al. [13]	16	199	UHPFRC	SF	2	CB or S
Osman and Fam [2]	35	115, 119	UHPFRC	POM	1–3	CB & RC

*Extracted as reported in [22].
**The 'f_c' column in this table indicates the mean concrete cylinder compressive strength.
***CB: Concrete breakout, RC: Radial Cracks, S: Splitting, StF: Steel Failure.

3.2 Effect of Embedment Depth

Effective embedment depth is one of the main factors governing the ultimate tensile capacity of cast-in anchors and is directly included in the CCD model (Eqs. (1) and (2)). Figure 3 illustrates the normalised ultimate loads from 177 pull-out tests conducted on NC and NRC plotted against the embedment depth as reported for each test campaign. The normalisation removes the influence of compressive strength and allows the effect of anchor embedment depth on ultimate capacity to be isolated. Predictions from the CCD model are also highlighted, as defined by Eq. (1). It is shown that the increased embedment depth leads to higher tensile capacity and that the CCD predictions capture the overall behaviour very well across the dataset.

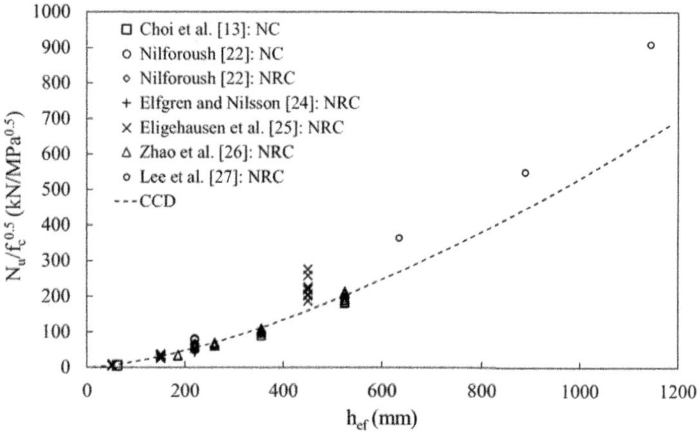

Fig. 3. Normalised tensile capacity of cast-in anchors in NC as a function of h_{ef}

Figure 4 shows the normalised tensile capacity of anchors embedded in UHPFRC. Similar trends to NC can be observed, with an exponential increase in tensile capacity with increasing embedment. It should be noted that for the presented tests embedment depths up to 80 mm were investigated. While the CCD model generally provides safe predictions for anchors in UHPFRC, with only a few exceptions that are unsafe, it tends to be quite conservative for most of the safe cases. Similarly, the model proposed in Choi et al. [13] appears to be unsafe for lower dosages of steel fibres, however it captures the steeper increase in capacity with embedment depth. In addition to steel fibres, recent studies have focused on the inclusion of alternative fibres such as polyoxymethylene (POM) fibres [2]. As shown in Fig. 5 three fibre contents were tested and all test data remain consistently below analytical predictions. This pronounced overestimation highlights the limitations of applying models developed for NC to cast-in fasteners in fibre-reinforced substrates. It is evident that POM fibres do not provide the same enhancement in tensile anchorage capacity as the steel counterparts. This is likely due to the limited effectiveness of POM fibres to crack bridging and lower modulus of elasticity.

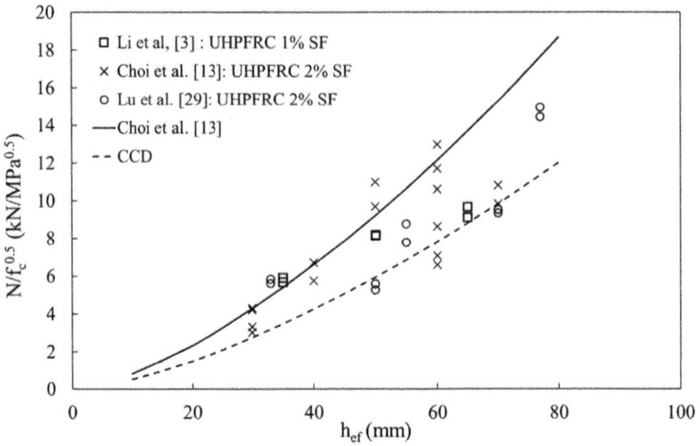

Fig. 4. Normalised tensile capacity of cast-in anchors in UHPFRC with steel fibres as a function of h_{ef}

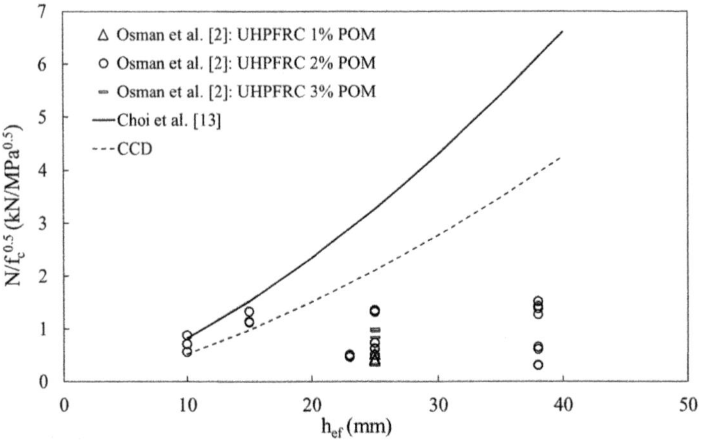

Fig. 5. Normalised tensile capacity of cast-in anchors in UHPFRC with POM fibres as a function of h_{ef}

3.3 Effect of Bearing Pressure

Under tensile loads cast-in anchors undergo a complex three-dimensional stress state beneath the anchor head. In practice, this region undergoes a combination of compressive, tensile, and shear stresses, with significant stress concentrations which is difficult to capture in analytical models. As a result, anchor head geometry is not directly accounted for in design models. To account for the influence of head size, ACI [11, 31] limits the bearing stress to approximately 15 times the concrete compressive strength [21, 22, 32].

To investigate the effect of anchor head, the bearing pressure (σ_b) is commonly employed [10] as determined using Eq. (6):

$$\sigma_b = N_u / A_h \tag{6}$$

where, N_u is the ultimate tensile load carried by the anchor and A_h is the effective bearing area under the fastener head.

Smaller head sizes can often lead to stress concentrations promoting localised crushing, reducing the embedment depth and therefore the tensile capacity of the connection [5, 30]. In Fig. 6 the normalised bearing stress is plotted against anchor embedment depth. Anchors in NC and NRC consistently show the highest ratios, often exceeding the reference value of 15. HPC shows lower ratios, while UHPFRC remains consistently below 5, allowing full concrete cone development. Notably, the UHPFRC tests focused on shallow embedment depths. Even with 1% fibre content, including POM fibres, bearing stress ratios decline possibly due to stress redistribution around the anchor head. However, as h_{ef} increases, NRC also exhibits reduced bearing stress, indicating that current design equations may not fully capture the behaviour of deep embedment cases.

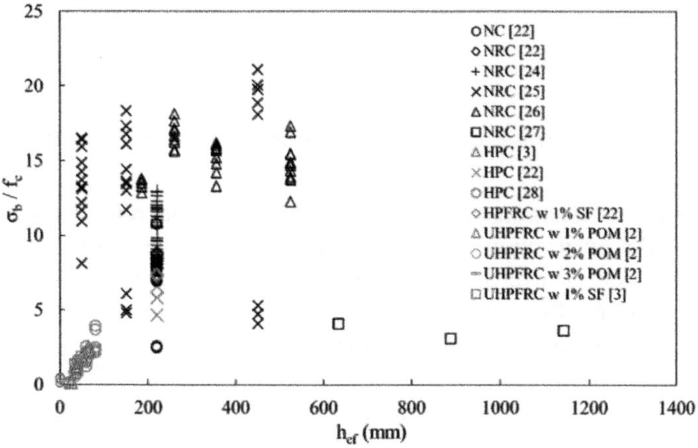

Fig. 6. Normalised bearing stress (σ_h/f_c) for collated data set. NRC: Normal strength Reinforced Concrete without fibres; SF: Steel Fibres; POM: POlyoxyMethylene fibres

3.4 Effect of Member Thickness

Slabs with depth to embedded depth ratio (H/h_{ef}) less than two are classified as thin elements whereas slabs with ratios greater than two are considered deep slabs [22]. For thin elements edge and thickness effects are more pronounced. As Fig. 7 shows, thin slabs have wide scatter in test results likely due to the truncation of the concrete breakout cone by the slab's bottom surface which causes splitting failure which is also in agreement with earlier studies [22, 33]. This reduction in effective breakout volume is not adequately accounted for by the CCD method, and current data are insufficient to confirm if fibre reinforcement can fully mitigate this limitation.

In deep slabs, a full breakout cone can develop, generally resulting in safe predictions. For UHPFRC in such configurations, however, the CCD model tends to yield conservative estimates. This is likely due to the fibres' ability to bridge cracks and provide post-cracking resistance, allowing anchors to carry more load than expected in conventional concrete. The scatter seen in the deep slab regime suggests that, while the CCD model gives a conservative baseline, the actual capacity gains depend on factors like fibre orientation, distribution, and mix design. Nonetheless, the overall trend confirms that UHPFRC can reliably exceed CCD predictions in deep slabs.

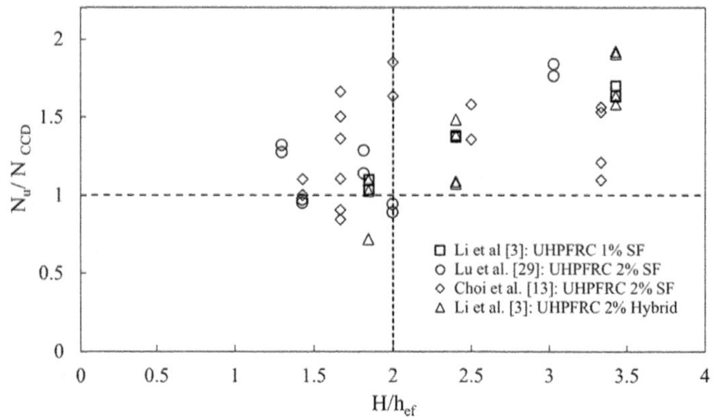

Fig. 7. Ultimate tensile capacity to CCD prediction for varying slab thickness

In Fig. 8 pull out tests on NC slabs with and without fibre reinforcement are examined, with similar trends observed to UHPFRC data. This reveals that the CCD method systematically results in overconservative predictions for deep NC slabs, even in unreinforced specimens. For thin NC slabs the CCD predictions exhibit significant variability with test data appearing on both sides of the reference line.

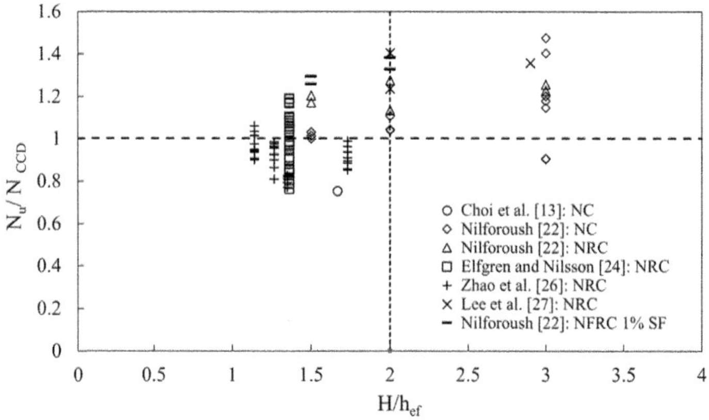

Fig. 8. Ultimate Tensile Capacity to CCD Prediction for varying slab thickness

4 Conclusion

This paper presented a literature review on cast-in anchors in UHPFRC along with a design assessment of the current codes and analytical models based on a collated experimental database. The following conclusions were drawn:

1. Cast-in anchors in UHPFRC deep slabs exhibit higher ultimate tensile capacities than those predicted by the CCD model, owing to improved mechanical properties and mitigating potential cracking during the early stages of loading. The behaviour of thin slabs is notably more complex with greater scatter in CCD predictions as concrete breakout can be accompanied by splitting failure mode.
2. Fibre content is a critical parameter in determining the ultimate capacity of connections, with higher fibre volumes leading to increased capacities; however, limited experimental evidence is available.
3. For substrates with non-steel fibres, analytical predictions require further recalibration. This highlights the need for further research to develop accurate design equations specifically tailored to fibre types.
4. UHPFRC exhibits lower bearing stress under headed anchors compared to NC, especially at lower embedment depths. This indicates that higher capacities can be achieved with UHPFRC even when using small headed anchors, as the reduced bearing stress helps to avoid pull-out failure of the anchor and instead promotes concrete breakout failure. The interaction between the concrete cone and the fibres contributes to enhancing the overall connection performance especially in applications requiring superior performance at reduced embedment depths.

Acknowledgements. The authors gratefully acknowledge the financial support provided by Aston University for funding the first author's doctoral studies.

References

1. Obayes, O., Gad, E., Lee, J., Pokharel, T., Abdouka, K.: Assessment of the tensile behaviour of cast-in headed anchors with void formers in early age concrete. Constr. Build. Mater. **317**, 125851 (2022). https://doi.org/10.1016/J.conbuildmat.2021.125851
2. Osman, H., Fam, A.: Pull-out strength of cast-in versus post-installed connectors in thin ultra-high performance concrete panels with polyoxymethylene fibers. J. Build. Eng. **103**, 112078 (2025). https://doi.org/10.1016/j.jobe.2025.112078
3. Li, F., et al.: Pullout behavior of studs in ultra-high performance concrete with steel fibers and novel structural fibers. Structures **44**, 405–417 (2022). https://doi.org/10.1016/J.ISTRUC.2022.08.017
4. Delhomme, F., Roure, T., Arrieta, B., Limam, A.: Pullout behavior of cast-in-place headed and bonded anchors with different embedment depths. Mater. Struct./Materiaux et Constructions **49**(5), 1843–1859 (2016). https://doi.org/10.1617/s11527-015-0616-4
5. Di Nunzio, G., Muciaccia, G.: Cast-in-place fasteners under tensile loading: a critical review. Structures **41**, 1532–1545 (2022). https://doi.org/10.1016/j.istruc.2022.05.059
6. Petersen, D., Lin, Z., Zhao, J.: Behavior and Design of Cast-in-Place Anchors under Simulated Seismic Loading Final Report (Volume II) Reinforcement for Headed Anchors (2013). https://nees.org
7. Akeed, M.H., et al.: Ultra-high-performance fiber-reinforced concrete. Part III: Fresh and hardened properties. Case Stud. Constr. Mater. **17**, e01265 (2022). https://doi.org/10.1016/j.cscm.2022.e01265
8. Zaha Hadid Architects. https://www.zaha-hadid.com/heydar-aliyev-center. Accessed 11 May 2025

9. Zaha Hadid Architects. https://www.zaha-hadid.com/changsha. Accessed 11 May 2025
10. Di Nunzio, G., Muciaccia, G.: A combined concrete cone breakout-pull-out model for cast-in-place fasteners (2025). https://doi.org/10.1002/suco.202400926
11. American Concrete Institute (ACI) Committee: ACI 318-19_Building Code Requirements for Structural Concrete and Commentary (2019). https://doi.org/10.14359/51716937
12. Lee, J.H., Choi, E., Cho, B.S.: Shear failure mode and concrete edge breakout resistance of cast-in-place anchors in steel fiber-reinforced normal strength concrete. Appl. Sci. **10**(19), 1–19 (2020). https://doi.org/10.3390/app10196883
13. Choi, S., Joh, C., Chun, S.C.: Behavior and strengths of single cast-in anchors in Ultra-High-Performance Fiber-Reinforced Concrete (UHPFRC) subjected to a monotonic tension or shear. KSCE J. Civ. Eng. **19**(4), 964–973 (2015). https://doi.org/10.1007/s12205-013-0246-8
14. European Committee for Standardization (CEN): EN 1992-4:2018_ Eurocode 2 – Design of concrete structures: Part 4 – Design of fastenings for use in concrete (2018)
15. Fuchs, W., Eligehausen, R., Breen, J.E.: Concrete Capacity Design (CCD) approach for fastening to concrete. ACI Struct. J. **92**(1), 73–94 (1995). https://doi.org/10.14359/1533
16. European Technical Approval Guidelines (ETAG) 001: Metal anchors for use in concrete – Annex C: Design methods for anchorages (1997)
17. Mallée, R., Fuchs, W., Eligehausen, R.: Design of fastenings for use in concrete – The CEN/TS 1992-4 provisions (2013). https://doi.org/10.1002/9783433602737
18. Stehle, E.J., Sharma, A.: Concrete cone breakout behavior of anchor groups in uncracked concrete under displacement-controlled cyclic tension load. Eng. Struct. **246**, 113092 (2021). https://doi.org/10.1016/j.engstruct.2021.113092
19. ACI Committee 349: ACI 349.2R-07_Guide to the Concrete Capacity Design (CCD) Method – Embedment Design Examples (2007)
20. Karmokar, T., Mohyeddin, A., Lee, J.: Predictive models for concrete cone capacity of cast-in headed anchors in geopolymer concrete. Eng. Struct. **285**, 116025 (2023). https://doi.org/10.1016/j.engstruct.2023.116025
21. Karmokar, T., Mohyeddin, A., Lee, J.: Tensile behaviour of cast-in headed anchors in ambient-temperature cured geopolymer concrete. Eng. Struct. **266**, 114643 (2022). https://doi.org/10.1016/j.engstruct.2022.114643
22. Nilforoush, R.: Anchorage in concrete structures – numerical and experimental evaluations of load-carrying capacity of cast-in-place headed anchors and post-installed adhesive anchors. Ph.D. dissertation (2017)
23. European Organisation for Technical Assessment (EOTA): Mechanical fasteners for use in concrete, EAD 330232-01-0601, pp. 1–122, December 2019
24. Elfgren, L., Nilsson, M.: Load-carrying capacity of cast-in headed studs in concrete structures – influence of surface reinforcement, Technical report 2009:1192221:1, Luleå University of Technology, 183 p. (2009)
25. Eligehausen, R., Bouška, P., Červenka, V., Pukl, R.: Size effect of the concrete cone failure load of anchor bolts. In: Bažant, Z.P. (ed.) Proceedings 1st International Conference on Fracture Mechanics of Concrete Structures (FRAMCOS 1), pp. 517–525 (1992)
26. Zhao, G.: Tragverhalten von randfernen Kopfbolzenverankerungen bei Betonbruch (in German). Ph.D. dissertation, University of Stuttgart, 204 p. (1993)
27. Lee, N.H., Kim, K.S., Bang, C.J., Park, K.R.: Tensile-headed anchors with large diameter and deep embedment in concrete. ACI Struct. J. **104**(4), 479–486 (2007)
28. Li, F., et al.: Tensile behavior of stud connectors in high strength concrete. Adv. Struct. Eng. **24**(16), 3677–3690 (2021). https://doi.org/10.1177/13694332211029731
29. Lu, K., Xu, Q., Wang, M., Yao, Y., Wang, J.: Anchorage performance of bolt connection embedded in thin UHPC members. Structures **34**, 1253–1260 (2021). https://doi.org/10.1016/j.istruc.2021.08.060

30. Di Nunzio, G., Muciaccia, G.: A literature review about the head-size effect on the capacity of cast-in anchors, pp. 1–12 (2020). https://doi.org/10.21012/FC10.239783
31. American Concrete Institute (ACI) Committee 349: ACI 349-01_Code Requirements for Nuclear Safety Related Concrete Structures (2001)
32. Ožbolt, J., Eligehausen, R., Periškić, G., Mayer, U.: 3D FE analysis of anchor bolts with large embedment depths. Eng. Fract. Mech. **74**(1–2), 168–178 (2007). https://doi.org/10.1016/j.engfracmech.2006.01.019
33. Nilforoush, R., Nilsson, M., Elfgren, L.: Experimental evaluation of influence of member thickness, anchor-head size, and orthogonal surface reinforcement on the tensile capacity of headed anchors in uncracked concrete. J. Struct. Eng. **144**(4), 1–14 (2018). https://doi.org/10.1061/(asce)st.1943-541x.0001976

Low-Velocity Impact Behavior of Auxetic Cementitious Cellular Composites (ACCCs)

Jinbao Xie(✉) and Branko Šavija

Delft University of Technology, Delft 2628CN, The Netherlands
J.Xie-1@tudelft.nl

Abstract. Auxetic cementitious cellular composites (ACCCs) exhibit promising mechanical properties under static loading conditions, including high fracture resistance and effective energy dissipation. However, their performance under impact loading remains largely unexplored. In this study, two ACCCs—designated P25 and P50—featuring different aspect ratios were designed using additive manufacturing (AM)-assisted casting and evaluated under low-velocity impacts using a Schmidt hammer with consistent impact energy. Impact resistance was assessed based on energy absorption, localized damage, crack propagation, and peak reaction force. In addition to single-impact testing, multiple impacts were applied until specimen failure, with performance compared against a reference specimen incorporating circular holes. Strain distribution during impact was captured using Digital Image Correlation (DIC) with a high-speed camera. A numerical model accounting for strain rate effects was developed to simulate the impact behavior of the ACCCs. The results reveal that the ACCCs significantly outperformed the reference design in terms of impact resistance, showing reduced localized damage, increased contact stiffness, and enhanced energy absorption under multiple impacts. This improved performance is attributed to the auxetic behavior, which pulls more material into the impact zone, enhancing energy dispersion and minimizing localized damage, thereby preserving the overall structural integrity. Among the two designs, P50 demonstrated superior impact resistance due to its enhanced auxetic behavior, which engages more ligaments in energy dissipation and further reduces localized damage. Given the widespread availability of cementitious materials, this study highlights the potential of ACCCs as lightweight, high-performance protective structural materials for impact mitigation in infrastructure applications.

Keywords: Auxetic cementitious cellular composites · Additive manufacturing · Impact loading · Energy absorption · Impact model · Strain rate effect

1 Introduction

Auxetics are materials that exhibit a negative Poisson's ratio [1]. Xu et al. [2] developed a cellular structure with periodic elliptical perforations using a fiber-reinforced cementitious composite as the base material. Under uniaxial compression, this structure demonstrated a distinctive auxetic response, characterized by a negative Poisson's

M. Posani and S. Singh (Eds.): RYS 2025, RILEM Bookseries 67, pp. 796–810, 2026.
https://doi.org/10.1007/978-3-032-14652-6_57

ratio. The auxetic behavior in the elliptical-shaped ACCC was primarily attributed to crack-bridging mechanisms within the cementitious matrix [2, 3]. Owing to their auxetic properties, ACCCs display favorable mechanical characteristics under static loading conditions, such as large deformation capacity and effective energy dissipation. Despite these advantages, no research has been conducted on their impact resistance. Given the widespread availability of cementitious materials, ACCCs are promising candidates for lightweight, high-performance protective structural materials for impact mitigation in infrastructure.

The novelty of this study lies in the evaluation of ACCCs' impact resistance based on their responses under low-velocity impacts, including rebound value, energy absorption, localized damage in the impact zone, crack propagation, and peak reaction force during impacts. Additionally, a numerical model incorporating strain rate effects was developed to simulate the impact behavior of ACCCs, providing further insight into their impact mitigation mechanisms. Given the widespread availability of cementitious materials, lightweight and high energy-absorbing ACCCs have potential as sacrificial protective components in infrastructure—such as building walls, columns, and bridge piers—by dissipating low-velocity impact energy and mitigating localized damage. Furthermore, the proposed ACCCs can contribute to the development of next-generation shock-absorbing Engineered Materials Arresting Systems (EMAS) [4, 5], where high-energy absorbing materials are installed at runway ends to reduce the impact of aircraft overruns.

2 Experimental Tests

2.1 Specimen Fabrication

The compressive deformation behavior of ACCCs was strongly affected by their geometric configurations. In our previous work [3], we found that elliptical ACCCs with chiral sections—specifically specimens P25 and P50—exhibited a negative Poisson's ratio (i.e., auxetic behavior) under quasi-static compression. This auxetic response resulted from cracking that triggered section rotation. The designations P25 and P50 refer to a 25% and 50% increase, respectively, in the major axis length compared to the reference specimen P0, which has circular holes. Based on these findings, P25 and P50 were selected for further evaluation under impact loading, while P0 served as the control specimen for comparison. The geometric parameters of P0, P25, and P50 are presented in Fig. 1. The aspect ratio of the holes in ACCCs—defined as the ratio of the major axis to the minor axis—was greater in P50 than in P25, indicating that P50 offers less compressible space. The quasi-static test results for the three specimens, previously presented in the earlier study [3], are illustrated in Fig. 2.

The specimens were produced using an "indirect printing" technique combined with additive manufacturing-assisted casting. Initially, a 4 × 4 cell structure was fabricated using an Ultimaker 2+ Fused Deposition Modeling (FDM) 3D printer with ABS (Acrylonitrile Butadiene Styrene) as the printing material. The ABS templates were first placed in a cardboard box. After being mixed in a 1:1 ratio, Poly-Sil PS 8510—a two-part silicone rubber—was then poured into the box. Following a room-temperature curing process lasting at least one hour, the silicone rubber solidified and was removed from

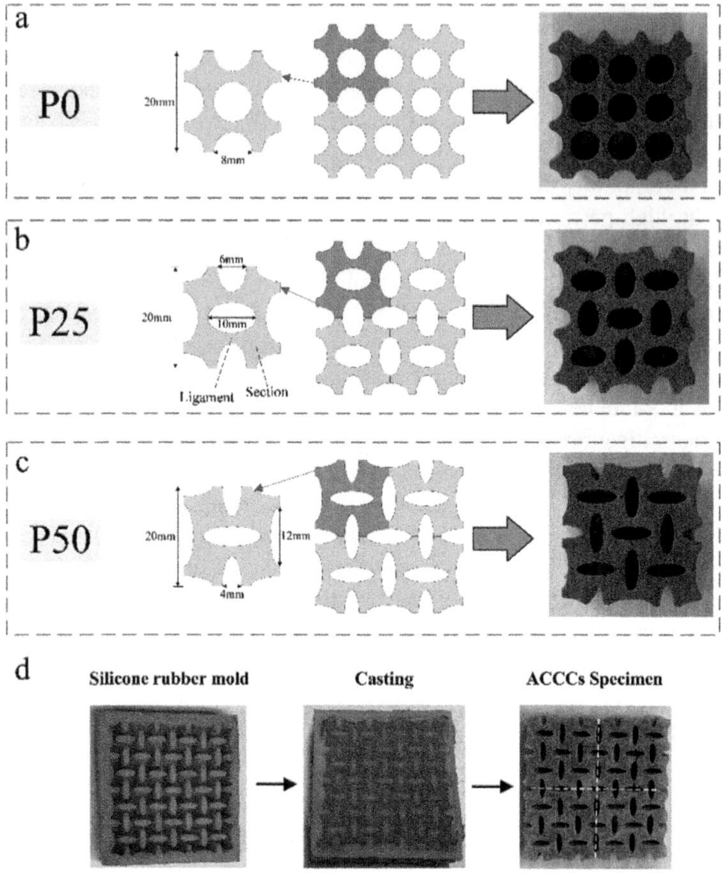

Fig. 1. Specimen fabrication.

the box. Cementitious materials were subsequently cast into the silicone rubber molds (see Fig. 1d). After setting, the cementitious structures were easily demolded due to the flexibility and reusability of the silicone molds.

The mixture composition of the ACCCs, as shown in Table 1, consisted of CEM I 42.5 N cement, fly ash, fine sand (with a grain size of 125–250 μm), water, a polycarboxylate-based superplasticizer, a viscosity modifying agent (VMA), and Polyvinyl Alcohol (PVA) fibers. The base material used was a fine-grained fiber-reinforced mortar, with PVA fibers incorporated at a volume fraction of 2%. To improve fiber dispersion, methylcellulose powder was added as a VMA. The superplasticizer used was MasterGlenium 51 from BASF (Germany), which ensured the required workability. A water-to-binder ratio of 0.46 was selected to facilitate the casting of the cementitious mix into the small molds, ensuring adequate flow and filling capacity. The mixing and curing procedures of specimens were consistent with those described in the earlier study [2, 3, 6, 7]. Upon completion of curing, the specimens were cut using a precision machine, dividing each 4 × 4 cell unit into four smaller 2 × 2 cell specimens. The cutting lines are indicated in white in Fig. 1d.

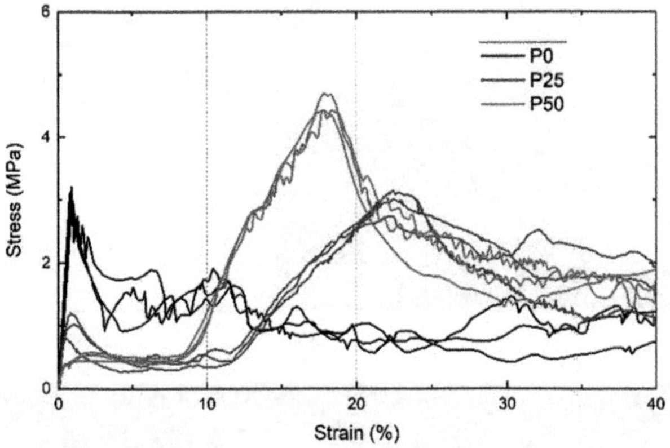

Fig. 2. Quasi-static tests results of the three specimens under uniaxial compression [3]

Table 1. Mix ratios of ACCCs (kg/m^3).

Cement	Fly ash	Sand (125 μm to 250 μm)	Water	Superplasticizer (Glenium 51)	VMA	PVA Fiber
453	535	370	450	1.58	0.29	25.6

2.2 Impact Test Setup

To evaluate cellular geometries with small dimensions under impact loading, a test was conducted using a Schmidt hammer OS-120PM, which simulates a drop-weight impact test with controlled specific impact energy, as illustrated in Fig. 3. The Schmidt hammer OS-120PM is commonly employed to assess the compressive strength of low-strength mortars through low-velocity impacts. Its working principle is based on the balance of energy between the elastic energy stored in its internal springs and the energy absorbed or dissipated during impact within a specimen. As depicted in Fig. 3a, upon contact with the specimen, the energy transferred from the hammer increases over time, reaching a maximum value at the point of peak compression just before the impactor begins to rebound. The energy level then decreases once the hammer loses contact with the specimen, and this final stabilized value corresponds to the energy absorbed by the specimen.

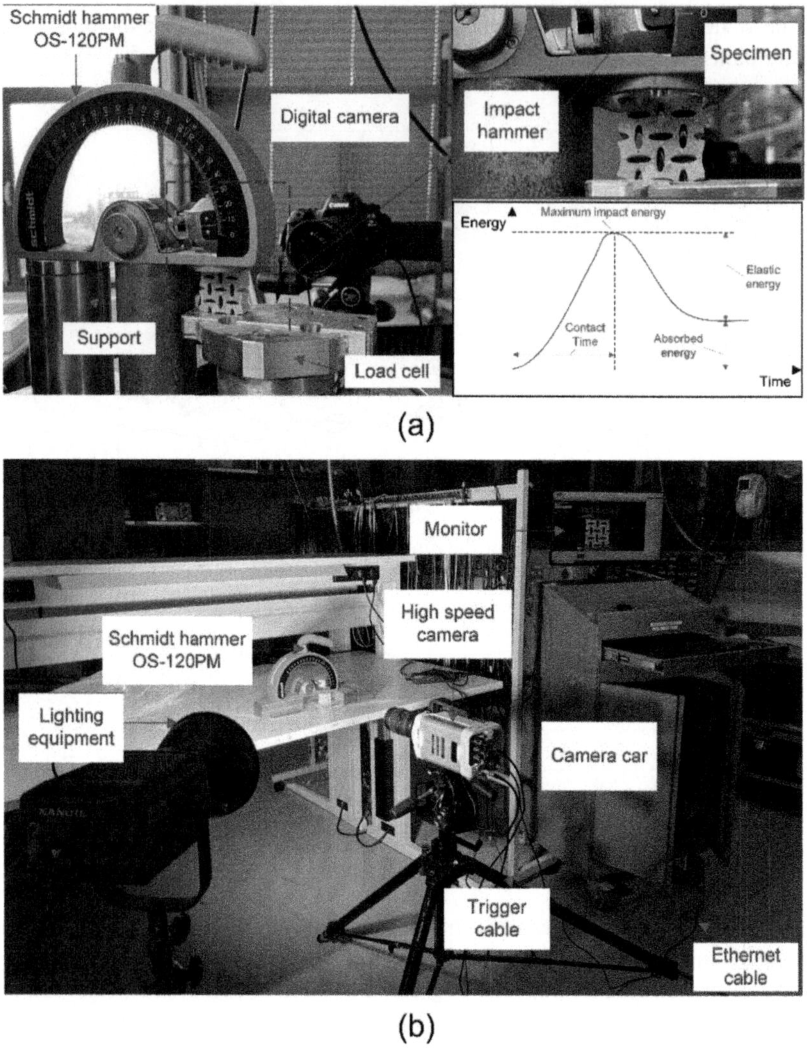

Fig. 3. (a) Experimental setup for multiple impact tests, (b) impact test setup using high-speed camera.

For the Schmidt hammer OS-120PM, each impact delivers 0.833 J of kinetic energy to the specimen via its impact tip. Before impact, the internal spring is compressed to a fixed initial position, denoted as $x_0 = 75$. After striking the specimen, the impact tip rebounds to a final position, denoted as *R*, which is recorded as the rebound value. This value represents the amount of elastic energy returned to the hammer and serves as an indicator of the specimen's ability to recover elastically. A higher rebound value suggests that the specimen returns a greater portion of the input energy, indicating better elasticity and reduced local damage or plastic dissipation. Conversely, when the rebound value approaches zero, it implies that nearly all of the impact energy has been absorbed by the specimen through localized failure and inelastic deformation, leaving no recoverable energy to return to the hammer. Neglecting minor losses due to heat and sound, the absorbed energy E can be estimated using Eq. (1).

$$E = \frac{x_0 - R}{x_0} * E_0 \tag{1}$$

where E_0 is the initial kinetic impact energy. For the Schmidt hammer OS-120PM, $E_0 = 0.833$ J.

Figure 3a presents the experimental setup used for conducting multiple impact tests on the specimens. Each specimen was placed on the load cell of the Instron 8874 testing machine, which recorded the reaction force generated during each impact. A digital camera was employed to document the crack patterns that developed on both the front and back sides of the specimens following each impact. To ensure proper alignment, supports were installed so that the top surface of the specimen aligned precisely with the bottom surface of the Schmidt hammer's impact tip (as shown in Fig. 3a). During the initial impact, the impact tip was positioned to strike the center of the specimen's top surface. For all subsequent impact tests, precise alignment was maintained to ensure that the hammer tip consistently contacted the same location as in the initial impact. To ensure the reliability and repeatability of the results, a minimum of three specimens was tested for each geometric configuration.

Following a similar setup to the impact test illustrated in Fig. 3a, a high-speed camera was employed to capture the strain distribution of ACCCs during the brief duration of the impact, as shown in Fig. 3b. The camera used was the Photron FASTCAM Mini AX200, which records at a resolution of 1 megapixel (1024 × 1024 pixels) with a frame rate of 6400 frames per second (fps). It was connected to a camera car via an Ethernet cable and operated through a computer. Additional lighting was used to ensure proper illumination of the specimens during impact loading. The testing procedure began with initiating the recording on the computer controlling the high-speed camera. The Schmidt hammer was then triggered to apply impact, and the recording was stopped using a trigger cable. The system operated in "end" trigger mode with a total recording duration of 3.41 s, meaning it captured the 3.41 s immediately preceding the trigger event. From this recorded sequence, a short segment of the impact event was selected and extracted for subsequent Digital Image Correlation (DIC) analysis. Analysis was limited to the initial impact, as visible damage was observed in the specimens prior to the application of a second impact.

3 Results and Discussion

3.1 Impact Resistance

Figure 4 compares the impact responses of the three specimens in terms of peak reaction force, rebound value, and energy absorption under multiple impacts. As shown, the rebound value decreases progressively for all specimens with increasing impact number, attributed to the accumulation of local damage in the impact zone and plastic dissipation throughout the structure. Conversely, energy absorption increases with the number of impacts, reflecting the inverse relationship with rebound behavior due to ongoing material degradation. Among the specimens, P50 exhibits the highest impact endurance, withstanding up to 11 impacts before local failure occurs in the impact zone. P25 subsequently withstands seven impacts prior to local failure, whereas P0 endures only five impacts before complete damage develops in the impact zone.

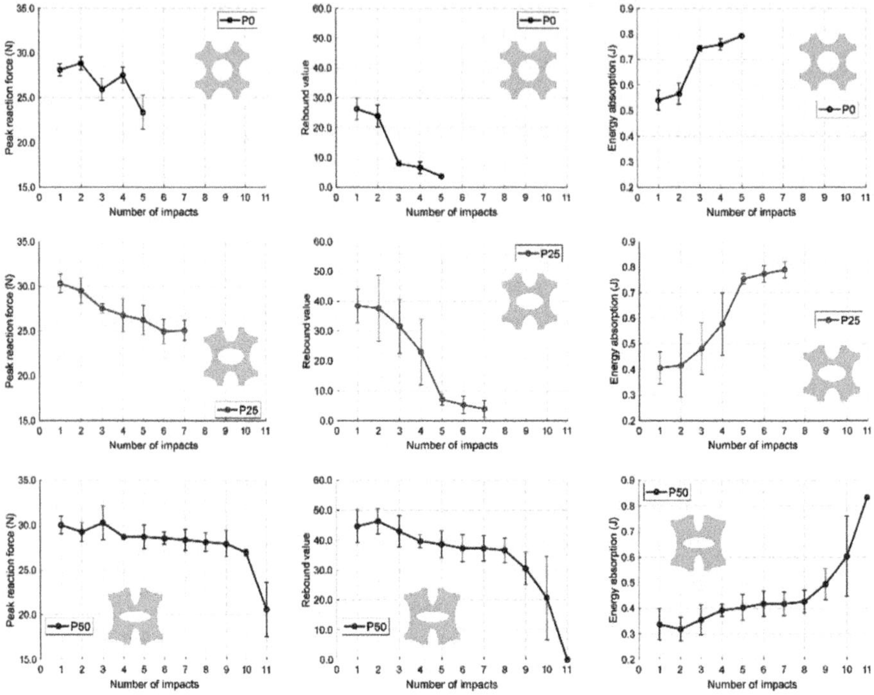

Fig. 4. Results of multiple impact tests conducted on the three specimens.

As shown in Fig. 4, from the onset of impact until failure in the impact zone, P50 exhibits the slowest progression of local damage, whereas P25 shows a faster rate, and P0 displays the most rapid development of local damage. After two impacts, P0 begins to exhibit a rapid rate of local damage progression. Similarly, P25 also shows a swift rate of local damage after two impacts, though slower than that of P0. In contrast, P50 maintains a significantly slower rate of damage development compared to both P0 and P25. In terms of accumulated energy absorption upon failure, P0 records the lowest value among the three specimens, with 3.4 J after five impacts. ACCC specimens P25 and P50 demonstrate higher energy absorption capacities before failure. Specifically, P25 reaches 4.2 J after seven impacts, while P50 achieves the highest accumulated energy absorption of 5 J after 11 impacts.

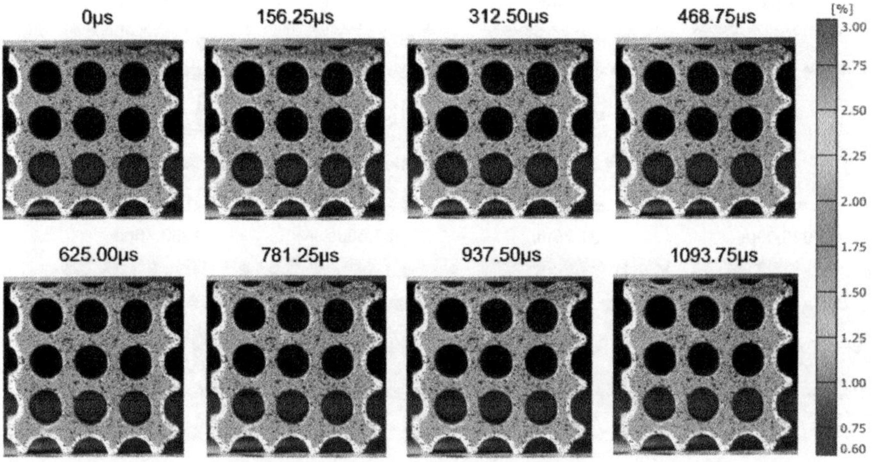

Fig. 5. Major strain of P0 during the first impact

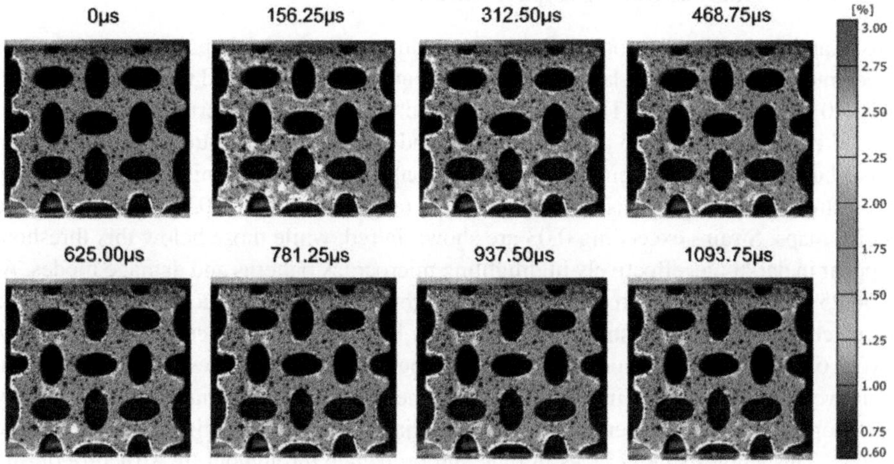

Fig. 6. Major strain of P25 during the first impact

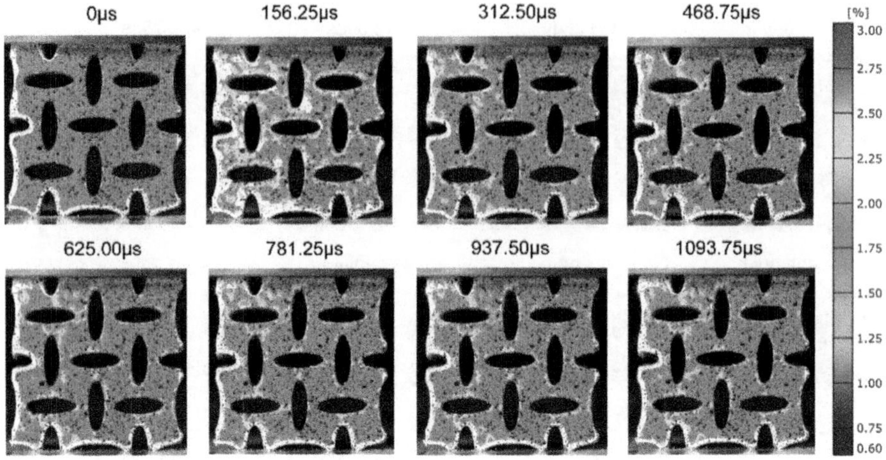

Fig. 7. Major strain of P50 during the first impact

3.2 Strain Analysis During Impact Duration

To examine the distribution of major strain during the initial impact, the state of the specimens captured in the last frame before contact was considered the initial condition (i.e., 0 μs in Figs. 5–7). The time interval between frames, determined by the high-speed camera, was 156.25 μs. Figures 5, 6 and 7 present the evolution of major strain distribution across all specimens during the contact period under initial impact loading. To better visualize these distributions, a strain range of 0.006 to 0.03 was applied in the color maps. Strains exceeding 0.03 are shown in red, while those below this threshold appear in dark blue, effectively highlighting microcrack patterns and damage modes. At 156.25 μs, the sudden force from impact slightly affects the accuracy of DIC. As the contact time progresses, strain concentrations in the ligaments increase and intensify. After 625.00 μs, the variations in strain distribution among the three specimens remain relatively small. Notably, in the ACCC specimens P25 and P50, major strain primarily develops along the ligaments that connect adjacent holes, indicating a ductile failure mode. This suggests that cracks initiate and propagate throughout the structure during impact, allowing for widespread energy dissipation. According to our previous study [2], the auxetic behavior in ACCC specimens results from lateral inward movement of unit cells due to rotation and inward folding under uniaxial compression. Under impact loading, the ligaments experiencing significant strain function as joints, enabling section rotation and folding, which in turn triggers auxetic behavior. This facilitates the engagement of surrounding material in the impact zone, enhancing the structure's ability to distribute impact energy effectively. Specifically, specimen P50 displays a broader and more uniform distribution of major strain across multiple ligaments compared to P25, which explains its superior resistance to multiple impacts, as shown in Fig. 4. In contrast, specimen P0 exhibits concentrated strain in the central ligaments near the impact point, indicating a brittle damage mode. Here, cracks form immediately near the impact site, with little further propagation, limiting energy dissipation. Due to its structural design, P0 does not support auxetic deformation through section rotation, resulting in poor energy transfer and localized damage. Finally, the inherent brittleness and heterogeneity of cementitious materials lead to noticeable strain concentrations around the holes in all specimens as a result of impact-induced stress localization.

$$\upsilon = -\frac{\varepsilon_x}{\varepsilon_y} \tag{2}$$

The variations in compressive and lateral strain during the initial impact were analyzed using DIC, as shown in Fig. 8. In standard uniaxial compression tests, specimens are placed between two plates and subjected to uniform compressive deformation, allowing for straightforward calculation of Poisson's ratio using Eq. (2). However, the impact scenario in this study involves localized loading, which results in non-uniform compressive deformation. Therefore, the compressive and lateral strains were indirectly used to assess each specimen's Poisson's ratio during the impact event. In Fig. 8, the vertical distance between two red markers on each specimen was used to measure compressive strain (strain in the y-direction). Similarly, lateral strain (strain in the x-direction) was determined by measuring the horizontal displacement between two red markers. All specimens exhibit an increase in compressive strain during impact, which eventually stabilizes at the end of the contact duration. The ACCC specimens show higher compressive strain compared to specimen P0, reflecting their greater deformability. Notably, specimen P25 reaches a maximum compressive strain of 1.64%, whereas P50 only reaches 0.26%. This difference is attributed to P25 having more compressible void space and less cementitious matrix material. In Fig. 8, P0 consistently exhibits positive lateral strain, indicating structural expansion during the impact. Because the impact on P0 is localized, the overall structural deformation is minimal, resulting in the smallest lateral strain magnitude. In contrast, the ACCC specimens show negative lateral strain, indicating contraction during the impact. Specifically, P25 initially exhibits contraction with a peak negative strain of 0.05%, then transitions to expansion (positive strain) after approximately 156.25 μs. Meanwhile, P50 maintains negative lateral strain throughout, reaching a peak contraction of 0.54%. According to Eq. (2), P50 demonstrates greater lateral strain and lower compressive strain compared to P25, which indirectly suggests a higher magnitude of negative Poisson's ratio—indicating more pronounced auxetic behavior.

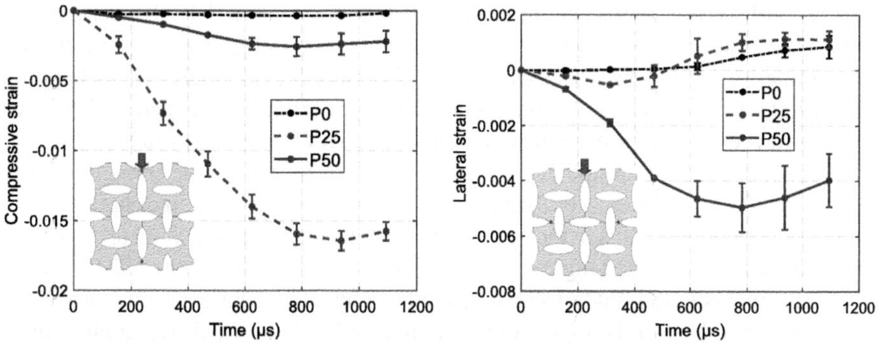

Fig. 8. DIC analysis of compressive and lateral strain evolution during the first impact.

4 Numerical Modeling

4.1 Simulation of ACCCs Under Impact Loading

The Concrete Damage Plasticity (CDP) model, incorporating strain rate effects, was developed using the Explicit Dynamics module in ABAQUS to simulate the impact behavior of ACCC specimens under dynamic loading. The ACCC specimen was discretized using CPS3 elements—three-node linear triangular plane stress elements that are free from hourglass effects. Table 2 presents the CDP model parameters for the ACCC specimens, where ρ, ν, E_0 denote the density, Poisson's ratio, and initial elastic modulus of the cementitious matrix, respectively. Material parameters related to the tensile and compressive behavior of the cementitious matrix, obtained from static tests in a previous study [6], are provided in Table 3. The bottom plate in the model, representing a steel plate consistent with the experimental setup, was assigned an elastic modulus of 206 GPa, a Poisson's ratio of 0.3, and a density of 7800 kg/m^3. As shown in Fig. 9, surface-to-surface contact interactions were defined between the impact hammer and the cementitious specimens. Frictional effects between the specimen and hammer tip were neglected. To prevent excessive interpenetration or overclosure in the normal direction, a hard contact interaction was applied between the specimen and the hammer tip. The same surface-to-surface contact method was also used between the specimen and the bottom plate. Additionally, the bottom surface of the steel plate was coupled to a reference point, which was constrained in all degrees of freedom.

In this study, the Dynamic Increase Factor (DIF) for tensile behavior is found to be more pronounced than that for compression, as the auxetic effect in ACCCs primarily results from tensile cracking within the ligaments. According to existing literature on the strain rate sensitivity of concrete under tension, the DIF exhibits a bilinear relationship with strain rate when plotted on a logarithmic scale. Notably, there is no significant increase in DIF for strain rates below 10^{-6} s^{-1}, and a noticeable change in slope is observed at a strain rate of approximately 1.0 s^{-1} [8]. The following summarizes the DIF for cementitious materials under tensile loading at high strain rates, as reported in existing literature. At an experimental strain rate of approximately 157 s^{-1}, concrete subjected to tension demonstrated a DIF of around 7.0 [8]. The SHCC (Strain-Hardening Cementitious Composite) experiments revealed a pronounced strain rate sensitivity in tensile strength, yielding a DIF of 6.7 within the measured strain rate range of 140 s^{-1} to 180 s^{-1} [9]. The enhancement in tensile strength observed in SHCC under dynamic loading conditions is comparable to the strain rate effects reported for conventional and high-performance concretes. Specifically, the tensile strength DIF reached approximately 6.0 for conventional concrete, while for high-performance concrete (HPC) and fiber-reinforced ultra-high-performance concrete (UHPC), the values ranged from 3.8 to 5.3 at a strain rate of around 100 s^{-1} [10]. The DIF for the tensile strength of plain geopolymer (GP) matrices was found to be approximately 6.5, as measured using a gravity-driven split-Hopkinson tension bar (SHTB) at strain rates reaching up to 300 s^{-1} [11]. With a unit cell side length of 20 mm and a ligament thickness of 2.0 mm, the ACCC specimens in this study were subjected to impact by a hammer tip at a velocity of 1582.8 mm/s. Due to the high strain rate involved, a DIF of 8.5 was applied in the numerical model. This value was determined by calibrating the simulation results against experimental observations. In contrast, the strain rate sensitivity of the elastic modulus has a significantly smaller influence [12, 13]. Therefore, the influence of strain rate on the elastic modulus was negligible in this study.

Table 2. CDP model parameters for ACCCs.

ρ_{aux} (kg/m^3)	E_0 (MPa)	ν	σ_{b0}/σ_{c0}	K_c	ψ	ε_{ec}	Viscosity Parameter
1870	3997	0.2	1.16	0.667	35	0.1	0.001

Table 3. Tensile behavior and compressive behavior parameters.

Tensile behavior		Compressive behavior	
Yield stress (MPa)	Displacement (mm)	Yield stress (MPa)	Inelastic strain
2.358	0	8.376	0
1.283	0.0519	12.273	0.0158
1.671	0.180	12.027	0.0367
1.477	0.265	10.984	0.0765
1.235	0.505	10.586	0.0889
0.959	0.669	10.060	0.1264
0.641	0.962	9.948	0.1557
0.334	1.551	10.720	0.2030
0.193	1.995	11.333	0.2312

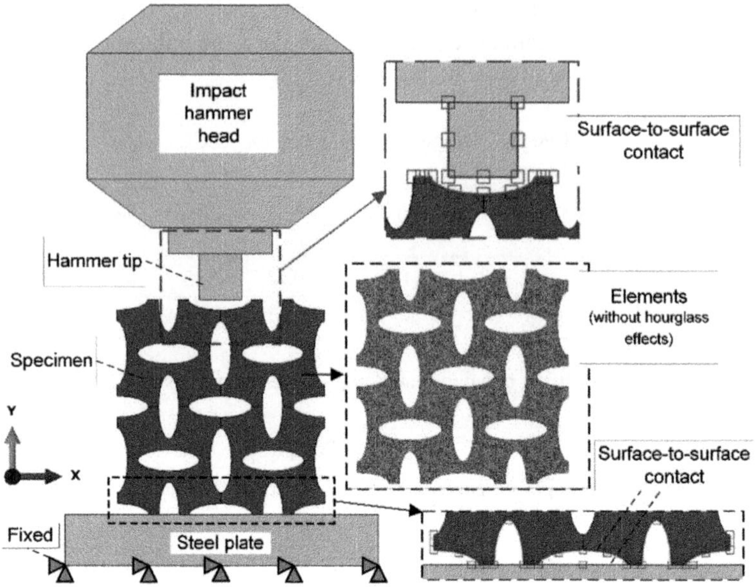

Fig. 9. Simulation of ACCCs under impact loading conditions.

4.2 Modelling Results and Discussion

In Fig. 10a, the simulation results align well with the experimental data regarding the energy absorption of the three specimens during the initial impact, thereby validating the model. In the experiments, the energy absorption of specimens P0, P25, and P50 during the initial impact was 0.541, 0.407, and 0.337, respectively. In the simulations, these values were 0.550, 0.495, and 0.356, respectively. Figure 10b illustrates the variation in internal energy of the three specimens during the initial impact in the model. When the artificial strain energy is negligible in the model, the internal energy can primarily be considered as the sum of plastic dissipated energy (Fig. 10c) and elastic strain energy (Fig. 10d). Upon contact with the impact hammer, all three specimens experience a rapid increase in elastic strain energy, followed shortly by a rise in plastic dissipated energy. As a result, the hammer's kinetic energy decreases as it is transferred into the internal energy of the specimens, as depicted in Fig. 10e. Compared to P0, specimens P25 and P50 exhibit a quicker rise in elastic strain energy and a delayed onset of plastic dissipation following impact. The internal energy of the specimens peaks when the hammer's kinetic energy is fully transferred and reduced to zero, at which point both elastic and plastic energy components reach their maximum values. Subsequently, a portion of the elastic strain energy is transferred back to the hammer through the reactive contact force and resulting acceleration, causing the internal energy of the specimens to decrease. This process leads to the hammer gradually disengaging and rebounding with reversed velocity. Eventually, the internal energy of the specimens stabilizes as the hammer separates and its kinetic energy levels off, leaving behind plastic dissipated energy and a smaller amount of elastic strain energy. It is noteworthy that P25 and P50 stabilize at lower levels of plastic dissipated energy compared to P0. Specifically, P50 shows a slower rate of energy dissipation and reaches the lowest stabilized value among the three. As shown in Fig. 10f, the total energy remains the same in the model throughout the initial impact.

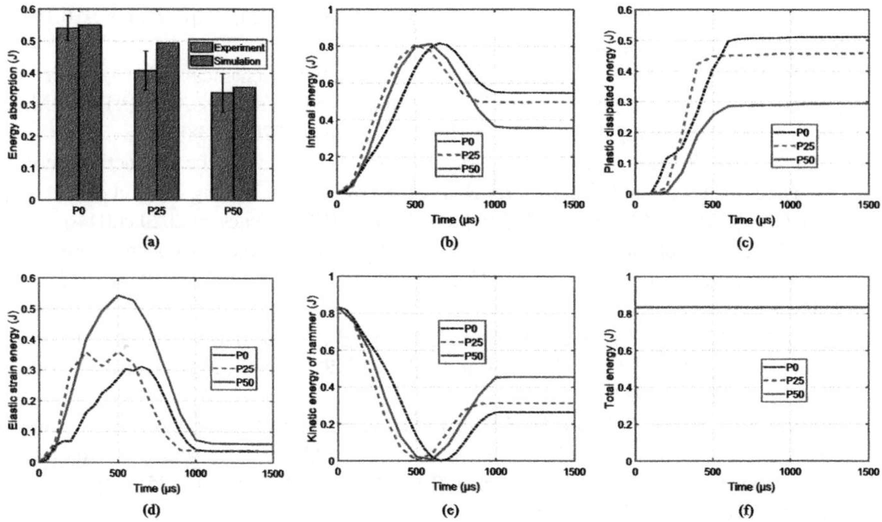

Fig. 10. Energy evolution in the model during the first impact test, (a) energy absorption after the first impact, (b) internal energy, (c) plastic dissipated energy, (d) elastic strain energy, (e) kinetic energy of hammer, (f) total energy.

5 Conclusions

ACCC specimens display a pronounced distribution of major strain along the ligaments across a wide region of the structure, a behavior attributed to their auxetic characteristics. This strain dispersion facilitates the engagement of surrounding material into the impact zone, promoting a more effective transfer of impact energy throughout the structure. As a result, contact stiffness is enhanced and localized damage within the impact zone is mitigated, helping to preserve overall structural integrity. In contrast, for P0 specimens, major strain is confined primarily to the central ligaments near the point of impact.

Compared to the reference specimen (P0), ACCC specimens exhibit reduced localized damage during the initial impact due to their auxetic behavior. This leads to higher peak reaction forces and greater elastic strain energy per impact. Throughout the multiple impacts, ACCC specimens accumulate local damage at a slower rate, enabling them to maintain higher peak reaction forces and endure a greater number of impacts before failure. This allows for higher total energy absorption prior to structural failure.

Among the ACCC specimens, P50 demonstrates superior impact resistance compared to P25, owing to its more pronounced auxetic behavior resulting from a higher aspect ratio. This configuration generates greater bending moments, enabling more ligaments to participate in energy dissipation through rotation-induced plastic deformation. As a result, P50 more effectively reduces localized damage.

References

1. Evans, K.E.: Auxetic polymers: a new range of materials. Endeavour **15**, 170–174 (1991). https://doi.org/10.1016/0160-9327(91)90123-S
2. Xu, Y., Zhang, H., Schlangen, E., Luković, M., Šavija, B.: Cementitious cellular composites with auxetic behavior. Cem. Concr. Compos. **111**, 103624 (2020). https://doi.org/10.1016/j.cemconcomp.2020.103624
3. Xu, Y., Schlangen, E., Luković, M., Šavija, B.: Tunable mechanical behavior of auxetic cementitious cellular composites (CCCs): experiments and simulations. Constr. Build. Mater. **266**, 121388 (2021). https://doi.org/10.1016/j.conbuildmat.2020.121388
4. Ketabdari, M., Toraldo, E., Crispino, M., Lunkar, V.: Evaluating the interaction between engineered materials and aircraft tyres as arresting systems in landing overrun events. Case Stud. Constr. Mater. **13**, e00446 (2020). https://doi.org/10.1016/j.cscm.2020.e00446
5. Barri, K., et al.: Multifunctional nanogenerator-integrated metamaterial concrete systems for smart civil infrastructure. Adv. Mater. **35**, 2211027 (2023). https://doi.org/10.1002/adma.202211027
6. Xie, J., Xu, Y., Meng, Z., Liang, M., Wan, Z., Šavija, B.: Peanut shaped auxetic cementitious cellular composite (ACCC). Constr. Build. Mater. **419**, 135539 (2024). https://doi.org/10.1016/j.conbuildmat.2024.135539
7. Xie, J., Xu, Y., Meng, Z., Liang, M., Zhou, Y., Šavija, B.: Impact behavior of auxetic cementitious cellular composites (ACCCs) architected through additive manufacturing (AM) assisted casting: experiment and modelling. Constr. Build. Mater. **471**, 140692 (2025). https://doi.org/10.1016/j.conbuildmat.2025.140692
8. Malvar, L.J., Ross, C.A.: Review of strain rate effects for concrete in tension. ACI Mater. J. 95 (1998). https://doi.org/10.14359/418
9. Mechtcherine, V., Millon, O., Butler, M., Thoma, K.: Mechanical behaviour of strain hardening cement-based composites under impact loading. Cem. Concr. Compos. **33**, 1–11 (2011). https://doi.org/10.1016/j.cemconcomp.2010.09.018
10. Nöldgen, M., Millon, O., Thoma, K., Fehling, E.: Hochdynamische Materialeigenschaften von Ultrahochleistungsbeton (UHPC). Beton- und Stahlbetonbau **104**, 717–727 (2009). https://doi.org/10.1002/best.200900038
11. Trindade, A.C.C., Heravi, A.A., Curosu, I., Liebscher, M., de Andrade Silva, F., Mechtcherine, V.: Tensile behavior of strain-hardening geopolymer composites (SHGC) under impact loading. Cem. Concr. Compos. **113**, 103703 (2020). https://doi.org/10.1016/j.cemconcomp.2020.103703
12. Schuler, H., Mayrhofer, C., Thoma, K.: Spall experiments for the measurement of the tensile strength and fracture energy of concrete at high strain rates. Int. J. Impact Eng **32**, 1635–1650 (2006). https://doi.org/10.1016/j.ijimpeng.2005.01.010
13. Thomas, R.J., Sorensen, A.D.: Review of strain rate effects for UHPC in tension. Constr. Build. Mater. **153**, 846–856 (2017). https://doi.org/10.1016/j.conbuildmat.2017.07.168

Methodology for the Selection of Process Parameters in 3D Printing Cementitious Materials: Multicriteria Analysis and Impact on Quality

Khristel Rodríguez[1(✉)], María De Los Ángeles Ortega Del Rosario[1,2,3], Melany Medina[1], Ana Sofia Guimarães[4], and Manuel Jesus[4]

[1] Research Group in Design, Manufacturing and Materials (DM+M), Faculty of Mechanical Engineering, Universidad Tecnológica de Panamá, Panama City 0819-07289, Panama
{khristel.rodriguez,maria.ortega}@utp.ac.pa

[2] Centro de Estudios Multidisciplinarios en Ciencias, Tecnología (CEMCIT-AIP), Panama City 0819-07289, Panama

[3] Sistema Nacional de Investigación (SNI), Clayton City of Knowledge Edif. 205, Panama City 0819, Panama

[4] LFC Research Group, CONSTRUCT Unit, Faculty of Engineering (FEUP), University of Porto, Rua Dr. Roberto Frias, 4200-465 Porto, Portugal

Abstract. The selection of process parameters plays an important role in determining the quality and performance of 3D-printed cementitious structures, as well as the overall efficiency of the construction process. However, the proper choice of these parameters varies depending on the application, materials, and geometric complexity involved. Furthermore, the diverse quantitative and qualitative metrics used to assess these parameters often complicate direct comparisons.

This study introduces a systematic methodology for selecting process parameters using three multicriteria decision analysis methods (MCDM): TOPSIS (Technique for Order of Preference by Similarity to Ideal Solution), VIKOR, and AHP (Analytic Hierarchy Process). TOPSIS and VIKOR are used to rank parameter sets by their closeness to an ideal or compromise solution, while AHP derives criteria weights through pairwise comparisons. The three rankings are integrated with the Borda count technique to obtain a robust final ordering. An illustrative case study shows how the framework balances printability, dimensional accuracy, geometric complexity, and mechanical performance. The approach is generic and will be validated with experimental data in future work.

The proposed methodology evaluates parameters such as printing speed, nozzle diameter, and environmental conditions like humidity and temperature to identify combinations that balance multiple performance criteria, including mechanical properties, printability, geometric complexity, and dimensional accuracy of printed elements.

This framework offers a replicable approach for researchers seeking to improve the consistency and quality of 3D-printed structures based on cementitious materials by highlighting the most influential parameters for adjustment. The findings contribute to advancing additive manufacturing in construction by

M. Posani and S. Singh (Eds.): RYS 2025, RILEM Bookseries 67, pp. 811–820, 2026.
https://doi.org/10.1007/978-3-032-14652-6_58

providing a structured strategy for process parameter selection, supporting the broader adoption of 3D printing technologies in the field.

Keywords: additive manufacturing · process parameters · cementitious materials · multicriteria decision analysis · TOPSIS · construction technology

1 Introduction

Additive manufacturing (AM), commonly known as 3D printing, has emerged as a transformative technology in the construction industry, enabling the creation of complex structures with enhanced design flexibility and material efficiency [1, 2]. This advancement has prompted extensive research aimed at improving the selection of process parameters to improve mechanical properties, printability, and dimensional accuracy of printed elements [3, 4].

A critical challenge in AM for cementitious materials is the selection of process parameters that directly influence material performance and build quality. Parameters such as printing speed, layer height, and nozzle diameter significantly impact both the structural integrity and feasibility of large-scale printing [5]. However, the most suitable configuration of these parameters varies based on specific applications, material compositions, and geometric complexities of the printed structures [4, 6]. Multicriteria decision-making (MCDM) methods such as TOPSIS (Technique for Order of Preference by Similarity to Ideal Solution), AHP (Analytic Hierarchy Process), and VIKOR have demonstrated effectiveness in various additive manufacturing applications by systematically evaluating and prioritizing relevant parameters [5, 6].

This study proposes an integrated methodology using these three MCDM approaches, coupled with the Borda count technique, to establish a systematic and replicable framework for selecting the most suitable values for parameters. The approach seeks to balance critical criteria and addresses existing gaps by explicitly focusing on printability and dimensional accuracy alongside mechanical properties, thus enhancing decision-making strategies for broader adoption of AM in construction.

2 Methodology

The methodology proposed in this study combines three MCDM methods to systematically identify the top-ranked combinations for 3D printing of cementitious materials. The combined use of these methods ensures a balanced and comprehensive evaluation, accounting for multiple performance criteria including mechanical properties, printability, dimensional accuracy, and geometric complexity. A final integrative step using the Borda count technique consolidates rankings from each MCDM approach, yielding robust parameter recommendations (Fig. 1).

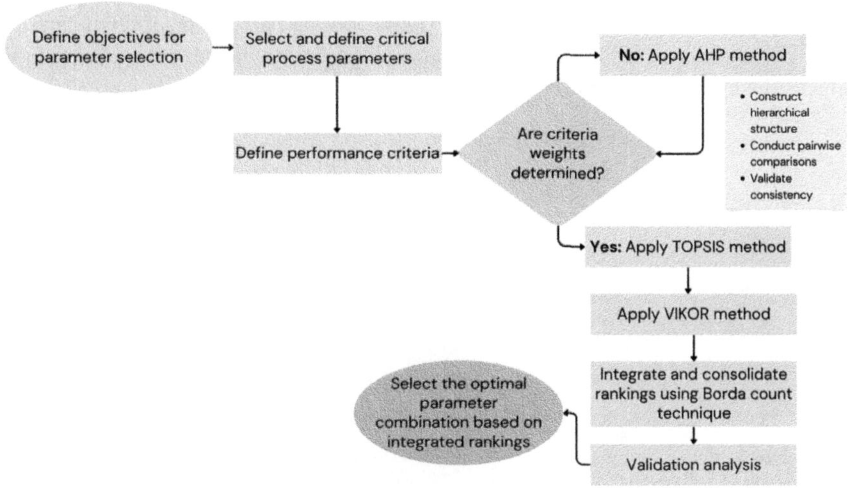

Fig. 1. Methodology for parameter selection in 3D printing of cementitious materials.

2.1 Description of MCDM Methods

Analytic Hierarchy Process (AHP). The AHP method organizes complex decisions into hierarchical structures of goals, criteria, and sub-criteria. Experts perform pairwise comparisons using a standardized numerical scale, assigning relative importance from 1 (equal importance) to 9 (extreme importance). The consistency of judgments is validated through the Consistency Ratio (CR), ensuring logical coherence of the assigned weights. The resulting criteria weights form the foundational input for subsequent ranking methodologies [7] (Fig. 2).

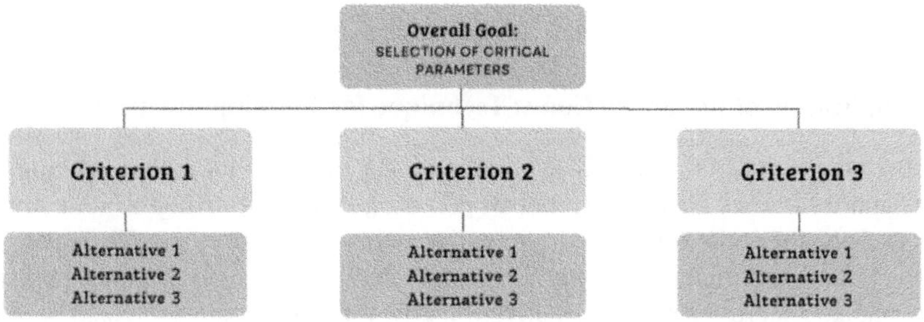

Fig. 2. Hierarchical structure for criteria weighting using AHP.

Technique for Order of Preferences by Similarity to Ideal Solution (TOPSIS). TOPSIS ranks alternatives based on their relative distances to ideal and negative ideal solutions. Initially, a decision matrix is normalized and weighted using criteria derived from AHP. Distances from each alternative to both ideal (optimal performance) and negative ideal (least optimal performance) points are computed. The alternative closest to the ideal and farthest from the negative ideal solutions achieves the highest rank [8] (Fig. 3).

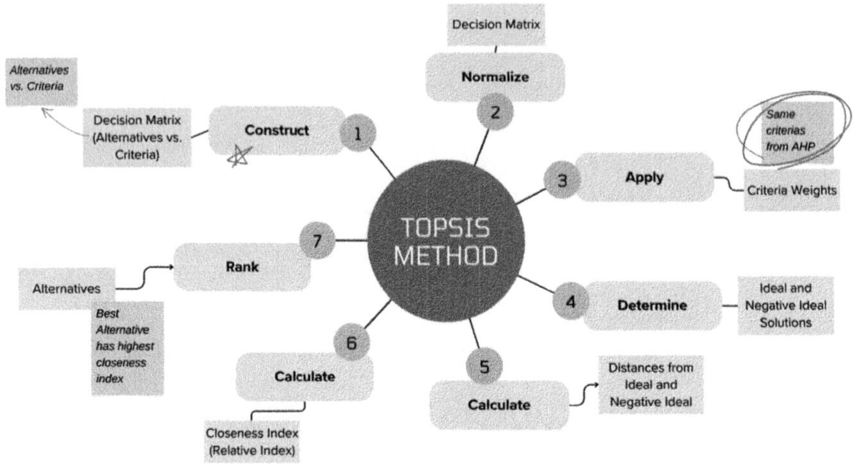

Fig. 3. Step-by-step procedure for alternative selection using TOPSIS.

VIKOR Method. VIKOR identifies suitable compromise solutions when balancing conflicting criteria. Similar to TOPSIS, alternatives are assessed based on closeness to an ideal solution. However, VIKOR introduces metrics for group utility and individual regret, offering balanced alternatives with acceptable trade-offs. The optimal alternative is determined by the lowest VIKOR index, representing the most suitable compromise among all criteria considered [9] (Fig. 4).

2.2 Application of the Borda Count Technique for Ranking Integration

To integrate results from the MCDM methods, the Borda count method is employed. In this approach, each alternative receives points based on its ranking position obtained from each method; the top-ranked alternative gains the most points, decreasing progressively for lower rankings. Points from individual rankings are summed, generating a consolidated score for each alternative. Finally, alternatives are re-ranked according to these cumulative scores, resulting in a comprehensive and robust decision-making outcome [10].

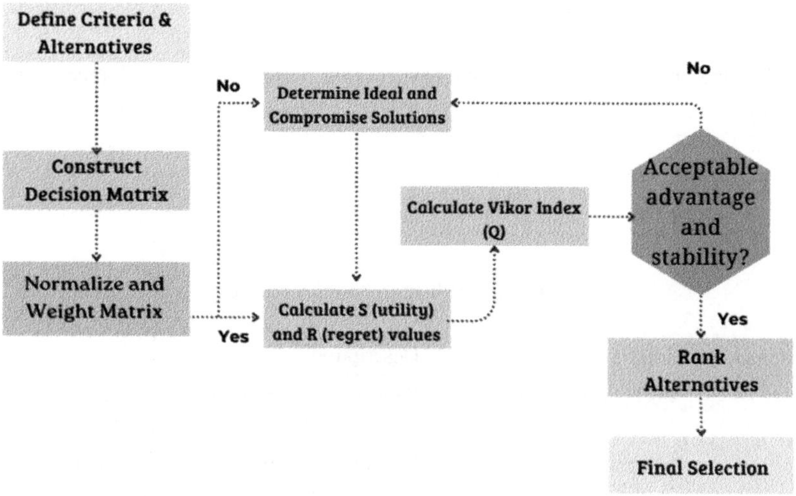

Fig. 4. Procedure for alternative selection using VIKOR

3 Results and Discussion

This section presents the outcomes of applying the proposed methodology to the selection of process parameters for 3D printing of cementitious materials. A case of study is included to demonstrate the methodology in practice. The results of each MCDM method are discussed independently, followed by the integrated results obtained through the Borda count technique.

3.1 Case Study: Evaluation of Three Parameter Sets

Note: The case study is illustrative and does not use laboratory data; scores reflect plausible tendencies reported in the literature and expert judgment. Its purpose is to demonstrate how the methodology works end-to-end. The spreadsheet implementation was verified with a R script to ensure identical rankings.

Three hypothetical parameter sets (A, B, and C) were evaluated using the proposed methodology. These sets varied in printing speed, nozzle diameter, temperature and humidity. The characteristics are shown in Table 1. The evaluation considered four criteria: mechanical performance, printability, dimensional accuracy, and geometric complexity.

Each combination was scored on the four criteria using expert judgment, previous experimental results, and observed behavior during fresh-state extrusion and post-print inspection (Table 2).

The scores assigned to each parameter set, were based on qualitative assessment informed by prior studies and expert judgment. The hypothetical values are consistent with ranges reported by Wang et al. [1], but were not derived from actual experiments. The scoring scale ranges from 1 (low performance) to 9 (high performance), as shown in Table 3.

Table 1. Parameter sets for case study

Set	Printing Speed (mm/s)	Nozzle diameter (mm)	Humidity (%)	Temperature (°C)
A	40	20	65	25
B	30	15	70	27
C	50	20	60	23

Table 2. Evaluation criterias selected

Abbreviation	Criteria	Description
C1	Mechanical performance	Expected compressive and flexural strength based on material formulation and printing configuration
C2	Printability	Stability of the fresh-state material during extrusion and its ability to retain geometry without collapse
C3	Dimensional accuracy	Degree to which the printed element matches the intended geometry, based on alignment, layer thickness, and over-extrusion
C4	Geometric complexity	Capacity to reproduce shapes with varying levels of curvature or detail without deformation

Table 3. Decision matrix of alternatives (raw scores)

Criteria	A	B	C
C1	8	6	7
C2	6	8	7
C3	5	9	7
C4	6	6	8

This matrix was normalized for use in the TOPSIS and VIKOR methods. The AHP method was used to derive weights for each criterion through pairwise comparison.

The weights assigned in Table 4 reflect the priority given to mechanical performance and printability in the context of structural applications for 3D printed cementitious materials. The consistency ratio (CR) of the AHP matrix was within acceptable limits ($CR < 0.1$).

All steps (normalization, weighting, distance calculations, and Borda aggregation) can be executed in a standard spreadsheet or with compact code in Python, R, or MATLAB; the computational effort grows linearly with the number of alternatives and criteria.

Table 4. Criteria weights derived from AHP and TOPSIS

Criteria	Weight
C1	0.35
C2	0.30
C3	0.20
C4	0.15

3.2 Results from MCDM Methods

The normalized scores of each alternative for each criterion were first introduced into the decision matrix. The weights of the criteria were obtained through AHP based on pairwise comparisons. These weights were applied in the TOPSIS and VIKOR methods. The final results are shown in Fig. 5.

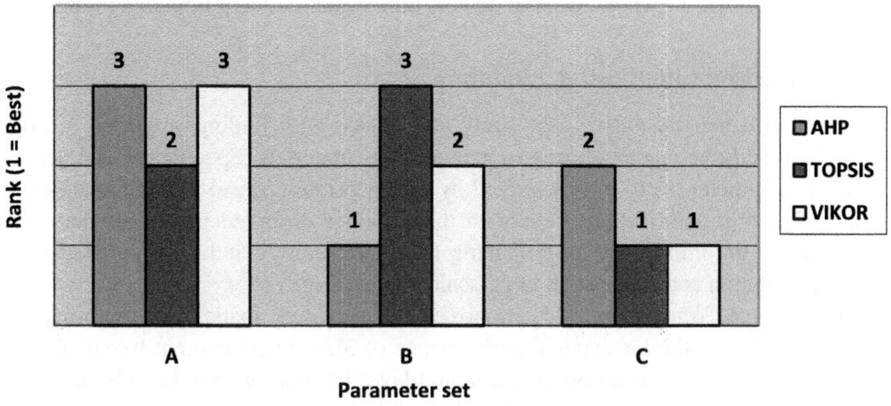

Fig. 5. Final results from MCDM methods for the case study

The results highlight differences in the criteria prioritization:

- AHP favored Set B due to better printability and geometric consistency.
- TOPSIS and VIKOR both ranked Set C first, as it offered strong mechanical strength with acceptable dimensional precision.

3.3 Integrated Results Using Borda Count

To integrate the rankings, the Borda count technique was applied. Each method awarded 3 points to the 1st place, 2 to the 2nd, and 1 to the 3rd. The totals were shown in Table 5.

Set C was identified as the most suitable combination, performing consistently well across all methods.

Table 5. Application of Borda count technique to results in the previous chart

Set	AHP	TOPSIS	VIKOR	Total score	Final Rank
A	1	2	1	4	3rd
B	3	1	2	6	2nd
C	2	3	3	8	1st

3.4 Sensitivity and Uncertainty Analysis

To assess robustness, we perturbed the inputs and repeated the full ranking pipeline. Criteria weights were varied by $\pm 10\%$ around their AHP values and alternative scores were perturbed by ± 1 on the 1–9 scale. Across these scenarios, the top-ranked combination (set C) remained unchanged, while Sets A and B occasionally exchanged positions when higher emphasis was given to dimensional accuracy. These outcomes indicate that the integrated ranking is stable under moderate uncertainty in expert judgment and scoring. The framework can also accommodate fuzzy or interval versions of AHP/TOPSIS/VIKOR when decision makers prefer qualitative or bounded inputs.

3.5 Comparison with Existing Literature

The results of this illustrative case study are aligned with findings reported in prior research on additive manufacturing of cementitious materials. Specifically, the performance of parameter set C (characterized by higher printing speed and moderate environmental conditions) is consistent with the behavior observed in studies that have demonstrated the importance of balancing material rheology and print parameters to avoid deformation and collapse in large-scale 3D printing.

For instance, Wang et al. [11] proposed an innovative spatial path optimization method to enhance the mechanical performance of 3D printed, cement-based components, highlighting that printing direction and layer orientation significantly influence compressive and flexural strengths. Similarly, Li et al. [12] reviewed the properties of extrusion-based 3D printed cementitious materials, emphasizing that printing speed and nozzle diameter significantly affect both the extrusion quality and the mechanical integrity of printed elements. Moreover, Mortada et al. [13] proposed test methods for assessing the printability of concrete materials, emphasizing the critical role of flowability, setting time, and buildability in ensuring successful 3D printing processes.

These parallels suggest that the MCDM-based approach can approximate realistic decision patterns and support parameter configuration in research and development context.

4 Conclusion

This study presented a structured methodology for process-parameter selection in 3D printing of cementitious materials by combining AHP for criteria weighting, TOPSIS and VIKOR for alternative ranking, and a Borda count to integrate results. An illustrative case study showed how the framework resolves differences across methods and yields a consistent, top-ranked parameter set while balancing mechanical performance, printability, dimensional accuracy, and geometric considerations. A simple sensitivity check ($\pm10\%$ in weights; ±1 in scores) indicated stable rankings, suggesting robustness to moderate uncertainty.

This study has two main limitations. First, the case study uses hypothetical scores and therefore does not constitute experimental validation. Second, AHP-derived weights are inherently subjective; different decision-makers may produce different pairwise judgments. Future work will validate the framework with experimental datasets from 3DCP, elicit weights from multiple expert panels, and model uncertainty explicitly (for example, by Monte Carlo sampling of weights/scores or by fuzzy variants of AHP, TOPSIS, and VIKOR). The procedure is implementable in a spreadsheet or with short scripts in Python, MATLAB, or R, which facilitates adoption in research and practice.

References

1. Wang, Q., Han, S., Yang, J., Li, Z., An, M.: Optimizing printing and rheological parameters for 3D printing with cementitious materials. Autom. Constr. **169**, 105881 (2025). https://doi.org/10.1016/j.autcon.2024.105881
2. Akhoundi, B., Modanloo, V.: A multi-criteria decision-making analysis on the extrusion-based additive manufacturing of ABS/Cu composites. Int. J. Interact. Des. Manuf. **17**, 1995–2003 (2023). https://doi.org/10.1007/s12008-023-01342-4
3. Arrêteau, M., Fabien, A., El Haddaji, B., Chateigner, D., Sonebi, M., Sebaibi, N.: Review of advances in 3D printing technology of cementitious materials: key printing parameters and properties characterization. Buildings **13**, 1828 (2023). https://doi.org/10.3390/buildings13071828
4. Quah, T.K.N., Tay, Y.W.D., Lim, J.H., Tan, M.J., Wong, T.N., Li, K.H.H.: Concrete 3D printing: process parameters for process control, monitoring and diagnosis in automation and construction. Mathematics. **11**, 1499 (2023). https://doi.org/10.3390/math11061499
5. Zhuang, Z., Xu, F., Ye, J., Hu, N., Jiang, L., Weng, Y.: A comprehensive review of sustainable materials and toolpath optimization in 3D concrete printing. NPJ Mater. Sustain. **2**, 1–14 (2024). https://doi.org/10.1038/s44296-024-00017-9
6. Mian, S.H., Nasr, E.A., Moiduddin, K., Saleh, M., Abidi, M.H., Alkhalefah, H.: Assessment of consolidative multi-criteria decision making (C-MCDM) algorithms for optimal mapping of polymer materials in additive manufacturing: a case study of orthotic application. Heliyon. **10**, (2024). https://doi.org/10.1016/j.heliyon.2024.e30867
7. Kumar Mondal, J., Das, S., Kumar, R., Maity, M.: Experimental study on FDM 3d printed object & position analysis using multicriteria decision-making process. Mater. Today Proc. (2023). https://doi.org/10.1016/j.matpr.2023.02.292
8. Dev, S., Srivastava, R.: Statistical analysis and multi-criteria optimization of fused deposition additive manufacturing process for acrylic butadiene styrene parts. Proc. Inst. Mech. Eng. Part E **237**, 2496–2509 (2023). https://doi.org/10.1177/09544089221139607
9. Elloumi, A., Jerbi, A., Ben Amor, R., Souissi, S.: A Comparative study of gray relational analysis and VlseKriterijumska Optimizacija I Kompromisno Resenje approaches for enhancing mechanical properties and productivity in 3D-Printed copper-filled PLA parts. J. Elastomers Plast. **56**, 675–692 (2024). https://doi.org/10.1177/00952443241254939
10. Biswas, S., Bandyopadhyay, G., Mukhopadhyaya, J.N.: A multi-criteria framework for comparing dividend pay capabilities: evidence from Indian FMCG and consumer durable sector. Decis. Making Appl. Manage. Eng. **5**, 140–175 (2022). https://doi.org/10.31181/dmame0306102022b

11. Wang, B., et al.: Research on mechanical properties of a 3D concrete printing component-optimized path by multimodal analysis. Sustainability. **16**, 9388 (2024). https://doi.org/10.3390/su16219388
12. Li, Z., et al.: Fresh and hardened properties of extrusion-based 3D-printed cementitious materials: a review. Sustainability. **12**, 5628 (2020). https://doi.org/10.3390/su12145628
13. Mortada, Y., Mohammad, M., Mansoor, B., Grasley, Z., Masad, E.: Development of test methods to evaluate the printability of concrete materials for additive manufacturing. Materials. **15**, 6486 (2022). https://doi.org/10.3390/ma15186486

Stochastic Multi-hazard Vulnerability Assessment of Railway Masonry Arch Bridges Subject to Seismic and Flood Hazards

Carlos Cabanzo[1(✉)], Monica Santamaria[1], Mitsuyoshi Akiyama[2], Paulo B. Lourenço[1], and José C. Matos[1]

[1] Department of Civil Engineering, University of Minho, ISISE, ARISE, Guimarães, Portugal
carlos.cabanzo@civil.uminho.pt

[2] Department of Civil and Environmental Engineering, Waseda University, Tokyo, Japan

Abstract. Natural disasters are unavoidable and may lead to serious damage to transportation networks, often resulting in catastrophic losses. Masonry arch bridges (MABs) represent around 40% of the bridge stock in many European railway networks. In Portugal, MABs with larger spans and thinner piers were built during the early 20th century to improve the national network. These larger structures are expected to continue operating as a crucial part of the railway network without elevated maintenance costs. Moreover, some are located in regions with high seismic and flood hazards. The joint effects of these hazards in out-of-plane behavior are often overlooked, and their assessment may lead to identifying possible mitigation needs. The present research presents the vulnerability assessment accounting for the likelihood of both seismic and flood events to occur simultaneously on two large multi-span masonry arch bridges in Portugal. The results show the effects of the loss in stiffness due to asymmetrical scour in the out-of-plane failure mechanism of the masonry arch bridges. Fragility curves are obtained by considering structure-related uncertainties for different levels of intensity given by peak ground acceleration for the seismic hazard and flow discharge for the flood hazard. The resulting vulnerability curves, obtained by employing surrogate models, provide useful information for prioritizing assets considering their exposure to multiple hazards, thus aiding in maintaining the desired performance of the railway network.

Keywords: Local scour · modal pushover · out-of-plane behavior · multi-hazard fragility curves · surrogate models

1 Introduction

Masonry arch bridges (MABs) are one of the most common bridge typologies in most European railway networks. They are expected to continue in operation without elevated maintenance costs for the foreseeable future [1, 2]. Bridge management is vital to maintain a satisfactory performance level of the large stock of built infrastructures exposed to harsh environmental conditions, which have been increasing under the effects

M. Posani and S. Singh (Eds.): RYS 2025, RILEM Bookseries 67, pp. 821–835, 2026.
https://doi.org/10.1007/978-3-032-14652-6_59

of climate change, leading to serious deterioration processes that reduce their structural performance over time [3, 4]. Moreover, the occurrence of extreme natural events (*e.g.*, earthquakes, landslides, floods, scour, hurricanes, typhoons) has also caused sudden bridge collapses in the past. Among these natural causes, flood-induced scour and seismic actions have been identified as one of the predominant triggers of bridge failure worldwide [4].

In the context of MABs, seismic fragility curves have been developed employing nonlinear kinematic approaches [5] and finite element [6–10] and discrete element models for the analysis of both in-plane [5–7] and out-of-plane behavior [5, 8–10]. The seismic fragility curves are often defined in terms of peak ground acceleration (PGA) [6–8, 10] as an intensity measure for multiple performance limit states. Analogously, MABs are vulnerable to the effects of local scour due to their rigid behavior and shallow footing foundations, which exacerbate the effects of settlements [11, 12]. In this context, flood-related research has been developed for multi-span MABs, considering in-plane [13–15] and out-of-plane [11, 16, 17] failure modes, employing both analytical models based on kinematics [14, 15], limit state models [13] and finite element models [11, 16, 17]. Flood-fragility curves have been developed for masonry bridges employing flow discharge [13], flow velocity [18], foundation rotation [14, 15], and scour depth [2] as intensity measures.

Some research has been conducted assessing the combined effects of both seismic and flood hazards. However, most of the research has been focused on reinforced concrete bridges [19–23]. In MABs, research has proven that local scour might affect modal characteristics, both shapes, and frequencies, affecting their dynamic performance under seismic actions [24, 25]. Thus, the assessment of larger multi-span masonry arch bridges under both seismic and flood hazards is of the utmost relevance due to their vulnerability to out-of-plane behavior due to slender piers and larger span lengths [26].

When assessing bridge safety, the lack of information causes high uncertainties when representing the loads and the material properties. Thus, by explicitly considering the uncertainties and implementing fully probabilistic approaches, the lack of information can be mitigated [2, 8, 27]. However, when performing stochastic analyses, the complexity and size of the models increase the computational cost of the analysis. In literature, surrogate models have been widely employed as a mitigation strategy to the computational problem by replacing the evaluation of the numerical model [28, 29].

The present work aims to present a consistent framework to derive fragility curves associated with the out-of-plane failure mode of multi-span MAB under flood and seismic hazards using surrogate models for reducing computational costs. Fragility curves were derived based on code-based response spectra to describe the seismic hazard employing PGA as intensity measure and flow discharge for flood hazard. Transversal displacement was used as the engineering demand parameter for the out-of-plane failure. Finally, the resulting vulnerability curves consider the probability of occurrence of both events to occur simultaneously and the multi-hazard fragility curves for the collapse limit state, allowing the identification of assets with high vulnerability and aiding the management authorities in the prioritization of interventions for mitigating the associated risks.

2 Case Studies

2.1 General Description

In Portugal, MABs represent around 30% of the total stock in the railway network, where 10% account for larger bridges, most of which were built in the early 20th century [1]. During this period, MABs located over a water passage are often characterized by a larger principal arch supported by two main piers situated at the edges of the river, presenting a higher cross-section area. Moreover, these single-track bridges have a width at the top of around 4.8 m, which increases in a ratio of 1:25 with height, which adds to their slenderness since they are often between 30 and 50 m in height. For the piers supporting arches of equal span, the width at the thinnest part is 20% of the arch span, *s* (see Fig. 1). Arches present two geometrical configurations of the block units, and the external configuration presents a constant thickness of 1 m, forming a semicircular arch. However, the inner arch has a variable cross-section (see Fig. 1). The internal angle between the ends of the arch is 120 degrees.

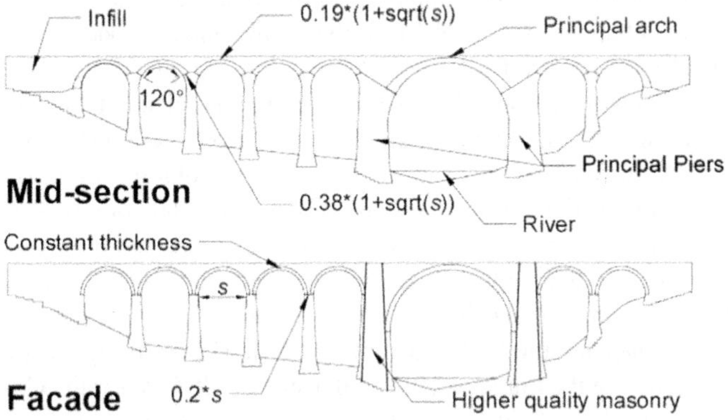

Fig. 1. Main geometrical characteristics of the Portuguese MABs over water passages of the early 20th century.

The materials employed during the construction of the bridges present different combinations of blocks and infills. Piers supporting the arches with equal spans present regular stone masonry for the exterior, and the infill is a mix of low-quality concrete and irregular masonry, with a higher quality regular masonry in the piers supporting the longest arch, as described in the technical documents and corroborated in the field tests [8, 30].

For this research, the Quinta Nova and Côa bridges were selected as representative structures of this typology. The former is located in southern Portugal, which presents high seismicity. It is part of the second most important railway line, known as *Linha do Sul*. The latter is in northern Portugal, part of the railway line *Beira Alta*. Both bridges have been the subject of experimental campaigns aimed at the determination of material and modal properties for the calibration of numerical models [8, 30]. Table 1 presents the main dimensions of the structural components of each case study.

Table 1. Principal geometrical characteristics of the main components of each case study.

Parameter	Quinta Nova bridge	Côa bridge
Number of spans	7	8
Span length [m]	30/12	38/20
Length [m]	174	238
Height [m]	31	56
Arch thickness crown [m]	1.25/0.85	1.40/1.05
Arch thickness ends [m]	2.50/1.70	2.80/2.10

2.2 Numerical Modeling

In the context of multi-span MABs, continuum numerical models are often employed to represent the global behavior, including out-of-plane effects [3, 8, 30]. A three-dimensional finite element model (FEM), utilizing the macro-modeling approach, was developed in DIANA FEA software [31], employing 3D quadrilateral elements. The mesh of each model was generated by guaranteeing regular elements in the three dimensions and at least six integration points along each section. Nonlinearity in the material was introduced by the total strain-based crack model with rotating crack orientation with exponential softening and parabolic hardening for tension and compression [31].

The mechanical properties of the materials and boundary conditions were retrieved and calibrated using the available information of the dynamic identification tests found in recent literature, using the elasticity modulus as the variable to achieve a good fit for the first three out-of-plane modal frequencies [8, 30]. The resulting elasticity modules for the masonry were in the range of 10–15 GPa and 20- 26 GPa for the Côa and Quinta Nova bridges, respectively. It was found that the abutments presented fixed behavior, while springs were employed to simulate the soil interaction at the foundation levels of the piers. Similar approaches have been widely used in scour analyses of MABs [11, 24]. Figure 2 presents the final numerical model, including the boundary conditions, mesh, and the first modal shape for both case studies.

For the present analysis, a modal pushover analysis was employed to determine the seismic capacity of the bridges. The pushover analysis was conducted through the DIANA software employing a mode-dependent equivalent lateral force applied over all nodes of the model [31]. The numerical solution of the nonlinear equations was obtained using the Modified Newton-Raphson iterative method with a convergence norm based on both force and displacement with a tolerance equal to 0.01.

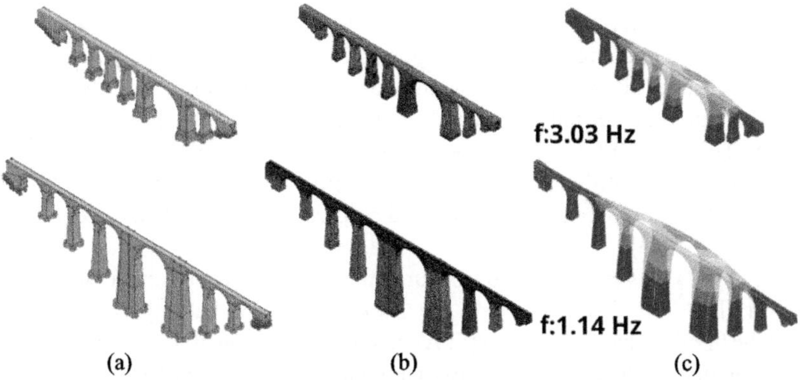

Fig. 2. Characteristics of the numerical model of the Quinta Nova (top) and Côa (bottom) bridges: (a) boundary conditions, in green the Winkler-type springs and in red the fixed supports; (b) mesh elements; (c) first mode, presenting the normalized transversal displacement with red and blue representing the maximum and minimum displacement, respectively.

3 Hazard Scenarios

3.1 Flooding Scenarios

Portuguese MABs are built considering a principal span length higher than the width of the stream, with very tall piers characterized by large cross-sections, as described in Sect. 2.1. Given these geometrical characteristics, only the foundations of the two main piers located at the riverbanks are likely to be affected by scour during flood hazard events. Scour at these piers is expected to result from flow acceleration and vortex formation caused by the obstruction to the flow. Since the bridge piers are not likely to be fully submerged due to the stream's size and the large cross-section of the piers, the scour process is anticipated to be similar to that typically observed at bridge abutments. There are several formulations for estimating the scour depth at abutments. Among them, the model in NCHRP 24-40 [32] developed abutment scour equations considering flow properties, sediment transport conditions, and a range of abutment types and configurations. Scour occurring when the abutment is located in or near the main channel, such as in the case of the MAB piers, is referred to as scour condition A. The equations consider abutment scour as a short contraction scour and apply a factor to amplify the scour depths to account for the large-scale turbulence that develops in the vicinity of the abutment [35]. Scour depth around an abutment, $y_{s,ab}$, can be determined as [32]:

$$y_{s,ab} = y_{max} - y_0 = (\alpha_A y_c) - y_0 \tag{1}$$

$$y_c = y_1 \left(\frac{q_{2c}}{q_1} \right)^{\frac{6}{7}} \tag{2}$$

where y_{max} is the maximum flow depth resulting from abutment scour, y_0 is the flow depth prior to scouring, α_A is the amplification factor for condition A of abutment scour, y_c is the flow depth, including live-bed or clear-water contraction scour, y_1 is the upstream flow depth, q_1 is the upstream unit discharge and q_{2c} is the unit discharge in the constricted opening. Since these equations describe a complex physical process through the use of simplified expressions derived from small-scale experiments, the predicted scour may be underestimated or overestimated. This epistemic uncertainty for the NCHRP 24-40 formula can be considered by multiplying the predicted scour $y_{s,ab}$ by a normally distributed model error λ_{sc} with a mean value equal to a bias factor representing the ratio between observed and predicted scour depth and a coefficient of variation (COV) [33].

From Eq. (2), it is evident that the scour depth heavily depends on the flow discharge of a river. The flow discharge for a given exceedance probability can be estimated through the flood frequency analysis (FFA) method, which consists of fitting a probability distribution function to the observations of annual peak flows recorded at stream gauging stations. In Portugal, this information is accessible free of charge through the National Information System for Water Resources (SNIRH) website [34]. In the case of the Côa bridge, located over the Côa River (Fig. 3a), the nearest stream gauge station is Castelo Bom (10P/01H). For the case of the Quinta Nova bridge, located over the Sado River (Fig. 3a), the nearest stream gauge station is Ponte Alvalade Sado (26G/05). Applying the FFA method, a Gumbel distribution is found to be a good fit for the peak flow discharges (Fig. 3b). Table 2 presents the flow discharges for the 10-, 100-, and 1000-year return periods at both locations. The scour depths are estimated for these three return periods using the parameters provided in Table 2 for the hydraulic analysis. Due to the uncertainties in these parameters, their probability distributions and COV values are adopted following Lagasse et al. [33].

Table 2. Variables for the computation of the scour depth of each case study.

Variables		Quinta Nova bridge		Côa bridge	
Name	Distribution	Mean	COV	Mean	COV
Flow discharge [m^3/s]	Gumbel	134.5	1.12	179.9	0.82
10-year		261.0		373.3	
100-year		561.0		631.4	
1000-year		886.0		884.9	
River width [m]	Normal	25.00	0.05	25.00	0.05
Longitudinal slope [-]	Lognormal	0.002	0.18	0.002	0.18
Manning coefficient [-]	Lognormal	0.075	0.05	0.075	0.05
Model error λ_{sc}	Normal	0.74	0.23	0.74	0.23

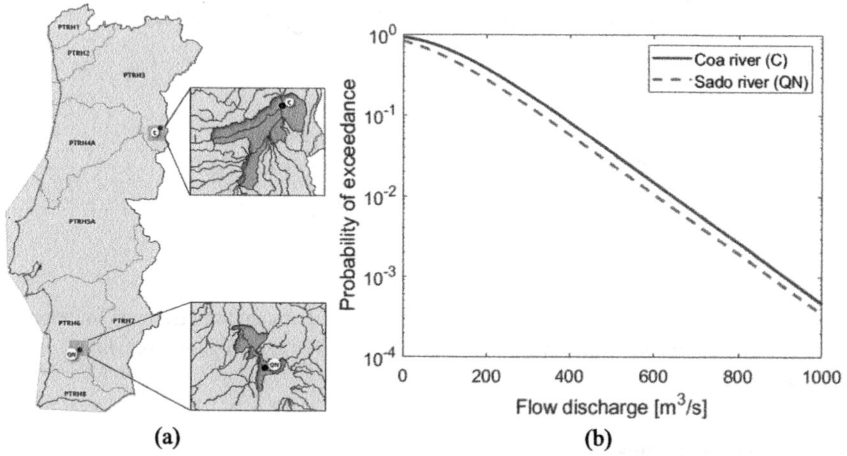

Fig. 3. (a) Water drainage basins at both bridge sites: Côa (C) bridge at Côa River and Quinta Nova (QN) bridge at Sado River; (b) flood-hazard curves at each bridge site.

The estimated scour depths for the three flood return periods are presented in Table 3. According to previous studies [19], the level of variability observed in the scour depth estimates does not significantly affect the seismic response of bridges. Consequently, scour depth is treated as a deterministic parameter in the seismic scenarios and is modeled by the removal of the soil springs around the piers. The scour shape has been widely discussed in the literature, with the most commonly adopted shape being triangular, defined by the soil's friction angle and the increasing water flow values [11, 12, 35]. Moreover, depending on the pier alignment with regard to the direction of the water flow, a symmetrical or asymmetrical scour configuration may occur. Since the water flow is not fully aligned with respect to the MAB piers' orientation, an asymmetric scour configuration is considered. Therefore, the scour shape for the removal of the soil springs is assumed to follow the geometrical characteristics illustrated in Fig. 4.

Table 3. Estimated scour depths for different flood return periods.

Flood return period	Gaussian distribution of the scour depth			
	Quinta Nova bridge		Côa bridge	
	Mean [m]	COV	Mean [m]	COV
10-year (S1)	3.25	0.258	5.03	0.257
100-year (S2)	5.15	0.257	6.9	0.258
1000-year (S3)	6.77	0.26	8.4	0.256

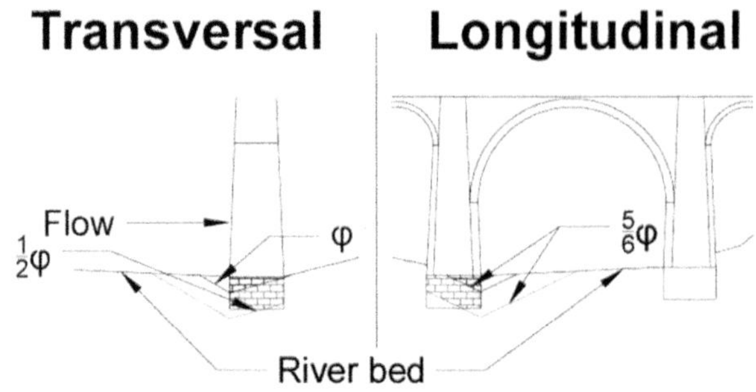

Fig. 4. The main geometrical characteristics of the scour shape and the exposed foundation are for the cases where the scour depth is lower (black dashed line) and higher (grey dotted line) than the foundation height.

3.2 Seismic Scenarios

Pushover analyses have been employed for the seismic assessment of MABs [6, 8–10, 36] due to the possibility of deriving fragility curves for different damage states based on the degradation of the stiffness of the pushover curves. The damage states (DS) define the physical condition of a structure limited by the performance limit states (PL). PL1-Damage limitation is associated with a predominantly elastic response up to 75% of the base shear; PL2-Significant damage represents the limit where the structure maintains its integrity, assumed as the point of 93% reduction of the stiffness of the pushover curve; and PL3-Near collapse is often associated with the collapse of the structure, represented by the endpoint in the pushover curve. These PLs have been successfully employed for the seismic assessment of multi-span MABs [9].

Each of the PL states will be compared with a seismic action represented by a demand spectrum, which can be associated with a given return period, T, and a given intensity in terms of PGA computed according to the Eurocode [37] based on the Portuguese seismic zones, Fig. 5a, and a correction factor. Equation (3) is used to account for the return period of the response spectrum, T, in relation with the reference return period, T_r, equal to 475 years, and k is a variable depending on the seismicity of the region (often assumed as 3). Finally, Fig. 5b presents the seismic hazard curves with the probability of exceedance of a seismic action in terms of PGA for a reference period of 50 years.

$$\left({}^{T_r}/_{T}\right)^{-1/k} \tag{3}$$

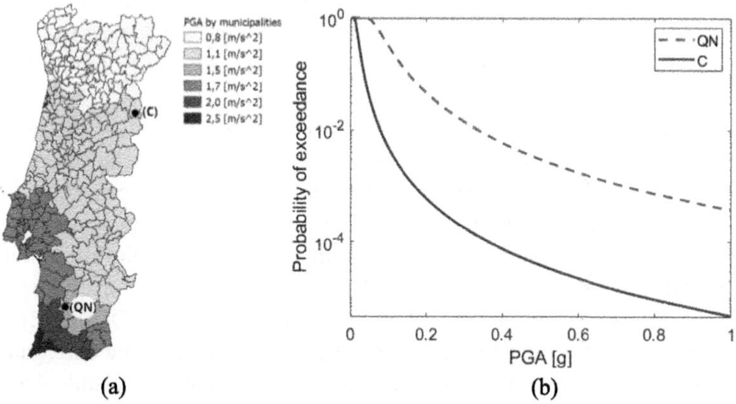

Fig. 5. (a) Seismic hazard map of Portugal with the location of the Quinta Nova-QN and Côa-C bridges; (b) Seismic hazard curve for the reference period of 50 years.

4 Vulnerability Assessment

For the definition of the multi-hazard curves, the process described in Fig. 6 was followed, which consists of an iterative process first selecting a given scour level, including the case of no scour. Then, an experimental design was defined based on the recommendations of the JCSS probabilistic model code [38], in which uncertainty is given by the introduction of independent lognormal variables associated with a mechanical property. In this case, the parameters are related to the elasticity modulus and compressive capacity of the different materials, *i.e.*, regular masonry, infill, and the combination of both [8, 30]. The sampling was generated in the standard Gaussian space to avoid the bias introduced by the scale and units of their associated mechanical parameters. Samples are distributed evenly by maximizing the Euclidean distance in the region with an equal probability of unexplored space. The number of samples is determined based on the number of variables and their joint probability function. For the present research, a probability of unexplored space of 10^{-2} was selected, yielding an experimental design of 50 samples [39].

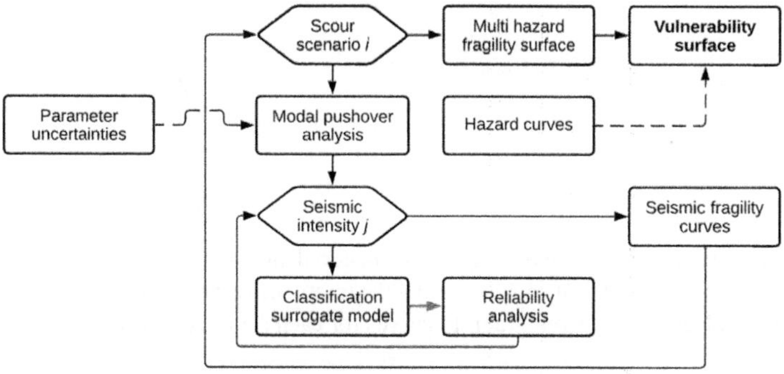

Fig. 6. Overview of the process for defining the multi-hazard vulnerability curves.

The experimental design is then employed in the numerical model to obtain the pushover curves associated with each scour level, allowing the comparison of each of the PL against the increasing intensity of the response spectra, *i.e.,* higher PGA, with each iteration.

The main limitation of assessing the reliability of a big structure is the large computational expense associated with the model evaluations required to achieve a reliable estimation [40, 41]. However, it can be reduced by employing a surrogate model to substitute evaluations of the numerical model in the reliability analysis. Support vector machines (SVG) are often used to assess structural systems due to their high-dimensional problems with limited and imbalanced datasets. The classification variation of the SVGs was implemented in this research for the reliability analysis through UQLab [42], employing the leave-one-out method for validation. For the training of the surrogate model, the intersection point between the performance limit state and the seismic demand was defined as the criteria for classifying a given sample into failure or success. Importance sampling was employed to obtain the system reliability [42, 43]. After concluding the iterative procedure, seismic fragility curves can be defined for each scour profile by correlating the maximum PGA with the exceedance probability of surpassing each PL state, which is described by a lognormal distribution. Finally, vulnerability is derived by combining each seismic fragility curve with both hazard curves, in this research limited to the PL3-Near collapse.

4.1 Fragility Analysis

Figure 7a presents the variation in the resulting pushover curves due to the effects of the uncertainties, not only affecting the capacity curve but also affecting the modal parameters that are employed for the transformation into a SDOF system.

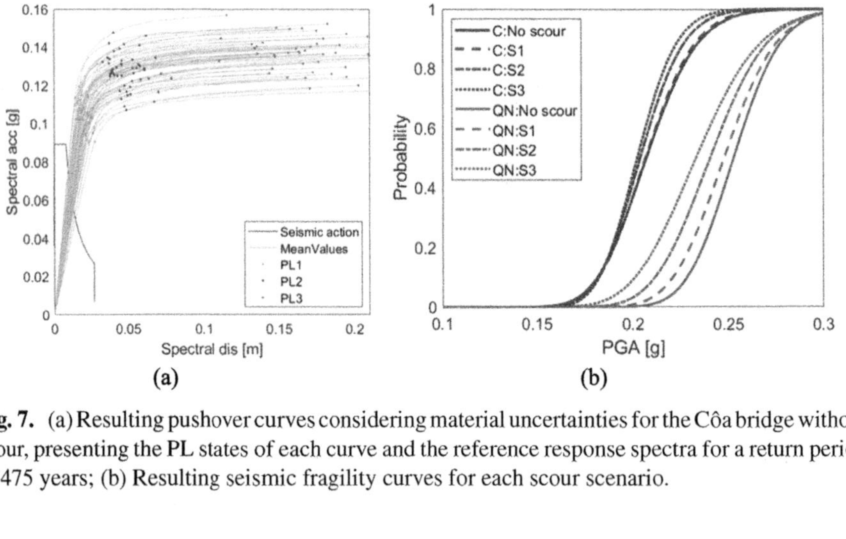

Fig. 7. (a) Resulting pushover curves considering material uncertainties for the Côa bridge without scour, presenting the PL states of each curve and the reference response spectra for a return period of 475 years; (b) Resulting seismic fragility curves for each scour scenario.

As expected, the resulting fragility curves (see Fig. 7b) present variation due to the lack of support of the pier in which the local scour is applied. It should be noted that for scenario S1, where the scour depth is lower than the foundation height (5m), the effects on the fragility curve are almost neglectable, which has been previously discussed in the literature [2, 12, 13]. From the fragility curves with no scour, it can be concluded that the bridges present a low probability of failure for the reference seismic actions according to the seismic zones in the Eurocode. The Quinta Nova bridge is in the zone with the highest reference PGA, around 0.2g. However, the bridge presents a low failure probability of collapse for that value of PGA. The Côa bridge presents higher fragility for lower values of intensity, which is expected due to its higher span and height and lower mechanical properties of the masonry.

Figure 8 shows the resulting fragility surfaces after accounting for the probability of a given scour depth to be a result of a given discharge intensity, as described in Sect. 3.1. It can be concluded that the Côa bridge presents higher fragility than the Quinta Nova bridge for similar ranges of both seismic and flood intensities.

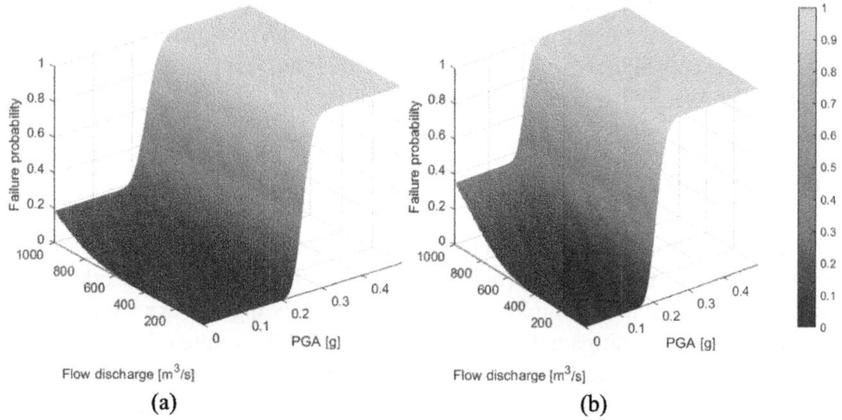

Fig. 8. Resulting multi-hazard fragility curves for the Quinta Nova (a) and Côa (b) bridges.

4.2 Vulnerability Surfaces

The vulnerability curves account for both the probability of exceedance of a given hazard intensity and the failure probability of the structure for that intensity. Figure 9 presents the vulnerability curves for both case studies, employing the reliability index as a performance indicator. The resulting surface is bimodal due to the shape of the hazard curves, which means that the highest vulnerability is associated with individual hazards. Likewise, there is an area near low-intensity values where the failure probability is neglectable. To better display relevant values of vulnerability, the curves were limited by a upper bound of failure probability of 10^{-5}, which corresponds to a reliability index of 4.3, a limit defined in the Eurocodes for structures with high human and economic losses associated with a reference period of 50 years [37].

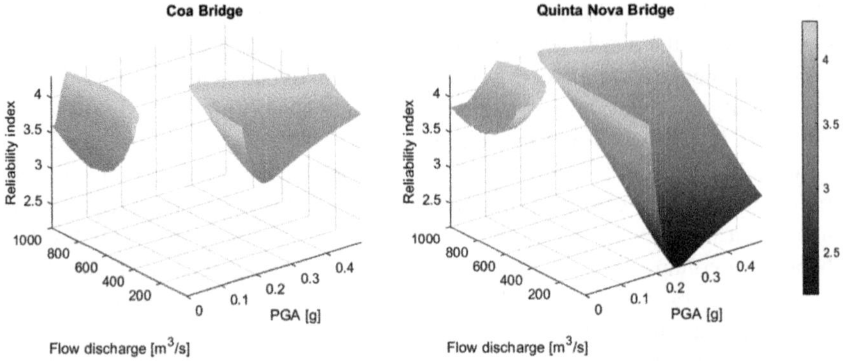

Fig. 9. Resulting multi-hazard vulnerability curves employing flow discharge and PGA as intensity measures.

The Quinta Nova bridge presents the highest vulnerability for a seismic action of around 0.25g, represented by a reliability index of approximately 2.2. Alternatively, the highest vulnerability for the Côa bridge, with a reliability index of 3.3, appears at a flow discharge of 700 m^3/s. From the vulnerability curves, it can be concluded that even if the Côa bridge presents higher fragility, it also presents lower exposure to seismic actions, resulting in an overall lower maximum vulnerability.

5 Conclusions

The proposed research has significant relevance for the transportation asset management industry for the identification of critical infrastructure in the railway network that might present high vulnerability to multiple hazards, thus allowing a better allocation of the available resources to mitigate the risks associated with the failure of assets.

Uncertainties were included to mitigate the limitations related to the availability of information on the mechanical properties of the different materials present in the structures. Moreover, in both cases, the mean values were derived by calibration of the first three modal shapes and frequencies based on dynamic characterization tests. Although uncertainties allow for the accuracy of the assessment to increase, they also heavily increase the computational cost. Machine learning algorithms were included to replace the numerical model and overcome the increase in the computational cost of performing the reliability analyses.

As expected, due to the scour depth eroding the soil around the foundation level, the structural capacity of the case studies in the out-of-plane behavior significantly decreased. However, higher effects were identified when the scour depth was higher than the foundation height [2, 11, 12, 44].

The findings show that fragility might not be enough to correctly prioritize assets in the network since the exposure heavily affects the resulting failure probability. In this study, it was determined that while the Côa bridge is the largest structure and the most susceptible to out-of-plane effects, it is in a region with low seismic hazard, resulting in a lower vulnerability with a reliability index of 3.4 in comparison with the obtained reliability index of 2.2 for Quinta Nova bridge, in both cases for a reference PGA between 0.2 g to 0.25 g. However, the Côa bridge presents higher vulnerability than the Quinta Nova bridge for flood hazard with a lower reliability index of 3.3 against 3.7, respectively, for a flow discharge between 700 m^3/s to 800 m^3/s.

Acknowledgment. This work is financed by national funds through FCT - Foundation for Science and Technology, under grant agreement 2022.09751.BD (https://doi.org/10.54499/2022.09751.BD) attributed to the 1st author. This work was partly financed by FCT/MCTES through national funds (PIDDAC) under the R&D Unit Institute for Sustainability and Innovation in Structural Engineering (ISISE), under reference UIDB/04029/2020 (https://doi.org/10.54499/UIDB/04029/2020), and under the Associate Laboratory Advanced Production and Intelligent Systems ARISE under reference LA/P/0112/2020.

References

1. da Silva, R.F.P.: Advanced nonlinear numerical simulation tools for in-service and retrofit-ting assessment of stone masonry railway arch bridges – experimental calibration and val-idation. Doctoral thesis, University of Porto. https://hdl.handle.net/10216/146901 (2022)
2. Baron, E.A., Mendoza Cabanzo, C., Bento, A.M., et al.: A study of stone arch bridge's flood reliability through a surrogate model approach. Struct. Infrastruct. Eng., 1–16
3. Deng, L., Wang, W., Yu, Y.: State-of-the-art review on the causes and mechanisms of bridge collapse. J. Perform. Constr. Facil. **30**, 04015005 (2016)
4. Proske, D.: Bridge collapse frequencies versus failure probabilities. Springer Inter-national Publishing, Cham (2018). https://doi.org/10.1007/978-3-319-73833-8
5. Manzini, C.F., Morandi, P., Borzi, B., et al.: Derivation of fragility curves for the seismic vul-nerability assessment of railway masonry arch bridges. In: Pellegrino C, Faleschini F, Zanini MA, et al. (eds.) Proceedings of the 1st Conference of the European Association on Quality Control of Bridges and Structures, pp. 893–902. Springer International Publishing, Cham (2022). https://doi.org/10.1007/978-3-030-91877-4_102
6. Tecchio, G., Donà, M., Saler, E., et al.: Fragility of single-span masonry arch bridges accounting for deterioration and damage effects. Eur. J. Environ. Civ. Eng. **27**, 2048–2069 (2023)
7. Homaei, F., Yazdani, M.: Seismic fragility, loss, and resiliency of old railway masonry arch bridges under near-field ground motion. Sustain. Resilient Infrastruct., 1–28 (2025)
8. Cabanzo, C., Mendes, N., Akiyama, M., et al.: Probabilistic framework for seismic performance assessment of a multi-span masonry arch bridge employing surrogate modeling techniques. Eng. Struct. **325**, 119399 (2025)
9. Gönen, S., Soyöz, S.: Reliability-based seismic performance of masonry arch bridges. Struct. Infrastruct. Eng. **18**, 1658–1673 (2022)
10. Shabani, A., Kioumarsi, M.: Seismic assessment and strengthening of a historical masonry bridge considering soil-structure interaction. Eng. Struct. **293**, 116589 (2023)
11. Tubaldi, E., Macorini, L., Izzuddin, B.A.: Three-dimensional mesoscale modelling of multi-span masonry arch bridges subjected to scour. Eng. Struct. **165**, 486–500 (2018)
12. Zampieri, P., Zanini, M.A., Faleschini, F., et al.: Failure analysis of masonry arch bridges subject to local pier scour. Eng. Fail. Anal. **79**, 371–384 (2017)
13. Mendoza Cabanzo, C., Santamaría, M., Sousa, H.S., et al.: In-plane fragility and parametric analyses of masonry arch bridges exposed to flood hazard using surrogate modeling techniques. Appl. Sci. **12**, 1886 (2022)

14. George, J., Menon, A.: Analytical fragility curves for displacement-based scour assess-ment of masonry arch bridges. Structures **46**, 172–185 (2022)
15. George, J., Menon, A.: A mechanism-based assessment framework for masonry arch bridges under scour-induced support rotation. Adv. Struct. Eng. **24**, 2637–2651 (2021)
16. Han, Y., Chun, Q., Gao, X.: Flood-induced forces and collapse mechanism of historical multi-span masonry arch bridges: the putang bridge case. Eng. Fail. Anal. **153**, 107564 (2023)
17. Di Marco, F., Tetougueni, C.D., Zampieri, P., et al.: Numerical analysis of masonry arch bridges subject to scour effect. In: Pellegrino, C., Faleschini, F., Zanini, M.A., et al. (eds.) Proceedings of the 1st Conference of the European Association on Quality Control of Bridges and Structures. Springer International Publishing, Cham, pp. 1335–1342 (2022). https://doi.org/10.1007/978-3-030-91877-4_152
18. Kosič, M., Anžlin, A., Bau, V.: Flood Vulnerability study of a roadway bridge subjected to hydrodynamic actions, local scour and wood debris accumulation. Water **15**, 129 (2022)
19. Yilmaz, T., Banerjee, S., Johnson, P.A.: Uncertainty in risk of highway bridges assessed for integrated seismic and flood hazards. Struct. Infrastruct. Eng. **14**, 1182–1196 (2018)
20. Biazar, S., Kameshwar, S., Balomenos, G.P.: Multi-hazard fragility modeling framework for bridges with shallow foundations subjected to earthquake, scour, and vehicular loading. Soil Dyn. Earthq. Eng. **178**, 108482 (2024)
21. Karatzetzou, A., Stefanidis, S., Stefanidou, S., et al.: Unified hazard models for risk assessment of transportation networks in a multi-hazard environment. Int. J. Disaster. Risk Reduct. **75**, 102960 (2022)
22. Argyroudis, S.A., Mitoulis, S.A.: Vulnerability of bridges to individual and multiple haz-ards-floods and earthquakes. Reliab. Eng. Syst. Saf. **210**, 107564 (2021)
23. Banerjee, S., Ganesh, P.G.: Seismic risk assessment of reinforced concrete bridg-es in flood-prone regions. Struct. Infrastruct. Eng. **9**, 952–968 (2013)
24. Scozzese, F., Tubaldi, E., Dall'Asta, A.: Damage metrics for masonry bridges under scour scenarios. Eng. Struct. **296**, 116914 (2023)
25. Sözen, Ş, Doğangün, A.: Evaluation of the effect of change in support conditions on the seismic behavior of a historical masonry bridge. J. Comput. Cult. Herit. **16**, 1–19 (2023)
26. Zampieri, P., Tetougueni, C.D., Pellegrino, C.: Nonlinear seismic analysis of masonry bridg-es under multiple geometric and material considerations: Application to an existing seven-span arch bridge. Structures **34**, 78–94 (2021)
27. Moreira, V.N., Fernandes, J., Matos, J.C., et al.: Reliability-based assessment of existing masonry arch railway bridges. Constr. Build. Mater. **115**, 544–554 (2016)
28. Tubaldi, E., Macorini, L., Izzuddin, B.A.: Identification of critical mechanical parameters for advanced analysis of masonry arch bridges. Struct. Infrastruct. Eng. **16**, 328–345 (2020)
29. Mahmoudi, S.N., Chouinard, L.: Seismic fragility assessment of highway bridges using support vector machines. Bull. Earthq. Eng. **14**, 1571–1587 (2016)
30. Neto, J.A.D.: Calibração de modelos dinâmicos de pontes ferroviárias em arco de alvenaria de pedra. Master of Science (MS), University of Porto (2015). https://hdl.handle.net/10216/80588
31. DIANA Finite Element Analysis (FEA) Software (2023). https://dianafea.com/
32. Ettema, R., Nakato, T., Muste, M.: Estimation of scour depth at bridge abutments. NCHRP 24-20, Washington, D.C.: Transportation research board (2010)
33. Lagasse, P., Ghosn, M., Johnson, P., et al.: Risk-Based Approach for Bridge Scour Prediction. Transportation Research Board, Washington (2013)
34. Departamento de Monitorização de Recursos Hídricos da Agência Portuguesa do Ambiente (APA). Sistema nacional de informação de recursos hídricos, snirh.apambiente. pt/
35. Ahamed, T., Duan, J.G., Jo, H.: Flood-fragility analysis of instream bridges – consideration of flow hydraulics, geotechnical uncertainties, and variable scour depth. Struct. Infrastruct. Eng. **17**, 1494–1507 (2021)

36. Shimpi, V., Sivasubramanian, M.V.R., Singh, S.B., et al.: Seismic vulnerability assessment and fragility curves for a multistorey gallery arch bridge. SN Appl. Sci. **3**, 662 (2021)
37. Eurocode 8: Design of structures for earthquake resistance. part 1: general rules, seismic actions and rules for buildings. https://eurocodes.jrc.ec.europa.eu/EN-Eurocodes/eurocode-8-design-structures-earthquake-resistance (2004)
38. Vrouwenvelder, T.: The JCSS probabilistic model code. Struct. Saf. **19**, 245–251 (1997)
39. Vořechovský, M.: Reliability analysis of discrete-state performance functions via adaptive sequential sampling with detection of failure surfaces. Comput. Methods Appl. Mech. Eng. **401**, 115606 (2022)
40. Soleimani, F.: Probabilistic seismic analysis of bridges through machine learning approaches. Structures **38**, 157–167 (2022)
41. Saraygord Afshari, S., Enayatollahi, F., Xu, X., et al.: Machine learning-based methods in structural reliability analysis: a review. Reliab. Eng. Syst. Saf. **219**, 108223 (2022)
42. Marelli, S., Sudret, B.: UQLab: a framework for uncertainty quantification in matlab. In: Vulnerability, Uncertainty, and Risk. Liverpool, UK: American Society of Civil Engineers, pp. 2554–2563
43. Tabandeh, A., Jia, G., Gardoni, P.: A review and assessment of importance sampling meth-ods for reliability analysis. Struct. Saf. **97**, 102216 (2022)
44. Scozzese, F., Ragni, L., Tubaldi, E., et al.: Modal properties variation and collapse assess-ment of masonry arch bridges under scour action. Eng. Struct. **199**, 109665 (2019)

Structural and energy retrofitting of masonry-infilled RC frames subjected to in-plane cyclic loading

Matthildi Monastiridou(✉), Szymon Cholostiakow, Christos Papakonstantinou, and Lampros Koutas

Department of Civil Engineering, University of Thessaly, Pedion Areos, GR-38334 Volos, Greece
mmonastiridou@uth.gr

Abstract. Textile-Reinforced Mortar (TRM) has proven to be an effective strengthening solution for masonry-infilled reinforced concrete (RC) frames. Recent studies attempted to integrate TRM strengthening with standard thermal insulation, with promising results. However, the so-far solutions make use of ordinary portland cement-based binders (OPC), the production of which is a major contributor to global CO_2 emissions. More sustainable mortars, such as those containing alkali-activated materials (e.g. geopolymers), can help address this issue, as recent studies have shown that they can offer similar or better performance than standard OPC mortars. This study aims to investigate the combination of structural and energy retrofitting of existing masonry-infilled RC frames. One-story, one-bay specimens were constructed at a 1:2.5 scale. Regarding the strengthened specimen, basalt fiber meshes embedded in metakaolin-based geopolymer mortar were used for structural retrofitting, while a conventional thermal insulation system with expanded polystyrene was used for energy retrofitting. The two specimens were subjected to in-plane cyclic loading.

Keywords: infilled frame · masonry infill · reinforced concrete · textile-reinforced mortar (TRM) · seismic retrofitting · energy retrofitting · in-plane loading

1 Introduction

Many existing residential buildings in seismic-prone European regions consist of reinforced concrete (RC) frames with masonry infills, constructed with minimal or no seismic design provisions and without strict energy efficiency regulations. The aging of materials and lack of maintenance further increase the vulnerability of these structures to seismic loads, negatively affecting, also, their energy performance [1].

Although infill walls are typically not considered in the design process, they can significantly influence the overall behavior of the structure. They require special attention due to their vulnerability to seismic actions and issues related to insufficient anchorage

M. Posani and S. Singh (Eds.): RYS 2025, RILEM Bookseries 67, pp. 836–845, 2026.
https://doi.org/10.1007/978-3-032-14652-6_60

to the surrounding RC elements. As a result, potential local failures in infilled frames may lead to partial or even total collapse of the structure [2, 3].

Textile-Reinforced Mortars (TRM) have proven to be an effective strengthening solution for RC frames with infills [4–7]. Recent studies have attempted to integrate conventional thermal insulation or capillary tube systems into TRM strengthening solutions [8–16]. These combined systems have shown promising results. However, the systems studied so far use cementitious mortars based on ordinary portland cement (OPC), the production of which contributes significantly to global CO_2 emissions [17]. More sustainable mortars, such as those containing alkali-activated materials (e.g., geopolymers) [18], can help address this issue. Recent studies suggest that these materials can offer similar or even superior performance compared to conventional OPC-based mortars [11, 19–23].

The aim of the present experimental study is to combine structural and energy upgrading of existing RC frames infilled with masonry walls in direct contact. The structural upgrade is achieved by jacketing with basalt fiber meshes embedded in a metakaolin-based geopolymer mortar, while the energy upgrade is provided by a conventional thermal insulation system with Expanded Polystyrene (EPS). In-plane cyclic loading tests were conducted on two masonry-infilled RC frames constructed at a 1:2.5 scale. Further details are provided in the following section.

2 Materials and Methods

The present experimental program aimed to assess the effectiveness of the proposed retrofitting approach for masonry-infilled RC frames subjected to in-plane cyclic loading. The first test was carried out on an unretrofitted specimen, hereafter referred to as the "control specimen (CON)", whereas the second test was conducted on an infilled RC frame that had been previously strengthened using a combined structural and energy retrofitting scheme (STR).

2.1 Infilled Frames

The infilled frames were designed to represent parts of existing structures built before 1985. Two identical one-story, one-bay masonry-infilled RC frames were constructed at a 1:2.5 scale, each with a span of 1570 mm and a height of 1000 mm (Fig. 1). The columns had rectangular cross sections measuring 100 x 160 mm. Their longitudinal reinforcement consisted of six 8 mm rebars. The upper beam had a rectangular cross section of 100 x 200 mm, with longitudinal reinforcement consisting of four 8 mm rebars. Transverse reinforcement for both columns and the upper beam comprised 6 mm closed stirrups at 80 mm spacing. The base beam had a cross-section of 260 x 250 mm and was adequately reinforced to remain uncracked during testing.

The infill walls consisted of a single-wythe, 60 mm thick fired clay brick masonry, placed eccentrically to contact the front face of the RC frame. Hollow clay masonry units with nominal dimensions of 60 x 90 x 140 mm (width x height x length) were used for their construction.

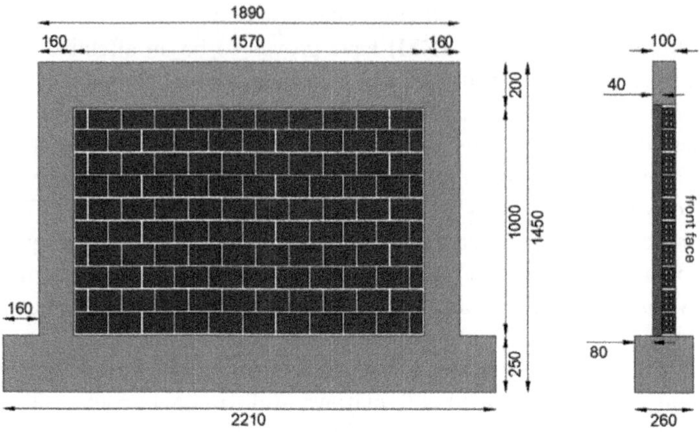

Fig. 1. Geometry of the masonry-infilled frames (dimensions in mm): a) front view, b) section view.

2.2 Material Properties

Concrete. The compressive strength of concrete was determined on the day of testing through compression tests performed on three 150 mm cubic specimens. The results are summarized in Table 1.

Table 1. Concrete compression test results (mean values).

Specimen ID	Compressive strength (MPa)	
	Base beam	Frame
CON	35.1	32.7
STR	37.9	33.6

Masonry. The average compressive strength of the clay bricks was determined by testing five randomly selected units, following the procedures outlined in EN 772-1 [24]. The compressive strength measured parallel to the perforations was 15.5 MPa, while the strength perpendicular to the perforations was 7.1 MPa.

The mortar used for the infill joints was a general-purpose masonry mortar with a mix ratio of 1:1:5 (cement:lime:sand). Its mechanical properties were determined according to EN 1015-11 [25] through flexural and compression tests conducted on the day of each frame's testing. The results are summarized in Table 2.

Table 2. Masonry mortar flexural and compression test results (mean values).

Specimen ID	Flexural strength (MPa)	Compressive strength (MPa)
CON	2.3	11.0
STR	2.3	10.2

Retrofitting Materials. For the strengthening of the masonry infills, commercially available basalt fiber meshes coated with epoxy resin were used (Fig. 2). The fiber rovings were arranged in two orthogonal directions, with an equal quantity of fibers in each direction. The mechanical properties, as provided by the manufacturer, are listed in Table 3.

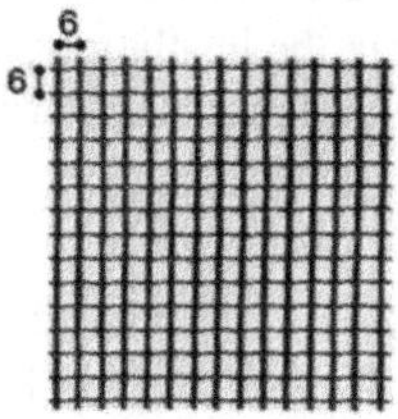

Fig. 2. Basalt textile reinforcing mesh.

Table 3. Mechanical properties of basalt textile.

Properties	
Mesh size	6 x 6 mm
Coating of fibre roving	Yes
Elastic modulus (uncoated fibers)	89 GPa
Total weight	250 g/m2
Weight distribution	50–50 %
Nominal thickness (per direction)	0.039 mm
Tensile strength (per direction)	1540 MPa 60 kN/m

The geopolymer mortar, used both as the matrix of the strengthening system and as an adhesive layer between the thermal insulation and the masonry in specimen STR, was developed at the Laboratory of Concrete Technology and Reinforced Concrete Structures (University of Thessaly). Designed to replace conventional binders based on ordinary Portland cement, the mortar exhibited similar mechanical properties. The maximum aggregate size was 1 mm, and polypropylene fibers were added at 1% per volume. Metakaolin was used as the precursor, and potassium silicate as the alkaline activator. Flexural and compressive strengths of the mortar were determined on the day of each frame's testing using three specimens measuring 40 x 40 x 160 mm. The results are presented in Table 4.

Table 4. Geopolymer mortar flexural and compression test results (mean values).

Specimen ID	Flexural strength (MPa)	Compressive strength (MPa)
STR	5.8	42.7

For the thermal retrofitting of specimen STR, 30 mm thick Expanded Polystyrene Insulation (EPS) panels with a thermal conductivity coefficient of 0.03 W/mK were applied. This thickness corresponds to 7.5 cm thick panels at full scale, which lies at the lower end of the range typically used in real practice. EPS panels are commonly used in energy retrofitting applications across Southern Europe, primarily due to their low cost. According to the manufacturer, the insulation panels have a minimum compressive strength of 0.02 MPa and a shear modulus of at least 1.0 MPa.

2.3 Strengthening Procedure

Specimen STR was strengthened on one side by applying the strengthening materials in a "sandwich" configuration (TRM layer – thermal insulation layer – TRM layer) on the front face of the frame. The TRM strengthening covered the entire exterior surface of the masonry infill, extended onto the surrounding RC members, and was anchored to the columns by means of a simple 180° wrap-around. In the second stage, EPS panels were installed and anchored to both the masonry and the frame with standard 6 mm diameter anchors. Finally, the second TRM layer was applied over the entire surface of the EPS panels, following the same procedure as for the first layer (Fig. 3).

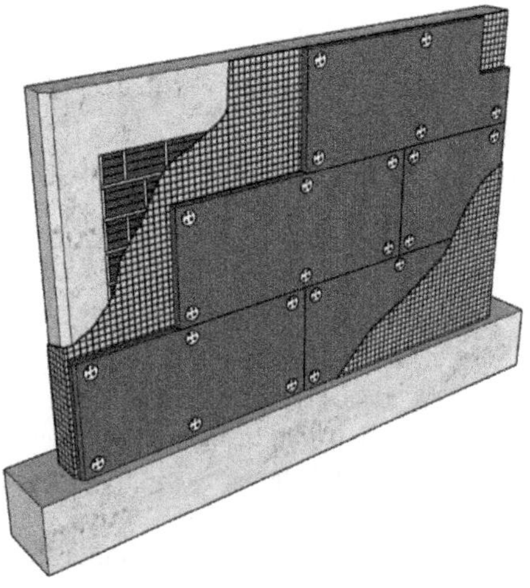

Fig. 3. Strengthening layout [26].

2.4 Test Setup

The specimens were tested within a rigid steel frame, to which the base of each specimen was anchored. Two steel guides with roller bearings were installed above each frame to prevent out-of-plane displacements. In-plane cyclic displacements of increasing amplitude were applied at beam height using a servo-hydraulic actuator with a capacity of 500 kN. The loading protocol for both specimens is illustrated in Fig. 4.

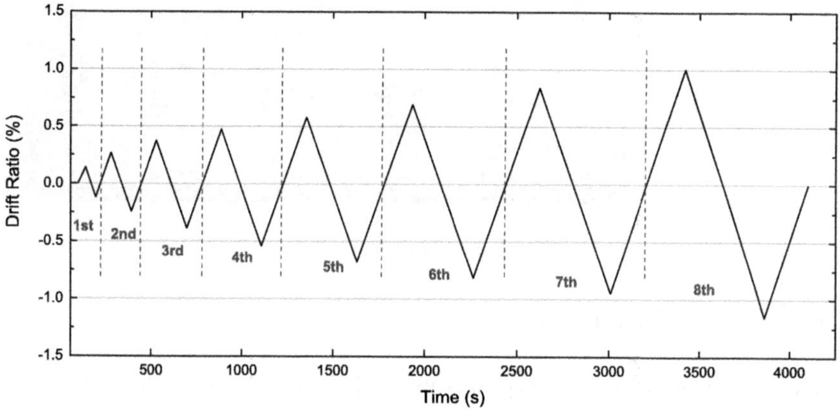

Fig. 4. Loading protocol.

In-plane displacement of the specimens was measured using an LVDT positioned at the beam axis level. Additionally, six LVDTs were placed on the back face of the specimens near the frame joints to capture the relative movement between the frame and the infill. One LVDT was used to measure the displacement of the base beam, and two wire gauges were installed across the diagonals of the infill to measure shear deformation (Fig. 5).

Fig. 5. Experimental setup.

3 Results and Discussion

The unretrofitted specimen (CON) was subjected to five loading cycles in total. CON reached a maximum load of 88.6 kN at a drift level equal to 0.71% in the push direction, and a maximum load of 81.6 kN at a drift level equal to 0.71% in the pull direction. The initial stiffness, as calculated for the first loading cycle, was equal to 30.6 kN/mm.

The retrofitted specimen (STR) was subjected to nine in-plane loading cycles. STR reached a peak load of 161.8 kN at a drift level of 0.99% in the push direction, and a maximum load of 144.6 kN at a drift level of 1.15%. The initial stiffness was equal to 37.7 kN/mm, showing 23.2% increase compared to the control specimen.

The hysteretic behavior of the control specimen (CON) in comparison to the retrofitted one (STR) is presented in Fig. 6. The main results are presented in Table 5.

Table 5. Main test results.

Specimen ID	Push direction		Pull direction		Initial Stiffness (kN/mm)
	Pmax (kN)	Drift Ratio (%)	Pmax (kN)	Drift Ratio (%)	
CON	88.6	0.71	81.6	0.71	30.6
STR	161.8	0.99	144.6	1.15	37.7

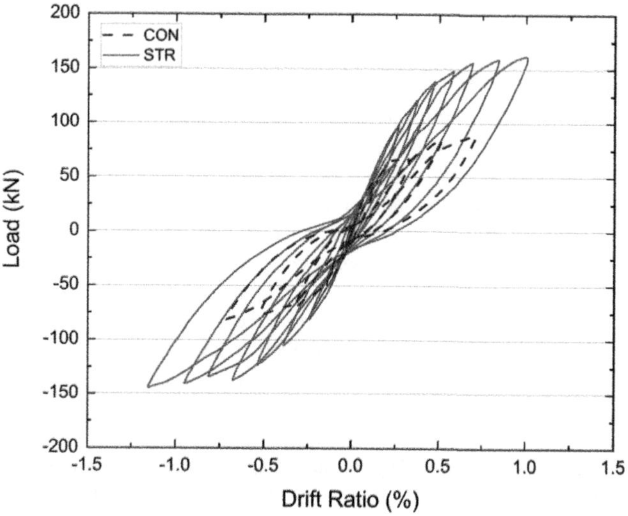

Fig. 6. Comparative response curves for the two specimens.

4 Conclusions

This experimental study investigated the effectiveness of a specific retrofitting system designed to simultaneously address both the structural and energy deficiencies of RC frame structures with infill walls. Two masonry-infilled RC frames were constructed at a 1:2.5 scale, one of which was retrofitted with two layers of Textile-Reinforced Mortar (TRM) and one layer of Expanded Polystyrene Insulation (EPS), arranged in a "sandwich" configuration. Both specimens were subjected to in-plane cyclic loading.

The experimental results demonstrated that this innovative combination of materials offers an efficient and sustainable solution for the simultaneous structural and energy upgrading of masonry infills. Specifically, the following conclusions were drawn:

- The retrofitting system led to an increase of 82.6% and 77.0% in terms of maximum load in Push and Pull direction, respectively.
- The retrofitted specimen, STR, showed 23.2% increase in initial stiffness, in comparison to the unretrofitted specimen, CON.

While the present study focused on the investigation of the experimental behavior of specimens retrofitted with the proposed strengthening scheme, future research effort could be directed towards the development of analytical tools capable of predicting the structural behavior of TRM-retrofitted frames when combined with thermal insulation panels.

References

1. Pohoryles, D.A., Bournas, D.A., Da Porto, F., Caprino, A., Santarsiero, G., Triantafillou, T.: Integrated seismic and energy retrofitting of existing buildings: A state-of-the-art review. J. Build. Eng., 105274 (2022)

2. Furtado, A., Rodrigues, H., Arêde, A., Varum, H.: Experimental evaluation of out-of-plane capacity of masonry infill walls. Eng. Struct. **111**, 48–63 (2016)
3. Mosalam, K.M., Günay, S.: Progressive collapse analysis of reinforced concrete frames with unreinforced masonry infill walls considering in-plane/out-of-plane interaction. Earthq. Spectra **31**(2), 921–943 (2015)
4. Koutas, L., Bousias, S.N., Triantafillou, T.C.: Seismic strengthening of masonry-infilled RC frames with TRM: experimental study. J. Compos. Constr. **19**(2), 04014048 (2015)
5. Sagar, S.L., Singhal, V., Rai, D.C.: In-plane and out-of-plane behavior of masonry-infilled RC frames strengthened with fabric-reinforced cementitious matrix. J. Compos. Constr. **23**(1), 04018073 (2019)
6. Koutas, L.N., Bournas, D.A.: Out-of-plane strengthening of masonry-infilled RC frames with textile-reinforced mortar jackets. J. Compos. Constr. **23**(1), 04018079 (2019)
7. Furtado, A., Rodrigues, H., Arêde, A., Melo, J., Varum, H.: The use of textile-reinforced mortar as a strengthening technique for the infill walls out-of-plane behaviour. Compos. Struct. **255**, 113029 (2021)
8. Gkournelos, P.D., Bournas, D.A., Triantafillou, T.C.: Combined seismic and energy upgrading of existing reinforced concrete buildings using TRM jacketing and thermal insulation. Earthq. Struct **16**(5), 625–639 (2019)
9. Gkournelos, P.D., Triantafillou, T.C.: Out-of-plane behavior of in-plane damaged masonry infills retrofitted with TRM and thermal insulation. J. Compos. Constr. **27**(6), 04023054 (2023)
10. Wang, F.: Experimental research on seismic performance of masonry-infilled RC frames retrofitted by using fabric-reinforced cementitious matrix under in-plane cyclic loading. Int J Concr Struct Mater **17**, 31 (2023)
11. Cholostiakow, S., Skyrianou, I., Koutas, L., Papakonstantinou, C.: Out-of-plane performance of structurally and energy retrofitted masonry walls: geopolymer versus cement-based textile-reinforced mortar combined with thermal insulation. Open Res. Europe **3** (2023)
12. Furtado, A., Rodrigues, H., Arêde, A., Varum, H.: A experimental characterization of seismic plus thermal energy retrofitting techniques for masonry infill walls. J. Build. Eng. **75**, 106854 (2023)
13. Christis, Z.C., Renos, V., Nicholas, K., Rogiros, I., Christiana, A.F., Stathis, B.: Seismic and energy upgrading of existing buildings—full-scale testing of retrofitted masonry-infilled RC frames. Earthquake Eng. Struct. Dynam. **52**(14), 4489–4517 (2023)
14. Bournas, D.A.: Concurrent seismic and energy retrofitting of RC and masonry building envelopes using inorganic textile-based composites combined with insulation materials: a new concept. Compos. B Eng. **148**, 166–179 (2018)
15. Baek, E., Pohoryles, D.A., Kallioras, S., Bournas, D.A., Choi, H., Kim, T.: Innovative seismic and energy retrofitting of wall envelopes using prefabricated textile-reinforced concrete panels with an embedded capillary tube system. Eng. Struct. **265**, 114453 (2022)
16. Kallioras, S., Bournas, D., Koutas, L., Molina, F.J.: Integrated seismic and energy retrofitting of masonry-infilled RC buildings with textile-reinforced mortar and thermal insulation: full-scale tests on a five-story prototype. J. Build. Eng. **98**, 111006 (2024)
17. Andrew, R.M.: Global CO2 emissions from cement production. Earth Syst. Sci. Data **10**(1), 195–217 (2018)
18. Davidovits, J.: Geopolymers: inorganic polymeric new materials. J. Therm. Anal. Calorim. **37**(8), 1633–1656 (1991)
19. Maras, M.M., Kose, M.M.: Structural behavior of masonry panels strengthened using geopolymer composites in compression tests. Iran. J. Sci. Technol. Trans. Civ. Eng. **45**, 767–777 (2021)

20. Longo, F., Cascardi, A., Lassandro, P., Aiello, M.A.: A novel composite reinforced mortar for the structural and energy retrofitting of masonry panels. Key Eng. Mater. **916**, 377–384 (2022)
21. Gkournelos, P.D., Azdejković, L.D., Triantafillou, T.C.: Innovative and eco-friendly solutions for the seismic retrofitting of natural stone masonry walls with textile reinforced mortar: In-and out-of-plane behavior. J. Compos. Constr. **26**(1), 04021061 (2022)
22. Arce, A., Kapsalis, P., Papanicolaou, C.G., Triantafillou, T.C.: Textile-reinforced alkali-activated mortar for in-plane shear capacity improvement of masonry before and after high temperature exposure. In International Conference on Structural Analysis of Historical Constructions, pp. 849–857. Springer Nature Switzerland, Cham (2023)
23. Cholostiakow, S., Koutas, L.N., Papakonstantinou, C.G.: Geopolymer versus cement-based textile-reinforced mortar: diagonal compression tests on masonry walls representative of infills in RC frames. Constr. Build. Mater. **373**, 130836 (2023)
24. CEN (European Committee for Standardization): Methods of test for masonry units —Part 1: Determination of compressive strength. EN 772-1. Brussels, Belgium: CEN (2011)
25. CEN (European Committee for Standardization): Methods of test for mortar for masonry—Part 11: Determination of flexural and compressive strength of hardened mortar. EN 1015-11. Brussels, Belgium: CEN (2006)
26. Cholostiakow, S.P., Koutas, L.N., Papakonstantinou, C.G.: Holistic approach to rehabilitation of substandard masonry infilled RC frames. In:1th International Conference on Fiber-Reinforced Polymer (FRP) Composites in Civil Engineering (CICE 2023), Rio de Janeiro, Brazil, 2023b, vol. 10 (2023)

Service Life and Durability

Ageing Chamber Tests of Cement Mortars Reinforced with Carbon Nanotubes

Eryk Goldmann(✉), Marcin Górski, and Barbara Klemczak

Faculty of Civil Engineering, Silesian University of Technology, Akademicka 5, 44-100 Gliwice, Poland
eryk.goldmann@polsl.pl

Abstract. The durability of construction materials is one of the most important properties related to their sustainability and longevity. With various degrading factors, including carbonation, chloride and sulphate attack, water ingress and cycles of temperature change, there is a wide area for improvement for modern construction materials. The use of carbon nanotubes in cement-based materials has the potential to create stronger, functional materials which combine better mechanical properties with functionality as a strain sensor or heating material. However, the question of the durability of cement nanocomposites under various factors is still not fully recognised. The presented research aims at assessing the resistance of cement mortars infused with different dosages of carbon nanotubes to rapid ageing cycles. Repeating cycles of high and low temperatures with wind and light exposure using an ageing chamber were employed to simulate the ageing process of cement mortar samples without other corrosive factors. A comparison of mechanical strength differences for samples exposed to rapid ageing and the control group was used to determine the effect of carbon nanotube addition and dosage on the materials' durability. The results show that a higher addition of carbon nanotubes leads to the smaller reduction of compressive strength after cycles in the ageing chamber with 0.2wt.% dosage of carbon nanotubes, resulting in a smaller compressive strength reduction than for the reference sample. The flexural strength of samples increased after exposure, which could result from re-occurring hydration caused by high temperatures and humidity in the ageing chamber.

Keywords: cement mortar · carbon nanotubes · ageing chamber · cement nanocomposites

1 Introduction

Carbon nanotubes are amongst the most widely researched carbon nanofillers in cement-based materials, along with graphene and carbon nanofibers [1]. The addition of these materials into the cement matrix impacts a variety of properties from early-age rheology [2], microstructure and mechanical strength [3, 4] up to electrical conductivity [5], which is the foundation for the creation of functional smart materials used as strain sensors [6], heating materials [7] or for energy harvesting [8, 9]. Amongst all of the abovementioned

M. Posani and S. Singh (Eds.): RYS 2025, RILEM Bookseries 67, pp. 849–858, 2026.
https://doi.org/10.1007/978-3-032-14652-6_61

properties of cement nanocomposites, however, not much attention is yet given to their durability, which could be a vital property for smart materials placed in various working conditions. Smart materials are meant to be used in varying conditions and might be exposed to temperature gradients and changing weather which could cause degradation and faster ageing.

Durability of cement-based materials can be considered from different points of view. Usually, the most researched topics in terms of durability are linked with the effects of corrosive media like chlorides or sulphates, or with ever-present carbonation. Ageing of materials combined with the influence of temperature changes and strong light is not considered as frequently. These factors are especially important for elements placed outside, like façade elements, walls or columns, which are exposed to various combinations of weather conditions. For cement nanocomposites used as smart materials, resistance to temperature changes and outside conditions might be especially important if they are to be used for energy harvesting or as strain sensors on infrastructure.

The presented research aims at assessing the effect of rapid ageing in an ageing chamber on the mechanical strength of cement mortars with the addition of carbon nanotubes, as a first step to the complex evaluation of the durability of cement nanocomposites. With possible improvement of the microstructure and mechanical strength of cement-based nanocomposites with carbon nanotubes a higher resistance to cycles of changing temperature was expected. Cycles of high temperature of 40 °C with strong light (visible spectrum, without UV exposition) and low temperature of −8 °C were used as exposition conditions. Flexural and compressive strength were tested for samples after exposition and compared with cement mortars kept in water. Additionally, SEM imaging was used in order to determine differences in the microstructure of mortars near the exposed surface.

2 Materials and Methods

Cement mortars used in tests were prepared in accordance with European standard PN-EN 196–1 [10]. Samples were composed of ordinary Portland cement CEM I 42.5N, standard sand, tap water and water suspension of multiwalled carbon nanotubes (MWCNT). Carbon nanotubes used in the research were industrial grade NANOCYL NC7000, which are one-dimensional carbon nanomaterial with 90% purity, a mean length of 1.5μm and mean diameter of 9.5nm. The SEM image of used carbon nanotubes is presented in Fig. 1.

Suspensions were prepared by first dosing naphthalene-based superplasticizer into 100g of deionised water. The superplasticizer-to-MWCNT ratio was constant 5:1 by mass. Then, without mechanical stirring, suspensions were sonicated using an ultrasonic homogeniser for 30 min. To avoid water evaporation, suspensions were placed in an ice bath, and the total time of sonication was divided into 30s intervals. Suspensions were added into the dry components along with tap water to achieve a water-to-cement ratio of 0.5. A sand-to-cement ratio of 2:1 by mass was used in order to improve the flowability of cement mortars without the addition of additional superplasticiser. Cement mortars were mixed in a planetary mixer using an automated procedure described in PN-EN 196–1 standard and placed in oiled moulds immediately after mixing. Samples of 4 ×

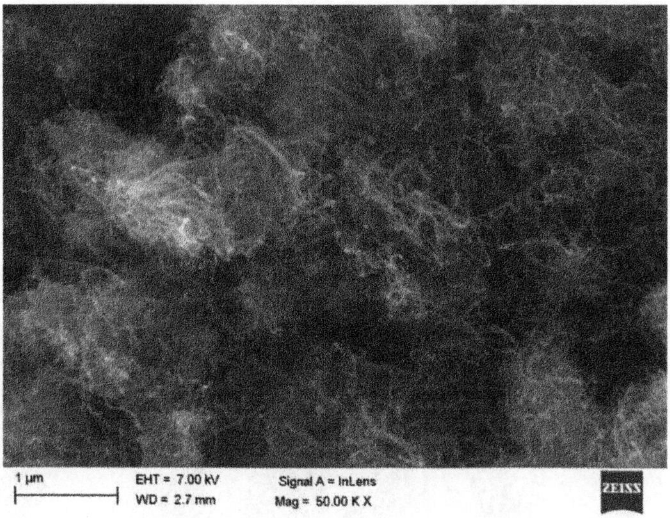

Fig. 1. Carbon nanotubes used in the study.

4 × 16 cm were casted and kept covered for 24 h. After the initial setting samples were demoulded and placed in water for 28 days. For each mix two sets of samples were prepared, one meant for exposition and the other as a reference. Reference samples were kept in water for the entire exposition time and beams intended for testing were securely attached to the walls of the ageing chamber. Composition of samples is given in Table 1.

Table 1. Composition of cement mortars used in tests.

Sample	CNT [wt.%]	w/c	Sand-to-cement ratio
2C CNT 0	0	0.5	2:1
2C CNT 0.05	0.05		
2C CNT 0.1	0.1		
2C CNT 0.2	0.2		

The ageing chamber used in tests is composed of four walls with mounting areas for samples. The main part of the chamber is placed on a rotating mechanism which allows for automated conduction of tests and simultaneous exposition to two different conditions at the same time. The chamber is equipped with three exposition parts called "sun chamber", "rain chamber" and "frost chamber" [11]. In the presents research only "sun chamber" and "frost chamber" were used. The "sun chamber" is equipped with 20 lamps with 8kW of power each capable of generating temperatures up to 65 °C and a system of additional lamps to simulate UV exposition. This chamber simulates exposition to sunlight and generates a combination of high temperature, visible light of 400–700 nm wavelength and UV radiation. The "frost chamber" is equipped with a system of air conditioning units coupled with condensers and is able to drop the temperature of exposed samples to −20 °C. In the presented research, each cycle consisted of 18 min of heating to 40 °C with strong light in a visible spectrum, without UV exposition and 18 min of cooling to −8 °C. The rotation of the chamber and preparation for exposition lasted additional 7 min. A total of 740 cycles were conducted before strength test of the mortars. The view of the chamber and its parts used in the tests is given on Fig. 2.

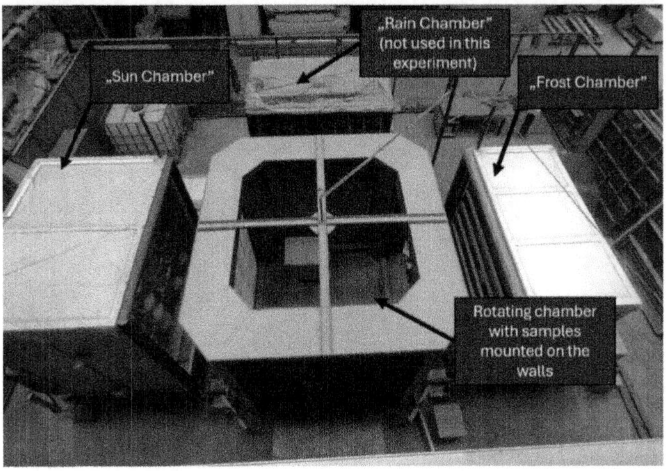

Fig. 2. Overview of the ageing chamber.

Mechanical strength of cement mortars was tested immediately after completion of exposition cycles in accordance with the procedure described in PN-EN 196–1 standard. For each mix, an exposed sample and a reference sample were tested to compare strength differences caused by exposure in the ageing chamber. Flexural strength was tested in a three-point bending test. Samples were placed on steel rolls with a diameter of 10 mm and spaced 100 mm from one another. A constant force was applied in the centre of the sample in rate of 0.05 kN/s until breakage. Samples from ageing chamber were placed in a way that caused the exposed surface to be in the tension zone. For compressive strength test parts of samples remaining after bending test were placed in the testing press in a way to assure an area loading of 40 × 40 mm and loaded at a rate of 2.4 kN/s until destruction. A total of three samples were tested in the flexural test and six in the compressive strength test for each of the mixes. Standard deviation in each of the tests did not exceed 5%.

SEM imaging was used to observe changes in microstructure of samples near the exposed surface. Splinters of cement mortar samples were carefully taken from the side of the exposed area and sputtered with a few μm layer of gold to improve electrical conductivity and tested with HR SEM Supra 35 microscope [12].

3 Results

Visual inspection of the samples revealed no significant differences between group exposed to changing conditions in the ageing chamber and control group. The results of strength test with the comparison between groups are presented in Table 2 and Figs. 3 and 4. Column "difference" in Table 2 represents the difference between mean mechanical strength of sample after exposition and its relative reference. There was no significant difference in the failure mode for any of the samples.

Table 2. Comparison of the mechanical strength of samples after exposure in the ageing chamber and their references.

Sample	ft [MPa]		Difference [%]	fc [MPa]		Difference [%]
	Ref	Exp		Ref	Exp	
2C CNT 0	7,61	10,35	+36,01	66,75	64,24	−3,76
2C CNT 0.05	7,25	9,69	+33,66	70,21	62,86	−10,46
2C CNT 0.1	7,51	8,31	+10,65	68,64	63,81	−7,06
2C CNT 0.2	7,74	9,13	+17,96	60,83	60,41	−0,72

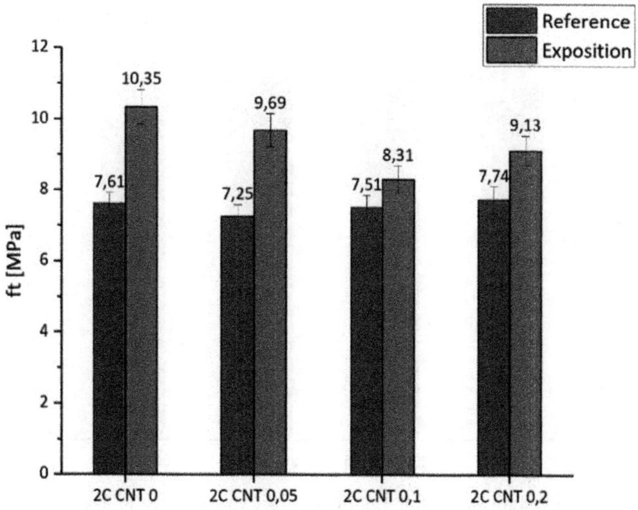

Fig. 3. Flexural strength of mortar samples after exposure in the ageing chamber and their references.

All of the tested samples had higher flexural strength after exposition when compared to the references. The highest improvement were noticed in samples 2C CNT 0 and 2C CNT 0.05, while samples with higher MWCNT content had significantly smaller increase in flexural strength. Reference samples had all relatively similar flexural strength with 2C CNT 0.2 having a slightly higher value than the sample without MWCNT addition. These differences were much higher for samples after ageing chamber cycles, where samples with MWCNT had significantly lower flexural strength, with the value decreasing for higher nanotubes content. This is an unexpected results as naturally, for cementitious materials exposed to temperature differences, it is expected for them to have lower mechanical strength after exposition due to microstructural damage and cracks appearing. A possible explanation for an increment in flexural strength after exposition could be a re-occurring hydration near the exposed surface. It is possible that some unhydrated cement grains underwent a hydration reaction under high temperature in the ageing chamber, which strengthened and densified an area of cement matrix close to the surface and increased flexural strength of samples. On the other hand however, there was not much additional moisture present in the chamber, since no rain exposition was employed in the tests. Therefore, only water present in the pores of samples could have reacted with unhydrated cement grains in a limited way. Under this theory observed differences in flexural strength for samples after exposition could be caused by a positive influence of carbon nanotubes on hydration during samples curing. In sample 2C CNT 0 there were more unhydrated cement grains left than in samples with MWCNT, which allowed for a higher strength increase for the sample without nanofiller.

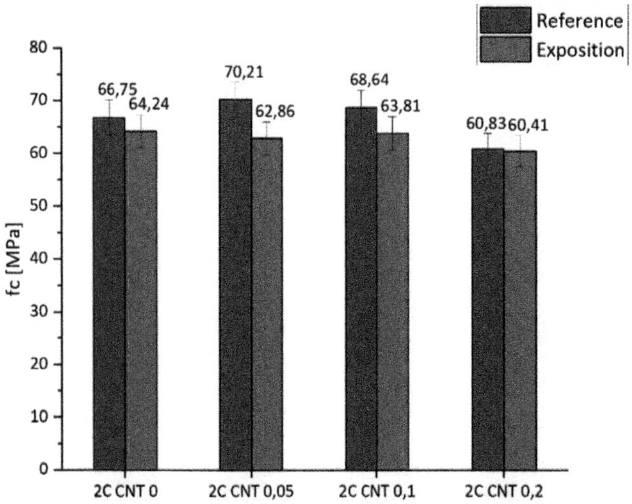

Fig. 4. Compressive strength of mortar samples after exposure in the ageing chamber and their references.

Compressive strength of samples was reduced after exposition for all of the tested samples with the highest reduction for sample 2C CNT 0.05 and the lowest for 2C CNT 0.2 which is a result typically expected for samples exposed to temperature variations. A different trend was observed compared to the flexural strength results. While for flexural strength, values for all samples were similar for the reference series and significant differences were observed for the series after cycles in the chamber, for compressive strength, the trends were opposite. For reference series compressive strength was higher for samples with low MWCNT content than for 2C CNT 0, while 2C CNT 0.2 was significantly lower. After exposition in the ageing chamber, the compressive strength was similar for samples 2C CNT 0, 2C CNT 0.05 and 2C CNT 0.1, while 2C CNT 0.2 remained significantly lower. While for all of the samples the compressive strength was lower after exposition, the increase in MWCNT content lead to lower relative reduction. At the same time, only sample 2C CNT 0.2 had lower reduction in compressive strength than the sample without the addition of MWCNT. The observed reduction in compressive strength could be caused by the influence of negative temperature in the ageing chamber. While high temperature was applied through strong light emission, low temperature was applied with air conditioning units and condensers, first of which influenced mainly the exposed surface of samples because of the application through lamps, while the latter could have influenced a larger part of the samples. This could have led to strengthening and reoccurring hydration in the surface area of the samples and internal cracking in the other parts. SEM images of tested samples from both series are presented in Fig. 5. Images were taken using shards carefully acquired from the exposed surface of samples (right column) and reference series (left column) to assess if there are visible differences in microstructure, which could support the theory of renewed hydration occurring in an ageing chamber under high temperature. The comparison shows that the microstructure of exposed samples appears to be more refined and less porous than the reference series. The amorphous C-S-H phase is denser, and no clear indication of other hydration products was observed. This observation could confirm that additional hydration occurred in samples exposed in the ageing chamber, causing densification of the microstructure. The highest increment on flexural strength was observed in samples 2C CNT 0 and 2C CNT 0.05 which is also visible in SEM images. Microstructure of these samples is denser than 2C CNT 0.1 and 2C CNT 0.2 which had a significantly smaller increase in flexural strength after exposition in the ageing chamber.

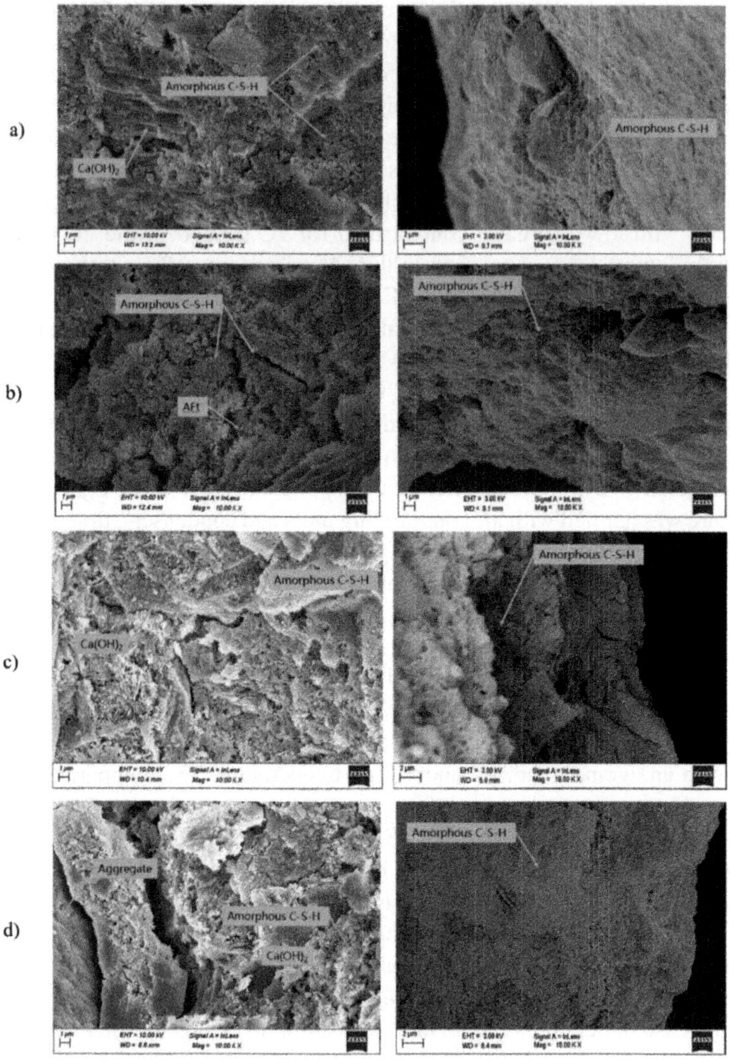

Fig. 5. SEM images of microstructure of cement mortars taken near the exposed surface reference (left) and after exposition (right) a) 2C CNT 0, b) 2C CNT 0.05, c) 2C CNT 0.1, d) 2C CNT 0.2.

4 Conclusions

Presented research focused on the assessment of the influence of repeated cycles of high temperature with light and low temperature in the ageing chamber on the mechanical strength of cement mortars reinforced with multiwalled carbon nanotubes. Flexural and compressive strength tests and SEM imaging were performed on samples after 740 cycles in the ageing chamber, and the reference series was kept in water for the same time.

Acquired results show that flexural strength increased for all of the tested mortars, which is an unexpected result but it could be an effect of high temperature and light exposure on the surface, which caused re-occurring hydration in the area near the exposed surface. Meanwhile the compressive strength was lower for all the tested samples exposed to cycles in the ageing chamber which could be explained by the influence of high temperature mostly on one of the surfaces of samples while low temperature applied through air conditioning units and condensers influenced more areas of each sample leading to cold induced damage of the microstructure and lowering of compressive strength.

In terms of mechanical strength values, flexural strength was similar for all samples in the reference series, while it was significantly lower for samples with MWCNT after cycles in the chamber, which could imply that more unhydrated grains of cement were left and underwent additional hydration in sample without the addition of carbon nanotubes. The increment of flexural strength was lower for all samples with MWCNT compared to 2C CNT 0, which could either imply a smaller influence of carbon nanotubes on possible reoccurring hydration or their positive influence on hydration in curing stage, which caused fewer unhydrated grains of cement to be left in mortars. It is possible that the addition of MWCNTs influenced hydration during initial setting and curing of samples, therefore covering cement grains with hydration products and reducing the effect of reoccurring hydration under chamber conditions. This theory however requires further material testing including an influence of MWCNTs on hydration dynamics, mineral composition and microstructure including tests on samples exposed to conditions similar to applied in the ageing chamber tests.

Compressive strength of all samples was lower after exposition and on a similar level, while for reference samples a significant increase was observed for samples with low MWCNT content compared to 2C CNT 0. The decrease in compressive strength was lowering as the dosage of MWCNT increased and for sample 2C CNT 0.2 was smaller than for 2C CNT 0, which could imply that it exhibited the highest resistance to cycles of varying conditions used in tests in terms of compressive strength.

Acquired results show interesting but unexpected mechanisms linked with the assumed exposure conditions. The influence of carbon nanotubes varied for flexural and compressive strength results after exposition. For flexural strength, the lowest amount of 0.05wt.% provided the best results, while for compressive strength the highest amount of 0.2wt.% performed the best. In case of assessing the resistance to cycles in the ageing chamber, it can be concluded that the amount of 0.2wt.% could be considered the best as it exhibited the lowest reduction in compressive strength compared to the reference sample. On the other hand, this amount had worse mechanical strength in terms of values, which, for a possible application in smart materials might not be the most important property. If tested mortars were to be used in façade elements used as energy harvesting materials, the mechanical strength could be considered as a secondary property, with potential electrical conductivity and durability being more important.

The presented research could be considered a first step in the assessment of the durability of cement mortars with carbon nanotubes. Additional exposition conditions focused on different aspects of real-life working conditions should be applied as a comparison. Moreover, in-depth material tests will be performed to further back up the theory about re-occurring hydration in exposed samples.

References

1. Goldmann, E., Górski, M., Klemczak, B.: Recent advancements in carbon nano-infused cementitious composites. Materials **14**, 5176 (2021). https://doi.org/10.3390/ma14185176
2. Goldmann, E., Cygan, G., Klemczak, B., Górski, M.: Rheological properties of carbon nanotube infused cementitious composites with various amounts of CNT. Buildings **13**, 2960 (2023). https://doi.org/10.3390/buildings13122960
3. Chen, J., Akono, A.-T.: Influence of multi-walled carbon nanotubes on the hydration products of ordinary Portland cement paste. Cem. Concr. Res. **137**, 106197 (2020). https://doi.org/10.1016/j.cemconres.2020.106197
4. Wang, X., Li, Q., Xu, S.: Failure modes of carbon nanotubes in cement composites under flexural load. Eng. Fract. Mech. **273**, 108740 (2022). https://doi.org/10.1016/j.engfracmech.2022.108740
5. Hong, G., Choi, S., Yoo, D.-Y., Oh, T., Song, Y., Yeon, J.H.: Moisture dependence of electrical resistivity in under-percolated cement-based composites with multi-walled carbon nanotubes. J. Mater. Res. Technol. **16**, 47–58 (2022). https://doi.org/10.1016/j.jmrt.2021.11.151
6. D'Alessandro, A., Tiecco, M., Meoni, A., Ubertini, F.: Improved strain sensing properties of cement-based sensors through enhanced carbon nanotube dispersion. Cem. Concr. Compos. **115**, 103842 (2021). https://doi.org/10.1016/j.cemconcomp.2020.103842
7. Nishat, F.M., Chakraborty, R., Oh, H.-J., Yoo, D.-Y., Mohapoo, R., Yeon, J.H.: Design and performance evaluation of an electrically heated concrete panel for sustainable winter maintenance. Dev. Built. Environ. **16**, 100259 (2023). https://doi.org/10.1016/j.dibe.2023.100259
8. Tzounis, L., Liebscher, M., Fuge, R., Leonhardt, A., Mechtcherine, V.: P- and n-type thermoelectric cement composites with CVD grown p- and n-doped carbon nanotubes: demonstration of a structural thermoelectric generator. Energy Build **191**, 151–163 (2019). https://doi.org/10.1016/j.enbuild.2019.03.027
9. Choi, K., Kim, D., Chung, W., Cho, C., Kang, S.-W.: Nanostructured thermoelectric composites for efficient energy harvesting in infrastructure construction applications. Cem. Concr. Compos. **128**, 104452 (2022). https://doi.org/10.1016/j.cemconcomp.2022.104452
10. CEN PN-EN 196-1:2016-07 methods of testing cement. In: Part 1: Determination of strength. Brussels, Belgium (2016)
11. Bochen, J., Ślusarek, J.: Analysis of resistance to accelerated aging of the top layer of small-sized concrete elements. A case study. Constr. Build. Mater. **264**, 120211 (2020). https://doi.org/10.1016/j.conbuildmat.2020.120211
12. Bialas, O., et al.: Laser assisted fabrication of mechanochemically robust Ti3Au intermetallic at Au-Ti interface. Eng. Sci. Technol. Int. J. **42**, 101413 (2023). https://doi.org/10.1016/j.jestch.2023.101413

Durability of LC3 Concrete Exposed to the Cuban Marine Environment

Alejandro Fernández Domínguez[1](✉), Fernando Martirena-Hernandez[2], Raúl González López[2], and Ernesto Díaz Caballero[3]

[1] Technological University of Havana, Havana, Cuba
alefingeniero@gmail.com
[2] Central University of Las Villas, Villa Clara, Cuba
[3] Research and Development for Construction Centre, Havana, Cuba

Abstract. The durability of concrete in marine environments is critical for long-term structural performance, particularly in tropical climates where chloride-induced corrosion is a primary concern. This study evaluates the long-term performance of limestone calcined clay cement (LC3) concrete compared to ordinary Portland cement (OPC) concrete under natural exposure conditions in Cuba's marine environment. Reinforced concrete elements were installed in a tidal zone and monitored over four years using durability indicators (DIs) such as carbonation depth, chloride profiles, air permeability, electrical resistivity, and chloride migration. Results demonstrated that LC3 concrete exhibited superior resistance to chloride penetration, lower air permeability, and higher electrical resistivity than OPC concrete, attributed to its refined pore structure and enhanced chloride-binding capacity. Despite LC3's slightly higher carbonation rate, projected carbonation depths after 50 years remained negligible for both concretes. Additionally, LC3 showed better resistance to surface wear, further mitigating chloride ingress over time. These findings highlight LC3 as a promising sustainable alternative for marine applications, offering improved durability and reduced environmental impact.

Keywords: LC3 concrete · marine environment · chloride penetration · durability indicators · carbonation · electrical resistivity · air permeability

1 Introduction

The primary cause of deterioration of reinforced concrete in marine environments is chloride-induced corrosion [1–4]. When the chloride concentration at the rebar reaches a critical threshold, it disrupts the passive protective layer, initiating corrosion [1, 3–7].

Chloride penetration through the concrete pore matrix occurs via various transport mechanisms, including migration, convection, capillary sorption, and diffusion, with the latter being dominant in most cases, especially in marine environments [7].

Currently, one of the most common strategies to mitigate this phenomenon is the use of supplementary cementitious materials (SCMs) in concrete mixtures [8] partially

M. Posani and S. Singh (Eds.): RYS 2025, RILEM Bookseries 67, pp. 859–872, 2026.
https://doi.org/10.1007/978-3-032-14652-6_62

replacing the Portland cement (PC). These materials densify the concrete matrix, reducing the size, quantity, and interconnectivity of pores, thereby enhancing the resistance to chloride ion penetration [9, 10].

One SCM introduced in recent decades is the combination of calcined clays (CC) and limestone (L), known as LC2. Among clays, those of the kaolinite group exhibit the highest reactivity when thermally activated [11]. Studies show that clays with at least 40% kaolin content, when calcined between 700 and 850 °C, possess sufficient pozzolanic activity to be used as SCMs [12]. The combination of PC with LC2 is known as limestone-calcine clay cement (LC3). [12].

The performance of LC3 concrete depends on the PC/LC2 and CC/L ratios, as well as the physical and chemical properties of each component [13]. Generally, concretes prepared with a 2:1 CC/L ratio and 50% PC, termed LC3–50, exhibit mechanical properties comparable to PC-based concrete [12, 14] and superior resistance to chloride-induced corrosion [13, 15]. This enhanced durability is attributed to a less interconnected pore matrix, which increases tortuosity and resistivity, hindering chloride ingress and electron movement after steel depassivation [16]. Additionally, LC3 systems form abundant alumina phases, enabling greater chloride binding as Friedel's salt, further reducing diffusion [17, 18].

The resistance of concrete to chloride penetration has been assessed in recent decades using various durability indicators (DIs) [19]. Chloride profiling, for instance, is used to determine the chloride diffusion coefficient, a key parameter for evaluating resistance to penetration [5, 20]. Other DIs employ an electric field to accelerate chloride migration and measure resistance [21]. Similarly, electrical resistivity correlates strongly with chloride resistance [19, 22]. Additional tests evaluate physical durability-related parameters, such as air permeability [23], or carbonation depth, which affects the release of bound chlorides [24, 25].

While most durability studies are conducted under laboratory conditions, long-term evaluation of structural elements under natural exposure conditions [26–29], are needed to account for time-dependence evolution of concrete properties, particularly for modern materials like LC3.

In 2014, reinforced concrete (RC) elements of equivalent strength grade, fabricated with PC and LC3, were installed in a tidal zone on Cuba's north-central coast. Over four years, their durability was monitored using DIs such as carbonation depth, chloride profiles, air permeability, surface resistivity, and chloride migration. This paper aims to evaluate the long-term durability performance of these PC and LC3 concrete elements.

1.1 Durability Indicators for Performance-Based Design

To achieve a performance-based durability design, durability indicators (DIs) must be applicable at various stages: prescribed during design, obtained during prequalification testing, incorporated into service-life models, tested during construction for quality control, and monitored during service life to assess structural condition [30]. Given the diversity of DIs, some are better suited to specific stages. For example, non-destructive, rapid tests are ideal for quality control, whereas more invasive, time-intensive tests often provide higher precision for prequalification and modeling.

Researchers have employed a wide range of DIs in recent decades, including carbonation depth, air permeability, electrical resistivity, and chloride migration.

Carbonation Depth. Carbonation is another mechanism triggering rebar corrosion [31, 32]. When CO_2 penetrates the concrete pore matrix in the presence of water, it reacts with pore solution alkalis, lowering pH and leading to rebar depassivation and corrosion [32, 33].

LC3 concretes exhibit reduced carbonation resistance due to: (i) a lower clinker content (fewer alkalis available to react with CO_2) and (ii) the pozzolanic reaction of the CC consumes all the produced alkalis [34].

The peak of carbonation rates occurs when relative humidity (RH) is between 55–75% [32]. In coastal marine environments where RH is over 85, chloride-induced corrosion prevails over carbonation.

However, because carbonation lowers the chloride threshold [34] and releases bound chlorides [24, 25], studying carbonation in such environments—particularly for LC3 concretes—remains relevant.

Air Permeability. The gas permeability of the concrete cover is a key DI for assessing resistance to chloride ions and CO_2 ingress [35]. Among available methods, the Torrent Permeability Tester (TPT) stands out [35], and is included in the Swiss Standard SIA 262/1 [30, 36].

The TPT-derived air permeability coefficient (kT) correlates strongly with porosity, pore connectivity [37] and other permeability tests [36, 38]. However, the TPT is highly sensitive to concrete saturation, requiring careful handling [38, 39].

Electrical Resistivity. Electrical resistivity is a critical DI, directly linked to chloride transport and concrete microstructure [40]. It is widely used for in situ quality monitoring and service-life modeling [41].

One of the most used test methods to evaluate concrete electrical resistivity is the Wenner Probe Method (WPM). This method consists in of using four contact electrodes spaced at equal distance and positioned on the concrete surface [41]. The equipment applies an alternating current (I) between the outer electrodes and measures the potential difference (V) between the two inner electrodes [41]. The electrical resistivity is then calculated as the V/I ratio [22].

Because the applied current flows through pore water, high resistivity suggests lower porosity and a pore matrix that is finer and more tortuous [42, 43].

Chloride Migration. Migration tests accelerate chloride penetration using an electric field, enabling rapid determination of the chloride diffusion coefficient form migration test (D_{RCM}).

The Ion Migration Test (IMT) was developed by SIMCO Technology as a modification of the Rapid Chloride Permeability Test (RCPT) by ASTM C1202 [44].

During the IMT, the current (I) passing from the upstream- to the downstream cell is measured, allowing the determination of the concrete conductivity (inverse of resistivity); where lower conductivity indicates higher resistance to chloride penetration. Good correlations between RCPT, WPM, chloride profiles and other DIs has been stablished in the literature [45–47].

2 Materials and Methods

The experimental program aimed to compare the durability performance of PC and LC3 concrete by analyzing cores extracted from reinforced concrete elements exposed to a tropical marine coastal environment.

2.1 Materials

Reinforced concrete (RC) elements (20 cm height) where cast between November 2013 and February 2014 at a precast concrete plant located 5 km form the shore. Two concrete mixes—M32 (PC-based) and M381 (LC3–50 based) —where used, which mix proportions are detailed in Table 1. Chemical compositions of the used PC and LC3 are detailed in Table 2. The RC elements remain at the plant on outdoor conditions until October of 2014.

Table 1. Mixes proportions

Mixture ID	Binder type	Binder (kg/m3)	water/ binder	Superplasticizer (lt/m3)	Manufacture date	28 days compressive strength (MPa)
M32	PC	360	0.47	2.08	Feb/2014	31.4
M381	LC3–50	360	0.47	4.04	Nov/2013	27.9

Table 2. Binder chemical composition

Cement type	SiO_2	Al_2O_3	Fe_2O_3	CaO	MgO
PC	21.20	5.79	2.70	62.64	1.22
LC3–50	27.33	4.58	4.58	49.80	1.30

2.2 Exposure Site

In early 2014, an exposure site was established at Punta Matamoros, Santa María Key, on the northern coast of Villa Clara Province, Cuba (Fig. 1). The precast reinforced concrete (RC) elements were installed at the site in the early October (Fig. 2).

Over the following four years, annual core extraction campaigns were conducted to evaluate long-term durability. The timing of each campaign (month/year) and corresponding exposure time (in years) are detailed in Table 3.

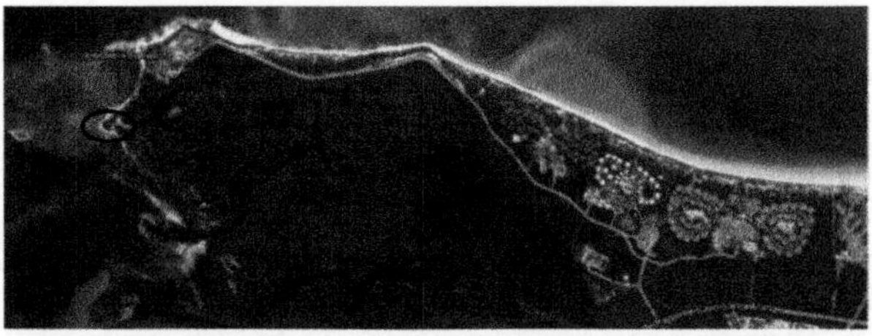

Fig. 1. Satellite picture of the location of the exposure site at Punta Matamoros, Santa Maria Key, Cuba.

Fig. 2. Exposure site at Punta Matamoros, Santa Maria Key, Cuba.

Table 3. Timeline for exposure and core extractions

Stage	Placement	1st extraction	2nd extraction	3rd extraction	4th extraction
Date	Oct-2014	Feb-2015	Apr-2016	Feb-2017	Feb-2018
Exposure time (years)	0	0.34	1.50	2.34	3.34
Carbonation depth		x	x	x	x
Air permeability		x	x	x	x
Total chloride profiles		x	x	x	
Surface resistivity			x	x	x
Ion migration		x	x	x	x

2.3 Test Methods

The extracted cores were evaluated using multiple durability indicator (DI) tests (Table 3). Air permeability with the TPT was performed on 150 mm diameter cores previously dried at 100 °C until the surface humidity was less than 5%. The actual humidity was controlled by an impedance-based instrument as suggested in [30]. Surface resistivity was measured by the WPM on the same cores after a 24-h immersion in a $Ca(OH)_2$ saturated solution. Carbonation depth was measured on 30 mm slices, split by the diameter into two halves, taken from the top surface of 100 mm diameter cores, using a phenolphthalein indicator. Also in 100 mm cores, the total chloride (acid soluble) content was determined at three depths from the surface (10, 20 and 30 mm).

Finally, two 30 mm height (*L*) slices taken from the middle of the same 100 mm diameter cores were submitted to the Ion Migration Test (IMT), a modification of the ASTM C-1202 [44] developed by SIMCO Technologies inc. In the IMT, after a conditioning similar to the stablished in [44], the sample was mounted between two cells filled with a 0.3M NaOH solution. The upstream cell was also filled with a 0.5 NaCl. A 20V voltage (*V*) was then applied for more than 60 h while monitoring the current through the specimen. The minimum current (I) measured after the initial hours indicated that Cl^- ions had reached the downstream cell and steady-state flow was achieved. This I value was then used to calculate the concrete conductivity (σ) using Eq. (1).

$$\sigma = (I\,x\,L)/(V\,x\,A) \tag{1}$$

where:

A: area of the transverse surface of the concrete core.

3 Results and Discussion

The results of carbonation depth, surface resistivity and air permeability performed on the concrete cores at the different exposure times are shown in Table 4. Also, the minimum current determined from the IMT and the conductivity calculated with (1) are shown on Table 4. Finally, chloride concentrations at 10, 20 and 30 mm during the same periods are depicted in Table 5.

Table 4. Carbonation depth, air permeability and surface resistivity measured at cores extracted annually during almost four years of exposure at Punta Matamoros.

Mixture	Exposure time (years)			
ID	0.34	1.50	2.34	3.34
	Carbonation depth (mm)			
M32	4.8	5.0	5.7	5.7
M381	1.9	2.1	2.2	3.2
	Air permeability (kT values)			
M32	0.017	0.015	0.021	0.05
M381	0.001	0.004	0.0042	0.022
	Surface resistivity (kΩ*cm)			
M32		11.1	11.7	10.9
M381		19.9	40.8	44.7
	Minimum current (mA)			
M32	35.36	97.7	152	209.5
M381	24.18	24.3	35.1	29.96
	Conductivity (S/m)			
M32	0.015	0.0195	0.0289	0.04
M381	0.004	0.0046	0.0067	0.0057

Table 5. Total chloride content (% wt. Concrete) at 3 depths from the top surface measured at cores extracted from 2015 till 2017

Mixture	Depth	Exposure time (years)		
ID	(mm)	0.34	1.50	2.34
M32	10	0.32	0.24	0.66
	20	0.23	0.30	0.51
	30	0.21	0.19	0.36
M381	10	0.29	0.23	0.44
	20	0.23	0.28	0.42
	30	0.08	0.11	0.27

3.1 Carbonation Depth

Carbonation depth (CD) was higher for M32 samples than for M381, though LC3 systems are expected to carbonate faster as explained in Sect. 1.1 This discrepancy could be attributed to the fact that M32 samples were manufactured three months earlier than M381.

In Fig. 3, CD is plotted against the square root of exposure time at Punta Matamoros. The data show a good linear fit, and as expected, the carbonation rate of M381 concrete is higher than that of M32. The projected carbonation depths in 50 years are 9 mm and 10.6 mm for M32 and M381, respectively. These values suggest that carbonation-induced corrosion and chloride ion unbinding will be negligible in both concrete types under these conditions.

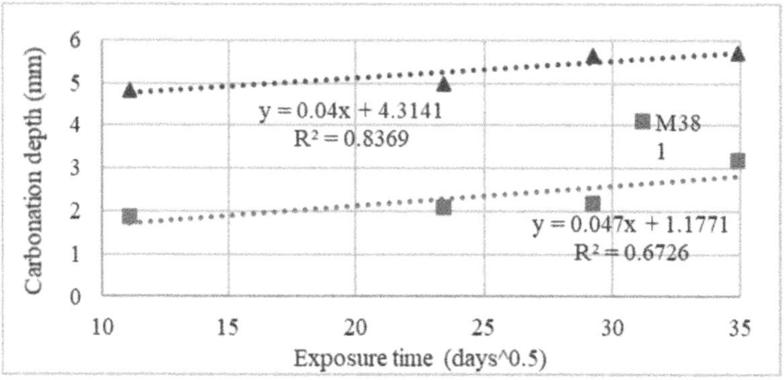

Fig. 3. M32 and M381 carbonation depth against the square root of exposure time at Punta Matamoros

3.2 Chloride Profiles

Table 4 and Fig. 4 show the total chloride profiles for M381 and M32 concretes obtained at 3 different exposure times. As can be noted, M381 consistently exhibited lower chloride concentrations than M32 at all depths and exposure periods, confirming the better resistance to chloride ingress reported in the literature [13, 14, 17, 18]. It should be noted that the difference in chloride content between concretes becomes higher with the exposure time. This issue will be addressed in the next sections.

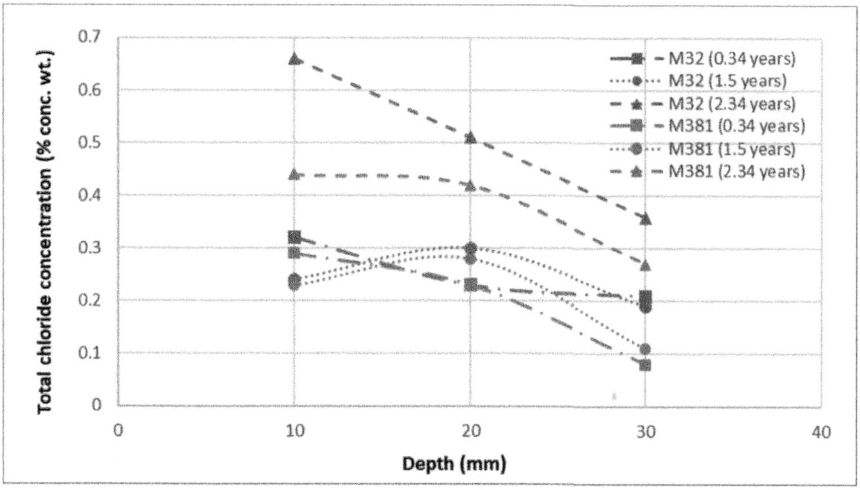

Fig. 4. Total chloride profiles for M381 and M32 from 2015, 2016 and 2017 cores

3.3 Air Permeability

In terms of air permeability, the kT values for M381 were consistently one order of magnitude lower than those for M32 at every measurement, demonstrating the finer pore structure of LC3 systems compared to PC-based concrete.

The observed increase in the air permeability coefficient over time indicates that surface erosion exerts a stronger influence on this parameter than the ongoing hydration of both concretes in this very humid environment. At the fourth measurement, a significant increase in kT was observed for both concrete types, attributed to an increase in the surface wear of the concrete elements (Fig. 5 and Fig. 6).

Concrete surface wearing in tidal zones is related to various aspects, such as the leaching of ettringite (which is soluble in the presence of chlorides), erosion, and salt crystallization [48]. Notably, the LC3 elements exhibited much better resistance to this phenomenon than PC elements because the finer pore matrix of LC3 concrete limits seawater penetration, thereby reducing surface wear damage.

This phenomenon is also influencing the chloride concentration inside the concrete elements, since chloride diffusion is governed by the chloride surface concentration. This fact, plus the better performance against surface wear of LC3 concrete over OPC, could be the explanation behind the increasing difference over time of chloride content between samples.

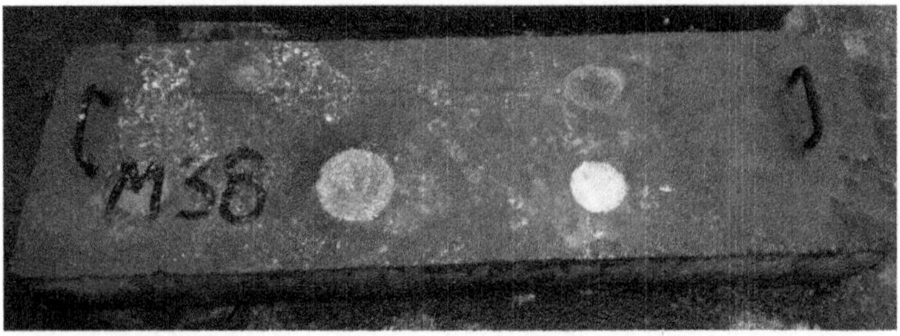

Fig. 5. Surface wearing of M381 reinforced concrete element after more than 3 years at the exposure time at Punta Matamoros

Fig. 6. Surface wearing of M32 reinforced concrete element after more than 3 years at the exp sure time at Punta Matamoros

3.4 Surface Resistivity

Electrical resistivity measurements obtained using the WPM showed values 2–4 times higher for LC3 concrete compared to PC elements (Table 4 and Fig. 7). These findings align perfectly with the trends observed in the previously analyzed durability indicators (DIs), higher resistivity for LC3 correlates with a lower air permeability and a lesser chloride ingress.

In terms of time evolution (Fig. 7), OPC concrete presents a very stable value from the beginning as expected, since clinker phases undergo maximum reaction at an early age (by 7 days) and there is a slowdown in their contribution with time [15]. In the case of the LC3 concrete, the resistivity values show a high increase from the first till the second measurement that tends to stabilize over the third one. This behavior could be showing an increase in the pore refinement until at least three years after casting because of continuous hydration of LC3 components. These results, are consistent with the findings by Zunino and Scrivener [49].

It can be discarded the influence of surface wearing on thess results, since the test is done on the lateral surface of the cores.

3.5 Ion Migration Test

As can be noted on Table 4 and Fig. 7, the conductivity of M32 concrete at every year is much higher than M381, proving as expected, a much higher tortuosity of the pore matrix of the latter.

Contradictions arrive when the analysis is on time evolution. M381 conductivity remains stable through the years, with a light trend to increase. On his side, M32 values increase at a high ratio after the second measurement. This behavior of the conductivity is contrary to what theoretically should happen, since the continuous hydration of the concrete, especially in a very humid environment should lead to a decrease in porosity and hence a decrease in conductivity [47, 50]; especially in LC3 concrete, where the pozzolanic reaction continues at least until the third year [49].

However, these results don´t differ from the findings from Van Noort et al. [47], who demonstrated, from large database analysis, that for concretes with lower resistivities, high and more variable values of D_{RCM} (and hence of concrete conductivity) could be obtained; while at high resistivities, lower and more stable values of D_{RCM} could be found. This issue will be addressed in a more comprehensive way in future papers.

Since the IMT is performed on slices taken from the medium part of the cores, it can also be discarded the influence of surface wearing and also of chlorides from sea water on the conductivity values measured.

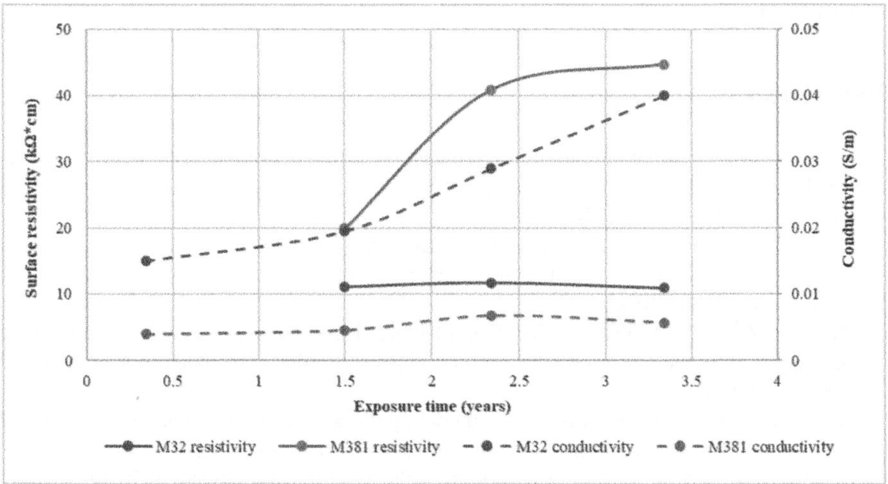

Fig. 7. Surface resistivity and conductivity of M32 and M381 concrete cores after more than 3 years of exposure at Punta Matamoros

4 Conclusions

This study provides a comprehensive evaluation of the durability performance of LC3 and OPC concretes exposed to a tropical marine environment over four years. Key findings include:

- LC3 concrete consistently exhibited lower chloride concentrations at all depths compared to OPC, confirming its superior resistance to chloride ingress due to a denser and tortuous pore matrix.

- While LC3 showed a higher carbonation rate, the projected carbonation depth after 50 years (10.6 mm for LC3 vs. 9 mm for OPC) represent a minimal risk for carbonation-induced corrosion and chloride unbinding in such high-humidity environments.
- LC3's finer pore structure was reflected in its significantly lower air permeability and concrete conductivity, alongside higher electrical resistivity, which correlates with reduced chloride diffusion and enhanced long-term durability.
- LC3 demonstrated superior resistance to surface wear caused by tidal action, reducing chloride ingress and maintaining structural integrity over time. Surface wear influenced chloride diffusion by increasing chloride concentrations at the concrete surface.
- LC3's resistivity increased over time, suggesting ongoing pore refinement, whereas OPC's conductivity unexpectedly increased, warranting further investigation.

These results underscore LC3's potential as a sustainable and durable alternative to OPC in marine environments, offering improved durability performance alongside a reduced carbon footprint that aligns with global efforts to mitigate CO_2 emissions in construction.

References

1. Arreola-Sanchez, M., et al.: Reinforced concrete structure performance in marine structures: analyzing durability indexes to obtain more accurate corrosion initiation time predictions. Materials **14**(24), 7662 (2021)
2. Nogueira, C.G., Yoshio, L., Zacchei, E.: Deterministic and probabilistic approaches for corrosion in RC structures: a direct proposed model to total service life predictions. Case Stud. Constr. Mater. **18**, e01913 (2023)
3. Szweda, Z., et al.: Analysis of the effect of protective properties of concretes with similar composition on the corrosion rate of reinforcing steel induced by chloride ions. Materials **16**(10), 3889 (2023)
4. Zhao, R., Li, C., Guan, X.: Advances in modeling surface chloride concentrations in concrete serving in the marine environment: a mini review. Buildings **14**(6), 1879 (2024)
5. Gonzalez, J.R., et al.: Response variability of diffusion models applied to chloride concentration profiles from concrete structures. J. Mater. Civ. Eng. **36**(1), 04023493 (2024)
6. Dudi, L., Krishnan, S., Bishnoi, S.: Numerical modeling for predicting service life of reinforced concrete structures exposed to chloride. J. Building Eng. **79**, 107867 (2023)
7. Xu, Z., Ye, G.: Understanding chloride diffusion coefficient in cementitious materials. Materials **16**(9), 3464 (2023)
8. Barbara Lothenbach, K.S., Hooton, R.D.: Supplementary cementitious materials. Cement and Concrete Research (2010)
9. Dhandapani, Y., Santhanam, M.: On the correlations between different chloride transport parameters and their role in service life estimation. Sustain. Resilient Infrastruct. **8**(2), 240–255 (2023)
10. Ribeiro, D.V., et al.: Effects of binders characteristics and concrete dosing parameters on the chloride diffusion coefficient. Cement Concr. Compos. **122**, 104114 (2021)
11. Martirena-Hernandez, J.F., Scrivener, K.: Hacia la sostenibilidad en la producción de cemento en América Latina: cemento en base a arcillas calcinadas y caliza (LC3). (2019), FICEM

12. Karen Scrivener, F.M., Shashank, B., Soumen, M.: Calcined clay limestone cements (LC3). Cement and Concrete Research (2017)
13. Sharma, M., et al.: Limestone calcined clay cement and concrete: a state-of-the-art review. Cem. Concr. Res. **149**, 106564 (2021)
14. Avet, F., S.L., Scrivener K.: Concrete performance of limestone calcined clay cement (LC3) compared with conventional cements. In: Advances in Civil Engineering Materials (2019)
15. Dhandapani, Y., et al.: Mechanical properties and durability performance of concretes with Limestone calcined clay cement (LC3). Cem. Concr. Res. **107**, 136–151 (2018)
16. Dhandapani, Y., Santhanam, M.: Assessment of pore structure evolution in the limestone calcined clay cementitious system and its implications for performance. Cement Concr. Compos. **84**, 36–47 (2017)
17. Pillai, R.G.: Service life and life cycle assessment of reinforced concrete systems with limestone calcined clay cement (LC3). Cement and Concrete Research (2021)
18. Santhanam, M., et al.: Perspectives on durability of blended systems with calcined clay and limestone. In: Calcined Clays for Sustainable Concrete, Springer. pp. 581–591 (2020)
19. Andrade, C., et al.: Testing and modelling chloride penetration into concrete. Constr. Build. Mater. **39**, 9–18 (2013)
20. Reichert, T.A., Balestra, C.E.T., de Medeiros-Junior, R.A.: Coupled model on the variation of diffusion coefficient and surface chloride concentration in reinforced concrete structures over time. Constr. Build. Mater. **411**, 134715 (2024)
21. Szweda, Z., et al.: Comparison of standardized methods for determining the diffusion coefficient of chloride in concrete with thermodynamic model of migration. Materials **16**(2), 637 (2023)
22. Robles, K.P.V., Yee, J.-J., Kee, S.-H.: Electrical resistivity measurements for nondestructive evaluation of chloride-induced deterioration of reinforced concrete—a review. Materials **15**(8), 2725 (2022)
23. Sakai, Y.: Correlations between air permeability coefficients and pore structure indicators of cementitious materials. Constr. Build. Mater. **209**, 541–547 (2019)
24. Shen, X.-H., et al.: Numerical study of carbonation and its effect on chloride binding in concrete. Cement Concr. Compos. **104**, 103402 (2019)
25. Guo, B., et al.: Effect of natural carbonation on chloride binding behaviours in OPC paste investigated by a thermodynamic model. J. Build. Eng. **49**(104021), 2022 (2022)
26. Mohammed, T.U., Yamaji, T., Hamada, H.: Chloride diffusion, microstructure, and mineralogy of concrete after 15 years of exposure in tidal environment. ACI Mater. J. **99**(3), 256–263 (2002)
27. Baroghel-Bouny, V., et al.: Aging and durability of concrete in lab and in field conditions–pore structure and moisture content gradients between inner and surface zones in RC structural elements. J. Sustain. Cement-Based Mater. **6**(3), 149–194 (2017)
28. Thomas, M.D., Bremner, T., Scott, A.C.: Actual and modeled performance in a tidal zone. Concrete Int. **33**(11) (2011)
29. Nanukuttan, S.V., et al.: The performance of concrete exposed to marine environments: predictive modelling and use of laboratory/on site test methods. Constr. Build. Mater. **93**, 831–840 (2015)
30. Bjegović, D., et al.: Test methods for concrete durability indicators. Performance-based specifications and control of concrete durability: state-of-the-art report RILEM TC 230-PSC, pp. 51–105 (2016)
31. Lollini, F., Redaelli, E.: Service life of reinforced concrete structures made with blended cements and exposed in urban environment. Struct. Concr. **23**(2), 694–705 (2022)
32. Stefanoni, M., Angst, U., Elsener, B.: Corrosion rate of carbon steel in carbonated concrete–a critical review. Cem. Concr. Res. **103**, 35–48 (2018)

33. Sahani, K., Khadka, S.S., Sahani, S.K.: Influence of corrosion on lifespan of reinforced concrete structures: a comprehensive review. Kathmandu Univ. J. Sci. Eng. Technol. **18**(1) (2024)
34. Cabrera, E., et al.: Preliminary results on corrosion rate in carbonated LC 3 concrete. In: Proceedings of the International Conference of Sustainable Production and Use of Cement and Concrete: ICSPCC 2019. Springer (2020)
35. Neves, R., Branco, F., de Brito, J.: About the statistical interpretation of air permeability assessment results. Mater. Struct. **45**, 529–539 (2012)
36. Ebensperger, L., Torrent, R.: Medición "in situ" de la permeabilidad al aire del hormigón: status quo. Revista ingeniería de construcción **25**(3), 371–382 (2010)
37. Katpady, D.N., et al.: Durability assessment of blended concrete by air permeability. Int. J. Concrete Struct. Mater. **12**, 1–11 (2018)
38. Romer, M.: Effect of moisture and concrete composition on the Torrent permeability measurement. Mater. Struct. **38**, 541–547 (2005)
39. Kucharczyková, B., Misák, P., Vymazal, T.: Determination and evaluation of the air permeability coefficient using torrent permeability tester. Russ. J. Nondestr. Test. **46**, 226–233 (2010)
40. Zhou, R., et al.: Assessment of electrical resistivity and oxygen diffusion coefficient of cementitious materials from microstructure features. Materials **14**(12), 3141 (2021)
41. Lencioni, J.W., Medeiros-Junior, R.A.: Analysis of different parameters in the electrical resistivity test of concrete. Int. J. Civil Eng. **19**(1), 27–38 (2021)
42. Azarsa, P., Gupta, R.: Electrical resistivity of concrete for durability evaluation: a review. Adv. Mater. Sci. Eng. **2017**(1), 8453095 (2017)
43. Koleva, D., et al.: Electrical resistivity and microstructural properties of concrete materials in conditions of current flow. Cement Concr. Compos. **30**(8), 731–744 (2008)
44. Material, A.S.O.T.: Standard test method for electrical indication of concrete's ability to resist chloride ion penetration (2022)
45. Dhanya, B., Santhanam, M.: Performance evaluation of rapid chloride permeability test in concretes with supplementary cementitious materials. Mater. Struct. **50**, 1–9 (2017)
46. Huang, K.-S., Yang, C.-C.: Using RCPT determine the migration coefficient to assess the durability of concrete. Constr. Build. Mater. **167**, 822–830 (2018)
47. Van Noort, R., Hunger, M., Spiesz, P.: Long-term chloride migration coefficient in slag cement-based concrete and resistivity as an alternative test method. Constr. Build. Mater. **115**, 746–759 (2016)
48. Chen, W.-F., Liew, J.R.: The Civil Engineering Handbook. CRC Press (2002)
49. Zunino, F., Scrivener, K.: Microstructural developments of limestone calcined clay cement (LC3) pastes after long-term (3 years) hydration. Cem. Concr. Res. **153**, 106693 (2022)
50. Audenaert, K., Yuan, Q., De Schutter, G.: On the time dependency of the chloride migration coefficient in concrete. Constr. Build. Mater. **24**(3), 396–402 (2010)

How to Protect Our Façade Coating Mortars? A Glance at Different Approaches to Improve Surface Protection

Jéssica Bersch[1(✉)], Angela Masuero[2], Denise Dal Molin[2], and Inês Flores-Colen[1]

[1] CERIS, Instituto Superior Técnico, Universidade de Lisboa, Av. Rovisco Pais 1, 1049-001 Lisboa, Portugal
jessica.d.bersch@tecnico.ulisboa.pt

[2] NORIE/PPGCI, Universidade Federal do Rio Grande do Sul, Av. Osvaldo Aranha 99, Porto Alegre 90035-190, Brazil

Abstract. This study addresses the challenge of protecting building façades by evaluating lime-based coating mortars treated with thirteen surface finishes, including silicate paints and commercial hydrophobic, oleophobic, and photocatalytic products. The goal was to identify the most effective surface protection solutions by assessing multiple performance aspects, particularly self-cleaning and anti-graffiti capabilities. A straightforward assessment framework was established, focusing on key surface properties to support decision-making. Initial evaluations included measurements of gloss, chromatic coordinates, static water contact angle, and roughness. Subsequently, specimens were stained with methylene blue and exposed to ultraviolet light for 145 h to evaluate self-cleaning performance via color change monitoring. A graffiti removal was then conducted, using silver spray paint and steam cleaning, with the cleaning efficiency assessed using an established scale. At 60°, gloss was highest for the anti-graffiti coating, though all values were generally low. Overall, color compatibility was preserved. Hydrophobic behavior was observed in silicate and hydrophobic treatments, while most photocatalytic coatings showed hydrophilic properties. The lime-based photocatalytic paint demonstrated the best self-cleaning performance, and, as expected, anti-graffiti coatings were most effective in graffiti removal. Some non-photocatalytic coatings also led to color changes under light, and one photocatalytic treatment provided partial resistance to graffiti. The findings underscore the need for additional research on rain exposure and long-term durability. At this stage, selecting a suitable coating depends on environmental exposure and anticipated degradation agents.

Keywords: Façade · Mortar Surface · Hydrophobic · Oleophobic · Photocatalytic

M. Posani and S. Singh (Eds.): RYS 2025, RILEM Bookseries 67, pp. 873–887, 2026.
https://doi.org/10.1007/978-3-032-14652-6_63

1 Introduction

Building façades represent both symbolic and architectural expressions within the built environment, contributing to community well-being and physical comfort [1]. Additionally, façades are critical to the performance and durability of buildings, as they are continuously exposed to degrading climatic agents. They play a key role in protection, particularly by preventing water ingress [2]. The deterioration of façades primarily results from exposure to air and rain, further intensified by soiling due to biological and non-biological deposits [3].

Degradation mechanisms can trigger chemical reactions and mechanical stresses in façade materials, affecting properties such as porosity, surface texture, layering, and internal microstructure [4]. Consequently, cladding materials are decisive in the degradation patterns and service life of façades [5]. In this context, applying preventive measures to renders can reduce the incidence of anomalies, minimize the need for invasive interventions, and promote durability [6, 7].

Among the protective strategies for mortar renders, photocatalytic, hydrophobic, and oleophobic treatments are particularly noteworthy. Photocatalytic coatings offer promising potential for façade preservation by enabling self-cleaning and depolluting effects [8] and may also suppress biological growth [9]. Hydrophobic treatments prevent the ingress of water and harmful ions [10]. Combining hydrophobic and oleophobic properties can further enhance resistance to graffiti vandalism [11].

Photocatalysis involves activation by sunlight, producing reactive species such as hydroxyl radicals (•OH), which, in conjunction with oxygen (O_2), degrade pollutants through redox reactions [12, 13]. The removal of surface impurities is facilitated by high wettability [14], often driven by photoinduced superhydrophilicity [15]. In contrast, hydrophobic treatments require external pressure to allow water penetration [16], thereby reducing moisture content, freeze-thaw damage, and heat flux: factors that can preserve thermal conductivity and mitigate moisture risks [17]. Oleophobic coatings serve as barriers that prevent graffiti paints from penetrating the substrate [18], and their low surface energy enables easier cleaning [19].

Despite the variety of available treatments, few studies offer a comparative analysis of these protective mechanisms for mortar surfaces or suggest strategies for improving their long-term performance. Therefore, this research investigates thirteen surface finishes for lime-based mortar specimens, aiming to determine which mechanisms offer the best overall protection for façades. The study addresses reference mortar, silicate paints, and a range of commercial products, including photocatalytic, hydrophobic, hydrophobic-oleophobic, and sacrificial anti-graffiti coatings. In addition to ranking their performance, the research also aims to identify limitations that should be explored in future studies to achieve a more comprehensive classification.

2 Materials and Methods

2.1 Materials

Thirteen surface-protected mortar configurations were studied, as summarized in Table 1, together with their corresponding reference codes. The combinations between mortar types and surface coatings were selected according to the intended use of each product and the manufacturers' application guidelines. All the materials were commercially available.

Table 1. Studied mortars and their respective surface coatings.

Code	Mortar	Surface Coating
M1PhotLime	1: monolayer render	Lime-based photocatalytic paint
M1PhotLimeE	1: monolayer render	Lime-based photocatalytic paint + extra coat
M1Sil1	1: monolayer render	Silicate paint 1
M1Sil1Phot1	1: monolayer render	Silicate paint 1 + photocatalytic coating 1
M1Sil2	1: monolayer render	Silicate paint 2
M1Sil2Phot1	1: monolayer render	Silicate paint 2 + photocatalytic coating 1
M1Sil2Phot2	1: monolayer render	Silicate paint 2 + photocatalytic coating 2
M2Ref	2: finishing render	Reference (uncoated mortar)
M2Hyd1	2: finishing render	Hydrophobic coating 1
M2Hyd2	2: finishing render	Hydrophobic coating 2
M2HydOleo	2: finishing render	Hydrophobic and oleophobic coating
M2Phot2	2: finishing render	Photocatalytic coating 2
M2AG	2: finishing render	Sacrificial anti-graffiti coating

The mortars (M1 and M2) were dry-mix formulations composed of natural hydraulic lime, siliceous and calcareous aggregates, and admixtures. Mortar 1 (M1) was designed as a monolayer render with improved crack resistance, while mortar 2 (M2) was a two-layer system, with the outer layer serving as the final decorative finish. M1 was coated primarily with mineral-based paints, with or without added photocatalysts or photocatalytic layers. In contrast, M2 was treated with suspensions and solutions evidently intended for protection, having a transparent aspect, and not attributing decorative, opaque colors to the surface.

All photocatalytic coatings (Phot) were titanium dioxide-based (TiO_2). Since Phot1 was not suitable for direct application to the mortar surface, it was only applied over silicate paints used as primers. The potassium silicate paints (Sil) contained inorganic pigments and various admixtures; for Sil1, the presence of organic polymers (less than 5% by weight) was declared by the manufacturer.

Hydrophobic coating 1 (Hyd1) consisted of silane and siloxane resins in a solvent-based solution, while Hyd2 was based on polysiloxanes. The hydrophobic and oleophobic coating (HydOleo) was a water-based dispersion of fluoropolymers. The sacrificial anti-graffiti coating (AG) was an aqueous emulsion of polymeric waxes recommended for removal using hot pressurized water.

2.2 Production of Specimens

Two specimens, each measuring 5 x 5 x 2 cm, were prepared for each combination listed in Table 1. These specimens were cut from larger panels and coated following the manufacturers' recommended application rates, as summarized in Table 2.

Except for Phot1, which required three cross-applied coats, all coatings were applied in two layers. For the M1PhotLimeE configuration, an additional coat, supplied with the lime-based photocatalytic paint, was considered to investigate whether it would enhance photocatalytic efficacy. In the case of the HydOleo coating, the second coat was applied while the first coat was still wet, as per manufacturer guidance. Figure 1 shows the prepared specimens.

Table 2. Surface coating consumption and application details.

Surface Coating	Consumption/Coverage	Observations
PhotLime	5–8 m^2/l (2 coats)	Diluted in 5–10% water
Sil1, Sil2	8–10 m^2/l per coat	-
Phot1	7–10 m^2/l (3 coats)	-
Phot2	100 g/m^2	-
Hyd1	250 g/m^2 per coat	Manufacturer recommends 250–800 g/m^2
Hyd2	2–8 m^2/l per coat	-
HydOleo	50 g/m^2 per coat	50–200 g/m^2 per coat recommended for low-porosity substrates
AG	50–80 g/m^2 per coat	Recommended range: 30–150 g/m^2

Fig. 1. Surface-protected mortar specimens.

2.3 Testing Methodology

Figure 2 presents an overview of the testing methodology, summarizing the framework established to evaluate the performance of different surface protection systems. When applicable, statistical analyses were performed using the software Statistica 7 (StatSoft).

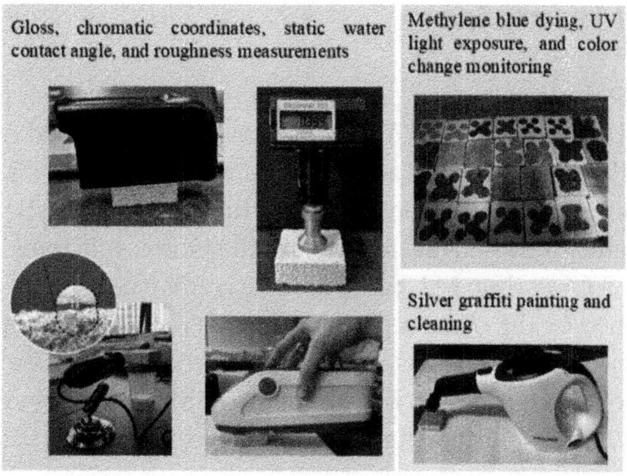

Fig. 2. Visual summary of performed tests – assessment framework.

Surface roughness was assessed using an Elcometer 223, with five measurements taken per specimen. Gloss was measured using a PCE-PGM 100 glossmeter at 20°, 60°, and 85° geometries, also with five measurements per sample. To evaluate surface wettability, static water contact angles were determined by photographing 4.0 μl droplets on the surfaces. For each of the thirteen configurations, the average angle was calculated based on at least ten droplets using the DropSnake plugin in ImageJ software.

Initial color characterization was performed in the CIEL*a*b* color space using a Konica Minolta CM-26d spectrophotometer, with five readings per specimen. After initial characterization, the self-cleaning ability of each coating was evaluated through a methylene blue degradation test. A 0.5 g/l aqueous solution of methylene blue (methylthioninium chloride, $C_{16}H_{18}ClN_3S$) was prepared following Silva et al. [6]. 50 μl of the solution was applied to five areas of each surface at least 8 days after painting. As previously noted by Bersch et al. [20], the staining patterns varied, probably due to differences in contact angle, water absorption potentials, and evaporation rates.

Specimens were then exposed to two 8W ultraviolet (UV-A) lamps (365 nm) for 145 h. Color changes (ΔE) were monitored over time by comparing the chromatic coordinates of the stained specimens before and after exposure, as calculated using Eq. (1) [21], where L* represents lightness, a* extends from green to red, and b* from blue to yellow [8].

$$\Delta E = \left[(\Delta L*)^2 + (\Delta a*)^2 + (\Delta b*)^2\right]^{1/2} \tag{1}$$

After the self-cleaning test, the same specimens underwent anti-graffiti evaluation. A silver spray paint was applied, following the methodology of Gil et al. [22], and removal was attempted using a low-pressure (3 bar) water steam jet. The cleaning process involved two rounds of 12 passes, alternating brushing directions orthogonally while steam was applied. Cleaning performance was classified according to the scale by García and Malaga [19]: CN 0 (complete removal, 100%); CN 0.5 (isolated, small residues); CN 1 (pinpoint residues); CN 1.5 (laminar residues with > 90% cleaning efficiency); CN 2 (laminar residues); CN 3 (visible residues with 30–85% efficiency); CN 4 (surface only slightly affected, <30% efficiency); CN 5 (no cleaning effect).

3 Results and Discussion

Gloss results for the thirteen surface-protected mortar configurations are shown in Fig. 3. According to EN ISO 2813:2014 [23], all surfaces were classified as matt (gloss < 10 glass units (GU)). Parracha et al. [24] also reported very low gloss (< 2.5 GU) for External Thermal Insulation Composite Systems (ETICS) finished with acrylic and silicate paints.

At 20°, gloss values reflected the influence of the mortar substrate: M1-based samples averaged around 1.5 GU, while M2-based surfaces remained closer to 1; this behavior is not directly explained by the other surface properties measured in this research. At 85°, rougher surfaces yielded lower gloss. In contrast, smoother ones showed higher values (roughness is presented later in Fig. 6). Although an 85° geometry is recommended for low-gloss surfaces, in the present study, it primarily captured surface roughness rather than coating type.

The 60° measurements exhibited more variability and proved most helpful to distinguish between protection systems. Analysis of Variance (ANOVA) with a 95% confidence level revealed significant differences in gloss at 60° among the studied surface-protected mortars. Corroborated by the Tukey test, presented in Table 3, hydrophobic treatments showed the lowest gloss, with no significant differences between M2Hyd1 and M2Hyd1. At the same time, the anti-graffiti coating (M2AG) achieved the highest gloss, in conjunction with M1SilPhot2. Except for M2AG, all the other products applied over M2Ref reduced its final gloss. In addition, apart from M1SilPhot2, photocatalytic configurations led to intermediate gloss values. The application of Phot1 onto M1Sil2 did not cause significant differences in gloss, unlike its behavior when applied to M1Sil1.

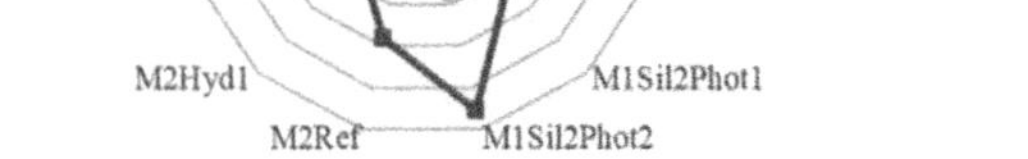

Fig. 3. Gloss of the 13 surface-protected mortar surfaces.

Table 3. Tukey test for the gloss values of the surface-protected mortar surfaces at a 60° geometry - **** indicates homogeneous groups, with a maximum 5% difference.

Code	Mean	1	2	3	4	5	6
M2Hyd2	1.086	****					
M2Hyd1	1.152	****					

(continued)

Table 3. (*continued*)

Code	Mean	1	2	3	4	5	6
M1Sil2	1.483		****				
M2HydOleo	1.563		****	****			
M1Sil2Phot1	1.580		****	****			
M2Phot2	1.774			****	****		
M1Sil1	1.802			****	****		
M1PhotLime	1.887				****	****	
M1PhotLimeE	1.903				****	****	
M2Ref	1.905				****	****	
M1Sil1Phot1	2.086					****	
M1Sil2Phot2	2.759						****
M2AG	2.934						****

Figure 4 shows the chromatic coordinates measured after preparing the surface-protected mortars. One-way ANOVA (5% significance value) showed the significant influence of the different types of surface-mortar configurations in all three coordinates: lightness (L*), a*, and b*.

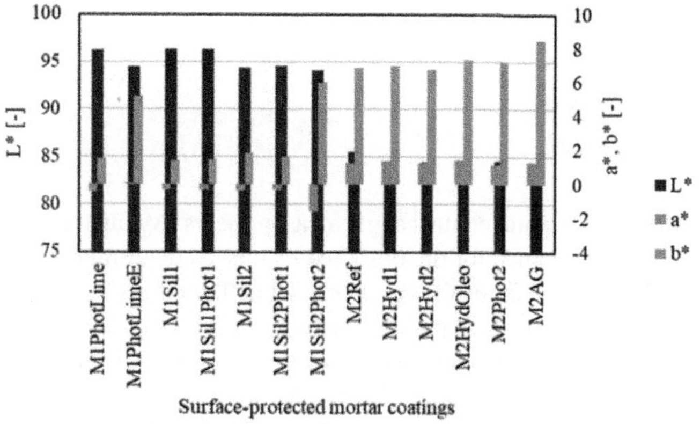

Fig. 4. Chromatic coordinates of the as-prepared 13 surface-protected mortar surfaces.

M1Sil2Phot2 reduced a* and increased b* compared to M1Sil2, as confirmed by the Tukey test. The additional coat in M1PhotLimeE increased a* and b*, reducing its whiteness. Except for these two configurations, typically no significant differences were observed for a* and b* among M1 surface-protected mortars. Generally, no differences were observed for M2 too, except for the b* coordinate of M2AG, which was the highest among all, indicating yellowing.

Figure 4 suggests that L* values dominated across all samples, indicating surfaces that tended toward a white appearance, particularly in M1 configurations coated with mineral paints. The Tukey test for L* is shown in Table 4. M2AG and M2Hyd1 were the least white surfaces, in contrast to M1PhotLime, M1Sil1Phot1, and M1Sil1. Considering the absolute values, however, the color differences introduced by the protective coatings over their previous silicate paint or reference substrate generally supported good aesthetic compatibility with the substrates [25].

Table 4. Tukey test for the lightness of the as-prepared surface-protected mortar surfaces - **** indicates homogeneous groups, with a maximum 5% difference.

Code	Mean	1	2	3	4	5	6
M2AG	83.292	****					
M2Hyd1	83.616	****					
M2HydOleo	84.262		****				
M2Hyd2	84.460		****				
M2Phot2	84.527		****				
M2Ref	85.524			****			
M1Sil2Phot2	94.033				****		
M1Sil2	94.305				****	****	
M1PhotLimeE	94.420					****	
M1Sil2Phot1	94.517					****	
M1PhotLime	96.258						****
M1Sil1Phot1	96.290						****
M1Sil1	96.324						****

Figure 5 displays the water contact angle measurements. Again, the different surfaces yielded statistical significance for the observed results, as determined by ANOVA with a 5% p-value. The Tukey test results are presented in Table 5.

Aside from M1PhotLime, all photocatalytic surfaces were hydrophilic, especially when applied over the silicate paints, which, by themselves, yielded the second-highest contact angles, surpassed only by M2Hyd1 and M2Hyd2. M2HydOleo and M2AG yielded values similar to those of the uncoated M2Ref and M1PhotLime, indicating a minimal influence on surface wettability.

Studying transparent coatings applied over glass specimens and considering only varying airflow velocities, Huang et al. [26] identified that wettability impacts dust deposition. Superhydrophobic coatings ($\theta > 150°$) had the highest resistance to dust deposition, followed by hydrophobic ($90° < \theta < 150°$), hydrophilic ($10° < \theta < 90°$), and superhydrophilic finishes ($\theta < 10°$ [27, 28]) [26]. Therefore, water contact angles may impact not only the cleaning activity but also how dirt adheres to the surfaces.

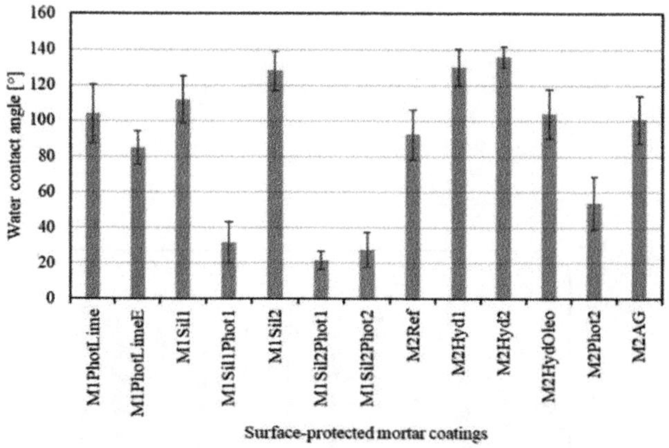

Fig. 5. Water contact angle of the 13 surface-protected mortar surfaces.

Regarding roughness, ANOVA again showed a significant influence of the different surface coatings on the results. As shown in Fig. 6, higher variability in roughness was observed for M1Sil2. M1PhotLime, M1PhotLimeE, M1Sil1, and M1Sil1Phot1 resulted in the smoothest surfaces. Since these configurations were all obtained with M1, it may suggest that Sil2 was applied with a lower thickness than PhotLime and Sil1 to the mortar substrate. This is reinforced by the fact that Sil2 configurations, with and without an additional photocatalytic layer, yielded results with no significant difference compared to those of M2, which exhibited uniform roughness regardless of the applied coatings (Table 6).

Table 5. Tukey test for the water contact angles of the as-prepared surface-protected mortar surfaces - **** indicates homogeneous groups, with a maximum 5% difference.

Code	Mean	1	2	3	4	5	6	7
M1Sil2Phot1	21.51	****						
M1Sil2Phot2	27.59	****						
M1Sil1Phot1	31.43	****						
M2Phot2	53.74		****					
M1PhotLimeE	84.97			****				
M2Ref	92.13			****	****			
M2AG	100.79			****	****	****		
M2HydOleo	103.91				****	****		
M1PhotLime	103.95				****	****		
M1Sil1	111.95					****	****	
M1Sil2	128.06						****	****
M2Hyd1	130.10							****

(continued)

Table 5. (*continued*)

Code	Mean	1	2	3	4	5	6	7
M2Hyd2	135.88							****

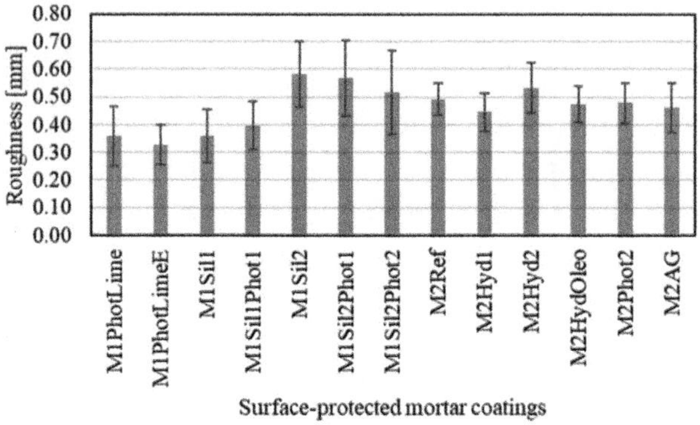

Fig. 6. Roughness of the 13 surface-protected mortars.

Microroughness can increase dirt adhesion [29], but overall, moderate roughness may lead to a better balance between photocatalytic activity and durability for TiO_2-based surfaces [30]. Comparison with contact angle data suggests that roughness did not directly dictate wettability, which is likely influenced primarily by the surface coating; lower surface energies resulted in larger water contact angles and lower wettability [31]. Unfortunately, there is still no consensus on how roughness and surface energy influence soiling processes [26].

Table 6. Tukey test for the roughness of the as-prepared surface-protected mortar surfaces - **** indicates homogeneous groups, with a maximum 5% difference.

Code	Mean	1	2	3	4
M1PhotLimeE	0.327	****			
M1Sil1	0.359	****	****		
M1PhotLime	0.359	****	****		
M1Sil1Phot1	0.397	****	****	****	
M2Hyd1	0.446	****	****	****	****
M2AG	0.462		****	****	****
M2HydOleo	0.473		****	****	****
M2Phot2	0.478		****	****	****

(*continued*)

Table 6. (*continued*)

Code	Mean	1	2	3	4
M1Sil2Phot2	0.518			****	****
M2Ref	0.523			****	****
M2Hyd2	0.533			****	****
M1Sil2Phot1	0.570				****
M1Sil2	0.582				****

Table 7 presents the ΔE values over 145 h of UV exposure. Examining the results (145 h), ANOVA reveals a significant influence of the different surface coatings at a 95% confidence level. In general, all coatings increased ΔE, including those not explicitly designed for photocatalytic action, such as hydrophobic and anti-graffiti products, suggesting they may facilitate surface-level dye degradation due to their composition.

According to Tukey test results (Table 8), M1PhotLimeE exhibited the most effective self-cleaning performance, followed by M1PhotLime and M1Sil1Phot1, with similar results. M1Sil2Phot1 also showed strong dye degradation, highlighting the efficacy of Phot1 when applied over a silicate base. Phot2 did not show the same effectiveness as Phot1, since it did not lead to statistically significant improvements when compared to M2Ref or M1Sil2.

To support the findings in Table 7, Fig. 7 presents the Energy Dispersive X-ray Spectrometer (EDS) report obtained from a Scanning Electron Microscopy (SEM) analysis of a selected area from M1PhotLime. The weight concentration of the titanium element on the surface is 10.34%, which could indicate the amount of photoactive material within the paint.

Table 7. Color change (ΔE) during methylene blue degradation over 145 h.

Code	0 h	1.5 h	4 h	18 h	15 h	145 h
M1PhotLime	0.00	1.61	1.98	10.83	12.24	22.09
M1PhotLimeE	0.00	5.24	6.52	8.79	9.29	27.13
M1Sil1	0.00	3.28	7.39	6.72	6.92	9.21
M1Sil1Phot1	0.00	3.71	5.09	7.34	8.61	18.59
M1Sil2	0.00	2.88	3.30	5.16	4.94	11.04
M1Sil2Phot1	0.00	2.44	7.13	8.92	8.33	13.54
M1Sil2Phot2	0.00	1.37	1.65	2.66	3.50	7.94
M2Ref	0.00	0.79	0.97	1.43	1.63	2.62
M2Hyd1	0.00	1.29	1.95	3.59	3.69	5.77
M2Hyd2	0.00	1.80	2.34	4.00	4.54	6.35
M2HydOleo	0.00	2.99	3.99	5.90	6.10	8.68

(*continued*)

Table 7. (*continued*)

Code	0 h	1.5 h	4 h	18 h	15 h	145 h
M2Phot2	0.00	1.16	1.36	2.13	2.32	3.60
M2AG	0.00	3.15	3.97	5.96	7.04	12.09

Four configurations showed partial graffiti removal: M1Sil2Phot2 (CN 4), M2Phot2 (CN 4), M2HydOleo (CN 4), and M2AG (CN 3). However, only M2AG (sacrificial anti-graffiti) achieved a cleaning efficiency of more than 30%. All other systems were classified as CN 5; thus, they had no effective cleaning.

Figure 8 illustrates M2AG specimens before and after graffiti cleaning. As a sacrificial product, its removal is also expected to affect other surface properties, such as gloss and contact angle, and should be reapplied after cleaning.

Table 8. Tukey test for color change (ΔE) at 145 h of methylene blue degradation - **** indicates homogeneous groups, with a maximum 5% difference.

Code	Mean	1	2	3	4	5	6	7
M2Ref	2.62	****						
M2Phot2	3.60	****						
M2Hyd1	5.77	****	****					
M2Hyd2	6.35	****	****					
M1Sil2Phot2	7.94		****	****				
M2HydOleo	8.68		****	****	****			
M1Sil1	9.21		****	****	****			
M1Sil2	11.04			****	****	****		
M2AG	12.09				****	****		
M1Sil2Phot1	13.54					****		
M1Sil1Phot1	18.59						****	
M1PhotLime	22.09						****	
M1PhotLimeE	27.13							****

Element name	Atomic concentration	Weight concentration
Carbon	13.69%	5.02%
Oxygen	47.05%	22.99%
Calcium	26.25%	32.13%
Titanium	7.07%	10.34%
Palladium	2.26%	7.33%
Gold	3.68%	22.19%

Fig. 7. SEM-EDS for M1PhotLime.

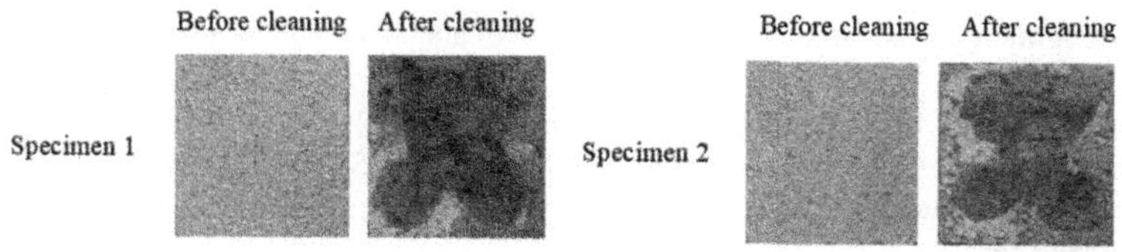

Fig. 8. Graffiti removal for M2 protected with a sacrificial anti-graffiti coating (M2AG).

4 Conclusion

This study evaluated surface performance for protective coatings applied to lime-based mortars for façade rendering. The evaluation framework included measurements of gloss, chromatic coordinates, static water contact angles, and surface roughness, as well as evaluations of self-cleaning and anti-graffiti performance.

For gloss, the 60° geometry provided the clearest differentiation between protective coatings. The sacrificial anti-graffiti resulted in the highest gloss, but all configurations remained in the matt category. Overall, the coatings preserved the initial chromatic coordinates, demonstrating good color compatibility with the mortars. Surface wettability was primarily determined by the type of coating rather than by surface roughness. Most photocatalytic coatings (excluding the lime-based version) were hydrophilic (20°–40°), while silicate and hydrophobic treatments yielded hydrophobic behavior. Roughness was lowest for coatings with lime-based photocatalytic and one silicate paint; other surfaces were similar.

The lime-based photocatalytic coatings showed the best performance in methylene blue degradation under UV exposure, followed by a photocatalytic solution over silicate paint. The self-cleaning assessment excluded water interaction, which could influence cleaning due to hydrophilic effects; only light exposure was considered. Some non-photocatalytic coatings also showed dye degradation, raising questions about potential chemical interactions under light exposure.

In general, photocatalytic and hydrophobic products were not effective in graffiti removal, except for the hydrophobic-oleophobic coating and one photocatalytic solution (which showed limited effectiveness in self-cleaning). As expected, the best performance was obtained for the sacrificial anti-graffiti (>30% efficiency).

These results underscore the complexity of selecting optimal façade protection. The best solution depends on context-specific degradation risks: anti-graffiti coatings may be ideal for vulnerable lower-level walls, while photocatalysts can be advantageous in polluted environments if they also ensure water resistance.

Future research should include water exposure and durability assessments to understand how rain and runoff affect soiling and cleaning. Long-term aging, including UV exposure, freeze-thaw cycles, and other aggressive agents, is essential to support decision-making. A complete façade system application could benefit the assessment, especially if it enables understanding of the impacts of constructive details, different heights, and orientations. Furthermore, a deeper study on the compositions of the coatings must be carried out, at least to investigate why non-photocatalytic products can also degrade organic dye.

Acknowledgments. The authors acknowledge the financial support of the Foundation for Science and Technology (FCT) through the project UIDB/04625/2025 of the research unit CERIS, the grant 2023.05316.BD (DOI: https://doi.org/https://doi.org/10.54499/2023.05316.BD) and the ReCoveRing project. Additional thanks are extended to UFRGS, CAPES, and CNPq. The authors also gratefully acknowledge the materials provided by Levenger, Mapei, Neuce, Secil, Steikos, and Graphenstone.

References

1. Prieto, A., Oldenhave, M.: What makes a façade beautiful? J. Facade Des. Eng. **9**, 21–45 (2021)
2. de Souza, A.L.R., de Lima Meneses Romeiro, T.R., Bauer, E.: Influence of the exposure degree on the degradation of facades of buildings in Brasília—Brazil. Buildings **14**, 133 (2024)
3. Marie, I.: Diagnostic approach for quantifying the soiling level and the aging of limestone façade. J. Build. Eng. **7**, 292–299 (2016)
4. Bauer, E., Souza, A.L.R.: Failure patterns associated with facade zones and anomalies in the initiation and propagation of degradation. Constr. Build. Mater. **347**, 128563 (2022)
5. Barrelas, J., Dias, I., Silva, A., de Brito, J., Flores-Colen, I., Tadeu, A.: Impact of environmental exposure on the service life of façade claddings—a statistical analysis. Buildings **11**, 615 (2021)
6. Silva, A.S., et al.: Evaluating the effectiveness of self-cleaning products applied on external thermal insulation composite systems (ETICS). J. Coat. Technol. Res. **19**, 1437–1448 (2022)
7. Bergamonti, L., et al.: Hydrophobic and photocatalytic treatment for the conservation of painted lecce stone in outdoor conditions: a new cleaning approach. Appl. Sci. **14**, 1261 (2024)
8. Diamanti, M.V., Luongo, N., Massari, S., Lupica Spagnolo, S., Daniotti, B., Pedeferri, M.P.: Durability of self-cleaning cement-based materials. Constr. Build. Mater. **280**, 122442 (2021)
9. Guerra, F.L., et al.: The effectiveness of spraying nTiO_2 on coating mortars in historical buildings aimed at reducing fungal growth. Buildings **13**, 2751 (2023)
10. Wang, L., et al.: Investigation on the effects of polyaniline/lignin composites on the performance of waterborne polyurethane coating for protecting cement-based materials. J. Build. Eng. **64**, 105665 (2023)
11. Lettieri, M., Masieri, M., Pipoli, M., Morelli, A., Frigione, M.: Anti-graffiti behavior of oleo/hydrophobic nano-filled coatings applied on natural stone materials. Coatings **9**, 740 (2019)
12. Hot, J., Dasque, A., Topalov, J., Mazars, V., Ringot, E.: Titanium valorization: from chemical milling baths to air depollution applications. J. Clean. Prod. **249**, 119344 (2020)
13. Fujishima, A., Rao, T.N., Tryk, D.A.: Titanium dioxide photocatalysis. J. Photochem. Photobiol., C **1**, 1–21 (2000)
14. Castro-Hoyos, A.M., Rojas Manzano, M.A., Maury-Ramírez, A.: Challenges and opportunities of using titanium dioxide photocatalysis on cement-based materials. Coatings **12**, 968 (2022)
15. Munafò, P., Goffredo, G.B., Quagliarini, E.: TiO_2-based nanocoatings for preserving architectural stone surfaces: an overview. Constr. Build. Mater. **84**, 201–218 (2015)
16. Silva, L., Flores-Colen, I., Vieira, N., Timmons, A.B., Sequeira, P.: Durability of ETICS and premixed one-coat renders in natural exposure conditions. In: Delgado, J. (ed.) New Approaches to Building Pathology and Durability. Building Pathology and Rehabilitation, pp 131–158. Springer, Singapore (2016)

17. Chen, X., De Kock, T., Cnudde, V., Van Den Bossche, N.: The influence of material characteristics and environmental circumstances on the efficiency of water repellent treatments of historic facades. Constr. Build. Mater. **386**, 131610 (2023)
18. Moura, A.R., Flores-Colen, I., De Brito, J.: Anti-graffiti products for porous surfaces: an overview. In: Charola, A.E., Rodrigues, J.D. (eds.) Hydrophobe VII - 7th International Conference on Water Repellent Treatment and Protective Surface Technology for Building Materials, pp 225–233. LNEC, Lisbon (2014)
19. García, O., Malaga, K.: Definition of the procedure to determine the suitability and durability of an anti-graffiti product for application on cultural heritage porous materials. J. Cult. Herit. **13**, 77–82 (2012)
20. Bersch, J.D., et al.: Desempenho individual e combinado de produtos fotocatalítico e anti-graffiti em revestimento de Fachada. In: de Brito, J., et al., (eds.) Livro de atas. 5° Congresso Luso-Brasileiro de Materiais de Construção Sustentáveis. Congresso Construção 2024, pp 1804–1815. CERIS, Lisbon (2024)
21. ISO - International Organization for Standardization: EN ISO/CIE 11664-4:2019: Colorimetry - Part 4: CIE 1976 L*a*b* Colour Space. ISO, Brussels (2019)
22. Gil, B.C., Borsoi, G., Parracha, J.L., Dionísio, A., Veiga, R., Flores-Colen, I.: Effectiveness and durability of anti-graffiti products applied on ETICS: towards a compatible and sustainable graffiti removal protocol. Environ. Sci. Pollut. Res. **30**, 65160–65176 (2023)
23. CEN - European Committee for Standardization: EN ISO 2813:2014: Paints and varnishes - Determination of gloss values at 20 degrees, 60 degrees and 85 degrees. CEN, Brussels (2014)
24. Parracha, J.L., et al.: Performance parameters of ETICS: correlating water resistance, bio-susceptibility and surface properties. Constr. Build. Mater. **272**, 121956 (2021)
25. Pondelak, A., et al.: Efficiency of novel photocatalytic coating and consolidants for protection of valuable mineral substrates. Materials **12**, 521 (2019)
26. Huang, Z.-S., et al.: Experimental investigation of the anti-soiling performances of different wettability of transparent coatings: superhydrophilic, hydrophilic, hydrophobic and superhydrophobic coatings. Sol. Energy Mater. Sol. Cells **225**, 111053 (2021)
27. Marmur, A.: Hydro- hygro- oleo- omni-phobic? Terminol. Wettab. Classif. Soft Matter **8**, 6867 (2012)
28. Si, Y., Guo, Z.: Superhydrophobic nanocoatings: from materials to fabrications and to applications. Nanoscale **7**, 5922–5946 (2015)
29. Dantas, S.R.A., de Oliveira Romano, R.C., Vittorino, F., Loh, K.: Effects of surface roughness and light scattering on the activation of TiO_2 on mortar photocatalytic process. Const. Build. Mater. **270**, 121421 (2021)
30. Bersch, J.D., Picanço Casarin, R., Maia, J., Masuero, A.B., Dal Molin, D.C.C.: TiO_2-based mortars for rendering building envelopes: a review of the surface finishing for sustainability. Sustainability **15**, 16920 (2023)
31. Klein, N.S., Bachmann, J., Aguado, A., Toralles-Carbonari, B.: Evaluation of the wettability of mortar component granular materials through contact angle measurements. Cem. Concr. Res. **42**, 1611–1620 (2012)

Point Cloud-Based Damage Detection for Reinforced Concrete Structures by Deep Learning

Kazuma Shibano[1(✉)], Toma Tsubota[1], Moeka Mukai[1], Hiromu Tanaka[2], Hikaru Umezawa[2], Ninel Alver[3], and Tetsuya Suzuki[4]

[1] Graduate School of Science and Technology, Niigata University, Niigata, Japan
kazuma3267@gmail.com
[2] Faculty of Agriculture, Niigata University, Niigata, Japan
[3] Department of Civil Engineering, Faculty of Engineering, Ege University, Bornova, Turkey
[4] Institute of Agriculture, Niigata University, Niigata, Japan

Abstract. This study investigates the factors affecting crack detection accuracy using three-dimensional point clouds acquired by Terrestrial Laser Scanners (TLS). The research focuses on understanding how measurement conditions—including scanning distance, incidence angle, and point cloud density—influence the ranging results by TLS. In the crack detection procedure, PointNet architecture is employed for binary crack classification, with various input feature combinations evaluated including 3D coordinates, RGB values, intensity data and local geometric features derived from eigenvalue decomposition. Experiments were conducted on a reinforced concrete headworks structure constructed in 1976, using both TLS and Handheld Laser Scanner (HLS) measurements for ground truth validation. Results demonstrate that local geometric features significantly improve detection performance, achieving improvement of F1 score compared using only 3D coordinates. The study reveals that point cloud density, determined by scanning distance and incidence angle, critically affects ranging accuracy in the depth direction. Optimal neighborhood radius for local feature extraction varies between planar and curved surfaces. These findings provide practical guidance for improving crack detection accuracy in civil engineering applications using point cloud data and deep learning methods.

Keywords: Terrestrial Laser Scanner (TLS) · Point cloud · Crack · PointNet · Local feature

1 Introduction

Point clouds are increasingly utilized due to the development of 3D reconstruction techniques, such as Structure from Motion (SfM), laser scanners, etc. [1]. Three-dimensional reconstruction technology is being explored for application in maintenance and operational fields within civil engineering, due to its characteristics of acquiring shape information [2]. Tasks for geometric quality assessment and defect detection are performed by

M. Posani and S. Singh (Eds.): RYS 2025, RILEM Bookseries 67, pp. 888–898, 2026.
https://doi.org/10.1007/978-3-032-14652-6_64

point cloud obtained using a device corresponding to required resolution [3]. Accuracy and resolution of Terrestrial Laser Scanner (TLS) and photogrammetry are millimeter order and centimeter order, respectively, among three-dimensional ranging technologies. Therefore, examples of detected object by TLS are limited to spalling of bridge piers, cracks and deflection, in contrast to examples of that by photogrammetry are deflection, deformation of façade, or pothole on pavement.

Crack detection of concrete structures is a key challenge for long-term period. In terms of resolution and required density of data, TLS is mainly used for crack detection. Detection accuracy is influenced by distance between Area of Interest (AOI) and TLS, and incident angle since laser spot size and shape are changed [4]. Most existing studies compare profiles of point clouds by TLS and designed surfaces assuming that concrete surface is flat. However, in-service structures exhibit surface roughness due to exposed aggregates, wearing and construction variations. There is no established method for obtaining ground truth for point clouds, as TLS ranging is recognized as the most accurate.

Several technical challenges arise when processing data using point clouds. Point clouds represent unstructured data, unlike images or voxels data. Therefore, convolution which is defined for grid data cannot be performed, making it difficult to apply models designed for images. For this reason, it is necessary to use deep learning models specifically designed for point cloud data. Designed models for point clouds are proposed as PointNet [5], graph convolutional neural network (GCNN) [6], Point Transformer [7] and so on. Feature extraction for point cloud data format needs a definition of neighborhood. When crack detection procedure is defined as a segmentation problem, it needs to quantify the relationship of any given point to its surroundings. For example, in the case of an image, a kernel such as 2x2 is prepared, and features are obtained by convolution and pooling. However, since point clouds are not grid data, a point within a sphere of radius is defined as the neighborhood, or the nearest points are defined as the neighborhood, and features are extracted. Setting the neighborhood area is considered important, nevertheless there is little discussion about this, and detection accuracy tends to be set as the important indicator. Moreover, batch size also is a target for a similar discussion. The batch size can adjust the proportion of cracks in a batch, and is an important indicator that contributes to resolving imbalanced data.

These issues of crack detection using TLS point clouds are summarized. TLS-based crack detection faces accuracy challenges stemming from distance and incident angle variations that alter laser spot characteristics. While most research assumes flat concrete surfaces, actual structures exhibit roughness due to aggregates and construction variations. The performance of crack detection is compound by the lack of established ground truth methods since TLS itself represents the most accurate ranging technique. Additionally, the unstructured characteristics of point cloud data prevent the application of traditional convolution-based image processing models. Specialized deep learning architectures and careful consideration of neighborhood definition and batch size parameters are necessary to address inherent imbalance data issues. In this study, HLS (Handheld Laser Scanner) is used to obtain the ground truth of the concrete surface shape and the TLS point cloud is compared with it. The distance and incidence angle, as well as the nearest-neighbor distance, which indicates the point cloud density, are then investigated.

In crack detection using deep learning of point clouds, a comparison is made with PointNet, which adds color information from RGB capturing and local geometric parameter performed along with distance measurement and local features. The selection of local features is performed by considering the definition of the neighboring point cloud.

2 Point Cloud-Based Crack Detection

2.1 Detection Model, PointNet

Using TLS point clouds, crack detection is attempted by targeting cracks with sufficient width to obtain depth information. This study employs PointNet [5] (see Fig. 1), which was proposed for handling three-dimensional point clouds in deep learning. Three-dimensional point clouds, unlike images or voxel data, do not have order or grid structure. PointNet uses a network that combines Shared MLP and (Max-)Pooling as a Symmetric Function to handle unordered data. A Symmetric Function refers to a function where the output does not change even if the order of input data changes. Shared MLP is a method that applies the same MLP (Multi-layer Perceptron) in the channel direction for each point. For an input of $(\boldsymbol{p}_1, \boldsymbol{p}_2, \ldots, \boldsymbol{p}_i, \ldots, \boldsymbol{p}_j, \ldots, \boldsymbol{p}_N)$, since Shared MLP is applied at each point, the output becomes $(f(\boldsymbol{p}_1), f(\boldsymbol{p}_2), \ldots, f(\boldsymbol{p}_i), \ldots, f(\boldsymbol{p}_j), \ldots, f(\boldsymbol{p}_N))$. Each MLP is defined as Eq. (1).

$$\forall_i \epsilon \{1, 2, \ldots, L\} : h_i = ReLU(BN(W_i h_{i-1} + b_i)), \tag{1}$$

where, $h_0 \epsilon \mathbb{R}^{d_0}$ denotes input data, $W_i \epsilon \mathbb{R}^{(d_i \times d_{i-1})}$ denotes weights, b_i denotes $b_i \epsilon \mathbb{R}^{d_i}$, *ReLU* is activate function (Rectified Linear Unit) and BN is batch normalization.

In Max-Pooling, the features of each point are aggregated. Pooling is applied to the entire point cloud for each channel. By using the maximum value as the pooling function, since the maximum value across the entire point cloud does not change even if the order of points changes, the output becomes independent of the order of points.

PointNet architecture is trained on TLS point cloud data for binary crack classification. The network is optimized using the Adam optimizer with an initial learning rate of 0.001. The batch size varies between 10, 20, and 50 to evaluate its impact on learning stability and classification performance. Training is performed for 100 epochs, with early stopping applied based on validation loss to prevent overfitting. When signs of overfitting are observed during training, the maximum number of epochs is reduced to 10 to avoid excessive fitting to the training data. To mitigate class imbalance between crack and non-crack, a class-weighted cross-entropy loss function is adopted. The input to the network consisted of either 3D coordinates alone, 3D coordinates with intensity, RGB values or 3D coordinates supplemented with local geometric features such as third eigen value.

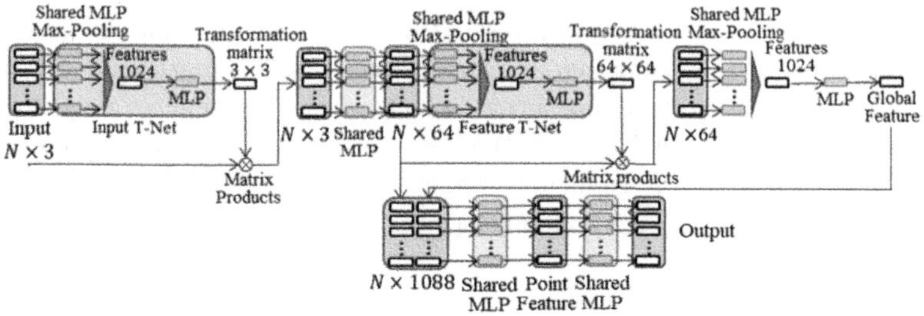

Fig. 1. Architecture of PointNet.

2.2 Extraction of Local Feature

Extraction of local geometric features of point clouds is performed by principal components based on eigenvalue decomposition [8]. The third eigenvector is used to detect normal. Local geometric features are generated by combining eigenvalues. In this study, the third eigenvalue is calculated and used as a local geometric feature. The procedure for calculating local features is as follows. First, obtain the neighboring point cloud within a sphere of radius r at an arbitrary point. Then, perform eigenvalue decomposition using the neighboring point cloud. Covariance tensor is calculated as Eq. (2).

$$\Sigma_i = \frac{1}{N}\sum\nolimits_{n \in P_i^N} (p_n - \overline{p})(p_n - \overline{p})^T \tag{2}$$

where P^N denotes the set comprising the N nearest neighbors of P_i, and $\overline{p} = \underset{n \in P_i^N}{\mathrm{med}}\, p_n$ is the centroid. The eigenvalues $\lambda_1 \geq \lambda_2 \geq \lambda_3 \geq 0$ and eigenvectors e_1, e_2, e_3 of the covariance tensor are shown in Fig. 2.

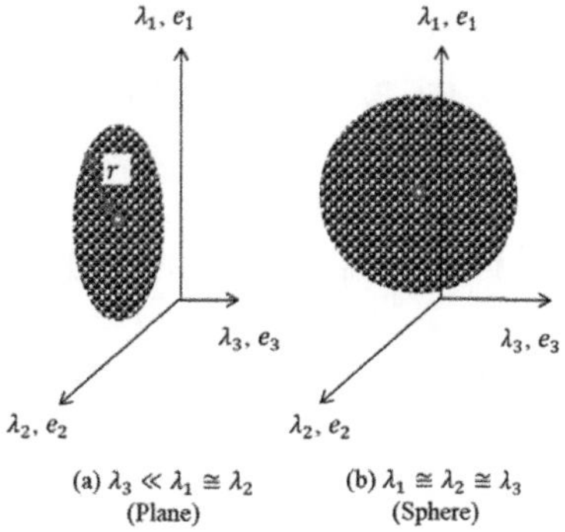

Fig. 2. Calculation of local geometric features of point clouds.

3 Material and Measurement Methods

3.1 TLS and HLS Target Reinforced Concrete Structure

TLS and HLS measurements were conducted on a reinforced concrete headworks constructed in 1976. TLS measurements were performed targeting the entire structure from installation positions with different distances and incident angles. The point cloud used for the study was the pier section on the left bank side where numerous cracks were observed (see Fig. 3(a)). Four cases were set to investigate the impact of incident angle on TLS ranging accuracy: curvature surface with large incident angle (Case 1), curvature surface with frontal positioning (Case 2), planar surface with frontal positioning (Case 3) and planar surface with large incident angle (Case 4). The TLS used was a FARO® Focus S150 Laser Scanner. The technical specifications of TLS and HLS are shown in Table 1.

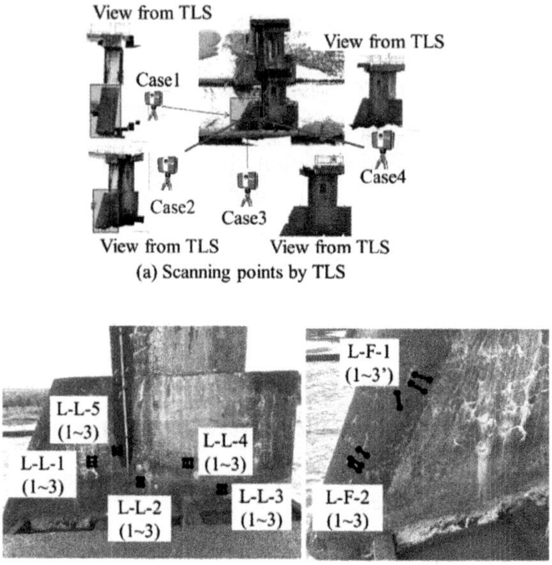

(a) Scanning points by TLS

(b) Scanning points for planar (left) and curvature (right) surface by HLS

Fig. 3. Scanning points.

Table 1. Technical specifications of TLS and HLS.

Parameter	TLS	HLS
Range	0.6–150 m @90%	85 mm ± 20 mm from sensor
Wavelength	1,550 nm	650 nm (1 mW)
Beam divergence	0.3 mrad (1/e)	-
Beam diameter	2.12 mm (1/e)	-
Laser spot size	6.3 mm @10 m	0.9 mm × 1.5 mm
Resolution (ranging direction)	1 mm (range accuracy)	10 μm
Resolution (planar direction)	15 arcsec (horizontal and vertical)	0.1 mm

(*continued*)

Table 1. (*continued*)

Parameter	TLS	HLS
Nearest point distance	3.1 mm @10 m	-

Since comprehensive HLS measurement is difficult, scanning points for planar and curvature surface are shown in Fig. 3(b). HLS measurement and data processing are shown in Fig. 4. In HLS measurement, scanning is performed at 0.1 mm intervals over 150 mm length, making it possible to acquire 1,500 data points. While HLS measurement provides the distance to the target as raw data, since the installation position is not necessarily flat, the distance from a reference line assuming an undeformed surface to the target is considered as deformation. Simply obtaining a linear regression line using all data cannot detect the undeformed surface, so the reference line is determined using RANSAC line detection, which is robust against data containing outliers [9]. At first, RANSAC randomly selects samples from the data to create a model, then evaluates the consistency with the remaining data. Models that match with many points are repeatedly selected, and the best-fitting model is output as the final estimation result. TLS point clouds corresponding to the HLS scanning range are extracted, and the TLS point clouds are linearly interpolated and resampled to 1,500 data points. To evaluate the degree of divergence between the TLS point clouds and the distance from the reference line obtained by HLS, Eq. 3 shown in the following equation is calculated.

$$NRMSE = \sqrt{\frac{\sum_{i=1}^{n}\left(y_i - \widehat{y_i}\right)^2}{n}} / \sigma \tag{3}$$

where, y_i is the distance from the reference line obtained by HLS (mm), $\widehat{y_i}$ is the roughness of TLS (mm), n is the data size (1,500), and σ indicates the standard deviation.

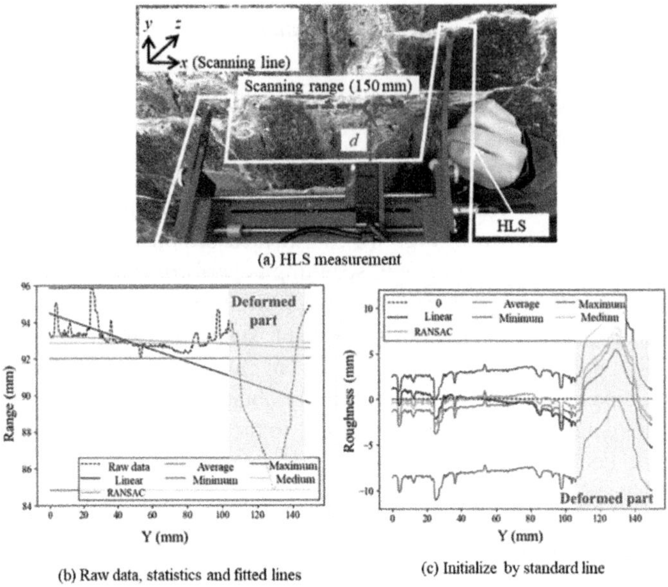

Fig. 4. HLS measurement and data processing.

4 Results and Discussion

4.1 Evaluation of Ranging Measurement Accuracy of TLS Point Cloud by HLS

The distance measurement accuracy of TLS point cloud is affected by the surface characteristics of the material, atmospheric conditions, the distance between the target and the scanner, and the incidence angle due to the constraints of the TLS installation position [10]. In this section, the effect of the measurement conditions on the point cloud density is investigated from aspects of the relationship between the distance from the target to the scanner, which can be quantitatively evaluated, the incidence angle, and the distance between the nearest points. The distance between the nearest points indicates the distance to the nearest point at each point. This distance is considered to correspond to the detectable crack width. Figure 5 shows the relationship between the measurement conditions and the distance between the nearest points. Figure 5(a) shows that the distance between the nearest points tends to increase as the distance between the target and the scanner increases, however this relationship is not clear and is indicated to be influenced by other factors. Figure 5(b) shows that the distance between the nearest points shows a positive correlation with the incidence angle ($R = 0.589$), and the distance between the nearest points is larger than the measurement condition value (3.1 mm) when the incidence angle is 70 to 90°. This suggests that when using a TLS scan, which has the smallest distance between the object and the scanner and the smallest angle of incidence, a point cloud with a distance of 3.1 mm or less between nearest points can be obtained, making it possible to detect cracks with a width of approximately 3.1 mm.

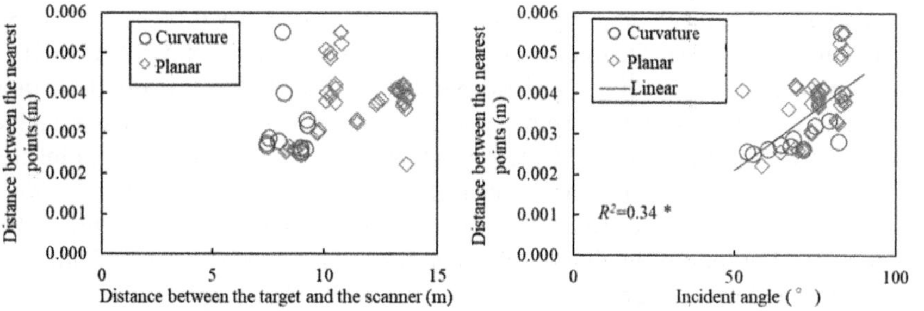

(a) Relationship between distance between the nearest points and distance between the target and the scanner (b) Relationship between distance between Incident angle

Fig. 5. Relationship between the measurement conditions and the distance between the nearest points.

Figure 6 illustrates the relationship between point cloud density and ranging accuracy of TLS point cloud. Figure 6(a) shows the acquisition status of a point cloud on a curvature surface. It can be seen that there are areas on the curvature surface where the point cloud density is high and areas where it is low due to a change of incident. It can be seen that in areas where the point cloud density is high, the TLS point cloud tracks the distance from the HLS reference line (NRMSE = 0.96). In areas where the point cloud density is low, the distance measurement accuracy in the depth direction decreases (NRMSE = 1.00).

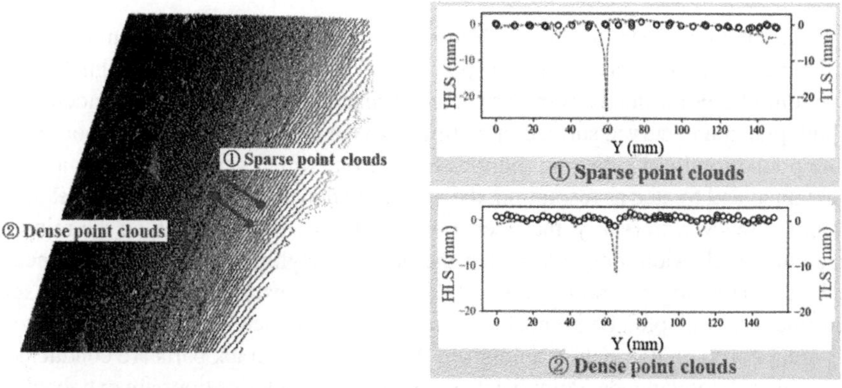

(a) Acquisition status of a point cloud on a curvature surface (b) Ranging accuracy of TLS point cloud

Fig. 6. Relationship between point cloud density and the ranging accuracy of TLS point cloud.

4.2 Local Geometric Feature and Crack Detection by PointNet

Regarding the radius r that defines the neighborhood point cloud, the radius r that minimizes the NRMSE is calculated for each measurement point in the comparison between TLS and HLS. The candidates for radius are 0.001, 0.003, 0.005, 0.008, 0.01, 0.03, 0.05, 0.08 and 0.1 m. Figure 7 shows the distribution of optimal radius for defining the neighborhood point cloud. On a planar surface, the frequency is highest at 0.008 m, and on a curvature surface, the frequency is highest at 0.05 m. The optimal NRMSE is not obtained with a radius of 0.005 m or less on either surface. This is because if the radius is too small, it is not possible to secure a neighborhood point cloud for calculating local features. In addition, the reason why the radius that defines the neighborhood point cloud on a planar surface is smaller than that on a curvature surface is that it is not possible to approximate the optimal linear if a radius that is too small is used on a curvature surface. If a wide neighborhood point cloud is obtained for a planar surface, it is difficult to fit a good linear line due to the unevenness.

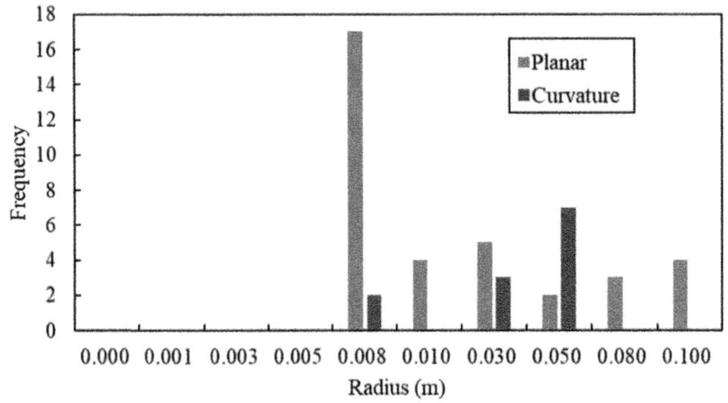

Fig. 7. Distribution of optimal radius to define neighbor point clouds.

In the previous section, it is confirmed that accurate distance measurement in the depth direction is achievable when the point cloud density is sufficiently high. For crack detection, TLS point clouds from Cases 2 and 3 characterized by minimal incident angles on both planar and curved surfaces are utilized. Assuming the nearest neighbor distance in the point cloud to be the minimum distance from which geometric information can be obtained and treated as resolution, the minimum detectable crack width is set at least 3.1 mm. However, in reality, the laser footprint differs at each point, so the minimum detectable crack width does not match the nearest neighbor point distance. Since the crack part in the original training and validation data are insufficient for achieving stable performance, available TLS point cloud data for another concrete structure and in internet [11] are incorporated into the training set. Annotations of crack parts are conducted on RGB-colored point clouds. Point batches centered on crack locations are extracted. The coordinate values (x, y, z) and point cloud information are normalized to the range of 0 to 1 for each batch.

To evaluate the effects of different input feature combinations, four cases are compared as shown in Table 2. In the case that only XYZ coordinates are used (Case 1), F1 score of 0.639 is obtained, with recall of 0.738 and precision of 0.564. It is suggested that a lot of cracks are successfully identified, although some false positives are included. In the case that intensity values are added (Case 2), recall increased to 0.837, however precision dropped significantly to 0.483, resulting in an overall F1 score of 0.612, indicating a tendency toward over-detection. Despite adding expandable variable, clear improvement of detection accuracy is not confirmed. It is indicated that intensity values are not important to detect crack on variable surface due to other influencing factors such as color, moisture, roughness and material properties [10]. The inclusion of RGB information (Case 3) improves recall of 0.828 and yields an F1 score of 0.636. Although the number of input variables is increased, no modification is made to the model complexity, potentially resulting in insufficient capacity to effectively process the additional information. The best performance is achieved in Case 4, where local geometric feature, third eigenvalue, is combined with XYZ coordinates. This configuration yielded the highest F1 score of 0.667, with precision and recall of 0.591 and 0.765, respectively. These results suggest that incorporating local geometric features significantly improves the balance between crack detection sensitivity and specificity.

Table 2. Crack detection results by PointNet using TLS data.

Case	F1 score	Precision	Recall
Case1: XYZ (Max epoch: 100, batch: 20)	0.639	0.564	0.738
Case2: XYZ + intensity (Max epoch: 10, batch: 10)	0.612	0.483	0.837
Case3: XYZ + RGB (Max epoch: 10, batch: 20)	0.636	0.517	0.828
Case4: XYZ + local geometric feature (Max epoch: 100, batch: 50)	0.667	0.591	0.765

5 Conclusions

This study investigated the influence of measurement conditions and input feature combinations on crack detection accuracy using TLS point clouds and PointNet deep learning architecture. Key findings include:

1. Point cloud density, determined by scanning distance and laser incidence angle, critically affects TLS ranging accuracy. Incidence angles exceeding 70–90° result in nearest-neighbor distances larger than 3.1 mm, limiting crack detection capability. Optimal scanning conditions (minimal distance and incidence angle) enable detection of cracks.
2. HLS measurements provide effective ground truth for evaluating TLS point cloud accuracy, revealing that high point cloud density regions achieve better ranging accuracy (NRMSE = 0.96) compared to low-density areas (NRMSE = 1.00).
3. Optimal neighborhood radius for local geometric feature extraction varies significantly between surface types: 0.008 m for planar surfaces and 0.05 m for curvature surfaces. Radius smaller than 0.005 m proves insufficient for reliable feature calculation regardless of surface geometry.
4. Local geometric features derived from eigenvalue decomposition significantly improve crack detection performance compared to coordinate-only or RGB-enhanced inputs. The combination of XYZ coordinates with local geometric features achieves the highest F1 score of 0.667, demonstrating improved balance between detection sensitivity and specificity.

These findings provide emphasizing the importance of measurement condition control and appropriate feature selection for deep learning models processing unordered point cloud data.

Acknowledgments. The authors are grateful to M. Mori, S. Kawabe, T. Kanamori, and Y. Kimura of the National Agriculture and Food Research Organization (NARO) for their cooperation in HLS measurement. This research was supported by the Research and implementation promotion program through open innovation grants (JPJ011937) from the Project of the Bio-oriented Technology Research Advancement Institution (BRAIN) (Project ID: 25132859).

References

1. Ma, Z., Liu, S.: A review of 3D reconstruction techniques in civil engineering and their applications. Adv. Eng. Inform. **37**, 163–174 (2018)
2. Walsh, S.B., Borello, D.J., Guldur, B., Hajjar, J.F.: Data processing of point clouds for object detection for structural engineering applications. Comput.-Aided Civil Infrastruct. Eng. **28**(7), 495–508 (2013)
3. Kim, M.K., Wang, Q., Li, H.: Non-contact sensing based geometric quality assessment of buildings and civil structures: a review. Autom. Constr. **100**, 163–179 (2019)
4. Stałowska, P., Suchocki, C., Rutkowska, M.: Crack detection in building walls based on geometric and radiometric point cloud information. Autom. Constr. **134**, 104065 (2022)
5. Qi, C.R., Su, H., Mo, K., Guibas, L.J.: Pointnet: deep learning on point sets for 3d classification and segmentation. In: Proceedings of the IEEE Conference on Computer Vision and Pattern Recognition, pp. 652–660 (2017)
6. Li, D., Lu, C., Chen, Z., Guan, J., Zhao, J., Du, J.: Graph neural networks in point clouds: a survey. Rem. Sens. **16**(14), 2518 (2024)
7. Zhao, H., Jiang, L., Jia, J., Torr, P.H., Koltun, V.: Point transformer. In: Proceedings of the IEEE/CVF International Conference on Computer Vision, pp. 16259–16268 (2021)
8. Hackel, T., Wegner, J.D., Schindler, K.: Contour detection in unstructured 3D point clouds. In: Proceedings of the IEEE Conference on Computer Vision and Pattern Recognition, pp. 1610–1618 (2016)
9. Fischler, M.A., Bolles, R.C.: Random sample consensus: a paradigm for model fitting with applications to image analysis and automated cartography. Commun. ACM **24**(6), 381–395 (1981)
10. Oytun, M., Atasoy, G.: Effect of terrestrial laser scanning (TLS) parameters on the accuracy of crack measurement in building materials. Autom. Constr. **144**, 104590 (2022)
11. Stałowska, P., Suchocki, C.: TLS data for cracks detection in building walls. Data Brief **42**, 108247 (2022)

The Effect of Combined Carbonation and Chloride Action on Concretes with a Combination of Cements CEM II 42.5N and CEM II 32.5N

Eleni Mariolopoulou[1,2](✉) and Kosmas Sideris[2]

[1] Department of Civil and Environmental Engineering, Politecnico Di Milano, Milan, Italy
eleni.mariolopoulou@polimi.it

[2] Department of Civil Engineering, Democritus University of Thrace, Xanthi, Greece

Abstract. Concrete structures are designed to have a service life of many decades but in many cases their durability is affected by external factors, undermining their safety. A lot of studies have focused on the carbonation of concrete and chlorides penetration on the structures. However only a few have combined both. This study aims to investigate the impact of continuous cycle exposure to air-water conditions on concrete specimens, helping us understand the problems that occur in constructions with alternating exposure, such as the splash zone of marine structures. The experimental program involves two mixtures with a combination of CEM II/A-M 42.5N and CEM II/B-M 32.5N and other two with CEM I 42.5R cement. All formulations were subjected to a series of tests to further investigate the effect of concrete synthesis on the durability factors. After exposure to accelerated carbon dioxide and/or chloride environment, the carbonation depth and chloride penetration were experimentally measured on the specimens. The results indicate that the mixtures with CEM I present lower carbonation and chloride penetration depths. Chlorides presence and cement type strongly influence carbonation, and results differ depending on whether the concrete is in dry-wet cycles or solely submerged in water.

Keywords: Durability · Concrete · Accelerated Carbonation · Chloride Penetration

1 Introduction

Concrete in general is a very durable material. All over the world, there are structures that have survived 2.000 years without significant interventions and will survive hundreds more. The Pantheon in Rome and the numerous bridges built during the Roman Empire are among them. However, nowadays, many concrete structures of the last century have deteriorated prematurely [1]. According to most standards, structures are expected to have a lifespan of at least 100 years. In reality, repair and renovation are often necessary after no more than 30 years [2]. Carbonation and chloride action are the two of the main causes of degradation in concrete structures. In marine constructions, where both

M. Posani and S. Singh (Eds.): RYS 2025, RILEM Bookseries 67, pp. 899–908, 2026.
https://doi.org/10.1007/978-3-032-14652-6_65

of them are present, the most severely deteriorated segment of the wharves is the tidal zone of the piles [3]. In this zone with constant dry-wet cycles, carbonation changes on concrete microstructure and pH, can potentially have a direct impact to the chloride diffusion into concrete [4].

The last decade, more and more researchers have start to look into the combined degradation mechanisms of CO_2 and Cl, highlighting the urgent need to understand their effects. Saeki et al. [5] focused on the investigation of chloride ions immobilization in cement hydrates observing the fixation effect of AFm. Since then, many different experimental programs have been presented, comparing mixtures with and without fly ash [6, 7], studying structural lightweight aggregate concrete [8], evaluating blended cements [9] or corelating exposure to service-related crack width [10]. Cement types I and V have also been studied focusing on the ratio of water-soluble chloride to the acid-soluble chloride content on the carbonated or non-carbonated test series [11]. The contribution of Tassos et al. [12] investigate mixtures with type I and II cements, but it is solely focused on the carbonation after exposure to open air conditions. Actually, no research has ever been made with the purpose to analyze the combined carbonation and chlorides action to concretes containing commonly used type II cements.

This paper presents the outcome of combined exposure to CO_2 and Cl conditions for specific concretes with type I and II cements. The research is focused on four mixtures of two strength classes that have been prepared with three different cements. The dry-wet alterations are imitated by five circles of exposure. Various experimental tests are investigated, and the results are corelated. The analysis intends to highlight the importance of the concrete composition and the effects of the type of cement in the combined deterioration mechanisms of CO_2 and Cl.

2 Experimental Program

During this research, four concrete mixtures were produced, two of C25/30 and two of C30/37 strength class according to EN 206–1. For each strength class, a combination of cements CEM II/A-M 42.5N and CEM II/B-M 32.5N was used in one mixture and CEM I 42.5R in the other. Retarder and polycarboxylate superplasticizer were also added, to achieve the desired workability and setting time. The mixing ratios and slump test measurements for every mixture are shown analytically in Table 1. The mixtures with cement CEM I 42.5R were used for reference.

A total of 16 specimens were prepared for each mixture (Fig. 1). Of these, 4 were cubic specimens of 150x150x150 mm and 3 cylindrical specimens of 100x200mm. All were maintained in a curing chamber ($T = 20 \pm 2$ °C and $RH > 95\%$) for 28 days. Moreover, 9 cubic specimens with dimensions of 100x100x100 mm were also made to be used for the measurement of carbonation and chlorides penetration depth. Of them, 6 were maintained for 3 days and the other 3 for 7 days at the curing chamber with the abovementioned conditions. In this study, the term carbonation refers to the process where carbon dioxide reacts with calcium hydroxide and other hydration products in the cement matrix, forming calcium carbonate and lowering the pore solution pH, which may depassivate the reinforcing steel.

The cubic specimens of 150x150x150mm were used to measure the compressive strength at the age of 28 days. The EN 12390–18 standard [13] was followed for the

Table 1. Mixing ratios and properties of the concrete mixtures.

kg/m3	C25/30 - CEM II/A-M 42.5N + CEM II/B-M 32.5N	C25/30 - CEM I 42.5R	C30/37 - CEM II/A-M 42.5N + CEM II/B-M 32.5N	C30/37 - CEM I 42.5R
CEM I 42.5R	0	300	0	330
CEM II/A-M 42.5N	100	0	130	0
CEM II/B-M 32.5N	200	0	200	0
Crushed stone sand	473	480	473	473
River/sea sand	567	575	567	567
Gravel	378	383	378	378
Coarse	473	480	473	473
Water	165	165	165	165
Superplasticizer	0.80%	0.80%	1%	1%
Retarder	0.20%	0.20%	0.30%	0.30%
W/C	0.55	0.55	0.5	0.5
Slump (mm)	170	155	200	170

evaluation of the chloride diffusion coefficient D_{nssm} of the mixtures. The Eq. (1) was used according to NT Build 492 Standard [14]. For this, the 3 cylindrical specimens of 100x200mm were cut into 4 equal pieces each. The top and the bottom part of the cuts were excluded of the tests. Form the remaining 6 pieces, 3 were selected to satisfy the thickness of 50 ± 2 mm. These remained in $Ca(OH)_2$ solution for 18 ± 2 h before the test.

$$D_{nssm} = \frac{0.0239(273+T)L}{(U-2)t}\left(x_d - 0.0238\sqrt{\frac{(273+T)L\,x_d}{U-2}}\right) \quad (1)$$

where:

D_{nssm}: non-steady-state migration coefficient, × 10^{-12} m^2/s;
U: absolute value of the applied voltage, V;
T: average value of the initial and final temperatures in.
the anolyte solution, °C;
L: thickness of the specimen, mm;
x_d: average value of the penetration depths, mm;
t: test duration, hour.

The methods followed to investigate the carbonation depth in concrete are in accordance with the standards of the Fib 2010 regulation. For each mixture, the 3 cubic specimens of 100mm edge that were maintained for 7 days, were used as reference.

When they were removed from the curing chamber, they remained in a laboratory environment until the age of 28 days. After, they were placed in the accelerated carbonation chamber (T = 20 °C, RH = 56% and CO_2 = 2%) for 28 days.

For the study of the alternating carbonation and chloride penetration conditions, the 6 cubic specimens of 100mm edge of each mixture were used. After being maintained in the abovementioned curing chamber for 3 days, these specimens were removed from the chamber and got divided into 2 groups (B and C). This categorization was made to facilitate the planning of CO_2 – Cl rotations. Each mixture had 3 specimens in every group. The specimens of group B were exposed only to cycles of accelerated carbonation conditions while the ones of group C were exposed to cycles of both chlorides and accelerated carbonation conditions.

a) b)

Fig. 1. a) Production of the specimens. b) Final specimens of two mixtures.

On each cycle, the specimens of group C were placed in chlorides for 1 week while the ones of group B remained in laboratory environment. After 1 week, the samples were removed from the chlorides and placed in laboratory environment to dry for 3 days. After 3 days, both groups B and C were transferred from the lab environment to the abovementioned accelerated carbonation chamber for 2 weeks. Two weeks later, the specimens were removed, and the cycle was complete. In our experiment 5 cycles were completed. The specimens of group C were coated with a layer of resin, apart from one side so that the penetration of CO_2 and chloride agents could be unhindered. The exposure to chlorides was carried out using a tank containing water and NaCl = 10%. The carbonation depth in each specimen was measured by applying the phenolphthalein solution.

3 Results and Discussion

3.1 Compressive Strength

The compressive strength of all mixtures was measured at the age of the 28 days. The final strength (Table 2) was obtained from the average of the values of the 3 cubic 150 mm specimens of each mixture and converted from kN to MPa.

All the results obtained were fully expected. In both C25/30 and C30/37 strength classes, it is observed that the reference mixtures with CEM I 42.5R cement have higher compressive strength than the ones with the combination of CEM II/A-M 42.5N and CEM II/B-M 32.5N cement. The CEM I 42.5R is a Portland cement that is used to produce high-strength concrete. It shows fast hydration, very high early age strength, high final strengths, and relatively low porosity. Therefore, it is consequential to present higher compressive strength than the composite Portland cements contained in the other mixtures.

Table 2. Compressive strength of concrete mixtures

	C25/30 - CEM II/A-M 42.5N+ CEM II/B-M 32.5N	C25/30 - CEM I 42.5R	C30/37 - CEM II/A-M 42.5N+ CEM II/B-M 32.5N	C30/37 - CEM I 42.5R
Specimen n. 1	960	1170	1120	1340
Specimen n. 2	980	1110	1140	1310
Specimen n. 3	960	1110	1120	1360
Average (kN)	966,67	1130,00	1126,67	1336,67
Compressive strength (MPa)	42,96	50,22	50,07	59,41

3.2 Chloride Penetration Measures

For the measurement of the depth of chloride penetration, the EN 12390–18 standard [13] was followed. After the end of the procedure, the 3 cylindrical specimens of 50mm width for each mixture were cracked in half and sprayed with $AgNO_3$. This solution reacts with chloride ions to form a mixture of silver oxide and silver chloride that changes the color of the concrete surface, creating a distinct area that correspond to the chloride penetration depth. The recording of the depth was made along the diameter every 10mm except from the two external points. A total of 7 values was obtained from each sample and the 10mm distances from which they were taken were modified from $xd1 = - 30$ mm to $xd7 = 30$ mm, having as zero point the middle of the specimen. For all mixtures, the specimens remained in the device for 24 h. All measurements are presented in detail at Fig. 2 and Table 3 below.

It can be observed that in the reference mixtures with CEM I 42.5R, the D_{nssm} coefficient is lower than the mixtures with the combination of CEM II42.5N and CEM II 32.5N cement. More specifically, for the strength class C25/30, the mixture with the combination of cements shows an increase in the chloride penetration coefficient by 11.22% compared to the reference mixture. Respectively, for the C30/37 category, an increase of 25.18% is observed. These results are considered logic and reasonable as the increase is due to the reduced porosity that CEM I 42.5R cement presents as a

Portland cement, compared to the composite Portland CEM II cements. CEM I cements are hydrated more quickly, the hydration products are formed sooner and the pores of the concrete close in a shorter time than CEM II cements. Therefore, it is more difficult for chlorides to penetrate the pores and for this reason the reference mixtures of our research present a lower D_{nssm} coefficient.

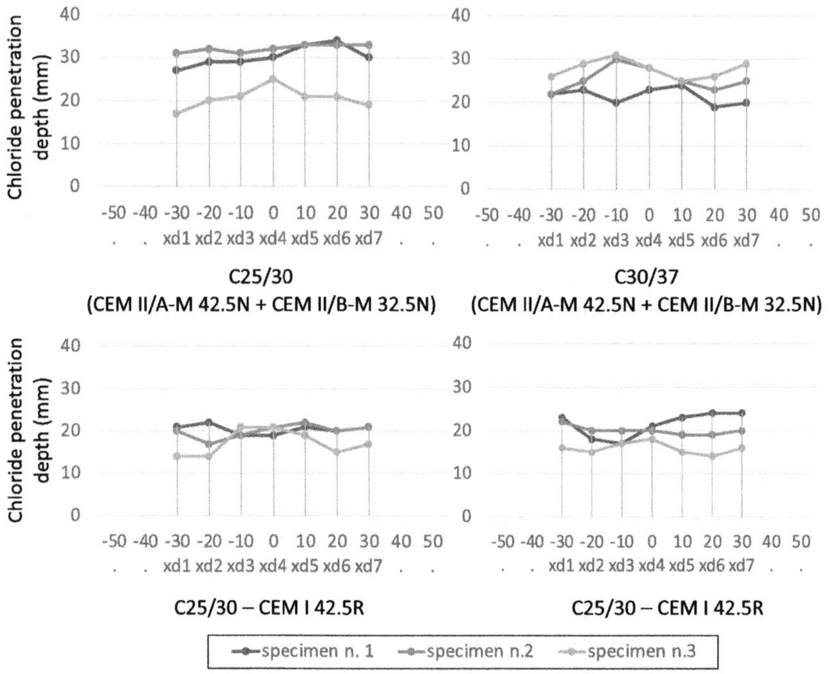

Fig. 2. Chloride penetration depth on every point of the three specimens for every mixture.

Table 3. Chloride penetration coefficient (x10^{-12} m^2/s) of the concrete mixtures.

x10^{-12} m^2/s	C25/30 - CEM II/A-M 42.5N + CEM II/B-M 32.5N	C25/30 - CEM I 42.5R	C30/37 - CEM II/A-M 42.5N + CEM II/B-M 32.5N	C30/37 - CEM I 42.5R
Dnssm	18,74	16,85	17,12	13,66

3.3 Carbonation Depth

The carbonation depth was measured according to the EN 14630 Standard [15]. The measurements were taken at various points along the sample sides except from the one with the label of the mixture name. On every mixture the 3 specimens where measured and the average depth was calculated. The results are presented in Table 4.

Table 4. Accelerated carbonation depth (mm) of the concrete mixtures.

(mm)	C25/30 - CEM II/A-M 42.5N+ CEM II/B-M 32.5N	C25/30 - CEM I 42.5R	C30/37 - CEM II/A-M 42.5N+ CEM II/B-M 32.5N	C30/37 - CEM I 42.5R
carbonation depth	8,77	5,42	7,42	3,81

It is noticed that in both strength categories the reference mixture with CEM I 42.5R presents a lower carbonation depth than the mixtures with the combination of cements CEM II 42.5N and CEM II 32.5N. The C25/30 mixture with the combination of cements has a higher depth by 3.35mm (61.81%) than the reference mixture. A big increase is also seen in the C30/37 mix with a combination of cements which has a carbonation depth 3.61mm (94.75%) greater than the reference mixture. The reduced carbonation of the reference mixtures is due to the rapid production of hydration products in the concretes with CEM I, as in the case of chloride penetration analyzed above. The result is lower porosity, which hinders carbonation. Thus, it is logical that the two mixtures with CEM II 42.5N and CEM II 32.5N, which have higher porosity, are more carbonized and precent bigger carbonation depth.

3.4 Carbonation Depth After Alternating Exposure Conditions

The measurement of the carbonation depth was made after the end of 5 exposure cycles in chlorides and accelerated carbonation conditions, as explained above. The EN 14630 Standard [15] was followed, and the average of the measurements in every concrete mixture is shown in Fig. 3. The samples that were exposed only to accelerated carbonation are indicated as CO_2 and the ones that were exposed to alternating conditions of accelerated carbonation and chlorides and are indicated as $CO_2 + Cl$. As analysed in chapter 2, the specimens exposed only to carbonation conditions are the 100mm edge cubic specimens that constituted group B and the specimens that were alternately exposed to carbonation and chloride conditions are the 100mm edge cubic specimens that constituted group C.

From the values, when we correlate the measurements of the group B and C on every mixture, it is noticeable that depths in the CO_2 samples are higher than the $CO_2 + Cl$ samples in all mixtures. Specifically, in the mixtures of C25/30 strength class, the CEM II 42.5N and CEM II 32.5N $CO_2 + Cl$ specimens present a carbonation depth lower than that of the ones of CEM II 42.5N and CEM II 32.5N CO_2 by 9.44mm (62.64%). Similarly, the CEM I 42.5R $CO_2 + Cl$ specimens have lower depth than the ones of CEM I 42.5R CO_2 by 8.76mm (89.75%). For the C30/37 strength class, the specimens CEM II 42.5N and CEM II32.5N $CO_2 + Cl$ present a carbonation depth reduced by 12mm (85.71%) compared to the CEM II 42.5N and CEM II 32.5N CO_2 ones, and the CEM I42.5R $CO_2 + Cl$ also decreased by 8.14mm (89.06%) compared to CEM I42.5R CO_2. Therefore, the specimens that were exposed to alternating environmental conditions have lower carbonation depth than those that were exposed only to carbonation conditions.

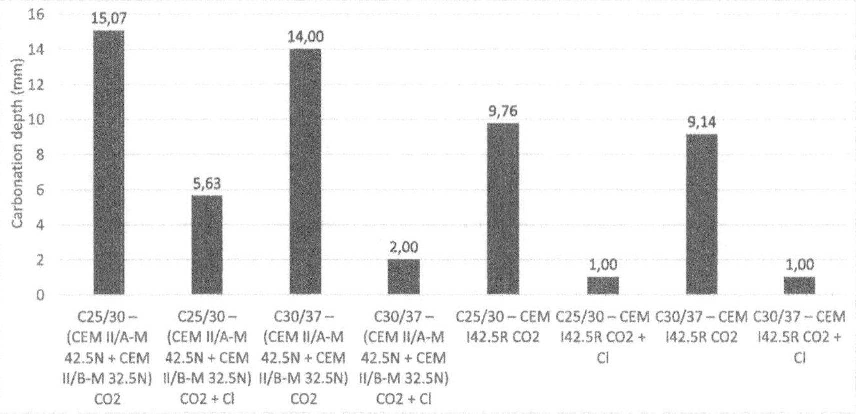

Fig. 3. Carbonation depth of concrete mixtures after exposure cycles.

When we correlate the mixtures by strength class with different type of cements taking into consideration the group B and C specimens separately, it is observed that in the C25/30 strength class, the mixture with the combination of cements that was exposed to CO_2 present an increased carbonation depth by 54.4% compared to the mixture with CEM I, while when it comes to specimens exposed to combined action the increase is 463%. Accordingly, in the C30/37 category, the mixture with the CEM II combination has an increase in carbonation depth of 53.17% compared to the mixture with CEM I, and in the combined exposure the increase is 100%.

It is also worth noting that the C30/37 mixture with combination of CEM II 42.5 and CEM II 32.5 cements, present lower durability than C25/30 with CEM I 42.5R as its carbonation depth is increased almost by 43,44% when exposed only to CO_2 (group B specimens) and 100% in combined actions of CO_2 and Cl (group C specimens). Once again, all mixtures with CEM I have lower carbonation depth than the corresponding mixtures containing CEM II cements. As we have already mentioned, this is normal, and it is a result of the nature of CEM I cements that form hydrate products at a very quick rate compared to composite Portland cements.

In this study, prolonged saturation during imbibition limited CO_2 diffusivity and thus slowed carbonation. However, carbonation is known to proceed most rapidly at intermediate humidity (~50–70% RH), where sufficient moisture enables the reaction while pores remain open for gas diffusion [16, 17]. Under such wetting–drying conditions, both chloride ingress and carbonation may be accelerated simultaneously.

4 Conclusions

The paper presents the experimental results obtained by the exposure of C25/30 and C30/37 mixtures with CEM I 42.5R, CEM II 42.5N and CEM II 32.5N cements to carbonation and chloride environments. The composition of concrete, the type of cement it contains, the porosity and the compressive strength, appear to affect the depth of carbonation and chloride penetration. Concretes with CEM II present higher chloride penetration compared to those containing CEM I, while concretes with strength class C30/37 perform lower chloride penetration than the ones of C25/30 with the same type of cements, because of the lower w/c ratio used in the production of the C30/37 mixtures.

Moreover, mixtures with CEM I cement have reduced carbonation depth by half in both strength categories compared to the corresponding mixtures with the CEM II cement combinations. Additionally, when concrete is cyclically exposed to combined carbonation and chlorides, it presents a carbonation depth half than the one it has after exposed exclusively to CO_2 regardless of the cement it contains. Concretes that are exposed to alternating environmental conditions have lower carbonation depth than those that were exposed only to carbonation conditions.

In general, the type of cement has great impact to the penetration of CO_2 and chlorides. Concretes that are considered durable with high strength can present higher carbonation and chloride depths than concretes of lower strength class that contain CEM I. This highlights the importance of the right selection of cements regardless the strength class that the concrete will be, in order to accomplish its purpose in the best possible way and take advantage of its characteristics to the fullest.

References

1. Liu, R., Jiang, L., Xu, J., Xiong, C., Song, Z.: Influence of carbonation on chloride-induced reinforcement corrosion in simulated concrete pore solutions. Constr. Build. Mater. **56**, 16–20 (2014)
2. Torrent, R.J., Neves, R.D., Imamoto, K.-I.: Concrete Permeability and Durability Performance: From Theory to Field Applications, 1st edn. CRC Press, Boca Raton (2021)
3. Suh, K., Mullins, A., Sen, R., Winters, D.: Effectiveness of fiber-reinforced polymer in reducing corrosion in marine environment. ACI Struct. J. **104**, 76–83 (2007)
4. Malheiro, R., Camões, A., Meira, G.R., Amorim, M.T., Castro-Gomes, J., Reis, R.: Chloride penetration into carbonated concrete. In: Proceedings of the 7th International Conference on Safety and Durability of Structures ICOSADOS 2016, UTAD, Portugal (2016)
5. Saeki, T., Andrade, C., Kropp, J.: Effect of carbonation on chloride penetration in concrete. In: Third International RILEM Workshop on Testing and Modelling Chloride Ingress into Concrete, pp. 381–394, RILEM Publications SARL (2004)
6. Zhang, D., Shao, Y.: Effect of early carbonation curing on chloride penetration and weathering carbonation in concrete. Constr. Build. Mater. **123**, 516–526 (2016)
7. Malheiro, R., Camões, A., Meira, G., Amorim, M.T., Castro-Gomes, J., Ferreira, R.M.: Behaviour of cementitious matrices subjected to the combined action of chloride ions and carbonation. Revista Materia **23**(3), Article e-12162 (2018)
8. Bogas, J.A., Real, S.: A review on the carbonation and chloride penetration resistance of structural lightweight aggregate concrete. Materials **12**(20), 3456 (2019)
9. Hren, M., Bosiljkov, V., Legat, A.: Effects of blended cements and carbonation on chloride-induced corrosion propagation. Cem. Concr. Res. **145**, 106458 (2021)
10. Al-Ameeri, A., Rafiq, M.I., Tsioulou, O.: Combined impact of carbonation and crack width on the chloride penetration and corrosion resistance of concrete structures. Cement and Concrete Composite **115**, 103819 (2020)
11. Lee, M.K., Jung, S.H., Oh, B.H.: Effects of carbonation on chloride penetration in concrete. ACI Mater. J. **110**(5), 559–566 (2013)
12. Tassos, C., Sideris, K., Chatzopoulos, A., Pistofidis, N., Chaniotakis, E.: Influence of cement type on carbonation of concrete mixtures. In: MATEC Web of Conferences, **163**, 05005 (2018)

13. EN 12390–18. Testing hardened concrete - Part 18: Determination of the chloride migration coefficient, CEN (2021)
14. NT Build 492. Chloride migration coefficient from non-steady-state migration experiments, Nordtest methods (1999)
15. EN 14630. Products and systems for the protection and repair of concrete structures - Test methods - Determination of carbonation depth in hardened concrete by the phenolphthalein method, CEN (2006)
16. RILEM TC 281-CCC. RILEM Technical Committee 281-CCC: carbonation of concrete with supplementary cementitious materials. Materials and Structures **53**, 140 (2020)
17. Thiel, C., Rechenberg, J., Gehlen, C.: Influence of environmental conditions on carbonation of concrete – review and discussion of current knowledge. CivilEng **3**(4), 877–898 (2022)

Uniaxial Tensile Stress Relaxation and Cracking Behavior of SHCCs Incorporating Blast Furnace Slag and Fly Ash

Faizudin Hafiz Zadah(✉) and Yao Luan

Saitama University, Saitama, Japan
hafiz.z.f.152@ms.saitama-u.ac.jp

Abstract. This study investigates the early-age (7 days) tensile stress relaxation and cracking behavior of slag- and fly ash-based strain-hardening cementitious composites (SHCCs) under sealed curing. Mechanical properties were characterized via compressive strength, Young's modulus, and uniaxial tensile tests, while stress relaxation was assessed under both pre-cracking (70% and 90% of the first cracking stress) and post-cracking (0.6% and 1.0% tensile strain) conditions. Crack formation and evolution were monitored using digital image correlation (DIC). Compared to fly ash-based SHCCs, slag-based SHCCs exhibited higher compressive strength and stiffness due to faster hydration and the development of a denser, more viscoelastic matrix. Both mixtures displayed ductile, strain-hardening behavior; however, slag-based SHCCs showed slightly superior tensile performance with higher strength and finer multiple cracking. Stress relaxation in slag mixes was more pronounced under both pre- and post-cracking conditions, particularly at higher stress levels, likely due to enhanced microcrack formation and fiber debonding. In contrast, fly ash-based SHCCs exhibited greater relaxation at lower stress levels. For both materials, the most significant stress decay occurred within the first 30 min, followed by a slower, stable phase. DIC analysis confirmed minimal change in existing crack geometry during relaxation, but revealed the initiation of new microcracks, especially in slag-based specimens, contributing to greater deformation and stress decay. These findings enhance the understanding of time-dependent tensile behavior in SHCCs and support their application in structural repair systems requiring long-term deformation control.

Keywords: Strain-Hardening Cementitious Composites (SHCC) · Slag · Fly Ash · Uniaxial Tension · Multiple Fine Cracking · Stress Relaxation

1 Introduction

Strain-hardening cementitious composites (SHCC) are a class of fiber-reinforced cement-based materials characterized by exceptional tensile ductility. This ductility is achieved through the development of multiple fine and tightly spaced microcracks rather than a single dominant crack, a behavior made possible by fiber bridging mechanisms [1]. As a result, SHCC exhibits pseudo-ductile and strain-hardening responses under

M. Posani and S. Singh (Eds.): RYS 2025, RILEM Bookseries 67, pp. 909–923, 2026.
https://doi.org/10.1007/978-3-032-14652-6_66

uniaxial tensile loading. These unique mechanical properties make SHCC highly suitable for applications demanding both enhanced durability and significant deformation capacity, whether in new construction or the repairment of deteriorated structures.

As the number of structures requiring repair continues to increase globally, repair applications, such as patch repairs, often involve restraint conditions where the autogenous and drying shrinkage strains of the repair material are restricted by the existing substrate concrete [2]. This mismatch leads to the development of internal tensile stresses within the repair material. Among the most influential factors governing these restrained tensile stresses are time-dependent properties, particularly stress relaxation and creep [3].

Beushausen and Alexander investigated the failure mechanisms and tensile stress relaxation in bonded concrete overlays subjected to differential shrinkage [4]. Their study revealed that tensile relaxation relieved approximately 40–50% of the tensile stresses in the overlay, highlighting the critical role of relaxation in enhancing the serviceability and long-term performance of bonded concrete overlays. In previous studies that examined the stress relaxation behavior of cement mortar and concrete specimens with varying water-to-binder (w/b) ratios and curing ages, relaxation was found to significantly reduce tensile stresses [5, 6]. According to these studies, approximately 80% of the ultimate stress relaxation in cement mortar occurred within the first 12 h, while in concrete, about 74% of the total relaxation was observed within the first 5 h. These findings underscore the importance of early-age relaxation in mitigating restrained tensile stresses in cement-based materials.

With the growing use of SHCC as patch repair materials, understanding their time-dependent behavior is crucial. In a series of tensile creep tests on notched and dumbbell-shaped SHCC specimens, Boshoff et al. concluded that tensile strain and crack width increase significantly over time, primarily due to fiber pull-out rather than fiber creep, with the average crack width being more influenced by the duration of loading than by the applied tensile creep load [7, 8]. Ratnayake et al. [9] simulated SHCC creep in practical repair scenarios and observed substantial stress decay within the first four hours, attributed to crack widening under sustained loading and the initiation of new cracks. This was followed by a slower, gradual reduction in stress. The study also reported that crack widening under stress relaxation was significantly less pronounced than under constant loading, indicating that SHCC can maintain long-term resistance to water and chemical ingress when used as a repair material. A similar conclusion was drawn in another study [10].

Tensile stress relaxation, defined as stress decay under constant strain, plays a more critical role than tensile creep under restrained conditions, as the latter involves strain development under constant load. However, studies on this phenomenon remain scarce, especially in both pre-cracking and post-cracking scenarios.

This study investigates the tensile stress relaxation behavior of SHCC under both pre-cracking and post-cracking conditions. Stress relaxation tests were conducted at defined stress and strain levels to simulate different stages of structural response. In addition, digital image correlation (DIC) was used to monitor crack formation and evolution under sustained tensile strain, providing insights into time-dependent deformation and cracking mechanisms.

To examine the role of binder composition, SHCC mixtures incorporating fly ash (FA) and ground granulated blast furnace slag (BFS) were compared. FA is widely used in SHCC to enhance ductility and reduce cement content [11]; however, its relatively low pozzolanic reactivity at early ages limits its effectiveness at high replacement levels; an important consideration when aiming for substantial reductions in CO_2 emissions. Conversely, blast furnace slag (BFS) exhibits higher early-age reactivity compared to fly ash. In addition, previous studies [12] have reported enhanced durability and improved resistance in aggressive environments for slag-based systems, making BFS a suitable choice for use at higher replacement ratios. By comparing these two binder systems, the study aims to clarify how binder selection influences tensile stress relaxation and crack evolution in SHCC in restrained structural conditions.

2 Experimental Program

2.1 Materials and Mix Preparation

Materials. High early-strength Portland cement, with a density of 3.14 g/cm^3, conforming to Japanese Industrial Standards (JIS), was used as the primary binder. Two types of supplementary cementitious materials (SCMs) were incorporated: Type II fly ash (hereinafter FA or fly ash) with density of 2.36 g/cm^3 and fineness of 4,340 cm^2/g, and ground granulated blast furnace slag (hereinafter BFS or slag) with density of 2.91 g/cm^3 and fineness of 4,140 cm^2/g. Silica sand No. 7, with a maximum particle size of 0.3 mm, was used as the fine aggregate. A polycarboxylate-based superplasticizer (SP) was added to achieve consistent workability across mixtures. Polyvinyl alcohol (PVA) fibers with a density of 1.30 g/cm^3, length of 12 mm, diameter of 40 μm, tensile strength of 1,560 MPa, and Young's modulus of 41 GPa, were incorporated as reinforcement in the SHCC mixtures.

Mix Proportions. Two SHCC mixtures were prepared: one incorporating 55% fly ash, a commonly used dosage in previous studies [11], and another with 70% slag, which has been reported to exhibit superior tensile performance compared to fly ash at 7 days [15]. Both mixes used a constant water-to-binder (w/b) ratio of 0.26 and a PVA fiber volume fraction of 2%. The dosage of superplasticizer was adjusted for each mix to ensure comparable flowability. The detailed mix proportions are presented in Table 1.

Table 1. Mix proportions

Mix	C	W	FA	BFS	SS	SP
	(kg/m^3)					(%)
Fly ash	569	329	696	0	455	1.8
Slag	404	351	0	944	458	0.8

C: Cement, W: Water, FA: Fly ash, BFS: Blast furnace slag, SP: Superplasticizer, SS: Silica sand

Mixing and Casting Procedures. All materials were combined in a mechanical mixer following a standardized sequence to ensure uniform dispersion of fibers and homogeneous paste. During mixing, the mortar was manually checked by hand to confirm that no clumps remained. Once the mixture was confirmed to be fully blended, the flowability was assessed using the mortar flow table test, following the procedures outlined in JIS R 5201. The purpose of this test was to verify that a consistent and similar flowability was achieved in all mixes by adjusting the SP dosage, while also preventing bleeding, fiber segregation, or excessive stickiness. After quality confirmation, the fresh SHCC mortar was cast into plastic molds and subjected to sealed curing for seven days. Sealed curing was carried out in an environmental chamber maintained at a temperature of 21 ± 3 °C and a relative humidity of 60%. The surface of each mold was covered with plastic film to prevent any moisture exchange with the environment. Sealed curing was selected over water curing to better simulate practical application conditions, particularly for SHCC used in repair scenarios.

2.2 Compressive Strength and Young's Modulus Tests

Cylindrical specimens with dimensions of 50 mm in diameter and 100 mm in height were cast and seal-cured for 7 days. For each mixture, compressive strength and static Young's modulus were evaluated using a universal testing machine on three specimens, and the average value was reported. Axial strain during compressive loading was measured with a compressometer securely attached to each specimen to accurately capture deformation throughout the test. The static Young's modulus was determined from the linear portion of the stress-strain curve, based on the applied compressive load and the corresponding axial strain.

2.3 Uniaxial Tension and Stress Relaxation Tests

Dog-bone-shaped specimens were prepared for each SHCC mixture and seal-cured for 7 days prior to testing.

Uniaxial Tensile Test. The uniaxial tensile test was performed in accordance with the JSCE recommendations [13]. Five specimens per mixture were tested under direct tension to evaluate tensile performance. The tests were conducted using an automatic loading machine at a constant displacement rate of 0.20 mm/min until tensile failure. Elongation and applied load were measured using two linear variable differential transducers (LVDTs) mounted symmetrically on either side of the specimen's central gauge length (80 mm). The average first cracking stress, tensile strength, and strain capacity were obtained from these tests and used as reference parameters for subsequent relaxation testing.

Stress Relaxation Test. The stress relaxation tests were performed using a manually operated tensile loading machine in a room maintained at 21 ± 3 °C and 60% relative humidity. The tensile load was gradually applied by slowly tightening a nut located on the plate at the top of the machine until the desired stress or strain level was reached. Once the target condition was achieved, the nut was firmly locked in place to maintain the applied deformation and prevent any release of tension during the sustained loading phase. The corresponding stress at this point was defined as the initial tensile stress for the relaxation test. The SHCC specimens were tested under two distinct conditions: pre-cracking, at 70% and 90% of the average first-cracking stress obtained from the uniaxial tensile test, and post-cracking, at imposed tensile strains of 0.6% and 1.0%. The specimen naming and corresponding details are provided in Table 2.

Table 2. Test conditions and specimen labels for SHCC stress relaxation

SHCC	Pre-cracking condition		Post-cracking condition	
	70%	90%	0.6%	1.0%
Flay ash	FA07	FA09	FA06	FA10
Slag	BFS07	BFS09	BFS06	BFS10

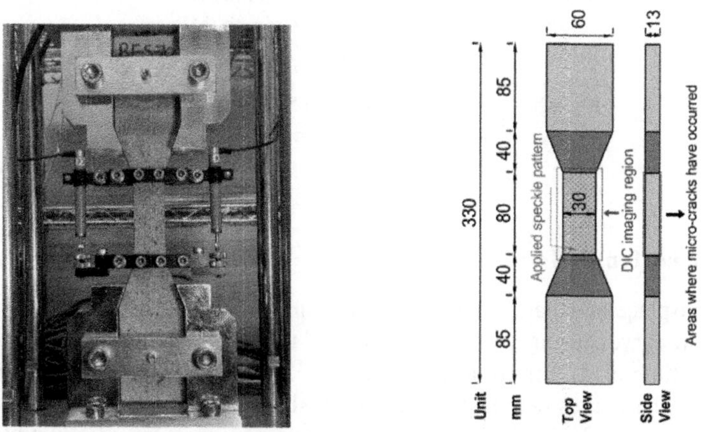

(a) (b)

Fig. 1. Tensile relaxation test procedure, (a) test setup and specimen mounting, (b) dimensions of specimen and application of speckle patterns

At each of these points, the load was sustained for 12 h, during which the decay in load under constant strain was recorded. A load cell and displacement transducers were connected to a data logger and computer system for continuous monitoring of load and displacement throughout the relaxation period. The resulting load decay curves were used to evaluate the stress relaxation behavior of each SHCC mixture.

A speckle pattern suitable for DIC analysis was applied to the central portion of each specimen using black and white spray paint. During testing, digital images were automatically captured at 60-s intervals using a Hozan L-836 camera. Subsequently, digital image correlation (DIC) analysis was performed in MATLAB using the Augmented Lagrangian-based DIC method (AL-DIC) [14] to evaluate surface strain fields, crack initiation, and crack width development over time during the relaxation process.

Figure 1a and Fig. 1b show the tensile relaxation test setup and application of speckle pattern, respectively.

3 Results and Discussion

3.1 Compressive Strength and Young's Modulus.

The results of compressive strength and Young's modulus show that slag exhibited marginally higher values than fly ash in both tests, as shown in Table 3. The relatively lower compressive strength and Young's modulus observed in the fly ash mixture can be attributed to the low pozzolanic reaction of fly ash at early ages. Despite its lower blending ratio compared to slag (55% vs. 70%), the fly ash exhibited a lower degree of reaction and resulted in coarser porosity.

Table 3. Compressive strength and Young's modulus

SHCC	Young's modulus (GPa)		Compressive strength (MPa)	
	Average	S. D	Average	S. D
Fly ash	20.2	0.4	59.75	2.06
Slag	21.5	0.7	66.04	4.24

3.2 Uniaxial Tensile Test

The load-displacement data obtained from the uniaxial tensile tests were converted into stress-strain curves, using the cross section of the tensile gauge at the center of the specimen (30 mm × 13 mm), as shown in Fig. 2. Each mixture was tested using five specimens. The curves demonstrate that both slag- and fly ash-based SHCCs exhibited clear strain-hardening behavior, which is typical sign of pseudo-ductility in these composites. This behavior is characterized by the continued increase in stress after the initial cracking, enabled by the formation of multiple fine cracks and effective fiber bridging. As expected in SHCCs, variability was observed among the five specimens in terms of ultimate tensile strain. This variation is a commonly reported phenomenon and is typically attributed to the presence of micro-defects, heterogeneities, or minor differences in fiber distribution and orientation within the matrix. From the stress-strain curves, three key mechanical parameters were extracted and averaged for each mix: (1) first cracking stress, (2) ultimate tensile stress, and (3) ultimate tensile strain. The first cracking stress corresponds to the stress level at which the initial crack forms. This is typically identified by either a sudden drop in stress or a deviation from the linear elastic regime into a nonlinear response. This value was further used to define two pre-cracking conditions (70% and 90% of the first cracking stress) for subsequent relaxation tests. The ultimate tensile stress represents the peak load the composite can carry after first cracking, generally occurring during the strain-hardening phase. In this stage, multiple fine cracks are formed, and the embedded fibers effectively bridge these cracks, enabling continued load-bearing capacity. The ultimate tensile strain indicates the end of the strain-hardening phase and the onset of strain-softening, occurring at the peak tensile stress.

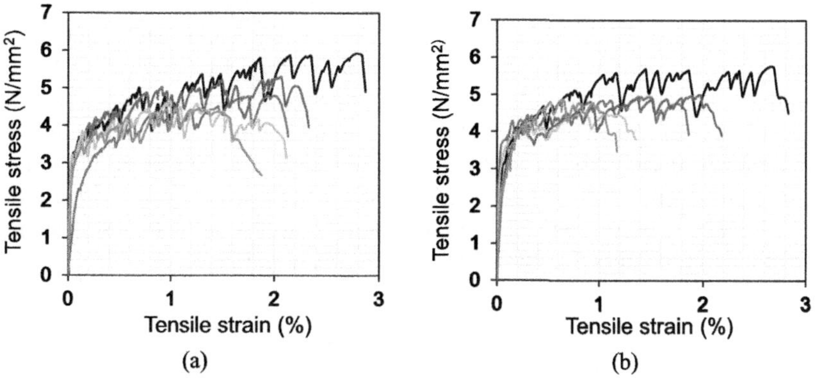

Fig. 2. Stress-strain curves obtained from uniaxial tensile test, (a) BFS, (b) FA

Table 4 presents the average values of the three tensile parameters for both slag- and fly ash-based SHCCs. Slag-based SHCC exhibited higher first cracking stress and ultimate tensile stress compared to its fly ash counterpart. This trend aligned with the compressive strength results and can be attributed to the higher early-age reactivity of slag, which leaded to a denser matrix and stronger bond between the fibers and matrix. These factors improved the ability of the composite to transfer stress and resist cracking.

Table 4. Tensile parameters from uniaxial tensile test

Tensile parameters	Fly ash		Slag	
	Average	S. D	Average	S. D
First cracking stress (N/mm^2)	2.63	0.27	2.87	0.39
Ultimate tensile stress (N/mm^2)	5.02	0.46	5.03	0.58
Ultimate tensile strain (%)	1.54	0.79	1.81	0.70

While both SHCCs exhibited pseudo-ductile behavior, the slag-based mix demonstrates superior tensile strain capacity compared with fly ash-based case with a small margin. This improvement aligns with the crack-width histogram from microscopic observations (Fig. 3), where slag-based specimens exhibited a higher proportion of fine cracks ($<10\ \mu m$). This is likely due to enhanced fiber-matrix bonding and greater fiber-bridging from the higher early-age reactivity of slag, leading to more effective crack control and greater energy dissipation under tensile loading. Similar findings were reported by Hafiz Zadah and Luan [15], observing that high volume slag-based SHCCs (above 70%) exhibited greater ductility than the mixes with 55% fly ash at early ages.

3.3 Stress Relaxation Results

As a preliminary trial, tensile stress relaxation tests were conducted on one slag-based and one fly ash-based SHCC specimen under a constant tensile strain of 0.5% for 116 h. Significant stress relief was observed during the first half hour, after which the rate of stress decay gradually slowed and stabilized. For the slag-based SHCC, the remaining stress was 68.2% of the initial stress at 12 h and further decreased to 63.2% at 116 h. Similarly, the fly ash-based SHCC retained 69.8% of the initial stress at 12 h and 64.7% at 116 h, as shown in Fig. 4. Due to the limited number of available test setups, the duration of each subsequent stress relaxation tests was limited to 12 h in order to shorten the overall testing schedule. Since the difference between this duration and the 116-h value is only about 5%, it is considered that this did not cause a significant underestimation of stress relaxation within the short-term range.

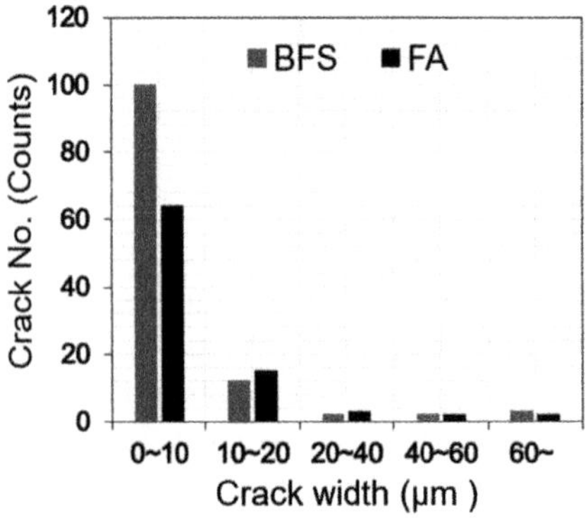

Fig. 3. Crack width distribution

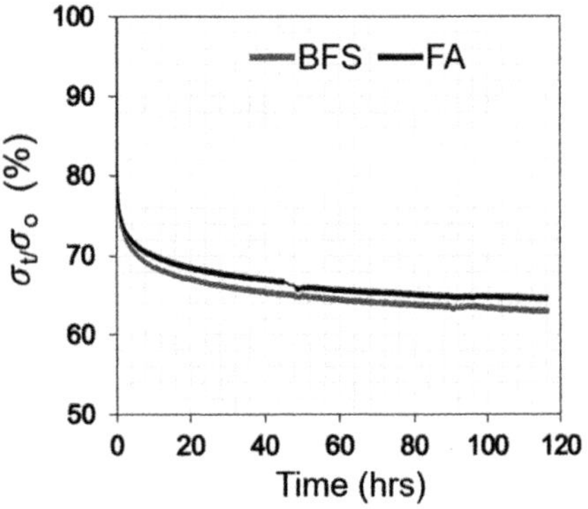

Fig. 4. Stress relaxation at 0.5% strain

As described in the experimental program, stress relaxation was evaluated under two conditions: (1) the pre-cracking condition, with initial applied stresses at 70% and 90% of the average cracking stress; (2) the post-cracking condition, with the initial applied strains of 0.6% and 1.0%.

Pre-Cracking Condition. Figure 5 presents the normalized stress relaxation behavior of the fly ash- and slag-based SHCCs at two predefined stress levels, corresponding to 70% and 90% of the average first-cracking stress. The normalized stress was calculated by dividing the stress at each time point by the initial stress value. At these loading levels, no visible cracks were observed, and the corresponding stress-strain data confirmed that the specimens remained within the elastic range.

Both plots showed a significant stress drop within the first half hour (over 60% of total loss for slag-based and over 50% for fly ash-based SHCCs), followed by a gradual, stable decay. Previous studies on both SHCC and non-SHCC materials [5–9] have reported similar patterns of substantial stress relaxation during the initial hours, followed by slower and more stable decay. This behavior may be attributed to the initiation of invisible microcracks due to the high initial stress-to-strength ratio and the viscoelastic nature of SHCC materials. At the stress levels of 70% and 90%, the slag-based SHCCs exhibited greater stress relaxation and overall stress decay compared to the fly ash-based counterparts. This difference may be attributed to the denser microstructure and higher elastic modulus of the slag-based matrix at early ages, which result in higher initial stress under strain-controlled conditions and thus a higher stress-to-strength ratio. In SHCC systems, such elevated stress levels can activate microcracking and fiber bridging mechanisms earlier, promoting stress redistribution through localized damage and interfacial slip. As a result, greater stress relaxation is observed in the slag-based mixtures. Furthermore, the increased stress decay observed at higher initial stress levels may also reflect more pronounced time-dependent evolution of interfacial friction and bond behavior in the stiffer slag matrix.

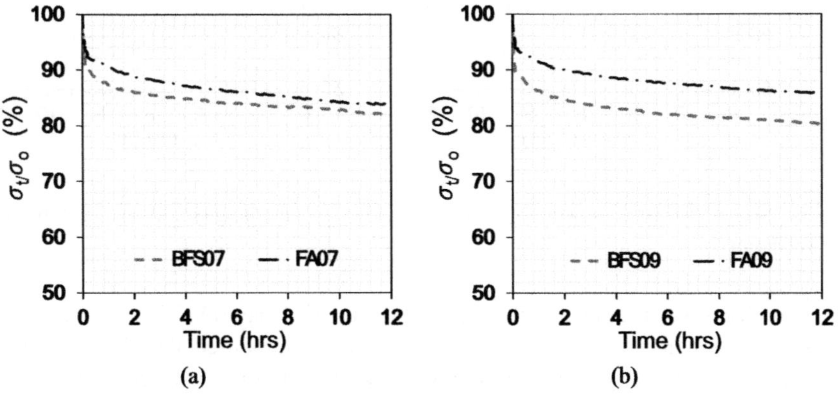

Fig. 5. Normalized stress relaxation of SHCCs under pre-cracking conditions: (a) 70%, (b) 90%

In contrast, the fly ash mix displayed the opposite trend, with the specimen under lower initial stress (FA07) showing more relaxation than that under higher stress (FA09). The reason for this unexpected behavior remains unclear and requires further investigation. In both the slag and fly ash cases under the pre-cracking condition, the ultimate stress loss appears to be primarily influenced by a more pronounced initial stress drop. Following this early stage, all specimens exhibited a similar trend characterized by a slow, stable, and gradual stress decay.

Post-Cracking Condition. Figure 6 shows the normalized stress relaxation behavior of the SHCC specimens over a 12-h period at the two predefined strain levels of 0.6% and 1.0%. Both fly ash- and slag-based SHCCs exhibited a large and rapid drop in tensile stress at the beginning of the relaxation test, with around 60–70% and 80–90% of total loss occurring in the first half hour, respectively. However, the rate of stress decay gradually slowed over time, eventually reaching a steady and gradual relaxation phase. Beushausen et al. [5] studied stress relaxation in plain cement mortars and found that approximately 80% of the total relaxation occurred within the first 12 h. Similarly, Kizito et al. [6] reported that about 74% of the total relaxation in concrete occurred within the first 5 h. This rapid early-age stress decay may be attributed to the formation of microcracks resulting from a high initial stress-to-strength ratio. Comparable trends were observed in tensile creep tests on SHCC materials. For instance, Ratnayake et al.

[9] simulated SHCC creep in practical repair scenarios and recorded substantial stress decay within the first four hours followed by a slower and more gradual stress reduction. The behavior may be due to the new microcracks initiation, viscoelastic behavior of the SHCC, high stress-to-strength ratios, and fiber pull-out mechanisms.

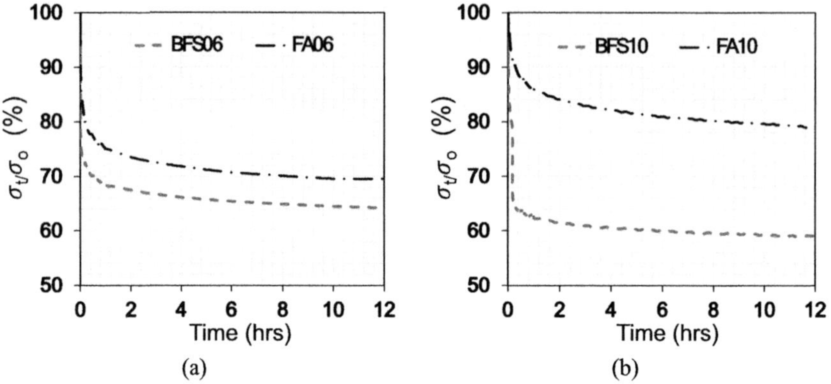

Fig. 6. Normalized stress relaxation of SHCCs under post-cracking conditions: (a) 0.6% strain, (b) 1.0% strain

In both cases of the post-cracking condition, the slag-based SHCCs exhibited greater stress relaxation than their fly ash-based counterparts. Slag-based SHCCs tend to develop a denser matrix and higher strength, which may lead to higher internal stresses under sustained loading and potentially promote the formation of more distributed microcracks. In the post-cracking phase, stress relaxation is strongly influenced by the behavior of the fiber-matrix interface near the cracks. Under sustained loading, the increased internal stress in the slag matrix may intensify interfacial slip and partial fiber debonding at the crack interfaces. This progressive degradation of the fiber-matrix bond reduces the load transfer efficiency across the cracks, thereby accelerating stress relaxation in the slag-based mixtures. Notably, the formation of new microcracks during prolonged loading may further contribute to the observed stress relaxation in slag-based systems [16, 17]. Furthermore, stress relaxation in slag-based mixtures increased with higher strain levels, as seen in BFS10, which exhibited more stress decay than BFS06. This trend aligns with the strain-hardening nature of SHCC, where higher strain levels correspond to higher stress states, resulting in more active interfacial damage and time-dependent redistribution. In contrast, the fly ash-based mixtures showed the opposite trend, as FA06 experienced greater stress decay than FA10. A similar pattern was observed under pre-cracking conditions, where stress relaxation increased with higher initial stress in the slag-based mixtures but decreased in the fly ash-based ones. This reverse trend in fly ash mixtures may be associated with early microcrack formation or differences in interface quality, though further investigation is needed to clarify the underlying mechanism.

Figure 7 and Fig. 8 present the strain fields and crack patterns of the slag-based (BFS06) and fly ash-based (FA06) SHCCs, obtained using digital image correlation (DIC) at a tensile strain level of 0.6%. The results are shown at three time points of 0,

6, and 12 h from the start of the relaxation test. In both cases, no measurable changes were detected in the width, spacing, or propagation of existing cracks over the relaxation period. However, a new microcrack was initiated and evolved at the early stages in both materials. While both SHCC materials exhibited the formation of a new microcrack, the crack observed in BFS06 was more pronounced than that in the FA06. This indicates enhanced potential for time-dependent crack initiation and development in the slag-based mixture, contributing to its higher degree of stress relief during relaxation. Such time-dependent tensile strain and reduction of stress, triggered by delayed microcrack formation under sustained tensile loading, has also been reported by Boshoff et al. [16], who attributed the phenomenon to time-dependent fiber pull-out or fiber creep. These findings are consistent with the tensile stress relaxation results, reinforcing the correlation between time-dependent cracking and stress decay in slag-based SHCCs.

Figure 9 and Fig. 10 illustrate the evolution of strain fields and crack patterns in slag-based (BFS10) and fly ash-based (FA10) SHCCs, respectively, under a tensile strain of 1.0%. In BFS10, although no significant changes were observed in the width, length, or propagation of existing cracks during the 12-h relaxation period, a new crack was detected approximately 12 min after the start of the test, followed by a stable crack pattern without further new formation. This new crack coincided with a secondary stress drop in the relaxation curve of BFS10, after which the stress decay became gradual and steady, as shown in Fig. 6. Sahar et al. [17] reported that the formation of new cracks and time-dependent fiber pull-out are significant contributors to the time-dependent deformation of SHCC under sustained loading. In contrast, for the FA10 specimen, no visible evolution in crack width, length, or spacing was observed, and no new cracks developed during the relaxation period. Interestingly, the lower-strain FA06 specimen exhibited new crack formation, as seen in Fig. 8. This behavior aligns with the relaxation data measured by displacement transducers, suggesting that unlike slag-based mixtures, fly ash-based SHCCs may exhibit greater relaxation at lower strain levels. This reverse trend in the FA mixture may be related to early microcrack formation or local differences in fiber-matrix interface quality. Additionally, heterogeneity in fiber distribution or early-age porosity may have facilitated earlier interface degradation in FA06. However, these hypotheses remain speculative and require further investigation.

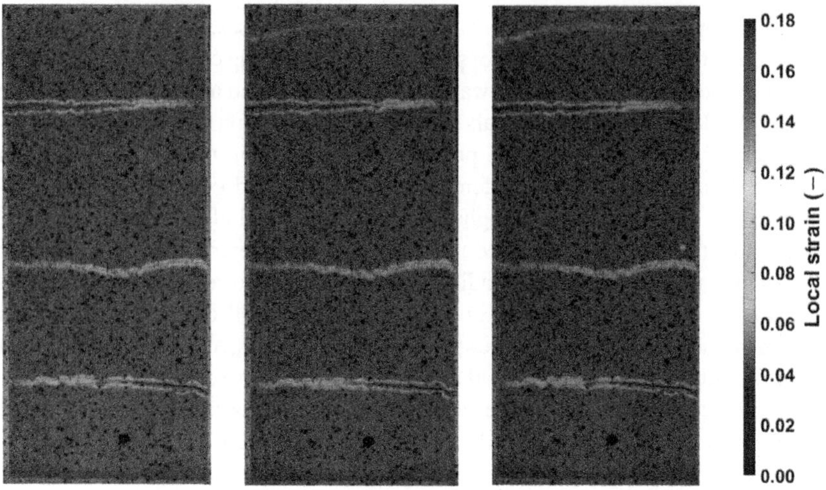

Fig. 7. Strain fields and crack patterns of BFS06 at 0, 6, and 12 h at 0.6% strain.

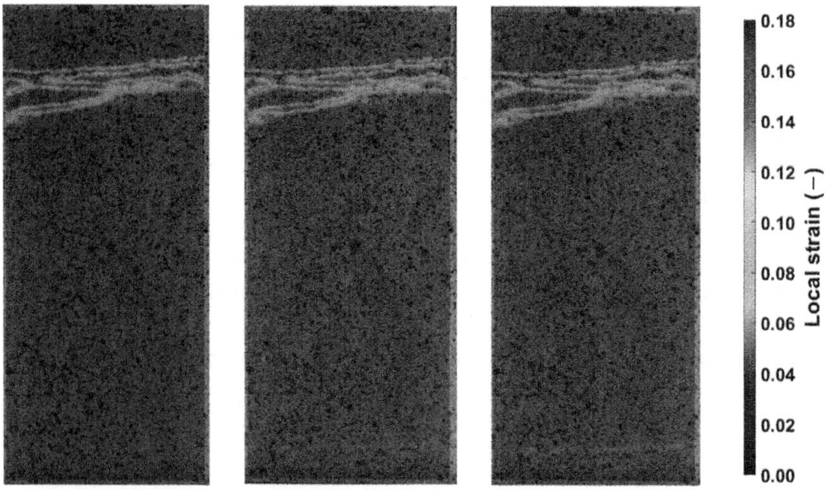

Fig. 8. Strain fields and crack patterns of FA06 at 0, 6, and 12 h at 0.6% strain.

While DIC provides valuable insights into visible cracking behavior, it should be noted that its spatial resolution may limit the detection of sub-microscopic damage or interfacial slip, which could also contribute to relaxation. On the other hand, utilizing advanced microstructural analysis techniques, such as X-ray micro-computed tomography (X-ray μCT) or scanning electron microscopy (SEM), to detect early microcrack formation and evaluate differences in interface quality may yield additional insights and elucidate the fundamental mechanisms governing the material's behavior.

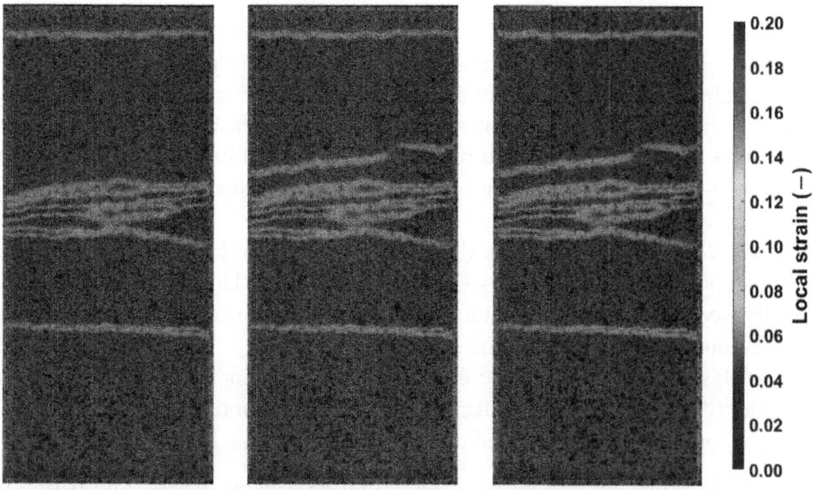

Fig. 9. Strain fields and crack patterns of BFS10 at 0, 6, and 12 h at 1.0% strain.

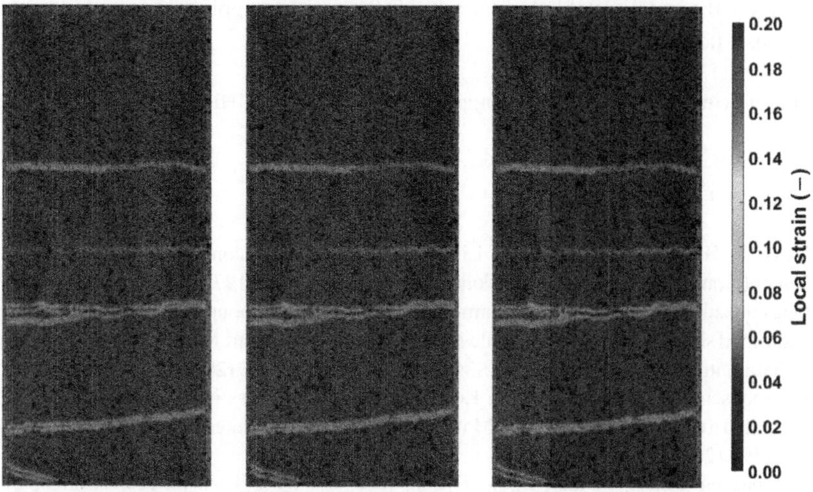

Fig. 10. Strain fields and crack patterns of FA10 at 0, 6, and 12 h at 1.0% strain.

4 Conclusions

This study investigated the uniaxial tensile stress relaxation and cracking behavior of slag- and fly ash-based SHCCs under sealed curing, with the aim of understanding their time-dependent tensile response and microcrack evolution mechanisms. Mechanical and time-dependent performance were evaluated through compressive, tensile, and stress relaxation tests under both pre- and post-cracking conditions. Crack evolution was monitored via DIC. The key findings are summarized as follows:

1) Slag- and fly ash-based SHCCs exhibited similar compressive strength and Young's modulus, with slag showing marginally higher values.

2) Both slag- and fly ash-based SHCCs exhibited ductile and strain-hardening behavior, while the slag-based mix showed slightly improved tensile performance. This may be attributed to a denser and more viscoelastic C-S-H matrix that contributed to enhanced fiber-matrix interaction and more extensive fine cracking prior to failure.
3) Substantial stress relaxation occurred within the first half hour in all SHCCs (approximately 50–70% for fly ash- and 60–90% for slag-based mixes), followed by a slower, more stable phase thereafter.
4) Slag-based SHCCs consistently exhibited greater stress relaxation than fly ash-based SHCCs under both pre- and post-cracking conditions. However, the fly ash-based mix showed greater relaxation at lower stress levels, indicating a reverse trend and a distinct time-dependent relaxation behavior.
5) DIC analysis showed that while existing cracks remained largely stable in width and intensity, new microcracks occasionally initiated, more frequently in slag-based SHCCs, contributing to increased deformation and stress decay.

While this study highlights the importance of time-dependent SHCC properties influenced by material type, further microstructural investigations using techniques such as X-ray μCT and SEM are required to better understand the underlying mechanisms and identify the most suitable binder system for real-life applications under restrained conditions, such as patch repair.

Acknowledgements. This work was supported by JSPS KAKENHI Grant Number 22K04257.

References

1. Yang, E.-H., Wang, S., Yang, Y., Li, V.C.: Fiber-bridging constitutive law of engineered cementitious composites. J. Adv. Concr. Technol. **6**(1), 181–193 (2007)
2. Ghasemzadeh, F., Harsini, I., Mohammadi, B.: Behavior of bonded concrete overlays under restrained shrinkage. In: First Middle East Conference on Smart Monitoring, Assessment and Rehabilitation of Civil Structures (SMAR2011), Dubai, UAE (2011)
3. Beushausen, H., Alexander, M.G.: Localised strain and stress in bonded concrete overlays subjected to differential shrinkage. Materials and Structures/Materiaux et Constructions **40**(2), 189–199 (2007)
4. Beushausen, H., Alexander, M.G.: Failure mechanisms and tensile relaxation of bonded concrete overlays subjected to differential shrinkage. Cem. Concr. Res. **36**(10), 1908–1914 (2006)
5. Beushausen, H., Masuku, C., Moyo, P.: Relaxation characteristics of cement mortar subjected to tensile strain. Mater. Struct. **45**(1), 1181–1188 (2012)
6. Kizito, N.M., Beushausen, H., Moyo, P., Arito, P.: An experimental study on the creep and relaxation behaviour of concrete. In: Second International Conference on Advances in Cement and Concrete Technology in Africa (ACCTA 2016), Dar es Salaam, Tanzania (2016)
7. Boshoff, W.P.: Cracking behavior of strain-hardening cement-based composites subjected to sustained tensile loading. ACI Mater. J. **111**(5), 553–560 (2014)
8. Boshoff, W.P., Adendorff, C.J.: Effect of sustained tensile loading on SHCC crack widths. Cement Concr. Compos. **37**(1), 119–125 (2013)
9. Ratnayake, K.A.S.D., Leung, C.K.Y.: A novel load-sharing system to simulate the creep of strain-hardening cementitious composites (SHCCs) in practical situations. Materials **17**(10), 1–19 (2024)

10. Ratnayake, K.A.S.D., Leung, C.K.Y.: Cracking behaviour under creep in strain-hardening cementitious composites (SHCC) applied as a protective layer on reinforced concrete flexural elements. In: 11th International Conference on Fracture Mechanics of Concrete and Concrete Structures (FraMCoS-11), International Association for Fracture Mechanics of Concrete and Concrete Structures, pp. 1–8 (2024)
11. Li, V.C., Kanda, T.: Engineered cementitious composites for structural applications. ASCE J. Materials in Civil Engineering **10**(2), 66–69 (1998)
12. Wang, T., et al.: Consideration of coupling of crack development and corrosion in assessing the reliability of reinforced concrete beams subjected to bending. Reliab. Eng. Syst. Saf. **233**, 1–20 (2023)
13. JSCE. Recommendations for design and construction of high-performance fiber reinforced cement composite with multiple fine cracks (HPFRCC). Concrete Library 127 (2007)
14. Yang, J., Bhattacharya, K.: Augmented Lagrangian digital image correlation. Exp. Mech. **59**(2), 187–205 (2019)
15. Hafiz Zadah, F., Luan, Y.: Study on strain-hardening cementitious composites with high volume slag: uniaxial tensile performance and chloride migration. In: 47th JCI Technical Conference **47**(1), 1164–1169 (2025). Japan Concrete Institute (JCI), Morioka, Japan
16. Boshoff, W.P., Zijl, G.P.A.G.V.: Time-dependent response of ECC: characterization of creep and rate dependence. Cem. Concr. Res. **37**(5), 725–734 (2007)
17. Sahar, U.-U., Tsubaki, T.: Modelling of time-dependent behavior of SHCC by using multi-layered microplane model. Cement Science and Concrete Technology **69**, 580–587 (2015)

Hygrothermal Behaviour

Assessment of the Hygrothermal Performance of Prefabricated Concrete Panel Materials for Modular Construction

Raquel Sarmento(✉), Henriqueta Teixeira, Joana Silva, and Maria da Glória Gomes

CERIS, Department of Civil Engineering, Architecture and Georesources, Instituto Superior Técnico, Universidade de Lisboa, Av Rovisco Pais, 1049-001 Lisbon, Portugal

Abstract. The increasing housing shortage in Portugal, alongside the industrialisation of the construction sector, has driven growing interest in modular construction. This innovative method offers faster building times, cost efficiency, and reduced waste. However, despite its advantages, aspects such as the thermal performance of modular systems have often been overlooked in previous studies. This highlights the need for a deeper understanding of how modular construction materials can be optimised to address these challenges.

This research focuses on the hygrothermal behaviour of various concrete compositions through the investigation of their performance under different environmental conditions, with an emphasis on relative humidity and thermal conductivity. Experimental tests were conducted to provide critical insights into how different concrete compositions react to temperature and moisture fluctuations. The results revealed that lower water content in concrete leads to a decrease in thermal conductivity.

The results of this research can help optimize the design of prefabricated concrete panels to enhance thermal efficiency and hygrothermal performance. This would ensure more stable and comfortable indoor environments while minimizing energy consumption in modular buildings. By aligning material performance with sustainability goals, the research aims to support the broader objective of creating environmentally responsible, energy-efficient housing solutions.

Ultimately, this work seeks to advance the understanding of concrete's hygrothermal behaviour, contributing to the development of sustainable practices and ensuring modular construction achieves a balance between performance, occupant comfort, and environmental sustainability.

Keywords: Hygrothermal Behaviour · Thermal Performance · Prefabricated Concrete Sandwich Panels · Modular Construction

1 Introduction

Portugal faces a growing housing shortage driven by insufficient political intervention, urban population growth, and the expansion of short-term rentals, all contributing to rising house prices. Coupled with relatively low-income levels, these factors have

M. Posani and S. Singh (Eds.): RYS 2025, RILEM Bookseries 67, pp. 927–936, 2026.
https://doi.org/10.1007/978-3-032-14652-6_67

made affordable housing increasingly inaccessible [1, 2]. Modular construction offers a promising solution by industrialising the building process, reducing construction time and costs, improving quality control, and minimising waste and environmental impact [3]. However, significant financial, regulatory, technical, and cultural barriers remain [4, 5].

Given the technical and regulatory challenges still surrounding modular construction, understanding its thermal performance is crucial—particularly as national and European standards continue to evolve. Improving thermal behaviour can further reduce environmental impact and energy consumption, which is especially relevant in the context of climate change [6]. Factory-controlled environments also enable greater precision in building envelope construction, potentially reducing thermal bridging and enhancing energy efficiency compared to traditional on-site methods.

In modular construction, the use of Prefabricated Concrete Sandwich Panels (PCSPs) is common, with numerous studies highlighting their effectiveness in achieving thermally efficient façades [7]. These panels typically consist of two concrete wythes connected by shear connectors—often made of steel—and a thermal insulation core, as seen in Fig. 1.

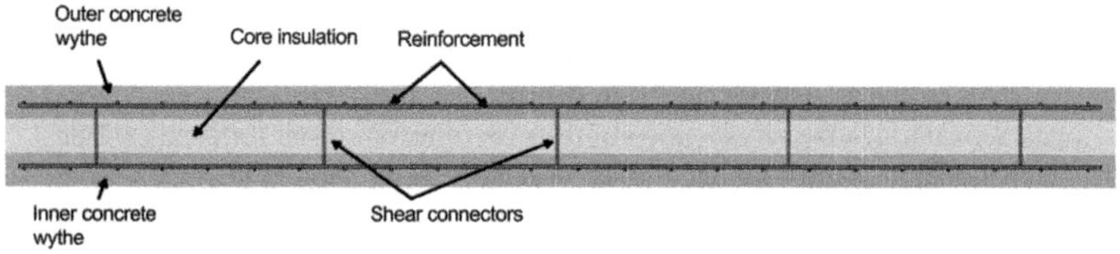

Fig. 1. Representation of a Prefabricated Concrete Sandwich Panel.

The common thermal reoccurring problem of PCSP depicted in the literature is the thermal bridge effect due to the steel sheer connectors, which can be solved with the use of Fiber Reinforced Polymer (FRP) due to its low thermal conductivity [7–10] or by staggering the shear connectors in order to provide a longer path for heat to travel, increasing the thermal resistance [11]. Although thermal bridging at the steel shear connectors in PCSPs is widely documented in the literature, it is important to acknowledge that other types of thermal bridges—such as those occurring at module junctions, corners, or interfaces with other construction elements—can also significantly affect the overall thermal performance of modular buildings and should be equally considered [12]. The composition of the used concrete, including aggregate type, water-to-cement ratio, and admixtures, may also significantly influence its thermal conductivity and moisture behaviour [13]. Similarly, the type and thickness of thermal insulation can impact not only thermal resistance but also moisture buffering and drying potential [14, 15].

Despite the growing body of research on PCSPs, a comprehensive hygrothermal assessment of these systems remains lacking. It is therefore essential to evaluate how the composition of the concrete layers affects the overall hygrothermal behaviour of the panels, especially in relation to risks such as mould growth and freeze-thaw cycles, which may compromise durability and occupant comfort [16–18].

In this context, the present study aims to investigate the hygrothermal properties of different concrete compositions developed within the scope of the "R2UTechnologies – Modular Constructions" project [19]. An experimental campaign will be conducted to characterise the relevant thermal and moisture-related properties under controlled conditions.

2 Experimental Campaign

In this study, the hygrothermal performance of different concrete compositions was assessed only through sorption curve analysis due to time constraints. Since the concrete compositions were developed within the framework of the "R2UTechnologies – Modular Constructions" project [19], the detailed composition of some concretes mixes is not available. Therefore, their thermophysical properties are evaluated solely based on their hygrothermal behaviour under varying relative humidity (RH) conditions, considering their water-to-cement (w/c) ratio (Table 1), and knowing that all samples use Portland cement. Furthermore, as for the aggregates, all concrete samples have coarse and fine gravel, the first column of Table 1, referring to CP, 25RCP, 30CC, 30CC15LFF, and 30BFA15LFF, include coarse and fine sand in their compositions, while the rest only fine sand.

Table 1. Different types of concrete under study.

Concrete ID	w/c	Concrete ID	w/c	Concrete ID	w/c
CP	0.45	BF1	0.45	BF8	0.46
25RCP	0.60	BF3	0.36	V1	0.42
30CC	0.64	BF5	0.37	V2	0.39
30CC15LFF	0.80	BF6	0.39		
30BFA15LFF	0.82	BF7	0.53		

For the hygroscopic tests, measurements were performed at different moisture content levels, following the procedures defined in ISO 12570 [20] and ISO 12571 [21]. The water content of a material in contact with air can vary from absolute zero (under 0% RH) to the critical moisture content reached after prolonged exposure to saturated conditions. In this study, RH setpoints of 15%, 35%, 50%, 75%, and 95% were selected to cover a broad range of environmental conditions.

Although ISO 12571 [21] recommends conducting measurements between 35% and 95% RH to characterise the hygroscopic region, an additional point at 15% RH was included to explore material behaviour under extremely dry conditions. Including this point might provide further insight into the lower boundary of moisture uptake [22].

As stated in ISO 12571 [21], before initiating sorption measurements, each sample was dried in an oven at 60 ± 5 °C until constant mass was achieved, as shown in Fig. 2. Constant mass is defined as a mass variation of no more than 0.1% across three consecutive weighings taken at 24-h intervals.

Fig. 2. Concrete samples undergoing oven drying at 60 ± 5 °C until achieving constant mass, in accordance with ISO 12571 [21].

Following this drying phase, the sorption curve measurements were initiated. The specimens were placed in a climatic chamber (Fig. 3) maintained at a constant temperature of 23 ± 0.5 °C and exposed to each RH setpoint until mass stabilisation was observed, indicating equilibrium conditions.

Fig. 3. Samples placed in a climatic chamber at 23 ± 0.5 °C and exposed to different relative humidity setpoints for sorption curve measurements.

At each RH level, thermophysical properties were also measured, including thermal conductivity (λ) [W/(m·K)], volumetric heat capacity (c_p) [$\times 10^6$ J/(m^3·K)], and thermal diffusivity (α) [m^2/s] using the transient method with the Isomet 2114 device [23]. Since the samples were rigid, a surface (planar) probe was used. Each measurement lasted approximately 30 min at a controlled temperature of 20 ± 3 °C. The test procedure followed ASTM D5334-22 [24], ASTM D5930-17 [25], and ISO 22007-2 [26] standards. However, as transient methods such as the one employed may present limited reliability under certain conditions, particularly in heterogeneous or moisture-sensitive materials, complementary validation using steady-state heat flux measurements should be considered in future work to reinforce the robustness of the results.

The method evaluates the thermal response of each material to a controlled heat pulse generated by an internal electrical resistance within the probe. A surface (planar) probe was employed due to its suitability for testing rigid materials such as concrete. Given the small cylindrical geometry of the specimens (Fig. 4), measurements were taken at multiple locations—specifically, the top surface, the bottom, and once more on the top surface near the centre—to account for potential variability across the specimen and to ensure representative results. Measurements were conducted closer to the centre in order to minimise edge-related heat losses.

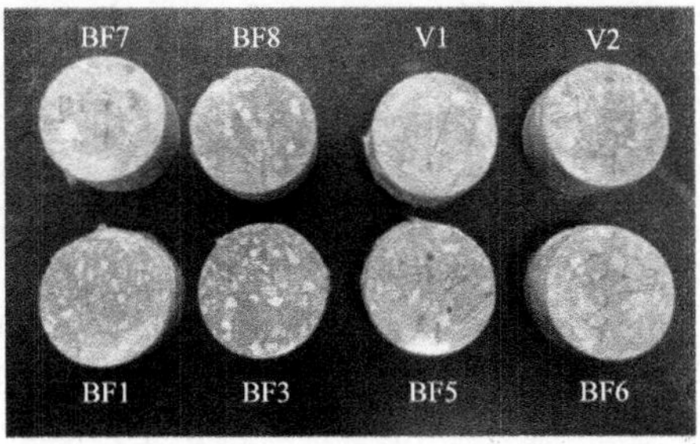

Fig. 4. Cylindrical concrete specimens used for thermal property measurements.

To further reduce thermal exchange with surrounding surfaces, specimens were placed on an insulating board during testing, as shown in Fig. 5. Additionally, each sample was wrapped in cling film to prevent moisture exchange with the surrounding environment during measurements.

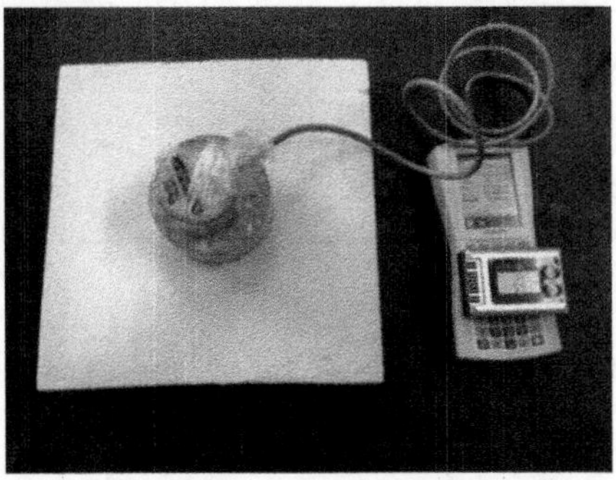

Fig. 5. Experimental setup for thermal property measurements using the Isomet 2114 device with a surface probe [23].

Each thermophysical property was measured three times, and the final reported value corresponds to the average of these three measurements. The results obtained from these measurements will be presented and discussed in the following chapters.

3 Results

This section presents the results obtained from the sorption experiments conducted on the concrete samples.

The sorption curves presented in Fig. 6 illustrate the relationship between equilibrium moisture content and relative humidity (RH) at a constant temperature. The measurements were carried out in accordance with ISO 12571 [21], and the values for moisture content by volume (Ψ) [m3/m3] were determined following ISO 12570 [20]. The selected RH setpoints—15%, 35%, 50%, 75%, and 95%—were chosen to capture both the lower and upper bounds of moisture adsorption. The results reflect the moisture uptake capacity of each sample and highlight differences in sorption behaviour.

As shown in Fig. 6, the sample of concrete 30CC15LFF exhibits the lowest sorption curve, indicating that it absorbs and retains less water vapour from the environment than the other concrete samples. This behaviour is unexpected, given its relatively high water-to-cement (w/c) ratio of 0.80, which would typically correspond to increased porosity and, consequently, greater moisture uptake.

On the other end of the spectrum, sample BF7 displays the highest sorption curve, despite having a moderate w/c ratio of 0.53, which is not the highest value among the tested mixtures.

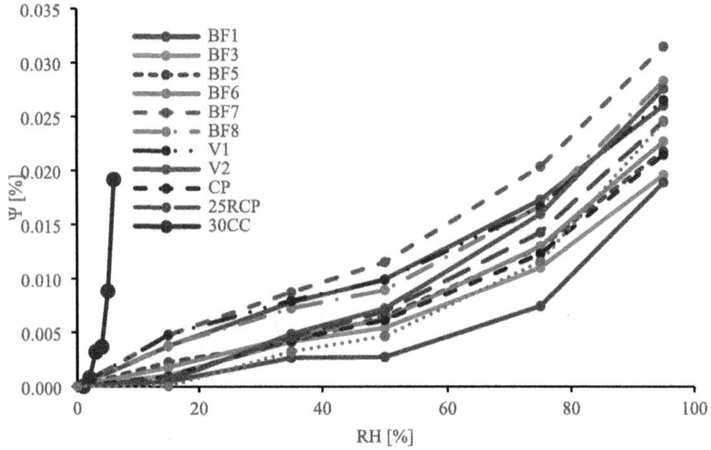

Fig. 6. Sorption curve of moisture content (Ψ) [m^3/m^3] as a function of RH [%] at constant temperature.

These results suggest that moisture uptake cannot be explained by the w/c ratio alone. Other factors, such as the type and dosage of additives or fillers, aggregate dimensions, the degree of hydration, pore structure, or curing conditions, play a significant role in defining the hygroscopic behaviour of the concrete. This highlights the complexity of moisture transport in cementitious materials [13].

The results for thermal conductivity (Fig. 7), generally exhibit the expected trend of increasing values with rising relative humidity. As moisture is absorbed into the concrete's pore structure, its thermal conductivity increases, indicating a decline in its insulating capacity.

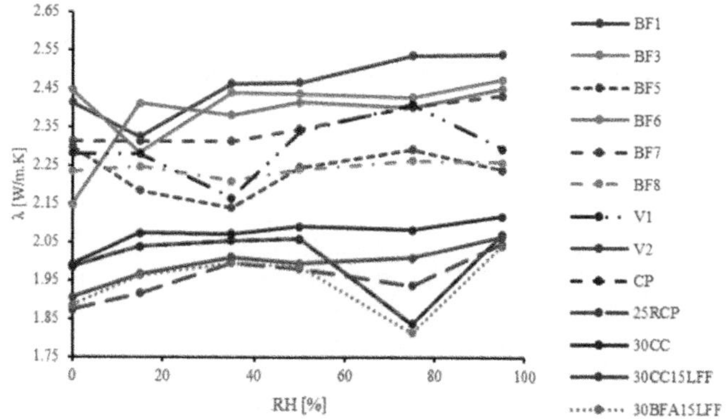

Fig. 7. Sorption curve of thermal conductivity λ [W/(m•K)] as a function of RH [%].

In Fig. 8, it can be observed that the volumetric heat capacity of the concrete samples increases with rising RH. This trend indicates that the materials' ability to store thermal energy is enhanced by the presence of moisture. Among the samples, 25RCP exhibits the highest sensitivity to humidity changes, as reflected by its notably elevated values of volumetric heat capacity.

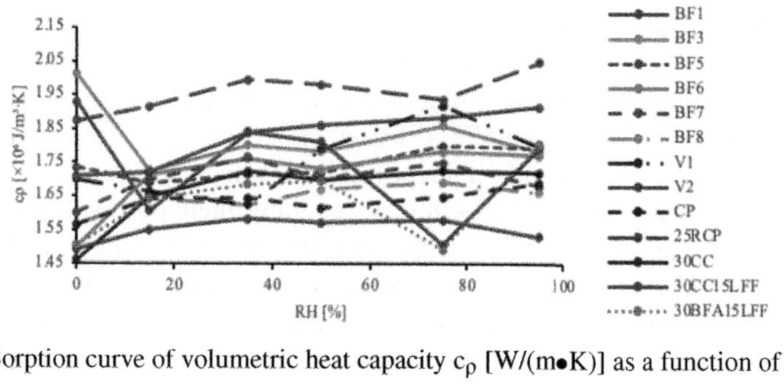

Fig. 8. Sorption curve of volumetric heat capacity c_p [W/(m•K)] as a function of RH [%].

Figure 9 does not reveal a clear trend in thermal diffusivity across the different relative humidity levels. The values fluctuate for all samples, showing both increases and decreases without a consistent pattern, indicating an unstable or non-monotonic behaviour. The observed variability could also suggest that the microstructural response of each concrete composition to moisture is non-linear, depending on factors such as porosity, pore connectivity, and aggregate-matrix interaction. It is also possible that minor experimental variability or sensitivity of the transient method at certain moisture levels contributes to this irregular behaviour, particularly given the relatively small size of the specimens. Further investigation, including more RH setpoints and complementary microstructural characterisation techniques, such as mercury intrusion porosimetry (MIP), may be necessary to better understand the underlying causes of these fluctuations.

In conclusion, the pore network plays a critical role in the uptake and retention of moisture, which directly influences heat transfer processes and thermal storage within the material. A more detailed understanding of these microstructure should be pursued in future works to support the interpretation of thermal behaviour.

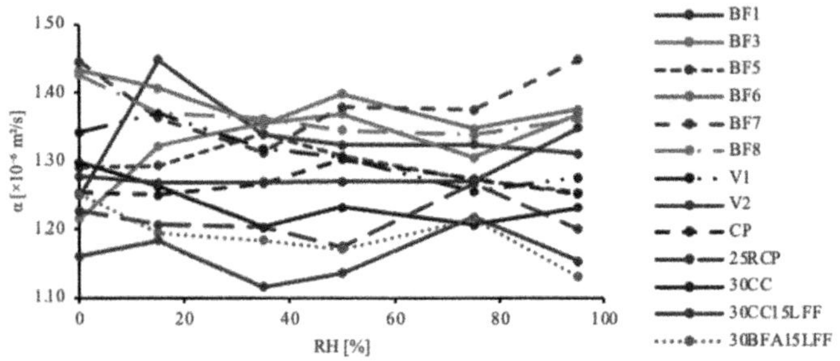

Fig. 9. Sorption curve of thermal diffusivity α [m^2/s] as a function of RH [%].

4 Discussion

This study assessed the hygrothermal behaviour of different concrete compositions by analysing their sorption characteristics and key thermophysical properties under varying RH conditions. The findings demonstrate that moisture uptake, as reflected by the sorption curves, does not correlate solely with the w/c ratio. For instance, although sample 30CC15LFF exhibited the highest w/c ratio, it showed the lowest sorption capacity, whereas sample BF7, with a lower w/c ratio, displayed the highest sorption capacity. These results highlight the multifactorial nature of moisture transport in cementitious materials, where factors such as additive dosage, aggregate dimensions, pore structure, and curing conditions significantly influence hygroscopic behaviour.

The thermal conductivity of all samples generally increased with RH, indicating reduced insulating performance due to moisture absorption. Similarly, volumetric heat capacity also increased with RH, confirming that the presence of moisture enhances the material's thermal energy storage capacity. Among the samples, 25RCP exhibited the highest sensitivity to humidity variations in this respect.

In contrast, thermal diffusivity did not exhibit a consistent trend across the RH range. The irregular fluctuations observed suggest complex interactions between moisture and the concrete microstructure, potentially influenced by by variations in pore connectivity, matrix-aggregate interfaces, and microstructural heterogeneity. These inconsistencies may also be partially attributed to limitations inherent in the transient measurement method. Steady-state methods, where heat flux can be directly measured, should be considered to obtain more reliable and comprehensive conclusions. Further studies involving a broader RH range and complementary characterisation techniques are recommended to clarify the mechanisms governing thermal diffusivity in moist concrete.

Overall, the results underscore the importance of incorporating moisture-dependent thermophysical behaviour when evaluating the performance of concrete in energy-efficient and durable construction applications. In particular, understanding these parameters is critical for the accurate development of hygrothermal simulations of PCSP, as they enable more reliable predictions of moisture transport and thermal behaviour under real-world boundary conditions. Such simulations can support the early identification of potential issues such as condensation, thermal bridging, and long-term durability concerns within the building envelope.

References

1. Lorga, M., Januário, J.F., Cruz, C.O.: Housing affordability, public policy and economic dynamics: an analysis of the city of Lisbon. J. Risk Financ. Manage. **15**(12), 560 (2022)
2. Santos, A.D., Castro, H.: Housing and setting constraints: the Portuguese evidence. Sustainability **14**(18), 11720 (2022)
3. Hořínková, D.: Advantages and disadvantages of modular construction, including environmental impacts. IOP Conf. Ser. Mater. Sci. Eng. **1203**(3), 032002 (2021)
4. Parisi, L., Donyavi, S.: Modular momentum: assessing the efficacy of modular construction in alleviating the UK housing crisis. Discov. Appl. Sci. **6**(10), 548 (2024)
5. Khan, A.A., Amirkhani, M., Martek, I.: Overcoming deterrents to modular construction in affordable housing: a systematic review. Sustainability **16**(17), 7611 (2024)
6. Zhou, A., Wong, K., Lau, D.: Thermal insulating concrete wall panel design for sustainable built environment. Sci. World J. **2014**, 279592 (2014)
7. Pozo-Lora, F.F., Sorensen, T.J., Al-Rubaye, S., Maguire, M.: State-of-the-art and practice review in concrete sandwich wall panels: materials, design, and construction methods. Sustainability **17**(8), 3704 (2025)
8. Tawil, H., Tan, C.G., Sulong, N.H.R., et al.: Mechanical and thermal properties of composite precast concrete sandwich panels: a review. Buildings **12**(9), 1429 (2022)
9. Li, G., Li, X., Fang, C., Liu, R.: Thermal performance of precast concrete sandwich walls with a double-layer insulation system. Constr. Build. Mater. **472**, 140785 (2025)
10. Woltman, G., Noel, M., Fam, A.: Experimental and numerical investigations of thermal properties of insulated concrete sandwich panels with fiberglass shear connectors. Energ. Build. **14**, 22–31 (2017)
11. Bida, S.M., Abdul Aziz, F.N.A., Jaafar, M.S., et al.: Thermal resistance of insulated precast concrete sandwich panels. Int. J. Concr. Struct. Mater. **15**(1), 41 (2021)
12. Sorensen, T., Dorafshan, S., Maguire, M.: Thermal evaluation of common locations of heat loss in sandwich wall panels. In: Proceedings of the 17th International Conference on Cold Regions Engineering, pp. 173–184 (2017)
13. Gomes, M.G., et al.: Thermophysical properties under different hygroscopic conditions of an innovative composite concrete pre-walls system. Constr. Build. Mater. **307**, 124938 (2021)
14. Cascione, V., et al.: A review of moisture buffering capacity: from laboratory testing to full-scale measurement. Constr. Build. Mater. **200**, 333–343 (2019)
15. Zhang, M., et al.: Moisture buffering phenomenon and its impact on building energy consumption. Appl. Therm. Eng. **124**, 337–345 (2017)
16. Lee, S.-T., et al.: Effect of freeze-thaw cycles on the performance of concrete containing water-cooled and air-cooled slag. Appl. Sci. **11**(16), 7291 (2021)

17. Xue, Y., et al.: The moisture distribution in wall-to-floor thermal bridges and its influence on mould growth. UCL Open Environ. **4** (2022)
18. Zhou, X., Derome, D., Carmeliet, J.: Hygrothermal modeling and evaluation of freeze-thaw damage risk of masonry walls retrofitted with internal insulation. Build. Environ. **125**, 285–298 (2017)
19. R2UTechnologies – Modular Construction Homepage. https://r2utechnologies.pt. Accessed 20 May 2025
20. European Committee for Standardization: EN ISO 12570:2000/A2:2018 - Hygrothermal performance of building materials and products - Determination of moisture content by drying at elevated temperature (2000)
21. European Committee for Standardization: EN ISO 12571:2021 - Hygrothermal performance of building materials and products - Determination of hygroscopic sorption properties (2021)
22. Ranefjärd, O., Strandberg-de Bruijn, P.B., Wadsö, L.: Hygrothermal properties and performance of bio-based insulation materials locally sourced in Sweden. Materials **17**(9) (2024)
23. L. Applied Precision, ISOMET 2114 Thermal properties analyzer user's guide, Version 120712
24. Advancing Standards Transforming Markets: ASTM D5334-22 - Standard test method for determination of thermal conductivity of soil and soft rock by thermal needle probe procedure (2022)
25. Advancing Standards Transforming Markets: ASTM D5930-17 Standard test method for thermal conductivity of plastics by means of a f line-source technique (2017)
26. International Organization for Standardization: ISO 22007-2:2022 - Plastics - Determination of thermal conductivity and thermal diffusivity - Part 2: Transient plane heat source (hot disc) method (2022)

Effective Water Vapor Permeability Estimation of Earthen Materials from Dynamic Vapor Sorption/Desorption Test

Habib Jalili[1,2(✉)], Tariq Ouahbi[1], Said Taibi[1], Sarah Nassar[2], Joanna Eid[1,2], and Ichrak Hamrouni[1]

[1] Université Le Havre Normandie, Normandie Univ, LOMC, UMR 6294 CNRS, 76600 Le Havre, France

[2] R&D Department, AI Environment, FACEA Group, Noisy le Grand, France

Abstract. Cup method based on the EN ISO 12572 is the widely used method to measure the water vapor permeability in porous building materials. The cup test is categorized into two ranges: dry and wet conditions. This technique evaluates the vapor pressure gradient across the sample, and the entire cup is assessed by monitoring its weight over time. While the external environment is precisely regulated by climate chamber, appropriate solid desiccants or saturated salt solutions are used to control the interior of cup. Dynamic vapor sorption (DVS) method provides distinct advantages because of the highly reproducible data over the wide range of the relative humidity as a function of time. DVS method was utilized to measure sorption hysteresis and equilibrium water content and kinetic of sorption. In this paper, we applied the macroscopic model to analyze the dynamic vapor sorption in an earthen brick with taking into account the sorption hysteresis. The effective water vapor permeability calculated by the optimization method in our model and compared against the standard cup method test according to EN ISO 12572. The results showed that applied model is able to predict the water vapor permeability of earthen brick as a function of the relative humidity and as well as the resistance factor. The error in the simulation results was below 10% within the relative humidity in the range of 20% and 95%.

Keywords: Dynamic vapor sorption · hysteresis model · water vapor permeability · adsorption/desorption · parameter estimation

1 Introduction

Among classical building materials, raw earth materials, such as compressed earth bricks, adobe, cob, and rammed earth, are experiencing increasing use each year due to their environmental sustainability. Earthen building materials are classified as a good hygroscopicity range, with a Moisture Buffer Value (MBV) in the range of 1–2 according to the MBV classification based on the Nordtest method [1]. In fact, due to their breathability and high thermal mass, raw earth building materials can help stabilize hygrothermal

M. Posani and S. Singh (Eds.): RYS 2025, RILEM Bookseries 67, pp. 937–954, 2026.
https://doi.org/10.1007/978-3-032-14652-6_68

indoor conditions and maintain constant fluctuations in temperature and humidity, particularly during the summer months. The absorption or desorption rate of water vapor in earthen building materials varies depending on the environmental conditions, like temperature, pressure, and humidity [2]. The moisture accumulated in the materials can have detrimental effects for several applications. Moisture fosters mold growth [3], breaks polymer chains, which in turn can reduce the material strength, and leads to the corrosion of metals [4]. Moisture can accelerate the corrosion process in metals, leading to rust formation and eventual structural failure. The moisture degradation mechanisms are a challenge to a wide range of industries including food packing [5], microelectronics [6], airbags, cultural heritage conservation [7]. In such scenarios, prior knowledge of the moisture uptake and outgassing properties of materials not only allows for the selection of suitable materials but also the predictions the future performance. Thus, a complete understanding of moisture behavior and the ability to predict long term outgassing behavior are crucial [8].

Historically, vapor sorption-diffusion and predicting the moisture uptake in polymeric and non-polymeric materials were limited to a small range of moisture or humidity levels. Specifically, prior modeling efforts have been limited to single relative humidity (RH) step studies [9]. Depending on the material type and composition, moisture transfer can be modeled in multiple ways. Fick's law is the simplest and most straightforward way to represent vapor diffusion via gradients of vapor pressure. Various driving potential have been considered for modeling the vapor transfer in porous materials, such as vapor pressure, vapor concentration, capillary pressure, saturation degree, and relative humidity [10].

The characterization of the hygrothermal properties of building materials such as earthen bricks (compressed bricks) allow a better understanding of the material, obtaining its thermal and hydric properties, which are used as input data of the various vapor transfer models [11–13]. The hydric properties are the moisture storage capacity, the vapor and liquid permeabilities [14, 15]. The vapor permeability is measured by the cup method based on EN ISO 12572 [16]. Although other alternative methods have been developed and studied over the years. Unlike thermal properties (thermal conductivity, heat capacity) which can be determined after relatively fast characterization campaigns (a few days) [17], the characterization of moisture properties (moisture storage capacity, vapor and liquid permeability) can be very long (few months) [18]. To overcome these constraints, an alternative solution is to use inverse methods to quickly estimate these hydric parameters. This method consists in estimating parameters of a problem from experimental observations. In practice, this is done by calculating the sum of the mean square error between the numerical solution obtained from a mathematical model and the observed data using a minimization algorithm [19]. Sharma et al. [20] tested a dynamic Triple-Mode sorption model to analyze vapor diffusion in various materials. To evaluate the model's performance, they conducted isotherm experiments to study dynamic moisture sorption and desorption in two markedly different materials: a polymeric material (Sylgard-184) and a ceramic aluminosilicate (Zircar RS-1200), across different temperature conditions. By fitting the model to experimental data obtained from dynamic vapor sorption (DVS) measurements, they reported key parameters such as vapor diffusivity, desorption rate, Langmuir capacity etc. Further, Castonguay et al. [21]. Studied the

Triple-Mode model in the framework of the multi-material system. For this purpose, various materials were parameterized using the model, which are then used in examples to demonstrate the model's ability to capture complex sorption processes in multi-material systems. Guivier et al. [22], investigated various methods and compared them against the standard cup method for measuring the water vapor permeability. They investigated water transport phenomena in PLA flms coated with nanometric layers of chitosan and nanocelluloses, utilizing a combination of techniques including dynamic vapor sorption (DVS) and long-term water vapor adsorption–diffusion experiments (back-face measurements) to understand the influence of each layer on the behavior of multilayer film. They highlighted the limitations of standard cup method, especially when dealing with materials containing ultrathin nanocelluloses layers.

Within this context, the present contribution explores the material-specific sorption/desorption behavior in a wide range of relative humidity and studying the simplified hysteresis model in predicting the moisture behavior of the earthen materials under realistic environmental conditions. In this study, the sorption and diffusion of water vapor in earthen bricks were investigated using two methods: water vapor permeation assessment (cup method), dynamic vapor sorption (DVS). In general, DVS test is used for sorption isotherm and subsequently, for moisture capacity of a material. Our objective is to apply the inverse approach to DVS experimental data for estimating the effective water vapor permeability in building materials.

2 Materials and Experiments

2.1 Material Information

The earthen compressed bricks utilized in the current study is made of natural silt GO available in large quantities in Gonfreville l'Orcher, Le Havre Normandy, France [23]. The soil composition consists of approximately 67% sand (particle sizes ranging from 0.06 mm to 2 mm), about 25% silt (2 μm to 60 μm), 0% clay (particles < 2 μm), and 8% gravel (particles > 2 mm).Silt GO was mixed with water and compacted in a Proctor mold using a static press to achieve a target dry density of 2359 kg/m^3 and a moisture content of 10.52% (g/g), respectively. A compacted cylindrical brick with a diameter of 10 cm and a thickness of 1.2 cm was prepared. The brick was subsequently cut into smaller specimens for experimental testing. Selected samples were oven-dried in an autoclave at 100 °C for 48 h to use in dynamic vapor sorption/desorption (DVS) tests. The details of the selected samples used for DVS test are summarized in Table 1.

To carry out the water vapor permeability test, the compacted material was cut into small cylindrical specimens with dimensions slightly smaller than those of the test cups (see Section II.2 for details). Two specimens were used for this purpose, and their characteristics are summarized in Table 2. Two methods have been used for evaluating the dry density of the bricks including the Kerdane oil and geometrical methods. Sample dimensions (**a x b x c**) were measured using high precision calipers from Swiss Precision Instruments. The geometrical method for sample density was computed utilizing sample weight and measured sample volume. In addition, Kerdane oil method includes the measurement of the sample total volume by immersed weighting using non-wetting Kerdane oil [17]. In fact, the Kerdane test starts with measuring the wet weight of 1 cm3 of raw earth sample. Subsequently, the sample is immersed in oil to fill all pores and replace air with oil. Afterward, it is removed and wiped with absorbent paper to remove excess oil from its outer surfaces. Then, the weight is measured. Using a basket suspended from the bottom of the balance, the sample is immersed in a container of Kerdane oil, and the submerged weight is measured. Finally, the sample is dried in an oven and its dry weight is measured which allow to determine the density and porosity.

Table 1. Dimensions and physical properties of selected samples in dry state.

	a (cm)	b (cm)	c (cm)	Volume (cm^3) a	Dry density (gr/cm^3)		Dry mass (gr)	Void ratio (%)	Porosity (%)
					Geometrical method	Kerdane oil method			
Brick 1	1.00	1.00	1.50	1.50	1.44	1.70	2.16	55.90	35.86
Brick 2	0.90	1.10	1.50	1.49	1.53	1.72	2.28	54.57	35.30
Brick 3	1.30	1.10	1.40	2.00	1.69	1.72	3.39	54.26	35.17
Brick 4	2.00	1.50	1.50	4.50	1.70	1.72	7.65	53.84	35.00

a All the bricks are assumed to be parallelepiped shape, each with a volume of a × b × c

Table 2. Dimensions of the samples used for Water vapor permeability test.

	Diameter (mm)	Thickness (mm)	A (m^2)	Dry weight (gr)
Wet cup	88.45	13.28	0.006141371	153.13
Dry cup	89.83	12.96	0.006334502	149.46

2.2 Experiments

2.2.1 Dynamic Vapor Sorption/Desorption Test

All experiments reported in this study were carried out using the GraviTest 9000-Series device, developed by GINTRONIC [24], which is equipped with 0.1 mg resolution balance. The instrument has relative humidity ranges from 10 to 95%, various temperatures from 15 to 45 °C and controlling air velocity inside the chamber with 1 to 4 m/s. The GraviTest device comes with standard test cups, making it suitable for conducting water vapor permeability tests according to EN ISO 12572. However, for the DVS test, the device is not programmed to automatically apply relative humidity steps. Therefore, the humidity conditions had to be manually adjusted at each step. DVS test is performed immediately after the samples are removed from autoclave and placed into the device.

This is done quickly to avoid any adsorption from the ambient humidity. Both adsorption and desorption have been performed. The device is set to a temperature of 23 °C, an air velocity of 1 m/s, and a time interval of 10 min for each measurement step. The water content is defined as Eq. (1):

$$w = \frac{m - m_0}{m_0} \quad \text{(Eq.1)}$$

where m_0, m, and w are the initial dry mass, instantaneous moist mass, and the mass uptake or water content in gr /gr, respectively.

2.2.2 Standard Cup Method Based on the EN ISO 12572

Cup method test is a classical method to measure the water vapor permeability, mostly referred in building physics to the EN ISO 12572 [16]. In this study, both dry and wet cup methods were employed to approximate the water vapor permeability through cup testing (see Fig. 1(a) and (b)). This test is also conducted using the GraviTest apparatus, the same device employed for the DVS test, as shown in Fig. 1(a). The dry cup method utilized silica gel as the interior desiccant, while the wet cup test involved water. After the preparation of the samples and cups, the device is adjusted in 50% relative humidity with 23 °C, and 1 m/s as air velocity in chamber. The quantity of water vapor that passed through the samples, denoted as Q_{exp}, is determined from the slope of the graph plotting the variation of the mass (Δm) against time(t). The trend of the Δm as a function of time shows the linear for building materials. Subsequently, the water vapor permeability is then calculated using the Eq. (2) [25]:

$$\delta_{p,exp} = \frac{Q_{exp} L_0}{A \Delta P_v} \quad \text{(Eq.2)}$$

where ΔP_v is the pressure difference between the two exposed faces of the cylindrical samples. It is related to the relative humidity differences and saturated pressure. Saturated pressure at 23 °C is 2813.102 Pa and $\Delta P_v = P_{sat}(RH_{exterior} - RH_{interior})$. . It should be noted that Eq. (2) is applied under the assumption that there is no air layer resistance on the top and bottom surfaces of the samples [26].

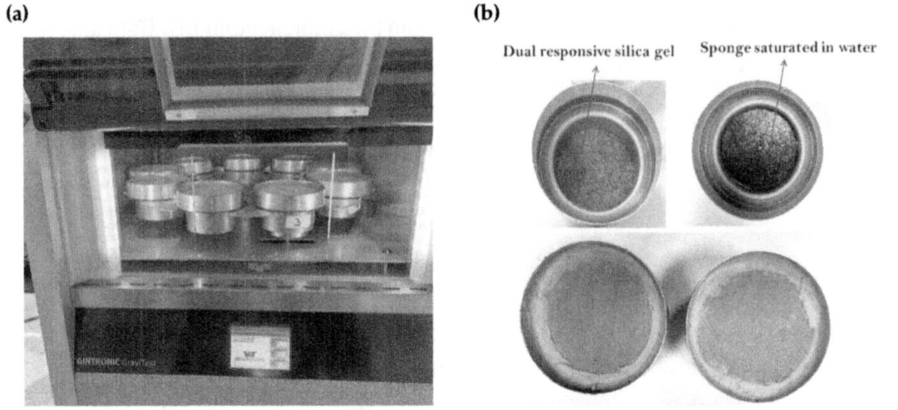

Fig. 1. **(a)** Measurement scheme of the DVS approach and water vapor permeability by the GraviTest instrument, (b) Prepared standard cups for water vapor permeability test for wet and dry cup test, dual responsive silica gel is showing the yellow before adsorption of water vapor and become green after saturated with water vapor.

3 Simplified Hysteresis Model

In DVS experiments, the process can be macroscopically described in terms of the relative humidity in the ambient gas (air-water mixture) as a driving potential. Water vapor transport in microporous and soluble solids is governed by a mass balance equation and a definition for the diffusive flux, generally given by Fick's Law. Fick's law is often expressed in terms of a gradient in concentration, vapor pressure, relative humidity as it allows for the mass balance equation, Eq. 3 is one-dimensional isothermal model of vapor transfer with the relative humidity as a driving potential [12]:

$$\begin{gathered} \xi(t)\frac{\partial\varphi}{\partial t} = \frac{\partial}{\partial x}\left(D(\varphi)\frac{\partial\varphi}{\partial x}\right), x \in (0, L), t > 0 \\ D(\varphi) = \frac{\delta_{air}}{\mu(\varphi)}P_{sat}, \mu(\varphi) = 1 + ae^{-b\varphi} \\ \overline{\varphi}(t) = \frac{1}{L}\int_0^L \varphi(x, t)dx \end{gathered} \tag{Eq.3}$$

$$\begin{gathered} \xi(\varphi, t) = \begin{cases} \xi_{ads}, if\,\overline{\varphi}(t) < \varphi_b(t), (adsortpion) \\ \xi_{des}, if\,\overline{\varphi}(t) \geq \varphi_b(t), (desorption) \end{cases} \\ \text{Initial condition: } \varphi(x, 0) = 0, x \in (0, L) \\ \text{Boundary conditions: } \varphi(0, t) = \varphi_b(t), t > 0 \end{gathered} \tag{Eq.4}$$

$$\varphi(L, t) = \varphi_b(t), t > 0 \tag{Eq.5}$$

where $\boldsymbol{\varphi}$ is humidity, $\boldsymbol{\delta_{air}}$ (kg/ (m.s. Pa)) is vapor permeability in the ambient at 23 °C with the value of 1.88×10^{-10} (kg/ (m.s. Pa)), $\boldsymbol{P_{sat}}$ (Pa) is saturated vapor pressure, at 23 °C with value of 2813.102 (Pa) and $\boldsymbol{\xi(\varphi)}$ (kg/m^3) is moisture capacity of material. The parameter of $\boldsymbol{\mu}$ is the resistance factor and is related to the vapor permeability of a material by $\boldsymbol{\delta_{p,material}} = \boldsymbol{\delta_{air}/\mu(\varphi)}$. The system of Eq. (4) shows the simplified approach for hysteresis modeling in vapor sorption/desorption process, it does not concern the exact mechanism by moisture hysteresis happens but rather focus on the modeling of it. In this relation, storage coefficient ($\boldsymbol{\xi}(\boldsymbol{\varphi}, \boldsymbol{t})$) is governed by dynamic regime of sorption/desorption process based on the spatial average as defined by $\overline{\boldsymbol{\varphi}}(\boldsymbol{t})$**.** Equation 5 shows the initial and boundary conditions used in modeling of the sorption/desorption process (see Fig. 2).

Initially, the material is dry, and the entire system starts with zero relative humidity. The external RH condition is subsequently set to various relative humidity over 766500 s (12775 min), allowing the system to approach equilibrium at each relative humidity step. The time dependent Dirichlet boundary conditions ($\boldsymbol{\varphi}_{\boldsymbol{b}}(\boldsymbol{t})$) are set at the following RH values: 10%, 20%, 30%, 40%, 60%, 80%, and finally 95% then followed by a return sequence to observe the hysteresis behavior, as shown in Fig. 2. The domain of the study focuses on Brick 1 and Brick 4 as representative samples. A one-dimensional simulation domain is considered, with a thickness of 1 cm for Brick 1 and 2 cm for Brick 4.

Sorption /desorption isotherm curves are generally modelled using existing prediction models in the literature such as the Guggenheim-Anderson-de Boer (GAB) model [27]. After fitting and obtaining the optimized isotherm parameters which serve as an input for vapor transfer model.

$$w(\varphi) = \frac{w_m G_{GAB} K\varphi}{(1 - K\varphi)(1 - K\varphi + G_{GAB} K\varphi)} \quad \text{(Eq.6)}$$

$$\xi = \frac{\partial w(\varphi)}{\partial \varphi} \quad \text{(Eq.7)}$$

where $\boldsymbol{w}(\boldsymbol{\varphi})$ is the water content (gr/gr) and $\boldsymbol{w_m}$, $\boldsymbol{G_{GAB}}$, and $\boldsymbol{K}$ are the fitting parameters of the GAB model. Note that the GAB model obtained for both adsorption and desorption process. Thus, derivation of each is referred to $\boldsymbol{\xi_{ads}}$ and $\boldsymbol{\xi_{des}}$.

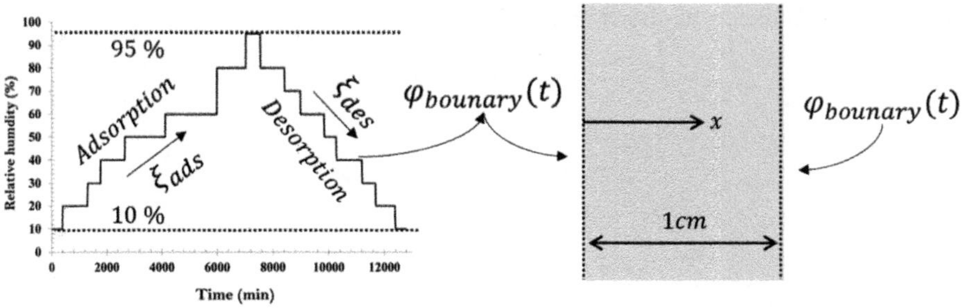

Fig. 2. Schematic of the simplified hysteresis model, a time-varying relative humidity is applied at two side sample.

4 Inverse Approach and Numerical Solution

An inverse problem, as opposed to a direct problem, is a problem in which one tries to determine the causes of a phenomenon from the experimental observations of its effects. The resolution of an inverse problem supposes a knowledge of the direct problem resulting from the modeling of the physical phenomenon. Thus, knowing the evolution of the system, it is possible to recover the parameters of the system defined in the direct problem [28]. In our study, the function of interest, i.e., is the observable water content of the material $\boldsymbol{w}(\overline{\boldsymbol{\varphi}})$. The inverse problem consists in finding the parameters of resistance factor $\boldsymbol{\mu}(\boldsymbol{\varphi})$,including a and b in Eq. 3, from minimizing the difference between the experimental data of water content and the estimated numerical values ($\boldsymbol{w}(\overline{\boldsymbol{\varphi}})$). The optimization problem can be solved by defining the objective function and minimizing it which is defined as the sum of square error between experimental measurements and estimated values, as follows:

$$\text{Objective function } = min\left(w(\overline{\varphi}) - w_{experimetnal}\right)^2 \quad \text{(Eq.8)}$$

Various optimization algorithms are available within the software environment to achieve a reliable solution. Among these, the Nelder–Mead method was employed to estimate the model parameters. It is a derivative-free, direct search algorithm used for nonlinear optimization problems. The objective function was defined as a variable within the software, and a minimization strategy was selected in the solver settings. Since the simulation is time-dependent, the sum of the objective function values over time is used as the optimization target in COMSOL Multiphysics. Additionally, the initial-based solution setting was applied to ensure consistent convergence behavior throughout the simulation. The numerical solution and optimization method is implemented in COMSOL Multiphysics software (version 6.2) based on discretizing the differential equation system by the finite element method. Strategy has followed in the software to estimate the parameters as shown in Fig. 3.

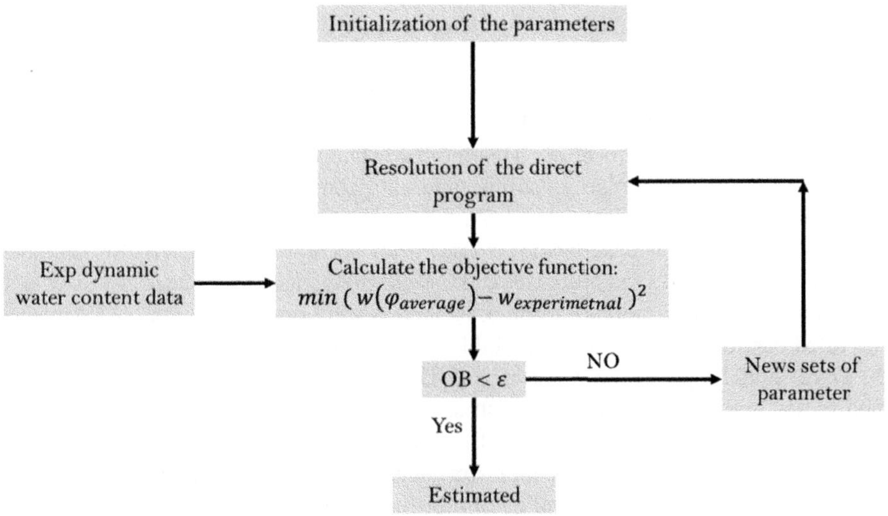

Fig. 3. Strategy implemented for minimizing the objective function and estimation of water vapor permeability.

5 Results and Discussions

5.1 Results of Dynamic Vapor Sorption/Desorption Test

The dynamic vapor sorption results for four earthen bricks- information on the bricks have been previously listed in Table 1–with different dimensions are presented in Figs. 4(a) and (b). Up to 30% relative humidity, all samples exhibited a similar trend. However, beyond this point, differences between the smaller and larger dimensions became noticeable. Since all samples were extracted from the same compacted brick, their average properties are the same, such as porosity, with only their dimensions varying. It is observed that the water content increases as the specimen size decreased; on the graph of Fig. 4(a), specimens with the smaller dimension consistently show higher water content than the bigger ones. The differences between the steady state of water content depends on the penetration depth of the bricks. As a result, the deeper the vapor penetration, the greater the amount of moisture that can be adsorbed within the specimen [29]. In earth-based materials with relatively small thickness, water vapor can rapidly permeate the entire volume, leading to faster saturation. In contrast, thicker specimens, Brick 4 in our case, exhibit that water content increases more slowly than smaller samples (Brick 1), as illustrated in Fig. 4(b). Consequently, varying the dimensions from Brick 1 to Brick 4 reveals an inverse relationship between the average water content and both material thickness and the time required to reach equilibrium.

Equilibrium water content is considered when the water content, at each applied relative humidity, reaches steady state conditions. Equilibrium water content is plotted against relative humidity, as shown in Fig. 5(a). As a result, the shape of sorption/desorption isotherms are significantly influenced by both the surface polarity and the pore structure of the material. This indicates that the interaction between surface characteristics, such as polarity, and the internal geometry, specifically pore size distribution and geometry, plays a key role in the resulting hysteresis behavior. A thorough understanding of these factors is crucial for accurately characterizing the material's adsorption/desorption processes and for optimizing its performance across various applications. The adsorption/desorption isotherms were modelled by the GAB method and fitted parameters are reported in Table 3. There is a very good agreement between the GAB modelling and the experimental data. The results from the fitting the GAB models are used to calculate the moisture capacity, which is defined as the slope of the sorption isotherm curves. Figure 5(b) shows the sorption/desorption isotherms and moisture capacity for BTC 1. The moisture capacity and sorption isotherm are essential parameters for macroscopic modeling of moisture uptake in porous materials [30].

Table 3. Optimized parameters for the GAB model.

Adsorption				
BTC	w_m	G_{GAB}	K	R-squared
1	0.0068	2.5743	0.8349	0.9988
2	0.0073	2.4332	0.8157	0.9986
3	0.0066	2.7557	0.8074	0.9989
4	0.0055	3.4683	0.7993	0.9988
Desorption				
BTC	w_m	G_{GAB}	K	R-squared
1	0.0102	2.7757	0.7391	0.9992
2	0.0112	2.7359	0.7098	0.9993
3	0.0101	2.975	0.6975	0.9997
4	0.0084	3.5583	0.6876	0.9999

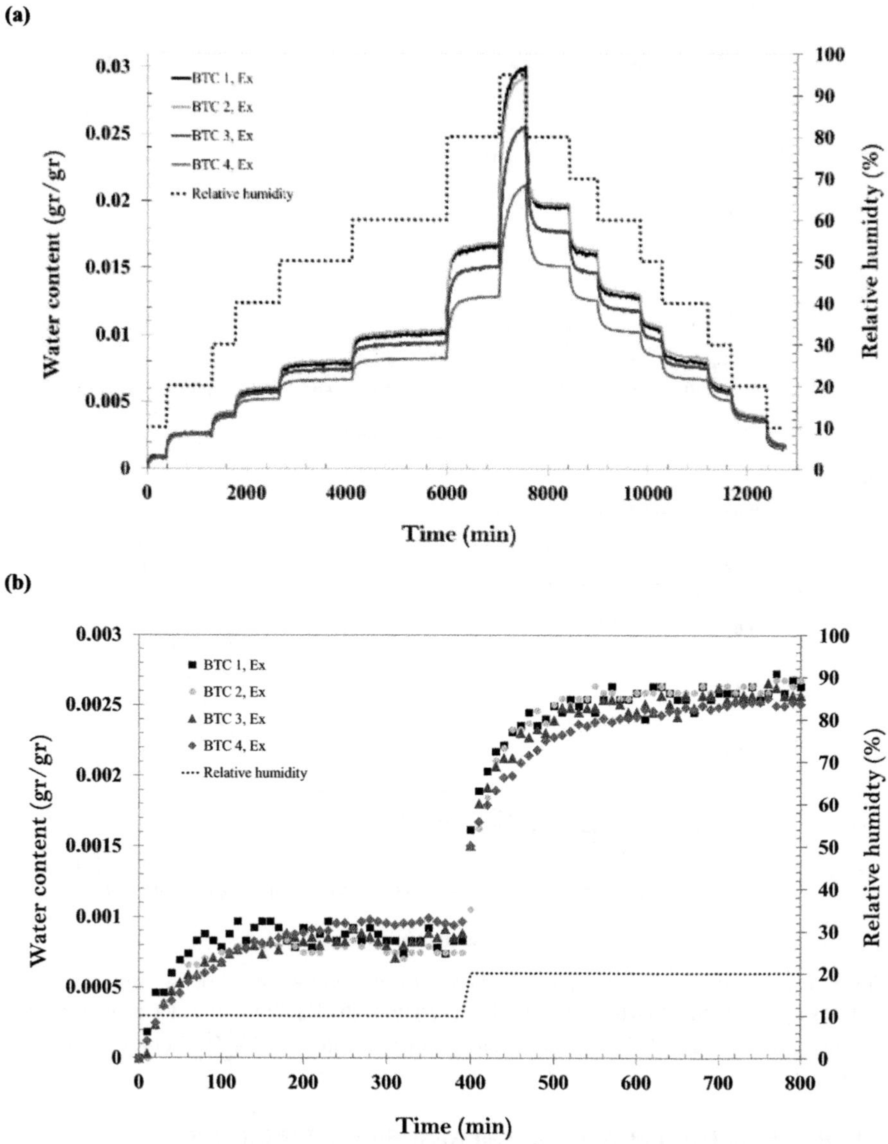

Fig. 4. (a) Typical water uptake (in gr/gr) vs. time (in minutes) for earthen materials ranging from smaller dimensions (1 cm × 1 cm × 1 .5 cm for BTC 1) to larger dimensions (2 cm × 1.5 cm × 1.5 cm for BTC 4); (b) The same measurements within a relative humidity range of 10–20%, with highlighted experimental data points recorded every 10 min.

(a)

(b)

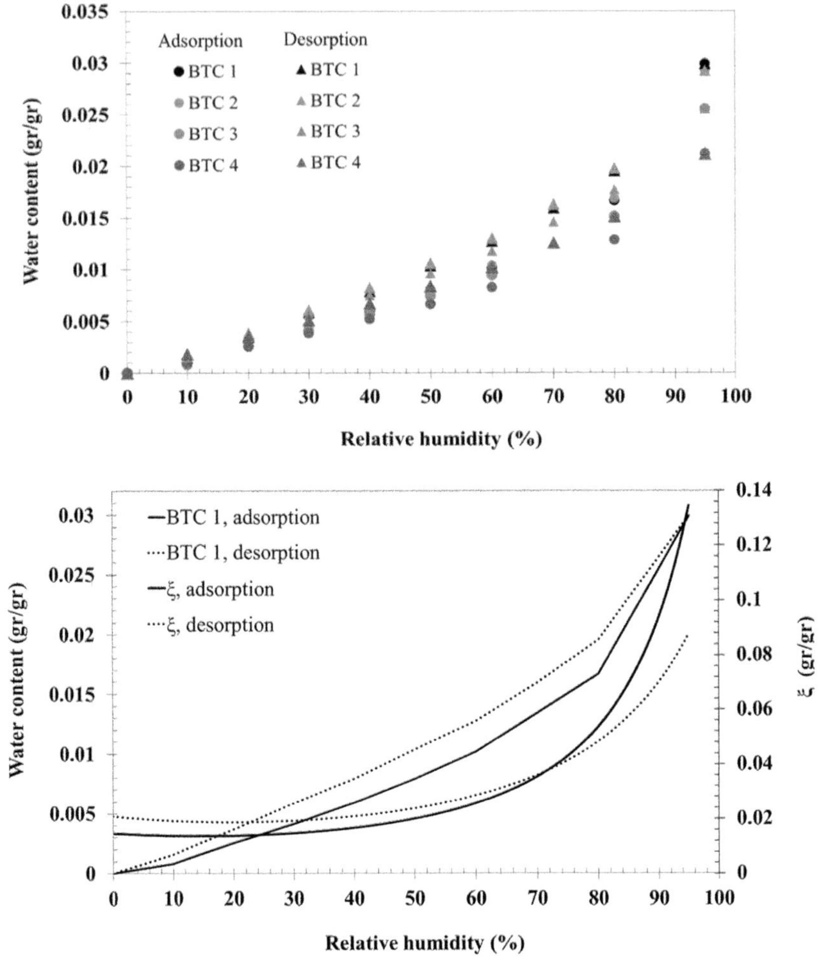

Fig. 5. **(a)** Adsorption and desorption isotherms for the four earthen bricks. Solid lines represent adsorption, while dotted lines indicate desorption, **(b)** Equilibrium water content for both sorption/desorption for BTC 1 and calculated moisture capacity ($\frac{\partial w(\varphi)}{\partial \varphi}$) from the fitted GAB model.

5.2 Results of Dry and Wet Cup Test According to EN ISO 12572

As previously explained in section II.2, the method used involves monitoring the mass change of the samples over time, each 10 min. Figure 6 illustrates transient and steady-state regimes observed during the water vapor permeability tests. At the beginning of the transient phase, a decrease in mass was recorded for the dry cup test, while an increase in mass was observed for the wet cup experiment. This behavior is attributed to the initial moisture conditions of the samples. In the dry cup test, the sample is initially quasi-saturated, meaning its internal humidity is higher than the ambient vapor pressure gradient imposed by the test. Consequently, the sample releases moisture to reach equilibrium, resulting in an initial mass loss. In contrast, during the wet cup test, the imposed vapor pressure gradient is higher than the sample's initial humidity, leading to a gain in mass as the sample absorbs moisture to equilibrate.

Figure 6 demonstrates a clear linear relationship between mass change and time under steady-state conditions, with a correlation coefficient of $R^2 = 0.99$ for the dry cup test and $R^2 = 0.97$ for the wet cup test. Therefore, the slope of the curve corresponds to the $\boldsymbol{Q_{exp}}$, in Eq. (2). Note that this approach is primarily suited for materials with high permeability. Using the slope values obtained from the mass–time plots in Fig. 6 (converted from g/h to kg/s), along with the sample thickness and surface area provided in Table 2, the water vapor permeability was calculated to be 2.29 10^{-11} for the wet cup test and 3.52 10^{-12} for the dry cup test.

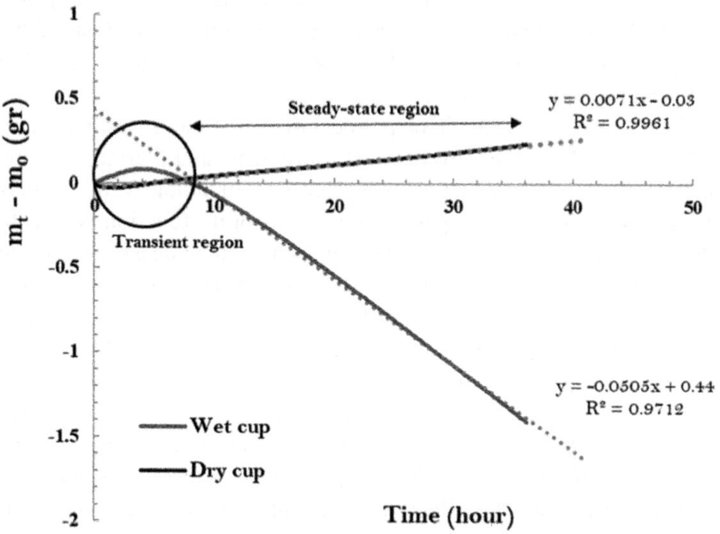

Fig. 6. Mass change over time for both samples used for the wet cup and dry cup tests.

5.3 Estimating the Resistance Factor and Effective Water Vapor Permeability

A numerical study was carried out to analyze the resistance factor and to optimize the fitting parameters" a" and" b" in Eq. 3. A wide range of resistance values was examined during parametric study to guide the initial selection of these parameters. The resistance factor was varied from 2 to 30, based on values reported in the literature [31], and to minimize simulation costs. Following this initialization, the inverse method was applied to estimate the parameters b and a that best fit the experimental data points. Two bricks have participated in numerical study as a representative of four bricks, one smaller (BTC 1) and one bigger dimension (BTC 4).

Figure 7(a) presents a comparison between experimental data and the model simulations, along with the predicted resistance factors. It shows the complete moisture uptake curves, together with the simulated results for two bricks (BTC 1 and BTC 4) at 23 °C. The right-hand y-axis in Fig. 7(a) illustrates the relative humidity variations over time, which were applied as boundary conditions in the simulation. The relative humidity changes start from 0.1, with only a single step increase of 10% during the sorption process (from 60% to 80%), while for desorption and outgassing, the initial condition is set to zero relative humidity.

As a result, at low relative humidity (RH = 10%), both during adsorption and desorption, a significant deviation in moisture content and outgassing was observed for both cases. This deviation is illustrated by the relative errors presented in Fig. 7(b), where the relative error reached approximately 60% for both bricks during both the sorption and desorption processes. Our model predictions are in agreement with the experimental data within the error margin ($\leq$ 10%) except for relative humidity of 10% in sorption/desorption conditions. However, optimized resistance factors for BTC 1 and BTC 4 are plotted in Fig. 7(c) and (d), and estimated water vapor permeability is calculated from the relation of the resistance factor discussed in Eq. 3 as $\boldsymbol{\delta_p} = \boldsymbol{\delta_{air}}/\boldsymbol{\mu}(\boldsymbol{\varphi})$ where $\boldsymbol{\mu}(\boldsymbol{\varphi}) = 1+\boldsymbol{ae^{-b\varphi}}$. A comparison has been made to quantitative analysis of the effective water vapor permeability which was obtained as function of relative humidity and cup method test performed in Section V.2.

Figure 8 presents the water vapor permeability (δ_p) values estimated from DVS test, along with standard dry and wet cup measurements and literature comparison for earthen materials [31]. To have an estimate that is comparable with dry cup and wet cup, an average of the estimated water vapor permeability function is calculated from 10% to 50% RH and from 50% to 95% RH. The results indicate a strong dependency of permeability on both RH range and sample size. Standard dry cup and wet cup methods based on the EN ISO 12572 yield values of 3.52×10^{-12} and 2.29×10^{-11} kg/(m·s·Pa), respectively, confirming that permeability increases with higher RH and that the RH-dependent method used in this study captures this nonlinearity more effectively. For the BTC 4 as a bigger sample during the wet cup conditions (average of WVP between 50% and 95% RH), it is overestimated and five times bigger than the standard wet cup test. However, as a result, smaller samples in DVS test can work effectively in inverse methods to capture the effective water vapor permeability function with the dependency of relative humidity. Still the minimum error between this technic and cup method was 96% (compared with average BTC 1). This error comes from the vapor transfer phenomena that includes both vapor and liquid transfer. The effective water vapor estimated is combined with the vapor transfer and liquid transfer coefficient as explained in details in reference [12].

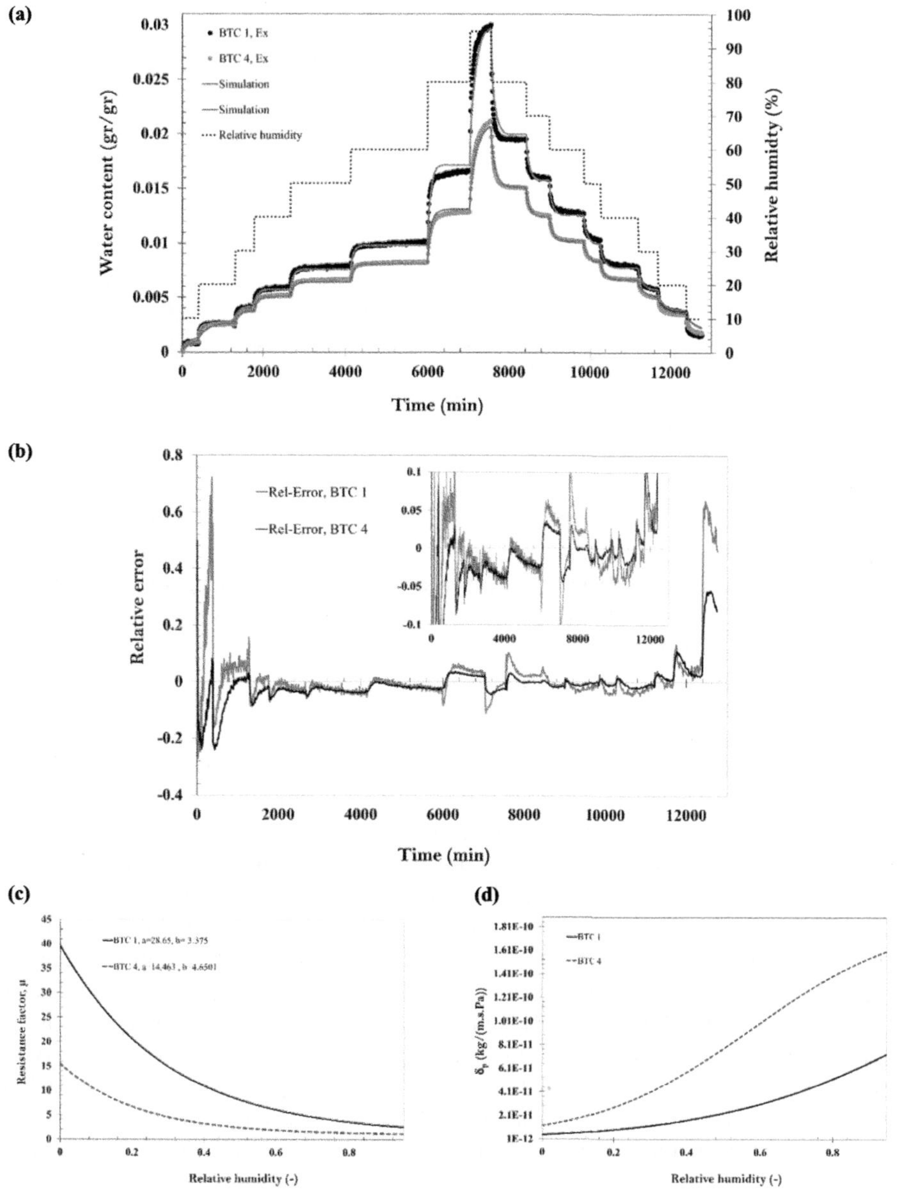

Fig. 7. (a) Experimental and simulated water content for BTC 1 and BTC 4. (b) Relative error between the simulation results and experimental data with subfigure highlighted the relative error with the range of 10%. (c) Resistance factor estimated from the optimized parameters, BTC 1, a = 28.65, b = 3.375 and for BTC 4, a = 14.463 and b = 4.6501. (d) Water vapor permeability (δ_p) calculated from the optimized parameters of resistance factor ($\delta_p = \frac{\delta_{air}}{1+ae^{-b\varphi}}$)

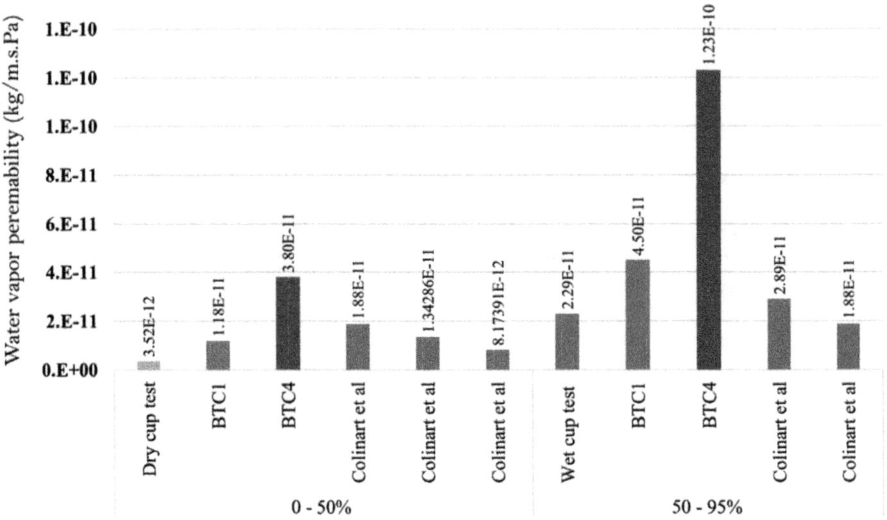

Fig. 8. Water vapor permeability estimated from inverse approach for both BTC 1 and BTC 4, during the wet (ranges from RH 50%-RH 95%) and dry cup conditions (ranges between RH 10% and RH 50%) compared against the standard wet cup test and dry cup test according to the EN ISO 12572 [16] and literature [31]

6 Conclusion

Moisture uptake rates and capacity of the earthen materials, and moisture prediction in materials are important phenomena that can dictate the lifetime and moisture performance of a material. This study applied a simplified hysteresis model that can follow the sorption and desorption lines by altering the moisture capacity function instead of the scanning curve methods. This study examined earthen bricks with varying dimensions, from smaller to larger scales, while considering DVS tests and effective water vapor permeability. The inverse method was applied to estimate the effective water vapor permeability from a DVS test and compared against the well-known cup method test according to EN ISO 12572. This macroscopic isothermal diffusion model can be applicable for sorption/desorption modeling of earthen materials and to any other types of building materials. Future work will focus on expanding our modeling approach in three-dimensional simulations and addressing multi-dimensional cases. And improving the model for estimation of both liquid diffusion and water vapor permeability from DVS test.

References

1. Posani, M., et al.: Low-carbon indoor humidity regulation via 3D-printed superhygroscopic building components. Nat. Commun. **16**, 425 (2025). https://doi.org/10.1038/s41467-024-54944-1
2. Benmahiddine, F., Cherif, R., Bennai, F., Belarbi, R., Tahakourt, A., Abahri, K.: Effect of flax shives content and size on the hygrothermal and mechanical properties of flax concrete. Constr. Build. Mater. **262**, 120077 (2020). https://doi.org/10.1016/j.conbuildmat.2020.120077

3. Boardman, C.R., Glass, S.V., Lepage, R.: Dose-response simple isopleth for mold (DR SIM): a dynamic mold growth model for moisture risk assessment. J. Build. Eng. **68**, 106092 (2023). https://doi.org/10.1016/j.jobe.2023.106092
4. Du, C., Wang, Y., Li, B., Xu, M., Sadrizadeh, S.: Grey image recognition-based mold growth assessment on the surface of typical building materials responding to dynamic thermal conditions. Build. Environ. **243**, 110682 (2023). https://doi.org/10.1016/j.buildenv.2023.110682
5. García-Alvarado, M.A., Pacheco-Aguirre, F.M., Ruiz-López, I.I.: Analytical solution of simultaneous heat and mass transfer equations during food drying. J. Food Eng. **142**, 39–45 (2014). https://doi.org/10.1016/j.jfoodeng.2014.06.001
6. Jalili, H., Raad, M., Fallah, D.A.: Numerical study on the mixing quality of an electroosmotic micromixer under periodic potential. Proc. Inst. Mech. Eng. C J. Mech. Eng. Sci. **234**, 2113–2125 (2020). https://doi.org/10.1177/0954406220904089
7. Lawrence, M.G.: The Relationship between relative humidity and the dewpoint temperature in moist air: a simple conversion and applications. Bull. Amer. Meteor. Soc. **86**, 225–234 (2005). https://doi.org/10.1175/BAMS-86-2-225
8. Or, D., Lehmann, P., Shahraeeni, E., Shokri, N.: Advances in soil evaporation physics—a review. Vadose Zone J. **12**, 1–16 (2013). https://doi.org/10.2136/vzj2012.0163
9. Challansonnex, A., Casalinho, J., Perré, P.: Non-Fickian diffusion in biosourced materials: experimental determination of the memory function using minute samples. Constr. Build. Mater. **224**, 560–571 (2019). https://doi.org/10.1016/j.conbuildmat.2019.07.013
10. Jalili, H., Ouahbi, T., Eid, J., Taibi, S., Hamrouni, I.: Exploring historical perspectives in building hygrothermal models: a comprehensive review. Buildings **14**, 1786 (2024). https://doi.org/10.3390/buildings14061786
11. Rode, C., Hansen, P.N., Hansen, K.K.: Combined heat and moisture transfer in building constructions (1990). https://api.semanticscholar.org/CorpusID:136307513
12. Künzel, H.M.: Simultaneous Heat and Moisture Transport in Building Components: One- and Two-Dimensional Calculation Using Simple Parameters. IRB Verlag, Stuttgart (1995)
13. Hagentoft, C.-E., et al.: Assessment method of numerical prediction models for combined heat, air and moisture transfer in building components: benchmarks for one-dimensional cases. J. Therm. Envelope Build. Sci. **27**, 327–352 (2004). https://doi.org/10.1177/1097196304042436
14. M. Krus, Moisture transport and storage coefficient of porous mineral building materials, 1996
15. Fang, A., Chen, Y., Wu, L.: Transient simulation of coupled heat and moisture transfer through multi-layer walls exposed to future climate in the hot and humid southern China area. Sustain. Cities Soc. **52**, 101812 (2020). https://doi.org/10.1016/j.scs.2019.101812
16. 14:00-17:00, ISO 12572:2016, ISO (2024). https://www.iso.org/standard/64988.html. Accessed 11 Jan 2024
17. Hamrouni, I., Jalili, H., Ouahbi, T., Taibi, S., Jamei, M., Mabrouk, A.: Thermal properties of a raw earth-flax fibers building material. Constr. Build. Mater. **423**, 135828 (2024). https://doi.org/10.1016/j.conbuildmat.2024.135828
18. Hamrouni, I., Ouahbi, T., El Hajjar, A., Taibi, S., Jamei, M., Zenzri, H.: Water vapor permeability of flax fibers reinforced raw earth: Experimental and micro-macro modeling. Eur. J. Environ. Civ. Eng. **27**, 3020–3039 (2023). https://doi.org/10.1080/19648189.2022.2123857
19. Roussey, C., Perré, P., Casalinho, J., Colin, J.: Inverse analysis of oxygen diffusivity in oak wood using the back-face method: application to cooperage. Wood Sci. Technol. **56**, 219–239 (2022). https://doi.org/10.1007/s00226-021-01325-2
20. Sharma, H.N., Harley, S.J., Sun, Y., Glascoe, E.A.: Dynamic triple-mode sorption and outgassing in materials. Sci. Rep. **7**, 2942 (2017). https://doi.org/10.1038/s41598-017-03091-3

21. Castonguay, S.T., et al.: Modeling diffusion and types I-V sorption of water vapor in heterogeneous systems. Chem. Eng. Sci. **275**, 118695 (2023). https://doi.org/10.1016/j.ces.2023.118695
22. Guivier, M., Chevigny, C., Domenek, S., Casalinho, J., Perré, P., Almeida, G.: Water vapor transport properties of bio-based multilayer materials determined by original and complementary methods. Sci. Rep. **14**, 50 (2024). https://doi.org/10.1038/s41598-023-50298-8
23. El Hajjar, A., Ouahbi, T., Hamrouni, I., Taibi, S., Eid, J., Hattab, M.: Experimental characterization of desiccation cracking mechanisms in clayey soils. Eur. J. Environ. Civ. Eng. **27**, 3380–3401 (2023). https://doi.org/10.1080/19648189.2022.2136249
24. Gravitest (n.d.). https://www.gintronic.ch/en/
25. Casnedi, L., Cappai, M., Cincotti, A., Delogu, F., Pia, G.: Porosity effects on water vapour permeability in earthen materials: experimental evidence and modelling description. J. Build. Eng. **27**, 100987 (2020). https://doi.org/10.1016/j.jobe.2019.100987
26. Colinart, T., Glouannec, P.: Accuracy of water vapor permeability of building materials reassessed by measuring cup's inner relative humidity. Build. Environ. **217**, 109038 (2022). https://doi.org/10.1016/j.buildenv.2022.109038
27. Nagata, G.A., Costa, T.V., Perazzini, M.T.B., Perazzini, H.: Coupled heat and mass transfer modelling in convective drying of biomass at particle-level: model validation with experimental data. Renew. Energ. **149**, 1290–1299 (2020). https://doi.org/10.1016/j.renene.2019.10.123
28. Kouchade, A.C., Passard, J., Perre, P.: Inverse method using TransPore 1-D code for the mass-diffusivity determination in wood. **6** (2018)
29. Cunningham, M.J.: Effective penetration depth and effective resistance in moisture transfer. Build. Environ. **27**, 379–386 (1992). https://doi.org/10.1016/0360-1323(92)90037-P
30. Phung, T.A., Le Guern, M., Boutouil, M., Louahlia, H.: Hygrothermal behaviour of cob material. In: Reddy, B.V.V., Mani, M., Walker, P. (eds.) Earthen Dwellings and Structures, pp. 345–356. Springer, Singapore (2019). https://doi.org/10.1007/978-981-13-5883-8_30
31. Colinart, T., et al.: Preliminary results of interlaboratory tests on hygrothermal properties of CEB. In: Beckett, C., Bras, A., Fabbri, A., Keita, E., Perlot, C., Perrot, A. (eds.) Second RILEM International Conference on Earthen Construction, pp. 458–465. Springer Nature Switzerland, Cham (2024). https://doi.org/10.1007/978-3-031-62690-6_47

Mycelium-Based Panels for Sustainable Construction: Evaluating Hygrothermal Performance

Carlotta Basso(✉) and Michele Libralato

Polytechnic Department of Engineering and Architecture, University of Udine, Udine, Italy
carlotta.basso98@gmail.com

Abstract. Construction industry is increasingly prioritizing sustainability through life cycle assessment, aiming to evaluate its environmental impact at every stage. The cradle-to-cradle approach emphasizes the use of materials designed with end-of-life reuse or biodegradability in mind. Mycelium, the root-like structure of fungi, offers a solution to these challenges. It can be cultivated on agricultural waste, obtaining panels made of organic residues, made more compact and durable. The aim is to obtain mycelium-based products that are fully biodegradable, in line with the principles of circular economy. To assess the feasibility of mycelium in construction, heat and moisture transfer (HMT) simulations can be utilized. A preliminary study is carried out on three types of commercial mycelium panels prepared with different substrates. The samples are tested using a Dynamic Vapour Sorption Analyser. The moisture sorption curve, the Moisture Buffer Value (MBV) and the water vapour resistance factor are estimated, using the dynamic vapour sorption measures. Finally, the material properties are used to quantify the effects in WUFI Plus: a small office room is considered as a case study, and the effect of the panels is evaluated on the temperature and relative humidity of the room and on the HVAC system consumption.

Keywords: Mycelium · moisture · biobased · WUFI · DVS

1 Introduction

Construction sector is one of the larger contributors to global greenhouse emissions and energy consumption, not only due to the operation of buildings, but also to the embodied carbon of the used materials. As a result, the study of the Life Cycle Assessment of building materials became a relevant tool for the climate change mitigation [1–3]. The Cradle-to-Cradle approach has brought renewed relevance to materials – even those with lower performance – that enable the reuse of waste from other production processes and can be safely disposed of in the environment due to their biodegradable nature.

In this context, a significant role is played by materials made from chipped agricultural waste bound together by the mycelial network of fungi. This type of compostable material production process makes it possible to achieve mechanical properties suitable

M. Posani and S. Singh (Eds.): RYS 2025, RILEM Bookseries 67, pp. 955–963, 2026.
https://doi.org/10.1007/978-3-032-14652-6_69

for use in packaging and building panels. These panels similar to wood fiber boards, providing thermal insulation, or, by leveraging the distinctive aesthetic of mycelium, serve as visible sound-absorbing elements. In both cases, these materials remain biodegradable and may be sensitive to the environmental conditions typically found in buildings. One method to anticipate such issues is through hygrothermal analysis of the materials and the simulation of coupled heat and moisture transport [4–7].

Using the same hygrothermal models it is possible to evaluate also the impact on the indoor environment and the passive control of air humidity. This approach is also studied with the Moisture Buffer Value a simple parameter that can provide an evaluation of the dynamic moisture absorption and desorption from the air [8–11]. In this work, three mycelium-based composites used to produce commercial acoustic panels are studied. Their hygrothermal properties are evaluated experimentally and then used to simulate the moisture buffering effect in a small office room.

2 Materials and Methods

Mycelium-based composites are obtained by inoculating a sterilized substrate (usually chipped plant byproducts) in a mold with mycelium. The fungi grow the hyphae structure in few days, binding the substrate together, then the inactivation phase stops the growth of the hyphae. The material, removed from the mold, is heated in a stove and dried, until it gets its mechanical properties. After this treatment the fungi stops its growth, and the resulting panels can be used as building material.

For this work, three commercially available sound adsorbing acoustic panels are considered. The substrate for the growth of the mycelium is composed of hemp and cotton fibers obtained as industrial byproducts. After the forming process the panels two of the three panels are covered with a coating and with paint.

2.1 Description of Test Procedures

Three samples of mycelium-based composites are cut from three different panels into squares of about 5x5cm. The samples are described in Table 1 and shown in Fig. 1.

The practical Moisture Buffer Value (MBV) [12], the isotherm of desorption and the water vapour resistance factor μ of the three samples are estimated with the ProUmid Vsorp Basic, a dynamic vapour sorption analyzer (DVS) at the Thermal Systems Lab of the University of Udine.

The Moisture Buffer Value is evaluated from the dynamic measures performed by the DVS. The samples are prepared sealing all the surfaces but one with aluminum tape. The NORDTEST procedure is followed with the exception of the exposed surface of the samples which is less than 0.01 m^2. Then, the samples are exposed to the controlled environment of the DVS and after reaching equilibrium with the air at 23 °C and 50% RH, they are exposed to cycles of 8 h at 75% RH and 16 h at 33% RH at 23 °C for five days. The MBV values are calculated according to Eq. 1.

$$MBV = \frac{m_{75\%} - m_{33\%}}{A \cdot \Delta RH} \quad (1)$$

where $m_{75\%}$ and $m_{33\%}$ are the masses of the samples at the end of the 75% and 33% relative humidity steps respectively, expressed in g, A is the surface of the sample in m^2 and ΔRH is the relative humidity difference between 75% and 33% expressed in %.

The μ value is estimated from the dynamic measurements of the adsorption phase, from a fitting procedure. The value of the vapour resistance factor have been estimated using the procedure presented in [13], based on the analytical solution of the non-steady state diffusion of a species in a plane sheet, reported in the corresponding section of [14]. The samples, covered with aluminum tape on all the surfaces, except for one, are considered as one dimensional samples with one dimensional Fickian diffusion, with the same behavior of a plane sheet. From the fitting of the equation to the measured dynamic values of the absorption from 0% to 50% RH allowed to estimate the diffusivity and then the permeability of the samples.

The three samples have different thickness, composition and coating. Due to the commercial nature of the products, the exact composition and the coatings are unknown to authors.

Table 1. Description of the samples studied

Identifier	Description	Thickness (cm)	Sample surface (cm^2)
S1	No coating	3.5	25
S2	With coating	3.5	28
S3	With coating	1.5	30

Fig. 1. Samples of mycelium based composite panels.

This procedure does not allow for a clear separation between the behavior of the base material and that of the coating. Therefore, the measured properties should be considered representative of the combined response of the composite and its coating, even when used in the whole building simulation.

Finally, the material properties of the sample S2 are used to perform a whole building simulation of a room with and without the panels, evaluating the moisture buffering effect on the internal environmental conditions. The density of the material is 153 kg/m^3 while the missing thermal properties are taken from the "Wood-Fibre Insulation Board" material available in the WUFI Plus material database with 0.042 W/m · K of thermal conductivity and 1400 J/kg · K of specific heat capacity.

2.2 Whole Building Hygrothermal Simulation

The simulations are performed using WUFI Plus, the walls are modelled with heat and moisture transport model to capture the interaction between the room air and the wall materials.

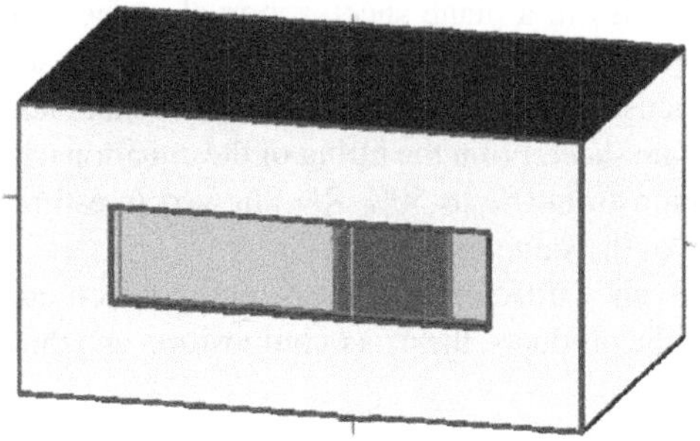

Fig. 2. Simulated office room geometry.

The simulated office room, shown in Fig. 2, is a 24 m^2 room facing the external environment from just one wall. The other walls face other rooms with the same internal conditions (adiabatic surfaces). The external wall has one window of 4 m^2 and the opposite wall has a wooden door. The walls are made of 20 cm autoclaved aerated concrete bricks covered in 2 cm of lime plaster, while the external wall is externally insulated with 15 cm of mineral wool. The used weather file has been measured in 2004 in (Italy) and is provided within the WUFI Plus database. The surface heat transfer resistance values are set to 0.04 m^2K/W for the exterior surfaces and to 0.13 m^2K/W to the internal surfaces.

The office is occupied from Monday to Friday, from 9:00 to 17:00 by 4 people. The infiltration is set to 0.2 ACH. The heating and cooling systems are modelled as ideal systems, and the air temperatures are kept between 20 °C and 27 °C during occupation hours.

Two simulations are performed, a baseline simulation, with the internal surface of vertical walls covered in lime plaster, while the second simulation is performed covering the internal surface with 3 cm of the S2 mycelium composite properties. The total composite exposed surface is about 53 m^2.

The building is simulated for one year with 30 days of preheating since the aim of the study is a comparison between two cases.

3 Results

The tests and simulations performed using the estimated material properties allowed to show that the panels have good moisture buffering properties with effects on the environmental conditions.

3.1 Hygrothermal Properties

Figure 3 shows the desorption isotherms for the hygroscopic region. The points at 33%, 50% and 75% relative humidity are measured and are used to fit the GAB curves to be used in WUFI Plus. The fitting parameters have been calculated using the least squares method. The curves show values of moisture content typical for biobased materials. The samples S2 and S3 have similar values, while S1 shows lower values of moisture content that could be caused by different substrate compositions or the different coatings.

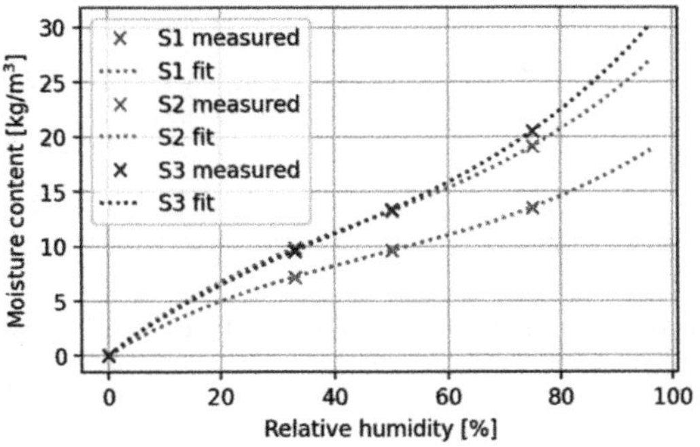

Fig. 3. Isotherms of absorption of the three samples and the GAB fitting used in the simulations.

Table 2 lists the estimations of μ and MBV for the three samples. The vapour resistance factors μ are of the same order of magnitude and the values are similar to the ones of wood fiber panels found in WUFI Plus. The *MBV* are between "good" and "excellent" Moisture Buffer Value classes.

Table 2. Estimated properties of the composites studied.

Sample	Vapour resistance factor μ [-]	*MBV* [g/(m^2 · %$_{RH}$)]
S1	3.2	2.9
S2	3.0	3.0
S3	8.6	2.0

It is worth noting that sample S3 has a thickness of 1.5 cm, which is less than half that of samples S1 and S2. This reduced thickness could explain the lower MBV value, even though the isotherm shows higher moisture content.

3.2 Simulation Results

Among the three samples, sample S2 is used for simulations. The panels are expected to reduce the relative humidity oscillation amplitude and improve comfort avoiding extreme values. A small effect is found also on the temperature values (Fig. 4) with a maximum absolute difference of 0.5 K. Figure 4 shows a week without heating or cooling, where the temperature is comprised between the two thermostats. In this case, the effects are negligible and the influence on heating and cooling load is not relevant.

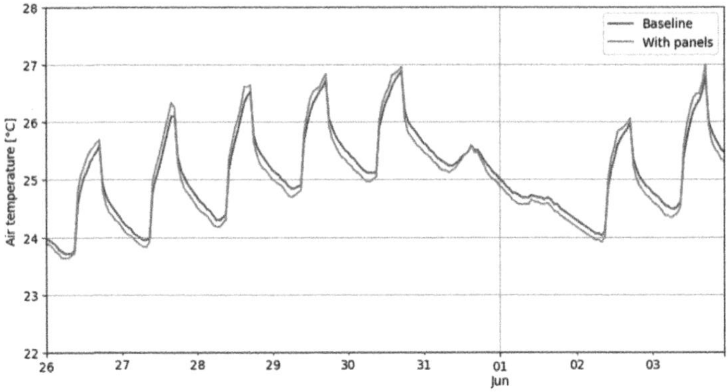

Fig. 4. Hourly dry bulb air temperature of the simulated office room during a week without heating or cooling.

On the other hand, relative humidity, as expected, shows larger deviations with a maximum of 8%. Figure 5 shows a week of August with high humidity peaks, and the presence of the panels reduces the peaks of 5% RH, closer to comfort conditions. Temperatures and vapour pressure are reported to exclude the effect of temperature on air. The peak in temperature is caused by the shutdown of the HVAC system.

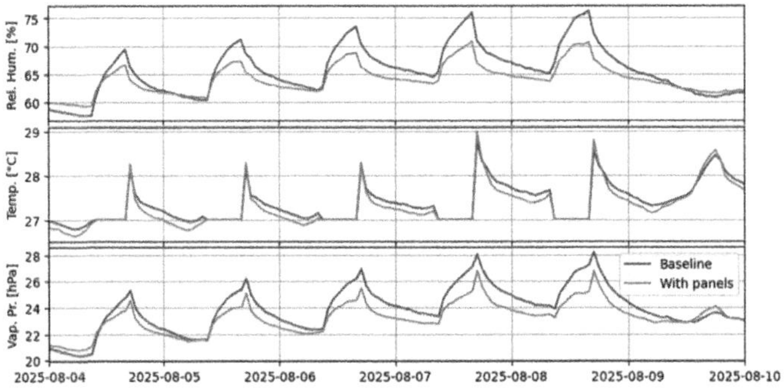

Fig. 5. Simulated hourly relative humidity, dry bulb air temperature and vapour pressure in the simulated office room during a summer week without heating or cooling.

Figure 6, on the other hand, shows the effects of the panels during days with low humidity values and low temperatures, during January. The effect of the panel is to lower the relative humidity during the occupied hours and to increase it during the non-occupied hours, as expected from materials with high *MBV*.

Finally, a relevant effect of the panels is found on the internal surface of the external wall. This variable presents a maximum difference of 8.5% RH. During the cold months

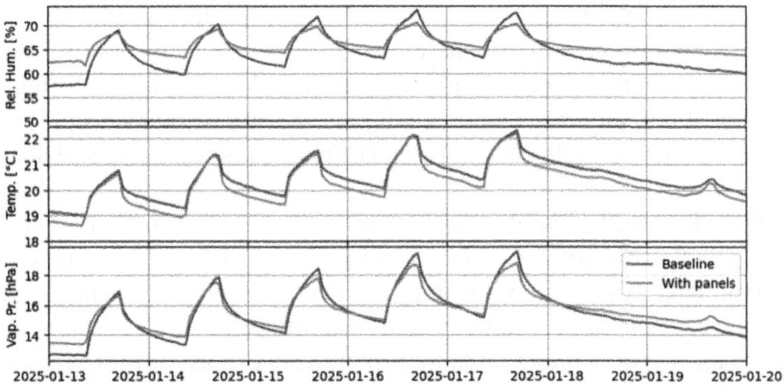

Fig. 6. Simulated hourly relative humidity, dry bulb air temperature and vapour pressure in the simulated office room during a week without heating or cooling.

the lower surface temperature of the external wall might be subject to mould growth conditions, above 80% RH, During the simulated period the maximum surface RH is found in the first week of October, shown in Fig. 7. On the 2nd and 3rd of October RH is higher than 80% for a few hours for the baseline simulation, while the simulation with panels is always on the safe side, below 77%. This behavior could be exploited in moisture critical environments to reduce the dehumidification demand and moisture related risks.

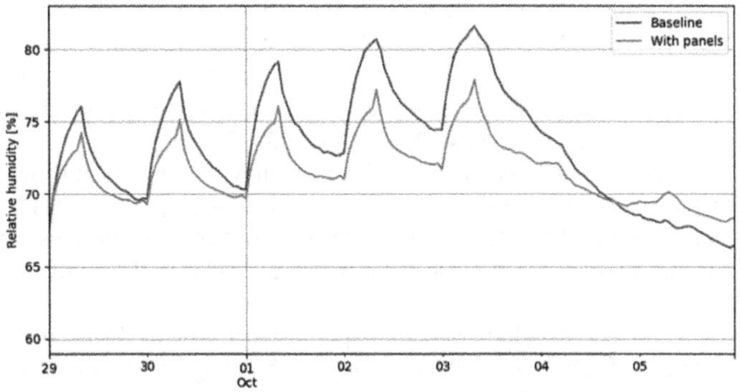

Fig. 7. Relative humidity at the internal surface of the external wall during the week with higher values.

4 Discussion and Conclusions

The moisture buffer values of the mycelium-based composites are between 2 and 3 g/(m^2 · $\%_{RH}$). These values are higher than the ones measured for spruce boards (1.2 g/(m^2 · $\%_{RH}$)) or birch panels (1) in [12], while are comparable with clay mortar (2.30), climatic plaster (2.57 g/(m^2 · $\%_{RH}$)) and lime plaster (2.62 g/(m^2 · $\%_{RH}$).) measured in [15]. The fact that some traditional non-bio-based material presents higher MBV underlines the relevance of having the possibility to compare the performance of materials with a parameter when designing the moisture response of the building envelope.

Whole building hygrothermal simulations is then a valuable tool that allows us to estimate the effectiveness of a material, nevertheless, the simulations present many limitations, for example, when considering the one of which is neglecting moisture hysteresis, that could influence indoor conditions and risk assessments [16] or the fact that the moisture loads are constant and could be not representative of realistic moisture loads [15].

The effect of the panels on the heating and cooling loads are negligible, while the effect on the air relative humidity and the surface relative humidity is more relevant (8% RH) and could lead to small comfort improvements and mould growth risk reduction, confirming the capacity of reducing the high humidity peaks.

In conclusion, the presented simulations show that the moisture buffering capabilities of mycelium-based composites, a circular bio-based material obtained from agriculture byproducts, could improve comfort and health conditions of a small office with high occupation and low air exchanges.

References

1. Mouton, L., Allacker, K., Röck, M.: Bio-based building material solutions for environmental benefits over conventional construction products–Life cycle assessment of regenerative design strategies (1/2). Energ. Build. **282**, 112767 (2023)
2. Le, D.L., Salomone, R., Nguyen, Q.T.: Circular bio-based building materials: a literature review of case studies and sustainability assessment methods. Build. Environ. **244**, 110774 (2023)
3. Ben-Alon, L., Loftness, V., Harries, K.A., Hameen, E.C.: Life cycle assessment (LCA) of natural vs conventional building assemblies. Renew. Sustain. Energy Rev. **144**, 110951 (2021)
4. Elsacker, E., Vandelook, S., Brancart, J., Peeters, E., De Laet, L.: Mechanical, physical and chemical characterisation of mycelium-based composites with different types of lignocellulosic substrates. PLoS ONE **14**(7), e0213954 (2019)
5. Danovska, M., et al.: Numerical and experimental study on the impact of humidity on the thermal behavior of insulated timber walls. Proc. Build. Simul. Appl. BSA **2019**, 91–99 (2020)
6. Jin, Q., Zhang, Z., Chen, J.: A study on the thermal performance of Pleurotus Ostreatus/Straw mycelium composites and its application in building envelopes. J. Build. Eng. **92**, 109646 (2024)
7. Alaneme, K.K., et al.: Mycelium based composites: a review of their bio-fabrication procedures, material properties and potential for green building and construction applications. Alex. Eng. J. **83**, 234–250 (2023)
8. Rode, C., et al.: Moisture Buffering of Building Materials. Technical University of Denmark DTU (2006)
9. Posani, M., et al.: Low-carbon indoor humidity regulation via 3D-printed superhygroscopic building components. Nat. Commun. **16**(1), 425 (2025)
10. Gentile, V., Libralato, M., Fantucci, S., Shtrepi, L., Autretto, G.: Enhancement of the hygroscopic and acoustic properties of indoor plasters with a Super Adsorbent Calcium Alginate BioPolymer. J. Build. Eng. **76**, 107147 (2023)
11. Gentile, V., Autretto, G., Libralato, M., Fantucci, S., Serra, V.: Modeling moisture buffering of innovative plasters from material properties to room scale. J. Build. Eng. **118**, 114983 (2026). https://doi.org/10.1016/j.jobe.2025.114983
12. Qin, M., Hou, P., Wu, Z., Wang, J.: Precise humidity control materials for autonomous regulation of indoor moisture. Build. Environ. **169**, 106581 (2020)

13. Gentile, V., Autretto, G., Libralato, M., Fantucci, S., Serra, V.: (Unpublished manuscript) Modeling Moisture Buffering of Innovative Plasters from Material Properties to Room Scale
14. Crank, J.: Section 4.3 "Non-steady state". The Mathematics of Diffusion. Oxford University Press (1975)
15. Larcher, M., Leonardi, E., Troi, A., Stefani, A., Nerobutto, G., Herrera-Avellanosa, D.: Assessing the impact of moisture buffering properties of materials on indoor environmental quality: a study on a recycled material plaster. Build. Environ. **267**, 112170 (2025)
16. Libralato, M., De Angelis, A., D'Agaro, P., Cortella, G., Qin, M., Rode, C.: Damage risk assessment of building materials with moisture hysteresis. J. Phys. Conf. Ser. **2069**(1), 012043 (2021). IOP Publishing

In-Situ Hygric Characterization of Historic Brick Masonry Using Karsten and Pleyers Tubes

Florian Verbruggen[1(✉)], Minting Chen[1], and Evy Vereecken[1,2]

[1] Department of Civil Engineering, Section of Building Physics and Sustainable Design, KU Leuven, Heverlee (Leuven), Belgium
florian.verbruggen@student.kuleuven.be, minting.chen@kuleuven.be
[2] Monuments and Monumental Decoration Lab, Royal Institute for Cultural Heritage (KIK-IRPA), Brussels, Belgium

Abstract. This study evaluates the accuracy and reproducibility of the Karsten tube and Pleyers tube methods for determining the capillary absorption coefficient of ceramic bricks and stones. Seven different brick and stone types with varying hygric properties were tested. For each type, five individual bricks were subjected to three repeated Karsten tests to assess inter- and intra-brick variability. The coefficient of variation (CV) for intra-brick measurements ranged from 5.9% to 10.6%, indicating good reproducibility. Inter-brick variability, however, was more pronounced, with CVs up to 29.3%, highlighting significant material heterogeneity within some brick types.

To improve measurement accuracy, a custom-designed ring was developed that ensures a consistent and circular effective contact area, eliminating the need to manually measure the radius and reducing uncertainty in the derived capillary absorption coefficient.

A comparison between the Karsten and Pleyers methods was conducted on the same brick for each type, with the bricks dried in between tests to ensure identical starting conditions. The results from both methods were next compared to a reference value obtained from a standard water uptake test. The findings show that the Pleyers tube does not yield more accurate results than the Karsten tube including analytical description and is more challenging to handle. Given its ease of use, broad applicability, and sufficient accuracy, the Karsten method is preferred.

Overall, this research supports the use of the Karsten tube, especially when combined with the ring, as a robust in-situ method for evaluating the capillary absorption coefficient of masonry materials.

Keywords: Karsten Tube · Pleyers Tube · Capillary absorption coefficient · non-destructive measurements · moisture properties

1 Introduction

Historic buildings represent a vital part of our cultural heritage and contribute significantly to the identity of cities and regions. Often protected due to their historical and architectural value, they carry the narratives of the past and are considered indispensable elements of the built environment. As such, they must be preserved for future generations.

M. Posani and S. Singh (Eds.): RYS 2025, RILEM Bookseries 67, pp. 964–975, 2026.
https://doi.org/10.1007/978-3-032-14652-6_70

In the case of listed heritage buildings, any intervention that may compromise their cultural value is strictly restricted. This principle, known as passive conservation, prohibits alterations that could visually or structurally damage the historic fabric [1]. Masonry is one of the most prevalent construction techniques in heritage buildings and structures. However, due to conservation regulations, it is often not permitted to extract samples destructively for laboratory-based characterization of masonry properties.

Brick masonry is particularly vulnerable to various forms of deterioration, such as frost damage (as shown in Fig. 1), salt crystallization, and mould growth. These forms of damage are often related to capillary moisture transport within the masonry. Such degradation processes can weaken the structural integrity of a building and pose a serious threat to the long-term preservation of heritage assets [2]. Therefore, understanding the mechanisms by which moisture is absorbed, transported, and retained in masonry materials is essential for effective conservation.

Fig. 1. Frost damage on the perimeter wall of the Arenberg Park, Kardinaal Mercierlaan, Heverlee (Belgium).

In many cases, internal insulation is the preferred solution for renovation purposes, as external insulation can alter or obscure the building's historic appearance. However, internal insulation introduces additional hygrothermal risks, such as condensation and mould growth, especially if not properly assessed. To mitigate these risks, a detailed hygrothermal risk assessment is necessary. Such a risk assessment relies on numerical simulations that model the heat and moisture transport within the building envelope. The reliability of these simulations is highly dependent on the availability of accurate and material-specific input data [3].

Traditional methods for determining the hygric properties of masonry materials typically involve destructive laboratory tests. In the context of heritage conservation, obtaining such samples is often not feasible. As a result, promising in-situ measurement techniques, such as the Karsten tube and Pleyers tube, are being explored as non-destructive alternatives. These methods enable the direct assessment of moisture-related properties of bricks on-site, preserving the material integrity and aligning with the principles of passive conservation.

2 Methods and Materials

To establish a benchmark, standard water uptake tests [4] will initially be conducted on a selection of brick samples. These standardized laboratory tests serve as a reference, enabling a direct comparison with the results obtained through in-situ measurement techniques. The primary objective of this study is to perform a precise and comprehensive comparison between in-situ methods and traditional laboratory tests.

Seven distinct types of bricks and stones were selected for this study, each differing in their physical characteristics. This diversity allows for the assessment of the performance and reliability of the various in-situ measurement techniques under different material conditions. The selected brick types are presented in Fig. 2 and are identified by a unique letter code.

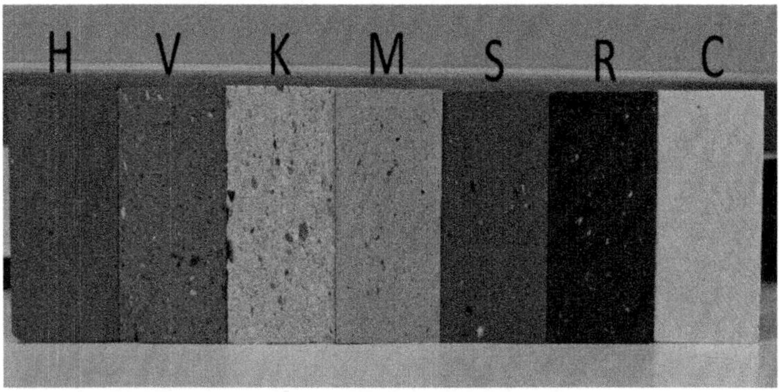

Fig. 2. Selection of distinct brick types identified by unique letter codes.

2.1 Karsten Tube Method

Currently, the most commonly used simplified method to estimate the capillary absorption capacity of a material is the $\Delta_{(5-15)}$-method. This approach involves determining the increase in moisture content between two fixed time intervals—typically 5 and 15 min—during a Karsten tube test. Importantly, this method does not take into account the three-dimensional nature of moisture transport into the material, as it simplifies the absorption process to a one-dimensional analysis. This simplification makes the method especially practical for rapid or in-situ assessments, albeit at the cost of precision.

2.1.1 Analytical Description

To enable a more general physical interpretation and comparison of Karsten tube results, we drew upon the sharp front approximation theory proposed by Smettem et al. [5] and referred to the computational approach outlined by Hendrickx [6], which we slightly modified for the current study [7]. Based on these methods, the results obtained from the Karsten tube were converted into the widely used physical parameter, the capillary absorption coefficient, A_{cap} [kg/(m^2·s)]. The analytical procedure is detailed as follows. Neglecting gravity, the analytical equation for the three-dimensional water flow from the source was derived using the sharp front approximation, as shown in Eq. (1). In this model, the wetted front is assumed to be circular rather than parabolic, as depicted in Fig. 3.

$$V_{abs} = \pi R_e^2 A_{cap}\sqrt{t} + \frac{\pi R_e \gamma A_{cap}^{2}}{\theta_{cap}} t \tag{1}$$

where:

V_{abs} is the absorbed water volume (mL); R_e is the effective water source radius (mm), defined as the actual contact radius between the Karsten tube and the brick, excluding the sealant; A_{cap} is the capillary absorption coefficient (kg/(m^2·s)); θ_{cap} is the capillary saturated volume moisture content (m^3/m^3); t is the time (s); γ is a constant, empirically determined as 0.75 [5].

The parameter θ_{cap} is an intrinsic material property and should remain constant during the test. It is defined as the ratio of the absorbed water volume V_{abs} to the total wetted volume V_{wet}, see Eq. (2):

$$\theta_{cap} = \frac{V_{abs}}{V_{wet}} \tag{2}$$

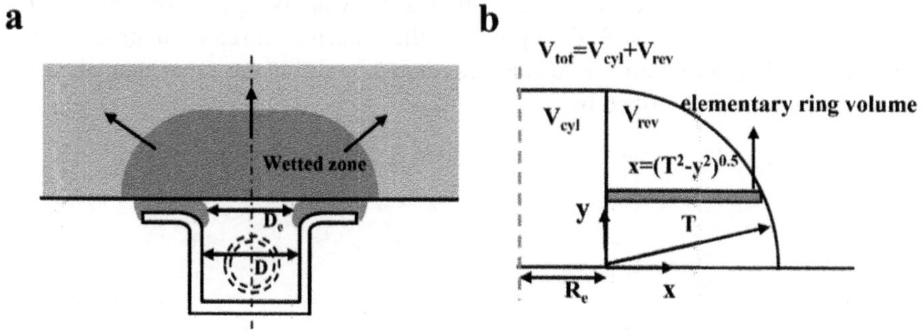

Fig. 3. Wetted zone [6]: (a) Horizontal section through the contact zone of the Karsten tube, illustrating the approximate geometry of the wetted zone based on the sharp front approximation, (b) Geometry of the wetted volume (a section containing both the axis and a radius), with an indication of an elementary ring volume used for integrating the total volume of the revolution solid, Vrev.

As shown in Fig. 3b, the total wetted volume V_{wet} consists of the sum of the volume of a cylinder (V_{cyl}) and a revolution solid (V_{rev}) which is generated by the rotation of a quarter-circle with the penetration depth as its radius. While the penetration depth is a variable, within the integral calculus used to determine the volume, it is treated as a constant, denoted as T. This parameter defines the relationship between the variables x and y. By eliminating x, the integral is computed as a summation of elementary ring volumes, each with a rectangular cross-section of height dy and width $\sqrt{T^2 - y^2}$, where y varies from 0 to the penetration depth T.

$$Vring = \pi\left(T^2 - y^2 + 2Re\sqrt{T^2 - y^2}\right)dy \tag{3}$$

$$V_{rev} = \int_0^T \left[\pi\left(T^2 - y^2 + 2Re\sqrt{T^2 - y^2}\right]dy\right] \tag{4}$$

$$V_{\text{wet}} = V_{\text{cyl}} + V_{\text{rev}} = \pi R_e{}^2 T + \frac{1}{6}\pi T^2(3\pi R_e + 4T) \tag{5}$$

When T is replaced by a variable penetration depth x(t), this formula expresses the wetted volume as a function of time. Considering that the penetration depth x(t) is equivalent to the wetted front radius $R_{wet}(t)$, Eq. (5) can be rewritten as:

$$V_{\text{wet}} = V_{\text{cyl}} + V_{\text{rev}} = \pi R_e{}^2 R_{wet}(t) + \frac{1}{6}\pi R_{wet}(t)^2(3\pi R_e + 4R_{wet}) \tag{6}$$

2.1.2 Procedure

The determination of the capillary absorption coefficient A_{cap} follows a systematic procedure, as illustrated in Fig. 4 [6]. First, the absorbed water volume V_{abs} and the radius of the wetted front R_{wet} are recorded over time. Using these measurements, the capillary moisture content θ_{cap} is calculated at each time step by applying Eqs. (6) and (2). The average value of θ_{cap} across all data points is then taken as a representative material parameter. Finally, the parameter A_{cap} is determined by fitting the experimental data to the theoretical model described by Eq. (1).

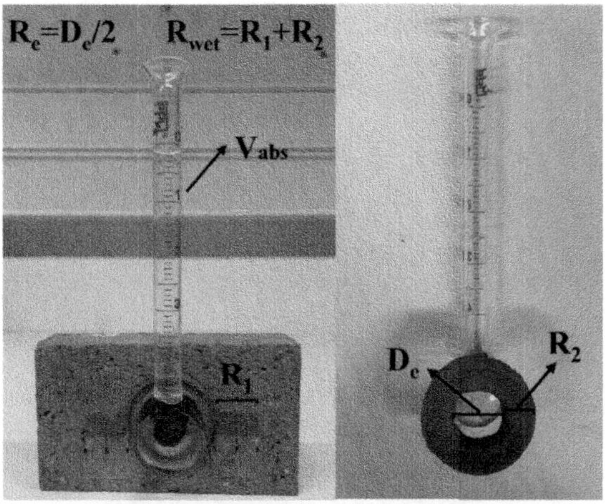

Fig. 4. Parameter Measurement.

2.1.3 Practical Observations

(1) **Invisible Moisture Front**

During testing, the lateral moisture front cannot be directly observed due to obstruction by the sealing compound. It only becomes visible once it extends beyond the seal, at which point the radius R_{wet} can be measured and used in standard calculations.

(2) **High Capillary Absorption Coefficients**

Materials with high capillary absorption coefficients such as calcium silicate bricks (C-brick) and Vandersanden Robusta bricks (R-brick) with $A_{cap} > 0.50\ kg/(m^2 \cdot s)$, absorb water rapidly, causing the water level in the graduated Karsten tube to drop quickly. This makes accurate reading more difficult and requires more frequent refilling, which may further affect precision.

(3) **Surface Irregularities and Cracks**

The Karsten test should not be performed on surfaces with visible cracks or irregularities. Cracks can accelerate water ingress, leading to an overestimation of the material's capillary absorption and resulting in unreliable measurements.

2.2 Pleyers Tube Method

Alternatively, a Pleyers tube could be used (Fig. 5), which includes an outer ring via which the three-dimensional effect of the water absorption (that occurs via the inner tube) can be minimized or precluded [8].

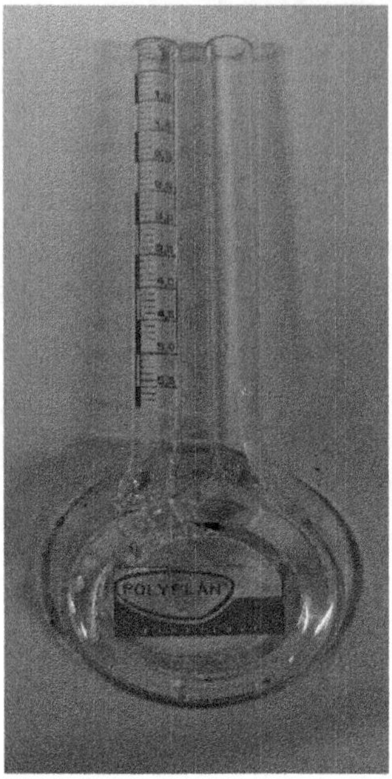

Fig. 5. Pleyers tube

The Pleyers tube method is conceptually similar to the Karsten tube test but introduces a key refinement. In the Pleyers setup, two concentric water cups are used, each connected to a separate measuring tube. The inner water cup is attached to a calibrated tube—much like in the Karsten test—while the outer water cup surrounds the inner one and is connected to an uncalibrated tube. The purpose of this outer cup is to saturate the surrounding area and thus prevent lateral, three-dimensional water flow from the inner cup. This setup ensures that water absorbed from the inner cup is transported strictly in one dimension, eliminating the need for corrections related to three-dimensional moisture transport, which are required in the Karsten method. This can be described analytically using Eq. (7).

$$m_w\left[\frac{kg}{m^2}\right] = A_{cap}\left[\frac{\mathrm{kg}}{\mathrm{m}^2\sqrt{\mathrm{s}}}\right]\sqrt{t}[\sqrt{\mathrm{s}}] \tag{7}$$

2.2.1 Practical Observations

(1) **Difference in Absorption Rates**

The uncalibrated tube typically absorbs water faster than the calibrated one, due to three-dimensional moisture spreading. In contrast, the calibrated tube measures controlled, one-dimensional uptake. To ensure equal pressure on the test surface, the uncalibrated tube must be regularly refilled and kept at the same level as the calibrated tube.

(2) **High Capillary Absorption Coefficients**

The same practical observances apply as with the use of the Karsten tube.

(3) **Communicating Vessels Effect**

In bricks with very open pore structures, the two tubes may behave as communicating vessels—keeping the water levels perfectly equal. This indicates internal hydraulic connectivity and makes reliable measurement impossible.

(4) **Limitations in On-Site Application**

The Pleyers apparatus has a diameter of 70 mm, which exceeds the height of standard bricks. As a result, proper placement on a single brick without overlap or leakage is often not feasible, limiting in-situ use.

(5) **Surface Irregularities and Cracks**

The same practical observances apply as with the use of the Karsten tube.

3 Results and Discussion

3.1 Karsten Tube Error Analysis

During the measurement campaign, three variables (R_e, R_{wet} and V_{abs}) are recorded simultaneously as shown in Fig. 4. To determine which parameter has the greatest influence on the final result, an error analysis is performed. The possible reading error is assumed to be equal to the smallest unit of measurement, which is $\Delta R_e = 1$ mm, $\Delta R_{wet} = 1$ mm and $\Delta V_{abs} = 0.1$ ml. To assess the impact of these uncertainties on the experimental results, a local perturbation was conducted for the sensitivity analysis, evaluating how variations within these error ranges influence the outcomes while keeping other conditions constant. The results of this analysis are presented in Fig. 6. The figure shows that the measurement error in R_e increases rapidly, whereas the error in R_{wet} grows more gradually. In comparison, V_{abs} contributes the least to the overall uncertainty.

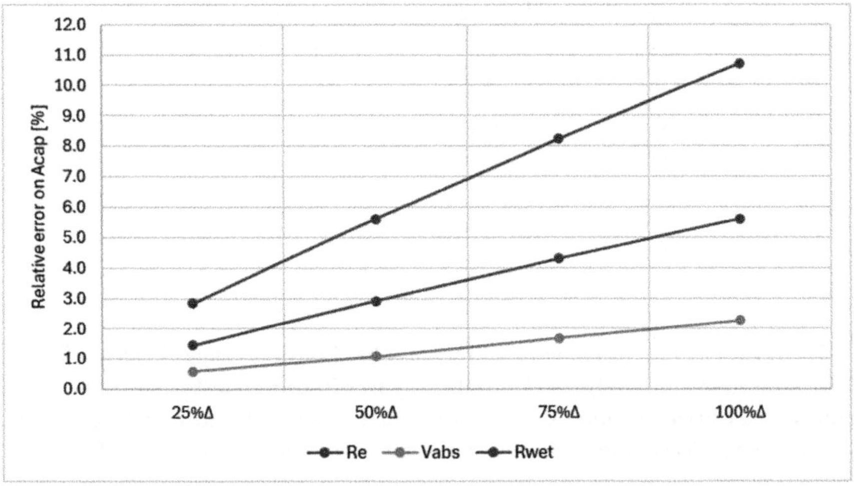

Fig. 6. The variation in error factors affecting the experimental results of the Karsten tube

To minimize error propagation, a technique is required that ensures an accurate reading of the effective radius. For this purpose, a specialized ring was designed by the Building Physics Laboratory at KU Leuven. This ring eliminates the need to measure the effective diameter D_e, as it guarantees an optimal and perfectly circular shape. Additionally, the ring fits precisely into the water cup of the Karsten tube without obstructing water flow. The usage of the ring is illustrated in Fig. 7.

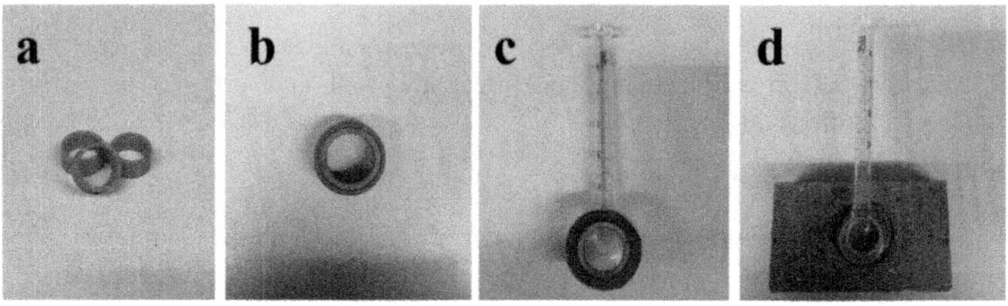

Fig. 7. Application process of the Ring on the Karsten Tube: (a) Ring with a diameter of 21 mm, (b) Sealant applied around the ring, (c) The sealant-coated ring attached to the Karsten tube, (d) Karsten tube with the ring fixed to the brick surface.

3.2 Repeatability of the Karsten Tube and Material Variability in Bricks

Seven brick types with varying hygric properties were tested. For each type, five individual bricks were subjected to the Karsten tube, with three repeated measurements per brick. Bricks were fully dried between tests to ensure consistent starting conditions. This setup allowed evaluation of both intra-brick (repeatability) and inter-brick (material variability) differences. The results are shown in Table 1.

Table 1. CV of Capillary Absorption Coefficients: Intra- and Inter-Brick Variability

Type brick	$CV_{intrabrick}$ [%]	$CV_{interbrick}$ [%]
H	9.3 ± 4.3	9.5
V	5.9 ± 2.6	29.3
K	6.2± 2.6	15.6
M	8.6 ± 5.3	2.4
S	7.3 ± 4.9	12.6
R	10.6 ± 5.4	3.6
C	8.3 ± 5.9	6.6

The coefficient of variation (CV) for intra-brick results ranges between 5.9% and 10.6%, indicating relatively good repeatability of the Karsten test on the same brick. In contrast, inter-brick variation shows a wider range from 2.4% to 29.3% depending on the brick type. These results suggest that differences between individual bricks of the same type can vary significantly. When inter-brick variability is high, such as for brick type V (29.3%), it indicates that the material is less homogeneous. Nevertheless, the relatively low intra-brick CV in such cases (e.g., 5.9% for type V) confirms that the test method itself remains stable and reproducible, even when material variation increases.

3.3 Pleyers Tube Error Analysis

To assess the impact of reading errors on the final result, it is essential to identify which parameters require manual reading. In the Pleyers tube, this is limited to the graduated measurement tube. Although the effective contact diameter would typically need to be read as well, this is avoided, similar to the Karsten tube, by using a custom-designed ring developed by the Building Physics Laboratory at KU Leuven. This ring ensures a constant, perfectly circular effective diameter, thereby enhancing measurement accuracy (Fig. 8).

Fig. 8. Rings for the Pleyers tube

The potential reading error on the water level corresponds to the smallest division on the graduated tube, i.e., 0.1 l/m^2, and directly affects the accuracy of the calculated capillary absorption coefficient A_{cap}. Equation (8) expresses the absolute error on A_{cap} as a function of this reading uncertainty.

$$\Delta A_{cap} = \sqrt{\left(\frac{\partial A_{cap}}{\partial m_w}\Delta V_{abs}\right)^2} = \frac{1}{\sqrt{t}}\Delta V_{abs} \tag{8}$$

3.4 Comparison of the Pleyers Tube, Karsten Tube, and Standard Water Uptake Method

For consistency, the test was conducted on the same brick that had previously been used for the Karsten test. Between both tests, each brick was thoroughly dried to ensure that every measurement started from a comparable initial condition. By performing both tests on the same specimen, the testing methods can be directly compared without the influence of inter-brick material variability. Additionally, a standard water uptake test was conducted for one sample on the same brick for each type. This allows the results of the Pleyers and Karsten tests to be compared against a common reference value. The results are shown in Fig. 9.

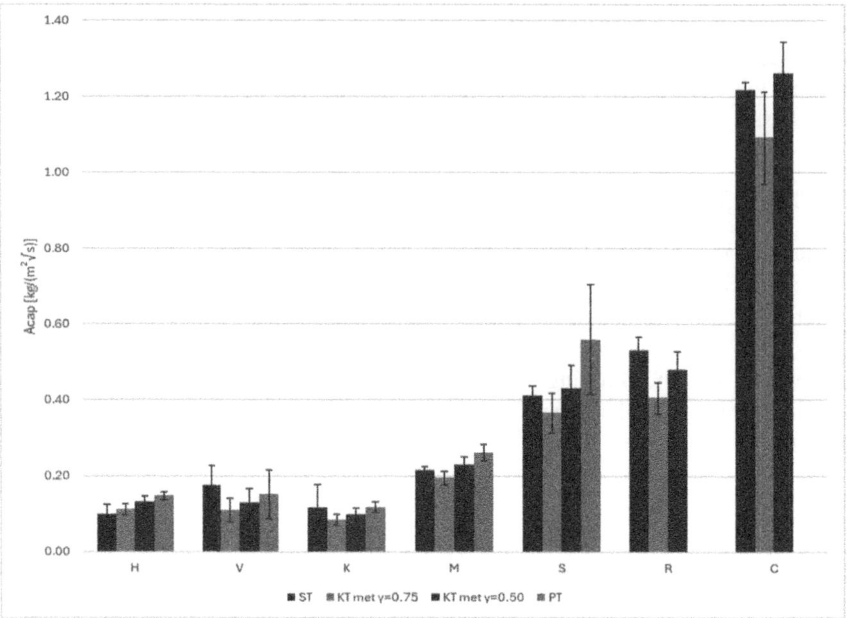

Fig. 9. Comparison of the Pleyers Tube, Karsten Tube, and Standard water uptake method

Figure 9 shows that the Pleyers tube does not yield better results than the Karsten test when compared to the reference values from the standard water uptake test. Since the Pleyers tube is more difficult to handle and does not offer additional accuracy, the Karsten tube is preferred. Its smaller diameter makes it easier to attach and allows for broader applicability across different types of bricks.

4 Limitations of External Influences

In in-situ measurements, prior drying of the material is not possible. As a result, external influences on the material under investigation must be taken into account. Environmental factors such as precipitation, humidity, and temperature can significantly affect both the progress and the outcome of the measurements [9, p. 14], [9, p. 83], [9, p. 88], [10] and [11].

The main external factor influencing the capillary absorption coefficient in measurements is driving rain, as it leads to an increased initial moisture content of the masonry. A second important factor is the temperature of the test water used. In contrast, the relative humidity has only a limited effect compared to these two factors [7].

5 Conclusion

The results of the conducted tests demonstrate that the Karsten method is a reliable and reproducible technique for determining the capillary absorption coefficient of bricks. Analysis of intra- and inter-brick variations shows that measurements on the same brick are relatively consistent, with coefficients of variation ranging between 5.9% and 10.6%. In contrast, variations between different bricks of the same type can be more pronounced, indicating material inhomogeneity. Error analysis reveals that the measurement uncertainty in the effective radius (R_e) has the most significant impact on the final result. To minimize this error, a specially designed ring was developed that eliminates the need for manual measurement and ensures a constant, circular contact radius.

When comparing the Karsten tube with the Pleyers tube, the Pleyers tube does not provide a noticeable improvement in accuracy, despite being more difficult to handle. Additionally, the smaller diameter of the Karsten tube allows for easier attachment and wider applicability across different brick types. Both test methods were benchmarked using a standard water uptake test performed on the same brick, enabling a direct comparison without inter-brick variability influencing the results.

In summary, the Karsten method offers the most advantages for determining the capillary absorption coefficient of masonry, thanks to its simplicity, reproducibility, and accuracy. The use of the specially developed ring further optimizes the test setup by reducing the margin of error.

References

1. Vlaamse overheid: Agentschap Onroerend Erfgoed, "Onroerend erfgoed: een toelichting", Brussels (2014)
2. Buildwise: Vocht in gebouwen: Bijzonderheden van opstijgend vocht. Buildwise, Brussels (2014)
3. Dang, X., Vereecken, E. Janssen, H., Roels, S.: Impact of semi-empirical methods implemented in heat, air, and moisture (HAM) models on predicted wind-driven rain (WDR) loads and hygrothermal responses. Elsevier (2024)
4. EN ISO 15148:2003, Hygrothermal performance of building materials and products - Determination of water absorption coefficient by partial immersion (ISO 15148: 2002), Brussel: European committee for Standardization (2002)
5. Smettem, K.R.J., Parlange, J.Y., Ross, P.J., et al.: Three-dimensional analysis of infiltration from the disc infiltrometer: 1. A capillary-based theory Water Resources Research (1994)
6. Hendrickx, R.: Using the Karsten tube to estimate water transport parameters of porous building materials: the possibilities of analytical and numerical solutions. Mater. Struct. (2013)
7. Verbruggen, F.: Hygrothermal characterization of historic masonry based on in-situ testing: Karsten and Pleyers tube. KU Leuven, Belgium (2025). (in Dutch)
8. Haindl, K., Schöner, T., Zirkelbach, D., Fritz, C.: Was ist bei Karsten & Co. Zu beachten. Fraunhofer Institut für Bauphysik, IBP, Holzkirchen (2016)
9. Stelzmann, M.: Entwicklung und Validierung eines Verfahrens zur Untersuchung des Schlagregenschutzes von Fassaden denkmalgeschützter Bestandsgebäude. Technische Universität Dresden: Institut für Bauklimatik, Dresden (2020)
10. Hall, C., Hoff, W.D., Skeldon, M.: The sorptivity of brick: dependence on the initial water content. Building Science Laboratory, Department of Building, UMIST, Manchester M60 1QD, UK (1983)
11. Vandevoorde, D., Cnudde, V., Meynen, V., de Bouw, M., Janssens, K.: Influence of environmental conditions on water adsorption measurements performed by non-destructive in situ applicable techniques. Built Heritage 2013 Monitoring Conservation Management (2013)

Cultural Heritage

Carbonation Pathways of Mayan Engineering: Mineralogical Insights into Lime-Based Mortars from the Muyil Pyramid

Md. Montaseer Meraz[1], Pavan Akula[1(✉)], Francisco Javier Castaneda[2], and O. Burkan Isgor[1]

[1] Oregon State University, Corvallis, OR 97333, USA
pavan.akula@oregonstate.edu
[2] Grupo Calidra, 01210 Mexico City, Mexico

Abstract. The Mayans' skill in crafting highly durable lime-based mortars from natural in-situ materials highlights the sophistication of their engineering. Characterization of such mortars can provide valuable insights into the phases that contribute to long-term physicochemical performance. This study presents Mayan mortar's preliminary mineralogical characterization data collected from the Muyil pyramid (built around 300 B.C.) in Quintana Roo state of Mexico. Specifically, X-Ray Diffraction (XRD), and Thermogravimetric analysis (TGA) of the mortar indicate that mortar primarily comprises lime and calcareous soil. The strength gain is attributed mainly to the binding of soil particles from calcium carbonate precipitation. Representative modern mortar samples cast with lime and in-situ soil sourced from the Muyil pyramids indicate that the calcium carbonate polymorphs present in the soil played a critical role in the kinetics of carbonation. The high amounts of Aragonite and calcite in the in situ soil significantly enhance the carbonation kinetics and contribute to the strength improvement. Preliminary results from mineralogical characterization indicate that Aragonite and calcite act as nucleation sites, facilitating the improved kinetics of lime carbonation.

Keywords: Mayan mortar · Ancient mortar carbon mineralization · In-situ resource utilization · Carbonation kinetics

1 Introduction

In Mesoamerica, the Maya developed a lime-based building material and successfully utilized it to construct a variety of structures that have been standing for thousands of years. These structures symbolize the Mayan's understanding of geology and skill, demonstrating a remarkable durability by surviving in hot, humid, aggressive environmental conditions [1].

The utilization of calcined lime was initially introduced during the Neolithic period, approximately 10,000 BCE [2]. The technology became widespread across various continents and cultures. Archaeological evidence indicates that the use of burnt lime emerged in Mesoamerica between 1100–1200 BCE [3]. To produce lime plasters, the limestone

M. Posani and S. Singh (Eds.): RYS 2025, RILEM Bookseries 67, pp. 979–991, 2026.
https://doi.org/10.1007/978-3-032-14652-6_71

was burned to create quicklime, which was subsequently combined with water to create slaked lime. However, the distinctiveness of the lime plasters in ancient Maya constructions is attributed to the incorporation of $CaCO_3$-rich local material called sascab and an organic extract from tree bark known as Chukhum [4–6].

Another distinguishing feature of the Mayan mortar that sets it apart from other civilizations of the same historic times is the level of sophistication that it possesses. While the Egyptians utilized binders based on gypsum and calcite, the Anatolians utilized a straightforward lime formulation [7]. On the other hand, Roman mortar incorporated volcanic ash to improve its durability. This is where the Mayan strategy was essentially distinct. Rather than utilizing geological accelerators, they proficiently used biological analogs [8]. Rodriguez-Navarro et al. [9] found that the Mayan mortars from the Copan region had intracrystalline characteristics inside the calcite matrix, influencing the hardness and durability of the material.

Limited studies have explored the secrets of Mayan mortar and hypothesized the mechanisms that contributed to their durability [6, 9–13]. Although the exact mechanisms remain under investigated, these studies suggest that the use of in-situ materials (e.g., soil, and organic resin extracts from Chukum trees) enhanced the properties of these materials. This has garnered significant scholarly interest. In previous studies, considerable attention was directed towards understanding and characterizing the properties of Mayan plasters; however, the characteristics of the mortar and the mineralogical kinetics influencing the engineering performance and durability remain under investigation.

In this study, mortar samples collected from the Muyil pyramids were analyzed using advanced mineralogical characterization techniques, including X-ray Dif-fraction (XRD) and Thermogravimetric Analysis (TGA). Representative modern mortar mixtures were prepared using lime and locally sourced soils from the region surrounding the pyramids and evaluated for both engineering properties and mineralogical evolution. Preliminary results reveal key mechanisms governing the reaction kinetics and phase transformations, offering insights into the long-term performance of these ancient lime-based materials.

2 Methodology

2.1 Materials

The mortar sample (Fig. 1) was collected from the Muyil pyramid (Fig. 1) (Zona Arqueológica de Muyil pyramid built between 600 and 900 CE) [14] by the permission of INAH Center Quintana Roo (Instituto Nacional de Antropología e Historia, National Institute of Anthropology and History) of the Mexican government. The collected sample appears to be a solid lump with an uneven, coarse surface roughness and is light gray in color with darker grey spots. The dimensions of the sample was measured using a slide caliper at three different points along both the length and width to account for shape irregularities (Table 1).

In-situ soil from regions adjacent to the pyramid and traditional lime from Mexico were used for preparing the replicate mortar mixture. The soil has a clay content of 2.05% and the silt content was found to be 0.5.% from the hydrometer test according to

ASTM D7928-21 [15]. The specific gravity of soil was 2.59. The moisture content of the soil was 1.11%.

The sand used in this study was collected from local supplier. The Fineness Modulus of the sand was 2.60. The specific gravity and water absorption of the sand was 2.63 and 0.4% respectively.

Fig. 1. Appearance of the collected Sample.

Table 1. Dimensions of the samples.

Sample ID	Length (mm)			Average Length (mm)	Width (mm)			Average Width (mm)
Mayan Sample	11.99	14.81	11.71	12.84	9.52	13.23	9.87	10.87

2.2 Methods

Mix Design of Representative Modern Mortar

The mix design adopted in this study for the representative modern mortar is summarized in Table 2. The control mix (Mix C) consisted of commercially available sand conforming to ASTM C778-21 [16] and traditional high-calcium lime (comprising ~90% CaO and ~10% $CaMg(CO_3)_2$). The Mix A was prepared using in-situ soil collected from the vicinity of the Mayan pyramid site and traditional high-calcium lime, without any sand replacement. All components listed in Table 2 were dry-mixed for two minutes using a Hobart mixer. Water was then added gradually and mixed for an additional three minutes to ensure homogeneity.

Cubic specimens measuring 50 mm × 50 mm × 50 mm were cast for compressive strength testing. Two distinct curing regimes were employed to assess both engineering and mineralogical performance. Normal Curing (NC) condition comprises of curing at 23 ± 2 °C and 55% Relative Humidity (RH). Similarly, the Accelerated Curing condition comprised of curing at 20% CO_2, 25 °C, and 60% RH in a controlled carbonation chamber. The accelerated curing was only intended to increase carbonation as deep as possible through the samples rather than being a representative carbonation exposure.

Table 2. Mix proportions by volume.

Mix ID	Mayan Soil	Mexican Lime	Sand	w/s (Water to binder Ratio)
C (Control)	–	1	3	1.35
A	3	1	–	1.35

Experimental Framework

The full testing matrix is shown in Table 3. For each condition and age, three replicate specimens were tested for compressive strength, X-ray diffraction (XRD), and thermogravimetric analysis (TGA), with depth profiling conducted to capture mineralogical gradients.

Table 3. Experimental framework.

Mix ID	Curing Regime	Compressive Strength	Target Depths for XRD/TGA (mm)
Mix C	1, 28 days NC; 28 days AC	✓	1, 3, 6, 11, 16, 21
Mix-A	1, 28 days NC; 28 days AC	✓	1, 3, 6, 11, 16, 21

NC – Normal Carbonation Curing, AC – Accelerated carbonation curing (20% CO_2) condition, ✓ - carried out with three replicates.

Compressive strength testing was conducted in accordance with ASTM C109/C109M-20 [17]. Cube specimens (50 mm) were tested at 7- and 28-days following demolding. Additional samples were cured for 28 days under accelerated carbonation conditions. Testing was performed using a calibrated universal testing machine equipped with hardened steel bearing platens. The loading rate was maintained at 2.4 ± 0.2 kN/s (equivalent to 0.9 ± 0.1 MPa/s) until failure. Peak load at failure was recorded and used to compute compressive strength.

XRD analysis was conducted using a Bruker D2 diffractometer. Approximately 7 g of powder from each depth was manually ground using an agate mortar and pestle to preserve crystallinity and prevent contamination. The powdered samples were scanned across a relevant from 5° to 75° 2θ range to identify both primary and minor crystalline phases.

TGA was performed using a TGA-550 instrument. Approximately 40 mg of each ground sample was placed in a platinum crucible and heated from 22 °C to 900 °C at a constant rate of 20 °C/min under flowing air or nitrogen. Mass loss was recorded to quantify phase decomposition, particularly for portlandite, calcite, and other carbonate-bound phases.

3 Results

3.1 Mayan Mortar Characterization

XRD

Figure 2 represents the XRD data of the Mayan Mortar and Mayan Soil. The Mayan mortar sample was collected from the Muyil Pyramid (Fig. 1), as described in the Materials section. Additionally, the Mayan soil, used in experimental mixes, was also sourced directly from the same site to maintain material authenticity. The data shows that the mortar does not contain any residual lime or portlandite, indicating complete carbonation. In addition, calcite and aragonite, calcium carbonate polymorphs, were the two dominating phases identified in the sample. The XRD of the in-situ soil shown in Fig. 2 also indicates the presence of calcite and aragonite with different recorded intensities for both phases. The presence of aragonite in both the soil and the mortar indicates that the soil was used in making the mortar. The higher intensity of calcite and aragonite peaks in the mortar sample indicates that the mortar has undergone carbonation over time. Similar characteristics, such as the use of in-situ materials (e.g., tree bark extracts) have also been observed in the lime plasters from other Mayan structures [9]. Mineralogically, Calcite is hexagonal in shape and exhibit a rhombohedral cleavage [18]. It is also more stable at room temperature and pressure. On the other hand, the aragonite is orthorhombic in shape and exhibits a prismatic cleavage [18]. Aragonite is more stable at high temperatures and pressures. Aragonite has a higher density and hardness than Calcite [19].

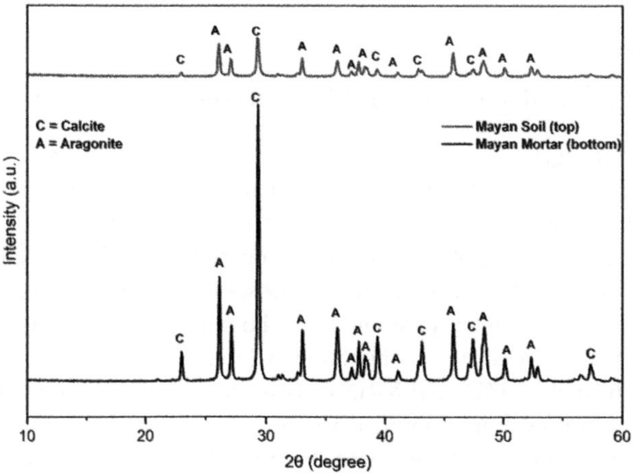

Fig. 2. XRD pattern of the Mayan Mortar (Sample 1) and Mayan Soil.

TGA

Figure 3 shows the TGA and DTG plots for the mortar sample (Mayan Sample. The DTG shows a peak at 800 °C (approximately), corresponding to the decarboxylation peak of calcium carbonate. The absence dehydroxylation peaks typically observed around 400 °C for portlandite, further validates carbonation of lime in them mortars. This data agrees with the XRD results indicating that calcium carbonate as the prominent phase in the mortar.

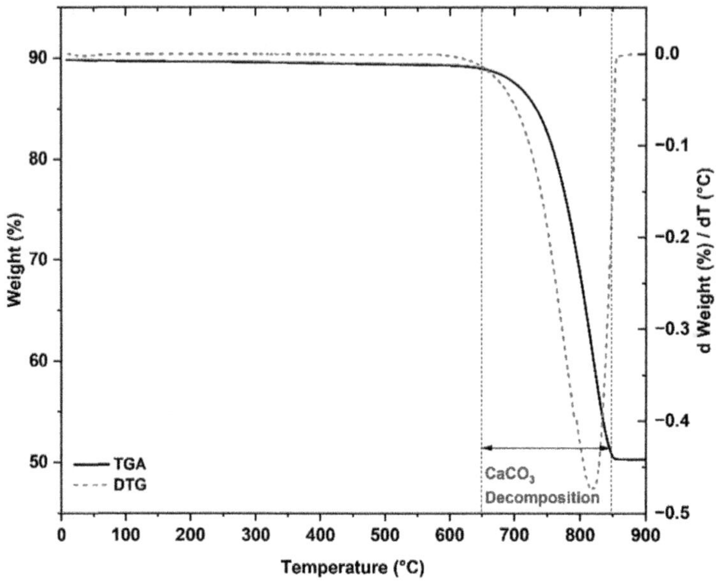

Fig. 3. TGA and DTG curve of the Mayan mortar (Mayan Sample).

3.2 Representative Modern Mortar Characterization

Compressive Strength

The compressive strength of the mixes is depicted in Fig. 4, which demonstrates that Mix-A consistently exhibits a higher compressive strength than Mix-C (Control mix) throughout all curing ages. In the normal curing condition, Mix-A attained 1.07 and 2.89 MPa at 7 and 28 days, respectively, representing an 89.3% and 16.5% increase over Mix-C. The enhancement in strength can be attributed to the higher consumption of portlandite in Mix-A during carbonation, which ultimately led to greater $CaCO_3$ content, as confirmed by TGA and XRD analysis in the following section. Furthermore, Mix-A achieved a maximum compressive strength of 8.44 MPa after 28 days under accelerated carbonation conditions, representing an 88% improvement over the control mix (Mix-C). The type of aggregate used also affected the compressive strength of the mixes, as Mayan soil was used as aggregates for mix-A, which likely provided possible nucleation for carbon sequestration for mix-A. As a result, a greater amount of Portlandite was converted into Calcite, which in turn contributed to the observed variation in compressive strength. The results indicate that accelerated carbonation yields a greater degree of carbonation product in Mix-A compared to Mix-C.

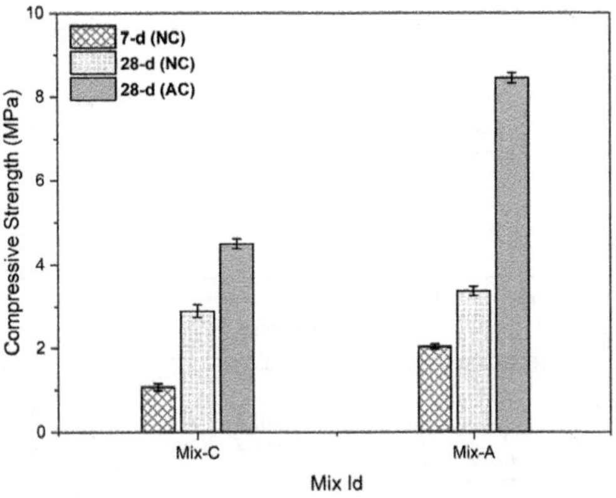

Fig. 4. Compressive Strength of the mixes.

Mineral Characterization

After 1-day Curing

Figures 5(a) and 5(b) show the DTG plots for Mix-C (control mix) and Mix-A at 1 day of curing and at selected depths. The plots show two distinct peaks at 400 °C (approximately) and 700 °C, respectively. The former corresponds to the decomposition of portlandite, and the latter is a characteristic decarboxylation peak from calcium carbonate. Irrespective of the depth, portlandite and Calcium Carbonate are observed in all samples. However, the amount of portlandite is higher for Mix-C as compared to Mix-A. For Mix C, the total amount of portlandite increases from 2.71 mg at a depth of 1 mm to 3.24 mg at 16 mm (Fig. 5(c)). On the other hand, the mix-A had a gradual increase of portlandite, increasing from 1.97 mg at 1 mm to 2.59 mg at 16 mm. These results suggest that carbonation is more pronounced in Mix-C. The lower portlandite content in Mix-A, despite the presence of carbonate phases, implies that pre-existing calcium carbonate in the soil may serve as nucleation sites, thereby promoting early-stage carbonation reactions. The presence of such nucleation sites effectively accelerates the carbonation process more rapidly. As a result, a higher amount of portlandite was consumed by Mix-A.

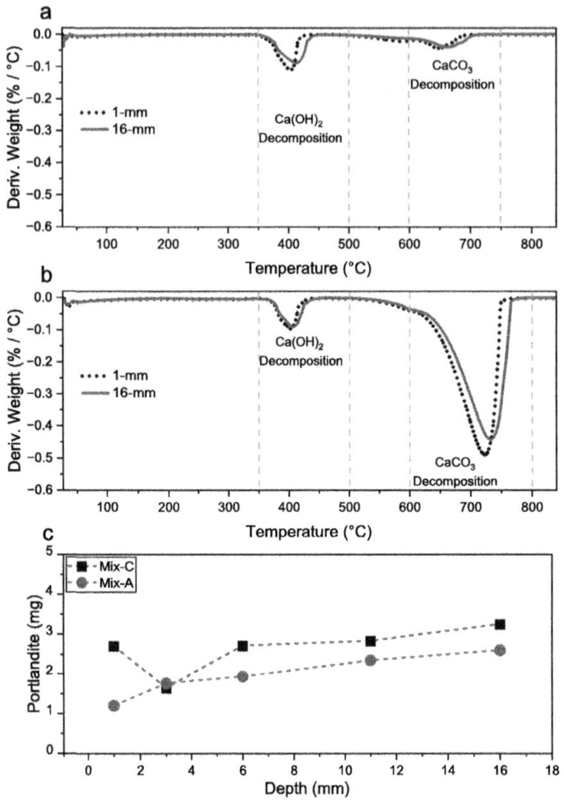

Fig. 5. TGA data at selected depths from the surface (a) Mix-C (NC) at 1 day of curing, (b) Mix-A (NC) at 1 day of curing, and (c) Amount of Port landite for Mix-A (NC) and Mix-C (NC)

Figures 6 (a) and (b) shows the XRD at 1 day for Mix C and A, respectively. The data indicates the presence of quartz, portlandite, and calcite (Fig. 5(a)). The intensity of the portlandite peak increased whilst the intensity of the Calcite peak decreased with depth, as evidenced by the peaks at $2\theta = 18.27°$ and $2\theta = 29.5°$, respectively (Fig. 6 (a)). In Mix-A, the presence of calcite and aragonite peaks was significantly more pronounced (Fig. 6 (b)) than in the control mix. The intensity of the portlandite peak ($2\theta = 18.27°$) increases with depth,, while the intensity of the Calcite peak ($2\theta = 29.5°$) decreases with increased depth, (Fig. 6b). The finding agrees well with the thermogravimetric analysis data, validating the reduction in portlandite and increase in calcite with increase in depth.

After 28-day Curing

Figures 7(a) and 7(b) show the DTG curves for the control mix and Mix-A after 28 days of curing under normal carbonation conditions. Figures 7(c) and 7(d) show the corresponding curves for samples subjected to accelerated carbonation. The control mix (Mix C) under normal carbonation exhibited the most prominent portlandite decomposition peaks between 350 °C and 450 °C (Fig. 7 (a)), indicating a higher residual portlandite content. In contrast, Mix-A showed significantly attenuated peaks in this region (Fig. 7 (b)), consistent with a lower initial content of portlandite. A similar trend was observed under accelerated carbonation- the portlandite decomposition peak in Mix-A remained minimal (Fig. 7 (d)), while a pronounced mass loss between 600 °C and 800 °C indicated increased formation of calcite. The control mix under accelerated carbonation exhibited a slight reduction in the intensity of the portlandite decomposition peak relative to the normal carbonation condition (Fig. 7 (c)). Collectively, the reduction in DTG peak intensity between 350 °C and 450 °C under accelerated carbonation confirms the enhanced conversion of portlandite to carbonate phases, in line with previous studies.

Quantitative estimates of portlandite content at different depths are presented in Figs. 7(e) and 7(f). Under normal carbonation, the control mix showed an increase in portlandite from 1.19 mg at 1 mm to 2.42 mg at 16 mm depth, while Mix-A exhibited a lower but gradual increase from 1.03 mg to 1.44 mg over 16 mm (Fig. 7 (e)). This depth-dependent increase in unreacted portlandite is attributed to the unidirectional diffusion of CO_2 from the exposed surface, with deeper regions remaining less carbonated. In the accelerated carbonation samples (Fig. 7f), the control mix showed a scattered and less consistent trend in portlandite content, increasing slightly from 2.62 mg at a depth of 1 mm to 2.42 mg at 16 mm, while, Mix-A had significantly lower portlandite cotent throughout the 16 mm profile, ranging from 0.21 mg to 1.32 mg. These results demonstrate that long-term curing (simulated through accelerated carbonation) shows a higher amount of carbonation in Mix A as compared to Mix C. This further validates the nucleation effect of calcium carbonates in the soil.

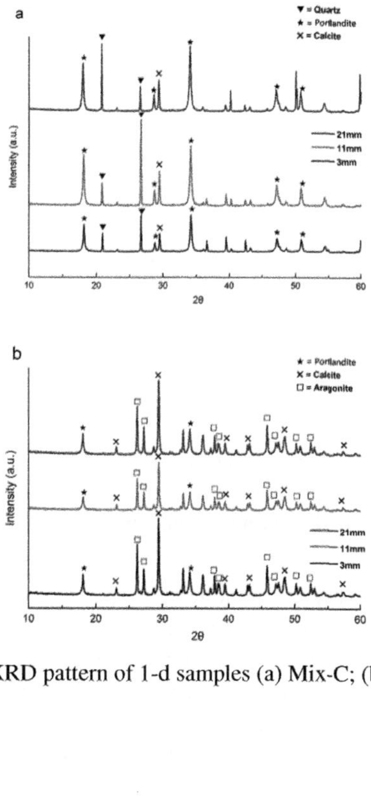

Fig. 6. XRD pattern of 1-d samples (a) Mix-C; (b) Mix-A

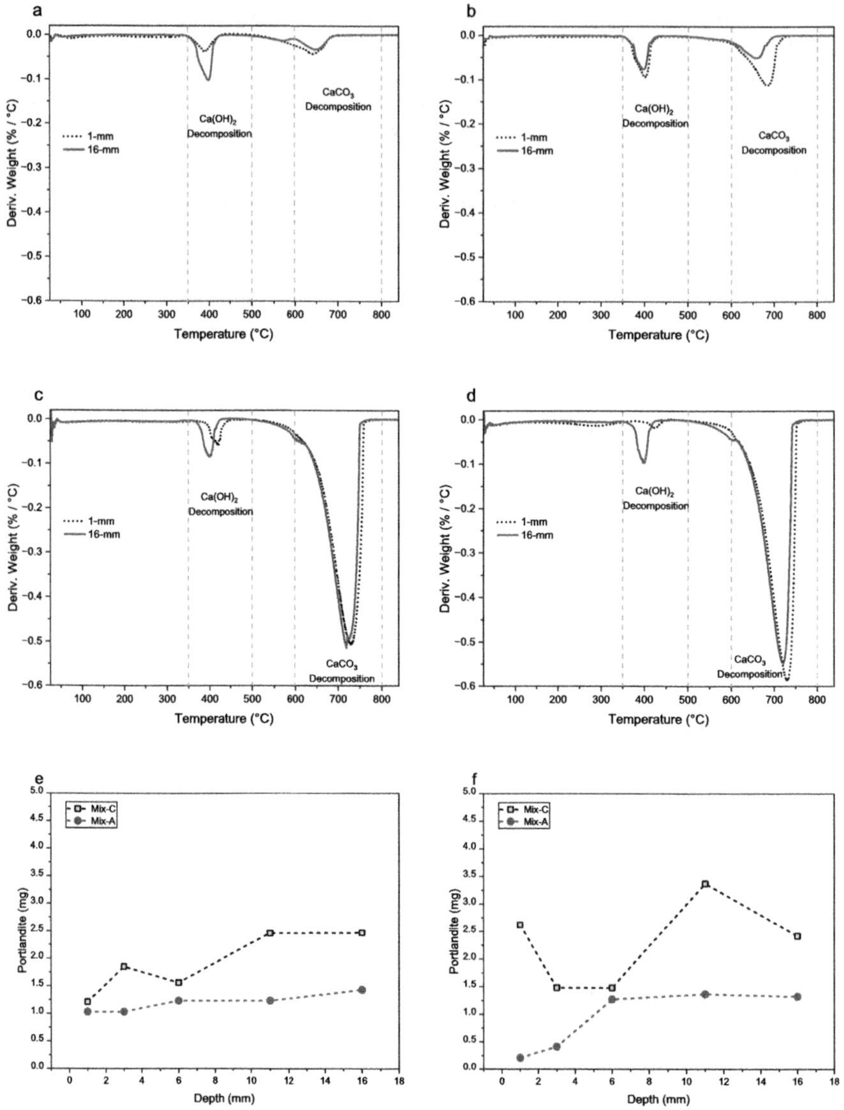

Fig. 7. (a) DTG of mix-C (NC) at 28-d, (b) DTG of mix-C (AC) at 28-d, (c) DTG of mix-A (NC) at 28-d, (d) DTG of mix-A (AC) at 28-d, (e) Amount of portlandite for mix-C and mix-A in NC condition, (f) Amount of portlandite for mix-C and mix-A in AC condition.

Figure 8 presents the X-ray diffraction (XRD) patterns of the 28-day samples for the control mix and Mix-A under both normal and accelerated carbonation conditions. For the control mix, diffraction peaks corresponding to quartz, portlandite, and calcite were observed under both curing regimes (Figs. 8a and 8b). Under normal carbonation, the intensity of the portlandite peak at $2\theta = 18.27°$ increased with depth, while the calcite peak at $2\theta = 29.5°$ showed a corresponding decrease, indicating a depth-dependent carbonation gradient (Fig. 8a). Under accelerated carbonation, the intensity of the calcite peak was notably higher, whereas the portlandite peaks at $2\theta = 18.27°$ and $2\theta = 46.8°$ were significantly attenuated (Fig. 8b), consistent with increased carbonation. These trends align with the thermogravimetric data.

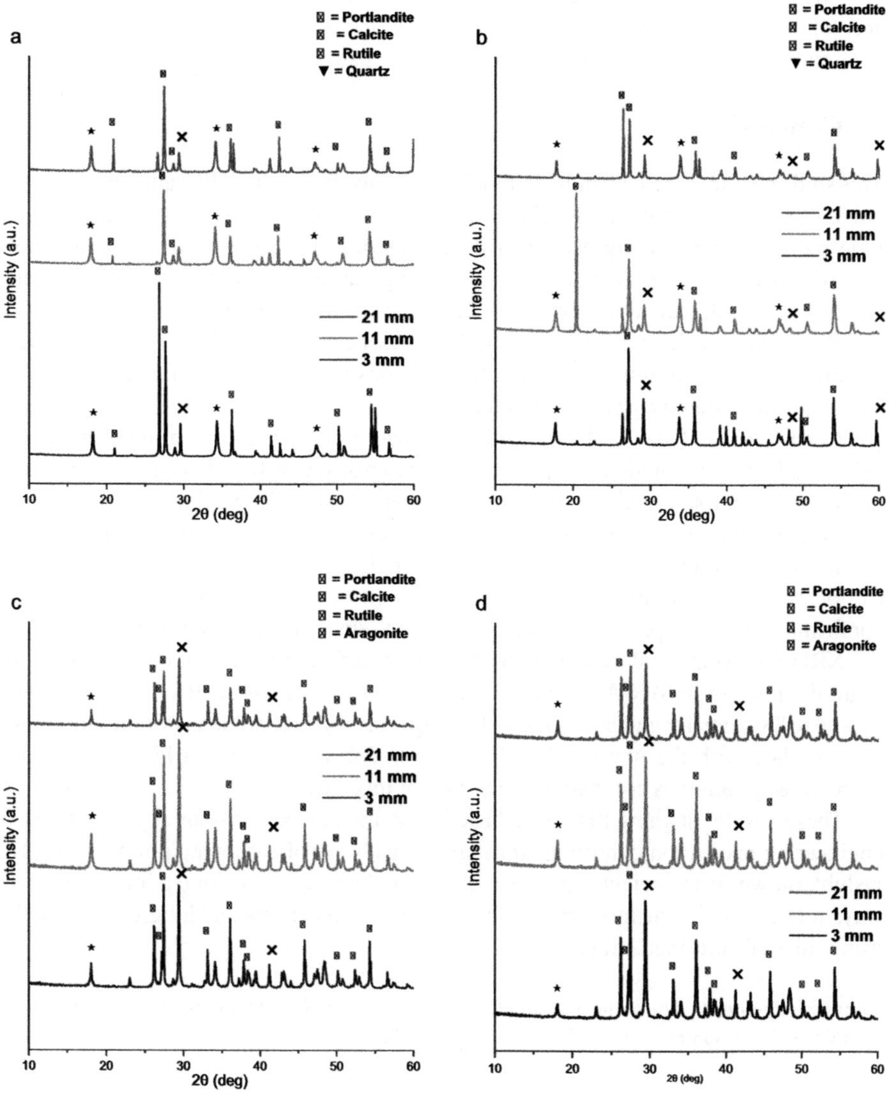

Fig. 8. XRD pattern of 28-d samples (a) Mix-C (NC); (b) Mix-C (AC); (c) Mix-A (NC); (d) Mix-A (AC)

In the case of Mix-A, the XRD profiles revealed the presence of portlandite, calcite, and Aragonite (Figs. 8c and 8d). Notably, the intensities of the calcite and Aragonite peaks were more pronounced in Mix-A than in the control mix, particularly under accelerated carbonation. Similar to the control mix, portlandite peak intensity increased with depth while calcite peak intensity diminished (Fig. 8c). Under accelerated carbonation, the portlandite peak at $2\theta = 18.27°$ was markedly reduced (Fig. 8d), indicating greater carbonation. Overall, Mix-A exhibited substantially lower portlandite peak intensities and higher carbonate phase intensities (calcite and Aragonite) than the control mix, reflecting enhanced carbonation of the lime–soil matrix.

4 Conclusion

This study characterized the mortar samples collected from Muyil Pyramid and representative mixtures. The mineralogical and mechanical evolution observed in representative lime mortars incorporating Mayan soil demonstrates the significant role of naturally occurring carbonate polymorphs in determining carbonation pathways. The presence of aragonite in the ancient mortars suggests the incorporation of local soil comprised of aragonite and calcite in the mortars. The presence of such calcium carbonate polymorphs provides preliminary evidence that the constituents acted as effective nucleation substrates that altered carbonation kinetics.

At early ages (1 day), Mix-A exhibited lower portlandite content than the control (Mix C), despite initial identical portlandite contents, suggesting that calcium carbonate in the soil promoted nucleation both at the surface and deeper into the matrix. The relatively higher compressive strength of the mixes can be attributed to the high intrinsic reactivity of the traditional Mexican lime used in this study. After 28 days of accelerated carbonation, portlandite content in Mix-A was up to 90% lower compared to Mix-C, confirming rapid and preferential transformation due to nucleation effects.

XRD and TGA analyses also validated that carbonation was more complete and spatially uniform in Mix-A. The higher initial compressive strengths obtained in Mix-A—up to 88% greater under accelerated curing—are not solely due to Portlandite content but correlate with the improved carbonation kinetics facilitating a higher amount of calcite precipitation as compared to the control mix (Mix C).

These results suggest that lime-based mortars can be substantially improved by engineering nucleation conditions through the addition of carbonate polymorphs. This insight, drawn from ancient practice and validated through modern testing, opens pathways for designing lime mortars with improved early reactivity and long-term durability using mineral-informed strategies.

Acknowledgement. The authors would like to acknowledge Lic. Margarito Molina, Director of the INAH Center Quintana Roo, for supplying the Mayan mortar sample.

References

1. S. Institution: Precolumbian cements: a study of the calcareous cements in prehispanic mesoamerican building construction / by David S. Hyman. Smithsonian Institution. https://www.si.edu/object/siris_sil_999343. Accessed 23 May 2025

2. The beginnings of pyrotechnology, Part II: production and use of lime and gypsum plaster in the Pre-Pottery Neolithic near east. J. Field Archaeol. **15**(2). https://www.tandfonline.com/doi/abs/10.1179/009346988791974501. Accessed 23 May 2025
3. Prehispanic Maya burnt lime industries: previous studies and future directions. Anc. Mesoam. https://www.cambridge.org/core/journals/ancient-mesoamerica/article/abs/prehispanic-maya-burnt-lime-industries-previous-studies-and-future-directions/E0F1BF97202100037EA005F55D0FCE0C. Accessed 23 May 2025
4. Ancient Maya burnt -lime technology: cultural implications of technological styles - ProQuest. https://www.proquest.com/openview/12f2088861131c38d9ce5154df34cad0/1?cbl=18750&diss=y&pq-origsite=gscholar. Accessed 23 May 2025
5. Morris, E.H., Charlot, J., Morris, A.A.: The Temple of the warriors at Chichen Itzá, Yucatan (1931). https://agris.fao.org/search/en/providers/122376/records/6474731a2d5d435c424f1312. Accessed 06 May 2025
6. Magaloni, D., Pancella, R., Fruh, Y., CaÑetas, J., CastaÑo, V.: Studies on the Mayan mortars technique. MRS Online Proc. Libr. OPL **352**, 483 (1995). https://doi.org/10.1557/PROC-352-483
7. Reller, A., Wilde, P.-M., Wiedemann, H.G., Hauptmann, H., Bonani, G.: Comparative studies of ancient mortars from Giza, Egypt, and Nevali ÇOri, Turkey. MRS Online Proc. Libr. OPL **267**, 1007 (1992). https://doi.org/10.1557/PROC-267-1007
8. Pachta, V., Stefanidou, M., Konopisi, S., Papayianni, I.: Technological evolution of historic structural mortars. J. Civ. Eng. Archit. **8**(7) (2014). https://doi.org/10.17265/1934-7359/2014.07.005
9. Rodriguez-Navarro, C., Monasterio-Guillot, L., Burgos-Ruiz, M., Ruiz-Agudo, E., Elert, K.: Unveiling the secret of ancient Maya masons: biomimetic lime plasters with plant extracts. Sci. Adv. **9**(16), eadf6138 (2023). https://doi.org/10.1126/sciadv.adf6138
10. Lowland Maya lime plaster technology: a diachronic approach - ProQuest. https://www.proquest.com/openview/95d5005933606c0b1bedc69962a31fa9/1?cbl=2026366&pq-origsite=gscholar. Accessed 06 May 2025
11. Guasch-Ferré, N., et al.: Polysaccharide remains in Maya mural paintings: is it an evidence of the use of plant gums as binding medium of pigments and additive in the mortar? STAR Sci. Technol. Archaeol. Res. **5**(2), 200–220 (2019). https://doi.org/10.1080/20548923.2020.1720377
12. Littmann, E.R.: Ancient mesoamerican mortars, plasters, and stuccos: the use of bark extracts in lime plasters. Am. Antiq. **25**(4), 593–597 (1960). https://doi.org/10.2307/276642
13. Villaseñor, I., Graham, E.: The use of volcanic materials for the manufacture of pozzolanic plasters in the Maya lowlands: a preliminary report. J. Archaeol. Sci. **37**(6), 1339–1347 (2010). https://doi.org/10.1016/j.jas.2009.12.038
14. Witschey, W.R.T.: Recent investigations at the Maya Inland Port City of Muyil (Chunyaxche), Quintana Roo, Mexico. Mexicon **10**(6), 111–117 (1988)
15. ASTM International. Standard Test Method for Particle-Size Distribution (Gradation) of Fine-Grained Soils Using the Sedimentation (Hydrometer) Analysis. ASTM D7928 (2021)
16. ASTM International. Standard Specification for Standard Sand. ASTM C778 (2021)
17. ASTM International. Standard Test Method for Compressive Strength of Hydraulic Cement Mortars. ASTM C109/C109M (2020)
18. Druzhbin, D., Rashchenko, S.V., Shatskiy, A., Crichton, W.A.: New high-pressure and high-temperature CaCO3 polymorph. ACS Earth Space Chem. **6**(6), 1506–1513 (2022). https://doi.org/10.1021/acsearthspacechem.2c00019
19. Hu, H., et al.: Bioprocessing-inspired water-mediated crystallization and densification of strong nacre-like aragonite ceramic from hydrated amorphous precursor. Adv. Funct. Mater., 2421628. https://doi.org/10.1002/adfm.202421628

Development of Correlations for Physico-Mechanical Properties of Sandstones Present in India (Vindhyan and Marwar Basins)

Narupa Devi Singh(✉) and Swathy Manohar

Department of Civil Engineering, Indian Institute of Technology Bombay, Mumbai, India
narupadevisingh@gmail.com

Abstract. The use of sandstones, a stratified sedimentary rock, have been integral in construction practices due to their durability and aesthetic appeal in India. Historically used in heritage structures, sandstones continue to be employed in modern applications such as cladding, tiles, masonry, and dimension stones. With India hosting 27 sedimentary basins that house a wide variety of sandstones, this study investigates the physical and mechanical properties of various sandstone types from Vindhyan Basin and Marwar Basin, namely: Agra Red, Teak Wood, Kandla White Beige, Jodhpur Beige, Jodhpur Brown, Mandana, Bansipaharpur, Gwalior Mint, etc., sourced from Rajasthan and Madhya Pradesh. A series of standardized tests, such as Ultrasonic Pulse Velocity (UPV), Rebound Hammer and Karsten Tube test were performed to characterize the sandstones. Key properties evaluated include specific gravity, porosity, compressive strength, bulk specific gravity, water absorption, and water permeability. The results revealed substantial variations influenced by geological and environmental conditions, such as burial depth and mineralogical composition. For instance, Mandana and Jodhpur Beige sandstones exhibited higher compressive strength and lower porosity, suggesting greater durability, while Teak Wood sandstone displayed higher porosity and water absorption, indicating its susceptibility to environmental deterioration. The study also establishes empirical correlations linking physical and mechanical properties, hence it can aid in property predictions especially in situations where certain tests cannot be conducted such as heritage conservation, where non-destructive evaluation is critical, and for modern construction, where rapid property estimation is essential. The comprehensive documentation of sandstone properties and developing correlations specific to Indian geological contexts can facilitate performance predictions. This study contributes to the sustainable use of sandstones in construction and heritage preservation.

Keywords: Stones · Sedimentary basins · Sandstones · Correlating properties · Characterization

M. Posani and S. Singh (Eds.): RYS 2025, RILEM Bookseries 67, pp. 992–1010, 2026.
https://doi.org/10.1007/978-3-032-14652-6_72

1 Introduction

Sandstones have been used in the construction industry for ages; the proof shows in heritage structures such as Rashtrapati Bhawan, Delhi, Amer Fort, Jaipur, Rajasthan, Rani Ki Vav, Patan, Gujarat, Agra Fort, Agra, Uttar Pradesh, Khajuraho Group of Temples, Chhatarpur District, Madhya Pradesh (Kaur et al. 2019 and Priya 2010). Recently, sandstones have been mainly used as cladding, tiles, aggregates, dimension stones, etc. Sandstones are readily available around India due to its 27 sedimentary basins with various formations consisting of multiple types (Directorate General of Hydrocarbons 2025). Vindhyan and Marwar basins are two of India's major basins regarding sandstone quarrying (Agarwal 2025). Vindhyan basin spreads across four states: Rajasthan, Madhya Pradesh, Uttar Pradesh and Bihar, while Marwar basin spreads across Rajasthan. Since Rajasthan is India's largest sandstone producing state which gives approximately 90% of the country's total production, the sandstones present in this region are majorly from the Vindhyan and Marwar basin (Agarwal 2025). Additionally, Rajasthan also houses Jaisalmer Basin which produces sandstones such as Jaisalmer Yellow sandstone which were used in heritage structures in the Golden City of Jaisalmer. Vindhyan Basin also contains limestone, shales and siltstones and it covers approximately 166 400 km^2 (Verma et al. 2020). It is divided into main supergroups which are Son Valley (eastern) and Chambal Valley (western). Further, the basin is divided into upper and lower Vindhyan which are stratigraphically subdivided into four groups namely, from oldest to youngest, Semri supergroup is in the lower Vindhyan and Kaimur, Rewa and Bhander supergroups are in upper Vindhyan Basin (Verma et al. 2020). Marwar Basin contains sandstones, carbonate rocks and shales and covers approximately 100 000 km^2 (Zhang et al. 2024 and George 2020). It is divided into three groups namely, from oldest to youngest, Jodhpur, Bilara and Nagaur groups. Jodhpur group is further divided into Pokaran, Sonia and Girbhakar formations, Bilara groups is subdivided into Dhanapa, Gotan and Pondlo formations and Nagaur group is subdivided into Nagaur and Tunklian formations (Xu et al. 2022).

As sandstones are naturally formed, they may contain various impurities which affects their properties. They primarily comprise quartz, feldspar, lithic fragments and heavy metals. Factors such as burial depth in the basin, environmental conditions, mineralogy and age of formation influence the physical and mechanical properties of sandstones. Past studies on physical properties of various sandstones in the Vindhyan Basin shows that; Itwa sandstone, which is also referred to as Asan sandstone from Lower Rewa Region, is approximately 3000–5000 m thick. These sandstones are fine to very fine-grained and grey was observed being the predominant colour with small sections of green and pink pigmented sandstones embedded into it. They were found to be dominated by the mineral quartz (Verma et al. 2020). As for Sagar sandstones which are a part of Upper Rewa Group, there are two sections: one which has coarse-grained sandstones of grains larger than 2 mm and the other section which has coarse to fine-grained sandstones that are subrounded to angular. These sandstones are red in colour due to their siliceous and ferruginous material content (G. K. Singh et al. 2023). Other major sandstones, which are quarried and used extensively since time immemorial, present in this basin are Bansipaharpur and Dholpur sandstone, which is a part of Bhander Group, which was used in the construction of Agra Fort, Hawa Mahal, Qutub Minar, etc. (Kaur et al. 2019).

In modern time it is used in construction of Ram Mandir at Ayodhya, New Parliament, New Delhi and series of Swaminarayan temple and Akshardham temple. This sandstone was reported to have high amounts of quartz, small amounts of aluminium and calcium oxides and they are fine grained (Guha Roy et al. 2023). These properties would suggest a strong and durable material. Past studies on physical properties of various sandstones in the Marwar basin shows that; Sonia formation contains fine to medium grained red sandstones, Girbhakar sandstone is coarse, angular, micaceous with interlayers of shale and Nagaur sandstones are fine to coarse grained and are red-brown (Zhang et al. 2024 and Xu et al. 2022). It can be clearly seen that even though these sandstones are formed in one region, in a single basin, there are large variations among their appearance and physical properties.

Mechanical properties of a material become important when designing various structural components. This refers to the material's compressive strength, tensile strength, modulus of elasticity, UPV, etc. Variation of these properties would significantly affect the usage of the material, for example structural elements would require higher compressive strength, UPV, etc., whereas non-structural elements like cladding may not require a high compressive strength, tensile strength, etc. Variations between the mechanical properties of some Indian sandstones can be seen in Table 1.

Table 1. Mechanical Properties of Indian Sandstones.

Sandstone Name	Sedimentary Basin	Compressive Strength (MPa)	Tensile Strength (MPa)	Ultrasonic Pulse Velocity (km/sec)	Modulus of Elasticity (GPa)	References
Dholpur Sandstone	Vindhyan	44.50	6.28	3.12	12.00	Guha Roy et al. (2023)
Chunar Sandstone	Vindhyan	79.30	–	3.43	–	A. Singh et al. (2023)
Bhagwanpur Sandstone	Vindhyan	66.22	12.27	–	6.30	Chaudhary et al. (2022)
Jabalpur White Sandstone	Satpura	68.00	11.18	3.41	20.00	Guha Roy et al. (2017)
Jabalpur Red Sandstone	Satpura	76.00	10.22	2.85	21.00	Guha Roy et al. (2017)
Majher Khad Sandstone	Himalayan Foreland	14.46–25.75	0.96–2.15	–	–	Sahu et al. (2024)
Dharot Sandstone	Himalayan Foreland	51.56–88.25	9.57–12.98	–	–	Sahu et al. (2024)
Betamcherla Sandstone	Cuddapah	76.95	–	2.17	–	T. D. Singh et al. (2023)

Correlations are important to understand relationships between various properties as well as property prediction in cases where tests cannot be done or where there are time constraints, because these properties govern the performance of the stones. Correlations are developed by studying the properties of the material and incorporating them in a regression analysis to achieve the best fit of the data. The correlations can be affected by the mineral composition of the material, grain structure and size, cementation of minerals (lithification), burial depth, degree of compaction of sediments and diagenesis. Those characteristics will vary from different types of sandstones found in various locations; hence empirical equations formed will only be valid for those types of sandstones studied. Various studies have been done for sandstones, like in Jining, China and Qom, Iran. Relationships for the sandstones found in Jining were linear fits between porosity and density, porosity and water absorption, and P-wave velocity and compressive strength. Nonlinear fits (fitting curves) were reported between porosity and P-wave velocity, and porosity and compressive strength (Shi et al. 2021). For the sandstones in Qom, the study differed by considering the properties at different levels of saturation, as it was found that the properties varied significantly with this change. It was reported that the majority of loss of compressive strength was at 50% saturation level, and P-wave velocity decreased with higher degrees of saturation, this is because of the obstruction of the wave propagation due to the moisture (Abdi et al. 2024). This study presents a novel contribution by systematically evaluating a broad spectrum of Indian sandstones from the Vindhyan and Marwar basins using both physical and mechanical characterization techniques, and by establishing specific empirical correlations between these properties. Unlike previous studies that either focused on a limited set of sandstones and/or properties (e.g., sandstones in Jining, China or Qom, Iran), this work develops property correlations between multiple types of sandstones, which are uniquely tailored to explore non-destructive and destructive testing methods to achieve various properties such as specific gravity, bulk specific gravity, etc., of Indian sandstones. By integrating multiple test methods, including techniques such as UPV and rebound hammer, the study facilitates accurate property prediction in scenarios where destructive testing is impractical, such as heritage conservation.

2 Materials and Methods

2.1 Materials

Most sandstones in India are from Rajasthan, which houses a large part of the Vindhyan Basin and the entire Marwar Basin. Multiple sandstones from these basins were collected and studied, eleven from Rajasthan and two from Madhya Pradesh. The details are presented below. Three samples of each type of sandstone were considered for each test, and the average was computed.

Agra Red Sandstone was sourced from Sarmathura, Dholpur, Rajasthan. It is found in the Lower Rewa formation of the basin. It is shown in Fig. 1a. Teak Wood Sandstone was sourced from Badi, Khatu, Nagaur, Rajasthan. It is shown in Fig. 1b below. Kandla White Beige Sandstone was sourced from Sukhpura, Bijoliya, Bhilwara, Rajasthan. It is shown in Fig. 1c. Jodhpur Beige Sandstone was sourced from Dhanroop Mines, Bujawar, Jodhpur, Rajasthan and is shown in Fig. 1d. Jodhpur Brown sandstone was sourced from

Jodhpur, Rajasthan and is shown in Fig. 1e. Mandana Sandstone was sourced from Kota, Rajasthan and is shown in Fig. 1f. Bansipaharpur Sandstone was sourced from Bharatpur, Rajasthan and is shown in Fig. 1g. Gwalior Mint Sandstone was sourced from Gwalior, Madhya Pradesh and is shown in Fig. 1h. Sagar Black Sandstone was sourced from Sagar District, Madhya Pradesh. Bundi-Bijoliya Sandstone was sourced from Bundi and Bijoliya, Bhilwara Districts of Rajasthan. Dholpur Sandstone was sourced from Dholpur, Rajasthan. Karauli Sandstone was sourced from Karauli, Rajasthan. Bharatpur Sandstone was sourced from Bharatpur, Rajasthan.

Fig. 1. (a) Agra Red Sandstone, (b) Teak Wood Sandstone, (c) Kandla White Beige Sandstone, (d) Jodhpur Beige Sandstone, (e) Jodhpur Brown Sandstone, (f) Mandana Sandstone, (g) Bansipaharpur Sandstone and (h) Gwalior Mint Sandstone.

2.2 Methods

The physical and mechanical properties studied are qualitative estimation of durability via the Ultrasonic Pulse Velocity (UPV) test and Rebound Hammer, Water Absorption, Water Permeability via Karsten Tube test, Open Porosity, Bulk Specific Gravity, Specific Gravity via Helium Gas Pycnometer test, and Compressive Strength tests. The methods/ standards used for these tests are described below. These tests were done on three 50 mm cube specimens, and the average of the results was taken.

Ultrasonic Pulse Velocity (UPV)

This test was conducted in accordance to IS 516 Part 5 section 1, 2018 code. Out of the 3 ways, direct, semi-direct and indirect, the direct method was done as it gives accurate results since each surface of the specimen was accessible. The oven dried specimen was cooled and the centre of each side of the cubes was marked, a sealant (grease) was applied to the transmitter and receiver and placed on two opposite sides of the specimen. This was repeated 3 times and an average was taken.

Rebound Hammer

This test was conducted in accordance with IS 516 Part 5, 2020 code. It was done on 2 sides of the cube specimen three times each; then the average was taken. The rebound hammer was held 90° upright and then pressed down on the centre of the surface of the cube.

Water Absorption and Bulk Specific Gravity

This test was conducted in accordance with ASTM C97. The specimens were oven dried and then weighed. Later, they were submerged with mineralized water for 48 h, then the submerged and saturated weights were measured and the water absorption and bulk specific gravity values were computed.

Water Absorption via Karsten Tube Test

This test was conducted in accordance with Rilem Test Number II, 4 code. Karsten tubes were attached to the cube specimens on one side. Water was filled into the tube and the water absorbed by the material at different time intervals was recorded.

Open Porosity

This test was conducted in accordance with BS EN 1936 2006 code. The specimens were oven dried and then weighed. They were then placed under vacuum for twenty-four (24) hours, after which water was immersed in the desiccator and kept for twenty-four (24) hours. The vacuum was then removed by letting air into the system and kept for another twenty-four (24) hours after which the submerged and saturated weights were measured and the open porosity was computed.

Specific Gravity

This test was conducted by Helium Gas Pycnometer test, whereby powdered samples passing 75 μm were filled into the cell and pressure was added to the expansion chamber to determine the volume. Then the sample was weighed and the specific gravity was computed.

Compressive Strength

This test was conducted in accordance with IS 1121 Part 1, 1974 code. Three oven-dried specimens were given a uniform loading rate of 140 kg/cm^2 or 572 N/s until failure. The average compressive strength was computed by the stress at which the sandstones failed.

Mapping of Sedimentary Basins and Sandstones' Sources

Mapping was done utilizing QGIS Version 3.38.1 software. Georeferencing was done from the map of sedimentary basins in India from the Ministry of Natural Gas and Petroleum source. The map of Vindhyan and Marwar basins is presented in Fig. 2. It also shows the location of the sources of the sandstones in this study.

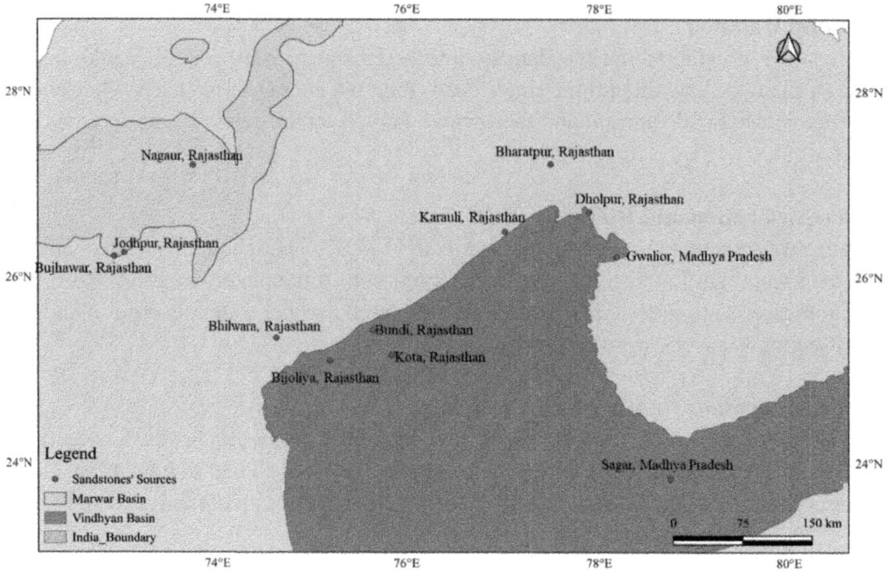

Fig. 2. Map of the sources of the sandstones.

3 Results and Discussions

3.1 Physical Properties

It is essential to understand a material's physical properties as they determine the material's ability to be used for various applications. Their porosity, specific gravity, bulk specific gravity and water absorption are the physical properties explored below and the values are presented in Table 2. The results of water absorption by Karsten tube tests are presented in Table 3. Figure 3 shows the variation of the above mentioned properties in the different sandstones.

It can be seen that Kandla White Beige sandstone, Mandana sandstone and Jodhpur Brown sandstone have the highest specific gravities and lowest open porosities and water absorption. This means the pore structure may contain heavier particles, smaller pores or less connected pores. Whereas, Teak Wood sandstone and Bansipaharpur sandstone have lower bulk specific gravities and higher open porosities and water absorption. This is because the material is more porous and may contain larger pores. The mineralogy also significantly affects these properties; if there are clay minerals present, it would quickly absorb water and cause swelling which can further deteriorate the sandstone. Teak Wood, Gwalior Mint and Jodhpur Beige sandstones have the highest water permeabilities whereby in less than 120 min (2 h), 5 ml of water was absorbed whereas for other sandstones, they did not absorb 5 ml of water even after 360 min (6 h). This shows that Teak Wood, Gwalior Mint, and Jodhpur Beige sandstones have connected and small pore sizes, therefore capillary absorption is higher due to the strong surface tension. The other sandstones appear to have larger and less connected pores, which cause high capillary resistance due to the weak capillary forces.

Table 2. Physical Properties of the Sandstones.

Sandstone Name	Specific Gravity	Bulk Specific Gravity	Standard	Open Porosity (%)	Standard	Water Absorption (%)	Standard
Agra Red	2.793	2.2	ASTM C97	12.9	BS EN 1936 2002	4.0	ASTM C97
Teak Wood	2.767	2.0	ASTM C97	20.8	BS EN 1936 2002	6.3	ASTM C97
Kandla White Beige	2.817	2.5	ASTM C97	13.1	BS EN 1936 2002	1.1	ASTM C97
Jodhpur Beige	2.782	2.3	ASTM C97	9.2	BS EN 1936 2002	2.1	ASTM C97
Jodhpur Brown	3.027	2.5	ASTM C97	6.2	BS EN 1936 2002	2.5	ASTM C97
Mandana	3.051	2.6	ASTM C97	2.9	BS EN 1936 2002	1.1	ASTM C97
Bansipaharpur	3.064	2.2	ASTM C97	9.5	BS EN 1936 2002	4.0	ASTM C97
Gwalior Mint	2.815	2.3	ASTM C97	7.6	BS EN 1936 2002	2.6	ASTM C97
Sagar Black	–	2.3	ASTM C97	–	–	1.7	ASTM C97
Bundi-Bijoliya	–	2.6	ASTM C97	3.5	IS 1124	1.1	ASTM C97
Dholpur	–	2.3	ASTM C97	6.5	IS 1124	2.9	ASTM C97
Karauli	–	2.2	ASTM C97	10.0	IS 1124	3.1	ASTM C97
Bharatpur	–	2.3	ASTM C97	12.5	IS 1124	3.15	ASTM C97

Table 3. Water Absorption by Karsten Tube Test of the Sandstones.

Time (minutes)	Water Absorbed (ml)							
	Agra Red	Teak Wood	Kandla White Beige	Jodhpur Beige	Mandana	Jodhpur Brown	Bansipaharpur	Gwalior Mint
1	0.07	0.60	0.03	0.25	0.08	0.07	0.17	0.30
2	0.15	1.00	0.03	0.37	0.08	0.07	0.17	0.40
3	0.18	1.47	0.03	0.53	0.08	0.07	0.22	0.55
4	0.22	1.83	0.03	0.67	0.08	0.10	0.23	0.67
5	0.23	2.17	0.03	0.80	0.10	0.13	0.23	0.82
10	0.27	3.27	0.05	1.27	0.10	0.17	0.30	1.60
15	0.28	3.13	0.05	1.63	0.10	0.23	0.37	2.18
20	0.35	3.80	0.05	2.03	0.10	0.25	0.43	2.63
25	0.42	4.30	0.07	2.37	0.10	0.30	0.47	3.03
30	0.43	4.25	0.07	2.70	0.10	0.27	0.52	3.40
40	0.53	5.00	0.07	3.25	0.10	0.33	0.62	4.20
50	0.62	NA	0.07	3.78	0.08	0.37	0.70	4.80
60	0.68	NA	0.08	4.17	0.12	0.43	0.77	5.00
120	1.05	NA	0.10	5.00	0.17	0.63	1.13	NA
180	1.50	NA	0.12	NA	0.17	0.80	1.60	NA
360	3.03	NA	0.13	NA	0.23	1.47	2.60	NA

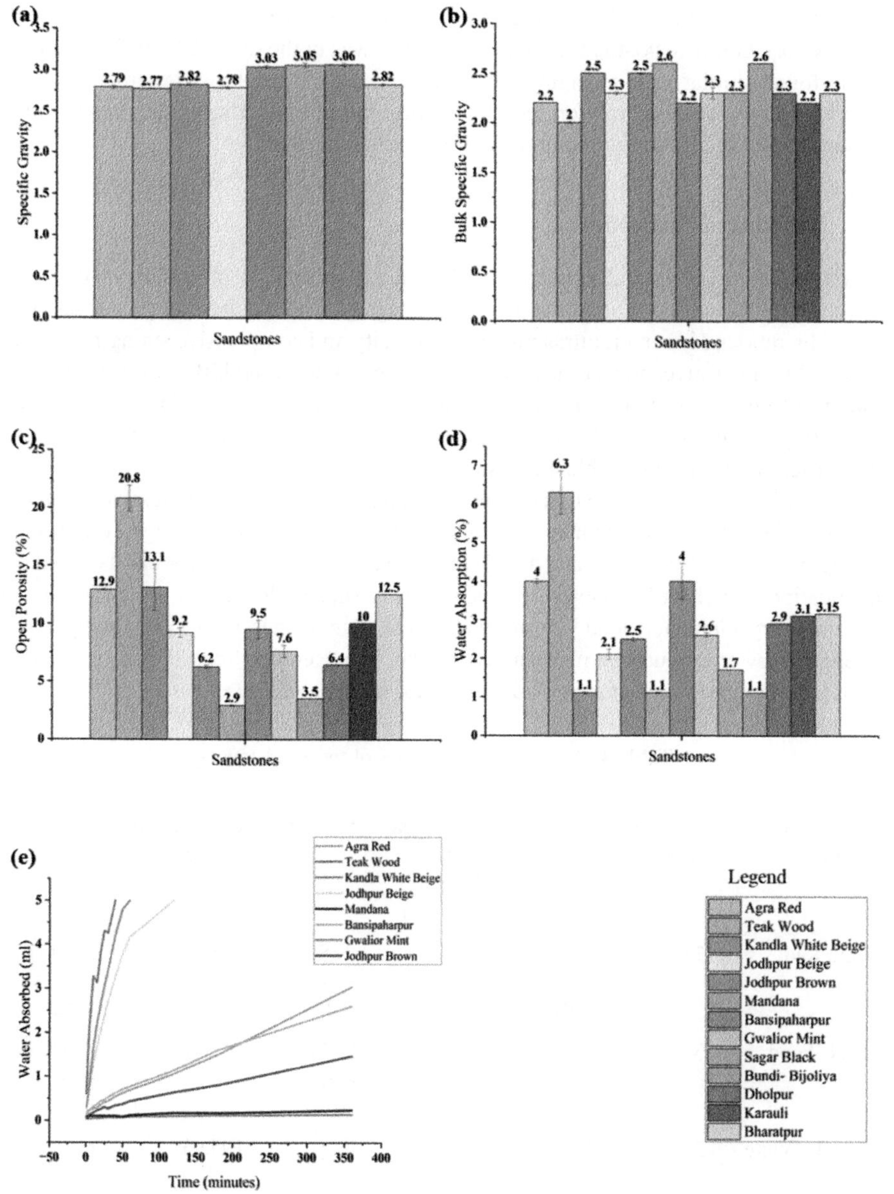

Fig. 3. (a) Variation of specific gravity, (b) Variation in bulk specific gravity, (c) Variation of open porosity, (d) Variation of water absorption, (e) Variation in water absorbed with time of the sandstones studied.

Singh et al. (2023), Guha Roy & Singh (2023), Chaudhary et al. (2022), Sahu et al. (2024) and Guha Roy et al. (2017), explored the properties of various types of sand stones across India such as Dholpur, Bhagwanpur, Chunar, Jabalpur White, Jabalpur Red, Majher Khad and Dharot sandstones. The results showed porosity of 2.6 to 21% and this study had a similar finding whereby the porosity is in the range 2.9 to 20.8%. Specific gravity variation of sandstones in previous studies was in the range 2.02 to 2.66 and this study found similar values ranging from 2.77 to 3.11. These properties are affected by various factors while it was formed, its position in the basin (burial depth/exposed to environment) and its seasoning process after being quarried.

3.2 Mechanical Properties

It is essential to understand a material's mechanical properties as they indicate the material's strength (indirectly its durability) and integrity. The mechanical properties were tested by rebound hammer, ultrasonic pulse velocity and compressive strength test. The rebound number gives the surface hardness of the material, the UPV contributes to the integrity (presence of pores, microcracks and fissures) of the material and compressive strength gives the load up to which the material can withstand before failure. These properties can be seen in Table 4 and variation is shown in Fig. 4.

Sagar Black and Bundi-Bijoliya sandstones have the highest compressive strengths. Kandla White Beige, Jodhpur Beige and Mandana sandstones have high compressive strengths, rebound numbers and UPV. This means these sandstones can be used for applications which will be subjected to heavy loads, such as floor slabs, dimension stones and masonry elements (load bearing walls), etc. Sandstones with lower compressive strengths, lower rebound number and UPV, like Agra Red and Teak Wood, etc., can be used for other purposes such as claddings, tiles, aggregates, etc.

Table 4. Mechanical Properties of the Sandstones.

Sandstone Name	Rebound Number	Standard	Ultrasonic Pulse Velocity (km/sec)	Standard	Compressive Strength (MPa)	Standard
Agra Red	34.11	IS 516 Part 5, 2020	2.8	IS 516 Part 5, 2018	52.84	IS 1121 Part 1, 1974
Teak Wood	33.72	IS 516 Part 5, 2020	3.6	IS 516 Part 5, 2018	47.99	IS 1121 Part 1, 1974
Kandla White Beige	32.67	IS 516 Part 5, 2020	5.0	IS 516 Part 5, 2018	105.93	IS 1121 Part 1, 1974
Jodhpur Beige	35.78	IS 516 Part 5, 2020	4.8	IS 516 Part 5, 2018	77.51	IS 1121 Part 1, 1974
Jodhpur Brown	41.06	IS 516 Part 5, 2020	3.6	IS 516 Part 5, 2018	41.40	IS 1121 Part 1, 1974

(continued)

Table 4. (*continued*)

Sandstone Name	Rebound Number	Standard	Ultrasonic Pulse Velocity (km/sec)	Standard	Compressive Strength (MPa)	Standard
Mandana	49.61	IS 516 Part 5, 2020	4.6	IS 516 Part 5, 2018	83.29	IS 1121 Part 1, 1974
Bansipaharpur	44.67	IS 516 Part 5, 2020	3.4	IS 516 Part 5, 2018	45.92	IS 1121 Part 1, 1974
Gwalior Mint	42.00	IS 516 Part 5, 2020	3.9	IS 516 Part 5, 2018	42.91	IS 1121 Part 1, 1974
Sagar Black	–	–	–	–	129.00	ASTM C170
Bundi-Bijoliya	–	–	–	–	139.00	ASTM C170
Dholpur	–	–	–	–	86.00	ASTM C170
Karauli	–	–	–	–	55.50	ASTM C170
Bharatpur	–	-	–	–	106.50	ASTM C170

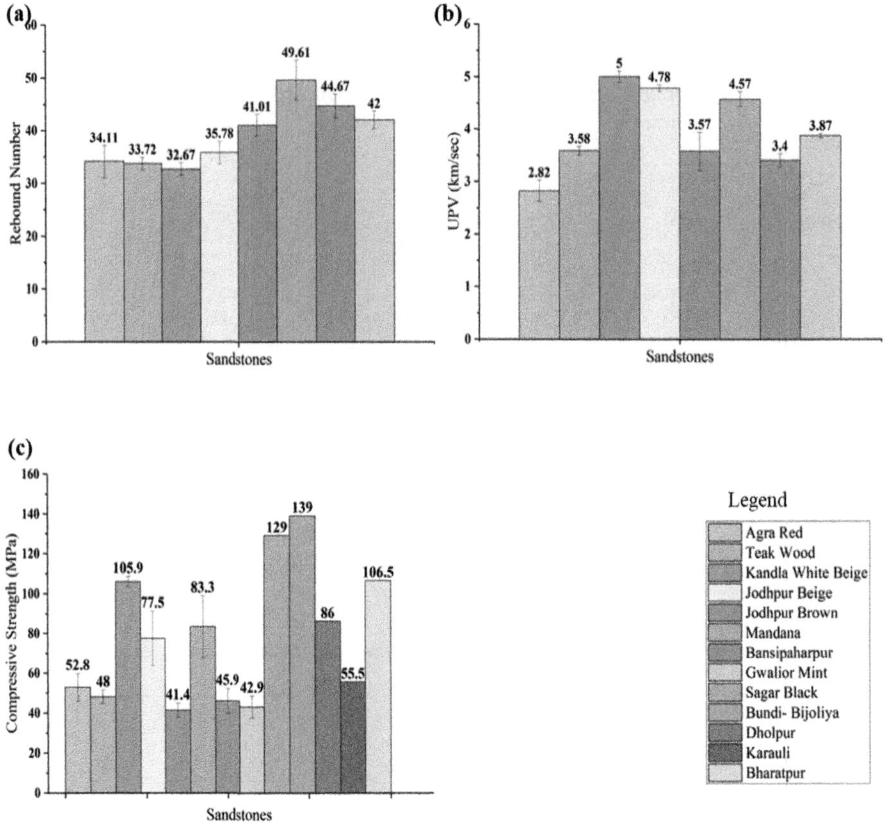

Fig. 4. (a) Variation of rebound number, (b) Variation of ultrasonic pulse velocities and (c) Variation of compressive strength of the sandstones studied.

From Table 1, compressive strength variation in various sandstones across 4 basins in India is 14 to 88 MPa, whereas the variation of the sandstones in this study ranged from 41 to 139 MPa. This variation may be due to the age of formation; the older the sandstone is, the compaction may result in higher compressive strength. Burial depth and environmental conditions also affect the compressive strength. The deeper the sandstone is buried, the more compact it may be due to pressure. Weathering conditions also have an effect; groundwater can cause dissolution of grains and can affect the porosity of the sandstones.

From Table 1 ultrasonic pulse velocity range is 2.17 to 3.43 km/sec whereas this study found UPV Values of around 2.8 to 5.0 km/sec. This may be due to the porosity of the sandstones, duration of seasoning and weathering effects. Similarly, the other properties vary due to various factors.

3.3 Correlations Between Physical and Mechanical Properties

Correlations are very specific to the sandstone it was developed for; this is due to the change of properties of sandstones because of factors like geological location, environmental conditions, burial depth, mineralogical composition, degree of compaction, lithification, diagenesis, etc. Hence, these correlations can only be used for the sandstones in this study. They can be in the form of linear or nonlinear fits and they have a R^2 value which tells the accuracy of the fit. The R^2 value is also called the coefficient of determination and it measures the model's fit to the data set. The closer the R^2 value is to 1, the more accurately the data set fits the curve. Whereas, the closer the R^2 value moves to 0, the accuracy reduces. To achieve a higher coefficient of determination, points which fall out of the fit are marked as outliers and are not included in the analysis.

The correlations can be seen in Fig. 5 below. In Fig. 5a, it shows Specific Gravity plotted against UPV and it has a nonlinear curve. This curve has an R^2 value of 0.999, which means the accuracy is very high. There are four outliers which are Kandla White Beige, Jodhpur Beige, Jodhpur Brown and Bansipaharpur sandstones and are marked as red points. From the correlation, it is understood that when specific gravity increases, so does the UPV. This is due to the density increase in the sandstones which causes the wave to pass through the material faster as there are less voids, hence increasing UPV.

Figure 5b shows the relationship between Open Porosity and UPV which is a nonlinear fit and has an R^2 value of 0.979. This is very close to 1, hence accuracy is achieved. The outlier which is the red point is Kandla White Beige sandstone. As seen in the graph, the UPV increases when the open porosity decreases. This is because less porosity means less discontinuity for the wave to pass through, hence an increase of UPV.

Figure 5c shows Water Absorption against UPV and it has a nonlinear fit with an R^2 value of 0.922. There are no outliers in this correlation which means it can be used for the eight sandstones: Agra Red, Teak Wood, Kandla White Beige, Jodhpur Beige, Jodhpur Brown, Mandana, Bansipaharpur and Gwalior Mint. Similarly to the correlation between open porosity and UPV, water absorption and UPV have a similar relationship whereby UPV decreases when water absorption increases. Since P-waves discontinue when in contact with air (pores or voids) or water, the time will increase, resulting in lower UPV.

Figure 5d shows the relationship between Rebound Number and UPV which has a nonlinear fit with an R^2 value of 0.301. The accuracy of this correlation is quite low. More data sets can be added to this correlation to improve the accuracy. The outliers marked as red points are Kandla White Beige and Jodhpur Beige sandstones. From the correlation, when the rebound number increases, the UPV increases. This is because the rebound number indicates surface hardness, which is usually high for a denser, less porous material. The wave will pass through such a material fast, resulting in higher UPV.

Figure 5e is the plot between Compressive Strength and UPV which has a nonlinear fit with an R^2 value of 0.988. The accuracy is high and there are no outliers in this correlation plot. Compressive strength increases when UPV increases. When the material is of good quality with high integrity, the UPV is high because of less porosity, microcracks or fissures in the material, this contributes to its higher compressive strength.

Figure 5f is the plot between Specific Gravity and Open Porosity which has a nonlinear fit with an R^2 value of 0.674. The accuracy of this correlation is not very high, and the outlier marked in red is Bansipaharpur sandstone. Open Porosity increases when specific gravity decreases. This is because the material's mass decreases when more pores are there for a fixed volume, so the specific gravity decreases when porosity increases in the system.

Figure 5g is the plot between Water Absorption and Open Porosity which has a linear fit with an R^2 value of 0.864. The accuracy is high and the outlier marked in the red point is Kandla White Beige. These two parameters increase uniformly, because a material with more connected pores will allow more water to be absorbed at a faster rate. However, this does not necessarily mean the material may retain the water for a long time.

Figure 5h is the plot between Water Absorption and Rebound Number which has a linear fit with an R^2 value of 0.962. The accuracy is high and the outliers marked in the red points are Kandla White Beige and Bansipaharpur sandstones. Water absorption decreases when rebound number increases. This is because the harder the material's surface is, the more compact and less porous it is. Hence, less water would be able to penetrate through the material, that forms this relationship.

Figure 5i shows the plot between Open Porosity and Rebound Number which has a nonlinear fit with an R^2 value of 0.994. The accuracy is high and there are no outliers. A higher rebound number means a harder surface, this is possible when there is denser, less porous material, so it correlates to lower open porosity.

Figure 5j shows the plot between Compressive Strength and Open Porosity which has a nonlinear fit with an R^2 value of 0.952. The accuracy is high but there are 6 outliers which are Kandla White Beige, Mandana, Bansipaharpur, Gwalior Mint, Jodhpur Brown and Bharatpur sandstones. Compressive strength is found as decreasing with increasing porosity. This is because pores are the weak links in a material which cannot resist loads, so they contribute to a lower material strength.

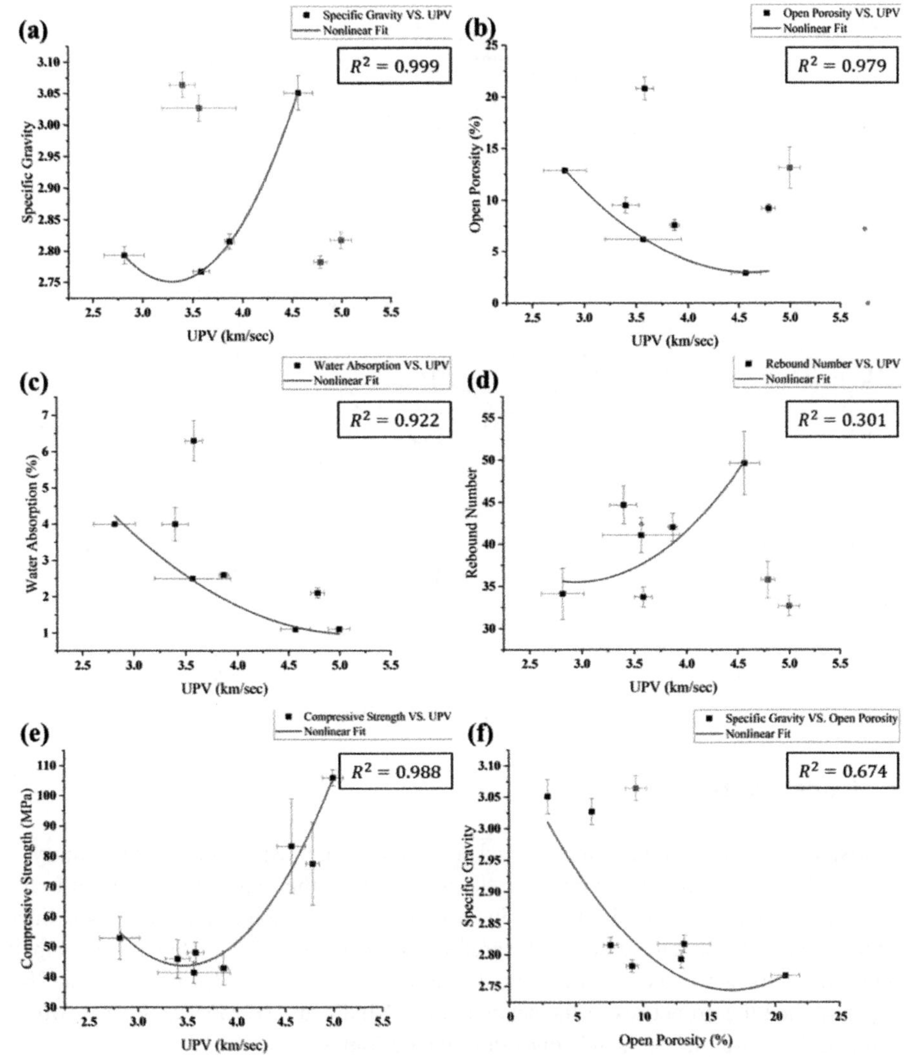

Fig. 5. Correlation between (a) Specific Gravity and UPV, (b) Open Porosity and UPV, (c) Open Porosity and UPV, (d) Rebound Number and UPV, (e) Compressive Strength and UPV, (f) Specific Gravity and UPV, (g) Water Absorption and Open Porosity, (h) Water Absorption and Rebound Number, (i) Open Porosity and Rebound Number and (j) Compressive Strength and Open Porosity. The red points are outliers which were excluded to increase the value of the coefficient of determination.

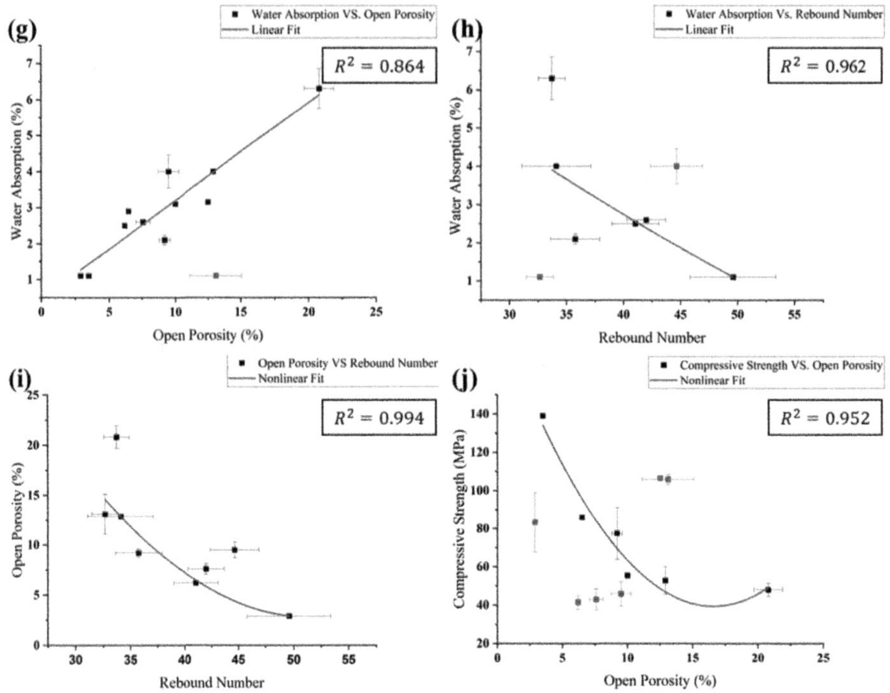

Fig. 5. (*continued*)

4 Conclusions

Sandstones are in abundance in India and are available in most regions of the country. It is clear that there is large variability in the properties of the sandstones found in India, specifically Vindhyan and Marwar Basins which are undocumented and unknown. Since applications of these sandstones, and therefore their performance, are dependent on their properties, it is necessary to understand the variabilities of the sandstones in India. This paper studied the properties of the sandstones in Vindhyan and Marwar Basins, where some of the most quarried sandstones in India are found.

Porosity and pore structure of the sandstones have effects on the other properties of the material. Kandla White Beige, Jodhpur Beige, Mandana and Bundi-Bijoliya sandstones are seen to have the lowest open porosities and highest compressive strengths, hinting to their durable and strong nature. While, Agra Red, Bansipaharpur and Jodhpur Beige sandstones have the highest open porosities and lowest compressive strengths. This hints that these sandstones are suitable for non-structural elements such as claddings, tiles, etc. Hence, these material properties contribute to a sustainable use of sandstones in mentioned basins.

The developed correlations are specific to the sandstones in this study excluding the outliers due to factors mentioned above like burial depth, environmental conditions, mineralogy, etc. These correlations contribute to the understanding of the relationships between the properties. They were developed to be used in cases where only NDT tests are possible, or there is none to limited material where all the tests cannot be conducted, various properties can be achieved from only knowing one, from a combination of these correlations.

Acknowledgements. The authors would like to thank the Centre for Development of Stones (CDOS) and Mr. Mukul Rastogi, CEO, CDOS, for their contribution of data for properties of Sagar Black, Bundi-Bijoliya, Dholpur, Karauli and Bharatpur sandstones and for providing samples for testing purposes. The authors also thank Dr. Pradeep Agarwal for his valuable comments and insights in this study.

References

Abdi, Y., Khanlari, G.R., Jamshidi, A.: Correlation between mechanical properties of sandstones and p-wave velocity in different degrees of saturation. Geotech. Geol. Eng. **42**(1), 665–674 (2024). https://doi.org/10.1007/s10706-018-0721-6

Agarwal, P.: Rajasthan - the land of exotic stones. Laghu Udyog Bharati, New Delhi (2025)

American Society for Testing and Materials. Standard Test Method for Compressive Strength of Dimension Stone (ASTM C170 2014). https://store.astm.org/

American Society for Testing and Materials. Standard Test Methods for Absorption and Bulk Specific Gravity of Dimension Stone (ASTM C97 2018) (2018). https://store.astm.org/

British Standards Institution. Natural Stone test methods - Determination of real density and apparent density and of total and open porosity (BS EN 1936 2006) (2006). https://knowledge.bsigroup.com/

Bureau of Indian Standards. Method of Test for Determination of Water Absorption, Apparent Specific Gravity and Porosity of Natural Building Stones (IS 1124 1974) (1974). https://law.resource.org/pub/in/bis

Bureau of Indian Standards. Methods of Test for Determination of Strength Properties of Natural Building Stones (IS 1121 1974) (1974). https://law.resource.org/pub/in/bis

Bureau of Indian Standards. Hardened Concrete —Methods of Test (IS 516 2018) (2018). https://archive.org/details/gov.in

Bureau of Indian Standards. Hardened Concrete Methods of Test (IS 516 2020) (2020). https://archive.org/details/gov.in

Chaudhary, V., et al.: Physico-mechanical characteristics of vindhyan sandstone, India. J. Inst. Eng. (India): Ser. D **103**(1), 295–302 (2022). https://doi.org/10.1007/s40033-021-00301-1

Directorate General of Hydrocarbons. Sedimentary Basins of India (2025). https://dghindia.gov.in/

George, B.G.: Geology of the neoproterozoic - early cambrian marwar supergroup, Rajasthan: A synthesis. In: Proceedings of the Indian National Science Academy, vol. 86, issue 2, pp. 1057–1068. Indian National Science Academy (2020). https://doi.org/10.16943/ptinsa/2019/49712

Guha Roy, D., Singh, T.N.: Effect of heat-treatment on the pure- and mixed-mode fracture properties of a homogeneous sandstone. Geomechan. Energy Environ. **33** (2023). https://doi.org/10.1016/j.gete.2022.100430

Guha Roy, D., Singh, T.N., Kodikara, J., Das, R.: Effect of water saturation on the fracture and mechanical properties of sedimentary rocks. Rock Mech. Rock Eng. **50**(10), 2585–2600 (2017). https://doi.org/10.1007/s00603-017-1253-8

Kaur, G., et al.: Vindhyan Sandstone: a Crowning Glory of Architectonic Heritage from India. Geoheritage **11**(4), 1771–1783 (2019). https://doi.org/10.1007/s12371-019-00389-8

Priya, T.L.: Protecting the tangible and intangible heritage of Rani ki Vav: a unique subterranean step well in Gujarat. Adv. Mater. Res. **133–134**, 1057–1064 (2010). https://doi.org/10.4028/www.scientific.net/AMR.133-134.1057

Sahu, A., Pandey, M., Singh, J., Singh, S.: Physico-mechanical properties estimation of upper siwalik sandstone using field (MASW) and laboratory test, Himachal Himalaya. India Article J. Geol. Soc. India (2024). https://doi.org/10.17491/jgsi/2024/173840

Shi, Z., Zhang, W., Wang, Z.: Correlation of physical and mechanical properties of Jurassic sandstone in Jining, Shandong province (2021). https://doi.org/10.1007/s12517-021-07655-6/ Published

Singh, A., Verma, A.K., Kumar, A., Singh, C.S., Roy, S.K.: Effect of temperature on physico-mechanical properties of chunar sandstone, Mirzapur, U.P., India. J. Min. Sci. **59**(1), 53–62 (2023a). https://doi.org/10.1134/S1062739123010064

Singh, G.K., Rai, A.K., Singh, A.K.: Diagenesis, facies and palaeocurrent analysis of Upper Rewa Sandstone around Sagar, Central India. J. Palaeogeog. **12**(4), 546–563 (2023b). https://doi.org/10.1016/j.jop.2023.07.003

Singh, T.D., Prakash, S., Manohar, S.: Assessment of the performance and compatibility of acrylic polymer and silane based consolidants on deteriorated heritage masonry units subjected to salt weathering. J. Build. Eng. **77** (2023c). https://doi.org/10.1016/j.jobe.2023.107490

Verma, A., Shukla, U.K.: Heterolithic lower rewa sandstone of the neoproterozoic rewa group, Vindhyan Basin, U. P., India: an example of tidal point bar. Precambrian Res. **350** (2020) https://doi.org/10.1016/j.precamres.2020.105932

Xu, H., Meert, J.G., Pandit, M.K.: Age of the Marwar Supergroup, NW India: a note on the U–Pb geochronology of Jodhpur Group felsic volcanics. Geosci. Front. **13**(1) (2022). https://doi.org/10.1016/j.gsf.2021.101287

Zhang, J., et al.: Quartz trace element geochemistry and internal morphology as proxies for provenance characterization: Results from the Marwar basin, NW India. Precambrian Res. **409** (2024). https://doi.org/10.1016/j.precamres.2024.107436

Environmental Impact Assessment

Micro-Structured Design of Binder-Based Absorbers for Carbon Dioxide (CO2) Uptakes

Mohamad Reza Sadeghi[1], Ignacio Peralta[4], Victor Fachinotti[2], Fadi Aldakheel[3], and Antonio Caggiano[1](✉)

[1] DICCA, University of Genova, Genova, Liguria, Italy
antonio.caggiano@unige.it
[2] CIMEC (UNL/CONICET), Santa Fe, Argentina
[3] IBNM, Leibniz Universität Hannover, Hannover, Germany
[4] Laboratorio de Flujometría, Facultad Regional Santa Fe, Universidad Tecnológica Nacional, Santa Fe, Argentina

Abstract. This work introduces the key aspects to optimize binder-based microstructures using machine learning techniques, with the primary goal of enhancing their uptake properties, particularly the capacity to absorb carbon dioxide (CO_2). The utilization of CO_2 in concrete and its components is currently achieved by converting gaseous CO_2 into solids (carbonates) through a process known as mineral carbonation. To bind CO_2 in solids as carbonates, two essential components are required: i) a supply of cations to form solid carbonate minerals, and ii) a supply of alkalinity that enables the rapid conversion of CO_2 into carbonate ions. Most common carbonate formed is calcium carbonate ($CaCO_3$): i.e., calcium serves as the cation. Alternatively, systems rich in magnesia also show potential for CO_2 curing, resulting in the formation of hydroxy-magnesites and/or other magnesium-carbonate hydrates. The proposed approach encompasses several key numerical steps to design micro-structured concrete absorber based on machine-learning procedures, such as data collection and model training, rendering microstructural variables compatible with machine-learning algorithms, selection of appropriate machine-learning algorithms, model validation, optimization, and prediction. By enabling the exploration of a vast range of microstructures and mineral carbonation processes, this approach offers an avenue to engineer concrete materials with superior CO_2 absorption capacities.

Keywords: CO2 Uptake · Concrete · Machine Learning · LCA · Carbonation

1 Introduction

Concrete and cement production are among the largest industrial sources of carbon dioxide (CO_2), responsible for about 8% of global emissions [1]. Cement manufacturing emits CO_2 both through energy consumption and the chemical calcination of limestone. With rising infrastructure demands, reducing the carbon footprint of concrete is critical. Innovative materials, such as CO_2-absorbing micro-structured foamed concrete, offer a path to mitigate emissions and transform concrete into a carbon sink. This

M. Posani and S. Singh (Eds.): RYS 2025, RILEM Bookseries 67, pp. 1013–1021, 2026.
https://doi.org/10.1007/978-3-032-14652-6_73

study focuses on designing sustainable concrete capable of actively capturing CO_2 by using a novel approach both experimentally and numerically through machine learning techniques. Recent advancements in sustainable construction materials have opened new possibilities for carbon mitigation. One promising innovation is micro-structured foamed concrete that is specifically engineered to absorb atmospheric CO_2 (2021) [2]. By enhancing the surface area and porosity of the concrete matrix, this material actively captures and stores carbon over time, potentially transforming traditionally high-emission construction materials into net carbon sinks. This study aims to explore and validate the performance of such sustainable concrete through a dual approach: experimental testing to assess physical properties and carbon uptake, and computational modeling using advanced machine learning techniques to predict and optimize CO_2 absorption under varying environmental and material conditions. This integrative method not only supports the development of greener infrastructure but also contributes to the broader goal of climate change mitigation through material innovation.

2 Materials and Methods

In this study, different types of concrete composites will be developed by incorporating various components, including nano-powders, recycled concrete aggregates (RCA), ordinary Portland cement (OPC), and biochar. The main objective will be to compare the CO_2 uptake performance of each composite material (Table 1).

After a standard curing period of 28 days under water, the concrete specimens will be placed inside a controlled CO_2 chamber. The samples will be exposed to a concentrated CO_2 environment for a specified duration to simulate accelerated carbonation (Fig. 1).

To accurately assess the CO_2 capture capacity and carbonation behavior of each concrete composite, several advanced characterization techniques will be utilized (Fig. 2):

a) Thermogravimetric Analysis (TGA): TGA will measure the mass change of the samples as they are heated, allowing precise quantification of the amount of CO_2 absorbed by the material during carbonation.
b) Scanning Electron Microscopy (SEM): SEM will provide high-resolution images of the concrete microstructure, revealing physical changes such as pore structure refinement and the formation of carbonation products at the microscopic level.
c) X-ray Diffraction (XRD): XRD will be used to identify and confirm the presence of crystalline carbonate phases (such as calcium carbonate), offering direct evidence of chemical carbonation within the material.
d) Phenolphthalein Test: This visual test will be performed by spraying phenolphthalein indicators on freshly broken concrete surfaces. Areas that remain colorless will indicate carbonated zones, while pink coloration will indicate non-carbonated areas, allowing for the measurement of carbonation depth.

By combining these methods, a comprehensive evaluation of both the extent and mechanisms of CO_2 sequestration across different concrete types will be achieved.

Table 1. Components and CO2 uptake.

Concrete Type	CO_2 Uptake Capacity (%)	Key Findings	Reference
OPC Concrete	~ 20% of cement mass over 27 days	Standard OPC concrete under accelerated carbonation can absorb about 20% of the cement mass in CO_2 over 27 days	[3]
RCA Concrete	~25–30% higher than natural aggregates	RCA concrete shows approximately 25–30% greater CO_2 uptake compared to conventional natural aggregate concrete, due to its higher porosity	[4]
Nano-Powder Concrete	~15–18% increase compared to plain cement	Addition of nano-TiO_2 increases the carbonation rate by 15–18% compared to non-modified cement pastes	[5]
Biochar Concrete	~10–15% of concrete mass	Incorporating biochar enhances CO_2 uptake up to 10–15% of the concrete mass, depending on the porosity and loading of biochar	[6]

(a) (b) (c) (d)

Fig. 1. (a) Dispersing iron oxide nano-powder (Fe2O3) using an ultrasonic disperser to achieve homogeneous mixing within the concrete mixture; (b) Measuring the slump of the Fe2O3 concrete sample; (c) Measuring the slump of the OPC concrete sample; (d) Specimens immersed in water for 28 days for hydration curing.

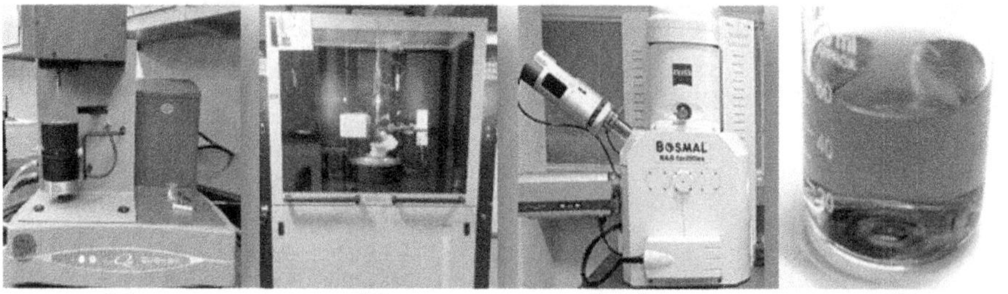

(a) (b) (c) (d)

Fig. 2. Experimental methods: (a) is the TGA device; (b) is the XRD device; (c) is the SEM device; (d) is phenolphthalein solution.

3 Numerical Assessments

3.1 Meso-Scale and NRGene Code

The numerical modeling in this study will employ NRGene, a microstructure generation tool developed by [7]. NRGene can produce three-dimensional representations of porous concrete structures, providing detailed information about the geometry and porosity distribution within a sample (Fig. 3).

Using this code, a 3D cubic sample can be digitally created, incorporating internal voids and solid voxels. This enables the prediction of key features such as the number of pores, size distribution, and air inclusions inside the material.

The generated virtual microstructures will serve as the basis for simulating and analyzing the carbonation behavior numerically, thereby reducing the reliance on purely experimental methods.

$$h_{\max} = \frac{D_1}{n_{\mathrm{ref}}} \qquad V_{\mathrm{RVE}}^{\max} = \frac{V_{\mathrm{por}}^{\max}}{p}$$

$$n_{\mathrm{vox}} = \mathrm{ceil}\left(\frac{L_{\mathrm{RVE}}}{h_{\max}}\right) \qquad V_{\mathrm{por}}^{\max} = \sum_{q=1}^{Q} V_q^{\max}$$

$$h_{\mathrm{vox}} = \frac{L_{\mathrm{RVE}}}{n_{\mathrm{vox}}} \leq h_{\max} \qquad V_q^{\max} = n_q \frac{\pi}{6} D_q^3$$

Fig. 3. Left figure is the voxelized porous model of sample and right pictures are the related equations (see Fachinotti et el. [6]).

3.2 Experimental Data of Foamed Concretes

To validate the effectiveness of the NRGene code in simulating realistic microstructures of foamed concretes, nine distinct types of Portland Cement–Carbonated Foamed Concrete (PC-CFC) samples by [8] will be examined. These variations are systematically designed by altering key parameters such as:

- Foam Volume: Adjusting the volume of foam introduced into the mix (e.g., 4, 6, and 8 L),
- Carbonation Pretreatment Duration: Varying the duration of carbonation pretreatment (e.g., 60, 90, and 120 min).

Each combination of foam volume and pretreatment time results in a unique PC-CFC variant, culminating in a total of nine different sample types. These samples are prepared based on the methodology proposed by Ta et al. (2023) [8] in their study on CO_2-foamed concretes for heat insulation and carbon storage.

Subsequently, the NRGene code is applied to generate three-dimensional digital models of these samples, capturing detailed internal geometries, including pore distribution and connectivity. The successful replication of these microstructures confirms the NRGene code's capability to accurately model the complex geometries inherent in PC-CFCs (Fig. 4).

Fig. 4. 3D model of PC-CFCs of (from left to right) 4, 6, and 8 L of foam contents.

3.3 Carbonation Equation

To implement the phenomenon of carbonation within a finite element method (FEM) framework, a special strategy is adopted. The carbonation process is simulated by adapting the classical heat transfer equation to be representative of the diffusion-reaction of carbon dioxide.

By mathematically converting the thermal diffusion equation into a form analogous to CO_2 transfer, the carbonation process is treated similarly to a heat diffusion problem. This trick allows the simulation of CO_2 penetration and concentration profiles over time within the concrete microstructure, enabling a realistic numerical prediction of carbonation fronts and depths (2015) [9] (Fig. 5).

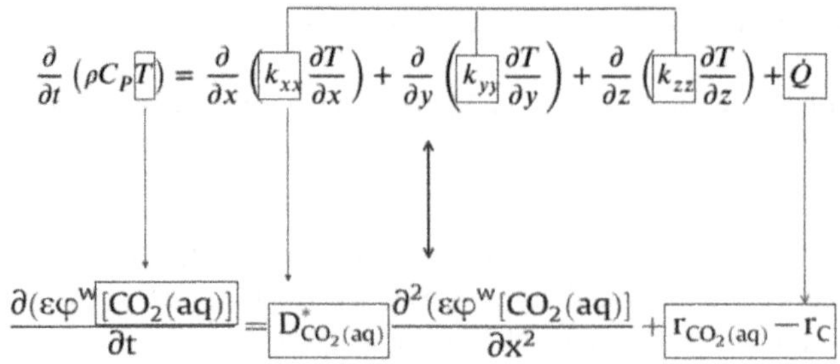

Fig. 5. Heat Transfer equation (top) and Carbonation equation (bottom).

In the first equation, T is the temperature that will assume the role of the CO_2 concentration, K_{xx}, K_{yy}, and $K_{zz,}$ will assume the diffusion coefficient D_{CO2}, and dot(Q) will be the carbonation and dissolution rate ($r_{CO2()}$ - r_C), respectively.

3.4 Machine Learning

This study applies inverse design powered by machine learning to optimize concrete compositions for CO_2 uptake. Instead of predicting behavior from known inputs, inverse design starts from the desired performance (high CO_2 capture) and works backward to identify optimal material properties such as density and porosity.

Using trained models, we aim to efficiently find mixtures that enhance CO_2 sequestration, significantly reducing experimental cost, time, and material waste.

Kong et al. (2024) [10] propose a data-driven inverse mix design framework for alkali-activated materials (AAMs), enabling mix proportioning based on target properties. The inverse design approach predicts input parameters from desired outputs like workability and compressive strength, bypassing traditional trial-and-error methods. They construct forward prediction models using XGBoost trained on a comprehensive literature-based dataset. A Gaussian Mixture Model (GMM) generates synthetic mix designs, capturing a broad design space. This framework (Fig. 6) directly outputs viable mix designs from performance requirements, offering high efficiency and flexibility.

Validation tests confirm that the proposed inverse design method produces sustainable, high-performance mixes with reduced environmental impact.

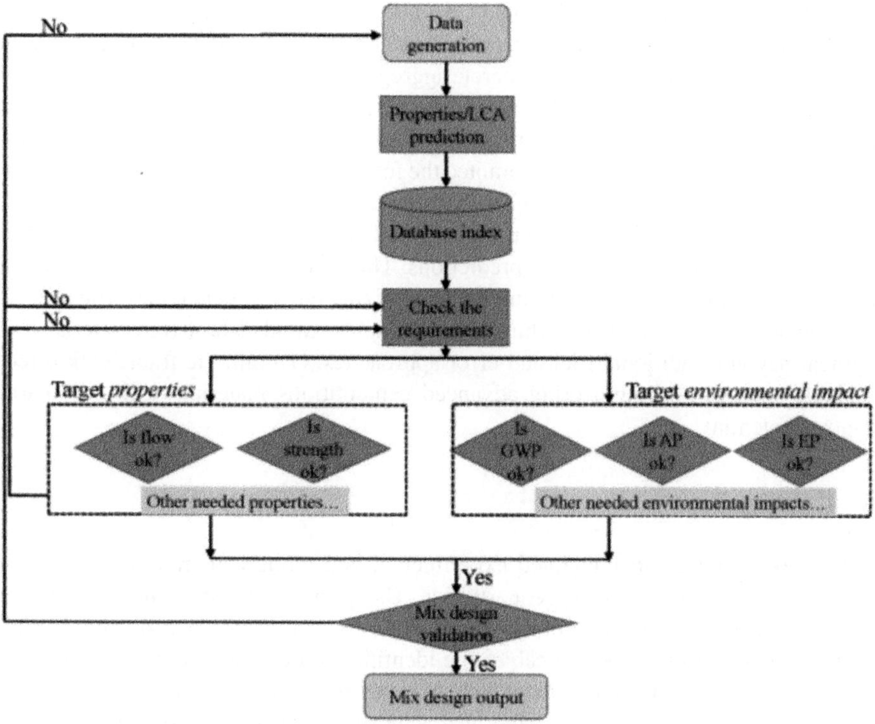

Fig. 6. Workflow of predicting the desired AAM mixture [10]

4 Life Cycle Assessment (LCA)

After identifying the best-performing materials for CO_2 uptake, a Life Cycle Assessment (LCA) will be conducted to evaluate the carbon footprint associated with each component. By analyzing the total CO_2 emissions during the production, use, and end-of-life phases, we can ensure that the selected material not only captures CO_2 efficiently but also contributes to zero or near-zero net carbon emissions. To achieve this, the Ecoinvent database (2016) [11] will be employed to provide comprehensive inventory data related to raw material extraction, manufacturing, transportation, and disposal. This database allows for a detailed environmental evaluation of each component, helping to identify materials that maximize CO_2 uptake while minimizing life cycle emissions. This step is essential to ensure that the final material selection aligns with sustainability goals and represents a genuinely climate-friendly solution.

5 Results and Discussions

Machine learning-based inverse design successfully identified optimal concrete microstructures for enhanced CO_2 uptake.

By starting from a desired outcome—high CO_2 absorption—the model determined ideal combinations of porosity, density, and material components. Among the nine

Portland Cement–Carbonated Foamed Concrete (PC-CFC) variants studied, those with higher foam volumes and specific pretreatment times demonstrated superior performance. The inclusion of recycled concrete aggregates, nano-powders, and biochar significantly influenced carbonation behavior. Thermogravimetric Analysis (TGA) validated the CO_2 sequestration efficiency, revealing distinct trends based on material formulation. Calcium-rich compositions promoted the formation of calcium carbonate ($CaCO_3$), while magnesia-enhanced mixes led to hydroxy-magnesite phases.

These results align with known mineral carbonation mechanisms and support the reliability of the machine learning predictions. The observed differences underscore the importance of microstructural tuning for maximizing environmental performance.

This study illustrates that machine learning enables rapid, data-driven material development beyond traditional trial-and-error approaches. Overall, the framework offers a scalable strategy for engineering advanced cementitious materials with high carbon capture potential.

6 Conclusion

This work proposes an integrated experimental and numerical framework to design concretes with enhanced CO_2 sequestration. Using material testing and 3D modeling with NRGene, the carbonation behavior of different composites is evaluated. Machine learning-based inverse design enables the identification of optimal material properties while reducing cost and time. A Life Cycle Assessment ensures that the selected materials not only capture CO_2 effectively but also achieve low overall carbon emissions. This combined methodology offers a promising route toward developing sustainable, climate-resilient construction materials.

Acknowledgments. The authors gratefully acknowledge the financial support from the EU Commission through the MSCA under GA ID: 101086440, which has enabled the realization of the BEST (Bio-based Energy-efficient Materials and Structures for Tomorrow) project: https://doi.org/10.3030/101086440.

References

1. Kazemian, M., Shafei, B.: Carbon sequestration and storage in concrete: a state-of-the-art review of compositions, methods, and developments. J. CO2 Utilization **70**, 102443 (2023)
2. Park, B., Choi, Y.C.: Investigation of carbon-capture property of foam concrete using stainless steel AOD slag. J. Clean. Prod. **288**, 125621 (2021)
3. Hameed, R., Seo, J., Park, S., Amr, I.T., Lee, H.K.: CO_2 uptake and physicochemical properties of carbonation-cured ternary blend portland cement-metakaolin-limestone pastes. Materials (Basel). **13**(20), 4656 (2020). https://doi.org/10.3390/ma13204656.PMID:33086553; PMCID:PMC7603273
4. Huang, K., Li, A., Xia, B., et al.: Prediction on CO_2 uptake of recycled aggregate concrete. Front. Struct. Civ. Eng. **14**, 746–759 (2020). https://doi.org/10.1007/s11709-020-0635-2
5. Lopez-Arias, M., Moro, C., Francioso, V., Elgaali, H.H., Velay-Lizancos, M.: Effect of nanomodification of cement pastes on the CO2 uptake rate. Constr. Build. Mater. **404**, 133165 (2023)

6. Mesnage, M., Omnée, R., Colin, J., Ramezani, H., Jeong, J., Raymundo-Piñero, E.: Porous biochar for improving the CO2 uptake capacities and kinetics of concrete. Cement Concr. Compos. **157**, 105932 (2025)
7. Fachinotti, V.D., Peralta, I., Toro, S., Storti, B.A., Caggiano, A.: Automatic generation of high-fidelity representative volume elements and computational homogenization for the determination of thermal conductivity in foamed concretes. Mater. Struct. **56**(10), 179 (2023)
8. Ta, X., Zhang, Y., Wan, Z., Shi, P., Zhou, J.: Study on preparation and performance of CO_2 foamed concrete for heat insulation and carbon storage. Materials (Basel). **16**(7), 2725 (2023). https://doi.org/10.3390/ma16072725.PMID:37049022;PMCID:PMC10095924
9. Kashef-Haghighi, S., Shao, Y., Ghoshal, S.: Mathematical modeling of CO2 uptake by concrete during accelerated carbonation curing. Cem. Concr. Res. **67**, 1–10 (2015)
10. Kong, Y.K., Kurumisawa, K., Tokoro, C., Li, Z., Chu, S.H.: Data-driven inverse mix design for sustainable alkali-activated materials. J. Sustain. Cement-Based Mater. **13**(12), 1857–1878 (2024)
11. Wernet, G., Bauer, C., Steubing, B., et al.: The ecoinvent database version 3 (part I): overview and methodology. Int. J. Life Cycle Assess. **21**, 1218–1230 (2016). https://doi.org/10.1007/s11367-016-1087-8

Author Index

M. Posani and S. Singh (Eds.): RYS 2025, RILEM Bookseries 67, pp. 1023–1026, 2026.
https://doi.org/10.1007/978-3-032-14652-6

GPSR Compliance

The European Union's (EU) General Product Safety Regulation (GPSR) is a set of rules that requires consumer products to be safe and our obligations to ensure this.

If you have any concerns about our products, you can contact us on ProductSafety@springernature.com

In case Publisher is established outside the EU, the EU authorized representative is:

Springer Nature Customer Service Center GmbH
Europaplatz 3
69115 Heidelberg, Germany

Batch number: 10755284

Printed by Printforce, the Netherlands